Toxicology and Risk Assessment

Toxicology and Risk Assessment: A Comprehensive Introduction

Second Edition

Edited By

HELMUT GREIM

Technical University of Munich, Munich, Germany

ROBERT SNYDER

Rutgers University and EOHSI, USA

WILEY

This edition first published 2019
© 2019 John Wiley & Sons Ltd

Edition History

Toxicology and Risk Assessment: A Comprehensive Introduction, First Edition, Wiley 2008.

All rights reserved. No part of this publication may be reproduced, stored in a retrieval system, or transmitted, in any form or by any means, electronic, mechanical, photocopying, recording or otherwise, except as permitted by law. Advice on how to obtain permission to reuse material from this title is available at http://www.wiley.com/go/permissions.

The right of Helmut Greim and Robert Snyder to be identified as the authors of the editorial material in this work has been asserted in accordance with law.

Registered Offices
John Wiley & Sons, Inc., 111 River Street, Hoboken, NJ 07030, USA
John Wiley & Sons Ltd, The Atrium, Southern Gate, Chichester, West Sussex, PO19 8SQ, UK

Editorial Office
The Atrium, Southern Gate, Chichester, West Sussex, PO19 8SQ, UK

For details of our global editorial offices, customer services, and more information about Wiley products visit us at www.wiley.com.

Wiley also publishes its books in a variety of electronic formats and by print-on-demand. Some content that appears in standard print versions of this book may not be available in other formats.

Limit of Liability/Disclaimer of Warranty

In view of ongoing research, equipment modifications, changes in governmental regulations, and the constant flow of information relating to the use of experimental reagents, equipment, and devices, the reader is urged to review and evaluate the information provided in the package insert or instructions for each chemical, piece of equipment, reagent, or device for, among other things, any changes in the instructions or indication of usage and for added warnings and precautions. While the publisher and authors have used their best efforts in preparing this work, they make no representations or warranties with respect to the accuracy or completeness of the contents of this work and specifically disclaim all warranties, including without limitation any implied warranties of merchantability or fitness for a particular purpose. No warranty may be created or extended by sales representatives, written sales materials or promotional statements for this work. The fact that an organization, website, or product is referred to in this work as a citation and/or potential source of further information does not mean that the publisher and authors endorse the information or services the organization, website, or product may provide or recommendations it may make. This work is sold with the understanding that the publisher is not engaged in rendering professional services. The advice and strategies contained herein may not be suitable for your situation. You should consult with a specialist where appropriate. Further, readers should be aware that websites listed in this work may have changed or disappeared between when this work was written and when it is read. Neither the publisher nor authors shall be liable for any loss of profit or any other commercial damages, including but not limited to special, incidental, consequential, or other damages.

Library of Congress Cataloging-in-Publication Data:

Names: Greim, Helmut, editor. | Snyder, Robert, editor.
Title: Toxicology and risk assessment : a comprehensive introduction / edited by Helmut Greim, Technical University of Munich, Munich, Germany, Robert Snyder, Rutgers University and EOHSI, USA.
Description: Hoboken, NJ : Wiley, 2019. | Originally published. 2008. | Includes bibliographical references and index. |
Identifiers: LCCN 2018009797 (print) | LCCN 2018010582 (ebook) | ISBN 9781119135920 (pdf) | ISBN 9781119135937 (epub) | ISBN 9781119135913 (cloth)
Subjects: LCSH: Toxicology. | Health risk assessment.
Classification: LCC RA1211 (ebook) | LCC RA1211 .T635 2018 (print) | DDC 615.9/02–dc23
LC record available at https://lccn.loc.gov/2018009797

Cover Design: Wiley
Cover Images: © zffoto/Shutterstock;
© chromatos/Shutterstsock; © Mopic/Shutterstock

Set in 10/12pt TimesLTStd by SPi Global, Chennai, India
Printed in Singapore by C.O.S. Printers Pte Ltd

10 9 8 7 6 5 4 3 2 1

Contents

List of Contributors	xv
Short CVs of Authors	xix
Preface	xxix
Dedication	xxxi
List of Abbreviations	xxxiii

1 Introduction to the Discipline of Toxicology — 1
Helmut Greim and Robert Snyder
- 1.1 Introduction — 1
- 1.2 The Risk Assessment Process — 2
  - 1.2.1 Hazard Identification — 2
  - 1.2.2 Dose Response and Toxic Potency — 3
  - 1.2.3 Exposure Assessment — 4
  - 1.2.4 Risk Characterization — 6
- 1.3 Toxicological Evaluation of New and Existing Chemicals — 6
  - 1.3.1 General Requirements for Hazard Identification and Risk Assessment — 7
  - 1.3.2 General Approach for Hazard Identification and Risk Assessment (for details see Chapter 2.1) — 12
  - 1.3.3 Toxicological Issues related to Specific Chemical Classes — 13
  - 1.3.4 Existing Chemicals — 14
  - 1.3.5 Test Guidelines (Chapter 4.1) — 14
  - 1.3.6 Alternatives to Animal Experiments (Chapter 4.5) — 15
  - 1.3.7 Evaluation of Mixtures (Chapter 2.7) — 15
  - 1.3.8 Evaluation of Uncertainties — 16
  - 1.3.9 The Precautionary Principle — 16
  - 1.3.10 The TTC Concept — 17
  - 1.3.11 Classification and Labeling of Chemicals — 17
- 1.4 Summary — 18

2 Principles in Toxicology — 21
- 2.1 General Concepts of Human Health Risk Assessment — 21
Paul T.C. Harrison, Philip Holmes, and Ruth Bevan
  - 2.1.1 Introduction — 21
  - 2.1.2 Principles of Risk Assessment — 21
  - 2.1.3 Application of Risk Assessment in Setting Exposure Standards and Limits — 30

	2.1.4	The Wider Importance of Risk Assessment	32
	2.1.5	Summary	32
2.2	Toxicokinetics		34

Johannes G. Filser

	2.2.1	Definition and Purpose	34
	2.2.2	Absorption, Distribution, and Elimination	35
	2.2.3	Toxicokinetic Models	47
	2.2.4	Summary	61
2.3	Biotransformation of Xenobiotics		61

Wolfgang Dekant

	2.3.1	Introduction	61
	2.3.2	Phase I Reactions	64
	2.3.3	Phase II Reactions	74
	2.3.4	Factors that Influence the Biotransformation of Xenobiotics	80
	2.3.5	Role of Bioactivation in Toxicity	87
	2.3.6	Interactions of Reactive Intermediates formed during the Biotransformation	90
	2.3.7	Summary	94
2.4	Cytotoxicity		98

Daniel Dietrich

	2.4.1	Introduction	98
	2.4.2	The Cell	102
	2.4.3	Cellular Targets of Toxins	106
	2.4.4	Mechanisms underlying Cell Death	115
	2.4.5	Summary	120
2.5	Toxicogenetics		122

Lesley A. Stanley

	2.5.1	Introduction	122
	2.5.2	Toxicogenetics and Toxicogenomics	123
	2.5.3	Genotype and Phenotype	123
	2.5.4	The Role of Xenobiotic-metabolising Polymorphisms in Susceptibility to Toxic Agents	127
	2.5.5	Study Numbers and Effect Size	135
	2.5.6	Summary	139
2.6	Receptor-mediated Mechanisms		140

Jens Schlossmann and Franz Hofmann

	2.6.1	Introduction	140
	2.6.2	Ligand–Receptor Interactions	141
	2.6.3	Biological Consequences of Ligand–Receptor Interactions	142
	2.6.4	Receptor Signal Transduction	144
	2.6.5	Summary	147
2.7	Mixtures and Combinations of Chemicals		148

Hermann Bolt

	2.7.1	Introduction	148
	2.7.2	Types of Mixed Exposures	148
	2.7.3	Types of Joint Actions and their Role in Safety Evaluation	149
	2.7.4	Role in Safety Evaluation	151
	2.7.5	Summary	158

2.8		Chemical Carcinogenesis: Genotoxic and Non-Genotoxic Mechanisms	159
Thomas Efferth and Bernd Kaina			
	2.8.1	Introduction	159
	2.8.2	Mechanisms of DNA Damage, Repair, and Carcinogenesis	160
	2.8.3	Cancer Development	171
	2.8.4	Non-Genotoxic Mechanisms of Carcinogenesis	181
	2.8.5	Implications of Initiation and Promotion for Risk Assessment	184
	2.8.6	Summary	185
2.9		Threshold Effects for Genotoxic Carcinogens	186
Helmut Greim			
	2.9.1	Introduction	186
	2.9.2	Development of Cancer due to Genotoxic Carcinogens	187
	2.9.3	Cellular Reaction to DNA Damage	187
	2.9.4	Examples of Dose-dependent Reactions in Cases of Genotoxicity	193
	2.9.5	Summary	193
2.10		Reproductive Toxicology	195
Horst Spielmann			
	2.10.1	Introduction	195
	2.10.2	Characteristics of Reproductive Toxicology	196
	2.10.3	Adverse Effects on Female and Male Fertility	197
	2.10.4	International Test Methods in the Field of Reproductive Toxicology	199
	2.10.5	Pre- and Postnatal Toxicology	201
	2.10.6	Effects of Drugs and other Chemicals on Lactation	207
	2.10.7	Endocrine Disrupters	208
	2.10.8	Summary	208
2.11		Ecotoxicology: More than Wildlife Toxicology	209
Peter Calow and Valery E. Forbes			
	2.11.1	Introduction	209
	2.11.2	Protection Targets	210
	2.11.3	Necessary Information	211
	2.11.4	Risk Assessment	216
	2.11.5	Fast-track Approaches	217
	2.11.6	Summary	218

3 Organ Toxicology — **221**

3.1		The Gastrointestinal Tract	221
Michael Schwenk			
	3.1.1	Introduction	221
	3.1.2	Structure and Function	221
	3.1.3	Fate of Xenobiotics in the GI Tract	223
	3.1.4	Toxicology	226
	3.1.5	Summary	231
3.2		The Liver	231
Jan G. Hengstler			
	3.2.1	Introduction	231
	3.2.2	Structure and Function	232
	3.2.3	Toxicology	237
	3.2.4	Myths of the Liver	247
	3.2.5	Summary	248

3.3		The Respiratory System	248
Florian Schulz			
	3.3.1	Introduction	248
	3.3.2	Structure	249
	3.3.3	Function	252
	3.3.4	Protective Systems	254
	3.3.5	The Respiratory Tract as a Target for Toxicity	256
	3.3.6	Respiratory Allergy and Asthma	261
	3.3.7	Lung Cancer	262
	3.3.8	Test Systems to Detect the Toxic Effects of Inhaled Materials	263
	3.3.9	Summary	263
3.4		The Nervous System	264
Gunter P. Eckert and Walter E. Müller			
	3.4.1	Structure and Function of the Nervous System	264
	3.4.2	The Nervous System Site of Attack for Toxins	273
	3.4.3	Clinical Signs and Symptoms induced by Neurotoxins	279
	3.4.4	Summary	281
3.5		Behavioral Neurotoxicology	283
Andreas Seeber			
	3.5.1	Introduction	283
	3.5.2	Exposure Assessment	283
	3.5.3	Methods	284
	3.5.4	Neurobehavioral Effects in Humans	291
	3.5.5	Summary	295
3.6		The Skin	296
Brunhilde Blömeke			
	3.6.1	Structure	296
	3.6.2	Function	298
	3.6.3	Toxicology of the Skin and the Anterior Segment of the Eye	300
	3.6.4	Summary	309
3.7		Kidney and Urinary Tract	311
Helmut Greim			
	3.7.1	Introduction	311
	3.7.2	Anatomy and Function	311
	3.7.3	Toxicology	313
	3.7.4	Summary	319
3.8		The Hematopoietic System (Bone Marrow and Blood)	320
Robert Snyder			
	3.8.1	Introduction	320
	3.8.2	Hematopoiesis	321
	3.8.3	The Bone Marrow Niche	323
	3.8.4	Toxicological Features of Circulating Blood Cells	323
	3.8.5	Leucocytes (White Blood Cells)	326
	3.8.6	Platelets (Thrombocytes)	328
	3.8.7	Impairment of Bone Marrow Function	328
	3.8.8	Mechanisms by which Chemicals can Induce Leukemia	330
	3.8.9	Summary	330

3.9		The Immune System	331

Peter Griem

	3.9.1	Introduction: the Innate and Specific Immune System	331
	3.9.2	Antigen Recognition	333
	3.9.3	Activation of T and B lymphocytes	334
	3.9.4	Immunologic Tolerance	336
	3.9.5	Sensitization and Allergy	337
	3.9.6	Risk Assessment of Immunotoxic Effects	342
	3.9.7	Chemical-induced Autoimmunity	345
	3.9.8	General Immunostimulation by Chemicals	346
	3.9.9	Chemical Immunosuppression	348
	3.9.10	Summary	350
3.10		The Eye	351

Ines Lanzl

	3.10.1	Introduction	351
	3.10.2	Structure and Function of the Eye	351
	3.10.3	Routes of Delivery of Xenobiotics to the Eye	354
	3.10.4	Specific Toxicology of the Eye	355
	3.10.5	Summary	363
3.11		The Cardiovascular System	364

Helmut Greim

	3.11.1	Structure and Function	364
	3.11.2	Toxicology	366
	3.11.3	Summary	371
3.12		The Endocrine System	372

Gerlinde Schriever-Schwemmer

	3.12.1	Introduction	372
	3.12.2	Structure and Function	373
	3.12.3	Foetal Development of the Hypothalamus–Pituitary–Gonad Axis	381
	3.12.4	Testing of Sexual Function in Toxicology	383
	3.12.5	Hazard Identification and Risk Assessment of Endocrine Disruptors	386
	3.12.6	Summary	386

4	**Methods in Toxicology**		**389**
	4.1	OECD Test Guidelines for Toxicity Tests *in vivo*	389

Rüdiger Bartsch

	4.1.1	Introduction	389
	4.1.2	Requirements for *in vivo* Tests	390
	4.1.3	Acute Toxicity	392
	4.1.4	Skin and Eye Irritation	394
	4.1.5	Skin Sensitization	397
	4.1.6	Toxicity after Repeated Dosing	398
	4.1.7	Reproductive Toxicity	401
	4.1.8	Other Test Guidelines	406
	4.1.9	Other Regulatory Bodies	407
	4.1.10	Summary	407

4.2	Genotoxicity		408
4.2A	*In vitro* Tests for Genotoxicity		408
Hans-Jörg Martus			
	4.2A.1	Introduction	408
	4.2A.2	Bacterial Test Systems	409
	4.2A.3	Test Systems employing Mammalian Cells	411
	4.2A.4	Cell Transformation Assays	420
	4.2A.5	Xenobiotic Metabolism	420
	4.2A.6	Summary	421
4.2B	Mutagenicity tests in vivo		422
Ilse-Dore Adler and Gerlinde Schriever-Schwemmer			
	4.2B.1	Introduction	422
	4.2B.2	Chromosomal Mutations in Somatic Cells	424
	4.2B.3	Gene Mutations in Somatic Cells	427
	4.2B.4	Chromosome Mutations in Germ Cells	429
	4.2B.5	Gene Mutations in Germ Cells	433
	4.2B.6	Summary	435
4.3	Assessment of the Individual Exposure to Xenobiotics (Biomonitoring)		440
Thomas Göen			
	4.3.1	Introduction	440
	4.3.2	Prerequisites for Carrying Out Biomonitoring	442
	4.3.3	Examples of Biomonitoring of Special Substance Groups or Special Biomonitoring Parameters	447
	4.3.4	Summary	449
4.4	Epidemiology		450
Kurt Ulm			
	4.4.1	Introduction	450
	4.4.2	Measures to Describe the Risk	450
	4.4.3	Standardization	452
	4.4.4	Types of Epidemiological Studies	454
	4.4.5	Statistics	456
	4.4.6	Meta-analysis	457
	4.4.7	Bias, Confounding, Chance, Causality	459
	4.4.8	Summary	460
4.5	Alternatives to Animal Testing		461
Thomas Hartung			
	4.5.1	Introduction	461
	4.5.2	The Birth of Doubt in Animal Experiments	461
	4.5.3	Early Successful Alternatives	463
	4.5.4	The Replacement of Animal Tests is Possible	463
	4.5.5	Validation of Alternative Methods: Animal Welfare must not Trump Patient and Consumer Safety	466
	4.5.6	How Reliable are Animal Tests?	467
	4.5.7	The Animal Test Ban for Cosmetics in Europe as an Engine of Change	467
	4.5.8	"Toxicological Ignorance": the European REACH Program as a Driver for Alternative Methods	468
	4.5.9	Outlook	469
	4.5.10	Summary	471

4.6		Omics in Toxicology	472

Laura Suter-Dick

	4.6.1	Introduction	472
	4.6.2	Concept of Toxicogenomics	472
	4.6.3	Technology Platforms	475
	4.6.4	Bioinformatics and Biostatistics	480
	4.6.5	Applications of Toxicogenomics	481
	4.6.6	Summary	484
4.7		Introduction to the Statistical Analysis of Experimental Data	486

György Csanády

	4.7.1	Introduction	486
	4.7.2	Descriptive Statistics	488
	4.7.3	Error Propagation	491
	4.7.4	Probability Distribution	492
	4.7.5	Inferential Statistics	496
	4.7.6	Regression Analysis	504
	4.7.7	Probit Analysis	506
	4.7.8	Experimental Designs	507
	4.7.9	Statistical Software	508
	4.7.10	Summary	509
4.8		Mathematical Models for Risk Extrapolation	510

Jürgen Timm

	4.8.1	Introduction	510
	4.8.2	Basic Approach of Linear Extrapolation	512
	4.8.3	Some Special Methods of Linear Extrapolation	514
	4.8.4	Consideration of Time Aspects	515
	4.8.5	Models of Carcinogenesis	518
	4.8.6	Assumptions and Limits of Extrapolation in Mathematical Models	521
	4.8.7	Summary	523

5 Regulatory Toxicology — 525

5.1		Regulations on Chemical Substances in the European Union	525

Werner Lilienblum and Klaus-Michael Wollin

	5.1.1	Introduction	525
	5.1.2	Current Legislation in the EU	526
	5.1.3	Risk Issues and Some Definitions in Terms of Chemical Substances	528
	5.1.4	International Co-operation and Harmonization Supported and Implemented by the EU	531
	5.1.5	EU Legislation on Chemical Substances and Their Uses	535
	5.1.6	Legislation on Chemical Substances in the Environment	552
	5.1.7	Summary	556
5.2		Regulations Regarding Chemicals and Radionuclides in the Environment, Workplace, Consumer Products, Foods, and Pharmaceuticals in the United States	557

Dennis J. Paustenbach

	5.2.1	Introduction	557
	5.2.2	Occupational Health Regulations	558
	5.2.3	Food and Drug Regulations	561
	5.2.4	Environmental Regulations	563

	5.2.5	Consumer Product Regulations	568
	5.2.6	Radionuclides Regulations	569
	5.2.7	Governmental Agencies on Human Health	570
	5.2.8	Centers for Disease Control	571
	5.2.9	Litigation is Nearly as Effective as Regulation in the United States	571
	5.2.10	Summary	572
5.3	The Concept of REACH		573

Jörg Lebsanft

	5.3.1	Introduction	573
	5.3.2	Historical Development	574
	5.3.3	Substances, Mixtures and Articles	574
	5.3.4	The Main Elements of REACH	575
	5.3.5	Allocation of Responsibilities and Administration of REACH	580
	5.3.6	Downstream Users	582
	5.3.7	Outlook	583
	5.3.8	Summary	583

6 Specific Toxicology — 585

6.1	Persistent Halogenated Aromatic Hydrocarbons		585

Heidrun Greim and Karl K. Rozman

	6.1.1	Introduction	585
	6.1.2	Polychlorinated Dibenzodioxins and Dibenzofurans	586
	6.1.3	Polychlorinated Biphenyls	588
	6.1.4	Dichlorodiphenyltrichloroethane	589
	6.1.5	Hexachlorobenzene	590
	6.1.6	Physico-chemical Properties	591
	6.1.7	Toxicity	592
	6.1.8	Mechanisms of Action	600
	6.1.9	Metabolism	603
	6.1.10	Enzyme Induction	603
	6.1.11	Kinetics	604
	6.1.12	Summary	606
6.2	Metals		607

Andrea Hartwig and Gunnar Jahnke

	6.2.1	General Aspects	607
	6.2.2	The Importance of Bioavailability	608
	6.2.3	Acute and Chronic Toxicity as well as Carcinogenicity	609
	6.2.4	Toxicology of Selected Metal Compounds	611
	6.2.5	Summary	624
6.3	Toxicology of Fibers and Particles		625

Paul J.A. Borm

	6.3.1	Introduction	625
	6.3.2	Particle Toxicology: Basic Concepts	626
	6.3.3	Particle Properties	633
	6.3.4	Nanoparticles: A Special Case?	636
	6.3.5	Special Particle Effects	637
	6.3.6	Summary	640

| | 6.4 | Principles of Nanomaterial Toxicology | 641 |

Thomas Gebel

	6.4.1	Introduction	641
	6.4.2	Toxicology	643
	6.4.3	Summary	648
6.5	Endocrine Active Compounds		649

Volker Strauss and Bennard van Ravenzwaay

	6.5.1	Introduction	649
	6.5.2	Thyroid Hormone Affecting Compounds	651
	6.5.3	Sex Hormones	654
	6.5.4	Low-dose, Non-monotonic Dose-effect Relation and Additive Effects	659
	6.5.5	Summary	659
6.6	Assessment of Xenoestrogens and Xenoantiandrogens		661

Helmut Greim

	6.6.1	Introduction	661
	6.6.2	Modes of Action and Testing	662
	6.6.3	A Weight of Evidence Approach and Future Improvements	666
	6.6.4	Limited Evidence for Endocrine Disruption	667
	6.6.5	Summary	667
6.7	Solvents		671

Wolfgang Dekant

	6.7.1	Introduction	671
	6.7.2	Toxicology of Selected Solvents	672
	6.7.3	Hydrocarbons	677
	6.7.4	Aliphatic Alcohols	680
	6.7.5	Summary	685
6.8	Noxious Gases		686

Kai Kehe and Horst Thiermann

	6.8.1	Introduction	686
	6.8.2	Airborne Systemic Poisons	686
	6.8.3	Respiratory Tract Irritants	692
	6.8.4	Irritant Gases	694
	6.8.5	Summary	698
6.9	Fragrance Materials		699

Anne Marie Api

	6.9.1	Introduction	699
	6.9.2	Evaluation of Toxicity	699
	6.9.3	A Tiered Approach to the Risk Assessment of Fragrances	700
	6.9.4	Summary	702
6.10	Pesticides		703

Roland Alfred Solecki and Vera Ritz

	6.10.1	Introduction	703
	6.10.2	Toxicity of Selected Pesticidal Active Substances	706
	6.10.3	Fungicides	709
	6.10.4	Insecticides	711
	6.10.5	Substances of Biological Origin	713
	6.10.6	Insect Growth Regulators	714
	6.10.7	Other Pesticidal Active Substances	715

	6.10.8	Regulatory Toxicology of Pesticidal Active Substances	716
	6.10.9	Toxicological Endpoints	717
	6.10.10	Classification and Cut-off Criteria	718
	6.10.11	Human Health Risk Assessment	718
	6.10.12	Summary	720
6.11	Polycyclic Aromatic Hydrocarbons		722

Heidrun Greim and Hermann Bolt

	6.11.1	Introduction	722
	6.11.2	Physico-chemical Properties	722
	6.11.3	Occurrence in the Environment and in the Workplace	723
	6.11.4	Toxicity	724
	6.11.5	Mechanisms of Action	729
	6.11.6	Evaluation by National and International Organisations	730
	6.11.7	Toxicity Equivalency Factors for PAHs	731
	6.11.8	Summary	732
6.12	Diesel Engine Emissions		733

Heidrun Greim

	6.12.1	Introduction	733
	6.12.2	Contents of Diesel Engine Emissions	734
	6.12.3	Toxicokinetics	735
	6.12.4	Toxicity	736
	6.12.5	Mechanisms of the Carcinogenic Effects of Diesel Engine Emissions	738
	6.12.6	The Exposure of Humans to Diesel Engine Emissions	740
	6.12.7	Evaluation by International Organizations	740
	6.12.8	Summary	741
6.13	Animal and Plant Toxins		742

Thomas Zielker

	6.13.1	Introduction	742
	6.13.2	Animal Toxins	742
	6.13.3	Plant Toxins	751
	6.13.4	Summary	759

Glossary of Important Terms in Toxicology — **761**

Index — **769**

List of Contributors

Dr. Ilse-Dore Adler   Neufahrn, Germany

Dr. Anne Marie Api   Science Fellow, Human Health Sciences, Research Institute for Fragrance Materials, Inc., New Jersey, USA

Dr. Rüdiger Bartsch   Karlsruhe Institute of Technology (KIT), Institute for Applied Biosciences (IAB), Department of Food Chemistry and Toxicology, Freising, Germany

Dr. Ruth Bevan   IEH Consulting Ltd, Rectory Farm, Marston Trussell, Market Harborough, Leicestershire, UK

Professor Dr. Brunhilde Blömeke   Trier University, Faculty VI Department of Environmental Toxicology, Trier, Germany

Professor Dr. Hermann M. Bolt   IfADo Leibniz Research Centre for Working Environment and Human Factors, Dortmund, Germany

Professor Dr. Paul J.A. Borm   Born Nanoconsult Holding BV, CV Meerssen, The Netherlands

Professor Dr. Peter Calow   Humphrey School of Public Affairs, University of Minnesota, Minneapolis, MN, USA

Professor Dr. György Csanády   Formerly: GSF Research Center for Environmental Health, Institute of Toxicology, Oberschleißheim, Germany

Professor Dr. Wolfgang Dekant   University of Würzburg, Department of Toxicology, Würzburg, Germany

Professor Dr. Daniel Dietrich   University of Konstanz, Faculty of Biology, Human and Environmental Toxicology, Konstanz, Germany

Professot Dr. Gunter P. Eckert   Justus Liebig University, Institut for Nutritional Sciences, Gießen, Germany

Professor Dr. Thomas Efferth   Johannes Gutenberg University, Institute of Pharmacy and Biochemistry, Mainz, Germany

Dr. Johannes Filser   Helmholtz Zentrum München, Institute of Molecular Toxicology and Pharmacology, Oberschleißheim, Germany

Professor Dr. Valery E. Forbes University of Minnesota, Department of Ecology, Evolution and Behaviour, St. Paul, MN, USA

Professor Dr. Thomas Gebel Federal Institute for Occupational Safety and Health, FB4, Dortmund, Germany

Professor Dr. Thomas Göen Friedrich-Alexander-Universität Erlangen-Nürnberg, Arbeits-, Sozial- und Umweltmedizin, Erlangen, Germany

Dr. Heidrun Greim Karlsruhe Institute of Technology (KIT), Institute for Applied Biosciences (IAB), Department of Food Chemistry and Toxicology, Freising, Germany

Professor Dr. Helmut Greim Technical University of Munich, Toxikology and Environmental Hygiene, Freising, Germany

Dr. Peter Griem Symrise AG, Director Toxicology, Holzminden, Germany

Professor Dr. Paul T.C. Harrison IEH Consulting Ltd, Rectory Farm, Marston Trussell, Market Harborough, Leicestershire, UK

Professor Dr. Thomas Hartung Johns Hopkins University, Bloomberg School of Public Health, Baltimore, MD, USA

Professor Dr. Andrea Hartwig Karlsruhe Institute of Technology (KIT), Institute for Applied Biosciences (IAB), Department of Food Chemistry and Toxicology, Karlsruhe, Germany

Professor Dr. Jan G. Hengstler IfADo Leibniz Research Centre for Working Environment and Human Factors, Dortmund, Germany

Professor Dr. Franz Hofmann Technical University of Munich, Institute of Pharmacology and Toxicology, München, Germany

Philip Holmes IEH Consulting Ltd, Rectory Farm, Marston Trussell, Market Harborough, Leicestershire, UK

Dr. Gunnar Jahnke Karlsruhe Institute of Technology (KIT), Institute for Applied Biosciences (IAB), Department of Food Chemistry and Toxicology, Karlsruhe, Germany

Professor Dr. Bernd Kaina University Medical Center Mainz, Institute of Toxicology, Obere Mainz, Germany

Dr. Kai Kehe Sanitätsakademie der Bundeswehr, Abt. Medizinischer ABC-Schutz, München, Germany

Professor Dr. Ines Lanzl Department of Ophthalmology, Klinikum Rechts der Isar, München, Germany

Dr. Jörg Lebsanft Federal Ministry for the Environment, Nature Conservation, Building and Nuclear Safety, Bonn, Germany

Dr. Werner Lilienblum Formerly: Niedersächsiches, Landesgesundheitsamt, Hannover, Germany

Dr. Hans-Jörg Martus Novartis Institutes for BioMedical Research, Muttenz (Basel), Switzerland

Professor Dr. Walter E. Müller Goethe University Frankfurt, Department Pharmacology Biocenter, Frankfurt, Germany

Dr. Dennis J. Paustenbach Natural Resources and Health Sciences Division Cardno, San Francisco, CA, USA

Dr. Bennard van Ravenzwaay BASF SE, GV/T – Z470, Experimental Toxicology and Ecology, Ludwigshafen, Germany

Dr. Vera Ritz The German Federal Institute for Risk Assessment (BfR), Department Pesticides Safety, Berlin, Germany

Professor Dr. Karl K. Rozman Formerly: University of Kansas Medical Center, Department Pharmacology, Toxicology, Therapeutics, Institute of Pharmacy, Kansas City, USA

Professor Dr. Jens Schlossmann University of Regensburg, Department of Pharmacology and Toxicology, Institute of Pharmacy, Regensburg, Germany

Dr. Gerlinde Schriever-Schwemmer Karlsruhe Institute of Technology (KIT), Institute for Applied Biosciences (IAB), Department of Food Chemistry and Toxicology, Freising, Germany

Dr. Florian Schulz Fraunhofer Institute for Toxicology and Experimental Medicine ITEM, Chemical Risk Assessment, Hannover, Germany

Professor Dr. Michael Schwenk Tübingen, Germany

Professor Dr. Andreas Seeber Institut für Arbeitsphysiologie an der Universität Dortmund, Dortmund, Germany

Professor Dr. Robert Snyder Rutgers University, Ernest Mario School of Pharmacy, EOHSI - Toxicology, Piscataway NJ, USA

Dr. Roland Alfred Solecki The German Federal Institute for Risk Assessment (BfR), Department Pesticides Safety, Berlin, Germany

Professor Dr. Horst Spielmann Zerbster Strasse Berlin, Germany

Dr. Lesley A. Stanley Consultant in Investigative Toxicology, Linlithgow, West Lothian, UK

Dr. Volker Strauss BASF SE, Experimental Toxicology and Ecology, Ludwigshafen, Germany

Professor Dr. Laura Suter-Dick The University of Applied Sciences Northwestern Switzerland, Institute for Chemistry and Bioanalytics, Muttenz, Switzerland

Professor Dr. Horst Thiermann Institut für Pharmakologie und Toxikologie der Bundeswehr, München, Germany

Professor Dr. Jürgen Timm University of Bremen, FB3, Institute of Statistics, Bremen, Germany

Professor Dr. Kurt Ulm Technical University of Munich, Institut of Medical Informatics, Statistics and Epidemiology, München, Germany

Dr. Klaus-Michael Wollin Niedersächsisches, Landesgesundheitsamt, Hannover, Germany

Professor Dr. Thomas Zilker Franziskanerstrasse München, Germany

Short CVs of Authors

Adler, Ilse-Dore Ilse-Dore Adler studied biology, chemistry and geography in Berlin and Tübingen to become a high school teacher. She graduated in Berlin (1965) and did her PhD thesis work (1965–1968) in Heidelberg at the Institute of Human Genetics on the topic of the genetic effects of caffeine on the germ cells of mice. After a post-doctoral period (1970–1971) at the Children's Cancer Research Foundation, Harvard Medical School, Boston, USA, she joined the Institute of Mammalian Genetics of the GSF, Neuherberg, Germany (now Helmholz-Zentrum München) and worked there on germ cell mutagenesis topics until her retirement in 2004. She served as Secretary and then President of the EEMS, was Adjunct Professor (1989–2004) at the Texas University Medical School, Galveston, USA) and served as advisor and reviewer of scientific projects of NIEHS.

Api, Anne Marie Anne Marie Api is Vice President, Human Health Sciences at the Research Institute for Fragrance Materials, Inc. (RIFM). Dr. Api has used her advanced knowledge of fragrance ingredient safety to establish a quality record of managing fragrance ingredient safety at RIFM. She has authored over 100 scientific publications and is a member of numerous scientific organizations. She received for the 2018 Toxicology Forum Philippe Shubik Distinguished Scientist Award.

Bartsch, Rüdiger Dr. Rüdiger Bartsch is certified expert in toxicology (DGPT) and has worked for more than 20 years in the scientific secretariat of the Permanent Senate Commission for the Investigation of Health Hazards of Chemical Compounds in the Work Area of the German Research Foundation. His main work is the derivation of occupational exposure levels for industrial chemicals, focusing on the underlying toxic mechanism. Dr. Bartsch is a biologist and received his PhD from the Technical University of Munich.

Bevan, Ruth Dr. Ruth Bevan has considerable expertise in toxicology and human health risk assessment in areas connected with environmental or occupational exposure to chemicals. She has published on a broad range of environment and health issues, notably in the field of occupational cancer burden and biomonitoring of environmental exposures (including via consumer articles, drinking water, food and air).

Blömeke, Brunhilde Brunhilde Blömeke received her PhD in biology and started her academic career as a molecular toxicologist. At present she holds the position of Professor for Environmental Toxicology at the University of Trier, Germany. Her primary research interest is focused on the molecular mechanisms of chemicals and the development of *in vitro* methods for their detection and risk assessment. She serves as member of several advisory committees, including the DFG

Permanent Senate Commission for the Investigation of Health Hazards of Chemical Compounds in the Work Area (German MAK commission).

Bolt, Hermann   Professor Dr. Hermann Bolt studied medicine and biochemistry at the universities of Cologne and Tübingen, and started his academic career in the Institute of Toxicology at the University of Tübingen. In 1979 he became head of the Section of Toxicology at the Institute of Pharmacology, University of Mainz and in 1982 Professor of Toxicology and Occupational Medicine at the University of Dortmund with the position of Director of the Institute of Occupational Health of the Leibniz Research Center for Working Environment and Human Factors. Dr. Bolt is a general toxicologist with major research experience in the metabolism and toxic mechanisms of steroids and industrial chemicals. Dr. Bolt has served in many national and international committees and chaired the Scientific Committee on Occupational Exposure Limits of the General Directorate Employment of the European Commission.

Borm, Paul   Paul Borm holds degrees in biochemistry (MSc) and pharmacology (PhD), and has focused his academic career (1984–2004) on inhalation toxicology and risk assessment. Since 2011 Paul Borm has been CSO and shareholder of Nano-Imaging (Aachen), as well as managing director of Nanoconsult. He supports both start-ups in business development and advises major international companies with regard to risk management and product stewardship. In addition he holds a professorship at the University of Düsseldorf and teaches medical imaging.

Calow, Peter   Peter Calow is a professor at the Humphrey School of Public Affairs, University of Minnesota and has held previous professorial positions at the universities of Nebraska-Lincoln (USA), Sheffield (UK), and Roskilde (Denmark). His current research is on the interface between science and public policy, with special interest in the better use of risk assessment in environmental policy and regulation. He has written more than 300 articles and edited/authored more than 20 books.

Csanády, György András   György András Csanády (1958–2011) studied chemistry at the Eötvös-Loránd-University, Budapest, where he received his PhD. After research positions in Austria and the chemical industry at the Institute of Toxicology, North Carolina he joined the Institute of Toxicology at the federal Research Center in Munich, Germany. His research activities focused on the development of physiological based toxicokinetic models for the risk assessment of chemicals. He held the position of Associate Professor at the Technical University of Munich.

Dekant, Wolfgang   Dr. Wolfgang Dekant has been Professor of Toxicology at the Department of Pharmacology and Toxicology of the University of Würzburg since 1992. He is involved in international toxicology research projects as well as advisory services and collaborates with the relevant authorities and organizations, including the European Union, the World Health Organization, the US Environmental Protection Agency, the Federal Environment Agency, and federal ministries in Germany. He is also responsible for numerous scientific publications as an author and editor.

Dietrich, Daniel   Professor Daniel Dietrich, PhD, ERT, FATS, studied biochemistry and biology at the University of Zurich and ETH Zurich, gaining a Master of Science (1984) and a PhD in toxicology (Institute of Toxicology ETH Zurich, 1988). He is head of the Institute of Human and Environmental Toxicology at the University of Konstanz, elected external expert "Life Sciences for Human Well-being" for the European Parliament, ad hoc expert for the High Level Group of the European Commission's Scientific Advice Mechanism, and Academic Advisor for the European Risk Forum.

Eckert, Gunter Peter Professor Dr. Gunter Peter Eckert is full-professor for nutrition in prevention and therapy at the Justus-Liebig-University of Giessen. He studied food chemistry and environmental toxicology, and holds a PhD in pharmacology. He is a trained pharmacologist at the Goethe-University in Frankfurt, where he is responsible for teaching toxicology.

Efferth, Thomas Professor Dr. Thomas Efferth is chair of the Department of Pharmaceutical Biology, Institute of Pharmacy and Biochemistry, Johannes Gutenberg University, Mainz, Germany. He is honorary professor at the Northeast Forestry University, Harbin, and at the Zhejiang Chinese Medical University, Hangzhou, China. Moreover, he is visiting professor at the Zhejiang University of Science and Technology, Hangzhou, China. He has published over 490 PubMed-listed papers in the field of cancer research, pharmacology, and natural products, and is editor-in-chief of the journal *Phytomedicine*. He is scientific advisory board member of the German Pharmaceutical Society and several other institutions.

Filser, Johannes G. Johannes G. Filser, PhD, studied biochemistry at the University of Tübingen; habilitation thesis at the University of Mainz in 1987. His main research topics at the Institutes of Toxicology at the Universities of Tübingen and Mainz and later on at the federal Research Center (former GSF) were toxicokinetics and metabolism of xenobiotics in laboratory animals and humans. Between 1993 and 2012 he served as associate professor at the Technical University of Munich.

Forbes, Valery Valery Forbes is Dean of the College of Biological Sciences at the University of Minnesota, a position she has held since July 2015. A marine biologist by training, Dr. Forbes' research aims to improve the science underlying risk assessments of chemicals and other stressors on ecological systems.

Gebel, Thomas Professor Dr. Thomas Gebel is head of the toxicology unit at the German BAuA, the Federal Institute for Occupational Safety and Health. He deals with the toxicological evaluation of workplace chemicals with respect to both hazard classification and occupational exposure limit setting. He is associate professor for environmental toxicology at the Technical University of Dortmund.

Göen, Thomas Professor Dr. Thomas Göen is a chemist and expert in the human metabolism of hazardous substances as well as analytical procedures for the determination of such substances and their metabolites in biological matrices. He is a member of the DFG Senate Commission for the Investigation of Hazardous Compounds in the Work Area and the Human Biomonitoring Commission of the German Federal Environment Agency. He is also engaged in the organization of the German External Quality Assessment Scheme, which offers the proficiency control of laboratories worldwide for about 200 biomonitoring parameters.

Greim, Heidrun Heidrun Greim has a PhD in chemistry and has been involved in toxicology for many years. As head of the scientific secretariat of the GDCH Advisory Committee on Existing Chemicals and the DFG Permanent Senate Commission for the Investigation of Health Hazards of Chemical Compounds in the Work Area (German MAK Commission), the focus of her work is the hazard as well as the risk assessment of chemicals and the regulatory toxicology.

Greim, Helmut Helmut Greim is Professor Emeritus at the Technical University of Munich. After receiving his MD in Pediatrics he joined the Institute of Pharmacology of the Freie University at Berlin. In 1955 he moved to the Institute of Toxicology at the University of Tübingen and after holding a Research Associate Professorship at the Department of Pathology at the Mount

Sinai School of Medicine, New York between 1970 and 1973 he became director of the Institute of Toxicology of the Gesellschaft für Strahlen- und Umweltforschung in Munich, Germany in 1975 and between 1987 and 2003 was Director and Chairman of the Institute of Toxicology of the Technical University of Munich. He has served as member or chairman of many national and international committees such as the Health Effects Institute, Boston, the Health and Environmental Safety Institute, Washington DC, the Research Institute of Fragrance Materials (Woodcliff NJ, chair 2000–2008), Scientific Committee on Health and Environmental Risks (chair 2004–2012), the Risk Assessment Committee of the European Chemicals Agency and the Commission for the Investigation of Health Hazards of Chemical Compounds in the Work Area of the German Research Foundation (chair 1992–2007). Dr. Greim's major research interest has been the metabolism and mechanisms of toxic chemicals, alternative methods to animal testing as well as training in toxicology.

Griem, Peter Dr. Peter Griem studied biochemistry in Berlin and Tübingen, Germany and did his PhD in the area of immunotoxicology at Düsseldorf University, completing postgraduate training in toxicology (DGPT, ERT, DABT). After working with FoBiG GmbH, Wella AG, and Clariant AG Dr. Griem joined Symrise AG in 2011 as Head of Toxicology. Throughout his professional career the interaction of chemicals with the immune system has stayed one of his focal points of interest.

Harrison, Paul Professor Paul Harrison has an impressive depth and range of experience in the field of toxicological risk assessment, having for many years investigated the linkages between environmental quality and human health. His principal areas of expertise are dust and fiber toxicology, indoor air quality, endocrine disruption and chemical risk assessment. He has held positions on numerous expert groups and committees and has published many scientific papers and articles on a broad range of toxicological issues.

Hartung, Thomas Thomas Hartung, MD, PhD, is the Doerenkamp-Zbinden-Chair for Evidence-based Toxicology with a joint appointment for Molecular Microbiology and Immunology at Johns Hopkins Bloomberg School of Public Health, Baltimore. He holds a joint appointment as Professor for Pharmacology and Toxicology at University of Konstanz, Germany. He also is Director of Centers for Alternatives to Animal Testing (CAAT). CAAT hosts the secretariat of the Evidence-based Toxicology Collaboration, the Good Read-Across Practice Collaboration, the Good Cell Culture Practice Collaboration, the Green Toxicology Collaboration and the Industry Refinement Working Group. As PI, he heads the Human Toxome project funded as an NIH Transformative Research Grant. He is the former Head of the European Commission's Center for the Validation of Alternative Methods, Ispra, Italy, and has authored more than 500 scientific publications.

Hartwig, Andrea Andrea Hartwig is Full Professor and Chair of the Department of Food Chemistry and Toxicology at the Karlsruhe Institute of Technology, Karlsruhe, Germany. Her main research area focuses on the impact of carcinogenic metal compounds, including metal-based nanoparticles, as well as essential trace elements on genomic stability, with special emphasis on DNA damage induction and DNA damage response systems. Since 2007 she has been chair of the Permanent Senate Commission for the Investigation of Health Hazards of Chemical Compounds in the Work Area (German MAK commission).

Hengstler, Jan G. Jan G. Hengstler studied medicine, began his research work at the institute of Toxicology in Mainz, became Professor of Molecular Toxicology at the Leipzig University, and currently is director of the Leibniz Research Center in Dortmund. His research interests include liver physiology and hepatotoxicity and he was among the first to introduce systems modeling and functional imaging to better understand mechanisms in toxicity.

Hofmann, Franz   Professor Dr. Franz Hofmann studied medicine at the Universities of Heidelberg, Munich and Berlin. He started his academic career at the Institute of Pharmacology at the University of Heidelberg, worked between 1973 and 1975 at the Department of Biological Chemistry, in the Medical School at the University of California, Davis, USA and after returning to pharmacology in Heidelberg he became Chairman and Professor of Physiological Chemistry at the University of the Saarland in 1985. Between 1990 and his retirement in 2007 he served as chairman and Professor of Pharmacology and Toxicology at the Technical University of Munich. His research activities focused on the function, distribution and mechanisms of the cAMP- and cGMP-dependent protein kinases. He also paid special attention to cardiac and smooth muscle function, and carried out experiments to understand the role and function of the cAMP- and cGMP-dependent protein kinases in these tissues.

Holmes, Philip   Philip Holmes has over 40 years' experience as a toxicologist and chemical risk assessor, having worked in pharmaceutical and agrochemical development, academia, and consultancies. During this time he has studied a wide range of chemical classes in relation to the risk posed to humans from occupational or environmental exposure, as well as the application of toxicological and epidemiological data to the socioeconomic analysis of chemical regulation options.

Jahnke, Gunnar   Gunnar Jahnke studied food chemistry and did his PhD thesis in the field of metal toxicology and DNA repair in the group of Andrea Hartwig at the Technical University of Berlin, Germany. Since 2007 he has been a member of the scientific secretariat of the Permanent Senate Commission for the Investigation of Health Hazards of Chemical Compounds in the Work Area (German MAK Commission). He is a European Registered Toxicologist.

Kaina, Bernd   Professor Bernd Kaina obtained his PhD in genetics in 1976. He completed his postdoctoral training at the Institute of Genetics in Gatersleben (Germany), the Department of Molecular Biology in Leiden (The Netherlands), the German Cancer Research Center in Heidelberg and, as a Heisenberg fellow, at the Department of Genetics at the Nuclear Research Center in Karlsruhe, Germany. In 1993 he obtained a full professorship at the Institute of Toxicology of the University in Mainz, and since 2004 has acted as director of the Institute. His working fields are DNA repair and damage signaling, regulation of cell death, and mechanisms of carcinogenesis.

Kehe, Kai   Colonel (MC) PD Dr. Kai Kehe received his doctorate in medicine from the Technical University of Munich in 1991 and was a postdoctoral fellow at the Bundeswehr Institute of Pharmacology and Toxicology and the Walther-Straub Institute of Pharmacology and Toxicology, Ludwig-Maximilians-University of Munich. Dr. Kehe specializes in pharmacology and toxicology, and is lecturer and assistant professor in pharmacology and toxicology at the Ludwig-Maximilians-University of Munich. In 2016 he earned a Master of Business Administration for medical doctors from the University of Neu-Ulm, Germany. Dr. Kehe is currently head of the Medical CBRN Defense Division at the Bundeswehr Medical Academy in Munich, Germany.

Lanzl, Ines   Professor Dr. med. Ines Lanzl is currently teaching ophthalmology at the Medical School of the Technical University of Munich and is practicing ophthalmology in the city of Prien, Bavaria, Germany. Her research and clinical focus is on a multidisciplinary approach to ophthalmic pathology with a special interest in ocular surface, immunology, and perfusion as well as increasing disease awareness and patient compliance.

Lebsanft, Jörg Jörg Lebsanft studied biochemistry at the University of Tübingen and received a PhD at the former Gesellschaft für Strahlen- und Umweltforschung, Institute of Toxicology, Munich. Since 1988 he has worked in the chemicals unit of the Federal Ministry for the Environment, Nature Conservation and Nuclear Safety, Bonn, Germany. From 1999 to 2002 he was seconded to the European Commission, DG Environment, Brussels and from 2007 to 2012 he worked for the European Chemicals Agency in Helsinki, Finland.

Lilienblum, Werner Dr. Werner Lilienblum, a chemist by training, received his PhD at the University of Marburg, Germany, became a certified toxicologist (Fachtoxikologe DGPT) at the University of Göttingen in 1983 and is a EUROTOX Registered Toxicologist. He headed the State Authority for Occupational Safety and the Environment, Hannover and Hildesheim, Germany. Since retirement he has been a toxicology consultant and independent researcher. He served as a member of the Scientific Committee for Consumer Safety of the European Commission, sat on many scientific national boards, and served as a member of the German Society of Toxicology.

Martus, Hans-Joerg Hans-Joerg Martus was trained in biology and toxicology at the University of Mainz, Germany and Harvard University, USA. Currently he is the Global Head of Genetic Toxicology and Photosafety and a Project Toxicologist at the Novartis Institutes for BioMedical Research in Basel, Switzerland. He is the author of multiple publications and book chapters, and teaches various educational programs.

Müller, Walter E. Dr. Walter E. Müller is Professor Emeritus at the Department of Pharmacology, Biocenter Goethe University Frankfurt. He got his PhD in pharmacology at the University of Mainz and became Associate Professor at the Central Institute of Mental Health Mannheim, Heidelberg University in 1983. From 1997 to 2013 he was Full Professor of Pharmacology at the Biocenter University Frankfurt. He has received several professional awards, including the Fritz Külz Preis of the German Pharmacological society, the award in Psychopharmacology of the AGNP, and the felicitation lecture at the Neurocon 2017 meeting. He was awarded fellowship of the Amercan College of Neuropsychopharmacology, and he is an honorary member of the Austrian Society of Biological Psychiatry. He has published more than 600 papers and book chapters.

Paustenbach, Dennis Dennis Paustenbach has been the President of ChemRisk for more than 20 years. About 5 years ago, ChemRisk merged with Cardno, an Australian environmental services firm. Dr. Paustenbach has a BS in chemical engineering from the Rose-Hulman Institute of Technology (Terre Haute), an MS in industrial hygiene and toxicology from the University of Michigan (Ann Arbor) and a PhD in toxicology from Purdue University. He did postgraduate work at the Wright-Patterson Air Force Base and the Harvard School of Public Health. He has authored more than 250 peer-reviewed publications, about 50 book chapters, and two books on risk assessment that are used by many graduate schools of public health and medicine. His specialty is environmental and occupational toxicology.

Ritz, Vera Dr. Vera Ritz has worked at the German Federal Institute for Risk Assessment since 2006 and heads the Steering and Overall Assessment Biocides unit in the Department of Safety of Pesticides. Her expertise is the toxicological evaluation of biocides and plant protection products as a European Registered Toxicologist. Her background is a diploma in biology and a PhD in toxicology and genetics.

Rozman, Karl Karl Rozman (1945–2017) studied organic and pharmaceutical chemistry at the University of Innsbruck and received his PhD in 1973. He then worked in the Institutes of Ecological Chemistry and of Toxicology of the former Gesellschaft für Strahlen- und Umweltforschung, first in Munich and since 1974 in the branch at the Albany Medical College in Alamogordo, New Mexico. In 1981 he joined the Department of Phamacology, Toxicology and Therapeutics of Kansas Medical Center in Kansas City, where he was appointed professor in 1986. His main field of work was the toxicology of halogenated hydrocarbons such as TCDD.

Schlossmann, Jens Jens Schlossmann studied chemistry at the Universities of Tübingen and Munich. For his diploma thesis he worked at the Max-Planck-Institute for Biochemistry (Martinsried) in the Department of Cell Biology. In 1990 he joined the Institute of Physiological Chemistry of Munich University and between 1995 and 2007 worked at the Institute of Pharmacology and Toxicology of the Technical University of Munich investigating substrate proteins of cGMP-dependent protein kinase. In 2007 he was appointed Professor for Pharmacology and Toxicology at the Institute of Pharmacy of the Regensburg University, Germany. His research fields comprise NO/cGMP signaling mechanisms in cardiovascular, renal and immunological functions, and he also studies the role of cyclic pyrimidinic nucleotides. He is editor and reviewer of several scientific journals.

Schriever-Schwemmer, Gerlinde Gerlinde Schriever-Schwemmer studied biology and mathematics with a focus on human genetics and did her PhD thesis in the field of immunology and carcinogenicity at the DKFZ, German Cancer Research Center, Heidelberg. Thereafter she worked in the field of mammalian mutation research in the group of Ilse-Dore Adler at the Helmholtz Center, Munich, the German research center for environmental health. Since 2000 she has been a member of the scientific secretariat of the Permanent Senate Commission for the Investigation of Health Hazards of Chemical Compounds in the Work Area (German MAK Commission).

Schulz, Florian Dr. Florian Schulz holds a PhD in biochemistry. During his doctoral thesis in the Institute of Toxicology at Hannover Medical School he was engaged with the cellular mode of action of bacterial toxins. Afterwards Dr. Florian Schulz joined the Department of Chemical Risk Assessment at the Fraunhofer Institute for Toxicology and Experimental Medicine in Hannover, Germany. Since then he has worked as a scientist in the field of inhalation toxicology, including the derivation of occupational exposure limit values for particles and fibers.

Schwenk, Michael Michael Schwenk studied biochemistry and medicine at the University of Tübingen, Germany. He specializes in pharmacology, toxicology and environmental medicine, and is Member of IUPAC Division on Chemistry and Human Health.

Snyder, Robert Robert Snyder is a trained chemist and biochemist and received his PhD at the College of Medicine, Syracuse, NY. After a postdoctoral fellowship in the Department of Pharmacology, University of Illinois College of Medicine he joined the Department of Pharmacology at the Jefferson Medical College, Philadelphia, being finally promoted to Professor of Pharmacology. Between 1981 and 2010 he held the Professorship of Toxicology at Rutgers, State University of New Jersey. During this time he served as Professor and Chairman of Pharmacology and Toxicology, Director of the Health Effects Assessment Division at the New Jersey Institute of Technology, and Director of the Division of Toxicology of the Environmental and Occupational Health Institute. Dr. Snyder's main research interest is the metabolism and toxic mechanism of benzene. He has been the chief organizer of the International Symposia on Benzene 1995 in Piscataway, NJ, 1998 in Ottawa, 2005 and 2009 in Munich, Germany, and at the New York Academy of Sciences in 2012. He served as member of several national advisory committees (USEPA, FDA, NAS/NRC) and was President of the American College of Toxicology.

Solecki, Roland Alfred Dr. Roland Alfred Solecki is a biologist and toxicologist by training and Head of Department of Pesticide Safety in the German Federal Institute for Risk Assessment. He is involved in the toxicological testing and human health risk assessment of pesticides. Dr. Solecki was a World Health Organization panel member in the JMPR and is currently a member of the Scientific Committee of EFSA.

Spielmann, Horst Horst Spielmann is Professor for Regulatory Toxicology at the Freie Universität Berlin, Germany. His book *Drugs in Pregnancy and Lactation*, with various co-authors, was first published in 1987 in German and has now appeared in seven German editions, three English editions, one edition each in Russian and Chinese. He is one of the international promoters of alternative test methods in toxicology and an Honorary Member of the following societies: Japanese Society for Alternatives to Animal Experiments (2007), European Society for Toxicology In Vitro (2012), Society for Dermopharmacy (2013) and the Chinese In Vitro Science Academic Committee (2017).

Stanley, Lesley Dr. Lesley Stanley is a toxicologist with over 30 years' experience in assessing the effects of chemicals on human health. Since May 2005 she has been a freelance consultant in investigative toxicology, advising clients in academia, government and industry on experimental strategy and assisting with literature reviews, grant applications and report/manuscript preparation. Her previous experience includes six years as a Senior (latterly Principal) Lecturer in Biomedical Science at De Montfort University, Leicester, UK as well as at the University of Oxford, the Medical Research Council Toxicology Unit and the National Institute of Environmental Health Sciences, North Carolina, USA.

Strauss, Volker Volker Strauss is certified specialist of clinical pathology. He has 15 years of toxicology expertise in the pharmaceutical and chemical industry. Presently he is Senior Scientist in the Department of Experimental Toxicology and Ecology, BASF SE, Ludwigshafen, Germany.

Suter-Dick, Laura Laura Suter-Dick is European Registered Toxicologist and Professor for Molecular Toxicology in the School of Life Sciences at the University of Applied Sciences Northwestern Switzerland. She acquired more than 20 years of research experience in the pharmaceutical industry before moving to academia in 2012. Her research included the fields of toxicogenomics and molecular toxicology, applying *in vivo* assays, new technologies, and alternative *in vitro* methods.

Thiermann, Horst Colonel (MC) Professor Dr. Horst Thiermann studied medicine. He started his career in the Bundeswehr Hospital, Munich, Germany in the departments of anaesthesiology and surgery. Thereafter, he moved to the Bundeswehr Institute of Pharmacology and Toxicology. He specialized in Pharmacology and Toxicology at the Walther-Straub-Institute of Pharmacology and Toxicology, Ludwig Maximilians-University, Munich in 1996. In 2002, he completed his advanced studies of clinical pharmacology at MDS Pharma Services, Höhenkirchen-Siegertsbrunn. Since November 2006 he has been director of the Bundeswehr Institute of Pharmacology and Toxicology. In January 2012 he was appointed Professor at the Technical University of Munich.

Timm, Jürgen Jürgen Timm is Professor for Mathematics and Applied Statistics at the University of Bremen, where he founded the Centre of Competence for Clinical Trials and the masters program Medical Biometry/Biostatistics. He has served as leading biostatistician in hundreds of biomedical projects, as a biometrical expert for the federal and state government, and as a temporary WHO advisor. For 20 years he managed the University of Bremen as Rektor (President).

Ulm, Kurt After studying mathematics and information science at the Technical University of Munich Kurt Ulm received a PhD in statistics and later on became Professor at the Institute of Medical Statistics and Informatics. He spent about a year at the University of Washington at Seattle, USA. Dr. Ulm serves as member of the Permanent Senate Commission for the Investigation of Health Hazards of Chemical Compounds in the Work Area (German MAK Commission).

van Ravenzwaay, Bennard Bennard van Ravenzwaay is Associate Professor for Toxicology at the University of Wageningen, the Netherlands. At present he is Senior Vice President of the Department of Experimental Toxicology and Ecology, BASF SE, Ludwigshafen, Germany and Chairman of the Scientific Committee of the European Centre for Ecotoxicology and Toxicology, Brussels, Belgium.

Wollin, Klaus-Michael Klaus-Michael Wollin studied chemistry at the Technical University of Dresden and the University of Rostock. In 1980 he received his doctorate from the University of Rostock, where he worked at the Institute of Public Health from 1979 to 1989. From 1990 to 2004 he headed the Department for Risk Assessment of Contaminated Sites at the State Agency of Ecology Hildesheim-Hannover and 2006 moved to the Centre of Health and Infection Control at the Lower Saxony Agency of Public Health Hannover, Germany. He is a certified toxicologist (DGPT, ERT). He is a member of the German federal advisory bodies Committee on Hazardous Substances (AGS) and Human Biomonitoring Commission, and of the pool of experts on Rapid Risk Assessment at the EU's SCHER. His research focuses on health effects from environmental pollution.

Zilker, Thomas Thomas Zilker studied medicine at the University of Munich, where he received his MD. He specialized in internal medicine, endocrinology, and environmental medicine. After serving as Assistant Medical Director of the Department of Clinical Toxicology at the Technical University of Munich he directed this institution for almost 20 years.

Preface

About 40 years ago the need for trained toxicologists in the German chemical industry prompted Professors Herbert Remmer and Helmut Greim to organize a 3-year toxicology training program for 20 chemists. Using this experience the German Society of Pharmacology and Toxicology developed criteria to receive a certificate designating the "Fachtoxikologe" and initiated a broad training program to provide the information required. Later on the criteria to become a "Certified Toxicologist" were developed by the European Society of Toxicology. It became obvious that a textbook was needed to accompany the classroom work to meet the needs of the students. The first book[*] was published in the German language in 1995 and subsequently in Italian.[**] When time came for a new edition, the publishers, who were interested in expending the market, suggested that a new edition, which could service a broader representation of the community of scholars in toxicology, should be written in English. The editors, Helmut Greim and Robert Snyder, decided to prepare a completely new book to ensure that recent achievements in toxicology were covered and each chapter produced by the faculty contained essential knowledge for a toxicologist or anyone interested in understanding the basics of our discipline. In the meantime the German book was updated and published in 2017.[***] We now present the second edition of the English textbook. Apart from two, all chapters have been rewritten, mostly by new authors, and chapters have been added to cover new areas of toxicological relevance, including general concepts of human health risk assessment, threshold effects for genotoxic carcinogens, the endocrine system, principles of nanomaterial toxicology, pesticides, fragrances, and diesel engine emissions. Since an understanding of the regulations of dangerous materials has become increasingly important, in addition to the US regulations the corresponding EU regulations and the concept of REACH are covered.

This book is intended for people with a broad range of toxicological interests, including both practical and science-based subjects. References at the end of each chapter allow the reader to go beyond this book into more detailed information.

The authors and editors hope that the book proves useful to all users and provides information at a level that will enable them to understand the basic principles of toxicology and to successfully study our discipline.

There are two famous admonitions in toxicology. The first, by Paracelsus, "the dose makes a poison", appears in the introduction. The second has been credited to any of several of our colleagues: "Toxicology can be learned in two lessons, each 10 years long."

---

[*] Toxikologie. Eine Einführung für Naturwissenschaftler und Mediziner, H. Greim and E. Deml (eds), Wiley-Verlag Chemie, Weinheim, 1995.
[**] Tossicologia, H. Greim and E. Deml (eds), Zanichelli, Bologna, 2000.
[***] Das Toxikologiebuch: Grundlagen, Verfahren, Bewertung, Wiley-VCH, Weinheim 2017.

We specifically thank Heidrun Greim and Isabel Schaupp for handling the cumbersome and time consuming final editing during the proof-reading process in cooperation with Hari Sridharan, the Production Editor. They all did an excellent job, which is highly appreciated.

Dedication

This book is dedicated to Herbert Remmer (1919–2003) and John Doull (1922–2017). Herbert Remmer was Professor and Director of the Institute of Toxicology at the University of Tübingen. In Germany, together with Dietrich Henschler in Würzburg he converted toxicology from a mere observational discipline to a research-based branch of medicine and sciences.

John Doull was Professor of Pharmacology and Toxicology at the University of Kansas. In 1981 in his article "The Discipline of Toxicology" in *Fundamental and Applied Toxicology* he questioned "Is it desirable for toxicology to be viewed as a scientifically rigorous discipline? The answer is clearly yes. One of the most important reasons is that if we expect to recruit the bright students to the discipline of toxicology we must strive to be scientifically rigorous and objective."

The contributions of both these men as educators, authors, and editors served to build a solid and stable base for the discipline of toxicology. Their work established a standard of excellence for generations of scientists to come. They are sorely missed.

List of Abbreviations

5-CSRTT   5-choice serial reaction time task

2,4-D   2,4-dichlorophenoxyacetic acid

2,4-DB   4-(2,4-dichlorophenoxy)butyric acid

2,4,5-T   2,4,5-Trichlorophenoxyacetic acid

AAF, 2-AAF   2-acetaminofluorene

AAS   Atomic absorption spectrometry

ACD   Allergic contact dermatitis

ADH   Alcohol dehydrogenase

ADI   Acceptable (allowable) daily intake

ADME   Absorption, distribution, metabolism, excretion

ADP   Adenosine phosphate

Ah   Aryl hydrocarbon

AHH   Aryl hydrocarbon hydroxylase

AhR, AHR   Ah receptor

ALARA   As low as reasonably achievable

ALDH   Aldehyde dehydrogenase

AMH   Anti-Muellerian hormone

AML   Acute myelogenous leukemia

ANFT   2-amino-4-(5-nitro-2-furyl)thiazole

AOEL   Acceptable operator exposure level (for applicators of pesticides)

AP   Apurin

AP-1   Activator protein 1

Apaf-1   Apoptosis protease activating factor-1

APC   Antigen-presenting cells
APS   Adenosine-5-phosphosulfate
AR   Androgen receptor
ARE   Antioxidant/electrophile response element
ARfD   Acute reference dose
ARNT   Aryl hydrocarbon receptor response element
AT   Acetyl transferase
ATE   Acute toxicity estimates
ATM   Ataxia-telangiectasia mutated (kinase)
ATP   Adenosine triphosphate
ATR   Ataxia telangiectasia and Rad3-related (kinase)
ATRIP   ATR-interacting protein
AUC   Area under the curve
BALF   Bronchoaleveolar lavage fluid
B[a]P   Benzo[a]pyrene
BARS   Behavioral assessment and research system
Bax   bcl-2-associated X protein (expressed by p53)
BBN N-butyl-N-(4-hydroxybutyl) nitrosamine
BER   Base excision repair
BMD   Benchmark dose
BMDL   Benchmark dose level
BPA   Bisphenol A
BrdUrd, BrdU   Bromodesoxyuridine
BSEP   Bile salt export pump
BUN   Blood urea nitrogen
CA   Carboanhydrase
CAD   Caspase activated DNase
cAMP   Cyclic adenosine monophosphate
CAR   Constitutive androstane receptor
CASE   Computer automated structure evaluation
CAT   Catalase

CANTAB   Cambridge neurophysiological test automated battery

CDK   Cyclin-dependent kinase

CEO   Cyano ethylene oxide

cGMP   Cyclic guanosine monophosphate

CHK, Chk   Checkpoint kinase

CHMP   Committee for Medicinal Products for Human USE (of EMA)

CK   Creatine kinase

CKMB   Creatine kinase primarily in myocardial cells

CLP   1. Classification, Labelling and Packaging (of Substances and Mixtures); 2. Common lymphoid progenitor

CLRTAP   Convention on long-range transboundary air pollution

CMP   Common myeloid progenitor

CPT   Continuous performance test

CRP   C-reactive protein (parameter for systemic inflammatory processes)

CSE   Chronic solvent-induced encephalopathy

cSNP   Coding SNP (single nulcleotide polymorphism)

CTBP   Cytosolic T3-binding protein

CTL   Cytotoxic T-lymphocytes

CYP   Cytochrome P450

Cys   Cysteine

DAG   Diacylglycerol

DDE p,p'-dichlorodiphenyl dichloroethene

DDR   DNA-damage response

DDT   2,2-bis(chlorophenyl)-1,1,1-trichloroethane

DEHP   Di(2-ethylhexyl)phthalate

DES   Diethylstilbestrol

DHT   Dihydrotestosterone

DIGE   Difference gel electrophoresis

DISC   Death-inducing signalling complex

DMAP   Dimethylaminophenole

DMSO   Dimethylsulfoxide

DMT-1   Divalent metal transporter-1

DNA   Desoxyribonucleic acid

DNEL   Derived no effect level

DPRA   Direct peptide reactivity assay

DRE   Dioxin responsive element

DROSHA Double-stranded RNA-specific endoribonuclease

DSB   Double-strand break

DTH   Delayed type hypersensitivity

EAA   Excitatoric amino acid

EAC   Endocrine-active compound

EBV   Epstein–Barr virus

EC   European Commission

ECG   Electrocardiography

ECHA   European Chemicals Agency

ED   Endocrine disrupter

ED$_{50}$ Effective dose causing the expected effect in 50% of exposed individuals

EE   Ethinylestradiol

EFSA   European Food Safety Authority

EGF   Epidermal growth factor

ELISA   Enzyme-linked immunosorbent assay

EMA, EMEA   European Medicines Agency

EMP   Erythromyeloic progenitor

EMS   Ethyl methanesulfonate

ENTIS   European Network of Teratology Information Service

EoBP   Eosinophil/basophil progenitor

EOGRTS Extended one-generation reproductive toxicity study

EPA   Environmental Protection Agency (USA)

ER   Estrogen receptor, endoplasmic reticulum, excess risk

ESTR   Expanded single tandem repeat (assay)

EU   European Union

EURATOM European Atomic Energy Community

FAD   Flavin adenine dinucleotide

FADD Fas-associated protein with death domain

FANFT N-[4-(5-nitro-2-furyl)-2-thiazolyl]formamide

FAO Food and Agriculture Organization

Fapy Formamidopyridimidine

FasR Fas-receptor (CD95)

FDA Food and Drug Administration (USA)

FELS Fish early life stage

FISH Fluorescense *in situ* hybridization

FMN Flavin mononucleotide

FMO Flavin-dependent monooxygenase

FNT 2-(4-(5-nitro-2-furyl)-2-thiazolyl)hydrazin

FOB Functional observation battery

FPG Formamidopyrimidine DNA glycosylase

FSH Follicle-stimulating hormone

FS-OOH Fatty acid hydroperoxides

G6PD Glucose-6-phosphate dehydrogenase

GABA receptor Gamma-aminobutyric acid receptor

GALT Gut-associated lymphoid tissue

GAP GTPase-activating protein

GBP Granular biopersistent particles

GC Guanylyl cyclase

GCP Good clinical practice

GD Gestation day

GDP Guanosindiphosphate

GEF Guanine nucleotide exchange factor

GGR Global genomic repair

GHS Globally Harmonized System for classification and labelling of chemicals

GI Gastrointestinal

GLP Good laboratory practice

Glu Glutamic acid

Gly Glycine

GMP Granulocyte-macrophage progenitor, good manufacturing practice
GnRH Gonadotropin-releasing hormone
GPCR G-protein-coupled receptor
GPMT Guinea pig maximization test
GPT Glutamate pyruvate transaminase
GSEC Genetic susceptibility to environmental carcinogens
GSH Glutathione (reduced)
GSSG Glutathione (oxidized)
GST Glutathione-S transferase
GTP Guanosine triphosphate
GW Gestation week
GWAS Genome wide association studies
Hb Hemoglobin
HC5 Hazardous concentration 5%
HCB Hexachlorobenzene
HCBD Hexachloro-1,3-butadiene
hCG Human choriogonadotropin
HDI Hexamethylene-diisocyanate
HGF Hepatocyte growth factor
HGPRT Hypoxanthine-guanine phosphoribosyl transferase
HL Half-life (time)
HLA Human leukocyte antigen
HNPCC Heriditary non-polyposis colon cancer
hOGG Human 8-oxoguanine-DNA-glycosilase
HPG axis Hypothalamus–pituitary–gonades axis
HPRT Hypoxanthine-phosphoribosyl transferase
HPT Hypothalamus–pituitary–thyroid axis
HR Homologous recombination
HRE Hormone-responsive element
HRIPT Human repeated insult patch test
HPTE 2,2-bis(*p*-hydroxyphenyl)-1,1,1-trichloroethane
IARC International Agency for Research on Cancer

ICH	International Conference on Harmonization of Technical Requirements for Registration of Pharmaceutical for Human Use
ICL	Interstrand crosslinks
ICRP	International Commission for Ray Protection
IFRA	International Fragrance Association
Ig, IG	Immunoglobulin
IL	Interleukin
ILO	International Labour Organization
ILSI	International Life Science Institute
i.m.	Intramuscular
INR	International normalized ratio; (Parameter to estimate blood clotting)
i.p.	Intraperitoneal
IP_3	Inositol triphosphate
IPCS	International Programme on Chemical Safety of WHO
IPDI	Isophorone-diisocyanate
IQ	Intelligence quotient
ISO	International Organization for Standardization
IUPAC	International Union of Pure and Applied Chemistry
i.v.	Intraveneous
JECFA	Joint FAO/WHO Expert Committee on Food Additives
JMPR	Joint FAO/WHO Meeting on Pesticide Residues
JRC	Joint Research Center of EU (Ispra, Italy)
LC	Liquid chromatography
LC_{50}	Lethal concentration at which 50% of animals die
LD_{50}	Lethal dose at which 50% of animals die
LDH	Lactate dehydrogenase
LDL	Low density lipoprotein
LH	Luteotropic hormone
LLNA	Local lymph node assay
LMPP	Lymphomyeloic progenitor
LO(A)EC	Lowest observable (adverse) effect concentration
LO(A)EL	Lowest observed (adverse) effect level

LPS   Lipopolysaccharide
MA   Mycoplasma arthritidis
MAD   Mutual acceptance of data (between OECD member states)
MAO   Monoamine oxidase
MAP kinases   Mitogen-activated protein kinases
MCL-1   Anti-apoptotic survival factor
MCT   Monocarboxylate transporter
MDI   4,4-methylenediphenylisocyanate
MEHP   Monoethylhexylphthalate
MEP   Megacaryocyte-erythro progenitor
MEST   Mouse ear swelling test
MGMT O^6-metylguanine-DNA-methyltransferase
MHC   Major histocompatibility complex
MIC   Methylisocyanate
MK2   Mitogen-activated protein kinases
MLP   Multilymphoid progenitor
MMR   Mismatch repair
MNU   *N*-Methyl-*N*-nitrosourea
MOE   Margin of exposure
MOS   Margin of safety
MPP   Multipotent progenitor
mPTP   Mitochondrial permeability transition pore
MPTP   1-methyl-4-phenyl-1,2,3,6-tetrahydropyridine
MRL   Maximum residue limit
mRNA   Messenger RNA
MS   Mass spectrometry
MTD   Maximum tolerated dose
MWM   Morris–Water maze
MXC   Methoxychlor
NADH   Nicotinamide dinucleotide (reduced NAD)
NADPH   Nicotinamide dinucleotide phosphate (reduced NADP)
NAG   *N*-acetyl-glucosaminidase

NAPQI	*N*-acetyl-*p*-benzochinonimin
NASH	Nonalcoholic steatohepatitis
NAT	*N*-acetyltransferase
NCTP	Neurobehavioral core test battery
NER	Nucleotide-excision repair
NES	Neurobehavioral evaluation system
NESIL	No expected sensitization induction level
NF-kappaB pathway	Nuclear factor kappaB; pro-inflammatory pathway
NHEJ	Non-homologous end joining
NIEHS	National Institute of Environmental Health Sciences (USA)
NIS	Sodium iodide symporter
NMDA	*N*-Methyl-d-aspartate
NNK	Nicotine-derived nitrosamine ketone, 4-(methylnitrosamine)-1-(3-pyridyl)-1-butanone
NO	Nitrogen monoxide
NO(A)EC	No observed (adverse) effect concentration
NO(A)EL	No observed (adverse) effect level
NOD	Non-obese diabetic
NOGEL	No observed genotoxic exposure level
Noxa	Apoptotic regulator
NP	4-nonylphenol (branched); nasopharyngeal
OATP	Organic anionic transport protein
OECD	Organisation of Economic Co-operation and Development
OP	4-*tert*-octylphenol
OR	Odds ratio
OTIS	Organization of Teratology Information Services
PA	Procainamide
PAH	Polycyclic aromatic hydrocarbon
PAMP	Pathogen-associated molecular pattern
PARP-1	Poly(ADP-ribose)-polymerase 1
PBDE	polybrominated diphenylether
PBPK Model	Physiologically based pharmacokinetic model

PCB Polychlorinated biphenyls
PCDD Polychlorinated dibenzo-*p*-dioxins
PCDF Polychlorinated dibenzofurans
PCNA Proliferating cell nuclear antigen
PCP Pentachlorophenol
PCR Polymerase chain reaction
PDGF Platelet-derived growth factor
PEC Predicted environmental concentration
PEPCK Phosphoenolpyruvate carboxykinase
PET Positron emission tomography
PFOS Perfluerooctanesulfonic acid
PGS Prostaglandin synthase
PHAH Polyhalogenated aromatic hydrocarbons
PhIP 2-amino-1-methyl-6-phenylimidazo-[4,5-b]pyridine
PIC Prior informed consent (EU Regulation, Rotterdam Convention)
PKC Proteinkinase C
PLNA Popliteal lymph node assay
PL-OOH Phospholipid hydroperoxides
p.o. Per os
POD Point of departure; peroxidase
POE Polyethoxylated
Polβ DNA-polymerase β
POP Persistent organic pollutant
PPAR Peroxisome proliferator-activated receptor
PPD *p*-Phenylenediamine
ppm Parts per million
PPP Plant protection product
PR Progesterone receptor
PRR Pattern recognition receptor
PS(L)P Poorly soluble (low toxicity) particles
PT Prothrombin time
PTT Partial thromboplastin time

Puma   Apoptotic regulator

PXR   Pregnane X receptor

RAC   Risk Assessment Committee (ECHA)

Rb   Retinoblastoma (protein)

REACH   Registration, Evaluation, Authorization and Restriction of Chemicals (EU)

RGS   Regulator of G-protein signalling

RIFM   Research Institute for Fragrance Materials (New Jersey, USA)

RISC   RNA-induced silencing complex

RMO-Analysis   Risk management options analysis

RNA   Ribonucleic acid

RNAP   RNA polymerase

ROAT   Repeated open application test

ROS   Reactive oxygen species

RPA   Replication protein A

RR   Relative risk

α-SMA   α-smooth muscle actin

rSNP   Regulatory single nucleotide polymorphism

RTK   Receptor-tyrosine kinase

SAF   Sensitization assessment factor

s.c.   Subcutaneous

SCCS   Scientific Committee on Consumer Safety (EU)

SCE   Sister chromatid exchange

SCOEL   Scientific Committee on Occupational Exposure Limits (EU)

SEAC   Committee for Socio-Economic Analysis (ECHA)

SERM   Specific estrogen receptor modulators

SHBG   Steroid hormone binding globulin

SIEF   Substance Information Exchange Forum (REACH)

SMR   Standardized mortality ratio

SNP   Single nucleotide polymorphism

SPE   Pyrogenic streptococcal exotoxins

SPES   Swedish Performance Evaluation System

SR   Steroid receptor

srSNP   Structural RNA SNP (Single - nucleotide polymorphism)

SSBR   Single-strand break repair

SSD   Species sensitivity distribution

ST   Sulfotransferase

STAR Protein   Steroidogenic acute regulatory protein

SULT   Sulfotransferase

T3   Triiodothyronine

T4   Tetraiodothyronine

TB   Tracheobronchial

TBG   Thyroxin binding protein

TBT   Tributyltin

TCDD   2,3,7,8-Tetrachlorodibenzo-*p*-dioxin

TCDF   2,3,7,8-Tetrachlorodibenzofuran

TCR   Transcription coupled repair

TDI   1. Tolerable daily intake; 2. Toluene diisocyanate

TE   Toxicity equivalence

TEF   Toxicity equivalence factor

TETRAC   Tetraiodothyroacetic acid

TFT   Trifluorothymidine

TG   1. 6-thioguanine; 2. Thyreoglobulin; 3. Test guideline

TGF   Transforming growth factor

THC   Tetrahydrocannabinol

TIMES-SS   Times metabolism simulator platform used for predicting skin sensitization

TK   Thymidine kinase

TME   Terrestrial model ecosystem

TMR   Test method regulation

TNF   Tumour necrosis factor

TOPKAT   Toxicity prediction computer-assisted technology

TPA   12-*O*-tetradecanoylphorbol-13-acetate

TPO   Thyroid peroxidase

TR   Thyroid hormone receptor

TRH   Thyrotropin-releasing hormone

TRIAC   Triiodothyroacetic acid

TSH   Thyroid-stimulating hormone

TTC   Threshold of toxicological concern

TUNEL   Terminal deoxynucleotide transferase-mediated dUTP nick-end labeling

UDGP   Uridine diphosphate glucose

UDPGT, UGT   UDP-glucuronosyl transferase

UDP   Uridine diphosphate

UDS   Unscheduled DNA synthesis

UGT, UDPGT   UDP-glucuronosyl transferase

UN   United Nations

UNECE   United Nations Economic Commission for Europe

UNEP   United Nations Environmental Programme

UNG   Uracil-*N*-glycosylase

UPDRS   Unified Parkinson's Disease Rating Scale

UROD   Uroporphyrinogen decarboxylase

UR   Unit risk

UTP   Uridine triphosphate

VICH   International Cooperation on Harmonization of Technical Requirements for Registration of Veterinary Medicinal Products

VLDL   Very low density lipoprotein

VOC   Volatile organic compounds

VTG   Vitellogenin

WHO   World Health Organization

XO   Xanthine oxidase

XP   Xeroderma pigmentosum

ZEA   Zearalenone

1

Introduction to the Discipline of Toxicology

Helmut Greim and Robert Snyder

"In all things there is a poison, and there is nothing without a poison. It depends only upon the dose whether a poison is poison or not."

Paracelsus (1493–1541)

1.1 Introduction

> The discipline of toxicology is concerned with the health risks of human exposure to chemicals or radiation. According to Paracelsus' paradigm toxicology is charged with describing the adverse effects of chemicals in a qualitative sense, and with evaluating them quantitatively by determining how much of a chemical is required to produce a given response. Taking these two together, we can describe the intrinsic properties of an agent (hazard identification) and we can estimate the amount of the chemical required to produce these properties (risk characterization).

Humans may be exposed to chemicals in the air, water, food, or on the skin. The external dose at which a chemical exerts its toxic effects is a measure of its potency, i.e. a highly potent chemical produces its effects at low doses. Ultimately, the response to the chemical depends upon duration and route of exposure, the toxicokinetics of the chemical, the dose–response relationship, and the susceptibility of the individual. To characterize the risk, the dose–response can be evaluated and the exposure at which the chemical will produce adverse effects identified. It is obvious from this that risk characterization comprises three elements:

- hazard identification, i.e. a description of the agent's toxic potential

- evaluation of the dose response, including information on the concentration above which the agent induces toxic effects to identify the no observable adverse effect level (NOAEL)
- exposure assessment to understand the concentration of the agent in the relevant medium, time, and routes of human exposure.

It is necessary to establish toxicological profiles of each chemical, either pre-existing or newly developed, to ensure that it can be utilized safely either by the public or under specific conditions of use such as in the workplace. Toxicological evaluations may take different forms for new and existing chemicals. In the case of newly developed drugs, pesticides or new chemicals a stepwise procedure starting from structure–activity evaluation and simple *in vitro* and *in vivo* short-term tests and proceeding to life-time testing in experimental animals. Depending on the hazardous potential of the agent, studies can be extended to evaluate toxicokinetics and the toxic mode of action. For existing chemicals the available information is collected and a risk assessment based on exposure data, knowledge of the dose–response relationship, and the mode of action is performed.

The parameters that determine toxic potential and potency are discussed in the following chapters. Here they are briefly discussed to indicate their importance for the risk characterization process, which is presented in detail in Chapter 2.1.

1.2 The Risk Assessment Process

1.2.1 Hazard Identification

Chemicals induce local and/or systemic effects such as embryotoxicity, hepatotoxicity, neurotoxicity, etc. after absorption from the gastrointestinal tract, through the skin or via the lungs. Reactivity, solubility, and metabolism of the chemical, its metabolites, and their distribution within the organism determine the target organ of the critical effects.

Acids or bases can be directly acting agents which cause irritation or corrosion at the site of exposure such as skin, mucous membranes of the eye, the gastrointestinal tract or the respiratory system. However, most chemicals induce systemic effects such as embryotoxicity, hepatotoxicity, neurotoxicity, etc. after absorption from the gastrointestinal tract, through the skin or via the lungs. Depending on exposure concentration and time of exposure, acute or chronic effects may result. Acute intoxication usually occurs in response to large doses. Chronic effects are seen after repeated exposure, during which time the chemical reaches critical concentrations at the target organ, e.g. liver, kidney or central and peripheral nervous system. Histopathological and biochemical changes have been the major parameters used to detect organ toxicity. Increasing availability of sensitive methods in analytical chemistry and molecular-biological approaches including toxicokinetics and the various "omics" (Chapter 4.6) have significantly improved the understanding of the mechanisms by which cellular and subcellular functions are impaired and how the cells respond to toxic insults. This results in a better understanding of toxic mechanisms, species differences and the consequences of exposures at high and low concentrations over different times.

Exposure to some chemicals, such as 2,3,7,8-tetrachlorodibenzo-p-dioxin (TCDD), can result in retention and long-lasting effects even after a single high exposure (Chapter 6.1). This is because TCDD is lipophilic and not well metabolized, which results in very slow elimination. The consequence is accumulation in adipose tissue. In humans the half-life of excretion is about 8 years. In laboratory animals and humans TCDD induces tumors in various organs. Since TCDD does not induce DNA damage or mutations, the carcinogenic effect is considered to have a threshold, i.e. there are doses below which no adverse effects will be observed.

Induction of sensitization and of allergic responses by sensitizing agents are also considered to require to reach a threshold dose, although at very low doses and the NOAELs of these effects are rarely known. When establishing acceptable exposure standards thresholds are not considered to be a property of the dose–response curves for genotoxic carcinogens because so far any genotoxic event is considered irreversible. (For more detailed discussion of this concept see Chapters 2.8 and 2.9.)

1.2.2 Dose Response and Toxic Potency

The Paracelsian admonition teaches us that the occurrence and intensity of toxic effects are dose dependent. This paradigm addresses the concept of threshold effects, which implies knowledge of the dose–response relationship. Animal or human exposure is usually defined as the dose, e.g. mg of the chemical/kg body weight/day. This daily dose may result from oral, inhalation or dermal exposure or as a sum thereof. The external dose leads to a specific internal dose, which depends on the amount absorbed via the different routes and the distribution to the critical target (tissue, organ). Absorption rates via the different routes can vary significantly, although oral and inhalation exposure usually lead to the highest internal dose. For example, about 50% of cadmium in inhaled air, e.g. in tobacco smoke, is absorbed in the lung, whereas cadmium absorption from the gastrointestinal tract is about 10%. Ultimately, it is the dose which reaches the cellular target over a given time period that results in the toxicological response. No toxic effects will be seen if the dose is below the NOAEL, whereas effects increase with increasing exposure. The dose–response curve may be expressed using a variety of mathematical formulas. Using the semilogarithmic form of the dose–response relationship the curve is sigmoidal in shape and varies in slope from chemical to chemical. Thus, if the curve is shallow a doubling of the dose results in a small increase in effects, whereas effects increase several-fold when the slope is steep (see Figure 1.1). The log of the dose is plotted on the abscissa (X axis) and increases toward the right. The location of the curve on the abscissa is a measure of the potency of the chemical.

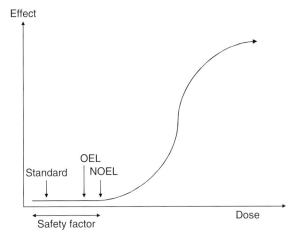

Figure 1.1 Dose–response curve showing log dose on the X axis and % response (effect) on the Y axis. The figure illustrates the location of regulatory values such as the NOEL (NOAEL), occupational exposure levels (OELs), and environmental standards such as acceptable daily intake (ADI). Note that a doubling of dose in the lower or upper part of the S-shaped curve results in small increases in effects, whereas it is much more prominent in the steep part.

1.2.3 Exposure Assessment

> **Since toxic effects are dose dependent, knowledge of the extent and duration of exposure is an integral part of the risk assessment process. Exposure defines the amount of a chemical to which a population or individuals are exposed via inhalation, oral, and dermal routes. Animal or human exposure is commonly defined by mg of the chemical/kg body weight per day.**

Toxicologists are concerned with exposure to any chemical, by any route, which may lead to adverse health effects. Exposure defines the amount of a chemical to which a population or individuals are exposed via inhalation, oral, and dermal routes. Animal or human exposure is commonly defined by mg of the chemical/kg body weight per day. According to the general principle of toxicology the consequences of human or environmental exposures depend on the amount and duration to which these individuals or populations are exposed. Thus, exposure assessment or prediction of exposure is an ultimate requirement for risk assessment and to decide whether regulations are needed. Since occupational exposure is regular and repetitive it can easily be measured in the air of the workplace by use of personal monitoring equipment or by biomonitoring.

Exposure of the general population is more difficult to assess. It usually is a combination of the presence of the compounds in indoor/outdoor air, drinking water or food, or use of products that contain the chemical (aggregate exposure). Moreover, frequency, duration, and site of exposure, and the concentration and weight of the substance in the products need to be considered. Children represent a special case of exposure. They may be exposed to chemicals that are released from articles such as toys during mouthing, via skin contact or from ingestion of contaminated dust or soil. Exposure can be modeled based on data such as information on frequency of mouthing, migration rates of the specific compound from the toy during mouthing, and absorption rates from the oral cavity and gastrointestinal tract.

The rate of absorption through the skin will also determine the internal exposure (body burden) of the chemical. Use of these parameters to assess exposure is plagued by many uncertainties, which often lead to overestimation of the actual exposure.

External exposure does not necessarily correlate with internal exposure. Therefore, risk assessment of internal exposure either requires knowledge of the dose response of internal exposure versus adverse effects or information to what extent external and internal doses correlate. Exposure assessment is even more complicated when mixtures of chemicals are the source of exposure (Chapter 2.3).

Ultimately, it is the dose that reaches the cellular target over a given time period that results in the toxicological response. Thus, the toxic potency of a chemical is the product of the interrelated external, internal, and target doses that result from the multiple pathways and routes of exposure to a single chemical (aggregate exposure). In the case of existing chemicals an appropriately designed program to measure the chemical in different media will provide the necessary information.

The measurement of external dose is either done on collected samples, i.e. food, or by direct measurement, i.e. in the ambient air. When collected samples are used representative sampling and appropriate storage conditions as well as accurate and reproducible measurement techniques are essential. This also applies to biomonitoring programs.

For new chemicals such data are not available and cannot be provided, so modeling of exposure is the only option.

In the EU Technical Guidance Document on Risk Assessment Part II the following core principles for human exposure assessment for new and existing chemicals and biocides are listed:

- Exposure assessments should be based upon sound scientific methodologies. The basis for conclusions and assumptions should be made clear and be supportable, and any arguments developed in a transparent manner.
- The exposure assessment should describe the exposure scenarios of key populations undertaking defined activities. Such scenarios that are representative of the exposure of a particular (sub)population should, where possible, be described using both reasonable worst-case and typical exposures. The reasonable worst-case prediction should also consider upper estimates of the extreme use and reasonably foreseeable other uses. However, the exposure estimate should not be grossly exaggerated as a result of using maximum values that are correlated with each other. Exposure as a result of accidents or from abuse shall not be addressed.
- Actual exposure measurements, provided they are reliable and representative for the scenario under scrutiny, are preferred to estimates of exposure derived from either analogous data or the use of exposure models.
- Exposure estimates should be developed by collecting all necessary information (including that obtained from analogous situations or models), evaluating the information (in terms of its quality, reliability, etc.), and thus enabling reasoned estimates of exposure to be derived. These estimates should preferably be supported by a description of any uncertainties relevant to the estimate.
- In carrying out the exposure assessment the risk reduction/control measures that are already in place should be taken into account. Consideration should be given to the possibility that, for one or more of the defined populations, risk reduction/control measures which are required or appropriate in one use scenario may not be required or appropriate in another (i.e., there might be subpopulations legitimately using different patterns of control which could lead to different exposure levels).

Biomonitoring of exposure (Chapter 4.3) is the best tool to determine actual individual exposure by determination of the chemical or its metabolites in blood, critical organs, urine or exhaled air. It allows determination of:

- the amount of a chemical taken into the organism by all routes (aggregate exposure)
- the metabolic fate of the chemical, its persistence in the organism, its rate of elimination, and from that the total body burden at the time of measurement
- the amount of the chemical and/or metabolites that reaches the target organs.

This procedure is also helpful to evaluate whether an environmental exposure such as increased indoor air concentrations or contaminated dust or soil, which might be ingested by children, actually leads to an increased body burden.

Reactive compounds may react with macromolecules like proteins or DNA. The latter does not automatically lead to a genotoxic effect because mutations usually occur at much higher doses. Thus, DNA adducts are markers of exposure and do not necessarily indicate effects.

Biomonitoring of effects determines the changes in a cellular function such as enzyme activity.

1.2.4 Risk Characterization

> **The sensitivity of analytical chemistry has advanced to the point where infinitesimally small amounts of chemicals can be detected and identified in the various media that characterize our environment. Detection of a chemical *per se* does not mandate that a toxicological effect in exposed people will be observed. Since the dose determines the poisonous effect, effects only occur when exposure exceeds the NOAEL.**

The risk assessment process requires differentiation between reversible and irreversible effects. The dose–response curves for chemicals that induce reversible effects display a region below which no effects are observed. The highest dose at which no effects are seen is called the no observable adverse effects level (NOAEL). The point at which effects become observable is called the lowest observable adverse effect level (LOAEL). The term "adverse" requires evalution whether the effect identified is adaptation or actually adverse. Note that a threshold is not the equivalent of an NOAEL since it describes concentration or exposure where the slope of the dose–response curves changes.

If damage is not repaired the effect persists and accumulates upon repeated exposure. In such cases a NOAEL cannot be determined and every exposure can be related to a defined risk. Reversibility depends on the regenerative and repair capacity of cells, subcellular structures, and macromolecules during and after exposure. Epithelial cells of the intestinal tract or the liver have a high regenerating capacity and rapidly replace damaged cells by increased cell replication. The highly specialized cells of the nervous system lose this capacity during natal and postnatal development. Consequently, damaged cells are not replaced, at least in the adult.

For chemicals that induce reversible effects, the NOAEL of the most sensitive endpoint is determined and compared with the human exposure to describe the margin of exposure (MOE) (or margin of safety, MOS). If the NOAEL is derived from animal experiments a MOE of 100 or greater is desirable (see Section 1.3.2).

A MOE of at least 10 is sufficient if the NOAEL is derived from human data (see Figures 1.1 and 1.3). These factors can be modified when data on the toxicokinetics or specific sensitivities indicate that the MOE can be reduced or needs to be increased.

The covalent binding of genotoxic mutagens and carcinogens to DNA is considered an irreversible event despite the availability of repair processes. Although there is increasing knowledge about DNA-repair mechanisms, the role of tumor-suppressor genes and apoptosis, their interactions and dose responses are not sufficiently understood to conclude whether genotoxic effects exhibit a threshold at low exposure or even a NOAEL (see Chapters 2.8 and 2.9). So far, the general agreement remains that the potency of genotoxic carcinogens increases with increasing dose so that the risk at a given exposure is determined by linear extrapolation from the dose–response data obtained from experimental studies in animals or from data obtained from humans (Chapter 4.8).

1.3 Toxicological Evaluation of New and Existing Chemicals

The various toxic effects that chemicals may exert and the different applications for which chemicals are designed require in-depth understanding of the cause and effect relationship, i.e. knowledge of the chemical and the specific organs upon which it impacts. As a result toxicologists tend to focus on specific organs, specific applications (e.g., pesticides or drugs), specific compounds like metals or solvents, or specific effects like carcinogenicity. Chapters in this book are devoted to specific organ toxicity and the specific effects of compounds such as carcinogenicity and mutagenicity. Institutions that document the toxicological, including epidemiological, data for hazard and risk assessment are given in Table 1.1.

Table 1.1 International institutions that publish documentations on chemicals.

Institution	Contact
Agency for Toxic Substances and Disease Registry (ATSDR)	http://www.atsdr.cdc.gov/
American Conference of Governmental Industrial Hygienists (ACGIH)	ACGIH Threshold Limit Values (TLVs) - ChemSafetyPro.COM
Canadian Centre for Occupational Health and Safety	http://www.ccohs.ca/
Dutch Expert Committee on Occupational Standards (DECOS)	https://www.gezondheidsraad.nl/en/publications
Environmental Protection Agency (EPA)	http://www.epa.gov/
European Centre for Ecotoxicology and Toxicology of Chemicals	http://www.ecetoc.org/
Health and Safety Executive (HSE)	http://www.hse.gov.uk/
International Programme on Chemical Safety (INCHEM)	http://www.inchem.org/ International Agency for Research on Cancer (IARC) Summaries and Evaluations Concise International Chemical Assessment Documents (CICADS) Environmental Health Criteria (EHC) monographs Health and Safety Guides (HSG) JECFA (Joint Expert Committee on Food Additives) Monographs and evaluations JMPR (Joint Meeting on Pesticide Residues) Monographs and evaluations. OECD Screening Information Data Sets (SIDS)
MAK Commission (German Research Foundation)	http://onlinelibrary.wiley.com/book/10.1002/3527600418/topics
National Institute for Occupational Safety and Health (NIOSH)	http://www.cdc.gov/niosh/homepage.html
Occupational Safety and Health Administration (OSHA)	http://www.osha.gov/
SCOEL – EC Scientific Committee on Occupational Exposure Limits	SCOEL recommendations
The Nordic Expert Group	http://www.nordicexpertgroup.org//

1.3.1 General Requirements for Hazard Identification and Risk Assessment

> **Toxicological evaluation of chemicals requires knowledge of the health consequences of acute, subchronic and chronic exposure via routes relevant to the common use of the chemical. Therefore, all elements of risk assessment (hazard identification, dose response, exposure, and the risk) have to be evaluated.**
>
> **Organ specificity and other relevant endpoints like fertility, pre- and postnatal toxicity or carcinogenicity, their dose response, and determination of the NOAEL can only be identified by appropriate repeated dose studies in animals. The use of *in vitro* testing can contribute important pieces of information, but so far cannot replace whole animal experimentation.**

To obtain sufficient information on the hazardous properties of a chemical requires investigation of:

- acute, subchronic, and chronic toxicity (oral, inhalation, dermal)
- irritation (skin, mucous membranes, eye) and phototoxicity
- sensitization and photosensitization
- genotoxicity (*in vitro* and *in vivo* methods)
- carcinogenicity (lifetime studies)
- reproductive toxicity
- toxicokinetics
- mode and mechanism of action.

In all studies information on the dose response of effects is essential to identify the slope of the dose–response curves, possible thresholds, NOAEL, LOAEL, and maximal tolerated dose (MTD). For most of the relevant tests guidelines have been proposed (e.g., see the Organization for Economic Cooperation and Development (OECD) Guidelines, Chapter 4.1).

Acute Toxicity, Subchronic Toxicity, and Chronic Toxicity

Acute toxicity studies describe toxic effects assessed after a single administration of the chemical to rodents and are primarily aimed at establishing a range of doses in which the chemical is likely to produce lethality. After dosing, the animals are observed over a period of one to two weeks to determine immediate or delayed effects. It is possible to plan studies in which other endpoints are examined as well.

Having established the lethal dose range the chemical may be examined for effects produced upon repeated administration. The common practice in such **repeated dose studies** is to treat animals each day for a few weeks or months. These studies usually include rodents, but larger species such as dogs, and in the case of new drugs, monkeys or apes, may be employed. The animals must be observed for effects on general, as well as specific, organ toxicity. At the termination of these studies the animals are usually examined for gross and microscopic pathology.

Chronic studies, usually in rodents, involve treatment of animals for several months up to a lifetime. Their intent is to examine the likelihood of the development of pathology after long-term exposure to low levels of chemicals and in the case of lifetime studies are focused on cancer.

There is an ongoing discussion regarding the extent to which *in vitro* studies and consideration of structure–activity relationships provide sufficient information to waive repeated *in vivo* exposure studies (Chapter 4.5). From a toxicological point of view it has to be stressed that this discussion is primarily concerned with cost reduction and protection of animals. It is necessary to ensure that in this climate protection of human health and the environment do not become secondary considerations.

In vitro studies allow identification of hazardous properties of substances, but only those that can be detected by the specific test system. When the dose response in the *in vitro* test system is known, toxicokinetic modeling may predict the dose response at the specific target *in vivo*. Even when the test system has a metabolic capacity, its appropriateness must be verified in intact organisms. Consequently, identification of all relevant endpoints, their dose response, thresholds, and NOAELs can only be determined in the intact animal by repeated dose studies. In the absence of such information hazard identification is incomplete and there is no basis for appropriate assessment of the risk of human exposure.

Irritation and Phototoxicity

Dermal irritation of compounds is evaluated by studies in animals and humans prior to testing for sensitization. These are usually performed by using a single occluded patch under the same conditions as applied when testing skin sensitization. Phototoxicity and photoallergic reactions have to be expected when compounds show significant absorption in the ultraviolet range (290–400 nm). Using the test strategy for irritation, an additional patch site is irradiated immediately after application of the test substance or after patch removal. Phototoxicity can also be tested by validated *in vitro* tests, such as uptake of Neutral Red by 3T3 cells. If such a test is negative further *in vivo* testing may not be necessary.

Sensitization and Photosensitization

For detection of the sensitizing potential of products the choice of a relevant animal is crucial (Chapter 3.9). However, in many cases animal models may be inappropriate for detection of a sensitizing potential so most dermatologists prefer studies in humans. An acceptable alternative may be studies with non-human primate species like cynomologus or rhesus monkeys. Generally, the Buehler guinea pig test and the local lymph node assay (LLNA) in mice are used in the preclinical testing program. The LLNA received great attention because it is the only reliable test for screening compounds that cause sensitization via routes other than the skin. So far the test has been successfully applied to determine relative potencies of contact allergens and has been reported to closely correlate with NOAELs established from human repeat patch testing. When the animal data indicate a weak contact sensitizing potential human skin sensitizing testing is conducted, usually by a human repeated-insult patch test (HRIPT). In any case, detection of antibodies in the serum during the studies using specific ELISA methods or bioassays to measure antibodies may be appropriate.

Genotoxicity

Test systems and test strategies to evaluate possible genotoxicity of a compound are described in detail in Sections 4.2.1 and 4.2.2. Generally, a bacterial mutation assay and an *in vitro* cytogenetic assay are performed. The results are usually verified by the mouse bone marrow micronucleus test, a reliable and widely used test system that detects aneugens as well as clastogens. Chemicals that yield positive responses to these tests frequently do not undergo further development. However, those which appear to lack genotoxicity may be carried forward and evaluated in lifetime carcinogenicity studies in rodents.

Carcinogenicity

The design of carcinogenicity studies is similar to that of chronic studies. At least three adequately spaced doses are tested, the highest dose being the MTD. Usually relatively large doses/concentrations are used to maximize the chance of finding a possible increase in tumor incidence in relatively few animals (50 per sex/dose). If thought necessary additional animals are included for investigations at 12 and/or 18 months. Rats and mice are used because of their relatively short lifespan of about 2 years and available information on their susceptibility to tumor induction, physiology, and pathology. A large historical database on tumor incidence in most strains and tissues exists, which is important in view of the large variability of tumor incidence in untreated animals and among different strains. Although the tumor incidence of the control group of the specific study is of major relevance, the historical control data of the specific strain used in the specific laboratory can be helpful to evaluate the incidences in the controls, especially when they deviate from the expected values. The incidence of spontaneous and substance-induced tumors increases in older animals, so it is necessary to terminate the study after a defined period to avoid the impact of different lifespans on the interpretation of the data.

Since it is difficult to predict the MTD, severe toxicity may occur at this dose. This requires specific consideration when interpreting the results. In such cases the metabolism of the animal may be overwhelmed and/or detoxifying mechanisms, such as GSH levels or DNA repair, may no longer be operative. It should also be recognized that a number of tumors are species-specific and may not be relevant for humans. One example is the α-urinary globulin-induced kidney tumors of the male rat. Moreover, rodents are more sensitive to compounds that disturb thyroid hormone metabolism and are also much more sensitive to compounds that induce peroxisome proliferation in the liver. When such mechanisms have been properly demonstrated the resulting tumors are of low relevance to humans. In any case, the incidence and type of tumors found have to be evaluated by experts and the underlying mechanisms as well as possible high-dose effects, such as overwhelmed metabolism of the test compound, have to be taken into account in the final judgment of whether a substance is considered to be carcinogenic to humans.

Toxicity for Reproduction and Development

Studies to evaluate reproductive and developmental effects may only be needed if there are indications that the chemical, or critical metabolites, can reach the embryo and/or fetus and could cause teratological, feto-toxic, or developmental effects. In such cases tests such as a reproduction/developmental toxicity screening test (OECD 421), a combined repeated dose toxicity study with a reproduction/developmental toxicity screening test (OECD 422), or the appropriate standard tests to evaluate effects on reproduction (one-generation reproduction toxicity, OECD 415) and prenatal developmental study (OECD 414) may be performed.

Toxicokinetics

> **Toxicokinetics describe the absorption, distribution, metabolism, and elimination (ADME) of a chemical in humans, experimental animals or cellular systems. Of specific importance for the interpretation of animal studies and for the extrapolation of hazards between species is the comparative information on the exposure and the dose that reaches the critical target (Chapter 2.2).**

A chemical may enter the body via food, air or the skin. The amount absorbed depends on the concentration in the different media, on physical-chemical parameters such as solubility in water and fat, stability, and the route of exposure (Figure 1.2).

Upon inhalation or skin penetration the compound directly enters the circulation and is distributed into the organs. When absorbed from the gastrointestinal tract the chemical enters the liver via the portal vein. The epithelial cells of the gut wall and the liver demonstrate a large capacity for metabolizing chemicals so that a compound may be extensively metabolized by this "first-pass effect" before entering the (cardiovascular) systemic circulation. Larger molecules, e.g. the glucuronosyl-conjugates, can be excreted via the biliary system into the duodenum, where the conjugates may be hydrolyzed so that the original compound is reabsorbed and re-enters the liver. This process is defined as **enterohepatic circulation**. Inhalation or dermal exposure to a chemical or intravenous or intraperitoneal injection may result in different effects than after oral exposure because of the first-pass effect.

After entering the cardiovascular system the chemical or its metabolites distribute to the organs, where they can accumulate, such as in fat or bones, or are further metabolized. Reactive metabolites will interact with tissue components and may induce cellular damage. This "tissue dose", i.e. the concentration of a chemical or its metabolite at the critical target over a given time, is

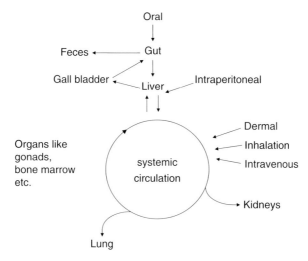

Figure 1.2 *Routes of exposure and systemic distribution of a compound within the organism. After oral ingestion the compound reaches the liver, where it can be extensively metabolized. Upon inhalation or dermal exposure and intravenous application the compounds reach the circulation without major metabolism.*

an important factor that helps to explain the correlation between internal exposure and external (environmental) exposure in relation to toxicity. By comparing tissue doses in different species at similar exposures it also helps us to understand species differences in the sensitivity to chemicals, as well as interindividual variations.

The chemical or its more water-soluble metabolites are primarily excreted via the kidneys or the biliary system. Volatile compounds may be exhaled. The great variety of processes observed during absorption, metabolism, distribution, and excretion cannot be predicted by modeling or by *in vitro* experiments without confirmatory data from animals and humans.

Mode and/or Mechanism of Action

> **Identification of the modes or mechanisms by which a chemical induces toxicity and the dose–response relationship are essential to understand species specificities, species differences, sensitive populations or the interpretation of data regarding threshold or non-threshold effects. They also help to evaluate the relevance of the toxic effects derived from experimental animals for humans. Whereas a toxic mechanism is often not known in detail, modes of action, which can be described in a less restrictive manner, are undergoing consideration for inclusion in the risk assessment process.**

There is an array of mechanisms by which chemicals, or any other stressors like heat or radiation, can lead to toxicity. They may be differentiated as follows:

Physiological changes are modifications to the physiology and/or response of cells, tissues, and organs. These include mitogenesis, compensatory cell division, escape from apoptosis and/or senescence, inflammation, hyperplasia, metaplasia and/or preneoplasia, angiogenesis, alterations in cellular adhesion, changes in steroidal estrogens and/or androgens, and changes in immune surveillance.

Functional changes include alterations in cellular signaling pathways that manage critical cellular processes such as modified activities for enzymes involved in the metabolism of chemicals such as dose-dependent alterations in Phase I and Phase II enzyme activities, depletion of cofactors and their regenerative capacity, alterations in the expression of genes that regulate key functions of the cell, e.g. DNA repair, cell cycle progression, post-translational modifications of proteins, regulatory factors that determine rate of apoptosis, secretion of factors related to the stimulation of DNA replication and transcription, or gap–junction-mediated intercellular communication.

Molecular changes include reversibility or irreversibility of changes in cellular structures at the molecular level, including genotoxicity. These may be formation of DNA adducts and DNA strand breaks, mutations in genes, chromosomal aberrations, aneuploidy, and changes in DNA methylation patterns.

As indicated in Chapter 4.6, data derived from gene expression microarrays or high-throughput testing of agents for a single endpoint become increasingly available and need to be evaluated for suitability for use in the hazard and risk assessment process. As long as the information is not related to functional changes their relevance is poor and there is the possibility of overinterpreting the effects observed. High-throughput data on specific endpoints may aid in the identification of common mechanisms of multiple agents.

Mechanistic information is most relevant for the evaluation and classification of carcinogens. If the carcinogenic effect is induced by a specific mechanism that does not involve direct genotoxicity, such as hormonal deregulation, immune suppression or cytotoxicity, the detailed search for the underlying mode of action may allow identification of a NOAEL. This can also be considered for materials such as poorly soluble fibers, dusts and particles, which induce persistent inflammatory reactions as a result of their long-term physical presence that ultimately lead to cancer.

1.3.2 General Approach for Hazard Identification and Risk Assessment (for details see Chapter 2.1)

Before starting any evaluation, structural alerts and physical-chemical parameters like water/lipid solubility and volatility need to be identified as well as the purpose of the hazard identification. To screen for specific effects such as relative cytotoxicity, mutagenicity or hormonal effects simple *in vitro* tests may be appropriate. This allows identification of specific wanted or unwanted effects and by that selection of useful compounds for further studies or their elimination.

For a more detailed evaluation the stepwise procedure usually starts with the determination of the LD_{50} to determine acute toxicity and the evaluation of genotoxicity by an *in vitro* bacterial test system (Ames test) and cytogenicity in mammalian cells. Positive results are verified *in vivo*, usually by the mouse bone marrow micronucleus test. For structural alerts or questionable results the compound needs further evaluation by additional tests, including studies on toxicokinetics or potential genotoxic mechanisms.

The information so far collected provides information on the reactivity of the test compound, its absorption and distribution in the organism, and possibly on critical targets. This allows a decision on whether or not the database is appropriate for further testing by repeated dose studies in animals for 28 and 90 days, which depending on the outcome and intended use of the chemical are followed by a 6-month or lifetime study to evaluate potential effects upon long-term exposure, including carcinogenicity.

Introduction to the Discipline of Toxicology 13

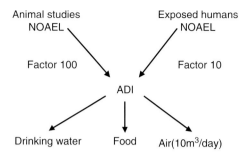

Food: 90 % of ADI
Air: 10 % ADI
Drinking water: 1 % of ADI (usually detection limit)

Figure 1.3 *Assessment of the acceptable daily intake (ADI)/tolerable daily intake (TDI) and limit values for drinking water, food and air.*

This information finally allows appropriate risk assessment for potential human exposure or the setting of acceptable exposure limits for risk management. For example, when the detailed toxicological evaluation can exclude genotoxic and carcinogenic effects the NOAEL in long-term studies in experimental animals can be determined. This NOAEL is the starting point to set the acceptable daily intake (ADI), which is usually 100-fold below the NOAEL (Figure 1.3). This factor considers a 10-fold difference between the sensitivity of the experimental animals and humans, and another factor of 10 to take into account possible interindividual differences among the human population. If the NOAEL is derived from studies in humans the interspecies factor is not necessary and the factor of 10 to cover possible individual differences within the population is applicable only. These factors can be reduced if specific information is available to conclude that the species–species or intraspecies differences are less than 10. From the ADI permissible concentrations in food, drinking water, consumer products, indoor and outdoor air, and other environmental compartments may be established.

1.3.3 Toxicological Issues related to Specific Chemical Classes

Jurisdictions and regulatory agencies around the world have established a variety of guidelines for risk assessment and permissible exposure standards for chemicals in the workplace, the home, and the general environment. Regulatory decision-making depends upon the estimation of health risks from chemical exposure.

Health risks of chemicals designed for specific applications, e.g. consumer products, drugs or pesticides, must be assessed when people are exposed in the many types of environment in which people can be found. Therefore all elements of risk assessment (hazard identification, dose response, exposure, and the risk) have to be thoroughly evaluated.

Data requirements for **new and existing chemicals** usually depend on annual production rate and the extent of human exposure. When there is considerable exposure regulatory requirements demand an extensive toxicological evaluation of the potential adverse effects of the specific chemical and the likelihood of their expression under the conditions of use or exposure and the definition of the MOE or the health risk under defined conditions of exposure (Chapters 5.2 and 5.3).

For **drugs** special emphasis must be placed on efficacy, therapeutic index, potential side effects, and the effects of overdosage (Chapter 5.1).

14 *Toxicology and Risk Assessment*

For **pesticides** the relative impacts of the chemical on the target versus on people is a critical requirement. Thus, the NOAEL for people must be established and an acceptable daily intake (ADI) must be determined because of the possibility of contamination of food and other consumer products with the pesticide, and the margin of safety needs to be established (Chapter 6.10).

Exposures to **chemicals in the workplace** is, accordingly to law, controlled by the Occupational Safety and Health Administration (OSHA) in the United States and by the Chemicals Law Act (1992) in Europe. Various governmental and non-governmental institutions are involved in setting occupational exposure standards. Since the institutions publish the complete toxicologically relevant information and a justification for the proposed limit value, these documentations are valuable sources for the toxicological database of the compounds. Institutions that publish these documents are listed in Table 1.1.

1.3.4 Existing Chemicals

In 1992 the European Commission estimated that about 100,000 chemicals were in use. They are produced in quantities ranging from less than a ton to several million tons produced per year. Except for drugs and pesticides, the data requirement for existing or new chemicals has not been regulated. Although it is the responsibility of the producer and downstream user to release safe products, there are high-volume products with a relatively small database. Several programs have been launched to obtain knowledge at least for compounds with high annual production rates. In the United States the Environmental Protection Agency (EPA) has initiated an HVP program. In an international cooperation the OECD has launched the International Council of Chemical Associations (ICCA) program, which evaluates and documents the available information on environmental and human health hazards and risks for about 1000 chemicals. In Europe, Risk Assessment Reports under the Existing Chemical Program of about 150 compounds are being produced and Registration, Evaluation and Authorization of Chemicals (REACH) regulation started in 2008.

REACH (Chapter 5.3)

REACH regulation in the European Union identifies substances of hazardous properties and evaluates the risks of human and environmental exposure. The regulation became effective by 2008. It is the responsibility of the producer or downstream user to provide the necessary information to the relevant agency. The extent of toxicological information largely depends on the annual production rate of a chemical. As long as there is no indication of a specific risk the chemicals will be registered for the intended use. Special attention will be paid to carcinogens, mutagens, and reproductive toxins (CMR compounds) and to other serious toxic effects as well as chemicals that show bioaccumulation, persistence, and toxicity (BPT compounds) in the environment. The specific use of such compounds needs to be authorized.

1.3.5 Test Guidelines (Chapter 4.1)

For reproducibility and acceptance by regulatory agencies standardized study protocols for each test have been developed. Moreover, only qualified laboratories are accredited to perform tests for regulatory purposes and they must adhere to good laboratory practice (GLP) guidelines. These guidelines describe how to report and archive laboratory data and records. GLP guidelines also require standard operating procedures (SOPs), statistical procedures for data evaluation, instrumentation validation, materials certification, personnel qualification, proper animal care, and independent quality assurance (QA). GLP regulations have been developed by the US Food and Drug Administration (FDA) and by the OECD.

REACH requires the use of test methods as described in the Commission Regulation (EC) No. 440/2008 or any methods based on internationally recognized scientific principles. Use of the data for classification and labeling of chemicals according to the Globally Harmonized System (GHS) is described in the European Chemicals Agency (ECHA) Guidance on the Application of the classification, labelling and packaging (CLP) of substances and mixtures, Version 5.0 (July 2017). For pharmaceuticals the International Conference on Harmonisation of Technical Requirements for Registration of Pharmaceuticals for Human Use (ICH) has produced a comprehensive set of safety guidelines to uncover potential risks (www.ich.org, Safety Guidelines). These include repeated dose toxicity testing, toxicokinetics and pharmacokinetics, genotoxicity studies, carcinogenicity studies, and immunotoxicology studies. They refer in part to OECD guidelines for test descriptions and include guidance for result interpretation. Guidelines prepared by the US FDA or the European Medicines Agency (EMA) are generally in agreement with the ICH guidelines.

1.3.6 Alternatives to Animal Experiments (Chapter 4.5)

In recent years the 3R (refinement, reduction, and replacement) strategy for substituting animal experiments has received increasing attention. This intends to preferentially use *in vitro* studies and consideration of structure–activity relationships to waive and reduce studies in laboratory animals. Most success in the development of alternative methods is in local toxicity and acute toxicity testing, when the effects of the chemical, not its possible metabolites, are determined. The standardized *in vitro* mutagenicity/genotoxicity tests include some metabolizing capacity. Since these tests are frequently afflicted by false-positive results there is a need to verify their outcome by appropriate *in vivo* testing. *In vitro* studies allow identification of hazardous properties of substances, but only those that can be detected by the specific test system. Even when the test system has a metabolic capacity, its appropriateness must be verified in intact organisms. As indicated before, toxicokinetic modeling may predict the external or internal dose response in an intact organism. Consequently, identification of all relevant endpoints, their dose response, thresholds, and NOAELs can only be determined in the intact animal by repeated dose studies. In the absence of such information hazard identification is incomplete and without information on the dose–response relationship obtained from these studies there is no basis for appropriate assessment of the risk of human exposure. Thus, in case of the methodologies for long-term testing for systemic and reproductive toxicity and carcinogenicity, which consume the highest number of animals, validated alternatives are lacking. In spite of this Regulation (EC) No. 1223/2009 requires validated alternative methods, in particular *in vitro* replacement methods for the safety evaluation of cosmetic substances and products.

1.3.7 Evaluation of Mixtures (Chapter 2.7)

Humans and their environments are exposed to a wide variety of substances. The potentially adverse effects of substances when present simultaneously have been analyzed in several reviews and documentations. Most recently the available scientific literature has been analyzed by SCCS/SCHER/SCENIHR 2011. The general conclusions are that chemicals with common modes of action will act jointly to produce combination effects that are larger than the effects of each mixture component applied singly (dose/concentration addition). However, effects only occur when the concentrations of the individual compounds are near or above their zero effect levels (NOAELs). For chemicals with different modes of action (independently acting), no robust evidence is available that exposure to a mixture of such substances is of health or environmental concern if the individual chemicals are present at or below their zero effect levels. If no mode of action information is available, the dose/concentration addition method should be preferred. Prediction of possible interaction requires expert judgement and hence needs to be considered on a case-by-case basis.

1.3.8 Evaluation of Uncertainties

Derivation of the ADI, derived no effects level (DNEL), NOAEL, and risk assessments includes uncertainties. For example, the NOAEL may not be a real NOAEL for statistical reasons in that too few animals have been used in the specific experiment. Alternatively, the NOAEL may be rather conservative because the next higher dose, which determines the LOAEL of a weak adverse effect, is 10-fold higher. Usually this uncertainty is covered by applying assessment factors that build in a margin of error to be protective of the population at risk. In case of ADIs or DNELs the uncertainty factor of 100 covers the uncertainties of inter- and intra-individual differences unless toxicodynamic and/or toxicokinetic information allows its reduction. Whereas the experts who have performed the risk assessment are usually aware of uncertainties, the risk manager tends to use the numbers as such, with the consequence that any exposure even slightly higher than the ADI or DNEL is not considered to be acceptable. To acknowledge these uncertainties statistical analysis may be used to characterize and weight the different assumptions from various components of the risk assessment process, such as dose response, emissions, concentrations, exposure, and valuation. This will improve understanding of the uncertainties, thereby allowing better mean estimates of risk and of the risk for different individuals and populations. For uncertainty analysis see European Food and Safety Authority (EFSA, 2006) and ECHA (2010).

1.3.9 The Precautionary Principle

The precautionary principle is a measure to enable rapid response in the case of a possible danger to human, animal or plant health, or to protect the environment. In particular, where scientific data do not permit a complete evaluation of the risk this principle may, for example, be used to stop distribution or order withdrawal from the market products likely to be hazardous. Since this description allows various interpretations a more precise definition is given in the Communication from the Commission of 2 February 2000. There it is outlined that the precautionary principle may be invoked when a phenomenon, product or process may have a dangerous effect, identified by a scientific and objective evaluation, if this evaluation does not allow the risk to be determined with sufficient certainty. The Commission specifically stresses that the precautionary principle may only be invoked in the event of a potential risk and that it can never justify arbitrary decisions.

Other states use slightly different definitions, for example the Canada definition is as follows:

> "The precautionary principle is an approach to risk management that has been developed in circumstances of scientific uncertainty, reflecting the need to take prudent action in the face of potentially serious risk without having to await the completion of further scientific research."

All definitions of the precautionary principle refer to the Rio Conference on Environment and Development (Principle #15 of the June 1992 Declaration).

1.3.10 The TTC Concept

The threshold of toxicological concern (TTC) is a concept to establish a level of exposure for chemicals, regardless their chemical-specific toxicity data, below which there is no appreciable risk to human health. The concept is based on knowledge of structural alerts, the amount of a specific chemical in a product, and the daily human exposure. The TTC as applied to foods is defined as a nominal oral dose which poses no or negligible risk to human health after a daily lifetime exposure. At a mean dietary intake below the level of the TTC, toxicology safety testing is not necessary or warranted. By that, the TTC concept can contribute to a reduction in the use of animals for safety tests. The TTC concept may also represent an appropriate tool to evaluate or prioritize the need for toxicological testing. There is ongoing discussion on its general applicability for the safety assessment of substances that are present at low levels in consumer products such as cosmetics or for impurities or degradation products.

1.3.11 Classification and Labeling of Chemicals

Criteria for the classification and labeling (C&L) of chemicals have been developed by several national and international agencies. For the most recent guidance see ECHA (2017). It is based on the Globally Harmonized System (GHS) of Classification, Labelling and Packaging of Chemicals, which has been developed for worldwide harmonization in the evaluation of chemicals. In the health hazard part the ECHA guidance addresses all endpoints of toxicological relevance as well as the presence of such chemicals in mixtures. Since the CLP criteria only consider the toxic potential not the potency, consideration of high-dose effects that are irrelevant to humans may lead to unnecessary C&L. For example, the hazard-based concept for classification of carcinogens is based on qualitative criteria and reflects the weight of evidence available from animal studies and epidemiology. The mode of action and potency of a compound are either not taken into account or at best are used as supporting arguments. The advancing knowledge of reaction mechanisms and the different potencies of carcinogens at least triggered a discussion for a re-evaluation of the traditional concept.

The proposals for restriction and authorization require estimation of cancer risk at a given exposure, which is estimated by a linear or sublinear extrapolation from the high-dose effects observed in animals to the usually lower human exposure. However, the EFSA recommended avoidance of this extrapolation because of the inherent uncertainties. Instead, the MOE between a benchmark dose, or the T25 calculated from a carcinogenicity study in animals, and human exposure should be determined. A MOE of 10,000 or more is considered to be of minor concern. The advantage is that neither a debatable extrapolation from high to low doses is performed nor are hypothetical cancer cases calculated.

> **The systems for classification of carcinogens used by various national or international institutions were developed in the 1970s. So far classification has been based on qualitative criteria only, and reflects essentially the weight of evidence available from animal studies and epidemiology. More recently some institutions have included mechanistic considerations, dose response, and exposure.**

Classification is usually based on the certainty with which a carcinogenic potential for a chemical can be established.

Generally three categories, the definitions of which slightly differ, are used:

- human carcinogens
- animal carcinogens, reasonably anticipated to be human carcinogens
- not classifiable because of inadequate data.

The International Agency for Research of Cancer (IARC, 2006) and the OECD proposed to use data on the carcinogenic mechanism and potency in decision-making on whether carcinogenicity is likely or not likely below a certain dose. The German Committee on occupational exposure limits (the MAK committee) and the Scientific Committee to set Occupational Exposure Limits (SCOEL) of the General Directorate Employment of the European Commission apply information on carcinogenic mechanisms and potency as criteria for a revised classification. The US EPA recommended consideration of the mode of action and has published a modified concept for classification. These activities in part originate from the recognition that one can distinguish between mechanisms of carcinogenicity caused by non-genotoxic and genotoxic carcinogens. Thus, it is possible to identify a NOAEL for non-genotoxic carcinogens, provided there is sufficient information on the primarily non-genotoxic mechanism. The American Conference of Governmental Industrial Hygienists (ACGIH, 1997) has used a concept that considers carcinogenic potency for classification since 1995.

1.4 Summary

Toxicology is charged with describing the adverse effects of chemicals in a qualitative sense, and with evaluating them quantitatively by determining how much of a chemical is required to produce a given response. Fundamental to understanding toxicology are the definitions of hazard, potency, dose response, exposure, and their integration into the risk assessment process. Since humans or organisms in the environment can be exposed via different routes, the concentrations in the different environmental compartments are a prerequisite for appropriate risk assessment. This needs to be specifically recognized, since the sensitivity of measurements in analytical chemistry has advanced to the point where infinitesimally small amounts of chemicals can be detected and identified in the various media of the human environment. In understanding the principles of toxicology it is obvious that the presence of a chemical does not necessarily imply a health hazard. Since the dose makes the poison, effects only occur when exposure exceeds the NOEL. This applies tor chemicals that induce reversible effects. If damage is not repaired the effect persists and accumulates upon repeated exposure. In such cases every exposure represents a defined risk, which needs to be quantified.

There is an array of testing procedures to determine the hazardous properties of a chemical, such as acute, subchronic, and chronic toxicity, irritation and phototoxicity, sensitization and photosensitization, genotoxicity, carcinogenicity, or toxicity to reproduction. Information on the toxicokinetics and mechanisms of the toxic effects helps to evaluate the relevance of the findings for humans. More recent methodologies, like toxicogenomics or high-throughput testing of agents for a single endpoint, will become increasingly available and may improve hazard identification and aid in the identification of common mechanisms of multiple agents.

In summary, toxicology describes the intrinsic properties of an agent (hazard identification) by applying conventional and substance-specific test procedures to estimate the amount of the chemical required to produce these effects (risk characterization).

Reference

Paracelsus, *Selected Writings*, edited by J. Jacobi, translated by N. Guterman, page 95, Princeton University Press, Princeton, NJ (1951).

Further Reading

ECHA 2010: Guidance on the information requirements and chemical safety assessment, European Chemicals Agency: http://echa.europa.eu/web/guest/guidance-documents/guidance-on-reach.
EFSA 2006: Guidance of the Scientific Committee on a request from EFSA related to uncertainties in dietary exposure assessment. *The EFSA Journal*, **438**, 1–54.
EFSA 2005: Opinion of the Scientific Committee on a request from EFSA related to a harmonized approach for risk assessment of substances which are both genotoxic and carcinogenic, European Food and Safety Authority: https://www.efsa.europa.eu/de/efsajournal/pub/28.
Greim H and Albertini R, eds: *The cellular response to the genotoxic insult*. The question of threshold for genotoxic carcinogens. RCS Publishing 2013.
IARC: Monographs on the evaluation of carcinogenic risks to humans, Geneva.
Kroes R, Kleiner J and Renwick A (2005): The threshold of toxicological concern concept in risk assessment. *Toxicological Sciences*, **86**, 226–230.
Opinions of Scientific Committees of DG SANCO: http://ec.europa.eu/health/scientific_committees/environmental_risks/opinions/index_en.htm#id9

- SCCS/SCHER/SCENIHR (2011) Toxicity and Assessment of Chemical Mixtures.
- SCCS/SCHER/SCENIHR (June 8, 2012) Use of the Threshold of Toxicological Concern (TTC) Approach for Human Safety Assessment of Chemical Substances with focus on Cosmetics and Consumer Products.

The Globally Harmonized System for Hazard Communication. In: *Hazard Communication*. Occupational Safety & Health Administration, United States Department of Labor, 10 June 2009.
US EPA *Assessing Health Risks from Pesticides. Pesticides: Topical and Chemical Fact Sheets*. http://www.epa.gov/pesticides/factsheets/riskassess.htm.

2
Principles in Toxicology

2.1 General Concepts of Human Health Risk Assessment

Paul T.C. Harrison, Philip Holmes, and Ruth Bevan

2.1.1 Introduction

Early approaches to chemical risk assessment were rather simple subjective assessments of the level of risk that a particular substance might pose to people or the environment, often based on little more than the best judgement of an individual knowledgeable in the field. Now, however, a range of more sophisticated approaches are available to estimate risk, combining available evidence on the intrinsic hazard of the material with the likely scenarios under which humans (or other target organisms) might be exposed to the material.

In this chapter, we explain the underlying principles of chemical risk assessment in relation to the safeguarding of human health, focusing particularly on how these principles are used to derive health-based exposure standards.

2.1.2 Principles of Risk Assessment

When discussing risk assessment it is necessary to clearly define what is meant by 'risk' and distinguish it from the term 'hazard'. Normally – and throughout this chapter – risk is considered a function of the nature of the hazard posed by any agent and the degree of exposure. Hazard refers to the intrinsic harmful property of a substance – for a chemical this will be its inherent toxicity.

> **For a material to pose a risk it must have intrinsic hazardous properties and there must also be the potential for exposure to occur; the degree of risk is determined by the potency of the hazard and the extent of exposure.**

Thus, the assessment of risk posed by any chemical or mixture requires information on both the intrinsic toxicological qualities of the chemical (or chemicals present in a mixture) and the potential for it to cause harm at relevant levels and routes of exposure.

A number of special terms and acronyms are used in risk assessment. Some of the more common terms are listed in Table 2.1.

> **Risk assessment traditionally involves three basic steps:**
> - *hazard identification*
> - *hazard characterisation*, including dose–response assessment
> - *exposure assessment*.
>
> An additional process, *risk characterisation*, is then applied to integrate the findings.

Any attendant uncertainties relating to the first three steps of the risk assessment process are integrated into the risk characterisation step. It is this information that is drawn on for the related, but distinct, process of defining the necessary risk management measures (RMM) required to ensure safe use of the material.

Qualitative and Quantitative Methods

Historically, risk assessment entailed a largely qualitative description of the nature of the risk and the use of semi-quantitative terminology to describe the extent of the risk posed, e.g. by use of terms such as 'minimal' or 'severe'. However, such descriptions may be of limited value where the objective is to establish an exposure standard that can be used to decide if a given exposure is acceptable or not, or to define monitoring requirements or justify enforcement action. Therefore, risk should, where possible, be described in quantitative (numerical) terms.

Normally risk assessments incorporate both qualitative (descriptive) and quantitative (numerical) elements.

Hazard Identification

> **The first step of risk assessment, hazard identification, involves identifying the source of risk that is capable of causing adverse effects together with a qualitative description of the nature of those effects.**

The specific potential adverse (toxic) effects of a substance can be identified through consideration of its chemical and physical properties in conjunction with critical endpoint data.

A critical endpoint is an observable outcome in a whole organism, such as a clinical sign or pathologic state, that is indicative of an adverse outcome that can result from exposure to the material in question.

For hazard identification, critical endpoints are generally identified from experimental toxicological studies that most closely match the route and duration of exposure for which the risk assessment is being carried out. A wide variety of studies and analyses are used to support hazard identification, including toxicokinetics and toxicodynamics (the study of how the body handles, interacts with and eliminates toxins) and mode of action analysis.

Hazard Characterisation

> **Hazard characterisation involves a quantitative or semi-quantitative evaluation of the nature of the adverse effects that occur following exposure to the hazard (in this case a particular chemical substance).**

Table 2.1 Terminology used in risk assessment.

Occupational exposure level (OEL)	OELs are used to provide standards or criteria against which measured exposure levels may be compared in order to ensure that, as far as the current state of knowledge permits, control is adequate to protect health, or for designing new plants and processes to ensure that they are engineered in such a way that exposures can be controlled at levels that will not damage health.
Margin of safety (MoS)	The relationship (ratio) between the POD value (e.g. NOAEL/LOAEL) for the relevant effect(s) and the dose or concentration to which humans are exposed. Values below 100 (where the POD is based on animal data) are usually taken to indicate a need for more comprehensive risk evaluation.
Point of departure (POD)	The critical dose level, usually a NOAEL or LOAEL, derived from the key (most relevant/authoritative) study.
No observed adverse effect level (NOAEL) and lowest observed adverse effect level (LOAEL)	The 'no' or 'lowest' observed adverse effect level in any experimental study.
Benchmark dose (BMD)	The dose at which a defined level of response occurs (by convention 5 or 10%, depending on the nature of the effect). It is an alternative to the use of a NOAEL/LOAEL as a POD.
T(D)10/T(D)25	For non-threshold carcinogens the BMD may be described as a T(D)10 or T(D)25, indicating the calculated dose for a 10% or 25% tumour incidence, respectively, using actual dose response data or modelled data assuming, for example, a linear dose–response relationship.
Assessment factors	Deriving an OEL from health effects data invariably requires the application of assessment factors to account for uncertainty in the process, particularly in relation to various extrapolations and assumptions that need to be made, for example in using animal data to predict human risk.
Derived no-effect level (DNEL)	The derived no-effect level for threshold substances is calculated for REACH purposes according to ECHA guidance (Chapter R.8). This guidance gives various default values for the assessment factors that are to be applied, differentiating between workers and the general public and for different routes of exposure. An airborne DNEL for workers may be taken as equivalent to an OEL.
Derived minimal effect level (DMEL)	The derived minimal effect level is similar in principle to a DNEL but is calculated differently and applied to substances with non-threshold effects, such as genotoxic carcinogens. A DMEL for workers may be taken as equivalent to an OEL.
Mode of action (MoA)	The mode of action of a substance indicates whether a threshold or non-threshold approach should be used in establishing the DNEL/DMEL and/or OEL. For non-threshold substances a so-called 'risk-based' OEL is set.
Acceptability and tolerable risk	For threshold effects, a risk is usually considered acceptable if the ratio DNEL:exposure (the risk characterisation ratio) is less than 1. For non-threshold substances such as genotoxic carcinogens, risk levels ranging between 10^{-3} and 10^{-6} for workers and/or the general public may be considered 'tolerable'.

Table 2.1 (Continued).

Indicative occupational exposure limit value (IOELV)	This is a health-based limit conventionally established only for substances for which it is possible to establish a threshold or a no-effect level considered to be protective of health. They are, in essence, OELs recommended by SCOEL.
Binding occupational exposure limit value (BOELV)	The binding occupational exposure limit value may be formally established by the Council and European Parliament in cases where an appropriate limit of exposure can be identified based on risk as well as socio-economic impact assessment. BOELVs primarily apply to non-threshold substances covered by the EC Carcinogens and Mutagens Directive (CMD), and are also recommended by SCOEL.
Weight of evidence (WoE)	Weight of evidence is an approach involving the assessment of the strengths, weaknesses and relative weights of different pieces of information. It requires expert judgement and is influenced by a variety of factors, including data quality and consistency, nature and severity of effects, and relevance. Reliability, relevance and adequacy for purpose must always be taken into account in the WoE approach.

During this process, available data are assessed to develop a weight of evidence (WoE) argument linking exposure to the material and the likelihood and severity of any adverse effect (the critical endpoint). Most often the critical endpoint is identified using the dose–response relationship shown by one pivotal toxicity study (i.e. that which most closely resembles exposure conditions under assessment and that is considered to be of a suitable quality in terms of study design and reporting) in the most sensitive test species (e.g. rat, mouse, rabbit or dog) considered of relevance to humans. As many chemicals may show a number of different dose–response relationships for different endpoints, the one that occurs at the lowest dose is usually selected as the critical endpoint for risk assessment purposes. This practice is based on an assumption that if the 'critical' effect is prevented, then no other effects of concern will occur. It is also necessary to characterise the target human populations since responses to exposure may vary across subpopulations and life stages.

> **For the majority of chemicals it is possible, on the basis of mechanistic understanding and available experimental data, to define a threshold exposure level below which the critical effect is no longer discernible.**

Such a threshold may be defined as the no observed adverse effect level (NOAEL) or, if utilising benchmark dose modelling to define the dose response, the benchmark dose (BMD) at which a defined level of response is predicted. The threshold dosage is generally used as the point of departure (POD) (also referred to as the reference point) for extrapolating the degree of risk; this involves the application of various adjustment factors to address uncertainties (see *Application of Assessment Factors*, this chapter) and ensure an adequately precautionary outcome. If the dataset available for hazard assessment does not allow definition of a 'no effect' threshold then, provided understanding of the toxic mechanism is strong enough to support the theoretical existence of a threshold, an alternative approach is to identify the lowest test dosage at which an adverse effect is apparent. This is termed the lowest observed adverse effect level (LOAEL) and may be used as the POD, although, because of the implicitly greater degree of uncertainty, additional adjustment factors are normally incorporated into the risk assessment process.

For most chemicals, the POD is expressed as the dosage or concentration to which the test organism is exposed, although in some circumstances other metrics, such as dose per unit surface area, might be adopted.

For a few types of toxic effect (e.g. cancer, mutation and sensitisation), the underlying mechanism may be such that it is not theoretically possible to establish a threshold below which an adverse effect can be discounted, that is, any exposure to the chemical might elicit some degree of response. The classic example of such a non-threshold chemical is one that causes cancer through mechanisms involving direct damage to the DNA, with genetic mutation a possible consequence. Such **genotoxic carcinogens** include aflatoxin, benzene and benzo[a]pyrene.

> **Chemicals for which no threshold can be established are treated differently from threshold chemicals in the risk assessment process, for example using various models to predict the scale of any impact at low levels of exposure (see *Carcinogens and Thresholds of Effect*, this chapter).**

In recent years the supposed absence of a threshold for a genotoxic carcinogen has increasingly been disputed (see Chapters 2.7 and 2.8). Also, although it is true to say that while theoretical principles and general understanding may lead to the conclusion that the human dose response to a substance has a threshold, determining where that threshold lies is often a very challenging endeavour. This is particularly the case with respiratory sensitisers, for example. It can reasonably be argued that all substances (including genotoxic carcinogens) are likely to have some threshold below which no relevant biological effect will occur – the problem always is determining where this threshold lies.

Human Data

> **Various sources of human data may be available for risk assessment purposes, including case reports, epidemiological studies, occupational studies and clinical research; case and field studies may also provide valuable information. Each of these information sources has associated advantages and limitations.**

Epidemiological studies provide a statistical evaluation of the relationship between exposure of a defined human population to a substance and the occurrence of adverse health effects. These studies assess real-world exposure scenarios, with relevant mortality and morbidity endpoints being measured. However, epidemiology studies have several significant inherent limitations. In particular, **causality** can be difficult to demonstrate, effects due to **confounders** are difficult to eliminate and accurate measurements of individual **exposures** are difficult to obtain. Also, such studies are potentially costly and time consuming, especially if assessing chronic health outcomes such as cancer.

Occupational studies can provide valuable information linking specific exposure levels with the appearance (or absence) of adverse health effects, but they may be subject to confounding due, for example, to the 'healthy worker effect': working populations usually exhibit better health and lower overall death rates than the general population because severely ill people are largely excluded from employment.

The most robust data for risk assessment come from statistically controlled human clinical studies. However, such studies are unlikely to address chronic effects and are rare as there are significant ethical concerns associated with any kind of human testing.

Animal Data

> **As human data are often unavailable, data from studies on experimental animals (e.g. rats, mice, rabbits and dogs) are frequently utilised. Such studies have the advantage that they can be designed, controlled and conducted to address specific gaps in knowledge and in accordance with strict test criteria (e.g. OECD guidelines for the testing of chemicals).**

Animal studies can provide valuable information on the mode of action (MoA) and/or adverse outcome pathway (AOP) for the material being assessed. However, as responses of humans and other species to a given chemical exposure may be substantially different (both physiologically and behaviourally), there are always inherent uncertainties associated with extrapolating results from an animal species to humans. A further common limitation of animal studies is that dose–response relationships are usually examined over much higher concentration ranges than would be expected to occur for human exposure. Extrapolation to lower doses is therefore required, leading to further uncertainty; the implications of dealing with uncertainty within the context of risk assessment and regulation are discussed more fully in *Margins of Safety and Acceptability of Risk* (this chapter).

In vitro, in silico *and 'Omics' Data*

> **A number of *in vitro* models, *in silico* (computational) tools, and so-called 'omics' technologies have been developed to investigate toxicological processes at different levels of biological organisation. Using *in vitro* models to accurately predict what will occur *in vivo* is difficult and, when interpreting results, it is vital that relevance to the *in vivo* situation is established.**

In vitro models can provide extremely reliable (repeatable), rapid and relatively inexpensive approaches to identifying MoA/AOPs, and detailed time– and dose–response relationships can be carried out to define concentration-dependent transitions in the mechanisms of toxicity. Also, comparison of results from *in vitro* studies using animal and human cells can help inform extrapolations from animals to humans and provide evidence of the relevance of animal study findings to humans. However, a major limitation of all *in vitro* models is that they have a reduced level of integrity compared with the *in vivo* situation, as interactions with neighbouring cells or remote tissues, and the operation of homeostatic (compensatory) mechanisms will not be the same as would occur in an intact organism. In addition, the types of toxicological endpoints that can be measured with *in vitro* models are limited (e.g. behavioural effects cannot be mimicked).

The use of computational or *in silico* tools to predict responses following exposure to a chemical can help provide essential information contributing to the understanding of MoA/AOPs. These tools include (quantitative) structure activity relationships ((Q)SARs) that are generally based on the assumption that a chemical's structure can be used to predict its physicochemical properties and reactivities, leading to an understanding of potential toxicity.

> **(Q)SARs are typically used in combination with other, desk-based methods such as read-across. In read-across, structural elements of the chemical being assessed are compared with those of another chemical having a similar structure and an established hazard profile, providing insight into the toxic potential of the new chemical.**

Various tools have been developed to assess biochemical changes associated with a MoA/AOP at the level of DNA/RNA (transcriptomics), proteins (proteomics) and the whole metabolome (metabolomics). Such biochemical changes can act as biomarkers of exposure in humans and animals for use in dose–response modelling.

'Omics' technologies also offer a means by which interspecies differences and relevance to humans might be investigated, including the consideration of human variability (age differences, inter-ethnic differences, polymorphisms) as well as the investigation of patterns of gene transcripts, proteins, and metabolites within an AOP (see Chapter 4.6). However, understanding the relevance of the observed changes at a biological or toxicological level is challenging, and their use for many substances and scenarios is limited, particularly when dealing with novel chemical structures.

Carcinogens and Thresholds of Effect Conventionally, carcinogens with no threshold of effect are referred to as **genotoxic carcinogens** and those with a threshold as **non-genotoxic carcinogens**. It is now recognised, however, that a distinction can be made between 'primary' and 'secondary' genotoxic carcinogenicity, where the 'secondary' effect has a measurable threshold.

Genotoxic Carcinogens

> **As primary (DNA-reactive) genotoxic carcinogens are considered not to have a threshold of effect, it is normally assumed that increases in DNA damage occur in a linear fashion with increasing dose. It is therefore theoretically possible that any level of exposure may result in a mutation, leading to increased risk of cancer, although this risk is likely to be extremely small.**

Under EU Directive 2004/37/EC on the protection of workers from the risks related to exposure to carcinogens or mutagens at work, it is stated that where a risk assessment shows a risk to worker health or safety, the workers' exposure must be prevented. This may be achieved by replacing the carcinogen or mutagen with a less hazardous substance (substitution) or by using a closed process. However, if these options are not technically possible then risk must be lowered by ensuring that exposure of workers is 'reduced to as low a level as is technically possible'. Similarly, as complete prevention of exposure of the general population to carcinogens is not always feasible, a widely accepted approach to reduce risk from genotoxic carcinogens is to ensure that levels are **as low as reasonably achievable** (ALARA). This may mean preventing exposure entirely or identifying a practical 'minimal risk level' through application of an appropriate uncertainty factor to an identified POD.

Genotoxic carcinogens that act through indirect mechanisms (so-called secondary genotoxic carcinogens) are considered to have a threshold of effect related to the MoA, and linear extrapolations for cancer risk are thus inappropriate for these substances.

Non-genotoxic Carcinogens

> **Non-genotoxic carcinogens are carcinogenic substances for which there is no evidence that genotoxicity is the primary biological mechanism. For these materials it is accepted that there is a threshold dose below which no effect occurs.**

Many of these substances induce tumours as a secondary consequence to a primary toxic response for which a threshold dose (i.e. a NOAEL) can be identified. In such cases, exposures below which no appreciable risk to human health will occur can be estimated by the application of appropriate uncertainty factors to the threshold POD (discussed further in *Risk Characterisation*, this chapter).

Exposure Assessment

> **Defined as 'the quantitative or semi-quantitative evaluation of the exposure of humans and/or the environment to risk sources from one or more media', exposure assessment includes the direct exposure of workers and consumers and/or indirect exposure of the general public via the environment.**
>
> **Estimates of exposure and risk derived from the use of models are subject to a large degree of uncertainty associated with the accuracy and applicability of the various assumptions, simplifications, techniques and data utilised.**

Exposure assessment takes into account the magnitude, frequency and duration of existing or future exposures of a defined population.

A variety of approaches exist for quantifying human exposure. Direct methods involve measurements of exposure taken at the point of contact or uptake at the moment it occurs (e.g. by personal monitoring and biomonitoring). Biomarkers of exposure – such as measured levels of chemical metabolites in urine – can give an indication of whether exposure has occurred and, in some cases, also the extent of exposure. Where direct measurement is not possible, exposure may be estimated indirectly through the extrapolation of exposure estimates from other measurements and available data (e.g. environmental monitoring data, questionnaires, work diaries and exposure models).

Exposure Models   Exposure assessment accounts for possibly the greatest level of uncertainty in the overall risk assessment process. It is therefore important to assess the quality of measurements reported and to use statistical techniques in the analysis of the data that take account of possible measurement errors. Due to the relative scarcity of actual exposure data (be it in relation to environmental, dietary, consumer product, occupational, aggregate or cumulative exposures), a wide variety of exposure models have been developed to help meet the exposure assessment needs of regulators, industry and academia.

Risk Characterisation

> **Risk characterisation, the final step in the risk assessment process, can be defined as 'the quantitative or semi-quantitative estimate, including attendant uncertainties, of the probability of occurrence and severity of adverse effect(s)/event(s) in a given population under defined exposure conditions based on hazard identification, hazard characterisation and exposure assessment'.**

In practice, human health risk characterisation is usually accomplished by comparing the POD values (e.g. NOAEL/LOAEL) for the relevant effect(s) with the dose or concentration to which humans are known or anticipated to be exposed. Historically, it has been common to express the relationship between the two terms as the margin of safety (MoS), i.e. the ratio of the predicted exposure to the POD. The magnitude of, and level of certainty in, the MoS is used to form a basis for risk management, with certainty being informed by consideration of the assumptions, limitations and uncertainties from each component of the risk assessment. MoS values less than 100 are normally interpreted as indicating a need for a more comprehensive evaluation of risk. However, such a simple procedure may not fully take into account differences in susceptibility between humans and animals, or within individual animal species or humans. A further pragmatic approach (discussed earlier) is to seek to control exposure to levels as low as reasonably practicable (ALARP); this may be supplemented by providing information on the MoS that exists at anticipated human exposures.

Another approach is use of the POD (NOAEL or LOAEL, or other) established from experimental studies to determine an exposure level in humans that is considered 'acceptable' by society. This is based on the premise that the POD in an experimental animal is an appropriate starting point for the derivation of an exposure level that would be safe for humans. It is done through application of so-called assessment factors (sometimes termed uncertainty factors or safety factors – see *Application of Assessment Factors*, this chapter) to the POD value. The derived 'acceptable' exposure level may be variously referred to – depending on jurisdiction and the type and nature of exposure – as acceptable daily intake (ADI), tolerable daily intake (TDI), tolerable concentration (TC), minimal risk level (MRL), reference dose (RfD) or reference concentration (RfC). Acceptable exposures for workers are generally termed permissible exposure levels (PELs), occupational exposure levels (OELs) or recommended exposure levels (RELs).

Application of Assessment Factors

> **The process of extrapolating findings from the high dosages used in experiments to low (human-relevant) levels is predictive in nature and inherently uncertain. Risk assessors make use of assessment factors not only to account for limitations in datasets but also to address inherent uncertainties in the extrapolation process itself.**

Historically, regulatory authorities have addressed uncertainty pragmatically, for example by applying a default 100-fold factor to a NOAEL from a chronic (lifetime) animal study to determine the acceptable exposure for humans (based on an assumed 10-fold inter-species variability in sensitivity and 10-fold intra-species variability in sensitivity). Additional factors might be included because of limitations in datasets (e.g. from 1- to 100-fold, on a case-by-case basis) when extrapolating from short- to longer-term durations or if using a LOAEL rather than a NOAEL as the POD. Further adjustment might be applied to address the severity of effect and reservations on dataset quality, based on professional judgement. Where specific toxicokinetic and/or toxicodynamic data are available a combination of chemical-specific and default values may be used.

Guidance on the **R**egistration, **E**valuation, **A**uthorisation & restriction of **Ch**emicals (REACH; Regulation (EC) No 1907/2006) published by the European Chemicals Agency (ECHA) provides a table of default assessment factors, the purpose of which is to provide a standardised approach to the calculation of derived no effect levels (DNELs) and derived minimal effect levels (DMELs) in the REACH Registration process (see *Margins of Safety and Acceptability of Risks*, this chapter). This reflects the current pragmatic approach to the use of assessment factors but can be criticised on the basis of lack of flexibility and a certain lack of detail in the guidance provided.

2.1.3 Application of Risk Assessment in Setting Exposure Standards and Limits

Occupational Exposure Limits

> **Occupational exposure limits (OELs) have been a feature of industrialised society for over 50 years, with the earliest limits being established during the 19th and early 20th centuries for agents such as carbon monoxide and quartz dust. By the 1920s, OELs had been proposed for at least 30 chemicals.**

Whilst there is now a globally recognised classification system for the classification of chemical hazards (the Globally Harmonised System, GHS), OELs are established and enforced at the level of individual national or trans-national jurisdictions. In the EU, it is the primary responsibility of the Scientific Committee on Occupational Exposure Limits (SCOEL) to recommend 'health based' OELs where 'a review of the total available scientific data base leads to the conclusion that it is possible to identify a clear threshold dose/exposure level below which exposure to the substance in question is not expected to lead to adverse effects'. 'Risk-based' OELs may be recommended for substances that are genotoxic, carcinogenic or respiratory sensitisers where, at least according to current paradigms, it is not possible to define a threshold of effect any level of exposure however small, might carry some finite risk. It is the responsibility of the European Commission (EC) to set risk-based OELs, a process which requires consultation with interested parties.

SCOEL advises that OELs are principally used 'to provide standards or criteria against which measured exposure levels in existing workplaces may be compared in order to ensure that, as far as the current state of knowledge permits, control is adequate to protect health'.

Margins of Safety and Acceptability of Risk

OELs may be established using human and/or animal data and are intended to be protective under realistic workplace exposure conditions (e.g. by mandating controls on the maximum exposure during a working day or on peak short-term exposures).

> **In deriving an OEL, regulatory bodies have to consider what level of risk is societally acceptable. Various approaches may be adopted for this, such as the previously discussed concept of MoS, where, depending on the endpoint being considered, a low MoS may suggest the need for further evaluation. However, in some circumstances, and in some jurisdictions, there may be the need to include consideration of aspects such as practicality and achievability, not just risk.**

Hence, depending on the particular socioeconomic, legislative and political environment, different regulatory regimes may reach somewhat differing conclusions regarding what constitutes an appropriate OEL for the same chemical.

Since 2010 in the EU, following the implementation of REACH, the concepts of DNEL for threshold effects and DMEL for non-threshold effects have been applied; the latter is intended to represent an exposure level where the likelihood that an adverse effect would occur in a population is considered to be sufficiently low as to be of essentially no concern. Unlike the traditional OEL setting process for which expert judgment was frequently drawn upon to decide on an appropriate level of safety depending on the severity and likelihood of an adverse effect occurring and the technical (feasibility) constraints with regard to exposure control, DNEL or DMEL setting for workers under REACH is a somewhat mechanical process that involves applying assessment factors to the POD, mostly involving the use of various pre-defined default values (see also *Application of Assessment Factors*, this chapter).

In REACH, DNEL values derived for threshold effects and DMEL values for non-threshold effects are used to compare against estimated human exposures to decide on the acceptability of a risk. For threshold effects, a risk is considered acceptable if the ratio DNEL:exposure (termed the risk characterisation ratio, RCR) is <1. For non-threshold substances (such as genotoxic carcinogens), the approach under REACH is less clear cut since the regulations do not specifically define an 'acceptable' risk and in its guidance ECHA states simply that ' … based on these experiences, cancer risk levels of 10^{-5} and 10^{-6} could be seen as indicative tolerable risks levels when setting DMELs for workers and the general population, respectively'.

A somewhat different approach is taken by SCOEL, which, under Council Directive 80/1107/EEC as amended by Council Directive 88/642/EEC, is tasked with developing proposals for either binding or indicative occupational exposure limit values (BOELVs or IOELVs, respectively); the former refer to situations where an exposure threshold for the health effect cannot be reliably identified. If there is a threshold, health-based OELs are recommended while risk-based OELs may be established for non-threshold changes (particularly genotoxicity, carcinogenicity and respiratory sensitisation). For a substance, SCOEL establishes assessment factors on a case-by-case basis, whereas under REACH default values are commonly used. As a consequence, the derived safety margins may differ significantly.

Principles of Precaution

The precautionary principle is now widely referenced in situations where decisions need to be made based upon risk assessment, particularly in relation to chemical-related issues.

> **The term 'precautionary approach' – reflecting the idea that society should seek to avoid environmental damage by careful forward planning and preventing harmful activities – was first coined in the London Declaration, the Ministerial Declaration for the Second International Conference on the Protection of the North Sea (25 November 1987).**

In the European Union (EU), the principle was recognised by Article 191 of the 1992 Maastricht Treaty and in 2000 the EC issued a Communication on the subject which stated:

> 'The precautionary principle applies where scientific evidence is insufficient, inconclusive or uncertain and preliminary scientific evaluation indicates that there are reasonable grounds for concern that the potentially dangerous effects on the environment, human, animal or plant health may be inconsistent with the high level of protection chosen by the EU'.

The statement also noted that the precautionary principle should be used by decision-makers when deciding on risk management policy but should not be confused with use of assessment factors by scientists during risk assessment.

Although it is now widely, although not universally, accepted that the precautionary principle can be of value when deciding on risk policy, there remains debate as to its meaning and, specifically, when and how to apply it in practice. The ongoing wider controversy surrounding the meaning and implementation of the precautionary principle for practical risk management is perhaps not surprising given the diverse views held by various stakeholders with interests in a particular decision and also the diversity of definitions and interpretations of the principle. The OECD has recognised that shifting the burden of proof involves practical and conceptual problems due primarily to the nature and treatment of risk and uncertainty. The precautionary principle is thus perhaps best interpreted as simply the need to favour the side of risk aversion during risk assessment.

The Issue of Feasibility and Achievability

> **The object of any risk assessment – in the context of chemical hazards – should be to define, in a transparent manner based upon scientific evidence, the nature and extent of risk that is posed by the use of the substance in particular scenarios so as to inform risk management decisions; this latter step being influenced by practical, political and economic considerations.**

Each risk management decision will need to consider a wide range of additional issues, including, for example, societal aspirations, social effects, economic costs, feasibility, availability of resources and availability (or not) of alternatives. If a risk is identified, potential options for risk management may range from a minor adaptation of a process or intended use, or taking measures to increase the level of protection provided to workers by reducing their exposure, through to the complete banning of the substance.

Final risk management decisions are often informed by a socioeconomic cost–benefit analysis (CBA), which may consider factors such as technical feasibility, societal and economic issues, ethical and cultural values, and the legislative and political situation, as well as scientific considerations.

An important factor influencing the successful implementation of risk management decisions is that the 'acceptable' level derived through risk assessment should be realistic and able to be substantiated. Thus, the feasibility of achieving a desired exposure level needs to be considered and it must be possible to measure exposure in the workplace or in the environment to a sufficiently high degree of accuracy to demonstrate whether or not control has been achieved. In the case of non-threshold chemicals the magnitude of the technical challenge may be greater, given that larger assessment factors are likely to be applied to derive an 'acceptable' level of risk. In such circumstances, use of the ALARA concept – where the objective is to achieve an acceptable risk by adopting reasonably achievable control measures in a manner that balances risks and benefits – remains appropriate.

2.1.4 The Wider Importance of Risk Assessment

It is essential to maintain stakeholder confidence that transparent and consistent approaches to risk assessment are adopted worldwide. Without this, widely differing standards or statutory limits for chemicals, particularly in relation to occupational, ambient air pollution or food and water safety, will continue to be derived and applied in the various jurisdictions around the world. Such a situation is unsatisfactory and confusing, not just for businesses that manufacture or use chemicals, but also for workers and the wider population.

2.1.5 Summary

The degree of risk resulting from exposure to a hazardous substance is determined by the potency of the hazard and the extent of exposure. Risk assessment traditionally involves hazard identification, hazard characterisation (including dose–response assessment) and exposure assessment. Risk characterisation then integrates the findings, providing a quantitative or semi-quantitative estimate of the probability of occurrence and severity of adverse effects in a given population under defined exposure conditions.

For the majority of chemicals it is possible to define a threshold exposure level below which harmful effects do not occur. Chemicals for which no threshold can be established are treated differently in the risk assessment process, for example using models to predict the scale of any impact at low levels of exposure. Primary genotoxic carcinogens are considered not to have a threshold of effect and it is normally assumed that DNA damage occurs in a linear fashion with increasing dose, meaning that any level of exposure may theoretically result in a mutation, leading to increased risk of cancer. For non-genotoxic carcinogens there is no evidence that genotoxicity is the primary mechanism and it is accepted that there is a threshold dose below which no effect occurs.

Human data are used in the risk assessment process whenever available, for example from case reports, epidemiological studies, occupational studies and clinical research. However, as human data are often not available, data from studies on experimental animals are frequently utilised. Such studies have the advantage that they can be purposely designed, controlled and conducted to address specific gaps in knowledge. In addition, a number of *in vitro* models, *in silico* (computational) tools and 'omics' technologies have been developed to investigate toxicological processes, but when interpreting results it is important to establish their relevance to the *in vivo* situation.

The process of extrapolating findings from the high dosages used in experiments to low (human-relevant) levels is inherently uncertain and requires the use of (multiplicative) assessment factors to account for limitations in datasets and to address uncertainties in the extrapolation process itself. The precautionary principle, which is essentially the need to favour the side of risk aversion during risk assessment, is used by decision-makers when deciding on risk management policy and is not to be confused with use of assessment factors by scientists during the risk assessment process.

In deriving OELs, regulatory bodies have to consider what level of risk is societally acceptable. Various approaches may be adopted for this, and in some jurisdictions there may be the need to include consideration of practicality and achievability.

The object of any risk assessment is to define, in a transparent manner based upon scientific evidence, the nature and extent of risk that is posed by the use of the substance in particular scenarios to inform risk management decisions, the latter being influenced by practical, political and economic considerations.

Further Reading

Bevan, J. and Harrison, P.T.C. (2017) Threshold and non-threshold carcinogens: a survey of the present regulatory landscape. *Regulatory Toxicology & Pharmacology*. Available at: http://dx.doi.org/10.1016/j.yrtph.2017.01.003.

Bolt, H.M. (2008) The Concept of 'Practical Thresholds' in the Derivation of Occupational Exposure Limits for Carcinogens by the Scientific Committee on Occupational Exposure Limits (SCOEL) of the European Union. *Genes and Environment*, **30**(4), 114–119. Available at: https://www.jstage.jst.go.jp/article/jemsge/30/4/30_4_114/_pdf.

EC (2000b) *Communication from the Commission on the precautionary principle*. COM(20001)1 Brussels, 02.02.200. European Commission.

ECHA (2012a) *Guidance on information requirements and chemical safety*. Part E: Risk Characterisation. Version 2.0 November 2012. European Chemicals Agency, Helsinki, Finland. Available at: http://echa.europa.eu/documents/10162/13632/information_requirements_part_e_en.pdf.

ECHA (2012b) *Guidance on information requirements and chemical safety assessment*. Chapter R.8: Characterisation of dose [concentration] response for human health. Version: 2.1 November 2012. European Chemicals Agency, Helsinki, Finland. Available at: http://echa.europa.eu/documents/10162/13632/information_requirements_r8_en.pdf.

ECHA (2012c) *Guidance on information requirements and chemical safety assessment*. Chapter R.14: Occupational exposure estimation. Version: 2.1 November 2012. European Chemicals Agency, Helsinki, Finland. Available at: http://echa.europa.eu/documents/10162/13632/information_requirements_r14_en.pdf.

ECHA (2012d) *Guidance on information requirements and chemical safety assessment*. Chapter R.15: Consumer exposure estimation. Version 2.1 October 2012. European Chemicals Agency, Helsinki, Finland. Available at: http://echa.europa.eu/documents/10162/13632/information_requirements_r15_en.pdf.

EFSA (2014) Modern methodologies and tools for human hazard assessment of chemicals. *EFSA Journal*, **12**(4), 3638. Available at: http://www.efsa.europa.eu/en/efsajournal/doc/3638.pdf.

Fryer, M., et al. (2006) Human exposure modelling for chemical risk assessment: a review of current approaches and research and policy implications. *Environ. Sci. Policy*, **9**, 261–274.

IGHRC (2004) *Guidelines for good exposure assessment practice for human health effects of chemicals*. Report cr10, MRC Institute for Environment and Health, Leicester. Available at: http://www.iehconsulting.co.uk/IEH_Consulting/IEHCPubs/IGHRC/cr10.pdf.

IPCS (2000) *Human Exposure Assessment*. Environmental Health Criteria Monograph no. 214. International Programme on Chemical Safety, WHO, Geneva.

NRC (2009) *Science and Decisions: Advancing Risk Assessment*. National Research Council, National Academies Press, Washington, DC.

SCHER, SCCP, SCENIHR (2009) *Risk assessment methodologies and approaches for genotoxic and carcinogenic substances*. Scientific Committee on Health and Environmental Risks (SCHER), Scientific Committee on Consumer Products (SCCP), and Scientific Committee on Emerging and Newly Identified Health Risks (SCENIHR), European Commission Directorate-General Health & Consumer Protection. Available at: http://ec.europa.eu/health/ph_risk/risk_en.htm.

SCOEL (2013) *Methodology for the derivation of occupation exposure limits*. Key Documentation (version 7) June 2013. Scientific Committee on Occupational Exposure Limits, European Commission DG Employment, Social Affairs & Inclusion. Available at: http://www.google.co.uk/url?sa=t&rct=j&q=&esrc=s&source=web&cd=1&ved=0CCEQFjAA&url=http%3A%2F%2Fec.europa.eu%2Fsocial%2FBlobServlet%3FdocId%3D4526%26langId%3Den&ei=fG-FVNTPHsOrU76Og4gC&usg=AFQjCNEGbqUBE71ITtMcGMJcbSexCWdICw&bvm=bv.80642063,d.d24.

US EPA (undated) *Human Health Risk Assessment*. Available at: https://www.epa.gov/risk/human-health-risk-assessment

US EPA (2005) *Guidelines for Carcinogen Risk Assessment*. Available at: http://www.epa.gov/raf/publications/pdfs/CANCER_GUIDELINES_FINAL_3-25-05.PDF.

2.2 Toxicokinetics

Johannes G. Filser

2.2.1 Definition and Purpose

> **The object of toxicokinetics is the investigation of the absorption, distribution, and elimination of toxicants and their toxicologically relevant metabolites as functions of dose (the amount administered) and time.**

In toxicokinetic studies, concentrations of the administered substance and its relevant metabolites in body fluids, organs, and excrement are determined in a time-dependent manner.

If substances are volatile, concentration–time courses are monitored in the exhaled air. *In vitro*, concentration–time courses are examined in organs, cells and cell fractions, and distribution and binding studies are carried out.

For toxicological studies, experimental animals are used. Since the probability of detection of a toxic or carcinogenic effect is a function of the dose, very high doses of the toxicant are used so that even with small numbers of animals adverse effects can be seen. The interpretation and extrapolation of the obtained dose–response curves to low doses and concentrations relevant in the human environment requires knowledge of the ultimate active chemical species (the parent compound or a metabolite) and of its tissue burden as a function of dose and time. Therefore, toxicokinetic studies must be carried out with experimental animals, and very wide concentration ranges of the toxicant must be used. On the other hand, to allow appropriate extrapolation of the dose–response curves to humans, it is also necessary to know in this species the systemically available dose of the ultimately active substance. For ethical reasons, toxicokinetic studies in humans can be carried out only to a very limited extent. Therefore, the toxicokinetics in organ and tissue fractions from experimental animals and humans are compared in additional *in vitro* studies. For the extrapolation to humans of the results of toxicokinetic studies with experimental animals *in vivo* and *in vitro*, and with human tissue *in vitro*, physiological toxicokinetic models that take into account physiological and biochemical species differences are increasingly used.

2.2.2 Absorption, Distribution, and Elimination

> **Absorption means uptake of a substance into the lymph and bloodstream. Distribution comprises both the transport of the substance with the circulating blood and its accumulation in organs and tissues. Elimination describes the removal of the substance from the organism. It covers two processes, the biotransformation (metabolism) of the substance into other products (metabolites) and the excretion from the organism (Figure 2.1).**

Once formed, metabolites can also be distributed via the bloodstream and finally eliminated. The basic physiology of these processes is summarized below.

Passage through Membranes

> **During the processes of absorption, distribution, and elimination, substances must pass through biological membranes.**

Biological membranes separate morphological and functional differing entities (e.g., cells, nuclei, mitochondria, Golgi apparatus, endoplasmic reticulum). They consist primarily of a phospholipid bilayer in which proteins are floating. These proteins serve to maintain the membrane structure but also function as enzymes. Carriers that transport substrates through the membrane span the entire membrane. Some proteins form "pores" that permit water to cross the membrane. The biological membrane is selectively permeable. Lipophilic and small polar molecules can easily diffuse through it. Others require specific transport mechanisms.

> **In decreasing order of importance mechanisms for the transport of substances through a biological membrane include passive diffusion, convective transport, active transport, facilitated diffusion, and endocytosis.**

36   Toxicology and Risk Assessment

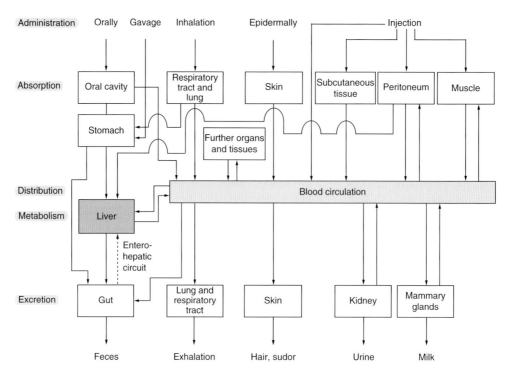

Figure 2.1 *Routes taken by substances in the organism: absorption, distribution, and elimination (metabolism and excretion).*

Passive Diffusion   Passive diffusion through membranes is very important for the transfer of substances and takes place in both directions. To be able to be transported by passive diffusion, the permeating substance must be in solution. The rates of diffusion are directly proportional to the concentrations on the two sides of the membrane and to the surface area of the membrane, and are inversely proportional to the thickness of the membrane. At equilibrium, the amounts of substance diffusing per time unit through the membrane from the two sides are the same, but the concentrations on the two sides of the membrane need not necessarily be the same: the concentrations of a substance which accumulates in individual compartments of cells and tissues are determined by hydrophilic and hydrophobic interactions between the substance and its biological surroundings. Lipophilic substances, for example, accumulate in the lipids of an organism. A measure of the maximum possible accumulation of a substance in the various tissues is given by the partition coefficients. They are thermodynamic constants that express the ratio of the concentrations of a substance in two phases in equilibrium. In Table 2.2 the partition coefficients muscle:blood and fat:blood for some hydrocarbons and the partition coefficients blood:air for their vapors are shown.

Ionized organic substances (e.g., quaternary ammonium compounds), which cannot normally diffuse through biological membranes, can nonetheless be absorbed from the gastrointestinal tract. It is assumed that these compounds combine with endogenous polyionic substances to form ion pairs, which are externally uncharged and so can diffuse passively through the membranes. Subsequently, the complexes are thought to dissociate. Such "ion pair transport" would also be a passive diffusion process.

Table 2.2 *Partition coefficients for some hydrocarbons (muscle:blood and fat:blood) and their vapors (blood:air) (data for human tissues).*

Substance	Partition coefficient		
	Muscle:blood	Fat:blood	Blood:air
n-Pentane	1.84	104	0.38
n-Hexane	6.25	130	0.80
n-Heptane	5.25	162	2.38
Benzene	2.23	51.5	7.37
Toluene	2.31	63.7	15.1
p-Xylene	1.61	51.9	38.9

Values were taken or calculated from a collection of blood: air and tissue:air partition coefficients (Meulenberg et al., Toxicol. App. Pharmacol. 165: 206–216). Partition coefficients fat:blood and muscle:blood were obtained by dividing the corresponding tissue:air partition coefficients by the respective blood:air partition coefficients.

Convective Transport Convective transport or "filtration" is understood to mean the permeation through membranes of substances dissolved in water, which flow with the water through pores of 7–10 Å diameter. This route is especially important for ionized substances, for small hydrophilic molecules such as urea, and for substances such as the lower alcohols with properties like those of water. In general, spherical and thread-like molecules with molecular weights up to about 150 and 400, respectively, can cross biological membranes by convective transport. Convective transport is another kind of passive transport; the driving force is mainly the difference in hydrostatic pressure between the two sides of the membrane. In addition, osmotic pressure plays an important role and, with charged particles, also the charge on the walls of the pores.

Active Transport When a substance is transported through the membrane with the help of a carrier molecule in an energy-requiring process, the mechanism is described as active transport. A carrier is a membrane-bound enzyme, generally an adenosine triphosphatase, which binds the substance more or less specifically and transfers it to the other side of the membrane, where it is released.

Being a process catalyzed by an enzyme, active transport obeys saturation kinetics (see *Saturation Kinetics*, this chapter). This means that an increase in the dose is associated with an increase in the rate of transport only within a certain dose range. Active transport can also take place against a concentration gradient as it involves energy consumption.

Among the enzyme-catalyzed processes is exchange diffusion, an energy-consuming process in which Na^+ is actively transported out of the cell in exchange for K^+, which enters passively. This sodium pump is responsible for reducing intracellular low Na^+ in exchange for extracellular K^+.

Facilitated Diffusion Facilitated diffusion is also carrier-mediated and saturable. However, in this case the concentration gradient is the driving force. The energy reserves of the cell are not used.

Endocytosis Endocytosis is an active transport mechanism by which larger or smaller aggregates are moved into the cell. The aggregate is enclosed in the cell membrane and incorporated by the cell in a vesicle. In the cytoplasm, the vesicle membrane is broken down and the aggregate released. In this way even solid objects such as plastic particles can be taken up.

Absorption

Of most toxicological significance for the uptake of chemicals by the human organism is absorption via the respiratory tract, the gastrointestinal tract, and the skin. In animal studies other routes also play a role.

Respiratory Tract

Foreign compounds (xenobiotics) can be inhaled not only when they are in gaseous form but also when they are particles, solid, or liquid (aerosols), suspended in the air.

Gaseous substances can be absorbed in the whole respiratory tract and most especially in the alveoli. The alveoli are surrounded by a dense capillary net, which is perfused with the blood from the pulmonary circulation. The gas exchange between the alveolar space and the capillary blood proceeds extremely rapidly because the separating alveolocapillary membrane is less than 1 μm thick. This membrane consists of the alveolar epithelium, an interstitium with elastic fibers, and the capillary endothelium. Consequently, when neglecting any uptake in the upper airways, the inhalation and exhalation processes can be described by considering only the mass transfer between air and capillary lung blood (Figure 2.2). The amount of a gaseous substance taken up from the capillary blood per time unit (v_{alvi}) is equal to the product of the alveolar ventilation (Q_{alv}) and the concentration of the substance in the inhaled air (c_{air}):

$$v_{alvi} = c_{air} \cdot Q_{alv} \tag{2.1}$$

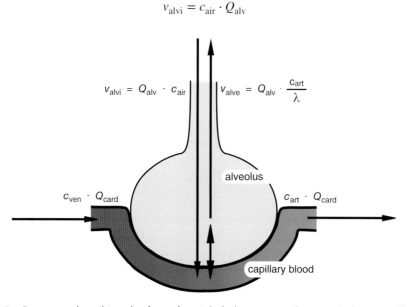

Figure 2.2 Processes describing the fate of an inhaled gaseous substance in lung capillaries. c_{art}, concentration in oxygen-rich (arterial) blood; c_{air}, concentration in the inhaled air; c_{ven}, concentration in oxygen-poor (venous) blood; λ, partition coefficient blood:air; Q_{alv}, alveolar ventilation; Q_{card}, blood flow through the lung; v_{alve}, rate of alveolar elimination (exhalation); v_{alvi}, rate of absorption in the capillary blood (inhalation); ↕, immediate distribution between oxygen-rich capillary lung blood and alveolar air according to λ.

The alveolar ventilation at rest is about 120 ml/min and 5 l/min in a rat of 250 g body weight and an adult human, respectively.

The amount of gaseous substance exhaled per time unit into the alveolar space (v_{alve}) is equal to the product of Q_{alv} and the concentration of the substance in the alveolar air that is given by the ratio of the substance concentration in the oxygen-rich blood (c_{art}) leaving the lung capillaries to the partition coefficient blood:air (λ):

$$v_{alve} = Q_{alv} \cdot c_{art}/\lambda \qquad (2.2)$$

The quantity of the substance entering the lung capillaries per time unit (v_{ven}) is equivalent to the product of the blood flow through the lung (Q_{card}) and the concentration of the substance in the oxygen-poor blood (c_{ven}) entering the lung capillaries:

$$v_{ven} = Q_{card} \cdot c_{ven} \qquad (2.3)$$

Q_{card} is identical to the cardiac output – the total volume of blood pumped by the ventricle per time – which is about 83 ml/min in a rat (250 g body weight) and 6.2 l/min in a human (70 kg body weight) at rest.

The absorbed amount of the substance that leaves the lung capillaries per time unit (v_{abs}) in the oxygen-rich blood and enters the residual body is equivalent to the product of Q_{card} and c_{art}:

$$v_{abs} = Q_{card} \cdot c_{art} \qquad (2.4)$$

Considering all four processes and taking into account that the amounts entering the capillaries equal those leaving them per time unit, Equation 2.5 is obtained:

$$Q_{card} \cdot c_{ven} + c_{air} \cdot Q_{alv} = c_{art} \cdot Q_{alv}/\lambda + c_{art} \cdot Q_{card} \qquad (2.5)$$

$Q_{card} \cdot c_{ven}$ is zero during the first inhalation process when there is still none of the xenobiotic in the organism. Under this condition the absorption of the inhaled substance into the oxygen-rich blood occurs with maximum rate (v_{absmax}). From Equations 2.4 and 2.5 v_{absmax} is:

$$v_{absmax} = c_{air} \cdot [Q_{alv} \cdot \lambda \cdot Q_{card}/(Q_{alv} + \lambda \cdot Q_{card})] \qquad (2.6)$$

From this equation it may be seen that for large values of λ, v_{absmax} becomes the product of only c_{air} and Q_{alv}, and at small values of λ, it becomes independent of Q_{alv}, resulting in the product of c_{air}, λ and Q_{card}. Only at high values of λ is the maximum rate of absorption of gaseous substances limited by the alveolar ventilation. If λ is small, estimation of v_{absmax} only from the alveolar ventilation yields values which are too high (Figure 2.3). During intensive physical exercise, which leads to an increase especially in Q_{alv}, the v_{absmax} for substances with small λ increases less than that for those with large λ.

With the exposure duration at constant c_{air}, the concentration of the substance in the oxygen-poor venous blood (c_{ven}), which flows from the heart to the lung, increases to a plateau determined not only by λ but also by the rate of metabolic elimination of the substance. The effectiveness of metabolic elimination can be very different for different substances. During the time to the plateau of c_{ven} the absorption rate v_{abs} decreases from the initial v_{absmax} and then remains constant until the end of exposure.

The absorption of inhaled particles depends strongly upon their sizes and the kind of breathing (nose or mouth). Particles larger than about 10 μm are almost exclusively filtered by the nose under conditions of nose breathing. At mouth breathing, even large particles reach the bronchi. Generally, particles can penetrate into the alveoli when they are smaller than 2-3 μm. Some of the

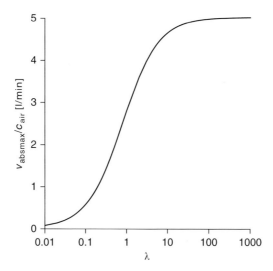

Figure 2.3 Ratio of the maximum rate of absorption (v_{absmax}) of an inhaled gaseous substance to the concentration of the substance in the inhaled air (c_{air}) as a function of the blood:air partition coefficient (λ) in persons at rest. Only when λ is larger than about 10 is v_{absmax} given by the product of the alveolar ventilation (5 l/min at rest) and c_{air}. At small values of λ, v_{absmax} can be much smaller than this product.

particles that are deposited in the tracheobronchial tree are transported by the ciliated epithelium into the throat, where they are swallowed and can then be absorbed in the gastrointestinal tract. Soluble particles can be absorbed directly through the epithelium of the respiratory tract into the bloodstream.

Oral Cavity and Gastrointestinal Tract

> **In humans, the absorption of toxic substances through the oral cavity and from the gastrointestinal tract takes place when individuals have eaten contaminated food or accidentally ingested the toxin. It is also relevant for toxicological studies with laboratory animals to which a xenobiotics is administered via food or by gavage.**

In humans, uncharged and sufficiently lipophilic substances can readily enter the systemic circulation through the oral mucosa, which is well-supplied with blood vessels and has an effective surface area of about 0.02 m² in adults. The administration of chemicals into the stomach by means of a tube (gavage) is a standard method in animal studies. As a rule, substances are taken up rapidly through the mucous membrane of the stomach, which has an acid milieu, and that of the small intestine, which is weakly alkaline, and transferred with the blood via the portal vein to the liver, the main metabolic organ of the body. However, some substances, especially some heavy metals, are only poorly absorbed from the gastrointestinal tract. Substances that are not stable in acid, such as epoxides, can be rapidly hydrolyzed in the stomach, resulting in a reduced absorption. Many compounds are subjected to a first biotransformation in the intestinal walls and, even more so, in the liver (**first-pass effect**). Not until they have passed through the liver can they, or at least the fraction which remains unmetabolized, enter the blood of the systemic circulation.

Because of the first-pass effect, usually only part of the dose of substances administered orally becomes available systemically. This phenomenon is described as reduced bioavailability. The **bioavailability** expresses the fraction of the dose that reaches the systemic blood circulation. When the substance is absorbed through the lungs or the skin, the bioavailability is usually not or only slightly reduced.

Skin

> **The rate of absorption of a substance through the skin (surface area about 300 cm² for a rat of 250 g and 1.8 m² for an adult man of 70 kg body weight) depends on the physicochemical properties of the substance and the state of the skin. The stratum corneum, the outermost layer of the skin, made up of stacked interconnected dead corneocytes embedded in a lipophilic matrix, forms the main barrier and represents the rate-determining step for the passage of xenobiotics through the skin.**

There are four possible routes of percutaneous diffusion of substances: through the cells (transcellular) or between the cells (intercellular) of the stratum corneum, through the excretory ducts of the sebaceous and sweat glands (transglandular), and along the hair shafts through the hair follicles (transfollicular). The last two possibilities are routes through pores in the skin. As the pores account for only 0.1% of the surface area of the human skin, these routes are of less importance than those through the stratum corneum. Of these, hydrophilic compounds prefer the transcellular and lipophilic substances the intercellular route. In general, the hydro- or lipophilicity of a substance and its molecular size determine its ability to penetrate the skin. Additionally, the permeability of the skin can be altered by the xenobiotic, for example, as a result of swelling or defatting.

Water penetrates the stratum corneum only very slowly. Other substances may penetrate relatively rapidly, especially lipophilic substances such as organophosphates and polychlorinated biphenyls. For most gaseous substances, percutaneous absorption is only a few percent of that from inhalation. High rates of transdermal uptake are found for vapors of some volatile amphiphiles which have very high partition coefficients between the tissue and the gas phase. For example, under conditions of whole body exposure the vapors of 2-butoxyethanol and of dimethylformamide are taken up more rapidly through the skin than by inhalation.

The rate of absorption through the skin (i.e., the stratum corneum) may be described in a simplified way in terms of Fick's first law:

$$J = k_p \cdot \Delta c \tag{2.7}$$

J is designated as the flux. It expresses the quantity of substance that penetrates the skin per unit area and time. Δc is the difference between the concentrations of the substance on the two sides of the stratum corneum and k_p is the permeability constant (dimension: distance/time). Normally the concentration under the stratum corneum is assumed to be negligible. This yields the generally used equation:

$$J = k_p \cdot c \tag{2.8}$$

Here, c is the concentration of the substance on the surface of the skin. The value of k_p depends on the properties of the substance and the state of the stratum corneum.

Intravascular, Subcutaneous, Intramuscular, and Intraperitoneal Injection

> **In animal studies, toxicants are often injected into a blood vessel (intravascular), directly under the skin (subcutaneous), into the muscles (intramuscular), or into the abdominal cavity (intraperitoneal).**

Intravascular, mostly intravenous, injection of a substance bypasses the absorption phase so that the administered dose is systemically available at once.

Substances administered by subcutaneous or intramuscular injection are transferred relatively slowly into the blood so that, although the bioavailability is the same, the initial concentrations in the blood are lower than after intravascular injection. Substances injected intraperitoneally are absorbed through the peritoneum, a membrane that consists of a thin epithelium covering a layer of well-vascularized connective tissue and lines the abdominal cavities (peritoneum parietale) and covers the abdominal viscera (peritoneum viscerale). In humans and rats the peritoneum has surface areas of about 1 m^2 and 500 cm^2, respectively. The movements of the intestines distribute the injected substance rapidly over the peritoneum. Lipophilic substances are absorbed very rapidly. After passage through the peritoneum, the absorbed substance can be transported in the lymph ducts or bloodstream in two directions: via the lymph ducts and blood vessels of the membrane covering the walls of the abdomen (peritoneum parietale) the absorbed substance is transported to the heart and so directly into the systemic circulation, and via the blood vessels of the membrane covering the peritoneum viscerale the absorbed substance is transported through the portal vein to the liver. Because of this passage through the liver, after intraperitoneal application a first-pass effect may be seen.

Distribution

> **The distribution of a substance within the organism takes place via the bloodstream.**

At continuous uptake, substances accumulate in the organism until **steady state** is reached, that is, until the concentration-dependent rate of elimination (see below) equals the rate of uptake. The length of time required until steady state depends on the physicochemical and biochemical properties of the substance and can be very different for different substances; if the substance has a high affinity for poorly vascularized tissues and if it is slowly or not metabolized the time until steady state can be very long. Lipophilic substances accumulate in the adipose tissue, and heavy metals such as strontium or lead with chemical properties similar to those of calcium are stored in bones in the form of poorly soluble salts. Many substances can bind reversibly to plasma and tissue proteins.

After a single dose or a brief exposure, redistribution of a substance between tissues can take place. The substances are first distributed in the well-perfused organs such as brain, heart, liver, lungs, spleen, and kidneys. The accumulation in tissues that are less well or poorly perfused (muscle and adipose tissue, respectively) takes place more slowly and is still continuing when the substance concentrations in blood and well-perfused organs are already declining due to end of exposure (Figure 2.4a). When a substance is absorbed continuously at a constant rate, as it is during exposure to a constant concentration in the inhaled air or during continuous infusion under constant conditions, the tissues become saturated with the substance in sequence. Then, after the end of exposure or infusion, redistribution does not take place. Instead the concentrations decrease in all tissues, first more rapidly in the blood and in the better perfused organs. Finally,

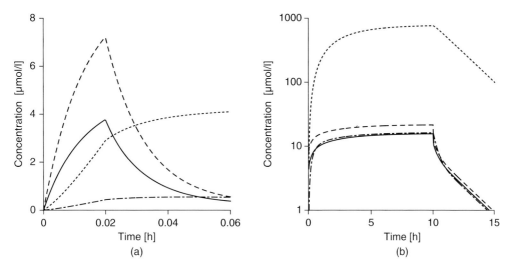

Figure 2.4 Modeled concentration–time courses of inhaled styrene in various tissues of a rat during and after exposures to styrene vapor at a concentration of 100 ppm in the inhaled air. (a) Duration of inhalation exposure 0.02 h; post-exposure period 0.04 h; linear plot. (b) Duration of inhalation exposure 10 h; post-exposure period 5 h; half-logarithmic plot. ———, oxygen-poor (venous) blood; – – –, richly-perfused organs; - — - —, moderately perfused tissues (e.g., muscles); - - - -, adipose tissue. Curves were calculated with a physiologically-based toxicokinetic model for styrene (Csanády et al., Arch. Toxicol. 68: 143–157).

in the terminal elimination phase, the concentrations decrease parallelly in all tissues. This phase is often determined by the slowest of the processes, which release the substance into the blood. With well-metabolized lipophilic substances, for example styrene, this is generally the release of the substance from the poorly perfused adipose tissue that serves as a storage organ (Figure 2.4b).

Elimination

The two elimination routes are conversion of substance into metabolites and excretion of the unchanged substance.

Metabolism Metabolism or biotransformation is generally catalyzed by enzymes. It can take place in all organs and tissues. For the biotransformation of most xenobiotics, however, the liver plays the quantitatively most important role.

Metabolites are other substances than the initial one; they have their own chemical and toxicological properties and their own toxicokinetic profiles. Frequently it is a metabolite that is considered to be the cause for the adverse effect of a substance.

Saturation Kinetics

Enzyme-catalyzed metabolism obeys saturation kinetics which, in the simplest case, can be described by Michaelis–Menten kinetics.

The Michaelis–Menten equation expresses the rate of an enzyme-mediated substance conversion (dc/dt; dimension: concentration/time) as a function of the concentration of the substrate (c)

at the enzyme, the maximum change of substance concentration per time (S_{max}), and the Michaelis constant (K_m):

$$dc/dt = (S_{max} \cdot c)/(K_m + c) \quad (2.9)$$

K_m equals that value of c at which half of S_{max} is attained and is a measure of the affinity of the substrate for the enzyme: the smaller the value of K_m, the larger is the affinity. In the low concentration range where c is much smaller than K_m, the reaction obeys **first-order kinetics**. This means that dc/dt is directly proportional to c, as may be seen from Equation 2.9a, when $c \ll K_m$:

$$dc/dt \cong (S_{max}/K_m) \cdot c \text{ (when } c << K_m) \quad (2.9a)$$

Here the proportionality factor is identical with the ratio S_{max}/K_m.

With increasing c, direct proportionality is no longer found; dc/dt increases more slowly than c and reaches 91% of S_{max} when c is 10 times the K_m. Further increases in c only have a small effect on dc/dt. When S_{max} is attained, dc/dt obeys **zero-order kinetics**. In a zero-order reaction, dc/dt is independent of c and is constant:

$$\lim_{c \to \infty} dc/dt = S_{max} = \text{constant} \quad (2.9b)$$

$$c \to \infty$$

The value of S_{max} is proportional the total concentration of the enzyme involved.

The simultaneous presence of other substances which can interact with the enzyme can reduce the rate of conversion. The most important example of such enzyme inhibition is **competitive inhibition**, in which a second substance competes with the substrate for the binding site of the enzyme. This process causes a reduction in the value of dc/dt to below that for the substrate on its own. The properties of the enzyme itself are not changed; however, the apparent K_{mih} values obtained are higher than the real K_m value determined in the absence of the inhibitor. In Figure 2.5, the relationship between dc/dt and c is shown for Michaelis–Menten kinetics in the absence and presence of a competitive inhibitor.

Excretion

> **Excretion takes place mainly via the kidneys, the intestinal tract, and the lungs, but the skin and mammary glands may also be involved.**

Kidneys

> **Substances below a certain molecular size are excreted via the healthy kidney; in humans these are mostly substances with molecular weights below 300.**

Many substances are converted by Phase I and Phase II reactions into more water-soluble compounds, which are excreted in the urine. Both passive and active transport processes can be involved in the excretion from the blood plasma via the kidney into the urine. In the passive processes, the average rate of excretion with the urine (dimension: amount/time) is directly proportional to the concentration of the free substance in the blood plasma (c_P). This is also the case for active processes if the concentration of the substance is markedly smaller than the K_m value

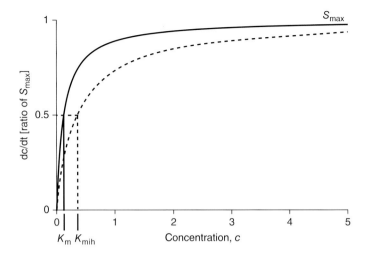

Figure 2.5 Rate of conversion of a substrate as a function of its concentration: Michaelis–Menten kinetics in the presence and absence of a competitive inhibitor (linear plot). ———, without inhibitor; – – –, with inhibitor; K_m, Michaelis constant: substrate concentration at half maximum conversion rate; K_{mih}, apparent K_m in the presence of a certain concentration of the inhibitor; S_{max}, maximum change of substance concentration per time.

of the active enzyme-mediated transport process (see above). The proportionality factor between the rate of urinary excretion and c_P is called the **renal clearance** (Cl_{ren}). To determine Cl_{ren}, c_P is measured in the middle of a relatively short urine collection period. In addition, the quantity of substance eliminated with the urine during this period (Δt) is determined. It is the product of the measured urine volume (V_{ur}) and the concentration of the substance in the urine (c_{ur}). Thus, Cl_{ren} is given by:

$$Cl_{ren} = (V_{ur} \cdot c_{ur})/(\Delta t \cdot c_P) \qquad (2.10)$$

In toxicokinetics, clearance (Cl) is given the dimension volume/time. It links a rate (v, dimension: amount/time), with a concentration (c, dimension: amount/volume):

$$Cl = v/c \qquad (2.11)$$

Cl is a constant over the concentration range in which the elimination obeys first-order kinetics and thus v is directly proportional to c. When saturation kinetics applies, the clearance becomes smaller with increasing concentration because the rate of the active process tends gradually towards a maximum rate and thus becomes independent of the concentration. The concentration dependence of the clearance when kinetics according to Michaelis–Menten apply is given by (cf. Equations 2.9 and 2.11):

$$Cl = V_{max}/(K_{mapp} + c) \qquad (2.12)$$

where V_{max} is the maximum rate (dimension: amount/time) and the concentration K_{mapp} is called the apparent Michaelis constant. (The abbreviation K_{mapp} demonstrates that it is not the true enzyme-specific K_m).

Generally, a clearance is a measure of the elimination of a substance. Clearances can be defined not only for the kidney but for any eliminating organ or tissue and for the sum of such organs and tissues. As the clearance is related to a concentration, in all cases the medium in which the concentration was determined, mostly plasma or blood, must be given. Frequently, metabolic clearance is determined; this parameter enables the calculation of the rate of metabolism in a certain organ (e.g., liver) or the whole organism in association with the concentration of the substance in the medium. The sum of all clearances from the organism, if related to the concentration in a single medium (e.g., plasma), is the total clearance (Cl_{tot}).

Intestinal Tract

> **The intestinal tract includes the small and large intestines. Substances that are excreted with the feces can have entered the intestinal tract from the liver via the bile or have been excreted directly through the intestinal membrane.**

Whereas low molecular weight substances are mostly excreted via the kidneys (see above), substances with molecular weights greater than 300 are mostly transferred into the intestines with the bile. The bile is actively secreted by the parenchymal cells of the liver. To be excreted with the bile, substances must not only be of a certain minimum size but must also contain a polar group. Both conditions may be fulfilled after the substance has undergone Phase II conjugation reactions. Of most quantitative significance is the conjugation with glucuronic acid, which is catalyzed by the enzyme UDP-glucuronyltransferase. The concentration of xenobiotics in the bile can be a multiple of that in plasma and so high concentrations of toxic substances can be found in the hepatobiliary system. Some of the substances that enter the intestines with the bile are metabolized by bacterial enzymes. For example, glucuronic acid conjugates can be cleaved by bacterial β-glucuronidases. The substances so released can be absorbed and transferred via the portal vein to the liver and, after conjugation with glucuronic acid, are excreted with the bile back into the intestine, where they are released again. This is the so-called **enterohepatic circulation** (Figure 2.1).

Substances can enter the intestines directly through the intestinal mucosa, as has been demonstrated, for the cardiotonic glycoside, digoxin, or mercury.

Lung

> **Via the lung volatile substances and gases can be eliminated by exhalation.**

Inhaled gases, vapors and volatile metabolites can be excreted by respiration. The rate of alveolar elimination (v_{alve}; Equation 2.2) depends not only on Q_{alv} and λ but also on the concentration of the substance in the oxygen-rich blood (c_{art}) which flows from the lung via the heart into the arteries of the body. It is obtained by reformulating Equation 2.5 to:

$$c_{art} = (c_{ven} \cdot Q_{card} + c_{air} \cdot Q_{alv})/(Q_{card} + Q_{alv}/\lambda) \qquad (2.13)$$

In its turn c_{art} depends on the rate of alveolar absorption from the air (v_{alvi}; Equation 2.1), v_{alve}, Q_{card} and c_{ven}, the concentration of the substance in the oxygen-poor blood which enters the lung capillaries (Equation 2.3 and Figure 2.2). The extent of metabolic elimination has a considerable influence on c_{ven} and, consequently, also on v_{alve}.

The ratio of the amount of a volatile substance eliminated by exhalation to that eliminated metabolically is substance-specific and concentration-dependent if saturation kinetics occur. It can have very different values. Some substances (e.g., 1,1,1-trichloroethane) are mainly exhaled unchanged; others (e.g., styrene inhaled at concentrations of less than 200 ppm) are eliminated predominantly by the metabolic route.

Skin

> **Perspiration is also an elimination route.**

The final product of normal protein catabolism, urea, and a number of drugs have been shown to be eliminated by this route. Arsenic and thallium can be readily detected in hair because they accumulate in the skin and its appendages.

Mammary Glands

> **The elimination of xenobiotics with contaminated breast milk can result in an internal exposure of the baby.**

Passive diffusion is the main mechanism by which xenobiotics enter a mother's milk, the pH of which (6.6) is slightly lower than that of the organism (7.4). Not only many medicines (e.g., antibiotics) but also everyday drugs such as nicotine, ethanol and environmental substances such as heavy metals, polyhalogenated biphenyls, polychlorinated dibenzodioxins, and polychlorinated dibenzofurans (PCDD/PCDF) are eliminated with the milk. In particular, the elimination of the slowly metabolized, highly lipophilic PCDD/PCDF in milk fat has the effect of markedly reducing the body burden of these substances in the breast-feeding woman. Conversely, in breast-fed babies higher PCDD/PCDF levels have been determined than in those who were not breast-fed.

2.2.3 Toxicokinetic Models

> **Toxicokinetic models describe the fate of substances (absorption, distribution, elimination) by mathematical functions. Mostly compartment models or physiologically based toxicokinetic models are used.**

In both kinds of model, open compartments are defined, which are characterized by their volumes and current concentrations of substance. The number of compartments in a model depends on the physicochemical and biochemical properties of the substance investigated and on the problems that are dealt with.

Compartment Models

> **In compartment models (Figures 2.6 and 2.13), the compartments are usually imaginary entities that do not need a physiological basis.**

In the low concentration range, invasion and elimination processes generally obey first-order kinetics. A kinetic is first order when a rate is directly proportional to a concentration. It is

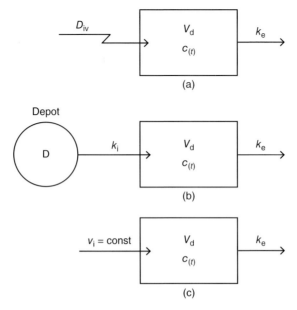

Figure 2.6 One-compartment model for the description of the toxicokinetics of a substance administered by various routes and eliminated according to first-order kinetics. (a) intravascular injection, (b) absorption from a depot after extravascular administration, and (c) inhalation of a substance at constant concentration in the air or continuous intravascular infusion. $c_{(t)}$, concentration at time t; D_{iv}, intravascular dose; D, dose; k_e, elimination rate constant; k_i, invasion rate constant (first-order kinetics); V_d, apparent volume of distribution; v_i, constant invasion rate.

therefore also described as linear kinetics. In such a process (absorption, distribution, metabolism, or excretion) an exponential function of the general form

$$c_{(t)} = A_1 \cdot e^{-k \cdot t} + A_2 \qquad (2.14)$$

describes the substance concentration in a time-dependent manner. The concentration in the compartment at any time t is given by $c_{(t)}$. A_1 and A_2 are constants, with their sum being the initial substance concentration $c_{(0)}$ at time $t = 0$. k is a rate constant with the dimension time^{-1}. If deviations from first-order kinetics are observed, like those always seen at higher concentrations in saturable processes, the kinetics are said to be non-linear (see *Saturation Kinetics*, this chapter). To indicate the direction of a process in a particular tissue, its rate (v; dimension amount/time) is given a positive sign if the amount of substance in the compartment is increasing and a negative sign if it is decreasing.

The One-compartment Model

> **The one-compartment model is the most straightforward toxicokinetic model. It represents the whole body as a single compartment. The one-compartment model (Figure 2.6) may be used when the distribution processes are much more rapid than the absorption and elimination processes.**

In the section below, the reader will be familiarised with the basic principles in toxicokinetics exemplified by the one-compartment model.

Apparent Volume of Distribution The absorbed substance is often not evenly distributed in the organism but concentrated in certain depots (e.g., adipose tissue, protein binding) according to its physicochemical properties. The use of a factor (V_d) makes it possible to associate the concentration $c_{(t)}$ measured in a defined body fluid at a certain time point t with the amount of substance $N_{(t)}$ present at this time in the body:

$$N_{(t)} = V_d \cdot c_{(t)} \tag{2.15}$$

V_d has the dimensions of a volume and is called the apparent volume of distribution. Usually, it is a purely mathematical quantity. In the one-compartment model, in which it is assumed that the distribution processes take place so rapidly that they have no effect on the toxicokinetics, the value of V_d is constant (which is not the case in models that have more than one compartment). V_d can be calculated following intravascular administration of the substance from the dose D_{iv} (the administered amount; see Figure 2.6a), and the measured (or extrapolated) concentration $c_{(0)}$ at time point $t = 0$:

$$V_d = D_{iv}/c_{(0)} \tag{2.16}$$

Because most substances are not distributed homogeneously in the various tissues, the size of V_d depends on the medium in which c is measured. For example, if the substance-specific partition coefficient blood:blood plasma (P_{BP}), which can be determined *in vitro*, is not equal to 1, then the value of V_d obtained when c is measured in blood plasma (c_P) is different from that obtained when c is measured in blood (c_B). The apparent volume of distribution when the concentration of the substance is measured in plasma (V_{dP}) is given by:

$$V_{dP} = D_{iv}/c_{P(0)} \tag{2.17}$$

Considering that in the one-compartment model there is no distribution phase, that is

$$c_{B(t)} = c_{P(t)} \cdot P_{BP} \tag{2.18}$$

the apparent volume of distribution related to the concentration in the blood (V_{dB}) is given by:

$$V_{dB} = D_{iv}/(c_{P(0)} \cdot P_{BP}) \tag{2.19}$$

or, considering Equation 2.17,

$$V_{dB} = V_{dP}/P_{BP} \tag{2.20}$$

In a few cases, V_{dP} is identical to the volume of blood plasma, which accounts for about 4 % of the body weight of an adult human. This applies, for example, to heparin or Evan's Blue macromolecules, which can penetrate neither the vessel walls nor the cell membranes of the erythrocytes and which are therefore distributed only in the intravascular plasma volume. For ethylene oxide, an epoxide, and the solvents ethanol, ethylene glycol, and acetone, V_{dP} is slightly larger than the volume of the total body water, which is about 60% of body weight in adult men.

Many xenobiotics bind to plasma proteins, which reduces the concentration of the free substance in the plasma. This phenomenon can result in very large values for V_{dP} because it is related to the concentration of the free substance. In general, the ratio of free to bound substance is constant over a wide concentration range. However, with increasing concentration the binding sites at the plasma proteins are increasingly occupied and finally the fraction of the total substance that is free also increases. In such cases, V_{dP} becomes concentration dependent, having a smaller value at high concentrations than at lower ones.

Intravascular Administration, Elimination, Total Clearance, and Half-life The easiest toxicokinetic challenge is the elimination kinetics of an intravascularly, as a single-dose (D_{iv}) administered substance which is distributed much more rapidly than it is eliminated, and for which the elimination obeys linear kinetics and so the amount eliminated per time unit (dN_e/dt) is directly proportional to the amount of the substance in the compartment, which is given by the product of its apparent volume of distribution V_d with the actual concentration $c_{(t)}$ (Figure 2.6a). Consequently, the rate of elimination is described by:

$$dN_e/dt = -k_e \cdot V_d \cdot c_{(t)} \qquad (2.21)$$

The proportional factor k_e is the rate constant (dimension: time^{-1}) for the elimination process and the product of k_e with V_d represents the total clearance of elimination from the organism (Cl_{tot}):

$$Cl_{tot} = k_e \cdot V_d \qquad (2.22)$$

The value of Cl_{tot} can be calculated not only from Equation 2.22. There are also other possibilities (see below).

Since dN_e/dt equals the product of the concentration change in the compartment with V_d, Equation 2.21 can be rewritten as:

$$dc_{(t)}/dt \cdot V_d = -k_e \cdot V_d \cdot c_{(t)} \qquad (2.21a)$$

Canceling V_d in the differential Equation 2.21a and solving it for $c_{(t)}$ we obtain:

$$c_{(t)} = c_{(0)} \cdot e^{-k_e \cdot t} \qquad (2.23)$$

The constant $c_{(0)}$ stands for the initial concentration when the whole dose D_{iv} is still present in the organism. Consequently, Equation 2.23 can also be expressed by considering Equation 2.16 as:

$$c_{(t)} = (D_{iv}/V_d) \cdot e^{-k_e \cdot t} \qquad (2.23a)$$

Concentration–time curves corresponding to the function given in Equation 2.23 are shown in Figures 2.7a and 2.8. Plotting $c_{(t)}$ in the log scale versus t in the linear scale (half-logarithmic plot) this function yields a straight line with a slope of $-k_e/\ln 10$ (Figure 2.7b). After taking the logarithm of Equation 2.23, k_e can be expressed as:

$$k_e = \ln (c_{(0)}/c_{(t)})/t \qquad (2.24)$$

Substituting $c_{(0)}/2$ for $c_{(t)}$ in Equation 2.24 yields the elimination half-life ($t_{1/2}$) at which the initial concentration has been reduced to one half:

$$t_{1/2} = \ln 2/k_e \approx 0.693/k_e \qquad (2.25)$$

This equation demonstrates that $t_{1/2}$ depends only on k_e and is independent of the concentration of the substance in the organism in the exposure range in which the elimination follows first-order kinetics. The value of $t_{1/2}$ gives the time span required to halve a given $c_{(t)}$.

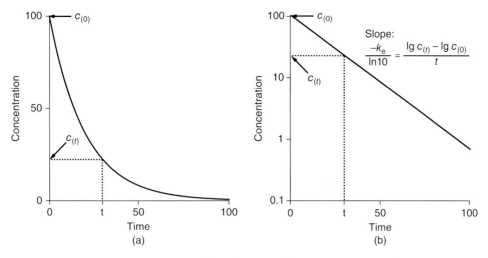

Figure 2.7 Concentration–time course of a substance administered intravascular, as a single dose and eliminated according to first-order kinetics as shown by the one-compartment model: (a) linear plot and (b) half-logarithmic plot. $c_{(0)}$, $c_{(t)}$, concentration at time 0 and time t, respectively; k_e, elimination rate constant; ln 10, factor for the conversion of logarithm base 10 into natural logarithm.

Extravascular Administration and Bateman Function The most straightforward model to describe the absorption of a substance administered as a single oral, intraperitoneal, intramuscular, epicutaneous or subcutaneous dose (D) is shown in Figure 2.6b. It is assumed that the substance moves quantitatively in only one direction, from the application site into the compartment that represents the organism. In analogy to Equation 2.23, the quantity of substance $N_{i(t)}$ at the application site (intestinal tract, peritoneum, muscle, skin or subcutaneous tissue) is given at any time by:

$$N_{i(t)} = D \cdot e^{-k_i \cdot t} \tag{2.26}$$

Because the substance is taken into the compartment from the application site, the rate constant for the invasion process is designated as k_i.

The derivative of Equation 2.26 with respect to time yields the rate of disappearance of the substance at the application site:

$$dN_{i(t)}/dt = -k_i \cdot D \cdot e^{-k_i \cdot t} \tag{2.27}$$

If the substance is taken up quantitatively (no first-pass effect!), the rate of absorption into the organism is the same as the rate of disappearance of substance at the application site. Thus, taking into account the volume of distribution V_d, the increase in the concentration ($c_{(t)}$) of the absorbed substance in the compartment is given by:

$$dc_{(t)}/dt = +(k_i \cdot D/V_d) \cdot e^{-k_i \cdot t} \tag{2.28}$$

Integrating Equation 2.28 over time and considering that $c_{(0)}$ at time point $t = 0$ is zero, Equation 2.29 is obtained:

$$c_{(t)} = D/V_d \cdot [1 - e^{-k_i \cdot t}] \tag{2.29}$$

This function describes a curve, as shown as a dotted line in Figure 2.8. If there is no elimination from the organism ($k_e = 0$ in the model), the concentration of the absorbed substance increases rapidly at first and finally reaches a plateau when the whole dose has been taken up.

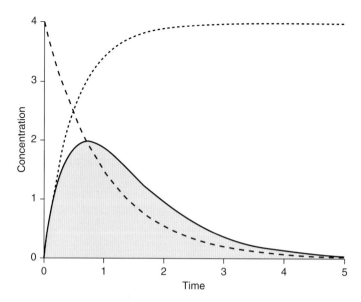

Figure 2.8 Concentration–time courses of a substance during absorption and elimination obeying first-order kinetics as shown by the one-compartment model. — — —, elimination after intravascular injection (Equation 2.23a); - - - -, invasion from a depot without elimination (Equation 2.29); ——, interaction of invasion and elimination (Equation 2.31, Bateman function); shaded area: area under the Bateman curve.

In general, substances are not only taken up by the organism but also eliminated ($k_e > 0$). The elimination of a substance begins simultaneously with its absorption. In the one-compartment model (Figure 2.6b) the concentration change per unit time is written when both processes, absorption (Equation 2.28) and elimination (Equation 2.21a), are considered by:

$$dc_{(t)}/dt = +(k_i \cdot D/V_d) \cdot e^{-k_i \cdot t} - k_e \cdot c_{(t)} \tag{2.30}$$

Given that the initial concentration in the body $c_{(0)}$ is zero, the solution of this differential equation yields a function describing the substance concentration in the one-compartment organism in a time-dependent manner which is called "Bateman function" (Figure 2.8):

$$c_{(t)} = (D/V_d) \cdot [k_i / (k_e - k_i)] \cdot [e^{-k_i \cdot t} - e^{-k_e \cdot t}] \tag{2.31}$$

The half-life of the final concentration–time course depends on that expression $e^{-k \cdot t}$ in the Bateman function that has the smaller value of k. The value of the expression with the smaller k approaches zero more slowly than the value of that with the larger k. In the rare cases where the terminal elimination phase is determined by k_i and not by k_e, one speaks of "flip-flop" kinetics. In such cases the absorption is slow when compared to the elimination.

Area under the Concentration–Time Curve and Bioavailability The probability of chronic damage and the extent of such damage are often correlated with the area under the concentration–time curve (AUC) of the active substance in blood or plasma.

The AUC (dimension: concentration · time) for the period between $t = 0$ and $t = \infty$ is given by:

$$\mathrm{AUC}_0^\infty = \int_0^\infty c_{(t)} \cdot \mathrm{d}t \qquad (2.32)$$

For concentration ranges in which linear kinetics apply, $c_{(t)}$ is generally given by a function of the form:

$$c_{(t)} = A_1 \cdot e^{-k_1 \cdot t} + A_2 \cdot e^{-k_2 \cdot t} + \ldots + A_n \cdot e^{-k_n \cdot t} \qquad (2.33)$$

A_1 to A_n (dimension: amount/volume) are concentration constants and k_1 to k_n (dimension: time^{-1}) are rate constants (cf. Equations 2.14 and 2.31).

For such a function, AUC_0^∞ is given by:

$$\mathrm{AUC}_0^\infty = A_1/k_1 + A_2/k_2 + \ldots + A_n/k_n \qquad (2.34)$$

According to Equations 2.33 and 2.34, AUC_0^∞ obtained following intravascular administration (Equation 2.23a) is given by

$$\mathrm{AUC}_0^\infty = D_{\mathrm{iv}}/(k_\mathrm{e} \cdot V_\mathrm{d}) \qquad (2.35)$$

The AUC_0^∞ of the Bateman function (shaded grey in Figure 2.8) is given by

$$\mathrm{AUC}_0^\infty = (D/V_\mathrm{d}) \cdot [k_\mathrm{i}/(k_\mathrm{e} - k_\mathrm{i})] \cdot [1/k_\mathrm{i} - 1/k_\mathrm{e}] \qquad (2.36)$$

which can be cancelled to

$$\mathrm{AUC}_0^\infty = D/(k_\mathrm{e} \cdot V_\mathrm{d}) \qquad (2.35\mathrm{a})$$

Obviously, the AUC_0^∞ of the Bateman function is identical with the AUC_0^∞ following intravascular administration when the same doses are given. This is because the extravascularly administered dose D was presupposed to be taken up completely (see also below).

From Equations 2.35, 2.35a and 2.22 it can be seen that Cl_tot can also be obtained from the administered dose and AUC_0^∞:

$$Cl_\mathrm{tot} = D_\mathrm{iv}/\mathrm{AUC}_0^\infty \qquad (2.37)$$

and, if D is completely taken up:

$$Cl_\mathrm{tot} = D/\mathrm{AUC}_0^\infty \qquad (2.37\mathrm{a})$$

Because the value of V_d depends on the medium in which the substance concentration is measured (see above), this also holds true for Cl_tot. A simple way of determining Cl_tot in blood (Cl_totB) or blood plasma (Cl_totP) is based on the experimental determination of $\mathrm{AUC}_{0\,\mathrm{iv}}^\infty$ in blood ($\mathrm{AUC}_{0\,\mathrm{ivB}}^\infty$) or blood plasma ($\mathrm{AUC}_{0\,\mathrm{ivP}}^\infty$) after intravenous injection of a dose D_iv. The intravenous administration has to be chosen because only by this method is a complete bioavailability ensured.

If the partition coefficient P_BP is known, Cl_totP and Cl_totB can be interconverted:

$$Cl_\mathrm{totB} = Cl_\mathrm{totP}/P_\mathrm{BP} \qquad (2.38)$$

When determining Cl_tot via an experimentally determined $\mathrm{AUC}_{0\,\mathrm{iv}}^\infty$, no use of a toxicokinetic model is required. It is emphasized again that Cl_tot is a concentration-independent constant only if the elimination kinetics is linear.

After extravascular administration of a substance, often only a fraction (F) of the administered dose reaches the systemic blood circulation (e.g., because of a first-pass effect). The bioavailability expresses this fraction. For example, under conditions of first-order kinetics the bioavailability F of an orally administered dose (D_{po}) can be calculated from the corresponding dose-normalized $AUC_{0\ po}^{\infty}$ and from the dose-normalized $AUC_{0\ iv}^{\infty}$ determined after intravenous injection of a dose (D_{iv}) in a second experiment:

$$F = (AUC_{0\ po}^{\infty}/D_{po})/(AUC_{0\ iv}^{\infty}/D_{iv}) \tag{2.39}$$

F is a dimensionless number. When $F = 1$ a substance is completely bioavailable and there is no first-pass effect (applicable for first-order kinetics).

Continuous Administration The one-compartment model for continuous administration is shown in Figure 2.6c. During continuous administration of a substance, for example during intravascular infusion or exposure to a constant concentration of a gas in the inhaled air, the absorption of the substance into the compartment representing the organism can be modeled as a process obeying zero-order kinetics: the substance enters the compartment with a constant absorption rate v_i (dimension: amount/time). When the elimination from the organism obeys first-order kinetics (Equation 2.21) the change in the concentration in the organism in a time-dependent manner is given by:

$$dc_{(t)}/dt = + v_i/V_d - k_e \cdot c_{(t)} \tag{2.40}$$

Solving this differential equation for $c_{(t)}$ and considering that $c_{(0)} = 0$ yields:

$$c_{(t)} = [v_i/(k_e \cdot V_d)] \cdot [1 - e^{-k_e \cdot t}] \tag{2.41}$$

A plot of the concentration–time curve as given by this function is shown in Figure 2.9. With increasing time $c_{(t)}$ approaches a plateau concentration (c_{ss}) at which steady state is attained, that is, the elimination rate ($k_e \cdot V_d \cdot c_{ss}$; see Equation 2.21) is exactly equal to the constant rate of absorption (v_i). Thus the change in concentration of the substance in the organism at the steady-state concentration (c_{ss}) is equal to zero:

$$dc_{ss}/dt = 0 \tag{2.42}$$

Inserting Equation 2.42 into Equation 2.40 and solving for c_{ss} yields:

$$c_{ss} = v_i/(k_e \cdot V_d) \tag{2.43}$$

This equation shows that the plateau concentration c_{ss} (the maximum accumulation of a substance) is determined only by the ratio of v_i to Cl_{tot}. As may be seen from Figure 2.9, increasing v_i by a factor of 2 doubles the value of c_{ss}. The time until c_{ss} is attained does not depend on v_i; it is determined only by k_e. After four elimination half-lives, about 94% of c_{ss} has been attained. In the elimination phase after the end of administration, the course of the reduction in concentration is also given by an exponential function with $t_{1/2} = \ln 2/k_e$ (Figure 2.9).

Repeated Administration The simplest case of repeated administration via a depot (e.g., oral or intraperitoneal doses; bioavailability $F = 1$; linear kinetics) is shown schematically in Figure 2.10. Regular administration of the dose D with a constant time interval τ between doses is equivalent to a constant dose rate D/τ. In analogy to Equation 2.43, the mean concentration $\bar{c}_{ss}$ which is attained at steady state is given by:

$$\bar{c}_{ss} = (D/\tau)/(V_d \cdot k_e) \tag{2.44}$$

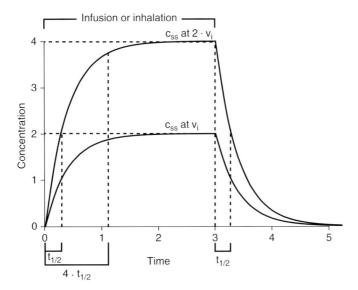

Figure 2.9 Concentration–time courses of a substance during continuous absorption (e.g., by intravenous infusion or inhalation) and thereafter as shown by the one-compartment model (linear plot). Elimination follows first-order kinetics. c_{ss}, concentration at equilibrium; $t_{1/2}$, half-time; v_i, invasion rate.

Thus, just as in the case of continuous administration, the concentration of the substance accumulates to a maximum value. Whereas during constant continuous administration of a substance a constant c_{ss} is attained, during repeated administration this concentration varies between a maximum (c_{ssmax}) and a minimum (c_{ssmin}) value. The mean concentration $\bar{c}_{ss}$ is not identical with the arithmetic mean of these two values because the concentration–time course obeys first-order kinetics. The value of $\bar{c}_{ss}$ can be calculated from the area under the concentration–time curve from $t = 0$ to $t = \infty$ obtained after administration of a single dose ($\text{AUC}^{\infty}_{0\,\text{SD}}$). When steady state is reached under repeated dosing conditions, the area $\text{AUC}^{t+\tau}_{t}$ under a dose interval τ is identical with $\text{AUC}^{\infty}_{0\,\text{SD}}$ (see Figure 2.10). Therefore $\bar{c}_{ss}$ is given by:

$$\bar{c}_{ss} = \text{AUC}^{t+\tau}_{t}/\tau = \text{AUC}^{\infty}_{0\,\text{SD}}/\tau \tag{2.45}$$

Saturation Kinetics For a first-order process, AUC^{∞}_{0} is proportional to the administered dose (cf. Equations 2.37 and 2.37a) provided that the administration route is not changed or the bioavailability F is equal to 1. In toxicological studies, however, it is often observed that as the dose is increased AUC^{∞}_{0} increases more than in proportion with the administered dose. AUC^{∞}_{0} is then no longer given by Equation 2.34 and derived equations. Often concentration–time curves are observed like the two upper ones shown in Figure 2.11 for administration by intravenous injection. Such curves are characteristic for a process that obeys the saturation kinetics of the kind shown for enzyme-catalyzed metabolism.

In such cases, the rate of elimination is usually described according to a Michaelis–Menten equation. In the one-compartment mode, the saturation kinetics of a metabolic elimination can then be allowed for by replacing the concentration-independent Cl_{tot} (given as the product of k_e and V_d in Equation 2.21a) by the concentration-dependent clearance presented in Equation 2.12.

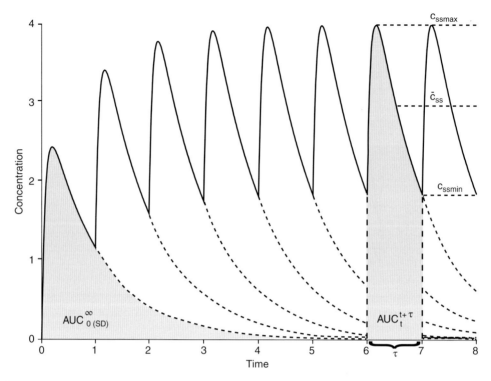

Figure 2.10 Concentration–time courses of a substance administered repeatedly at constant intervals (e.g., oral or intraperitoneal administration) with complete absorption and with elimination, both of which obey first-order kinetics as shown by the one-compartment model (linear plot). ———, concentration–time course resulting from multiple dosing; – – –, theoretical concentration–time courses for each dose; shaded area: areas under the curves following the first dose and a dose given when steady-state conditions are reached; $\mathrm{AUC}_{0(SD)}^{\infty}$, area under the concentration–time curve from $t = 0$ to $t = \infty$ for a single dose; $\mathrm{AUC}_{t}^{t+\tau}$, area under the concentration–time curve for a dose interval at steady state; $\bar{c}_{ss}$, mean concentration at equilibrium; c_{ssmax}, maximum concentration during a dose interval at equilibrium; c_{ssmin}, minimum concentration during a dose interval at equilibrium; τ, dose interval.

According to Equation 2.21a, the obtained rate of elimination is then expressed by:

$$dc_{(t)}/dt \cdot V_d = -[V_{max}/(K_{mapp} + c_{(t)})] \cdot c_{(t)} \qquad (2.46)$$

To determine $c_{(t)}$ as a function of t from this differential equation, iterative calculations must be used.

Frequently it is not the absorbed substance itself which is of toxicological relevance but a metabolite. The toxicokinetics of such a metabolite can also be described by the processes given above. Its rate of formation (corresponding to the rate of absorption) can be determined from the rate of metabolic elimination of its precursor.

The Two-compartment Model

Linear Kinetics and Elimination from the Central Compartment When, after intravascular injection, the concentration–time course of the substance is plotted on a half-logarithmic scale, it is often possible to recognize a first phase (the α-phase in Figure 2.12) in which the concentration

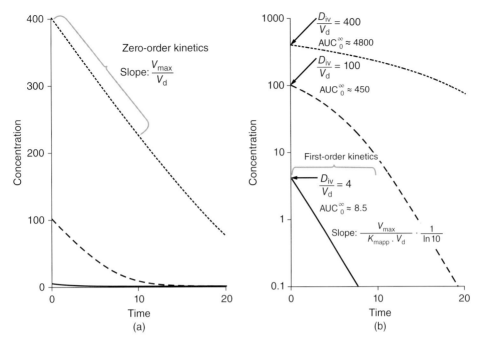

Figure 2.11 Concentration–time courses of a substance administered as single variously sized dose and eliminated according to saturation kinetics as shown by the one-compartment model: (a) linear plot and (b) half-logarithmic plot. AUC_0^∞, area under the concentration–time curve from $t = 0$ to $t = \infty$; D_{iv}, intravenous dose administered; K_{mapp}, apparent Michaelis constant; V_d, apparent volume of distribution; v_{max}, maximum elimination rate; ln 10, factor for the conversion of logarithm base 10 into natural logarithm.

in blood or plasma decreases rapidly, mainly as a result of distribution of the substance between organs and tissues. This distribution phase is followed by an elimination phase (the β-phase in Figure 2.12) which is characterized by a slower decrease in concentration that obeys an exponential function. The whole time course of the concentration changes is described by:

$$c_{(t)} = A_1 \cdot e^{-\alpha \cdot t} + A_2 \cdot e^{-\beta \cdot t} \tag{2.47}$$

where α and β are two rate constants.

Such a concentration–time course cannot be reflected by the one-compartment model, which is based on the assumption that distribution is instantaneous. Therefore the model must be extended by including a further peripheral or deep compartment which takes into account the distribution phase.

Figure 2.13 shows a frequently used open two-compartment model with a central compartment and a peripheral compartment. First-order kinetics describe the elimination from the central compartment and the distribution process between both compartments. In this model, the changes in amount per time unit $dN_{1(t)}/dt$ in the central and $dN_{2(t)}/dt$ in the peripheral compartment are described by the following differential equations:

$$dN_{1(t)}/dt = -(k_{12} + k_e) \cdot V_1 \cdot c_{1(t)} + k_{21} \cdot V_2 \cdot c_{2(t)} \tag{2.48}$$

and

$$dN_{2(t)}/dt = -k_{21} \cdot V_2 \cdot c_{2(t)} + k_{12} \cdot V_1 \cdot c_{1(t)} \tag{2.49}$$

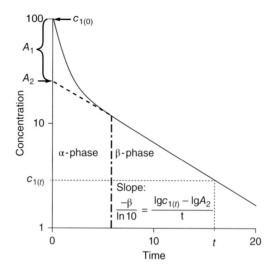

Figure 2.12 Biphasic concentration–time curve as shown by the two-compartment model for first-order kinetics: concentration–time curve in the central compartment after administration of a single dose of a substance into the central compartment (half-logarithmic plot). ———, concentration in the central compartment; — — —, backwards extrapolation of the β-phase curve to time 0; - - - - -, end of the α-phase and the beginning of the β-phase; $c_{1(0)}$, initial concentration $(A_1 + A_2)$; $c_1(t)$, concentration at time t; β, rate constant of the β-phase; ln 10, factor for the conversion of logarithm base 10 into natural logarithm.

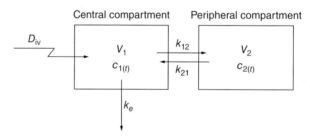

Figure 2.13 Two-compartment model for the description of the toxicokinetics of a substance administered as a single intravascular injection, distributed from the central into the peripheral compartment and eliminated from the central compartment, in each case according to first-order kinetics. $c_{1(t)}$, concentration in the central compartment at time t; $c_{2(t)}$, concentration in the peripheral compartment at time t; D_{iv}: intravenous dose; k_{12}, k_{21}, rate constants of the processes of distribution between the central and peripheral compartments; k_e, elimination rate constant; V_1, V_2, volumes of the central and peripheral compartments, respectively.

where $c_{1(t)}$ is the concentration of the substance in the central compartment with volume V_1 and $c_{2(t)}$ that in the peripheral compartment with volume V_2. The rate constants k_{12} and k_{21} describe the processes of distribution between the two compartments and k_e is the elimination rate constant for the central compartment. After a dose D_{iv} of the substance has been injected into the central compartment, which generally represents the blood and the well-perfused organs (lung, brain, heart, liver, spleen, kidneys), the substance is immediately evenly distributed in this compartment.

From here it enters the peripheral compartment, which generally represents the less well-perfused tissues (muscle, skin, adipose tissue). There, the accumulation of the substance is determined by the rates of distribution between the two compartments ($V_1 \cdot k_{12} \cdot c_{1(t)}$ and $V_2 \cdot k_{21} \cdot c_{2(t)}$). These rates are not constant because they are determined by the concentrations $c_{1(t)}$ and $c_{2(t)}$, which change with time.

In the β-phase, the ratio of the concentrations $c_{2(t)}$ to $c_{1(t)}$ is constant and the elimination from both compartments has the same half-life of $t_{1/2} = \ln 2/\beta$.

The area under the concentration–time curve, AUC_0^∞, may then be calculated from Equations 2.33, 2.34 and 2.47 as

$$\text{AUC}_0^\infty = A_1/\alpha + A_2/\beta \tag{2.50}$$

The constants α and β (dimensions: time^{-1}) are complex functions of the rate constants k_{12}, k_{21} and k_e (see Figure 2.13):

$$\alpha = 0.5 \left[(k_{12} + k_{21} + k_e) + \left[(k_{12} + k_{21} + k_e)^2 - 4 \cdot k_e \cdot k_{21} \right]^{0.5} \right] \tag{2.51}$$

$$\beta = 0.5 \left[(k_{12} + k_{21} + k_e) - \left[(k_{12} + k_{21} + k_e)^2 - 4 \cdot k_e \cdot k_{21} \right]^{0.5} \right] \tag{2.52}$$

The apparent volume of distribution V_d related to the central compartment is not constant in the two-compartment model, unlike in the one-compartment model, but increases in the time course of the α-phase until the β-phase begins. In the β-phase, the apparent volume of distribution ($V_{d\beta}$) may be determined from the intravenous dose (D_{iv}), the resulting area under the concentration–time curve in the central compartment $\text{AUC}_{0\,iv}^\infty$, and the rate constant β:

$$V_{d\beta} = D_{iv} / (\text{AUC}_{0\,iv}^\infty \cdot \beta) \tag{2.53}$$

The total clearance (Cl_{tot}; related to the concentration in the central compartment) equals the expression $D_{iv}/\text{AUC}_{0\,iv}^\infty$ (see Equation 2.37). It is independent of the toxicokinetic model. If the value of V_1 is known, the elimination rate constant k_e can be calculated in various ways:

$$k_e = V_{d\beta} \cdot \beta / V_1 = Cl_{tot}/V_1 = D_{iv}/(\text{AUC}_{0\,iv}^\infty \cdot V_1) \tag{2.54}$$

Appropriate equations can be derived for the calculation of the constants $A_1, A_2, \alpha, k_{12}, k_{21}$ and the function for the time-dependence of V_d.

Physiologically-based Toxicokinetic Models

In physiologically-based toxicokinetic models, compartments represent organs, tissues or groups of tissues with their actual volumes. All rates are described by means of physiological, physiochemical, and biochemical parameters. For instance, the transport of substances between the tissues takes place via the bloodstream and the accumulation in them is dependent on tissue:blood partition coefficients and on tissue-specific metabolic parameters.

A straightforward model for exposure of an animal or a human to a lipophilic gaseous substance is exemplified in Figure 2.14. Gas exchange between atmosphere and lung blood is modeled as described in detail earlier (see heading *Lung*, above). The organism is subdivided into several compartments representing organs and tissues. Adipose tissue (the storage tissue) and liver are represented by own compartments. Metabolism follows saturation kinetics according to Michaelis and Menten and takes place exclusively in the liver. The compartment "richly perfused organs" summarizes primarily lung, brain, kidney, spleen, heart and intestines, the compartment "moderately perfused tissues" represents mainly muscle and skin. The scarcely perfused bones and

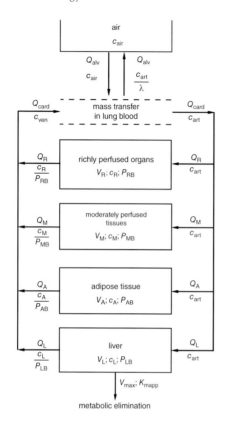

Figure 2.14 *Straightforward physiologically-based toxicokinetic model for a lipophilic gaseous substance enabling the description of absorption by inhalation, distribution by the blood flow in tissues and organs, metabolic elimination obeying saturation kinetics in the liver and exhalation. Equations: (a) concentration of substance in oxygen-poor venous blood and (b) concentration of substance in oxygen-rich arterial blood (obtained by reformulating Equation 2.5). Mass balance equations giving the mass changes in (c) the richly-perfused organs, (d) the moderately perfused tissues, (e) adipose tissue, and (f) the liver. c_{air}, concentration of substance in the air at time t; c_{art}, c_{ven}, concentrations of substance in the oxygen-rich arterial blood leaving the lung and in the oxygen-poor venous blood entering the lung at time t; c_R, c_M, c_A, c_L, concentrations of substance in richly-perfused organs, moderately perfused tissues, adipose tissue, and liver at time t; K_{mapp}, apparent Michaelis constant related to the concentration of substance in the liver; λ, substance-specific partition coefficient blood:air; P_{RB}, P_{MB}, P_{AB}, P_{LB}, substance-specific partition coefficients richly-perfused organs:blood, moderately-perfused tissues:blood, adipose tissue:blood, and liver:blood; Q_{alv}, alveolar ventilation; Q_{card}, cardiac output (equals the blood flow through the lung); Q_R, Q_M, Q_A, Q_L, blood flows through the richly-perfused organs, the moderately-perfused tissues, the adipose tissue, and the liver; V_R, V_M, V_A, V_L, volumes of richly-perfused organs, moderately-perfused tissues, adipose tissue, and liver; V_{max}, maximum rate of metabolism.*

cartilage are disregarded. The model is a so-called perfusion-limited model, that is, it is assumed that the substance in each tissue is always in equilibrium with the blood leaving the tissue. Differential equations describe the mass changes in the compartments (Figure 2.14); computer programs are available for the solution of these equations.

Physiologically-based toxicokinetic models have the advantage over the classical compartment models because they permit knowledge of the fate of a substance in individual tissues as, for instance, in a tissue which is the target of an adverse effect. They are useful for the extrapolation of tissue burdens between species (species scaling) when sufficient anatomical, physiological and biochemical information is available. For these reasons, the use of such models is continuously growing. However, there are also certain disadvantages: physiologically-based toxicokinetic models require a large number of experimental *in vivo* and *in vitro* data, which are essential for model calibration and validation.

2.2.4 Summary

Toxicokinetics is fundamental to the understanding of the quantitative relationships between the amount of a substance administered or taken up and its toxic effects. It describes by means of mathematical functions the processes of the absorption of a substance into an organism, its distribution and accumulation in various organs and tissues, and its elimination from the organism. Usually, substances are absorbed via the lung, the skin or the gastrointestinal tract. In the latter case, a first-pass effect may occur due to metabolism of the substance in the liver before it enters the systemic blood circulation. The most common absorption mechanisms include diffusion through biological membranes and active transport catalyzed by enzymes located in cell membranes. The absorption phase can be circumvented experimentally by direct injection of the substance into the bloodstream. Elimination processes are also mostly mediated by enzymes and so are subject to saturation kinetics, which becomes especially relevant at the high substrate doses or concentrations used in animal studies. Many toxicants only become harmful when they are metabolically activated. Some absorbed substances are excreted unchanged. Excretion of xenobiotics takes place mainly via the kidneys, the intestinal tract, and the lungs, but the skin and mammary glands may also be involved. For the mathematical analysis of experimentally determined changes in substance concentration with time, compartment models and physiologically-based toxicokinetic models are used. The most frequent toxicokinetic expressions are defined by means of the one-compartment model, which is presented in detail. Additionally, an introduction is given into the basic principles of the two-compartment model and a physiologically-based toxicokinetic model.

Further Reading

Gibaldi M, Perrier D: Pharmacokinetics, 2nd edition, Marcel Dekker (1982).
Ritschel WA, Kearns GL: Handbook of Basic Pharmacokinetics … Including Clinical Applications, 7th edition, APhA Publications (2009).
Reddy M, Yang RS, Andersen ME, Clewell HJ: Physiologically based Pharmacokinetic Modeling: Science and Applications, Wiley (2005).

2.3 Biotransformation of Xenobiotics

Wolfgang Dekant

2.3.1 Introduction

In the absence of efficient systems for the elimination of fat-soluble chemicals, the long-term intake of low doses of a chemical may result in accumulation. The excretion

> of non-volatile chemicals with low molecular weight is mainly carried out with urine, therefore a sufficient water solubility of the chemical in question is a prerequisite for rapid elimination. On the other hand, lipophilic chemicals are well absorbed into the body due to their good penetration through biological membranes and are rapidly distributed. Therefore, lipophilic chemicals must be efficiently converted into water-soluble and thus excretable compounds to prevent accumulation.

The term biotransformation describes the metabolic transformations of lipophilic xenobiotics in the organism, which leads to the formation of water-soluble products. An example of the important role of biotransformation in excretion and detoxification is provided by 2,3,7,8-tetrachlorodibenzodioxin (TCDD). TCDD is metabolically stable and very lipophilic. Because of the lack of biotransformation, TCDD accumulates in adipose tissue due to its slow elimination, with an elimination half-life of more than seven years. On the other hand, even very lipophilic chemicals can be excreted quickly in the form of water-soluble metabolites if these can be metabolized efficiently.

Biotransformation reactions are mostly enzymatically catalyzed; only a few, mostly highly reactive, chemicals are converted by a direct reaction with water (hydrolysis) or with other body-specific nucleophiles such as the thiol-containing tripeptide glutathione without enzymatic catalysis.

For systematization, the large number of enzymatic reactions that contribute to the conversion of fat-soluble xenobiotics into water-soluble products in mammals is divided into two classes (Table 2.3):

- Enzymes of Phase I reactions catalyze the oxidation, reduction or hydrolysis of lipophilic xenobiotics and introduce functional groups or release them (functionalization).
- Enzymes of Phase II reactions couple xenobiotics with functional groups with an endogenous and highly water-soluble substrate (conjugation) (Reaction 2.1).

Due to the conjugation in Phase II, a high water solubility of the Phase I metabolite and thus rapid elimination is achieved. Since the excretion of many Phase II metabolites can be further accelerated by active transport through biological membranes, the active transport of polar chemicals in excretion media such as urine is also referred to as Phase III of the biotransformation. An example of a Phase III reaction is the active transport of glutathione conjugates of a chemical from the liver cell through the canalicular membrane into bile.

Functionalization describes the introduction of polar groups such as –OH or hydrolysis of esters, ethers, and substituted amines to expose functional groups such as –OH, –SH, –NH$_2$, or –COOH. The presence of these functional groups permits the coupling of the Phase I metabolite to endogenous molecules to form a water-soluble conjugate through Phase II reactions. An example of the interaction of Phase I and Phase II reactions is the metabolism of benzene to phenyl glucuronide (Reaction 2.2).

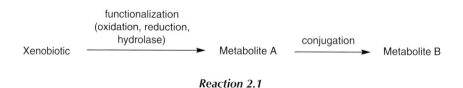

Reaction 2.1

Reaction 2.2

In the first step, benzene is oxidized to phenol in a Phase I reaction. Since phenol is largely undissociated in the physiological pH range, the coupling with the readily water-soluble glucuronic acid leads to a very polar Phase II metabolite present as an anion under physiological pH values. Due to the very good solubility in water, conjugation products such as glucuronides are usually excreted much faster than the Phase I metabolites.

> **The highest concentrations of biotransformation enzymes are normally found in the liver of mammals, as the liver processes absorb food components from the gastrointestinal tract.**

The portal vein transports blood, which contains many nutrients, practically quantitatively from the gastrointestinal tract to the liver. The high permeability of the membranes of the liver cells to the blood capillaries results in a very rapid intake of chemicals dissolved in the portal blood in the liver. Metabolites of xenobiotics formed in the liver can either be returned to the bloodstream or excreted with the bile into the intestine.

Enzymes of biotransformation are also present in other organs of mammals, such as in the intestine, in the kidney, or in the lungs. The extent of the biotransformation reactions occurring in these organs is dependent on the pathway of absorption but is often quantitatively lower than that in liver. Organ-specific expressed enzymes of biotransformation can be responsible for organ-specific toxic effects for chemicals that are subject to bioactivation.

The enzymes of the biotransformation are not uniformly distributed to all cell types present in an organ. Enzymes that catalyze Phase I reactions are often bound to the membranes of the endoplasmic reticulum. When the organ is homogenized, the endoplasmic reticulum forms small vesicles, which can be separated from other cell parts by centrifugation. These membrane vesicles are referred to as **microsomes**, which are often used as an *in vitro* system for the investigation of biotransformation reactions.

> **After the addition of appropriate cofactors, microsomes still have the enzymatic capacity to metabolize chemicals. Addition of microsomes is often used in studies in *in vitro* systems, in bacteria or cultured mammalian cells, in order to introduce a capacity for biotransformation. Examples are the Ames test or studies on the genotoxicity of chemicals in mammalian cell systems, such as the HGPRT test.**

In addition to the cellular enzymes of the biotransformation found in virtually all mammals, biotransformation reactions can also be catalyzed by microbial enzymes (anaerobic bacteria in the microflora of the intestine). Due to the large amount of bacteria in the intestine, microbial reduction can make an important contribution to the biotransformation of chemicals.

The enzymes of the biotransformation of chemicals usually have low turnover rates. These are often orders of magnitude lower than the turnover rates of enzymes that are specialized in the biosynthesis and degradation of endogenous chemicals. However, the low turnover rates are compensated by high enzyme concentrations, especially in the liver. Moreover, enzymes of the biotransformation of xenobiotics usually have a broad substrate specificity. For example, the cytochrome P450 enzyme 2E1 oxidizes both ethanol, chloroform or dichloromethane, as well as short-chain saturated and unsaturated aliphatic hydrocarbons and mononuclear aromatics such as benzene. However, with endogenous substrates such as testosterone or prostaglandins, certain cytochrome P450 enzymes show a high degree of regio- and stereoselectivity.

2.3.2 Phase I Reactions

Cytochrome P450 Enzymes

Monooxygenases such as cytochrome P450 and the flavin-dependent monooxygenases are the most important enzyme systems for the oxidation of chemicals. Monooxygenases transfer an oxygen atom formed from the cleavage of molecular oxygen (O_2) to the xenobiotic, the other oxygen atom is reduced to water using NADPH as a cofactor. The oxygen atom used for the oxidation is either introduced into C–H bonds or added to double bonds and free electron pairs. For the oxidation of xenobiotics monooxygenases NADPH is consumed, with subsequent stoichiometry (Reaction 2.3):

> **The cytochrome P450-dependent monooxygenases are very important enzymes for mammalian biotransformation. In mammals, cytochrome P450 enzymes are found in almost all tissues, but the highest activities of these enzymes are usually in the liver.**

Cytochrome P450 enzymes are integrated into the phospholipid matrix of the endoplasmic reticulum. Cytochrome P450 enzymes consist of the NADPH-dependent cytochrome P450 reductase and the heme and iron-containing enzyme cytochrome P450. The phospholipids of the endoplasmic reticulum play an important role in the function of cytochrome P450 enzymes since they allow an interaction between the cytochrome P450 and the reductase. The reduced form of cytochrome P450 enzymes has a divalent iron atom (Fe^{2+}) in the active site. This property was previously used to determine cytochrome P450 enzymes since Fe^{2+}-containing heme protein forms a complex with carbon monoxide with maximum UV absorption at 450 nm; hence the term cytochrome P450. Maximum UV absorption at 450 nm is observed only in functional cytochrome P450 enzymes. After denaturation, cytochrome P450 enzymes show absorption at 420 nm, the same as other heme iron proteins. The transfer of oxygen to a substrate involves several steps and takes place by one-electron transfer reactions.

$$\text{Substrate (RH)} + O_2 + \text{NADPH} + H^+ \longrightarrow \text{Oxidated substrate (ROH)} + H_2O + \text{NADP}^+$$

Reaction 2.3

Table 2.3 Biotransformation enzymes.

Phase I	Phase II
Cytochrome P450	Glucuronyl transferases
Flavin-dependent monooxygenases	Sulfo transferases
Alcohol dehydrogenases	Glutathione S-transferases
Aldehyde dehydrogenases	N-Acetyl-transferases
Esterases	Epoxide-hydrolase
Amidases	Amino acid conjugation

More than 50 different human cytochrome P450 enzymes have been purified over the last 30 years and have been characterized in terms of structure, amino acid composition of peptide chains, and substrate specificities. Of the more than 50 human cytochrome P450 enzymes, approximately 20 may catalyze the oxidation of xenobiotics. The other cytochrome P450 enzymes in mammals catalyze specific steps in the biosynthesis of endogeneously important molecules such as steroids, bile acids, and vitamin D.

Purified cytochrome P450 enzymes have different molecular weights, organ distributions, and substrate specificities. Common to all cytochrome P450 enzymes is a hydrophobic amino acid sequence at the N-terminus for the anchoring of the enzyme in the membrane of the endoplasmic reticulum. The individual enzymes differ by amino acid sequences and by the spatial structure of the substrate binding sites. The differences in the spatial structure of the substrate binding site cause differences in the substrate specifics of the individual cytochrome P450 enzymes (Table 2.4). Homologs of human cytochrome P450 enzymes are found in virtually all mammalian species.

> **According to the extent of the consistency of the amino acid sequence, the individual cytochrome P450 enzymes are divided into families (Arabic numbers) and subfamilies (letters followed by an Arabic number). At present, at least 10 families and 60 different subfamilies can be distinguished from one another, and these are expressed differently in different organs and species (Table 2.4).**

In contrast to the many enzymes of cytochrome P450, only one form of cytochrome P450 reductase is known. The concentration of this enzyme in the endoplasmic reticulum is only about 10% of the concentration of cytochrome P450.

The oxidation of a foreign chemical via cytochrome P450 enzymes involves the binding of the substrate to the enzyme, the binding of molecular oxygen to the heme protein, several electron transfer steps, and, finally, the incorporation of oxygen into the substrate. In many cases, rearrangement reactions occur, followed by detachment of the final product of the reaction from the active center of the enzyme.

An insertion of oxygen takes place in the case of the aliphatic hydroxylation catalyzed by cytochrome P450 enzymes (Reaction 2.4) and in the case of oxidative dealkylation (Reaction 2.5).

Cytochrome P450 enzymes also catalyze the oxidation of alkenes and alkynes (Reaction 2.7 and 2.8) with epoxides as intermediates or terminal products. For example, ethene is metabolically converted to ethene oxide. Because of its high volatility, ethene oxide can be exhaled after exposure to ethene (Reaction 2.6). Epoxides are formed during the cytochrome P450-catalyzed oxidation of olefins in a multistage reaction. In the first step, a very short-lived intermediate is

Table 2.4 Location in tissues and preferred substrates of cytochrome P450 enzymes involved in the biotransformation of xenobiotics.

Cytochrome P450	Tissue	Selected substrates
CYP1A1	Several	Benzo[a]pyrene
CYP1A2	Liver	Aflatoxin$_{B1}$
		Coffeine
		Heterocyclic aryl amines
		Phenacetine
CYP2A6	Several	Coumarine
		Nicotine
		N-Nitroso diethylamine
CYP2C8	Liver	Tolbutamide
	Throat	R-Mephenytoin
CYP2D6	Liver	Bufuralol
	Throat	Debrisoquine
	Kidney	Sparteine
CYP2E1	Liver	Carbontetrachloride
	Throat	Ethanol
	Leucocytes	Dimethylnitrosamine
CYP3A4	Gastrointestinal tract	Aflatoxins
	Liver	Cyclosporine
		Nifedipine
		Testosterone

$$R-CH_2-CH_2-CH_3 \xrightarrow[\text{cytochrome P450}]{O_2} R-CH_2-CHOH_2-CH_3$$

Reaction 2.4

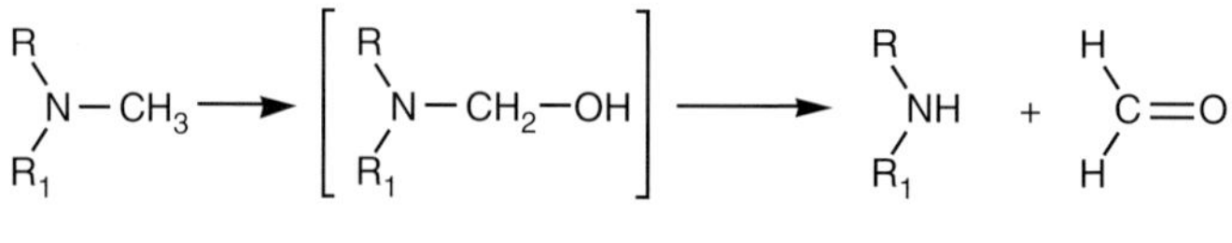

$$\begin{array}{c} R \\ \diagdown \\ N-CH_3 \\ \diagup \\ R_1 \end{array} \longrightarrow \left[\begin{array}{c} R \\ \diagdown \\ N-CH_2-OH \\ \diagup \\ R_1 \end{array} \right] \longrightarrow \begin{array}{c} R \\ \diagdown \\ NH \\ \diagup \\ R_1 \end{array} + \begin{array}{c} H \\ \diagdown \\ C=O \\ \diagup \\ H \end{array}$$

Reaction 2.5

formed, which can be converted into an epoxide by ring closure. However, the degree of epoxide formation from olefins is dependent on structure; the intermediates formed in the first step are frequently rearranged in the presence of migratory atoms in the molecule, such as chlorine atoms (Reaction 2.7).

Chloral is therefore formed as a stable and analytically detectable product in the oxidation of trichloroethene by cytochrome P450. Alkynes are also converted as intermediates into highly reactive ketenes (Reaction 2.8), which react rapidly with water to form epoxides.

Cytochrome P450 enzymes can also catalyze further reactions, for example the oxidation of siloxanes or heteroatoms such as nitrogen or sulfur (parathion to paraoxone), oxidation of carbon–heteroatom double bonds, and ester cleavages and dehydrogenations (Figure 2.15 and Reactions 2.9 and 2.10).

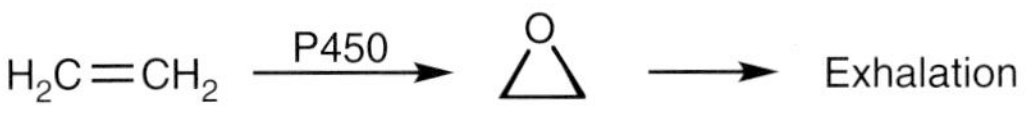

Reaction 2.6

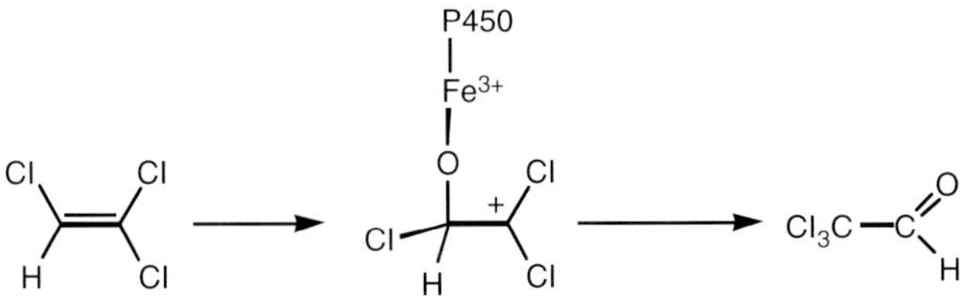

Reaction 2.7

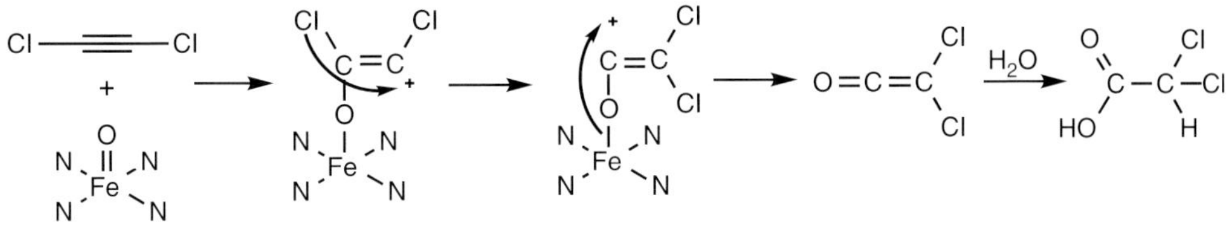

Reaction 2.8

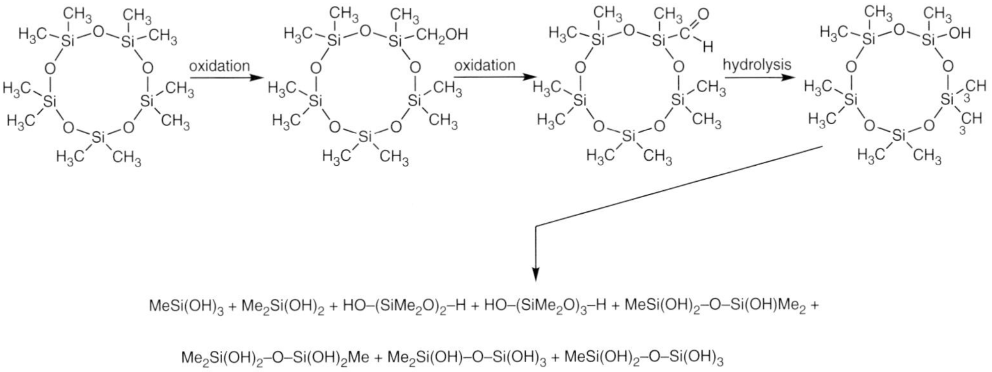

Figure 2.15 *Oxidative biotransformation of decamethylcyclopentasiloxane by cytochrome P450. The ring-cleavage reactions and the degradation of the siloxane chain are carried out by hydrolysis without enzymatic participation.*

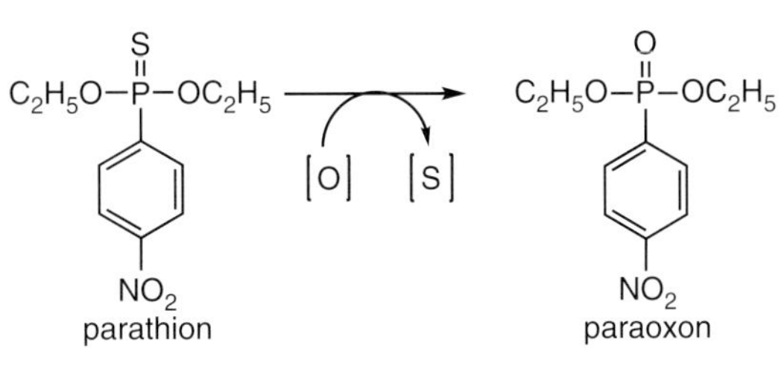

parathion → paraoxon

Reaction 2.9

Reaction 2.10

(valproic acid → CYP P450 → β-oxidation product COSCoA)

Oxidations: Flavin-dependent Monooxygenases

> The endoplasmic reticulum contains a second family of monooxygenases. Flavin-dependent monooxygenases (FMOs) are flavoproteins containing flavin adenine dinucleotide (FAD). FMOs often catalyze the conversion of substrates in competition with cytochrome P450. However, FMOs contain neither heme groups nor metals and have a high activity in humans and pigs, and only a comparatively low activity in rats and mice. In contrast to cytochrome P450, FMOs transmit two electrons in one reaction step (Figure 2.16).

FMOs are found in high concentrations in extrahepatic tissues such as the kidney or the lungs. FMOs (FMO 1 to 5) also exist in several families with different organ and species distribution.

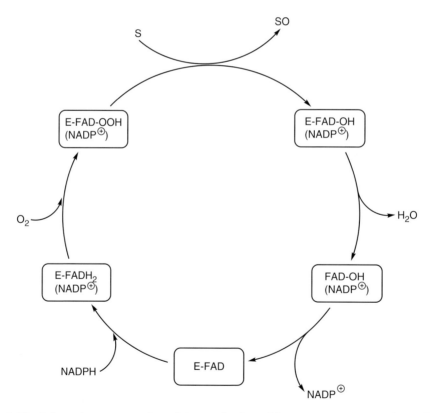

Figure 2.16 Schematic representation of the mechanism of flavin monooxygenase-dependent oxidation of xenobiotics. E-FAD, flavin-dependent monooxygenase; S, substrate.

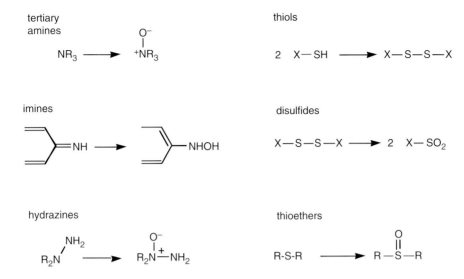

Figure 2.17 Examples of oxidations catalyzed by flavin-dependent monooxygenases.

In contrast to cytochrome P450 enzymes and many enzymes of the Phase II metabolism, FMOs are only slightly inducible (see *Induction of Metabolic Enzymes*).

FMOs prefer to oxidize soft nucleophiles (Figure 2.17) such as sulfur, nitrogen, selenium, and phosphorus compounds with very low substrate specificity.

Oxidations: Prostaglandin Synthase and other Peroxidases

The formation of prostaglandins takes place in a multistep reaction. The starting point is arachidonic acid, which is converted into prostaglandin G2 (PGG2) in the first step. This reaction is catalyzed by a cyclooxygenase activity. In the second step, a peroxidase activity catalyzes the reduction of the hydroperoxide function to the alcohol prostaglandin H2 (PGH2). Both reactions are mediated by prostaglandin synthase, an enzyme containing heme and iron (Figure 2.18).

In the first step of the conversion of arachidonic acid, the cyclooxygenase reaction, a peroxy radical (ROO) is formed as intermediate. This can serve as a source for reactive oxygen species or oxidize xenobiotics with double bonds to epoxides. The cyclooxygenase activity may transform diol metabolites of polycyclic aromatic compounds to the corresponding diol epoxides.

Xenobiotics can also serve as reducing substrates for the subsequent reaction of the hydroperoxide in prostaglandin H2. The xenobiotics are oxidized to one radical in one-electron steps.

Prostaglandin synthase is present in high concentrations in the microsomal fractions of the skin and kidney, in the brain, and in thrombocytes.

> **Prostaglandin synthase is often associated with the metabolism of xenobiotics, but since P450-dependent monooxygenases often form identical oxidation products from xenobiotics, the contribution of prostaglandin synthase to the biotransformation of a certain foreign chemical often cannot be well estimated. Prostaglandin synthase may be responsible for the oxidation of xenobiotics in extrahepatic tissues, which have low cytochrome P450 activities.**

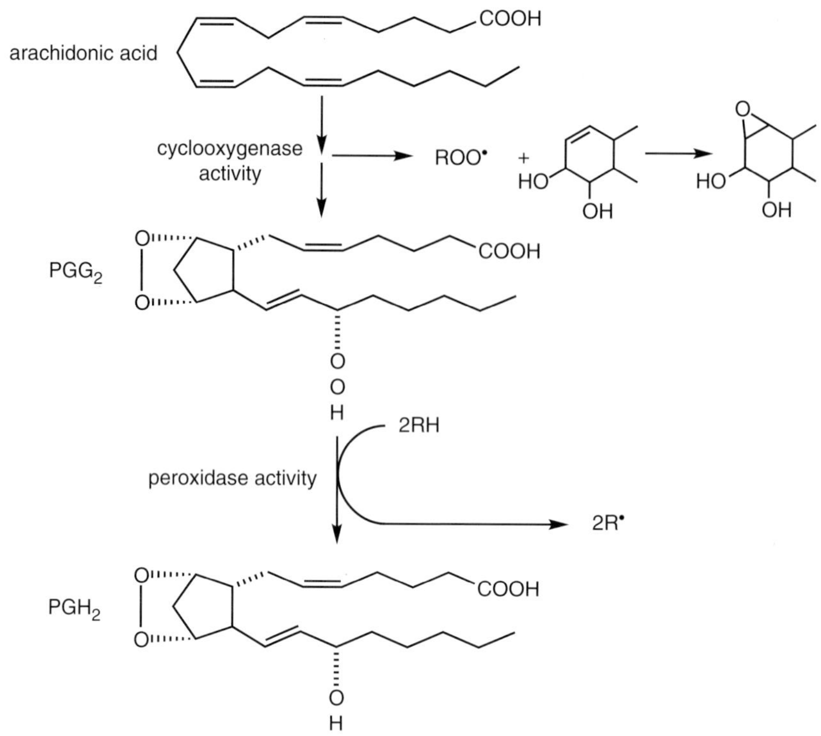

Figure 2.18 Cooxidation of xenobiotics by prostaglandin synthase. The conversion of arachidonic acid into prostaglandin H2 is catalyzed in two steps by the enzyme prostaglandin synthase. In both steps, xenobiotics can serve as cosubstrates and be oxidized by the reactive oxygen species formed.

$$R-\overset{O}{\underset{\|}{C}}-O-R_1 + H_2O \longrightarrow R-\overset{O}{\underset{\|}{C}}-OH + HOR_1$$

Reaction 2.11

Prostaglandin synthase-dependent reactions might explain the carcinogenic effect of some steroid hormones in certain target organs such as the kidneys. A link between bioactivation by this enzyme and the transplacentar carcinogenicity of certain synthetic hormones such as diethylstilböstrol is also discussed here.

Esterases and Amidases

Unspecific esterases and amidases are widely distributed enzymes in mammalian tissues and hydrolyze ester and amide functions both in xenobiotics and in endogenously present esters such as acetylcholine. The reaction produces carboxylic acids and alcohols (Reaction 2.11) or amines or ammonium ions (Reaction 2.12). Normally, esters are hydrolyzed faster than amides, and the cosubstrate is water.

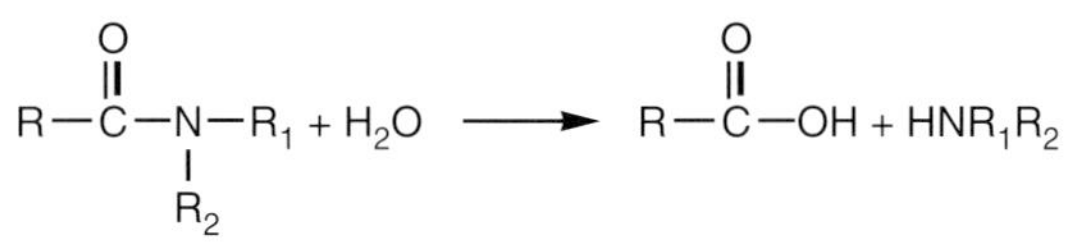

$$R-\underset{\underset{R_2}{|}}{\overset{\overset{O}{\|}}{C}}-N-R_1 + H_2O \longrightarrow R-\overset{\overset{O}{\|}}{C}-OH + HNR_1R_2$$

Reaction 2.12

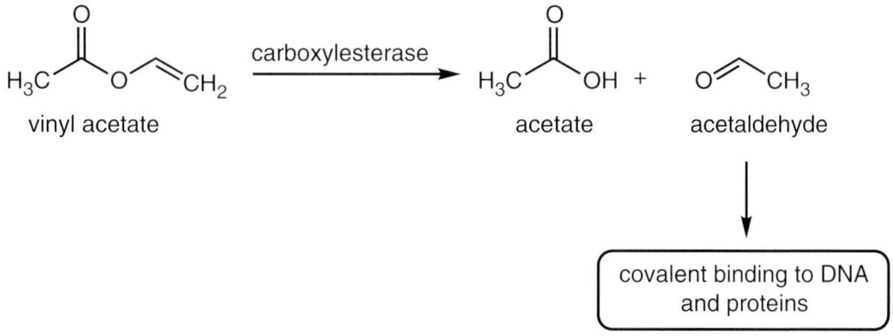

Figure 2.19 Biotransformation of vinyl acetate by carboxylesterase.

There are both membrane-bound (microsomal) and soluble (cytosolic) esterases and amidases. Cytosolic esterases often have a high conversion rate for endogenously present esters, and many of these esterases are serine hydrolases. In contrast, membrane-bound esterases and amidases usually catalyze the hydrolysis of xenobiotics. A toxicologically relevant and esterase-catalyzed reaction is the hydrolysis of vinyl acetate by esterases of the nasal mucosa of the rat (Figure 2.19). The acetaldehyde formed by the esterase-catalyzed reaction from vinyl acetate is responsible for the local toxicity of a vinyl acetate ionization on nasal tissue in the rat after inhalation.

The regulation of enzymes with esterase and amidase activities in the organism and their organ distribution are complex. There are also many different enzymes (e.g., carboxylesterases, cholinesterases, paraoxonases, β-glucuronidases) with often overlapping substrate specificity and complex regulation of their expression.

Systems for the Oxidation and Reduction of Alcohols, Aldehydes and Ketones

> **Xenobiotics and their metabolites often contain –OH, aldehyde, and ketone groups. These functional groups can also be oxidized or reduced by various enzyme systems. The most important of these enzymes are alcohol dehydrogenase, aldehyde reductases, ketone reductases, and aldehyde oxidases. These enzymes are found primarily in the cytosol of mammalian cells. NAD+ is the most significant cofactor in the oxidation reactions catalyzed by these enzymes; cofactors for reduction reactions are NADH or NADPH.**

Alcohol dehydrogenase (ADH) is the most important enzyme in this group and catalyzes the oxidation of ethanol to acetaldehyde and of methanol to formaldehyde. This cytosolic enzyme, which is mainly found in the liver, catalyzes the oxidation of most of the absorbed ethanol in humans after consumption of alcoholic beverages.

Aldehydes are oxidized to carboxylic acids by the use of NAD+ as a cofactor by *aldehyde dehydrogenase* (Reaction 2.13). In mammals, there are various aldehyde dehydrogenases with

72 *Toxicology and Risk Assessment*

$$CH_3OH \xrightarrow{ADH} H\underset{H}{\overset{O}{\|}}C \longrightarrow H\underset{OH}{\overset{O}{\|}}C$$

Reaction 2.13

different substrate specificities. Aldehyde dehydrogenases are cytosolic and mitochondrial, and microsomal enzymes and more than 20 different enzymes are known; their expression is dependent on genetic factors. Many of the aldehyde dehydrogenases involved in the metabolism of xenobiotics have specific endogenous substrates such as retinal or γ-amino butyraldehyde. The oxidation of aldehydes to the corresponding carboxylic acids is very efficient, acetaldehyde formed from ethanol has a half-life of only a few minutes in humans.

Carbonyl-reducing Enzymes Carbonyl reduction of aldehydes and ketones to their corresponding hydroxy derivatives is the counterpart to oxidative pathways and plays an important role in the Phase I metabolism of many endogenous (biogenic aldehydes, steroids, prostaglandins, reactive lipid peroxidation products) and xenobiotic (pharmacologic drugs, carcinogens, toxicants) compounds.

Carbonyl-reducing enzymes are grouped into two superfamilies, the aldo-keto reductases (AKR) and the short-chain dehydrogenase/reductases (SDR).

The AKR superfamily (www.med.upenn.edu/akr) represents a growing superfamily of NADP(H)-dependent oxidoreductases, which accept many structurally different substrates of endogenous and exogenous origin. This protein superfamily includes, amongst others, several hydroxysteroid dehydrogenases (EC 1.1.1.x), some of which have previously been designated dihydrodiol dehydrogenases (EC 1.3.1.20).

The SDRs (http://sdr-enzymes.org) comprise a wide range of enzymes involved in the metabolism of steroids, sugars, aromatic hydrocarbons, and prostaglandins and xenobiotics.

An example for the competing role of carbonyl-reducing enzymes versus cytochrome P450 enzymes is the tobacco-specific nitrosamine 4-(methylnitrosamino)-1-(3-pyridyl)-1-butanone (NNK, nicotine-derived nitrosamine ketone) which is carbonyl reduced by AKR1B10, AKR1C1, AKR1C2, and AKR1C4 from the AKR superfamily and carbonyl reductase 1 (CBR1) and 11β-hydroxysteroid dehydrogenases type 1 (11βHSD1) from the SDR superfamily.

Reductive Reactions of Biotransformations

Reductive reactions of biotransformations are particularly important for nitro and azo compounds and are often catalyzed by enzymes from the bacterial microflora of the gut.

> **The contribution of the intestinal flora to the metabolism of xenobiotics is often underestimated. Due to the large quantity of bacteria in the intestine, bacterial reductions for certain chemicals make a significant contribution to the metabolism of this chemical. A toxicologically relevant example of bacterial cleavage reactions in the intestine is the reductive cleavage of azo dyes such as Congo red with the release of aromatic amines (Figure 2.20).**

By reductive biotransformation, aniline is finally formed from aromatic nitro compounds such as nitrobenzene by stepwise reduction via nitrosobenzene and phenylhydroxylamine

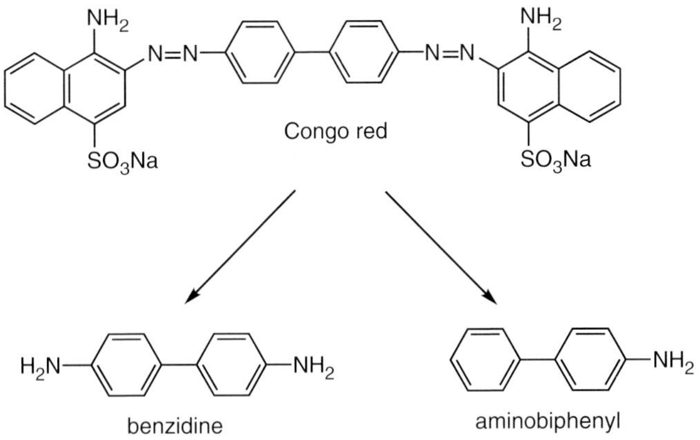

Figure 2.20 *Reductive cleavage of the azo dye Congo red by bacterial enzymes of the intestinal microflora.*

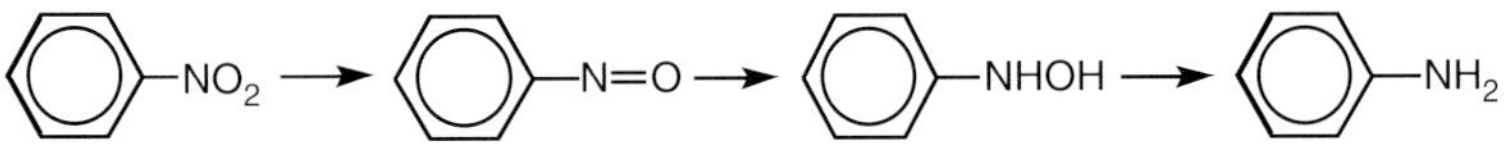

Reaction 2.14

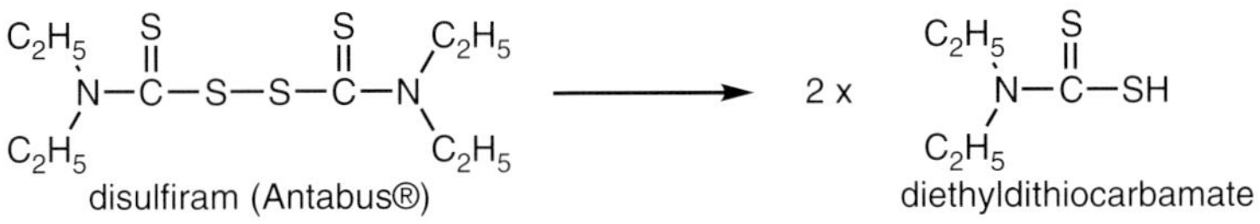

Reaction 2.15

(Reaction 2.14). Reduction reactions can also be catalyzed by cytochrome P450 at low oxygen concentrations. In cytochrome P450-catalyzed reductions, xenobiotics consume the reduction equivalents provided by the NADPH-dependent P450 reductase and can be reduced in both one- and two-electron steps.

Disulfides can also undergo metabolic reduction reactions (Reaction 2.15):

A reductive biotransformation can also take place with halogenated alkanes such as carbontetrachloride (see Reaction 2.24) or with some fluoroalkanes such as 1,1-dichloro-2,2,2-trifluoroethane (Reaction 2.16). For this chemical, reduction appears to be catalyzed by a cytochrome P450 enzyme.

Epoxid Hydrolases

> **Epoxide hydrolases catalyze the trans-addition of water to aliphatic and aromatic epoxides to form diols.**

Microsomal epoxide hydrolases are adjacent to the cytochrome P450 enzyme system in the cell. The low spatial distance to cytochrome P450 promotes the detoxification of the reactive epoxides formed by cytochrome P450. Microsomal epoxide hydrolases catalyze the hydrolysis

Reaction 2.16

Reaction 2.17

of aromatic and aliphatic epoxides formed by cytochrome P450 enymes to the corresponding trans-1,2-dihydrodiols.

The highest activity of epoxide hydrolases is found in the liver. The only cofactor for microsomal epoxide hydrolases is water; other low molecular weight cofactors are not necessary for enzymatic activity (Reaction 2.17).

Microsomal epoxide hydrolases are found in the microsomal fraction of many organs. In organs with different cell types, the distribution of the enzyme between the individual types is heterogeneous. In addition to the microsomal epoxide hydrolase, there are also at least three cytosolic epoxide hydrolases.

2.3.3 Phase II Reactions

> In the Phase II reactions, xenobiotics or their Phase I metabolites are coupled with a substrate provided by the intermediary metabolism. Phase II metabolic reactions are therefore biosynthetic reactions and require energy to perform the reaction. This applied energy usually serves for the biosynthesis of reactive co-substrates (UDP-glucuronic acid, 3'-phospoadenosine-5'-phosposulfates, acetyl-coenzyme A) as reaction partners or, in special cases, for the activation of the foreign chemical (formation of acetyl-coenzyme thioesters). Therefore, the ability of the organism to biosynthesis the co-substrates may influence the extent of Phase II reactions and result in dose-dependent changes in biotransformation.

In addition to the reactions described below, the organism is capable of further conjugation reactions (conjugation with phosphate, taurine, thiosulfate). Phase II reactions also include the formation of less readily water-soluble chemicals by biotransformation, such as the acetylation of amines or the methylation of thiols.

Glucuronidation

Glucuronidation is one of the most important conjugation reactions and serves to convert endogenous and exogenous chemicals with functional groups to very readily water-soluble end products. As a cofactor, the enzymatic reaction requires activated glucuronic acid (UDP-glucuronic acid) (Reaction 2.18). The acceptor molecule must provide a functional group for coupling.

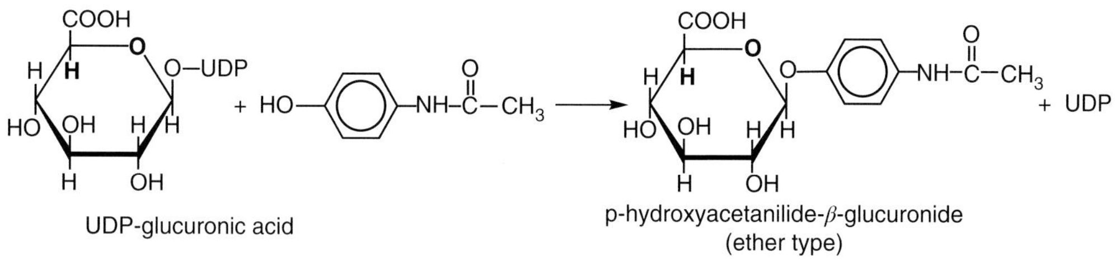

$$\text{UDP-glucuronic acid} + \text{HO-}\langle\text{phenyl}\rangle\text{-NH-C(=O)-CH}_3 \longrightarrow \text{p-hydroxyacetanilide-}\beta\text{-glucuronide (ether type)} + \text{UDP}$$

Reaction 2.18

Activated glucuronic acid as cofactor is produced by several coupled enzymatic reactions (Figure 2.21). Examples of different functional groups as substrates for glucuronidation reactions are listed in Table 2.5. It should be noted that ester glucuronides are chemically reactive, therefore the formation of ester glucuronides from certain drugs such as diclophenac is a bioactivation reaction (see Table 2.5) which can lead to the formation of protein adducts of the drug and thus to hypersensitivity reactions.

Bilirubin, steroid hormones, and fat-soluble vitamins are known as endogenous substrates for UDP-glucuronyltransferases. Many xenobiotics are also conjugated by this enzyme. The conjugation reaction of glucuronic acid can take place with alcohols, carboxylic acids, thiols, amines, and hydroxylamines (Table 2.5).

UDP-glucuronyltransferases are membrane-bound enzymes that occur in many organs. High concentrations are found in the liver, kidneys, and intestinal epithelium. Like cytochrome P450

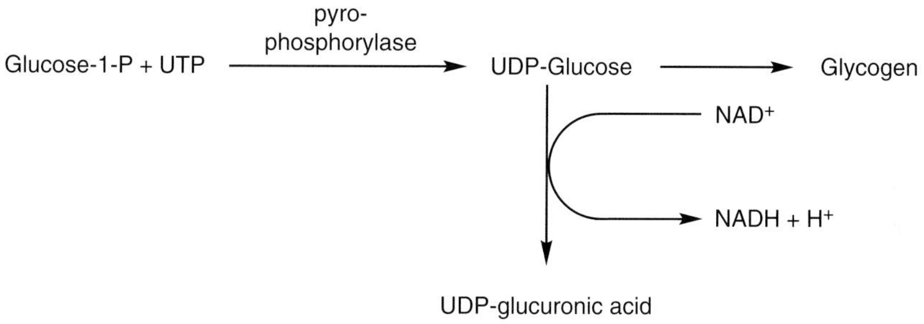

$$\text{Glucose-1-P} + \text{UTP} \xrightarrow{\text{pyrophosphorylase}} \text{UDP-Glucose} \longrightarrow \text{Glycogen}$$

$$\text{UDP-Glucose} \xrightarrow{\text{NAD}^+ \rightarrow \text{NADH} + \text{H}^+} \text{UDP-glucuronic acid}$$

Figure 2.21 Biosynthesis of UDP-glucuronic acid (activated glucuronic acid). UTP, uridine triphosphate; P, phosphate; UDP, uridine diphosphate.

Table 2.5 Examples of different classes of glucuronides.

Type of glucuronides	Chemical class of xenobiotic	Examples
O-Glucuronide		
Ether type	Alkohol	Trichloro ethanol
Ester type	Carbonic acid	o-Aminobenzoic acid
N-Glucuronide	Carbamate	Meprobamate 2-Naphtylamine
	Aromatic amine	Sulfadimethoxine
	Sulfon amide	
S-Glucuronide	Aromatic thiol	Thiophenol
	Dithiocarbaminic acid	Diethyl dithiocarbamate
C-Glucuronide	1,3-Dicarbonyls	Phenylbutazone

enzymes, UDP-glucuronyltransferases are also divided into families which have different, partially overlapping substrate specificities. More than 20 different UDP-glucuronyltransferases are known in humans. The concentration of UDP-glucuronyltransferases, similar to that of cytochrome P450, is increased by various inductors such as phenobarbital, 3-methylcholanthrene, and 2,3,7,8-tetrachlorodibenzo-p-dioxin (see *Induction of Metabolic Enzymes*).

Sulfatation

In mammals, the sulfation of hydroxyl groups is another important conjugation reaction. This reaction is catalyzed by sulfotransferases, a group of enzymes present in the liver, kidney, gastrointestinal tract, and lung. These enzymes couple the sulfate group to the xenobiotic. An activated intermediate (activated sulfate, 3′-phosphoadenosine-5′-phosphosulfate, PAPS) serves as reaction partner. Activated sulfate is produced in the cell from sulfate and ATP. In the first step of biosynthesis, an ATP sulphurylase links the starting materials to adenosine 5′-phosphosulfate (APS) (Reaction 2.19). In the second step, a kinase forms PAPS from the adenosine 5′-phosphosulfate (Reaction 2.20). Due to the close coupling of the two enzymes, the total reaction is rapid.

The inorganic sulfate necessary to form the activated sulfate is provided by the degradation of sulfur-containing amino acids, its availability is therefore dependent on nutritional factors and is limited. If large amounts of xenobiotics have to be sulfated, sulfation can no longer take place due to the lack of this cofactor. Therefore, the rate of excretion and biotransformation of sulfated endogenous chemicals can be influenced by the cofactor deficiency.

Sulfotransferases are also a large group of enzymes that can be divided into two different classes. The membrane-bound sulfotransferases in the Golgi apparatus of the cell usually catalyze the sulfation of endogenous molecules such as the amino acid tyrosine in proteins, whereas the soluble enzymes involve the conjugation of xenobiotics (e.g., phenol, Figure 2.22) and low molecular weight endogenous molecules (e.g., steroid alcohols). In humans, about 20 different sulfotransferases have been identified. In the enzymes responible for the biosynthesis of PAPS, genetic defects can occur which can cause interindividual differences in the excretion kinetics of xenobiotics in humans. In addition, pronounced species differences in the expression of individual sulfotransferases enzymes are described.

The most important consequences of a conjugation of xenobiotics molecules with the sulfate moiety is that the sulfate conjugate is dissociated at physiological pH, resulting in a reduction or loss of biological activity as well as an increase in their water solubility and thus an accelerated elimination. Sulfation competes with the glucuronidation; at low substrate concentrations, many xenobiotics are preferably sulfated.

$$SO_4^{2-} + ATP \xrightarrow{\text{sulfurylase}} APS + \text{pyrophosphate}$$

Reaction 2.19

$$APS + ATP \xrightarrow{\text{APS-phosphokinase}} PAPS + ADP$$

Reaction 2.20

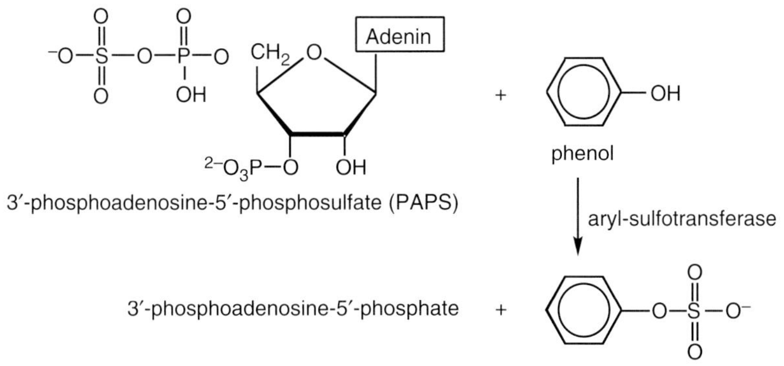

Figure 2.22 Sulfotransferases catalyze the sulfation of phenol. The cofactor PAPS provides activated sulfate.

> Sulfates of xenobiotics are usually released into the blood from the liver cell because of their lower molecular weight and thus rapidly reach the excretory organ kidney. There, sulfates are only poorly reabsorbed as molecules dissociate at physiological pH values and are therefore rapidly excreted. In contrast, because of the higher molecular weights, glucuronides are preferably released from the liver into the bile and, after cleavage in the intestine by bacterial glucuronidases, can be resorbed from the gut and thus be subjected to an enterohepatic circulation. Therefore, sulfation usually results in a faster elimination of xenobiotics as compared to glucuronidation.

Amino Acid Conjugation

Metabolically formed carboxylic acids can be conjugated with different amino acids. These reactions result in the formation of a peptide bond between the carboxylate group of the xenobiotic and the amino group of the amino acid. The most important substrates for amino acid conjugation are aromatic carboxylic acids such as benzoic acid, arylacetic acids, and substituted acrylic acids. In contrast to most glucuronidation and sulfation reactions, the conjugation of xenobiotics with amino acids converts the xenobiotic into a more reactive form by forming a coenzyme-A ester (Reaction 2.21).

The formation of amino acid conjugates is a coupled reaction catalyzed by different enzymes. In the first step, the carboxylic acid is converted into a coenzyme-A thioester by catalysis of ATP-dependent acid-CoA ligases. This thioester transfers its acyl group to the amino group of the

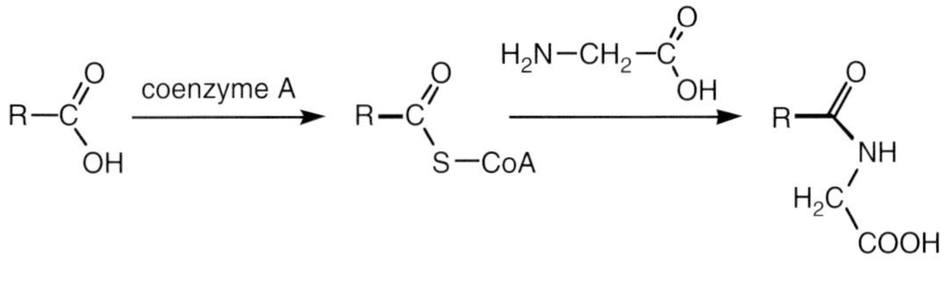

Reaction 2.21

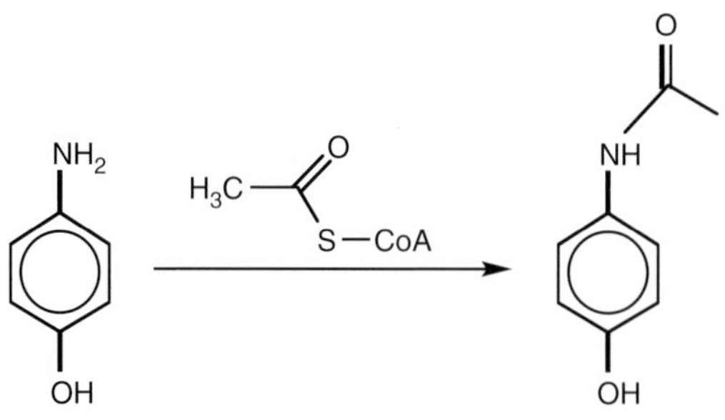

Reaction 2.22

amino acid. The reaction is catalyzed by an N-acetyl transferase. Both the ligase and the N-acetyl transferase are soluble enzymes present in the liver and kidney that occur in several forms. The amino acid conjugation of aromatic carboxylic acids competes with the glucuronidation of these compounds, therefore xenobiotics in different species are often excreted in different proportions as glucuronides or as amino acid conjugates. Frequent amino acid substrates are glycine or glutamine.

Acetylation

N-acetylation is a significant metabolic pathway for aromatic amines, α-amino acids, hydrazines, and sulfonamides. Acetyl-coenzyme-A (Reaction 2.22) serves as cosubstate. The enzymes of N-acetylation are located in the mitochondria, microsomes, and cytosol.

> **In most species, several different forms of N-acetyltransferases occur; genetic differences in the activity of N-acetyltransferases are observed in humans.**

In many cases, however, the acetylation of amino groups does not lead to the formation of more water-soluble products. Introduced acetyl groups are frequently removed by amidases. This hydrolysis introduces complex equilibria between free amines and the corresponding amides.

Glutathione S-transferases

Glutathione S-transferases are a group of enzymes that catalyze the first step in the formation of mercapturic acids (N-acetyl cysteine derivatives) from xenobiotics. Several cytosolic and and one membrane-bound glutathione S-transferase are known. Glutathione S-transferases are found in very many organs and are present in high concentrations in the liver, stomach, and intestine as well as in testes. Soluble glutathione S-transferases consist of at least 10 isoforms; the individual isoforms are dimers, which differ in the composition of their subunits. Like the other enzymes of biotransformation, glutathione S-transferases have a broad and partially overlapping substrate specificity. As a cofactor, glutathione S-transferases use the tripeptide glutathione (γ-glutamylcysteinyl-glycine), which is present in high concentrations (up to 10 mM in the liver cell).

Glutathione S-transferases accelerate the spontaneous reaction of electrophiles with glutathione, and a low spontaneous reaction rate with glutathione is observed with all substrates.

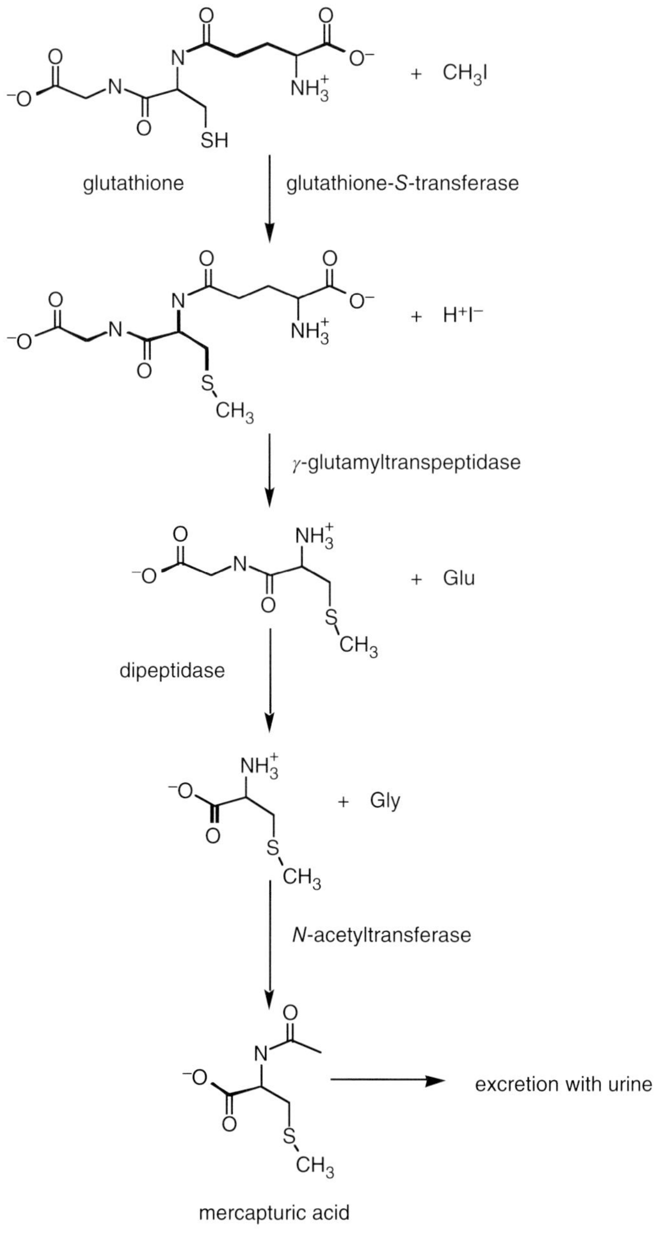

Figure 2.23 Metabolic formation of a glutathione S-conjugate and degradation of the S-conjugate to the precipitatable mercapturic acid using the example of methyl iodide.

During conjugation of a xenobiotic with glutathione, a thioether is formed. For example, S-methyl glutathione is formed from methyl iodide. The glutathione conjugates formed are further processed in the organism to the final excretory products, the mercapturic acids, which appear in the urine (Figure 2.23). In the first step of this reaction, the glutathione conjugate is converted by the enzyme γ-glutamyl transpeptidase to the corresponding cysteinyl glycine conjugate, which is further cleaved by dipeptidases to the corresponding cysteine conjugate. Cysteine conjugates of xenobiotics are ultimately metabolized by an N-acetyl transferase specific for cysteine conjugates to the mercapturic acid excreted with urine.

> **Conjugation with glutathione is the most important detoxification reaction for electrophilic compounds and metabolically formed electrophiles. Many toxic compounds are eliminated as mercapturic acid derivatives with the urine or as glutathione conjugates with the bile. Glutathione transferases also catalyze the detoxification of reactive oxygen species. In this reaction, glutathione is oxidized to glutathione disulfide.**

Besides their role as enzymes, glutathione S-transferases can play an important role in the storage of xenobiotics such as certain metals and endogenous compounds as well as in protein transport.

Methylation

Methylation of amines and thiols is a widespread metabolic reaction. Like N-acetylation, the transfer of a methyl group to an amine or a thiol does not result in an increase, but in a reduced water solubility. In addition to amines and thiols, phenols and metals such as mercury and arsenic can be enzymatically methylated. The cofactor for the methyltransferases catalyzing the reactions is S-adenosyl methionine, which acts as a methyl group donor. Methyl transferases are also a large group of very different enzymes, which often have endogenous substrates and therefore play important roles in the intermediary metabolism.

> **Genetic polymorphisms have a significant impact on the capacity of individual individuals for methylation of a xenobiotic. For example, a genetically determined reduced capacity of individual patients for the methylation of the immunosuppressive mercaptopurine (Figure 2.24) may cause an increased sensitivity of these individuals to the immunosuppressive effects of the mercaptopurine. This results in an increased susceptibility to infection in these patients at dosages of mercaptopurine well tolerated by other patients.**

2.3.4 Factors that Influence the Biotransformation of Xenobiotics

The dose of the chemical and the exposure pathways have a great influence on the extent and pathways of biotransformation. Chemicals that are intensively metabolized in the liver can be metabolized quantitatively via liver-specific pathways after oral intake, while the contribution

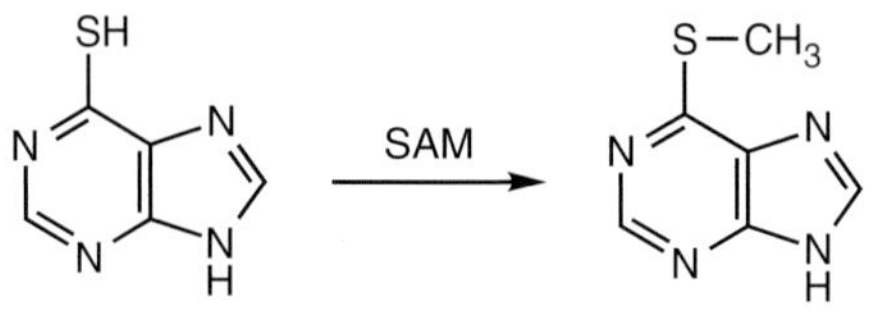

Figure 2.24 Deactivation of mercaptopurine by S-methylation.

of other organs to the biotransformation is small. When quantitative biotransformation of a xenobiotic in the liver after oral uptake occurs, only low concentrations of the xenobiotic can be reached in other organs. This is termed the hepatic first-pass effect. In the case of intravenous administration or after inhalation, disposition and bioavailability of such chemicals in other organs may be widely different.

> **Some enzymes of the foreign metabolism have a high affinity, but a low capacity for certain xenobiotics. These enzymes may be rapidly saturated after high doses of a xenobiotic, resulting in dose-dependent transitions in biotransformation pathways with enzymes with low affinity but high capacity increasingly participating in biotransformation. A good example of dose-dependent metabolism is acetaminophen (N-acetyl-p-aminophenol), which is metabolized mainly to a sulfate conjugate after low doses. After high doses, the availability of activated sulfate as cofactor for this reaction becomes limited and biotransformation reactions switch to glucuronide formation or bioactivation to a quinone imine.**

The availability of a chemical in an organ depends on the exposure pathway and the chemical properties of water solubility or lipophilicity. These two variables influence the distribution to blood and the influx of chemicals in the cell. The binding to plasma proteins also significantly influences the distribution of a chemical and its intercellular concentrations. Both intramolecular and extracellular proteins can bind xenobiotics. This binding is not specific (e.g., the binding of many compounds to serum albumin) and often has a great effect on the availability of the foreign chemical in the cell and the excretion kinetics.

In addition, diet, gender, age, and genetic predisposition can influence the activity of individual metabolic enzymes.

In rodents, the capacity to metabolize xenobiotics shortly before and after birth is low and the expression of individual cytochrome P450 enzymes differs widely. In human fetuses and in newborns, the capacity for biotransformation is low for some xenobiotics; corresponding enzyme activities develop with different time sequences only in the first 6 months of life. At an advanced age, the capacity of the body to metabolize xenobiotics is reduced in humans as well as in rodents.

The level of sex hormones is an important factor in the regulation of cytochrome P450 enzymes, since the expression of certain cytochrome P450 enzymes is under the control of these hormones. Male rats show two to three times higher monooxygenase activity than female rats; by administering male sex hormones, certain cytochrome P450 activities can be increased in female rats. In humans, differences in the cytochrome P450 profile have also been described between males and females.

The nutritional status of animals and humans may have a significant impact on bioactivation and detoxification. The lack of trace elements (calcium, copper, iron, magnesium, and zinc) and proteins in the diet leads to a reduction in the activity of cytochrome P450 and thus to a reduced metabolism of xenobiotics. Nutritional deficiencies, however, also affect the capacity of various protective mechanisms (glutathione, antioxidants) and the availability of cofactors for Phase II reactions. A diet rich in unsaturated fatty acids increases the concentrations of cytochrome P450 in the liver. Food restriction reduces some enzyme activities; on the other hand, malnutrition increases the concentration of certain cytochrome P450 enzymes. Diseases can also influence the metabolic fate of xenobiotics by directly changing the enzyme activities due to liver damage, co-substrate deficiencies and changes in the intestinal flora.

Induction of Metabolic Enzymes

The repeated administration of certain xenobiotics leads to an increased activity of enzymes of the biotransformation (examples in Table 2.6). This is often due to an increased synthesis or reduced degradation of the enzyme. This process is referred to as enzyme induction, and corresponding chemicals as enzyme inductors.

Induction by xenobiotics mainly increases the activity and quantity of membrane-bound enzymes. Enzyme induction has a specific time course for each inductor, is subject to saturation kinetics, and is often specific. After the enzyme inductor is discontinued, enzyme concentrations decrease within the characteristic time periods to the initial activities present before administration of the enzyme inductor.

Different inductors may increase enzyme concentration factors from three-fold to more than 100-fold. Certain inducers can specifically induce individual enzymes such as certain cytochrome P450 enzymes, other inducers increase the expression and thus also the activities of several enzyme systems (such as cytochrome P450 and glucuronyltransferases).

In addition to medicinal products or xenobiotics, food ingredients and excess consumption of alcoholic beverages as well as diseases can result in enzyme induction. The cytochrome P450 enzyme 2E1, which metabolizes many xenobiotics and solvents (benzene, trichloroethene, dichloromethane), is induced by chronic alcohol intake in the liver. This enzyme is also affected by disease states such as diabetes; sugar is increasingly formed by acetone, which acts as an enzyme inducer.

In addition to an increase in the activity of enzymes in the metabolism of xenobiotics, the administration of inductors of the biotransformation can also induce other, often very specific changes in the liver and in other organs. These are dependent on the type of the inductor. Interestingly, individual agents resulting in enzyme induction do not affect all organs to the same extent. Alcohol induces cytochrome P450-2E1 in the liver in the rat, but no increase in the concentration of this enzyme can be detected in other organs. In addition to the inductor, the extent of enzyme induction also depends on the organ and on the type, gender, and age of the test animals.

Enzyme induction may result in accelerated biotransformation of substrates for this enzyme. For chemicals which are bioactivated by these enzymes, the equilibrium concentration of reactive intermediates is increased by the accelerated biotransformation. This may increase toxicity.

Table 2.6 Enzyme inductors and enzyme systems influenced by their administration.

Inductor	Induced enzymes
2,3,7,8-tetrachlorodibenzo-*p*-dioxin	Cytochrome P450 1A1, glucuronyl transferases
Ethanol	Cytochrome P450 2E1
Phenobarbitone	Cytochrome P450, epoxide hydrolase, glucuronyl transferases
trans-stilbenoxide	Epoxide hydrolase
3-Methylcholanthren	Cytochrome P450, glucuronyltransferases
Perfluordecanoic acid	Cytochrome P450, UDP-glucuronyltransferases
β–Napthoflavone and xenobiotics that reduce glutathione concentrations in cells	Glutathione S-transferases, UDP-glucuronyl transferases, microsomal epoxide hydrolase

Table 2.7 Mechanisms of the induction of cytochrome P450 enzymes by xenobiotics.

Cytochrome P450	Inducing agent	Mode of action
1A1	Tetrachlorodibenzo-dioxin	Increase in transcription due to interaction with the Ah receptor
1A2	Methylcholanthrene	Increase in stability of mRNA
2B1, 2B2	Phenobarbitone	Increased transcription due to interaction with the CAR receptor
2A6, 2B6, 2C9	Bile acids, polychlorinated biphenyls, phenobarbitone	Increased transcription due to interaction with the PXR receptor
2C19		
2E1	Ethanol, acetone	Protein stabilization
3A1	Dexamethasone	Increase of transcription, independent of glucucorticoid receptor
3A1	Triacetyl oleandomycine	Protein stabilization due to inhibition of proteolytic enzymes
4A1	Clofibrate	Increased transcription due to interaction with the PPARα-receptor

> **Some xenobiotics, such as allyl isothiocyanate, can selectively induce Phase II enzymes and thus improve the capacity for detoxification of xenobiotics and endogenous toxicants; this induction may explain the anticarcinogenic effect of food with a high content of Phase II inducers, such as broccoli.**

The mechanisms of enzyme induction at the molecular level are manifold and often include a receptor-mediated increase in gene expression (Table 2.7). Due to the diverse mechanisms, enzyme inductors can also have very different structures.

The molecular mechanism of enzyme induction is well described by tetrachlorodibenzo-p-dioxin (TCDD). Dioxin and other planar molecules such as polycyclic aromatic hydrocarbons bind with high affinity to a cytosolic receptor. This receptor was named the Ah receptor (Ah = aryl hydrocarbon) because of its affinity for polycyclic aromatic compounds, but TCDD has a higher order affinity for this receptor compared to polycyclic aromatics. After binding of the TCDD to the receptor and interaction with further proteins, the TCDD receptor complex migrates into the nucleus, binds to DNA, and results in an activation of the DNA transcription and thus to increased protein synthesis (Figure 2.25). Cytochrome P450 1A1 and certain UDP-glucuronyltransferases are prominent enzymes under the control of the Ah receptor.

Barbiturates, which are strong inducers of a variety of foreign metabolic enzymes, interact with the constitutive androstane receptor (CAR) and the Pregnan X receptor (PXR), which also control the expression of many enzymes of biotransformation.

> **Enzyme induction may accelerate the degradation of xenobiotics and can therefore affect both the pharmacologically active profiles of drugs as well as the toxicity of xenobiotics.**

For example, the chemopreventive effect of isothio cyanates is attributed to an induction of glutathione transferases. Enzyme induction may require an increased dosage for low-therapeutic drugs to achieve the desired pharmacological effect, as the accelerated degradation of the drug

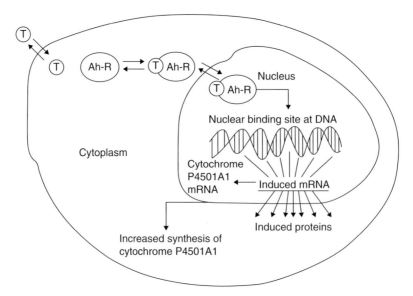

Figure 2.25 *Mechanism of the receptor-mediated action of dioxins such as TCDD (T, 2,3,7,8-tetrachlorodibenzo-p-dioxin). After binding the TCDD to the Ah receptor (Ah-R), this complex enters the nucleus and binds to DNA. As a result of binding, the expression of various proteins such as cytochrome P450 1A1 is increased.*

may prevent adequate blood levels at normal doses (Figure 2.25). This can be demonstrated by the example of the interaction of barbiturates and oral anticoagulants of the coumarin type such as dicumarol. Coumarins have a low therapeutic index that requires an adjusted individual dosage of the coumarin derivative to achieve optimal blood levels for the anticoagulant effect. If the optimal blood levels are only slightly below the threshold, the anticoagulant effect is no longer present (prothrombin time, shown on the right in seconds in Figure 2.26). If optimal blood levels are only slightly exceeded, hemorrhage occurs as a result of the increased pharmacological effect. If a well-adjusted patient is given a barbiturate in addition to the coumarin derivatives, the degradation of the coumarin derivative is accelerated and the anticoagulative effect is lost. The co-administration of enzyme inducers with other medicinal drugs can therefore lead to drug interactions.

Inhibition of Biotransforming Enzymes

In addition to an increase in the activity of biotransforming enzymes, a reduction in capacity for the biotransformation of xenobiotics can be observed after administration of various xenobiotics under certain nutritional circumstances and diseases states.

Chemicals that interfere with protein elimination and/or biosynthesis of functional groups in biotransformation enzymes (e.g., heme in cytochrome P450 enzymes) lead to a reduction in the concentrations of these enzymes and thus also to a decrease in the capacity for biotransformation. Consumption of cofactors can also influence enzymatic reactions, for example exhaustion of the available sulfate supply can lead to reduced sulfation (see above).

In chemical mixtures, the individual components can interfere with biotransformation. If a component has a very high affinity for a particular enzyme, it is preferably metabolized by the

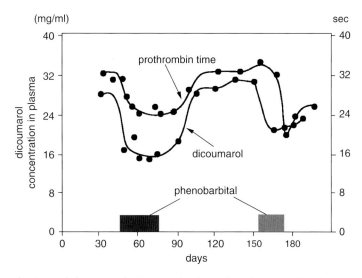

Figure 2.26 Induction of drug metabolism as the basis for pharmacokinetic interactions in drug therapy.

enzyme. The reaction of components of the mixture with lower affinity to this enzyme is then prevented or reduced. These xenobiotics can then be metabolized by other mechanisms and thus lead to potentially altered toxic effects compared to the administration of the single chemical.

> **In the enzymatic reaction of xenobiotics, reactive intermediates are frequently formed which also react with components of the enzyme itself (e.g., with the nitrogen atoms in the heme of the cytochrome P450). Such reactions lead to a chemical modification of the active site of the enzyme and thus usually to an irreversible inhibition. Since the active site of the enzyme is altered by covalent binding, enzymatic activity can only be restored by resynthesizing the enzyme.**

Irreversible inhibition of cytochrome P450 enzymes can occur as a result of the catalytic oxidation of ethins, secondary amines, or thioamides and also contribute to drug interactions (Table 2.8).

Table 2.8 Pharmacokinetic interactions: inhibition of metabolism by drug interactions.

Enzyme	Drug	Consequence
Xanthine oxidase	Purine analogs and allopurinol	Increased cytotoxicity
Aldehyde dehydrogenase	Ethanol + disulfiram + metronidazole	Increased concentrations of acetaldehyde with major effects on circulation
Cytochrome P450	Oral anticoagulants (phenprocoumon) + cimetidine + phenylbutazone + chloramphenicol	Increased bleeding
Cytochrome P450	Oral antidiabetics (tolbutamid) + dicoumarol + chloramphenicol	Hypoglycemia

Genetically Determined Species and Individual Differences

Many species differences in biotransformation and toxicokinetics are due to the lack or reduced activity of enzymes of biotransformation in certain animal species. Because of the different activities of the biotransformation enzymes, the metabolic profile of a chemical in different species may be very different. Thus, bioactivation reactions and the toxicity of xenobiotics may be highly species-dependent and certain species may be extremely sensitive to biotransformation-mediated toxic effects. For example, administration of acetamido fluorene to rats induces a high incidence of cancer, while guinea pigs are virtually resistant to the tumor-inducing effect of acetamido fluorene. The reason for these differences is probably the different metabolism (Figure 2.27).

Acetamidofluorene is hydroxylated in the rat via a P450-dependent mechanism and subsequently sulfated. These sulfates serve as a transport form for an electrophilic intermediate stage into the bladder that decomposes to reactive nitrenium ions. In the guinea pig, acetamidofluorene is oxidized on the aromatic ring system, and the formation of nitrenium ions does not take place.

Dogs have very low N-acetyl transferase activity and therefore cannot acetylate many substrates. Polymorphisms (interindividual differences in the activity of enzymes with genetic basis) in the acetylation of selected substrates have also been observed in humans, rabbits, and mice. These polymorphisms are based on genetically determined enzyme deficiencies and correlate

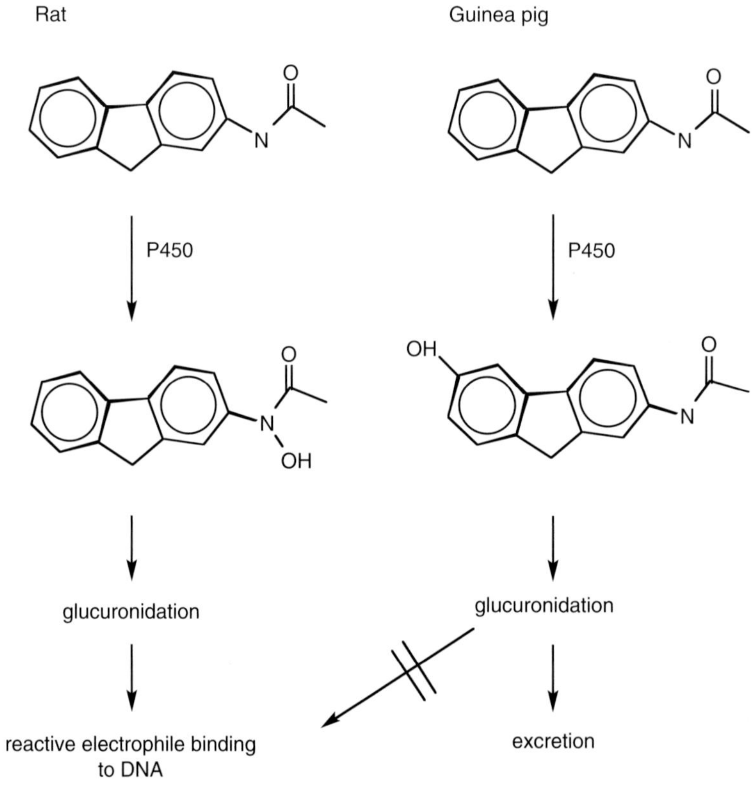

Figure 2.27 Metabolism of acetamido fluorene in rat and guinea pig. In the rat, acetamido fluorene is oxidized by a cytochrome P450 enzyme at the nitrogen atom and the hydroxamic acid formed can be converted into an instable glucuronide. In contrast, guinea pigs undergo hydroxylation on the aromatic ring system. This reaction does not resulting in an unstable glucuronide.

with sensitivity to toxic effects. In humans, slow and fast acetylators are present. This distinction is based on their capacity to acetylate the tuberculostatic isoniazid. Individuals who are poorly acetylated are particularly sensitive to the most important toxic effect of isoniazid: nerve damage.

> **Individual differences in foreign metabolism are mainly observed in humans. In the case of normally used laboratory animals, which are mostly inbred strains, this phenomenon is of little importance.**

For example, extreme differences in the extent of N-oxidation of caffeine in humans have been found. These differences are due to interindividual variations in the expression of a particular cytochrome P450 enzyme (1A1). Many other enzymes of biotransformation are also subject to a genetic control in humans and thus show interindividual different activities. The influence of such polymorphisms is particularly well investigated with respect to the drug metabolism since such polymorphisms can influence the desirable and undesirable effects of drugs. The genetically determined interindividual differences require a patient-specific dosage of drugs that are mainly metabolized by polymorphously expressed enzymes.

2.3.5 Role of Bioactivation in Toxicity

> **During biotransformation, many xenobiotics are transformed to metabolites with higher chemical reactivity as compared to the parent compounds. This process is called bioactivation or metabolic activation.**

When reactive intermediates formed in the metabolism react with cellular macromolecules, they alter their structure and often their function. As a result, a large number of toxic effects can be triggered. Apart from bioactivation reactions, however, reactions which efficiently convert the reactive intermediates into non-toxic products occur in the mammalian cell. Therefore, the equilibrium concentration of toxic intermediates and the toxic effects of many chemicals depend on the final balance between metabolic activation and deactivation.

> **In order to understand bioactivation reactions as the basis for toxic effects, knowledge of the chemical mechanisms of these reactions as well as knowledge of the enzymes involved in the reactions, their distribution in the organism, and their regulation are necessary. Bioactivation reactions are catalyzed by both Phase I and Phase II enzymes. Often, the metabolic activation of xenobiotics is the prerequisite for their toxic effect.**

Some examples of enzymatic bioactivation reactions and the resulting toxic effects are given in Table 2.9.

Bioactivation reactions can also be responsible for toxic effects on the human fetus and transplacental carcinogenic effects. For example, thalidomide, which when ingested during pregnancy induces malformations of the limbs in the fetus, is metabolized by the fetus into a reactive intermediate. Its interaction with specific proteins leads to the characteristic malformations in children whose mothers had regularly taken thalidomide during pregnancy.

88 Toxicology and Risk Assessment

Table 2.9 Examples of bioactivation reactions and the toxic effects they trigger.

Xenobiotic	Bioactiviation-dependent toxicity	Reactive intermediate formed
Acetamido fluorene	Bladder cancer	Nitrenium ion
Aflatoxin$_{B1}$	Liver tumors	Epoxidation
Benzen	Leukemia	Formation of benzochinone
Dimethyl nitrosamine	Tumors in several organs	α-Hydroxylation followed by rearrangement to a carbonium ion
Carbon tetrachloride	Liver necrosis	Formation of trichloromethyl radicals
Vinyl chloride	Liver tumors	Epoxidation

The process of bioactivation is complex. Often, different enzymatic reactions are involved or the precursors of the reactive metabolites undergo bioactivation reactions in different organs. Because of efficient detoxification at low doses, low equilibrium concentrations of reactive intermediates or switches in metabolic pathways at high doses, formation of toxic metabolites at low doses can occur without adverse effects. Endogenous chemicals which can also be enzymatically converted to more reactive intermediates, for example cholesterol, are epoxidized during biosynthesis of bile acids.

> **Electrophiles often form during bioactivation reactions. As reactive intermediates, methyl cations, oxiranes, ketenes, acyl chlorides, quinones, and chemically exotic species have been identified. The toxic effects of such compounds are due to non-selective alkylations and acylations of cellular macromolecules.**

Cytochrome P450 catalyses the bioactivation of many xenobiotics into electrophiles; however, other Phase I responses can be mediated as FMOs and also Phase II reactions may contribute to the formation of electrophilic metabolites. For example, chloroform is oxidized by cytochrome P450 2E1 by insertion of an oxygen atom into the C–H bond of the unstable trichloromethanol; phosgene is then formed as a reactive metabolite. Phosgene binds to terminal amino and thiol groups in proteins (Reaction 2.23).

Phase II enzymes can also catalyze bioactivation reactions. Enzymatic coupling with glutathione plays an important role in the bioactivation of chlorinated aliphatic and olefinic compounds. With 1,2-dihaloalkanes, glutathione reacts enzymatically by substitution of a halogen atom. The glutathione conjugates thus formed have a structural analogy to mustard gas (dichlorodiethyl sulfide). They cyclize to an electrophilic episulfonium ion. This electrophile then alkylates nucleophilic macromolecules such as DNA (Figure 2.28). Such alkylation reactions are responsible for the toxicity and carcinogenicity of 1,2-dibromoethane.

$$HCCl_3 \xrightarrow{P\text{-}450} [HOCCl_3] \xrightarrow{-HCl} \underset{Cl}{\overset{O}{\underset{\|}{C}}}_{Cl} \longrightarrow \text{covalent binding} \longrightarrow \text{toxic effect}$$

Reaction 2.23

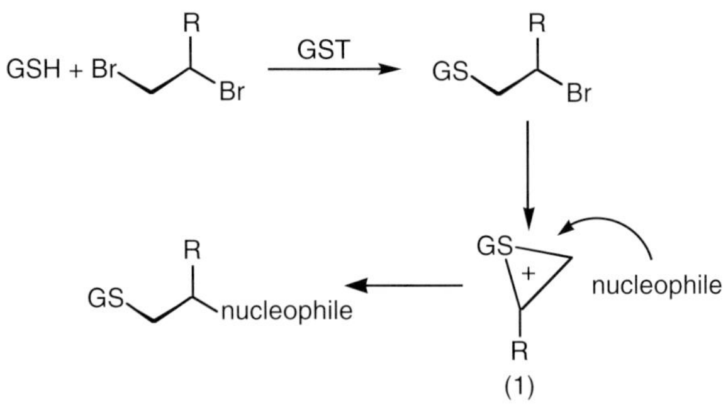

Figure 2.28 Bioactivation of 1,2-dibromoalkanes by glutathione conjugation. GSH, glutathione; GST, glutathione S-transferase. The formed episulfonium ion (1) reacts with nucleophilic macromolecules.

> **Bioactivation reactions may also result in free radicals. Neutral (e.g., the trichloromethyl radical) as well as positively (paraquat cations) and negatively (nitroradical anions) charged radicals can be generated (Table 2.10).**

The formation of radicals can occur by oxidative or reductive reactions. Bioactivation by one-electron reductions can be catalyzed by the NADPH-dependent cytochrome P450 reductase, nitroreductase or xanthine oxidase. The initiation of toxic effects by the metabolic formation of radicals is well studied, especially in the case of carbon tetrachloride. This potent hepatotoxicant is converted to the trichloromethyl radical by enzymatic reduction (Reaction 2.24). This radical reacts with fatty acids in lipid membranes and leads to lipid peroxidation (see *Interactions with Lipids*).

Important for the toxic effect of radicals is the initiation of radical chain reactions. Radicals such as the trichloromethyl radical are highly reactive, have short half-lives, and react with a variety of partners. Cleavage of endogenous chemicals under formation of radical species also plays an important role in the toxic effects of high-energy radiation.

The formation of reactive oxygen species can also be catalyzed by enzymatic reactions. The reduction of molecular oxygen occurs by a stepwise electron transfer through a series of intermediates to water, catalyzed by the enzymes and coenzymes of the respiratory chain. During the stepwise reduction, toxic oxygen species such as the hydroxyl radical are formed (Figure 2.29).

Table 2.10 Examples of xenobiotics that are metabolized to radicals.

Xenobiotic	Radical formed	Toxicity
Carbon tetrachloride	$\cdot CCl_3$	Liver necrosis
Paraquat	Paraquat radical cation and reactive oxygen	Lung fibrosis
Daunomycin	Daunomycin radical and O_2^{-}	Cardic muscle damage
Nitrofurantoin	RNO_2^{-}, O_2^{-}	Pulmonary edema

$$Cl_4C + e^- \xrightarrow{\text{P450 reductase}} {}^{\bullet}CCl_3 + Cl^-$$

Reaction 2.24

90 Toxicology and Risk Assessment

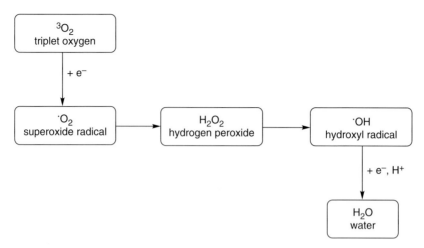

Figure 2.29 *Formation of active oxygen species by stepwise one-electron reduction of molecular triplet oxygen.*

The formation of reactive oxygen species without an interaction of xenobiotics plays an important role in non-specific infection control as well as in the pathogenesis of many diseases (arteriosclerosis, chronic polyarthritis).

The metabolism of certain xenobiotics also linked to the formation of reactive oxygen metabolites. Various stable radicals, for example semichinone radical anions, can transfer electrons to molecular oxygen. This produces the superoxide radical anion ($\bullet\, O_2^-$). The intermdiate during the reduction of molecular oxygen can be further reduced and thus act as a catalyst in the formation of superoxide radical anions. The most important reactive oxygen species is the hydroxyl radical because of its high chemical reactivity.

> **Oxidative stress caused by oxygen radicals plays an important role in the toxicity of quinones (e.g., used as cancer therapeutics), bis-pyridylium herbicides such as paraquat, some transition metals (Cr, Ni), and ionizing radiation. Oxidative stress occurs when the balance between oxidants and antioxidants shifts in favor of the oxidative side and detoxification processes are overwhelmed.**

2.3.6 Interactions of Reactive Intermediates formed during the Biotransformation

> **Reactive intermediates formed by biotransformation reactions from xenobiotics can react with both low molecular weight and macromolecular targets. These reactions may cause detoxification of the reactive intermediates formed or the initiation of acute and chronic toxic effects.**

Detoxifications

The reaction of electrophiles with the nucleophile water present in high concentration in the cell is the simplest example of a detoxification reaction. Many electrophilic intermediates are sensitive to hydrolysis and react quickly with water present in the cell. This reaction almost always results in detoxication and leads to the formation of excitable reaction products.

The most important function of glutathione in the biotransformation of xenobiotics is the detoxification of toxic electrophiles and radicals. Formation and degradation of glutathione conjugates are described in detail above.

Electrophiles react with the nucleophilic sulfur atom of the glutathione molecule to form glutathione S-conjugates; this reaction can proceed spontaneously or be enzymatically catalyzed. Spontaneously, only "soft electrophiles" (according to the concept of Pearson) react with the "soft nucleophile" glutathione. For the conjugation of "hard electrophiles" with glutathione, enzymatic catalysis is necessary. Glutathione also plays an important role in the detoxification of reactive oxygen species. Selenium-dependent glutathione peroxidases are important enzymes for the detoxification of hydrogen peroxide.

Various antioxidants in the cell also reduce reactive oxygen radicals to molecular oxygen or water. The most important lipid-soluble antioxidant in the cell is α-tocopherol, whose integration into lipid membranes prevents damage to membrane components by radicals. The OH radical is reduced to water by α-tocopherol: the superoxide radical anion is converted to hydrogen peroxide and peroxy radicals, resulting in conversion of α-tocopherol to a less reactive radical. In contrast to α-tocopherol, ascorbic acid is an essential water-soluble antioxidant in the cytosol of the cell.

Interactions with Cellular Macromolecules

> **Reactive intermediates react in the cell with lipids, proteins, and DNA. Enzyme inhibition, structural changes in enzymes and lipid membranes, destruction of organelles, and ultimately cell death may occur as a consequence. The reaction of cellular intermediates with DNA components leads to the formation of mutations during DNA replication and can influence gene expression. The alkylation of DNA is an important process in the induction of tumors.**

Electrophilic intermediates react with nucleophilic macromolecules to form a covalent bond. This results in an adduct from the metabolite and the macromolecule.

Interactions with Proteins Nucleophilic sulfur and nitrogen atoms in proteins are alkylated or acylated by electrophilic metabolites; reaction partners are the amino acids cysteine, histidine, valine, and lysine. Since free sulfur and nitrogen atoms in enzymes are important for catalytic activity, alkylation and acylation often lead to the inactivation of enzymes. The active centers of the enzymes that catalyze the formation of reactive metabolites can also be alkylated by these. This "suicide reaction" often results in inhibition of the enzymes. An example is the alkylation of the heme in cytochrome P450 (see above). If the alkylated proteins have important functions for the cell (such as the thiolate-containing enzymes of the mitochondrial respiratory chain), the functional loss resulting from the alkylation leads to cell death. Through alkylation and acylation, xenobiotics are bound to proteins or their spatial structures are altered. Therefore, under certain conditions, alkylated or acylated proteins are detected by the immune system as foreign and trigger an allergic reaction after sensitization. Many undesirable drug reactions, such as sudden liver failure, are due to the formation of modified proteins by drug metabolites. Such events are very rare, but often a reason for discontinuing the use of appropriate drugs. Therefore, clarification of a possible bioactivation to form reactive intermediates with the ability to bind to proteins has become an important part of the development of new drugs.

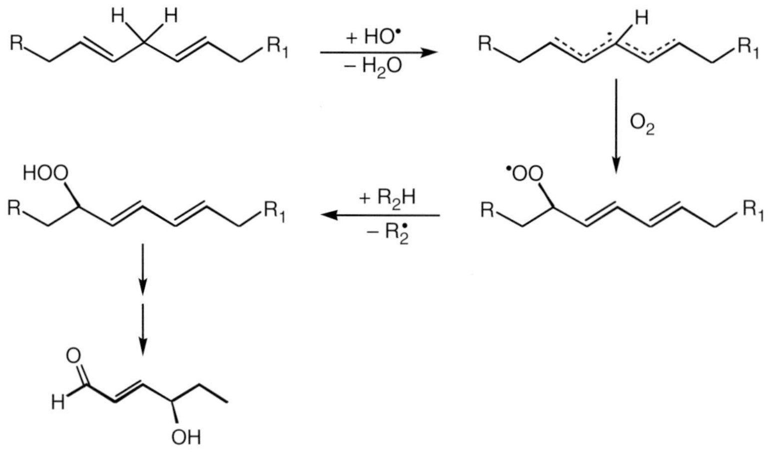

Figure 2.30 *The reaction of radicals with unsaturated fatty acids in biological membranes and the initiation of lipid peroxidation. The disintegration of the unsaturated fatty acids produces α,β-unsaturated carbonyl compounds such as 4-hydroxy-2-trans-hexanal.*

Radical and reactive oxygen species can oxidize thiolates in enzymes to the corresponding disulfides or induce the formation of carbonyl groups in proteins. The resulting conformational change also leads to changes in enzyme activity and denaturation of the proteins.

Interactions with Lipids Radicals formed from xenobiotics and reactive oxygen species can abstract hydrogen atoms from lipids. This reaction with the unsaturated fatty acids of the lipid-essential components of biological membranes leads to lipid peroxidation. The abstraction of hydrogen atoms results in fatty acid radicals, which can then react with molecular oxygen to form peroxy radicals and hydroperoxides (Figure 2.30). By these triggered radical chain reactions, C–C bonds can be cleaved in the fatty acids and these are broken down to short chain lengths (C2 to C6, e.g. 4-hydroxynonenal) and biological membranes desintegrate. This results in a loss of compartmentalization of the cell and ultimately cell death. In lipid peroxidation, liberated α,β-unsaturated aldehydes such as 4-hydroxynonenal can further intensify toxic effects due to their electrophilic properties and reactions with thiolates,.

Electrophiles can also react with nucleophilic centers in lipid membranes. The acylation and alkylation of the amino groups of phosphatidylethanolamine from lipid membranes has also been observed. The importance of this formation of phospholipid adducts for toxic effects is, however, little defined.

Interactions with DNA

> **The alkylation of DNA bases by reactive intermediates often leads to a change in the three-dimensional structure of the DNA and can lead to the incorporation of a false base into the daughter strand during DNA replication and thus to the generation of a mutation. DNA changes by reactive intermediates represent the first biochemically definable step in chemical carcinogenesis.**

Only xenobiotics with electrophilic structures such as alkyl sulfates, nitrogen and sulfur mustard, dichloroplatinum complexes such as cis-platinum, lactones and imines used in tumor chemotherapy do not require metabolic activation as a prerequisite for reaction with DNA bases. In contrast, many chemical carcinogens require bioactivation to react with DNA. The most important reaction partners for electrophilic metabolites in DNA are the nitrogen and oxygen atoms of purine and pyrimidine bases (Figure 2.30). DNA alkylation in the cell is, however, a step which proceeds only in very low yield. Even potent carcinogens such as dialkylnitrosamines alkylate only one of 10^7–10^8 bases in the DNA. The extent of the alkylation of oxygen and nitrogen atoms in individual bases is strongly dependent on the chemical structure and electrophilicity of the reactive intermediate. Hard electrophiles preferentially attack the enolized oxygen atoms (hard nucleophiles) while softer electrophiles react preferentially with exocyclic amino groups and nitrogen atoms in the rings (Figure 2.31). For many electrophilic metabolites, guanosine is the preferred target in DNA. Planar structures such as polycyclic aromatics and aflatoxin B1 often lead to a high yield of DNA adducts since the planar parts of the molecules can be intercalated into the DNA double strand.

Examples of the structures of reaction products of "classical" chemical carcinogens with DNA bases as they are present in target organ DNA are given in Figures 2.32 and 2.33.

Radicals can also damage DNA. Both the sugar-phosphate framework and DNA bases can be oxidatively modified. Breakdown of the sugar-phosphate framework can lead to DNA strand breaks or apurinic sites in the DNA. After the reaction of OH radicals with DNA bases, the most important reaction product is 8-hydroxy deoxyguanosine (Figure 2.34).

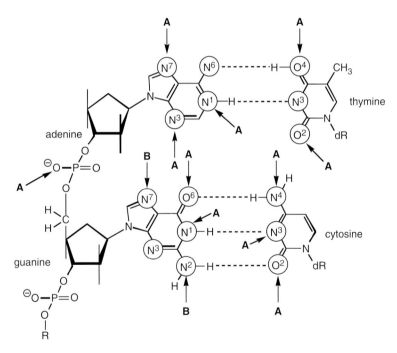

Figure 2.31 Formation of covalent bonds between electrophiles (A, B) and DNA. Electrophiles can react with all nucleophilic centers of the DNA (oxygen and nitrogen atoms of purines and pyrimidines, hydroxyl in phosphate groups). The preferential reaction of certain structures in DNA can be explained by the concept of hard and soft acids and bases. Hard electrophiles (acids, A) such as methyl cations react preferentially with hard nucleophiles (bases), such as the oxygen atoms of guanosine, and soft electrophiles (B) react preferentially with soft nucleophiles such as the exocyclic amino group (N6) Or the N7 in guanosine.

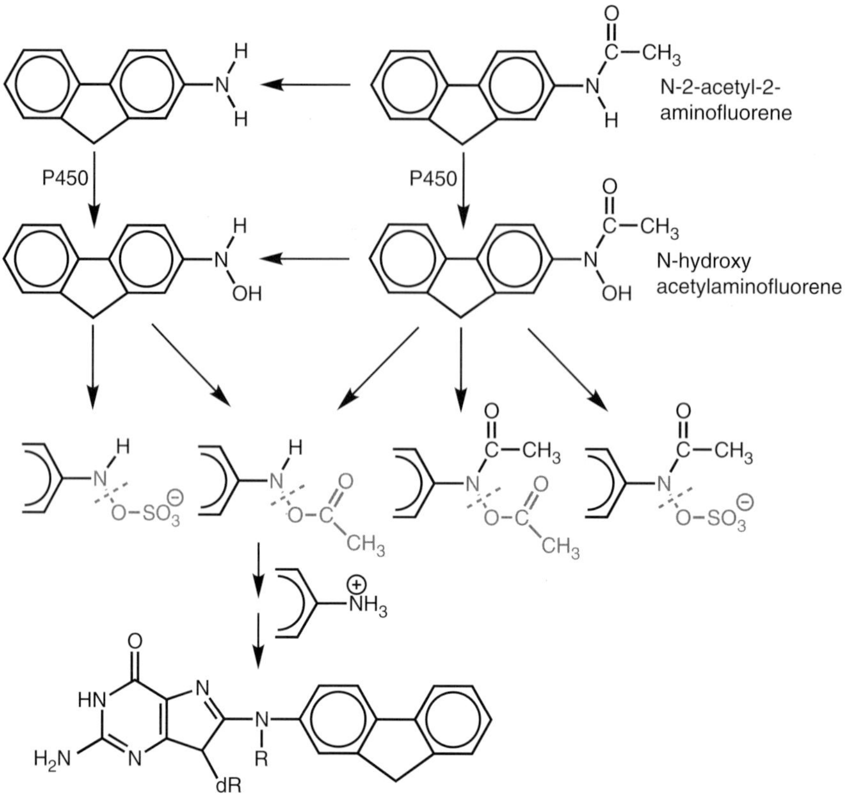

Figure 2.32 Bioactivation and DNA binding of N2-acetylaminofluorene. R = H, CH$_3$–CO.

Since reactive oxygen species are generated in the processes of cellular energy production and cannot be completey detoxified, a large number of DNA changes are found in mammalian cells even without an interaction with xenobiotics. In addition to reactive oxygen metabolites, other endogenous processes such as the formation of formaldehyde as an intermediate in biosynthetic reactions can lead to DNA modifications. The frequency of such DNA modifications is well above the extent of the DNA binding of xenobiotics such as polycyclic aromatics even in highly exposed individuals such as smokers (Table 2.11). Since DNA alterations induced by the action of radicals are high due to the measurable equilibrium concentrations of reactive oxygen species in the cell, cells have developed specific enzymes for the repair of oxidized DNA bases. These recognize oxidized DNA bases, remove them from the DNA strand, and replace them with the correct bases.

DNA alterations due to the action of radicals are discussed as biomarkers for the aging process. It is probable that a reduction in the concentration of antioxidant chemicals occurring with increasing age is responsible for an age-enhanced oxidative modification of DNA and proteins.

2.3.7 Summary

Lipophilic chemicals are only excreted from the mammalian body. An important function of the biotransformation is therefore to convert lipophilic chemicals into hydrophilic molecules, which are more readily excreted.

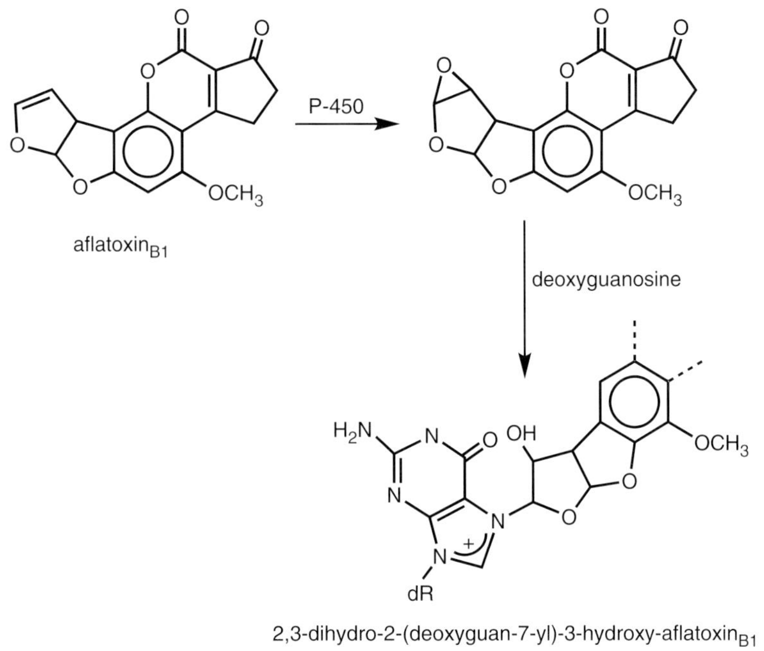

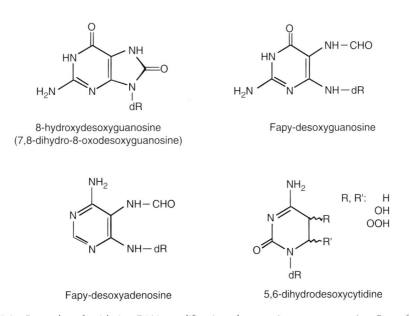

Figure 2.33 *Metabolic epoxidation of aflatoxin$_{B1}$ and reaction of the resulting aflatoxin$_{B1\text{-}8.9}$ oxide with deoxyguanosine. The guanosine derivative formed represents a prämutagene DNA change.*

Figure 2.34 *Examples of oxidative DNA modifications by reactive oxygen species. Fapy, formamidopyrimidine; dR, deoxyribose. Such alterations can also be prämutagen DNA lesions.*

Table 2.11 Concentrations of selected DNA adducts (from endogenous and exogenous sources) in human tissues.

Modified DNA constituent	Tissue	Concentration per 10^7 bases
Endogenous factors		
7,8-Dihydro-8-oxo-2'deoxyguanosin	Leucocytes	12 ± 7
5-Hydroxy-2'-deoxycytidin	Leucocytes	10 ± 5
5-Hydroxy-2'-deoxyuridin	Leucocytes	7 ± 6
2'-Deoxyuridin	Leucozytes	20 ± 15
7,8-Dihydro-8-oxo-2'deoxyadenosin	Leucocytes	230
5-(Hydroxymethyl)-2'-deoxyuridin	White blood cells	$2\,300 \pm 480$
O^6-Methyl-2'-deoxyguanosin	Lung	0.25 ± 13
7-Methyl-2'-deoxyguanosin	Lymphocytes	14
8-Hydroxy-6-methyl-1,N^2-propan-2'-deoxyguanosin	Liver	6
8-Hydroxy-1,N^2-propan-2'-deoxyguanosin	Liver	10
1,N^6-Etheno-2'-deoxyadenosin	Liver	0.7 ± 0.4
3,N^4-Etheno-2'-deoxycytidin	Liver	2.8 ± 0.9
Exogenous factors		
DNA-adducts derived from polycyclic aromatics in smokers	Lymphocytes	0.3–0.2
Guanosin adduct of 4-aminobiphenyl in smokers	Bladder	2–3

The biotransformation of xenobiotics takes place mainly in the liver, intestine, kidney, lungs, and to a lesser extent in other organs of the body. In addition, the intestinal flora also contribute to biotransformation, in particular by hydrolysis and reduction of xenobiotics. The ability of the organism to metabolize and eliminate xenobiotics with different structures is remarkable. This is due to the fact that the enzymes of biotransformation frequently have a broad substrate specificity.

Biotranformation is divided into two phases. In the reactions of Phase I, that is, oxidation, reduction, and hydrolysis, functional groups are inserted in the parent compound. In the reactions of Phase II, functional groups are coupled with functional groups formed during the Phase I biotransformation, that is, with glucuronic acid, sulfonate residue, carboxylic acids, methyl groups, amino acids, and glutathione. The products of the Phase II reaction are generally water-soluble. Exceptions are the products of acetylation and methylation.

The biotransformation does not necessarily lead to biologically inactive compounds. For example, certain drugs are converted into the actual active drug and certain xenobiotics into electrophilic toxic metabolites by the biotransformation. These can cause irreversible toxic effects by direct reaction with the DNA or with functionally important proteins and lipids. Indirect effects are also possible, for example by depletion of the glutathione or NADPH pool. Radical metabolites can also form reactive oxygen species due to redox cycling.

The enzymes of the biotransformation are compartmentalized in the cell. The P450-dependent monooxygenases, which are very important for biotransformation and which catalyze many oxidation reactions, are located in the membranes of the endoplasmic reticulum and the nuclear membrane of the cells, just like the glucuronosyltransferase and a form of the epoxide hydrolase. A part of the oxidation products formed can thus be coupled with glucuronic acid or, in the case of epoxides, converted into vicinal dihydrodiols near their site of formation. Other enzymes of the foreign metabolism are present in the cytosol, for example the sulfotransferases and most glutathione S-transferases.

Enzymes of biotransformation, such as P450-dependent monooxygenases and UDP-glucuronosyltransferases, can be induced by xenobiotics, including plant ingredients. These chemicals are mostly also substrates and induce increased biosynthesis or, in certain cases, a reduced degradation of these enzymes. The enzyme induction leads to an increased elimination rate of the xenobiotic.

Enzymes of biotransformation differ in their occurrence, specificity and activity between species and within individuals of one species. For example, the activity of individual biotransformation enzymes in humans can vary by a factor of 100. In contrast, the enzyme activity of inbred strains of the laboratory rat or mouse shows only slight differences. Knowledge of the differences in the metabolism of enzymes in experimental animals and humans helps in the extrapolation of results obtained in animal experiments on humans and thus in the assessment of the health risk of xenobiotics to humans.

Further Reading

Anders, M.W., 2008. Chemical toxicology of reactive intermediates formed by the glutathione-dependent bioactivation of halogen-containing compounds. *Chem Res Toxicol* **21**, 145–159.

Badenhorst, C.P., van der Sluis, R., Erasmus, E., van Dijk, A.A., 2013. Glycine conjugation: importance in metabolism, the role of glycine N-acyltransferase, and factors that influence interindividual variation. *Expert Opin Drug Metab Toxicol* **9**, 1139–1153.

Chai, X., Zeng, S., Xie, W., 2013. Nuclear receptors PXR and CAR: implications for drug metabolism regulation, pharmacogenomics and beyond. *Expert Opin Drug Metab Toxicol* **9**, 253–266.

Darnell, M., Weidolf, L., 2013. Metabolism of xenobiotic carboxylic acids: focus on coenzyme A conjugation, reactivity, and interference with lipid metabolism. *Chem Res Toxicol* **26**, 1139–1155.

Garattini, E., Terao, M., 2012. The role of aldehyde oxidase in drug metabolism. *Expert Opin Drug Metab Toxicol* **8**, 487–503.

Guengerich, F.P., 2001. Common and uncommon cytochrome P450 reactions related to metabolism and chemical toxicity. *Chem Res Toxicol* **14**, 611–650.

Hoffmann, F., Maser, E. 2007. Carbonyl reductases and pluripotent hydroxysteroid dehydrogenases of the short-chain dehydrogenase/reductase superfamily. *Drug Metab Rev* **39**, 87–144.

James, M.O., Ambadapadi, S., 2013. Interactions of cytosolic sulfotransferases with xenobiotics. *Drug Metab Rev* **45**, 401–414.

Jin, Y., Penning, T.M. (2007). Aldo-keto reductases and bioactivation/detoxication. *Annu Rev Pharmacol Toxicol* **47**, 263–292.

Kalgutkar, A.S., Obach, R.S., Maurer, T.S., 2007. Mechanism-based inactivation of cytochrome P450 enzymes: chemical mechanisms, structure-activity relationships and relationship to clinical drug-drug interactions and idiosyncratic adverse drug reactions. *Curr Drug Metab* **8**, 407–447.

Kodama, S., Negishi, M., 2013. PXR cross-talks with internal and external signals in physiological and pathophysiological responses. *Drug Metab Rev* **45**, 300–310.

Korzekwa, K., 2014. Enzyme kinetics of oxidative metabolism: cytochromes P450. *Methods Mol Biol* **1113**, 149–166.

Pandey, A.V., Sproll, P., 2014. Pharmacogenomics of human P450 oxidoreductase. *Front Pharmacol* **5**, 103.

Runge-Morris, M., Kocarek, T.A., Falany, C.N., 2013. Regulation of the cytosolic sulfotransferases by nuclear receptors. *Drug Metab Rev* **45**, 15–33.

Stingl, J.C., Bartels, H., Viviani, R., Lehmann, M.L., Brockmoller, J., 2014. Relevance of UDP-glucuronosyltransferase polymorphisms for drug dosing: A quantitative systematic review. *Pharmacol Ther* **141**, 92–116.

Wu, T.Y., Khor, T.O., Lee, J.H., Cheung, K.L., Shu, L., Chen, C., Kong, A.N., 2013. Pharmacogenetics, pharmacogenomics and epigenetics of Nrf2-regulated xenobiotic-metabolizing enzymes and transporters by dietary phytochemical and cancer chemoprevention. *Curr Drug Metab* **14**, 688–694.

Zanger, U.M., Schwab, M., 2013. Cytochrome P450 enzymes in drug metabolism: regulation of gene expression, enzyme activities, and impact of genetic variation. *Pharmacol Ther* **138**, 103–141.

2.4 Cytotoxicity

Daniel Dietrich

2.4.1 Introduction

The term "cytotoxic" may have different meanings. Frequently this term is used to indicate that a toxic chemical(s), defined here as "toxin(s)", induces apoptosis or necrosis. In the present chapter the effects of chemicals are also called cytotoxic when they disturb cellular functions in excess of normal physiological fluctuations, i.e. when chemicals damage the cell but do not cause death immediately. Nevertheless, the consequences may be fatal for the organism in the short or long term.

Toxin Targets and Consequences for the Tissue

> **Generally all cellular functions and structures may be the targets of toxins.**

Elucidation of the mechanisms of action of toxins is closely connected with the progress of biochemistry and cell biology during the past 80 years. On one hand, knowledge of the physiologic functions of various cell structures is a necessary prerequisite for the analysis and understanding of the interaction of toxins with these structures. On the other hand, basic discoveries, such as elucidation of the function of nerve membranes, cell division, DNA repair or mitochondria, were only possible by studying an experimentally caused disturbance of these functions by the use of toxins. The extent of the damage depends on the function and central importance of the affected cell constituents, as well as on the possibility of damage recognition and its subsequent repair by the cell (Table 2.12). Repair often occurs via the degradation and resynthesis of damaged cellular components, albeit inhibited or damaged repair in itself can constitute one of the many mechanisms leading to toxin-mediated cytotoxicity.

Irreversible damage of essential functions inevitably causes cell death (Table 2.12). If the necrotic areas are small or only single-cell necrosis is present, most tissues are capable of replacing the affected cells either by true regeneration or initally by fibroblast-like cells, thereby maintaining the anatomical functionality of the tissue/organ. However, cell regeneration is not possible for all cell types, e.g. the nerve cells of the brain cannot be replaced owing to their inability to divide. When larger areas of tissue are damaged, many organs replace the necrotic areas with connective tissue, resulting in scarring as is the case in liver cirrhosis, for example. The scarring process results in a more or less structurally repaired organ, but with greatly diminished functional capacity.

> **Cell loss by cytotoxicity can be repaired, with the exception of tissues/organs where cells no longer replicate (brain, ovaries). In the case of continued cytotoxicity repair is primarily structural and not functional, resulting in scarring.**

Frequently, however, the disturbance induced by toxic chemicals does not result in destruction of the cell, but only in a temporary adjustment of cellular metabolic pathways to a new steady state. These disturbances, also termed tissue or cell adaptation, may be manifested as a change in cell size (atrophy or hyperplasia) or in the storage of cellular contents such as the accumulation of triglycerides in liver cells (Chapter 3.2). In this context it is important to note that the impairment

Table 2.12 Criteria for the assessment of the meaning and possible consequences of cell damage for the cell and the whole organ/organism.

- Meaning and function of the cellular target (qualitative)
- Extent of damage (quantitative)
- Possibility of damage repair
- Reversibility of damage
- Possibility of tissue regeneration (structural and functional)

of cellular functions that are not involved in the survival of the cell may have serious consequences for the total organism. Examples are the disturbance of the function of the nerves or the heart muscle.

Structure–Activity Relationships

> **The toxicity of xenobiotics and natural toxins or their metabolites is due to their high structural specificity for distinct cellular targets (covalent or non-covalent interactions) or high chemical reactivity (e.g., covalent binding, radical formation).**

A characteristic feature of many natural toxins, whether of geologic, incineration, animal, plant, bacterial, fungal or cyanobacterial origin, is that they disturb specific cellular functions by highly selective interaction with distinct components of cells owing to their specific structural features (Table 2.13). The mode of interaction of the majority of natural toxins is non-covalent, albeit some, e.g. aristolochic acid or aflatoxins, can be metabolically activated to protein and DNA binding, while microcystins and okadaic acid bind to the catalytic subunit of the ser/thr protein phosphatases covalently and reversibly, respectively, and thus inhibit the dephosphorylation of proteins and enzymes critical for all cell proliferation and signal transduction processes. Natural toxins inhibit or activate enzymes, interfere with the synthesis, storage or secretion of autocoids, hormones, and neurotransmitters, activate or block receptors, block or inhibit transporters, inhibit the transfer of electrons within the mitochondrial respiratory chain, or interact with components of the cytoskeleton.

Similarly, synthetic chemicals exert their toxic actions via a disturbance of distinct cellular functions (Table 2.14). Some chemicals and their metabolites react non-specifically as well as specifically with various cellular structures. Electrophilic metabolites bind covalently to nucleophilic positions in proteins and lipids as well as RNA and DNA. Radicals that arise during the metabolism of a compound may undergo addition or abstraction reactions with cellular molecules (Chapters 2.3 and 2.8). However, just as is the case for natural toxins, synthetic chemicals can inhibit or activate enzymes, interfere with the synthesis, storage or secretion of autocoids, hormones, and neurotransmitters, activate or block receptors, block or inhibit transporters, inhibit the transfer of electrons within the mitochondrial respiratory chain, or interact with components of the cytoskeleton.

> **Thus, it is irrelevant whether a toxin is of natural or of synthetic origin, especially as a huge number of synthetic toxins have been derived from the basic structure of a corresponding natural toxin. Of much greater importance is the structure per se, which primarily determines whether or not it can activate, bind (reversibly or covalently), block, or inactivate cellular components (receptors, enzymes, skeletal components, organelles, proteins, RNA, DNA, etc.).**

Table 2.13 Disturbance of specific cell functions by animal and plant poisons.

Poison	Occurrence	Type of compound	Target
Acrylamide	Food processing of starch-containing foods	Amides	Neurotoxicity presumed to stem from inhibition of kinesin-based fast axonal transport, alteration of neurotransmitter levels, and direct inhibition of neurotransmission
Anatoxin-a(S)	Cyanobacteria: *Anabaena*	Organophosphate	Neurotoxicity resulting from the inhibition of acetylcholinesterases
Aflatoxin	Mould: *Aspergillus*		Epoxide metabolite of parent aflatoxin binds to protein and DNA
Amanitin	Mushroom: *Amanita*	Octapeptide	Inhibition of RNA-polymerases, in particular DNA-dependent RNA-polymerase II
Antimycine A	Bacteria: Streptomyces griseus		Inhibition of the mitochondrial electron transport at cytochrome b/c1
Aristolochic acid	Plant: *Aristolochia clematitis*	Carboxyclic acid	Formation of DNA and protein adducts
Atractyloside	Plant: *Atractylis gumnifera*		Inhibition of the ATP/ADP exchange at the inner mitochondrial membrane
Bisphenol F	Plant: sweet mustard (*Sinapis alba*)		Reversibly binds to eukaryotic estrogen receptors agonistically and androgen receptors antagonistically
Botulinus toxin	Bacteria: *Clostridium botulinum*	Protein (150 kD)	Inhibition of the release of acetylcholine from presynaptic neurons
Cholera toxin	Bacteria: *Vibrio cholerae*	Protein (87 kD)	ADP-ribosylation of the stimulatory G-protein of the adenylate cyclase causing increased c-AMP production
Colchicine	Plant: meadow saffron (*Colchicum autumnale*)	Alkaloid	Inhibition of the aggregation of the microtubules affecting mitotic spindle formation
Curare	Plant: *Strychnos* and *Chondrodendrons*	Alkaloids	Inhibition of neural-muscular signaling due to the blockage/occupation of cholinergic receptors

Table 2.13 (Continued).

Poison	Occurrence	Type of compound	Target
Digitoxin	Plant: foxglove (e.g., *Digitalis purpurea*)	Glycoside	Inhibition of Na$^+$/K$^+$-ATPase
Dioxins, furans, polycyclic aromatic hydrocarbons (PAH)	Incineration process residues found in kaolinite clays		Binding to the Ah-receptor, initiation of gene transcription
Diphtheriae toxin	Bacteria: *Corynebacterium diphtheriae*	Protein	Inhibition of the protein synthesis by ADP-ribosylation of the elongation factor (EF-2)
d-Limonene	Plant: citrus fruit (e.g., *Citrus limon*)	Terpene	Reversibly binds to male rat specific α2u-globulin, thereby leading to lysososmal overload with subsequent tubular necrosis and renal tumors
Microcystin (MC)	Cyanobacteria: *Microcystis aeruginosa, Planktothrix aghardii*, etc.	Cyclic heptapeptide (1 kD)	First reversible then covalent binding of MC to the catalytic subunit and thus irreversible inhibition of ser/thr protein phosphatases (PPP)
Phalloidin	Mushroom: death cap (*Amanita phalloides*)	Heptapeptide	Stabilization of actine filaments
Ricin	Seeds of the castor oil plant, *Ricinus communis*	Lectin (carbohydrate binding protein (60–65 kDa)	Type 2 ribosome-inactivating protein
Rotenone	Roots of several members of *Fabaceae*	Fucocoumarine-derivative	Mitochondrial electron chain complex 1 inhibitor
Tetrodotoxin	Fish: *Tetraodontiformes* (including pufferfish, porcupinefish, ocean sunfish, and triggerfish)		Inhibits the firing of action potentials in nerves by binding to the neuronal voltage-gated sodium channels and blocking the passage of sodium ions

For most chemicals, it is not clear which of the numerous possible cellular targets of reactive chemicals and metabolites are responsible for their overt toxic action. An answer to this question requires an analysis and toxicological assessment of the various targets. This has not yet been done sufficiently, but has increasingly become simpler due to the use of ever-advancing sophisticated computational toxicology that compares structures with potential cellular targets or mechanisms of toxicity. The comparison of numerous structures with numerous toxicologically relevant endpoints and thus activities is called "read-across". This allows not only iterative determination of the most likely mechanism of toxic activity of a given compound, but also, due to continuous learning and increasing database size, the prediction of the toxicity of a new structure and thus foresight of potential toxicity.

Table 2.14 Disturbance of specific cell functions by chemicals.

Chemical	Target
Acrylamide	Neurotoxicity presumed to stem from inhibition of kinesin-based fast axonal transport, alteration of neurotransmitter levels, and direct inhibition of neurotransmission
Dioxins, furans as well as coplanar and monoortho polychlorinated biphenyls (PCBs)	Binding to the Ah-receptor, initiation of gene transcription
Fluoracetate	Fluoracetate is converted to fluoroacetyl-CoA, which condenses with oxaloacetate to form fluorocitrate. This inhibits the enzyme aconitase and thereby the citric acid cycle
Disulfotone, Thiometone, Parathion (organophosphate insecticides), Sarin, Tabun, VX (chemical warfare agents)	Neurotoxicity derived from the inhibition of acetylcholinesterases
Hydrocyanic acid	Inhibition of the mitochondrial electron transport due to covalent binding to cytochrome C oxidase
Iodoacetate	Inhibition of glycolysis via the carboxylation of the SH group of the glyceraldehyde-3-phosphate dehydrogenase
Trimethylpentane, unleaded gasoline, 1,4-dichlorobenzene, isophorone, pentachloroethane, hexachloroethane, dimethylmethyphosphonate, tetrachloroethylene	Reversibly binds to male rat specific $\alpha 2u$-globulin, thereby leading to lysososmal overload with subsequent tubular necrosis, regenerative proliferation and renal tumors

In Chapter 3 some important targets of toxic chemicals are described. As this topic requires basic knowledge of the structure and function of the cell, an introduction to cell biology is given first.

2.4.2 The Cell

More than 200 different cell types can be found in the various organs of the body. They may exist separately or form closely connected cell layers such as the epithelial cells of the skin. Their morphology and that of the organs in which they are found are the result of the process of differentiation, which prepares them for the specific physiological functions they must perform, e.g. production of mechanical energy by muscle cells, absorption of nutrients by the epithelial cells of the intestine, or reception of light signals by the neural epithelium of the retina of the eye.

While cells contain the same genome, differentiation results from the fact that the various cell types only express distinct parts of the total genetic information.

Nevertheless, human and animal cells show similar structural elements, i.e. the outer membrane, the cell nucleus, and the **cytoplasm**, which is a colloidal fluid containing numerous cell organelles. The structural elements of the cell are described briefly in the following section (Figure 2.35).

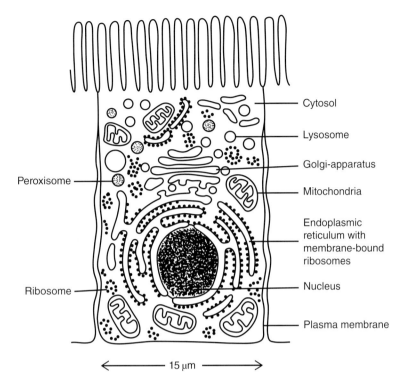

Figure 2.35 Schematic drawing of a cell. Reproduced from B. Alberts et al., Molecular Biology of the Cell, 4th edition, Garland Science, New York & London, 2002, with permission.

The Membrane System of the Cell

Membranes enclose the cell and subdivide it into various compartments.

Cell membranes consist of a double layer of polar lipids containing proteins (Figure 2.36). The **lipid bilayer** is a diffusion barrier for many compounds. Both non-polar molecules, such as O_2, N_2 or benzene, and small polar molecules, such as H_2O, urea and CO_2, can diffuse through the lipid phase of the cell membrane. Membrane permeability is minimal for larger uncharged polar molecules, such as glucose, whereas the membrane is almost impermeable to charged molecules (see *The Cell*, this chapter). Some polar molecules, e.g. certain inorganic ions, sugars, amino acids and numerous metabolites, which are formed in intermediary metabolism, can traverse the cell membrane due to the presence of highly specific transport proteins, e.g. organic anion transporters (OATs), organic cation transporters (OCT), organic anion transporting polypeptide transporters (OATPS), sodium–glucose transporters (SGLT), glucose transporters (GLUT), etc.

The transport of small non-polar molecules through the membrane may be driven by the concentration gradient of the chemical on either side of the barrier. Transporting charged chemicals against a concentration gradient often requires the cell to utilize mechanisms which require energy in the form of adenosine 5′-triphosphate (ATP). As a consequence of selective permeability and active transport processes cells develop characteristic environments within their membrane-enclosed compartments. The creation of specific reaction areas in the cell is a prerequisite for the well-ordered course of cellular metabolism. To give an indication of the

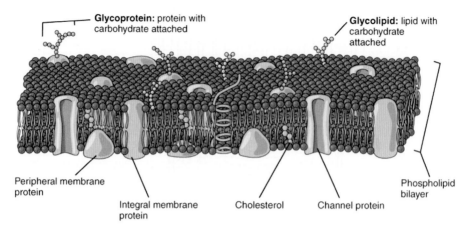

Figure 2.36 Schematic depiction of a cell membrane. The phospholipid bilayer contains (glyco)proteins. Source: OpenStax 2016, https://commons.wikimedia.org/wiki/File:0303_Lipid_Bilayer_With_Various_Components.jpg. Used under CC by 4.0 https://creativecommons.org/licenses/by/4.0/deed.en.

Table 2.15 Comparison of ion concentrations inside and outside a typical mammalian cell.

Ion	Intracellular concentration (mmol × l^{-1})	Extracellular concentration (mmol × l^{-1})
Na$^+$	5–15	145
K$^+$	140	5
Mg^{2+}	30	1–2
Ca^{2+}	1–2 $\leq 10^{-7}$ (free/unbound)	2–5

Modified from B. Alberts et al., Molecular Biology of the Cell, 4th edition, Garland Science, New York & London, 2002.

large differences in the composition of extracellular and intracellular fluids, the concentrations of some inorganic ions are listed in Table 2.15. The concentration gradients between the cytosol and the outer liquid and between the cytosol and distinct intracellular organelles are maintained by energy-dependent transport processes, e.g. mediated by the Na$^+$/K$^+$-ATPases and Ca^{2+}-ATPases.

The differential conductivity of membranes for Na$^+$, K$^+$ and Cl$^-$ is responsible for the generation of the electric membrane potential. As a basic feature of living cells, the membrane potential is of central importance for the transport and distribution of many molecules. Besides control of the transport of molecules, the cell membrane also plays an essential role in the reception and forwarding of the signals of many messenger molecules, such as hormones, which are indispensable for coordination of different tissues of the body. Proteins, peptides, steroids, amino acids, and fatty acid derivatives with hormonal activity are either produced by special gland tissues of the body and reach the target cells via the bloodstream, or are synthesized close to their place of action, often in the same tissue. It is even possible for cells to secrete growth factors for their own stimulation. Such mechanisms are termed autocrine and play an important role in several malignant tumors. Many hormones bind to specific receptor glycoproteins and lipids within the membranes of their target cells and set off an ordered sequence of cellular reactions, resulting in specific adjustment of cell metabolism.

Components of the Cytoplasm

> The cytoplasm consists of the cytosol and the organelles of the cell.

Many important reactions in cellular metabolism occur within the cytosol, which is the fluid portion of the cytoplasm, e.g. the synthesis of fatty acids and cytosolic proteins, metabolic pathways such as glycolysis and the pentose phosphate pathway, and ribosomal protein synthesis.

The cytosol contains a complex network of protein filaments called the **cytoskeleton**, which consists of actin filaments, intermediary filaments, and microtubules. The actin filaments and the microtubules contribute to the shape as well as to the stability and placement of surface structures of the cell. Microtubules are also components of the mitotic spindles, which move the chromosomes during cell division. Furthermore they form the centrioles (cylindrical organelles), which serve as the start of the spindle during mitosis.

The cytosol contains several cell organelles that are surrounded by a membrane: the **endoplasmic reticulum** (ER), the **Golgi apparatus** (GA), **lysosomes**, **peroxisomes**, and **mitochondria**. With the exception of the mitochondria these cell organelles develop from the ER. The ER is a network of tubules and stacks of flattened cisternae that are continuous with the outer nuclear membrane. Functionally, the ER is the site of the synthesis of lipids, membrane proteins and secretory proteins. The latter are synthesized by the so-called rough ER (rER), which is covered with ribosomes.

The GA consists of stacks of disc-shaped cisternae from which small vesicles bud and then may fuse with the plasma membrane and mediate secretory processes. In the GA oligosaccharides of glycoproteins, which were formed in the ER, achieve their final structure.

Lysosomes are small membranous organelles containing hydrolytic enzymes such as peptidases, lipases, and phophatases, and are the location of the controlled intracellular digestion of cellular components.

Peroxisomes are cell organelles in which several oxidative reactions requiring consumption of molecular oxygen occur. These reactions form toxic H_2O_2 and superoxide anions ($O_2^{\bullet -}$), which are largely detoxified owing to the high concentrations of catalase and Co/Zn superoxide dismutase.

Mitochondria differ significantly from other cell organelles. They are surrounded by two membranes and contain their own DNA. The outer membrane is relatively permeable. Mitochondria can grow and divide; for this they need both their own genetic information and that of the cell nucleus. The mitochondria perform several vital reactions of cellular metabolism such as the **citric acid cycle**, **anaplerosis**, the **degradation of fatty acids**, and **gluconeogenesis**. Furthermore, mitochondria play an essential role in maintaining the energy balance of most cells. In the degradation of energy-rich substrates, such as carbohydrates and fats, hydrogen atoms (reducing equivalents) are transferred to nicotinamide adenosine dinucleotide (NAD) and a flavoprotein in the citric acid cycle. These hydrogen atoms and their electrons are fed into the electron-transport chain of the inner mitochondrial membrane. The electron-transport chain consists of a sequence of oxidation–reduction (redox) systems through which substrate electrons pass until they reduce molecular oxygen to water (Figure 2.37). In this process the considerable energy of the oxyhydrogen is fractionated and finally used for the synthesis of energy-rich ATP molecules. This process is termed **oxidative phosphorylation**.

The **cell nucleus** is enclosed by the nuclear envelope, a double membrane with numerous pores. Compounds may exchange between the nucleus and the cytoplasm through these pores.

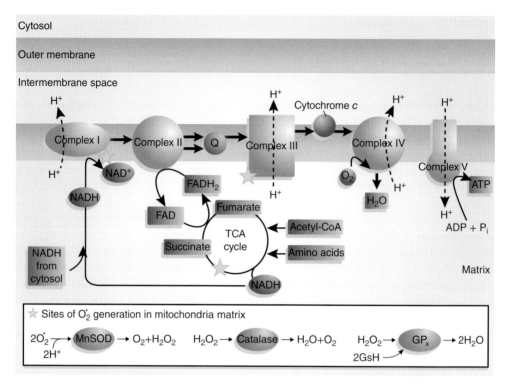

Figure 2.37 The electron transport chain of mitochondria is the major source of free radicals in the cell. Because of electron leak, free radicals react with oxygen (O_2) to generate superoxide radicals (O_2^-). The major sites of generation include the iron-sulfur clusters of complex I, coenzyme Q associated with complex III, and components of the tricarboxylic acid cycle, including alpha-ketoglutarate dehydrogenase. Superoxide radicals are dismutated by manganese superoxide dismutase in the mitochondrial matrix to generate O_2 and hydrogen peroxide (H_2O_2). H_2O_2 is then converted to H_2O by either catalase or glutathione peroxidase (GPx), which uses glutathione (GSH). Aging is associated with increased mitochondrial production of H_2O_2, leading to oxidative damage and mitochondrial DNA mutations. Schriner et al. found that the lifespan of mice can be substantially expanded and age-associated disease reduced by overexpressing catalase in mitochondria. Source: Reproduced with permission of Springer Nature.

The sequence of the four bases of deoxyribonucleic acid (DNA) in the nucleus determines the genetic information, which is packaged into segments of hereditary factors termed the genes. DNA covered with proteins is termed chromatin and represents the basic structure of the chromosomes. The somatic cells of the human body contain 46 chromosomes. The information from small DNA sequences is transcribed (**transcription**) into so-called **messenger-ribonucleic acid** (mRNA) and subsequently translated (**translation**) into the amino acid sequence of distinct proteins at the ribosomes. On their differentiation some cell types lose the nucleus (erythrocytes, thrombocytes) and thus cannot divide or synthesize RNA.

2.4.3 Cellular Targets of Toxins

Toxins react with very different components of the cell. Although only single mechanisms of toxicity will be discussed, it is important to remember that in almost every case toxins act simultaneously on several key functions of the cell. The simultaneous disturbance of several cell functions can cause an additive or, in rare cases, even a synergistic increase in toxicity. For example, this may be the case when cellular structures and repair processes are affected simultaneously.

Cell Membrane

Toxins may affect membrane fluidity and the function of membrane proteins.

Cellular membranes contain many important metabolic enzymes that deliver their products specifically into distinct compartments of the cell. Toxins may affect both the permeability of cellular membranes and the activity of the enzyme proteins in the membranes. The disturbance may be caused by radical attack as well as covalent bonding of a reactive metabolite or by a non-covalent interaction of the toxin with membrane components, e.g. lipid phase or proteins. In the case of lipid peroxidation the membrane is damaged and can become leaky (see *Lipid Peroxidation*, this chapter). A further mechanism of damage involves the oxidation of functional thiol groups of membrane proteins (see *Disturbance of Redox Systems*, this chapter).

Depending on their physicochemical characteristics lipophilic and amphipathic chemicals may concentrate in cellular membrane systems and affect the package density and mobility of the membrane lipids. Chemicals may increase or decrease the mobility of the fatty acid chains and, as a consequence, the membrane may become more rigid or fluid and the activity of distinct membrane enzymes and transport proteins may change. Very high concentrations of lipophilic chemicals can cause a complete disintegration of cellular membranes.

Historically, the narcotic action of various narcotics and solvents was thought to be due to their interaction with the lipid phase of the plasma membrane of neurons. This notion was supported by the correlation of the narcotic effectiveness of these lipophilic chemicals with their distribution coefficient in octanol/water or a phospholipid–cholesterol/water mixture (Figure 2.38). Recent studies, however, show that the narcotic action is most likely caused by a direct interaction

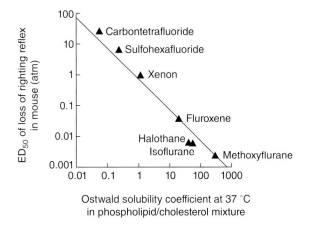

Figure 2.38 Comparison of the narcotic effect and the solubility of chemicals in artificial lipid membranes, which show similar phospholipid/cholesterol ratios as a neuron membrane. Modified according to K.W. Miller, Physical theories of general anaesthesia, Weekly Anesthesiology Update, 1979, 2, 1–8.

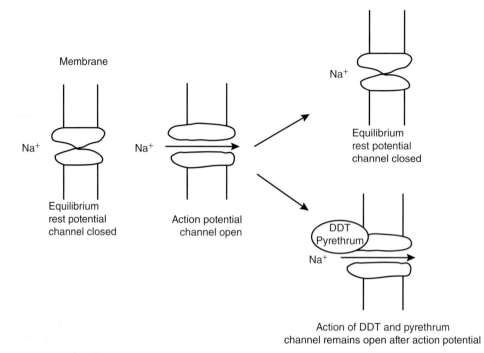

Figure 2.39 *Effect of DDT and pyrethrum on neuronal sodium channels. Under physiological conditions sodium channels are closed immediately after initiation of an action potential. After binding the insecticides the sodium channels remain open for a longer period. Note: natural toxins, e.g. brevetoxin and ciguatoxin, have the same effect.*

between lipophilic chemicals and the hydrophobic part of channel proteins in the membrane. However, in the latter case and as exemplified below, the distinction between true narcosis and specific neurotoxicity becomes fuzzy. Indeed, a direct interaction with membrane proteins underlies the action of DDT, pyrethroids and some other insecticides (Figure 2.39), and also some marine biotoxins, e.g. brevetoxin or ciguatoxin. These chemicals bind to a hydrophobic section of voltage-gated sodium channels and prevent channel closure after an action potential. (see "Inhibition of transporters", and Figure 2.26, this chapter). Consequently, the chemicals cause over-excitability at low concentrations and paralysis at high concentrations.

Lipid Peroxidation

> **Oxidative attack on unsaturated membrane lipids results in lipid peroxidation, as a result of which ethane and pentane are released and can be determined in exhaled air.**

In contrast to the interactions of chemicals with membranes described above, components of the membrane may be severely damaged during lipid peroxidation. Oxidative destruction of unsaturated membrane lipids has been studied primarily in the liver (Figure 2.40). In a monooxygenase-dependent reaction CCl_4 is cleaved homolytically to yield a chlorine ion and a trichloromethyl radical (CCl_3^-) in the liver cell; the free radical can react with unsaturated lipids, via H abstraction, to produce a lipid radical and $CHCl_3$. The weakly bound hydrogen of the

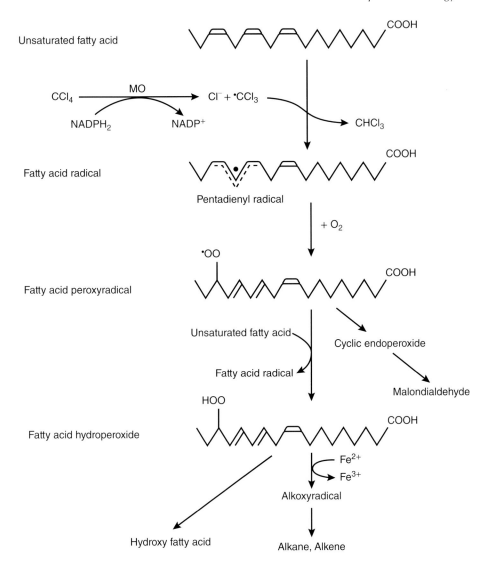

Figure 2.40 *Schematic representation of lipid peroxidation induced by CCl_4. MO, monooxygenase.*

methylene group of the divinylmethane grouping of multiple unsaturated lipids is particularly prone to H abstraction, allowing generation of a pentadienyl radical, which bonds with oxygen. It is likely that the ultimate location of the free electron is at the second oxygen atom. The result of the uptake of the oxygen by the isolated double bond is the energetically driven formation of a conjugated series of double bonds. The lipid peroxy radicals can now react with unsaturated fatty acids to generate lipid hydroperoxides and new fatty acid radicals via a chain reaction. The final products of the destruction of membrane lipids include inter alia malonic dialdehyde, alkanes, and alkenes. These products can be easily determined and are used for the detection of lipid peroxidation.

Lipid peroxidation is not only produced by strong oxidants but may take place in principle whenever the redox balance of the cell is disturbed (see *Disturbance of Redox Systems*, this chapter). The evidence that a chemical induces lipid peroxidation does not necessarily mean that this process plays a central role in its cytotoxic action. Often it is not possible to decide whether lipid peroxidation is a primary cause or only a consequence of cell damage. In the case of CCl_4, however, lipid peroxidation appears to be causally connected with the toxicity of the halocarbon.

Disturbance of Redox Systems

Disturbance of cellular redox homeostasis may lead to cytotoxicity by various means.

The cell contains various redox systems such as the hydrogen transferring coenzymes NAD, NADP and riboflavin, the substrate-product pairs lactate/pyruvate, acetaldehyde/acetic acid, and ascorbic acid/dehydroascorbic acid, the glutathione system, and protein thiols. Enzymes catalyzing one-electron transitions can change the redox state of these paired molecules. Both the concentration and the relative proportions of oxidized and reduced coenzymes and substrates determine the redox state of the cell. Its continued disturbance has serious effects on the vitality of cells, as has been demonstrated by numerous *in vitro* studies (Table 2.16). The significance of some of the following mechanisms for toxicity *in vivo* has not been sufficiently clarified yet.

The disturbance of cellular redox homeostasis induced by the redox cycling of chemicals, so-called **oxidative stress**, has gained particular interest. During redox cycling, catechols, quinones, nitroaromatics or aromatic azo dyes are reduced due to one-electron transitions and may subsequently react with oxygen, producing a superoxide radical (Figure 2.41). The superoxide radical is cytotoxic and genotoxic. These effects, however, are most likely not due to the superoxide radical but to other toxic oxygen species derived from superoxide, such as

Table 2.16 Possible consequences of oxidative stress.

- Decrease in the GSH/GSSG ratio
- Lipid peroxidation
- Oxidation of protein thiols and loss of the activity of (enzyme) proteins
- DNA strand breaks

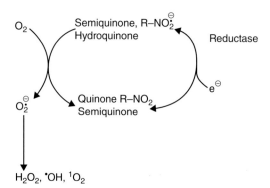

Figure 2.41 Formation of the superoxide radical and further reactive oxygen species by redox cycling of chemicals. Examples of reductases are NADPH-cytochrome P450 reductase, aldehyde reductase, and ketone reductase. $R-NO_2$ represents an aromatic nitro compound.

H_2O_2, $OH^·$, and $^1O_2$. The cell protects itself against reactive oxygen species and their cellular products such as lipid peroxides by means of enzymes such as superoxide dismutase, catalase, and glutathione peroxidase as well as antioxidants such as β-carotene and α-tocopherol.

If these systems are overwhelmed, the reactive oxygen species may damage cell components and thus the cell directly by an oxidative attack or indirectly by rendering the cell more vulnerable to other oxidants due to the disturbance of important redox systems.

The cellular targets of reactive oxygen species are lipids, proteins, and DNA. Lipid peroxidation, as exemplified by CCl_4, has been described above. Similarly, the reactive trimethyl radical may also directly react with the double bonds of multiple unsaturated lipids and thereby initiate the destruction of lipids. In proteins, thiol groups are preferred targets of toxic oxygen species. Frequently enzymes have reduced SH groups, which are essential for their catalytic activity. Thus, the oxidation of thiol groups is often accompanied by a loss of protein function (Figure 2.42). The inhibition of the catalytic activity of Ca^{2+} transport proteins of the plasma membrane and the endoplasmic reticulum may therefore play a role in cell damage because these carrier proteins contribute to the maintenance of a low concentration of cytoplasmic calcium. Furthermore, non-oxidative mechanisms, such as thiol acylation and arylation by reactive chemicals, can produce a loss of protein function (see *Mechanisms underlying Cell Death*, this chapter) (Figure 2.42). Finally, reactive oxygen species can produce cytotoxicity and arrest cell growth by reacting with DNA, resulting in oxidatively damaged DNA base lesions, e.g. 8-hydroxyguanine (8-OH-Gua), 5-hydroxyhydantoin (5-OH-Hyd), and **DNA-strand breaks**.

H_2O_2 is detoxified by either catalase or glutathione peroxidase. GSSH generated during the glutathione peroxidase reaction is in turn reduced to GSH by the enzyme glutathione reductase at the expense of $NADPH_2$, and the formation of $NADP^+$. When the formation of $NADPH_2$ becomes rate limiting, the cellular concentrations of $NADP^+$ and GSSG increase.

Damage to both redox systems can contribute to cytotoxicity in several ways. Some important cellular functions are very sensitive to changes in the GSH/GSSG ratio. If the portion of oxidized glutathione increases, protein synthesis and cell division may be inhibited. The cell protects itself against a disturbance of thiol redox status by the active export of GSSG from the cell.

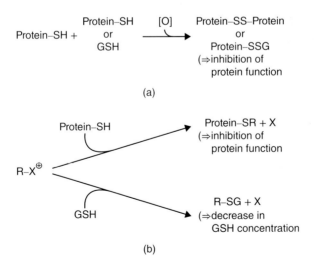

Figure 2.42 Reactions of (a) reactive oxygen species [O] and (b) electrophilic chemicals ($R-X^+$) with glutathione (GSH) and protein thiols (protein-SH).

112   Toxicology and Risk Assessment

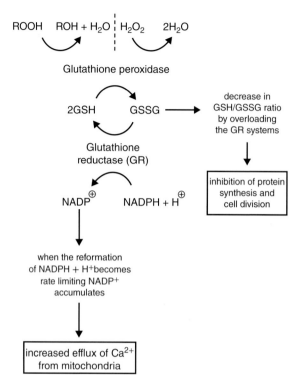

Figure 2.43 *Potential consequences when systems metabolizing H_2O_2 and hydroperoxides are overwhelmed. ROOH, organic hydroperoxides; ROH, alcohols; GR, glutathione reductase.*

Under extreme conditions this can lead to a decrease in glutathione concentration. Moreover, the concentration of reduced glutathione can decrease under a critical value by a change in the thiol redox ratio that can result from the extensive conjugation of electrophilic chemicals or their metabolites with GSH (Figure 2.43). This can cause a loss of vitality or cell death, since the capacity of the GSH-system is no longer sufficient to inactivate reactive, electrophilic chemicals and toxic oxygen species as well as their by-products.

Another factor which may contribute to cell damage is an increase in the concentration of oxidized pyridine nucleotides in the cytosol and the mitochondria. Elevation of the intramitochondrial oxidized pyridine nucleotides, at the expense of reduced pyridine nucleotides, may result in an excessive efflux of mitochondrial Ca^{2+} into the cytosol, a key event which may lead to cell death (see also *The Role of Disturbed Calcium Homeostasis in Cell Death*, this chapter).

Inhibition of ATP Formation

Inhibitors of ATP formation cause arrest of cellular metabolism.

Most of the cells of the body form their ATP via both **glycolysis** and **mitochondrial oxidative phosphorylation**. In principal, chemicals can inhibit glycolysis as well as mitochondrial ATP synthesis (Tables 2.13 and 2.14). Cyanide poisoning is a well-known example of a blockage of mitochondrial electron transport. Cyanide binds to the central Fe^{3+} of the last component of the

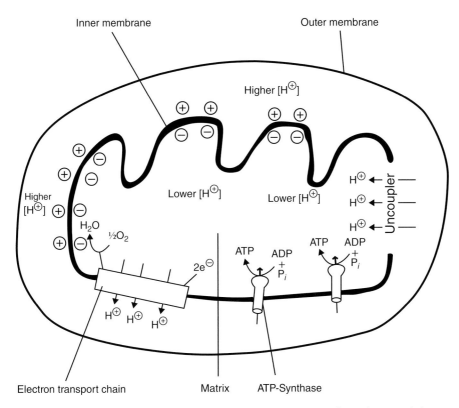

Figure 2.44 Schematic and simplified drawing of the chemiosmotic hypothesis and the action of uncouplers. During mitochondrial electron transport H+ ions are pumped through the inner mitochondrial membrane, which is basically impermeable to protons, generating an electrochemical proton potential. This potential is the driving force for the synthesis of ATP (proton-motive force). Uncouplers, e.g. rotenone inhibiting complex 1, increase the permeability of the inner mitochondrial membrane for protons and thus cause the breakdown of the electrochemical proton potential and a disturbance of the ATP synthase.

respiratory chain, cytochrome C oxidase (Complex IV), thereby inhibiting the electron flow to oxygen.

A second important mechanism for damage to the mitochondrial ATP generating system is the uncoupling of ATP synthesis from electron transport (Figure 2.44). According to the chemiosmotic hypothesis the energy of the redox reactions that accompany electron transport produces an electrochemical proton potential at the mitochondrial inner membrane, an essential prerequisite for this is that the inner membrane of the mitochondria is impermeable to protons and hydroxyl ions. The so-called proton-motive force that is generated is used for the synthesis of ATP. **Uncouplers**, e.g. 2,4-dinitrophenol, hydrophobic salicylanilide S-13 or hindered phenol SF 6847, cause a passive transport of protons through the inner membrane and thereby decrease the electrochemical proton potential. With the loss of the proton-motive force, the mitochondria lose their capability to synthesize ATP.

Total inhibition of the cellular energy supply, derived from glycolysis and mitochondrial ATP synthesis, for a sufficient period of time, causes the death of the cell. All energy-dependent cellular metabolic reactions and active processes that are responsible for the maintenance of homeostasis in the cell will cease. In contrast, partial inhibition of the energy supply will initially cause a decrease in cellular metabolism, but not necessarily cell death, and thus a decrease in cell proliferation capability. A consequence of the latter, however, is a decreased regenerative repair capacity and thus generally tissue/organ functionality.

Inhibition of the Synthesis of Macromolecules

> **Inhibition of the synthesis of macromolecules can have various causes.**

DNA, RNA, and protein synthesis are complex processes that proceed only in the presence of metabolic energy, protein factors, transfer-RNA (t-RNA), ribosomes, enzymes, and various anabolic pathways. Hence, it would be expected that disturbances in macromolecular synthesis are a frequent effect of cytotoxic chemicals. A decreased synthesis of macromolecules, however, may cause an altered steady state of cellular metabolism and may not necessarily lead to cell death. Even temporary complete inhibition of protein synthesis may not cause cell death. For example, cells may readily survive a short-term treatment with the antibiotic cycloheximide despite the inhibition of protein synthesis.

Inhibition of Critical Enzymes involved in Signal Transduction

> **Inhibition of critical enzymes involved in signal transduction can result in cellular structural damage and dysfunction or overt cytotoxicity.**

Toxins, e.g. microcystins (MCs) or okadaic acid (OA), enter the cell via either specific organic anion polypeptide transporters, as is the case for MCs, or membrane diffusion, as in the case for OA. Both OA and MCs bind to the catalytic subunit of ser/thr protein phosphatases (PPPs, Figure 2.45), whereby the enzymatic activity of the PPPs are inhibited. In contrast to OA, which inhibits PPPs reversibly, the methyldehydroalanine (Mdha) moiety of MCs allows the covalent binding to protein phosphatases by Michael addition of a proximate PPP cysteine residue. As PPPs are critical for the phosphorylation/dephosphorylation balance for numerous structural proteins as well as proteins involved in signal transduction, nearly complete PPP inhibition results in the hyperphosphorlyation and thus depolymerization of microtubuli, cytokeratins, and actin, the consequence of the latter being a disintegration of cellular structure and cell necrosis. Hyperphosphorylation of key regulators of cell proliferation, apoptosis DNA-damage repair and differentiation, e.g. p53 retinoblastoma protein, will result in a functionality of cells due to dysregulated signal transduction and loss of key controlling elements of cell homeostasis.

Inhibition of Transporters

> **Inhibition of transporters results in functional impairment and dysregulation of cellular homeostasis, leading to overt cytotoxicity.**

Some naturally occurring toxins have multiple actions on the transition states of voltage-gated sodium channels (NaV), leading to a shift in voltage dependence, inhibition of inactivation, increase in mean open times, and subconductance states. This leads to uncontrolled Na^+ influx

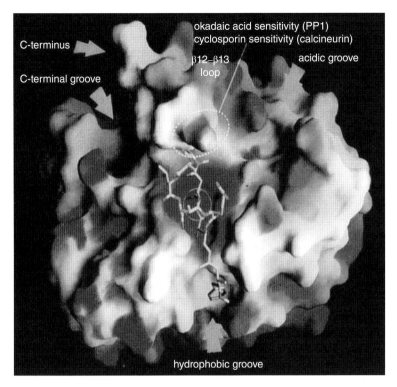

Figure 2.45 Molecular surface of ser/thr protein phosphatase 1, colored according to electrostatic potential. The microcystin molecule is drawn with nitrogen atoms in blue, oxygen atoms in red, and carbon atoms in yellow. The printed version is in the black and white, but the colors are shown in the electronic version. Source: Reproduced with permission of Springer Nature.

and depolarization of neurons followed by a persistent inactivated state and blockade of nerve conduction (Figure 2.46), and an increase in intracellular Ca^{2+}. Activation of NaV by these toxins (brevetoxin, ciguatoxin) has been shown in immune cells, inducing cell proliferation, gene transcription, cytokine production, and apoptosis.

2.4.4 Mechanisms underlying Cell Death

> **Apoptosis is an active process leading to cell death via an ordered sequence of events.**

The cellular events that ultimately lead to cell death are an area of active research and are as yet incompletely understood. In principle, cell death may be caused by the accumulation of injurious events, resulting in loss of the activity of many cellular functions. Alternatively, cell death may be the consequence of active processes. The first scenario may contribute to distinct forms of necrosis; the second may be the case in apoptosis, which is an active and often physiological form of cell death having characteristic morphologic features achieved via specific biochemical mechanisms. Apoptosis plays an important role in the elimination of embryonic tissue, the regulation of the size of organs, and the elimination of precancerous cells. Chemically-induced cell death

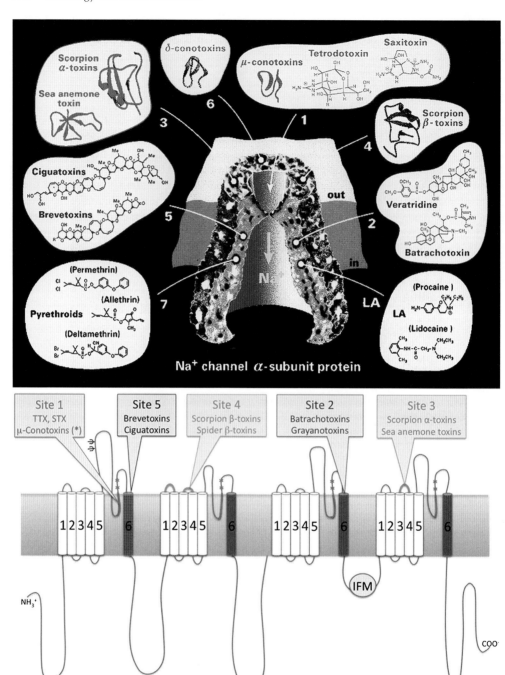

Figure 2.46 Inhibitory binding sites of voltage-gated sodium channel for different toxins, including CTX, BTX, TTX, and STX. Receptor binding sites for neurotoxins are denoted with numbers. Receptors of the lipid-soluble sodium channel modifiers, such as alkaloid toxins (e.g., veratridine, grayanotoxin, batrachotoxin – receptor site 2), marine cyclic polyether toxins (e.g., brevetoxins and ciguatoxins receptor site 5), synthetic pyrethroids (e.g., permethrin, deltamethrin – receptor site 7), and local anesthetics (LA, e.g., lidocaine), are located within the hydrophobic protein core. Receptor binding sites of water-soluble polypeptide toxins, such as scorpion α- and β-toxins (receptor sites 3 and 4), and δ-conotoxins (receptor site 6), are located at the extracellular side of the channel protein. The external vestibule of the ion-conducting pore, at the centre of the protein, contains receptor site 1, which binds the sodium channel blockers μ-conotoxins, tetrodotoxin (TTX), and saxitoxin (STX). The part of the channel responsible for fast inactivation is the short, highly conserved intracellular linker that connects domains III and IV (lower part of the figure). The three hydrophobic amino acids Ile, Phe, and Met (IFM motif) are the key sequence necessary for fast inactivation. Reprinted with permission from ALTEX, 30(4), 487–545 (2013). The printed version is in the black and white, but the colors are shown in the electronic version.

←———————————————————————————————————

resulting from either necrosis or apoptosis depends on the dose of the cytotoxic chemical used, e.g. diethylnitrosamine induces apoptosis in the liver at low doses, whereas it causes necrosis at higher doses.

Two general pathways leading to apoptosis have been described. One originates from specific "death receptors" at the cell membrane and the interaction with their ligands (Figure 2.47). The other involves the mitochondrial compartment and the release of cytochrome C (Figure 2.47).

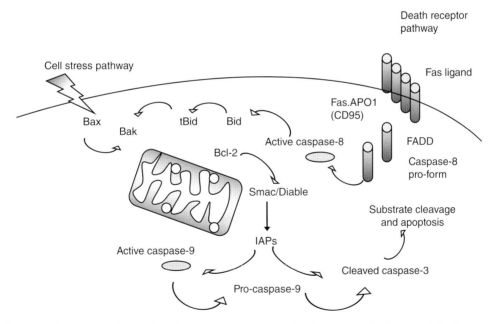

Figure 2.47 Schematic drawing of two general pathways leading to apoptosis, one originating from "death receptors" such as the CD 95 receptor recruiting and activating caspases. This process is mediated by adapter proteins such as Fas-associated death domain (FADD). Similarly, the other pathway leads to the activation of caspases but it originates from disturbed mitochondrias releasing cytochrome C. The cytochrome C forms together with caspase 9, ATP, and apoptosis protease activating factor-1 (Apaf-1), an apoptosome that catalyses the activation of the caspase.

In both pathways caspases, i.e. cysteine-containing proteases, which are characterized by their ability to split peptides or proteins at aspartate residues, play a central role. Various death receptors that belong to the tumor necrosis factor (TNF) receptor gene superfamily have been characterized. Fas, otherwise known as CD95 or APO1, is a cell surface receptor that belongs to the death receptor family. Ligand binding to Fas causes three of these receptor molecules to form trimers, which can then attract Fas-associated death domain (FADD), which in turn binds caspase 8 via the so-called death effector domain (DED). The result is activation of the caspase, which then activates other caspases downstream, ultimately leading to activation of executor caspases such as caspase 3, which cleave essential cellular proteins, leading to the death of the cell (Figure 2.47). Distinct cytotoxic chemicals may induce apoptosis via this pathway. One example is bleomycin, which induces the expression of the CD 95 receptor and its ligand in a hepatoma cell line.

In the other apoptotic pathway the mitochondria of the cell are compromised. Cytochrome C, a component of the electron transport chain that is localized on the outside of the inner mitochondrial membrane, is released into the cytoplasm (Figure 2.47) and participates, with ATP, in the apoptosis protease activating factor 1 (Apaf-1)-mediated activation of caspase 9. The result is the onset of a caspase cascade ultimately leading to cell death. If the loss of cytochrome C from mitochondria is massive, necrosis, but not apoptosis, is induced, owing to an interruption of the mitochondrial electron transport chain and consequently of ATP production.

During the apoptotic process various pro-apoptotic and anti-apoptotic proteins play a regulatory role. In this context the members of the bcl-2 multigene family have attracted special attention. Thus, bcl-2 and bcl-x_L exert anti-apoptotic activity, whereas bax, bad, and bcl-x_s promote apoptosis. Most likely the relative concentrations of anti-apoptotic and pro-apoptotic proteins determine whether the cell lives or dies. The binding of bcl-2 or bcl-x_L to bax results in mutual inactivation of their biological activities.

Morphology of Necrosis and Apoptosis

> **The morphologic appearance of apoptosis and necrosis is very different.**

Apoptosis shows very distinct morphologic changes in the nuclear area. In particular the condensation of the chromatin, which has been termed pyknosis, is prominent; other cell organelles remain largely intact. In general, a progressive contraction of cellular volume occurs. Subsequently apoptotic bodies are formed, which undergo phagocytosis by other cells such as macrophages.

In contrast, necrotic cells demonstrate increased cell volume and marked changes of many cell organelles, including swelling of the ER and the mitochondria. The cristae (invaginations) of the mitochondria vanish; flocculent and crystalline materials consisting of denatured proteins and hydroxyapatite accumulate in the mitochondria. In contrast to apoptosis, the nucleus usually does not show significant changes and only in special cases does caryolysis, i.e. disintegration of the nucleus, take place. The final stage of necrosis is characterized by widespread disintegration of cellular membranes and the appearance of organelles in the extracellular fluid. The results are prompt immunologic reactions and inflammation, which do not occur in apoptosis since the plasma membrane remains intact.

The Role of Disturbed Calcium Homeostasis in Cell Death

> **A non-physiologic increase in cytosolic calcium concentration can either cause necrosis or contribute to apoptosis.**

Under physiologic conditions the calcium concentration in the cytosol amounts to only 50–300 nmolar versus about 1 mmolar in the plasma. This high concentration gradient is maintained by various mechanisms (Figure 2.48):

1. Ca^{2+}-ATPases and Na^+/Ca^{2+}-exchange transport systems, which continuously pump calcium out of the cell
2. transport proteins, which catalyze the uptake of calcium into intracellular compartments, i.e. endoplasmic reticulum and mitochondria
3. binding of calcium to molecules such as the polar head groups of phospholipids.

The free calcium concentration in the cytoplasm plays a central role in the regulation of many essential cell functions such as growth, differentiation, secretion, change of cell shape, and transfer of hormonal signals. Exact control of calcium homeostasis is, therefore, of vital importance for the cell. As already described in *Cell Membrane* (this chapter) transport proteins can be targets of toxic chemicals. For example, calcium transport proteins contain functional SH groups that can be oxidized as well as alkylated or arylated. The loss of the ability to pump calcium either out of the cell and/or into intracellular storage sites becomes critical under conditions where other mechanisms cause an increase in cytoplasmic calcium. In this context, it is important to remember that damage to the cell membrane results in increased influx of calcium into the cell and oxidation

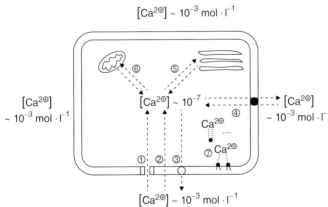

① Calcium channel

② Permeability of the cell membrane

③ Ca^{2+}-ATPases (transport proteins)

④ Na^+/Ca^{2+} exchange systems (transport proteins)

⑤ Sequestration into the ER by Ca^{2+} ATPases as well as release from the ER

⑥ Release and uptake by mitochondria (energy-dependent uptake systems as well as Ca^{2+}/H^+ exchange systems

⑦ Binding to proteins and anions as well as to the polar head groups of membrane lipids

Figure 2.48 Schematic drawing of the processes which affect or regulate the concentration of free calcium in the cell. 1, Calcium channels; 2, permeability of the cell membrane; 3, Ca^{2+}-ATPases (transport proteins); 4, Na^+/Ca^{2+} exchange systems (transport proteins); 5, sequestration into the ER by Ca^{2+}-ATPases as well as release from the ER; 6, release and uptake by mitochondria (energy-dependent uptake systems as well as Ca^{2+}/H^+ exchange systems); 7, binding to proteins and anions as well as to the polar head groups of membrane lipids.

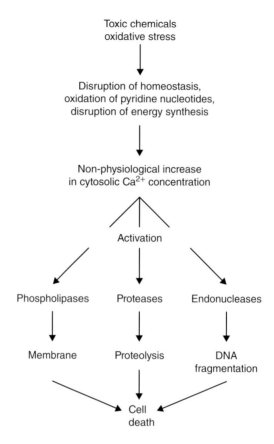

Figure 2.49 Possible contribution of calcium-activated catabolic processes to cell death.

of pyridine nucleotides (*Disturbance of Redox Systems*, this chapter) leads to increased efflux of calcium from the mitochondria. Furthermore, inhibition of ATP synthesis decreases the activity of transport proteins. All of these mechanisms result in unphysiological increases of cytoplasmic calcium concentration, which induce cell death (Figure 2.49). Thus, Ca^{2+}-stimulated activation of phospholipase A_2 enzymes leads to the formation of lysophosphatides and release of arachidonic acid. The altered turnover of phospholipids and the formation of toxic lysophosphatides can compromise the integrity of the plasma membrane. Activation of proteases by calcium can lead to the degradation of essential cell structures, e.g. proteins of the cytoskeleton are substrates of calcium-activated proteases. Last but not least, calcium-dependent endonucleases may contribute to cell death by the fragmentation of DNA.

2.4.5 Summary

All cell structures and functions may be targets of cytotoxic chemicals. Damage may be due to covalent or non-covalent interaction with the toxin. The consequences of the damage depend on the structure and function of the affected cellular component and the cell's ability to repair the damage. Important cellular targets of toxins are membrane systems, the generation of metabolic energy, the synthesis of critical macromolecules, control of intracellular signalling, and the control of redox homeostasis. The plasma membrane and intracellular membrane systems create specific reaction compartments due to their selective permeability, which is a prerequisite of the ordered metabolism of the cell. Damage to the integrity of the cell membrane results in serious consequences because enzymes may be inhibited, the fluidity of the lipid phase of the membrane may be altered, and lipid peroxidation may ensue. An extended disturbance of the synthesis of ATP or macromolecules is not compatible with life; whereas a brief interuption of these central cell functions will only cause a transient alteration in the steady state of the metabolic pathways of the cell. A disturbance of the redox homeostasis of the cell can be caused by the redox cycling of chemicals. The resulting inactivation of enzyme proteins due to the oxidation of their functional thiol groups may lead to cell death.

Cell death induced by cytotoxic chemicals may take the form of necrosis or apoptosis, depending on the cellular context and the biologically available concentration of the toxin. Necrosis is characterized by an increase in cell volume, marked changes in cell organelles and ultimately a disintegration of cell membranes. Due to the emergence of cell constituents in the extracellular fluids, immunological reactions and inflammation occur. In contrast, apoptosis is an active and well-ordered process, which may be induced by the action of "death receptors". In apoptosis cell organelles remain largely intact. There is a progressive contraction of cellular volume and a condensation of the chromatin of the nucleus. The resulting apoptotic bodies are phagocytized and immunological reactions do not occur.

Further reading

Artal-Sanz M, Tavernarakis N. Proteolytic mechanisms in necrotic cell death and neurodegeneration. *FEBS Lett.* **579**: 3287–96, 2005.

Ashton-Rickardt PG. The granule pathway of programmed cell death. *Crit Rev Immunol.* **25**: 161–82, 2005.

Blaise GA, Gauvin D, Gangal M, Authier S. Nitric oxide, cell signaling and cell death. *Toxicology* **208**: 177–92, 2005.

Brautigan DL. Protein Ser/Thr phosphatases – the ugly ducklings of cell signalling. *FEBS J.* **280**(2):324–45, 2013.

Carmody RJ, Cotter TG. Signalling apoptosis: a radical approach. *Redox Rep.* **6**: 77–90, 2001.

Cronin M, Dietrich DR. Computational toxicology to support predictions of ADMET properties and as a promising tool for 21st century risk assessment. *Comp Toxicol.* **1**: 1–2, 2017.

Daneshian M, Botana LM, Decheraoui Botttein M-Y, Buckland G, Campás M, Dennison N, Dickey RW, Diogène J, Fessard V, Hartung T, Humpage A, Leist M, Molgó J, Quilliam MA, Rovida C, Suarez-Isla BA, Tubaro A, Wagner K, Zoller O, Dietrich DR. A roadmap for hazard monitoring and risk assessment of marine biotoxins on the basis of chemical and biological test systems – a t4 report. *ALTEX.* **30**(4): 487–545, 2013.

Dietrich DR, Swenberg JA. The presence of alpha 2u-globulin is necessary for d-limonene promotion of male rat kidney tumors. *Cancer Res.* **51**: 3512–21, 1991.

Dietrich DR, Höger S. Guidance values for microcystins in water and cyanobacterial products (blue-green algal supplements): a reasonable or misguided approach? *Toxicol Appl Pharmacol.* **203**(3): 273–89, 2005.

Dietrich DR, Hengstler JG. Highlight report: From bisphenol A to bisphenol F and a ban of mustard due to chronic low dose exposures? *Arch Toxicol.* **90**(2), 489–91, 2016; doi: 10.1007/s00204-016-1671-5.

Goldberg J, Huang H, Kwon, Y, Greengard P, Nairn AC, Kuriyan J. Three-dimensional structure of the catalytic subumit of protein serine/threonine phosphatase-1. *Nature.* **376**: 745–53, 1995.

Lee AG. How lipids affect the activities of integral membrane proteins. *Biochim Biophys Acta.* **1666**: 62–87, 2004.

Los DA, Murata N. Membrane fluidity and its roles in the perception of environmental signals. *Biochim Biophys Acta.* **1666**: 142–57, 2004.

Orrenius S, Zhivotovsky B, Nicotera P. Regulation of cell death: the calcium-apoptosis link. *Nat Rev Mol Cell Biol.* **4**: 552–65, 2003.

Owuor ED, Kong AN. Antioxidants and oxidants regulated signal transduction pathways. *Biochem Pharmacol.* **64**: 765–70, 2002.

Pinkoski MJ, Green DR. Apoptosis in the regulation of immune responses. *J Rheumatol Suppl.* **74**: 19–25, 2005.

Segal-Bendirdjian E, Dudognon C, Mathieu J, Hillion J, Besancon F. Cell death signalling: recent advances and therapeutic application. *Bull Cancer.* **92**: 23–35, 2005.

Xu JJ, Diaz D, O'Brien PJ. Applications of cytotoxicity assays and pre-lethal mechanistic assays for assessment of human hepatotoxicity potential. *Chem Biol Interact.* **150**: 115–28, 2004.

Zimber A, Nguyen QD, Gespach C. Nuclear bodies and compartments: functional roles and cellular signalling in health and disease. *Cell Signal.* **16**: 1085–104, 2004.

2.5 Toxicogenetics

Lesley A. Stanley

2.5.1 Introduction

> **Individuals vary in their responsiveness to toxic agents. Slow acetylators have increased susceptibility to isoniazid-induced neuropathy and hydralazine-induced systemic lupus erythematosus, while normal therapeutic doses of the antihypertensive drug debrisoquine can cause sudden collapse due to a catastrophic drop in blood pressure in CYP2D6 poor metabolisers. These and other observations led to the initiation of the discipline now known as toxico- (or pharmaco-) genetics.**

The fact that individuals differ in their responsiveness to toxic agents has been known for a very long time. The phenomenon of favism, for example, was first described by the Ancient Greeks. Favism is the most common human enzyme deficiency and describes the phenomenon whereby haemolysis occurs in certain individuals on ingestion of fava beans due to a deficiency of the enzyme glucose-6-phophate dehydrogenase. The defect is sex-linked, being transmitted from mother (usually a healthy carrier) to son (or daughter, who would be a healthy carrier too). It is the most common human enzyme deficiency, affecting some 6 million people, and exists in more than 400 genetic variants. Agents that can cause haemolysis in glucose-6-phophate dehydrogenase-deficient individuals include antimalarial drugs (primaquine, pamaquine, pentaquine and chlorophine), sulfonamides, nitrofurans (e.g. furadantin), paracetamol, naphthalene, certain vitamin K derivatives, acetylsalicylic acid and probenecid (Benemid). The favism-inducing toxins in beans are believed to be divicine and isouramil, the aglycone moieties of vicine and convicine.

Modern interest in the role of xenobiotic-metabolising enzymes in individual differences in susceptibility to the adverse effects of drugs and chemicals began with several serendipitous observations, including the fact that so-called 'slow acetylators' had increased susceptibility to isoniazid-induced neuropathy and hydralazine-induced systemic lupus erythematosus and the dramatic effects of the antihypertensive drug debrisoquine in an individual who subsequently turned out to be what is now called a CYP2D6 poor metaboliser. A laboratory worker studying the metabolism of the drug took a normal therapeutic dose and collapsed due to a catastrophic drop in blood pressure. These and other observations led to the initiation of the discipline now known as toxico- (or pharmaco-) genetics.

2.5.2 Toxicogenetics and Toxicogenomics

> **Toxicogenetics and toxicogenomics consider the role of an individual's genetic makeup in determining the effects of xenobiotics to which an individual is exposed. Toxicogenetics is the study of genetic variation in specific candidate genes relevant to a particular toxic response and is based on an existing paradigm such as the proposed mechanism of action of a carcinogen. Toxicogenomics is the study of genetic variation across the whole genome in an unbiased manner. An understanding of toxicogenetics/toxicogenomics will facilitate the evaluation of interindividual variability in pharmacological and toxic responses and permit the identification and protection of susceptible populations.**

Elaboration of toxicogenetics is relatively inexpensive and easy to perform with current technologies, and because it arises from a biological hypothesis the biological consequences of the results obtained are usually quite readily interpretable. This is the way that the study of genetic influences on individual susceptibility to toxic agents started back in the 1960s. Toxicogenomics involves the simultaneous study of many genetic variants (several million markers have been identified in the human genome), and is by definition an unbiased approach, but its disadvantage is that the functional consequences of the observations made are usually unknown.

2.5.3 Genotype and Phenotype

> **Toxicogenetic variation is determined by the individual genes, each of which is found at a specific chromosomal locus. A polymorphism is conventionally defined as a sequence variant found at a particular locus in at least 1% of the population.**

Toxicogenetic variation is determined by the genes that encode proteins responsible for the metabolic processing of, and response to, xenobiotics. Each gene is found at a specific locus. At each locus there may be one or more possible sequence variants, or alleles. The existence of more than one possible allele at a particular locus is called a polymorphism, and loci where multiple alleles exist are described as polymorphic. A polymorphism is defined as a sequence variant that is present in at least 1% of the population. Humans, being diploid, have two copies of each locus (i.e. they carry two alleles of each gene), and the combination of alleles carried by an individual defines his or her genotype at that locus. *If both copies are the same, the individual is said to be homozygous at this locus; if two different alleles are present, the individual is said to be heterozygous.*

Many polymorphisms affect the amino acid sequence of a protein, but some affect the non-coding regulatory regions of genes. Polymorphisms which affect the amino acid sequence may be detected at the level of either the nucleotide sequence or the amino acid sequence. Alternatively, those which affect function may be detected directly by measuring the function of the protein (e.g. its enzyme activity). *This is the phenotype*; it is a function of the underlying DNA sequence but may also be influenced by regulatory, physiological and environmental factors as well as by pathological conditions. Polymorphisms in non-coding regions may affect the level of expression of a particular protein and can be detected at the nucleotide sequence level or by looking at mRNA or protein expression. The complete absence of a gene may also be classified as a polymorphism.

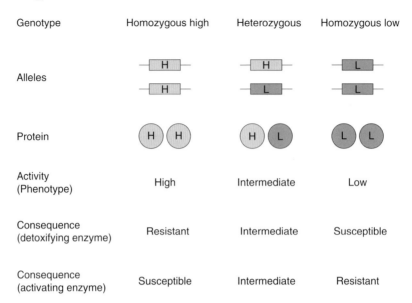

Figure 2.50 *This figure illustrates the possible consequences of polymorphism in a hypothetical xenobiotic-metabolising enzyme that has both a high-activity and a low-activity variant. In the case of an individual who is homozygous for the high-activity variant, two high-activity alleles will be present in the genome (genotype) and these will be transcribed and translated to form entirely high-activity enzyme molecules. Thus, the individual's liver will have high enzyme activity (phenotype). If the enzyme in question is a detoxifying enzyme, the individual is predicted to have reduced susceptibility to toxicity, whereas if it is a metabolic activating enzyme the individual may have increased susceptibility. The reverse is the case for an individual who is homozygous for the low-activity variant while a heterozygous individual has one high- and one low-activity allele, a mixture of high- and low-activity enzyme molecules in the liver, and hence an intermediate phenotype.*

The consequences of polymorphic variation at a hypothetical allele are illustrated in Figure 2.50.

Genotyping

> **Genetic characteristics, i.e. the allelic composition, define the genotype. Genotyping involves examining the altered nucleotide sequences of polymorphic variants in the DNA itself, whereas phenotype refers to the biological expression of the genotype. Phenotyping involves measuring biological functions, e.g. enzyme activity. Genotyping involves examining the altered nucleotide sequences of polymorphic variants in the DNA itself and is usually carried out using polymerase chain reaction (PCR)-based methods. Genotyping is becoming increasingly popular as a means of evaluating toxicogenetic variation; however, it is important to remember that it is the phenotype that determines function and will mediate any observed effects on susceptibility.**

Genotyping involves examining the altered nucleotide sequences of polymorphic variants in the DNA itself. This approach traditionally involved cloning and sequencing the gene of interest from different individuals, but recent technological developments mean that genotyping is now usually undertaken using PCR-based methods.

Single Nucleotide Polymorphisms

> **A single nucleotide polymorphism (SNP) is a variation in a single nucleotide that occurs at a specific position in the DNA. SNPs are present at a frequency of approximately 1/300 bp in the human genome. Most SNPs are functionally silent; in humans, only about 2000 mis-sense or non-synonymous SNPs cause an actual amino acid change and may affect protein function. An individual may be identified by his/her SNP profile and this profile may be used to identify subgroups with increased or decreased responsiveness to either beneficial (therapeutic) or adverse (toxic) effects of chemicals.**

The vast majority (99.5%) of the human genome is common to everybody; the differences between individuals are mainly SNPs. For example the base C may appear in most individuals, whereas in a minority of individuals, the position is occupied by base A. SNPs are estimated to account for as much of 90% of human genetic variation. On average there is about one SNP every 300 bases, meaning that an average-sized gene of 25 kb may contain up to 80 SNPs, but they are not evenly spread through the genome: there are SNP hot spots where the density of SNPs is much higher than in other regions, whereas 99% of SNPs are found in regions which do not contain genes, such as intergenic regions.

The changes caused by SNPs may be silent, harmless or latent. Most SNPs are located in non-coding/non-regulatory regions of the genome and are functionally silent; it is currently estimated that only about 2000 cause an actual amino acid change. These mis-sense or non-synonymous SNPs alter the amino acid sequence of the cognate protein and may affect its function, whereas synonymous SNPs do not have any effect on coding information, although if located within a regulatory region they can affect responses to drugs and toxic chemicals.

Most SNPs are biallelic, that is, there are only two possible variants (usually a major and a minor variant), so they are relatively easy to type using automated methods. In addition, they exhibit a fairly low rate of recurrent mutation, meaning that they act as stable markers of human evolutionary history. They are usually detected by PCR-based methods.

Currently, the favoured high-throughput method for SNP detection is direct DNA sequencing. The fact that PCR products are usually quite short (a few hundred nucleotides) means that they are easy to sequence, and recent advances in sequencing technology and automation have made it possible to sequence the products of hundreds of PCR reactions simultaneously. As a consequence, PCR and direct sequencing is now the most common method used to detect SNPs.

Haplotypes

> **A haplotype is a combination of alleles which are physically linked (in linkage disequilibrium) along a chromosome; in other words, multiple loci that are inherited as a unit. For any given genetic region an individual has one haplotype from each parent. The study of haplotypes provides increased statistical power to detect associations compared with single marker approaches, meaning that less genotyping effort is required to evaluate the role of common genetic variants.**

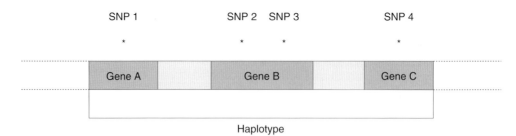

Figure 2.51 *Relationship between SNPs, genes and haplotypes. A haplotype is a section of chromosome which is inherited as a block. It may contain one or more genes and/or a variety of SNPs which are co-inherited because of linkage disequilibrium.*

SNPs that are separated by a large distance are typically not very well correlated because recombination occurs in each generation, mixing the allele sequences of the two chromosomes. Haplotypes can be assigned to both of the chromosomes in a diploid cell, so for any given region (apart from on the sex chromosomes in males) there is one haplotype from the mother and one from the father (Figure 2.51).

In practice, linkage disequilibrium means that it is not necessary to type all 10 million potential SNPs in order to identify genetic risk factors – a few hundred thousand to a million haplotypes can provide the same information. The study of haplotypes therefore provides increased statistical power to detect associations and map disease mutations compared with single marker approaches. This approach means that less genotyping effort is required to evaluate the role of common genetic variants.

Phenotyping

> **Phenotyping is a function of the underlying DNA sequence but may also be influenced by regulatory, physiological and environmental factors as well as by pathological conditions.**
>
> **Phenotyping involves measuring biological functions such as enzyme activity. This will identify variations in activity due to either the existence of different variants of an enzyme or differences in the level of expression of a uniformly active enzyme. If the observed variability is due to differences in expression, this may also be estimated by determining the level of expression at the protein or mRNA level.**

Genotyping is becoming increasingly popular as a means of evaluating toxicogenetic variation because of the advent of high-capacity automated PCR methods. However, it is important to remember that it is the phenotype which determines function and will mediate any observed effects on susceptibility. Genotype is determined by the DNA sequence and is unchanging, whereas phenotype may change during life due to developmental changes in gene expression and, potentially, xenobiotic-induced effects such as induction or enzyme inhibition. Traditional methods for phenotyping xenobiotic-metabolising enzymes involve measuring enzymatic activity using diagnostic substrates. The advantage of this approach is that it directly addresses the actual function of the enzyme and does not require assumptions about genotype-phenotype correlations, but it has become less popular in recent years because it is more labour-intensive than genotyping.

The genotype can be determined by looking at any cell type because all of an individual's cells, except for red blood cells and germ cells, contain the same genomic DNA sequence. This means that it is not necessary to examine genotype in the target tissue in which toxicity is expressed, and that more accessible surrogate tissues can be used.

Many investigators prefer to use mRNA-based techniques such as TaqMan® real-time PCR, a simple and rapid method to quantify the expression of a particular mRNA, rather than using labour-intensive protein-based approaches. However, the results of these studies must be examined critically because differences in mRNA levels may not be reflected at the protein, and hence functional, level.

Correlating Genotype and Phenotype

> **In order to use genotype as an indicator of phenotype it is necessary to demonstrate a clear correlation between genotype and phenotype. In some, but by no means all, cases a clear correlation between genotype and phenotype has been established.**

The drawback of genotyping methods in toxicogenetics is that they provide only an indirect measure of the actual activity of the enzyme or protein of interest. To interpret the results it is necessary to make the assumption that in a given individual the phenotype is a direct function of the genotype. In order to justify this assumption it is necessary to demonstrate a clear correlation between genotype and phenotype. In some cases, such as the aromatic amine-metabolising enzyme NAT2, a clear correlation between genotype and phenotype has been established. However, in other cases the relationship between genotype and phenotype is more complex.

2.5.4 The Role of Xenobiotic-metabolising Polymorphisms in Susceptibility to Toxic Agents

> **Polymorphisms exist among both Phase I and Phase II xenobiotic-metabolising enzymes. Where polymorphisms exist, a star number is allocated to each polymorphic variant, for example *CYP2C9*2* represents cytochrome P450, family 2, subfamily C, member 9, polymorphic variant 2. The basis of naming of other xenobiotic-metabolising enzyme polymorphisms is similar, except that it is simpler where subfamilies do not exist, for example *NAT2*4* represents *N*-acetyltransferase, family 2, polymorphic variant 4.**

Polymorphisms exist among both Phase I and Phase II xenobiotic-metabolising enzymes. The CYP gene family, for example, is very diverse, comprising 57 genes in 18 gene families. Where polymorphisms do exist, a star number is allocated to each polymorphic variant. Thus, for example, the designation *CYP2C9*2* represents:

- cytochrome P450
- family 2
- subfamily C
- member 9
- polymorphic variant 2.

The basis of naming of other xenobiotic-metabolising enzymes is similar, except that it is simpler where subfamilies do not exist; thus, for example, *NAT2*4* represents:

- *N*-acetyltransferase
- family 2
- polymorphic variant 4.

This is the main rapid acetylator variant of NAT2.

Various metabolic polymorphisms are known to have a significant impact on individual susceptibility to toxic and carcinogenic compounds, although the literature in this area can often be contradictory. In this chapter, rather than presenting a long list of polymorphisms which may or may not be involved in susceptibility to chemically-induced disease processes and toxicity, a few of the better characterised examples are presented.

Interindividual Variability in Susceptibility to Industrial Chemicals

Acrylonitrile

> **CYP2E1-mediated metabolism of acrylonitrile generates a reactive epoxide metabolite, cyanoethylene oxide (CEO), which undergoes further metabolism to generate cyanide. Both acrylonitrile and CEO can cause glutathione depletion; CEO also reacts with DNA and is mutagenic, suggesting that this metabolite may be responsible for the carcinogenic effects of acrylonitrile. Individuals with the Val^{104} variant of GSTP1 have an increased level of *N*-(cyanoethyl)valine adducts compared with the wild type (Ile^{104}), possibly due to altered affinity of the enzyme for electrophilic substrates. A trend towards higher adduct levels in individuals has also been observed in individuals with at least one copy of the A316G variant of *CYP2E1*, which is associated with lower levels of oxidative metabolism.**

Relatively few well-characterised examples of acute toxicity affected by polymorphic xenobiotic-metabolising enzymes currently exist. One of the best examples to date is acrylonitrile, an industrial solvent which is, in addition to being carcinogenic (IARC classifies it as 'possibly carcinogenic to humans'), acutely toxic. Acrylonitrile is metabolised via oxidative and reductive routes. The glutathione-dependent detoxification pathway leads via the primary metabolites *S*-cyanoethylglutathione and *S*-methylglutathione to the mercapturic acids *N*-acetyl-*S*-cyanoethylcysteine and *N*-acetyl-*S*-cyanomethylcysteine, the final urinary excretion products. The alternative oxidative pathway of acrylonitrile metabolism, mediated by the isozyme CYP2E1, generates a reactive epoxide metabolite, CEO, which is further metabolised to cyanide (Figure 2.52). Both acrylonitrile and CEO also react with tissue thiols, leading to glutathione depletion. The acute toxicity of high doses of acrylonitrile is thought to be gated by the depletion of hepatic glutathione, and this has led to the recommended use of *N*-acetylcysteine as an antidote to acrylonitrile poisoning.

There is significant inter- and intraspecies variability in the acute toxicity of acrylonitrile. Humans are known to have a higher level of CYP2E1-mediated oxidative metabolism of acrylonitrile than rodents, although inducing agents such as acetone can be used to upregulate CYP2E1 activity in rats for use in experimental studies, and this makes it difficult to extrapolate directly from the results of animal experiments to the effects of acrylonitrile in humans. In addition, humans have an active epoxide hydrolase pathway which leads to the generation of cyanide from CEO whereas this is not the case in rodents. Oxidative metabolism, which leads to the formation of cyanide, seems to be much less important in animals than it is in humans. The acute toxicity of

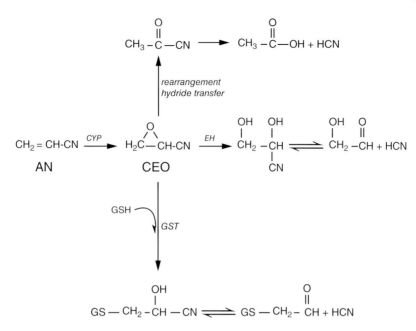

Figure 2.52 A proposed scheme showing the role of cytochrome P450 and epoxide hydrolase enzymes in the metabolism of acrylonitrile to cyanide. Reproduced from Wang, H., Chanas, B., and Ghanayem, B.I. (2002) Cytochrome P450 2E1 (CYP2E1) is essential for acrylonitrile metabolism to cyanide: Comparative studies using CYP2E1-null and wild-type mice. Drug Metab Dispos **30**: 911–917 with the permission of the publisher.

acrylonitrile in animals seems to be mediated by glutathione depletion whereas in humans acute toxicity is largely determined by the metabolic formation of cyanide.

The formation of an *N*-(cyanoethyl)valine adduct at the *N*-terminus of haemoglobin can be used as a biomarker of blood levels of acrylonitrile. By measuring levels of *N*-(cyanoethyl)valine in relation to acrylonitrile exposure, higher adduct levels have been demonstrated in individuals with at least one copy of a CYP2E1 promoter variant (A316G). This may be associated with individual variation in the inducibility of CYP2E1. Slower CYP2E1-mediated metabolism of acrylonitrile in some individuals could lead to lower levels of metabolic activation (to CEO and cyanide) and increased blood levels of the parent compound, reflected in higher levels of the biomarker.

Initially, studies on individual cases of acrylonitrile intoxication provided clues about the role of metabolic polymorphisms in the detoxification of acrylonitrile in humans. When the effects of acute acrylonitrile exposure were compared in two individuals, one with low and one with high glutathione *S*-transferase (GST) activity, the individual with low activity experienced headache, nausea and vomiting. Furthermore, the level of hydrocyanic acid in his blood was within the lethal range, although fortunately he recovered following treatment with the antidote, *N*-acetylcysteine. This was consistent with the hypothesis that, particularly in individuals with low GST activity, toxicity is gated by glutathione depletion. If insufficient glutathione is available for conjugation (or the activity of GST is too low), free acrylonitrile may become available to enter the CYP2E1-mediated oxidative pathway, leading to toxicity.

Biomonitoring studies have suggested a role for the GST isozyme GSTP1-1 in the human metabolism of acrylonitrile. Individuals with the wild type form of GSTP1 (*GSTP1*A*) appear to have a lower level of *N*-(cyanoethyl)valine adducts (i.e. they detoxify acrylonitrile more efficiently) compared with those carrying a polymorphism at codon 104 of this gene (*GSTP1*B* and *GSTP1*C*). This was thought to be due to altered affinity of the enzyme for electrophilic substrates: the wild type enzyme efficiently conjugates glutathione to electrophilic substrates (i.e. it has a low K_m and a high K_{cat}/K_m ratio) whereas the variants have higher K_m and lower K_{cat}/K_m values. This would be expected to lead to less efficient detoxification in the individuals who carry variant forms of GSTP1 and potentially to increased susceptibility to acrylonitrile toxicity. Furthermore, multiplex analysis has implicated the combined genotypes of CYP2E1, GSTM1 and mEH4 in urinary excretion and the formation of acrylamide and glycidamide adducts in individuals exposed to acrylamide in the workplace.

These results suggest that the biomonitoring of exposure to acrylonitrile should be supported by genotyping and illustrate the way in which the combination of toxicogenetic data with pharmacokinetic modelling and biomonitoring of actual exposures could, in the future, be used to identify and protect susceptible subgroups within the exposed subpopulation.

Individual Susceptibility to Bladder Carcinogens

> **Polymorphisms in drug metabolism may make it possible to identify individuals with increased or decreased susceptibility to certain types of tumour. The simplest paradigm is susceptibility to bladder cancer induced by exposure to aromatic amines. Bladder cancer is associated with occupational exposure to aromatic amines, and studies in the 1980s and 1990s indicated that slow acetylation is a contributory risk factor in occupationally-induced bladder cancer. Hepatic activating and detoxifying enzymes compete for aromatic amines, generating metabolites which are subsequently delivered to the bladder, and it is believed that individuals who are slow for NAT2, rapid for CYP1A2, GSTM1 null and rapid for NAT1 are likely to have the highest risk of developing bladder cancer.**

Aromatic amines such as 4-aminobiphenyl are used in various industrial processes and found in cigarette smoke. Bladder cancer is associated with occupational exposure to aromatic amines in the rubber, textile, dye and chemical industries. Smoking is also a major risk factor (approximately 66% of bladder cancers in Western countries are attributable to cigarette smoking), probably due to the presence of aromatic amines as well as polycyclic aromatic hydrocarbons and nitrosamines in cigarette smoke.

Carcinogenic aromatic amines are metabolised by *N*-acetylation. Individuals may be classified as rapid or slow acetylators depending upon the variant of NAT2 they carry. As mentioned above, acetylation by NAT2 was one of the first examples of pharmacogenetic variation recognised.

One of the reasons that the role of the NAT2 polymorphism in mediating the effects of xenobiotics is relatively well understood is that convenient methods exist for measuring the activity of this enzyme both *in vitro* and *in vivo*; PCR-based methods are also available for the determination of NAT2 genotype. There are now known to be at least 65 genetic variants of NAT2, but seven key SNPs have been identified and found to identify most of the variant alleles. Indeed, studies using panels of two, three, four or all seven SNPs, in combination with a proposed tag SNP which has been proposed as an activity biomarker, have demonstrated that genotyping for four SNPs (G191A, T341C, G590A and G857A) is sufficient for the prediction of phenotype (as measured using sulphamethazine) in a range of ethnic populations.

The availability of convenient genotyping methods and the clear correlation between genotype and phenotype for this enzyme have facilitated epidemiological analysis of the NAT2 polymorphism. The deduction of NAT2 phenotypes from genotypes is based upon the assumption that

Table 2.17 *Epidemiological analysis of the* NAT2 *polymorphism in bladder cancer.*

Exposure	Number of slow acetylators/total in study	
	Patients	Controls
Dye factories	22/23 (95.7%)	
Urban	74/111 (66.6%)	118/207 (57%)
Industrial	44/62 (71%)	
No exposure	83/127 (65.4%)	26/59 (44.1%)

Summarised from Cartwright, R.A., Glashan, R.W., Rogers, H.J., Ahmad, R.A., Barham-Hall, D., Higgins, E., and Kahn, M.A. (1982). Role of N-acetyltransferase phenotypes in bladder carcinogenesis: a pharmacogenetic epidemiological approach to bladder cancer. *Lancet* **2**: 842–845 and Risch, A., Wallace, D.M., Bathers, S., and Sim, E. (1995). Slow N-acetylation genotype is a susceptibility factor in occupational and smoking related bladder cancer. *Hum Mol Genet* **4**: 231–236.

rapid and slow alleles are codominant; this means that as well as rapid acetylators (with two rapid alleles) and slow acetylators (with two slow alleles), it is also possible to identify heterozygotes (i.e. people with one rapid and one slow allele) as an intermediate group. There is further heterogeneity in the slow acetylator phenotype because each SNP has a distinct effect on the activity and stability of the corresponding protein.

In an influential study published in the 1980s, it was found that 22/23 (95.7%) dye factory employees (or ex-employees) with bladder cancer were slow acetylators, whereas only ~60% of controls were slow acetylators (Table 2.17). Although subsequent studies found somewhat lower slow acetylator frequencies in bladder cancer patients, the observation that slow acetylators have an increased risk of bladder cancer is now well established and has been confirmed by meta-analysis. This effect seems to be specific to smokers: NAT2 slow acetylators are especially susceptible to the adverse effects of smoking on bladder cancer risk whereas slow acetylation does not increase bladder cancer risk among subjects who have never smoked.

The role of NAT2 should not be considered in isolation, since susceptibility to bladder cancer is also modulated by interaction with other xenobiotic-metabolising enzymes. The metabolic activation of aromatic amines is illustrated in Figure 2.53. The first step in this process is hepatic *N*-hydroxylation by the CYP isozyme CYP1A2. The resulting *N*-hydroxyl aromatic amines form glutathione, glucuronide or sulphate conjugates (enzymatically catalysed by GSTs, UDP-glucuronyltransferases (UGTs) and sulfotransferases (SULTs), respectively). These soluble metabolites are transported to the bladder via the systemic circulation and may then be taken up by transitional epithelial cells and hydrolysed, either under acidic conditions in the urine itself or as a result of intracellular hydroxylase activity, regenerating the original *N*-hydroxylamine.

> **While individuals can readily be classified as rapid or slow acetylators according to their NAT2 genotype/phenotype, the genotype/phenotype correlation for NAT1 is less clear. The other NAT isozyme (NAT1) metabolises *P*-aminobenzoic acid as well as endogenous substrates such as *p*-aminobenzoyl glutamate, is expressed in a number of tissues including the urothelium and is thought to play a key role in cellular homeostasis. Most of the polymorphic variants of NAT1 result from one or more SNPs, although in some cases small deletions are observed.**

Like NAT2, NAT1 can metabolise aromatic amines, but it is as yet unclear what the consequences of this polymorphism are for bladder cancer susceptibility. It is currently believed that

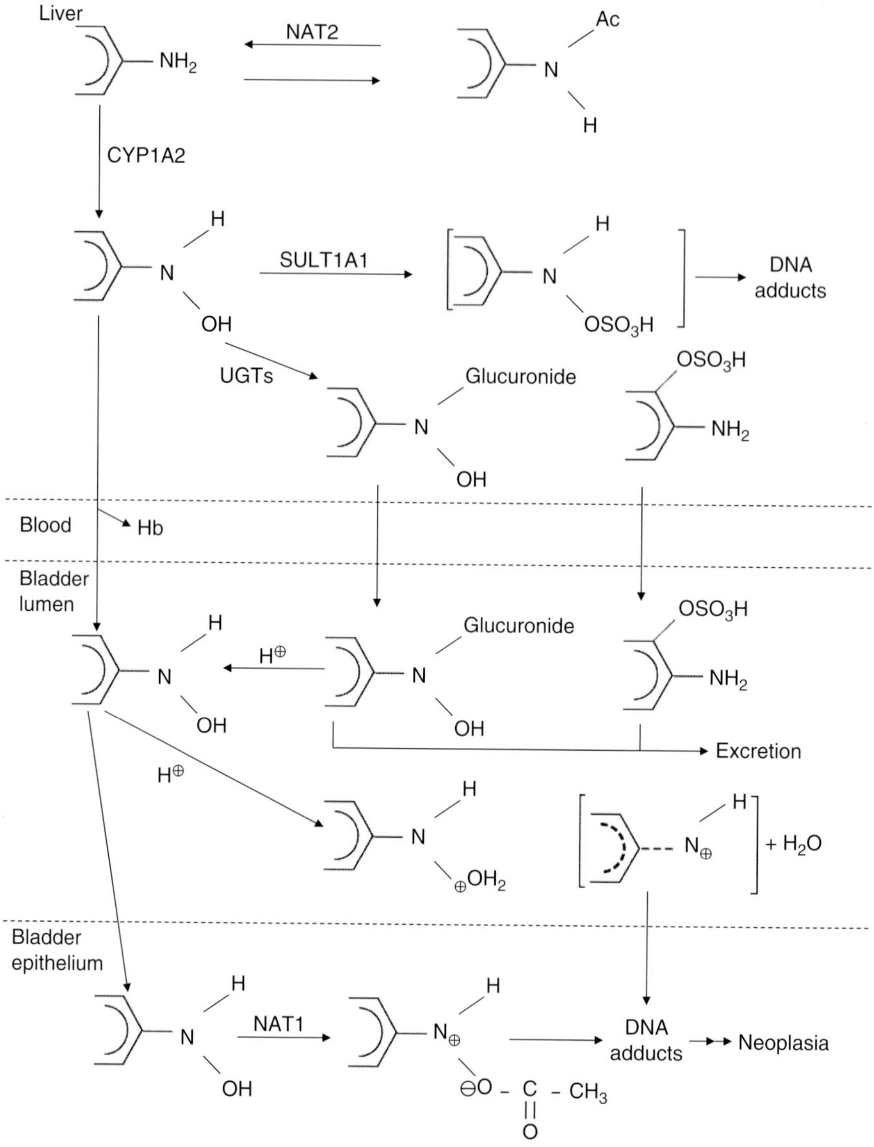

Figure 2.53 *Pathways of aromatic amine metabolism in the liver and bladder. Reproduced from Kadlubar, F.F. and Badawi, A.F. (1995) Genetic susceptibility and carcinogen-DNA adduct formation in human urinary bladder carcinogenesis. Toxicol Lett **82–83**: 627–632 with the kind permission of the late Dr. Fred Kadlubar and the publisher.*

NAT1 *O*-acetylates *N*-hydroxy aromatic amines *in situ* in the transitional epithelium, forming highly reactive *N*-acetyoxy esters which are able to bind directly to DNA, potentially leading to the initiation of carcinogenesis. This illustrates the way in which the same enzyme may mediate detoxification in some circumstances and metabolic activation in others: NAT1-mediated *N*-acetylation is a first-pass detoxification route for aromatic amine carcinogens in the skin whereas in the colon NAT1 mediates metabolic activation of the *N*-hydroxy aromatic amine metabolites of the same aromatic amines.

The difficulty in relating *NAT1* genotype to phenotype arises because NAT1 phenotype is influenced by factors other than SNPs in the coding region (there is, for example, evidence that NAT1 is subject to transcriptional regulation via the glucocorticoid receptor), and because the distribution of NAT1 activity (at least in red blood cells) is unimodal rather than bi/tri modal as for NAT2. There is currently no epidemiological evidence for a link between NAT1 genotype and bladder cancer.

The GST isozymes GSTM1 and GSTT1 may also contribute to the risk of bladder cancer. Polymorphisms in these genes take the form of null alleles designated *GSTM0* and *GSTT0*; individuals may carry two, one or no copies of each of these genes. Case-control studies and meta-analyses indicate that the risk of bladder cancer is increased by ~50% in homozygous *GSTM1* null individuals, and that this may be associated with increased accumulation of DNA damage in the transitional epithelium. In addition, carriers of the *GSTP1* A313G polymorphism appear to have increased risk of bladder cancer.

Bladder cancer has historically been considered to be one of the few clear-cut examples of a tumour type that is consistently associated with a particular polymorphism in xenobiotic metabolism. In this scenario, activating and detoxifying enzymes in the liver compete for the aromatic amine substrate, leading to the generation of metabolites that are subsequently delivered to the bladder. Low levels of hepatic *N*-acetylation (NAT2) allow *N*-hydroxylation by CYP1A2 to predominate. Following further metabolism by GSTs and transport to the bladder, hydrolysis regenerates a chemically reactive *N*-hydroxyl aromatic amine that is susceptible to *O*-acetylation by NAT1, leading to the initiation of carcinogenesis. The hypothesis based upon this scheme is that individuals who are slow for NAT2, rapid for CYP1A2, GSTM1 null and rapid for NAT1 are likely to have the highest risk of developing bladder cancer.

This attractive scenario has recently been called into question by a large, well-conducted study in which there was no difference in the proportion of slow acetylators between bladder cancer cases (64.1%) and controls (64.0%). This may trigger a re-evaluation of the role of the NAT2 polymorphism in bladder cancer susceptibility; however, it should be noted that the proportion of slow acetylators in both case and control populations in this study was high compared with that normally seen in Caucasian populations. One possible reason for the absence of a clear association between slow acetylation and bladder cancer in recent studies may be better control of occupational exposure to aromatic amines, meaning that fewer individuals are exposed to bladder carcinogens in the workplace, and that those who are exposed receive lower doses. Thus, while NAT2 genotype may lack utility as a predictive marker of bladder cancer risk in the population as a whole or among exposed workers (although it has recently been suggested that an 'ultra-slow' NAT2 phenotype may exist and be associated with bladder cancer susceptibility), there is currently no reason to reject mechanistic explanations of susceptibility to bladder carcinogenesis based on relative rates of hepatic *N*-acetylation and *O*-hydroxylation combined with *O*-acetylation in the transitional epithelium.

Individual Susceptibility to Colorectal Cancer

The determination of colorectal cancer susceptibility involves both genetic and environmental/lifestyle factors. Heterocyclic amines generated during the cooking of red meat are thought to mediate the carcinogenic process in the colon. The main CYP isozyme involved in the Phase I metabolism of heterocyclic amines is CYP1A2. The key Phase II metabolites implicated in colorectal cancer induction are the *N*-acetoxy derivatives. These may be generated by the action of NAT1 and NAT2. In theory, the

> **rapid NAT2 phenotype should confer an increased risk of colorectal cancer due to the enzyme's capacity to *O*-acetylate the products of Phase I metabolism, especially when combined with the rapid CYP1A2 and regular consumption of well-done red meat; however, no clear link has been established between NAT2 genotype and colorectal cancer susceptibility.**

Attempts to elucidate the role of metabolic polymorphisms in colorectal cancer further illustrate how difficult it is to obtain definitive answers concerning the roles of individual xenobiotic-metabolising polymorphisms in cancer susceptibility. An individual's risk of colorectal cancer is modified by both genetic and environmental/lifestyle factors, including a first-degree family history of colorectal cancer, overeating, physical inactivity, a high intake of red meat, alcohol use, smoking and a low intake of vegetables.

Heterocyclic amines are generated during chemical reactions that occur during the cooking of red meat. Thus, people who frequently consume cooked meat are exposed to heterocyclic amines on a regular basis. These are thought to mediate the carcinogenic process in the colon; haem iron and nitrate/nitrite intake may also play a role. The heterocyclic amine 2-amino-1-methyl-6-phenylimidazo-[4,5-b]pyridine (PhIP) is considered to be of particular significance with respect to colorectal cancer because it is usually the predominant heterocyclic amine found in cooked meat, it is mutagenic in bacterial and mammalian cell-based assays, and it is a colon carcinogen in rats.

Heterocyclic amines such as PhIP are subject to both ring- and *N*-oxidation mediated by CYPs (Figure 2.54); the *N*-hydroxy metabolites are direct-acting mutagens whereas the ring hydroxylated metabolites are not mutagenic. The main CYP isozyme involved in the Phase I metabolism of heterocyclic amines is CYP1A2; indeed meta-analysis suggests that the *CYP1A2*1C* and *CYP1A2*1F* variants of CYP1A2 could be risk factors for colorectal cancer, at least in Asian populations. The parent heterocyclic amines and their Phase I metabolites also undergo Phase II metabolism catalysed by GSTs, UGTs, SULTs and NATs. The key metabolites implicated in colorectal cancer induction are the *N*-acetoxy derivatives, which may be generated by the action of NAT1 and NAT2. Phase I metabolism of heterocyclic amines is primarily hepatic, but *O*-acetylation can occur in either the liver or in the colon.

The pathway of metabolism of heterocyclic amines suggests that the rapid NAT2 phenotype would be likely to confer an increased risk of colorectal cancer, especially when combined with rapid CYP1A2 and regular consumption of well-done red meat. Phenotyping studies in the early 2000s suggested that this was the case, at least in smokers. It has also been suggested that the *NAT1*10* variant of NAT1, which is associated with high acetylation activity in colon tissue, may be a risk factor. However, the results of studies using genotyping alone have been contradictory, especially when the studies in question are small, and meta-analysis of 40 studies has not indicated any clear link between acetylation genotypes and colorectal cancer susceptibility.

The difference between aromatic amine carcinogenesis, where the rapid acetylator phenotype confers protection, and heterocyclic amine carcinogenesis, where the rapid acetylator phenotype is, if anything, a risk factor, is a function of the chemistry of the two classes of amines. Heterocyclic amine carcinogens exhibit steric hindrance of the exocyclic amine moiety. This makes them resistant to *N*-acetylation and means that they are more likely to undergo *O*- but not *N*-acetylation, whereas aromatic amine carcinogens can undergo both reactions.

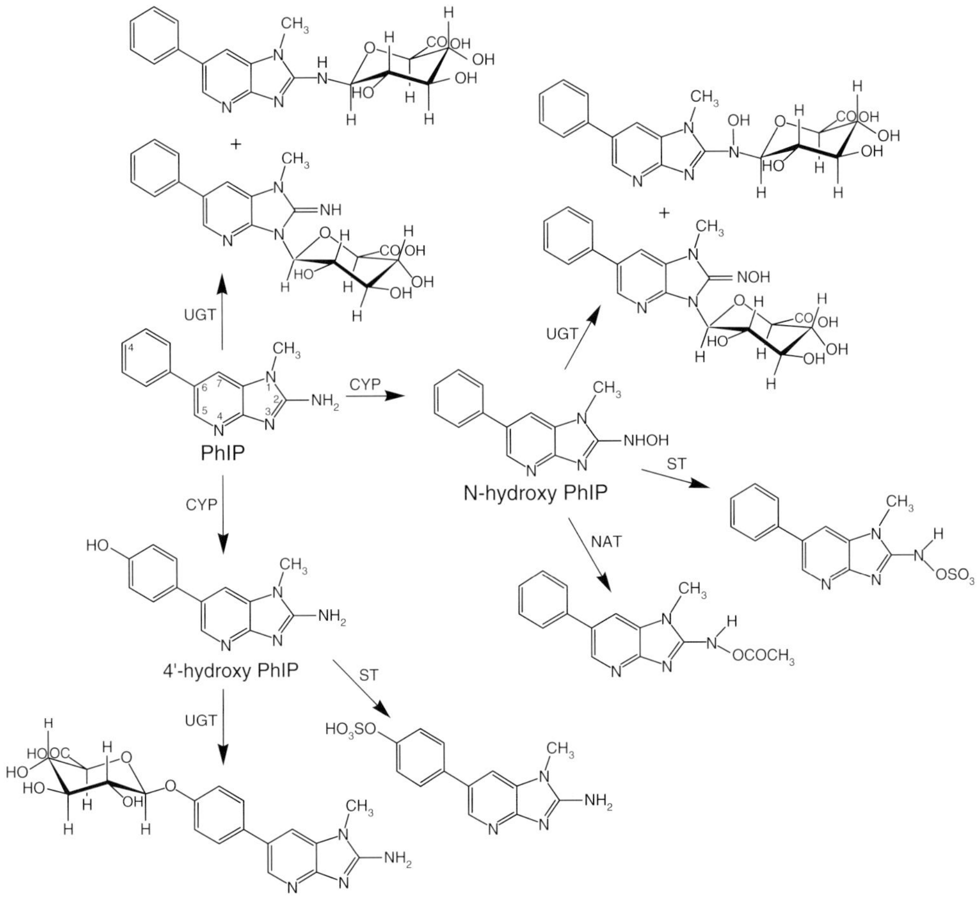

Figure 2.54 Pathways of heterocyclic amine metabolism. Reproduced from Gooderham, N.J., Murray, S., Lynch, A.M., et al. (2001). Food-derived heterocyclic amine mutagens: variable metabolism and significance to humans. Drug Metab Dispos **29**: 529–534 with the kind permission of Professor Nigel Gooderham and the publisher.

> It is evident that the determination of colorectal cancer susceptibility is multifactorial, involving both genetic and environmental factors. There is some evidence that the risk of colorectal cancer is increased in individuals who are current or ex-smokers, prefer red meat well done and have the rapid phenotypes of both CYP1A2 and NAT2. There is also evidence that GSTM1 and GSTT1 null genotypes confer additional risk for colorectal cancer.

2.5.5 Study Numbers and Effect Size

> The confusion over the relative roles of different polymorphic xenobiotic-metabolising enzymes in modulating susceptibility to colorectal cancer illustrates the importance of

> **adequate study size in toxicogenetic studies. The majority of studies on polymorphic variation and risk of toxicity or disease have, for scientific and resource reasons, been quite small (a few hundred subjects, at most) and have concentrated on one or two key genes. In case-control studies the magnitude of the effect is usually expressed as an odds ratio; if the odds ratio deviates from 1.0, the risk is either increased (odds ratio > 1) or decreased (odds ratio < 1). The multifactorial nature of common diseases means that odds ratios for specific variants are usually low, in the range of 1.5 to 3.0; large studies and/or meta-analysis are required to reveal such small increases in risk. The best approach is now to concentrate effort on large studies such as the international project on Genetic Susceptibility to Environmental Carcinogens (GSEC), which considers tens of thousands rather than hundreds of cases and controls.**

Redundancy within xenobiotic-metabolising pathways means that, if the activity of one enzyme is reduced, potential toxic agents may be directed down other pathways. For example, in smokers, various enzymes, including GST variations M, T and P (GSTM, GSTT, GSTP, see *Disturbance of the Redox System*, this chapter), NAT1 and NAT2, may exert combined effects on DNA adduct formation and cancer susceptibility. In addition, it is now clear that metabolism is not the only phase of drug disposition which is subject to polymorphic variation: polymorphisms in nuclear receptors and drug transporters probably also exert toxicogenetic effects, and our understanding of how these genes determine susceptibility to toxic agents is much less advanced than in the case of xenobiotic-metabolising enzymes, partly because the size of the epidemiological studies required in order to understand the roles of all the possible polymorphic variants of these proteins and variation in drug response was considered prohibitive before about the year 2000.

Unlike so-called Mendelian diseases such as sickle cell anaemia and cystic fibrosis, in which alterations in a single gene explain the vast majority of occurrences, common diseases such as cancer are likely to be influenced by multiple genes, each with a relatively small effect, whose actions in concert with each other and with environmental influences cause clinical disease.

Methodological limitations meant that the epidemiology studies of the 1960s to the 1980s tended to focus on the postulated effects of polymorphisms in individual xenobiotic-metabolising enzymes on susceptibility to adverse effects such as cancer and many preliminary studies aimed at the identification of risk factors were relatively small, with only 100–300 cases and controls. This number is sufficient to detect common polymorphisms which double risk, but will not detect rare polymorphisms or those which cause less than a doubling in risk. Much larger studies and/or meta-analysis are required to reveal small increases in risk. It is also important to note that polymorphic variants may be present at different frequencies within ethnic groups, for example ~55% of Caucasians but only ~10% of Japanese individuals are NAT2 slow acetylators. This means that the implications of the identification of a particular variant may differ depending on the ethnic makeup of the exposed population. It may not always be possible to extrapolate risk assessments from one ethnic group to another.

The best approach is now considered to be to concentrate effort on large studies such as the international project on Genetic Susceptibility to Environmental Carcinogens (GSEC), which was initiated in 1996 with the aim of undertaking pooled data analysis on polymorphisms in genes known to affect metabolic susceptibility. Its aims were to quantify gene frequencies in healthy populations by ethnicity and geographical distribution and then to study the association between each gene frequency and cancer. By working together, the investigators hoped to be able to pool a large enough number of subjects to study the effects of multiple gene polymorphisms on cancer risk. Contributions were invited from investigators who had published case-control studies on Phase I and II polymorphisms up to June 1999. As of February 2008, it involved 185 investigators, 304 studies and 124,456 subjects (53,072 cancer cases and 71,384 controls) who had been genotyped for a range of polymorphic xenobiotic-metabolising enzymes. As well as providing information about the geographic and ethnic distribution of known polymorphisms in these genes, this provided a resource which could be used to evaluate the roles of multiple metabolic polymorphisms in disease susceptibility. The data are now being used to further our understanding of susceptibility to a range of neoplastic conditions, including colorectal cancer, lung cancer, oral and pharyngeal cancers, gastric cancer, bladder cancer and breast cancer. These studies have highlighted the difficulties involved in assessing the true consequences of variation in genes whose individual effects are of borderline significance, leading to the establishment of guidelines for the evaluation of cumulative epidemiological evidence.

Genome-wide Association Studies

Candidate gene approaches such as those which led to the identification of the roles of NAT2 in bladder cancer and GSTP1 in acrylonitrile toxicity are limited in that they only look at selected genes and so may miss other important toxicogenetic factors. The effects of individual polymorphisms are often subtle because susceptibility to environmentally-induced disease is multifactorial, being determined by numerous genetic factors as well as environmental exposures. With the advent of high throughput methods for SNP analysis it has become possible to look at thousands of markers across the human genome in so-called genome-wide association studies (GWASs) without making prior assumptions as to which markers might confer susceptibility.

> **The goal of a GWAS is to test the links between subtle changes in the DNA sequences of individuals within a population and a trait (or phenotype) by assaying the majority of common SNPs across the entire genome. Its advantage over the candidate gene approach is that all genes are considered equally, with no influence from preconceived ideas, while its disadvantage is that it does not provide functional information.**

A GWAS, also known as a whole genome association study, is an examination of genetic variation across the genome, designed to identify genetic associations with observable biological phenomena such as health conditions. In human studies, this might include traits such as blood pressure or weight, or susceptibility to specific diseases or conditions such as serious adverse drug reactions. The goal of a GWAS is to test the links between subtle differences in the DNA sequences of individuals within a population and a trait (or phenotype) by assaying the majority of common SNPs across the entire genome. Its advantage over the candidate gene approach is that all genes are considered equally, with no influence from preconceived ideas. This falls under the heading of functional genomics, defined as the study of relationships between particular genotypes and specific phenotypes (essentially genotype–phenotype correlation at the genomic level). The GWAS approach has recently been used to indentify some 15 loci associated with bladder cancer susceptibility and four genetic regions associated with colon cancer. However, the limitation of this approach in toxicogenomics lies in the attempt to associate particular SNPs with exposures, since prior exposure to toxic chemicals may not have been recorded, and there is no opportunity to conduct family studies because it is unethical to expose other family members to a toxic chemical in order to test an association.

The advantage of the GWAS approach is simultaneous, unbiased testing of millions of SNPs while its disadvantage is that it does not provide functional information for the implicated loci, so when interpreting a GWAS study it is important to ask two questions:

- Is any association detected real?
- Are the results actually of any use?

Collaborative Programmes

GWAS require large populations to retain adequate power to detect any significant effect. Over the last couple of decades, several international collaborations have been established in order to address this issue. This includes the establishment of biobanks to collect and store biological samples for ongoing and future research (Table 2.18).

Table 2.18 Collaborative programmes.

Project	Methodology	Aim	Results
SNP consortium, 1999–2001	SNP map of the human genome	Improving individual drug therapy	1.7 mio SNPs identified, 1.5 mio mapped
International HapMap Project, 2002–2009	Multi-country haplotype analysis	Prevention and treatment of 40 common diseases in 11 populations	30–50 individuals per population investigated (trios: 2 parents, 1 adult child)
1000 Genomes Project, 2008–2015	Largest public catalogue of human variation and genotype data	Characterisation of ≥95% of variants with a frequency of ≥1% in the five major human populations	HapMap project extended by 2500 individuals of 27 populations
Environmental Genome Project, extension of Human Genome Project, ongoing	SNP characterisation	Characterisation of toxicologically relevant functions like DNA repair, apoptosis, cell-cyclic control, drug metabolism	92,486 SNPs of 647 genes of possible relevance in 95 individuals
The UK Biobank, ongoing	Biological samples of 500,000 individuals aged 40–69 years investigated and stored for further studies	Characterisation of the role of genetic factors, lifestyle and environmental exposures in the major diseases	Data linked to health records and blood biochemistry Genotyping undertaken on all participants Data made available for health research in the public good

For details on the International HapMap and 1000 Genomes Projects see http://www.1000genomes.org/home; for the Environmental Genome Project, see http://egp.gs.washington.edu/; for the UK Biobank, see http://www.ukbiobank.ac.uk/.

2.5.6 Summary

> **Individual variation in responses to toxic chemicals is often caused by underlying genetic variation. Polymorphisms in xenobiotic metabolism and its regulation clearly contribute to susceptibility to environmental toxins and carcinogens. Genetic testing is becoming widely accepted but attitudes to it are often based on a simplistic view of what toxicogenetics can actually achieve. It is important to remember that it is the phenotype that determines function and will mediate any observed effects on susceptibility.**
>
> **The characterisation of SNPs and haplotypes is currently in progress; we do not yet have sufficient data for use in risk assessment, but significant developments are anticipated once the baseline data are all in.**

The emerging field of toxicogenetics promises to enhance our understanding of susceptibility to toxic agents in ways that could not have been imagined even a few years ago. Genetic testing is now becoming more widely accepted but attitudes to it are often based on a simplistic view of what toxicogenetics can actually achieve. For example, no current test can predict outcomes with 100% accuracy and rare variants may have large effects, but explain only a small proportion of variability within the population.

The availability of high-throughput PCR-based methods has made it possible to examine large numbers of samples for many different polymorphisms simultaneously, generating huge volumes of data that should be of great value in informing risk assessment as long as the limitations of the available methods are taken into account. Large population studies with multiple markers have the potential to explain a large proportion of variability, but smaller combined effects are harder to detect consistently. Whole genome sequencing for each individual may allow susceptibility to be predicted, at least to a limited extent, as well as making personalised therapy a realistic prospect, but it poses practical problems of cost-effectiveness and data handling for the healthcare provider as well as ethical and legal issues.

The strength of the toxicogenetic approach is that it makes it possible to examine susceptibility to xenobiotics in human populations, thus answering concerns relating to differences in susceptibility between humans and animals. However, the nature of toxicogenetic analysis, including the fact that, for ethical reasons, chemicals may not be administered to humans at known toxic doses and invasive methods may not be used to obtain relevant tissue samples, means that a degree of variability and uncertainty will always be present.

In conclusion, therefore, polymorphisms in xenobiotic metabolism and its regulation clearly contribute to susceptibility to environmental toxins and carcinogens. The effects of individual genes are often subtle but specific combinations of multiple genotypes, found in specific subpopulations, may have a marked effect on susceptibility. The current status of research is that the characterisation of SNPs, haplotypes and their distribution is currently in progress. We do not yet have sufficient data to allow this information to be used in the risk assessment process, but significant developments are anticipated once the baseline data are all in.

Further Reading

Abecasis, G.R., Altshuler, D., Auton, A., et al. (2010). A map of human genome variation from population-scale sequencing. *Nature* **467**(7319): 1061–1073.

Daly, A.K. (2004). Development of analytical technology in pharmacogenetic research. *Naunyn Schmiedebergs Arch Pharmacol* **369**(1): 133–140.

Economopoulos, K.P. and Sergentanis, T.N. (2010). GSTM1, GSTT1, GSTP1, GSTA1 and colorectal cancer risk: a comprehensive meta-analysis. *Eur J Cancer* **46**(9): 1617–1631.

Figueroa, J.D., Ye, Y., Siddiq, A., Garcia-Closas, M., et al. (2014). Genome-wide association study identifies multiple loci associated with bladder cancer risk. *Hum Mol Genet* **23**(5): 1387–1398.

Garcia-Closas, M., Malats, N., Silverman, D., et al. (2005). NAT2 slow acetylation, GSTM1 null genotype, and risk of bladder cancer: results from the Spanish Bladder Cancer Study and meta-analyses. *Lancet* **366**(9486): 649–659.

Huang, Y.F., Chiang, S.Y., Liou, S.H., et al. (2012). The modifying effect of CYP2E1, GST, and mEH genotypes on the formation of hemoglobin adducts of acrylamide and glycidamide in workers exposed to acrylamide. *Toxicol Lett* **215**(2): 92–99.

Ioannidis, J.P., Boffetta, P., Little, J., et al. (2008). Assessment of cumulative evidence on genetic associations: interim guidelines. *Int J Epidemiol* **37**(1): 120–132.

Manolio, T.A., Brooks, L.D. and Collins, F.S. (2008). A HapMap harvest of insights into the genetics of common disease. *J Clin Invest* **118**(5): 1590–1605.

Manolio, T.A., Weis, B.K., Cowie, C.C., et al. (2012). New models for large prospective studies: is there a better way? *Am J Epidemiol* **175**(9): 859–866.

Pesch, B., Gawrych, K., Rabstein, S., et al. (2013). *N*-acetyltransferase 2 phenotype, occupation, and bladder cancer risk: results from the EPIC cohort. *Cancer Epidemiol Biomarkers Prev* **22**(11): 2055–2065.

Peters, U., Jiao, S., Schumacher, F.R., et al. (2013). Identification of genetic susceptibility loci for colorectal tumors in a genome-wide meta-analysis. *Gastroenterology* **144**(4): 799–807 e724.

Qin, X. P., Zhou, Y., Chen, Y., et al. (2013). Glutathione *S*-transferase T1 gene polymorphism and colorectal cancer risk: an updated analysis. *Clin Res Hepatol Gastroenterol* **37**(6): 626–635.

Thorisson, G.A. and Stein, L.D. (2003) The SNP Consortium website: past, present and future. *Nucleic Acids Res* **31**(1): 124–127.

Wu, K., Wang, X., Xie, Z., Liu, Z. and Lu, Y. (2013a). *N*-acetyltransferase 1 polymorphism and bladder cancer susceptibility: a meta-analysis of epidemiological studies. *J Int Med Res* **41**(1): 31–37.

Wu, K., Wang, X., Xie, Z., Liu, Z. and Lu, Y. (2013b). Glutathione *S*-transferase P1 gene polymorphism and bladder cancer susceptibility: an updated analysis. *Mol Biol Rep* **40**(1): 687–695.

Zhang, L., Zhou, J., Wang, J., et al. (2012). Absence of association between *N*-acetyltransferase 2 acetylator status and colorectal cancer susceptibility: based on evidence from 40 studies. *PLoS One* **7**(3): e32425.

Zhao, Y., Chen, Z.X., Rewuti, A., et al. (2013). Quantitative assessment of the influence of cytochrome P450 1A2 gene polymorphism and colorectal cancer risk. *PLoS One* **8**(8): e71481.

2.6 Receptor-mediated Mechanisms

Jens Schlossmann and Franz Hofmann

"Dosis facit venenum"

2.6.1 Introduction

> **Receptors are components of an organism which bind molecules of diverse chemical structures. These molecules are ligands that activate or inhibit the receptor function and thereby elicit a physiological response. Ligands that activate a response are agonists; those that block the response are antagonists.**

The classical concept of receptors was developed by J.N. Langley and Paul Ehrlich. Receptors can be considered to be locations within the body with which drugs can react to provoke a biological response. Some drugs act in a generalized way without reacting with a specific biological site, e.g. osmotic diuretics increase urine flow by increasing the osmotic pressure within the kidney tubule, thereby preventing the reabsorption of water. Most drugs, however, act at localized sites to produce specific responses. It was recognized in the mid-19th century that there must be some relationship between their chemical structures and the responses that they induce. As early as 1878 J.N. Langley suggested that atropine and pilocarpine altered the secretion of saliva in dogs by reacting with physiological substances within the gland, and in 1905 in his report on the effects of nicotine and curare on muscles he described their locus of action as a "receptive substance". Independently, Paul Ehrlich, in his pioneering studies in immunology, suggested that the specificity of immunological interactions required an interaction of specific chemical structures to form antibodies. He formulated the sentence "Corpora non agunt nisi fixate" (Substances do not act unless bound).

Today we know that receptors are rather large proteins located at specific sites on or within cells, at which chemicals (agonists) react to produce responses. When other chemicals react at the receptor to inhibit the effect of the agonists we call then antagonists. The agonist is also often called the ligand. The receptor–ligand interaction follows the law of mass action and its kinetics are similar to the Michaelis–Menten equilibrium except that the products of the Michaelis–Menten type of interaction are metabolites whereas interactions of the agonist at the receptor usually do not result in a change of chemical structure of the agonist. The reversible binding of an inhibitor to the agonistic receptor site leads to competitive antagonism. The allosteric inhibition of the agonist action results in non-competitive inhibition. These effects on the receptor action can be measured and defined by a dose–response curve. Therefore, only agonists or antagonists in sufficient concentrations based on the dose–response curve can modulate receptor-mediated functions.

2.6.2 Ligand–Receptor Interactions

> **Classes of receptors are hormone, neurotransmitter, growth factor and cytokine receptors, ion channels, carriers, enzymes, transcription factors, structural proteins, lipids, mRNA or DNA. Most ligands bind to protein receptors. The specific binding of a ligand at its receptor is a prerequisite for its action and is based on the law of mass action. The interaction between a ligand and its receptor can trigger a cascade of events.**

The term receptor usually refers to receptors which are membrane bound, or cytosolic and respond to hormones, neurotransmitters, growth factors or cytokines. As suggested above, there are a few non-receptor-mediated mechanisms (e.g., acid–base neutralization, osmosis, high or low temperature, alteration of the redox potential of a cell, exhaustion of essential cellular molecules).

The ligand–receptor interaction is based on the law of mass action, which was first described by A.J. Clark. The kinetic analysis is derived from the proposal of Michaelis and Menten for the kinetics of enzymatic reactions. The traditional receptor theory was developed for the ligand binding to membrane-bound hormone and neurotransmitter receptors. These receptors were thought to be metabolically inactive. Therefore, this interaction was mathematically described as a bimolecular reaction:

$$L \text{ (ligand)} + R \text{ (receptor)} = L\text{–}R \text{ (ligand–receptor complex)}$$

Based on this theory, the receptor itself is inactive and ligand binding leads to the active conformation, which transduces the signal. The ligand binding is determined by binding curves from

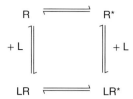

Figure 2.55 Two-state model of ligand–receptor (L–R) interaction.

which the amount of bound ligand at a given concentration can be calculated. Rearrangement of the above equation to

$$[L] \times [R]/[L-R] = k_{-1}/k_{+1} = K_D$$

yields the dissociation constant K_D, which defines the affinity of the ligand to the receptor. The value of the K_D is often between 0.1 and 1000 nM. A smaller K_D value represents a higher receptor affinity.

In the second half of the 20th century a receptor model evolved which provides a better understanding of ligand–receptor mechanisms. The basis of this model is an inactive conformation of the receptor R and an active conformation R* (Leff, 1995) (Figure 2.55).

The receptor itself changes between the R state and the R* state. Binding of the ligand to its receptor can induce different reactions:

a) The ligand binds with high affinity to the active conformation R* and shifts the equilibrium between R and R* towards R*. This increases receptor signaling and is characteristic of receptor binding by an agonist.
b) The ligand binds with identical affinity to R and R*. It does not change the existing equilibrium between inactive (R) and active (R*) receptors, but occupation of the binding site prevents binding of the endogenous ligand and, thereby, activation of receptor signaling. This type of binding is characteristic of an antagonist. (Note that some antagonists react covalently with, and inactivate, the receptor.)
c) The ligand binds with slightly higher affinity to the R* than the R conformation and prevents binding of the endogenous ligand. This type of binding is characteristic of a partial agonist.
d) The ligand binds with high affinity to the R conformation and shifts the equilibrium towards the R conformation. This type of binding is characteristic of an inverse agonist.

Thus, we can define the following types of ligands based on binding affinity: an agonist, an antagonist, a partial agonist, and an inverse agonist.

2.6.3 Biological Consequences of Ligand–Receptor Interactions

> **Toxic agents can act as agonists or antagonists of receptors. The effect of ligand binding on receptor activation is termed efficacy. Various types of agonists can be differentiated. Full agonists elicit a maximal tissue response upon receptor binding. In contrast, partial agonists elicit a submaximal response.**

Agonistic Effects

The receptor occupancy theory has been useful for helping to interpret the results of studies of ligand–receptor interactions. The theory suggests that the magnitude of the response is directly related to the number of receptor sites occupied. Thus, in the absence of ligand there is no response, when 50% of the receptors are occupied we observe 50% of maximal activity, and when all receptors are occupied there is maximal response. These responses over a range of doses gives rise to the dose–response curve in its various forms. Empirical examination of receptors often demonstrates that it is not necessary for all receptors to be occupied by agonists to elicit a maximal response, which suggests the presence of spare receptors.

The agonistic effect of a ligand is measured by the dose–response curve seen in Figure 2.56, which permits the measurement of the maximal response of the ligand, the dose needed to evoke 50% of a maximal response (ED_{50}), and the dose required to produce any other fractional response. In most situations, the ED_{50} is numerically equal to the K_a, the concentration constant that gives 50% of the maximal response. This value does not need to be identical with the K_D value (Figure 2.56). The availability of these values permits us to compare the potencies of different ligands. Some full agonists are capable of eliciting a maximal response at low occupancy of the available receptor binding sites because the investigated response is already maximal when, for example, 20% of the receptors are activated (see Figure 2.56). This finding usually is caused by an unopposed receptor signaling within a cell.

Antagonistic Effects

Various subclasses of antagonists can be distinguished by dose–response curves. The inhibitory concentration of an antagonist is standardized by an ID_{50} value or, if the receptor is known, by a K_i value. Antagonists which bind reversibly to the agonist binding site on R and R* are competitive antagonists. Raising the concentration of the agonist at a given concentration of a competitive antagonist restores the maximal tissue response because the agonist competes with the antagonist for the binding site of R*. In contrast, the maximal response is reduced by an antagonist that binds irreversibly to the agonist receptor interaction site. Furthermore, the maximal response of

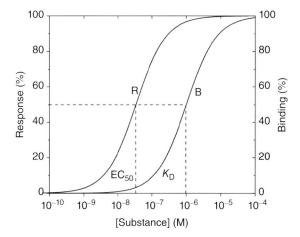

Figure 2.56 Dose–response curve demonstrating the correlation between the binding (B) and response (R) of a substance at 90% spare receptor action ($K_D = 1\ \mu M$; $EC_{50} = 50\ nM$). If the receptor is known, EC_{50} is replaced by K_a.

the agonist is also diminished by a non-competitive antagonist, which binds to a second site on the receptor that interacts allosterically with the agonist receptor interaction site.

> **Replacement of a physiological agonist by a competitor depends on its relative affinity to the receptor and its concentration. For example, replacement of the physiological ligand from the receptor by a compound of 1000-fold lower affinity requires a compound concentration that is 1000-fold higher than that of the physiological ligand. This demonstrates that information on the relative binding affinities of the compounds and their concentration in the organism are important criteria to evaluate possible effects of compounds.**

When investigating the dose–response relationship for toxicity it is possible to select from a variety of endpoints. Historically, determination of lethality involved treating groups of animals with different doses of a chemical, plotting the dose–response curve, as discussed above, and determining the dose which killed 50% of the animals, i.e. the LD_{50}. Although the method was highly accurate and precise for a given group of animals of the same strain, the data could not always be transferred to other strains of mice, rats, or other species. Furthermore, it required many animals. More practical methods are available today which provide a reasonable estimate of the lethality of chemicals and require few animals.

When studying drugs, it is often important to estimate the potential for adverse effects by comparing the effective dose range with the range of doses likely to cause some form of toxicity. In the past, a term called the therapeutic index was used and defined as the ratio LD_{50}/ED_{50} in animal studies. Obviously, the higher the value, the safer the drug. However, one must be certain that there is no overlap between the efficacy and lethality curves. An alternative would be to use the ratio LD_{01}/ED_{99} (or a value of LD as low as can be determined). In the clinical situation, toxicity endpoints such as nausea, rash, headache, etc., i.e. effects which may be unpleasant but not necessarily life threatening, are preferably used.

2.6.4 Receptor Signal Transduction

> **The receptors are sensing elements, which induce endogenous signal transduction pathways. They are divided into several classes:**
> 1. **G-protein coupled receptors**
> 2. **ion channels**
> 3. **enzymes**
> 4. **nuclear receptors/transcription factors (Figure 2.57).**

G-protein-coupled Receptors

G-protein-coupled receptors (GPCR) are heptahelical receptors located at the plasma membrane. They represent a large family of about 500 proteins and are expressed by about 5% of all invertebrate genes. Many ligands act via GPCRs. In addition to known and well-characterized GPCRs, a group of GPCRs exist that have no defined ligands and functions. These GPCRs are identified as orphan receptors. GPCRs can acts as monomers but increasing evidence supports the notion that homomers and heteromers of GPCRs have significant impact on GPCR signaling.

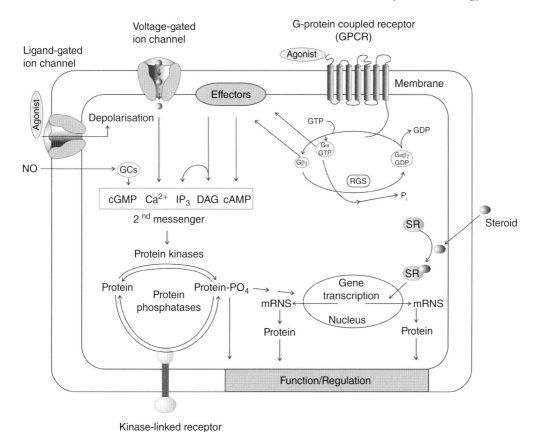

Figure 2.57 Summary of receptor-mediated cellular regulation by G-protein-coupled receptors (GPCR), voltage-gated ion channels, ligand-gated ion channels, guanylyl cyclases (GCs), kinase-linked receptors and steroid receptors (SR). NO, nitric oxide; cGMP, cyclic guanosine monophosphate; IP_3, inositol trisphosphate; DAG, diacylglycerol; cAMP, cyclic adenosine monophosphate; RGS, regulator of G-protein signaling.

Upon ligand binding, a trimeric G-protein is activated. G-proteins consist of three subunits, a Gα-subunit that binds and hydrolyses GTP, a β-subunit, and a γ-subunit (Wettschureck and Offermanns, 2005). The rate of GTP hydrolysis can be accelerated by RGS proteins (regulator of G-protein signaling). The G-proteins are separated into the different Gα-protein classes: Gα_s, Gα_i, Gα_q/α_{11} and Gα_{12}/α_{13}. Activation of these different G-proteins leads to diverse cellular signals. Gα_s stimulates adenylyl cyclase, which synthesizes the second messenger cyclic adenosine monophosphate (cAMP), Gα_i inhibits adenylyl cyclase, Gα_q/α_{11} stimulates phospholipase C and, thereby, the synthesis of the second messengers inositol trisphosphate (IP_3) and diacylglycerol (DAG), and Gα_{12}/α_{13} activates the small GTP-binding protein Rho.

Ion Channels

Ion channels comprise cation-selective or non-selective channels (e.g., for Na^+, Ca^{2+}, K^+) and anion channels (e.g. for Cl^-), which can be divided into ligand-gated and voltage-gated channels. For example, a ligand-gated channel is the nicotinic acetylcholine receptor which upon acetylcholine binding is permeable to sodium ions. The voltage-gated calcium channels are activated by depolarization. Ion channels are large proteins consisting of a channel protein with several trans-membrane segments and often of several additional subunits. The activity of these channels is altered by phosphorylation/dephosphorylation through kinases/phosphatases and by intracellular signals such as Ca^{2+} and ATP.

Enzymes as Receptors

Tyrosine- and serine/threonine-phosphorylation by kinases is a common theme in intracellular signaling cascades. Kinase-linked receptors mediate the actions of a variety of signaling molecules, including growth factors (e.g., epidermal growth factor, EGF), cytokines (e.g., erythropoietin), and hormones (e.g., insulin). An important class of these receptors are the tyrosine kinases (RTK). RTK are integrated into the plasma membrane, are often dimeric, and comprise a large extracellular ligand-binding and an intracellular effector domain. Intracellular autophosphorylated tyrosine allows the binding of the SH2 domain of an adaptor protein, which activates a signaling cascade, usually leading to gene transcription. Guanylyl cyclase-linked receptors synthesize the second messenger cyclic guanosine monophosphate (cGMP). They mediate the actions of nitric oxide (NO) and that of distinct peptides, e.g. the atrial natriuretic peptide.

Nuclear Receptors

These are intracellular proteins which act as transcription factors regulating nuclear-based transcription mechanisms. Their ligands include steroid hormones (e.g., gluco- and mineralo-corticoids and sex hormones), thyroid hormones, vitamin D and retinoic acid, and lipid lowering and anti-diabetic drugs. The nuclear receptors are assembled into three independent domains: the carboxy-terminal domain binds to the ligand, the central domain mediates the specific interaction with hormone-responsive elements in the nuclear DNA located upstream from the regulated gene, and the amino-terminal variable domain modulates the receptor function and allows the binding of co-transcription factors. The nuclear receptors act as dimers, but they are additionally modulated by other regulating molecules. Activation of the nuclear receptors by the different ligands induces or represses specific gene transcription and, thereby, regulates protein synthesis. The selection of the interacting co-transcription factor depends on the tissue and the conformation of the ligand-bound transcription factor. This conformation may vary with the ligand as best exemplified by the specific estrogen receptor modulators (SERM). The physiological effects of the nuclear receptors are delayed but occur within hours or days.

Recently, the efficacy of xenoestrogens (e.g., bisphenol A, *o,p'*-DDE, polychlorinated biphenyls or dieldrin) on estrogen receptors and its ecological effects were debated. *In vitro* tests showed that the potency of xenoestrogens compared to 17-β-estradiol differs by more than 1×10^{-6}. Therefore, it is very unlikely that xenoestrogens are able to induce estrogenic effects in humans. However, it is not established whether xenoestrogens act differently on the diverse estrogen receptors ERα or ERβ. Furthermore, xenoestrogens altered the expression of estrogen receptor ERα. Additive effects of xenoestrogens were also reported. In conclusion, mediation of toxic effects of xenoestrogens by occupancy of the estrogen receptors are unlikely, if extensive exposures are excluded.

The duration of a receptor signal is regulated at several levels. Desensitization mediated at the level of transcription, translation and degradation alters long-term receptor function. A short-term change of receptor activity is induced by phosphorylation through kinases and through endocytosis of membrane-bound receptors. Furthermore, subunits and cofactors of receptors are regulators of receptor function. Therefore, it is essential to define each receptor function in its given physiological background.

2.6.5 Summary

Receptors are components of an organism that bind molecules of diverse chemical structures. These molecules are ligands that activate or inhibit the receptor function and thereby elicit a physiological response. Ligands that activate a response are agonists; those that block the response are antagonists. The receptor–ligand interaction follows the law of mass action and its kinetic follows mostly the Michaelis–Menten equilibrium. The reversible binding of an inhibitor to the agonistic receptor site leads to competitive antagonism. The allosteric inhibition of the agonist action results in non-competitive inhibition. These effects on the receptor action can be measured and defined by a concentration–response curve. Therefore, only agonists or antagonists in sufficient concentrations based on the concentration–response curve can modulate receptor-mediated functions.

Further Reading

Albini A, Rosano C, Angelini G, Amaro A, Esposito AI, Maramotti S, Noonan DM, U. Pfeffer U. Exogenous hormonal regulation in breast cancer cells by phytoestrogens and endocrine disruptors. *Current Medicinal Chemistry* (2014) **21**, 1129–1145.

Catterall WA. Molecular properties of sodium and calcium channels. *Bioenerg Biomembr* (1996) **28**, 219–230.

CSTEE, Report of the Working Group on Endocrine Disrupters of the Scientific Committee on Toxicity, Ecotoxicity and the Environment 1999, 17–20.

Delfosse V, Grimaldi M, Cavaillè V, Patrick Balaguer P, William Bourguet W. Structural and functional profiling of environmental ligands for estrogen receptors. *Environ Health Perspect* (2014) **122**, 1306–1313.

González-Maeso J. Family a GPCR heteromers in animal models. *Front Pharmacol.* (2014) **5**, 226.

Gronemeyer H, Gustafsson JA, Laudet V. Principles for modulation of the nuclear receptor superfamily. *Nat Rev Drug Discov* (2004) **3**, 950–964.

La Rosa P, Pellegrini M, Totta P, Acconcia F, Maria Marino M. Xenoestrogens alter estrogen receptor (ER) a intracellular levels. *PLOSOne* (2014) **9**, e88961.

Leff P. The two-state model of receptor activation. *Trends Pharmacol Sci* (1995) **16**, 89–97.

NTP-CERHR Monograph on the Potential Human Reproductive and Developmental Effects of Bisphenol A. (2008) NIH Publication No. 08 – 5994: Appendix A, p40f.

Rasmussen TH, Nielsen F, Andersen HR, Nielsen JB, Weihe P, Grandjean P. Assessment of xenoestrogenic exposure by a biomarker approach: application of the E-Screen bioassay to determine estrogenic response of serum extracts Environmental Health: A Global Access Science Source *(*2003) **2**, 12.

Routledge EJ, White R, Parker MG, Sumpter JP. Differential effects of xenoestrogens on coactivator recruitment by estrogen receptor (ER) and ERb. *J Biol Chem* (2000) **275**, 35986–35993.

Schlessinger J. Cell signaling by receptor tyrosine kinases *Cell* (2000) **103**, 211–125.

Schlossmann J, Feil R, Hofmann F. Insights into cGMP signalling derived from cGMP kinase knockout mice. *Front Biosci* (2005) **10**, 1279–1289.

Wettschureck N, Offermanns S: Mammalian G proteins and their cell type specific functions. *Physiol Rev* (2005) **85**, 1159–1204.

Xue L, Rovira X, Scholler P, Zhao H, Liu J, Pin J, Rondard P. Major ligand-induced rearrangement of the heptahelical domain interface in a GPCR dimer. *Nat Chem Biol* (2015) **11**(2), 134–140.

2.7 Mixtures and Combinations of Chemicals

Hermann Bolt[1]

2.7.1 Introduction

> **Toxicity testing and risk assessment have traditionally focused on single chemicals. However, humans are exposed to complex and ever-changing mixtures and combinations of chemicals in the air they breathe, the water and beverages they drink, the food they eat, the surfaces they touch and the consumer products they use. Insight into the health consequences of such complex exposures is urgently needed to bring the safety evaluation of chemicals and real life into harmony. To this end a conceptual framework that defines three types of joint action (viz. dissimilar and similar joint action, and interaction) has been introduced.**

In real life, humans are exposed simultaneously or sequentially to large numbers of chemicals via multiple routes. Given this reality, insight into the health consequences of such complex exposures is needed to judge whether current approaches to the safety evaluation of chemicals offer adequate protection to humans. A key question concerning the toxicity of a chemical mixture is whether the toxic effects of a mixture can be predicted on the basis of the toxic effects and underlying mechanisms of toxicity of the individual components. To answer this question a conceptual framework that defines three types (modes) of joint action (viz. dissimilar and similar joint action, and interaction) has been introduced. In addition to information on the toxicity and mechanism of action of the individual chemicals in a mixture, knowledge about the mode of joint action of the chemicals is often critically important for properly assessing the safety (the potential health risk) of a mixture.

The consistent use of terms that describe mixed exposures is also a key issue for understanding the toxicological consequences of such exposures.

In this chapter, the various types of mixed exposures and joint action are defined, designs for mixture toxicity studies are briefly described, and methods for the safety evaluation of mixed exposures are discussed.[2]

2.7.2 Types of Mixed Exposures

> **In the absence of internationally harmonized terminology, types of mixed exposures should always be defined. The major types are mixture, simple mixture, complex mixture, combination of chemicals, specified combination of chemicals and cumulative exposure. These terms are described in this section.**

A *mixture* is any combination of chemicals characterized by simultaneity of exposure to the constituents as a result of their joint occurrence. A *simple mixture* is a mixture consisting of a

---

[1] This chapter is based on the contribution by V.J. Feron and D. Jonker in the previous edition.

[2] Combined exposure to initiators (mutagens), promotors, converters, and/or co-carcinogens in chemical carcinogenesis, as well as sequential exposure to toxicants and antidotes, are not discussed in this chapter because such combined exposures are specific areas in toxicology that have been extensively studied and are hardly recognized as examples of combination toxicology.

relatively small number of chemicals, say ten or less, the composition of which is qualitatively and quantitatively known (e.g., a cocktail of three pesticides or a combination of two medicines taken simultaneously). A *complex mixture* is a mixture that consists of tens, hundreds, or thousands of chemicals, and its composition is qualitatively and quantitatively not fully known (e.g., diesel exhaust, welding fume, drinking water).

A *combination* is any combination of chemicals regardless of whether or not they occur as a mixture. In a *specified combination* all components are known. An example of a specified four-component combination is exposure to the air contaminants sulphur dioxide and formaldehyde, the food additive butyl hydroxyl toluene and the pain-killer acetyl salicylic acid (aspirin). This example involves various independent exposures, via different routes, which may, but do not necessarily fully, overlap in time.

Cumulative exposure refers to exposure to multiple chemicals and the cumulative risk from such exposures.

2.7.3 Types of Joint Actions and their Role in Safety Evaluation

> **Three types of joint (or combined) action of chemicals in a mixture or combination have been defined: dissimilar joint action, similar joint action and interaction. Dissimilar and similar joint action are non-interactive: the chemicals in the mixture or combination do not affect each others' toxicity. With dissimilar joint action the chemicals in the mixture act independently. With similar joint action the chemicals in the mixture induce similar effects because they act similarly regarding primary physiological processes. With interaction one chemical influences the toxicity of one or more other chemicals by direct chemical–chemical interaction (a "new" chemical is formed), toxicokinetic interaction (alteration in metabolism or disposition of chemicals) or toxicodynamic interaction (interactive effects at target sites). The type of joint action in combination with the margin of safety (the margin between the actual exposure level and the no observed adverse effect level) of the individual chemicals in the mixture determines to a high degree whether or not exposure to a mixture is of health concern.**

The toxicity of mixed exposures can be estimated when the toxicity of the individual chemicals as well as types of joint action of the constituents are known.

Joint (or combined) action describes any outcome of exposure to multiple chemicals, regardless of the source or spatial or temporal proximity. Three different types of joint action are distinguished:

Dissimilar joint action (or independent joint action, or simple independent joint action, or response addition) is non-interactive, namely, the chemicals in the mixture do not affect each others' toxicity. In other words, the chemicals act independently, so that the body's response to one of the chemicals is the same regardless of whether or not the other chemicals are present. The modes of action and probably the nature and the site of the toxic effects differ among the chemicals in the mixture. An important characteristic of independent joint action is that a chemical is assumed not to contribute to the (adverse) effect of the mixture when that chemical is present at a level below its own individual (adverse) effect threshold. For example, in a 4-week toxicity study, rats were exposed to a combination of eight different chemicals (sodium metabisulphite, Mirex, Loperamide, metaldehyde, di-*n*-octyltin dichloride, stannous chloride, lysinoalanine and potassium nitrite) that were arbitrarily chosen regarding target organ and mechanism of action. The study produced no (convincing) evidence for increased risk from exposure to the combination when each chemical was administered at or below its own no observed adverse effect level (NOAEL).

Similar joint action (or simple similar action, or dose addition, or concentration addition) is also non-interactive. The chemicals induce similar effects because they act similarly in terms of mode of action and primary physiological processes (uptake, metabolism, distribution, elimination). They differ only in their potency. The toxicity of the mixture can be calculated using summation of the doses of each individual chemical after adjusting for the differences in potency. The addition of doses implies that toxicity can be expected if the summed dose is high enough to exceed the threshold of toxicity of the mixture, even when the dose level of each individual chemical is below its own effect threshold. For example, in a 4-week toxicity study, rats were exposed to a combination of four different but similarly acting nephrotoxicants (tetrachoroethylene, trichloroethylene, hexachloro-1:3-butadiene and 1,1,2-trichloro-3,3,3-trifluoropropene). Kidney effects of the mixture were seen at dose levels not showing renal toxicity of the individual compounds. Thus, the study provided support for the assumption of dose additivity for mixtures of similarly acting systemic toxicants under conditions of concurrent, repeated exposure at dose levels below the toxicity thresholds of the individual constituents.

Interaction is characterized by one chemical influencing the toxicity of one or more other chemicals in the mixture, resulting in an effect/response to the mixture that deviates from that expected from (dose or response) addition. Interactive joint actions can be less than additive (commonly termed antagonistic, inhibitive, infra- or subadditive) or greater than additive (synergistic, potentiating, supra-additive). Antagonistic and synergistic interaction indicate in which direction a response to a mixture differs from what is expected under the assumption of additivity.

An interaction may be based on direct chemical–chemical interaction, toxicokinetic interaction or toxicodynamic interaction.

Direct chemical–chemical interaction means that one chemical directly reacts with another, resulting in a "new" chemical that might be more or less toxic than the original chemicals. In cases of decreased toxicity, this mechanism is one of the common principles of antidote treatment. An example of increased toxicity is the formation of carcinogenic nitrosamines in the stomach through the reaction of non-carcinogenic nitrite (from food or drinking water) with secondary amines (e.g., from fish proteins).

Toxicokinetic interactions involve alterations in the metabolism (biotransformation) or disposition of a chemical. These alterations are often divided in changes in absorption, distribution, metabolism, and excretion. Essentially, toxicokinetic interactions alter the amount of the toxic agent(s) reaching the cellular target site(s) without qualitatively affecting the toxicant–receptor site interaction. With respect to their potential toxicological consequences at low doses, interactions in the process of metabolism (enzyme induction or inhibition) are considered most relevant. The toxicological impact of an alteration in absorption, distribution or excretion is expected to be small at low dose levels. *Toxicodynamic interactions* involve interactive effects occurring at or among target sites. Interactions at the same site resulting in antagonism have been described as receptor antagonism (e.g., the antagonistic effect of oxygen on carbon monoxide). Interactions resulting from different chemicals acting on different sites and causing opposite effects have been described as functional antagonism (e.g., opposing effects of histamine and noradrenaline on vasodilatation and blood pressure). Another type of toxicodynamic interaction is alteration of a tissue's response or susceptibility to toxic injury (e.g., immunomodulation, depletion of protective factors, and changes in tissue repair or haemodynamics).

Finally, an example of synergistic interaction is combined exposure to non-effective doses of 2,3,7,8-tetrachlorodibenzo-*p*-dioxin (TCDD) and the synthetic glucocorticoid hydrocortisone (HC) induced cleft palate, an irreversible structural change, in all exposed mouse embryos. This marked synergistic interaction was explained by mutual upregulation of the receptors which mediate the biological activity of TCDD and HC (the aryl hydrocarbon receptor and the glucocorticoid receptor, respectively).

2.7.4 Role in Safety Evaluation

Approaches for the safety evaluation (risk assessment) of mixtures and combinations of chemicals will be discussed in detail in the last section of this chapter. Nevertheless, the consequences of using the aforementioned types of joint action as guidance in the process of the safety evaluation are briefly discussed here because insight into these consequences is also of relevance for designing mixture toxicity studies.

Whether a mixed exposure constitutes a safety concern mainly depends on two key factors: the type of joint action (similar or dissimilar joint action, or synergistic or antagonistic interaction) and the margin of safety (the margin between the actual exposure level and the NOAEL) of the individual constituents of the mixed exposure. Moreover, experience has shown that *as a rule* exposure to mixtures at *non-toxic exposure levels* of the individual chemicals in the mixture is of no health concern. However, there are exceptions to this rule. When a mixture consists of chemicals with a *similar mode of action* (and the same target organ) *dose addition* is to be expected and should be taken into account in the safety evaluation of such a mixture. Furthermore, since *synergistic interaction* can never be fully excluded, its likelihood of occurrence (to be estimated on the basis of data on the mechanism of toxicity of the individual chemicals) and its toxicological relevance in the light of *actual exposure levels* should be estimated on a case-by-case basis. If a case of potential health risk emerges, a wide variety of methods for hazard identification and risk assessment is available. Mixtures with very small (or no) margins of safety for the individual chemicals (e.g., heavily polluted outdoor air containing ozone, nitrogen dioxide, sulphur dioxide, ammonia, and ultrafine particles with the lung as a common target organ) are to be regarded as priority mixtures because adverse effects due to combined exposure are likely to occur. When, on the other hand, the margins of safety for the individual chemicals of a mixture are large, say a factor of 5 or 10 or greater, additive or interactive adverse effects are unlikely to occur, and thus such mixtures hardly need any further consideration.

Designs for Toxicity Studies of Mixtures or Combinations

> **Factors that influence study design include the number of chemicals in a mixture or combination, the availability of a mixture for testing it in its entirety, the extent to which the toxicity of a mixture or combination needs to be characterized in terms of dose–response relationship or departure from additivity, and available resources. Major designs include whole-mixture studies, full factorial designs, fractional factorial designs, the additivity approach, the isobolographic method, effect/response-surface analysis, and mechanistic methods including physiologically-based toxicokinetic modelling. In this section, the various designs are briefly discussed, including their strengths and limitations.**

Given the diversity of exposure scenarios, the questions to be addressed in toxicity studies of mixtures and combinations of chemicals vary widely and consequently there is no "one size fits all" design. Factors that influence study design and methods to analyse the results include the number of chemicals in a mixture or combination, the availability of a mixture for testing it in its entirety, the extent to which the toxicity of a mixture or combination needs to be characterized in terms of dose–response relationship or departure from additivity, and available resources.

Major designs include whole-mixture studies, full factorial designs, fractional factorial designs, the additivity approach, the isobolographic method, effect/response-surface analysis, and mechanistic methods, including physiologically-based toxicokinetic modelling.

In a *whole-mixture study*, a given mixture or combination is examined as a whole; the mixture may be viewed as a "single chemical". This approach has been used for complex mixtures with poorly characterized but stable composition (e.g., industrial effluents, welding fumes) and for specially designed mixtures with completely characterized composition. Obviously, whole-mixture studies do not provide information about possible interactions between components of the mixture, nor do they enable identification of the most toxic components or fractions.

A *full factorial design* is a design in which each dose level (concentration) of each chemical is combined with each dose level (concentration) of every other chemical in the mixture or combination. The number of groups in a factorial design is m^k where k is the number of chemicals and m is the number of dose levels (concentrations) of each chemical. The simplest form of a factorial design is a *2 × 2 design*, which measures the effects/responses to the control situation (concentration zero for both chemicals) to one dose level of each of two chemicals and to the same dose levels of these two chemicals combined. This design can provide, for example, evidence that the chemicals interact antagonistically (when the response to the two chemicals combined is smaller than that to the individual chemicals). The number of treatment groups in a full factorial design increases so drastically with the number of chemicals that such a design is unfeasible when a mixture or combination contains more than four or five chemicals, especially for costly studies in experimental animals. However, the number of groups can be limited by considering only part of the possible combinations by using a so-called *fractional factorial design.* A illustrative example of the use of such a design is the application of a fractional two-level factorial design to examine the toxicity of a combination of nine different chemicals (widely varying in chemical nature) in a 4-week toxicity study in rats. In this study, a full factorial design would have required 2^9 (= 512) treatment groups; instead, the 1/32 fraction (=16 groups exposed to various carefully selected combinations of the individual chemicals) plus four groups exposed to different dose levels of the entire combination were used. The investigators were able to identify cases of non-additivity as well as the components responsible for adverse effects of the combination on specific endpoints.

The *additivity approach* uses the dose–response curves of the individual chemicals to calculate the expected response for a given combination (or mixture), under the assumption of additivity. Then the predicted response is compared with the experimentally obtained response to that combination. For a combination of k chemicals, the number of treatment groups would amount to $(m \times k) + 1$ (m dose levels of k chemicals + the combination) which is much less than the m^k groups required for a full factorial design. A disadvantage of this approach is that it can only detect whether the response to the whole combination or mixture deviates from additivity; it cannot identify the chemicals which cause the interaction.

The *isobolographic method* is the classical approach to determine whether or not two chemicals interact. An isobologram is a graphical representation of the joint effect of two chemicals in which the doses of the chemicals A and B are given on the x and y axis, respectively, and the experimentally determined dose combinations of these two chemicals, which all elicit the same effect, are plotted and then connected by a line, the *iso-effect line* or the *isobole* (Figure 2.58). The experimentally determined line is then compared with the theoretical iso-effect line based on the assumption of additivity. Differences between these lines indicate departure from additivity.

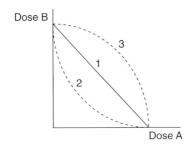

Figure 2.58 Isoboles of two chemicals indicating different types of joint action. On the axes are the doses of chemicals A and B that produce an equal effect (e.g., 60% increase in liver weight relative to body weight). Isobole 1 represents dose additivity, isobole 2 supra-additivity or synergism, and isobole 3 infra-additivity or antagonism.

Isoboles below the line of additivity indicate synergism (in the presence of chemical A, less of chemical B is required to generate the specified response than would be the case under the assumption of additivity), those above it indicate antagonism. A major disadvantage of the isobolographic method is its extensive data demand. Also, because of its graphical nature and the use of perpendicular axes, isobolograms are unsuitable for mixtures of more than three chemicals.

Effect/response surface-analysis yields a statistically based mathematical relation between the doses of each chemical in a mixture and the effect parameter. This method is much easier to perform but conceptually more difficult to interpret than isobolographic methods. The advantage of effect/response-surface analysis is that it includes all data points obtained and does not require a dose-effect equation for each individual chemical per se. In a study on the cytotoxicity to nasal epithelial cells of mixtures of aldehydes (formaladehyde, acrolein, and crotonaldehyde), effect-surface analysis revealed that two- and three-factor interactions were significantly different from zero, although the magnitude of the interaction was relatively small compared with the effect of single aldehydes. Besides the above empirical methods, *mechanistic methods* can be used to study the toxicity of mixtures or combinations of chemicals. For example, physiologically-based toxicokinetic modeling approaches are potentially useful for predicting quantitatively the consequences of interactions, and to conduct extrapolations (high to low dose, route to route, binary to multichemical mixtures) of the occurrence and magnitude of interactions from laboratory *in vitro* or *in vivo* studies to human real-life exposure scenarios.

Safety Evaluation

> **The safety of a mixture (or combination) of chemicals can be evaluated by approaching the mixture as:**
>
> - a single entity
> - a number of fractions
> - a number of individual constituents.
>
> The safety evaluation based on the individual constituents can be approached by considering :
>
> - all constituents
> - the "top *n*" constituents
> - the "pseudo top *n*" constituents.

154 Toxicology and Risk Assessment

Then, depending on the type of joint action of the chemicals in a mixture (or combination), the safety evaluation can be performed using:

- the dose additivity/Hazard Index method
- the response additivity/Hazard Index method, and/or
- the Hazard Index/weight-of-evidence method.

The safety of a mixture (or combination) of chemicals can be evaluated by approaching the mixture as a single entity, as a number of fractions or as a number of individual constituents (Figure 2.59).

Mixture as a Single Entity Once the NOAEL of a mixture has been established, the possibility of recommending an exposure limit can be examined, using methods similar to those for individual substances. It is common practice to accept toxicity data on a mixture as an entity, although the intention may not always be to recommend exposure limits and to set standards. If a mixture has not been examined toxicologically but another mixture with a reasonably similar composition has, one may consider basing the safety evaluation on data for the similar mixture. In the simplest situation, the data can be adopted directly for the mixture concerned. In more complex situations, sophisticated techniques (such as pattern recognition, principle component analysis, and partial

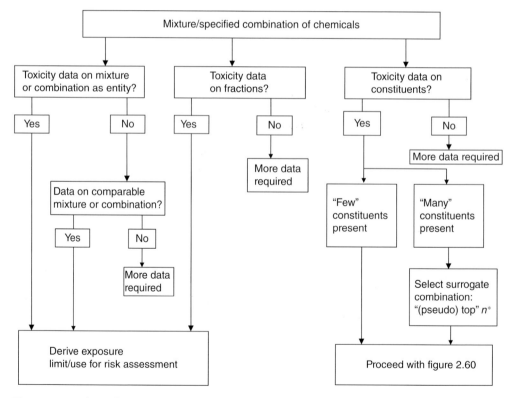

Figure 2.59 *Flow chart for the safety evaluation of mixtures (or combinations) of chemicals. *"(pseudo) top n": see text for explanation.*

least square projections to latent structures) can be used to compare mixtures physico-chemically and toxicologically. Such techniques, for example, have been used to compare the dioxin composition of sludge and to determine the toxicity of diesel exhaust fumes.

Mixture as a Number of Fractions If a mixture cannot be tested in its entirety, it may be possible to divide it into fractions according to its chemical or physical properties, and to study the toxicity of these fractions. The data obtained may be useful for risk assessment and there is also, at least theoretically, still the option of calculating an exposure limit through conversion and extrapolation of the NOAEL for the most harmful fraction. However, when taking this approach, one needs to be prepared for changes in toxicity due to chemical changes that fractionation may cause. Fractionation has been successfully applied to diesel exhaust fumes; the particulate matter fraction appeared to be entirely responsible for the respiratory tract inflammation and cancer discovered in animal studies, and neither the free constituents nor the gaseous constituents adsorbed to particles contributed to these effects.

Mixture as a Number of Individual Constituents The third possibility to evaluate the safety of a mixture (or combination) of chemicals is to take the individual constituents as the starting point. Depending on the number of chemicals present in a mixture, the safety evaluation can be approached by considering all constituents, the "top n" constituents or the "pseudo top n" constituents.

For mixtures (or combinations) consisting of a small number of chemicals, say ten or less, the safety evaluation should include *all individual constituents*; they should all be considered in detail (see the section *Type of Joint Action as the Basis for Safety Evaluation*). However, for mixtures or combinations consisting of a large or very large number of constituents the evaluation has to be reduced to manageable proportions. In such cases, the evaluation should focus on the most risky chemicals, viz. the chemicals that constitute the highest health risks based on toxic potential and (estimated) exposure. The joint toxicity of the most risky chemicals is assumed to represent that of the entire mixture or combination, or at least to closely approximate it.

At what number of constituents does the safety evaluation require a surrogate mixture or combination? Ten has been suggested because risk assessment was considered to become too difficult and too complex above this number. However, it is more appropriate to decide about the precise number case by case and therefore n has been introduced. Above n, a surrogate combination (of no more then n constituents) has to be selected and assessed for combined toxicity. In case of a not overly complex situation, it is best to select the "top n". In case of "very many" constituents, the "pseudo top n" is the best option: first placing each constituent in a (chemically or physiologically) related class followed by identification of the n most risky classes. Then for each class a representative chemical (a true chemical or a pseudo-chemical representing the fictional average of a class) is identified. Once the "top n" or "pseudo top n" constituents have been identified, they will be approached as a simple (defined) mixture or combination. Toxicity testing should first involve two or three chemicals and later deal with a gradually increasing number and thus a gradually increasing n. The combination issue must be realistic and the evaluation must extend to determining which health-protection measures are advisable. Working in this way, the number n can be determined, incorporating both feasibility and scientific soundness. Experience has shown that $n = 1$ is sometimes sufficient: in the case of polycyclic hydrocarbons, benzo[*a*]pyrene could represent the total. A similar approach has successfully been used for more then a decade (in the United States) to identify the priority substances (the "top n" or "pseudo top n" chemicals) that are released from hazardous waste sites. A slightly different, though in principle comparable, procedure has recently been developed for the safety evaluation of natural flavour complexes

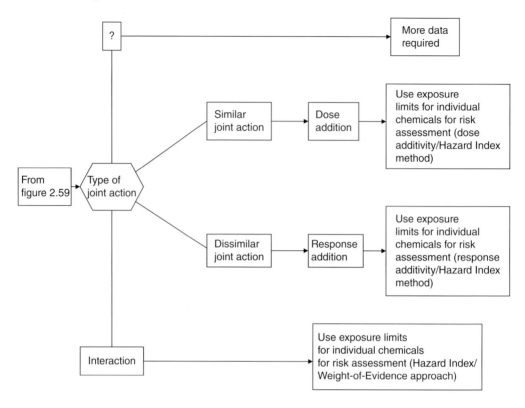

Figure 2.60 Flow chart for the constituent-based safety evaluation based on type of joint action.

obtained from botanical sources and intended to be used as flavouring substances for foods and beverages.

Type of Joint Action as the Basis for Safety Evaluation Figure 2.60 shows a flow chart for the safety evaluation based on individual constituents and type of joint action. The following question has to be answered: is the type of joint action expected to be similar or dissimilar, or interaction? If this question cannot be answered more data are required to perform a safety evaluation.

In the case of *similar joint action* dose addition can determine the effect of the combined exposure using empirically obtained exposure-effect data. To assess the potential health risk of such combined exposures the dose additivity/Hazard Index (HI) method should be used. This method involves summation of the ratios of the actual exposure concentration to the established exposure limit value (hazard quotients) for each individual chemical. If the sum exceeds unity, the combined exposure is considered of health concern. For example, an atmosphere contains 400 ppm acetone (exposure limit is 500 ppm) and 150 ppm sec-butyl acetate (exposure limit is 200 ppm). Both chemicals are central nervous system depressants. Application of the dose additivity/HI method shows that the sum exposure exceeds 1 ($400/500 + 150/200 = 0.8 + 0.75 = 1.55$), indicating an exposure situation of concern.

Another method based on dose additivity is the toxic equivalency factor (TEF) method, which was developed as a procedure to assess the toxicity of complex mixtures of polychlorinated dibenzo-*p*-dioxins and dibenzofurans. The TEF method assumes a common mode of action. In this method, the dose of each component is normalized against the dose of one of the components, usually the most potent and best-studied one (called the index chemical), to derive a relative potency for each component. The relative potencies are then summed to estimate the toxicity of the combination or mixture.

In the case of *dissimilar joint action*, response addition is the designated method for determining the combined effect. As long as the various exposure concentrations do not exceed the respective exposure limits, there is adequate protection. For example, an atmosphere contains 0.04 mg/m^3 of the neurotoxicant lead (exposure limit is 0.05 mg/m^3) and 0.9 mg/m^3 of the irritant sulphuric acid (exposure limit is 1 mg/m^3). Application of the response additivity/HI method shows that the exposure situation is non-hazardous because the HI (the hazard quotient) of both substances is smaller than 1. viz. 0.8 (0.04/0.05) for lead and 0.9 (0.9/1) for sulphuric acid. Only if at least one of the components has a hazard quotient that exceeds unity is exposure to the combination of health concern. Indeed, here the exposure limits of the individual chemicals constitute the reference point. There is one (important) exception: response addition in the case of substances of which the effect has no zero risk exposure level (e.g., genotoxic carcinogens). If the effect measure is an incidence (or likelihood) and the substances act dissimilarly without interaction but have similar health effects (e.g., different types of cancer), the combined effect can be calculated through response addition. In that case, cancer (not the type of cancer) is the reference point, and combined exposure to the two substances, each below its own recommended exposure limit, may exceed the acceptable cancer risk level. If indeed the acceptable cancer risk level is exceeded, the exposure concentrations of the individual chemicals (carcinogens) must be reduced to maintain the level of protection.

In the case of *interaction* it may or may not be possible to draw (semi-)quantitative conclusions depending on the data. Since neither the dose additivity/HI method nor the response additivity/HI method takes into account interactions, these methods cannot be used as such. A method that takes into account both possible synergistic or antagonistic interactions is the so-called HI/weight-of-evidence (HI/WoE) method. With this method, a WoE classification is used to estimate the joint actions (additivity, antagonism, synergism) for all pairs of chemicals in a simple mixture or specified combination, based on information for the individual chemicals:

$$HI_A = HI \times UF_I^{WoE}$$

where HI_A is the HI adjusted for the uncertainty for interactions, UF_I is a "fixed" (chosen) uncertainty factor for interactions, for example 10, and WoE is the weight-of-evidence score (ranging from -1 for the highest possible confidence in significant antagonistic interaction to $+1$ for the highest possible confidence in significant synergistic interaction, with 0 as full confidence in the absence of any interaction). In the WoE score, several weighting factors are taken into account, such as mechanistic understanding and toxicological significance of the binary interactions, route, duration and sequence of exposure, and whether data are from *in vitro* or *in vivo* studies. The better the data set on the individual chemicals, the more precisely the joint action of the combined exposure can be predicted.

New Developments

Comparisons of observed and predicted responses of polycyclic aromatic hydrocarbons (as a model of a complex mixture) under different mathematical models have indicated that joint action may over-estimate mixture responses compared to what is actually observed. This points to a need for further refinement of current procedures. One new research avenue to more precisely deal with the problem of complex mixtures is the integration of transcriptomics data. In the absence of precise toxicological effect data, individual chemical-induced transcriptomics changes associated with toxicity may indeed help to better predict the toxic mixture potential (Labib et al., 2017). Thus, the integration of "omics" technologies into traditional mixture assessment procedures appears to be a promising research field on the crossroads of basic toxicology and regulatory risk assessment.

2.7.5 Summary

Definitions are given for the following types of mixed exposures: mixture, simple mixture, complex mixture, combination and specified combination of chemicals, and cumulative exposure. Three types of joint action of chemicals in a mixture or combination have been defined: dissimilar joint action, similar joint action and interaction. With respect to interaction, three types are often distinguished: direct chemical–chemical, toxicokinetic and toxicodynamic interaction. Given the wide diversity of exposure scenarios, the questions to be addressed in toxicity studies of mixtures (or combinations) of chemicals vary widely and consequently there is no "one size fits all" design for toxicity studies of mixtures. Major factors that influence study design as well as the method of analysis of the results include the number of chemicals in a mixture or combination, the availability of a mixture for testing it in its entirety, and the extent to which the toxicity of a mixture or combination needs to be characterized in terms of dose–response relationship or departure from additivity, and available resources. To evaluate the safety of mixtures, they can be approached as a single entity, a number of fractions or a number of individual constituents. The safety evaluation based on the individual constituents can be approached by considering all constituents, the "top n" constituents or the "pseudo top n" constituents. Depending on the type of joint action of the chemicals in a mixture (or combination) the safety evaluation can be performed using the dose additivity/HI method for similarly acting chemicals, the response additivity/HI method for dissimilarly acting chemicals, and/or the HI/WoE method for chemicals showing interactive toxicity. There are promising new developments in the integration of "omics" technologies into the risk evaluation of complex chemical mixtures that may help to further refine the evaluation processes.

Further Reading

European Commission (2012). Toxicity and assessment of chemical mixtures. European Commission, Scientific Committees on Health and Environmental Risks, Emerging and Newly Identified Health Risks, and Consumer Safety, Brussels, Belgium. Available online: http://ec.europa.eu/health/scientific_committees/environmental_risks/docs/scher_o_155.pdf.

Feron VJ , Bolt HM (1996). Combination toxicology. *Food Chem Toxicol* **34**, 1025–1185.

Feron VJ, van Vliet PW, Notten WRF (2004). Exposure to combinations of substances: a system for assessing health risks. *Environ Toxicol Pharmacol* **18**, 215–222.

Jonker D, Woutersen RA, Feron VJ (1996). Toxicity of mixtures of nephrotoxicants with similar or dissimilar mode of action. *Food Chem Toxicol* **34**, 1075–1082.

Jonker D, Freidig AP, Groten JP, de Hollander AEM, Stierum RH, Woutersen RA, Feron VJ (2004). Safety evaluation of chemical mixtures and combinations of chemical and non-chemical stressors. *Rev Environ Health* **19**, 83–139.

Labib S, Williams A, Kuo B, Yauk CL, White PA, Halappanavar S (2017). A framework for the use of single-chemical transcriptomics data in predicting the hazards assiciated with complex mixtures of polycyclic aromatic hydrocarbons. *Arch Toxicol* **91**, 2599–2616.

Organisation for Economic Co-operation and Development (2001). Harmonised hazard classification criteria for mixtures. In: Harmonised Integrated Classification System for Human Health and Environmental Hazards of Chemical Substances and Mixtures, Report ENV/JM/MONO(2001)6, Paris, France.

Simmons JE (1995). Chemical mixtures and quantitative risk assessment, *Toxicology* **105**, 109–441.
Suk WA, Olden K, Yang RSH (2002). Application of technology to chemical mixtures research. *Environ Health Perspect* **110**, Suppl 6, 891–1036.
US Environmental Protection Agency (2003). *Framework for cumulative risk assessment, EPA/600/P-02/001F, Office of Research and Development*, National Center for Environmental Assessment, Washington DC.
Yang RSH (1994). *Toxicology of Chemical Mixtures*, Academic Press, San Diego, pp 720.

2.8 Chemical Carcinogenesis: Genotoxic and Non-Genotoxic Mechanisms

Thomas Efferth and Bernd Kaina

2.8.1 Introduction

Cancer is a multistep process that can be dissected into tumor initiation, tumor promotion and tumor progression (Figure 2.61). Tumor initiation is based on the induction of genetic mutations, and all initiators harbour a mutagenic potential. Tumor promotion rests on stimulation of cell

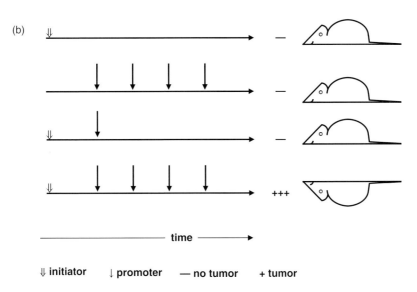

Figure 2.61 *Stageing of tumor formation on mice skin papilloma model. All DNA damaging agents can be considered to be initiators. Tumor promoters stimulate cell proliferation, notably of initiated cells. They also may prevent apoptosis, thus causing initiated cells to survive and propagate in the tissue. Tumor promoters must be applied repeatedly in order to provoke tumor promotion; their effect is reversible and their primary target is not the DNA. This is in contrast to initiators that exert irreversible genetic changes.*

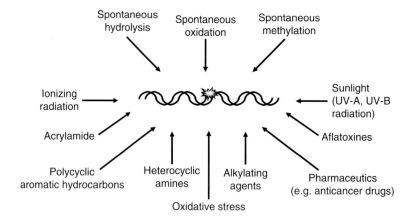

Figure 2.62 *Spontaneous and exogenously induced insults damaging DNA.*

division of initiated cells by activation of signaling pathways through tumor promoters. Tumor progression requires further genetic changes, which are induced by mutagenic initiators and often further triggered by genomic instability in benign tumors. Tumorigenesis results from altered expression of genes that regulate cell proliferation, cell–cell communication, cell migration, and cell adherence. The most important genes are proto-oncogenes and tumor suppressor genes. Other target genes of initiators are DNA repair genes, whose inactivation results in genomic instability. Alteration in their function or their expression may be caused by mutations, but change in expression can also be caused by epigenetic mechanisms via CpG methylation (formation of 5-methyl cytosine in gene promoter regions) that often results in gene silencing. Mutations, however, appear to be the predominant mechanism underlying the process of carcinogenesis.

Mutations arise from DNA damage that may occur spontaneously or is induced by exogenous insults, e.g. DNA adduct formation by reactive metabolites of carcinogenic chemicals or radiation (Figure 2.62). Therefore, it is important to consider the mechanism of DNA adduct formation, as well as the potency of metabolically activated carcinogens. Because DNA adducts are subject to repair, DNA repair is a major defence mechanism at the cellular level against cancer. However, mutations in DNA repair genes evoke genomic instability and predispose the affected individuals to tumor formation. Furthermore, during the process of DNA replication DNA damage may be erroneously processed by low-fidelity DNA polymerases (translesion polymerases such as DNA polymerase ζ) that would result in the generation of mutations. Thus, the DNA repair process may result in either removal of DNA damage or the fixation of altered DNA bases in the form of mutations.

2.8.2 Mechanisms of DNA Damage, Repair, and Carcinogenesis

> **DNA damage can result from loss of any of the bases, modification of base structure via adduct formation or oxidation by reactive oxygen and nitrogen species, impact of ultraviolet (UV) light, cross-linking of DNA strands, cross-linking between DNA and proteins, or by strand breakage. Xenobiotics and endogenous DNA damaging agents, such as reactive oxygen species, may be detoxified by either metabolic inactivation or antioxidant mechanisms such as superoxide dismutases, catalase, and glutathione-related enzymes.**

> DNA damage can be repaired by a number of sophisticated DNA repair pathways, including damage reversal, base excision repair, nucleotide excision repair, mismatch repair, non-homologous end joining, and homologous recombination. Alternatively, DNA damage can be inadvertently fixed by template switching and lesion bypass and by the activity of translesion DNA polymerases.

DNA Damaging Agents

Damage to DNA is not necessarily mutagenic unless it is fixed to form permanent changes of the genetic code (base substitutions, base deletions and insertions) that are starting points in the process of carcinogenesis. The main types of DNA damage are:

- **damage to DNA bases:** typical examples are O^6-methylguanine, 8-oxo-guanine, thymine glycol, fragmented bases resulting from oxidative stress, UV-light induced cyclobutan dimers, and 6-4-photoproducts; alkyladducts occur by alkylating agents and large, bulky adducts by polycyclic hydrocarbons
- **lesions at the DNA backbone** that are abasic sites and DNA single- and double-strand breaks.
- **cross-links:** bifunctional agents such as nitrogen mustard, cisplatinum, psoralen etc. form interstrand adducts between two complementary DNA strands and DNA protein adducts
- **aberrant non-duplex DNA forms** such as DNA bubbles, recombination structures, aberrant fork structures, which are generated during disrupted replication, and recombination or DNA repair.

Exogenous DNA Damaging Agents DNA lesions may result from the metabolic activation of xenobiotic carcinogens, which by virtue of their electrophilicity tend to form covalent adducts with DNA. Exposure to carcinogens may arise from release of industrial chemicals in the occupational setting or in the general environment. Furthermore, foods and beverages also contain compounds such as N-nitrosamines and heterocyclic amines that can damage DNA and which are potentially mutagenic.

Among the plethora of carcinogenic compounds five large classes with some of the main representatives are listed here (see also Figure 2.62):

- **polycyclic aromatic hydrocarbons**, e.g. benzo[a]pyrene, 3-methyl-cholanthrene, and benzanthracene
- **aromatic amines**, e.g. β-naphthylamine, 2-acetylaminofluorene, and *o*-toluidine
- **N-nitroso compounds**, e.g. N-nitrosodimethylamine and N-nitroso-N-methylurea
- **carcinogenic natural products**, e.g. aflatoxin B1, safrol, and aristolochic acid
- **cross-linking and alkylating agents**, e.g. sulfur and nitrogen mustards and ethylene oxide.

The most important chemical reactions leading to adduct formation of carcinogenic molecules with DNA are the transfer of alkyl-, arylamine-, or aralkyl-groups. Alkylated or aralkylated compounds are generated by the oxidation of carbon atoms, whereas the oxidation or reduction of nitrogen atoms lead to arylaminated substances.

Carcinogens form adducts with DNA bases. Alkylating agents frequently bind to exocyclic oxygen or ring nitrogen atoms, i.e. at the O^6 position of guanine and the N7 position of guanine, respectively. Arylaminated substances bind to nitrogen atoms at the N7 position. Subsequently, rearrangements can take place and C8-deoxyguanosine adducts can be generated. Polycyclic aromatic hydrocarbons form various adducts by binding to exocyclic nitrogen atoms of adenine or guanine. The formation of adducts can lead to the substitution of DNA bases.

Endogenous DNA Damage The hydrolysis of N-glycosidic bonds of bases in the DNA is frequently observed. It is estimated that about 30,000 purines are lost per day in the human genome. In comparison to purines, only 5% of the pyrimidines are affected. Apurinic/apyridinic sites (AP sites) are generated, which are cytotoxic and mutagenic since they block DNA replication, cause base misincorporation, and may lead to DNA strand breaks. The hydrolytic separation of exocyclic amino groups in cytosine and 5-methylcytosine leads to the formation of uracil and thymidine, respectively, which then occur as mismatches in the DNA. The rate of cytosine deamination is high, estimated to be as much as 500 per cell per day. In a comparable manner, adenine can be deaminated to hypoxanthine and guanine to xanthine.

Another cause of spontaneous DNA damage is oxidation by reactive oxygen species (ROS) and reactive nitrogen species (RNS), which are generated as by-products of oxidative metabolism, during the inflammation processes, and after ionizing radiation. The mitochondrial electron transport chain is not only the main source of ATP production by oxidative phosphorylation, but also the main source for the generation of ROS (H_2O_2 and $O_2^{\cdot-}$). The spatial proximity of ROS generation and mitochondrial DNA explains why oxidative damage is more frequently found in mitochondrial DNA than in nuclear DNA. During chronic inflammation, activated macrophages and neutrophilic leukocytes release NO, $O_2^{\cdot-}$, ·OH, and HOCl, which damage the DNA of neighboring cells. Fatty acid radicals, aldehydes, and other compounds are generated during lipid peroxidation. They cause etheno-adducts of pyrimidines and purines. The oxidation of bases, which occurs mainly at electrophilic carbon centers, leads to very mutagenic bases in DNA such as 8-oxo-guanine, formamidopyrimidine, and thymine glycol. The amount of 8-oxo-guanine formed spontaneously per cell was estimated in the range of 2,400 per day. Endogenous products of normal metabolism such as estrogens, heme precursors, amino acids, and glycol-oxidation products may also damage DNA.

Prevention and Repair of DNA Damage

> **The cell is equipped with several defense mechanisms to avoid mutations caused by endogenous and exogenous DNA damaging agents.**

There are several defense mechanisms that protect the cell from DNA damage:

- mechanisms of oxidative stress response for the avoidance of oxidative DNA lesions by endogenous or exogenous reactive oxygen species
- multiple control mechanisms of the DNA replication machinery to minimize the error rate, i.e. <1 misincorporation/10^6 nucleotides
- mechanisms to regulate cell cycle progression to guarantee error-free duplication and segregation of chromosomes during cell division; cell cycle checkpoints control this function
- DNA repair mechanisms: over 130 DNA repair genes have thus far been identified
- if DNA cannot be repaired after massive damage, cells undergo apoptosis, i.e. programmed cell death; hence, from the viewpoint of the entire organism, death of damaged cells can be looked upon as a protective mechanism
- DNA lesions can evoke transcriptional changes characterized by up- or down-regulation of DNA repair gene expression, which can lead to an elevated level of defense due to increased repair enzyme activity.

If mutations cause the activation of proto-oncogenes and inactivation of tumor suppressor genes, malignant transformation of cells can take place.

Antioxidant Mechanisms

> **ROS attack cell membrane lipids, generating DNA reactive compounds and damage DNA directly. Antioxidant response genes (superoxide dismutase, glutathione S-transferases, and catalases) counteract the deleterious effects of ROS.**

The interaction of ROS with lipids is deleterious, since a single ROS molecule can generate many toxic reaction products due to autocatalytic dispersion. Examples are hydrogen peroxide, peroxyradicals, alkoxyradicals and unsaturated aldehydes. The reaction of ROS with DNA is rated as most relevant for carcinogenesis. It is mainly based upon the induction of DNA strand breaks and base damage and the induction of certain cellular functions. Both events can finally lead to apoptosis. Organisms have, therefore, developed various protective mechanisms to fight oxidative stress induced by ROS and lipid peroxidation products, e.g. catalase, superoxide dismutases, and glutathione-associated proteins (glutathione reductase, glutathione peroxidase, glutathione S-transferases) as well as non-enzymatic molecules such as glutathione or α-tocopherol. Antioxidant enzymes protect against the development of cancer.

Various isoenzymes of **superoxide dismutase** (SOD) have been identified, e.g. iron-dependent FeSODs in the cytosol, mitochondria, and chloroplasts. The extracellular EC-SOD appears to be membrane-associated in the extracellular space. The copper- and zinc-dependent CuZnSOD of eukaryotes is localized cytoplasmatically. Moreover, a manganese-dependent SOD (MnSOD) has been observed in eukaryotes. SODs catalyze the dismutation of superoxide anions ($2O_2^-$) in the presence of protons ($2H^+$) to molecular oxygen (O_2) and hydrogen peroxide (H_2O_2).

Subsequently, hydrogen peroxide is transformed to water by **catalase**. **Glutathione S-transferases** (GSTs) catalyze the nucleophilic reaction of glutathione (GSH) to many hydrophobic xenobiotics, which are then less toxic and easier to excrete. Oxidized glutathione is returned to the reduced form by **glutathione reductase**. Furthermore, GSTs can participate in the repair of oxidative damage at membrane lipids and DNA.

Selenium-dependent **glutathione peroxidases** protect against lipid peroxidation by termination of the lipid peroxidation cascade through reduction of fatty acid hydroperoxides (FS-OOH) and phospholipid hydroperoxides (PL-OOH). Hydroxyl radicals are eliminated by a glutathione peroxidase-mediated reaction and are converted to water, while GSH is oxidized to GSSG. The cellular GSH level is regenerated by the NADPH-dependent glutathione reductase from GSSG. **Glucose-6-phosphate dehydrogenase**, which maintains the cellular NADPH level, can, therefore, also be considered an antioxidant enzyme.

DNA Repair

> **DNA repair is essential for maintenance of the genome. Most DNA repair processes are error-free and clearly protect against cytotoxicity of genotoxicants and cancer induced by chemical agents, ultraviolet light and ionising radiation. The main mechanisms of DNA repair encompass damage reversal by methylguanine-DNA methyltransferase (MGMT) and AlkBH proteins, which repair O6-alkylguanine and N1-methyladenine, respectively. Small lesions are repaired by base excision repair, while larger adducts are repaired by nucleotide excision repair (NER). Mismatches are subject to mismatch repair, and cross-links to cross-link repair that involves components of NER and recombination repair. DNA double-strand breaks are repaired by non-homologous end joining (NHEJ) and homologous recombination (HR). Both are complex pathways involving a different set of proteins. NHEJ, especially the backup pathway, is an error-prone repair process leading to gross chromosomal changes.**

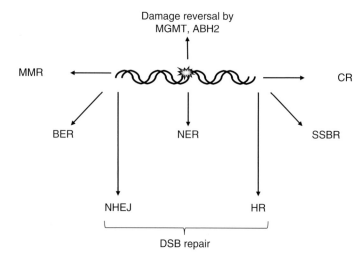

Figure 2.63 Mechanisms of repair of DNA damage. MMR, mismatch repair; CR, cross-link repair; BER, base excision repair; NER, nucleotide excision repair; SSBR, single-strand break repair; NHEJ, non-homologous end joining; HR, homologous recombination, DSB, double-strand break.

DNA has a special role as a critical target molecule compared to other macromolecules (lipids, proteins, RNA), since in case of damage it cannot simply be replaced. Therefore, the repair of DNA lesions is of eminent importance for life, and a plethora of different DNA repair pathways have been developed during evolution. The main mechanisms (Figure 2.63) are discussed below.

Damage Reversal

> **Damage reversal mechanisms are rapid repair processes that remove specifically highly pre-mutagenic and pre-toxic lesions. Among the best-known examples is the repair protein MGMT that clearly protects against cancer initiation and progression. Damage reversal is also mediated by alkB homologous proteins, notably ABH2.**

Alkylating agents produce N- and O-alkylated purines and pyrimidines as well as DNA phosphotriesters. O^6-alkylguanines (O^6-methylguanine and O^6-ethylguanine) are critical lesions, which cause GC to AT transition mutations (Figure 2.64). Likewise, O^4-methylthymine is critical, forming TA–CG mutations. These lesions are repaired by MGMT, also designated as alkyltransferase. The protein removes the methyl group from the O^6-position of guanine and, less efficiently, from the O^4 position of thymine, and transfers it to an internal cytosine residue. This transfer reaction leads to irreversible inactivation of MGMT. MGMT is, therefore, not an enzyme in the classical sense; it has been termed a suicide enzyme. Since O^6-methylguanine is a major promutagenic (Figure 2.64) and procarcinogenic lesion, MGMT clearly protects against cancer formation, which has experimentally been shown in MGMT overexpressing mouse models that are protected against cancer initiation and progression.

A second recently discovered damage reversal mechanism is represented by ABH proteins (ABH = alkB homologues of *E. coli*). At least three of these proteins are known to occur in human cells: ABH1, ABH2 and ABH3. ABH2 and ABH3 proteins belong to a superfamily of ketoglutarate- and Fe(II)-dependent dioxygenases that repair N1-methyladenine, N3-methylcytosine and N1-ethyladenine in a reversal reaction involving oxidative demethylation

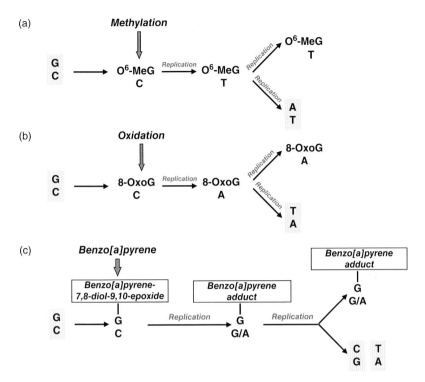

Figure 2.64 Mechanism of mispairing mutagenesis by the instructive lesions (A) O^6-methylguanine, (B) 8-oxo-guanine, and (C) the non-instructive bulky lesion induced by the active metabolite of benzo[a]pyrene, benzo[a]pyrene 7,8-diol-9,10-epoxide. Mutagenesis through bulky lesions requires low fidelity DNA polymerase action.

and restoration of the undamaged base. ABH3 repairs single-stranded DNA and even RNA. The repair mechanism contributes to cellular protection against alkylation damage. Conversely, MGMT-lacking mice are hypersensitive regarding cancer formation induced by methylating agents.

Base Excision Repair

> **Many modified bases in DNA are removed by base excision repair (BER). Although several mouse strains deficient in BER have been created, human disorders characterized by a clear defect in BER are not known. Nevertheless, it is believed that BER plays a crucial role in the body's defense against cancer, chronic inflammation, and malformations.**

"Small" lesions in DNA, i.e. non-bulky adducts, which do not cause clear distortion of the DNA helix, are repaired by BER (Figure 2.65). The process is highly specific because of recognition of lesions by specific enzymes, the DNA glycosylases. BER results in elimination of incorporated uracil (a base normally present in RNA, but not DNA), fragmented pyrimidines, N-alkylated purines such as N7-methylguanine and N3-methyladenine, 8-oxo-guanine, thymine glycol, and others. From these lesions, 8-oxo-guanine is presumably the most important oxidative DNA lesion. It is mutagenic because of its ability to mispair with adenine (Figure 2.64).

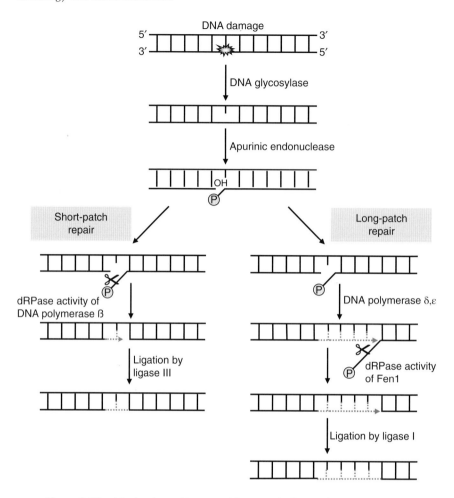

Figure 2.65 Mechanism of base excision repair. For explanation see text.

BER can be subgrouped into single nucleotide and long-patch BER. The steps of BER are as follows:

1. Damage recognition, base removal, and incision: First, **DNA glycosylases** recognize and remove a damaged base by hydrolysis of the N-glycosidic bond, thereby generating apurinic or apyrimidinic sites (**AP sites**). Subsequently, **AP endonucleases** cut the phosphodiester bond at the AP site, generating a DNA single-strand break.
2. Nucleotide insertion: **DNA polymerase-β** (Polβ) inserts a nucleotide at the AP site.
3. Depending on the nature of the base damage, **short-patch** or **long-patch repair** takes place. During long-patch repair **proliferation-dependent nuclear antigen (PCNA)** cuts out several nucleotides close to the lesion.
4. Strand displacement: During long-patch repair, up to 10 nucleotides are excised by endonucleases.
5. Ligation: **DNA ligases I and III** ligate the newly synthesized strand with the DNA. Ligase I is active in long-patch BER, ligase III in short-patch repair.

An important accessory protein in BER is XRCC1, which is a scaffold protein stabilizing ligase III. Another important protein is poly(ADP) ribosyl transferase (PARP-1), which binds to single-stranded DNA, including apurinic sites, becomes activated, and in turn poly(ADP) ribosylates itself and other proteins. This is supposed to have a regulatory role in the repair process, preventing the damaged DNA from uncoordinated cleavage of abasic sites.

Nucleotide Excision Repair

> **NER is a highly complex process for the repair of bulky DNA lesions. It discriminates between active and non-active genes in order to repair preferentially the DNA that is in the process of transcription. Several genetic diseases are characterized by defects in NER. Xeroderma pigmentosum displays a defect in global genomic repair, while Cockayne's syndrome is defective in transcription-coupled repair.**

A common feature of BER and NER is that the complementary DNA strand is used as a matrix for the new synthesis after elimination of the damage. Therefore, both pathways are error-free. BER and NER differ from each other in that:

- BER is used for small base adducts (**non-bulky lesions**), while NER is activated for larger damage (**bulky lesions**).
- About 30 nucleotides are replaced in NER, but only 1–10 in BER.
- In BER, the damaged base is removed as free base, whereas in NER the damaged site is cut out as part of a longer single-stranded fragment.

Bulky adducts are induced in guanine by N-acetyl-aminofluorene, benzo[a]pyrene, aflatoxin, heterocyclic amines, and cis-platinum. Psoralen and UV light induce thymine cross-links (CPDs) and 6–4-photoadducts that inhibit DNA replication. Error-prone DNA polymerases can bypass the lesion. This causes mutations, such as G → T transversions during DNA replication.

NER represents a complex pathway involving approximately 30 proteins. The main pathways are **global genomic repair (GGR)** and **transcription-coupled repair (TCR)**. GGR is independent of the transcriptional activity of affected gene sequences. This pathway removes lesions in transcriptionally active and inactive DNA regions with similar efficacy. It operates more slowly than TCR.

XP proteins contribute to GGR and TCR. The name XP derives from the UV-hypersensitive hereditary disease Xeroderma pigmentosum. There are seven XP subgroups (defective in the repair proteins XPA to XPG) plus XP variant defective in the translesion polymerase η. Mutations in *XP* genes cause extreme sensitivity to sunlight and development of skin cancer that may occur already early in life. For unknown reasons, Cockaynes syndrome patients defective in TCR do not develop cancer at high incidence. The patients, however, die early because of severe developmental disorders.

DNA repair by NER occurs in four steps:

1. DNA damage recognition: The protein complexes XPC-HR23B and RPA-XPA recognize DNA lesions such as 6 4-PPs or cis-platinum–DNA adducts which are characterized by distortion of the DNA. Another protein complex consists of the damaged binding proteins XPE and XPG. Binding of these proteins at the site of the lesion further distorts the DNA and invites repair.
2. DNA unwinding: A complex of the transcription factor TFIIH and nine other proteins binds to the damaged site. This causes unwinding of the DNA around the damaged site.
3. Excision of the DNA lesion: XPD induces incisions in the $3'$-position and the ERCC1-XPF complex in the $5'$-position flanking the damaged site. A 27–29 bp sequence containing the damaged site binds to the protein complexes and is removed.
4. Repair synthesis: The gap is filled by DNA polymerases POLδ and POLε and sealed by DNA ligase I and other accessory factors.

TCR is less well understood. DNA lesions in constitutive active (housekeeping) genes are more efficiently repaired than lesions in non-coding sequences. Within genes, the transcribed 5′-strand is more efficiently repaired than the non-transcribed 3′-strand. TCR only takes place in genes that are transcribed by RNA polymerase II. Block of transcription resulting from DNA damage provides the signal for repair, which is recognized by the Cockayne's syndrome protein CSB and it finally activates the other NER repair proteins. Lesions in RNA polymerase II-dependent ribosomal genes are scarcely repaired. The same is true for mitochondrial genes.

Mismatch Repair

> **With the finding that a subgroup of colon carcinomas is defective in mismatch repair (MMR), this repair pathway has attracted much interest as a tumor defense strategy. MMR not only removes misincorporated bases from the daughter strand after replication, it is also considered as a sensor of DNA damage involved in the process of DNA damage-triggered apoptosis. MMR defective cells are highly resistant to the killing effect of methylating carcinogens that further contributes to the high mutation frequency and genomic instability in the affected cells.**

MMR repairs single-base mismatches and small insertion and deletion mismatches of multiple bases. Base mismatches can occur spontaneously or by chemically induced base deamination, oxidation, and methylation as well as by replication errors. Therefore, MMR has also been termed replication error repair (RER). Examples of chemically induced base mismatches are alkylation-induced O^6-methylguanine matched with thymine, cis-platinum-induced 1,2-intrastrand cross-links, UV-induced 6–4-PPs, purine adducts of benzo[a]pyrene or 2-aminofluorene, and 8-oxo-guanine mispaired with adenine.

MMR can be divided into four steps:

1. DNA damage recognition: Mismatches are recognized by the dimeric proteins MSH2/MSH6 or MSH2/MSH3, which bind to the site of the lesion and distort the DNA.
2. Strand discrimination: Two mechanisms have been proposed. The molecular switch model claims that ADP binds to the MutSα complex, thereby recognizing the mismatch. The MutSα–ADP complex corresponds to the active state. By binding to the mismatch, an ADP → ATP transition and intrinsic ATPase activity is stimulated. This causes a conformational change and the docking of the MutLα-complex (MLH1-PMS2). In the hydrolysis-driven translocation model, the ATP hydrolysis induces a translocation of the MutSα complex.
3. Excision: After binding to the mismatch, MutSα associates with MutLα. The excision of the mismatched is mediated by exonuclease I.
4. Repair synthesis: The new synthesis is performed by POLδ.

DNA Double-Strand Break Repair

> **Double-strand breaks (DSBs) are considered to be the most lethal DNA lesions, therefore their repair is of utmost importance for cellular survival. In most cases DSB repair accurately replaces damaged DNA. However, a subpathway of DSB repair is error-prone, leading to the formation of chromosomal rearrangements. The cells may survive, but at the expense of chromosomal mutations that may ultimately result in cancer.**

DSBs occur as a result of ionizing radiation or exposure to DNA topoisomerase II inhibitors. DSBs can induce either genotoxic effects or apoptosis. DSBs are repaired via two pathways: error-free HR and error-prone non-homologous end-joining (NHEJ). NHEJ takes place in the prereplicative phase of the cell cycle (G0/G1) whereas HR requires an intact homologous DNA strand and therefore occurs in the S and G2 phases of the cell cycle.

NHEJ is a mechanism for repairing DSBs which permits direct ligation of the broken ends of DNA without the need for a homologous template to guide the joining process. There are two sub-pathways of NHEJ: D-NHEJ and B-NHEJ (or ALT-NHEJ). In the classical D-NHEJ pathway the first step is the binding of a heterodimer termed Ku70, Ku80 to the broken ends, which then facilitates binding of the cytalytic subunit of DNA-PK$_{CS}$. The whole complex is termed DNA-PK, which exerts kinase activity phosphorylating many proteins in order to facilitate DNA repair. Other proteins are required in order to complete repair, such as Artemis, which "cleans" the DNA ends, the scaffold protein XRCC4, and ligase IV, which finally catalyses the ligation. The B-NHEJ is less complex. It requires PARP-1, which recognizes the broken ends, XRCC1, and ligase III. This pathway obviously compedes with D-NHEJ. It is considered the most error-prone repair pathway leading to chromosomal aberrations. During HR the damaged chromosome attracts an undamaged strand of DNA with a homologous sequence which serves as a template for the repair. HR starts with nucleolytic resection of the DSB in 5′–3′ orientation. Two 3′ single-stranded DNA ends are generated. A protein called RAD52 binds to these DNA ends and helps to initiate the annealing of homologous strands of DNA. This is mediated by RAD51, which binds to single-stranded DNA and acts as a recombinase. RAD51 is responsible for the DNA strand exchange of damaged DNA with a homologous region of the undamaged DNA.

Homologous recombination with RAD51 being involved is also most important in repairing blocked replication forks resulting from bulky DNA lesions or abasic sites. If translesion synthesis does not occur, the replication fork may become unstable, leading to nuclease attack and the formation of one-ended DSB. If not repaired, this will give rise to a chromatid break. HR at blocked replication forks results in sister chromatid exchanges, which are supposed to be cytogenetic changes resulting from an error-free repair process.

Tolerance to DNA Damage

> **Bulky DNA lesions block DNA replication. Unless they are removed by NER, they may cause misreplication and mutations because of error-prone DNA polymerases that read over the lesion. Thus, bulky DNA lesions may go unrepaired and result in cytotoxicity or mutations. For many chemical carcinogens which form large adducts on DNA, such as benzo[a]pyrene and aflatoxins, error-prone translesion synthesis is the major pathway of mutagenesis, therefore error-prone polymerases are presumably crucially involved in carcinogenesis induced by genotoxins that form replication - blocking adducts in the DNA.**

Considering the high frequency at which DNA lesions occur, e.g. 30,000 depurinations per cell per day, only a small but nevertheless significant portion escape error-free DNA repair. Since persistant DNA damage impedes DNA replication and is cytotoxic, mechanisms in which DNA damage is tolerated have developed and are termed template switching and lesion bypass.

Template Switching Despite the inhibition of DNA synthesis on the damaged strand, DNA sysnthesis can take place to some extent on the non-damaged strand. The newly synthesized daughter strands serve as templates. After dissociation of both newly synthesized daughter strands, re-annealing to the original parental strands and semi-conservative replication takes place. Thereby, the damaged site can be bypassed and the replication machinery can proceed with normal DNA synthesis. Since the damaged site is bypassed, DNA synthesis continues in a error-free manner.

Lesion Bypass   In contrast to template switching, the damaged DNA strand is used in the lesion bypass mechanism. Nucleotides are incorporated opposite to the damaged strand (**translesion synthesis**). Then, DNA synthesis is extented, and the replication machinery continues DNA synthesis. An error-free lesion bypass takes place if the right nucleotide is incorporated. However, error-prone incorporation can also occur. Lesion bypass is performed by specific DNA polymerases, which use the damaged DNA strand as a template. Polη catalyzes error-prone DNA synthesis with adducts induced by benzo[a]pyrene (Figure 2.65) or GG-cisplatinum intrastrand adducts. The human **DNA polymerase I** (Pol I) represents an exception, since this enzyme does not follow the common Watson–Crick base pairing. Pol I frequently incorporates G instead of A at sites opposite to a T in undamaged DNA regions. As a consequence, DNA elongation truncates opposite to Ts (T-stop). The enzyme displays less catalytic efficacy opposite to a C. Hence, Polζ catalysis is cumbersome, and only short DNA sequences are synthesized. The enzyme promotes somatic hypermutation during immunoglobulin development. Somatic hypermutations in heavy and light antibody chains occur with 100,000-fold higher frequency than in other genes.

Halting Cell Cycle Progression for DNA Repair or Mitosis Termination

> **Halting cell cycle progression allows a cell to repair DNA damage. Defined cell cycle checkpoints (G1/M, G2/M, S/M) maintain quality control of DNA and, hence, guarantee genomic integrity. Upon DNA damage, master proteins (ATM, ATR, p53) induce cell cycle arrest by regulation of effector proteins (cyclin-dependent kinases, cyclins).**

Cell division is of the utmost importance in biology. Correct and error-free repair processing during cell cycling is crucial to a cell's life. Defined checkpoints serve the maintenance of genomic intregrity. These checkpoints allow halting of the cell cycle and DNA repair. Upon massive damage, the cell can undergo either apoptosis or permanent cell cycle arrest (also called **cellular senescence**).

ATM and **ATR** are very important sensor proteins that activate cell cycle arrest upon DNA damage (Figure 2.66). Both proteins switch on the G1/S and G2/M checkpoints by phosphorylation of Chk1, Chk2, and p53. Another checkpoint at the S boundary of the cell cycle is driven by the ATR-interacting protein (ATRIP). Therefore, ATM, ATR, and ATRIP are **master proteins** for checkpoint regulation. As a rule, the DNA damage-dependent kinase ATM becomes activated by DSBs while ATR becomes activated by blocked replication forks. The upstream sensor for DSBs is the MRN complex, a trimeric protein composed of mirin, Rad50, and nibrin (also known as NBS1). MRN is required in order for ATM to become activated. As a result it phosphorylates more than 500 different proteins, including p53. Thus the complex scenario of DNA damage response is evoked.

At the G1/M DNA damage checkpoint the cell cycle can be halted by wild-type p53 in early and mid G1 phase as a result of DNA damage. Mutations of p53 result in loss of the ability to control the cell cycle. At the G2/M DNA damage checkpoint DNA damage by oxidative stress or ionizing radiation leads to G2 arrest. At the replication checkpoint (S/M checkpoint) mitosis is inhibited when errors occur during DNA replication, i.e. after depletion of nucleotide pools or DNA damage by alkylating agents (e.g., methyl-methanesulfonate, MMS). If the replication

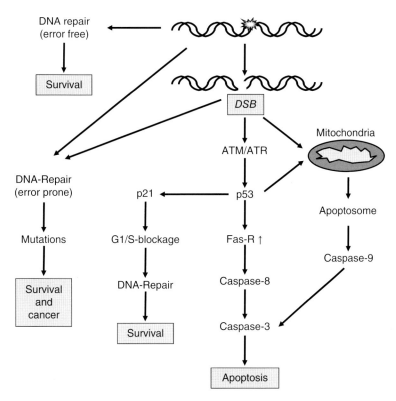

Figure 2.66 Survival and death pathways triggered by DNA lesions, notably DNA double-strand breaks (DSB). For explanation see text.

fork collides with a damaged site, replication is stopped, and the transition from S- to M-phase is inhibited.

2.8.3 Cancer Development

Multiple-Step Carcinogenesis

> **Persistant DNA damage, which has escaped the DNA repair defence mechanisms, may initiate tumor development. This process is traditionally divided into three steps: initiation, promotion, and progression. During the initiation phase, cells develop mutations in one or several critical genes. During tumor promotion, proliferative stimuli lead to clonal expansion, notably of the initiated cell. The progression phase is characterized by further accumulation of genetic changes that may be driven by chromosomal instability. Deregulated apoptosis and telomerase activity as well as neoangiogenesis support the carcinogenic process.**

The development of cancer takes place in several steps: initiation, promotion, and progression (Figure 2.61a). During **initiation**, DNA damage leads to replication errors and mutations such as base exchanges or frameshift mutations. If mutations appear in coding regions of genes that are relevant for cell growth and differentiation, malignant tumor formation can be initiated.

Initiation Several cellular processes are important for initiation:

- metabolic activation of carcinogens
- errors in DNA repair
- transfer of mutations to daughter cells by cell proliferation.

The initiation phase is irreversible. Nevertheless, not all initiated cells end up as tumors, since most damaged cells undergo apoptosis. An initiated cell is not yet a tumor cell, since it does not grow autonomously. Many DNA lesions can be accumulated during the life span of an organism without the generation of a tumor. Therefore, additional events are required in order to fully transform a cell.

Apart from the inactivation of DNA repair genes, **proto-oncogenes** are activated and **tumor suppressor genes** are inactivated. Furthermore, **epigenetic events** contribute to the transformation of cells from a benign to a malignant state. The minimal constellation for the development of cancer also consists of a constitutive expression of **telomerase** that is responsible for the immortalization of normal cells as a precondition for cancer development. Immortalization preceeds transformation. Telomerase circumvents the erosion of telomers at the chromosomal ends. The permanent shortening of telomeres during each replication round leads to senescence of healthy cells.

Promotion During the second phase of cancer development (promotion), initiated cells are stimulated to grow. Initiated cells start to replicate, and clonal cell expansion takes place. Chemical compounds, which do not damage DNA and which are not genotoxic, but stimulate the proliferation of initiated cells, are termed tumor promoters. Tumor development starts with pre-neoplastic precursors, many of which regress spontaneously via clonal expansion. Phenobarbital and tetradecanoyl phorbol acetate (TPA) are examples of compounds with cancer-promoting properties. TPA has structural similarities to diacylglycerole (DG), which is a signal molecule for the activation of protein kinase C (PKC). PKC stimulates cell growth. TPA can substitute for DG in the corresponding signal transduction pathway. While the signaling effect of DG is controllable, TPA (upon repeated application) permanently activates PKC, which stimulates continual tumor cell proliferation.

Carcinogens can be classified based on their carcinogenic function:

- **primary carcinogens** act directly without prior metabolic activation
- **secondary carcinogens (pre-carcinogens)** are activated by biotransforming enzymes, e.g. cytochrome P450 monooxygenases
- **cocarcinogens** boost the effect of carcinogens by induction of biotransformation enzymes
- **promoters** indirectly increase the carcinogenic effects by the stumulation of cellular proliferation and/or suppression of apoptosis.

Progression Cancer progression is characterized by the **invasion** of tumor cells into the neighboured normal tissues, the spread of **metastases**, and the **genetic instability** of the tumor. Genetic instability not only means the acquisition of point mutations (base substitutions) as indicated above, but also larger structural changes at the chromosomal level (translocations, amplifications, deletions etc.) as well as gain or loss of entire chromosomes (aneuploidy). The availability of nutrients and oxygen is an important condition for progressive tumor growth. Therefore, tumor cells excrete growth factors and other substances, which foster the generation of novel blood vessels in the tumor tissue (**neo-angiogenesis**).

Other Theories for Multistage Carcinogenesis The **classical three-step model of carcinogenesis** (initiation, promotion, progression) was launched in the 1950s and is still useful. Subsequently, further theories for multistage carcinogenesis have been developed. For example, as shown by

Vogelstein and colleagues, the **development of colon cancer** necessitates mutations of at least four genes on different chromosomes. Colon cancer represents a suitable model to study tumor progression since histologically distinguishable tumor stages can easily be separated from each other. The functional loss of the tumor suppressor p53 by deletion of the corresponding chromosomal locus on chromsomal 17 or by point mutations in the *TP53* gene causes a malignant colon carcinoma.

The **two-hit hypothesis** has been postulated by Knudsen for retinoblastoma, a childhood tumor of the eye. According to this hypothesis, tumor suppressor genes reveal a tumorigenic effect if both alleles are inactivated by deletion or mutation. If one allele of the *RB* (*retinoblastoma*) gene on chromosome 13 has already been inactivated in the germ cells, a hereditary predisposition for retinoblastoma can be observed. This mutation is recessive. In somatic cells, an inactivation of the second *RB* allele causes the development of retinoblastoma. Loss of heterozygosity (LOH) occurs when a mutated allele is inherited from both parents, from the paternal and the maternal chromosome. This also leads to the onset of retinoblastoma.

Mutator Phenotype

> **A mutator phenotype is generated by mutations in genes responsible for the maintenance of genetic integrity. It has been postulated to occur in normal or premalignant cells as an early event during carcinogenesis.**

In normal cells, DNA replication and chromosomal segregation are processes of highest precision. In cell culture, about 10^{-10} single-base substitutions per nucleotide per cell division and 2×10^{-7} mutations in genes can be observed. Stem cells have even lower frequencies. The fidelity of these processes is also important for cells of the germ line. Curiously, tumor cells contain many thousands of mutations. This speaks against the assumption that rarely occuring, spontaneous mutations in normal cells are sufficient to explain the entire magnitude of genetic changes in tumor cells. Therefore, a mutator phenotype has been postulated as an early event during carcinogenesis. This phenotype is generated by mutations in genes, which are responsible for the maintenance of genetic integrity in normal cells. Mutations in genes encoding DNA polymerases and DNA repair enzymes increase the susceptibility of DNA for further mutations. They cause the erronous incorporation of bases or the insufficient repair of DNA lesions. This results in genetic instability of tumors at an early stage of carcinogenesis. Thus, the mutator phenotype is linked to mutations in certain target genes. Generally speaking, caretaker and gatekeeper genes have to be considered to fulfill such a function.

Caretakers are proteins whose encoded genes ensure the fidelity of DNA synthesis or proper and efficient DNA repair. Hereditary defects in **DNA repair genes** are suitable models to study mutator phenotypes. Examples are

- xeroderma pigmentosum: inherited mutations in *XP* genes, which repair UV-induced DNA lesions by nucleotide excision repair, increase the susceptibility for UV-induced skin cancer
- hereditary non-polyposis colon cancer (HNPCC) is associated with mutations in mismatch repair genes.

The **DNA polymerases** Polα, Polδ, and Polε are also caretakers. Polβ is associated with BER. Polγ is responsible for mitochondrial DNA synthesis. Polδ plays a major role in DNA replication and $3' \to 5'$ exonuclease activity. It can remove erronously incorporated non-complementary bases.

DNA helicases belong to caretaker genes too. They unwind double-stranded DNA before it is read by DNA polyerases. Hereditary mutations in DNA helicases cause syndromes with increased cancer susceptibility:

- **Bloom syndrome**: The DNA synthesis is defective, and many sister chromatid exchanges can be found. This is due to defects in the *BLM* gene and an insuffient repair of DNA double-strand breaks.
- **Werner syndrome**: Homozygous mutations in the *WRN (RecQL1)* gene disturb the repair of double-strand breaks and the maintenance of telomeres.

Gatekeeper genes regulate the cell cylce and apoptosis. Typical diseases of this group are:

- **Ataxia telangiectasia** (AT): Cells show defective DNA repair, which is due to mutations in the *ATM* gene (*ataxia telangiectasia mutated*). ATM is involved in cell cycle control (see above). AT patients frequently develop a lymphoma.
- **Li Fraumeni syndrome**: Heterozygous mutations in the tumor suppressor gene *TP53* are inherited over the germ line. If the remaining *TP53 w*ild-type allele is lost in somatic cells, mesenchymal or epithelials tumors develop in early ages.

Clonal selection is another important feature contributing to the mutator phenotype. Mutated cells obtain a selective growth advantage, which leads to the spread of this cell population. DNA lesions cannot become manifest without cellular proliferation. Clonal expansion drives tumor progression and causes further mutations and more aggressive cell types. The clonal selection of mutations in specific genes allows overcoming limitations in the host tissue set by restricted oxygen supply, missing growth factors etc. The mutation frequency and genetic instability increases with every selection round. At this stage, **heterogeneity** of cellular subpopulations, which is typical for tumors, emerges. Tumor progression is irreversible. The magnitude of mutations indicates that the fidelity of DNA replication may decrease to a degree where the limit of cell survival is reached. Indeed, the fraction of apoptotic cells in tumors is much higher than in normal tissues.

Chromosomal Instability

> **There are various types of chromosomal instability that occur spontaneously and are induced by chemical carcinogens:** *translocations*, *inversions*, *amplifications*, **and** *deletions* **are changes within one or even several chromosomes. The gain or loss of entire chromosomes can also occur. Changes in the number of chromosomes are termed** *aneuploidy*.

Translocations are characterized by the transfer of chromosomal material from one chromosome to another. Two patterns can be distinguished: **idiopathic translocations** and **specific translocations**. Idiopathic translocations occur by chance and vary from patient to patient regardless of the tumor type. Their number increases in the course of tumor progression. It is unclear to what extent they contribute to the aggressiveness of tumors. More interesting are **specific translocations**. They appear consistently in certain tumor types. They have a causative role for the development of these tumor types. They appear predominately in leukemia, lymphoma, and some sarcoma types, but only rarely in the most common solid epithelial tumors types. In contrast, idiopathic translocations can be found in solid epithelial tumors. Important specific translocations are:

- **t(8;14)(q24;q32)**, which activates the *c-MYC* proto-oncogenes in Burkitt lymphoma
- **t(14;18)(q32;q21)**, which activates the anti-apoptotic *BCL-2* gene in follicular lymphoma

- **t(9;22)(q34;q11)**, which generates a fusion protein consisting of the proto-oncogene *c-ABL* and the break point cluster (BCR) region. BCR/ABL is an important signal transducer in acute lymphoblastic and chronic myeloid leukemia.

Inversions Translocations originate from non-repaired DNA double-strand breaks. Frequently, this is due to a functional loss of proteins involved in double-strand break repair or cell cycle control, i.e. ATM, ATR, BRCA1, BRCA2, p53 etc. Very likely, non-homologous end joining of double-strand breaks is involved. Idiopathic and specific forms are also known for inversions. Inversions are characterized by the unhingement of chromosomal segments and the reinstallment in reverse orientiation.

Amplifications The **amplification** of proto-oncogenes increases the probability of a cancer cell to survive. Healthy cells recognize gene amplifications as genetic instability, and the tumor suppressor induces apoptosis in such cells. Since many tumors carry mutated TP53 genes, apoptosis is not induced. Instead, the amplified DNA sequence (**amplicon**) is further amplified in subsequent replication rounds. Gene amplifications are typical for tumors in the progression phase. Examples are the amplification of the epidermal growth factor receptor genes *HER1(EGFR/cErbB1)* and *HER2(neu/cErbB2)* in breast cancer and other tumor types.

Deletions Deletions are characterized by a loss of genetic material. The size of a deletion can vary considerably from single bases to long chromosomal parts. Small deletions can cause a shift of the open reading frame during transcription (frameshift mutations) and the translation of inactive or truncated proteins. Parts of genes, entire genes or several genes can be lost by larger deletions. An example represents the tumor suppressor gene *CDKN2A*, which codes for p16INK4a. Since p16 has an important role in the regulation of cell growth, the deletion of the coding gene causes uncontrolled growth of tumor cells.

Oncogenes and Tumor Suppressor Genes

> **The targets for carcinogen-induced mutations are proto-oncogenes and tumor suppressor genes. Mutations critical for carcinogenesis may also occur in DNA repair genes. Defects in DNA repair predispose to tumor formation similar to mutations in proto-oncogenes and tumor suppressor genes. Chemical carcinogenesis is a multistep process based on the accumulation of mutations in more than one critical target gene.**

Oncogenes Initially, oncogenes were identified in cancer-causing viruses. There are two types of tumor viruses:

- DNA viruses with linear, double-stranded DNA
- RNA viruses, which carrry a reverse transcriptase to rewrite RNA into DNA.

Viral oncogenes (v-onc) are viral genes that have a sequence similiar to cellular proto-oncogenes. Proto-oncogenes are DNA sequences involved in the regulation of cellular differentiation and proliferation. Mutations in proto-oncogenes may lead to "true" oncogenes which promote dedifferentiation and uncontrolled proliferation common to cancer. Upon invasion of a cell by RNA-based retroviruses, the viral reverse transcriptase generates DNA, which may be inserted into the DNA of the host cell. Transcription of the incorporated viral genome may lead to escape from regulated growth.

Alternatively, the reverse transcriptase-generated DNA may activate an otherwise latent oncogene to initiate excessive cellular proliferation. The net effect is transduction of a normal into a malignant cell.

The proteins encoded by proto-oncogenes (oncoproteins) are involved in the cellular regulation of proliferation, apoptotis, and differentation signals. These signals interact with the cell via surface rceptors that transmit the information to the nucleus via signal transduction. The consitutive activation of proto-oncogenes by mutation and genomic instability causes a permanent and uncontrollable activation of the corresponding signaling pathways. According to their cellular localization and biochemical features, oncoproteins can be classified as follows:

- growth factors (e.g., Sis, Int-1)
- growth factor receptors (e.g., Kit, Met, ErbB1/Her1, ErbB2/Her2)
- cytoplasmic protein tyrosine kinases (e.g., Bcr-Abl, Src)
- cell membrane-associates guanine nucleotide-binding proteins (e.g., H-Ras, K-Ras, N-Ras)
- soluble cytoplasmic serine threonine protein kinases (e.g., Raf, Mos)
- nuclear proteins (e.g., c-Myc, c-Fos, c-Jun, Ets)
- anti-apoptotic proteins (e.g., Bcl-2).

Tumor Suppressor Genes

> **In contrast to oncogenes, *tumor suppressor genes* protect from carcinogenesis. Whereas oncogenes act in a dominant manner, tumor suppressor genes are recessive.**

Mutations and deletions inactivate tumor suppressor genes. While proto-oncogenes are activated sporadically in somatic cells, the inactivation of tumor suppressor genes can occur sporadically in somatic cells as well as in the germ cells. The origin of familial tumors, as well as inherited diseases with predisposition for cancer, is frequently caused by mutated tumor suppressor genes. Examples of important tumor suppressor genes are shown in Table 2.19.

> **Of the plethora of oncogenes and tumor suppressor genes, the ras-oncogenes and the *TP53* tumor suppressor will be discussed in more detail here.**

Ras proto-oncogenes There are four different Ras proteins (H-Ras, N-Ras, K-Ras4A, and K-Ras4B), which share 90% homology. Ras proteins are associated with the inner side of the cell membrane. Their ability to transduce external signals from surface receptors is regulated by the binding of GDP, which inactivates RAS, and by GTP, which activates RAS. Activated

Table 2.19 Tumor suppressor genes.

Gene	Tumor	Hereditary syndrome
TP53	Diverse	Li Fraumeni syndrome
RB1	Retinoblastoma	Retinoblastoma
APC	Colon carcinoma	Familiar adenomatosis polyposis
hMSH2	Colorectal carcinoma	Hereditary non-polypous colon carcinoma
NF1	Fibroma	Neurofibromatosis type 1
NF2	Schwannoma, menningioma	Neurofibromatosis type 2
MEN1	Insulinoma	Multiple endocrine neoplasia 1
MEN2	Pheochromocytoma	Multiple endocrine neoplasia 2
WT1	Wilms' tumor	Wilms' tumor
VHL	Kidney carcinoma	von Hippel Lindau syndrome

Ras molecules bind and activate effector molecules. Mutations in *ras* genes occur at codon 12, 13, or 61. Approximately 30% of all human tumors carry *ras* mutations, which cause permanent uncontrollable activity. Aberrant *ras* activation induces carcinogenicity by various mechanisms, e.g. increased cell proliferation, loss of cell cycle control, decreased apoptosis, increased angiogenesis, invasion, and metastasation.

Another Ras-dependant signal transduction pathway represents the phosphoinositide-3′-OH-kinase (PI3K)-mediated activation of Akt and Rac. After activation by Ras, PI3K phosphorylates the signal molecules phosphate idylinositole-4,5-biphosphate to form phosphateidylinositole-3,4,5-triphosphate (PIP_3). PIP_3 activates Akt, which upregulates the anti-apoptotic transcription factor NF-κB.

Tumor Suppressor Gene **TP53**   *TP53* is a gene that codes for the transcription factor p53, which is a tumor suppressor protein. Mutations in *TP53* result in loss of p53 function. Roughly half of all human tumors contain mutations in the *TP53* gene. Other tumors switch off p53 by amplification of the *MDM2* gene, which is an inhibitor of p53. Thus, the inactivation of p53 promotes carcinogenesis. Various hypotheses, which complement one another, help to explain the anticarcinogenic action of this gene:

- Termed the "guardian of the genome", wild-type p53 protects the genome from carcinogenic mutations caused by genotoxic stress such as UV radiation, nutrional factors, or xenobiotic compounds. The strongest evidence for this hypothesis comes from *TP53* knockout mice that all develop tumors, and patients with inherited *TP53* germ line mutations, who develop a cancer-linked disease known as the Li Fraumeni syndrome.
- Wild-type p53 induces apoptosis, which is a strong deterrent to the development of cancer. Tumors with mutated, inactivated p53 are resistant towards apoptosis. Tumor cells with mutated p53 exhibit a survival advantage compared to tumors with wild-type p53, which explains the high frequency of *TP53* mutations among human tumors.

Tumor cells can grow more rapidly than blood vessels are formed, which supply them with oxygen and nutrients because they can function better than normal cells under relatively anerobic conditions. In normal cells hypoxia activates wild-type p53, which induces apoptosis. Because mutations in p53 inhibit apoptosis, tumor cells survive under hypoxic conditions.

Apoptosis

> **Upon treatment with chemical carcinogens, cells may die by apoptosis, necrosis, autophagy or may undergo "premature senescence". Apoptosis appears to be the major route of cell death triggered by DNA damage. It is executed via a specific cellular program, involving the death receptor and/or the mitochondrial pathway. A key player in DNA damage-triggered apoptosis is p53. Apoptosis is thought to eliminate genetically damaged cells and, therefore, counteracts carcinogenesis.**

The initial cellular reaction to DNA damage is cell cycle arrest. This provides the cell with time to repair lesions before they are transmitted to the next cell generation. At extreme high amounts of lesions, which superceeds the cellular repair capacity, the cell switches to **programmed cell death (apoptosis)**. The death of a damaged cell prevents its malignant deterioration. This keeps the entire organism in a healthy state. Many tumors show a reduced ability to induce apoptosis, i.e. by mutations in apoptosis-inducing genes or permanent activation of apoptosis-repressing genes.

The switch from DNA repair to apoptosis (see the simplified scheme in Figure 2.66) is regulated by the cellular energy balance. Long-running DNA repair consumes much energy in the form of ATP. This energy is not further available for ATP-dependent ion pumps, which maintain low concentration of Ca^{2+} ions intracellularly. Rising Ca^{2+} levels activate the apoptotic cascade. Since apoptosis is also an energy-consuming process, apoptosis causes the cleavage of ATP-dependent DNA repair enzymes, i.e. ATM, DNA-PK, and PARP, thus DNA repair processes are switched off. Since PARP inhibits p53 by poly(ADP)adenylation, the cleavage of PARP activates p53, which in turn induces apoptosis.

With time, high apoptosis rates lead to the selection of apoptosis-resistant cell clones. Apoptosis resistance and insufficient DNA repair cause an accumulation of mutations, an increase in genomic instability, and tumor progression.

Regulation of Apoptosis by p53   The half-life of p53 protein is usually short but it may be extended by stabilization, in the face of genotoxic stress. After DNA damage, p53 arrests the cell cycle in the G1 or G2 phase by transcriptional activation of cell cycle-regulating proteins. DNA lesions result in extension of p53 activity via the phosphorylation of p53 through ATR - CHK1, ATM - CHK2 or the HiPK2 kinase, and the inhibition of the binding of Mdm2 to p53, which gives rise to p53 stabilization. When DNA repair is completed Mdm2 binds again to p53, which leads to its degradation, and the cell cycle continues. However, if cellular repair capacity is saturated, apoptosis may proceed via activation of the *Bax* gene, direct interaction of p53 with mitochondria, degradation of Bcl-2, release of cytochrome C, and activation of the apoptosome, or by a non-mitochrondrial pathway by activating apoptosis-inducing death receptors. In this case, p53 phosphorylated by the HiPK2 kinase at serine 46 is crtically involved, stimulating the promoter of the death receptor gene *FAS*.

> **There are both intrinsic and extrinsic pathways to apoptosis. The intrinsic pathway is activated by mitochondrial membrane damage, whereas the extrinsic pathway involves the activation of death receptors localized at the outer cell membrane. Both pathways can be activated by genotoxic stimuli.**

Intrinsic Pathway of Apoptosis   The intrinsic pathway is initiated by **mitochondria** and proteins of the **Bcl-2 family**. They are mainly localized in the outer mitochondrial membrane, but to a lesser extent also in the endoplasmic reticulum and the nuclear membrane. The family is composed of a series of both anti-apoptotic, e.g. Bcl-2, and pro-apoptotic, e.g. Bax, proteins. The Bax protein forms pores in the mitochondrial membrane, which allow the release of **cytochrome c** from the mitochondria into the cytosol. Cytochrome *c* stimulates the formation of the **apoptosome**, a protein complex consisting of Apaf-1 and procaspase-9. Procaspase-9 is autocatalytically cleaved in the apoptosome to caspase-9, which in turn activates downstream caspases required for apoptosis. The permeabilization of the mitochondrial membrane activates further pro-apoptotic proteins and AIF, the apoptosis-inducing factor.

Caspases are cysteine aspartyl proteinases. **Initiator caspases** (caspase-8, -9, -10) cleave and activate downstram **executioner caspases** (caspase-3, -6, -7) to amplify pro-apoptotic signals. Executioner caspases break down many cellular substrates. Laminin cleavage leads to chromatin condensation and nuclear shrinkage. The degradation of some DNA repair enzymes has already been mentioned above. Caspase-3 finally degrades ICAD, the inhibitor of caspase-activated DNAse that in turn becomes activated and cleaves nuclear DNA in nucleosomal size fragments. The degradation of the cytoskeleton proteins actin and plectrin causes cellular fragmentation and the release of **apoptotic bodies**, which are attacked by phagocytes. Phosphatidyl serine, which is localized at the inner side of plasma membranes of healthy cells, relocates to the outer surface of apoptotic cells and provides a point of attack for phagocytosis. Furthermore, there is a caspase-independent pathway of cell death. It is termed as **autophagy** or **type II cell death**, and is regulated by death-associated proteins (DAPs).

Extrinsic Pathway of Apoptosis   This pathway is initiated by the binding of specific death ligands to death receptors. There are at least six death receptors and a number of death ligands.

The ligand called TRAIL binds to the functional receptors TRAIL-R1 and TRAIL-R2, which transduce signals for apoptosis induction. Interestingly, TRAIL also binds to non-functional receptors called "decoy receptors" that may antagonize TRAIL-induced apoptosis.

The major death receptor is FAS (alias APO-1 or CD95). This receptor is subject to upregulation by genotoxins via p53. It becomes activated by the Fas ligand. The formation of a multimer protein complex at the intracellular site of the death receptor is crucial for the transduction of apoptosis signals. After binding of a ligand trimerization of the death receptor occurs. The adapter molecule ***Fas-associated death domain protein*** (**FADD/Mort1**) as well as procaspase-8 and caspase-10 are involved in complex formation. This complex is termed ***death-inducing signal complex*** (**DISC**). Procaspase-8 activates itself by autocatalytic cleavage. Active caspase-8 can transduce the signals in two ways:

- Large amounts of caspase-8 are activated in the DISC. Caspase-8 cleaves procaspase-3, and caspase-3 degrades several target proteins by proteolysis.
- Small amounts of caspase-8 are activated in the DISC, and signal transduction is performed via the mitochondria. Caspase-8 cleaves another member of the Bcl-2 family, Bid. Cleavage of Bid allows translocation of Bax to the mitochondria that permeabilizes the mitochondrial membrane, and cytochrome c is released into the cytosol. The apoptosome is generated and further caspases such as caspase-9 and -3 are activated (Figure 2.66).

Carcinogen-Induced Immortalization of Cells

> **A property of carcinogen-transformed cells is their ability to propagate indefinitely. This phenotype is designated as immortalization. A main mechanism of immortalization rests on upregulation of the enzyme telomerase that provokes telomere lengthening. Usually, tumor cells exhibit high telomerase activity and long telomeres. Whether telomerase gets upregulated as a direct consequence of carcinogen exposure is an open question.**

Normal diploid cells only have a limited doubling capability. They pass over to cellular ageing (senescence) after 50–70 cell divisions. The number of cell doublings depends on the length of telomeres. **Telomeres** have a size of 5–10 kb and consist of repetitive sequences $(TTAGGG)_n$. They cover the ends of chromosomes to protect DNA. During every cell division the telomeres shorten by 50–100 bp since the DNA replication machinery cannot replicate chromosomal ends. This phenomenon has been termed the **end replication problem**. If the telomeres reach a critical length, p53 is activated and a permanent cell cycle arrest or apoptosis is induced.

Telomerase counteracts telomere shortening by new synthesis of repetitive TTAGGG sequences. The enzyme consists of a RNA component (TERC), which serves as template for the synthesis of telomere sequences, and a reverse transcriptase (TERT). The activity of telomerase is highly restrictive. Telomerase is mainly active during embryogenesis. In the adult organism, telomerase activity is found in germ cells, stem cells, and activated lymphocytes. Most other mitotically active tissues (gastrointestinal epithelial cells, skin cells etc.) do not show telomerase activity. A continued telomerase shortening happens as a consequence in these tissues. The telomere length remains constant in mitotically inactive tissues (brain, heart etc.). Factors such as oxidative stress, epigenetic chromatin alterations, and others accelerate telomerase shortening and thus contribute to faster ageing.

Telomerase is active in over 80% of human tumors. The expression of telomerase in normal cells causes cellular transformation. **Transformation** confers unlimited growth cap ability on cells. The inhibition of telomerase activity prevents carcinogenesis. Though transformed cells are immortalized, they are not cancer cells. Hence, telomerase is not an oncogenic factor. Nevertheless, **telomerase supports the malignant process of carcinogenesis**. Telomere shortening contributes to cancer initiation. Chromosomes with shortened telomeres show higher genetic instability than chromosomes with long telomeres. Furthermore, telomerase is expressed as a consequence of p53 inactivation. Telomerase activity also contributes to tumor progression. Since telomere shortening can be induced by chemical carcinogens, e.g. as a result of adduct formation in telomeric sequences and their misrepair, it can be a mechanism for chemical carcinogenesis.

Chromosome Segregation and Aneuploidy

> **Normal human somatic cells possess 46 chromosomes (diploid chromosome set). Germ cells are haploid and posses 23 chromosomes. Most tumors have more than 46 chromosomes; the numbers vary between 60 and 90. These abnormal chromosome numbers are termed** *aneuploidy*.

Aneuploid tumor cells are frequently more than diploid but less than tetraploid. The explanation is that hyperdiplopid sets of chromosomes derive from a duplication of a diploid set of chromosomes (tetraploidy), and a gradual loss of single chromosomes takes place subsequently.

Aneuploidy contributes to the development of an aggressive phenotype during tumor progression, especially by polysomy of chromosomes with activated oncogenes or mutated tumor suppressor genes. Although most tumors are aneuploid, there is a small fraction of diploid tumors. They reveal specific defects in DNA repair pathways. The mechanisms of aneupolidy are different from those which lead to the accumulation of DNA mutations. DNA repair defects lead to increased mutational rates in microsatellite sequences. This phenomenon is termed **microsatellite instability (MIN)**. By contrast, **chromosomal instability (CIN)** causes aneuploidy. It is possible that CIN accelerates loss of heterozygosity of tumor suppressor genes and duplication of chromosomes harbouring oncogenes. Hence, tumor cells may get a growth advantage during carcinogenesis.

Abnormal mitotic spindles: Aneuploid tumors possess supernumerous centromers. An amplification of chromosomes increases the probability for abnormal mitosis and erronous chromosomal segregation. STK15/Aurora2 represents a centromer-associated serine-threonine-kinase, which is overexpressed in aneuploid tumors. The coding gene can be amplified. Aurora kinases phosphorylate histone proteins and foster the condensation of chromosomes.

Mitotic spindle checkpoints: These checkpoints guarantee the correct adjustment of chromosomes during metaphase and their fixation to the mitotic spindle. Two checkpoints are known:

- During the G_2M checkpoint, microtubule-dependent processes are controlled, i.e. the separation of duplicated centromers during the G_2 phase and delay of the G_2M transition in the presence of tubulin poisons.
- The metaphase checkpoint controls the fixation of the mitotic spindle to the kinetochors. If single kinetochors are not fixed to the spindle, the separation of daughter chromatids is stopped to enale fixation.

Two mutations can be found in protein which regulate the mitotic checkpoints: BUB1, BUBR1, and MAD2.

Neoangiogenesis in Chemical Carcinogenesis

> The formation of new blood vessels in tumor tissues is termed *neoangiogenesis*. It is necessary for tumor growth and for metastasis. Neoangiogenesis is an essential element of tumor progression. Tumors and normal tissues require oxygen and nutrients.

Early in carcinogenesis tumor growth is limited to about 0.5–1 mm, until such time neoangiogenesis provides nearby blood vessels for a supply of oxygen and nutrients. The avascular phase terminates and the vascular phase begins when new blood vessels reach as far as 150–200 μm from the tumor. In the absence of neoangiogenesis the tumor undergoes apoptosis.

Tumor neoangiogenesis is regulated by various pro- and anti-angiogenetic factors. If the balance is switched to pro-angiogenic, it is activated by the following:

- Angiogenic oncoproteins, e.g. Ras, upregulate the expression of pro-angiogenic proteins such as vascular endothelial growth factors (VEGFs), fibroblast growth factors (FGFs), platelet-derived growth factor (PDGF), and epidermal growth factor (EGF). Oncoproteins can also downregulate angiogenesis inhibitors such as thrombospondin, endostatin, angiostatin, and tumstatin.
- Mutations in tumor suppressor genes, e.g. *TP53*. Mutated p53 downregulates the expression of thrombospondin and upregulates vascular endothelial growth factor A. Tumor cells with p53 mutations are selected under hypoxic conditions, since they are apoptosis-resistant and survive.

Tumors also usurp existing vessels in normal tissues by displacement of normal cells. This process is called **vessel cooption**. Whether neo-angiogenesis can be driven by chemical carcinogens, thus contributing to tumor progression, is still a matter of dispute.

2.8.4 Non-Genotoxic Mechanisms of Carcinogenesis

> In recent years, the relevance of non-genetic, so-called epigenetic mechanisms for carcinogenesis has been recognized. Aberrant methylation of cytosines and deacetylation of histones are central epigenetic mechanisms. Furthermore, agents have been unravelled that do not directly damage DNA but promote tumor development by diverse alternative mechanisms, e.g. stimulation of signaling pathways triggering cell cycle progression or inhibition of apoptosis of initiated cells. So-called non-genotoxic carcinogens, e.g. asbestos and classical tumor promoters (e.g. TPA), may, however, also indirectly damage DNA by endogenous radical formation. DNA damaging radical burst can also be provoked during inflammation and infection. Sustained oxidative stress together with a chronic stimulus of cell division is an important driving force of tumor formation.

Epigenetics

> Apart from genetic lesions, there are epigenetic mechanisms of carcinogenesis. Epigenetic alterations are base sequence unrelated genomic changes that are tansmitted to the next generation of cells. They are, however, not related to mutations by base substitutions or deletions and translocations, but rather affect the transcriptional cellular program. *Epigenetics* deals with the propagation of information by regulatory mechanisms of temporally and spatially different gene activities.

The adaptation of organisms to changing environmental conditions by adaptive and hereditary epigenetic alterations challenges the central dogma of biological evolution, which dictates that mutation and selection are the sole driving forces of evolution. Epigenetic mechanisms are **alterations of the chromatin structure** via posttranslational histone modification, nucleosome adjustment, and DNA chromatin complexes as well as **methylation of cytosines** in the DNA. DNA methylation is also involved in **imprinting**, where homologous genes are differentially expressed depending on their maternal or paternal inheritance. This is in contrast to classical Mendelian genetics.

About 4% of cytosines are methylated in mammalian DNA. The main function of DNA methylation represents the transcriptional regulation of gene expression during development, in different tissues and organs, for imprinting, and for the switch off of transposon elements. Generally, DNA methylation causes repression of gene expression. Typically, methylated promoter regions suppress gene transcription. In some cases, hypermethylation, which is associated with increased gene expression, occurs in non-promoter regions, The maintenance of different methylation patterns is important for the regulation of normal gene expression.

On the one hand, methylation of promoter regions with high CG contents (**CpG islands**) inhibits the binding of transcription factors. On the other hand, other proteins can bind to these DNA areas and displace the actual transcription factors. Additionally, methylation of CpG islands is coupled with the **deacetylation of histones**. Histones are "packaging" proteins, which wrap DNA to repetitive nucleosomal units and fold them to chromatin fibers of higher order. Histones are modified at the post-translational level via acetylation, methylation, phosphorylation, and ubiquitination. The acetylation is controlled by histone acetyltransferases and histone deacetylases. Acetylation is linked to nucleosomal rearrangement processes and transcriptional activation. Deacetylation causes transcriptional repression by chromatin condensation.

DNA methylation and histone deacetylation support each other in terminating gene transcription. The **DNA methyltransferases** DNMT1 and DNMT3b are involved in DNA methylation. S-adenosyl methionine serves as a source for methyl groups. DNMTs recruit histone deacetylases. After the binding of the methyl-CpG-binding protein, MeCP2, to methylated DNA, the SIN3 protein, which is a part of the deacetylase complex, binds to MeCP2. This leads to histone deacetylation of methylated DNA.

Hypomethylated CpG islands are a characteritic feature of many tumor types. DNA hypomethylation during carcinogenesis indicates that carcinogenic chemicals can alter methylation patterns of genes that contribute to the toxic effects of xenobiotics or are oncogenic. For example, cadmium inhibits DNMT1 and thereby causes hypomethylation. Arsenic induces hypomethylation of the *ras* oncogene. Altered DNA methylation and histone acetylation have been observed in nickel-induced carcinogenesis.

The genetic and the epigenetic models of carcinogenesis mutually supplement each other. The epigenome as well as the genome contribute to carcinogenesis by specific alterations in gene expression. Epigenetic changes fit to the classical multiple-step model of carcinogenesis consisting of initiation, promotion, and progression. Potentiation of genetic lesions by epigenetic alterations contributes to the explanation of the age-related increase in cancer incidence. Since epigenetic alterations accumulate during an individual's life span, they may lead to a transient erosion of DNA methylation patterns in specific genes involved in carcinogenesis.

Non-Genotoxic Carcinogens and Tumor Promoters

> The term *non-genotoxic carcinogens* is a designation for a heterogenous group of agents that attack DNA indirectly. These agents do not form DNA adducts. Their carcinogenic potential is highly variable; upon acute exposure it is usually lower than the potential of genotoxic carcinogens that act as tumor initiators. However, looking more closely at DNA damage, it becomes more and more evident that many (if not all) non-genotoxic carcinogens are able to damage DNA indirectly by generating ROS and RNS or by inhibition of DNA metabolizing enzymes, such as DNA repair proteins. Indirect induced DNA damage may occur not only in the affected cell, but also in the surrounding cell population (bystander effect). For example, asbestos and wood dust (nanoparticles) may provoke a chronic radical burst in lung macrophages that causes genetic damage in the neighboring cells. If these cells are stimulated to undergo DNA synthesis, mutations may be induced in cells occupying the inflammatory area. Thus, non-genotoxic carcinogens may drive tumor formation by genetic damage they induce indirectly. The effect may even be enhanced by chronic stimulation of cell propagation that supports fixation of mutations. Non-genotoxic carcinogens are, therefore, often at the same time tumor promoters and vice versa.

Since point mutations and chromosomal aberrations have the potential to activate proto-oncogens and inactivate/delete tumor suppressor genes, all DNA damaging agents should be considered to act as genotoxic carcinogens. Non-genotoxic carcinogens, some examples of which are given in Figure 2.67, do not directly bind to and damage DNA. However, even without any reactivity towards DNA, DNA damage may be provoked by them indirectly. The most common way in which this can occur is radical formation. During chronic inflammation, because of infections or exposure to particles or fibers, macrophages and granulocytes respond with an oxygen burst that is strong enough to kill infectious particles. ROS and RNS produced in inflammatory tissues may damage cellular DNA forming, among other lesions, 8-oxo-guanine, which leads to mutations (due to mispairing with adenine). Asbestos permanently activates lung macrophages, causing chronic inflammation and presumably DNA damage. Another example is the tumor promoter TPA that promotes cell division by stimulation of the PKC-driven ERK pathway targeting c-Fos/AP-1. However, TPA was also shown to be able to induce chromosomal aberrations. The mechanism is unknown; it is likely to be related to the oxygen burst induced by TPA in the cell. Some carcinogenic metals such as arsenic do not directly bind to DNA. However, they are able to inhibit DNA repair enzymes and, therefore, may increase the "spontaneous" level of DNA damage. Given the fact that several hundred 8-OxoG and upto ~30.000 apurinic sites are formed per cell per day, inhibition of DNA repair enzymes by carcinogenic heavy metals will clearly contribute to the endogenous genotoxic stress level. These examples demonstrate that a clear distinction between genotoxic and non-genotoxic carcinogens is often arbitrary.

It should be noted that agents that are potent in inducing point mutations are clearly positive as tumor initiators, whereas most non-genotoxic carcinogens rather act as tumor promoters. Therefore, agents and exposures that do not directly attack DNA and have a low potency to induce point mutations should be considered as tumor promoters. This group encompasses not only the

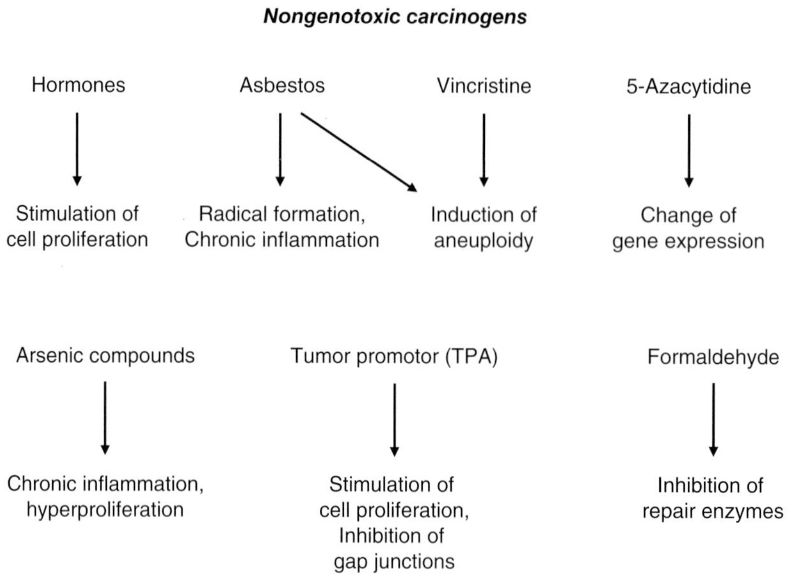

Nongenotoxic carcinogens

Hormones	Asbestos	Vincristine	5-Azacytidine
↓	↓ ↘	↓	↓
Stimulation of cell proliferation	Radical formation, Chronic inflammation	Induction of aneuploidy	Change of gene expression

Arsenic compounds	Tumor promotor (TPA)	Formaldehyde
↓	↓	↓
Chronic inflammation, hyperproliferation	Stimulation of cell proliferation, Inhibition of gap junctions	Inhibition of repair enzymes

Figure 2.67 Examples of non-genotoxic carcinogens and tumor promoters. TPA, tetradecanoyl phorbole acetate.

classical tumor promoters TPA, phenobarbital, and chlorinated hydrocarbons (DDT, TCDD) but also hormones (estradiol, diethylstilbestrol) and peroxisome proliferators. All they have in common is that they stimulate cell proliferation, which is required for fixation of mutations and the expansion of initiated cells.

Not all genotoxic agents are potent tumor initiators. A classical example is methyl methanesulfonate (MMS), which alkylates DNA to form predominantly N7-methylguanine and N3-methyladenine. These types of alkylating lesions are not mispairing. They give rise, however, to apurinic sites and therefore MMS is a very weak mutagen but rather a strong clastogen. Interestingly, MMS is quite potent in promoting tumor formation in mouse skin. Thus, given after a single dose of DMBA, MMS is as effective as TPA in tumor promotion. Both TPA and MMS induce chromosomal aberrations in skin keratinocytes. This has been taken to suggest that chromosomal aberrations together with stimulation of proliferation may drive tumor promotion, e.g. by loss of tumor suppressor genes. MMS provides an example to show that even clear genotoxic agents are not necessarily tumor initiators and, if they were applied on their own, would be only very weakly carcinogenic.

Should tumor promoters be considered as non-genotoxic carcinogens? If an agent has been experimentally proven not to be able to induce tumors on its own, it is, by definition, a tumor promoter and should not be considered as a "carcinogen". However, chronic application of a promoter may also cause tumor formation, although at a low frequency compared to the effect obtained if it is applied in combination with an initiator. Whether the slight tumorigenic effect of tumor promoters is due to fixation and expression of DNA lesions that were brought about by endogenous DNA damaging metabolites or to the indirect genotoxic effect of "promoters" (e.g., by endogenous radical formation) remains an open question.

2.8.5 Implications of Initiation and Promotion for Risk Assessment

Notably for the purposes of risk assessment it is important to distinguish between genotoxic and non-genotoxic carcinogens. In contrast to genotoxic carcinogens, the effect of non-genotoxic carcinogens (tumor promoters) is thought to be reversible and detectable only above a given dose threshold. Even if non-genotoxic carcinogens induce genetic changes indirectly, the level might be quite low and insufficient to induce cancer if the promoting agent was applied on its own. Therefore, for practical purposes most, if not all, non-genotoxic carcinogens should be considered as tumor promoters that exhibit a dose threshold in order to induce, upon chronic exposure, malignant transformation of cells.

2.8.6 Summary

> **Cancer-causing agents are termed carcinogens, and carcinogenic compounds that damage DNA are designated as being genotoxic. DNA alterations transform normal cells to tumor cells. Changes in RNA, proteins, and lipids are not causatively related to the development of cancer. Under certain circumstances they can, however, support malignant growth.**

The knowledge of molecular mechanisms of carcinogenesis has increased dramatically in recent years. Nevertheless, despite the sequencing of the human genome we do not understand the full range of molecular mechanisms of carcinogenesis. Apart from DNA alterations, transcriptional and translational mechanisms (alternative splicing), and post-translational modifications (phosphorylation, glycosylation, acetylation, methylation, ubiquitinylation etc.) must be taken into account in elucidating the spectrum of carcinogeneic and tumor-promoting mechanisms.

Although many carcinogenic and co-carcinogenic compounds have been identified in the past few decades and successful protection is now possible, personal risk assessment remains difficult to perform.

Pharmacogenetics (single nucleotide polymorphisms) and the novel "-omics" technologies (genomics, proteomics, metabonomics) may help to define patterns specifically linked to carcinogenic processes or individual risks. It is to be expected, however, that this new dimension of molecular analyses will raise even more questions than it can answer.

A new discipline on the horizon is systems biology. It attempts to integrate the vast amount of data generated by the "omics" technologies and to generate not only models to describe mechanisms and diseases, but also to develop predictive bioinformatic models for carcinogenic risk assessment at the level of each single individual.

Further Reading

Benigni R. (2005) Structure-activity relationship studies of chemical mutagens and carcinogens: mechanistic investigations and prediction approaches. *Chem Rev*, **105**, 1767–1800

Christmann M, Tomicic MT, Roos WP and Kaina B. (2003) Mechanisms of human DNA repair. *Toxicology*, **193**, 3–34.

Dixon K, Kopras E. (2004) Genetic alterations and DNA repair in human carcinogenesis. *Semin Cancer Biol*, **14**, 441–448.

Fahrer J, Frisch J, Nagel G, Kraus A, Dörsam B, Thomas AD, Reißig S, Waisman A, and Kaina B. (2015) DNA repair by MGMT, but not AAG, causes a threshold in alkylation-induced colorectal carcinogenesis. *Carcinogenesis*, **36**, 1235–1244.

Goodman JI, Watson RE. (2002) Altered DNA methylation: a secondary mechanism involved in carcinogenesis. *Annu Rev Pharmacol Toxicol*, **42**, 501–525.

Kim YI. (2004) Folate, colorectal carcinogenesis, and DNA methylation: lessons from animal studies. *Environ Mol Mutagen*, **44**, 10–25.

Klaunig JE, Kamendulis LM. (2004) The role of oxidative stress in carcinogenesis. *Annu Rev Pharmacol Toxicol*, **44**, 239–267.

Roos WP and Kaina B. (2006) DNA damage-induced cell dearth by apoptosis. *Trends in Mol Medicine*, **12**, 440–450.

Roos WP, Thomas AD and Kaina B. (2016) DNA damage and the balance between survival and death in cancer biology. *Nature Rev Cancer*, **16**, 20–33.

Valko M, Rhodes CJ, Moncol J, Izakovic M, Mazur M. (2006) Free radicals, metals and antioxidants in oxidative stress-induced cancer. *Chem Biol Interact*, **160**, 1–40.

Williams GM. (2001) Mechanisms of chemical carcinogenesis and application to human cancer risk assessment. *Toxicology*, **166**, 3–10.

2.9 Threshold Effects for Genotoxic Carcinogens

Helmut Greim[1]

2.9.1 Introduction

> **Animal experiments indicate that ineffective doses of genotoxic carcinogens do exist and that the dose–response curves of DNA adducts does not correspond with the resulting mutations. The dose–response curve of the mutations starts rising at higher concentrations than that of DNA adducts. The fact that the rate of spontaneous mutations does not increase at lower exposure concentrations also remains undisputed.**

There has been a general assumption that genetic changes are responsible for the development of cancer and that an ineffective concentration of genotoxic carcinogens (the no observed effect level, NOEL) cannot be derived. However, the increasing understanding of cellular defence mechanisms challenges the assumption that even the lowest exposure to a genotoxic agent can increase the risk of cancer.

Accordingly, the scientific expert committee of the European Food Safety Authority (EFSA, 2005) and the three scientific committees of the Directorate Generals of the European Commission (EC, 2009) assume that a dose exists below which genotoxic effects do not occur and, therefore, cancer cannot develop. Even the guidelines for Carcinogenic Risk Assessment by the US Environmental Protection Agency (EPA, 2005) also assume a non-linear dose–response curve. Furthermore, Calabrese (2011) examined the data underlying the concept of the linearity of the dose–response relationship for mutagenic effects and came to the conclusion that they could not substantiate the assumption of linearity. The latter was based on studies conducted on Drosophila, which were exposed to radioactive radiation where some of them showed a fall in the rate of spontaneous mutations. Nevertheless, the non-linearity was initially postulated for cancer caused due to radiation and was then adopted by the US National Academy of Sciences Safe Drinking Water Committee in 1977 for all genotoxic chemical carcinogens.

Since then, many descriptive *in vitro* and *in vivo* studies have either confirmed or refuted this concept. However, the examinations that showed an ineffective concentration of the genotoxic carcinogen are often called into question because statistical problems make it impossible

---

[1] This chapter is based on the publication by H. Greim and R. Albertini in *ASU International 2015* and in *Toxicol. Res.* 2015, **4**, 36–45.

to differentiate between the effects caused by low doses and the spontaneous rate of tumours. Even an experiment in 40,800 trout did not allow identification of an ineffective concentration. This shows that it is very difficult to prove the existence of an ineffective dose of gentotoxic carcinogens using such experiments.

2.9.2 Development of Cancer due to Genotoxic Carcinogens

The initial process of the development of cancer goes through the following stages (see Chapter 2.8):

- exposure of the target cell to DNA-reactive agents, and thereby to potentially mutagenic substances
- reaction with the DNA
- defective replication of the damaged section of a DNA or faulty repair of the DNA damage
- mutations in critical genes of the daughter cells.

Although the eventual development of cancer requires several further steps, it should be assumed that cellular defence mechanisms can influence the initial steps and thus prevent the occurrence of mutations at least up to a certain level of exposure.

Exposure of Target Cells

Toxicokinetic factors, e.g. absorption, distribution, metabolic activation and inactivation, initially determine if a potentially genotoxic substance reaches the target cell. Formaldehyde, a reactive substance produced endogenously through metabolism reacts with proteins and DNA, is a classic example. Sufficiently high doses of inhaled formaldehyde in animal experiments lead to mutations and tumours in areas of direct contact, i.e. upper airways. However, extensive experiments have shown that external formaldehyde exposure does not become systemically available and therefore does not add to the endogenous formaldehyde pool (IARC, 2006; Swenberg et al., 2011).

Reaction with DNA and DNA Damage

Mammalian cells are constantly exposed to DNA-reactive substances. They are produced due to either endogenous metabolism (e.g. acetaldehyde, ethylene oxide, formaldehyde or isoprene) or exogenous exposure (e.g. acetaldehyde from vinyl acetate or ethylene oxide from ethylene). In the case of ethylene oxide and acetaldehyde, it could be shown that the DNA adducts formed were comparable to the endogenous ones up to a certain level of exposure. This led to the conclusion that the cells did not have to deal with additional DNA damage in cases of lower levels of exposure.

2.9.3 Cellular Reaction to DNA Damage

Cells adopt several mechanisms to prevent DNA damage:

- **metabolic inactivation of the reactive agent**
- **protection of genetic material through membranes or proteins that intercept the reactive agents**
- **repair of the DNA lesions**
- **elimination of damaged cells through apoptosis or necrosis.**

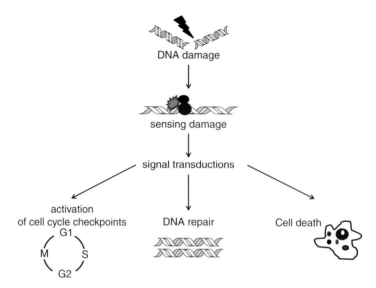

Figure 2.68 *The DNA damage response (DDR) system. Signals are transmitted after recognition of the damage. This activates enzymes through phosphorylation, which in turn slows down the cell cycle, activates DNA repair, or induces cell death programmes (apoptosis and necrosis) in cases of severe damage. Figure reproduced by permission of the Royal Society of Chemistry (RSC) from Chapter 4.2 by Surova and Zhivotovsky, in Greim and Albertini (2012).*

Based on the extensive literature available, it has been generally recognised that the metabolic inactivation of endogenous and exogenous genotoxic agents reduces the exposure of critical cellular structures like DNA, and that reactive agents are intercepted by membranes and proteins (see Chapter 2.4). Here various DNA repair mechanisms (homologous recombination, non-homologous end joining, nucleotide excision repair and mismatch repair) are described that are active depending on the cell cycle and also play a role in preventing DNA damage during DNA synthesis because the damage can be passed on to the daughter cells. The entire process of repair is regulated by the DNA damage response (DDR) system, which controls the repair, the rate of cell division and the mechanisms of apoptosis and necrosis. The individual mechanisms are briefly described below (Figure 2.68).

DNA Repair

> When DNA damage occurs, the process of DNA repair is activated and a reduction in the rate of cell division takes place. This allows more time for the repair, which thereby prevents the passing on of the damage to the daughter cells.

Base Excision Repair System According to Almaida and Sobol (2007) endogenous metabolism results in about 10,000 instances of DNA damage per cell per day through processes like oxidation and phosphorylation. Among these, oxidative damage caused due to the formation of reactive oxygen species, cytochrome P450 reactions and inflammation is of primary importance. Other causes of oxidative damage are intracellular metabolism of foreign substances including drugs, ionising radiation and X-rays. Of all the reactive oxygen species (ROS) released, the *OH radical is the most reactive and is responsible for most of the resulting about 10,000 oxidative DNA lesions per day. Since a normal cell contains only about 500 altered bases, the extent of the capacity of the base excision repair (BER) system becomes clear.

Most cases of oxidative DNA damage are repaired by the BER system. The damage is initially recognised by a glycosylase, which removes the altered base from the deoxyribose. The resulting apurinic site on the DNA strand is then cleaved by an endonuclease and the rest of apurinic part is eliminated by a polymerase and subsequently replaced by the corresponding intact nucleotide. The intact DNA strand is then reconstructed with the help of a DNA ligase. A number of glycosylases that are specific for bases altered due to oxidation are known to date; moreover, the activity of the individual glycosylases can be compensated by the others if one or the other is missing (Figure 2.69). There is evidence that intact DNA is capable of conducting electrical signals, which can be disrupted in cases of DNA damage – this attracts the enzymes responsible for DNA repair.

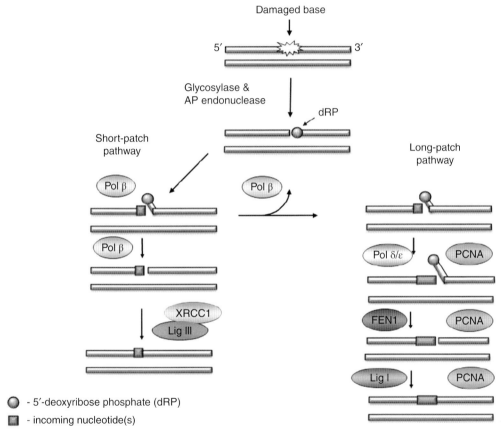

Figure 2.69 Base excision repair. A specific glycosylase eliminates the altered base once the damage is recognised. Polymerase β cleaves the DNA strand and removes the remaining abasic sugar phosphate. The corresponding intact nucleotide is then inserted into this gap and the single strand of DNA is reconstructed by XRCC1/ligase-III complex. When the polymerase β cannot cleave the DNA, multiple nucleotides get attached with the help of polymerase δ/ε generating a flap structure. These flaps are then removed by flap endonucleases after the intact nucleotide has been inserted, so that the ligase can assemble the strand again. Figure reproduced by permission of the Royal Society of Chemistry (RSC) from Chapter 3.3. by Dianov et al., in Greim and Albertini (2012).

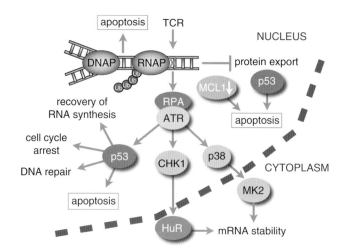

Figure 2.70 Cellular reaction in cases of blocked transcription (transcription coupled repair, TCR). When the RNA-polymerase (RNAP) stops at a damaged part of the DNA and cannot read the strand any further, several mechanisms are activated by RPA (replication protein A) and the stress kinase, ATR, through phosphorylation. Among these, the figure shows the activation of p53, which regulates RNA synthesis, cell cycle, DNA repair and apoptosis, as well as the activation of the kinase, CHK1 and p38, which finally lead to stabilisation of mRNA. The blocked RNAP is degraded through ubiquitylation. Protein synthesis continues while these processes are going on, until all the available free mRNA is used up. As a result, the anti-apoptotic survival factor MCL-1 that inhibits apoptosis is inactivated. Figure reproduced by permission of the Royal Society of Chemistry (RSC) from Chapter 4.3. by Ljungman, in Greim and Albertini (2012).

The information stored in the DNA is read by the RNA, which in turn regulates the synthesis of specific proteins in the proteasomes. This reading is controlled by the RNA polymerase complex. In the case of DNA damage, the process of reading is stopped at the site of the altered base and the process of transcription coupled repair (TCR) is activated. Simultaneously, stress kinases are activated to bring about an arrest of cell division by activating the tumour suppressor gene p53 till the damage is repaired and the synthesis of RNA starts once again. However, protein synthesis comes to a stop only when all the free RNA available in the cytoplasm has been used up. If this pool is completely exhausted, short-lived survival factors, like the anti-apoptotic factor MCL-1, are inactivated to initiate the mechanism of apoptosis that had been suppressed till then (Figure 2.70).

Even the BER system is under strict control. If the activity of its most important enzymes (XRCC1, ligase III and polymerase beta) exceeds the existing DNA damage, the enzymes are marked by ubiquitin chains, which aid their recognition and destruction by proteasomal enzymes. In cases of DNA damage, the ubiquitin ligases are inactivated.

ATM and ATR in the repair of DNA Damage Damages that disrupt DNA replication, like double-strand breaks (DSBs) and DNA cross-links, induce a cascade of repair mechanisms. The ataxia-telangiectasia mutated (ATM) protein is activated through DSBs, whereas the ATRad-3 related (ATR) protein is activated through single-strand breaks and by blocked DNA synthesis. Both are protein kinases, which activate different mechanisms that control DNA repair, rate of cell division and apoptosis (Figure 2.71). ATM activates exonucleases and endonucleases. ATR is activated by ATR-interacting protein (ATRIP) in cases of blocked DNA replication. The resulting

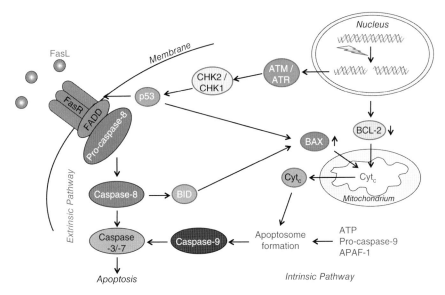

Figure 2.71 Apoptosis after double-strand breaks. See the text for details. Figure reproduced by permission of the Royal Society of Chemistry (RSC) from Chapter 4.1 by Kaina et al., in Greim and Albertini (2012).

complex binds to the replication protein A (RPA) DNA single-strand complex. This leads to phosphorylation of different proteins and thereby to homologous recombination, so that the replication blockade is removed. One of the phosphorylated proteins is histone 2AX, which can be detected microscopically using immunofluorescence and is used as a marker for DSBs resulting from genotoxicity and for their repair.

Apart from these, ATR and ATM control the cell cycle by means of checkpoint kinases 1 and 2 (Chk1 and Chk2). Chk1 is activated through ATR in cases of blockade of DNA replication, whereas Chk2 is activated through ATM in cases of DSBs. Both these enzymes phosphorylate the tumour suppressor p53 thereby preventing its degradation. This tumour suppressor activates p21, which inhibits the production of cyclin E/CDK2 complexes so that the cells do not enter the S-phase. This reduces both, the rate of cell division and DNA replication. Simultaneously p53 activates ERCC3, one of the DNA excision enzymes, which recognises and eliminates the damaged sections of DNA.

In this way, p53 plays a very important role in the repair of DNA damage because it inhibits the cell cycle through various mechanisms and activates DNA repair. The significance of the rate of cell division for the development of cancer has been recently emphasised again. Tissues with a higher rate of cell division have a higher rate of cancer than those with a lower one. The authors explain this by the high rate of DNA damage, which always occurs spontaneously during DNA replication.

Varying Persistence of DNA Adducts DNA adducts are formed by inhalative exposure to butadiene in mice. The half-life of the quantitatively most significant DNA cross-links (bis-N7G-butadiene) in the liver, kidneys and lungs is a few days, whereas the N7G-N1A-BD adducts and the 1,N6-HMHP-dA adducts persist and are possibly responsible for butadiene-induced genotoxicity and carcinogenicity. However, although specific DNA adducts persist longer or are eliminated only slowly, the exposures to butadiene were high and lower doses have not been examined. This could mean that the repair mechanisms could become saturated in cases of higher doses.

It is still not clear if long-term exposures lead to a reduction in the capacity of DNA repair as compared to acute exposures. One possibility that has been discussed is that the DNA repair proteins become restricted in their activity in cases of long-term exposure as a result of mutations caused by gradually accumulating DNA adducts.

Apoptosis and Necrosis In cases of severe or irreparable DNA damage, programmes that eventually lead to cell death are initiated. These can occur through the mechanism of apoptosis or programmed cell death, or through necrosis. Apoptosis is activated by the ATM- and ATR-induced stabilisation of p53. Interaction of p53 with the FasR/FADD receptor activates the caspase-8-triggered mechanism of apoptosis. In the case of mitochondrial mechanism, p53 activates Bax and this leads to an inhibition of Bcl-2. The release of cytochrome C from the mitochondria activates the Apaf monomers, which join together to form the apoptosome. This causes caspase 9 to activate caspases 7 and 3 (Figure 2.71). Both these mechanisms reverse the inhibition of the DNase, so that the DNA is split and nucleosomal fragments are formed.

Severe DNA damage also leads to an activation of PARP-1, which blocks mitochondrial respiration and causes a fall in the levels of NAD and ATP. ATP deficit leads to necrosis. Since the various mechanisms of apoptosis require energy, the process of apoptosis can occur only when mitochondrial respiration remains intact. The energy status of the cell seems to determine which process – apoptosis or necrosis – would occur. In the case of necrosis, the dying (necrotic) cells start swelling up, membranes become permeable, cytoplasmic contents are released and cell death occurs.

Epigenetic Mechanisms The most important processes of DNA repair (recognition of damage, excision of the DNA strand, elimination of the damaged base, DNA synthesis and attaching the repaired strand) are regulated by the XRCC1/ligase complex III and the so-called scaffold proteins. The enzyme proteins that are involved in this process are in turn regulated by epigenetic mechanisms like phosphorylation, acetylation, sumoylation (post-translational modification with small ubiquitin-related modifier (SUMO)), ubiquitylation and methylation. These post-translation mechanisms regulate the binding properties of the enzyme proteins, their turnover and their subcellular localisation and activity. The regulation of the most important enzymes involved in the BER system through ubiquitylation has been described above. Sumoylation is a post-translational modification with SUMO proteins. Hundreds of proteins involved in processes such as chromatin organisation, transcription, DNA repair, macromolecular assembly, protein homeostasis, trafficking and signal transduction are subject to reversible sumoylation. While the methylation of histones can cause specific chromosomal regions to be switched on or off, the process of methylation itself can be regulated by argonaute proteins. These proteins bind to microRNA and smaller sections of RNA, and thereby reduce the translation and the stability of mRNA. They regulate the activity of about 30% of all genes.

Intercellular Communication through Gap Junctions Individual cells within tissues are connected to each other by gap junctions, which permit the passage of small and hydrophilic molecules through them. It is assumed that a number of physiological processes are regulated by this exchange, and this in turn contributes to maintaining homeostasis. The channels are made of six connexin units, the synthesis of which is reduced in tumour cells as a result of an increase in the rate of methylation of connexin promoters. While genotoxic carcinogens do not seem to have an influence on the function of gap junctions, non-genotoxic substances like TCDD, i.e. tumour promoters, influence gap junctions by disturbing the balance between cell proliferation and cell death. The extent to which this disturbance in balance is counter-regulated (e.g. by increased synthesis of connexins) in cases of low exposures to such promoters has not yet been investigated.

2.9.4 Examples of Dose-dependent Reactions in Cases of Genotoxicity

> **The examples available so far are mainly applicable to the different dose–response curves between DNA adducts and mutations. This leads to the inference that DNA adducts are seen as markers of exposure, but are not necessarily considered genotoxic. This is true for the resulting mutations, which usually occur at higher exposures. This has been demonstrated in various examples.**

- Exposure to cyproterone acetate in transgenic big blue rats caused a linear increase in DNA adducts of between 25 and 75 mg/kg, and an increase in the rate of mutation from 75 mg/kg onwards. In earlier studies increased DNA and RNA synthesis and an increased rate of mitosis and liver enlargement occurred only at exposures to more than 40–100 mg/kg of cyproterone acetate. The authors therefore came to the conclusion that DNA adducts lead to mutations when an additional mitogenic effect occurs at higher doses, i.e. when the rate of cell division is stimulated.
- A corresponding parallelism of the dose–response curves between DNA adducts and the development of tumours has also been shown for vinyl chloride, where the DNA changes were detectable at lower doses as compared to the tumours.
- The NOAELs for the clastogenic effect in CD1 mice and Muta^{TM} mice which were treated with EMS (ethyl methanesulfonate) was 80 mg/kg in the case of CD1 mice and 25 mg/kg for the induction of mutations in bone marrow cells in the case of Muta^{TM} mice, while ethyl adducts were detectable on proteins and DNA at lower concentrations.
- A no observed genotoxic effect level (NOGEL) could be demonstrated, *in vitro* as well as *in vivo*, for the direct alkylans, methyl methanesulfonate, which correlates with an activation of the O^{6}MeG DNA methyltransferase (MGMT).
- In studies pertaining to the correlation of adducts with mutations after exposure to DNA-reactive substances like ethylene oxide or formaldehyde, mutations are induced only after higher exposures as compared to the adducts. This led to their inference that those adducts, which can be linearly followed up to the lowest exposure and then become lower than the spontaneous rate of adducts, should be seen as biomarkers of exposure. Mutations as biomarkers for genotoxicity are induced only by higher doses and correlate with the exposure only when they exceed the spontaneous rate of mutations.
- In 2-acetylaminofluorene-treated rats for a period of 16 weeks the cumulative dose of 0.125 mg/kg induced 0.6 per 10^8 specifically altered nucleotides for 1–3.1 per 10^8 altered nucleotides in the control group. This dose was taken as the NOAEL.

2.9.5 Summary

> **Normal cells are constantly exposed to DNA-reactive substances like ethylene oxide or reactive oxygen species (ROS), which are mostly produced endogenously. Additional**

> insults occur in the form of damage due to faulty repair and through exogenous exposure, e.g. due to ionising radiation, viral infections or reactive genotoxic chemicals. The cellular homeostasis, however, is maintained through various efficient mechanisms.

The cellular defence mechanisms include:

- Toxicokinetic factors, which prevent the reactive agent from reaching the DNA, e.g. metabolic inactivation, reaction with proteins, by restricting access to DNA with the help of membranes
- Enzymatic intracellular detoxification mechanisms, which keep the level of reactive agents like reactive oxygen species at physiological concentration
- An efficient system that does not allow the DNA damage to rise above the physiological equilibrium. This is ensured by repair of the constantly occurring DNA damage. The effect is reinforced by inhibition of cell proliferation. This makes more time available for the repair to take place and prevents the multiplication of the damaged genome through cell division.
- Elimination of severely damaged cells through apoptosis and necrosis.

In a state of equilibrium, a cell contains about 500 altered bases, which corresponds to approximately 1 lesion per 10^6 bases. However, about 10,000 bases become altered each day, which shows the high efficiency of this repair system. Since it is at least generally assumed that DNA damage caused by exogenous and endogenous substances are identical, it means that smaller additional exogenous insults are repaired and do not lead to mutations.

Additionally, in cases of severe irreparable DNA damage, so-called cell death programmes are induced, which eliminate such cells through apoptosis or necrosis.

It is therefore not plausible that low exposures to genotoxic chemicals lead to a disturbance of intracellular homeostasis, and thereby to an increase in the rate of spontaneous mutations and conversion to cancer cells.

Although many studies have examined the individual mechanisms responsible for the elimination of altered DNA and damaged cells, data about dose dependence with respect to the turning on or off of the individual reactions is not available in most cases. Thus, it is not possible to derive any clues about a concentration at which no effects occur or at which the system is saturated, i.e. pushed to the limits of its capacity. The latter in particular needs to be considered when interpreting results of carcinogenicity studies because in many cases effects are seen at higher exposures only. Since the cellular mechanisms described above could have become overburdened such effects may be irrelevant for human exposure and classification as a carcinogen may not be appropriate in case such exposures never occur.

Further Reading

Almaida KH, Sobol RW: A unified view of base excision repair: Lesion-dependent protein complexes regulated by post-translational modification. *DNA Repair* 2007; **6**: 695–711.

Barnes DE, Lindahl T: Repair and genetic consequences of endogenous DNA base damage in mammalian cells. *Annu Rev Genet* 2004; **38**: 445–476.

Calabrese EJ: Key studies to support cancer risk assessment questioned. *Environ Mol Mutagen* 2011; **52**: 595–606.

EC: Risk Assessment Methodologies and Approaches for Genotoxic and Carcinogenic Substances. JOINT opinion of the Scientific Committee on Health and Environmental Risks (SCHER), Scientific Committee on Consumer Products (SCPP) and Scientific Committee on Emerging and Newly Identified Risks (SCENIHR), Health and Consumer Protection Directorate, 2009.

EFSA: Opinion of the Scientific Committee on a Request from EFSA related to a harmonised approach for riskassessment of substances which are both genotoxic and carcinogenic. *EFSA Journal* 2005; **282**: 1–31.

EPA: Guidelines for Carcinogenic Risk Assessment, Washington DC, 2005.

Goggin M, Sangaraju D, Walker VE, Wickliffe J, Swenberg JA, Tretyakova N: Persistence and repair of bifunctional DNA adducts in tissues of laboratory animals exposed to 1,3-butadiene by inhalation. *Chem Res Toxicol* 2011; **24**: 809–817.

Greim H, Albertini RJ: *The cellular response to the genotoxic impact: The question of threshold for genotoxic carcinogens.* Issues in Toxicology, RSC Publishing, 2012.

Greim H, Albertini RJ: Cellular response to the genotoxic impact: The question of threshold for genotoxic carcinogens. *Toxicol Res* 2015; **4**: 36–45.

Izumi T, Wiederholt LR, Roy G, Roy R, Jaisval A, Bhakat KK, Mitra S, Hazra TK: Mammalian DNA base excision repair proteins: their interactions and role in repair of oxidative DNA damage. *Toxicology* 2003; **193**: 43–65.

Meisenberg C, Tait PS, Dianova II, Wrigth K, Edelmann MJ, Ternette N, Tasaki T, Kessler BM, Parsons JL, Kwon YT, Dianov GL: Ubiquitin ligase UBR3 regulates cellular levels oft he essential DNA repair protein APE1 andis required for genome stability. *Nucleic Acid Res* 2012; **40**: 701–711.

Moeller BC, Recio L, Green A, Sun W, Wright FA, Bodnar WM, Swenberg JA: Biomarkers of exposure and effect in human lymphoblasoid TK6 cells following [13C2]-acetaldehyde exposure. *Toxicol Sci* 2013; **133**: 1–12.

Parsons JL, Tait PS, Finch D, Dianova II, Allison SL, Dianov GL: CHIP-mediated degradation and DNA damage-dependent stabilization regulate base excision repair proteins. *Mol Cell* 2008; **29**: 477–487.

Preston RJ, Williams GM: DNA-reactive carcinogens: Mode of action and human cancer hazard. *Crit Rev Toxicol* 2005; **35**: 673–683.

Thomas AD, Jenkins GJ, Kaina B, Bodger OG, Tomaszowski KH, Lewis PD, Doak SH, Johnson GE: Influence of DNA repair on nonlinear dose-responses for mutation. *Toxicol Sci* 2013; **132**: 87–95.

Tomasetti C, Vogelsten B: Variation in cancer risk among tissues can be explained by the number of stem cell divisions. *Science* 2015; **347**: 78–80.

Williams GM, Duan J-D, Iatropulos MJ: Thresholds for the effects of 2-acetylaminofluorene in rat liver. *Toxicol Pathol* 2004; **32**: 85–91.

Williams GM, Kobets T, Duan JD, Iatropoulos MJ: Assessment of DNA binding and oxidative DNA damage by acrylonitrile in two rat target tissues of carcinogenicity: Implications for the mechanism of action. *Chem Res Toxicol* 2017; **30**: 1470–1480.

2.10 Reproductive Toxicology

Horst Spielmann

2.10.1 Introduction

> **Reproductive toxicology describes the impairment of female and male fertility, adverse effects during pregnancy and lactation, as well as impaired development in the next generation. Specific characteristics of reproductive toxicology are differences in the susceptibility of female and male germ cells and during pregnancy and lactation the simultaneous exposure of the mother and embryo or fetus or infant.**

Reproductive toxicology includes toxic effects on the entire reproductive system, starting with the impairment of female and male fertility, pre- and postnatal damage, and effects that will be apparent only in later generations. Due to the embryotoxic effects induced by thalidomide/Contergan, which were observed in Germany 60 years ago, originally the main focus in

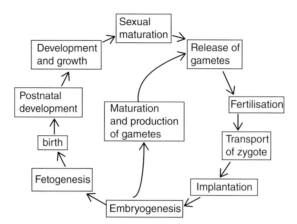

Figure 2.72 *The mammalian reproductive cycle. The effects of toxic agents on each stage of the reproductive cycle can be analysed in laboratory animal studies. This is more difficult in humans, where the generation time is 20–25 years.*

reproductive toxicology was on impaired development of the embryo or fetus. Only during the past 20 years have hormonally effective environmental chemicals, the so-called "endocrine disrupters", been evaluated more critically since they can affect female and male fertility in several animal species and possibly also in humans.

Experience from exposed human populations is used for hazard and risk assessment in reproductive toxicology after exposure to chemical or physical agents, and from testing in experimental animals according to international guidelines.

2.10.2 Characteristics of Reproductive Toxicology

Reproduction is essential to maintain the survival of species. Mammalian reproduction is characterized by a rather complex reproductive cycle that has allowed individual species to adapt to a continuously changing environment. Human reproduction is characterized by the same essential features and has allowed adult individuals around the globe to survive under quite complex environmental conditions. The mammalian reproductive cycle is in principle identical in all mammalian species, and the essential steps are depicted in Figure 2.72. To adapt to the environment, all species have developed species-specific deviations from the general pattern.

> **As a general rule, physical and chemical agents can adversely affect each of the essential steps of the mammalian reproductive cycle.**

Some essential features of reproductive toxicology are quite different from all other areas of toxicology, e.g.:

- adverse effects may only occur in the next generation/offspring
- adverse effects on the vital functions of all organs may indirectly induce adverse effects on reproduction.

> **Agents are termed "reproductive toxicants" when they are interfering with the reproductive cycle at a dose range that does not induce adverse effects in any other organ system.**

In the field of reproductive toxicology risk assessment is challenging due to the complexity and unusually long time frame/duration of the reproductive cycle, which covers at least two generations. To assess the potential of chemicals to interfere with reproduction, test methods in experimental animals must cover the following essential steps of reproduction:

- fertilization, e.g. fusion of oocyte and sperm resulting in a complete, diploid set of chromosomes
- normal cleavage divisions, implantation, intrauterine development, birth and postnatal development, including the period of lactation
- normal development of the offspring to fertile adult animals, which produce another healthy offspring.

For systematic reasons the essential steps of the reproductive cycle, which are important for female and male fertility, e.g. maturation and development of oocytes and sperm up to fertilization, will be described in this section separately from pre- and postnatal development.

2.10.3 Adverse Effects on Female and Male Fertility

Maturation of Oocytes (Oogenesis)

At birth the human ovary contains around 300,000 ovarian follicles and the numbers decrease during the 30 fertile years of a woman: between puberty and menopause ca. 400 mature oocytes will develop, while after menopause no follicles containing oocytes are left in the ovary.

During the four weeks of a normal ovarian cycle, only a single oocyte will develop into a primary follicle. Oocyte maturation is stimulated under the control of follicle-stimulating hormone (FSH) secreted by the pituitary. FSH also stimulates the maturation of the uterine mucosa to provide nutrition for the implanting preimplantation embryo that develops from the fertilized oocyte and migrates into the uterus. After ovulation growth of the uterine mucosa is stimulated by luteinizing hormone (LH) in the pituitary and, if the oocytes have not been fertilized, it is rejected/expelled every month with the uterine mucosa during the menstrual period.

> **The synchronized maturation cycles of ovary and uterine mucosa are controlled by gonadotropin-releasing hormone (GnRH) from the pituitary gland.**

The small proliferative cell population in the ovary and the intermittent nature of the proliferative stage make the ovary less susceptible to the disruptions of cell division. The most significant type of toxicants for female reproductive function are those that interfere with the dynamic endocrine balance required for folliculogenesis and oogenesis. Therefore, hormonal active chemicals that accumulate in the environment (endocrine disrupters) may be responsible for increased infertility in humans and several animal species.

Spermatogenesis

> **Spermatogenesis covers the maturation of sperm from immature progenitor cells.**

Spermatogenesis, which is dependent on sex hormones, starts at puberty and is localized in the tubuli of the testis. The stem cells of mammalian spermatogenesis, the spermatogonia, are attached to the wall of the seminiferous tubuli of the testis. At puberty these cells start to cleave into primary spermatrocytes, which migrate to the center of the seminiferous tubuli and contain a normal, diploid set of chromosomes. During the next step of maturation, the diploid chromosome set is reduced to a haploid set in secondary spermatocytes. In the subsequent stage spermatids continue to differentiate into mature sperm, which in contrast to the earlier stages are characterized by the neck and tail. Mature sperm are located in the center of the seminiferous tubuli and migrate to the epididymis, where they are stored and will leave the testis via the seminiferous tubuli.

Sertoli cells are also located in the tubuli of the testis and are attached to the wall. They are not only secreting hormones and proteins essential for the maturation of sperm but also provide the morphological substrate of the blood–testis barrier, which prevents the transfer of large and polar molecules into the seminiferous tubuli. Leydig cells, which are located in the interstitial tissue between the seminiferous tubuli, produce the male sex hormone testosterone under the control of LH produced by the pituitary gland.

In contrast to oocyte maturation, spermatogenesis is a continuous process and thus, in principle, men are fertile from puberty up to very old age.

The high rates of cellular division and metabolic activity associated with spermatogenesis are the basis for the high susceptibility to damage. During the duplication of genetic material and cell division, DNA is particularly vulnerable to damage. In addition, specific cellular proteins and enzymes are required for spermatogenesis. Therefore, chemicals that may cause DNA damage or inhibit the activity of specific proteins are of concern during spermatogenesis, e.g. reactive electrophilic chemicals such as alkylating agents and ionizing radiation.

Fertilization

> **In mammals, fertilization, the fusion of female and male gametes – oocyte and sperm – occurs in the fallopian tubes. At fertilization both oocyte and sperm contain only a haploid set of chromosomes (50% of "normal" somatic cells).**

The reduction of the diploid set (100%) of chromosomes to a haploid set (50%) during the two maturation divisions of each oocyte is termed **meiosis**, as are the two maturation divisions during spermatogenesis, when spermatocytes develop into spermatids (see above). The meiosis divisions of oocyte and sperm ensure that both parents, mother and father, contribute 50% of the chromosomes to the fertilized zygote and thus to the embryo. The sex of the embryo is controlled by the X and Y sex chromosomes; in humans XX individuals are female and XY are male. The basic physiological parameters controlling human fertilization are quite well established and during the past 30 years *in vitro* fertilization has become a standard therapy for many infertile couples.

Cleavage Divisions and Implantation

After fertilization the zygote undergoes cleavage divisions of preimplantation development while migrating from the ovarian tube to the uterus. During cleavage divisions all cells of the preimplantation embryo are not yet committed; they are omnipotent embryonic stem cells and can differentiate into all tissues of the embryo. After implantation the number of embryonic stem cells decreases while they differentiate into specific cells and tissues.

> **During early embryonic development, before implantation in the uterine mucosa, the embryo is rather insensitive to toxic agents, in particular since the embryonic stem cells forming the preimplantation embryo have a high capacity to replace damaged embryonic cells.**

When exposed to high doses of toxic agents, preimplantation embryos are not able to repair damaged cells and therefore die. At implantation into the uterine mucosa on days 6–7 after fertilization the early human embryo has reached the blastocyst stage and consists of 80–160 cells, among which only the **inner cell mass** still contains omnipotent embryonic stem cells.

Adverse Effects on Female and Male Fertility

The reproductive cycle (Figure 2.72) illustrates how different pollutants can affect fertility. The embryonic development of the body cells in the outer ring is shown and in the middle of it runs independently the so-called germ line, a specific phase in embryogenesis during which female and male germ cells are released in puberty. Whether offspring are capable of reproduction is shown in the daughter generation, and whether they are fertile is shown in the generation of grandchildren. Effects on fertility therefore can only be ruled out by studies on two generations.

Table 2.20 summarizes the main symptoms of reproductive harm as defined by the Organization for Economic Co-operation and Development (OECD) (2008).

> **Basically, all substances that affect the hormonal balance of the female and male sex hormones can affect the maturation of eggs and sperm, and also fertility. These are called endocrine disrupters (EDs) and can accumulate in the environment.**

2.10.4 International Test Methods in the Field of Reproductive Toxicology

To investigate the potential teratogenic effects caused by chemical or physical agents, morphological and functional criteria are analyzed in animal experiments. For this purpose, internationally harmonized, standardized studies must be conducted for pharmaceuticals according to the guidelines of the International Conference on Harmonization (ICH), in which the regulatory agencies for the drug safety of the major industrial nations (EMA (Europe), NIEHS (Japan) and FDA (USA)) are represented. For all other chemical substances, safety tests in animals must be carried out according to OECD test guidelines.

Reproductive Toxicity Tests for Medicinal Products and Medical Devices

For medicinal products, the ICH prescribes three trials in preclinical animal studies in the field of reproductive toxicology, the so-called segment 1, 2 and 3 studies, which examine the integrity of important parts of the reproductive cycle.

Table 2.20 Reproductive toxicity during different stages of life (OECD).

Reproductive and reproductive effects on female and male fertility include morphological and functional damage to the primary and secondary female and male sexual organs, as well as to the hormone systems that control them, which can interfere with the proper functioning of the following functions:

- beginning of puberty
- production and transport of female and male germ cells
- undisturbed expiration of the reproduction cycle
- male and female sexual behavior
- fertility
- pregnancy and childbirth
- lactation or lactation phase
- end of the fertile period (e.g. premature onset of the menopause).

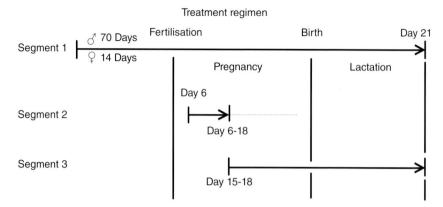

Figure 2.73 *Reproductive toxic testing of drugs: ICH segment 1–3 studies. The arrows show the treatment duration (days) of female and male animals in preclinical drug testing in the so-called segmental studies of ICH: in segment 1 a combined fertility, pre- and post-natal study is performed, in segment 2 a pure embryotoxicity study and in segment 3 a combined pre- and postnatal study. The results of these studies are recognized by the international regulatory agencies EMA (Europe), FDA (USA) and NIHS (Japan).*

As shown in Figure 2.73, male and female animals are treated before mating in a segment 1 study. Because of the different durations of germ cell maturation, the males are treated for 60 days and the females for 14 days before mating and also after birth until the end of lactation on day 21.

The segment 2 embryo toxicity study must be conducted in two species, rats and rabbits, since the human embryotoxic drug thalidomide (Contergan®) is embryotoxic in rabbits only but not in rats. The segment 3 study is a combined pre- and postnatal study that is performed from day 15 of pregnancy until the end of breastfeeding on day 21.

OECD Reproduction Toxic Test Methods

All substances and products that are neither drugs or medicinal products must be tested according to the OECD test guidelines. These include industrial and household chemicals, cosmetics, feed and food additives, pesticides and biocides. The results of these studies are accepted for regulatory purposes by the authorities of all 34 OECD member states. In the European Union (EU), the OECD test guidelines have been harmonized with the EU Registration, Evaluation, Authorization and Restriction of Chemicals (REACH) regulations.

Table 2.21 shows a comparison of the test methods of the ICH and OECD in the field of reproductive toxicology. Depending on the production volume, OECD testing requires an embryotoxicity study and one- or two-generation studies. In addition, the OECD has developed two reproductive toxic screening methods and an extended one-generation reprotox study (EOGRTS), with additional testing for behavioral disorders, immunological effects and fertility in the second generation. Moreover, a test for developmental neurotoxicity (DNT) has been developed specifically for neurotoxic pesticides and biocides.

The Importance of Reproductive Toxicity Test Methods for Risk Assessment

In preclinical drug testing, the results of the reproductive toxic ICH segment 1 to 3 studies are of great importance for the application of new drugs in humans. Since animal

Table 2.21 *International test guidelines for reproductive toxicology (ICH, OECD).*

Pharmaceuticals and medical devices (ICH, EMEA, US FDA)	All other chemicals (OECD, EU REACH, US EPA)

Segment 1 study
    Combined fertility
    Pre- and postnatal study
Segment 2 study
    Embryo toxicity study
Segment 3 STUDY
    Pre- and postnatal study
	OECD Test Guideline 414
	    Embryo Toxicity Study
	
	OECD Test Guideline 415
	    One-generation Reprotox Study
	OECD Test Guideline 416
	    Two-generation reprotox study
	OECD Test Guideline 443
	    Extended one-generation reprotox study
	OECD Test Guidelines 421 and 422
	    Reproductive toxic screening studies
	OECD Test Guideline 426
	    Neuro-embryotoxicity study

ICH, International Conference on Harmonization; EMEA, European Medicines Agency; US FDA, Federal Drug Administration; OECD, Organization for Economic Co-operation and Development; EU REACH, EU Registration, Evaluation, Authorization and Restriction of Chemicals; US EPA, US Environment Protection Agency.

studies have lost importance for drug counseling on risks in human pregnancy, it follows that the results of epidemiological studies on exposed women are now the basis of advice (www.embryotox.de).

The situation is different for all other chemicals, since data for only a few of them from epidemiological studies of exposed women or men are available and, therefore, only animal test data obtained according to the OECD test guidelines are available for risk assessment. These animal tests are also the basis for the classification and labelling of dangerous substances as carcinogenic, mutagenic and toxic to reproduction (CMR), and for their inclusion in the list of Substances of Very High Concern (SVHC) of the EU Chemicals Agency (ECHA) (see ECHA Candidate List of Substances of Very High Concern for Authorisation).

2.10.5   Pre- and Postnatal Toxicology

Principles of Drug Effects in Pregnancy

Experience during the past 60 years has proven that data from drug studies in pregnant animals are not very predictive for humans. The actual harmful potency in the therapeutic dose range in pregnancy can therefore only be determined in epidemiological studies on pregnant women, which are the basis for advising doctors and women on risks during pregnancy and lactation (Schaefer et al., 2014; www.embryotox.de).

Based on animal experiments, Wilson (1977) has established rules for the effects of drugs in pregnancy, which later proved to be valid also for human pregnancy. According to current knowledge these are:

First Principle: *In prenatal toxicology, the same dose–response relationships apply as in other pharmacology and toxicology studies (Figure* 2.74*)*

Low doses of a drug do not harm the embryo or the mother and only after exceeding a threshold dose do embryotoxic effects occur. At even higher doses, embryo lethality will occur and at even higher doses toxic effects may also occur in the mother.

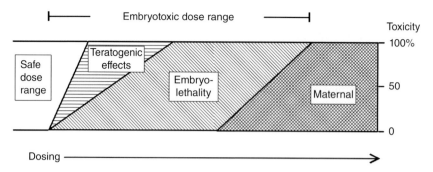

Figure 2.74 *The dose–response relationship in prenatal toxicology. Depending on the dose to which the mother is exposed, adverse effects are induced in the embryo/fetus in a dose-related manner. They increase from the non-toxic dose level to embryotoxic effects and embryo lethality at the highest dose levels, which finally will also induce maternal toxicity.*

Thus, increasing doses of an embryotoxic agent will increase the adverse effects in the embryo/fetus in a dose-related manner. Depending on the dose to which the mother is exposed adverse effects are induced and increase continuously from the non-toxic dose level at the lower end to embryo- and fetotoxicity effects and embryo lethality at the highest dose levels.

> **The effects of toxic agents in pregnancy are dose-related, as is usual in pharmacology and toxicology.**

At low doses neither embryo nor mother will be damaged. When increasing the dose, the embryotoxic/teratogenic range is reached first and malformations may be induced. At higher dose levels, the lethality range is reached, and the embryo/fetus will die and not survive to term. At the highest dose range maternal toxicity will be observed.

The embryotoxic risk to drugs and other chemicals is high when adverse effects are induced in the embryo/fetus at or even below the therapeutic dose range of the mother. This situation occurred in the thalidomide/Contergan® disaster since the drug had no side effects in adults and was specifically prescribed by doctors as a sleeping pill during pregnancy. It was therefore highly unexpected that a single dose of the drug taken in week four or five of pregnancy was able to induce severe limb malformations (phocomelia) in babies.

Second Principle: The sensitivity of the embryo to toxic agents depends on its genotype

During pregnancy the species-specific effects of drugs on humans and on experimental animals are determined by the genotype (hereditary traits) of the species. Moreover, embryotoxic effects in humans may vary due to genetic susceptibility. Therefore, genotyping can reveal polymorphisms and "sensitive" women can be treated with drugs posing a lower-risk.

Third Principle: The sensitivity of the embryo to toxic agents depends on its stage of development at the time of exposure (Figure 2.75)

In the pre-implantation phase before implantation in the uterus, the risk of malformation is low because the pluripotent embryonic stem cells can replace damaged cells and further development will proceed unaffected or the embryo will die. Development of a damaged embryo at this early stage into a malformed newborn is therefore most unlikely.

During the organogenesis phase, when the organs develop in humans around 15–60 days after fertilization, the embryo is particularly sensitive to toxic effects and malformations may be

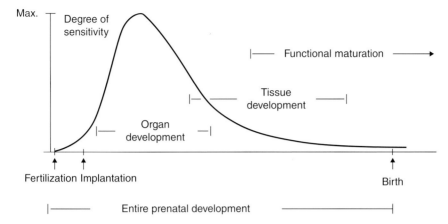

Figure 2.75 Sensitive phases of the embryo/fetus against toxic effects during pregnancy according to Wilson (1977).

induced (Table 2.23). In the fetal phase, during the development of tissues and the subsequent maturation of organ functions, the sensitivity decreases. At this stage (second and third trimester of pregnancy) toxic substances may cause organ dysfunction, e.g. alcohol interferes with brain maturation and fetal alcohol syndrome and some antihypertensive drugs may cause kidney failure.

Fourth Principle: *Embryotoxic agents may affect the morphological development of the embryo via different molecular pathways*

When Wilson formulated the rules 40 years ago, molecular mechanisms of action in toxicology were poorly understood. In the meantime, development-specific genes have been identified in all animal species and molecular receptors through which hormones and other agents will affect the expression of these genes. However, for Contergan® the embryotoxic mechanism in humans has so far not been conclusively identified.

Fifth Principle: *There are only a few final manifestations of abnormal development both in laboratory animals and in humans*

The following forms of development of the embryo will result after exposure to toxic agents during pregnancy:

- normal development due to the active repair capacity of the fast-growing embryo, defects are repaired; this is typical during early pregnancy but may also occur later
- intrauterine death of the embryo or fetus: miscarriage
- malformations of organs
- growth inhibition: general growth retardation or microcephaly
- disturbed organ function, e.g. brain, reproductive ability and immune competence (In humans disorders of brain development, behavioral problems, increased susceptibility to allergens and impaired fertility can be recognized only in later life and therefore a causal relation to prenatal exposure is difficult to prove.)
- tumors: diethylstilbestrol (DES) induced carcinomas of the vagina caused in daughters of exposed mothers (It has been proven that DES is a carcinogen in adults. DES is so far the only example of trans-placental carcinogenesis in humans.)

- germ cell mutation, which may cause damage in the next generation. (Genotoxic agents can cause damage to the germ line and affect the fertility of the next generation, which is detectable only in the second generation.)

Metabolism of Drugs and other Chemicals in Pregnancy

> **Drug metabolism in pregnancy is more complex than during any other period of life.**

The following factors influence the metabolism of drugs or other chemicals during pregnancy:

- pharmacokinetics and metabolism in the mother, who is exposed first
- placental transfer
- distribution in the embryo/fetus
- drug metabolism in the embryo/fetus
- excretion by the embryo/fetus
- excretion by the mother.

In the past it was assumed that the placenta is an effective barrier protecting the embryo against toxic effects induced by drugs and chemicals. Advanced analytical methods have shown that almost all drugs and chemicals as well as their metabolites will reach the embryo/fetus. The passage of toxic agents through the placenta is quite similar to transfer through the intestinal wall, since lipid-soluble materials will easily pass through this barrier, while a much smaller percentage of water-soluble, ionized materials will pass into the blood of the embryo/fetus.

> **According to current knowledge drugs that are readily absorbed after oral application will easily pass into the placenta and reach the embryo/fetus, while drugs that have to be injected will pass into the placenta at a significantly lower rate.**

Molecular weight also has a high impact on placental transfer of drugs and chemicals. Up to a molecular weight of 800 Da drugs will easily pass the placental barrier. Since the majority of drugs and their metabolites belong to this group, most of them will reach the embryo/fetus and may induce abnormal development.

On the other hand, the placenta is a tight barrier for complex and electrophilic molecules, which may, therefore, not reach toxic concentrations in the embryo/fetus, e.g. steroid and peptide hormones, insulin and even the mushroom toxin α-amanitine. The placental transfer of drugs with a high binding affinity to maternal serum proteins is low and, moreover, the drug metabolism activity of the placenta is also low and therefore does not significantly contribute to drug metabolism in pregnancy.

Embryonic and Fetal Toxicology

Drug Risks during Pregnancy It was a general concept that the embryo is well protected against damaging external effects while it develops in the uterus. That thinking changed in 1941 when Australian ophthalmologist Gregg described human rubella embryopathy. In the following years, the field of embryonic toxicology or teratology developed quite slowly. After Lenz in Germany and McBride in Australia described the thalidomide/Contergan® embryopathy in 1961, teratology developed as an independent science. Today, more than 50 years later, the risk of drug-induced human malformations can be quite effectively characterized:

- No drugs or chemicals have been detected that are embryotoxic in a similar way as thalidomide (Contergan®), e.g. the embryotoxic properties of retinoids were known from animal studies prior to their therapeutic use in humans.
- Malformation rates have not increased in the past 50 years despite the introduction of new drugs.
- Specific developmental disorders have been described after treatment with several groups of new drugs (see Tables 2.23a and 2.23b). However, even after treatment with one of these drugs, the risk of major malformations is less than 10% (prevalence 2–3%). High-risk exemptions include thalidomide, retinoids, alcohol and other drugs of abuse and combination therapy for severe epilepsy. In contrast, risk assessment during pregnancy is more difficult for occupational exposure to chemical and physical agents. Most importantly, new legislation, e.g. the Maternity Protection Act and occupational exposure limits, provides a framework that is sufficient to protect the mother and her unborn child.

Important Causes of Congenital Human Malformations

Epidemiological studies suggest that drugs and other chemicals, including food additives and recreational/lifestyle drugs, account only for 4–5% of malformations in humans (Table 2.22)

Table 2.22 summarizes estimates of the prevalence of abnormal development in humans 60 years after the thalidomide/Contergan® disaster. It can be seen that inheritable genetic disorders account for around 20% of malformations and that diseases of the mother, including infections, account for an additional 5–7%. In humans the percentage of malformations that may be induced by drugs and other chemicals is fairly low at about 4%.

About 20% of malformations are due to genetic diseases and about 3% are induced by diseases of the mother, including infections. The proportion of developmental disorders attributable to drugs or other chemicals fairly low at 2–4%, therefore, despite intensive research, in about 40–70% of congenital developmental disorders the causes remain unknown.

Table 2.22 Prevalence of causes of developmental disorders in humans (%).

Cause	Prevalence (% of all disorders)
Genetic diseases	8–20%
Chromosomal anomalies	3–10%
Morphological factors	Up to 3%
Anomalies of the uterus, twins	
Chemical and physical agents	2–4%
Alcohol, drugs and drugs of addiction, ionizing radiation, smoking, industrial and environmental toxicants	
Maternal diseases	Up to 3%
diabetes, epilepsy, thyroid malfunctions, phenyl ketonuria, infectious diseases (HIV, cytomegalia, listeriosis, ringworm, rubella, syphilis, toxoplasmosis, chickenpox, ZIKA)	
Multifactorial causes	Up to 49%
Combination of genetic and environmental factors	
Unknown causes	40–70%

According to Schaefer et al. (2014) and www.embryotox.de.

So far, there is no general hypothesis for the induction of prenatal development disorders, which may be induced by many different toxic agents. To help pregnant women and their doctors, prenatal counseling centers have been established in many countries. They conduct epidemiological studies on exposed pregnant women and they co-operate internationally via networks, e.g. in Europe the European Network of Teratology Information Services (ENTIS) and worldwide the Organization of Teratology Information Services (OTIS). Today, the experience of these centers forms the basis for advice on drug therapy in pregnancy, which is available on the Internet (www.embryotox.de).

Table 2.23a shows the results of epidemiological studies on exposure in the first trimester of pregnancy for human embryotoxic drugs and chemicals, e.g. alcohol, cocaine, diethylstilbestrol, methyl mercury, vitamin A and retinoids. Table 2.23b lists the most important fetotoxic drugs, which have been shown to affect the second and third trimester.

> Tables 2.22, 2.23a and 2.23b show that the proportion of congenital malformations caused by drugs and other chemicals has not increased over the past 50 years. At the same time, the result confirms that the methods developed for reproductive toxicity testing by the ICH and OECD are sufficiently sensitive that human embryos, fetuses and newborns are protected from drug and chemical risks.

Although this result is encouraging, it must be taken into account that in humans, due to the long generation time, there are no data on effects on grandchildren that would be comparable to animal experimental two-generation studies. Thalidomide/Contergan® is a special case since it does not cause any genetic damage. It is therefore encouraging that several of the thalidomide/Contergan® children have their own normal children.

Table 2.23a The most important human embryotoxic drugs in the first trimester of pregnancy Treatment with one of these first trimester medications rarely causes damage to the embryo. The risk of malformation, with the exception of thalidomide and retinoids (vitamin A derivatives), is less than 10%. A risk assessment for humans is not possible with most medicines due to missing epidemiological data, e.g. for antiepileptic drugs, which are needed to treat pregnant women.

Substance	Leading symptoms or predominantly affected organs
Androgenic hormones	Masculinization
Antimetabolites (cancer treatment)	Multiple malformations
Antiepileptic drugs	Spina bifida, heart, palate, urogenital system
Carbamazepine, phenobarbital primidon, phenytoin, trimethadione, valproic acid	Extremities, facial anomalies
Coumarin derivatives	Nose, extremities
Diethylstilbestrol	vaginal carcinoma
Lithium	Heart (Ebstein anomaly, very rare)
Misoprostol (abortion)	Moebius sequence, extremities
Penicillamin	Cutis laxa (rare)
Retinoids (vitamin A)	Ear, central nervous system, heart, skeleton
Thalidomide	Limbs, skeleton, heart
Vitamin A	(>25,000 IU/day) like retinoids low embryotoxic (in humans the risk is <1:1000)
Glucocorticoids	Cleft palate
Methimazole	Nasopharyngeal malformations
Trimethoprim/co-trimoxazole	Neural tube defects.

Table 2.23b *The most important fetotoxic drugs (second and third trimester) These drugs can harm the fetus in late pregnancy (second and third trimester) or during childbirth. Treatment will only rarely lead to developmental disorders with the specified symptoms. As described in detail, genetic predisposition, dose and treatment period are crucial.*

Substance	Leading symptoms or mainly affected organs
ACE inhibitors	Kidneys, oligohydramnios, anuria, joint contractures
Aminoglycoside antibiotics	Inner ear and kidneys
Androgenic hormones	Masculinization
Angiotensin II antagonists	Renal, oligohydramnios, contractures, cranial hypoplasia
Benzodiazepine	Respiratory depression, adjustment disorder, floppy infant syndrome
Coumarin derivatives	Cerebral hemorrhage
Ergotamine	Fetal hypoxia
Immuno-suppressants	Myelo-suppression (bone marrow)
Radioiodine	Thyroid hypoplasia or aplasia
Lithium	Floppy infant syndrome, hypothyroidism
Opioids/opiates	Withdrawal symptoms
Psychotropic drugs	Adjustment disorders, in SSRI serotonergic symptoms
Tetracycline	Yellow teeth
Cytostatic drugs	Bone marrow depression.

It is an important result that drugs, chemicals and environmental effects are responsible for only minor prenatal damage. In addition, the proportion of miscarriages due to diseases of the mother is low. Due to progress in the treatment of diabetes, diabetic women today enjoy a largely normal course of pregnancy, although until recently the affected mothers were infertile or had to expect a significant malformation risk.

It is quite disappointing that today more children are harmed by alcohol consumption during pregnancy than by an embryotoxic teratogenic drug. Of the many new drugs, only a few have been found to be embryotoxic or fetotoxic in humans.

2.10.6 Effects of Drugs and other Chemicals on Lactation

With modern chemical-analytical methods, it is possible to determine very low concentrations of most chemicals and their metabolites in breast milk. Fortunately, they reach toxic concentrations only in exceptional cases. Newborn and premature infants are particularly sensitive to many chemicals since their blood–brain barrier is not yet mature and, therefore, not able to prevent damage to the developing brain. Moreover, in newborns the functions of liver metabolism and renal excretion are still immature.

> **Most drugs and chemicals reach concentrations in breast milk that are below the therapeutic or toxic range for the infant.**

Unlike cow's milk, during the first days breast milk contains secretory immunoglobulin A, which is not digested by the "immature" gastrointestinal tract and is absorbed by the newborn during the first few weeks of life. This is important because the immune system of the newborn still cannot effectively produce antibodies. For the secretion of foreign substances into breast milk, substances with high fat solubility and low molecular weight (<200 Da) and low binding to maternal serum proteins are preferred.

Toxic, persistent organochlorine compounds that accumulate in the environment, e.g. pesticides, polychlorinated biphenyls and dioxins, and which accumulate in human adipose tissue are mobilized during breastfeeding and are enriched in breast milk. Although newborns are very sensitive to these toxic contaminants, epidemiological studies have shown that breastfeeding has no negative impact on early childhood development.

> **Today, expert breastfeeding commissions worldwide are urgently recommending breastfeeding because, despite earlier concerns, the benefits of breastfeeding are clearly demonstrated, especially the protection against infection by secretory immunoglobulins.**

2.10.7 Endocrine Disrupters

There is no doubt that synthetic chemicals can accumulate in the environment and due to their hormonal properties can have an influence on the reproduction of wild animal populations. These organochlorine compounds have been termed **endocrine disrupters** (EDs). Due to their toxic properties they are classified by the ECHA as Substances of Very High Concern (SVHC). The toxicological significance of these substances, which occur in the environment and also in human cells and tissues in very low concentrations, is currently causing some controversy. It has been proven that EDs are responsible for the sex shift in exposed fish and lower species living in water, e.g. snails. Several of these organochlorine compounds give positive results in hormone binding *in vitro* assays, but reproductive toxicity was only observed with some of them in exposed human populations. Another highly disputed consequence of endocrine disruption in humans is reduced sperm counts in men in industrialized countries. More definitive laboratory studies and risk assessment for a number of these chemicals indicate little or no potential for adverse effects in humans at environmentally relevant exposure levels. It is therefore a challenge for the scientific community to improve risk assessment for humans and the environment exposed to accumulating concentrations of EDs.

2.10.8 Summary

Reproductive toxicology covers all adverse effects on the reproductive cycle, including adverse effects on female and male fertility, and the development of the embryo, e.g. growth retardation, malformations and death of the embryo/fetus. Reproductive toxicology (RT) also includes the induction of adverse postnatal effects during the lactation period and impaired fertility and development of the following generations. To assess reproductive toxicology, it has to be taken into account that all stages of the mammalian reproductive cycle are controlled by the endocrine system. Tests in laboratory animals to assess the effects of drugs and other chemicals on reproduction and on pre- and postnatal development are therefore rather complex and time consuming.

For historic reasons the testing approach in reproductive toxicology is significantly different for drugs, for industrial chemicals and for pesticides. All drugs have to be tested in segment 1 to 3 studies in animals before they can undergo clinical testing in human patients. For pesticides reproductive toxicology testing is also quite extensive, since it includes developmental neurotoxicity studies and one- and two-generation studies in rodents. In contrast, reproductive toxicology testing requirements for industrial chemicals depend on the production volume. Thus most of the high production volume chemicals have been tested sufficiently for reproductive toxicology, while the vast majority of existing industrial chemicals have either not or only marginally been tested for RT. The EU chemicals policy REACH therefore focuses in particular on reproductive toxicology testing of approximately 30,000 existing chemicals with an annual production volume of more than 1000 tonnes.

During the past two decades, international clinical surveillance programs of pregnant women, who were exposed to drugs, have proven that the majority of drugs are safe during pregnancy and lactation. These studies have provided a list of drugs that may safely be used by women during pregnancy and lactation. In addition, epidemiological studies have proven that in general pesticides and industrial chemicals have no serious side effects on human pregnancy and lactation, while effects on fertility have to date not sufficiently been studied in women and men who are exposed either in the workplace or as consumers of finished products.

Further Reading

ICH: Technical Requirements for the Registration of Pharmaceuticals for Human Use. ICH Harmonized Tripartite Guidelines. Test Guideline S5R2 (Detection of Toxicity to Reproduction for Medicinal Products – Segment I, II and III studies and Toxicity to Male Fertility), International Conference on Harmonization ICH, Geneva. http://www.ich.org/fileadmin/Public_Web_Site/ICH_Products/Guidelines/Safety/S5_R2/Step4/S5_R2__Guideline.pdf.

OECD: Guidelines for Testing of Chemicals, Test Guideline (TG) 414 (Prenatal Developmental Toxicity Study), TG 415 (One-Generation Reproduction Toxicity Study), TG 416 (Two-Generation Reproduction Toxicity Study), TG 421 (Reproduction/ Developmental Toxicity Screening Study), TG 426 (Developmental Neurotoxicity Study), TG 443 (Extended One-Generation Reproductive Toxicity Study (EOGRTS)). Organization for Economic Co-operation and Development, Paris. http://www.oecd-ilibrary.org/environment/oecd-guidelines-for-the-testing-of-chemicals-section-4-health-effects_20745788.

OECD: Guidance Document 43 On Mammalian Reproductive Toxicity Testing and Assessment 2008. Organization for Economic Co-operation and DevelopmentOECD, Paris, 88 pp. http://www.oecd.org/officialdocuments/publicdisplaydocumentpdf/?cote=env/jm/mono%282008%2916&doclanguage=en.

OECD: TG (Test Guideline) 478: Rodent Dominant Lethal Test 2013. Organization for Economic Co-operation and Development, Paris, 13 pp. http://www.oecd.org/env/ehs/testing/OECD_TG478_Revision_Sept_2013.pdf.

REACH: EU Regulation (EC) No. 1907/2006 of the European Parliament and of the Council of 18 December 2006 concerning the Registration, Evaluation, Authorization and Restriction of Chemicals (REACH), 18 December 2006. http://eur-lex.europa.eu/legal-content/EN/TXT/PDF/?uri=CELEX:02006R1907-20161011&from=EN.

Schaefer, C., Peters, P.W.J, Miller, R.K. *Drugs During Pregnancy and Lactation – Treatment Options and Risk Assessment*. 3rd edition, E-Books ISBN: 9780124079014, Hardcover ISBN: 9780124080782, Academic Press, London & New York 2014 & www.embyotox.de.

Wilson, J.D. *Embryotoxicity of Drugs to Man; in Handbook of Teratology*, Vol. **1**. Eds. J.D. Wilson and C.J. Frazier, Plenum Press, New York 1977, pp. 309–355.

2.11 Ecotoxicology: More than Wildlife Toxicology

Peter Calow and Valery E. Forbes

2.11.1 Introduction

Ecotoxicology was first defined explicitly by Truhaut in the 1970s as "the branch of toxicology concerned with the study of toxic effects, caused by natural or synthetic pollutants, to the constituents of ecosystems, animal (including human), vegetable and microbial, in an integral context". It has subsequently been refined to: the description and understanding of impacts of industrial and agricultural chemicals on ecological systems. This

> includes the study of natural chemicals used in industrial processes, such as metals and oils, and synthetically produced chemicals that do not occur naturally in the environment, such as PCBs, detergents, most pesticides and most pharmaceuticals. Ecotoxicology has its roots in toxicology but the targets of interest are different: toxicology is concerned with impacts on individual human beings and ecotoxicology on ecological systems. So ecotoxicology is not just toxicology applied to wildlife (Calow and Forbes, 2006).

In what follows we first define the protection targets for ecotoxicology and demonstrate how they differ fundamentally from the protection targets of toxicology. We then consider the consequences of this in terms of the practicalities of carrying out ecotoxicological tests and then in using the information in assessing the risks posed by chemicals for the ecological targets.

2.11.2 Protection Targets

> **The focus for toxicology is on the individual whereas the focus for ecotoxicology is on collective groups of individuals of the same species (populations) and/or groups of individuals of different species living together (communities/ecosystems). Individuals may be removed from populations without impacts on population persistence, and similarly, some species may be lost from communities without impacts on community persistence or ecosystem processes. So ecotoxicology focuses on detecting those critical impairments that relate to the persistence of populations and communities through space and time.**

Toxicologists are interested in the extent to which chemicals impact the health and survival of individual human beings. Ecotoxicologists by analogy often refer to protection of ecosystem health. Yet this betrays a fundamental misunderstanding of the structure and organization of ecosystems. Within the human body, cells, organs and systems operate in concert for the well-being of the whole. By contrast there is no cooperation between the individuals and the species within ecosystems for the well-being of the whole. Natural selection favors selfishness, i.e. the ability of individuals and the genes that they contain to reproduce and spread at the expense of others. Cooperation of the kind seen between the constituents of organisms to ensure the survival of the individual is not an appropriate concept when referring to the different parts of ecosystems. Wherever cooperation appears to occur in nature, such as in the social insects, it can be explained in terms of selfish gene theory (Dawkins, 2006).

Objectively, the only characteristic that can be said to be important in ecological systems is persistence, i.e. of individuals in populations and species in communities and ecosystems. The persistence of populations depends on the survival and reproduction of individuals within them, but not all individuals need to survive and reproduce to ensure the persistence of the whole. Similarly, communities and ecosystems depend on the species within them, but again not all. It follows that whereas toxicologists are interested in detecting impairment of any aspect of the individual organism since it is likely to contribute to health and survival, ecotoxicologists ought to be interested only in detecting those critical impairments that relate to the persistence of populations and communities through space and time. In this context it should be noted that assessing persistence ought to involve more than simply estimating numbers of individuals or species. For populations, changes in genetic composition or age/size structure may also be of concern, and long-term persistence of communities seems to depend not only on the number of species they contain but also on their relative abundances and niches.

Although ecological entities do not have purpose in the sense that we do, it is the case that we have needs from ecosystems. These are referred to broadly as ecosystem services and come in the form of, for example, fish and forestry production, pollination, soil quality and natural flood defenses. Protecting these services is becoming a key feature of managing the risks that arise for ecosystems from human activities. So understanding the connections between the persistence criteria outlined above and ecosystem service delivery is important in informing what is measured in ecotoxicology (Forbes and Calow, 2013).

2.11.3 Necessary Information

> **Estimating the adverse effects and likely exposures from chemicals on ecological systems follows similar procedures as toxicology, but there are some complications. The assessment of effects involves standard tests but with species and multi-species systems supposed to be relevant for the protection of ecosystems. So plants, invertebrates and fish are more frequently used than laboratory rodents. Exposure assessments rarely consider doses but exposure concentrations, and here the challenges are not only in determining what these may be but also in how they relate to bioavailability and effects.**

Assessment of Effects

Some Principles Both toxicology and ecotoxicology are concerned with making predictions about effects on their targets from limited observations in the laboratory over limited time periods. In this context it is useful to make a distinction between dose/concentration–effect relationships that are measured at high concentrations over short periods (acute) and that usually involve lethal effects from those measured at lower concentrations over long periods (chronic) and that are usually measured as sublethal responses, including impacts on development, growth and reproductive performance.

In toxicological tests, exposure to the chemical toxicant is expressed in terms of dose, which typically may be precisely applied by one of several routes (e.g., oral, dermal, inhalation). By contrast, in ecotoxicological tests exposure to chemical toxicants is expressed in terms of environmental concentration relating to an appropriate environmental compartment (e.g., water, sediment, food). This important distinction also leads to some differences in the way that endpoints are expressed. For lethal effects, toxicologists often express the test result as an LD_{50}, i.e. the dose required to kill 50% of the test subjects. Ecotoxicologists in contrast express the equivalent test result as an LC50, i.e. the environmental concentration required to kill 50% of the test subjects. For sublethal endpoints toxicologists might express responses in terms of the no observed adverse effect level (NOAEL) or the lowest observed adverse effect level (LOAEL), where level refers to the applied dose. In contrast, ecotoxicologists refer to effects as no observed effect concentrations (NOECs) or lowest observed effect concentrations (LOECs).

An important difference between toxicology and ecotoxicology is that the former is generally concerned with extrapolating effects from a few standard test species to a single target (i.e., humans), whereas the latter involves extrapolating from a few laboratory test species to the many species in natural ecosystems. Toxicologists invariably focus on single species tests; ecotoxicologists, as well as using single species tests, may also use multi-species test systems, which are potentially important in picking up indirect effects due to species interactions (see *Multi-species Tests*, this chapter).

> In most cases, ecotoxicology is interested in adverse effects at the population level and above. So an important question is the extent to which changes that manifest themselves within and at the level of the individual organism (e.g., impairments of survival indicated by LC50s and of reproduction indicated by LOECs) do indeed have impacts at higher ecological levels. In principle the effects at lower levels of organization might either disappear at higher levels because of damping or be exacerbated at higher levels because of cascading effects.

Damping effects include the within-organism homeostatic responses that might be induced by the presence of chemical stressors, the ability of populations to persist (e.g., by increased recruitment) despite loss of individuals from them, and finally the ability of communities and ecosystems to maintain processes as a result of redundancy despite losses of species from them.

Cascading effects involve situations in which small changes in key metabolic processes within individuals, key individuals within populations or key species within communities lead to major and potentially catastrophic changes in the biological levels above. When cascading effects occur as a result of interactions between species they are referred to as indirect effects, and these are particularly challenging for ecotoxicology. This is because the ecotoxicological tests are most often on individual species and so may miss indirect effects due to, for example, a chemical destroying the food of the consumer rather than impacting the consumer directly as a toxicant. Other examples of indirect effects may involve interference with competitive interactions and removing predators. Clearly these examples illustrate that indirect effects can lead to positive as well as negative effects on species.

Ecotoxicology has barely started to take all these complications into account in assessing the effects that chemicals have on ecologically relevant endpoints but some progress has been made (Newman, 2014).

More Details on Test Systems

> Test systems have been developed for freshwater and marine organisms, and for terrestrial systems. These include single-species tests as well as multi-species tests that are representative of all major environmental compartments.

Key requirements of test systems are that they should be not only realistic but also replicable so that the results can be defended scientifically and in legal situations. For this reason tests are subject to standardization. The Organization for Economic Cooperation and Development (OECD) plays an important role in publishing standard test protocols. In addition, for regulatory purposes tests are carried out according to good laboratory practice (GLP).

Freshwater Systems Test systems are more developed for freshwater organisms than for other biota because rivers have been a major conduit of sewage and industrial wastes. In Europe, for example, the focus has been on test systems involving unicellular algae, invertebrate zooplankton (*Daphnia*) and fish as providing not only representative organisms but also representative feeding (trophic) groups: photosynthetic algae (primary producers) fed on by herbivorous zooplankton (primary consumers) fed on by predatory fish (secondary consumers). Short-term, acute tests seek to define the concentrations that cause a 50% reduction in population growth rates of the algal cells, in the number of individual daphnids moving in test vessels, and in survival in fish, each observed usually over periods of no more than 24–96 hours. Mobility is used for the daphnid because observing when individuals stop moving is relatively straightforward, and immobility is a precursor of death in these animals. For the algae and the daphnids the endpoint is expressed as an effect concentration and so should be referred to as an EC50 rather than LC50 (cf. above). Long-term, chronic tests (over several weeks) are carried out on daphnids and fish to determine effects on individual growth and reproductive performance. The algal tests do not readily fit into these kinds of classification: they are short term on a human scale but long term in the life of an algal cell and involve changes in the balance of cell division rates and cell death. They are in fact population responses, but in unicellular organisms under laboratory conditions.

All of the above tests represent open-water systems. There are some specific tests for organisms that inhabit the benthic sediments of lakes and rivers (e.g., midge larvae) but often the results from open waters are extrapolated to sediments on the presumption that sediment organisms are most exposed through the pore water (in the interstices of the sediments) that surrounds them. However, there is increasing evidence that the exposure route for sediment-dwelling organisms often also includes the food they eat (i.e., the sediment particles) and this is an area of active research.

Marine Systems There are some equivalent marine tests for open water and sediment organisms (e.g., involving mysid shrimps, sea urchin embryos and sediment-dwelling annelids; see Table 2.24), but often it is presumed that marine organisms will have similar sensitivities to their freshwater equivalents. Indeed there would seem to be no *a priori* physiological or ecological reasons for this not being the case. However, there are marine taxa that are not found in freshwaters at all, such as jellyfish, echinoderms and cephalopods, that could have unusual ecotoxicological features. This is used as an argument for treating these extrapolations, from observations on freshwater organisms to expectations for marine biota, with caution in marine risk assessments.

Terrestrial Systems For the terrestrial environment tests have been thoroughly developed to gauge the effects of plant protection products applied to crops on biota in abutting ecosystems such as field margins, sometimes on the biota intermingled with the crops (Table 2.25) and, of course, on the soil organisms. This is because agricultural application of such chemicals is an important source of deliberate exposure for terrestrial systems, and these products are designed to be toxic to plant pests. These tests include ones on beneficial arthropods, earthworms, springtails, plants and vertebrates.

Table 2.24 Marine "tox kits".

Because of their ability to form dormant cysts, rotifers and brine shrimps have been developed into "toxicant testing kits". They can be stored for long periods in the dormant stage without the need to maintain expensive cultures.

For brine shrimps hatching is initiated by adding cysts to seawater a few days before tests are to be performed. Freshly hatched larvae are rinsed and then put into wells in a multi-well tray, with the wells containing a series of replicated test solutions. The tray is covered to prevent evaporation and incubated at the desired temperature. After a prescribed time (usually 24 hours) live and dead larvae are counted and an LC_{50} calculated.

Two kinds of quality control can be applied. First, if more than 10% of larvae in the control well(s) die the test is considered invalid. Second, standard toxicants can be used to check endpoint fidelity.

Rotifer cysts are used in similar ways.

Table 2.25 Brief description of some laboratory ecotoxicity tests on beneficial (i.e., not pests) arthropods that live in and around crops.

Types	Exposure	Endpoint
Mites, spiders, various insects, including beetles, dipterans, hymenopterans	Natural foliage pretreated or glass slides onto which chemical has been applied	Most often mortality, occasionally emergence from eggs or pupae and fecundity

Table 2.26 The major classes of multi-species tests and their properties.

Description	Replication	Endpoint
Flasks or flow-through chemostats with cultures of bacteria and/or protozoans	Many possible	Abundance, diversity
Large flask, bench systems with periphyton and metazoans – may be constructed or naturally derived communities	Some	Relative abundance, production, respiration and nutrient uptake
Artificial streams ranging from small indoor to large outdoor systems	Occasional; system before treatment often used as control	Abundance of key species, diversity, production
Artificial ponds ranging from bowls to large outdoor systems	As above	As above
Enclosures of natural systems, e.g. tubes in lakes, field plots	As above	As above
Manipulation of whole ecosystems	Rarely	As above

Multi-species Tests As well as single-species tests, multi-species tests have been constructed as representative of all major environmental compartments. Depending on size and level of complexity they are referred to as micro-, meso- or macrocosm studies (Table 2.26). Very occasionally, and usually for agricultural applications, field or semi-field studies are carried out. On the one hand all these multi-species studies bring the advantage of more ecological realism, but drawbacks include cost, knowing what to measure and how to interpret it and, for -cosms, difficulties in setting up and maintaining complex systems. Moreover the complexity in itself may mean that the characteristics and responses of such systems are somewhat unique and not always easily translated to ecosystems in nature.

Exposure Assessments

> **As much as for toxicology, it is the dose that determines effect, and as in toxicology this depends on how much gets where in the organism and how long it remains there in toxic form. But there are added layers of complexity in ecotoxicology.**

The exposure concentration (EC) of a chemical might either be predicted (PEC) from models that take account of its release, distribution or fate (e.g., partitioning between air, water, soil and sediment) and persistence in toxic form in the natural environment. Alternatively, exposure concentration may be measured (MEC), for example in monitoring programs. Either way, neither PEC nor MEC need express the effective concentration for the targets since this will depend on other factors such as the extent to which the material present in the environment can enter the organism (bioavailability), e.g. it might be present in soils or sediments but bound so tightly to the materials there that it is not taken up by organisms (i.e., it is not bioavailable).

An important characteristic of chemicals is how they partition between water and organic phases, since this will determine likelihood of binding to organic particles in soil and sediments and/or accumulating in the bodies of organisms. The uptake of toxicants from their surrounding medium and food is referred to generally as bioaccumulation (bioaccumulation factor = concentration in organism/concentration in medium *and* food). A widely used technique for assessing the **bioaccumulation potential** of chemicals measures the partitioning of the chemical between octanol and water under standard conditions (i.e., the n-octanol-water partition coefficient or KOW).

Bioconcentration, on the other hand, refers more specifically to the capacity of aquatic organisms to take up toxicants from the surrounding water (bioconcentration factor = concentration in organism after specified time under standard conditions/concentration in surrounding water). In contrast, **biomagnification** refers to the increasing concentration of a toxic chemical in the tissues of organisms at successively higher trophic levels. Exposure through feeding is important because it opens the possibility of the biomagnification of toxicants. For example, there may be limited uptake of a toxicant in organisms at one trophic level, but feeders on these may eat many individuals and accumulate the toxicant in their tissues so increasing the level of exposure to animals that feed on them. Thus mercury may occur in very low concentrations in seawater and be absorbed by algae. These are eaten by invertebrates that are unable to excrete the mercury at a fast rate so it accumulates. The same goes for the fish that feed on them. Top predators such as sharks and swordfish may become exposed to dangerously high concentrations of the mercury. The commonness of biomagnification effects in nature is open to controversy, but biomagnification could be a potential source of problems for humans eating fish and game from nature.

The time course of the concentration of a toxicant at the site of toxic action in organisms, so-called **toxicokinetics** (TK), depends on the properties of the chemical and the physiological features of the exposed organisms that determine uptake, distribution, biotransformation and elimination of the chemical. These can be represented in toxicokinetic models and combined with an understanding of toxic action at the target sites, so-called **toxicodynamics** (TD), to provide a better understanding of the variability in toxic effects from the same chemicals across different species (Ashauer and Escher, 2010).

Finally, a feature that is of concern in both ecotoxicology and human health studies is the extent to which exposure to mixtures, rather than single substances, is important. Mixtures may be more complex and even less easily defined in ecosystems as compared with human exposures. In principle chemicals in these mixtures can act independently or in combination depending on their structure and concentration (European Commission Scientific Committees Opinion, 2012). Because the number and type of chemicals that may occur in mixtures is enormous they have to be considered on a case-by-case basis for ecological assessments.

A straightforward description of the principles and practice of fate and exposure assessment, as well as the types of models used, is given in van Leeuwen and Vermeire (2007).

2.11.4 Risk Assessment

> **Fundamentally, ecological risk assessment is similar to human health risk assessment in that it involves comparing likely exposure with likely effect concentrations. Both involve uncertainties especially concerned with extrapolation from observations in test systems to the real thing. Stepwise approaches leading to management decisions are a feature of both. However, the details of the steps and the role of expert judgement applied, differ between the two kinds of risk assessments.**

Risk assessment is concerned with predicting the extent to which exposure from a source of chemical pollution is likely to lead to adverse effects in the target (van Leeuwen and Vermeire, 2007). For both toxicology and ecotoxicology this amounts to comparing likely exposures with doses/concentrations known to cause adverse effects. The key challenges for assessing risks to human health and ecological systems from these data are therefore in terms of taking account of the uncertainties in using appropriate extrapolation techniques. As already mentioned the extrapolations involved in human health risk assessments are from observations on a few laboratory test species to human beings. On the other hand, the extrapolation for ecological risk assessment is from information on a few species to all those species in ecosystems that we seek to protect. For both there is usually the common challenge of extrapolating from effects observed at high concentrations over short time periods (the most frequent test situation) to effects of low concentrations over long time periods (often the most realistic exposure situation).

In toxicology predicted dose is compared with the NOAEL, and uncertainties are taken into account by considering if the difference is greater than a defined **margin of safety** (MOS). These MOSs can be explicitly defined but often it is left open to expert judgment to decide if the differences are big enough for safety.

For ecological risk assessments dealing with pesticides and herbicides a similar technique is used. The endpoint is compared with exposure in a toxicity exposure ratio (TER) and these have to exceed specified values for them to be considered acceptable. On the other hand for industrial chemicals PECs and/or MECs are compared with **predicted no-effect concentrations** (PNECs) to give a risk quotient (RQ = PEC or MEC divided by PNEC). The PNEC is calculated from whatever endpoints are available divided by uncertainty factors that are designed to take account of the extent of extrapolation that is involved. There is a lot of variability in the way this extrapolation is applied in ecological risk assessment, but one interpretation of the factors used within European Union legislation is summarized in Table 2.27.

Clearly the MOS approach allows expert judgment to be applied in weighing the difference between likely exposure and effects in each case whereas predefined uncertainty factors as used for RQs limit the scope for expert judgment, at least in principle. This appears paradoxical in the sense that there should be less uncertainty in extrapolating from effects in a few species to one than in extrapolating from few to many. On the other hand the more uncertainties of the latter case, representing the ecological situation, mean that expert judgments are themselves likely to be more variable and hence the need for a more restrictive and transparent procedure.

Sometimes sufficient data are available on the effects of a chemical on a large enough number of species to express variability between species in the form of a statistical distribution (**species sensitivity distribution**, SSD) and from this to predict that exposure concentration likely to have minimum effect (usually defined as affecting <5% of species) and using this as the PNEC. However, the sample of species that make up the SSD is rarely representative of the ecosystem(s) under consideration, the identity of the individual species and their role in ecosystem processes is

Table 2.27 Simplified version of extrapolation (assessment) factors suggested in the technical guidance currently used in EU legislation, but the allowances are our interpretation of the way the extrapolation factors have been compiled. Thus a factor of 10 allows for uncertainty between acute and chronic effects, a factor of 10 allows for uncertainty in interspecies chronic effects, and a factor of 10 allows for the uncertainty between laboratory single species tests to effects in multi-species or field tests. Extra factors might be added for other uncertainties, e.g. when using observations on freshwater biota to predict the effects on marine. After Forbes and Calow (2002a).

	Extrapolation Factors	Allowances
At least one acute L(E)C$_{50}$ from each of three trophic levels	1000	10
One chronic NOEC	100	10
At least three chronic NOECs from each of three trophic levels	10	10
Field data or model ecosystems	Reviewed on a case-by-case basis	

not taken into account, and sensitivity is measured by individual toxicological responses (ignoring other factors that determine species vulnerability). All of these contribute uncertainty to the interpretation of the SSD (Forbes and Calow, 2002b).

Much progress has been achieved during the last decade on the development and implementation of various mathematical and simulation models to extrapolate from observations on individuals and individual species to make predictions about population dynamics and biodiversity effects that are relevant to protecting the services people derive from ecosystems. These models provide a more quantitative and mechanistic approach to extrapolation than previous methods; they facilitate the integration of chemical exposure and effects, and, once developed, they can explore chemical impacts on relevant spatial and temporal scales more easily and at lower cost than empirical studies. More recent initiatives aim to use such mechanistic models to extend extrapolation to lower levels of biological organization to make use of high-throughput molecular data as well as to higher levels of organization to make linkages between ecological structures/processes and ecosystem service delivery more explicit (Forbes et al., 2017).

Like human health risk assessment, ecological risk assessment is usually performed using a tiered approach, with lower tiers designed to represent relatively realistic worst-case scenarios. If these give no cause for concern, then no more analyses are required. On the other hand if they do indicate likely risks then analyses involving more species and possibly "cosm" or field tests would be required. The idea is to get most information with minimum time, effort and hence costs.

2.11.5 Fast-track Approaches

> **With concerns about the time, cost and number of organisms used in ecotoxicological work there has been pressure for fast-track and short-cut techniques. Simplification as a general rule introduces uncertainties so these methods have to be treated with some caution.**

Examples of short-cut approaches similar to those used in human health risk assessment with attendant problems are models that link the structure of chemicals with their likely uptake and effect in organisms (so-called (Q)SARs) and *in vitro* systems (e.g., fish cell/tissue cultures) that are supposed to simulate whole organisms in responses to chemicals.

In-organism measures – on genes, enzymes, cells and tissues – have also been used as signals of exposure and effect in organisms sampled from circumstances where there was suspected pollution. These are broadly referred to as biomarkers and should be treated with some caution in an ecological context (Forbes et al., 2006). First, and similar to the concern in human health biomarkers (see Chapter 4.3), one effect might potentially be caused by various factors so the signal does not lead unequivocally to a conclusion about the actual cause. Second, it follows from the discussions above that even if the biomarker is signaling a specific exposure to a chemical, the effect may not be relevant in terms of harm to the ecological system concerned. Early warnings are often invoked as a rationale for taking these signals seriously, but that should be set against the costs of any management that follows and the likelihood of false positive results.

Another increasingly used method of fast tracking chemicals for restriction is based on the argument that those that are persistent (P), bioaccumulative (B) and have high toxicity (T) are likely to cause most problems for ecosystems. PBT criteria can be ascertained from simple and hence cheap methods and so can potentially identify problems for rapid action. However, it is important to recognize that all these criteria indicate potential to cause harm, i.e. are so-called hazard criteria (see Chapter 1), and may not always be good indicators of risk. For example, the PBT criteria say nothing about the amounts of a substance likely to escape into the environment and hence about likely exposure concentrations. Hence, management on the basis of PBT criteria should be modulated with an understanding of the costs of likely restrictions based on them.

2.11.6 Summary

Ecotoxicology aims to describe and understand how the structures and processes of ecological systems are affected by exposure to chemicals as a result of human activities. The responses of individual organisms are less interesting in this context than the responses of populations, communities and ecosystems. However, it is easier to make observations on individuals in test systems than on populations, communities and ecosystems in natural systems. Extrapolations from one level to another involve a number of uncertainties that are most often brought into consideration through the use of uncertainty factors. Because of the inherent complexity of ecological systems, methods promising short-cuts in assessments should be treated with caution. Ongoing initiatives to develop mechanistic effect models (Forbes and Galic, 2016) may provide an effective means of incorporating uncertainties and ecological complexities in a way that is scientifically robust yet practical.

Further Reading

R. Ashauer and B.I. Escher, Advantages of toxicokinetic and toxicodynamic modelling in aquatic ecotoxicology and risk assessment. *Journal of Environmental Monitoring,* **12**, 2056–2061, 2010.

P. Calow and V.E. Forbes, *Ecotoxicology, in* Encyclopedia of Life Sciences, John Wiley & Sons Ltd., Chichester, www.els.net, January 2006.

R. Dawkins, *The Selfish Gene – 30th Anniversary Edition*, Oxford University Press, Oxford, 2006.

European Commission Scientific Committees Opinion, *Toxicity and Assessment of Chemical Mixtures.* doi:10.2772/21444, 2012.

V.E. Forbes and P. Calow, Extrapolation in ecological risk assessment – balancing pragmatism and precaution in chemical controls legislation. *BioScience*, **52**, 249–257, 2002a.

V.E. Forbes and P. Calow, Species sensitivity distributions revisited: a critical appraisal. *Human and Ecological Risk Assessment*, **8**, 473–492, 2002b.

V.E. Forbes and P. Calow, Use of the ecosystem services concept in ecological risk assessment of chemicals. *Integrated Environmental Assessment and Management*, **9**, 269–275, 2013.

V.E. Forbes and N. Galic, Next generation ecological risk assessment: predicting risk from molecular initiation to ecosystem service delivery. *Environ Internat*, **91**, 215–219, 2016.

V.E. Forbes, A. Palmqvist and L. Bach, Use and misuse of biomarkers in ecotoxicology. *Environmental Toxicology and Chemistry*, **25**, 272–280, 2006.

V.E. Forbes, C.J. Salice, B. Birnir, R.J.F. Bruins, P. Calow, V. Ducrot, N. Galic, K. Garber, B.C. Harvey, H. Jager, A. Kanarek, R. Pastorok, S.F. Railsback, R. Rebarber and P. Thorbek, A framework for predicting impacts on ecosystem services from (sub)organismal responses to chemicals. *Environ Toxicol Chem*, **36**, 845–859, 2017.

C.J. van Leeuwen and T.G. Vermiere, *Risk Assessment of Chemicals: An Introduction*, 2nd edition. Springer, Dordrecht/The Netherlands, 2007.

M.C. Newman, *Fundamentals of Ecotoxicolgy, The Science of Pollution*, 4th edition. CRC Press, Boca Raton, Florida, 2014.

3
Organ Toxicology

3.1 The Gastrointestinal Tract

Michael Schwenk

3.1.1 Introduction

> The gastrointestinal (GI) tract forms a barrier between the intestinal lumen and the blood. It is the first organ that comes into contact with foodborne substances and controls digestion and absorption. Each of the sections of the GI tract has a unique physiological function and exhibits characteristic toxicological responses.

Ingested food passes from the mouth down the esophagus into the stomach, is then emptied into the small intestine and finally is transferred to the colon (large intestine). The lumen of the GI tract belongs to the "outside" of the organism. The GI tract has the function to digest and absorb nutrients and in addition to defend the organism against intruding toxicants and microorganisms. A single-cell layer of epithelial cells separates the contents of the GI lumen from the blood. Smooth muscle activity in the intestinal wall mixes and propels the intestinal content.

Toxicologically, the GI tract has several peculiarities: First, it may be exposed to higher (food-borne) toxicant concentrations than other organs. Second, it determines the extent of absorption of xenobiotics and is often involved in biotransformation. Finally, its microbiota has toxicological relevance. Cell necrosis and cancer belong to the GI-characteristic types of toxic damage. Beyond that, the networks of neurological/humoral regulations make the GI tract sensitive to neurotoxic agents. Its immunological defense system makes it sensitive to allergic reactions towards food ingredients.

3.1.2 Structure and Function

Structure

The GI tract extends as a hollow organ from the mouth to the anus and forms the barrier between the intestinal lumen and the blood. The barrier is composed of a single layer of epithelial cell types that differ between stomach, small intestine, and colon. The epithelial surface is protected by a mucus layer. The underlying tissue (mucosa and submucosa) is rich in supplying blood vessels, nerve fibers, regulatory endocrine cells, and varying numbers of immune cells (Figure 3.1). It is enclosed by muscle fibers (muscularis), which promote the GI contents (peristalsis). An outermost layer (serosa) provides mechanical stability. The small intestine has a very large surface due to numerous protrusions (villi) and invaginations (crypts) of the epithelial layer and a brush-like surface (brush border) on the luminal side of the epithelial cells. In humans the absorbing surface area corresponds to the size of a tennis court.

Function

After ingestion, degradation, and liquefaction of the food in the mouth, the food mash glides through the esophagus into the stomach. It is mixed there with gastric juice, which is produced by three specialized cell types: mucous cells (mucus), parietal cells (hydrochloric acid), and chief cells (pepsinogen). Secretions of the gastric mucosa are controlled by the vagus nerve, hormones (e.g., gastrin) and a number of local mediators, including prostaglandins, histamine, and neuroactive peptides.

The normal stomach epithelium is quite resistant to the irritant effects of the acidic gastric juice (which can be below pH 2), since mucous and bicarbonate secretions by mucous cells protect the cells. The gastric contents are emptied in portions into the small intestine, where they are mixed with inflowing bile and pancreatic juice. Bile salts act in the intestine as detergents by emulsifying nutritional fat in micelles, an essential step in lipid digestion. Pancreatic juice contains digestive enzymes which hydrolyze proteins (e.g., chymotrypsin) degrade triglycerides (lipase) and break down carbohydrates (amylase).

The brush border membrane carries digestive enzymes such as peptidases and disaccharidases that cleave predigested nutrients to absorbable small units. Substrate-specific membrane transporters of intestinal epithelial cells (enterocytes) move absorbable nutrient molecules, such as sugars, amino acids, fatty acids, nucleosides, vitamins, and minerals, from the intestinal lumen

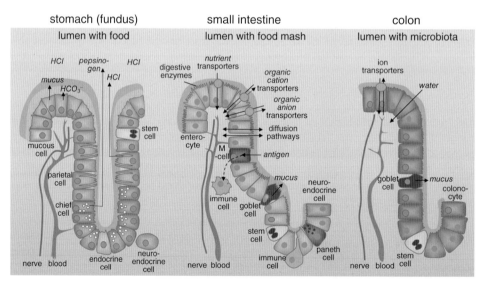

Figure 3.1 *Diversity of the GI barrier. Schematic presentation of the gastric, intestinal and colonic epithelia and some mucosal structures. The focus is on the respective epithelial cell types and their functions.*

into cells, often in an energy-dependent, accumulating process that involves a membrane transporter. The nutrient molecules pass through the cell body and are released through basolateral membrane transporters to the blood side.

The transit time for the passage of the food mash through the small intestine is about 3–8 hours. Undigested and unabsorbed components are delivered into the colon. It reabsorbs most of the approximately 8 liters of water that is daily secreted with the digestive juices. Disturbances of reabsorption will cause diarrhea with loss of water and electrolytes.

Microbiota

While the upper small intestine contains only few bacteria (100–1000 per ml), the colon contains some 10^{12} bacteria per ml. The bacteria decompose some undigested carbohydrates and synthesize some nutrients (e.g., vitamin K, folic acid). In the anoxic environment of the colon lumen, they produce toxic substances, such as ammonia, nitrite, phenols, endotoxine, and nitrosamines. These act as cofactors in various chronic liver disease and are a major cause of coma during hepatic failure, when they are not sufficiently metabolized and accumulate in blood.

Tissue Repair

Injuries of the barrier-forming epithelium are rapidly repaired by neighboring cells that migrate to the defective location and cover it (hours), followed by proliferation of stem cells (days). Deeper injuries will heal less readily, since they require reorganization and proliferation of many components, including blood capillaries. The normal lifetime of intestinal epithelial cells is about 6 days, that of stomach cells is much longer.

3.1.3 Fate of Xenobiotics in the GI Tract

Absorption of Xenobiotics

> **Absorption occurs mainly in the small intestine. Although xenobiotic molecules are mainly taken up by diffusion, involvement of membrane transporters often plays a significant role. The rate and extent of absorption depend on solubilization in the lumen, transport across the epithelial barrier, blood flow, and intestinal transit time.**

Solubilization in the GI tract is a precondition for absorption. Hydrophilic small xenobiotic molecules tend to be freely dissolved in the intestinal lumen. Moderately lipophilic substances are emulsified and solubilized through the detergent effect of bile. Highly lipophilic agents, such as liquid paraffin as well as compounds that are strongly bound to a matrix (e.g., soil), tend to remain poorly soluble in the GI lumen.

Absorption of an agent across the epithelial barrier can either occur through epithelial cells (transcytosis) or via molecular-sieve-like tight junctions that connect adjacent cells (paracellular route). It is assumed that the predominant mechanism for many xenobiotics is transcellular diffusion through enterocytes. Small molecular size, solubilization, and absence of an electric charge favor transcellular diffusion. Although the stomach plays only a minor role in absorption, it can absorb weak acids (non-ionized in acidic environment) whereas weak bases (ionized) will not be absorbed.

Energy-dependent membrane transporters of the ABC family are involved in absorption of organic molecules in the small intestine, but the relative importance compared to other uptake processes is often unclear. P-glycoprotein is a type of ABC transporter that transports a broad variety of xenobiotics out of intestinal epithelial cells back into the intestinal lumen; it can be induced by its substrates. The combined effect of absorbing and secreting intestinal organic transporters is a cause for drug–drug and drug–nutrient (e.g. statin–grapefruit) interactions.

There is a debate on the extent to which nanoparticles can be absorbed. Depending on their size, shape, composition, and surface features, possible routes of absorption are by endocytosis, pinocytosis, through leaky areas of the barrier (M cells or injured areas) or via the paracellular route. There are strategies to use synthetic nanoparticles as carriers to improve absorption of otherwise poorly absorbable drugs. For this purpose, suitable ligands for endocytosis are attached to the surface of engineered nanoparticles.

Concerning rates of absorption, the following rough rules may apply. Small xenobiotics with moderate lipophilicity (e.g., ethanol, dimethylnitrosamine or nitroglycerine) are rapidly absorbed in all segments of the GI tract. Xenobiotics with high lipophilicity (e.g., liquid paraffin) are water insoluble, which restricts their molecular access to the charged surface of the cell membrane; they are often absorbed less than 10% of the dose. Large water-soluble compounds (e.g., strophanthin G, atropine) are rejected by the lipophilic regions of the cell membrane and are often absorbed less than 10% of the dose. Macromolecules are believed not to be absorbed. Nevertheless, the highly toxic botulinum toxin, a protein with a molecular mass of 150,000 Da, may be absorbed in small amounts (0.01%), sufficient to produce neurotoxicity.

Biotransformation of Xenobiotics

> **The GI tract is one of the major xenobiotic-metabolizing organs. The respective enzymes are localized in epithelial cells. At the same time, some acid-catalyzed reactions in the stomach, as well as bacterial biotransformation in the intestinal lumen, contribute to formation of metabolites, some of which are toxic.**

Activities of cytochrome P450 (CYP) and Phase II enzymes are low in the esophagus and stomach, rise at the beginning of the small intestine, and then decrease gradually towards the colon. The enzymes are predominantly located in the villi and less often in the crypts. The activity of monoamine oxidase (MAO) in intestinal epithelial cells is higher than in liver cells. The activity of glucuronosyltransferases, sulfotransferases, acetyltransferases, and glutathione S-transferases is comparable with that in liver cells. CYP monoxygenases display only about 10% of the activity in liver tissue, but some isozymes can have activities comparable to those in liver cells (e.g., CYP3A4). The CYP isoenzyme pattern differs from that in the liver. Some forms of intestinal CYP enzymes and glucuronosyltransferases can be induced two–fold or more by food ingredients (e.g., ethanol, polyaromatic compounds). There are considerable variations depending on species, age, and nutrition.

In some cases, relevant biotransformation occurs during passage of a substance through enterocytes (intestinal first-pass effect). Thus, tyramine, which is abundant in fermented food (cheese, salami, herring) in amounts that would cause a life-threatening rise of blood pressure, is efficiently (>90%) detoxicated to *p*-hydroxyphenylacetaldehyde by MAO of intestinal epithelial cells (Figure 3.2). Likewise, many phenolic substances are conjugated with sulfate and glucuronic acid, often to an extent of about 50%. In general less than 10% of a dose is metabolized during absorption by intestinal CYP enzymes.

Some acid-catalyzed chemical reactions can take place in the acidic environment of the stomach. Notably, carcinogenic nitrosamines may be formed in the stomach by acid-catalyzed attachment of nitrite to secondary amines (Figure 3.3).

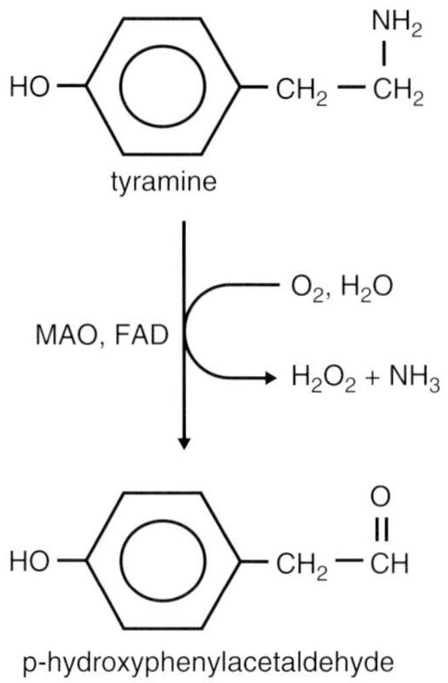

Figure 3.2 Inactivation of tyramine. The reaction is catalyzed by the enzyme monoamine oxidase (MAO) in intestinal epithelial cells; FAD is the coenzyme. The product hydroxyphenyl-acetaldehyde is biologically inactive and is further oxidized to p-hydroxyphenyl-acidic acid.

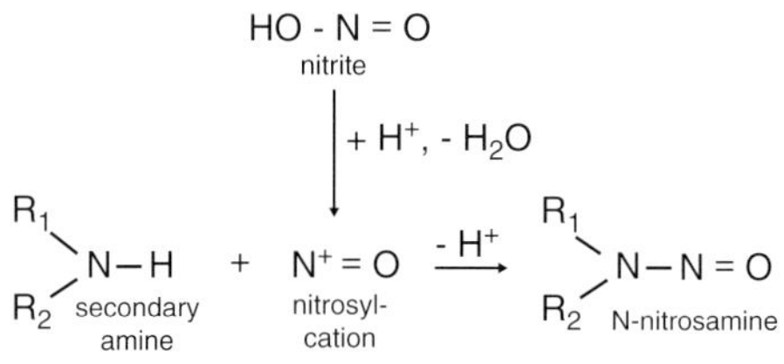

Figure 3.3 Nitrosamine formation in the stomach. The acidic environment of the stomach favors formation of nitrosamines from nitrite and secondary amines.

While biotransformations in the intestinal wall are oxidations (Phase I) and conjugations (Phase II) the microbiota in the colon with its anaerobic and substrate-deficient environment tend to use the opposite pathways: reduction of xenobiotics (e.g., azo compounds) and cleavage of conjugates (e.g., phenol glucuronide) provide energy sources for the microorganisms. Both bacterial reaction types may have adverse consequences for the host: chemical reductions may yield toxic products such as nitrite and ammonia. Cleavage of conjugates liberates non-conjugated parent compounds, which are usually more toxic and, being more lipophilic, can be reabsorbed. Glycosidase enzymes of colon bacteria are responsible for the toxicity of various glycosides in food, such as amygdalin in bitter almonds. They liberate hydrogen cyanide from the glycoside, which then exerts its toxic effects in the organism. Intestinal bacteria can form nitrosamines and other mutagens that are detectable in the feces and believed to contribute to cancer risk.

Enterohepatic Circulation of Xenobiotics

In the process of fat absorption, when chylomicrons are formed in enterocytes, some lipophilic foreign compounds (e.g., DDT) get bound to chylomicrons and are drained via the mesenteric lymph. The less lipophilic compounds pass via the portal vein to the liver. Many pharmaceutical drugs, metallo-organics, and organochloro-compounds are taken up into liver cells and from there secreted into bile either unchanged or in conjugated form. They are then delivered with bile into the intestine, from where they may be reabsorbed (either unchanged or after bacterial deconjugation). This enterohepatic circulation can prolong the biological half-life of some xenobiotics in the organism. Thus, about 90% of an oral dose of pentachlorophenol is eliminated into bile. Only 3% of this fraction appears in the feces and the rest is reabsorbed.

3.1.4 Toxicology

Disturbance of Function

> **The motility and digestive functions of the GI tract are regulated by the vagus nerve in combination with hormones (e.g., vasoactive intestinal polypeptide) and local mediators (e.g., histamine, neuroendocrine peptides). Toxic interference leads to disturbances of motility and water absorption/secretion, often without a morphological sign of pathologic damage.**

Organophosphates (e.g., parathione) inhibit acetylcholinesterase enzymes. This leads to a flooding of the respective synapses with the neurotransmitter acetylcholine that stimulates GI tract motility, salivation, and diarrhea. Atropine, the poison of the deadly nightshade, acts as an antagonist by blocking synaptic acetylcholine receptors. In severe poisonings with parathion or atropine, the life-threatening effects on the central nervous system usually dominate. Morphine is an agonist of endorphine receptors, which are present in the GI tract; in this way it reduces motility.

Even the highly toxic cholera toxin acts "solely" via functional disturbance. In cholera-infected individuals this toxin (a glycoprotein) is released from cholera bacteria into the intestinal lumen, from where it is taken up by endocytosis into intestinal epithelial cells. There, its components activate a regulatory protein (Gs), which then generates overshooting levels of cyclic AMP. This opens ion channels and induces excessive water secretions into the intestinal lumen. It leads to a life-threatening situation that is clinically managed by administrating water and electrolytes.

Lactose intolerance and fructose malabsorption are common causes for abdominal pain and diarrhea. Lactose intolerance is due to a (age-dependent) low intestinal lactase activity; symptoms occur after ingestion of milk products. Fructose malabsorption symptoms are a consequence of a low capacity of the intestinal fructose transporters in combination with a nutritional fructose overload (notably from soft drinks). In both cases, unabsorbed sugar will move into the lower intestine, dragging water with it (diarrhea) and favoring bacterial overgrowth.

In connection with silver nanoparticles that are used to protect textiles and other items from bacterial growth, there is some concern that accidental ingestion of silver nanoparticles might have a negative effect on the normal microbiota and subsequently on the GI tract.

Cytotoxic Effects

> **Ingested corrosives and cytotoxic substances damage epithelial cells of the GI tract. Agents that disturb the protective mucous layer or inhibit local blood flow may also result in tissue damage.**

When epithelial cells of the GI tract undergo necrosis (cell death), they scale off from the underlying tissue (erosion) into the GI lumen. If deeper tissue layers are involved, round-edged, bleeding defects are seen (ulcer).

Necrosis of the mucous membrane of mouth and GI tract may be caused by (accidental) intake of strong acids, bases or other corrosive chemicals. The necrosis does not hold on to predefined morphological borders. Acid necrosis often heals readily because acids precipitate tissue proteins, thus protecting the underlying tissue. In contrast, lye necrosis heals poorly because tissue is liquefied by alkali so that the destruction promotes into deeper tissue layers.

Heavy metal salts cause dose-dependent disorders in the GI tract. They block sulfhydryl, carboxy, and amino groups of cellular macromolecules and thus inhibit digestive enzymes, transporters, and healing processes. Moderate poisonings are associated with temporary malfunctions, such as malabsorption, intestinal cramps, diarrhea, and vomiting. Severe poisoning leads to massive mucosal cell death, ulceration, and complete destruction of areas of the mucous membrane.

Chemotherapeutics tend to exhibit side effects on the GI tract. They either inhibit vital cellular functions, leading to cell death, or inhibit proliferation of stem cells, so that dying cells will not be replaced. Destructions are especially observed in the small intestine, where the lifetime of intestinal epithelial cells is only a few days.

Gastric ulcerations and bleedings are common side effects of non-steroidal anti-inflammatory drugs (NSAIDs), inhibitors of prostaglandin synthesis. Lowered prostaglandin levels are accompanied by decreased formation of protective mucous, reduced local blood flow, and stimulated secretion of aggressive gastric acid. Ethanol-induced mucosal lesions are believed to result from a sequence of events involving decreased local blood flow, edema, insufficient oxygen supply, mediator release, and finally cell death. The experimental compounds cysteamine, propionitrile, and various toluene derivatives induce duodenal ulcers in experimental animals.

Immune Reactions

The intestinal mucosa possesses an elaborate immunological defense system called gut-associated lymphoid tissue (GALT) that is directed against penetrating microorganisms and food components. Undesirable immune responses in the GI tract can cause food allergies, mucosal inflammation, and autoimmune disease.

Even though the intestine forms a barrier against intruding pathogens and unwanted agents, antigens may pass the barrier and interact with the immune system, which forms a first line of immunological defense here. Thus, M cells in the epithelial lining support the passage of antigens from the GI lumen to antigen-presenting cells, lymphocytes migrate into the space between epithelial cells, IgA antibodies are secreted into the intestinal lumen, macrophages phagocytose intruding micro-/nanoparticles, and paneth cells secrete antibacterial substances. Both the innate immune system with its receptors for pathogen-associated molecular patterns (PAMP) that recognize viruses and microorganisms, as well as the adaptive immune system that develops a defense against non-self peptide sequences can undergo errors that may result in pathologic immune reactions. After consumption of an allergenic food component, the immune system may become sensitized. Re-exposure may then cause an allergic reaction with release of histamine, prostaglandins, leukotrienes, and cytokines followed by lymphocyte invasion and inflammation.

Food-induced allergy can be restricted to the GI tract with symptoms of a full, aching abdomen or may involve other organs, such as skin (rash), respiratory tract (asthma) or circulation (fall of blood pressure). Allergy-like symptoms are called "pseudoallergy" when initiated by foodborne mediators such as histamine or by ingested substances (e.g., salicylates) that induce mediator release from mast cells and basophils.

Celiac disease is a severe inflammatory reaction to gluten, notably the gliadin proteins, of wheat and other grain. According to present knowledge, the mechanism involves the following. Being poor substrates for proteases of the stomach, gliadins are emptied into the upper small intestine, where gliadin-derived peptides may get access to the tissue. They are deaminated by tissue transglutaminase, whereby adducts are formed. These are immunogenic, which means they induce proliferation of T cells and production of antibodies, both directed against the immunogenic agent, causing inflammation, autoimmune reactions, allergic response, and finally tissue damage. The mucosa of the affected areas shrinks, loses its ability to efficiently absorb nutrients, and becomes leaky. The risk for celiac disease is associated with the inherited major histocompatibility complex (MHC) type. The only effective therapy is to avoid any food that contains gluten or related proteins.

Cancer

> **While the colon is a predominant cancer location in many western countries, cancers in other sections of the GI tract are more common in other countries, such as mouth in India, esophagus in China, and stomach in Japan. These differences may in part be due to genetic factors, but observations suggest that lifestyle and exposure also play a role.**

It has been observed that Japanese immigrants to the USA have cancer locations similar to those found in Japan (mainly gastric cancer), but their descendants in the second generation had less gastric cancer and more colon cancer (like in the USA). From this and related observations it was assumed that lifestyle factors, such as carcinogenic ingredients in the food and country-specific viral infections, may play a role. Stomach cancer in Japan has been explained by the intake of nitrosamine-rich sea food and salt, colon cancer in the western world has been associated with low fiber and high-fat diets, cancer of the oral cavity in India with betel nut chewing, and cancer of the esophagus with zinc deficiency, alcohol, smoking, and consumption of unpeeled grain. However, a final causal relationship and mechanistic explanation is still missing in most cases.

Animal experiments suggest that the GI tract is sensitive to genotoxic chemical carcinogenesis. In the rat N-nitroso-N-methylbenzylamine produces cancer of the esophagus, N-nitroso-N-methyl-N'-nitroguanidin cancer of the stomach and 1,2-dimethylhydrazine cancer in the colon (Figure 3.4). The ultimate carcinogen of all three compounds is an alkyl cation, which reacts with DNA. In the case of the N-nitroso-N-methyl-N'-nitroguanidin the alkyl cation forms spontaneously, for N-nitroso-N-methylbenzylamine a CYP-dependent hydroxylation is involved, and with 1,2-dimethylhydrazine biotransformation to the reactive molecule proceeds through several steps (Figure 3.5). One of the chemical intermediates of 1,2-dimethylhydrazine metabolisms is methylazoxymethanol. A precursor of this compound occurs as a glycoside (cycasin) in the edible parts of the Asian sago palm. After consumption the glycoside is hydrolyzed by bacterial glycosidases of the colon. Cycasin is carcinogenic in normal rats, but induces no tumors in germ-free rats where methylazoxymethanol cannot be released from the glycosidic bond.

Alcohol consumption and smoking increase the risk of cancer in various sections of the GI tract, notably the esophagus and the colon. It is assumed that alcohol-associated esophagus cancer may be caused by a tumor-promoting effect. In general, cancer in the human stomach and colon is very common, but cancer in the small intestine is rare. The reason for this surprising difference is not known.

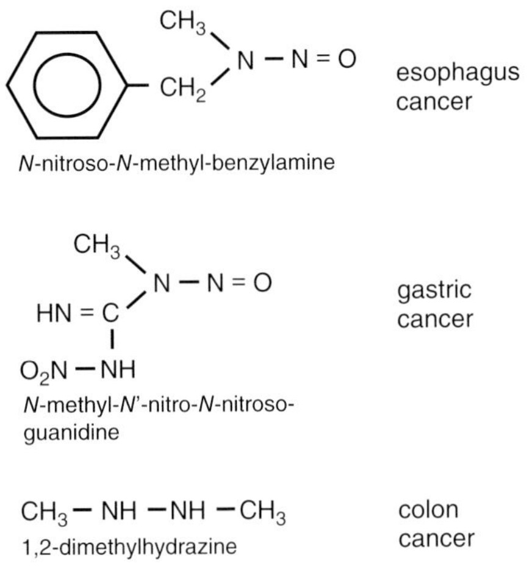

Organ Toxicology 229

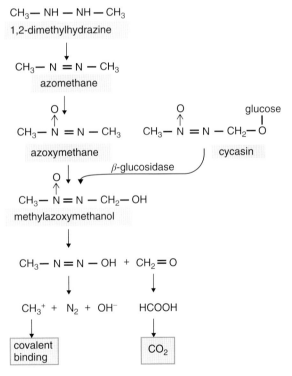

Figure 3.4 Some carcinogenic compounds which act on the GI tract. These compounds are metabolized and degraded to methyl cations, which react with the DNA in the cell nucleus.

Figure 3.5 Degradation of 1,2-dimethylhydrazine in the colon epithelium. Monoxygenases and possibly other enzymes are involved in the formation of methylazoxymethanol from 1,2-dimethlyhydrazine. Bacterial β-glucosidases are involved in its liberation from the naturally occurring glycoside cycasine. Methylazoxymethanol decomposes spontaneously and forms reactive methyl cations.

The difficulties of interpreting animal experiments with regard to their relevance for humans can be exemplified with the antioxidant butylated hydroxyanisole. Antioxidants usually inhibit the carcinogenic and toxic effects of various chemicals. However, when butylated hydroxyanisole is added in high concentrations of 2% to the feed of rats, it induces tumors in the forestomach, an organ that does not exist in humans. It is assumed that the tumorigenic action is due to local tissue irritation (tumor promotion). Because of the specific tumor location and the missing genotoxic activity of the compound, it has been argued that low concentrations of the antioxidant in food would not be associated with a cancer risk in humans.

Methods to Study Gastrointestinal Toxicology

A whole hierarchy of experimental methods is available, including whole animals, in situ perfused intestine, isolated perfused intestine, intestinal sacs, freshly isolated cells, and subcellular organelles in/from experimental animals. Each has its role in studying a specific toxic endpoint. Using such techniques, one should keep in mind that the intestinal musoca will easily be damaged during mechanical manipulations or oxygen deficiency. Endoscopic techniques in combination with biopsies are available to study in human intestinal absorption, biotransformation, motility, biliary secretion, and pathological tissue alterations. However, such human studies can be done only in small groups of individuals. Today toxicity studies are often performed in human cancer cell lines such as the colon cell line Caco-2. Such studies provide information about toxic mechanisms, but it is evident that complex toxic events of the GI tract with participation of neurons, immune cells, endothelial cells, epithelia, and microorganisms can hardly be imitated in assays with single cell types.

Table 3.1 Types of toxic damage in the gastrointestinal tract.

Type of damage	Compound	Source	Biological effect
Functional disturbances	Atropine	Deadly nightshade	Inhibition of motility/secretions
	Parathion	Insecticide	Stimulation of motility/secretions
	Cholera-toxin	Cholera infection	Stimulation of water secretion
	Morphine	Drug	Inhibition of motility
Erosions, ulcers	Alcohol, coffee	Luxury items	Stomach irritation
	Acid, lyme	Cleaning agents etc.	Necrosis in the affected GI section
	Many metal salts	Chemicals	Functional disturbance, necrosis
	Cytotoxic agents	Cytostatic drugs, toxins	Cell death, necrosis
	NSAIDS	Antiinflammatory drugs	Gastric ulcer
Hypersensitivity reactions	Allergens (e.g., food proteins)	Food ingredients	Allergic reaction (local or systemic)
	Pseudoallergens (e.g., salicylates)	Food ingredients	Allergy-like symptoms
	Immunogens	Food ingredients	Autoimmune disease
Cancer	Betel nut	Semi-luxury item	Oral cancer
	Cigarettes	Smoking	Esophagus cancer, colon cancer
	Alcohol	Semi-luxury item	Esophagus cancer
	Nitrosamines	Food or precursor	GI cancer?
	Vitamine C deficiency	Food	Gastric cancer?
	Low fiber intake	Food	Colon cancer?
	Dimethylhydrazine	Experimental	Colon cancer (rat)

3.1.5 Summary

The GI tract shows organ-specific reactivity to toxic compounds. Examples are compiled in Table 3.1. Considering the potentially high concentrations of foreign compounds in food, the GI tract exhibits an amazing resistance to toxic damage. It has the ability to detoxify and neutralize many compounds at the site of their entry. The importance of chemicals in human GI carcinogenesis is not clear at present.

Further Reading

Betton, G.R. (2013). A review of the toxicology and pathology of the gastrointestinal tract. *Cell Biol Toxicol*, **29**(5), 321–338.

Beyerle, J., Frei, E., Stiborova, M., Habermann, N., & Ulrich, C.M. (2015). Biotransformation of xenobiotics in the human colon and rectum and its association with colorectal cancer. *Drug Metab Rev*, 1–23.

Bryan, N.S., Alexander, D.D., Coughlin, J.R., Milkowski, A.L., & Boffetta, P. (2012). Ingested nitrate and nitrite and stomach cancer risk: an updated review. *Food Chem Toxicol*, **50**(10), 3646–3665.

Cerf-Bensussan, N. & Gaboriau-Routhiau, V. (2010). The immune system and the gut microbiota: friends or foes? *Nat Rev Immunol*, **10**(10), 735–744.

Ciecko-Michalska, I., Szczepanek, M., Slowik, A., & Mach, T. (2012). Pathogenesis of hepatic encephalopathy. *Gastroenterol Res Pract*, **2012**, 642108.

Coruzzi, G. (2010). Overview of gastrointestinal toxicology. *Curr Protoc Toxicol, Chapter 21*, Unit 21.1.

Gad, S.C. (2007). Toxicology of the Gastrointestinal Tract. CRC-Press.

Haiser, H.J. & Turnbaugh, P.J. (2013). Developing a metagenomic view of xenobiotic metabolism. *Pharmacol Res*, **69**(1), 21–31.

Li, Q. & Shu, Y. (2014). Role of solute carriers in response to anticancer drugs. *Mol Cell Therap*, **2**(15), 1–15.

Pang, K.S. (2003). Modelling of intestinal drug absorption: roles of transporters and metabolic enzymes (for the Gillette Review Series). *Drug Metab Dispos*, **31**(12), 1507–1519.

Wolfe, M.M., Lichtenstein, D.R. & Singh, G. (1999). Gastrointestinal toxicity of nonsteroidal antiinflammatory drugs. *N Engl J Med*, **340**(24), 1888–1899.

Zeilmaker, M.J., Bakker, M.I., Schothorst, R., & Slob, W. (2010). Risk assessment of N-nitrosodimethylamine formed endogenously after fish-with-vegetable meals. *Toxicol Sci*, **116**(1), 323–335.

3.2 The Liver

Jan G. Hengstler

3.2.1 Introduction

> **Functions of the liver include synthesis and secretion of proteins, secretion of bile, homeostasis of carbohydrate and lipid metabolism, as well as removal of bacterial fragments that enter the bloodstream and which first have to pass through the liver before they reach the general circulation. Importantly, the liver usually detoxifies endogenous compounds, such as ammonia, as well as xenobiotics. However, xenobiotics may also become metabolically activated.**

To fulfill its manifold tasks, the liver is strategically situated at the interface between the intestinal tract and the bloodstream. After oral ingestion, xenobiotics reach the intestinal tract, where they may be absorbed into the blood and reach the liver via the portal vein. Therefore, blood from the intestine reaches the general circulation only after passage through the liver, which thus allows for efficient hepatic detoxification of xenobiotics. On the other hand, the liver is also exposed to particularly high concentrations of compounds taken up by the oral route. As a result, hepatotoxicity of the parent compound or its metabolites is one of the most frequent reasons for drug withdrawal from the market. Identification of hepatotoxic compounds remains one of the major challenges in toxicology.

3.2.2 Structure and Function

Microstructure and Cell Types

> **Liver function and toxicity strongly depend on its microstructure. Venous blood from the stomach, gut and further gastrointestinal organs is collected in the portal vein. The latter enters the liver together with the liver artery, where it branches into smaller vessels. The final branches of the portal vein and liver artery reach the liver lobules, which are supplied with approximately 70% venous and 30% arterial blood (Figure 3.6A).**

Liver lobules are the smallest functional units of the liver (Figure 3.6A). Branches of the portal vein supply the lobules with blood, which enters the lobule in the periphery and flows along sheets of hepatocytes with a portal to central orientation. Hepatocytes are the parenchymal cells or "workhorses" of the liver (Figure 3.6A). Finally, the blood drains into the central vein. Branches of the central vein rejoin and form the liver vein, which drains into the inferior caval vein.

The sinusoids, microvessels of the liver lobules, are lined by sinusoidal endothelial cells (Figure 3.6B). The latter are flat cells with pores approximately 0.1 µm diameter. Xenobiotics or small proteins can enter via these pores to a 0.2–1.0 µm wide space between hepatocytes and endothelial cells. This so-called space of Dissé contains fibers of extracellular matrix, such as collagen, within a liquid environment.

> **Separation of this narrow compartment (space of Dissé) with almost standing liquid from the rapidly flowing blood in the sinusoids allows an efficient exchange of compounds between sinusoidal blood and hepatocytes.**

The space of Dissé contains a further cell type, namely the stellate cell (Figure 3.6B). Stellate cells (perisinusoidal or Ito cells) stretch relatively long, thin protrusions around the sinusoids. In healthy liver, one of their functions is to store vitamin A. After liver damage they play an important role in regeneration but they are also involved in the pathogenesis of liver fibrosis. Macrophages are located on the luminal (blood) side of the sinusoidal cells, also named Kupffer cells. Because of their interface to the gut, fragments of intestinal bacteria, such as lipopolysaccharides (LPS), may enter the sinusoids, where they are phagocytosed by Kupffer cells. The schematics of this arrangement of cells in Figure 3.6A,B are representative of those in current textbooks. However, it should be kept in mind that for didactic purposes they show a simplified architecture and do not correctly mirror the real spatial relationships. Using standard confocal or two-photon microscopes it became possible to reconstruct tissues with relatively little effort. The results illustrate that sinusoids and bile canaliculi form a seemingly chaotic intertwined network (Figure 3.6C,D). However, analysis of the three-dimensional structures revealed systematic order principles.

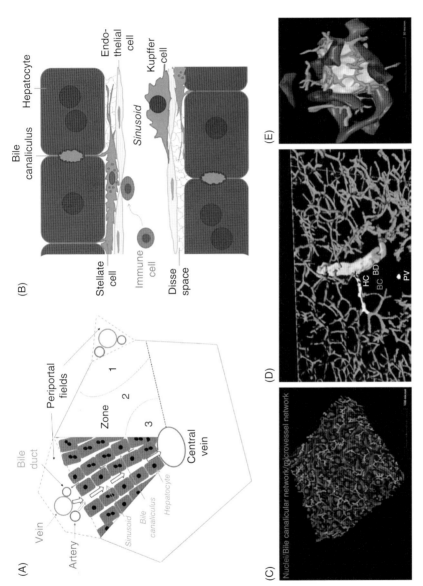

Figure 3.6 (A) Microarchitecture of the liver. The schematic shows a liver lobule, the smallest functional unit of the liver. The human liver consists of approximately 100,000 lobules. In the direction of the blood flow from periportal to pericentral three zones are differentiated: zone 1, periportal; zone 2, midzonal; zone 3, pericenral. (B) Cell types of the liver. (C) and (D) Confocal reconstructions of liver tissue to illustrate the three-dimensional structure. (C) The intertwined networks of sinusoids (red) and bile canaliculi (green) in a midzonal area of the liver lobule. The nuclei appear blue, while the cytoplasm is not visible. (D) The network of bile canaliculi (green) in a periportal region. The bile canaliculi join the so-called Hering channel (white) that connects to the larger periportal bile duct (yellow). Sinusoids and hepatocytes have not been visualized in this reconstruction. (E) Single hepatocytes (yellow) and their relationship to blood sinusoids (red) and bile canaliculi (green). (C–E) Reconstructions of confocal laser scans (from Hammad et al., 2014). The printed version is in black and white but the colors are shown in the electronic version.

> Approximately 25% of the surface of hepatocytes is in contact with a sinusoid and approximately 9% with a bile canaliculus (Figure 3.6E). Almost all hepatocytes adhere to this rule. This offers good conditions for an efficient exchange of compounds between blood and hepatocytes as well as hepatocytes and bile.

Bile Formation

> Bile salts are synthesized by hepatocytes and secreted into bile canaliculi. Bile canaliculi have their origin close to the central vein and are connected to the bile ducts in the periportal fields. The bile ducts rejoin to form the extrahepatic bile duct, which drains into the gall bladder and duodenum. The bile contains bile salts, cholesterol, bilirubin, phospholipids, glutathione and ions, and after secretion into the small intestine it plays a key role in the uptake of lipids.

Bile salts are formed by hepatocytes and secreted into bile canaliculi, where water follows an osmotic gradient. Bile canaliculi are formed from the membranes of hepatocytes and are not lined with endothelial cells (Figure 3.6B). Bile canaliculi originate close to the central vein and drain to the periphery, where they are connected to bile ducts (Figure 3.6A). In contrast to the bile canaliculi, bile ducts are lined by a specific cell type, namely cholangiocytes. The branches of the biliary tree rejoin to form the extrahepatic bile duct that drains into the gall bladder. The gall bladder is connected to the duodenum by a further duct and acts as a temporary store. It is present in humans and mice but not in rats.

Bile is a yellowish fluid that contains bile salts, cholesterol, bilirubin, phospholipids, glutathione and ions. After secretion into the small intestine, bile supports digestion, particularly absorption of lipids. Compromised bile secretion leads to steatorrhea and a lack of fat-soluble vitamins. A large fraction of xenobiotics are conjugated by UDP glucuronic acid, sulfate or glutathione (Phase II metabolism), followed by transport into the bile and excretion via the feces. Besides urinary excretion, biliary secretion represents the most relevant mechanism for excretion of xenobiotics. Excretion of compounds into the canalicular bile is achieved by active transport processes by carriers located at the apical (bile canalicular) membrane of hepatocytes. Further carriers at the basolateral (sinusoidal blood) side of hepatocytes are responsible for the uptake of compounds into hepatocytes. After uptake, they are further transported either into the canalicular bile or back to the sinusoidal blood (usually after Phase I and II metabolism) and further excreted via the urine. Carrier proteins present on hepatocyte membranes represent protein families that are often characterized by their affinity to specific substrates. The efflux carriers of hepatocytes are of particular relevance because their inhibition may lead to liver toxicity.

Metabolism of Xenobiotics and Endogenous Substrates

> The liver is the most active organ of xenobiotic and endogenous metabolism. It contains higher activities of xenobiotic-metabolizing enzymes than all other organs. In most cases, metabolism leads to detoxification and excretion of xenobiotics. In comparatively rare but highly relevant situations toxic intermediates are formed, usually as a consequence of Phase I metabolism.

The liver is the central organ of xenobiotic metabolism, with higher Phase I and II metabolizing activities than most other organs. The principles of xenobiotic metabolism, with functionalization by Phase I and conjugation by Phase II enzymes, have been explained elsewhere in this book. To understand the basis of hepatotoxicity, it should be kept in mind that metabolism by hepatocytes usually leads to detoxification and excretion. In some relatively rare cases toxic intermediates are formed, usually by Phase I metabolism. Importantly, Phase I enzymes are not homogeneously expressed in the hepatocytes of the liver lobule. Many Phase I enzymes, particularly members of the cytochrome P450 family, are expressed at much higher levels in the central region of the liver lobules (zone 3) (Figure 3.7). Therefore, the central region of the liver lobules is exposed to higher levels of reactive metabolites if the liver is exposed to compounds that are metabolically toxified by cytochrome P450 enzymes. This results in pericentral necrosis, a frequently observed damage pattern of the liver. In this case, the liver macroscopically shows a pattern of small light spots that correspond to the central necrotic lesions (Figure 3.7A).

The metabolic specialization of hepatocytes in different regions of the lobule is named zonation. In many cases, the advantage of metabolic zonation is obvious. An example is hepatic metabolism of ammonia, a toxic intermediate of protein metabolism. In lobular zones 1 and 2, ammonia is detoxified by enzymes of the urea cycle. These are high-capacity, low-affinity enzymes and therefore are adequate to remove the large amounts of ammonia that enter the lobule. The zonation of the rate-limiting enzyme of the urea cycle, carbamoyl phosphate synthetase-I, is shown in Figure 3.7B. To finally remove small amounts of ammonia that escape the urea cycle enzymes, glutamine synthetase is expressed in only one to two layers of hepatocytes around the central vein (Figure 3.7B). Glutamine synthetase detoxifies ammonia with low capacity but high affinity. Therefore, this inner ring of glutamine synthetase guarantees that only very low concentrations of ammonia enter the central vein and thereby the general circulation. Many of these spatially fine-tuned enzymatic processes exist, and together they are responsible for the amazing metabolic efficiency of the liver. In this context, the same cell type, e.g. the hepatocyte, can adopt different specializations. Hepatocytes in the periphery of the lobules are responsible for β-oxidation and synthesis of cholesterol and urea, while their centrilobular colleagues focus on lipogenesis, glycolysis, glutamine synthesis and metabolism by cytochrome P450. The pericentral expression pattern is thought to be mediated by specific cytokines, such as Wnt factors, released by the endothelial cells of the central vein, whereas the periportal differentiation depends on other cytokines, such as EGF, formed by other cells in the intestine and organism. Since these cytokines enter the lobules at its periphery, periportal hepatocytes may be exposed to higher concentrations. Zonation of the liver lobule can be disturbed by toxic compounds. After induction of hepatotoxicity, an altered zonation may persist for some time, even when the deteriorated hepatocytes have been replaced by new cells.

Regeneration

> **Besides its amazing detoxification capacity, the liver has a unique ability to regenerate. No other organ regenerates after toxic tissue damage as efficiently as the liver.**

The hepatotoxic compound carbon tetrachloride (CCl_4) destroys approximately 40% of the liver tissue at high doses (Figure 3.8). However, the original tissue architecture is restored in a tightly coordinated regeneration process. In the mouse, regeneration is completed within approximately eight days after intoxication. In humans, regeneration of larger liver damage may take several weeks. Early after induction of liver damage, inactive precursors of growth factors, e.g. of hepatocyte growth factor (HGF) in the extracellular matrix, are proteolytically activated. Moreover, sinusoidal endothelial cells are involved in sensing the damage and begin to secrete growth factors, such as HGF and Wnt factors. These cytokines stimulate hepatocytes to proliferate. They migrate into the necrotic lesions, whereby sinusoidal endothelial cells support orientation and serve as guide rails. Sinusoidal endothelial cells often survive exposure to hepatotoxic compounds because they express much less cytochrome P450 than hepatocytes. Moreover, immune cells,

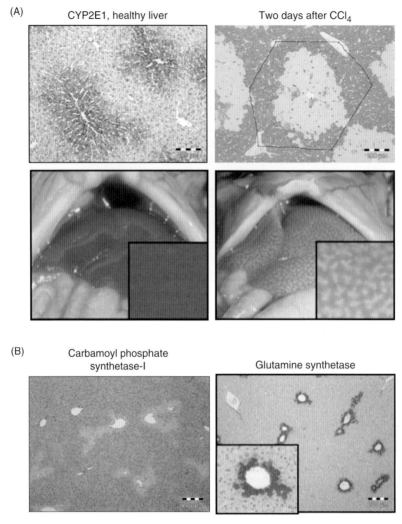

Figure 3.7 *Zonation of liver lobules. Some enzymes are preferentially expressed in specific zones of the liver lobule. An example is CYP2E1, which is located in a relatively broad pericentral region. (A) The pericentral zonation is typical for many members of the cytochrome P450 family. Therefore, compounds metabolically activated by cytochrome P450 cause pericentral necrosis (CCl_4-induced necrosis). Macroscopically, pericentral necrosis is characterized by its spotted pattern, whereby the necrotic regions appear pale. All images were obtained from mouse livers. In humans, zonation appears different (e.g. the CYP2E1 positive region is broader than in mice); however, the principles are similar. (B) Metabolic involvement in ammonia detoxification by zonation. The key enzyme of the urea cycle, carbamoyl phosphate synthetase I (CPSI), is expressed over the entire liver lobule, with the exception of a small region around the central veins. In this narrow CPSI-negative region, the ammonia detoxifying enzyme glutamine synthetase is expressed. These images represent the results of immunohistochemical stainings, where the presence of the stained enzymes is indicated by dark color (courtesy of Dr. Ahmed Ghallab).*

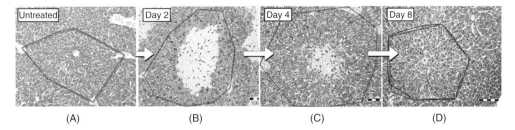

Figure 3.8 *Regeneration of pericentral liver damage caused by CCl_4. (A) Normal liver lobule with a central vein. (B) Approximately two days after administration of CCl_4 the necrotic lesion (pale in this image) reaches its maximal extension. (C) Partially completed regeneration at day 4. The small blue dots in the remaining necrotic region represent immune cell infiltrates that help with the removal of dead cells. (D) After eight days, the original state has been restored (from Hoehme et al., 2010).*

particularly neutrophils and macrophages, infiltrate the damaged liver tissue and remove dead cells. Together, these (and other) mechanisms guarantee rapid and perfect regeneration. However, a precondition of this process is that a critical fraction of sinusoidal endothelial cells survives intoxication. If sinusoidal endothelial cells are also destroyed, regeneration will be compromised and parenchymal tissue is replaced by scars (fibrotic tissue). The processes of toxic liver damage and regeneration have been understood relatively well and have been presented in spatio-temporal models. Moreover, two-photon microscopy allows the imaging of intact, living liver at cellular and subcellular levels. Videos showing the damage and regeneration processes of the liver can be seen at http://www.ifado.de/videos.

3.2.3 Toxicology

Cell Death

> **Relatively low doses of toxic chemicals may lead to adaptive responses without adverse effects. Higher doses may lead to reversible functional impairments. Examples are transient cholestasis or steatosis.**

In most cases, hepatotoxicity is a consequence of the cell death of hepatocytes, which is usually induced by necrosis or apoptosis. Necrosis is the most frequent mechanism of hepatocyte death induced by chemicals. Necrosis may exhibit various features depending on the mechanism of action of the responsible compound. Nevertheless, one commonality is compromised mitochondrial functions. Typically, the mitochondrial membrane potential decreases, leading to ATP depletion. This may already occur at relatively early stages as the result of a chain of events that finally leads to cell death. As a consequence of ATP depletion, numerous cell functions break down. Such characteristic "death sequences" can be observed by intravital two-photon microscopy. Intravital videos with mitochondrial membrane potential dyes show that the mitochondrial potential breaks down first; next, bile canaliculi may lose their capability to block the highly concentrated bile salts in their lumen from entering the cytoplasm of hepatocytes. Leakiness of bile canaliculi allows toxic bile to enter the hepatocytes, which causes irreversible cell death (http://www.ifado.de/videos). Subsequently, immune cells enter the necrotic region, first neutrophils, which represent the main fraction, followed by macrophages at later stages. Immune cells are required for the physiological regeneration process. However, over-activity of immune cells may also aggravate the initial damage. Typical features of necrosis include cell swelling, decay of the nucleus and leaking of hepatocyte contents into blood. Examples are the liver enzymes aspartate aminotransferase (AST) and alanine aminotransferase (ALT), which are routinely used to diagnose liver damage. Moreover, so-called damage-associated molecular patterns (DAMPs) are released from damaged hepatocytes and are distributed by diffusion. DAMPs may include fragments of nucleic acids and allow immune cells to find damaged regions of liver tissue.

> **In contrast to necrosis, apoptosis, i.e. programmed cell death, is energy dependent and requires intact mitochondrial function to generate ATP for activation of proteins. Typical morphological features include cell shrinkage, chromatin condensation and nuclear fragmentation. Finally, the apoptotic cell forms numerous apoptotic bodies that can be phagocytosed by macrophages. Since removal of apoptotic cells by macrophages is much more efficient, apoptosis leads to less inflammation than necrotic cell death.**

At the molecular level, two different mechanisms are known to induce apoptosis. The extrinsic mechanism can be induced by extracellular ligands, such as TNF-α or the fas ligand, to activate membrane receptors that recruit further proteins after activation. The resulting protein complex activates caspases, an enzyme family that degrades proteins and nucleic acids, which finally results in cell death. The second mechanism of apoptosis, also named the intrinsic pathway, induces pores to form in the outer mitochondrial membrane. This causes leakage of intermembrane proteins, such as cytochrome c, into the cytoplasm, which also activates caspases. Hepatotoxic chemicals often induce both necrosis and apoptosis. The form of cell death depends on the mechanism of action of the chemicals, as well as the dose.

Fibrosis and Cirrhosis

> **Liver fibrosis is the replacement of hepatocytes by fibrous connective tissue (Figure 3.9A,B). Liver fibrosis may progress to cirrhosis when fibrotic tissue increases resistance to blood flow through the liver. Progressed cirrhosis is often lethal because of hyperammonemic coma or bleeding from dilated veins in the stomach or esophagus.**

In liver cirrhosis, hepatocytes are replaced by scar tissue that reduces blood flow through the sinusoids. As a consequence, a shunt circulation develops. Vessels that bypass the cirrhotic liver, especially in the stomach and esophagus, may rupture, which may cause lethal bleeding. The second cause of death is hyperammonemic coma, caused by increased ammonia concentrations in blood. Ammonia increases since the cirrhotic liver loses its detoxification capacity. Because of its neurotoxicity, ammonia initially compromises cognitive functions and, at later stages, causes lethal coma.

Frequent causes of liver fibrosis are viral hepatitis and excessive consumption of alcohol. Since obesity became frequent in the general population, non-alcoholic steatohepatitis (NASH) plays an increasing role. Supernutrition initially leads to benign steatosis. However, the steatotic liver may be more susceptible to stress because of, for example, high doses of specific drugs. This may induce NASH, which frequently progresses to cirrhosis. Several drugs lead to an increased risk of liver fibrosis, which is accepted because better therapeutic alternatives are lacking. An example of this is rifampicin, which is used to treat tuberculosis. Typically, compounds that cause cirrhosis repeatedly induce cytotoxicity of hepatocytes over a longer period. A compilation of drugs causing fibrosis is given at https://livertox.nih.gov.

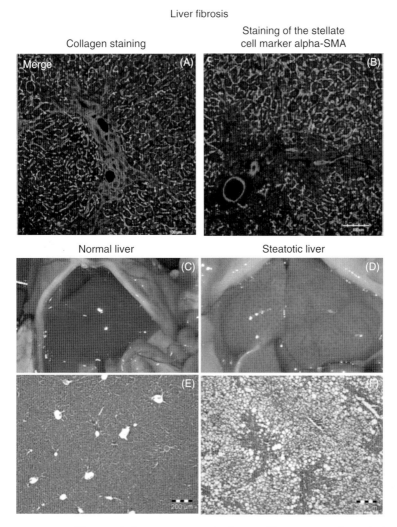

Figure 3.9 *(A, B) Liver fibrosis. Collagen staining (A) shows a fibrotic street through the liver tissue (red). Immunostaining of α-smooth muscle actin (α-SMA) visualizes some activated stellate cells (red) in the fibrotic street (B), indicated by white arrowheads. The network of bile canaliculi appears green, nuclei blue. (C–F) The steatotic liver appears larger and paler than healthy livers (C, D). Histologically, steatosis is characterized by numerous lipid droplets that appear white (F). All images represent mouse livers (courtesy of Dr. Cristina Cadenas and Dr. Seddik Hammad).*

> **Mechanisms causing liver fibrosis still represent a focus of intensive research. It is generally accepted that stellate cells (Figure 3.9B) play a key role in the pathophysiology of fibrosis.**

After induction of liver damage, a specific cell type, namely activated stellate cells, accumulates in the damaged region. These cells are formed by activation of quiescent stellate cells and represent a type of myofibroblast. Activated stellate cells support regeneration of acute liver damage. After liver damage induced by a single dose of a hepatotoxic compound, activated stellate cells are efficiently removed during the late regeneration phase and the original status of the liver tissue is restored. However, a different scenario is observed after repeated damage. In this case, more activated stellate cells are formed than removed. These then secrete an excess of extracellular matrix, e.g. collagen type I and III. Formation of fibrotic tissue increases to such a degree that antifibrotic mechanisms, such as proteases that digest extracellular matrix, can no longer compensate, finally leading to progressive fibrosis. Sinusoidal endothelial cells play an important role in the pathogenesis of fibrosis (Figure 3.6B). Experimental destruction of sinusoidal endothelial cells compromises liver regeneration and under these conditions even a single damage event may cause a scarring. Moreover, sinusoidal endothelial cells lose their fenestration in cirrhosis. This compromises the exchange of compounds between blood and hepatocytes, which further limits liver function.

Cholestasis

> In cholestasis, compounds appear in blood that are usually excreted via the bile, such as bile salts and bilirubin. The yellowish/brown bilirubin is deposited in the skin and eyes and is responsible for jaundice. Moreover, bilirubin is responsible for the dark color of the urine.

A typical cause of cholestasis is inhibition of hepatocyte efflux carriers. Numerous compounds inhibit the bile salt export pump (BSEP), which is located at the bile canalicular membrane of hepatocytes. As a consequence, excretion of bile salts into the bile canaliculi is reduced. This may lead to accumulation of bile salts up to cytotoxic concentrations in hepatocytes. A further relevant mechanism is obstruction of bile ducts. Such obstructions may be caused by some chemicals or by tumors.

A frequently used experimental system to study cholestasis in animal experiments is bile duct ligation, where the bile duct that connects the liver and the intestine is ligated. It may be expected that ligation of the bile duct leads to massive dilatation and rupturing of intrahepatic ducts. However, this is not the case because the epithelial cells of the bile ducts (cholangiocytes) reabsorb bile salts and water. Within a few days after obstruction, proliferation of cholangiocytes and corrugation of the inner membrane of bile ducts is observed. This increases the inner surface area and enhances the reabsorption capacity. After absorption of bile salts by the luminal side of cholangiocytes, they are excreted via the basolateral cholangiocyte membrane into the interstitium and reach the blood of the portal vein branches, which are in close proximity to the bile duct. Therefore, bile salts from the ducts recirculate into the liver lobule, where a fraction is absorbed at the basolateral hepatocyte membrane, transported into the bile canaliculi and the bile duct, from where it may recirculate again. This phenomenon has been named the cholehepatic shunt. The advantage of this shunt is that acute adverse effects, such as toxicity due to too high bile salt concentrations in the ducts, are avoided. The shunt leads to higher bile salt concentrations in blood and more excretion via the kidneys. A long-term effect of increased bile salt concentrations in blood is kidney damage.

A second toxicologically-relevant recirculation phenomenon is enterohepatic circulation. After biliary excretion, some compounds are reabsorbed in the intestine, reach the liver via the portal vein, are taken up by hepatocytes and again excreted into the bile. Bacterial enzymes, such as glucuronidases, are important in enterohepatic circulation since they may cleave Phase II metabolites and the resulting less hydrophilic products can be reabsorbed by the intestinal epithelium more readily, before the compound is again conjugated in the liver. Both shunt phenomena can massively influence pharmacokinetics and thereby toxicity of compounds.

Steatosis

> Fatty liver or steatosis is characterized by an increased content of lipids, particularly triglycerides, that significantly exceeds the lipid content of a normal liver of about 5% of the weight. Macroscopically, the steatotic liver appears enlarged and much paler than healthy livers (Figure 3.9C–F). In routine histology, steatotic liver tissue is characterized by round vacuoles since lipids are washed out by solvents during histological slide preparation procedures.

Overloading of hepatocytes with fat droplets can be so extreme that the nucleus is pushed to the margin of the cell. A frequent cause of steatosis is supernutrition, which leads to an excess of fatty acids that, after esterification, are stored in the form of triglycerides. Even if steatotic hepatocytes seem to be overloaded with triglycerides, they do not cause cytotoxicity and functional consequences to overall liver function are relatively small.

Chemicals can interfere with lipid metabolism. Particularly relevant is the inhibition of β-oxidation and inhibition of very low density lipoprotein (VLDL) formation. Fatty acids are transported into mitochondria, where they are metabolized to acetyl-CoA, which is further used for ATP synthesis in the citric acid cycle. Examples of drugs that can inhibit β-oxidation are amiodarone, tamoxifen and tetracycline. Decreased utilization of fatty acids after inhibition of β-oxidation leads to increased deposition of lipids in the form of triglycerides. Besides inhibiting β-oxidation, compounds such as amiodarone or tetracycline also inhibit utilization of triglycerides. Fatty acids derived from liver triglycerides are incorporated into VLDL, which leave the liver to supply peripheral tissues with fatty acids. Therefore, inhibiting both the utilization of fatty acids in mitochondria and export of fatty acids from hepatic triglyceride stores causes a particularly increased risk of steatosis. At least 20% of the population of industrialized nations have steatosis. This so-called benign steatosis is not considered to be a disease, but steatosis increases the risk of secondary diseases, such as the aforementioned steatohepatitis.

Porphyria

Some chemicals, such as polychlorinated biphenyls, dibenzodioxins and hexachlorobenzene, compromise heme biosynthesis in the liver. This leads to increased concentrations of heme precursors (porphyrins), which leave the liver and enter the blood. This may cause neurological disorders, psychoses and cramps, and fibrosis and cirrhosis may occur in the liver.

Inflammation

> Infiltration of immune cells into damaged liver tissue is of high relevance for removal of dead and damaged cells, and may support the regeneration process. However, some immune cell types may enhance the extent of liver damage, as described in section on idiosyncratic hepatotoxicity.

Numerous immune cell types are relevant in liver inflammation (hepatitis) and of these neutrophils and macrophages have been particularly well studied. After tissue damage, floating blood neutrophils can adhere to sinusoidal endothelial cells and infiltrate into the tissue. After adhesion, neutrophils become activated and release proteolytic enzymes and hypochlorous acid, which kill and degrade damaged cells. Next, macrophages, including liver tissue resident macrophages (Kupffer cells) and infiltrating macrophages from blood, help to phagocytose dead tissue. The complexity of the role of immune cells in liver damage and regeneration is impressive. For example, macrophages can secrete pro- and anti-inflammatory cytokines and even pro- and anti-fibrotic macrophages have been identified. A further important immune cell type is the natural killer cell. Their "kiss of death" may induce apoptosis of activated stellate cells, which has an antifibrotic effect. Stellate cells play an important transient role in regeneration of acute liver damage because they provide mechanical stability and secrete cytokines (e.g. hepatocyte growth factor, HGF) required for regeneration. However, after regeneration they have to be removed by other immune cells because their long-term presence would lead to fibrosis.

Examples of Hepatotoxic Compounds

Because of the high number of hepatotoxic compounds, this chapter will focus on only few representative examples. Acetaminophen (APAP), carbon tetrachloride (CCl_4), ethanol and allyl alcohol were chosen to illustrate key principles of hepatotoxicity. Further examples are summarized in Table 3.2. A comprehensive list is given at https://livertox.nlm.nih.gov/.

Acetaminophen and CCl_4

> **Acetaminophen (APAP) is one of the drugs most frequently used to treat pain and fever. Up to 4 g per day APAP is a safe dose for an adult; however, at higher doses, hepatotoxicity is frequently observed. APAP is a typical "threshold compound": while safe at therapeutic doses, toxicity occurs at only marginally higher doses. This threshold mechanism is due to the metabolism of the compound (Figure 3.10A).**

A large proportion of the parent compound, APAP, is conjugated with UDP-glucuronic acid or sulfate, which is then extensively (85–95% of the compound) excreted via the kidney. Notably, livers of toddlers have lower activities of UDP-glucuronosyl-transferases than adults, thus exhibiting lower conjugation and detoxification capacity compared to adults. Therefore, it is important to adjust APAP doses according to age and body weight. While adults may receive maximally 4 g per day, no more than 50–75 mg per kg per day should be given to toddlers.

Table 3.2 Examples of hepatotoxic compounds.

Damage type	Substance
Necrosis of hepatocytes; fibrosis after repeated doses	CCl_4, allyl alcohol, ethanol, acetaminophen, thioacetamide, α-amanitin (death cap), methotrexate
Steatosis	Valproic acid, amiodarone, tamoxifen, ethanol
Immune-medicated hepatitis	Halothane, diclofenac
Tumors	Aflatoxin B_1, androgens, arsenic compounds, vinyl chloride (hemangiosarcoma)
Cholestasis	Bosentan, chlorpromazine, cyclosporine A, estrogens
Damage to sinusoidal endothelial cells	Cyclophosphamide, pyrrolizidine alkaloids

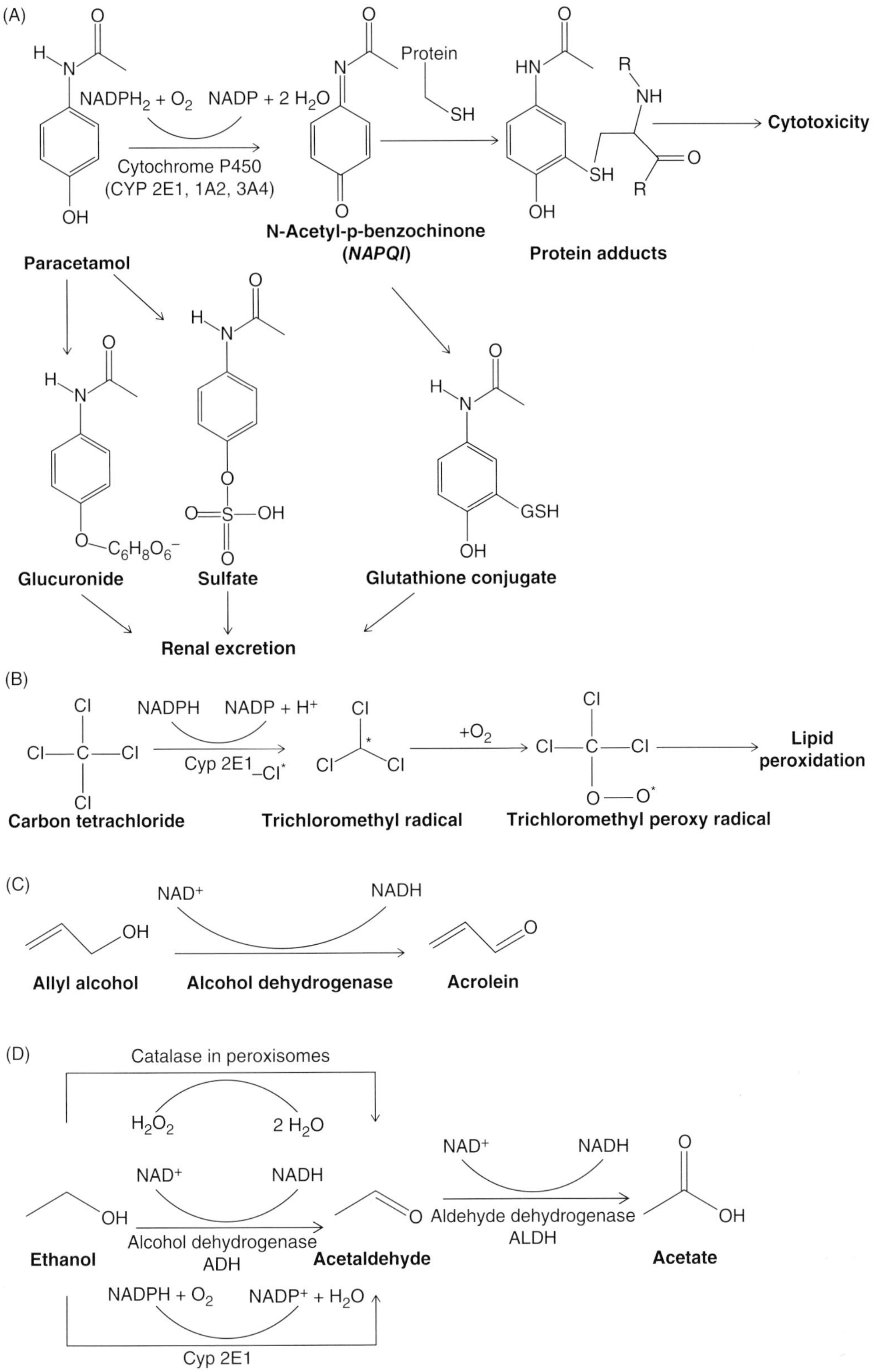

Figure 3.10 Metabolism of the hepatotoxic compounds (A) acetaminophen, (B) carbon tetrachloride (CCl_4), (C) allyl alcohol and (D) ethanol (courtesy of Dr. Regina Stöber).

Approximately 10% of the parent compound is metabolically activated to the toxic metabolite, N-acetyl-p-benzoquinone imine (NAPQI), by cytochrome P450. In humans, mostly CYP2E1 and, to a smaller degree, also CYP3A4 and CYP1A2 are responsible for catalyzing this reaction. In healthy livers, NAPQI is rapidly detoxified by conjugation with glutathione. Toxicity is observed only when NAPQI exceeds critical thresholds above which the glutathione stores of hepatocytes are depleted. The most effective therapy for APAP intoxication is a precursor of glutathione, N-acetyl cysteine, which can be administered intravenously or orally. After absorption into hepatocytes, cysteine is formed from N-acetyl cysteine, which is the rate-limiting amino acid in glutathione synthesis. When glutathione is depleted, NAPQI can covalently bind to proteins and compromise their functions. Although NAPQI binds to a wide spectrum of proteins, binding to mitochondria seems to be particularly critical for toxicity. The resulting loss of mitochondrial membrane potential may initially be reversible, but ultimately has irreversible consequences for the cell, such as leakage of toxic bile salts from canaliculi into the hepatocytes.

Similar to APAP, carbon tetrachloride (CCl_4) also has to be metabolically activated (mostly by CYP2E1) to exert its hepatotoxic effects (Figure 3.10B). Initially, the trichloromethyl radical ($CCl_3^{\bullet}$) is formed by reductive dehalogenization. $CCl_3^{\bullet}$ can react with oxygen to form the trichloromethyl peroxyl radical ($CCl_3OO^{\bullet}$). These radicals cause lipid peroxidation by withdrawing hydrogen from unsaturated fatty acids of phospholipids. This mechanism may ultimately also compromise mitochondrial functions. Both CCl_4 and APAP cause the characteristic pericentral necrosis since both require cytochrome P450 for metabolic activation. Despite different molecular initiating events, protein adducts in the case of APAP and lipid peroxidation for CCl_4, acute damage by both compounds is rapidly repaired and the original tissue architecture is perfectly restored. However, a completely different result is obtained if hepatotoxic doses are administered repeatedly. In this case, liver fibrosis is induced. This illustrates the amazing capacity of the liver to regenerate acute damage. By contrast, its vulnerability to repeated damage at high and prolonged exposure is much higher.

Allyl Alcohol Allyl alcohol is an industrially produced chemical required for the production of plastics and fire-resistant materials. It is metabolized to acrolein by aldehyde dehydrogenase, which is further metabolized to acrylic acid by alcohol dehydrogenase (Figure 3.10C). The reactive acrolein is responsible for hepatotoxicity because it binds to proteins. In contrast to APAP and CCl_4, allyl alcohol does not induce pericentral necrosis but damages the midzonal and periportal regions (zones 1 and 2). This is because ADH is not localized in the pericentral zone (in contrast to cytochrome P450). Because of the direction of blood flow from the periphery to the center of the lobule, hepatocytes in zones 1 and 2 may be exposed to higher concentrations of acrolein than in zone 3.

Ethanol

> **Ethanol activates several hepatotoxic mechanisms that together cause alcoholic steatosis and cirrhosis. The dose of ethanol that causes hepatotoxicity is well established. Drinking 40–60 g ethanol or more per day increases the risk of liver cirrhosis in men. Women should not drink more than 20 g ethanol per day. Probable reasons for the increased ethanol susceptibility of women are lower activity of alcohol dehydrogenase and a smaller volume of distribution compared to men. Almost 50% of all individuals drinking 200 g ethanol per day over a longer period of time will suffer from liver cirrhosis. A rule of thumb for estimation of ethanol doses: beer, wine and schnaps contain approximately 5%, 10% and 40% ethanol.**

More than 90% of ethanol consumed is metabolized in the liver. Three metabolic pathways of ethanol are particularly relevant for hepatotoxicity (Figure 3.10D). Most of the ethanol is oxidized to acetaldehyde and acetate by alcohol dehydrogenase and aldehyde dehydrogenase. This metabolic pathway contributes to hepatotoxicity by two mechanisms. Acetaldehyde is reactive and can form protein adducts. Acetate and NADH are required for synthesis of fatty acids, an excess of which supports the pathogenesis of steatosis. A toxicologically relevant polymorphic form of aldehyde dehydrogenase, namely ALDH2*2, leads to decreased enzyme activity. Therefore, alcohol consumption leads to higher concentrations of acetaldehyde, which may cause headaches. ALDH2*2 occurs only rarely in Caucasians; however, approximately 50% of Asians have this variant and it is responsible for alcohol intolerance ("flushing syndrome") in this population. The second metabolic pathway is mediated by cytochrome P450 2E1 (CYP2E1). CYP2E1 is inducible by ethanol and also catalyzes the formation of acetaldehyde. This reaction is toxicologically relevant because it forms reactive oxygen species. As already described, CYP2E1 is expressed in the center of liver lobules; therefore, mainly pericentral damage is induced. Alcohol dehydrogenase (metabolic path 1) has a Michaelis–Menten affinity constant (K_m) of 1 mM ethanol, compared to 10 mM for CYP2E1 (metabolic path 2), so that at higher ethanol exposures the CYP2E1 pathway predominates and thus the center of the lobule is affected. The third metabolic pathway is the reaction catalyzed by catalase in peroxisomes. This reaction reduces hydrogen peroxide to water, whereby ethanol serves as an electron donor. This reaction also generates acetaldehyde, although it is only responsible for the metabolism of at most 2% of the ethanol consumed.

> **The blood alcohol level can be roughly estimated by dividing the amount of alcohol ingested by the body's water content (females ≈ 50%, males ≈ 60%) and taking into account that due to metabolism about $0.15\%_0$ of the blood alcohol level decreases per hour.**

Further mechanisms of ethanol-induced hepatotoxicity are:

- Ethanol and acetaldehyde inhibit DNA binding of the transcription factor, PPARα, which is responsible for transcription of enzymes involved in the metabolism of fatty acids. This supports the pathogenesis of steatosis.
- Acetaldehyde inhibits integration of triglycerides into VLDL and thereby the export of lipids from the liver. This is a further mechanism that is involved in steatosis.
- Ethanol inhibits the activity of natural killer cells, which normally induce apoptosis of activated stellate cells.
- Ethanol increases concentrations of inflammatory cytokines by mechanisms which are still under investigation.

Concentration Dependent vs. Idiosyncratic Hepatotoxicity

A general assumption of preclinical toxicological studies is that identification of adverse effects is possible by dose escalation. This is the case for a vast majority of compounds; however, idiosyncratic hepatotoxicity may also occur. As the name infers, it is a relatively rare phenomenon such that out of 10,000 treated patients, only a small number will suffer from idiosyncratic hepatotoxicity. Nevertheless, idiosyncratic hepatotoxicity is one of the most frequent reasons why drugs have to be withdrawn from the market. This situation is particularly problematic because, on the one hand, patients suffer from drug-induced toxicity and, on the other hand, drugs have to be withdrawn from the market that are safe and efficient for the majority of patients. Therefore, research on the mechanisms responsible for idiosyncratic hepatotoxicity is of high relevance because in the future it may allow the identification of patients at risk. A current concept is that several risk factors have to coincide, including a low capacity of detoxification in combination with strong metabolic activation. Moreover, there is strong evidence of a genetic predisposition of the immune system. According to this concept, chemicals may bind cellular proteins that are presented to immune cells by helper cells. In a stimulating (e.g. inflamed) microenvironment, this may trigger a cytotoxic immune response. This concept is supported by genome-wide association studies that identified an association between specific HLA alleles and idiosyncratic hepatotoxicity. The situation is even more complex since these studies did not identify a single set of variants that was relevant to all compounds inducing idiosyncratic hepatotoxicity, rather different variants are associated with individual compounds. If it is possible to identify all relevant factors of idiosyncratic hepatotoxicity, this will be of high relevance for preclinical toxicity testing. It would mean that conventional studies based on the principle of dose escalation should be supported by additional tests that consider the susceptible genotype of idiosyncratic toxicity.

Liver Tumors

Strong genotoxic carcinogens alter the genetic information and simultaneously induce replacement proliferation by cytotoxicity. A typical example fulfilling these criteria is the mycotoxin, aflatoxin B_1.

Aflatoxin B_1 is formed by the mold *Aspergillus flavus*, which grows on grain or nuts if they are stored in a warm and humid atmosphere. Aflatoxin B_1 is metabolically activated by CYP3A4 to a reactive epoxide that binds to DNA, particularly to the N7 position of guanine. The resulting covalent DNA adducts cause mutations. Moreover, aflatoxin B_1 is cytotoxic to hepatocytes even at very low concentrations. Although hepatocellular cancer due to aflatoxin B_1 is rare in Europe, it represents the most frequent cancer type in Africa and southeast Asia because the warm climate facilitates growth of the mold.

Primary liver cancer can also be induced by carcinogens that primarily target sinusoidal endothelial cells or cholangiocytes. For example, the industrial chemical vinyl chloride damages sinusoidal endothelial cells in addition to hepatocytes and causes hemangiosarcoma. Thorotrast, a former contrast agent in medical radiography, is one of the best documented compounds that cause cholangiocarcinoma, tumors that originate from the cholangiocytes of bile ducts.

Increased Liver Weight

One of the parameters determined in chronic toxicity studies is organ weight in relation to body weight. Sometimes an increase in liver weight is the only finding from such studies. To interpret these findings, it is important to consider that the liver weight may vary in healthy animals. For example, rats fed ad libitum without stimulation of physical activity have a 20% higher relative liver weight compared to their companions on a controlled diet and underwent daily training on an exercise wheel.

In preclinical drug development, a small increase in relative liver weight without significant histological findings (e.g. necrosis, fibrosis, inflammation, steatosis, bile duct proliferation or cholestasis) or marked changes in clinical chemistry (e.g. no increase in blood liver enzymes levels) is interpreted as an adaptive and not an adverse effect. Similarly, induction of drug-metabolizing enzymes (e.g. CYP3A) without significant histology and clinical chemistry findings is considered to be adaptive. However, adaptive situations may decompensate at higher doses and lead to adverse effects. Therefore, possible hepatotoxicity should be monitored particularly carefully in further preclinical and clinical development if increased liver weight or liver enzyme induction have been observed. For establishment of occupational threshold values (e.g. MAK values), usually a more conservative procedure is chosen, particularly if, besides the finding of increased liver weight in animals, no further data of human relevance are available.

In this case, an increase in relative liver weight is used as point of departure for establishment of threshold values if it is more than 20% above controls or if it coincides with liver enzyme induction.

In vitro Systems and Interspecies Differences

> **Today, cultures of hepatocytes are well-accepted tools in pharmacology and toxicology. In the public debate, some politicians and non-governmental organizations demand that *in vivo* tests should be completely replaced by *in vitro* studies within relatively short periods of time. However, without an increased risk for humans, a complete replacement of animal experiments will only be possible when all relevant toxicity mechanisms are known and *in vitro* systems are available that allow their quantitatively correct analysis. Moreover, it will be necessary to model pharmacokinetics with sufficient accuracy. It is unlikely that all this will be achieved within the next 10 years.**

In recent years, hepatocyte *in vitro* systems have been developed that support the evaluation of hepatotoxic compounds. Commercially available cultured hepatocytes from human donors and numerous animal species allow the identification of compounds that induce or inhibit metabolizing liver enzymes. Inhibition or induction of liver enzymes in patients can be problematic because this may lead to drug interactions. However, the challenge remains that after identification of a concentration that alters enzyme activities *in vitro*, it still has to be clarified whether this concentration is (or is not) reached *in vivo*. Hepatocyte *in vitro* systems are routinely used to generate metabolites of a test compound because major metabolites of pharmaceutical drug candidates have to be considered in preclinical toxicity testing and to evaluate the human relevance of experimental studies for human exposure. Despite of some progress in the field of *in vitro* research, it is currently not yet possible to predict with sufficient accuracy the doses and blood concentrations of a test compound that will cause liver toxicity in humans under conditions of repeated exposure.

One frequently raised argument in the public debate is that animal experiments are not relevant for humans because of interspecies differences. It is correct that major differences in susceptibility have been observed between humans and experimental animals. Nevertheless, the argumentation that animal experiments do not allow conclusions for the human situation is not correct because modern toxicology has established tools to correctly interpret *in vivo* test results despite interspecies differences, as illustrated by the example of aflatoxin B_1 (AFB_1). This is a well-documented compound of major interspecies differences. Some mouse strains are more resistant to AFB_1-induced liver cancer than rats. As soon as the molecular initiating event (here DNA adducts of AFB_1) of the adverse effect (here hepatocellular cancer) is known, it can be used to compare hepatocytes of rats, mice and humans *in vitro*. In this case it has been shown that more AFB_1 DNA adducts are induced in cultured human and rat hepatocytes than in mouse hepatocytes. Later it became apparent that certain mouse strains are more resistant to AFB_1 because they detoxify the metabolically activated AFB_1 via a specific glutathione S-transferase isoenzyme more efficiently than rats and humans. Therefore, humans are as susceptible to AFB_1-induced hepatocellular cancer as rats and, unfortunately, not as resistant as mice. This example demonstrates that interspecies differences cannot be used as a generalized argument against *in vivo* studies. It rather shows that animal experiments should be accompanied by research to elucidate the toxic mechanisms and to compare them in the target cells of the respective animal species and of humans.

3.2.4 Myths of the Liver

Unlike any other organ, the liver is intertwined with myths. In the ancient legend, an eagle was sent to feed upon the liver of Prometheus. In order to be able to repeat the punishment Prometheus's liver regenerated within only one day. This legend is fascinating because it contains an element of truth, namely the outstanding regeneration capacity of the liver. This was already known in ancient Greece, which most probably stemmed from survivors of severe abdominal and liver injuries of ancient wars. In addition, ancient seers have been reported to predict the future from the abdominal organs of wild animals. This is, of course, irrational but the interesting core of this tradition is that seers analyzed scars of the liver. During periods of drought or deterioration of climate, wild animals had to deviate from their accustomed diet, which led to an increased risk of consumption of hepatotoxic plants and thus to liver fibrosis. Based on the liver fibrosis of wild animals, ancient seers predicted a dismal future for their own species.

3.2.5 Summary

The liver is responsible for numerous functions, including endogenous and xenobiotic metabolism. It expresses the highest levels and diversity of drug-metabolizing enzymes of all organs, secretes several serum proteins, stores glucose in the form of glycogen, plays a central role in lipid metabolism and produces bile that is required for absorption of lipids from the intestine. Xenobiotics absorbed from the gastrointestinal tract reach the liver via the portal vein. Many xenobiotics are metabolized in hepatocytes, after which they are either transported to the blood to be excreted by the kidneys or secreted into the bile and excreted via the feces. Hepatotoxic compounds can have a number of effects: they can impair metabolic processes leading, for example, to steatosis or porphyria, compromise sinusoidal blood flow (cirrhosis), block bile transport (cholestasis), and induce inflammation, as well as necrosis. Repeated damage of hepatocytes can cause cirrhosis, which is characterized by the replacement of normal liver tissue by fibrotic scars, leading to a destruction of liver microarchitecture. Potent liver carcinogens are characterized by their ability to simultaneously cause DNA mutations and induce cytotoxicity. Liver cirrhosis and viral hepatitis are strong risk factors of hepatocellular cancer.

3.3 The Respiratory System[1]

Florian Schulz

3.3.1 Introduction

> **The respiratory system of mammals facilitates oxygenating the blood via inhalation and removing excess carbon dioxide via exhalation. It acts by drawing air into the nasal passages and distributing the air through the tracheal and bronchiolar conducting airways to the alveoli, where gas exchange occurs. The adult human inhales about 20 m^3 of air daily. At maximum inhalation, the respiratory tract comprises a surface of approximately 140 m^2 and is thus the organ with the second largest surface within the human body after the small intestine (approximately 250 m^2). The surface is also considerably larger compared to the skin (approximately 1.75 m^2), the third organ in contact with the environment. Inhaled substances can deposit, retain and/or absorb at this large surface.**

---

[1] This chapter is an adapted and updated version of the original chapter entitled "The Respiratory System" by J. Pauluhn of the 1st edition of this book published in 2008.

The anatomy of the respiratory tract is designed to humidify and warm the inhaled air while filtering out dust particles and germs. Fine hairs cover the lining of the nasal epithelium and are able to capture larger particles before they reach the lungs. The gas exchange takes place within the widely branched alveoli. The fine structures of the alveolar septa consist of a highly specialized structural network including epithelial, interstitial, and endothelial cells. The alveolocapillary membrane is ≤ 0.5 µm thin and represents the blood–air barrier. This membrane serves as the basis for a highly efficient gas exchange between the lung and the blood capillaries over a large surface. On the other hand, the alveolocapillary membrane is coincidently an area vulnerable to inhaled toxic substances.

The respiratory tract is permanently exposed to potentially toxic inhaled substances present as gas, vapor, liquid (mist) or smaller solid dust particles, which may cause local inflammation or infection. The site of major deposition within the respiratory tract depends on particle shape and size, physicochemical properties, chemical reactivity and the blood:air partition coefficient. The most critical site in the lung is the alveoli because only a limited cell-based clearance system mediated by mobile macrophages is available. These cells allow clearance of cellular fragments, bacteria and deposited particles in this pulmonary region. Macrophages are able to phagozytose foreign matter, which is subsequently degraded and transported out of the alveoli.

Toxic substances reaching the alveolar region may also pass the blood–air barrier with subsequent distribution to other organs. This can be very rapid as all the blood returning from the lung to the left ventricle is subsequently distributed throughout the body. However, not all substances deposited and retained in the alveolar region pass the blood–air barrier. Accordingly, exposure to poorly soluble particles may lead to particle accumulation in the alveolar region and an overloaded macrophage-mediated clearance. Persisting particles may subsequently induce pulmonary damage. Likewise, the unsuccessful digestion of phagocytized poorly soluble particles within alveolar macrophages results in the discharge of inflammatory mediators and an excess formation of reactive oxygen and nitrogen species. The oxidative stress may lead to damage of adjacent epithelial cells as well as chronic inflammatory tissue reactions and lung tumors in the case of chronic exposure. Thus, persisting poorly soluble particles may induce diseases even when they are chemically inert.

3.3.2 Structure

> **The respiratory tract consists of the nasopharyngeal (NP), the tracheobronchial (TB), and the pulmonary (P) regions. Each region contains specific anatomical features and mechanisms of deposition and clearance.**

The **NP region** begins at the anterior nares and extends to the pharynx and larynx. The nasal passages are highly tortuous and in most animal species are lined with four distinct nasal epithelia. These include squamous, transitional, and pseudostratified respiratory epithelium in the anterior part of the main nasal chamber as well as the olfactory epithelium located in the dorsal region of the nasal cavity. The olfactory epithelium is the most metabolically active nasal epithelium and appears to be particularly vulnerable to metabolically induced lesions, e.g. the impact of acids released from esters via the action of esterases. Metabolism by the olfactory epithelium may contribute to provide or prevent access to inhaled substances to the brain. For example, metabolites of inhaled xylene may attain the brain by axonal transport while cadmium or mercury oxides presumably follow the same route without undergoing metabolism.

In the NP region, poorly soluble substances are subject to physical clearance by mucociliary transport to the pharynx and subsequent swallowing. By contrast, readily soluble substances are rapidly cleared into the bloodstream. The anatomical structures, the relative abundance and distribution of specific cell types as well as the airflow rate dependent on the surface dose of inhaled xenobiotics markedly vary between different species.

The **TB region** begins at the larynx and includes the trachea, the bronchial airways, and the terminal bronchioles (Figure 3.11). The TB region is both ciliated and equipped with mucus secreting cells. The velocity of mucus movement is slow in the lower airways and increases toward the trachea. After passing the NP region only a relatively small fraction of particles of all sizes deposits in the TB region (Figure 3.12). Deposition results from impaction at bifurcations and sedimentation as well as Brownian diffusion for very small particles. In contrast to nasal breathing, mouth breathing does not efficiently filter larger particles, resulting in an increased deposition in the TB region.

The **P region** with the alveolus as most important anatomical structure (Figures 3.11 and 3.13) is the functional gas exchange site. Each alveolus opens directly into an alveolar duct and alveolar sacs. Alveoli as well as alveolar ducts and sacs arising from one single conducting airway form a pulmonary acinus. A thin tissue barrier consisting of Type I and Type II alveolar pneumocytes facilitates highly efficient gas transfer over a large surface. A typical human alveolus contains approximately 32 Type I and 51 Type II pneumocytes, resulting in a surface area of 200,000–300,000 μm^2. By contrast, a typical rat alveolus contains only two Type I and three Type II pneumocytes, resulting in a surface area of about 14,000 μm^2.

Type I pneumocytes cover approximately 90% of the alveolar surface. Because of their large surface area relating to the cell mass these cells are vulnerable targets. **Type II pneumocytes** exhibit a cuboidal shape and are able to remove extravasated fluid from the alveoli. They are also metabolically active and produce surfactant. If Type I pneumocytes are damaged, Type II pneumocytes undergo an increased mitotic division and differentiation. The newly built cells can then replace the damaged Type I pneumocytes. The physiologically slow turnover of alveolar epithelium considerably accelerates after injury, and the number of Type II cells doubles within a short time. Collagen and elastin fiber producing mesenchymal interstitial cells preserve the integrity of the delicate alveolar septa. **Clara cells** are located in the terminal bronchioles and have a high content of xenobiotic-metabolizing enzymes.

Alveolar macrophages represent the numerically most variable lung cell type. Comparative studies indicate that the alveolar macrophage number is significantly higher in healthy human lungs compared to the lungs of other species. The macrophages contribute to a sterile environment within the lung by removing inhaled substances via phagocytosis and either subsequent enzymatic degradation in the phagolysosome or transport of non-degradable substances out of the lungs. Macrophages can leave the lungs via the mucociliary escalator of the TB branches toward the trachea or migrate into the lung interstitium and lymphatics.

The number of alveolar macrophages may increase by the immigration of interstitial macrophages. The latter are matured monocytes that enter the interstitium via the bloodstream. Thus, a variety of conditions influences the number of alveolar macrophages, potentially resulting in considerable fluctuations within a short time. These fluctuations in cell number may reflect different disease processes in the lungs.

Tight junctions between Type I pneumocytes of the alveolar epithelium facilitate the protein transfer across the epithelial barrier by passive diffusion through paracellular water-filled channels within the tight junctions. The permeability of the tight junction is strictly regulated and can dynamically change. Tight junctions are important to maintain cellular gradients originated from active transcellular mechanisms like endocytosis or pinocytosis.

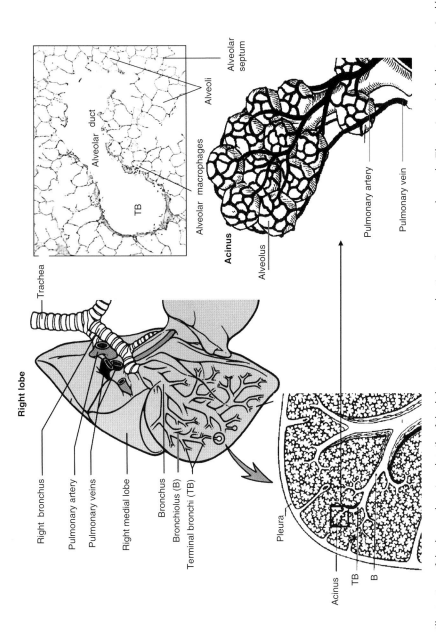

Figure 3.11 Illustration of the lung, showing details of the lobar structure, conducting airways, and vessels. The acinus includes a terminal bronchiole and its respiratory bronchioles, alveolar ducts, and alveolar sacs (alveoli). The histology photograph details these structural components in a trans-sectional manner.

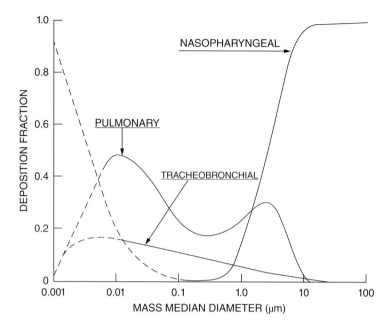

Figure 3.12 Regional deposition of inhaled airborne particles dependent on the aerodynamic diameter. This figure presents a generalization for regional deposition patterns in mammalian species.

The different epithelia and cell types of the respiratory tract, their localization and their function are summarized in Table 3.3.

3.3.3 Function

> **Gas exchange is the main function of the respiratory tract. The blood becomes oxygenated after inhalation and excess carbon dioxide is removed by exhalation. This requires highly specialized anatomical structures characterized by a thin membrane with a very large surface area, the alveoli (Figure 3.11).**
>
> **Efficient systems remove inhaled particles from the respiratory tract and protect the lung from the adverse effects of inhaled toxic substances. Further protective mechanisms and biochemical pathways prevent oxidative stress and proteolytic degradation of alveolar structures.**

Inhaled air passes from the nose to the nasopharynx. At this time, the glottis is open, with the epiglottis permitting airflow through the larynx into the trachea. However, during swallowing the epiglottis closes to cover the entrance of the glottis and to prevent food residues from entering the trachea. The highly branched conducting airways convey the air further to the acinus (Figure 3.11). The ciliated epithelium and mucus-producing cells within the airways allow the capture and transport of deposited particles to the oral cavity. Volatile substances are removed by convection.

The blood–air barrier separates the air from the vascular capillary bed and facilitates gas exchange. The efficacy of gas exchange depends on the integrity of this membrane and is diffusion limited. The transport of inhaled air in and out of the alveoli depends on lung mechanics

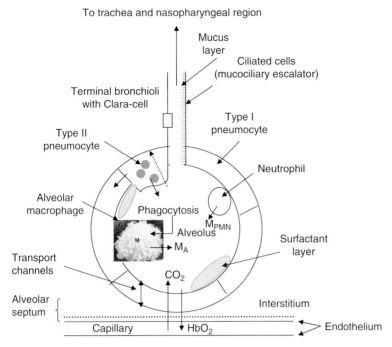

Figure 3.13 Diagrammatic representation of the terminal air space of the lung (terminal bronchial airways, alveolus with migratory cells, interstitium and vascular space). At the active air-exchange site, the blood–air barrier is formed by the capillary endothelium, the basal membrane and the Type I pneumocytes. The latter build the alveolar wall with its large surface. Type II pneumocytes are stem cells of Type I pneumocytes. They also synthesize, store and recycle surfactant, and generate an osmotic gradient to remove excess fluid from the alveolus. Interstitial cells stabilize the passive site. The surfactant layer is essential for the alveolar function as lack of surfactant leads to alveolar collapse (atelectasis). Clara cells are metabolically active and produce mucus as well as surfactant co-factors. Macrophages have a ruffled surface and clean the alveolar region from foreign matter by phagocytosis. Subsequently, macrophages move via the mucociliary escalator toward the larynx. This translocation is impaired by "overload" with particles, leading eventually to its "self-digestion" and further amplification of inflammatory stimuli, as illustrated in Figure 3.15. M_A derived from alveolar macrophages and M_{PMN} from neutrophilic granulocytes (PMN = polymorphonuclear neutrophils) are mediators either regulating early inflammatory responses or being directly cytotoxic. HbO_2 is oxygen reversibly bound to hemoglobin in red blood cells, the erythrocytes.

and is ventilation controlled. By contrast, the transport of gases from the blood to the alveoli is perfusion limited and controlled by the right ventricular cardiac output that regulates the blood flow through the lung capillary system. The alveolar membranes separate the inhaled air from the blood-containing capillaries, therefore the pressure is low in the alveoli and high in the capillaries. This pressure difference is maintained by the structural organization of specific cell packaging and physiological mechanisms that facilitate rapid and efficient gas exchange.

The pulmonary circulation carries approximately 70–80 cm^3 of blood, which is low in oxygen, per heartbeat from the right heart ventricle via the pulmonary veins to the lung. After the gas exchange, oxygen-rich blood reaches the left heart ventricle via the pulmonary arteries and subsequently the systemic circulation. Because of the high throughput, the lungs are exposed to substantial amounts of endogenous and exogenous substances carried via the blood circulation. The entire blood volume of the body passes the lung within one single circulation leading to the

Table 3.3 Localization and function of important epithelia and cell types of the respiratory tract.

Epithelium/cell type	Localization	Function
Olfactory epithelium	Nasal cavity	Odor perception, metabolism of xenobiotic substances
Ciliated epithelium (cylindric and cubic)	Airways	Mechanical clearance
Globlet cells, seromucous cells	TB region	Mucus production, stem cell
Clara cells (non-ciliated epithelium)	Bronchioles	Mucus production, metabolism of xenobiotic substances
Type I pneumocytes	Alveolar region	Blood-air barrier
Type II pneumocytes	Alveolar region	Surfactant production, stem cell, metabolism of xenobiotic substances
Alveolar macrophages	Alveolar region	Phagocytosis of microorganisms and particles
Endothelial cells	Blood capillaries	Blood–air barrier, inactivation of vasoactive, endogenous substances
Smooth muscles	TB region	Regulation of airway resistance
Supporting cells, elastic fibers	Connective tissue	Maintenance of structure and function
Neurons	Airways	Regulation of bronchial tonus, defensive reflex
Eosinophilic and polymorphonuclear granulocytes, mast cells	TB region	Mobile cell population infiltrating the lung after injury, may induce airway hyperreactivity and asthma

necessity to fulfill blood cleaning functions. These include filtration of microthrombi from the blood and maintenance of a protease transport system to eliminate proteases released from dead bacteria, neutrophils and lysosomes of alveolar macrophages. The most harmful among these enzymes is elastase, which may destroy structural elastin fibers leading to a disruption of the alveolar septum. Furthermore, the vascular endothelium of the lung efficiently clears vasoactive substances.

The lung must inevitably deal with inhaled toxic substances. In addition to a potent antioxidative system to protect from oxidative stress, several protective mechanisms at the various locations of the respiratory tract are present. In the nasal cavity, the larynx and the TB region, mucociliary clearance mechanisms remove deposited particles. By contrast, macrophages are the key players to remove foreign material from the alveolar region. The different epithelial cell types, barrier-maintaining Type I pneumocytes as well as secretory Type II pneumocytes, also contribute to the protective function. In addition, glutathione and proteins contained in the mucus and fluid of the conducting airways may inactivate water-soluble reactive vapors and gases.

3.3.4 Protective Systems

> **The protective functions of the respiratory tract are carried out by different mechanisms at different locations. Clearance and filtration mechanisms require secretory activity and the recruitment of phagocytotic cells to facilitate the trapping and removal of poorly soluble particles. Glutathione or proteins contained in the mucus and lining fluids of the conducting inactivate water-soluble reactive vapors and gases.**

Clearance from the pulmonary region occurs via a number of pathways and mechanisms. These involve both absorptive and non-absorptive processes, which may occur simultaneously or with temporal variations. In the NP and TB regions, the mucociliary transport preferentially removes poorly soluble substances toward the throat. The effective removal of insoluble particles from the NP region by this absorptive clearance mechanism may require 1 to 2 days.

By contrast, macrophage-mediated clearance of deposited particles dominates in the alveolar region. The macrophages move freely on the epithelial surfaces and phagocytize deposited material, which they contact by random motion or via directed movement facilitated by chemotactic factors. Inside the macrophages, foreign substances are degraded and detoxified. Furthermore, the activated macrophages release chemical messenger substances that attract further macrophages. Subsequently, macrophages are able to leave the lungs via the mucociliary escalator and thereby transport non-degradable substances out of the lungs.

Alveolar macrophage-mediated clearance is a slow process characterized by a monoexponential decay function. The clearance rate is approximately 1% per day for the rat, equivalent to an alveolar retention halftime of about 60–70 days. For humans, the alveolar retention halftimes are estimated to be up to 10-fold longer. A volumetric overloading of the macrophages with particles will result in restricted mobility of the macrophages and failure to move actively toward the mucociliary escalator. Poorly soluble substances, which are not efficiently cleared, retain within the lung and may translocate to other organs such as the lung-associated lymph nodes.

The large alveolar surface area promotes the absorption of inhaled substances by diffusion. The surface tension at the air–water interface produces forces that tend to reduce the area of the interface. Without the presence of surface-active material, the surfactant, as a counterpart, this would eventually lead to a collapse of alveoli. Surfactant also reduces the pressure gradient between the vascular system with high hydrostatic pressure and the alveolus with subatmospheric pressure. Thus, surfactant also prevents edema caused by extravasation of blood plasma into the alveolus.

Due to a high proportion of phospholipids, pulmonary surfactant is insoluble and floats on the surface of the alveolar cells. High-capacity lipid metabolism is required to maintain surfactant homeostasis. Macrophages mediate the surfactant turnover stimulated by the interaction of surfactant with various agents. Chemicals interfering with phospholipid metabolism may cause pulmonary diseases. As an example, phospholipidosis results from an inhibited phospholipid catabolism, leading to toxic levels of excess lipids in macrophages. A disturbed surfactant metabolism may eventually result in surfactant dysfunction, leading to a lung edema.

Proteases, such as elastases derived from inhaled dead bacteria or leachates of phagocytic cells, damage the structural proteins of the lung. Subsequently, this may cause a disruption of alveolar septa and lung emphysema. Protective mechanisms represent the removal of proteases with the mucus flow toward the larynx and inactivation of proteases by conjugation with the plasma protein α1-antitrypsin. The pulmonary blood flow and/or the lymphatic system subsequently removes the conjugated proteases. The plasma protein α2-macroglobulin further inactivates proteases by conjugation. The complexes are degraded in the liver.

Many enzymatic processes, especially those taking place in the mitochondria and lysosomes of phagocytic cells, generate highly reactive oxygen species (ROS) and reactive nitrogen species (RNS). These include superoxides, hydroxyl radicals and nitroperoxides. Some redox active transitional metals, e.g. Fe^{2+}–Fe^{3+}, may also produce reactive oxygen species by redox cycling via the Haber–Weiss reaction. The highly reactive metabolites may react with proteins, carbohydrates, lipids, and nucleic acids. ROS and RNS also contribute to an excessive consumption of reducing agents such as glutathione and NADPH, resulting in oxidative stress. Thus, they represent a considerable threat to the integrity of all cells.

Protection against oxygen radical attack in mammalian cells is mediated by a variety of antioxidant enzymes such as superoxide dismutase (SOD) and catalase as well as enzymes that increase the local concentrations of reduced glutathione. The enzyme glutathione peroxidase uses the reducing power of glutathione to convert hydrogen peroxide into water. Glutathione reductase recycles the oxidized glutathione via the co-enzyme NADPH as reducing agent. The pentose phosphate pathway in turn regenerates NADPH. In addition to these enzymatic systems, cells contain a variety of agents such as vitamin E, ascorbic acid and uric acid that interrupt chain reactions of free radicals or scavenge reactive oxygen metabolites. The relative abundance of these antioxidants differs between the lower and upper respiratory tract as well as across different mammalian species.

3.3.5 The Respiratory Tract as a Target for Toxicity

> **The distribution of inhaled gases, vapors and particles within the respiratory tract defines the pattern of toxicity, therefore potential damage to tissues depends on the surface dose. The presence of local enzyme systems may increase or decrease the toxicity of inhaled substances. In addition, an exceptional susceptibility of particular cell types may evolve at critical areas within the respiratory tract.**

Specific patterns with enhanced deposited doses at distinct locations of the respiratory tract imply that the initial dose is not distributed uniformly across the entire pulmonary epithelial surface. This is especially important for irritant vapors and particles that impair the tissue by direct contact. Highly reactive or water-soluble gases in moderate concentrations do not penetrate further than the nasal area. Accordingly, they are relatively non-toxic for the lower respiratory tract. However, associated with particles or vapors as "carriers" these gases may also penetrate to the deep lung and elicit toxic responses.

After passing the NP and TB regions, vapors and particles interact with the alveolar surfaces by sedimentation, diffusion, or interception. This may in turn lead to alveolar damage. A direct impairment of the blood–air barrier causes inflammation and an increased extravasation of plasma into the alveoli. Extensive damage may result in lung edema. Isocyanates, phosgene and chlorine are examples of irritant gases that may cause acute alveolar edema or death.

Poorly soluble particles such as crystalline silica are able to cause pulmonary fibrosis. The pathogenesis is determined by the size of the inhaled particles, which have to be small enough to reach the alveoli. Alveolar macrophages subsequently ingest the deposited particles. As part of the cytotoxic response to silica phagocytosis, the macrophage releases inflammatory cytokines and other substances, resulting in mitosis of fibroblasts and increased collagen synthesis. Epithelial cells react with an inflammatory response. In particular, within months of short-term exposure to silica a marked, self-perpetuating, and self-amplifying increase in inflammatory mediators was detected in the bronchoalveolar lavage fluid (BALF) corresponding to an inflammatory reaction (Figure 3.14). Thus, inhaled particles, which are phagocytized by alveolar macrophages, but are not transported out of the lung due to a cytotoxic reaction, may continuously induce sustaining inflammatory proliferative processes. This may lead to fibrosis (expressed as pneumoconiosis) and eventually tumor formation resulting from chronic oxidative stress and reactive cell proliferation (Figure 3.15).

The biochemical analysis of BALF parameters allows the detection of a toxic response in experimental animals. This technique is sensitive enough to obtain integrative, but not site-specific, information on the dose response or time course of the toxic response from initiation to

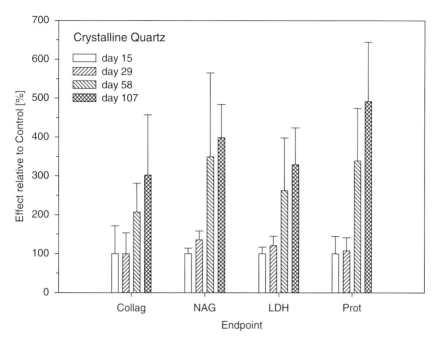

Figure 3.14 Time course of inflammatory responses in lungs of rats exposed to crystalline quartz dust by inhalation up to day 14. Over a post-exposure period of three months, the bronchoalveolar lavage was analyzed for the structural protein collagen (Collag), the lysosomal enzyme N-acetyl-glucosaminidase (NAG), the enzyme lactate dehydrogenase (LDH), and the total protein content.

termination. In addition, examination of BALF parameters facilitates the quantification of the relative potency of substances to induce alveolar damage. The endpoints shown in Figure 3.14 address changes in soluble collagen, the lysosomal enzyme β-N-acetyl-glucosaminidase (NAG), the enzyme lactate dehydrogenase (LDH), and total protein content. NAG reflects the disturbance in lysosomes due to an unsuccessful digestion of poorly soluble particles. LDH is indicative of local cytotoxicity and a changed protein content may serve as a marker of increased functional disturbance of the blood–air barrier.

Particles

In the average adult human, most particles larger than 10 μm in aerodynamic diameter deposit in the nose or oral pharynx and are unlikely to penetrate to tissues distal to the larynx. Very fine particles with a diameter of 0.01 μm and smaller are also trapped efficiently in the upper airways by diffusion. However, other particles may penetrate beyond the upper airways to a greater extent and deposit in the bronchial region, the terminal airways, and the alveoli (Figure 3.12).

Particles preferentially deposit where the airways are small and the velocity of airflow is low. These conditions exist in the smaller bronchi, the bronchioles, and the alveolar spaces. An airborne particle moves downward during sedimentation. While the gravitational force acts on the particle in downward direction, buoyancy and air resistance act on the particle in the opposite

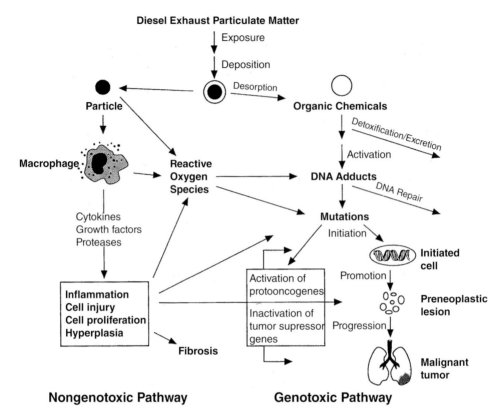

Figure 3.15 Diagrammatic representation of possible mechanisms for carcinogenesis induced by diesel exhaust particulate matter. The non-genotoxic pathway may generate reactive oxygen or nitrogen species, leading indirectly to genotoxicity so that both pathways overlap to some extent.

direction. Eventually, the gravitational force equilibrates with the sum of the buoyancy and the air resistance, resulting in a constant settling velocity of the particle. Sedimentation as a process of deposition becomes less important with decreasing particle size. Below an aerodynamic diameter of 0.5 μm diffusion processes increasingly determine the deposition. At a diameter of 0.5 μm, deposition is very low as sedimentation is minimal and diffusion is still less pronounced.

The clearance of deposited particles is an important aspect of lung defense. Rapid removal shortens the contact time available for potential tissue damage. In the upper airways, the mucociliary transport removes particles toward the pharynx, where particles are subsequently swallowed. In the lower respiratory tract, especially in the alveoli, clearance initiates with macrophage-mediated phagocytosis. Subsequently, the macrophages transport the particles via the ciliated bronchi and the mucociliary escalator to the pharynx. These clearance processes are particle size dependent. The fate of ultrafine particles (nano particles) may be very different from larger particles, even when they deposit at the same location. In contrast to larger particles, alveolar macrophages less efficiently phagocytose ultrafine particles. Consequently, ultrafine particles may interact more strongly with epithelial cells or penetrate more rapidly to interstitial sites and reach the endothelial cells of the blood capillaries. Thus, ultrafine particles are able to enter the blood circulation, resulting in their translocation to extrapulmonary tissues, provided the particles have not dissolved before.

Particle Overload Based on experimental data, phagocytized particles covering about 6% or more of the macrophage volume are assumed to impair macrophage motility and to retard the removal of particles. When phagocytoed particles attain 60%, the macrophages become immobile and are subject to cell death. The clearance function consequently ceases to exist (Figure 3.16).

The lung burden at which "overloading" starts to occur is approximately 1 mg ($\approx$1 µl) of particles per gram lung tissue. The excessive particle overload triggers a series of pulmonary events in small laboratory animals, as illustrated in Figure 3.15. These processes include the influx and activation of inflammatory cells, the inflammatory response, and subsequent tissue remodeling regulated by complex interrelationship of cytokines, chemokines and factors involved in maintaining lung homeostasis. These systems include the protease anti–protease system as well as the balance between the formation of ROS and the consumption of water- or lipid-soluble antioxidants. Chronic inflammatory responses commonly lead to a remodeling of the lung structures characterized by cell proliferation, focal fibrosis, localized emphysematous responses, and ultimately benign and/or malignant lung tumors.

The inflammatory potency increases with decreasing particle size. For ultrafine particles (<0.1 µm) the surface area – not the particle mass or particle volume – appears to be the crucial dose metric of exposure and toxic effect. On the other hand, ultrafine particles tend to build aggregates/agglomerates and the smaller agglomerate density compared to the material density may explain the stronger effect of ultrafine compared to fine particles as well. Besides the particle size, the extent of the toxic effect depends on geometric shape, crystal structure, and surface loading

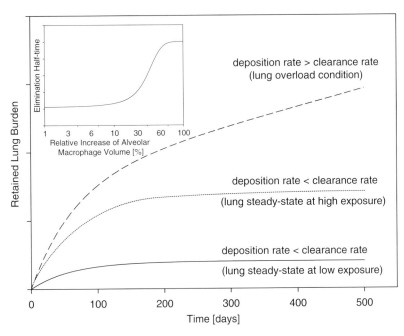

Figure 3.16 *Accumulation of poorly soluble particles in the rat lung during long-term inhalation exposure. Volumetric overloading of alveolar macrophages with particles causes an impaired macrophage-mediated clearance. Thereby, the macrophages fail to move actively toward the mucociliary escalator. Phagocytized particles covering about 6% or more of the macrophages volume are assumed to impair macrophage motility and on attaining 60%, the clearance function ceases to exist.*

properties as well as adherent particles like transitional metals, polycyclic aromatic hydrocarbons (PAH), or endotoxins.

The toxic response induced by poorly soluble particles in combination with an impaired clearance is determined at an early stage by the influx of inflammatory cells (neutrophilic granulocytes) into the alveolar space as well as histologically by inflammatory proliferative tissue changes.

Gases

> **The mechanisms affecting the transport and distribution of gases involve convection, diffusion, absorption, dissolution and chemical reactivity. Based on water solubility and reactivity three categories of gases are distinguished.**

Category I gases are highly water soluble and/or reactive. Thus, these gases are able to interact with the tissue or liquid film of the upper respiratory tract. The deposition is ventilation-dependent and the fraction exhaled is relatively low. Examples for Category I gases are hydrogen fluoride, chlorine, and formaldehyde as well as volatile organic acids and esters.

Category II gases are moderately water soluble and slowly metabolized in the tissues of the respiratory tract. Category II gases include ozone, sulfur dioxide, xylene and propanol.

Category III gases are insoluble in water and do not react with the extrathoracic and TB tissue and liquid film. Therefore, only relatively small doses are maintained in these regions. The uptake of Category III gases occurs predominantly in the pulmonary region and is rather perfusion than ventilation limited. Depending on the mechanism of gas uptake and pattern of exposure (intermittent or continuously), the estimated default retention of non-reactive gases is about 50%. Category III gases include halogenoalkanes, propellants and gases used for anesthesia.

The retention factors and the localization within the respiratory tract vary considerably between different gases and are dependent on gas concentration, duration of exposure, ventilation and physical activity. Therefore, default assumptions for gases represent only an approximation of the actual behavior. Among the most typical responses to reactive and irritant gases is protracted and sustainable damage to the conducting airways characterized by inflammation and overproduction of mucus. This leads to loss of airway patency and an impaired gas exchange. Consequently, the blood returning to the heart lacks oxygen and systemic hypoxemia may result when this blood enters the systemic circulation. The symptoms are similar to the interruption of O_2/CO_2 diffusion processes observed in alveolar edema.

Substance-specific Pulmonary Toxicity

Effects other than direct chemical reactivity may also damage the pulmonary region. These indirect effects are often mediated by cytotoxic factors liberated from alveolar macrophages upon phagocytosis (e.g. crystalline quartz). Additionally, physical factors may induce indirect effects such as substance-related changes in the surface tension of pulmonary surfactant (e.g. detergents, emulsifiers). The molecular weight may also be an influencing factor for acute inhalation toxicity. For instance, the acute toxic lethal potency of aerosols of polyalkene glycols increases with increasing molecular weight.

Some pharmaceuticals may also cause pulmonary toxicity. For instance, cationic amphiphilic drugs (CAD) induce storage disorders caused by phospholipids and/or mucopolysaccharides in a variety of animal species and in humans. CADs or CAD-phospholipid complexes are concentrated in lysosomes of macrophages and inhibit the intra-lysosomal breakdown of phospholipids. This results in an impaired metabolism of constituents of the surfactant system. Bleomycin is

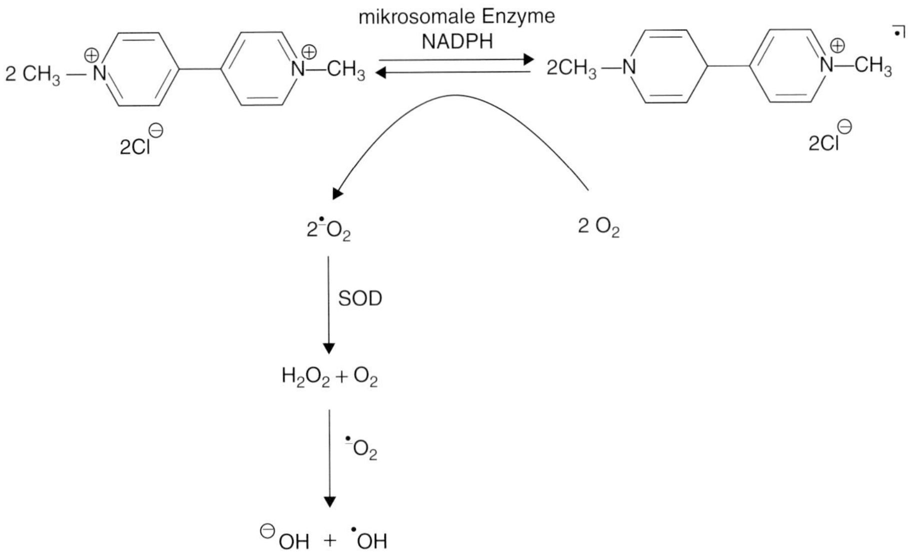

Figure 3.17 *Redox cycle of paraquat and formation of active oxygen species.*

a systemically administered chemotherapeutic agent that complexes dissolved iron ions (Fe^{2+}) thereby promoting the formation of ROS, which in turn stimulates the production of collagen and may lead to extensive fibrosis in the lung.

Some substances ingested with food are also able to induce lung injury. For instance, the herbicide paraquat (use prohibited in the European Union since 2007) may produce progressive and eventually fatal lung damage characterized by diffuse interstitial and intra-alveolar fibrosis. Paraquat enters the respiratory tract via the polyamine uptake system. The high susceptibility of the lung may be explained by subsequent accumulation in the lung. Once inside the cells, paraquat undergoes redox cycling with the concomitant formation of active oxygen species (Figure 3.17), depletion of NADPH and extensive lipid peroxidation.

Upon inhalation, some xenobiotics are either bioactivated by metabolically active cells of the nasal cavities (e.g. olfactory epithelium) and the lower respiratory tract (e.g. Clara cells in the bronchiolar epithelium) or cause site-specific cellular damage after deposition. Many soluble transitional metal ions may cause marked inflammatory responses in the lung upon inhalation as dust although they are essential nutrients at homeostatic concentrations. However, proteins capable of forming chelates protect the lung to some degree from metal-induced redox cycling. For example, Hg^{2+} and Cd^{2+} ions bind to metallothionein and ionic iron may build chelates with ferritin, transferrin and lactoferrin.

3.3.6 Respiratory Allergy and Asthma

Allergic asthma is a complex chronic inflammatory disease of the airways involving the recruitment and activation of many inflammatory and structural cells. These cells release inflammatory mediators, resulting in a typical pathological response. Studies with experimental animals may serve as models to investigate several features of asthma, including cellular infiltration in the lung, antigen-specific IgE production and Type 2 helper T cell-mediated immune response. The latter is characterized by elevated levels of typical cytokines seen upon allergen (hapten) sensitization and challenge. The number of mediators involved in the sensitization process and/or the development of a chronic inflammatory process in the mucosa of the lower airways, including airway remodeling, provides an impression of the complex network.

> **In contrast to an allergic reaction, airway hyper-reactivity is an exaggerated acute obstructive response of the airways to one or more non-specific stimuli. Hyper-reactivity is often associated with epithelial damage or disruption and even mild asthma.**

In experimental models used to identify potential allergens, animals are commonly sensitized by two consecutive treatments with the antigen (or hapten). Following a challenge exposition at a later point in time an allergen causes an asthmatic phenotype. However, secondary allergen challenges after a prolonged lapse of time reflect the human situation more closely. The route of exposure, application method and dose of allergen exposure determine the allergic phenotype. Short-term, high-level antigen exposure causes different airway lesions than chronic, low-level exposure. The methods used to assess immediate or delayed bronchoconstriction as well as the type and extent of induced airway inflammation and tissue remodeling affect the outcome of animal models involving allergen (hapten) sensitization.

The typical pathology of asthma is associated with reversible narrowing of the airways involving structural changes in the airway walls and extracellular matrix remodeling. Further features include abnormalities of bronchial smooth muscle, eosinophilic and neutrophilic inflammation of the bronchial wall as well as hyperplasia and hypertrophy of mucous glands. Asthmagenic compounds include many high-molecular-weight proteins (e.g. flour, animal dander, dust mite constituents, proteolytic enzymes used as detergents) and low-molecular-weight compounds (e.g. platinum salts, diisocyanates, organic anhydrides and reactive dyes). Experimental models have shown that dermal contact with such agents may induce asthma following subsequent inhalation exposure.

3.3.7 Lung Cancer

The primary cause of lung cancer in humans is cigarette smoking. In addition, several other substances, including many occupational agents, are causally associated with lung cancer. These include asbestos fibers and biopersistent man-made fibers, silica, metallic dusts (e.g. arsenic, beryllium, cadmium, chromium(VI) and nickel compounds), welding fumes, some wood dusts, coal tar, coke oven exhaust, diesel exhaust, radon and other radioactive materials, mustard gas, and vinyl chloride as well as industrial processes such as aluminum production and coke production.

The latency from first exposure to invasive carcinoma is usually in excess of 20 years. The pathogenesis of lung cancer is a multistep process involving the sequential accumulation of alterations in two types of genes: tumor suppressor genes and proto-oncogenes. In combination with increased growth-factor expression, these genetic changes result in an enhanced division of cells with a selective growth advantage over adjacent normal cells. Especially in the lung, genetic damage may result from oxidative stress and a derangement of antioxidant control mechanisms (Figure 3.15). Associated with a chronic inflammation driven sustained cell proliferation and lung remodeling, this may cause tumor promotion. Graded morphologic changes ranging from hyperplasia of the alveolar epithelium through adenoma to adenocarcinoma were accordingly observed in animal studies. These progressive changes are also reminiscent of the development of squamous-cell carcinomas in humans. Moreover, the acute inflammatory and proliferative response to high burdens of poorly soluble particles and the related enhanced oxidative stress appears to be a causative factor for tumor formation in rats. However, with the exception of asbestos and crystalline quartz, epidemiological evidence for poorly soluble particles as a causative agent for lung cancer in humans is lacking.

3.3.8 Test Systems to Detect the Toxic Effects of Inhaled Materials

> **Experimental models are designed to detect and quantify toxic hazards of inhaled substances. Standardized inhalation test systems are necessary to generate reliable experimental data used for regulatory purposes as well as to warrant the safe production and use of different materials.**

Inhalation studies allow the detection and quantification of toxic hazards caused by inhaled substances under highly standardized conditions and maximum pulmonary exposure. Competent authorities such as the Organization for Economic Cooperation and Development (OECD) provide internationally accepted protocols for inhalation toxicity studies to assure the reproducibility in testing laboratories. Nevertheless, results from inhalation studies have to be analyzed cautiously due to the large number of experimental variables. These include technical issues such as the generation of test atmosphere and special features of the animal model. Although inhalation exposure of experimental animals mimics the conditions in humans well and a route-to-route extrapolation is not necessary, not all generated results are relevant to real-life exposure conditions. The choice of a distinct animal model is usually based on guideline requirements and practical considerations such as available exposure technology, practicability and previous experience rather than consideration of the human relevance. Moreover, differences in routes of exposure associated with intermittent exposure regimens may result in dissimilar toxicokinetic behavior and substance-specific toxicodynamic differences. Being aware of these specific circumstances, inhalation studies provide a valuable tool to evaluate adverse effects and to derive health- or risk-based limit values.

3.3.9 Summary

The physiological function of the respiratory system is the enrichment of blood with oxygen and the removal of excess carbon dioxide. Within the alveoli, the presence of a thin membrane with a very large surface area facilitates an efficient gas exchange between inhaled air and the blood circulation. The surface area of this blood–air barrier exceeds the body surface of an individual several times. The thickness of the alveolocapillary membrane is limited by the requirements of gas diffusion and amounts to approximately 0.5 μm. The extensive volume change during inhalation also requires special structural organization. Furthermore, particular mechanisms protect the membrane, exposed daily to approximately 20 m^3 of air containing foreign substances, from injury.

Toxic substances may reach the lung via both the blood circulation and the inhaled air. The distribution of inhaled substances within the respiratory tract is non-homogeneous and depends on the physical and physicochemical substance properties. The respiratory tract exhibits several protective mechanisms to prevent foreign substances from entering the lower airways or to remove deposited particles. Inhaled biological particulate matter such as bacteria is (bio)chemically degraded. Any impairment or overload of these clearance mechanisms, including the macrophage-mediated elimination of particles in the alveoli and the mucociliary removal in the bronchi and NP region, may cause damage to health.

Animal models facilitate the assessment of pulmonary toxicity of chemicals and other substances. Nevertheless, the breathing pattern as well as the pulmonary architectural organization and cellular composition vary between different species to a certain degree. Therefore, species-specific differences have to be considered during extrapolation of data generated from animal studies to humans.

Further Reading

Bernstein DM (2007). Synthetic Vitreous Fibers: A Review Toxicology, Epidemiology and Regulations. *Crit Rev Toxicol* **37**, 839–886.

Brown JS, Wilson WE, Grant LD (2005). Dosimetric comparisons of particle deposition and retention in rats and humans. *Inhal Toxicol* **17**, 355–385.

Gardner DE, Crapo JD, McClellan RO (1993). *Toxicology of the Lung*, 2nd edition, Raven Press New York.

Gebel T, Foth H, Damm G, Freyberger A, Kramer PJ, Lilienblum W, Röhl C, Schupp T, Weiss C, Wollin KM, Hengstler JG (2014). Manufactured nanomaterials: categorization and approaches to hazard assessment. *Arch Toxicol* **88**, 2191–211.

Greim H, Borm P, Schins R, Donaldson K, Driscoll K, Hartwig A, Kuempel E, Oberdöster G, Speit G (2001). Toxicity of fibres and particles- report of the Workshop held in Munich, Germany, 26–27 October 2000. *Inhal Toxicol* **13**, 737–754.

Greim H, Utell MJ, Maxim D, Niebo R (2014). Perspectives on refractory ceramic fiber (RCF) carcinogenicity: comparison with other fibers. *Inhal Toxicol* **26**, 789–810.

IARC (International Agency for Research on Cancer) (2002). *Monographs on the evaluation of carcinogenic risks to humans, Man-Made Vitrous Fibres*, Vol **81**, Lyon, IARC Press.

Knaapen AM, Borm PJ, Albrecht C, Schins RP (2004). Inhaled particles and lung cancer. Part A: Mechanisms. *Int J Cancer* **109**, 799–809.

Maxim LD, Hadley JG, Potter RM, Niebo R (2006). The role of fiber durability/biopersistence of silica-based synthetic vitreous fibers and their influence on toxicology. *Regul Toxicol Pharmacol* **46**, 42–62.

McClellan RO, Henderson RF (1989). *Concepts in Inhalation Toxicology*, Hemisphere Publishing Corporation, New York, pp. 193–227.

Snipes MB (1989). Species comparison for pulmonary retention of inhaled particles. In: *Concepts in Inhalation Toxicology*, Hemisphere Publishing Company, New York, pp. 193–227.

3.4 The Nervous System

Gunter P. Eckert and Walter E. Müller

3.4.1 Structure and Function of the Nervous System

> **The nervous system (Figure 3.18), which consists of the peripheral nervous system (PNS) and the central nervous system (CNS), is an extremely complex network that provides the structural basis for electrochemical processes that are responsible for its multiple functions. These are cognition, sensation, memory, speech, locomotion, and unconscious hormonal and autonomic processes. The nervous system has an absolute requirement for mechanisms that enable it to transmit information within individual nerves and between nerve cells.**

Transmission of Neuronal Information

The principle function of a nerves is to transmit information to other nerves or various targets in the body (e.g. muscles, glands). The general structure of a nerve, which will be discussed below in detail, consists of the cell body from which projections extent which are categorized into dentrites and axons. Dentrites receive information whereas the longer axons send information

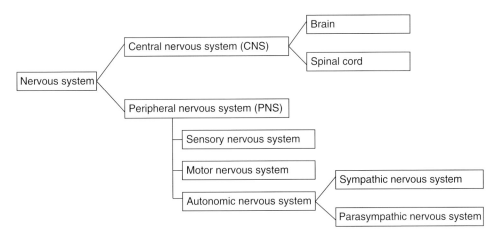

Figure 3.18 Schematic partition of the nervous system. The nervous system is divided into the central nervous system (CNS) and the peripheral nervous system (PNS). The CNS consists of the brain and spinal cord. The PNS consists of all other nerves and neurons that do not lie within the CNS. The PNS is divided into the somatic nervous system and the autonomic nervous system.

to other nerves or to target structures like muscles and glands. Information is transmitted from the cell body down the axon via an electrical signal generated by stepwise depolarization (action potential). Intercellular transmission of information is accomplished by the release of specific chemical mediators (neurotransmitters). The location of the chemical information transfer is called **synapse**.

In case of the motor neuron system (fig.3.18) the synapse lies between the foot of the axon of a motor neuron and the motor end-plates, specific structures at the muscle rich of specific receptors for the neurotransmitter released by the axon (in this case acrtylcholine). Upon release from the presynaptic terminal the neurotransmitter binds to these receptors, induces depolarization of the motor end-plate which finally activates the muscle to contract. Similar synapses control the interaction between nerves and different target structures.

For example, in the sympathetic nervous system, which is a subdivision of the autonomic nervous system, axons of nerves emerging from the spinal column reach the sympathetic ganglion clusters of nerves where they activate the cell bodies of the sympathetic nerves which extend to various organs of the body. Thus, we have preganglionic presynaptic fibers entering the synapse and transferring information, via chemical mediators, to postsynaptic postganglionic fibers, which then direct the activity of various muscles or glands.

The Peripheral Nervous System

The PNS consists of sensory neurons which transmit information from peripheral receptors to the CNS and motor neurons which transmit information from the CNS to the muscles and glands, otherwise termed effector organs. It is subdivided into the somatic sensory and motor systems and the autonomic nervous system (Figure 3.18). All spinal nerves contain both sensory and motor neurons. Sensory neurons transmit information from sensory organs, such as eyes, skin, ears or tongue to the CNS; the upper and lower motor neurons transmit information from the CNS to skeletal muscles.

The **autonomic nervous system** consists of sensory neurons and motor neurons that run between the CNS, particularly the hypothalamus and medulla oblongata, and various organs. It is responsible for monitoring and regulating the internal environment (Figure 3.19). In contrast to

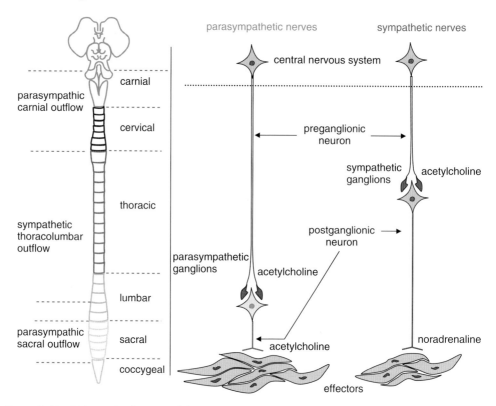

Figure 3.19 *Schematic diagram of the autonomic nerve system. The autonomic nervous system is commonly divided into two generally antagonistic subsystems: the sympathetic and parasympathetic nervous systems. It controls such vital functions as heart rate, dilation of the bronchioles, and dilation and constriction of the pupil; the digestive tract is controlled almost completely by autonomic mechanisms.*

the sensory-somatic system, the actions of the autonomic nervous system are mainly involuntary and use two groups of motor neurons to stimulate the effectors. Efferent axons leaving the CNS, except those innervating skeletal muscle, belong to the autonomic nervous system. The axons of lower motor neurons run without interruption to the neuromuscular junctions in skeletal muscles. However, the autonomic axons leaving the CNS make synaptic connections with peripheral neurons in ganglia, which in turn innervate the effector cells. The axons that form synapses in ganglion cells are called preganglionic autonomic fibers, while axons that innervate the effector cells are called postganglionic autonomic fibers.

The autonomic nervous system is subdivided into the **sympathetic nervous system** and the **parasympathetic nervous system**, based on the location of the cell bodies of the preganglionic autonomic axons in the CNS (Figure 3.19). The cell bodies which give rise to the preganglionic axons of the sympathetic division of the autonomic nervous system are located in the lateral horns of the thoracic and the upper two or three lumbar segments of the spinal cord. Synaptic transmission in the sympathetic ganglia and the effector organs is mediated by acetylcholine and norepinephrine (noradrenaline), respectively. Activation of the sympathetic nervous system generally primes the body to respond to stresses which might require rapid physical and/or mental activity and often demand increased aerobic metabolism, heart rate and muscular activity.

The preganglionic axons of the parasympathetic division emerge from the cranial and sacral region of the CNS. The synaptic transmissions in the parasympathetic ganglion and

Organ Toxicology 267

at the parasympathetic-innervated organs are also mediated by acetylcholine (Figure 3.19). Functionally, parasympathetic innervations are generally opposed to those of the sympathetic nervous system. For example, activation of parasympathetic nerves in the eyes causes miosis, whereas sympathetic stimulation causes mydriasis, parasympathetic nerves increase and sympathetic nerves decrease gastrointestinal activity, bronchial smooth muscles are contracted by the parasympathetic nerves, whereas they are relaxed by sympathetic nerves, etc. Both systems are controlled by the CNS.

The Central Nervous System

The **brain** and the **spinal cord** form the basic elements of the CNS (Figure 3.18). The **brain** receives sensory input mainly via spinal cord and cranial nerves, and uses most of its volume and computational power to process the various sensory inputs and to initiate appropriate and coordinated motor outputs as well as cognitive functions. The brain is an assembly of interrelated neural systems that regulate their own functions, as well as the functions of other body organs, in a dynamic, complex fashion, largely through intercellular chemical neurotransmission. Anatomically, the brain is divided into the forebrain, midbrain, and hindbrain (Table 3.4). These are composed of several specialized structures, each responsible for defined functions of the brain. The cerebral cortex, the limbic system, and the **diencephalon** all belong to the forebrain. The **cerebral cortex** consists of two hemispheres that represent the largest division of the brain. The specialized functions of a cortical region arise from the interplay between connections

Table 3.4 Divisions and macro-functions of the brain.

Parts of the brain	Location	Function
Brainstem	The brainstem is located at the juncture of the cerebrum and the spinal column. It consists of the midbrain, medulla oblongata, and the pons.	Alertness, arousal, breathing, blood pressure, contains most of the cranial nerves, digestion, heart rate, other autonomic functions, relays information between the peripheral nerves and spinal cord to the upper parts of the brain
Hindbrain (rhombencephalon)	The rhombencephalon is the inferior portion of the brainstem. It comprises the metencephalon, the myelencephalon, and the reticular formation.	Attention and sleep, autonomic functions, complex muscle movement, conduction pathway for nerve tracts, reflex movement, simple learning
Midbrain (mesencephalon)	The mesencephalon is the most rostral portion of the brainstem. It is located between the forebrain and brainstem.	Controls responses to sight, eye movement, pupil dilation, body movement, hearing
Forebrain (prosencephalon)	The prosencephalon is the most anterior portion of the brain. It consists of the telencephalon, striatum, diencephalon, lateral ventricle, and third ventricle.	Chewing, directs sense impulses throughout the body, equilibrium, eye movement, vision, facial sensation, hearing, phonation, intelligence, memory, personality, respiration, salivation, swallowing, smell, taste

from other regions of the cortex and non-cortical areas of the brain and a basic intra-cortical processing module of vertically connected cortical columns. Cortical areas called association areas process information from primary cortical sensory regions to produce higher cortical functions such as abstract thought, memory, and consciousness. The cerebral cortices also provide supervisory integration of the autonomic nervous system and integrate somatic and vegetative functions.

The **limbic system** is a term for an assembly of brain regions clustered around the subcortical borders of the underlying brain core to which a variety of complex emotional and motivational functions have been attributed. The main structures of the diencephalon are the thalamus and the hypothalamus. The **thalamus** lies in the center of the brain, beneath the cortex and basal ganglia and above the hypothalamus. The neurons of the hypothalamus are arranged into distinct clusters or nuclei. These nuclei act primarily to relay stimuli from the basal ganglia to associated regions of the cerebral cortex. Moreover, the thalamic nuclei and the basal ganglia exert regulatory control over visceral functions. The **hypothalamus** is the principal integrating region for the entire autonomic nervous system and regulates body temperature, water balance, intermediary metabolism, blood pressure, sexual and circadian cycles, secretion of the adenohypophysis, sleep, and emotion.

The **midbrain and brainstem** are considered as connection structures to the spinal cord and are primarily composed of the mesencephalon, pons, and medulla oblongata. These bridging areas of the CNS contain most of the nuclei of the cranial nerves, as well as the major inflow and outflow tracts from the cerebral cortices and spinal cord. They contain the reticular activating system, an important region of the gray matter linking peripheral sensory and motor events with higher levels of nervous system integration. The reticular activating system is essential for the regulation of sleep, wakefulness, and level of arousal, as well as for coordination of eye movement. These structures of the midbrain and the brain stem represent the points of central integration for coordination of essential reflexive acts, such as swallowing and vomiting, and those that involve the cardiovascular and respiratory systems. They also include the primary receptive regions for most visceral afferent sensory information.

The **cerebellum** arises from the posterior pons behind the cerebral hemispheres. In addition to maintaining postural regulation achieved by the proper tone of the musculature and providing continuous feedback during voluntary movements of the trunk and extremities, the cerebellum may also regulate visceral function and plays a significant role in learning and memory.

The spinal cord extends from the caudal end of the medulla oblongata to the lower lumbar vertebrae and transmits sensory information from the PNS, both somatic and autonomic, to the brain and routes motor information from the brain to various effectors, e.g. skeletal muscles, cardiac muscle, smooth muscle or glands. The spinal cord is divided into anatomical segments that correspond to divisions of the peripheral nerves and the spinal column (Figure 3.19). Within local segments of the spinal cord, autonomic reflexes can be elicited.

The blood–brain barrier

> **The blood–brain barrier (BBB) is a protective network that controls the blood flow to the brain. The BBB is composed of dense and tightly packed layers of endothelial cells around the brain blood vessels.**

The BBB regulates the access of physiological substances and xenobiotics to the CNS. It effectively impairs the transfer of charged, hydrophilic, and large molecular weight compounds from circulation to nervous tissue, but does not provide protection against small, lipid soluble compounds. Several transport mechanisms, like monocarboxylic acid transporters or organic anion transporters present in endothelial cells, mediate active transport through the BBB. Other transporters, like ABC transporters, mediate the excretion of compounds out of the endothelial cells back into the peripheral bloodstream. Leaking of the BBB makes the brain vulnerable to toxins. Areas that are not completely protected by the blood–brain barrier include those regions of the brain involved in neuroendocrine activity, such as the area postrema, hypothalamus, pineal, and places where the barriers are fenestrated, e.g. in autonomic ganglia, and motor and sensory nerve terminals. All these sites are potential points of entry for toxins.

Cells of the Nervous System

> **Neurons are electrically excitable and essential for signal transduction within the nervous system. They process and transmit information trough electrical and chemical signals. Neurons are structurally supported by glial cells, including astrocytes, oligodendrocytes, and microglia cells. In the PNS glia cells are called Schwann cells.**

Although the fine structure of the organelles of the CNS and PNS is not exclusive to these tissues, the interactions between cell types, such as synaptic contacts between neurons and myelin sheaths around axons, are unique. Diverse cell types are organized into assemblies and patterns such that specialized components are integrated into the nervous system. Neurons are cells within the nerve systems that integrate and conduct information, glial cells have supporting functions for the neurons, as well as immunological and signaling functions.

Neurons significantly differ from other cell types. Most neurons are highly branched cells that consist of cell bodies and projections, which are categorized into axons and dendrites (Figure 3.20). The **axons** represent projections of the nerve cell that are usually not branched. Parts of the axons, which transmit information from neurons to other nerve cells, organs or tissues, are surrounded by concentric myelin layers that have high electric impedance. The myelin layers are disconnected at the nodes of Ranvier, which occur at 1.5 mm intervals along the axon. During fast axonal transmission, electrical signals jump from one node of Ranvier to another. In non-myelinated axons, depolarization moves slowly down the axon. **Dendrites** are highly branched and are a point of communication to other neurons. They receive signals from other neurons, and integrate and transmit the information within the neuron.

Neurons of the adult brain are terminally differentiated cells that have not been thought to undergo proliferative responses to damage. However, neuronal stem cell proliferation as a natural means for replacement of neurons occurs to a small extent and in selected areas of the CNS only. Therefore, neurons of the CNS have evolved other adaptive mechanisms to maintain its functions following an injury. In the PNS damaged cells can be resubstituted. This requires surviving neurons, which expand their territory by axonal branching, overtaking territory vacated by dead neuronal cells. This process is ineffective in the CNS.

Other adaptive mechanisms provide the brain with considerable capacity for structural and functional modifications. The brain is highly resistant to damage or toxic effects and is functionally flexible. Dysfunction in selected brain areas can be compensated for by other areas of the brain. Another reason for the maintenance of function in the face of injury is the strong homeostatic protection of the brain milieu, which is partly based on the specialized vascularization of that organ.

Neurons are very active cells and have a high metabolic demand. Almost all neurons are highly branched. Long and energy-consuming routes of transport are necessary to provide the nerve terminals with nutrients and essential compounds. Purkinje cells of the cerebellum or motor neurons

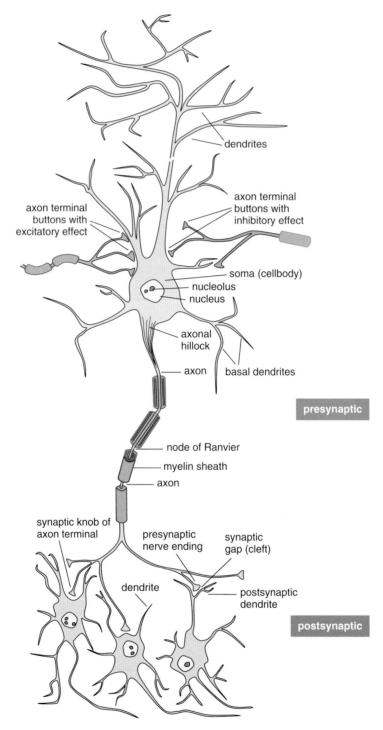

Figure 3.20 Schematic diagram of neurons. Neurons are a major class of cells in the nervous system. Their primary role is to process and transmit neural information.

in the PNS, which have very long axons, represent examples of highly effective systems for distributing metabolites between the cell body, the dendrites, and the axons of the neuron. Hence, neurons are especially vulnerable to chemicals which destroy the myelin layers or the cytoskeleton, or interfere with the supply of energy.

Neuronal cells express a high density of ion channels to reliably assure the transmission of action potentials from one neuron to another (Figures 3.21 and 3.22.). The maintenance of the ion gradient requires continuous and active ion transport during both normal and stimulated neurological activity. Glucose represents the main energy source for the brain and is used to generate adenosine triphosphate (ATP). ATP is generated during glycolysis and is even more effective during the oxidative phosphorylation within mitochondria. Neurons are very vulnerable to disturbances of the energy production. However, the vulnerability varies considerably, and is dependent upon on time, functional status, and cell type.

Other brain cell types include astrocytes, which assist in neurotransmission and neuronal metabolism by controlling the development of the nervous system in early developmental stages of life and sustain vascular, ionic, and metabolic homeostasis. Astrocytes synthesize components essential for the regeneration and maintenance of the nervous system. The **microglia** is part of the immune system in the CNS. **Oligodendrocytes** build up the myelin layers around long axons. In the PNS **Schwann cells** are responsible for this function.

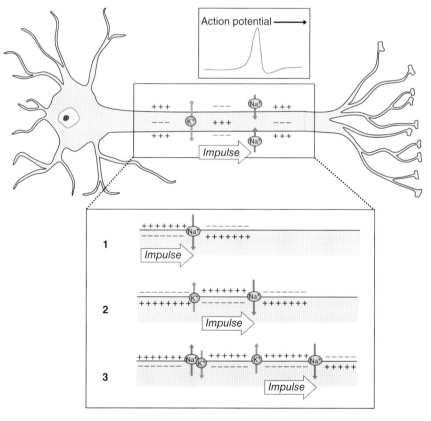

Figure 3.21 Schematic diagram of action potential transmission. The membrane voltage changes during an action potential as a result of changes in the permeability of the membrane to sodium and potassium ions.

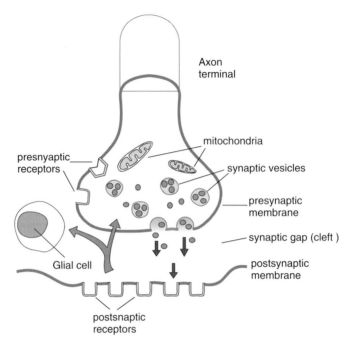

Figure 3.22 Schematic diagram of a synapse. Synapses are specialized junctions through which cells of the nervous system signal to one another and to non-neuronal cells such as muscles or glands.

The PNS contains two types of nerve fibers: thin, non-myelinated fibers that are embedded in Schwann cells and medullated fibers composed of Schwann cells enveloped by thick layers of lipids and proteins called the medullary nerve sheath. Myelin represents the main lipid component of oligodendrites and Schwann cells. Gliosis refers to the characteristic reaction of astrocytes after nerve damage, which involves the production of a dense fibrous network of neuroglia in areas of damage, resulting in scars in the CNS.

Neurotransmission

> **Neurotransmission describes the conduction of information between nerve cells or from nerve cells to other cells. Signals are transmitted (i) electrically within neurons and (ii) chemically from one neuron to another. The chemical transmission involves neurotransmitters, which are stored in synaptic vesicles. Synapses are the junctions between nerve terminals of different neurons. Upon electrical excitation, neurotransmitters are released in the synaptic cleft between two neurons. They diffuse to the postsynaptic site and bind to receptors, specialized proteins embedded in the postsynaptic membrane. Binding of neurotransmitters induces a biochemical response in the postsynaptic neuron, including as new action potentials.**

Nerve impulses elicit responses in neurons by liberating specific chemical neurotransmitters. The sequence of events involved in neurotransmission is of particular importance because pharmacologically active substances and neurotoxins are able to modulate the individual steps. The arrival of an action potential at the axonal terminals initiates a series of events that trigger the transmission of an excitatory or inhibitory impulse across the synapse. Synapses between neurons and smooth, cardiac, and skeletal muscles or exocrine glands are called neuro-effector junctions (Figures 3.21 and 3.22). The synaptic cleft separates the plasma membrane of the presynaptic and postsynaptic neurons. **Neurotransmitters** are synthesized by nerve cells and stored in vesicles at the presynaptic membrane (Figure 3.22). Neurotransmitters of the PNS include acetylcholine, norepinephrine (noradrenaline), epinephrine (adrenaline), serotonin, and histamine. Acetylcholine, serotonin, and histamine are also abundant in the CNS. However, a wider variety of chemical transmitters exists in the CNS.

Excitatory synapses are characterized by a densely thickened postsynaptic membrane and contain round vesicles with neurotransmitters such as glutamate and acetylcholine. In contrast, vesicles in inhibitory synapses are usually flattened and postsynaptic membranes of these synapses are only partly thickened. Inhibitory neurotransmitters of the CNS include gamma-aminobutyric acid (GABA) and glycine. The arrival of action potentials at a synapse causes the release of neurotransmitters out of the synaptic vesicles in the presynaptic cell (Figure 3.22). Released neurotransmitters diffuse across the synaptic cleft and bind to specific receptors in the plasma membrane at the postsynaptic site. Transmission may be modulated through inactivation of ion channels or internalization of receptors in the postsynaptic membrane, as well as binding of neurotransmitters to inhibitory presynaptic receptors. Neurotransmission is terminated by enzymatic degradation and/or re-uptake into the neuron or uptake of neurotransmitters into extra-neuronal cells (Figure 3.22).

Binding of neurotransmitters to specific receptors on the postsynaptic neuron generally induces signals that depolarize the postsynaptic membrane, tending to induce an **action potential**. The depolarization of the neuronal membrane due to opening of gated Na^+ channels is described in Figure 3.21. In resting neurons, non-gated K^+-channels are open, but the more numerous gated Na^+ channels are closed. The movement of K^+ ions outward establishes the inside-negative membrane potential characteristic of neuronal cells. Opening of gated Na^+ channels permits an influx of sufficient Na^+ ions to cause a reversal of the membrane potential. The impulse moves along the neuronal membrane, thousands of Na^+ channels open and that prolongs the action potential. Meanwhile, K^+ channels open and K^+ ions are excreted out of the cell, leading to a inside-negative characteristic of the neuronal cell. Finally, the Na^+/K^+ ion pump restores the original ionic conditions within the respected segment of the axon. The action potential spreads out over the whole neuron and is able to induce neurotransmitter release at a synaptic site of that particular nerve cell.

3.4.2 The Nervous System Site of Attack for Toxins

> **Neurotoxicology investigates adverse effects of various agents (drugs, chemicals, biologicals) on the structure or function of the nervous system. All parts of the CNS and the PNS are potential points of attack for toxins. The nervous system is especially vulnerable to toxins because it is very complex and passes through different development periods characterized by cellular migration, differentiation, and synaptic pruning. Neurotoxic effects are usually classified based on the structure of the impaired nerve. Hence, neuropathy, axonopathy or demyelination can be differentiated based on neurotoxic insults.**

Neurotoxins can be absorbed by inhalation, contact with the skin or ingestion. Absorbed compounds enter the circulation and are mainly metabolized in the liver, while sublingual, rectal or direct intravenous application avoids the initial liver passage. Exposure to neurotoxins mostly occurs through breathing, ingestion along with food or water, or contact through the skin. Uptake into the circulation from these sources will always be highly variable, and it is impossible to calculate intuitively the toxic dose unless suitable biomarkers are available to assess the body burden of the toxin. The parts of the nervous system that are affected by chemical substances depend on the distribution of the compounds in the body. Ionic and hydrophilic toxins mainly impair the PNS, since they do not cross the BBB. Lipophilic agents and substances that are carried by special transport mechanisms will enter the brain and affect the CNS. Toxicological effects in the CNS can vary extremely depending on brain region, cell type, and residence time. Thus, only a minor number of compounds are neurotoxic in general, and most toxins exhibit highly selective effects.

Compounds that harm endothelial cells, such as nitropropionate, dinitrobenzene or metronidazole, lead to a malfunction of the BBB, rendering it porous and permitting the entry of toxins to the CNS. Although not neurotoxic per se, the damage that they cause may ultimately result in neurotoxic effects.

Intoxication depends on the dose of compound and the duration of exposure. In terms of duration of exposure, acute and chronic intoxication that lead to reversible or irreversible effects have to be distinguished. For example, acute exposure of organic solvents causes a reversible CNS depression. Chronic exposure to small doses may result in peripheral nerve degeneration. Hence, it is important to distinguish between acute and chronic exposures resulting in acute and chronic effects.

Cellular Damage Resulting in Neurotoxicity

Impairment of neurons in defined regions of the nervous system results in characteristic disturbances grossly observed as symptoms of specific neuronal diseases. The resulting neuronal damage can be classified based on the subcellular target of the toxic agent.

Neuropathy The susceptibility of neurons to neurotoxins varies and only few chemicals affect all types of neuronal cells in the brain (Table 3.5A). Neuropathy is subdivided into cytoplasmic, nuclear, and postsynaptic neuropathies. Damage of the cell soma is followed by the loss of axons and dendrites. Neuropathies in the PNS are generally axonopathies and damage of neurons by chemicals can ultimately lead to cell death. The loss of neurons in the adult brain is virtually irreversible. A peripheral neuropathy may have its origin in the neuron, causing either cell death or dysfunction. There may be degeneration of the axon or disruption of neuronal or axonal function by damaging the myelin sheath. Alcoholism is the most common cause of a peripheral neuropathy. In contrast to the CNS, re-generation and re-innervation of surviving neurons is possible in the PNS.

Methyl mercury (MeHg) is the best-known neurotoxin associated with neuropathy, although the mechanisms of toxicity are poorly understood. The first well-documented mass poisoning by MeHg, from consuming contaminated fish, occurred in Minamata, Japan, 1956. The cerebellar granular cells are extremely sensitive to MeHg. Also, small neurons of the visual and the cerebellar cortex die after intoxication with MeHg, leading to central neuropathy. Symptoms include sensory deficits followed by ataxia and impaired motor coordination. Experimental data indicate that MeHg also causes peripheral neuropathies in rats, where accumulation in spinal ganglia leads to death of sensory neurons. However, this type of special peripheral neuropathy following MeHg intoxication has not been observed in humans. Even chronic exposure of low-dose MeHg over 20 years resulted in diffuse damage of the somatosensory cortex, bud did not damage the PNS.

1-methyl-4-phenyl-1,2,3,6-tetrahydropyridine (MPTP) damages dopaminergic neurons in the brain. The compound is a contamination produced during the synthesis of **MPPP** (1-methyl 4-phenyl 4-propionoxypiperidine). MPPP represents a designer drug that was

Table 3.5 Neurotoxins that cause structural damage in the nervous system.

Compound	Mechanism Clinical signs and symptoms
A Neuropathies	
Cyanate	Inhibition of mitochondrial cytochrome oxidase (complex IV); cell respiration failure Hypoxia
Carbon monoxide	High affinity to hemoglobin; carboxyhemoglobin formation; insufficient supply of oxygen Hypoxia
Methanol	Metabolic production of formic acid; destruction of visual nerves Blindness
MPTP	Active metabolite MPP$^+$; inhibition of mitochondrial respiration chain (complex I) Parkinson's syndrome Destruction of dopaminergic neurons
Hydroxydopamine	Generation of free radicals; oxidative stress; destruction of dopaminergic neurons Parkinson's syndrome
Lead	Replaces calcium; NMDA-receptor interactions; glutamatergic failure Encephalopathy
Triethyl lead	Excitatory compound; swelling of Golgi apparatus and endoplasmic reticulum; accumulation of lysosomes Delirium Loss of peroxisomes; necrosis specific for the limbic system (hippocampus)
Methyl mercury	High affinity to SH groups > inactivation of proteins > neuronal failure Encephalopathy
Trimethyl tin	Similar to triethyl lead; in contrast to triethyl tin no edema of the myelin layers Delirium
B Axonopathies	
n-hexane	Metabolites (γ-dicetones) interact with lysyl amino acids; destruction of neurofilamentes; impairment of intra-axonal transport mechanisms Sensor and motor malfunction
Carbon disulfide	Cross-linking of proteins; disturbances of neurofilamentes; dying back axonopathy Peripheral neuronopathy
Ethanol	Neuronal membrane interactions; membrane fluidity changes; disturbed ion permeability Encephalopathy
Organic phosphates	Inhibition of neuropathy target esterase; aging reaction; irreversible polyneuropathies Paralysis
2-Propenamide, acrylamide	Inhibition of retrograde axonal transport mechanisms; dying back axonopathy Neuronal dysfunction

(continued)

Table 3.5 (Continued)

Colchicin	Taxol disrupts microtubule polymerization; antimitotic; disturbances of neurofilaments
	Encephalopathy
Arsenic	High affinity to SH groups; inactivation of proteins; neuronal failure
	Polyneuropathy
C Demyelination	
Hexachlorophen	Membrane interactions; disturbed ion gradient; swelling of myelin; demyelination
	Ischemia (central)
Cuprizon	Copper chelator; enzyme inhibition; malfunction of oligodentrocytes; demyelination
	Ischemia (central)
Lead	Schwann's cell destruction; endoneuronal edema; no penetration into CNS
	Peripheral neuropathy
Triethyl tin	Membrane interactions; enhanced chloride and hydroxide exchange; myelin swelling
	Nausea; vertigo; convulsion
Tellurium	Squalenepoxidase inhibition; reduced cholesterol synthesis; reduced myelin production
	Nausea; vomiting

developed in the late 1970s as an alternative to heroin. MPTP results in an irreversible neurodegenerative syndrome, similar to that of Parkinsonism. The active metabolite of MPTP1-methyl-4-phenylpyridinum (MPP$^+$) is produced by the enzyme mono-oxidase B and is taken up through a active process involving the dopamine transporter in dopaminergic neurons. MPP$^+$ impairs mitochondrial function and thus induces neuropathy in dopaminergic neurons of the substantia nigra, an area of the brain important for motor coordination.

Trimethyl tin (TMT) derived organotin compounds are used in the chemical industry for organic synthesis. TMT is a potent biocide that selectively causes neuropathy in the limbic system, especially the hippocampus. TMT poisoning is characterized by a marked augmentation of excitability of hippocampal neurons but also cognitive deficits and epileptic seizures. TMT induces oxidative stress and inflammatory responses, finally resulting in neuronal death. TMT-induced neurodegeneration is used as an *in vivo* rodent model for neurodegeneration.

Axonopathy   Nerve damage classified as a central or peripheral distal axonopathy is produced by a variety of chemicals (Table 3.5). Distal axonopathy, also known as "dying-back" neuropathy, is typical of peripheral neuropathy resulting from metabolic or toxic derangement of the PNS. Distal swelling and degeneration of axons in the CNS and PNS is the most common response to neurotoxins. Axonopathy is typically gradual in onset, affecting first the distal regions of long axons and advancing slowly towards the nerve's cell body. Sensory loss tends to precede loss of motor function. The symptoms spread proximally as the axon "dies back" for as long as exposure lasts. Loss of ankle reflexes is an early indication of axonopathy. Depending on the severity of the damage, neurons may survive, but loss of neuronal connections occurs. If the noxious stimulus is removed, regeneration is possible depending on the duration and severity of the stimulus. Acrylamide and 2,5-hexanedione are prime examples of chemical toxicants that cause central-peripheral axonopathies characterized by distal axon swelling and degeneration.

Exposure to **acrylamide (ACR)** produces a distal auxopathy affecting both the CNS and PNS. ACR can be formed during the production of foods such as potato chips, bread or French fries. Carbohydrates and amino acids react at high temperature to so-called Maillard products. ACR is also used in laboratories for molecular biology. Recent experimental data suggests that ACR retards the slow retrograde axonal transport of specific neurofilaments and significantly reduces ATP levels. **2,5-Hexanedione (HD)** represents the toxic metabolite of the industrial solvents n-hexane and methyl-n-butyl ketone. The toxic γ-diketone induces a neuropathy that is characterized by a pathological process of neurofilament accumulation and degeneration of the axon. In humans, this pathology is manifested as a clinical peripheral neuropathy. Electrophysiological changes consist of decreased motor and sensory nerve conduction, with an increased distal latency period. Recent evidences indicate that the HD-mediated axon atrophy involves multiple cytoskeletal proteins that are cross-linked by covalent binding of diketones to lysine residues to form pyrrole derivatives.

Demyelination

Glial cells are non-neural support cells in the nervous system and are possible points of attack by neurotoxins. Oligodendrocytes of the CNS and Schwann cells of the PNS form layers around axons. Myelin that is produced by these glial cells helps to electrically isolate distinct axonal sections. Thus, damage to the myelin layer results in failures of nerve conduction. Myelin damage may involve swelling of myelin, primary demyelination, and secondary demyelination. The swelling of myelin may result from edema following disturbances in the myelin homeostasis or metabolic damage to glia. Functional consequences are disruption in nerve conduction and increased water content of the tissue. The swelling is partly reversible.

Primary demyelination occurs as a consequence of a direct attack of chemicals on myelin or myelin-producing cells resulting in loss of myelin or in myelin destruction (Table 3.5C). This damage is also partly reversible.

Secondary demyelination is an active process that ultimately leads to the destruction of axons. It occurs as a consequence of a traumatic destruction or toxic axonopathy, where myelin is digested by macrophages. Demyelinating neuropathies of the PNS result from damage to the Schwann cell or to the myelin sheath of the internodes. Examples are **diphtheria toxin** and **inorganic lead**. **Hexachlorophene** has also been implicated in myelin disruption. Recovery depends on the replication and activation of surviving Schwann cells. In regenerated axons the internodes tend to be shorter and myelin sheaths are thinner than in the axons prior to injury. Usually, the conduction velocity is slower in re-myelinated axons. Hexachlorophene also reaches the CNS and damages the sensitive long nerve fibers.

Effects on Neurotransmission

The transmission of information within the nervous system is extremely complex. Consequently, neurotransmission is especially susceptible to attacks by chemicals at several sites. Synapses are particularly sensitive to chemicals. Synaptic function requires the synthesis, storage, release, reaction, and inactivation of transmitters. Each step represents a possible point of attack. Reduced inactivation results in an excess of transmitters that causes continuous depolarization of the nerve cell. An alteration in synthesis, storage or release of neurotransmitters also profoundly disrupts neurotransmission. Compounds that are structurally related to transmitters interact with their receptors. Chemicals that activate receptors by mimicking transmitters may have agonistic effects. Antagonists occupy receptors and inhibit the binding of neurotransmitters to their receptors, thus preventing neuronal activation.

Organophosphate derivatives (OPDs) such as parathion (E 605) or dichlorphos (DDVP) have been used as pesticides in agriculture. E 605 was also used in household applications before it was banned due to a series of poisonings in the 1950s. OPDs such as sarin (GB) or tabun (GA) were also used in chemical weapons. Many organophosphates are potent neurotoxins, functioning by irreversibly inhibiting the enzyme acetylcholinesterase, which terminates cholinergic neurotransmission. Excessive acetylcholine causes continuous depolarization of the postsynaptic cholinergic receptors. After a short phase of initial hyper-activation, the neurotransmission completely breaks down, resulting in palsy.

The clinical signs of organophosphate poisoning are salivation, lacrimation, urination, defecation, gastrointestinal upset, and emesis. Poisoning with deadly doses leads to death within a very short time. Atropine acts as an antagonist and diminishes the effects of acetylcholine on postsynaptic muscarinic cholinergic receptors. Thus, high intra-muscular doses of atropine are used as an antidote to organophosphate poisoning. Since the half-time of atropine is much shorter than that of organophosphate, multiple doses have to be applied to overcome a poisoning.

Excitatory amino acids, such as domoic acid or kainic acid, act as false neurotransmitters at glutamatergic nerve cells. Activation of glutamate receptors results in excessive influx of calcium ions into nerve cells, which triggers the depletion of intracellular calcium stores in the endoplasmatic reticulum and mitochondria. The massive increase in the cytosolic imitates apoptosis and necrosis, and finally leads to neuronal death.

Domoic acid is a toxic algae amino acid produced, for example, by *Pseudo-nitzschia*, especially during marine algae blooms. The chemical can bioaccumulate in marine organisms that feed on phytoplankton, such as shellfish, anchovies, and sardines. Poisoning with gomoic acid in humans has been observed after consumption of contaminated shellfish and leads to so-called amnesic shellfish poisoning. Domoic acid acts as a neurotoxin, damaging mainly the hippocampus and the amygdaloidal nucleus, causing short-term memory loss and brain damage.

Kainic acid is also produced by algae and can be found in seaweed. The natural marine acid is a potent agonist at kainat receptors that represent a type of ionotropic glutamate receptors. Overstimulation by high doses can lead to neuronal death. Kainic acid is used in neuroscience to study neurodegeneration and epilepsy.

Glutamic acid, or its anionic form glutamate, is a non-essential proteinogenic amino acid. Many Asian dishes are characterized by the taste of glutamate, which comes from natural origins such as soy or fish sauce and acts a flavor enhancer. Components of flavor enhancers, such as monopotassium glutamate, calcium diglutamate, monoammonium glutamate, and magnesium diglutamate, have been implicated as a cause of Chinese restaurant syndrome. This syndrome is characterized by headache, flushing, sweating or sense of facial swelling and is supposed to occurs after eating in a restaurant serving Asian food. However, there is no evidence from scientific and clinical studies that glutamate is the causative agent and there is some doubt over whether a glutamate intolerance exists at all. The reported symptoms could also be related to other foodstuffs such as prawn or fish.

Other compounds like **muscimol** act as agonists on inhibitory receptors sensitive to the neurotransmitter GABA. **Strychnine** acts as an antagonist at glycine receptors and compensates for the inhibitory effect of glycine in spinal ganglions that results in seizures (convulsions).

Muscimol is a toxic psychoactive alkaloid present in the mushroom *Amanita muscaria*. It originates from decarboxylation of ibotenic acid. Both substances produce the same effects but muscimol is more potent. Muscimol specifically activates inhibitory $GABA_A$ receptors mediating sedative-hypnotic effects. Symptoms of poisoning generally occur within 1–2 hours after ingestion of the mushrooms, including drowsiness and dizziness, followed by a period of hyperactivity, excitability, illusions, and delirium.

Strychnine is a highly poisonous alkaloid present in the seeds of the strychnine tree (*Strychnos nux-vomica* L.). Strychnine acts as an antagonist at the inhibitory glycine receptor, a ligand-gated chloride channel in the spinal cord and the brain. Strychnine poisoning may result from inhalation, swallowing or absorption through eyes or mouth and can be fatal. Strychnine produces some of the most dramatic, terrifying, and painful symptoms that are manifested just a few minutes after exposure. Quickly after ingestion generalized muscle spasms occur as one of the first symptoms. Every muscle in the body will start to simultaneously contract. High doses result in respiratory failure, cardiac arrest, multiple organ failure, or brain damage that can lead to death within a short time. Lower doses provoke symptoms including seizures, cramping, hypervigilance, stiffness, and agitation.

Botulinum toxin (Botox) is produced by the bacterium *Clostridium botulinum* and related species. The toxic protein interacts with a fusion protein at the neuromuscular junction, preventing neurotransmitter vesicles from anchoring to the synaptic membrane to release acetylcholine. Thus, Botox interferes with nerve impulses and causes the paralysis of muscles. Botox is extremely toxic, with a lethal dose for humans of about 0.001 µg/kg body weight. Food poisoning with Botox is called botulism whereby the toxin is usually produced by *Clostridium botulinum* under anaerobic conditions, for example, in cans.

Tetrodotoxin (Tedox) represents a guanidine alkaloid produced by bacteria that colonize different marine animals, including the pufferfish. Tedox selectively inactivates sodium channels in nerve cell membranes, inhibits the transmission of action potentials and interrupts electrical signaling within nerves (Figure 3.21). Fish poisoning produced by consumption of members of the order *Tetraodon* forms one of the most violent intoxications caused by marine species. The gonads, liver, intestines, and skin of pufferfish can contain levels of Tedox sufficient to produce rapid and violent death. On the other hand, its concentration in muscles is low so that this meat can be eaten. Consumption of meat from the muscles results in a noticeable tingling.

Tedox poisoning leads to shortness of breath, numbness, tingling, light-headedness, paralysis, and irregular heartbeat. Due to its hydrophilic quaternary structure Todox does not cross the BBB barrier. Hence, the brain is not affected and patients remain conscious during paralysis.

3.4.3 Clinical Signs and Symptoms induced by Neurotoxins

> **The mature nervous system is remarkably vulnerable to toxin-induced damage. Since any damage potentially disrupts the extensive communication systems of the CNS, neurotoxins have the capacity to affect gait and posture, sensation, behavior and cognition, and produce a complex pattern of clinical signs and symptoms. These signs and symptoms may be expressed in the central, the peripheral and the autonomic nervous system. Poisoning with neurotoxins is often associated with pain, changes in the special senses of taste and smell, as well as with changes in visual acuity and hearing.**

Hypoxia describes a state of oxygen deficiency that causes an impairment of function. Reasons for hypoxia in the brain are mainly inadequate oxygen transport, the inability of the tissues to use oxygen, or a reduction in the oxygen carrying capacity of the blood. **Carbon monoxide** and **cyanide** are examples for poisons that produce a histotoxic hypoxia, which is defined as the inability of the tissues to use oxygen (Table 3.5A). Certain narcotics and alcohol also prevent oxygen use by the tissues (Table 3.5B). Symptoms of mild cerebral hypoxia include inattentiveness, poor judgment, memory loss, and a decrease in motor coordination. Brain cells are extremely sensitive to oxygen deprivation. Long lasting hypoxia can cause coma and even brain death. Whereas hypoxia is a general term denoting a shortage of oxygen, **ischemia** refers to a specific condition where there is shortage of blood supply to a particular organ. Since oxygen is mainly bound to hemoglobin in red blood cells, insufficient blood supply can cause tissues to become hypoxic, or, if no oxygen is supplied at all, anoxic, which will ultimately cause necrosis and cell death. The brain is especially sensitive to ischemia triggering a process in which proteolytic enzymes, reactive oxygen species, and other harmful chemicals damage and may ultimately provoke the death of brain tissue. Hypoxia or ischemia may cause **encephalopathy**, which manifests in neurological symptoms such as progressive loss of memory and cognitive ability, subtle personality changes, inability to concentrate, lethargy, and progressive loss of consciousness which depends on the type and severity of encephalopathy. Acute **encephalopathies** that are mild and that resolve within a few days are common. Numerous compounds including aluminum, cannabis, cocaine, domoic acid, lead, organic solvents, and trimethyl tin have been demonstrated to induce encephalopathies. The transition from the mild acute to chronic severe encephalopathy with associated loss of cognition and psychomotor function is relatively uncommon, but has been reported after acute severe poisoning by domoic acid, aluminum, cadmium, lead and following chronic abusive use of alcohol or organic solvents. Although acute signs usually rapidly recover persistent problems may have **serious adverse effects** which necessitate long-term psychiatric and psychological assessments and monitoring.

Cerebellar and basal ganglion dysfunctions result in **disorders of movement**, characterized by ataxia, intention tremor, and loss of coordination. This condition is best known as a feature of chronic exposure to mercury; however, overdose with a variety of potentially toxic drugs and chemicals such as MPTP, 5-fluorouracil, lithium, and acrylamide have also been implicated. Extrapyramidal syndromes such as Parkinsonism, dystonia, dyskinesia, and tics are relatively well known toxic syndromes. Tics are repeated, uncontrolled or impulsive actions that appear almost reflexive in nature. The toxic mechanisms are poorly understood but are usually reversible, although problems may recur many years after the original onset of the disorder. In most cases, the psychiatric abnormalities are relatively mild, but major cases of dementia and a Parkinsonism/dementia syndrome have been associated with aluminum toxicity, cerebellar ataxia with lithium overdose, and severe psychotic disorders related to use of psychedelic drugs such as lysergic acid diethylamide (LSD).

Common clinical symptoms of neurotoxins are the loss or changes in the perception of taste and smell. Although, organic solvents are often named the underlying mechanism are not yet fully understood. A perception of change in taste is common following the use of numerous therapeutic agents, but is usually reversible. The **loss of hearing** has been attributed to the abuse of organic solvents, e.g. toluene. It is more often associated with the use of well-known ototoxic drugs such as the aminoglycosides. Direct attack on the neuronal components of the **visual system** is less common. Mydriasis and miosis are obvious signs of exposure to cholinolytic and cholinomimetic agents, such as atropine or muscarinie, respectively (Table 3.5). The optic nerve can be damaged by toluene which causes demyelination or hexachlorophene which causes myelin deformation. Alcohol abuse is also associated with widespread damage to the neuronal components of the visual system, but it is suspected that the etiology is confounded by the poor nutritional status of many chronic abusers of alcohol Table 3.6.

Table 3.6 Assorted neurotoxins: mechanism of action and symptoms of intoxication.

Compound	Mechanism	Clinical signs and symptoms
Aluminum	Al^{3+} ions; disturbed cellular ion homeostasis	Encephalopathy
Atropine	Agonist for acetylcholine receptors; anticholinergic and parasympatholytic effects	For example, agitation, anxiety, central respiratory paralysis
Botulinum toxin (Clostridia toxins)	Affects moto- and vegetative neurons of the PNS; inhibition of exocytosis acetylcholine; flag of horizontally striped musculature/loss of vegetative functions	Palsy
Cocaine (alkaloid of *Erythroxylon coca*)	Inhibition of presynaptic neurotransmitter re-uptake; extended and accelerated effect on postsynaptic receptors	Paranoid psychopathy
Domoic acid (excitatory amino acid; produced by *dinoflagellate*)	Agonists at excitatory glutamate and glycine receptors	Headache, confusion, memory deficits
Glutamate	Agonist for excitatory glutamate receptors; overstimulation of glutamate receptors, e.g. increase of intracellular Ca^{2+} and H_2O levels; cell swelling	For example, apoplexy, epilepsy
Kainic acid (excitatory amino acid; produced by *dinoflagellate*)	Agonist for excitatory glutamate receptors; overstimulation of glutamate receptors, e.g. increase in intracellular Ca^{2+} and H_2O levels; cell swelling	For example, apoplexy, epilepsy
Tetanus toxin (Clostridia toxins)	Interneurons of the PNS; inhibition of exocytosis of inhibitory neurotransmitters; loss of inhibitory effects on moto neurons; continued contraction	Convulsion
Thallium	Tl^+ ions exchange K^+ ions, e.g. intra-cellular oxidation to Tl^{3+}; damage to mitochondria	Encephalopathy, polyneuropathy

3.4.4 Summary

The PNS and the CNS are extremely complex networks which provide the structural basis for electrochemical processes that are responsible for multiple functions such as cognition, sensation, memory, and speech. Neurons are the fundamental unit of the nervous system. The electrical charge across neuronal membranes allows the conduction of signals. For proper ion gradients, the supply of energy is essential for neurons. Hence, neurons are highly dependent on aerobic metabolism. Neurons are supported by glial cells, which build a supporting tissue in the nervous system. All parts of the nervous system are potential points of attack for toxins. The nervous system is especially vulnerable to toxins, since it is exceptionally complex and undergoes a prolonged period of development. Neurotoxic effects are usually classified based on the structure of the nerve cell that is impaired. Neurotoxins typically target the neuron, the axon or myelinating cells. Synthetic chemicals, as well as naturally occurring toxins, may interrupt the neurotransmission, including inhibition of cellular communication and inhibition of neurotransmitter re-uptake, or interfere with second-messenger systems. The mature nervous system is remarkably vulnerable to toxin-induced damage, and because any damage may disrupt the extensive communication systems that characterize the brain, neurotoxins have the capacity to affect gait and posture, sensation, behavior, and cognition, and produce a complex pattern of clinical signs and symptoms. Poisoning with neurotoxins is often associated with pain, changes in the special senses of taste and smell, as well as changes in visual acuity and hearing.

Further Reading

L. An, G. Li, J. Si, C. Zhang, X. Han, S. Wang, L. Jiang, K. Xie. Acrylamide retards the slow axonal transport of neurofilaments in rat cultured dorsal root ganglia neurons and the corresponding mechanisms. *Neurochem. Res.* **41**, 1000–1009 (2016).

F.E. Bloom. Neurotransmission and the central nervous system, in *Goodman & Gillman's the Pharmacological Basis of Therapeutics*, L.L. Brunton, L.S. Lazo, and K.L. Parker (Eds), McGraw Hill, New York, 2006.

S.C. Bondy, A. Campbell. Developmental neurotoxicology. *J. Neurosci. Res.* **81**, 605–612 (2005).

C.J. Ek, K.M. Dziegielewska, M.D. Habgood, N.R. Saunders. Barriers in the developing brain and neurotoxicology. *Neurotoxicology* **33**, 86–604 (2012).

S. Ekino, M. Susa, T. Ninomiya, K. Imamura, T. Kitamura. Minamata disease revisited: an update on the acute and chronic manifestations of methyl mercury poisoning. *J. Neurol. Sci.* **262**, 131–444 (2007).

S. Lee, M. Yang, J. Kim, S. Kang, J. Kim, J.C. Kim, C. Jung, T. Shin, S.H. Kim, C. Moon. Trimethyltin-induced hippocampal neurodegeneration: A mechanism-based review. *Brain Res. Bull.* **125**, 187–199 (2016).

R.M. LoPachin, T. Gavin. Toxic neuropathies: Mechanistic insights based on a chemical perspective. *Neurosci. Lett.* **596**, 78–83 (2015).

R.C. MacPhail. Principles of identifying and characterizing neurotoxicity. *Toxicol. Lett.* **64–65**, 209–215 (1992).

V.C. Moser, M. Aschner, R.J. Richardson, M.A. Philbert. Toxic responses of the nervous system, in *Cassarett & Doul's Toxicology*, C.D. Klaassen (Ed.), 8th edition, McGraw Hill, New York, 2013.

T. Narahashi, M.L. Roy, K.S. Ginsburg. Recent advances in the study of mechanism of action of marine neurotoxins. *Neurotoxicology* **15**, 545–554 (1994).

M.A. Philbert, M.L. Billingsley, K.R. Reuhl. Mechanisms of injury in the central nervous system. *Toxicol. Pathol.* **28**, 43–53 (2000).

F. Sánchez-Santed, M.T. Colomina, H. Herrero Hernández. Organophosphate pesticide exposure and neurodegeneration. *Cortex* **74**, 417–426 (2016).

L. Vecsei, G. Dibo, C. Kiss. Neurotoxins and neurodegenerative disorders. *Neurotoxicology* **19**, 511–514 (1998).

D.R. Wallace. Overview of molecular, cellular, and genetic neurotoxicology. *Neurol. Clin.* **23**, 307–320 (2005).

T.C. Westfall, D.P. Westfall. Neurotransmission: The autonomic and somatic motor nervous system, in *Goodman & Gillman's the Pharmacological Basis of Therapeutics*, L. Brunton, B. Chabner, and B. Knollman (Eds), McGraw Hill, New York, 2011.

D. Zemke, J.L. Smith, M.J. Reeves, A. Majid. Ischemia and ischemic tolerance in the brain: an overview. *Neurotoxicology* **25**, 895–904 (2004).

3.5 Behavioral Neurotoxicology

Andreas Seeber

3.5.1 Introduction

> **Neurotoxicity is the consequence of adverse effects on the structures and functions of the nervous system. Behavioural toxicology refers to the functional consequences of neurotoxicity and includes cognitive, executive and sensory effects, the personality domain, and peripheral nerve functions and electrophysiologically detectable effects, which are not elaborated here.**
>
> **To assess the variability of neurobehavioral effects certain methodological requirements have to be met.**

1. The exposure has to be measured using acute or chronic testing procedures and past exposures should be estimated.
2. The methodological approach should be supported by information on the underlying neurotoxic mechanism or the brain structures affected by the toxicant.
3. The design of a study and the statistical evaluation of the data should enable the identification of neurobehavioral effects as a function of exposure. Confounders such as age, education, verbal intelligence etc. that may affect the outcome of the study should be excluded.
4. Significant associations between exposure and neurobehavioral test results do not necessarily imply impairment of health. A careful check of criteria is necessary to evaluate whether an observed effect is adverse.

3.5.2 Exposure Assessment

> **For workplaces data on the exposure of a substance usually are air concentrations and represent time-weighted average (TWA) values of an 8-hour shift. If exposure is concentration (C) × time (T), the total exposure after repeated exposure would be $CT_1 + CT_2 + CT_3 \ldots = CT_{total}$, assuming that the biological effect depends on the total dose, including estimates of prior workplace exposures. Calculations of hygienic effects (HE) estimate exposure to mixtures of chemicals with regard to the limit values of the substances in the mixture. Although measures of exposure are usually expressed as mean values over an 8-hour day, it is appreciated that excursions resulting in high peak exposures could impact on the observed toxic effects.**

TWAs of air concentrations in workplaces or information from the biological monitoring of exposed workers provide data for investigations into associations between exposure and potential alterations of neurobehavioral function. Complete exposure assessment for all relevant substances over all periods of exposure is necessary to establish dose–effect relationships.

To estimate past exposures job exposure matrices are used that take into account technological changes in a plant during specific time periods at different work places. For example, in the European printing industry over a 20-year period exposures to toluene were reduced approximately 4-fold and in boat-building plants exposures to styrene were reduced at least by a factor of 3. Therefore, estimation of the individual lifetime weighted average exposure (LWAE) can only

Table 3.7 Exposure indices used in neurobehavioral studies.

HE, hygienic effect: Sum of TLV-weighted concentrations of the individual solvents

$$HE \text{ (job)} = \sum_{i=1}^{N} \frac{C_i}{TLV_i}$$

C_i = concentration of solvent i
TLV_i = threshold limit value of solvent i

CE, cumulative lifetime exposure for different jobs in an individual's working life

$$CE \text{ (person)} = \sum_{i=1}^{N} C_i t_i$$

C_i = concentration of the substance for the job i (ppm)
t_i = time in job i (years)

LWAE, lifetime weighted average exposure derived from CE

$$LWAE \text{ (person)} = \frac{1}{t} \sum_{i=1}^{N} C_i t_i$$

t = exposure time of the working life (years)

be accomplished after reviewing the work history and cumulative lifetime exposure (CE in ppm years) of the worker. LWAE and current information on workplace exposure should be considered independently when estimating total lifetime exposure. Table 3.7 shows how the expoure indices HE, CE and LWAE are calculated.

The HE is a summarized index of threshold limit value-weighted concentrations of the compounds in chemical mixtures, such as solvents. When searching for associations between HE and neurobehavioral effects, either equal neurotoxic actions or no interaction or potentiation between the compounds is assumed. HE values provide relative information on exposure depending on the threshold limit value for a substance. Usually only compounds exceeding 10% of the limit value are included in the calculation. HE values increase if threshold limit values are reduced and if the number of compounds in the mixture is elevated. HE < 1 indicates an exposure level below the summarized limit of the compounds in the mixture whereas HE > 1 indicates greater exposure. A comparison between studies of different countries or time periods requires information on corresponding exposure limits.

Establishment of NOAEL or LOAEL values in behavioural toxicology are facilitated when the broadest possible range of exposures to the chemical(s) in question are investigated.

3.5.3 Methods

> **Ideally, the relationship between behaviour and its underlying morphological and physiological mechanisms can be established. However, significant gaps exist between the biological and behavioural basis of neurotoxicity. Experimental or epidemiological studies in exposed people using neurobehavioral methods can be done side by side with morphological and physiological analyses. Estimating the manner in which these data are linked unambiguously is a challenge. However, it is possible to view the results as self-contained data obtained by a broad spectrum of standardized methods upon which cause and effect hypotheses can be conceptualized.**

Neuromorphological and Neuropathological Methods

Neuromorphological and neuropathological methods determine alterations of nerve cell bodies, axons, dendrites, neurofilaments, synaptic terminals and myelin sheaths. Histopathological and histochemical methods permit detection of pathological lesions in nerves cells and nervous tissue with respect to the types of cell injury, the time course over which damage appears, and the extent of injury. Neuroanatomical data characterize toxin-induced structural damage to central and peripheral nervous structures. Quantitative methods are available to measure how changes in the total number of specific types of neurons may be related to exposure to specific toxicants. The results of such studies can provide the basis for developing hypotheses aimed at linking these lesions to changes in behaviour.

Neurophysiological Methods

The primary electrophysiological procedures include studies of membrane ion cannels, synaptic transmission and analysis of senory-evoked potentials. The latter are intended to localize areas of neurotoxicological impact, and will be outlined in some detail.

The study of sensory functions by sensory-evoked potentials is a procedure to examine complex neural systems both in animals and humans. It includes the stimulation of sensory receptors or afferent nerves. Typical stimuli are light flashes or changes of visual patterns aimed at eliciting visual-evoked potentials, sounds intended to elicit auditory-evoked potentials, and stimulation of peripheral nerves with electrical impulses to generate somatosensory-evoked potentials.

The investigation of sensory reactions presents problems due to the small voltage of the evoked potentials. The data consist of positive and negative voltage deflections. Event-related potentials are commonly depicted as waveforms representing voltage against time. However, techniques must be used to overcome problems associated with the signal-to-noise ratio. Evoked potentials can be optimized by careful placement of the electrodes, adequate filtration and amplification of the potentials, and repeated measurements. Event-related potentials are commonly depicted as waveforms representing voltage against time. Interpretation of the data requires the use of averaging procedures and statistical analysis. Neural electrical activity can be detected from electrodes placed on the skull or on the surface of the cortex. The signals may arise from excitatory postsynaptic potentials, inhibitory postsynaptic potentials, and action potentials from an electrically active focus called a neural generator. Surface mapping techniques permit the localization of neural generators.

Examples of Sensory Evoked Potentials *Electroretinograms* are evoked responses from the cornea in response to visual stimulation. Negative and positive waves and oscillations of the waves are electrical characteristics, which can be interpreted as responses of rods and cones, of the middle retina or other retinal structures.

Flash-evoked potentials represent the cortical responses to visual stimulation mapped by use of surface recordings at the visual cortex and surrounding areas. Positive and negative waves can be associated, over their time course, with different origins in the cortical structure, e.g. a negative wave 40 ms after flash (N_{40}) is thought to be a sign for depolarization processes in the visual cortex lamina 4 following the thalamocortical input.

Brain stem auditory-evoked responses are composed of several peaks occurring between 10 and 1000 ms after stimulation. Early peaks are generated by the auditory nerve and the cochlear nucleus. Subsequent peaks seem to be generated by activity in several structures representing higher level processing of the auditory signal at the level of brain stem.

Lists of generators of visual-, auditory- or somatosensory evoked potentials have been offered to support hypotheses on the association between exposure to neurotoxicants and alterations of evoked potentials in neural structures.

Neurobehavioral Methods

Approach Variations of human behavior can be measured using standardized methods. Deviations from expected norms or patterns of tested behavior can have many causes. Chemical exposure is only one of the possible external sources for variations of behavior. Typical internal sources are the level of education, age, the status in the diurnal rhythm, physical conditions during measurement, as well as motivational or emotional states.

Evaluating the likelihood that a change in behavior is caused by exposure to a toxicant must include a comparison with behavioral variations having other causes. Investigation of the causes of behavioral changes can be studied using carefully designed investigations in which there are controls for confounding external and/or internal factors.

> **A variety of neuropsychological tests are available for neurobehavioral measurement. These include surveys to evaluate cognitive functions, executive functions, psychomotor functions, and personality or emotional changes.**

Cognitive functions involve the processing of learned information. Successful performance on tests of cognitive function require speed and accuracy and can be measured within a short period of time using either paper and pencil methods or by using computer administered exams.

- Receptive function tests measure the selection, acquisition, classification, and integration of information.
- Memory and learning tests measure the storage and retrieval of information.
- Thinking tests measure the mental organization and reorganization of information.
- Expressive functions and mental activity tests involve evaluation of the means by which information is acted upon and cognitive operations by which the level of consciousness is expressed.

Executive functions relate to management skills. If executive functions are impaired, the individual can no longer organize parallel tasks or maintain social relationships satisfactorily. Deficits in executive functions affect different aspects of behavior:

- A defective capacity for self-control or self-direction.
- A tendency to irritability and excitability, impulsivity, and rigidity.
- Difficulties in shifting ones attention.
- An impaired capacity to initiate activity.
- A decreased motivation and defects in planning and carrying out sequential activities.

It is difficult to test for clear distinctions between cognitive and executive functions.

Psychomotor functions deal with visible behavior. Motor activity represents the interface between behavior and the environment. Measurable endpoints in psychomotor tests include the time, the force, the speed, and the spatial boundaries of the response. Neurobehavioral tests in animals and infants rely on psychomotor responses because verbal reactions are limited. Compared to cognitive and executive examinations measured deficits, especially in animal psychomotor tests, can be associated with neural structures, e.g. the relationships between:

- loss of individual or integrated movements and specific cortical structures
- tremor, dystonic postures, turning behavior, and changed blinking rates and the basal ganglia
- nystagmus, loss of postural reflexes and uncoordinated movements and the cerebellum
- diminished grip strength or leg orientation and changed gait characteristics to peripheral nerves and neuromuscular junctions.

Personality or emotional changes directed preferably at emotional lability are typical signs of brain damage by neurotoxic agents. They are measured using standardized questionnaires or interviews directed at examining:

- Symptoms of impaired mental and physical health
- Tolerance to frustration
- Loss of emotional sensitivity and capacity for modulating emotional behavior
- Episodes of affective changes.

Examples of Cognitive and Psychomotor Tests

> **Cognitive, executive and psychomotor functions are measured using studies designed to show the impact of exposure to chemical toxicants compared to other factors which may influence the analyzed function. Among these are verbal intelligence, age and alcohol consumption. It is necessary to control for these factors because they may mask some of the effects of the toxicant.**

Tests such as the National Adult Reading Test (NART) are available to estimate verbal intelligence. It is a single-word, oral reading test. The words are short and irregular, not following normal grapheme-phoneme correspondences (e.g. ache, gauche). Previous familiarity with the words supports successful test performance and makes minimal demands on current cognitive capacity. The other widespread method is testing of vocabulary by a multiple-choice method. In a block of several nonsense words and one correct word the last one has to be marked without additional explanations and time limitation. The number of correct answered blocks of increasing difficulty gives the estimation of verbal abilities used as "pre-morbid" ability.

A typical attention test very often used in neurobehavioral studies is the Symbol Digit Substitution test. It belongs to the performance tests of the Wechsler Intelligence Scale. In the traditional paper–pencil version (Digit Symbol Test, DST) the subjects have to draw in a limited time symbols belonging to digits according to a code. The number of correct symbols serves as the score. The computer-administered version (Symbol Digit Substitution Test, SDST) requires a reversed order of actions. The subjects have to press the keys of digits for the corresponding symbols presented in random order. Translations from symbols to digits – not from digit to symbols – are required. Time and correct reactions are registered. The SDST version requires more than the DST version executive functions of cognitive flexibility. Figure 3.23 shows the use of SDST in a study on the effects of solvent mixtures in paint manufacturing. Raw data and the corrected data are presented taking into account lifetime weighted exposure to solvent mixtures, age, alcohol consumption measured by carbohydrate-deficient transferrin, and verbal intelligence measured by a vocabulary test. After control of the factors age, alcohol, and verbal intelligence a significant association between exposure and performance became apparent.

Meta-analyses on applications of DST and SDST in epidemiological studies of lead- or mercury-exposed workers indicate reproducible associations between biomonitoring data on exposure and test performance.

A typical example for measuring executive functions is the computer-administered test *Switching Attention*, which covers three subtasks. First, the subjects have to react compatibly on the left/right position of a square on the screen (block version), then compatibly on the left/right direction of an arrow (arrow version), and finally they have to react either in the block or in the arrow manner depending on the written instruction given just before the stimulus appears. Speed and errors are measured.

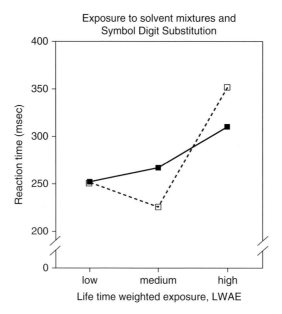

Figure 3.23 *Example for adjustment of data. Reaction time from the Symbol Digit Substitution Test in an epidemiological study with workers exposed to solvent mixtures in paint manufacturing. The open squares represent the raw data, the black squares the adjusted data. Derived from Seeber et al. Food Chem. Toxicol. 34, 1113–1120, 1996.*

Measurements of *simple and choice reaction times* represent attention tests with low cognitive demands on simple psychomotor stimulus response performances. Simple reaction tests include single visual or auditory stimuli and a simple response whereas choice reaction tests usually cover two or more stimuli (often four) with fixed corresponding responses. In these tests a first reaction time covers the processing from appearance of the stimuli up to the beginning of the response movement, a second movement time covers the process up to pressing the response key. Figure 3.24 shows the results of a choice reaction test (four stimuli with corresponding keys). From two experiments with 3 and 4 hours of styrene exposure, respectively, the results for constant 0.5 ppm and 20 ppm exposure are shown. In a special design the exposure conditions were combined with the time of day for carrying out the experiments. The exposure conditions do not provoke different reactions whereas the time delay of about 6 hours does this significantly. This validates the conclusion that 20 ppm styrene does not prolong the reaction or movement time.

A typical memory test also originating from the Wechsler Intelligence Scale is the *Test Digit Span*. In the original paper–pencil version a standardized series of digits is presented verbally and the response also has to be given verbally. The longest sequence of presented digits reproduced correctly – both forward and backward – is the score for the digit span. In the computer-administered version the sequence appears on the screen and the response is given on the keyboard. An adaptive mechanism regulates the presented sequences again forward and backward to find the individual optimum of the longest sequences for the score digit span.

Another often used memory test with higher portions of visual-spatial processing is the *Benton test*, which is also a paper–pencil or computer administered test. More or less complex figures are presented and the subject has to reproduce each one either by drawing or by marking the correct pattern out of similar patterns in a multiple choice response.

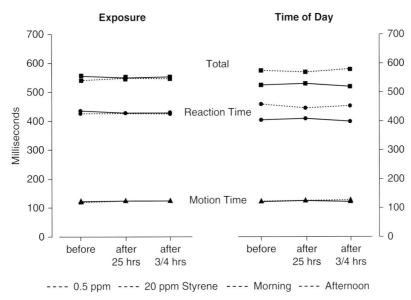

Figure 3.24 *Example for benchmarking of data. Results from a choice reaction test in experiments with two constant styrene exposures combined with two daytime conditions. The curves for exposure do not differ. According to Seeber et al. Toxicol. Lett. 151, 183–92, 2004.*

Measurements of *psychomotor functions* in humans are usually conducted with test batteries for manual dexterity. These include subtests for steadiness (holding a contact pen without contacting the boundary of a hole), line tracing (directing the contact pen on a pattern line without contacting the boundaries), aiming (directing the contact pen quickly over contact plates), tapping (touching contact plates as often as possible in a constant time), and peg board (manipulating pegs in small ports). Speed and precision are measured. The tests are performed subsequently with both hands.

Symptom Measurement and Personality Characteristics Symptom measurements are essential for neurobehavioral studies in humans. Mostly self-administered questionnaires are used with items covering chronic and/or acute neurotoxic effects. For example, EUROQUEST is a common questionnaire that investigates solvent effects. Chronic neurological problems, psychosomatic symptoms, mood lability, memory and concentration problems, tiredness, and sleeping problems as well as acute effects of solvents like irritations and systemic effects are listed. In addition, personality characteristics like environmental sensitivity, trait anxiety, and feeling healthy are asked for in order to measure confounding factors. The experienced symptoms have to be scaled between "seldom/never", "sometimes", "often", and "very often".

Table 3.8 summarizes typical symptoms identified by questionnaires in 28 neurobehavioral studies on solvent effects. The questionnaires were different, thus the concordance of symptoms is limited.

Figure 3.25 demonstrates the association of symptoms both to data on exposure and to a personality trait, particularly, for example, trait anxiety. The symptoms depend predominantly on individual levels of anxiety in a study at low dioxin and furan exposures whereas in the solvent mixtures study they depend additively on existing exposure and trait anxiety.

Table 3.8 Symptoms of occupational solvent exposure reported in 28 studies using no uniform questionnaires.

Symptom	Number of investigations asking for the symptom	Number of significant effects shown
Sleepiness	10	4
Tiredness	10	7
Physical exhaustion	7	5
Mental exhaustion, passivity	12	6
Tension, intolerance of stress	14	5
Headaches	17	6
Nausea	9	3
Dizziness	14	7
Breathing difficulties (including coughing)	8	5
Irritation of eyes, nose, airways	10	5
Feeling of drunkenness, confusion	10	5

Significant effects refer to comparisons between exposed and non-exposed groups independently on the level of solvent exposure.

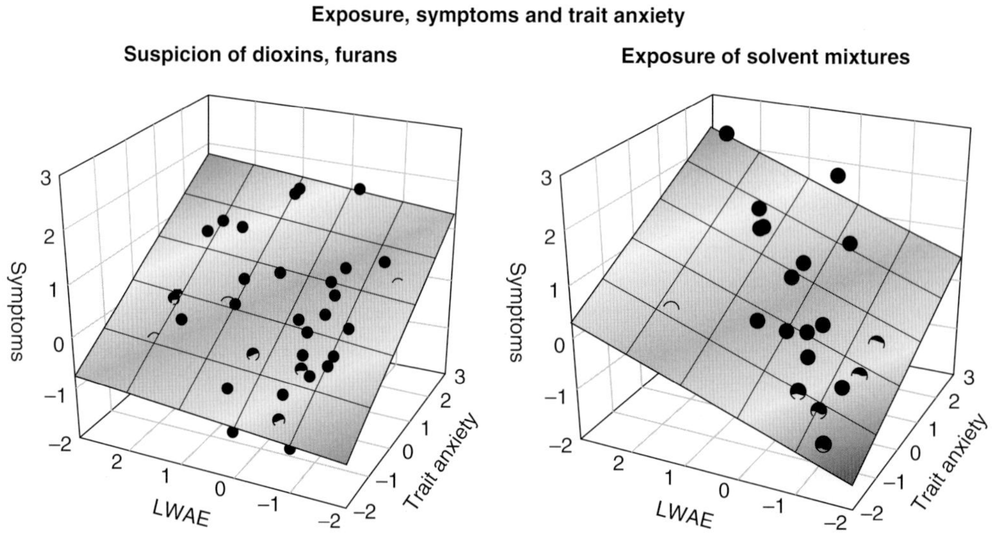

Figure 3.25 Example for additive effects. Standardized symptom scores related to standardized lifetime weighted average exposure and trait anxiety. Left: Results from persons assumed to be exposed to dioxins and furans without biological proof of exposure. Right: Results from exposed workers in paint manufacturing. According to Seeber et al. NeuroToxicology 21, 677–684, 2000.

Sensory Functions Increasingly sensory functions are measured in neurobehavioral studies. In particular, color discrimination and examination are used to evaluate chronic solvent effects.

Impaired *color vision* in the blue–yellow discrimination – classified as acquired dischromatopsia – is predominantly measured by the Lanthony Panel D-15d desaturated test. The D-15d consists of 15 moveable, colored caps chosen to represent perceptually equal steps of hue and to form a natural hue circle. Subjects must arrange the caps in order according to color. Scoring is done by the Color Confusion Index (CCI), which indicates the distance of the individual placements from the perfect cap replacement. In studies on solvent exposures (toluene, styrene and solvent mixtures) increased CCI related to elevated exposures could be shown.

Measurements of *auditory thresholds*, especially of high-frequency hearing loss, are part of epidemiological studies on solvent effects. Diminished auditory thresholds were shown repeatedly to be related to higher levels of solvent exposure and the interaction of solvent, especially toluene, and noise exposure.

Effects on *olfactory functions* are known for numerous chemicals. According to the regeneration cycle of the olfactory system reversibility of changes in the olfactory threshold and discrimination can be expected in most occupational exposures. Measurements of *standing stability* under different conditions that change the feedback system for the function are also included in neurobehavioral studies.

Neurobehavioral Test Batteries Various recent neurobehavioral test batteries have further developed the traditional paper–pencil tests by using fully computerized systems to reduce diversity. The Neurobehavioral Core Test Battery was recommended by a WHO working group. Computer-assisted test batteries such as the Swedish Performance Evaluation System (SPES) and the Neurobehavioral Evaluation System (NES) have mainly been used. The NES has been adapted to newer computer techniques and revised to meet specific demands in epidemiological research.

Table 3.9 lists examples of test applications using paper–pencil tests. The test principles are comparable with new computer-based approaches, and vocabulary tests are regularly included.

The advancements of the Neurobehavioral Evaluation System (NES) lead to the broad application of this computer-administered test battery in recent studies on neurobehavioral effects in humans. New developments in computer-based testing consider (a) standardized training procedures to compensate for missing PC experiences, (b) adaptive computer-based speech communication, and (c) adaptive task difficulties optimized for an individually adjusted level.

Evaluation of Neurobehavioral Effects To evaluate the effect of results of neurobehavioral studies evaluation criteria were developed. They discriminate between "substance-related" and "adverse" effects. Substance-related effects show statistical significant associations between exposure and a neurobehavioral variable. However, statistical significance does not mean that the result is important for health and well-being. Adverse effects implicate significance for health and well-being because an intolerable loss of capability or sense of well-being is observed. Important criteria for the transition from "substance-related" to "adverse" are the extent, number, concordance, severity and irreversibility of effects shown in different studies.

3.5.4 Neurobehavioral Effects in Humans

> **Toxic encephalopathy is a clinical disease pattern associated with chronic and high exposures to solvents, lead, and mercury. For solvent-induced encephalopathy three categories were defined that describe the transition from early effects characterized as reversible slight impairments in well-being and in cognitive and executive functions**

> up to severe mental deficits and multiple complaints in the final stage after long and intensive exposures.
> The differences between the exposure-related effects of solvents, lead and mercury can be hardly documented, whereas similarities are seen regarding attention, memory and psychomotor functions. For mercury a dominance of impaired psychomotor functions could be shown.

Solvents

Neurobehavioral methods have been used to document neurotoxic effects due to occupational and environmental exposures of single solvents and solvent mixtures. For occupational solvent exposures a well-defined disease pattern is *chronic toxic encephalopathy* (also known as painter's syndrome and solvent-induced psycho organic syndrome) with. In its severest form (Type 3) – recently very seldom observed - it is characterized as dementia with a marked global deterioration in intellect and memory, often accompanied by neurological signs or neuroradiological findings and not reversible. The weakest and reversible Type 1 form includes non-specific symptoms such as fatigue, memory impairment, difficulty in concentration, and loss of initiative without objective evidence of neuropsychiatric dysfunction. Between these

Table 3.9 Examples of methods in neurobehavioral test systems and their usage in neurotoxicological studies.

Tests	Number of inclusions in 16 test batteries	Number of applications in 210 studies
Cognitive functions		
Visuospatial design (block design)	11	31
Symbol digit/symbol digit substitute	12	51
Visual pattern processing		
Choice reaction		26
Trail making	3	13
Memory span forward/backward	13	37
Memory scanning		4
Visual retention (Benton)	8	14
Associate learning		13
Eye–hand coordination		4
Visuomotor coordination	6	19
Verbal reasoning		13
Continuous performance/attention	10	6
Executive functions		
Simple reaction time	9	39
Tracking tasks		3
Finger tapping/aiming	5	19
Vocabulary test (pre-morbid intelligence)		10

Numbers indicate the application in test batteries analysed until 1990 and in studies analysed until 1994. Before 1994 traditional paper–pencil tests were mainly used, rather than computer-administered tests.

forms, sustained personality or mood changes (Type 2a) and impairments of intellectual function (Type 2b) with questionable reversibility occur. Substances possibly inducing a chronic toxic encephalopathy are:

(a) the aliphatic hydrocarbons n-hexane and n-heptane
(b) the aromatic hydrocarbons benzene, toluene, styrene, and xylene
(c) the chlorinated aliphatic hydrocarbons dichloromethane, 1,1,1-trichloroethane, trichloroethylene, and tetrachloroethylene
(d) the ketones 2-butanone and 2-hexanone.
(e) the alcohols methanol, ethanol, and 2-methoxyethanol.

So far no substance-specificity of the effects has been identified, mostly because substance-specific neurobehavioral symptoms or performance deficits cannot be defined. This may be because (a) the mechanisms that induce behavioural impairments are unknown, (b) there is human intellectual and emotional capacity to compensate slight specific cognitive or emotional impairments or (c) the methodology is not available to identify distinct types of effects.

Some additional neurobehavioral data for the solvent *toluene* can be provided as a example. For high occupational exposures ($\gg$100 ppm) clear clinical manifestations were described with lists of symptoms experienced during work days (e.g. drunken feeling, headache, dizziness etc.) and after 6 months of exposure (nervousness, loss of consciousness, forgetfulness, headache etc.). Intentionally inhaling toluene vapors (misuse) has been associated with cognitive deficits in attention, memory, visuospational and complex executive functions, motor disorders, cerebellar dysfunctions (ataxic gait, tremor) as well as in brain stem and cranial nerve functions. In such cases hearing impairments and optic atrophy were also reported.

The regional distribution of toluene in the brains of rats and humans showed highest concentrations in regions with more white matter whereas the gray matter of the cerebral cortex and hippocampus had lower concentrations. This seems to reflect the low molecular weight, high lipid solubility, and lack of protein binding capability of toluene.

In the neurobehavioral literature on toluene some data on weak cognitive deficits below 40 ppm were published, but past exposures of the workers were higher by up to a factor of 5. Findings on deficits in color discrimination at this level and hearing loss at about 50 ppm were published. However, these data from cross-sectional studies were checked in a 5-year longitudinal study with repeated measurements. It was shown that current toluene exposures of 26 ppm and lifetime weighted average exposures of 45 ppm did not induce neurobehavioral effects.

Lead

Lead is one of the best-investigated neurotoxicants. It is known to be a toxic agent in the cognitive development of children through environmental exposure as well as impacting on the mental abilities of adults through environmental and occupational exposures.

High occupational risk operations are smelting, welding, and cutting of lead-containing materials, spray painting or scraping of lead paints. Moderate risks are associated with the activities of lead miners, solderers, plumbers, type formers, cable makers, lead glass workers, and automotive repair personal. Sources of environmental exposure to inorganic lead are air, soil, dust, and food. The emission of organic lead compounds (tetraalkyl, tetraethyl and tetramethyl lead) into the atmosphere was signifiacnt before their ban as gasoline additives. At present soil- and water-related exposures and exposure through aquatic organisms reaching the food chain dominate.

The well known neurotoxic effects of lead are:

- impaired development and function of oligodendrocytes
- abnormal myelin formation
- altered neurotransmitter release and receptor density
- abnormal myelin formation, dendritic branching pattern, and neurotrophic factor expression
- disruption of the blood–brain barrier and of thyroid hormone transport
- impaired neuropsychological functioning and lowered IQ.

Young woman living in lead-contaminated housing or who were poisoned with lead themselves as adolescents can pass lead on to a fetus. Strong correlations between maternal and umbilical cord blood lead levels prove the transfer of lead to the unborn child. The lead level in breast milk correlates with the maternal blood lead level. Pre- and perinatal blood lead levels of neonates, along with other factors (reduced body weight, circumference of the head, and length), are associated with disturbances in cognitive development.

Although such disturbances are considered to be the most sensitive endpoint of reproduction toxicology, there are results of prospective studies in independent cohorts with young children which do not provide a concordant pattern of results. These studies include consideration of confounders such as the maternal intellectual status, the socio-economic status of the family, and housing conditions (e.g. the Home Observation for Measurement of the Environment scale). In some studies of about 1200 children the association could not be shown by standardized test batteries for early cognitive and psychomotor development whereas in other studies with 3300 children associations between maternal-induced exposure data and impaired early development were shown up to the fourth year. With increasing age the postnatal individual exposure determines the strength of lead-induced mental impairments. Without lasting exposure other factors determine further development, especially educational conditions, and earlier lead impacts can be compensated.

Critical lead concentration in the pre- or perinatal blood for measurable effects evidence is about 100 µg lead/l blood with effects over 2 years whereas critical blood lead levels due to environmental exposure on later periods of the infantile development are less than 100 µg lead/l blood. The overall analysis of all data indicates that increasing blood levels from 50 to 200 µg lead/l is associated with increasingly impaired functions. An increase of 100 µg lead/l blood is associated with a decrease of up to 3 points in full-scale IQ. A threshold of blood lead levels under which adverse health effects are to be expected could not be derived. The critical discussion of these findings emphasizes the variety of different factors influencing the intellectual development of children, but the congruence of conclusions from different studies is convincing.

Mercury

The toxic effects of exposure to inorganic mercury have been known for centuries whereas the effects of organic mercury compounds only became apparent over the last few centuries. Organic mercury compounds arise after methylation of mercury by microorganisms in sediment, soil, and water, which enter the human food chain through plants and fish. However, the use of mercury in manufacturing products has decreased and a ban on mercury-containing products has reduced the occurrence of mercury as hazards in workplaces and other environments.

Because different rates of absorption and transfer through the blood–brain barrier, various mercury compounds have different neurotoxic potencies.

Acute inhalation of elemental mercury induces inflammation of the upper respiratory tract along with general malaise, headache, vomiting, nausea, fever, and chills leading to metal fume fever syndrome after very high exposure. After some hours neurological effects, including tremor or delirium, occur and persist whereas other symptoms diminish.

Chronic exposure to elemental mercury induces symptoms of "mercurialism", including fatigue, general weakness, loss of appetite, diarrhoea, mood changes, insomnia, and bilateral tremor. In servere cases progressive motor neuropathy with fasciculations of muscle and atrophy occurs associated with distal axonal degeneration and denervation atrophy in muscles.

Chronic exposure to organic mercury through contaminated fish and grain treated with methylmercury leads to decreased intellectual ability, loss of concentration and memory, emotional changes, and depression. In addition, cerebellar signs of ataxia and stumbling gait, incoordination, paresthesias, sensory loss, and other sensory impairments are noted. Prenatal exposure to methylmercury can result in brain damage in children. The neurophysiological effects have been verified by electroencephalography (EEG), somatosensory evoked potentials (SSEP), nerve conduction velocities (NCV), and electromyography (EMG).

In biological monitoring studies 100 µg Hg/g creatinine seems to generate no neuropsychological deficits. Later studies, partly in dental workers with contact to dental amalgam, revealed significant effects in tests measuring attention, perception, reasoning, motor speed and motor precision at exposure levels of about 25 µg Hg/g creatinine. In other studies effects on motor functions but not in cognitive functions at exposure levels of about 60 µg Hg/g creatinine were seen. These findings were supported by the meta-analysis of the available studies. Motor performance tests showed the strongest mercury effects. Less but still significantly affected was the performance shown in memory tests, while the performance in tests for attention was hardly affected at all.

3.5.5 Summary

Variations in human behavior can be observed and measured with standardized methods and deviations from the expected or predicted behavior patterns described. Sources inducing deviations from "normal" are multifarious, external and internal from person to person. Education, age, status in the diurnal rhythm, physical conditions during measurement as well as motivational or emotional states can induce behavioural variations to a measurable extent. In evaluating the neurotoxicity of a chemical the exposure-related variations of behavior have to be compared with variations induced by other conditions that also affect human behavior. In this way inhalation or another form of exposure to a substance can be seen as an assured source inducing significant variations in behaviour. Following this approach, the objective of studies in behavioural toxicology, especially in human neurobehavioral approaches, is to obtain data showing any relationship between exposure parameters and behaviour measurement parameters. The data should also show that confounding conditions of behavioral variations can be controlled statistically or by the study design. Before a toxicological interpretation is given to the test results, the methodical background of the data should always be analysed.

Further Reading

Chang LW and Slikker W (Eds.) Neurotoxicology. Approaches and Methods. Part II. Academic Press, San Diego, New York, 1995.

Chouanière D, Cassitto MG, Spurgeon A, Verdier A, Gilioli R. An international questionnaire to explore neurotoxic symptoms (EUROQUEST). *Environ. Res.* **73**, 70–72 (1997).

Feldman RG. *Occupational and Environmental Neurotoxicology*. Lippincott-Raven Philadelphia, New York, 1999.

Iregren A, Gamberale F, Kjellberg A. SPES: A Psychological Test System to Diagnose Environmental Hazards. *Neurotoxicol. Teratol.* **18** (4), 485–491 (1996).
Letz R. *NES3 user's manual*. Neurobehavioral Systems Inc., Atlanta, MA, 2000.
Lezak MD, *Neuropsychological Assessment*, 3rd edition, HDI Publishers, Oxford, 1995.

3.6 The Skin[1]

Brunhilde Blömeke

> **The skin and the anterior segment of the eye, with their appendages, represent the outermost boundary of an organism. The surface area of the skin is approximately 1.5–2 m^2 and its weight is 3.5–10 kg depending on body weight and size. The skin and the anterior segment of the eye have developed characteristics that enable them to endure direct exposure to the external environment. In this chapter the anterior part of the eye and its toxicology will be dealt with only to the extent necessary to understand its function as a boundary to the surrounding environment as effected by xenobiotics.**

3.6.1 Structure

Skin

> **The skin consists of three major components: epidermis, dermis, and subcutis.**

The epidermis is a stratified squamous epithelium. Its external surface hornifies and forms the stratum corneum (thickness 5–20 microns). Depending on location, age and gender the thickness of the epidermis varies between 80 and 160 microns. Up to four different cell layers can be distinguished depending on the body area. The cell layer above the basement membrane separating epidermis and dermis is the stratum germinativum, where cell proliferation takes place. As the epidermal cells migrate to the surface, forming the stratum granulosum and stratum lucidum, they undergo morphological and physicochemical changes, step by step losing their cell structure.

In the outermost part of the stratum corneum, cell structure is no longer discernible and the only histologically distinguishable feature is the desquamating stratum disjunctum. The turnover time for this process is 4 weeks. Other cell types of the epidermis are melanocytes, Langerhans cells, Merkel cells, and some lymphocytes mostly located in the basal cell layer. The epidermis also contains some nerve endings but no vasculature. The basal lamina separates epidermis and dermis from each other and is formed by extracellular components of both layers.

The thickness of the dermis, which is approximately 0.5–3 mm, cannot be measured accurately because it is contiguous with the subcutaneous tissue, which varies with anatomical site. The dermis consists of an upper papillary and a lower reticular layer not clearly separated. Common cell types occurring in the dermis are histiocytes (active macrophages), mast cells, and fibroblasts producing copious amounts of collagenous and elastic fibers. This structure is embedded in a gelatinous matrix, forming a thick network which provides firmness to the skin. The dermis also possesses a rich vasculature with a high rate of blood flow.

---

[1] This chapter is an adapted and updated version of the original chapter entitled "The Skin" by T.A. Rozman, M. Straube and K.K. Rozman from the 1st edition of this book published in 2008.

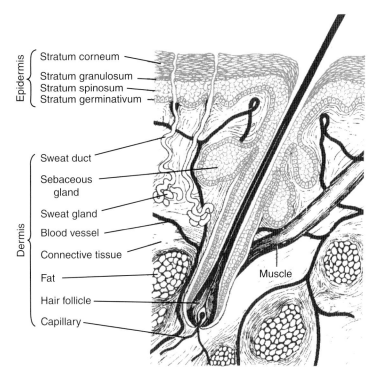

Figure 3.26 Cross-section of the skin. From Cassarett & Doull's Toxicology, 2001 (with permission from McGraw-Hill).

Appendages of the skin, such as hair follicles, sebaceous glands, which secrete oils, apocrine and eccrine glands, which secrete sweat, hair and nails, extend deep into the dermis. A cross-section of the epidermis and dermis is shown in Figure 3.26.

The subcutis consists of loose connective tissue in between lobes of fat cells and contains approximately 10–12% fat cells, sometimes forming a continuous layer of fatty tissue. The septa of the fat islets bring the nerve and blood supply to the dermis. Some collagen and elastic fibers also permeate into the reticular layer.

Anterior Segment of the Eye

The cornea and adjacent parts of the sclera form the external barrier which protects internal ocular structures.

The cornea is a transparent and avascular tissue with an anterior non-keratinizing epithelial cell layer. Below the cornea lies the stroma, which is also completely translucent and is composed of water, collagen, and glycosaminoglycans. At the inner surface an endothelial cell layer can be found. Descemet's membrane separates this endothelial cell layer from the substantia propria.

Anterior and posterior chambers containing the transparent aqueous humor are separated by the iris and delimited towards deeper eye structures by the lens and ciliary body. The iris itself is a highly vascularized tissue consisting mainly of connective tissue, containing a large number of pigmented cells. In addition, the mucosa is another non-keratinizing area of the human skin.

3.6.2 Function

Skin

> **The skin protects the organism from mechanical, chemical, and immunological insults as well as against dehydration, heat, and excessive light exposure.**

The surface of the skin (i.e. the stratum corneum) is covered by a lipid layer, which is produced by the sebaceous glands. Its pH of 5.7 is achieved by the secretion of sweat by sweat glands and other acidic components such as free fatty acids produced by resident flora. This combination is particularly suitable as defense against attacks by bacteria but also against some xenobiotics. Langerhans cells provide support for immune responses whereas melanocytes can diminish potential light-induced damage by producing UV-absorbing melanin.

Thermoregulation by sweating and diffusion of water through skin layers to the outside (perspiratio insensibilis) are both life-supporting functions of the skin. These functions can be carried out effectively only because the dermis and subcutis are perfused by quantities of blood 20 times higher than needed to provide nutrients.

Under the influence of light, 7-dehydrocholesterol is metabolized to cholecalciferol in the skin, which is the precursor of the active metabolites of vitamin D_3 in both liver and kidney required for calcium and phosphate absorption, deposition and mobilization. Another important function of the skin is communication with the external world. This occurs through organelles of different sensory qualities such as sensation of heat, cold, touch, and pain.

Subcutaneous fat serves as insulation against overt loss of heat. At the same time, it is also the site of deposition and storage of persistent lipophilic xenobiotics.

In addition to thymus and lymph nodes, the skin is also an important organ for the immune response and contains cells of the innate and adaptive immune system. The highest number of immune active cells can be found in the dermis, followed by a lower number in the epidermis.

Anterior Segment of the Eye

> **Accessory glands on the rim and posterior surface of the lids together with the lacrimal glands produce the lacrimal film.**

The cornea as well as the bulbar and palpebral conjunctiva is bathed in a continuously produced tear film consisting of three domains: a hydrophobic lipid layer, an aqueous layer and a mucoid layer, the latter acting as interface between the hydrophilic aqueous layer and the hydrophobic epithelial cell membranes of the cornea. Eyelids and lacrimal glands are a very efficient means of defense to prevent foreign bodies from injuring the cornea. If xenobiotics splash into the eye, lacrimation immediately dilutes and rinses the corneal sac and the cornea.

Dermal Absorption

> **Penetration through the skin is governed primarily by passive diffusion as defined by Fick's law. Primary determinants are the thickness of the stratum corneum, its moisture content and the temperature of the skin as well as the molecular weight and lipid solubility of the compound in question; in addition the properties of the vehicle are also important. Generally, amphiphilic molecules with low molecular weight (MW) penetrate the skin most readily whereas for large hydrophilic molecules it is almost completely impenetrable.**

The very thin stratum corneum consists of only a few layers of cornified, flattened, dead cells that are the actual rate-determining barriers in transdermal penetration. The fairly homogenous horny layer consists mainly of keratin and sphingolipids, in which hydrophilic and hydrophobic layers alternate. This unique structure and composition are ideally suited to prevent polar, ionized, and very apolar substances from entering the organism. The horny layer and subsequently the epidermis are most readily penetrated by compounds of mixed properties, i.e. having some polar as well as apolar characteristics, e.g. many suitably substituted corticosteroids.

MW is another important property, and is inversely related to penetration. Chemicals having MWs above 6000–7000 are generally unable to penetrate the skin.

Once a molecule crosses the horny layer diffusion takes place at a higher rate. The situation is comparable to a dam and a reservoir behind it, the stratum corneum playing the role of the reservoir and the epidermis playing the role of the outflow.

In addition to the properties of drugs and other chemicals, the condition of the skin itself is also of importance for transdermal penetration. Rate-limiting conditions are skin temperature and degree of hydration. The degree of hydration is of particular importance for enhanced penetration. Occlusive bandages greatly increase the moisture content of the stratum corneum, thereby enhancing transdermal penetration, often by a factor of 10 or more.

Partitioning of compounds and their vehicles (or solvents) into the lipids of the horny layer is a critical factor determining diffusion through the stratum corneum and epidermis (octanol/water partitioning ratio, K_{ow}). This effect is often used therapeutically for enhanced and controlled drug delivery through the skin (e.g. when propylene glycol is the vehicle). Some detergents and solvents, e.g. DMSO and DMF, render the horny layer more porous and penetrable by dissolving and removing sphingolipids from the stratum corneum.

The importance of the transdermal pathway as a potential route of entrance into an organism to cause toxicity is exemplified by monochloroacetic acid. It is an intermediate in the synthesis of numerous detergents and many other chemicals used in large quantities by industry. Monochloroacetic acid can cause lethal poisoning if as little as 6% of the skin is exposed for a short period of time.

Ocular Absorption

> **The lacrimal film and the different layers of the anterior segment of the eye are equally efficient barriers against direct entry of xenobiotics to the inner structures of the eye.**

If penetration through the cornea does occur, clearance into blood can occur via aqueous flow through the canal of Schlemm. Molecules that can penetrate the blood/central nervous system barrier may be secreted in the process of aqueous humor production.

Systemic absorption can also occur through the conjunctiva, in the nasolacrimal duct, and further down in the nose and epipharynx without exhibiting a first-pass effect in the liver. Absorbed compounds can be distributed via blood, partitioning into cells or interstitial fluid or sequestered in organs. For retinal toxicity to occur, systemic absorption is much more important than penetration through the cornea, aqueous humor, lens, ciliary body or vitreous body.

Excretory Function of the Skin

> **Cutaneous appendages (hair follicles, sebaceous glands, eccrine and apocrine sweat glands, hair, and nails) serve primarily excretory functions for compounds that cannot be eliminated from the organism otherwise.**

The excretory function of sweat glands for inorganic chemicals is well established. Sebaceous glands can reduce the body burden of lipophilic xenobiotics which are not otherwise amenable to biodegradation. Chemicals can also translocate into the epidermis by passive diffusion and are externalized in the process of cornification and desquamation. Other appendages, like hair and nails, may also serve as sites of accumulation and storage. For example, there are metals, especially heavy metals like arsenic, lead, cadmium, mercury, and thallium, that may be deposited in hair or nails in some instances, leading to topical toxicity in the form of loss of hair or nails. Some toxicants are deposited and stored in the dermis. The hair follicle, unlike the appendages cited above, provides an avenue for transdermal absorption. Thus, experimental animals with dense fur absorb material through the skin up to three or four times more rapidly than corresponding nude animals because the epidermal cells in the hair follicle do not cornify, i.e. do not generate, a horny layer. However, it is not possible to accurately predict the rate at which dermal absorption will take place in furry animals because of the large number of hair follicles, and as a result we cannot extrapolate from these data to predict dermal absorption in nude animals or in humans.

Metabolism in the Skin

> **The skin contains both Phase I and Phase II enzymes active in the metabolism of xenobiotic compounds.**

The skin contains most enzymes needed for the metabolism of glucose, lipids, and proteins, as well as for the synthesis of glycogen, lipids, and proteins. The epidermis appears to be the most active layer. The papilla of the hair follicle is one of the most metabolically active tissues in the body. The skin is a constantly regenerating tissue with a turnover rate of about 4 weeks. It requires a constant energy supply provided by ATP derived from different sources.

Xenobiotic-metabolizing enzyme activities are relatively low in the skin compared to other major organs, e.g. the liver. Under normal conditions approximately 2% of total enzymatic activity of the body can be found in skin. However, it has been shown that enzymatic activity in the fully induced state can be as high as 20% or more of the whole body, e.g. after exposure to TCDD, a well-known inducer of various cytochrome P (CYP) isozymes. Crude coal tar is another example of an inducer of the CYP system which is widely used in dermatological therapy. Phase II enzyme activity has been demonstrated in the epidermis. The presence of glucuronidase has also been documented in the stratum corneum, therefore many absorbed chemicals can mostly be metabolized before reaching the circulation. Hormones such as testosterone and estrone, and drugs such as corticosteroids are metabolized in the epidermis and only approximately two-thirds of a dose applied topically reaches the systemic circulation unchanged. The structures of the anterior segment of the eye also contain all major Phase I and Phase II enzymes.

Typical examples of diseases caused by enzyme activity or lack of it in the skin are male pattern baldness and xeroderma pigmentosum. In the first, increased 5-alpha reductase production in keratinocytes of genetically predisposed individuals leads to enhanced transformation of testosterone to Dihydrotestosterone (DHT) and as a consequence hair loss. Enzyme deficiency in xeroderma pigmentosum will be dealt with in *Skin Tumors*, this chapter.

3.6.3 Toxicology of the Skin and the Anterior Segment of the Eye

> **Because of its direct exposure to the environment, the skin is a readily available target for the toxic effects of xenobiotic chemicals, ultraviolet light, and extreme temperatures.**

In most industrialized countries half or more of reported occupational diseases are related to the skin. Direct injury in the form of burns or corrosions can be caused by alkali and acids as well as oxidizing agents. Most xenobiotics, however, are less aggressive, but penetration through the stratum corneum is a prerequisite for the exertion of toxicity. Many xenobiotics may also cause immune reactions.

Acidic and Alkaline Corrosion

> **Alkali, acids, some salts of metals, oxidizing agents, and different organic compounds are in many instances aggressive enough to cause cell death and even complete destruction of skin and/or cornea/conjunctivae, respectively.**

Injuries at times are severe enough that *restitutio ad integrum* [complete recovery] is not possible. Unlike mostly complete recoverable contact dermatitis, in such cases epithelial and other cells and cell layers will be replaced by scar formation. This is particularly damaging to the anterior segment of the eye, especially to the cornea. In severe cases the cornea becomes opaque due to vascularization and is completely impenetrable for light. Acids precipitate intra- and extracellular proteins, creating a barrier that prevents further penetration into deeper structures. Concentrated sulfuric acid denatures protein through dehydration and the development of heat. Alkali have a different mode of action with even more damaging consequences. They cause so-called colliquation necrosis by hydrolyzing peptide bonds of proteins, which can result in long-lasting and deepening lysis of tissues. Therefore, alkaline injury to the skin or eye may look less damaging at first than acid-based injury but delayed effects may lead to the development of a deep, poorly healing ulcer. Only immediate irrigation with copious amounts of water may diminish or prevent further destruction of the respective organs. Table 3.10 lists some compounds that have corrosive effects on the skin and eyes.

Irritation and Contact Dermatitis

> **Acute primary irritation is defined as a reversible inflammatory response of normal skin or the eye to a single topical injury by an agent of chemical, biological or physical origin. The mechanism provided is not attributed to an immunological response.**

Contact of an irritant with the skin or the eye resulting in reversible erythema and edema due to extravasation termed acute primary irritation. Patch testing is a common diagnostic procedure to reveal the causative agent in both irritant and allergic contact dermatitis by applying the suspected compound or, if unknown, a set of common allergens to the back of a patient.

Table 3.10 Compounds with corrosive effects on skin and eye.

Metals	Dichloromercury, pottasium dichromate
Acids	Hydrochloric acid, sulfuric acid chloroacetic acid, formic acid, oxalic acid
Alkalis	Sodium hydroxide, potassium hydroxide, ammonia
Phenols	Cresol, carbolic acid
Solvents	Trichloroethylene

Table 3.11 A few common contact irritants.

Poison ivy, poison oak
Detergents
Solvents
Adhesives
Soaps
Bleaches
Pharmaceuticals
Many other chemicals

In the evaluation of a potential irritant the bioassay usually involves an albino rabbit in which a comparison is made of the effects of applying the material to either intact or abraded skin. The skin is read after 24 and 72 hours and the degree of erythema and edema is recorded on a scale of 1 to 4. Testing of cumulative irritation (repeated application) may reveal the irritant potential of compounds not detectable in an acute test.

When measuring ocular irritancy in the eye of the albino rabbit, the test compound is administered to one eye of the animal, with the other eye serving as control.

Scores at 1, 24, 48 and 72 hours are assigned on the basis of damage inflicted upon the cornea, conjunctivae, and iris. The following criteria are applied: degree of opacity of the cornea, hyperemia of conjunctivae and iris, light reaction of the iris, and chemosis (swelling) of the conjunctivae.

Acute contact dermatitis arises from exposure to a primary irritant, such as those shown in Table 3.11. Old or damaged skin is particularly prone to contact irritation, which is characterized by intense redness, tenderness, heat, and swelling. In severe cases, oozing, blistering, and even necrosis may be observed. If exposure becomes chronic, lichenification (thickened skin), itching, and scaling might become the more predominant symptoms. At times irritant contact dermatitis might not be distinguishable from allergic contact dermatitis.

Allergic Contact Dermatitis

In cases of allergic contact dermatitis, exposure to a few molecules can produce an intense skin reaction. The underlying pathomechanism is a cell-mediated type IV hypersensitivity disorder.

The best-known example of this skin condition is chronic eczema suffered by masons working with cement treated with potassium dichromate, diluted alkali or organic solvents.

Allergic contact dermatitis is initiated by prior exposure to a small reactive molecule, called the hapten, which penetrates the stratum corneum and binds covalently to a carrier protein. The complex, termed the complete antigen, may then be incorporated via pinocytosis in the stratum spinosum into dendritic phagocytes called Langerhans cells for processing. Langerhans cells subsequently migrate to regional lymph nodes to present the processed antigen to the immune system in the lymph nodes. A complex interplay of cellular and humoral factors leads to the formation of a large number of specifically sensitized T-lymphocytes (skin sensitization). Renewed contact with the allergen at any site of the skin triggers an allergic reaction. There is ample evidence for the role of genetic predisposition in contact sensitization. Moreover, some haptens must be metabolically activated by Phase I and/or Phase II enzymes of the epidermis before forming the

Table 3.12 Groups of compounds that frequently give rise to allergic contact dermatitis.

Antiseptics and preservatives	Parabens
	Formalin
	Mercury compounds
	Chlorohexidine
	Dichlorophen
Dyes	Azo-compounds
	p-Phenylenediamine
Metal salts of	Nickel
	Chromium
	Cobalt
	Mercury
Antibiotics	Bacitracin
	Sulfonamides
	Aminoglycosides
Other therapeutic agents	Phenothiazines
	Benzocaine
	Idoxuridine
	Fluorouracil

hapten–carrier complex (for further details see Chapter 3.9). Symptoms of acute and chronic allergic contact dermatitis resemble those of contact dermatitis, the minute amount of substance needed to elicit the skin reaction being the only distinguishing feature. The number of recorded cases of allergic contact dermatitis has increased worldwide, mostly in industrialized countries, and can be provoked by a very large number of agents of chemical, plant or animal origin (Table 3.12).

Table 3.12 displays the most potent sensitizers. However, exposure to thousands of other compounds, including fragrances (see Chapter 6.9), may also lead to allergic contact dermatitis. In many instances the sensitized immune system reacts to substances that are similar, but not identical, to the contact allergen. Such cross-reactivity is called cross-sensitivity.

Toxic Epidermal Necrolysis (Lyell Syndrome)

> **Toxic epidermal necrolysis (TEN) is a severe, in some instances life-threatening, usually drug-induced immunological response resulting in detachment and sloughing of significant amounts of skin, mainly mucosae.**

TEN, which mostly occurs after oral ingestion, is characterized by severe bullous eruptions on skin and mucous membranes, fever, malaise, conjunctivitis, and general redness of the entire skin. Detachment of epidermis and mucous membranes from adjacent tissues causes skin to slough off in large pieces, leading to infections and causing life-threatening loss of body fluids. Mortality may range from 15% to 40%. Most frequently, it occurs after drug treatment when, in genetically susceptible individuals, the drugs or their metabolites accumulate in skin, resulting in an immune response resembling graft-versus-host disease. It has been proposed that T-lymphocytes and macrophages induce an inflammatory response leading to significant apoptosis of epidermal cells. The disease seems to be closely related to severe erythema multiforme and Stevens–Johnson syndrome.

Table 3.13 Phototoxic agents.

Psoralens
Polycyclic aromatic hydrocarbons
Tetracyclines
Sulfonamides
Chlorpromazine
Non-steroidal anti-inflammatory drugs (NSAIDs)
Porphyrins

Phototoxicity

Phototoxic reactions resemble severe sun burn (acute dermatitis). The distinguishing feature is a very rapid appearance of redness and blisters, sometimes within minutes of exposure to light.

The most frequently involved wavelengths of light in these disorders is UVA (320–400 nm) though UVB in some cases may also be effective. The endogenous and exogenous chemicals listed in Table 3.13 absorb UV light readily and cause phototoxic responses in the skin.

There are two mechanisms known to initiate phototoxicity. Some compounds, like chlorpromazine and tetracyclines, are excited to a higher energy state by the absorption of light and form cytotoxic free radicals, which can cause cell death. It is possible to take advantage of the light activation process for therapeutic purposes. For example, one can administer porphyrins, which can accumulate in neoplasms. When the tissue is irradiated by UV light of optimized wavelength the activation of the porphyrins results in detachment of tumor cells. Another mechanism of action in phototoxicity is characteristic of psoralens. After topical or systemic absorption, they intercalate with DNA, resulting in a covalently bound adduct between psoralens and pyrimidine bases. The resulting reaction products inhibit synthesis and repair of DNA, thereby diminishing cell proliferation. Thus, administration of 8-methoxypsoralen, followed by exposure to UVA light, can be used in the treatment of hyperproliferative conditions like psoriasis, eczemas, and cutaneous T-cell lymphomas.

Photoallergy

In contrast to phototoxicity, photoallergy is a true allergic reaction in that minute quantities of a photoallergen causes a type IV delayed hypersensitivity reaction.

Whereas in phototoxicity the activation of chemicals in skin cells via the absorption of UV light leads to reactive metabolites which cause cell death, in photoallergy the active metabolite acts as a hapten, which reacts with cell proteins to form photo antigens and generate allergic responses much like allergic contact dermatitis. Systemic treatment may lead to phototoxicity and in some instances also to photoallergy, but photoallergic agents usually act following direct application to the skin and exposure to UVA light (especially 310–340 nm). The rate of onset of photoallergy is slower than phototoxicity because it is a multistep process.

Table 3.14 A few typical photoallergens.

Chlorhexidine
Hexachlorophene
Chlorpromazine
Promethazine
6-Methylcumarin
Musk ambrette
Salicylates
Sulfanilamides
Thiourea
Triclosan
Topical antibiotics

Furthermore, photoallergy in general can also be caused by oral medication such as chlorpromazine and antidiabetics of the sulfonylurea type. Some photoallergens are listed in Table 3.14.

The most predictive animal model for photoallergy is the guinea pig. To distinguish between a phototoxic and a photoallergic response, testing is done on the backs of guinea pigs or human subjects by applying the test compounds in two symmetric rows. One side is uncovered 24 hours later and is then irradiated by UV light.

Readings are taken immediately after irradiation as well as 24, 48 and 72 hours later.

Allergic reaction on the irradiated side only is proof of true photoallergy.

Urticaria

> **The appearance of urticaria is comparable to a wheal or hive. It is a more or less circular red, spongy lesion caused by hyperemia and local edema. The underlying pathomechanism can be immune- or nonimmune mediated.**

Allergic urticaria is a type I IgE-mediated hypersensitivity disorder. The severity may vary considerably from a few hives with minimal redness and edema to severe, generalized forms. Other type I allergic diseases are rhinoconjuntivitis, bronchial asthma, and insect bites with or without urticaria. Anaphylaxis is the most severe and life-threatening condition attributed to a type I IgE-mediated hypersensitivity reaction. It has been described occasionally in the context of exposure to latex.

Millions of healthcare workers are exposed occupationally every day to latex-containing gloves with sensitization as a frequent consequence.

A common feature of non-allergic urticaria is the release of histamine and vasoactive peptides from mast cells as a result of a large number of physical, mechanical, and chemical causes. The urticarial lesion develops usually within minutes accompanied by itching, tingling, and burning of varying intensity. Non-allergic causes of urticaria include such diverse compounds as aspirin, azo dyes, benzoates and toxins of plant or animal origin. Quite common physical causes are cold or warm temperatures, as in cholinergic urticaria. In general, most recorded cases of acute urticaria lasting less than 6 weeks are the result of an allergic trigger; chronic urticaria lasting longer than 6 weeks is rarely due to an allergy.

Loss of Hair

> The hair follicle is a metabolically highly active tissue. Its proliferative activity is matched only by the bone marrow and some parts of the gastrointestinal tract.

Every hair follicle develops through three stages before maturation and shedding occur. These are growth phases collectively called anagene: the germinative phase, catagen and telogen. Hair loss occurs following the use of chemotherapeutic drugs such as antimetabolites or alkylating agents because proliferation is inhibited by these compounds; heavy metals, especially thallium can cause similar symptoms (Table 3.15). A large number of unrelated drugs and other chemicals may impair the mitotic activity of the hair matrix and cause its reduction or complete abolishment. Hair loss follows 2 weeks later and new growth does not commence for 2–4 months after cessation of treatment or other forms of exposure. Hair loss affects rapidly growing areas (scalp and beard) more than those of low growth rate (eye brows). Severe loss of hair is relatively rare in environmental toxicology and slight loss of hair can pretty much be reversed.

Accidental poisoning with the rat poison thallium sulfate may occur in households and lead to complete depilation. Large-scale poisonings in the industrial world resulting in hair loss have not been reported. However, many compounds, most of which are drugs, are suspected of causing loss of hair, usually without conclusive proof. One of the reasons for this may be that taking reliable hair counts is quite difficult because damaged hair broken off at the surface may be mistaken for shedding instead of toxicity to the matrix.

Chloracne (Halogen Acne)

> The label chloracne originates from the first observation of xenobiotic-induced acne to octa chloronaphthalene more than 100 years ago. The term halogen acne is more appropriate since bromine- and also to some extent iodine-containing compounds have been recognized as being comedogenic.

The basic lesion of chloracne is the comedone consisting of a dilated hair follicle containing surplus amounts of squamae, sebum, and bacteria. (corynebacteriumacnes). If open, the surface is usually dark due to pigment containing cell debris. If closed, it has a more yellowish appearance. Resident anaerobic bacteria present in everybody's skin find in such an environment plenty of nourishment and start multiplying.

At an early stage, excretory ducts of oil glands become thickened. If outflow is blocked, pressure builds up until the blocked hair follicle bursts and empties its contents into the surrounding

Table 3.15 Some agents causing hair loss.

Antimetabolites
Alkylating agents
Alkaloids like colchicine
Dixyrazine
Oral contraceptives
Anticoagulants
Thallium and its salts

Table 3.16 Some of the most potent chloracne-inducing agents.

Halogenated polycyclic aromatic hydrocarbons
Tar products
Lubricating oils
Plant oils in cosmetics

dermis. Propionibacterium acnes decompose sebum into free fatty acids, which are highly inflammatory. The result is an inflammatory lesion of various intensities, ranging from a red infiltrated papule to pustules. In severe cases cysts and tunnel systems may be formed in the dermis. Face, neck, and trunk are sites of strong proclivity. In severe poisoning, the whole body surface may be affected. Table 3.16 lists some compounds capable of inducing chloracne.

Polycyclic halogenated aromatic hydrocarbons such as dioxins, biphenyls, dibenzofurans, and azobenzenes usually have long half-lives and tend to accumulate over time. Therefore, typical acneiforme lesions may persist as long as 30 years after first appearance. The exact mechanism of action is not known but induction of hyperkeratinization and probably excess formation of sebum in the hair follicle is a common feature of all cases of chloracne.

In systemic poisoning with halogenated aromatic compounds severe conjunctivitis and hyperkeratosis in the Meibomian glands, which results in squamous cysts, may accompany the clinical picture. Topical agents with shorter half-lives without tendency towards accumulation cause less severe and less persistent lesions, with clearance (remission) occurring as early as within several weeks. The detailed formation mechanism of chloracne is still insufficiently studied. More recent information indicates that sebaceous gland atrophy might be a major feature of dioxin toxicity of the skin. This may be caused by a repression of lipid metabolism leading to accumulation of lipids in the sebaceous glands.

Bromism is another skin decease in which patients develop a skin condition called bromoderma, which is characterized by cuneiform papular eruptions in the face and on the hands. Until about 50 years ago it was a well-known disease entity because of the large number of bromide-containing drugs used in psychiatry. Recently, many cases of intoxication have been reported in workers exposed to solvents like n-propyl bromide and isopropyl bromide with manifestations of neurotoxicity and skin lesions. It was not clear how a compound with a half-life of 20 minutes to an hour could cause such severe toxicity until it was determined that bromide was the proximate toxicant and has a half-life of 12 days, having been liberated via a glutathione-mediated reaction. It is interesting to note that neurotoxicity and dermal toxicity usually occur together, most likely because both tissues are of ectodermal origin.

Disturbances of Pigmentation

> **Hyper- and hypopigmentation can be specific or non-specific responses of the skin to chemical, physical, and mechanical injury.**

Both endogenous and exogenous causes may change the pigmentation of the skin. Table 3.17 shows typical agents causing hyper- or hypopigmentation.

Endogenous hyperpigmentation is mostly due to increased melanin production and deposition in the basal cell layer or in tumor cells of epithelial origin. The formation and deposition of

Table 3.17 Agents causing hyper- or hypopigmentation.

I. Hyperpigmentation due to endogenous causes
  Ultraviolet light exposure
  Post-inflammatory changes (melanin and/or hemosiderin depositions)
  Hypoadrenalism
  Internal malignancy (acanthosis nigricans)
  Primary acquired melanosis (conjunctiva)
Hyperpigmentation due to chemicals
  Coal tar
  Anthracene
  Picric acid
  Psoralens
  Hydroquinone (high dose)
  Heavy metal salts of Hg, Ag, Bi, Pb, Au and As
  Beta-carotene
  Canthaxanthin (eye/skin)
Hyperpigmentation due to drugs
  Chloroquine
  Amiodarone
  Bleomycin
  Zidovudine
  Minocycline

II. Hypopigmentation due to endogenous causes
  Post inflammatory loss of pigment
  Vitiligo
Hypopigmentation due to chemicals (leukoderma)
  Hydroquinone (low dose)
  Benzyl, ethyl and methyl ethers of hydroquinone
  p-Butylphenols
  Mercaptoamines
  Phenolic germicides
  p-Butylcatechols
  Butylated hydroxytoluene

Modified from Table 9 in Casarett & Doull's Toxicology, 2001.

hemosiderin in the upper dermis or in poorly healing wounds is another mechanism of hyperpigmentation generation.

Chemically induced hyperpigmentation usually occurs as a result of accumulation of deposition of particles in different cellular elements and appendages of the skin or in mucous membranes. The epidermis most often remains unaffected, especially in cases of deposits of heavy metals. The anatomical site of deposition varies from compound to compound. Deposition of lead or mercury results in grayish-black or bluish discoloration of the gingiva. Silver depositions confer a bluish discoloration upon the skin and conjunctivae (argyrosis) in sun-exposed areas. Interestingly, metal salt deposition in skin is often accompanied and enhanced by stimulation of melanin production.

Loss of pigmentation in vitiligo is a disturbance of melanin production due to genetically determined enzyme defects of the tyrosine pathway. It is therefore not surprising that analogs of tyrosine (Table 3.13) are the most common compounds which cause depigmentation of the skin, leading to the well-known disease entity termed leukoderma. Most potent are substances with an alkyl group in the para position, e.g. methyl ether of hydroquinone (MBEH) or *p-tert*-butylphenol.

MBEH was the first compound to be recognized 50 years ago as the cause of leukoderma, a skin condition characterized by complete loss of pigmentation at the site of contact. Another common property of these compounds is their ability to form semiquinone radicals and thus to induce lipid peroxidation.

Skin Tumors

> **Most of the benign or malignant neoplasms of skin originate in the epidermis. Benign skin tumors such as papillomas, warts, fibromas, and hemangiomas are not life-threatening.**

In skin carcinogenesis, three types of malignant neoplasms are important: basal cell carcinoma, squamous cell carcinoma and melanoma. There are many different causes of skin cancer, ranging from excess exposure to sunlight to different sources of radioactive irradiation, including X-rays, as well as exposure to chemicals such as polycyclic aromatic hydrocarbons (PAHs) and arsenic. Two fundamental mechanisms appear to be operational, exemplified by PAHs and arsenic. PAHs require metabolic activation, via CYP isozymes, to epoxides, which form DNA adducts having long half-lives. Therefore, the accumulated adducts induce mutations in critical genes and/or inhibit the p53 tumor suppressor protein. Chronic irritation depletes the repair capacity of the skin, facilitating growth of malignant cells. UV light also has immunosuppressive effects that may contribute to increased survival of malignant cells by weakening defense mechanisms.

Skin cancer incidence is highest in the tropics in individuals with fair complexions at sites which are exposed to UV irradiation for long periods of time. *Xeroderma pigmentosum* best exemplifies the mechanism of action of sunlight-induced cancer.

In this disease repair of pyrimidine dimers is genetically deficient in affected individuals and they develop squamous cell carcinomas after minimal exposure to UV light. Tumorous melanocytes (melanoma) show an aggressive growth for the most part.

Scrotal squamous cell cancer in chimney sweeps was the first occupational cancer recognized as a result of chronic exposure to soot. Lip cancer of heavy pipe smokers is another example of chemically induced skin cancer. Coal tar, pitch and creosote, rich in polycyclic aromatic hydrocarbons, are also known skin carcinogens.

Coal tar in combination with UVB light used for the treatment of severe forms of psoriasis carries an additional risk of skin cancer.

Arsenic is another chemical carcinogen that causes skin tumors. High doses result in precancerous hyperkeratosis followed by the development of squamous cell carcinoma. Arsenic in drinking water has been shown to cause squamous cell carcinomas of the skin in populations living in areas with high soil content of arsenic. Unlike PAHs arsenic is not mutagenic and, hence, it is considered an epigenetic carcinogen, also called a promoter. In contrast to genotoxic agents the continuous presence of epigenetic carcinogens is required at the site of action because promotion is a reversible effect and irreversibility (cancer) does not occur until the stage of progression is reached.

3.6.4 Summary

Skin and the anterior segment of the eye protect the organism efficiently against toxic insults from chemical, biological, mechanical, and physical injuries. The stratum corneum and the tear film, if intact, are impenetrable to many xenobiotics. When this protective capacity is exceeded or destroyed, absorption occurs with the consequence of potential damage to adjacent cells and/or distant structures. Although the epidermis and cornea are not vascularized, the next layers beneath them have an extensive blood supply facilitating not only thermoregulation in the form of *perspiratio insensibilis* [insensible perspiration] but also quick removal of absorbed molecules and transport of substances from the circulatory system to the skin. Furthermore, the skin is a highly active and immunocompetent organ. Incidental to the lymphatic glands, the Langerhans cells of the epidermis and dermal dendritic cells play an important role in antigen presentation, resulting in an immune response to invading organisms or chemicals.

In addition, the skin and its appendages can have an excretory function, which in some instances is the only way to rid the organism of persistent xenobiotics, which are not readily metabolized.

Toxic insults to the skin and/or eye by chemical, physical, biological or mechanical agents cause first irritation, followed by an acute local inflammatory response when protective barriers are penetrated or destroyed. When the invader is a protein or hapten–protein complex, an allergic reaction at the local site or at a distant site may occur.

Although the skin is the most important organ for photochemical reactions, sunlight is the most potent environmental insult to the skin (especially UV). Historically, it was believed that exposure to sunlight is beneficial for the health of an individual. Only during the last century has it been recognized that excessive exposure to sunlight has detrimental effects ranging from the destruction of elastic elements of the skin to carcinogenesis. In some cases, a combination of light and the presence of a chemical in the epidermis are required to bring about either a phototoxic or photoallergic reaction, depending on the underlying pathology.

Chronic exposure to irritants or allergens produces chronic inflammation accompanied by tissue proliferation, scaling, and itching. The appendages of the skin are particularly prone to toxic responses, affecting mainly hair follicles and sebaceous glands. Chronic irritation in combination with aggravating factors such as chemical or physical carcinogens is the most common pathway for skin carcinogenesis. Testing of compounds for potential skin irritation involves application of material into the conjunctival sac of test animals. This procedure is standardized and if performed in at least two species has a reasonably good predictive value for humans. Conversely, for research into cancerogenous agents and influences regarding the skin, animal testing is inappropriate due to different circumstances, e.g. limited activation capacity. Skin and eyes are potential entry portals for xenobiotics and should be always taken into consideration when conducting risk assessments for chemicals.

Further Reading

S. Amisten, R. Neville, M. Hawkes, S.J. Persaud, F. Karpe, A. Salehi. An atlas of G-protein coupled receptor expression and function in human subcutaneous adipose tissue. *Pharmacol. Therap.* **146**: 61–93, 2015.

R.L. Bronough, H.I. Maibach. *Percuteneous absorption*. Marcel Decker, New York, 1989.

R.A. Clark, B. Chong, N. Mirchandani, N.K. Brinster, K. Yamanaka, R.K. Dowgiert T.S. Kupper. The vast majority of CLA+ T cells are resident in normal skin. *J. Immunol.* **176**: 4431–4439, 2006.

K. Fraser, L. Robertson. Chronic urticaria and autoimmunity. *Skin Therap. Lett.* (Review) **18**: 5–9, 2013.

U. Gundert-Remy, U. Bernauer, B. Blömeke, Döring, E. Fabian, C. Goebel, S. Hessel, C. Jäckh, A. Lampen, F. Oesch, E. Petzinger, W. Völkel, P.H. Roos. Extrahepatic metabolism at the body's internal-external interfaces. *Drug Metab. Rev.* **46**: 291–324, 2014.

A. Kubo, K. Nagao, M. Yokouchi, H. Sasaki, M. Amagai. External antigen uptake by Langerhans cells with reorganization of epidermal tight junction barriers. *J. Exper. Med.* **206**: 2937–2946, 2009.

A. Patzelt, J. Lademann. Chemical methods in penetration enhancement: Drug manipulation strategies and vehicle effects. In: *Percutaneous Penetration Enhancers*, N. Dragicevic-Curic, H.I. Maibach (Eds.), Springer-Verlag, Berlin, Heidelberg, 2015, 43–51.

J.H. Saurat, G. Kaya, N. Saxer-Sekulic, B. Pardo, M. Becker et al. The cutaneous lesions of dioxin exposure: lessons from the poisoning of Victor Yushchenko. *Toxicol. Sci.* **125**: 310–317, 2012.

T.J. Slaga, A.J.P. Klein-Szanto, R.K. Boutwell, D.E. Stevenson, H.L. Spitzer. Skin carcinogenesis, Mechanism and human relevance. *Progr. Clin. Biolog. Res.* **298**, 1989.

H. Thai, K.P. Wilhelm, H.I. Maibach (Eds.). *Marzulli and Maibach's Dermatotoxicology*, 7th Edition, CRC Press, Taylor & Francis Group, Boca Raton, 2008.

K. Yamagawa, K. Ichikawa. Experimentelle Studie ueber die Pathogenese der Epithelgeschwuelste. *Mitteilungen der Med.* Fakultaet der Kaiserl. Univ. Tokyo **15**: 295–344, 1915.

K. Yoshida, A. Kubo, H. Fujita, M. Yokouchi, K. Ishii, H. Kawasaki, T. Nomura, H. Shimizu, K. Kouyama, T. Ebihara, K. Nagao, M. Amagai. Distinct behavior of human Langerhans cells and inflammatory dendritic epidermal cells at tight junctions in patients with atopic dermatitis. *J. Allergy Clin. Immunol.* **134**: 856–864, 2014.

3.7 Kidney and Urinary Tract

Helmut Greim

3.7.1 Introduction

> **The kidney is mainly involved in maintaining the physiological homeostasis of the organism and in the excretion of metabolic waste. 99% of the water of the filtrated primary urine is reabsorbed, mainly in the tubular system, leading to an almost 100-fold increase in the concentration of compounds that are not or poorly reabsorbed, which explains the specific vulnerability of the kidney by toxic chemicals.**

The kidneys play a central role in maintaining cellular homeostasis. They regulate body fluid volume and the content and composition of electrolytes, and eliminate useless or harmful water-soluble substances via the urine, such as urea or water-soluble foreign compounds and their metabolites. These functions are the result of the high blood flow through the glomerulus. The abundant blood supply, together with the absorption of water and substances in the tubular system, may result in high concentrations of toxicants present at low concentrations in the blood.

Phase I and Phase II enzymes in the kidney have a high capacity of metabolic activation or inactivation of foreign compounds. Presentation of extra-renal formed toxic metabolites in the kidney mainly depends on their water solubility and chemical stability. Although glucuronides, sulfates, and glutathione conjugates usually are less toxic than their parent compounds, several of these conjugates become more toxic by kidney metabolism than the original products (Chapter 6.7).

3.7.2 Anatomy and Function

> **The anatomy and function of the kidney are focused on filtration of water and water-soluble substances in the glomerulus, absorption of water and essential substances out of the lumen of the tubular system as well as elimination of metabolic waste into the urine. The functional unit is the nephron.**

The kidney consists of an outer part, the cortex, and an inner part, the medulla. The shape of the medulla is pyramidal, the units are the papillae, with the outlets of the collecting ducts on each tip (papillary duct). The papillae project to the renal pelvis, which narrows towards the renal hilus. Here the renal artery enters the organ, the renal vein and the ureter leave it. Via the ureter the urine is transferred to the bladder, from which it is voided via the urethra.

Blood Filtration

The nephron (Figure 3.27), which is located in the cortex, consists of the renal corpuscle (or corpuscle of Malpighi) with the glomerulus and the Bowman's capsule, which filters the primary urine. In the renal tubule approximately 65–70% of filtered sodium, chloride, calcium, and water, 80–90% of filtered bicarbonate, phosphate, and urate, and essentially all of the filtered amino acids, glucose, and low molecular weight proteins are reabsorbed while others are transported through the tubular membranes by active processes.

The renal artery supplies the blood for filtration as well as for organ supply. Its branches, the afferent arterioles, enter the Bowman's capsules at the vasculary pole and branch again to form a ball of capillaries, the glomerulus. The capillaries and the inner layer of the Bowman's capsule constitute a semipermeable membrane through which the ultrafiltrate (primary urine) leaves the blood and passes into the lumen of the capsule of Bowman. The ultrafiltrate is practically free of cells and larger proteins, but contains all soluble substances of the blood plasma, including small molecules with a molecular mass up to 15,000 Da. According to loading and structure larger molecules up to a molecular mass of 70,000 can also appear in the ultrafiltrate (hemoglobin of molecular mass of 68,000 to about 3%). Normally, no proteins, except traces, appear in the urine because they are almost completely reabsorbed during their passage through the tubular system.

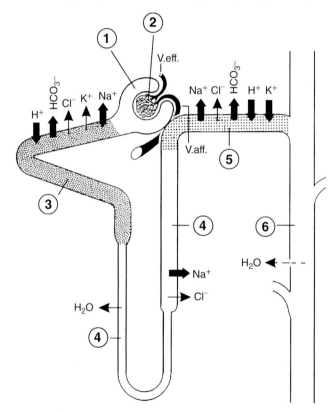

Figure 3.27 Structure and function of a nephron. 1, Bowman's capsule; 2, glomerulus with afferent and efferent arterioles; 3, proximal tubule; 4, loop of Henle; 5, distal tubule; 6, collecting duct. The efferent arterioles form a dense capillary network around the proximal and distal tubules and the loop of Henle, and regulate absorption of water and the different ions from the increasingly concentrating ultrafiltrate. Bold arrows indicate active transport, thin arrows passive transport.

The rate of the blood perfusion of the kidney is about 1 liter per minute, which corresponds to about one fourth of the volume that passes the heart per minute. Of this about 120 ml of ultrafiltrate per minute is produced. Thus, most of the blood leaves the Bowman's capsule at the urinary pole unfiltered via the efferent arterioles. The subsequent second capillary system of the kidney, the peritubular capillary net, encircles the tubules and takes up water and substances from the increasingly reduced volume of the ultrafiltrate.

During passage through the tubules 98–99% of the water previously filtered is reabsorbed. Most constituents of the ultrafiltrate are reabsorbed in the proximal convoluted tubule. Here active, adenosine triphosphate (ATP)-consuming processes selectively transport substances from the ultrafiltrate back to the blood, like sodium, bicarbonate, and phosphate ions, amino acids, and glucose. Urea, chloride ions, and water follow passively. About 60–80% of the water diffuses back in the proximal tubule. This passive process is primarily driven by ATP-dependent reabsorption of Na^+. It is obvious that any toxic effects in this part will heavily affect kidney function. Under physiological conditions the pH value of the blood is about 7.38, that of the voided urine about 6.0. Maintenance of this gradient is of great importance for the absorption of acids and bases. Any changes like increase (alkalosis) or decrease (acidosis) in the pH will affect their excretion and may result in toxic effects.

During the passage through the loop of Henle and the distal tubule, the absorption of water and sodium chloride continues. In the last part of the convoluted tubule and in the collecting ducts the fine adjustment in the concentration of electrolytes, the acid-and-base balance, and water takes place. The regulation of the electrolyte and water homeostasis is extremely complex and is partly controlled by hormones. For example, angiotensin II and vasopressin regulate blood flow and by that glomerular pressure.

The straight, ascending limb of the distal tubule closely joins the afferent arterioles near the corpuscle of Malpighi (glomerulus and capsule of Bowman). Here a complex feedback mechanism regulates the production of the ultrafiltrate. An increased volume of the solute or concentrations in the solute in the distal tubule reduces the blood flow in the afferent arteriole and by that reduces the glomerular filtration rate.

> **The human kidneys, with about two millions of nephrons, produce per day 1–2 l of final urine out of about 180 l of the ultrafiltrate. As a consequence, concentrations of compounds that are present in the ultrafiltrate and are not reabsorbed increase by 100-fold and may become toxic.**

Collection and Excretion of Urine

The final urine flows through the collecting ducts, the renal pelvis, and the ureter into the urinary bladder. Ureter and urinary bladder are muscular organs lined by a multilayered epithelium.

3.7.3 Toxicology

Intoxication of the Kidney

> **Intoxication of the kidneys mostly leads to impaired membrane function of the glomerulus or the tubular cells. Owing to the specific physiology of the kidneys, toxicity of chemicals may result from increasing concentrations in the lumen of the tubular system or in the tubular cells.**

Intoxication of the kidney usually impairs filtration and reabsorption. Some site-specific effects can be distinguished.

Glomerulus Although the glomerulus is the primary site of chemical exposure, disturbance of its function by non-drug chemicals is rare. Several mechanisms may be involved: impairment of ultrafiltration by vasoconstriction of renal arteriae (amphotericin), interaction with endothelial cell membranes (gentamycin), or direct cytotoxic effects on epithelial cells (cyclosporine). Such effects lead to impaired production of the ultrafiltrate. Another cause of glomerular injury is circulating immune complexes that become trapped in the glomerulus. These attract neutrophils and macrophages, which release reactive oxygen species and mediators for inflammation, contributing glomerulonephritis, the inflammation of the renal corpuscles. Deposition of antigen–antibody complexes against the glomerular basement membrane and/or thickening of the glomerular membrane will also impair filtration. A reduced ultrafiltrate production due to a reduced or less permeable filter surface or a decline in the filtration pressure causes a diminished excretion of metabolic waste. Accumulation of these compounds in the organism will result in azotemia.

Exposure to volatile hydrocarbons and organic solvents can enhance membrane permeability in the glomerulus. As a consequence, larger molecules like albumin and gamma-globulin, which are normally retained, pass to the ultrafiltrate and are excreted with the urine (proteinuria).

Tubular System Tubular atrophy and thickening of the basement membrane, especially in cells of the proximal segment, impair reabsorption of electrolytes, water, and substances. Intracellular intoxication can also occur by high concentrations of absorbed substances. Water reabsorption can lead to increasing and finally toxic concentrations of compounds in the tubular solute. This may lead to toxic effects in the surrounding epithelium or, by precipitation, to the formation of crystals and larger precipitates and by that to mechanical effects.

The proximal tubular system is preferentially affected by chemicals. Owing to high reabsorption rates chemicals can accumulate and reach toxic concentrations in the tubular epithelium. Such chemicals are the haloalkene S-conjugates, the alpha-$2_{\text{urinary}}(\alpha_{2u})$-globulin-bound chemicals such as limonene, cadmium, and mercury. Of the nephrotoxic drugs aminoglycosides, β-lactam-anitibiotics like penicillin, and the mycotoxin ochratoxin preferentially affect this part.

The loop of Henle and the distal tubular system are less affected. So far, only drugs like methoxyflurane or cisplatin or acidification are known to affect this area. Since this part controls the fine regulation of water and sodium uptake the clinical consequences of functional impairment are polyuria and loss of electrolytes.

As a consequence of tubular membrane damage, enzymes leak out and appear in the urine (enzymuria). Determination of specific enzymes allows identification of the site of the lesions. Enhanced urinary concentration of ligandin points to lesions in the proximal tubule, increased concentrations of lactate dehydrogenase indicate lesions in the distal tubule.

Functional disorders may occur in combination or in temporal sequence. After poisoning with halogenated alkenes, polyuria is observed at first, which in case of serious intoxication progresses to oliguria. Important factors are the dose and the time of exposure. A slight proteinuria will be without further consequences when the concentration of enzymes – traces of nearly all serum enzymes are normally present in the urine – is not exceeded dramatically. A severe proteinuria, on the other hand, leads to protein deficits, formation of edema, and renal failure.

Compounds Toxic to the Tubular System
Hydrocarbons

> Hydrocarbons affect the kidneys by different mechanisms. These comprise direct cytotoxicty, interaction with specific proteins (α_{2u}-globulin), and formation of reactive metabolites of mercapturic acids by β-lyase.

Cytotoxicity In humans kidney glomerulonephritis and tubular necroses were observed after exposure to petroleum hydrocarbons, like benzene, solvents, kerosene or diesel fuel (Table 3.18). These effects seem to be the direct interaction of the compounds with cellular membranes.

Table 3.18 *Nephrotoxic chemicals.*

Compounds	Mechanisms	Primary target	Relevance
Allyl chloride		Bowman's capsule	
Solvents	Increased permeability	Bowman's capsule, proximal tubule	
Unleaded gasoline, decaline, p-dichlorophenol, isophorone, limonene, methylisobutylketone	Formation of α_{2u}-globulin	Middle portion of proximal tubule	Male rat specificity
Trichloroethene, tetrafluoroethene, hexachloro-1,3-butadiene	β-Lyase-mediated formation of thioketene	Proximal tubule	Possible human carcinogens
Tetrachloromethane	Trichloromethyl radical metabolite	Proximal tubule	
Chloroform	Biotransformation to phosgene	Proximal tubule	
Ethylenglycol,	Precipitation of calciumoxalate	Proximal tubule	
Diethylenglycol	Increase in osmolality	Whole nephron	
Heavy metals	Various, see Chapter 6.2	Proximal tubule	
NTA	Ca^{**} extraction from membranes	Proximal tubule	
Aminoglycosides	Formation of undegradable phospholipids	Proximal tubule	Antibiotic drugs
Cisplatin	DNA cross-links	Loop of Henle	Cytostatic drug
N-nitroso compounds	Alkylating	Bladder	
2-(4-(5-nitrofuryl)-2-thiazolyl)hydrazine	Peroxidation by prostaglandin-H-synthase	Bladder, tubule	
2-Naphthylamine, 4-aminobiphenyl	Metabolism to aryl-hydroxylamines	Bladder	Contaminants in aniline production
Aromatic amines	Metabolism to aryl-hydroxylamines	Bladder	
Bladder stones	Mechanical irritation	Bladder	

NTA, nitrilotriacetic acid.

Halogenated alkanes and alkenes are basic substances in industrial production, and are used as solvents and for pest control. They are an important group among the kidney toxins and damage mainly the straight part of the proximal tubule. Allyl chloride enlarges the gap of Bowman's capsule. Some halogenated hydrocarbons exert a toxic potential in the liver as well as in the kidney. This applies to carbon tetrachloride (tetrachloromethane) and chloroform. The reactive metabolites are mainly generated in the liver and attain the kidney via the blood. In the case of chloroform the toxic metabolite phosgene is generated in the proximal tubule by cytochrome 450. It is detoxified by GSH conjugation and may react with various cell components.

α_{2u}-globuline-mediated effects The low molecular mass protein α_{2u}-globuline is formed in the male rat liver under the influence of testosterone. About 60% of the secreted protein is reabsorbed in the middle portion of the tubular system and is degraded within the tubular epithelium by lysosomal enzymes. The non-reabsorbed portion is excreted and constitutes the main component among the proteins found in the urine of this gender. In the aging male rat the content of α_{2u}-globulin declines and that of albumin rises, pointing to glomerulonephritis, which often spontaneously occurs in the aged rat. It entails necroses, regenerative growth, and kidney tumors.

In kidneys of male rats sex-specific kidney tumors occur upon exposure to volatile hydrocarbons like unleaded gasoline, decaline, p-dichlorophenol, isophorone, limonene, and methylisobutylketone. These compounds are metabolized in the liver, where the parent compounds and/or the reactive metabolites interact with alpha$_{2u}$-globulin and these complexes are transported to the kidneys. They are taken up by the epithelial cells of the middle portion of the proximal tubule and accumulate because the protein–α_{2u}-globulin complex cannot be degraded by lysosomal enzymes. This accumulation results in formation of hyaline droplets and the complexes can be verified by immunochemical staining. The resulting continuous cytotoxicity and reparative replication are responsible for the tumorigenic response.

α_{2u}-globulin is much less or not produced in other experimental animals and in humans, so that tumors induced by interaction of reactants with α_{2u}-globulin are considered to be species-specific for male rats and have no relevance to humans.

β-Lyase-mediated carcinogenicity An important mechanism in kidney toxicity is the bioactivation of GSH-conjugates, such as trichloroethylene or hexachlor-1,3-butadiene. Formation of such conjugates by GSH-transferases occur in the liver and the kidney. By means of γ-glutamyltransferases or dipeptidases, which split the γ-glutamyl- and the glycine residues, cysteine S-conjugates are formed in the intestine or in the kidney. These cysteine S-conjugates can be toxic by themselves or after further metabolism. Especially in the kidney, high activity of C–S-lyase (β-lyase) generates toxic metabolites, which alkylate macromolecules and ultimately lead to cancer (see Chapter 6.7).

Heavy Metals

> **Heavy metals particularly damage the proximale tubule.**

Many metals are reabsorbed by the epithelium of the proximale tubule. Among others these are cadmium, mercury, arsenic, chromium, platinum, and bismuth. The target can be limited, especially in cases of slight poisoning. Chromium preferentially affects the first portion of the convoluted tubule, mercury the straight part. With increasing mercury exposure the whole tubule becomes affected. Copper salts damage the epithelia of the loop of Henle and of the distal tubule as known from its use as fungicide. Mercury nephrotoxicity is the consequence of two mechanisms: The contracting effect on blood vessels increases blood pressure and the cytotoxic effect, which inhibits a series of enzymes and affects cell organelles. Probably the binding of mercury to the endoplasmic reticulum (ER) and the excretion into the lumen is a detoxifying process. The binding to lipoproteins in lysosomes or, in the case of cadmium and other heavy metals binding to metallothionein, is also protective.

Polyvalent Alcohols

Polyvalent alcohols damage the tubule in different ways.

Polyvalent alcohols such as the bivalent alcohols ethylene glycol and diethylene glycol damage the proximale tubule. They are used as solvents and antifreezing agents. Oxalate, the metabolite of ethylenglycol, and probably also intermediate products like glyoxylic acid, are nephrotoxic as well. During reabsorption of water, precipitations of calciumoxalate crystals occurs in the extracellular space, which obstruct the lumina and damage the epithelial cells mechanically, entailing urine retention (ischuria), excess pressure in the nephrons, and insufficient excretion of toxic waste from the blood. Diethylenglycol increases osmolarity in the kidneys and causes swelling and degeneration of epithelial cells. Formation of oxalate does not occur. Another example is the trivalent alcohol glycerol. After treatment with glycerol, nephrons situated near the surface of the kidney collapse.

Nitrilotriacetic Acid

In long-term repeated dose studies nitrilotriacetic acid (NTA) applied via feed or drinking water induced cytotoxicity, resulting in hyperplasia of the tubular epithelium and tumors in the kidney, the ureter, and the bladder. In the kidney the primary location of the carcinoma is the tubular cells. The lowest tumor-inducing dose is 100–140 mg/kg body weight (bwt).

NTA is readily absorbed from the gastrointestinal tract and excreted unchanged via the kidney. It has a high complexing activity for divalent ions. In the gut it preferentially binds zinc and NTA–zinc is the major complex excreted. Since there is little tubular reabsorption, NTA increases in concentration during passage through the tubular systems and finally precipitates at high exposures. At about 200 mg/kg bwt free NTA appears in the ultrafiltrate, which extracts Ca^{++} ions from the tubular epithelium and the epithelial cells of urinary system. Membrane damage, cytotoxicity, hyperplasia, and tumors are the result of reabsorbed zinc, extraction of calcium from the membranes as well as physical damage by precipitating NTA.

Other Chemicals

Among the anthropogenous substances in the environment there are pesticides and industrial chemicals with nephrotoxic potential. In pest control by soil fumigation hexachloro butadiene (HCBD) and the herbicides 1,1'-dimethyl-4,4'-bipyridinium (Paraquat) and 2,4,5-trichlorophenoxyacetic acid (2,4,5-T) are used. They inhibit the active elimination mechanism of organic ions in the proximal tubule and by that inhibit their own excretion. Some of the halogenated hydrocarbons, like polychlorinated biphenyls, dibenzodioxins and dibenzofurans, are not directly nephrotoxic. They induce enzymes, which metabolize foreign compounds and at high exposure may metabolically activate other toxicants.

Toxicology of the Urinary Tract

The urinary tract is a relatively insensitive target for toxic foreign compounds.

There is little information on the toxicity of foreign compounds in the ureter and the urethra. It may be that the short stay of the urine in these organs hinders any effective influence of toxic compounds.

Toxic effects on the urinary bladder by foreign compounds are practically unknown, but there are some potentially carcinogenic substances that may affect this organ (Table 3.18).

Carcinogenic Effects

Tumors of the Kidney

> **The formation of malignant neoplasms in the human kidney is uncommon, but there is a series of different tumors that are dependent on the type of the tissue and the age of the person. In recent years there has appeared increasing evidence that genetic defects play an important role, e.g. in the renal cell carcinoma a deletion on chromosome 3, segment p has frequently been found.**

There is little known about tumors in the human kidney induced by foreign compounds. Smoking is a factor of risk, but in this case the transitional epithelium of the renal pelvis seems to be affected rather than the renal parenchyma.

In contrast to the human situation, in experimental animals several chemicals are tumorigenic in the kidney. Among them are N-nitroso compounds, e.g. diethylnitrosamine and dimethylnitrosamine, with initiating potency in different target organs in rats and mice, prevalently in the liver but also in the kidney. Ethylnitrosurea produces tumors of the brain and the kidney. With methylnitrosurea the kidney is the preferential target of carcinogenesis.

The furan 2-(4-(5-nitrofuryl)-2-thiazolyl)hydrazine (FNT) causes tumors of the transient epithelium of the urinary bladder and tumors of the tubuli.

Naturally occuring substances like the mycotoxin aflatoxin and methylazoxymethanol, the aglycon of Cycasin, also induce tumors in experimental animals. In humans, after consuming flour from the Cycas fruit and peanuts contaminated with aflatoxin B_1 tumors of the liver but not of the kidney were observed.

Comparable to carcinogenesis in, for example, liver and skin, the steps of initiation and promotion can be distinguished. Besides the already mentioned initiating agents in kidney carcinogenesis, there are also promoting agents, at least experimentally, for example the plasticizer di(2-ethylhexyl)-phthalat) (DEHP).

Bladder Tumors

> **The bladder is a specific target for several occupational carcinogens.**

Since the end of the 19th century it has been known that 2-naphthylamine, which is an intermediate in aniline dyestuffs production, can give rise to tumors of the urinary bladder epithelium in humans, dogs, and other mammals (aniline cancer). This is one of the earliest examples of occupational carcinogenesis. Meanwhile other aromatic amines are known to be genotoxic carcinogens implicated in urinary bladder of mammals, such as 2-acetylaminefluoren (2-AAF) in mice. In humans exposure to aromatic amines results in tumors of the urinary bladder, besides nephrotoxicity (Table 3.18).

Analogous to the situation in the renal pelvis the ingredients of tobacco smoke and the roasting products of coffee beens are suspected to enhance the cancer incidence of the transient epithelium in the urinary bladder.

In animal experiments some nitrosamines, urea derivates, and nitrofurans have been identified as initiating agents. In particular the following substances have been found to be effective initiators: N-butyl-N-(4-hydroxybutyl)nitrosamine (BBN), N-methyl-N-nitrosurea (MNU), and N-[4-(5-nitro-2-fury)-2-thiazolyl]formamide (FANFT), which in the kidney is metabolized to 2-amino-4-(5-nitro-2-furyl)thiazol (ANFT). The 5-nitrofurans are metabolized to the ultimate carcinogens by peroxidation via prostaglandin-H-synthase in the urinary bladder, as shown also for benzidine. Enhanced amounts of N-nitroso compounds were found in the urine of bilharziasis (schistosomiasis) patients with or without cancer of the urinary bladder. Epidemiologic studies revealed that infections of the kidney and the urinary tract enhance the risk for urinary bladder cancer. Experimentally, FANFT in combination with *E. coli* induces tumors, preferentially of the renal pelvis. In animal studies FANFT or *E. coli* alone are ineffective regarding tumor development.

The promoting agents are saccharin and cyclamate. Promoting stimuli that may exert a modulating effect on tumorigenesis are the urine volume, pH value, and the time a compound remains in the urinary bladder. Such factors stimulate proliferation of epithelial cells and evoke hyperplastic changes. In addition, hypothermia or mechanical irritation by bladder stones stimulate proliferation. Calculi are produced experimentally by high doses of various chemicals, e.g. uracil. An important role is played by the urinary pH value. In an initiation–promotion experiment salts that increased the pH value of the urine promoted carcinogenicity. Those which lowered the pH value inhibited the carcinogenic process.

3.7.4 Summary

The kidneys maintain the physiological homeostasis of the organism by regulating the electrolyte and water balance. This takes place during blood filtration and reabsorption of essential amounts of water and salts. In the urine, which represents about 1% of the original ultrafiltrate, all water-soluble substances are eliminated. These can be those of excess, not useful for the body or toxic. Toxic effects are either the result of the high blood flow through the kidneys or the reabsorption of water, which increases the concentrations in the tubular system. There, concentrations of water-soluble compounds that are not reabsorbed increase 100-fold and may damage tubular epithelia or precipitate. Compounds that are reabsorbed may accumulate in the tubular epithelia and induce toxic effects. High urinary concentrations of toxic compounds may induce lesions in the epithelium of the urethra and the bladder.

Owing to its high metabolic capacity the kidney is able to metabolize foreign compounds to water-soluble and in some cases toxic derivates, which also induce toxic effects at various sites of the kidney, the urethra, and the bladder.

Further Reading

Barbier O, Jacquillet G, Tauc M, Cougnon M, Poujeol P: Effect of heavy metals on, and handling by the kidney. *Nephron Physiol* **99**, 105–110, 2005.

Brüning T, Bolt H: Renal toxicity and carcinogenicity of trichloroethylene: key results, mechanisms and controversies. *Crit Rev Toxicol* **30**, 253–285, 2000.

Endou, H: Recent advances in molecular mechanism of nephrotoxicity. *Toxicol Letters* **102–103**, 29–33, 1998.

Haschek WM, Rousseaux CG, Wallig MA: Kidney and lower urinary track. In: *Fundamentals of Toxicologic Pathology*, 2nd edition, Academic Press, New York, pp. 261–318, 2010.

Shitara Y, Horie T, Sugiyama Y: Transporters as a determinant of drug clearance and tissue distribution. *Eur J Pharm Sci* **27**, 425–446, 2006.

Van Vleet TR, Schnellmann RG: Toxic nephropathy: environmental chemicals. *Semin Nephrol* **23**, 500–5008, 2003.

Zalups RK, Ahmad S: Molecular handling of cadmium in transporting epithelia. *Toxicol Appl Pharmacol* **186**, 163–188, 2003.

3.8 The Hematopoietic System (Bone Marrow and Blood)

Robert Snyder

3.8.1 Introduction

> **Erythrocytes, various types of leucocytes, and thrombocytes arise in the bone marrow from stem cells, which undergo differentiation, proliferation, and maturation, and are released into the circulation. The toxicology of the blood and bone marrow involves two areas of concern: (1) the impact of drugs and chemicals on circulating blood cells and (2) the impact on cells developing in the bone marrow. Thus, hemolytic anemias are events that arise in the circulating blood as a result of factors such as genetics or chemicals that affect mature circulating erythrocytes. In contrast most cases of decreases in white blood cells, i.e. leucopenia, result from disturbances in the formation of specific types of leucocytes in the bone marrow.**

The bone marrow is the principal source of the cellular components of blood and maintains the proper levels of each cell type. For this purpose and the fact that the lifespan of each cell type differs, tens of billions new cells enter the circulation every day. Marrow is made up of fat through the hollow part of the bone with red marrow containing the blood cell-generating system largely at the ends of the long bones. Active marrow can also be observed in the bones of the spinal column, the ribs, the sternum, and in some pathological conditions, e.g. in the bones of the jaw and skull in the disease called thalassemia major. The total volume of the bone marrow approaches the size of the liver.

The circulatory system in adults normally contains about 5 l of blood. Each microliter contains 5 million erythrocytes. Therefore, at any given time there are 5×10^{12} erythrocytes in circulation. The lifespan of the red cell is about 120 days. Each day the marrow must supply 4.2×10^{10} new erythrocytes. There is a need to synthesize various white blood cell types and platelets as well. Thus, the marrow displays a very high capacity for generating blood cells. By the same token, interference with the normal functioning of the marrow may result in failure to properly generate cells, as in aplastic anemia, or in the uncontrolled production of aberrant cells, as in leukemia.

Table 3.19 contains a list of essential plasma proteins. Although not necessarily all of these are principally made in the bone marrow, they play a critical role in normal physiology of blood. Among these, albumin is essential for maintaining the osmotic pressure of the blood and it acts as a carrier for many drugs. The globulins have important immunological functions and also help to transport lipids. Other blood proteins include fibrinogen, which is converted to fibrin in the blood-clotting process, and transferrin and ferritin, both of which are essential for iron transport into the cells of the body.

Table 3.19 Normal components of blood.

Component	Normal amounts in circulating blood
Blood cells	
Erythrocytes (red blood cells)	4.5–5.5 million/µl
Total leucocytes (white blood cells)	5000–9000/µl
Neutrophils	60–70%
Eosinophils	2–4%
Basophils	0.5–1.0%
Lymphocytes	20–25%
Monocytes	3–8%
Thrombocytes (platelets)	
Plasma proteins	Approximate concentration (g/100 ml)
Total	7.3
Albumin	4.5
Globulins	2.5
Fibrinogen	0.3
Transferrin	0.25
Ferritin	0.015–0.3

This chapter will focus on the normal sequence of events in marrow function, the roles played by circulating cells in homeostasis, and on the adverse effects caused by exposure to various drugs and other chemicals leading to commonly observed hemopathies.

3.8.2 Hematopoiesis

> **The generation of blood cells in the bone marrow from the multipotent stem cell to the mature circulating cell involves differentiation to yield cells, which are morphologically and physiologically equipped to perform the functions of the mature cells. Differentiation drives the evolution of cells from the primitive state of the stem cell to the mature cell. Amplification involves a series of mitotic events which insure that there are a proper number of cells of each type released into the blood daily.**

The mechanism by which the bone marrow forms and supplies mature blood cells is termed hematopoiesis and is summarized in Figure 3.28. The process involves the sequential processes of differentiation from stem cells to progenitor cells to precursor cells, and, finally, to mature circulating cells. At any given stage the cells may undergo mitosis as well as differentiation. This complicates the identification of specific cell types in the bone based on morphological observations alone because at any given time a smear of bone marrow cells will offer mainly more advanced cells and very few of the early cell types.

Two approaches have been used to facilitate identification of cells in marrow. One is based on the ability to grow individual cell types in colonies by stimulating proliferation using specific growth factors. These have been termed colony-forming units (CFU) and are specifically named for the differentiation pathway in which they are found. Figure 3.28 provides an example of the differentiation pathway outlined in terms of identified CFUs. In addition, the development of antibodies, which react at specific cell surface receptors in bone marrow cells, provides

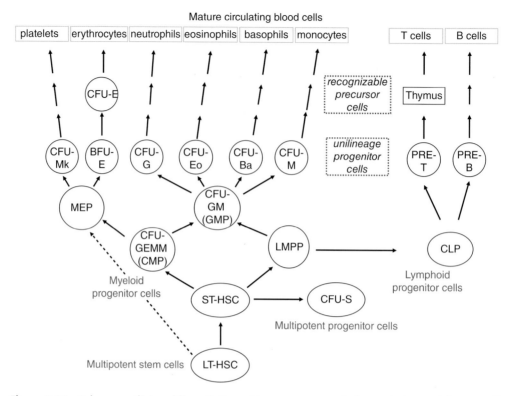

Figure 3.28 *Scheme outlining differentiation of bone marrow cells from pluripotential stem cells cells (CFU-S) through mature circulating blood cells. CFU, colony forming unit; S, stem cells; L, lymphocytes, G, granulocyte: MK, megakaryocyte; GEMM, granulocyte, erythrocyte, monocyte/macrophage, megakaryocyte.*

an additional approach to identification of bone marrow intermediate cells and provides a useful tool for determining differentiation, using changes in cell surface markers for specific antibodies. In this manner, all of the cells in Figure 3.28 can now be assesed using antibody staining.

The stem cell model argues that in the bone marrow there is a finite number of stem cells capable of replacing themselves by a mitotic process that also involves the release of a daughter cell that can progress through the differentiation/maturation scheme as influenced by specific growth factors. The first cell in the sequence has been termed the pluripotential, or multipotent, stem cell. The daughter cell (ST-HSC, CFU-S) is a progenitor cell capable of limited self-renewal and giving rise to the multilineage progenitor cells CFU-GEMM (colony assessment) or common myeloid progenitor (CMP, antibody assessment) and common lymphoid progenitors (CLP, antibody assessment). CFU-GEMM gives rise to all of the myeloid cells and CLP gives rise to lymphocytes. Further differentiation leads to unilineage progenitor cells for each of the circulating cell types. Although there may be some intermediary stages, ultimately the myeloid cells give rise to CFUs for each of the cell types in blood. Beyond the CFU stage morphological characteristics of the precursor cells are more readily apparent. The final stage is maturation from the last precursor cell to the mature circulating cell.

3.8.3 The Bone Marrow Niche

> The production of specific numbers of each cell type at various times and in different amounts is regulated by the collaborative community of cells in the bone marrow niche.

Niches are anatomical microenvironments that maintain the stem cell pool and regulate stem cell behavior. The niche is usually referred to as a reciprocal network of hematopoietic stem- and progenitor cells, and non-hematopoietic cells, which all interact and maintain tissue homeostasis and repair of the bone marrow. In recent years cellular components of the non-hematopoietic niche were shown to include mesenchymal stromal cells (MSC), both arterial and venous endothelial cells (EC), and neural cells. As indicated above, the hematopoietic stem cell (HSC) divides symmetrically or asymmetrically to produce quiescent HSCs and hematopoietic progenitor cells (HPCs), which in turn divide to produce the progenitor cells of the myeloid or lymphoid lineages. Whereas the proliferating HSCs play an important role in supporting routine blood production, the small pool of quiescent HSCs, which are primarily located in the endosteal and perivascular zones, are described as critical for long-term maintenance and for replenishing the active HSC pool.

Although there is limited understanding of all the cell types of the niche, specific MSCs fulfill different roles within the niche. These mesenchymal cells provide structural integrity by maintaining skeletal tissues, as well as providing support of HSCs. Different subpopulations of MSC progeny, such as osteoblasts, adipocytes, and so-called CXCL12-abundant reticular (CAR) perivascular cells, show distinct localizations within the bone marrow cavity, and have been shown to differentially maintain quiescent and activated HSCs.

The ECs not only deliver oxygen and nutrients; by specific angiocrine factors, such as for instance stem cell factor (SCF, cKIT ligand), but they also regulate HSC self-renewal, survival, and differentiation. Both arterial and sinusoidal venous ECs regulate HSCs and keep these in a quiescent state. HSCs are found close to hypoxic microenvironments, when the hypoxia-inducible factor-1α (HIF-1α) is stabilized and ATP is generated by anaerobic metabolism to improve survival. Angiopoetin 1 (Angpt1) triggers vascularization, increases oxygen supply and by that overcomes cell cycle quiescence and activates HSC to increase proliferation.

Due to their quiescent nature, HSCs are relatively resistant to cytotoxic agents, such as ionizing radiation and chemicals. As such, HSCs protect themselves by their limited cell cycle frequency against DNA damage responses and the possibilities of incomplete DNA repair, which render stem cells stress tolerant.

3.8.4 Toxicological Features of Circulating Blood Cells

Erythrocytes (Red Blood Cells)

> Anemia, which is a decrease in circulating functional erythrocytes, can result from inhibition of red cell development in the bone marrow, or as a result of hemolysis or inhibition of red cell function due to genetic defects or interactions with specific chemicals.

Erythropoiesis Erythrocytes are unique among cells in the body by virtue of both their morphology and their mission. Erythropoietin is a specific cytokine that directs differentiation of the erythroid cell line. Early recognizable precursor cells in the marrow, in sequence, are termed pronormoblasts, with several stages referred to as normoblasts, and reticulocytes. Each of these cell types is capable of synthesizing hemoglobin, the oxygen-carrying protein that defines red cell function. The final steps in maturation of the erythrocyte involve the loss of the nucleus at the stage described as the polychromatic normoblast. The resulting cell, the reticulocyte, continues to synthesize hemoglobin, is released into the circulation, and matures to the erythrocyte, at which point hemoglobin synthesis is complete.

The primary function of the erythrocyte is to transport oxygen from the lung to the tissues of the body. The red cell undergoes mechanical and oxidative damage during its 120 day lifetime. Eventually, as repair mechanisms fail, the cell is removed from the circulation, usually upon passage through the spleen, and the cell is lysed.

Hemoglobin Function Hemoglobin is a protein composed of four peptide chains (globins), two of which are α chains and two are β chains. Each chain carries a heme group composed of a mole of ferrous (Fe++) iron embedded in protoporphyrin IX and linked to an imidazole group on a histidine in the peptide chain (Figure 3.29). Hemoglobin can reversibly bind a molecule of O_2 to the Fe++. The extent to which hemoglobin is saturated with O_2, i.e. the degree to which the total number of heme groups are bound to O_2, is determined primarily by the oxygen tension. The air in the alveoli of the lung contains a higher concentration of O_2 than the blood circulating through blood vessels surrounding the alveoli. The concentration gradient drives the O_2 from the alveoli into the blood, where it combines with the heme groups of hemoglobin to form oxy-hemoglobin. During the course of circulation O_2 is released to tissues having a relatively low O_2 tension. Carbon dioxide is found in the tissues largely as bicarbonate, and is transported through the blood to the lung. Association of bicarbonate with oxy-hemoglobin enhances release of oxygen in the tissues whereas release of bicarbonate as CO_2 in the lung promotes oxy-hemoglobin formation. The presence of O_2 constantly oxidizes hemoglobin (HbFe^{++}) to methemoglobin (HbFe^{+++}). The level of methemoglobin in the blood is kept at about 1% by efficient NADH and NADPH dependent methemoglobin reductase systems. The key enzyme that provides NADPH is glucose-6-phosphate dehydrogenase (G6PD). G6PDH deficiency results in greater sensitivity to methemoglobin-inducing agents,

Hemolytic Anemias Hemolytic anemias are diseases characterized by a deficiency of effective erythrocytes resulting from lysis of abnormal red cells. Hemolytic anemias are usually the result of genetic defects in either the red cell membrane structure or the structure of hemoglobin. For example, hereditary spherocytosis is a hemolytic disease featuring genetic variants in which there are deficiencies of a number of erythrocyte membrane proteins leading to the production of small spherical red cells, which are trapped in the spleen and hemolyzed. Sickle cell anemia and the

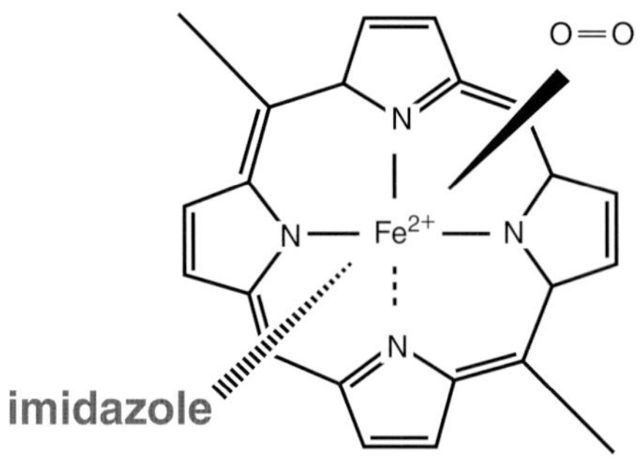

Figure 3.29 Heme group. Active component of hemoglobin, which transports oxygen from the lung to the tissues.

thalassemias are genetic diseases in which mutant globin chains are formed, resulting in ineffective red cells that are readily hemolyzed.

In some cases hemolysis results from the interaction of a chemical with normal or abnormal red cells.

Glucose-6-phosphate Dehydrogenase Deficiency   Reduced glutathione (GSH) is a critical factor in the maintenance of the red cell membrane. The oxidation of GSH to GSSG can be reversed by the NADPH-requiring glutathione reductase. NADPH is generated during the oxidation of glucose-6-phosphate by glucose-6-phosphate dehydrogenase (G6PD). People genetically deficient in G6PD fail to provide enough NADPH to meet the needs of glutathione reductase and subsequent membrane failure leads to hemolysis.

G6PD deficiency, which is an X-linked recessive disorder, is observed primarily in males. About 10% of African-American males demonstrate a relatively mild form of the disease, which has been called primaquine sensitivity because of its occurrence in black US servicemen treated with prophylactic doses of primaquine to protect against malaria during the Korean War. Favaism, which occurs among people who live in the Mediterranean region, is a more severe form of G6PDH deficiency disease in which people who eat fava beans (or inhale its pollen) may undergo hemolysis sufficiently severe to be fatal. Fava beans contain pyrimidine aglycones, which cause more excessive oxidation of GSH in G6PD-deficient erythrocytes than observed in primaquine sensitivity.

Carbon Monoxide Poisoning   Carbon monoxide binds to hemoglobin (Fe++) 200–300 times more avidly than oxygen. Carboxyhemoglobin formation results in a "cherry red" complexion with cyanosis. Low concentrations of CO lead to headaches, 40% carboxyhemoglobin results in impaired vision, tachycardia, and hyperpnea, and 60% can be fatal. Treatment involves increasing the ambient oxygen pressure by inhaling 100% oxygen or 95% oxygen:5% carbon dioxide. The role of carbon dioxide in the latter is to increase breathing frequency. In severe cases providing oxygen at hyperbaric pressures may be required to save the life of the exposed individual.

Methemoglobinemia

> **Hemoglobin in which the iron has been oxidized from the ferrous to the ferric state is called methemoglobin. It cannot bind and transport oxygen.**

During the normal course of oxygen transport an electron from the iron becomes partially associated with the oxygen, which assumes a superoxide-like structure, but returns to the iron upon the release of oxygen to the tissues. The process is not completely efficient, with the result that the electron may not always be restored to the iron. Under these circumstances the iron is converted from the reduced (ferrous) form to the oxidized (ferric) form. Hemoglobin with an oxidized iron is called methemoglobin, and cannot bind and transport oxygen. The stricken individual has a characteristic gray facial appearance at levels below 30% methemoglobin. Above that value the patient becomes cyanotic. Many nitrogenous compounds, sulfonamides, and quinones can induce the formation of methemoglobin. The cell normally maintains the level of methemoglobin at about 1% or less via the activity of an NADH-requiring cytochrome b_5 methemoglobin reductase. In the event of an overdose with a methemoglobin-forming agent additional reductase activity is required. A nascent form of methemoglobin reductase requires NADPH and its activity requires a reducing agent such as methylene blue.

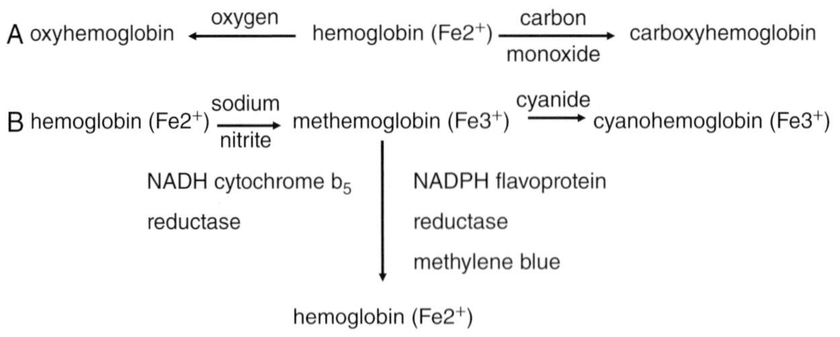

326 Toxicology and Risk Assessment

A oxyhemoglobin ⇌ (oxygen) hemoglobin (Fe2+) ⇌ (carbon monoxide) carboxyhemoglobin

B hemoglobin (Fe2+) →(sodium nitrite) methemoglobin (Fe3+) →(cyanide) cyanohemoglobin (Fe3+)

NADH cytochrome b₅ reductase | NADPH flavoprotein reductase methylene blue
↓
hemoglobin (Fe2+)

Figure 3.30 *A. Reactions of hemoglobin with oxygen and carbon monoxide. B. Methemoglobin reduction; methemoglobin reaction with cyanide.*

Cyanide is a mitochondrial poison that binds cytochrome oxidase in its Fe^{+++} configuration and prevents oxygen utilization. However, CN^- binds avidly to methemoglobin. Thus, in the event of cyanide poisoning, sodium or amyl nitrite can be administered to oxidize a small fraction of hemoglobin to methemoglobin, which acts as trap for the cyanide ion.

The interactions of hemoglobin, methemoglobin, carbon monoxide, and cyanide are summarized in Figure 3.30.

In the event of an overdose of cyanide, advantage can be taken of an enzyme called rhodanese, which is a normal blood component. Upon treatment with thiosulfate, rhodanese catalyzes a reaction with cyanide, the products of which are thiocyanate and sulfite. Thiocyanate is a detoxified form of cyanide.

3.8.5 Leucocytes (White Blood Cells)

> **The term leucocytes, or white blood cells, encompasses several cell types of differing morphology, with diverse physiological functions, i.e. protection against infection by neutrophils and macrophages, mediation of inflammatory reactions by eosinophils and basophils, T lymphocytes concerned with cell-mediated immunity, B lymphocytes which enhance immunity via the production of antibodies, and platelets, which are essential for blood clotting.**

The leucocytes are a heterogeneous group of cells that are distinguished from the erythrocytes primarily because they lack hemoglobin and all but the platelets contain a nucleus. They include the granulocytes (neutrophils, basophils, and eosinophils), the lymphocytes, and the thrombocytes (platelets), as well monocytes and macrophages.

Granulocytes

Granulocytes are named on the basis of their staining properties:

(a) Neutrophils comprise about 55% of leucocytes in the circulation. Normal neutrophil counts range from 3000 to 6000 cells/μl of blood. Their primary function is to engulf and destroy bacteria and other foreign material. Neutrophils are critically important in countering bacterial infections. People who exhibit neutropenia have a reduced capacity to resist infections. Agranulocytosis is a condition in which there is a severe reduction in neutrophils and is usually accompanied by high fevers indicative of infection.

Upon maturation of neutrophils in the bone marrow they require an array of stimuli to affect their release into the circulation. These may include endotoxin, androgens, glucocorticoids, and so-called CXC chemokines. Neutrophils leave the blood in response to a variety of stimuli, with a half-life of about 4–10 hours. They accumulate at sites of infection or inflammation, age, and die within 72 hours.

The bactericidal effect of neutrophils involves first engulfing the organism and then releasing lethal reactive oxygen species via the respiratory burst mechanism. The process is initiated via a cytochrome b-associated NADPH oxidase reaction leading to generation of superoxide, hydrogen peroxide, hydroxyl radicals, and singlet oxygen. The Haber–Weis reaction plays a key role in this sequence of events. In addition to the oxidative pathway neutrophils attack microorganisms by generating a series of proteins, such as the defensins and serprocidins, which alter surface protein activity and results in virucidal, fungicidal, and bactericidal activity.

(b) Eosinophils comprise about 3% of leucocytes or between 50 and 250 cells/µl of blood. They stain red when exposed to eosin, an acidic dye. Eosinophils are important in protecting against parasitic infections, but excessive accumulation of eosinophils in airways may further impair breathing in asthmatics.

(c) Basophils comprise less than 1% of leucocytes or about 15–50 cells/µl. They are concerned with generating allergic and hypersensitivity reactions. Their granules contain histamine, appear blue in the presence of a basic dye, and release histamine to initiate a variety of immediate hypersensitivity reactions such as asthma, urticaria, rhinitis, and possibly anaphylactic shock.

Lymphocytes

Lymphocytes comprise about 30% of circulating white cells or 1500–3000 cells/µl. The lymphocyte population may be divided into two types of cells, termed B and T cells. Both types arise in the bone marrow. B cells (about 10–15% of circulating lymphocytes) mature in the bone marrow but T cells (about 70–80%) migrate to the thymus to complete their maturation.

Lymphocytes are key components of the immune system. B cells are lymphocytes concerned with humoral immune response, whereas T lymphocytes function in cell-mediated immune response mechanisms. They react with antigens, which are macromolecules containing protein or polysaccharide structures, which may also have lipid characteristics. There are B cell and T cell receptors specific for designated antigen structures. Once the cells have taken up and processed the antigen they proceed to undergo mitosis to produce a clone of cells having the same antigen-responsive characteristics, which appear on the cell surface as so-called class II histocompatability molecules. These bind to helper T lymphocytes, which then generate lymphokines, which in turn stimulate the production of B cell receptors in a soluble form. The cells differentiate to plasma cells, which secrete the soluble receptors, which are now called antibodies and are capable of binding with and inactivating the antigens.

Alternative mechanisms by which T cells can act relate to the presence of either CD4 or CD8 cell surface glycoproteins. CD 8+ cells transmit cytotoxic molecules to infected cells, which induces apoptosis. CD 4+ cells combine with phagocytic cells, such as macrophages, to initiate an inflammatory response, which leads to the death of the antigen-containing cell. Thus, B and T lymphocytes play complex roles in support of the immune system.

The Monocyte/Macrophage System

Monocytes develop in the marrow and are released into the circulation, where they may function in phagocytosis, chemotaxis, and the killing of microorganisms. They have immunologic and secretory functions. Within a day of their entry into the circulation they leave the blood and migrate to various organs where they appear as in situ macrophages. Thus, there are pulmonary/alveolar macrophages in the lung and Kupfer cells in the liver, which perform the functions of macrophages in those organs.

Macrophages have a number of important physiological functions, but when activated they may release toxic mediators which have a deleterious effect on the surrounding tissues. Thus, among the many secretions emanating from macrophages there are cytokines, stress proteins, reactive oxygen species (which may be bactericidal or damage local tissues), tumerocidal mediators, fever-inducing pyrogens, bioactive lipids and oligopeptides, a variety of degradative enzymes, growth factors, and other materials. The ultimate impact will depend upon the physiological state of the local tissues and the effects of other agents.

3.8.6 Platelets (Thrombocytes)

> **Platelets are essential for initiating the process of blood coagulation. They are formed from a cell type called the megakaryocyte, which develops in the bone marrow in a fashion similar to other cells of the myeloid and lymphoid lines, but instead of maturing and entering the circulation they release small cells which lack a nucleus but are highly adapted to initiate blood coagulation.**

Upon damage to blood vessels collagen is released into the immediate circulation and in response platelets accumulate at the site of injury and release ADP, which enhances their ability to stick together and form a plug to halt blood loss. Fibrinogen adheres to the surface of the platelet plug and is converted to fibrin, the strong protein backbone of the clot, and other cells, such as erythrocytes, are trapped in the plug and reinforce the clot.

Thrombocytopenia is a decrease in blood platelets and can lead to hemorrhages. So-called "black and blue" marks, which appear as bruises after traumatic injury, are due to subdermal bleeding. In thrombocytopenia a form of excessive hemorrhage termed "thrombocytopenic purpura" can result in extensive bleeding and can be fatal. Chemicals and drugs which can cause aplastic anemia, such as benzene and a variety of anticancer alkylating agents, depress platelet production and can lead to purpura and other hemorrhages.

Platelets play a key role in the formation of the atherosclerotic plaque. High levels of blood cholesterol are associated with the accumulation of monocytes and macrophages in the vascular epithelium. They take up the cholesterol and become so fat laden that they are known as foam cells. Platelets are attracted to the site, smooth muscle cells proliferate, and the plaque is formed. The result is occlusion of the lumen of the blood vessel and thrombosis.

3.8.7 Impairment of Bone Marrow Function

> **The effects of many chemicals which impact on the bone marrow, such as benzene and anticancer alkylating agents, is to reduce the level of the various circulating white blood cells by interfering with the processes of differentiation and/or proliferation. Under some circumstances disturbances in differentiation and loss of control of proliferation lead to bone marrow cancers called leukemias.**

Depression of Bone Marrow Function

Drugs and chemicals may produce cytopenia by impairment of bone marrow function. Chronic exposure to benzene in the workplace has been know for over a century to cause decreases in circulating erythrocytes, leucocytes and/or thrombocytes. In the late 19th and early 20th centuries many cases of cytopenia and aplastic anemia were observed in factories where benzene exposure may have been as high as several hundred parts per million (ppm). In Western Europe and the United States benzene exposure is now controlled at low levels, e.g. 1–5 ppm, but there is evidence of continuing high exposures in developing countries. The mechanism of benzene-induced aplastic anemia appears to involve direct damage by benzene metabolites to bone marrow stroma and inhibition of both differentiation and proliferation of developing cells in the bone marrow.

The drugs that most frequently lead to impairment of bone marrow function are the antineoplastic agents. Both antimetabolites and alkylating agents are intended to inhibit cellular reproduction in cancer cells. Bone marrow cells are ideal targets for chemicals that inhibit proliferation because of their high mitotic rate. Thus, alkylating agents (e.g., mechlorethamine, melphalan, cyclophosphamide, chlorambucil, etc.) can produce decreases in all of the circulating cell types by virtue of their effects on developing cells in the bone marrow. By the same token antimetabolites (e.g., methotrexate, thioguanine, mercaptopurine, fluorouracil, cytarabine, etc.), which impair DNA synthesis in neoplastic cells, can cause similar disruptions of cell replication in the bone marrow.

In the Netherlands between 1974 and 1994 approximately 40 drugs, drug classes, or drug combinations were cited where there was a link between treatment and neutropenia, most of which was agranulocytosis. The drugs most frequently reported to cause agranulocytosis were dipyrone, mianserin, salazosulpha-pyridine, co-trimoxazole (a combination of trimethoprim and sulfamethoxazole), penicillins, cimetidine, thiouracils, phenylbutazone, and pencillamine.

Historically, many other drugs have been claimed to cause bone marrow depression, leading to aplastic anemia such as the antibiotic chloramphenicol, antithyroid thiourea derivatives such as propylthiouracil and methimazole, phenothiazines, non-steroidal anti-inflammatory agents such as phenylbutazone, sulfonamides, and many other drugs. Depression of the marrow usually begins with some form of leucopenia, i.e. neutropenia, thrombocytopenia, etc., and progresses to agranulocytosis or aplastic anemia. Although these tend to be rare events for any drug, the highest incidences have been observed with chloramphenicol and phenylbutazone. A higher frequency of aplastic anemia has been reported when aspirin, penicillin, acetopheneditin, phenytoin, streptomycin or sulfisoxazole was given with other drugs than when given alone.

Early forms of leucopenia or anemia can be reversed by withdrawing exposure to the offending agent. Some therapeutic intervention may be necessary. Damage sufficiently severe to result in aplastic anemia would require a bone marrow transplant to restore normal hematopoesis.

Exposure to benzene and many alkylating agents may result in myelodysplasia, a syndrome characterized by abnormal bone marrow cell morphology and chromosome damage to bone marrow cells. Exposure to benzene frequently results in myelodyplastic syndrome (MDS). A similar response may be observed in patients who have been treated with alkylating antineoplastic agents. Among these a significant percentage go on to acute myelogenous leukemia (see below).

Leucocytosis and Leukemias

Leucocytosis refers to increases in circulating mature leucocytes such as neutrophils. Non-neoplastic leucocytosis has been reported to result from treatment with steroids, β-agonists, lithium or tetracyclines.

Leukemias are cancers of the bone marrow manifested by excessive proliferation of transformed immature granulocytes or lymphocytes. Erythroleukemias have also been recorded. The most frequently observed forms of leukemias are referred to as acute or chronic myelogenous leukemia (granulocytic leukemias), acute or chronic lymphatic leukemias, Hodgkin's disease, non-Hodgkin's lymphoma, and multiple myeloma (a B cell leukemia arising from plasma cells).

Direct associations between exposure to chemicals and leukemogenesis are difficult to discern. Some authorities suggest that if a chemical can cause bone marrow damage any form of leukemia may result. Others point to benzene or the alkylating agents where the evidence demonstrates that acute myeloid leukemia is the form most frequently observed.

3.8.8 Mechanisms by which Chemicals can Induce Leukemia

Leukemias are bone marrow cancers and mechanisms of carcinogenesis in other organs can be invoked to attempt to explain the mechanism of leukemogenesis. Thus, covalent binding of carcinogens or their biological reactive intermediates, e.g. anticancer alkylating agents, to DNA can be envisioned as a trigger mechanism for initiating a mutagenic event. Thus, mutations in the *ras* family of oncogenes have been shown to be associated with some types of leukemia. Inhibition of topoisomerase II has been shown to be an effective chemotherapeutic strategy for some cancers, but frequently results in subsequent leukemia, probably because when the enzyme is inhibited at the cleavable complex stage of the reaction the double-stranded break created by the enzyme fails to re-anneal and the broken DNA chain represents a mutation. DNA repair is a critical mechanism for maintaining the integrity of DNA. Normally, in the absence of DNA repair, damage to DNA triggers cellular apoptosis, which is a mechanism aimed at insuring that mutated cells do not survive. If DNA repair fails and apoptosis is inhibited the stage is set for neoplastic transformation.

Epigenetic mechanisms of carcinogenesis have been proposed. These frequently involve hormone-mediated excessive cell proliferation. Protein-based carcinogenesis is more difficult to demonstrate because of the prevailing concept that there is no threshold dose for carcinogenesis. Demonstration of a protein-based mechanism of carcinogenesis is hindered because of the many potential protein targets for chemicals, which might influence control of cell physiology and proliferation.

Recent findings on altered homeostatic regulation of HSCs in the microenvironment of the bone marrow niche may lead to considerable improvement in our understanding of molecular mechanisms that contribute to myeloid neoplasia. Disruption of feedback signaling by myeloid progenitor cells, which may not represent targets of leukogenic transformation, leads to unregulated proliferation of quiescent HSCs. Consequently, recurrent cytogenetic abnormalities may emerge and lead to leukemic transformation.

Current studies of chemically induced leukemia are focused on the observation that remission of cancers following treatment for a variety of tumors with either alkylating agents or topoisomerase II inhibitors leads to so-called "second cancer," which is manifested as a form of acute myelogenous leukemia termed t-AML. Both treatments resulted in abnormalities in chromosomes 5 and/or 7. Patients with alkylating agent-associated t-AML developed myelodysplasia and experienced a latency period of 5–7 years between treatment and second cancer. Following topoisomerase II treatment the latency period was 1–3 years and no myelodyplasia was observed. Similar effects have not been observed with other forms of leukemia.

3.8.9 Summary

The anatomy of multi-organ animals requires a mechanism for the maintenance of homeostasis. The blood provides a system to maintain adequate cellular oxygen levels, body temperature, the pH of the cells of the body, etc. Furthermore, the cells of the body require anabolic nutrients and an avenue for the removal of the products of catabolism. The introduction of foreign agents, chemical, physical or biological, into the body requires routine or emergency response mechanisms, a prime example being the immune system. Furthermore, the bone marrow provides a milieu in which the system can continually revitalize itself. Clearly, the diverse mechanisms responsible for protection against disease are essential for life. The haematopoetic system, however, is also prone to diseases having a variety of etiologies including genetic impairments, nutritional deficiencies, traumatic injury, and the impact of chemicals or microorganisms. Cancers, i.e. leukemias, are the most insidious threat to life because all of the functions of the system which support the lives of the cells of the body are interrupted. Despite dramatic advances in the therapy of the leukemias, around the world they remain largely refractory to treatment and frequently fatal. The haematopoietic system has been the subject of intensive research for centuries. The discoveries of the circulatory system by Harvey, the microscope by Leeuwenhoek in the 17th century, and of oxygen and its biological functions by Priestly and Lavoisier in the 18th century opened the door to study of haematology, which continues to be an ever-expanding area of research into an understanding of the mechanisms by which the bone marrow and blood function in health and disease.

Further Reading

Calvi, L.M. and Link, D.C. The hematopoietic stem cell niche in homeostasis and disease. *Blood* **126**, 2443–2451, 2015.

Greim, H., Kaden, D.A., Larson, R.A., Palmero, C.M., Rice, J.M., Ross, D. and Snyder, R. The bone marrow niche, stem cells, and leukemia: impact of drugs, chemicals, and the environment. *Ann. N.Y. Acad. Sci.* **1310**, 7–31, 2014.

Hoffmann, R., Benz, E.J., Jr., Shattil, S.J., Furie, B., Cohen, H.J., Silberstein, L.E. and McGlave P. Hematology: *Basic Principles and Practice*. 3rd edition, 2584 pages, Churchill Livingstone, New York, 2000.

Larson, R.S. and Le Beau, M.M. Therapy related myeloid leukemia: A model for leukemogenesis in humans. *Chem.-Biol. Interactions* **153–154**, 187–195, 2005.

Mercier, F., Ragu, C. and Scadden, D. The bone marrow at the crossroads of blood and immunity. *Nat. Rev. Immunol.* **12**, 49–60, 2011.

Sanchez-Aguilera, A. and Mendez-Ferrer, S. The hematopoietic stem cell niche in health and leukemia. *Cell Mol. Life Sci.* **74**, 579–590, 2017.

Schreck, C., Bock, F., Grziwok, S., Oostendorp, R.A.J., and Istvanffy, R. Regulation of hematopoiesis by activators and inhibitors of Wnt signaling from the niche. *Ann. N.Y. Acad. Sci.* **1310**, 32–43, 2014. doi: 10.1111/nyas.12384.

Snyder, R. Benzene and leukemia. *Crit. Rev. Toxicol.* **32**, 155–210, 2002.

van der Klauw, M.M., Wilson, J.H., and Stricker, B.N. Drug-associated agranulocytosis: 20 years of reporting in The Netherlands (1974–1994). *Am. J. Hematol.* **57**, 206–211, 1998.

Vardiman, J.W. The World Health Organization (WHO) classification of the hematopoietic and lymphoid tissues: An overview with emphasis on myeloid neoplasms. *Chem.-Biol. Interactions* **184**, 16–20, 2010.

3.9 The Immune System[1]

Peter Griem

3.9.1 Introduction: the Innate and Specific Immune System

The immune system, its components and the multiple mechanisms controlling cell activation and inhibition may be affected by toxicants. General cytotoxic damage can

---

[1] The critical review of the manuscript by Debra L. Laskin, Dept. of Pharmacology and Toxicology, Ernest Mario School of Pharmacy, Rutgers, NJ, USA, is highly appreciated.

> weaken immune defense (immunosuppression) and cause reduced protection against infections and tumors. Partial damage of individual subpopulations or a general, non-specific immune stimulation may alter the immune homoeostasis, resulting in partial immunosuppression or an increased susceptibility to autoimmune or allergic reactions. The immune system can also specifically react against chemicals or self proteins altered by chemicals which may lead to allergic or autoimmune reactions.

During the course of evolution, the immune system evolved to protect organisms against the invasion, spread and pathogenic effects of viruses, bacteria, fungi, parasites and toxicants.

All living organisms maintain their integrity through a number of non-specific protection mechanisms, called the innate immune system. This comprises phagocytic cells that can engulf invading microbes, dead cells and cell debris as well as foreign (macro)molecules and soluble proteins, such as those of the complement system (capable of forming deadly pores in cell membranes), acute phase proteins and enzymes, e.g. proteases like lysozyme or single-stranded nucleases. With higher phylogenetic levels, the immune system becomes more and more complex and versatile.

At the level of vertebrate organisms, a second arm of the immune system has evolved, called the adaptive immune system. This arm uses antigen(Ag)-specific components, such as B and effector T lymphocytes. B cells produce highly Ag-specific immunoglobulins (Ig, aka as antibodies, Ab) which can neutralize toxins, aggregate foreign material and mark microbes or infected cells for elimination. Cytotoxic T lymphocytes (T_C) can kill infected cells and microbes, while helper T cells (T_H) stimulate B lymphocytes for Ab formation. Regulatory T lymphocytes (T_{Reg}) help terminate immune responses and thus prevent self-damage to the organism by overshooting immune reactions. Contrary to the immediate action of the innate immune system, the adaptive immune response requires several days to build up an effective cellular T cell and Ab response. The advantage of this response is its high specificity and effectiveness and that it builds an immunologic memory – in the form of memory T and B cells – after the first contact (primary immune response), allowing a much faster and more efficient secondary response upon a re-encounter with the same pathogen (e.g., protection from disease after vaccination against measles virus).

In vertebrate animals, including humans, the phagocytes have evolved into a plethora of specialized cells. In the blood, about two-thirds of white blood cells (leukocytes) are neutrophils. They share with eosinophils and basophils the presence of a lobed cell nucleus and large numbers of intracellular storage granules and are therefore called polymorphonuclear granulocytes. All of these immune cells are derived from pluripotent stem cells in the bone marrow by complex differentiation processes.

Neutrophils respond to chemotactic factors released by microbes or activated or damaged cells (e.g., leukotrienes, histamine, complement components) by leaving the blood vessel and migrating to the site of infection, where they destroy invading microbes by activating a respiratory burst (generation of reactive oxygen and nitrogen species), by degranulating vesicles containing, amongst others, pore-forming proteins, and by phagocytosis and degradation of the microbes. They recognize pathogen-associated molecular patterns (PAMP), such as bacterial mannanes, β-glucanes and lipopolysaccharides, with the help of genetically encoded pattern-recognition receptors (PRR) in their cell membrane. Activated macrophages secrete leukotrienes and cytokines that cause dilatation of small venules (visible as erythema in the skin) and lead to the attraction of leukocytes and flow of plasma into the tissue (perceivable as skin swelling). Taken together this process is clinically called a local inflammatory reaction.

Phagocytic cells are present in virtually all organs in the form of specialized tissue macrophages, such as alveolar macrophages in the lung, Kupffer cells in the liver, Langerhans cells in the skin, microglia cells in the brain and synovial macrophages in the joints. In addition, blood monocytes can migrate out of the blood vessels into tissues in the case of a local infection or inflammation. Besides secreting organ-specific proteins, macrophages form part of the innate immune defense and initiate inflammatory responses and the recruitment of cells of the specific immune system to the site of inflammation. In secondary lymphatic organs (spleen, lymph nodes, tonsils, appendix, Peyer's patches in gastrointestinal mucosa) macrophages differentiate into resident antigen-presenting cells (APC) where they present fragments of phagocytosed Ag to naïve T lymphocytes, allowing the latter to expand and differentiate into effector T cells.

3.9.2 Antigen Recognition

An Ag is a non-self structure, such as a microbial glycoprotein or other macromolecule, which can initiate an immune response. The recognition of Ag by the immune system is mediated by specialized receptors in the cell membrane of lymphocytes. The macromolecule is not recognized in its entirety, but Ag receptors bind to a small part of the Ag, which is called the antigenic determinant or epitope. The Ag receptors of B cells are membrane-bound Ab. Each Ab is Y-shaped with two identical Ag binding sites on the short arms. Each binding site has a three-dimensional structure that allows multipoint non-covalent binding to the epitope recognized. In case of protein Ag the epitope comprises about 15 amino acid side chains which can lie adjacent to each other in the polypeptide chain (linear epitope) or have a close spatial arrangement due to the native folding of the polypeptide chain (conformational epitope). Besides proteins, Ab can be directed against all kinds of structures, such as oligo- and polysaccharides, lipids, DNA and RNA as well as drugs and chemicals.

In contrast to B cells, T lymphocytes exclusively recognize protein Ag and are not able to bind with their Ag-specific T-cell receptor (TCR) to the native protein Ag, but need the help of APCs. The membrane-bound TCRs recognize peptide breakdown products of the protein Ag (about 8–25 amino acids long) that are bound to a special peptide-binding groove of a major histocompatibility complex (MHC) molecule on the APC membrane.

> **Complete Ag, e.g. viral glycoproteins, can initiate the activation of both B and T cells and usually have a molecular weight of >4000 Da. Many xenobiotic chemicals or drugs are too small to initiate an immune response. Some of them, nevertheless, can induce immune responses through binding covalently to a self protein of the organism, which turns this into an Ag. For some xenobiotics the covalent binding is brought about by conversion into reactive intermediate metabolites by, for example, cellular metabolism, autoxidation by air oxygen or photochemical reactions in the skin. Reactive chemicals that can bind to proteins and render them antigenic are called haptens.**

The MHC is a group of genes, located on chromosome 6 in humans, which encode two classes of MHC proteins (human forms are called HLA, for human leukocyte antigen). The MHC class I molecules are single transmembrane proteins comprising three immunoglobulin-like domains that associate with a soluble β_2-microglobulin (building the fourth domain in the spatial structure). MHC class I molecules are found on all nucleated cells in the body. The class II molecules are heterodimers of two transmembrane polypeptide chains (α and β chains), each having two immunoglobulin-like domains, so that their overall three-dimensional structure resembles that of a class I molecule associated with β_2-microglobulin. Class II molecules are expressed on monocytes, macrophages, dendritic cells, and B cells, all of which can act as APCs. Some other cell types, such as endothelial or epithelial cells, can be stimulated, e.g. by interferon(IFN)-γ, to transiently express MHC class II molecules.

The MHC genes code for an extremely polymorphic system, i.e. within the population exists a large number of genes (allels) encoding for slightly different polypeptide sequences. Differences affect primarily the amino acid side chains forming the peptide-binding groove of the MHC molecule and thus the range of peptides which they can bind and present to T lymphocytes. Due to these interindividual differences, the MHC molecules themselves constitute very powerful Ag that can lead to immune response-mediated rejection of transplanted organs.

> **Within the peptide-binding groove of MHC molecules two or three binding pockets (each accommodating one amino acid side chain) are responsible for anchoring the antigenic peptide to the MHC molecule. Only antigenic peptides having complementary amino acid side chains, which efficiently bind to the anchor pockets of a given MHC allel, will be presented to T cells. The APC cannot differentiate between self and non-self peptides and thus will present all peptides able to bind to its set of MHC molecules.**

The process of partial degradation of antigenic proteins by proteases into peptides is called Ag processing and occurs in APC. These cells can phagocytose Ag and the resulting intracellular phagocytotic vesicles are fused with lysosomes containing degrading enzymes, such as proteases, lipases, glycosidases and nucleases (then called phagolysosomes). MHC class II molecules are synthesized at the rough endoplasmic reticulum (ER) and are transported in vesicles to the phagolysosomes. They can also be recycled from the cell surface. During the protein degradation, peptides about 13–18 amino acid in length may bind to the class II molecules if they have suitable anchoring side chains, which saves them from further degradation. The peptide–MHC II complexes are finally transported to the APC surface where they are presented to T cells.

All nucleated cells of the body express about half a dozen different MHC class I molecules on the cell surface. The class I molecules bind peptides of intracellular proteins, which comprise normal cell proteins, mutated or abnormally expressed proteins as well as viral proteins in the case of an infection. These intracellular proteins are degraded by large multi-enzyme complexes (proteasomes). They cleave proteins into peptides of 8–10 amino acids in length and transport them into the ER lumen, where they can bind to newly synthesized MHC class I molecules, provided that they have the correct anchoring amino acid side chains. The peptide–MHC I complexes are then transported to the cell surface along the usual secretory pathway.

> **The two different classes of MHC I and II molecules and the compartmentalization of the Ag processing allows exogenous and endogenous Ags to be processed separately and presented to different T-lymphocyte subpopulations.**

3.9.3 Activation of T and B lymphocytes

T lymphocytes

The TCR is a heterodimer of an α and a β chain polypeptide (each consisting of two immunoglobulin-like domains) that jointly form the binding site for the MHC–peptide complex (similar to the Ag-binding site of an Ab). During the development of a T precursor cell, the genes for the α and β TCR polypeptide chains are formed by random fusion of three α and four β DNA segments out of a large number of segments available in the human genome. The DNA stretch of the newly combined gene segments codes for the MHC–peptide binding site of the TCR, i.e. the antigenic specificity, while the constant part of the gene encodes the Ig structure, the transmembrane sequence and the binding site for costimulatory molecules. This TCR specificity will be transferred to all daughter cells.

The mechanism of combining genetic building blocks into protein-coding genes is called somatic recombination und creates a practically unlimited number of TCR specificities, which constitutes the T-cell repertoire. The T-cell specificity comprises not only the antigenic peptide, but also the MHC allele that presents the peptide, which is called the MHC restriction of the T lymphocyte. A single TCR–MHC–peptide interaction may already cause an activation signal inside the T cell and the interaction of about ten TCRs with identical MHC–peptide complexes on the contacting APC may already be sufficient for a full T-cell stimulation. Full activation requires not only TCR binding to the peptide–MHC complex (signal 1), but also signals generated by the interaction of T cell and APC membrane proteins (signal 2) and by the release of cytokines by the APC (e.g., IL-1, IL-2, IL-12) into the cell–cell contact area (signal 3).

T lymphocytes can be subdivided into subpopulations identified by staining with monoclonal Ab recognizing certain surface markers (mostly transmembrane proteins). The most important subpopulations are the $CD4^+$ and $CD8^+$ T lymphocyte subsets.

> **$CD8^+$ T cells are cytotoxic T lymphocytes (CTL or T_C) which recognize peptides in context with MHC class I molecules. In the case of a viral infection, infected cells produce viral proteins in the cytoplasm and thus will present peptides of these on their MHC class I molecules on the cell surface. T_C recognize and kill the infected cells and thereby eliminate the infection.**
>
> **$CD4^+$ T cells are also called helper T cells (T_H) and recognize peptides in context with MHC class II molecules. Various cytokines released by different T_H cells stimulate B cells for Ab production and have a central role in initiating and controlling all specific immune responses.**

According to the set of cytokines produced, T_H cells can be subdivided into T_H1 cells that stimulate cytotoxic $CD8^+$ T cells (T_c) and macrophages through the release of IL-2, IFN-γ and lymphotoxin (TNF-β), and T_H2 cells, which stimulate Ab production by B cells via secretion of IL-4, -5, -6, -10 and -13. The cytokines released by T_H cells, as well as T_C cells and macrophages, determine which type of effector mechanism will be used against the Ag. An activated cell usually releases several different cytokines into the cell–cell contact site. The target cell expresses specific cytokine receptors on its cell surface that trigger intracellular signaling cascades upon cytokine binding, which, in turn, alter the activation status of the cell, e.g. inducing cell proliferation and/or differentiation, protein synthesis and secretion, or expression of cytokines and cytokine receptors. Cytokines are usually produced only by a small number of activated cells and will act only on cells contacting the producing cell (paracrine activity) or on the producing cell itself (autocrine activity). When released by a large number of cells, e.g. at the beginning of a virus infection, or when given therapeutically by injection, cytokines may produce systemic symptoms, such as fever, malaise, nausea, muscle and joint pain. A massive cytokine release, e.g. of TNF-α and IL-1β in septic shock, can be life-threatening.

B lymphocytes

As mentioned above, B lymphocytes can secrete Ab after having encountered their Ag and having received a cytokine signal from T_H cells. An Ig is composed of two heavy (large) polypeptide chains forming a Y-shaped spatial structure. At the upper ends, each heavy chain associates with a light (small) polypeptide chain, thereby forming two Ag binding sites. During B cell development the genes for the heavy and light chains are formed by well-controlled, stepwise somatic recombination of four and three, respectively, gene segments. Similar to the TCR, Ag specificity of the Ig is determined by the variable parts of both polypeptide chains, allowing the generation of a virtually unlimited number of Ag specificities (B-cell repertoire).

Newly formed B cells first express their Ig as a membrane-bound IgM. Ag binding at this early stage of development will lead to apoptosis in order to avoid the generation of autoreactive B cells (this process is called central tolerance formation). During further development into mature, naïve B cells membrane-bound IgD and further accessory membrane proteins are expressed.

The B cell internalizes Ag, e.g. a bacterial toxin, when it binds to its membrane Ig by endocytosis, then processes the Ag in phagolysosomes and presents the resulting peptides on its MHC class II molecules to T_H cells. Those T_H cells having TCRs with the correct specificity will release cytokines and stimulate proliferation of the B cell (clonal expansion) and its differentiation into Ab-secreting plasma cells and memory B lymphocytes.

The constant end of the heavy Ig chain determines its Ig class. Five Ig classes (or isotypes) exist: IgM, IgD, IgA, IgE and IgG. The Ig class determines the immunologic effector mechanism used to attack the Ag. For example, binding of IgG to surface proteins on bacteria will provoke their lysis and phagocytosis by T_C and natural killer (NK) cells that carry $F_c\gamma$ receptors on their surface, binding to the invariant part of the heavy IgG γ chain.

In naïve B cells (without Ag contact) membrane-bound IgM and IgD are formed by alternate splicing of mRNA. After Ag contact and T_H help, an Ig class switch is induced by somatic recombination of the Ag-specific heavy chain gene segment with a different segment for the constant part of the heavy chain (α, ε or γ gene segment). This process allows the production of soluble IgA, IgE and IgG of the same Ag specificity.

3.9.4 Immunologic Tolerance

> **While the specific immune system (B and T cells) must be able to react against virtually any non-self molecule in order to protect the integrity of the organism, it usually does not initiate immune reactions against self. This immunologic tolerance is acquired through contact with self proteins during development of B and T lymphocytes and needs to be actively maintained.**

During early stages of lymphocyte maturation in the primary lymphatic organs, namely the bone marrow and the thymus, Ag receptors are generated by somatic recombination irrespective of whether they react with self or non-self molecules. The development of mature T cells against self proteins is prevented in the thymus, where precursor T cells, so-called thymocytes, mature in a complex process. After generation in the bone marrow $CD4^-CD8^-$ (double negative) precursors differentiate into $CD4^+CD8^+$ (double positive) thymocytes. These will only be kept alive by cell–cell contacts and cytokine release from thymic epithelial cells if they have a low affinity for one of the host's MHC molecules (positive selection of thymocytes with correct MHC restriction). Autoreactive T cells with a high affinity for self peptide–MHC complexes are deleted by apoptosis (negative selection or clonal deletion).

To aid in this process, thymic medullary epithelial cells express proteins which are otherwise expressed exclusively in certain organs. A small percentage of the initially immigrated precursor cells are ultimately released from the thymus as mature, naïve T cells that travel through the secondary lymphatic organs in search of their Ag.

As not all self proteins are expressed in the thymus at sufficiently high concentrations to destroy all potentially autoreactive T cells, further mechanisms exist in the periphery to either destroy autoreactive T cells or render them functionally inactive (i.e., in a state of anergy). This may be accomplished when an autoreactive T cell recognizes its peptide–MHC complex (signal 1) without the accessory signals 2 and 3, e.g. because the APC is not stimulated by local inflammatory mediators.

3.9.5 Sensitization and Allergy

> **Hypersensitivity reactions are the most common immunotoxic reaction caused by chemicals. Mechanistically, the induction phase (primary immune reaction) in which sensitization is acquired precedes the elicitation phase (secondary immune reaction) in which re-encounter of the allergen causes a symptomatic allergic response. For skin sensitization (dermal contact with the allergen) validated and widely accepted predictive tests for hazard identification are available, while this is not the case for other exposure routes (inhalation, oral, parenteral).**

The **induction phase** is often symptomless, nevertheless it leads to the propagation of Ag-specific T cells, and eventually to Ab-producing B cells. In some circumstances a single contact is sufficient to induce sensitization (usually these are potent allergens, such as urushiol in poison ivy). In other cases repeated contacts over a long time period do not lead to sensitization (because the dose stays below the sensitization threshold) until the sensitization is facilitated by additional processes, e.g. contact on inflamed skin already containing activated immune cells.

In the **elicitation phase**, another contact with the allergen (accidentally or for diagnostic purposes in a patch test) induces a secondary reaction mediated by effector T cells or Ab accompanied by a broad range of symptoms, such as red skin, itching eyes, nasal discharge and, in severe cases, anaphylactic shock. Cell-mediated allergic reactions, e.g. allergic contact dermatitis, need one or a few days to develop as cells have to migrate into the contact site and are therefore referred to as delayed-type hypersensitivity (DTH) reactions. In contrast to DTH reactions, immediate-type hypersensitivity reactions develop within minutes and are mediated by IgE bound to $F_c\gamma$ receptors on the surface of mast cells or basophilic granulocytes that, upon Ag binding, release histamine and other vasoactive inflammatory mediators.

Dose–response relationships can be established for both the induction and elicitation reactions. Usually elicitation can be caused by (much) lower concentrations (doses). Thus, in risk assessment and management it needs to be decided if the naïve (non-sensitized) population is to be protected against acquiring sensitization or if an already sensitized subpopulation exists that is to be protected against the elicitation of allergic reactions. The following paragraphs describe methods allowing the characterization of induction thresholds and sensitizing potency, while elicitation will be addressed in *Elicitation Thresholds*, this chapter.

Animal Studies

Predictive tests with internationally harmonized and recognized OECD test guidelines are only available for the induction of skin sensitization (see Table 3.20 and IPCS,). Today's most widely

Table 3.20 Test methods identifying skin sensitizers.

Method	Principle	OECD test guideline*	Comment
In-vivo tests			
Local lymph node assay (LLNA) using tritium-labeled thymidine)	Measures DNA replication (proliferation of immune cells) in female CBA mice induced in the draining auricular lymph nodes after topical application onto the ears	TG 429	Threshold for a positive reaction is a stimulation index of 3.0; modification using additional UV irradiation for identification of photosensitizing substances is possible
LLNA (DA method)	Non-radioactive variant measuring adenosine triphosphate through bioluminescence as a measure of cell number increase in the auricular lymph node	TG 442A	Threshold for a positive reaction is a stimulation index of 1.8
LLNA (BrdU ELISA)	Non-radioactive variant measuring DNA replication through incorporation of 5-bromo-2′-desoxyuridine (BrdU) using a BrdU-specific ELISA	TG 442B	Threshold for a positive reaction is a stimulation index of 1.6
Guinea pig maximization test (GPMT; aka Magnusson–Kligman test)	Intracutaneous and epicutaneous induction followed by a epicutaneous challenge after which skin reddening and swelling are evaluated	TG 406	
Buehler test (epicutaneous sensitization test in guinea pigs)	Epicutaneous induction followed by a epicutaneous challenge after which skin reddening and swelling are evaluated	TG 406	Modification using additional UV irradiation for identification of photosensitizing substances
Mouse ear swelling test (MEST)	Elicitation test, e.g. in female Balb/c mice, measuring increase in ear thickness after challenge on the ear; prior immunization, e.g. by intradermal injection in Freund's complete adjuvant	No OECD guideline	

Table 3.20 (Continued).

Method	Principle	OECD test guideline*	Comment
In-vitro tests			
Direct peptide reactivity assay (DPRA)	Measures binding of chemicals to lysine or cysteine amino acid side chains in model peptides by high-performance liquid chromatography/mass spectroscopy (HPLC/MS)	TG 442C	Does not yet include metabolization in order to identify pro-haptens
KeratinoSens™ LuSens	Measures activation of the Keap1/Nrf2 signal transduction pathway using a luciferase reporter gene under ARE promotor control	TG 442D	
h-CLAT	Activation of the human monocytic tumor cell line THP-1 using flow cytometric measurement of the upregulation of surface markers CD86 and CD54	TG 442E	

*OECD test guidelines are available on the OECD website: http://www.oecd.org/env/ehs/testing/.
LLNA, local lymph node assay; h-CLAT, human cell line activation test; TG, test guideline.

used method is the local lymph node assay (LLNA) in mice. The animals receive the test substance topically on their ears for three days and increases in cell proliferation or cell count in the draining lymph node are measured by one of three guideline methods relative to a vehicle control group, i.e. the ratio is reported as a stimulation index (SI). When measuring DNA incorporation of radiolabeled thymidine, the test substance concentration causing an SI of 3 (called EC3) is used as a quantitative measure of the skin-sensitizing potency of the test substance.

Before the advent of the LLNA, skin-sensitization hazards were identified by studies in guinea pigs using two different protocols. In contrast to the LLNA that analyses the primary immune reaction caused by the test substance, the guinea pig tests measure the specific secondary immune response, i.e. the allergic reaction. In the maximization protocol (also named after its inventors, the Magnusson–Kligman method), the test substance is mixed with inactivated mycobacteria in mineral oil, called Freund's complete adjuvant, and injected subcutaneously. About 3 weeks later, a topical challenge with the test substance is performed and the local formation of edema and erythema is assessed. In the Buehler protocol the induction is accomplished by dermal administration of the test substance which renders the method somewhat less sensitive. The guinea pig methods are less suitable for a quantitative characterization of sensitizing potency than the LLNA.

In vitro Studies

No single *in vitro* test covers the complex interplay between cells that occurs in a primary immune reaction. Therefore, a combination of several tests is required to investigate the relevant molecular initiating events along the adverse outcome pathway for skin sensitization (OECD, 2012). *In vitro* assays can measure the skin permeation (OECD TG 428) and the protein binding of the substance, evaluated with synthetic heptapeptides containing either cysteine or lysine as electrophilic groups (OECD TG 442C).

Substances that can act as haptens often activate keratinocytes by altering cysteine side chains of the "sensor" protein Keap1 that will then dissociate from the transcription factor Nrf2, allowing it to translocate into the cell nucleus and activate transcription of genes, e.g. those coding for xenobiotic phase II metabolism enzymes, under control of the antioxidant/electrophile response element (ARE). The OECD TG 442D uses the human keratinocyte cell line HaCaT and a luciferase reporter gene under ARE-Nrf2 control, allowing cell activation to be measured by luminescence.

After activation, Langerhans cells in the skin will differentiate into dendritic cells while migrating to the nearest lymph node where they can present phagocytosed and processed Ag (e.g., haptenated proteins) to T lymphocytes. This activation could be measured using skin biopsies or isolated primary Langerhans cells or dendritic cells differentiated from blood monocytes. However, these methods are often difficult to standardize for reproducible results. Therefore, the OECD TG 442E uses the human leukemia cell line THP-1, which reacts after exposure to haptens with upregulation of CD86 and CD54 cell surface markers, which can easily be measured by flow cytometry using monoclonal Ab against the two marker proteins.

Another important step of the adverse outcome pathway would be the activation of naïve T cells *in vitro*. However, this has proven to be very difficult to establish *in vitro*. It was long thought that it is caused by a low frequency of T cells expressing a TCR recognizing the hapten or Ag in question. However, it seems more probable that regulatory T (T_{reg}) and dendritic cells act to suppress T-cell proliferation.

Human Studies

Tests in humans resemble the Buehler test design in that the subjects receive several patches prepared with test substance on a defined skin area during the induction period followed by a challenge patch about 2 weeks later using erythema and edema as end points. Currently, the human repeat insult patch test (HRIPT) is not used for hazard identification for ethical reasons. However, for substances that are designed for consumer skin contact, e.g. fragrance materials, the HRIPT is still used as a confirmatory assay to confirm concentrations of test substances predicted to cause no sensitization induction from the above-mentioned predictive studies.

For substances with a long-standing use in consumer products or in the workplace epidemiological data may be available showing existing sensitization in diagnostic patch tests. These data may be sufficient to justify classification and labeling as skin sensitizers according to globally harmonized system (GHS). Characterization of the sensitizing potency from epidemiological data is usually difficult.

Elicitation Thresholds

> **Dose–response relationships for the elicitation of allergic reactions that may be used for human risk assessment can only be established in allergic (sensitized) individuals.**

Dermatologists perform epicutaneous patch tests for diagnostic purposes in patients suspected of presenting allergy-mediated skin symptoms. In this context, contact allergy does not refer to a state of disease, but just describes that sensitization has taken place. The next step evaluates if the diagnosed sensitization is the cause of the skin symptoms presented to the dermatologist by the patient. After anamnesis additional patch tests or repeated open application tests (ROATs) may be required to elucidate whether a patient reacts to products used at the workplace, to household or cosmetic products or shoes, clothes, jewelry, etc. (see Table 3.21 for list of skin sensitizers). When this leads to the identification of the substance most likely causing the allergic skin symptoms, the patient is diagnosed with allergic contact dermatitis (ACD). It is often not possible to identify the original cause of the sensitization, which may have been the same product or an exposure to the same substance from a different source, e.g. plants in the garden or the handling of vegetables or spices. The sensitization might also have been caused by a structurally related substance, while the allergic reaction is caused by an immunologically cross-reactive substance.

Table 3.21 Examples of skin sensitizers.

Substance group	Examples
Pharmaceuticals	Antibiotics, such as bacitrazin, quinolones (gyrase inhibitors, e.g. ciprofloxacin), neomycin sulfate in antibiotic ointments; anesthetics, e.g. benzocaine, dibucaine, tetracaine; corticoids, e.g. budesonide
Biocides	Isothiazolinons, formaldehyde, dimethylfumarate, e.g. for treatment of clothes, shoes and furniture
Fragrance substances	Synthetic and plant-derived fragrance materials, e.g. cinnamal, hydroxycitronellal, isoeugenol, farnesol, Lyral® used in cosmetic and household products
Rubber chemicals (e.g., monomers)	N-Isopropyl-N'-phenyl-PPD, N-cyclohexyl-N'-phenyl-PPD, N,N'-diphenyl-PPD, diphenylguanidine, zinc dibutyldithiocarbamate, zinc diethyldithiocarbamate, mercaptobenzothiazole, tetramethylthiuram-monosulfide, -disulfide, disulfiram; for production of tires, belts, shoes, tubings, cables
Industrial chemicals	Ethylendiamine dihydrochloride, formaldehyde, paraphenylene diamine (also in permanent hair dyes)
Preservatives	Formaldehyde, bronopol (2-brom-2-nitro-1,3-propandiol), diazolidinyl urea, imidazolidinyl urea, quaternium-15 (other formaldehyde releasers), chloromethylisothiazolinone/methylisothiazolinone, methyldibromoglutaronitrile; used in medicinal drugs, cosmetics, colors, household products; thimerosal in antiseptic sprays, vaccines
Metals	Nickel (Ni(II) compounds), cobalt (Co(II) compounds), chromium (Cr(VI) compounds), e.g. in tools, coins, cement
Monomers	Diglycidylether in epoxy resins, acrylic acid esters (e.g., ethylacrylate)
Plant products	Balsam of Peru (tree sap of *Myroxylon balsanum* var. *pereirae*) as flavor in beverages and food, fragrances, medicaments and medical products. Colophony (sap resin of conifers or from talloil of paper production) in soldering, papier glueing, sport resins, violin bows, glues Sesquiterpene lactones (esp. compositae such as Arnica) from flowers and garden plants
Reactive dyes	Dispersion blue 106, e.g. as textile dyes
Surfactants	Cocoamidopropylbetain, e.g. in liquid soap, shampoo, shower gel
Animal products	Wool fat (lanolin) from sheep, e.g. in woolen clothes, cosmetics, medical products

Modified from: Davis, M., Hylwa, S.A., Allen, E.M. (2013). Basics of patch testing for allergic contact dermatitis. Semin. Cutan. Med. Surg. 32, 158–168.
PPD, p-phenylendiamine

In diagnostic epicutaneous tests typically dozens of substances in application systems, such as Finn chambers, are fixed on the skin of the back for 24 or 48 hours, followed by assessing the reactions 1, 2 or 3 days later. In these tests the concentrations of the test substances are usually high enough to identify an existing sensitization while minimizing false negative results, but low enough to avoid unspecific irritation reactions (false positive reactions) or even induce skin sensitization.

In order to characterize the concentration–elicitation relationship it is necessary to test different concentrations of the relevant allergen, often covering several orders of magnitude. As a result, the NOEC and LOEC value can be identified for each patient. From these data a benchmark concentration, the MET_{10} (minimum elicitation threshold causing a slight reaction in 10% of sensitized individuals) is calculated.

NOEL and LOEL values can also be determined in a ROAT. In this test patients daily apply a pre-determined amount of product onto a defined skin area. The product, e.g. a leave-on product such as a skin cream or a rinse-off product such as a liquid soap, contains the allergen in a very low concentration. When no symptoms are seen after 2–3 weeks, the product is exchanged with one containing a higher allergen concentration. This procedure is repeated until the patient develops slight allergic skin reactions or until the highest testable concentration has been reached. Again, the concentration that provokes an elicitation in 10% of the allergic individuals (ED_{10}) can be calculated. The advantage of the ROAT is the open application, which avoids a higher skin hydration by the occlusive patch test, and the use of a consumer-product-like matrix.

3.9.6 Risk Assessment of Immunotoxic Effects

The WHO International Program on Chemical Safety has developed structured decision trees to show how chemical-induced immunotoxic effects, including sensitization, autoimmunity, generalized immune-stimulation or immune-suppression, can be recognized and evaluated (IPCS, 2012). Existing data are evaluated according to their weight of evidence, consistence of effects, biological plausibility and possible data gaps.

Skin Sensitization

> **With regard to skin sensitizers the primary protection goal is usually the prevention of induction of skin sensitization.**

The development of a quantitative risk assessment methodology for skin sensitization began early in the 21st century. The goal is to derive safe use concentrations of chemicals with skin-sensitizing properties in order to prevent de novo sensitization of exposed individuals (consumers, workers, etc.). The point of departure for a quantitative risk assessment is the no-expected sensitization induction level (NESIL). This is the highest daily skin area exposure (expressed as mg or nmol substance per cm^2 of skin) that does not induce sensitization. The NESIL is often derived from a LLNA EC3 or from a confirmatory HRIPT. As for other toxicological assessments, certain safety factors (sensitization assessment factors) are applied to the NESIL to derive an acceptable exposure level (AEL given also in $mg/cm^2/day$). Exposure scenarios (assumed, measured or modeled) can then be evaluated through comparison with the AEL. The quantitative risk assessment of skin sensitizers was primarily developed by the cosmetic industry and the Research Institute on Fragrance Materials (www.rifm.org) and is employed in the safety assessment of perfume raw materials using an aggregated exposure model (see Safford et al., 2015).

The occurrence of sensitization via the dermal or other routes can never be prevented completely because of the high capability of the immune system to recognize all possible target structures, the evolutionary advantageous, genetically determined high interindividual variability in immune recognition, the heterogeneity of dermal exposures with regard to area concentration, exposed skin site and sensitization-favoring conditions caused by other factors, such as reduced barrier function and local inflammation. The prevalence of contact allergy, i.e. both clinically relevant and silent sensitization, in the non-selected general population is unknown. In dermatology patients the prevalence of contact allergy is often found to be in the single-digit percentage range. The fraction of the subgroup with contact allergy that have clinically relevant ACD is reported in the literature to be between 10% and 100%.

Respiratory Tract Sensitization

Of the different forms of sensitization of the respiratory tract, allergic bronchial asthma is receiving more attention than others, e.g. allergic rhinitis or alveolitis, because it occurs more often, is diagnosed more frequently and can lead to life-threatening episodes of allergy elicitation. While allergic asthma and allergic rhinitis are immediate-type hypersensitivity reactions mediated by allergen-specific Ab, such as IgE bound to mast cells, allergic alveolitis is a cell-mediated hypersensitivity conferring acute symptoms, such as breathing difficulties, subjective shortness of breath, cough and fever, that can lead in the long run to chronic lung fibrosis.

In contrast to the skin, the respiratory tract has no barrier, other than the bronchial mucous layer, against the entry of foreign substances into the body. Therefore, not only do small molecular weight chemicals bound to self proteins induce an immune response, but also substances of high molecular weight, such as foreign proteins or larger particles, such as plant pollen (see Table 3.22).

Currently there are no validated and internationally recognized test methods available for the identification of respiratory sensitizers. Therefore, classification of an agent as a respiratory sensitizer is typically based on cases reports in humans, e.g. for occupational exposure to isocyanates or organic acid anhydrides, or read-across evaluations that use data established for a structurally related substance. For diagnostic purposes the prick test is often employed. This involves testing a series of substances on the forearm by placing droplets of Ag solutions on the skin and then permeating part of it by stinging the skin with a small lancet, allowing larger molecules contact with mast cells. The developing urticarial reaction is assessed about 30 minutes later.

Several experimental approaches, e.g. in guinea pigs, mice or rats, are used to identify a respiratory sensitization potential for new compounds. While inhalation is obviously the most relevant exposure route for these tests, inhalation studies are technically demanding and costly; as a consequence other methods, such as intratracheal or intranasal instillation or intradermal injection are also used. For example, inhalation challenge with diisocyanates was reported to elicit allergic respiratory tract effects after dermal induction. Sensitization can be confirmed by measurement of Ag-specific Ab or T_H2-associated cytokines (e.g. IL-5) in serum, increased numbers of eosinophilic or neutrophilic granulocytes and cytokine concentrations in bronchoalveolar lavage fluid, or effects on ventilation rate or volume by whole-body plethysmography. Dose–response relationships can be established through varying airborne concentrations or applied doses of the allergen. Typically, no-effect concentrations (NOEC) that can be used for a quantitative risk assessment can only be derived from inhalation experiments. It seems that for identical concentration × time ($C \times t$) values, the ones with a higher exposure concentration for a shorter time period confer a higher risk of inducing sensitization.

Table 3.22 Examples of respiratory tract sensitizers.

Substance group	Substance name	Substance use
Metals	Beryllium	Production of fluorescent tubes and tools
	Cobalt and compounds	Glass, ceramic and metal industries
	Nickel and compounds	Smelting, electroplating, welding
	Platinum compounds, e.g. chloroplatinates	Platinum refineries, catalyst production
Anhydrides of organic dicarbonic acids	Phthalic anhydride and hydrogenated compounds	Use of epoxy resins in the electro and electronic industries
	Maleic acid anhydride	
	Trimellitic anhydride	
Isocyanates	Toluene diisocyanate	Production and converting of polyurethanes (foams, adhesives etc.)
	Methylene diphenylisocyanate, hexamethylene diisocyanate, isophorone diisocyanate	
Persulfates	Alkaline persulfates, diammonium peroxodisulfate	Production and use of bleaches, e.g. in hair bleaching products
Antibiotics	Cephalosporins, gyrase inhibitors, makrolides, penicillins, tetracyclines	Pharmaceutical industry and medical professions
Others	Glutardialdehyde	Disinfection, fixation
	1,2-Diaminoethane (ethylendiamine)	Resin production (diamine hardener)
	Piperazine	Pharmaceutical industry
Plant products	Cereals and feeds	Farming professions, transport and storage, flour mills
	Flour from wheat, rye etc.	Bakeries, pastry chops, flour mills
	Natural latex	Production of latex products, use of latex gloves, e.g. in medical and food-handling professions
	Hardwood, such as cedar, abachi, limba	Wood industries
	Flowers, fruits, vegetables, grass and tree pollen, soy beans, rhicinus proteins, herbs and spices, tea, coffee beans, weeds such as ambrosia	Farms, nurseries, plant shops, gardens, parks, food and feed industries
Animal products	Hair, bristle, feathers, horn, excrements, urine	Farming, forestry, animal contact, slaughterhouses
	Fish and sea food	Food and feed industries, gastronomy
	Mites (such as house dust mite, spider mite)	Farming, nurseries, animal keeping, transport and storage of natural products
Microorganisms	Bacteria, e.g. *Thermoactinomyces vulgaris, Saccharopolyspora rectivirgula*	Farming, nurseries, paper archives
	Molds, e.g. *Aspergillus* spp., *Penicillium* spp., *Cladosporium* spp.	Ventilation systems, wood, paper and food industries
Enzymes	Bromelain, cellulases, hemicellulases, papain, pepsin, subtilisins	Food, feed, pharmaceutical and laundry industry, bakeries, farming

Modified from: Technical Rules for Biological Agents, TRBS/TRGS 406 (2008), http://www.baua.de/de/Themen-von-A-Z/Gefahrstoffe/TRGS/TRGS-TRBA-406_content.html

Oral and Parenteral Sensitization

> While allergic symptoms from skin or respiratory tract sensitization manifest mainly at the site of contact with the allergen, systemic allergic symptoms can result from allergies caused by sensitization via the oral route (e.g., foodstuff and medicaments) or parenteral administration (e.g., injection of medicinal drugs).

Food allergies are often mediated by allergen-specific IgE bound to $F_c\gamma$ receptors on the surface of mast cells which upon binding of the allergen degranulate and release allergy mediators. This can cause not only gastrointestinal symptoms, such as nausea, vomiting and diarrhea, but can manifest also on the skin, e.g. as acute urticarial, atopic dermatitis, in the respiratory tract or even as systemic anaphylaxis, which is the result of a generalized release of inflammatory mediators and subsequent multi-organ reaction and cardiovascular collapse. These Ab-mediated allergies usually manifest quickly, while a similar gastrointestinal clinical picture may also be caused by a T-cell-dominated allergy one or a few days after allergen ingestion. The most common causes of oral allergies are foodstuffs, especially milk, egg, peanuts, hazelnuts and other nuts, fish, shellfish and seafood, soy, fruits, and vegetables. The prevalence of certain allergens depends on age and geographical culinary habits. Besides foodstuffs, insect venoms, e.g. from bees and wasps, and medicinal drugs, such as penicillin and other β-lactam antibiotics, quinolones, non-steroidal antiphlogistics like aspirin, ibuprofen, diclofenac, proton pump inhibitors and monoclonal Ab, can provoke systemic anaphylactic reactions in allergic persons.

For food allergies, which affect about 2–4% of the population, risk assessments usually focus on the elicitation of allergy in those already sensitized (e.g., see ILSI Food Allergy Task Force), and the occurrence and concentration of allergens in food products. Through combining data on elicitation thresholds and consumption data the yearly number of cases reaching a chosen severity level that will be provoked by a certain concentration of allergen can be estimated.

Some xenobiotics can also directly stimulate mast cells to release mediators in the absence of specific IgE. These reactions are called pseudoallergic or anaphylactoid reactions. Food additives which may provoke such reactions include food colorings, benzoates, salicylates, sodium glutamate, sulfites, biogenic amines, as well as injected iodinated x-ray contrast media and heparin or physical stimulants, such as cold. A range of different mechanisms cause these reactions and include direct release of histamine, classical and non-classical complement activation producing anaphylatoxins C3a and C5a, inhibition of cyclooxygenase causing an increased formation of bronchoconstrictors like cysteinyl-leukotrienes (e.g., in "aspirin asthma"), stimulation of peripheral or central nerve cells and genetic enzyme polymorphisms.

3.9.7 Chemical-induced Autoimmunity

> Autoimmunity and autoimmune diseases result from immune reactions against self molecules, cells, tissues or organs. For most autoimmune diseases the etiology is unknown and will, most likely, be a combination of intrinsic factors, such as genetic polymorphism, sex hormones, age, and extrinsic factors, such as certain infectious microbes, use of medicinal drugs and exposure against xenobiotics from tobacco, alcohol and illegal drugs.

Genetic predisposition for autoimmune diseases has been linked to polymorphisms of $F_c\gamma$ receptors, TGFβ, IL-1α, and IL-10. Other factors are estrogens (women are at a higher risk for many autoimmune diseases) and age (many autoimmune diseases develop after the third decade). The immunologic effector mechanisms leading to tissue or organ damage from autoimmune diseases are the same as those that protect the organism against invading pathogens or tumorigenic cells. However, the activated T lymphocytes and (eventually) autoantibodies (auto-Ab) produced are directed against self proteins, glycosylated proteins or polysaccharides.

Most autoimmune diseases are idiopathic, i.e. their molecular initiating events are unknown. Xenobiotic chemicals may contribute to their development through increasing or altering the spectrum of self peptides presented to the immune system, e.g. by causing an increased turnover of certain self proteins or altering their degradation during Ag processing. In addition, increased release of cytokines may lead to a stronger signal 2 in T-cell activation, which could make up for an insufficient signal 1 (presentation of self peptide), resulting in activation of autoimmune T cells. Hexachlorobenzene has been shown to directly stimulate macrophages and other cells of the innate immune system to produce an inflammatory signal that can cause a non-specific, polyclonal stimulation of T cells. Another mechanism may involve inhibition of apoptosis, which can cause survival of autoreactive cells that would normally be deleted. This mechanism is suspected of being involved in rheumatoid arthritis, Hashimoto thyroiditis and systemic lupus erythematosus (SLE).

Oxidative damage caused by free oxygen radicals and lipid peroxidation has also been discussed as potentially initiating autoimmune reactions. Auto-Ab specific for proteins and DNA altered by oxidation have been identified in patients with SLE, rheumatoid arthritis, type-I diabetes and autoimmune hepatitis. Oxidized self molecules could constitute neoantigens against which immune tolerance has not been established.

Exposure to xenobiotics can also cause autoimmunity but only after a considerable delay. Using mouse strains developing spontaneous autoimmunity, as well as non-autoimmune prone strains, it has been shown that pre- or perinatal exposure to the dioxin TCDD caused thymic atrophy, increased frequency of peripheral autoreactive T cells by interference with thymic negative selection, earlier onset of predisposed autoimmunity, and increased titers of auto-Ab and immune complexes in kidneys at adult age.

No validated and internationally recognized test methods are available for the detection of an autoimmune potential of chemicals. A number of autoimmunity-causing medicinal drugs and chemicals were identified from epidemiological investigations and clinical case reports. Additionally, studies of workers who may have been exposed to certain chemicals at much higher levels than the general population may help to identify an autoimmune potential of workplace chemicals, such as crystalline silica, vinyl chloride, some heavy metals and solvents (see Table 3.23).

Animal studies measuring an increased incidence or earlier onset of autoimmune diseases may also provide evidence of an autoimmune potential. Models employed to this end include, for example, genetically predisposed strains such as lupus-prone MRL mice and non-obese diabetic (NOD) mice (see IPCS, 2006). Findings that can lend support in a weight of evidence evaluation are altered immune cell frequencies, cytokine releases, weights of secondary lymphoid organs, or a relevant immune stimulation, e.g. measured in the popliteal lymph node assay (PLNA).

3.9.8 General Immunostimulation by Chemicals

> **Activation of immune cells is beneficial in fighting an infection, but needs to be controlled and shut down via feedback circuits to avoid damage to the organism.**

Table 3.23 Examples of autoimmune diseases associated with exposure against chemicals or other exogenous factors (modified from IPCS, 2012).

Autoimmune disease	Mechanism and target structure	Associated exogenous factors
Hemolytic anemia	Auto-Ab against haptenated membrane proteins on erythrocytes	Chlorpromazine, α-methyldopa, penicillins, dapsone, quinidine, non-steroidal antiphlogistics
Hepatitis	Autoreactive T cells and auto-Ab against haptenated hepatocyte proteins	Ethanol, halothane, IFN-α, lipid-lowering drugs
Myositis	Auto-Ab against intracellular components of muscle cells	Estrogens, L-tryptophan, UV radiation, heroin, penicillamine, different viruses and bacteria
Rheumatoid arthritis	Auto-Ab against connective tissue, immune cells reacting against immune complexes in joints	IFN-γ, organochlorinated pesticides, PCBs, quinidine, silica, tetracyclins, tobacco smoke
Scleroderma/systemic sclerosis	Auto-Ab against nuclear proteins, especially in skin, heart, lungs, kidney and gastrointestinal tract	Diphenylhydantoin, IL-2, silica, Spanish toxic oil, trichloroethylene, tryptophan, vinyl chloride
Systemic lupus erythematosus and lupus-like syndromes	Auto-Ab against DNA, nuclear proteins, membrane proteins of erythrocytes and thrombocytes; reactions against immune complex deposition	Aromatic amines, chlorpromazine, formaldehyde, hydralazine, IFN-γ, isoniazid, procainamide, silica, trichloroethylene
Thrombocytopenia	Auto-Ab against (haptenated) membrane proteins on thrombocytes	IFN-α, iodine, quinidine, rifampicin
Thyroiditis	Auto-Ab and autoreactive T cells against thyroid proteins	Iodine, lithium, PCBs, PBBs

A transient, local stimulation of the immune cells by chemical or biological adjuvants is used to achieve a high rate of immunizations from vaccinations. Unlike these valuable immunostimulatory effects, which are limited to B and T cells expressing the correct Ag receptors, an untoward general immunostimulation by a xenobiotic is not connected to a useful immune reaction or maintenance of immune homeostasis; rather it can favor unwanted allergic or autoimmune reactions or bias the immune system towards certain effector mechanisms so that infection incidence and/or severity can increase. For example, a chemical modulating the immune system towards T_H2 reactions may hamper the immune clearance of a viral infection because it is better fought with activated T_C cells from a T_H1 reaction.

Epidemiologic investigations, e.g. in controlled clinical trials or occupational studies investigating correlations between exposure and increased or decreased cellular or Ab-mediated immune reactions, autoimmunity or allergy carry the highest weight of evidence in identifying a general immunostimulatory effect. Animal studies may show enhanced hypersensitivity reactions or autoimmune reactions, or changes in infection models or tumor resistance assays. Other parameters that can lend support include altered Ab production, DTH reactions or T_H1–T_H2 modulation. After identification of a substance as causing general immunostimulation and establishing the dose–response relationship for the critical effect, a conventional risk assessment may be carried out using the NOEL as a point of departure.

Examples of chemicals that favor T_H2 reactions and thus the production of Ab against a test Ag, such as sheep erythrocytes, are mercury, cadmium, lead, malathione, aminocarb, deltamethrin, α-cypermethrin and nonylphenol.

Superantigens constitute another kind of general immunostimulant. These are proteins that can simultaneously bind to a MHC class II molecule outside the peptide-binding groove and to certain $V\beta$ chains of the TCR, thereby cross-linking both molecules and causing a polyclonal T-cell activation. Superantigens play a role in food poisonings caused by staphylococcal enterotoxins A, B, C, D and E, in septic shock caused by toxic shock syndrome toxin 1, Mycoplasma arthritidis, pyrogenic streptococcal exotoxins A, B, C or D, and in arthritis (rheumatic fever).

3.9.9 Chemical Immunosuppression

> **Similar to a general immunostimulation, undesirable immunosuppression can cause negative effects that may lead to increased incidence and severity of infections or cancer.**

Epidemiologic and clinical studies, as well as clinical case reports, have established an association between exposure to xenobiotics and immunosuppression measured as infection rate, primary Ab response to vaccination against influenza virus, secondary Ab response against tetanus toxoid or DTH reaction. In clinical studies higher incidences of opportunistic infections and neutropenia have been observed after high doses of IFN-α, azathioprine, cyclophosphamide or methotrexate. It is much more difficult to identify a weak immunosuppressive effect.

Examples of microbes leading to increased infection rates during states of immunosuppression are extracellular bacteria, such as *Streptococcus pneumonia* and *Haemophilus influenza*, associated with a decreased phagocytosis through neutropenia or inhibition of Ab production, intracellular bacteria, e.g. *Mycobacterium tuberculosis* and viruses, such as Herpes simplex virus, cytomegaly virus, associated with a reduced lytic ability of the phagocytes and reduced killing of infected cells by T_C. Tumors with increased incidence due to immunosuppression are virally-induced tumors, such as non-Hodgkin lymphomas, skin tumors, Kaposi sarcomas and Epstein–Barr virus induced B-cell lymphomas.

Laboratory animal studies can also be useful in identification of chemically induced immunosuppression, especially when their results are concordant between species and supported by functional immune parameters indicating a plausible mechanism. For more information on human and animal test methods see Table 3.24 and IPCS (1996).

Clinical experience with immunosuppressive drugs indicates that the immune system usually has a high capacity to regenerate after exposure has ended. However, irreversible effects may result from high doses and/or long-term exposure when hematopoietic stem cells are damaged. Damage to the developing immune system may also cause long-lasting effects. In industrialized countries levels of polyhalogenated aromatic hydrocarbons have been associated with increased infections and decreased vaccination efficiency. Children showing high levels of the DDT metabolite dichloro diphenyl dichloroethane, PCB or HCB had a higher incidence of ear infections. In rats, the immature immune system of unborn offspring was more sensitive than that of adult animals to the immunosuppressive effects of TCDD, diethylstilbestrole, diazepam, lead and tributyltinoxide. The immune system damage by TCDD and other halogenated aromatic hydrocarbons is mediated through binding to the arylhydrocarbon receptor (AHR), a transcription factor inducing transcription of genes with dioxin-responsive elements (DRE) in the promotor region. In susceptible mouse strains AHR ligands cause a direct suppression of T_H, T_C, B cells (including Ab production), hematopoietic stem cells in the bone marrow and

Table 3.24 Approaches for the experimental evaluation of immunosuppressive properties (modified from IPCS, 2012, 1996).

Endpoint	Parameter in specialized immuntoxicological tests	Coverage in standard subacute, subchronic or chronic animal studies	Possible coverage in (epidemiological) human studies
Non-functional tests			
Organ weight	Thymus, spleen, lymph nodes	Yes	No
Organ histopathology	Thymus, spleen, bone marrow, draining and distal lymph nodes, Peyer's patches	Yes	No
Plasma Ig concentrations	Total IgM, IgG, IgE or others in ELISAs	Yes	Yes
Leukocyte counts and phenotype	In blood or broncho-alveolar lavage; phenotyping using monoclonal antibodies and flow cytometry	Yes	Yes
Functional tests			
Phagocyte activity	*Ex vivo* phagocytosis of latex beads or SRBC, measurement of the respiratory burst	Can be added	Yes
NK cell activity	*Ex vivo* lysis of target cells, e.g. ^{51}Cr-labeled YAC lymphoma cells	Can be added	Yes, e.g. lysis of K562 tumor cells
Ag-specific antibody formation	*In vivo* synthesis of IgM and IgG (ELISA detection) against model Ag, e.g. tetanus toxid, ovalbumin, keyhole limpid hemocyanin, SRBC	Can be investigated in satellite group	Yes, e.g. immunization against diphtheria, tetanus or influenza vaccines
Proliferation stimulated by mitogens	*Ex vivo* proliferation of B or T cells following polyclonal stimulation with lipopolysaccharide, concanavalin A or phytohemagglutinin	Can be added	Yes
Mixed lymphocyte reaction	*Ex vivo* proliferation caused by allogeneic (MHC different) stimulator cells	Can be added	Yes
CTL test	Measurement of lysis of ^{51}Cr-labeled P815 mastocytoma cells following immunization of mice against P815	Can be investigated in satellite group	Yes, e.g. against influenza-infected cells

(*continued*)

Table 3.24 (Continued).

Endpoint	Parameter in specialized immuntoxicological tests	Coverage in standard subacute, subchronic or chronic animal studies	Possible coverage in (epidemiological) human studies
DTH test	Immunization and challenge (e.g., ear swelling test) against protein Ag, such as ovalbumin, listeria, BCG	Can be investigated in satellite group	Yes, e.g. challenge reaction against tetanus, diphtheria, streptococci or tuberculin Ag
Host resistance tests	Resistance against experimental infection with pathogenic microbes, such as *Listeria monocytogenes*, *Streptococcus pneumoniae*, *Candida albicans*, influenza virus, cytomegalo virus, *Trichinella spiralis* or syngeneic tumor cells, e.g. B16F10 melanoma or PYB6 fibrosarcoma in C57BL/6 mice, or MADB106 adenocarcinoma in Fischer 344 rats	Can be investigated in satellite group (high technical and safety requirements)	Not possible for ethical reasons

SRBC, sheep red blood cells.

increases of T_{REG} lymphocytes. In the fetal thymus, epithelial cells are altered, leading to higher apoptosis in thymocytes and a decreased immune reaction against influenza infections.

3.9.10 Summary

The immune system is composed of a large number of diverse cells and subsets that constitute the innate arm of the immune system, e.g. neutrophilic granulocytes, monocytes and macrophages, or to the Ag-specific arm, i.e. T and B lymphocytes. Immune cells develop from precursors in bone marrow and the thymus (primary lymphatic organs) and are found mainly in secondary lymphatic organs, such as spleen and lymph nodes, but patrol through every organ of the body in order to kill or inactivate invading pathogenic microbes, toxins and tumorigenic cells. This is achieved by a complex interplay of phagocytes, lymphocytes and soluble proteins, such as the complement system and acute phase proteins. The main immune effector mechanisms are Ag-specific T_C and Ab-producing B cells, which can neutralize toxins and kill microorgansims, virus-infected cells and tumorigenic cells. Xenobiotic chemicals can interact in different ways with the complex, feedback-controlled immune processes. Reactive chemicals or metabolites can covalently bind to protein and cause a specific T and B cell reaction to the altered protein, which can lead to allergic reactions, such as allergic contact dermatitis or allergic asthma. Other chemicals may lead to general immunostimulatory effects or modulate immune reactions, thereby favoring the development of autoimmune diseases or allergies. Some chemicals can also damage cells of the immune system or modulate differentiation or activation in such a way that a partial or general immunosuppression is caused, which may increase the incidence and severity of infectious diseases and the probability of tumor occurrence. Methodological concepts for the identification and toxicological evaluation of chemically induced allergic reactions, autoimmunity, immunostimulation and immunosuppression have been developed and employ a multitude of different *in vitro* and *in vivo* tests as well as structure–activity predictions.

Further Reading

ILSI (International Life Science Institute) Food Allergy Task Force, ILSI website.

IPCS (International Programme on Chemical Safety) (1996). Principles and methods for assessing direct immunotoxicity associated with exposure to chemicals. *Environmental Health Criteria 180*, World Health Organization, Geneva, http://www.inchem.org/documents/ehc/ehc/ehc180.htm.

IPCS (1999). Principles and methods for assessing allergic hypersensitization associated with exposure to chemicals. *Environmental Health Criteria 212*, World Health Organization, Geneva, http://www.inchem.org/documents/ehc/ehc/ehc212.htm.

IPCS (2006). Principles and methods for assessing autoimmunity associated with exposure to chemicals. *Environmental Health Criteria 236*, World Health Organization, Geneva, http://www.who.int/ipcs/publications/ehc/ehc236.pdf.

IPCS (2012). Guidance for Immunotoxicity Risk Assessment for Chemicals. IPCS Harmonization Project Document No. 10, World Health Organization, Geneva, http://www.inchem.org/documents/harmproj/harmproj/harmproj10.pdf.

OECD (2012). The Adverse Outcome Pathway for Skin Sensitisation Initiated by Covalent Binding to Proteins. Part 1: Scientific Evidence, ENV/JM/MONO(2012)10/PART1. Part 2: Use of the AOP to Develop Chemical Categories and Integrated Assessment and Testing Approaches, ENV/JM/MONO(2012)10/PART2. OECD Environment, Health and Safety Publications, Series on Testing and Assessment, No. 168, Paris, 2012.

Safford B et al. (2015) Use of an aggregate exposure model to estimate consumer exposure to fragrance ingredients in personal care and cosmetic products. *Reg Tox Pharmacol* **72**, 673–682.

3.10 The Eye

Ines Lanzl

3.10.1 Introduction

The eye and its adnexae represent a highly organized well-protected system, designed to gather and process visual information. The information processed within the rather small organ globe and along the visual system is important enough to occupy the largest portion of all sensory input in the human cortex. Considering its complex formation during fetal development its structure and function bear resemblances to such differentiated other organs such as skin, kidney and brain but have their own unique functions. An awareness of this is necessary to understand how the eye is affected by xenobiotics.

3.10.2 Structure and Function of the Eye

> **The eye is protected from the environment by the lids and tear film. The eye itself consists of three major segments: the anterior segment (conjunctiva, cornea, iris, lens), the posterior segment (sclera, choroid, retina, vitreous body) and the optic nerve (Figure 3.31). The outer antero-posterior diameter of an adult's globe measures about 24 mm.**

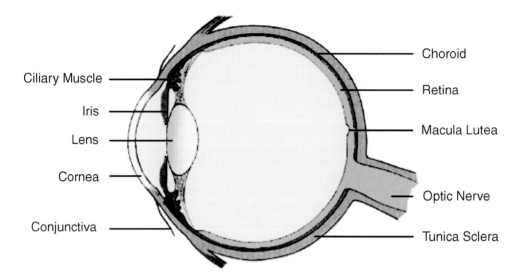

Figure 3.31 *Cross-section of the eye, Source: Courtesy of Chibret, Haar, Germany.*

The Lids

The outermost layer of the eye is created by the lids, which serve as a mechanical barrier to the outside and as a constant means of producing and spreading the tear film by blinking. The lids are comprised of skin, muscle, and sebaceous glands on the outside, and the tarsal plate, conjunctiva, and sebaceaous and accessory lacrimal glands on the inside.

The Anterior Segment

The Tear Film The tear film consists of an outer layer of lipid produced by the sebaceous glands of the eyelid and caruncle, a middle layer of aqueous fluid produced by the lacrimal and accessory lacrimal glands, and an inner layer of mucoprotein produced by conjunctival goblet cells. The conjunctival surface of the eyelids is adjacent to and in contact with the tear film layer. During blinking, this layer is distributed over the surface of the corneal epithelium, continually rewetting the entire surface.

The Conjunctiva The conjunctiva is a mucous membrane that lines the entire exposed surface of the eye from the lid margin up to the corneal limbus. In addition to acting as a physical barrier, it houses immune cells and its substantia propria is highly vascular. Its goblet cells produce mucin, which adsorbs to the glycoproteins coating the microvilli of corneal and conjunctival epithelial cells. This precorneal and preconjunctival mucin layer merges gradually with the overlying aqueous tear film to ensure complete wetting of the ocular surface.

The Corneal Limbus The conjunctival and corneal epithelia merge at the limbus, where the subtle transition from the nonkeratinized, stratified columnar epithelium of the conjunctiva to the nonkeratinized, stratified squamous epithelium of the cornea occurs. These two epithelial surfaces together cover the exposed portions of the eye and function as a barrier to chemical insults and to invasion of microorganisms.

The limbal region is of utmost importance for the maintenance of corneal transparency. First, intact conjunctival vascular arcades are necessary for nutrition of the anterior and peripheral parts of the cornea. The area also hosts corneal epithelial stem cells, which repopulate the corneal epithelium in a centripetal fashion.

If the limbus region is not intact, conjunctivalization of the corneal epithelium occurs, followed by opacification because of the following mechanism:

The conjunctival epithelium provides a source of cells to repopulate the corneal surface when the entire corneal epithelium has been denuded and the limbal stem cells have been destroyed by severe chemical injuries. After complete re-epithelialization of the cornea with conjunctival epithelium the structure of these cells changes by the process of transdifferentiation and the histology and biochemical function of the new epithelial surface increasingly resembles that of corneal epithelium. However, in cases of severe chemical injury, vascularization of the tissue occurs, with conjunctivalization of the corneal surface and thus loss of transparency.

The Cornea The cornea is a transparent and avascular tissue that forms the major refracting surface of the eye in conjunction with the tear film.

The corneal epithelium consists of several cell layers, is capable of rapid renewal, and needs to be intact for the regular spread of the tear film. If not intact it leaves terminal nerve endings free, which induce feeling of severe pain and increase tearing. The corneal epithelium has the unique function of providing a smooth optical surface and the transparency necessary to transmit images with minimal distortion. The chemically injured corneal epithelium may desquamate or become irregular and loses its clarity. When the integrity of the epithelium is compromised, exposure of the underlying stroma may result in an alteration of hydration that further compromises corneal transparency.

The corneal stroma is about 450 µm thick. The intricate parallel layering of its main constituents, collagen fibrils, ensures transparency. In case of damage this property is lost.

The inner corneal lining comprises endothelial cells stemming from neuroektoderm and therefore unable to regenerate. Nutrition of the avascular cornea is provided mostly by the aqueous humor produced by the endothelial cells of the anterior chamber and additionally by the tear film.

The Iris The iris is the anterior extension of the uvea that forms a mobile diaphragm between the posterior chamber and anterior chamber of the eye and is highly vascularized. The central aperture of the iris forms the pupil, through which light enters the eye.

The Anterior Chamber The anterior chamber represents an area between the posterior surface of the cornea, and the anterior surface of the iris and the pupillary portion of the lens, and peripherally by the trabecular meshwork, scleral spur, ciliary body, and iris root. These comprise the anterior chamber angle, where most of the aqueous humor contained in the anterior chamber is drained.

The Ciliary Body The ciliary body has three basic functions: production and removal of aquous humor, accommodation, and the production of vitreous mucopolysaccharides. The ciliary body represents the internal surface of the globe.

The Lens The lens is a transparent biconvex structure located behind the iris and the pupil and in front of the vitreous body. In adults the lens has a diameter of approximately 10 mm and is 4 mm thick. The lens is held in place and shaped according the accommodative needs by thin zonular fibres radiating into the ciliary body.

The lens continues to accumulate cells throughout life, such that the central portion of the lens becomes less pliable, more compact, sclerosed, and yellow. The increased density eventually reduces accommodation (zooming from far to near objects) and visual acuity. In addition to this sclerosis, the lens can become opaque because of a variety of other factors, including disorganization of fibre membranes or of lens proteins and accumulation of coloured, insoluble proteins in the lens as a result of numerous systemic or extrinsic factors.

The ciliary body processes, lens zonules and posterior lens capsule represent the border to the posterior segment of the eye.

The Posterior Segment

The Sclera The sclera shells the posterior segment of the eye. It forms five-sixths of the outer tunic of the eye, consists of dense fibrous tissue and contains openings and canals for the various vessels and nerves entering and exiting the globe. Its mechanical properties help contain the intraocular pressure and prevent deformations of the globe. It is relatively avascular and principally composed of collagen, with few fibrocytes and ground substance. Externally it is white, whereas internally it is slightly brown because of the presence of pigmented melanocytes.

The Choroid The choroid is a highly vascularized and pigmented tissue and is the extension of the ciliary body up to the optic nerve. The choroid is light to dark brown in color and sponge-like in appearance. It mostly consists of the blood vessels. Except for the central retinal artery that supplies the inner retina, the blood supply of the eye is derived from the ophthalmic artery.

The Retina The retina is a very thin and transparent membrane with a surface area of approximately 266 mm^2. It is loosely attached to the choroid via the pigment epithelium. The retina consists of two distinct layers: the neurosensory retina and the retinal pigment epithelium. These two layers extend over the inner surface of the eye from the optic nerve head to the retinal pigment epithelium, which continues into the pigmented epithelium of the ciliary body. The neurosensory retina extends into the nonpigmented epithelium of the ciliary body. The macular region is the neuronal zone close to the optical nerve head and provides the highest visual acuity.

The Vitreous Body The vitreous body is a clear, transparent, gel-like substance caused by a matrix composed of small-diameter collagen fibrils in hyaluronic acid that fills the eye between the lens and retina and contains 99% water.

The Optic Nerve

The optic nerve head is located slightly nasal to the posterior pole of the eye. Here, all the layers of the retina terminate, with the exception of the innermost nerve fiber layer. The ganglion cells of the retinal nerve fiber layers extend all the way to the lateral geniculate in the brain. From there the visual information is projected into the visual cortex.

3.10.3 Routes of Delivery of Xenobiotics to the Eye

> **Toxic compounds can enter the eye by two routes. The external route has to overcome the mechanical barriers of the lids, conjunctiva and/or cornea. The internal systemic route is via the bloodstream. Since the special feature of the eye is its transparent media, light toxicity has to be considered as well.**

Ocular Absorption

From the development of topical eye drops, we know that a substance cannot easily overcome the natural barriers of the anterior surface. Substances that come into contact with this surface may, however, penetrate the rather efficient barrier of the tear film and the cornea. The tear film is lipophilic on the outside and lipophobic on the inside due to its three-layer composition. The intact cornea presents a physical barrier to intraocular substance penetration. Active transport has not been demonstrated and substances appear to penetrate by diffusion. This process is characterized by the following factors: the rate of diffusion parallels drug concentration, the process is not saturable, is not retarded by metabolic inhibitors, and penetration will not occur against an electrochemical gradient.

Because of the limited time that substances in the lacrimal fluid are in contact with the cornea, the first barrier to penetration, the corneal epithelium, is the most important. Cell membrane lipids limit substance penetration. The greater the lipid solubility of a substance, the greater its diffusion into the corneal epithelium.

The corneal stroma is relatively acellular and made up primarily of collagen. It contains only 1% lipids. As a result, substances need to be water soluble to penetrate.

The third component of the cornea, the endothelium, is one cell layer thick. It represents a lipid barrier.

> **In order to penetrate the cornea, substances need to be both lipid and water soluble, which is called the differential solubility concept.**

Vasoconstrictors and vasodilators affect the blood flow in the conjunctiva and by decreasing or increasing conjunctival absorption alter both the amount and duration of substance in the tear film. These in turn will affect corneal and intraocular substance penetration.

Systemic Absorption

Lacrimal Drainage Systemic absorption and redistribution ultimately occur through blood vessels. The most obvious superficial ones in the eye are the conjunctival vessels. However, additionally the drainage of tears and substances through the canaliculi can be significant. Per blinking of the lids a volume of up to 2 µl may be pumped into the excretory lacrimal system. Ultimately the substances are reabsorbed by the nasal mucosa, thus bypassing the first-pass metabolism in the liver and entering the systemic blood flow in rather high concentrations. In addition, swallowing into the digestive system may occur. Again, from topical eye drop use it is known that, for example, β-blockers applied for glaucoma therapy may reach systemic therapeutic levels such as are used for blood pressure control.

Blood Flow Most of the blood flow to the eye stems from the inner carotid artery. Retinal vessels are end stream vessels, nourish the inner third of the retina and do not anastomose. The retinal vasculature resembles widely the vasculature in the central nervous system, including the blood–retina barrier, which is equal in its function to the blood–brain barrier. Choroidal vessels, which possess a fenestrated endothelium, nourish the outer two-thirds of the retina, including the photoreceptors. Thus, substances present in the bloodstream and able to penetrate the blood–brain barrier, such as methanol, reach and damage the eye and its delicate neuronal structures, the retina and optic nerve.

3.10.4 Specific Toxicology of the Eye

See also Chapter 3.6.

Chemical Injuries of the Eye

> **Chemical injuries are among the most urgent of ocular emergencies, often resulting in a dramatic decrease in visual acuity or loss of an eye. The prognosis for a burned eye**

> **depends not only on the severity of the injury but also on the rapidity with which therapy is initiated. Instant and copious long-term irrigation of the ocular surface, even if painful to the patient, is mandatory in order to reduce the contact of toxic substances. In general, alkali injuries are more damaging to the eye than those caused by acids.**

Chemical injuries are common in the chemical industry and laboratories, in machine factories, in agriculture, and in construction workers. They also are frequently reported from fabric mills, automotive repair facilities, and cleaning and sanitizing crews. With the work environment responsible for about 60% of chemical burns, roughly 35% of those injuries occur at home. Most of the eye injuries at home result from the use of lime and drain cleaners. Injuries caused by caustic chemicals are among the most severe; the prognosis for satisfactory recovery from these injuries is extremely poor.

Automotive battery acid burns have become increasingly more common and can be especially devastating when combined with the shrapnel resulting from explosion. These accidents typically occur in the colder months, almost always after dark, and usually involve young men. During recharging of a lead acid storage battery, which contains up to 25% sulfuric acid, hydrogen and oxygen produced by electrolysis form a highly explosive gaseous mixture. The most common causes of storage battery explosions are lit matches (used to see battery cells) and the incorrect use of jumper cables.

The deployment of automobile air bags involves the conversion of sodium azide to nitrogen gas, and is accompanied by the sudden release of an alkaline gas and powder. Ocular chemical injuries from air bag inflation must be considered dangerous and as great a threat to vision as other caustic injuries.

Alkalis

> **In alkali burns hydroxyl ions saponify the superficial cell membranes and intercellular bridges, which facilitates rapid penetration into the deeper layers and into the aqueous and vitreous compartments. Possible affected structures include almost the whole eye: lids, conjunctiva, cornea, sclera, iris, lens, and retina.**

Cell damage from alkaline agents depends on both the concentration of the alkali and the duration of exposure. The higher the pH, the greater the damage, with the most significant injuries occurring at a pH of 11 or higher. As the pH rises, destruction of the epithelial barrier becomes progressively more extreme. In the corneal stroma, alkali cations cause damage and necrosis by binding to the mucopolysaccharides and to collagen.

The most common alkalis involved in ocular injury are calcium hydroxide (lime), potassium hydroxide (potash), sodium hydroxide (lye), and ammonium hydroxide (ammonia). Of these, calcium hydroxide penetrates corneal tissue slowest because the calcium soaps are relatively insoluble, impeding further penetration of the agent. Potassium hydroxide penetrates corneal tissue faster. Sodium hydroxide penetrates even faster. Ammonium hydroxide passes through the cornea most rapidly because it destroys the epithelial barrier by saponification of the lipoidal cell walls and because it diffuses fastest through the stroma.

Acids

> **Acids quickly denature proteins in the corneal stroma, forming precipitates that retard additional penetration. Overall, the rates of penetration for acids at equivalent**

> concentrations and pH vary widely, with sulfurous acid penetrating more rapidly than hydrochloric, phosphoric, or sulfuric acids. Acid burns are usually limited to the anterior segment. The affected structures typically include lids, conjunctiva, and cornea.

Free hydrogen ions cause cellular necrosis and the strongest acids have the strongest necrotic potency. Weak mineral acids generally cause less severe ocular damage than alkalis. Corneal tissue has an inherent buffering capacity that tends to equilibrate local pH to physiological levels, but severe chemical injuries exhaust the cellular and extracellular buffering capacities, allowing extremes of pH that are incompatible with tissue survival.

Hydrofluoric acid is a strong inorganic acid (used in industry for cleaning and etching) that is particularly toxic because it has a complex mode of tissue injury. Along with necrosis from a high concentration of hydrogen ions, this agent causes cellular death because the fluoride anion binds calcium more quickly than the body can mobilize calcium. In addition, the fluoride blocks the Na-K ATPase of cell membranes, resulting in a fatal loss of potassium from the cells.

> *Severe chemical burns represent a devastating injury. If the conjunctiva, corneal limbal region, and cornea are severely damaged, opacification of the transparent structures may occur permanently. Even corneal transplants have a guarded prognosis. When the corneal epithelial stem cell area is damaged, neovascularization occurs and transparency of the transplant may not be ensured.*

Conjunctival Irritation and Inflammation

As described in Chapter 3.6, irritation not only of the lids, but also the conjunctiva may occur. An irritated conjunctiva has a large potential for swelling, even to the point when lid closure is no longer possible, which aggravates the problem by continued desiccation.

Contact inflammation is produced on the basis of a direct chemical effect on tissue; the resulting inflammation is not allergic in origin. Frequent defatting of the skin caused by excessive moisture also plays a role. Toxic reactions are much more common than allergic ones, accounting for about 90% of all reactions to topical ophthalmic medications. Toxic reactions include papillary conjunctivitis, follicular conjunctivitis, keratitis, pseudopemphigoid, and pseudotrachoma.

Contact Allergy

Contact inflammation of the eye may affect the lids, conjunctiva or cornea, or a combination of these, and can involve allergic or nonallergic (irritant/toxic) mechanisms.

Contact allergic reactions involve exposure to a sensitizing substance that is absorbed through the skin. Sensitization may take weeks to years to develop, depending on the ability of the hapten to act as a sensitizer, the amount applied and the duration, pre-existing lid or ocular disease, and individual susceptibility. In sensitized persons, contact allergic reactions can occur within 48–72 hours upon rechallenge, in keeping with type IV delayed-type hypersensitivity reactions.

Allergic contact dermato-conjunctivitis is the second most common type of drug reaction, accounting for 10% of reactions to topical ophthalmic medications.

In contact reactions of the eye and elsewhere, irritant and allergic mechanisms may coexist. Clinically, it is usually difficult to separate the two. Classically, a contact allergic reaction will begin in 48–72 hours, in keeping with type IV hypersensitivity reactions. An irritant reaction may begin within a few hours of contact or may occur only after prolonged use of a topical medication.

Although sometimes the offending agent may be obvious, often identification is difficult. Patch testing may provide the answer. A true allergic response will occur in 48–72 hours. An irritant can also cause a positive reaction, but this usually develops within a few hours and can be avoided by using lower doses of testing substances.

The acute lesions of contact allergic dermatitis/blepharitis resemble acute eczema, with erythema, vesicles, edema, oozing, and crusting. The chronic phase is characterized by dryness, crusting, fissuring, and thickening of the skin. Contact allergic conjunctivitis involves conjunctival injection and chemosis; there may be a papillary response and serous or mucoid discharge. Initially, the lower conjunctiva and lid are usually more affected; later, the entire conjunctiva and upper lid may also become involved. Itching can be prominent.

Several topical ocular drugs are known to act as sensitizers, including aminoglycosides (gentamicin, tobramycin, and neomycin), sulfonamides, atropine and its derivatives, topical antiviral agents (idoxuridine, trifluridine), topical anesthetics, echothiophate, epinephrine and phenylephrine, and preservatives (thimerosal, benzalkonium chloride). Other sensitizing agents include lanolin and parabens (cosmetics, skin creams, and lotions), nickel sulfate (jewelry), copper (colored lid shades), chromates (jewelry, leather products, fabrics, industrial chromated steel), and *p*-phenylenediamine (hair sprays, clothing, shoes). Rubbing the eyes after handling soaps, detergents, or chemicals may explain a localized ocular or periocular reaction after a more general exposure.

Lyell Syndrome

See Chapter 3.6.

Mucocutaneous Disease

Drug-induced Ocular Cicatricial Pemphigoid (Pseudopemphigoid): Drug-induced ocular cicatricial pemphigoid (pseudopemphigoid) includes a spectrum of disease ranging from a self-limited toxic form to a progressive immunologic form indistinguishable from true ocular cicatricial pemphigoid. Immunoglobulins bound to the basement membrane can be detected.

Stevens–Johnson Syndrome Stevens–Johnson syndrome (SJS) is an acute vesiculobullous disease characterized by systemic toxicity (fever, malaise, headache) and extensive cutaneous and mucous membrane involvement, including the conjunctiva. Precipitating factors include drugs (e.g., sulfonamides, penicillin) and infections, particularly HSV and *Mycoplasma*. Toxic epidermal necrolysis is considered to be a severe variant.

The clinical picture resembles that of ocular cicatricial pemphigoid. However, the major difference is that whereas the scarring in ocular cicatricial pemphigoid is chronic and progressive, the conjunctival scarring of SJS occurs as a result of the acute inflammatory episode and is self-limited. Clinical deterioration is a result of the subsequent tear deficiency and lid malpositions that result from the acute event, rather than chronic inflammation and progressive scarring.

Phototoxity and Thermal Effects

Cornea

> **The portions of the human body that are not protected by either melanin or increased thickness of the stratum corneum are the lips, the cornea, and the conjunctiva. Wavelengths of light shorter than 295 nm are absorbed and reflected by the corneal surface. Light with a peak near 288 nm induces fragmentation of epithelial nuclear protein with resultant cell death.**

One characteristic of ultraviolet injury is a latent period of several hours between exposure and effect. Corneal epithelial cells usually begin to die several hours after irradiation and are brushed off by the action of blinking or undergo spontaneous fragmentation. This exposes a number of bare nerve endings around each missing cell and results in excruciating pain. Ultraviolet damage may arise from artificial sources such as a welding arc, a close lightening strike, or an electrical flash caused by the breaking of a high-tension circuit. The amount of damage depends on the intensity of the light and the period over which it is delivered.

Photokeratitis is also common at high altitudes in the winter, when the reflectance from new, clean snow may be as high as 85%. The very bright light coming from below is not shielded by the eyebrows, forehead, or most headgear. This is most troublesome in high-altitude northern settings where there are few impurities in the air to absorb the short wavelengths of light. It is less troublesome in the desert where the reflectance from sand is only approximately 17%.

> **Changes caused by UV light in the corneal epithelium are reversible and heal completely. They are not like the effects of sunlight on the skin, which may lead to neoplastic changes.**

Conjunctiva Pterygium, a non-neoplastic hypertrophic growth of conjunctiva, occurs most often in persons who work in environments with a high surface reflectance of ultraviolet light. Those who live and work at latitudes of less than 30° have a much higher incidence of pterygium than those living at higher latitudes. Limbal stem cells seem to be activated by chromic ultraviolet light exposure to initiate this process. Climatic droplet keratopathy shares these same environmental associations.

Lens Transmittance of light through the ocular media rises rapidly from 400 to 442 nm. The decreased transmittance of shorter wavelengths is due primarily to absorption in the lens. This is an important consideration because sunlight peaks at 550 nm, and the large amounts of energy that are present in the shorter wavelengths of light are readily absorbed by the lens. These shorter wavelengths have been implicated in the formation of dark brown (brunescent) cataracts that occur in higher frequency in areas where the ultraviolet components of sunlight are most intense. It has been suggested that the short wavelength light results in lens damage by inducing chemical changes in proteins of the epithelial cells. This formation of toxic photo-products may inhibit growth and important metabolic activities. It has also been suggested that photo-oxidation of tryptophan may either remove this essential amino acid from its intended metabolic pathway or induce other effects.

> **Premature cataract formation is the ultimate clinical symptom of light-induced changes in the lens.**

Retina Damage occurs more commonly with short wavelengths of light near 441 nm falling on the retina and usually requires approximately 48 hours to induce cellular proliferation, with mitotic figures in the retinal pigment epithelium and in the choroid. Significant healing and regeneration may occur and visual acuity after a solar eclipse burn often improves. Short wavelengths of light reaching the retina have been found. Photochemical lesions are usually produced with light intensities several orders of magnitude below that needed to produce a direct thermal burn.

Thermal injury produces an intense central core of damage surrounded by edema. Nonlinear effects are caused by ultrashort exposure times and are produced by strong electric fields, acoustic signals, shock waves, and other phenomena generated by transient elevations in temperature gradients. Thermal effects are produced primarily by intense bright beams of light focused on the retina, such as the xenon photocoagulator or the argon or ruby laser.

Specific Retinopathies Solar retinopathy is a clinical entity that has been recognized for centuries. Synonymous terms include eclipse blindness, photoretinitis, photomaculopathy, and foveomacular retinitis. Solar retinopathy has been described in military personnel, sun bathers, religious sun gazers, solar eclipse viewers, and people under the influence of psychotropic drugs. Whereas this form of retinal insult is photochemical, it may be enhanced by thermal effects.

Although welder's maculopathy is the most common ocular injury associated with welding arc exposure, retinal damage has been reported to occur. Clinically, welder's maculopathy is similar to solar retinopathy. Clinical correlation between the use of the operating microscope and macular phototoxicity first was suggested in 1977 and has become a commonly recognized clinical syndrome associated with various anterior, posterior, and combined surgical procedures.

Cataract Formation

The solid mass of the lens consists of about 98% protein. These proteins undergo minimal turnover as the lens ages. Accordingly, on aging they are subject to the chronic stresses of exposure to light or other high-energy radiation and oxygen, and this damage is thought to be causally related to cataractogenesis. The term age-related cataract is used to distinguish lens opacification associated with old age from opacification associated with other causes, such as congenital and metabolic disorders or trauma.

Apart from irradiation-induced changes, alterations due to a change in lens metabolism by toxic substances is well known. Examples are organophosphates, naphthalene, cobalt, gold, silver, and lead. The most widely used iatrogenic substances are steroids, causing a characteristic posterior pole cataract. For other substances see Table 3.25.

Retinal Toxicity

> **Examples of chemicals that produce retinal toxicity are lead, methanol, mercury, n-hexane, naphthalene, organic solvents, and organophosphates.**
>
> **Representative medications that have recent reports of retinal toxicity are chloroquine and hydroxychloroquine, isotretinoin, sildenafil, vigabatrin, tamoxifen, and phenothiazines. A very small amount of damage (small fraction of 1%) may reveal a significant abnormality to the patient.**

Lead For over 100 years inorganic lead has been known to produce visual symptoms. Occupational lead exposure produces time- and concentration-dependent retinal alterations. At higher lead exposure levels the retina and optic nerve seem affected. At lower levels the rods, which are the retinal receptors for black and white information, and their neuronal pathway seem to be the target structure.

Methanol Methanol, which is widely used as a solvent, fuel source or antifreeze agent is quickly absorbed after any type of exposure. Since it easily crosses membranes it distributes well to tissues in relation to their water content.

Acute methanol poisoning induces permanent structural alterations in the retina and optic nerve, which results in blurred vision to complete loss of vision. Retinal ganglion cell edema and optic nerve head swelling may be observed. Methanol is oxidized in the liver to formaldehyde and then to formic acid.

Table 3.25 Site of action in the ocular system of select xenobiotics following systemic exposure (modified from Fox and Boyes, Caserett & Doulls Toxicology, 2001).

Toxicant	Cornea	Lens	Retina	Optic nerve
Acids	+			
Alkalis	+	+	+	
Acrylamide				++
Carbon disulfide			+	++
Chloroquine	+	+	+	+
Chlorpromazine	+	+	+	
Corticosteroids		++		+
Digoxin, digitoxin	+	+	++	+
Ethambutol			+	++
Hexachlorophene			+	+
Indomethacin	+	+	+	
Isotretinoin	+			
Lead	+	+	++	+
Methanol			++	++
Mercury, methyl mercury			+	
n-Hexane			+	+
Naphthalene		+	+	
Organic solvents			+	
Organophosphates		+	+	+
Styrene			+	
Sunlight and welding			+	
Tamoxifen	+		+	+
UV-light	+	+		

Formate is perceived as a mitochondrial poison that inhibits the oxidative phosphorylation of retinal photoreceptors, glial cells, and ganglion cells.

Organic Solvents Organic solvents are known neurotoxins. Their adverse effects on the retina are not well understood to date. However, color vision deficiencies have been noted in exposed workers.

Organophosphates Again, the neurotoxicity of organophosphates is well known, with the effect of exposure and retinotoxicity being still under investigation. Retinal degeneration has been described in some individuals chronically exposed to pesticides.

Chloroquine and Hydrochloroquine Hydroxychloroquine and chloroquine are taken by many patients on a chronic basis for rheumatoid disease. The incidence of retinopathy is very low when patients take less than 6.5 mg/kg/day of hydroxychloroquine or 3 mg/kg/day of chloroquine, and almost unknown within the first 5 years of usage at these dosage levels. These drugs accumulate in the melanotic retinal pigment epithelium (RPE) that lies just behind the retina, and gradually damage both the RPE and the overlying retina. If drug exposure continues, the retinal damage spreads and eventually both the center and the periphery of the retina can be destroyed. There is no known treatment for this retinopathy. Once recognized there will virtually always be some degree of permanent visual loss, and the damage may continue to progress long after stopping the drugs (sometimes for years, perhaps because of residual drug deposits in the body). The American

Academy of Ophthalmology therefore recommends that all individuals starting one of these drugs should have a complete baseline ophthalmologic examination within the first year of drug usage.

Isoretinoin Ocular side effects are dose related and probably the most frequent adverse reactions associated with these drugs. Retinal changes are rare compared with reports of blepharoconjunctivitis, subjective complaints of dry eyes, and transient blurred vision. Moreover, decreased ability to see at night after taking these agents may occur as early as a few weeks or after taking the drug for 1–2 years. Retinal dysfunction is probably due to the competition for binding sites between retinoic acid and retinol (vitamin A).

Sildenafil Sildenafil citrate, an oral therapy for erectile dysfunction (ED), is one of the best-selling prescription drugs in the world. This inhibitor of phosphodiesterase type 5 (PDE-5) is a unique class of drugs not previously used in humans. PDE-5 is responsible for the degradation of cyclic guanosine monophosphate (cGMP) in the corpus cavernosum. With increased levels of cGMP the smooth muscle in the corpus cavernosum is relaxed, allowing inflow of blood. Retinal side effects may occur because sildenafil, although selective for PDE-5, has a minor effect on PDE-6, an enzyme involved in light excitation in visual cells. The ocular side effects most commonly associated with sildenafil are a bluish tinge to the visual field, hypersensitivity to light, and hazy vision. These reversible side effects may last from a few minutes to hours, depending on drug dose. Visual changes are seen in approximately 3% of men taking the standard 50-mg dose, in 11% of men taking 100 mg and in 40% at a dose of 200 mg daily.

Vigabatrin Vigabrantin is a drug used in more than 50 countries for the treatment of refractory epilepsy. A 2% incidence of visual field abnormalities after 6 months of therapy is reported. The main ocular side effect is symptomatic or asymptomatic visual field constriction, which is usually bilateral and can progress to tunnel vision. More than 80% of these visual field defects appear to be irreversible.

Tamoxifen This antiestrogen is used primarily in the palliative treatment of breast carcinoma, ovarian cancer, pancreatic cancer, and malignant melanoma. An acute, debatable side effect, which is not well defined, may occur after only a few weeks of therapy, with any or all of the following: vision loss, retinal edema, retinal hemorrhage, and optic disc swelling. This may be a result of tamoxifen estrogenic activity, which may cause venous thromboembolism. These findings are reversible with discontinuation of tamoxifen. The typical crystalline retinopathy reveals striking white-to-yellow perimacular bodies. This finding has been reported for many medicines, including nitrofurantoin, canthaxanthine, and methoxyflurane. These occur most commonly after more than 1 year of therapy. with at least 100 g or more of the drug. There are, however, a number of cases of minimal retinal pigmentary changes occurring after a few months and only a few grams of tamoxifen. Loss of visual acuity in this chronic form is often progressive, dose dependent, and irreversible.

Phenothiazines Phenothiazines are used in the treatment of depressive, involutional, senile, or organic psychoses and various forms of schizophrenia. Some of the phenothiazines are also used as adjuncts to anesthesia, as antiemetics, and in the treatment of tetanus.

In patients on phenothiazine therapy for a number of years, a 30% rate of ocular side effects has been reported. If therapy continues over 10 years, the rate of ocular side effects increases to nearly 100%. The most significant side effects are reported with chlorpromazine and thioridazine, probably because they are the most often prescribed. The most common adverse ocular effect with this group of drugs is decreased vision, probably resulting from anticholinergic interference.

Retinopathy, optic nerve disease, and blindness are exceedingly rare at the recommended dose levels, and then they are almost only found in patients on long-term therapy.

Retinal pigmentary changes are most often found with thioridazine. A phototoxic process has been postulated to be involved in both the increased ocular pigmentary deposits and the retinal degeneration. The group of drugs with piperidine side chains (i.e., thioridazine) has a greater incidence of causing retinal problems than the phenothiazine derivatives with aliphatic side chains (i.e., chlorpromazine), which have relatively few retinal toxicities reported. The phenothiazines combine with ocular and dermal pigment and are only slowly released. This slow release has, in part, been given as the reason for the progression of adverse ocular reactions even after use of the drug is discontinued.

The patient should avoid bright light when possible. Sunglasses that block out UV radiation up to 400 nm are recommended.

Other Common Drugs Ibuprofen has caused visual evoked response alteration.

Indomethacin has been noted to cause retinal pigment epithelium (RPE) disturbances.

Desferrioxamine, used in the treatment of systemic iron overload, has been noted to cause retinal epithelial alterations.

Clofazimine, used in the treatment of dapsone-resistant leprosy, has been noted to cause a bull's eye maculopathy.

Methanol ingestion can cause visual field defects, retinal edema, and optic atrophy.

Toxic Optic Neuropathy

Toxic/nutritional optic neuropathy often presents as a painless, progressive, bilateral, symmetrical visual disturbance with variable optic nerve pallor. The patient may manifest reduction in visual acuity, loss of central visual field, and reduced color perception.

Toxic optic neuropathy may result from exposure to neuro-poisonous substances in the environment, ingestion of foods or from elevated serum drug levels. Nutritional deficiencies or metabolic disorders may also cause this disease. In most cases, the cause of the toxic neuropathy impairs the tissue's vascular supply or metabolism.

Among the common offenders is tobacco, which produces metabolic deficiencies as part of the systemic nicotine cascade. Alcohol (methanol as well as ethanol) produces its toxic effects through metabolic means. Chronic exposure typically leads to vitamin B_{12} or folate deficiency. Over time these deficiencies cause accumulations of formic acid. Both formic acid and cyanide inhibit the electron transport chain and mitochondrial function, resulting in disruption of ATP production and ultimately impairing the ATP-dependent axonal transport system.

Numerous other agents can produce toxic optic neuropathy, as can be seen in Table 3.25.

3.10.5 Summary

The eye is well designed by nature to protect itself against the insults of regular life activities. Lids and tear film constitute a potent barrier towards xenobiotics delivered externally. The blood–retina barrier partially shields the delicate neuronal retinal tissues from insults delivered systemically. However, all protective mechanisms have their limits and so external insults, e.g. from acids or alkalis, may lead to extremely devastating injuries of the eye, compromising visual acuity by destroying the cornea and its clarity as well as internal ocular structures.

The metabolism of the transparent lens is very vulnerable to systemically delivered insults. Disturbances result in cataract formation.

The retina and optic nerve may be considered as a specific part of the brain so that compounds toxic to the central nervous system are generally expected to affect the retina and the optic nerve as well. Because of the transparent optic media of the eye phototoxic damage to nonsuperficial structures is a unique feature of this organ.

Further Reading

Diamante GG, Fraunfelder FT. Adverse effects of therapeutic agents on cornea and conjunctiva. In Leibowitz HM, Waring GO (eds): *Corneal Disorders, Clinical Diagnosis and Management*, 2nd edition, Philadelphia, 1998.
Grant WM. *Toxicology of the Eye*, 4th edition, Springfield IL, Charles C. Thomas, 1993.
Kaufman PL. *Albert Alm Adler's Physiology of the Eye*, 10th edition, Mosby, St. Louis, Missouri, 2003.
Tasman W, Jaeger EA. *Duane's Ophthalmology*, Lippincott Williams & Wilkins, 2006.

3.11 The Cardiovascular System

Helmut Greim

3.11.1 Structure and Function

> **The heart is a muscular hollow organ that consists mainly of cardiac muscle, the myocardium. The heart is divided into chambers: the right and left *atrium*, and the right and left *ventricle*. The right atrium receives deoxygenated blood from the body via the superior and inferior *venae cavae* and from the heart itself during contraction, and passes it to the right ventricle. The left atrium receives oxygenated blood from the lungs and passes it to the left ventricle. The rate of the heart beat is primarily regulated by a pacemaker system consisting of the *sinuatrial node*, the *atrioventricular node*, both situated in the right atrium, and the *atrioventricular bundles* within the *interventricular septum*.**

The Myocardium

The heart, an organ mainly consisting of a single muscle, is the driving force for the pulmonary circulation (right half of heart) and for the greater systemic circulation (left half). The low-oxygen blood flowing back from the body through the veins enters the right half of the heart and is pumped through the lungs. The blood, now saturated with oxygen, flows back through the pulmonary veins to the left half of the heart from where it is transported back into the body through the aorta (Figure 3.32).

The pumping effect of the heart is produced by the rhythmic successions of contraction (**systole**) and dilatation (**diastole**) of the myocardium. The direction of the blood flow is determined by the opening and closing of valves (tricuspid and mitral valves).

The contraction of the myocardial cells is triggered by abruptly changing the direction of the electric potential at their outer membrane (80 mV, negative on the inner side) by means of a transmembrane ion flow (**action potential**). The action potential releases calcium ions from intracellular calcium reserves (**sarcoplasmatic reticulum**), which, in turn, initiate the contraction of the contractile elements actin and myosin in the myocardial cells. The ions flowing during the action potential phase regulate the filling level of the intracellular calcium reserves and consequently – together with other factors – the force of the contraction.

Organ Toxicology   365

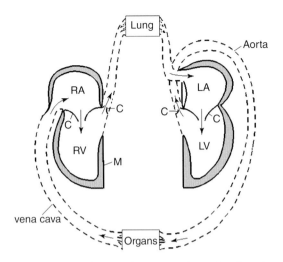

Figure 3.32 Diagram of cardiac circulation. The arrows show the direction in which blood flows. The left half of the illustration shows low-oxygen blood returning from the organs being pumped out to the lungs. The right half shows oxygen-saturated blood returning from the lungs being pumped out to supply the organs. M, myocardium; C, cardiac valves; RA, right atrium; RV, right ventricle; LA, left atrium; LV, left ventricle.

The System of Excitation and Conduction

> **The action potentials of the heart are normally produced in a specialized area (sinuatrial nodes), are then conducted via the atria to the atrioventricular nodes at the atrioventricular boundary; from here, they spread out into the entire myocardial tissue of the ventricles via a system of fibres specialized in conducting these stimuli (Figure 3.33).**

Any form of damage to this system of excitation and conduction, for example due to oxygen deficiency, necroses, scar formation, direct toxic impairment of the ion pumps and ion flows, results in cardiac arrhythmia, which can pose a threat to the pumping efficiency of the myocardium and thus a threat to life itself.

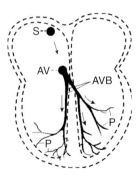

Figure 3.33 Diagram of the excitation system. The arrows indicate the direction in which the excitation process takes place. S, sinuatrial node, the site from which excitation normally starts; AV, atrioventricular node; AVB, atrioventricular bundles; P, Purkinje fibres.

The Coronary Blood Vessels

> **The supply of the heart with substrates and oxygen takes place via the coronary arteries, which branch out of the aorta and spread out to cover the entire cardiac organ. The walls of the heart are drained by the cardiac veins.**

After passing through the myocardial tissue the blood deprived of its oxygen is collected in the coronary veins and conducted to the right side of the heart. It is worth noting that the extraction of oxygen from the coronary blood by the heart is already very high in the state of rest when compared with the circulatory systems of other organs. As a greater amount of oxygen is required, for example, on physical or mental strain, the amount needed by the heart cannot simply be met by increasing the level at which oxygen is extracted from the coronary blood; an increase in blood flow must also take place. This makes it necessary for the coronary circulation to have a highly responsive and effective regulatory system. A pathological constriction of one or several coronary arteries therefore results in symptoms of oxygen deficiency in the heart. As a rule, primary toxic damage to the coronary arterial system has serious consequences for the myocardium and the system of excitation and conduction.

The Vascular System (Pulmonary and Systemic Circulation)

> **The vascular system is a closed system of tubes whereby blood permeates through all the tissues of the body. The pulmonary circulation serves the lung, the systemic circulation all other organs.**

The **pulmonary circulation** transports the deoxygenated blood from the heart to the lung and, after gaseous exchange of carbon dioxide and oxygen, transports the oxygenated blood back to the heart.

The **systemic circulation** transports oxygenated blood from the heart into the body and the deoxygenated blood back to the heart. It provides the tissues with oxygen, nutrients, which are absorbed from the alimentary tract, hormones, and enzymes and carries back waste products such as carbon dioxide or urea, which are eliminated via the lungs or the kidneys, respectively. The driving force is the contraction of the left ventricle during systole, which forcibly expels blood into the aorta and from there via the arteries, arterioles, and capillaries to the organs. Venules and veins carry the deoxygenated blood back to the right atrium.

The capillaries are thin-walled blood vessels situated between arterioles and venules. They are the site of the exchange of materials (oxygen, nutrients, carbon dioxide, and other waste products) between capillary blood and surrounding tissues.

3.11.2 Toxicology

Due to its constant activity and its rapid adaption to the need for increased blood flow, the heart muscle has a higher energy requirement than many other tissues. To meet this requirement cardiomyocytes have large numbers of mitochondria, which by oxidative phosphorylation produce and supply ATP. Myosin, a protein which binds with actin filaments to generate the contraction of the myofibrils contains ATPases, which hydrolyse ATP, releasing activated P, and activate cellular processes by phosphorylation. The ATPases are involved in muscle contraction and in the energy-driven pumping of ions including Na^+, K^+, and Ca^{++}. Thus, any interference with delivery or mitochondrial utilization of oxygen or the subsequent phosphorylation reactions will lead to disturbances of cardiac function.

> **In principle, each of the three functional cardiac systems (myocardium, excitation, and conduction) and the coronary arteries can be damaged by toxic substances.**

Damage to the Myocardium

Cadmium Cadmium reduces the **contractility** of the myocardium by blocking the calcium flow into the cells during the **action potential**. In addition, the membrane of the myocardial cell possesses an ion exchange system, which transports sodium and calcium. Cadmium principally competes with calcium at the binding site on the external surface of the cell membrane. By these mechanisms, cadmium reduces the flow of calcium into the heart muscle cell, which means that less calcium is available for contraction.

Cadmium also penetrates into the cell, reducing the available amount of adenosine triphosphate and phosphocreatine as well as the phosphorylation of the contractile protein. Furthermore, cadmium competes more powerfully than other divalent cations with calcium for those binding sites on the contractile protein, which finally triggers contraction.

Cobalt A considerable amount of experience has been gathered as regards the effect of cobalt on the human heart from a poisoning catastrophe. In the USA, Canada and Belgium during 1965/1966, cobalt was used as a foam-stabilizing additive in beer. This resulted in severe and sometimes fatal damage to the myocardium. Death occurred with the symptoms of severe cardiac insufficiency with circulatory collapse and acidosis. A similar intoxication occurred after industrial exposure to cobalt. The decisive effect is a disturbance of the myocardial energy metabolism, which is initiated by binding of cobalt to the disulfide groups of liponic acid. It was noted that, in the case of the catastrophe cited above, persons with insufficient nutritional intake were more liable to suffer from intoxication than those with a normal nutrition. This could be explained by the fact that cobalt binds to the sulfhydryl groups of food components (e.g., histidine and cysteine from proteins), so that binding to the endogenous disulfide groups is reduced.

Arsenic (Arsenic Trioxide) In 1900 a poisoning epidemic occurred with beer to which arsenic had been added. Around 6000 persons suffered, of which 70 cases were fatal. In nearly all cases heart failure was involved. As to whether the effects of the arsenic on the myocardium were the result of the capillary damage known to be produced by this poison or whether they were due to a direct effect on the myocardial cells remains open.

Arsines Poisoning through inhalation of arsines is not infrequent in the metal-processing industry. The most prominent feature of such toxic effects is a **haemolysis** (erythrocytic breakdown). However, additional damage to the heart also seems to take place. In persons who had died from arsine poisoning, a distension of the ventricles and, microscopically, a fragmentation of the myocardial fibres were found.

Lead Cases of acute to subacute lead poisoning have been described that resulted in death within a few days. Symptoms of severe acute **cardiac insufficiency**, which indicate damage to the myocardium, preceded death. The occurrence of severe **cardiac arrhythmia** also shows that, in addition to the myocardium, the system of excitation and conduction is also damaged by lead.

Antimony Cases of poisoning with antimony are no longer as frequent compared with earlier years. Antimony tartrate was used medically for a long time. Disturbances in repolarization, which indicate damage to the myocardium, predominated in larger groups of these patients.

Ethanol Approximately 1% of heavy alcohol drinkers fall prey to alcohol-produced damage to the myocardium (**cardiomyopathy**). In the light of widespread high alcohol consumption, this disease represents one of the most frequent forms of heart damage. As a rule, alcoholic cardiomyopathy develops after a high consumption of alcohol over the preceding 10 years. This alcohol-related damage should not be equated with the cardiomyopathy produced by vitamin B_1 deficiency, which occurs relatively often in alcoholics.

Among the acute effects of alcohol on the heart, the reduction in contractile force is to be emphasized, which is already extreme in isolated hearts at concentrations around 4% ethanol. In persons with healthy hearts, a single dose of 110 g ethanol reduces the stroke volume of the heart, resulting in a reduction of the contractile power of the myocardium.

Damage to the System of Excitation and Conduction

Halogenated Hydrocarbons Over recent years, a large number of halogenated hydrocarbons have produced cases of fatal poisoning, which were principally caused by damage to the system of excitation and conduction of the heart. However, accurate differentiation of the substances according to their toxicity, let alone differences in their active mechanisms, is not possible. Concerning the frequency at which fatal damage to the heart occurs, the following are to be cited in particular: **trichloroethylene** and **trichloroethane**, formerly used as narcotics and still processed in industry, and **dichlorodifluoromethane**, **trichlorofluoromethane**, **dichlorotetrafluoroethane** and **trichlorofluoroethane**, used as propellants for aerosols or as freezing agents.

In the United States in the 1960s it became fashionable for adolescents to inhale solvents for their intoxicating effects. As a result, within a relatively short time, over 100 fatal cases were reported. Fluorinated hydrocarbon propellants and trichloroethane were among the substances most frequently responsible for accidents of this kind. The mortalities occurred shortly after inhalation with such suddenness that fatal cardiac arrhythmia had to be regarded as being the probable cause of death.

Animal experiments demonstrated that halogenated hydrocarbons at low concentrations, at which they were by themselves not able to produce cardiac arrhythmia, intensified the arrythmogenic effect of adrenaline. It is not clear whether the phenomena described are produced by the direct action of the halogenated hydrocarbons on the excitation and conduction system or indirectly via induced reflex impulses of the sympathetic and parasympathetic nerves acting on this system.

In addition, halogenated hydrocarbons restrict the contractibility of the heart and reduce the cardiac output (per minute) and the blood pressure. With some of the substances, relatively low concentrations in the inhaled air are effective within a very short time, e.g. 1 % trichlorofluoroethane or 2.5 % dichlorotetrafluoroethane. However, in most cases, the levels of halogenated hydrocarbons in the environment are too low to have any major lasting effect and even at high levels (up to 15%) fatalities are rarely recorded. Among people who inhale these agents from closed bags to "get high", fatalities can result because the levels of these agents in the bags can reach 35–40%.

Cadmium Apart from blocking the flow of calcium into the cell, cadmium also inhibits excitation transmission in the atrioventricular node.

In workers who had been exposed to concentrations of 0.04–0.5 mg cadmium per m^3 inhaled air, a significantly disturbed excitation transmission was found.

Mercury Although mercury chloride and organic mercury compounds have effects on the heart similar to those of the highly toxic **digitalis glycosides**, cardiac involvement received little mention compared with the massive damage to the nervous system in the descriptions of epidemic poisoning with organic mercury compounds (in Minamata, Japan, and in Iraq). However, an electrocardiographic investigation in Iraq revealed changes in the electrocardiograms of all poisoned persons. Disturbed repolarization was found in all cases, indicating damage to the myocardium, but also a high percentage of different types of arrhythmias, so that toxic effects on the excitation and conduction system had to be assumed as well.

Barium Acute poisoning with barium chloride, inadvertently taken as a laxative, produced severe cardiac arrhythmias.

Damage to Coronary Arteries and other Blood Vessels

In relatively rare cases, toxic damage to coronary perfusion due to the short-term action of an environmental poison (**acute damage**) occurs. More frequently, an accelerated development of atherosclerosis of the coronary arteries under the long-term action of toxic substances in the environment (**chronic damage**) takes place. In such cases, additional factors are also involved as a rule (**high blood pressure**, **high blood cholesterol concentrations**).

In many cases, not only are the coronary arteries affected, but also other blood vessels, e.g. those supplying the brain, the kidneys or the muscles, so that circulatory disturbances can also occur as sequelae there, resulting in reduced cerebral performance, stroke, renal failure, and muscle pains, mostly as a consequence of oxygen deficiency.

Carbon Disulfide After many years' exposure to carbon disulfide (CS_2), increased blood pressure, angina pectoris, accelerated arteriosclerosis, and higher mortality from myocardial infarction can be observed in viscose workers.

In persons exposed to CS_2, an increase in the cholesterol concentration in their plasma has been found. In experimental animals, functional and morphological changes of the heart, including necrosis of the myocardium, have been attributed to both a direct effect of CS_2 on the heart and to increased incorporation of cholesterol and lipoproteins into heart vessels, leading to arteriosclerosis. In this context, a disturbed glucose metabolism caused by CS_2 is to be mentioned, as the metabolic disturbances occurring in diabetes also promote the formation of coronary sclerosis.

Compared with this chronic effect, the acute effects of CS_2 on the heart are of little importance. The concentrations at which acute effects have been obtained in animals are relatively high.

Nitrate Esters In human toxicology, cases of acute intoxication with organic nitro compounds are rare. Chronic poisonings, particularly with nitroglycerine and ethylene glycol dinitrate did, however, occur in the past with a relatively high frequency. Nitroglycerine is readily absorbed through the skin and has usually been taken up by this route where poisonings occur. Ethylene glycol dinitrate also readily penetrates the skin. However, as it is volatile, poisoning usually takes place via inhalation.

At higher doses, both substances, similar to the longer acting substances used in medicine, are responsible for a dilatation of all blood vessels, whereas at low doses mainly the veins are involved. At low doses, a congestion of blood in the veins of the lower extremities and in the abdominal cavity occurs, which results in less blood flowing back to the heart. The filling of the heart decreases and the volume of transported blood becomes smaller. By this means and due to the direct effect on the vessels, the blood pressure drops. The heart rate increases as a reflex. Orthostatic collapse may occur if the body is in an upright position. The characteristic **nitrate headache** is seen as the result of vascular distension and also occurs after administration of other vasodilators. In addition, there is a drop in the pressure of the cerebrospinal fluid, which also plays a part in causing headaches.

After intake of higher doses the organism develops tolerance (tachyphylaxis) to the described effects of organic nitrate esters within 1–2 days. When exposure to these substances is interrupted, the ability to react is restored after a relatively short time (about 2 days), so that the symptoms described above recur after re-exposure (Monday headache in exposed workers).

A different reaction can occur after contact with ethylene glycol dinitrate and/or nitroglycerine lasting for several years, shown in the form of withdrawal symptoms when exposure is discontinued. In such cases, symptoms of angina pectoris appear 2–3 days after termination of exposure. Fatal cases have also occurred. As sclerotic changes in the coronary arteries are not usually present, the term "non-arteriosclerotic ischaemic heart disease" has been coined. In such cases, spastic constrictions of the coronary arteries have been demonstrated, which were reversible after the administration of nitroglycerine.

Carbon Monoxide An increase of the haemoglobin carbon monoxide (HbCO) concentration to 5% causes a rise in blood flow through the coronary blood vessels similar to hypoxia; this finding is to be interpreted as a reaction to the reduced supply of O_2 to the myocardium. Correspondingly, physical exercise can disturb the relationship between the oxygen supply and the oxygen requirements of the myocardium at mean HbCO concentrations of only 5% and may lead to an angina pectoris attack.

High concentrations of HbCO in the blood reduce cardiac contractility and the pumping efficiency of the heart. The electrocardiogram indicates ischemia, as seen in cases with insufficient blood supply of the myocardium. Cardiac arrhythmias frequently occur.

Necroses of the myocardium are found in fatal CO poisoning. In addition to binding to haemoglobin, recent experimental findings indicate that the binding of CO to intracellular myoglobin and cytochromes plays an additional role. It has been found, for example, that the CO-myoglobin concentration is two to three times higher than the corresponding HbCO concentration. The affinity of the CO to cytochrome P450 is greater than that to haemoglobin.

Cigarette Smoking Epidemiologically, it has been proved that smoking promotes the development of atherosclerotic diseases of the heart and other organs. It seems that pipe and cigar smokers are less at risk than cigarette smokers. The explanation for this is probably that cigarette smokers inhale more frequently, so that the substances in the smoke are absorbed to a greater extent. Which ingredient or ingredients of the cigarette smoke are responsible for the development of **arteriosclerosis** is a matter of controversy. Lipid peroxidation caused by radicals probably plays a decisive role.

Besides the chronic effect, cigarette smoke also has acute effects, which can be of importance for a previously damaged heart. By means of increasing the HbCO content of the blood, smoking produces a **reduction in the oxygen supply,** which is significant for the heart with coronary sclerosis. Smoking mainly nicotine produces a rise in blood pressure and heart rate due to its effect on the vegetative nervous system and release of endogenous catecholamines.

Water Hardness There is no longer any doubt that there is an inverse correlation between the hardness of drinking water and the frequency of coronary heart disease. While it is certain that atherosclerotic coronary heart disease occurs more frequently in areas where the water is soft, the causal relationship is still doubtful.

Lead Patients with chronic lead poisoning frequently complain of palpitations, dyspnoea and precordial pains, which constitute **saturnine angina pectoris**, a set of pathological symptoms which have been known for a very long time. These are probably due to contractions of the coronary arteries.

The question as to whether long-term effects of lead increase the frequency of atherosclerotic coronary heart disease and other atherosclerotic diseases is to be considered. Epidemiological investigations and animal experiments indicate an enhancing effect of chronic lead poisoning on the development of arteriosclerosis, both in the coronary blood vessels and in the aorta.

Cadmium According to several authors there is an increased risk of death due to coronary sclerosis and other atherosclerotic diseases in persons chronically exposed to cadmium. However, the way in which cadmium can affect the coronary blood vessels is not known. As an increased blood pressure is one of the major factors in the development of atherosclerosis, the effect of cadmium on blood pressure may be involved.

In animal experiments, hypertension can be induced by long-term feeding with small to moderate amounts of cadmium, although this was not reproducible in all of the studies. This contradiction can mostly be explained because the dependence of blood pressure on cadmium dose actually has two phases. The hypertensive effect starts at a dose that is low compared with frequent human exposures, i.e. 0.01 ppm in the drinking water; it reaches a maximum at 1 ppm before dropping steeply, then rapidly turning to a hypotensive effect at 10 ppm.

Fibers and Particles Epidemiological studies have shown that ambient exposure to particles that display a mass mean diameter of 10 μm (PM_{10}) is related to both respiratory and cardiovascular mortality and morbidity (see Chapter 6.3). These cardiovascular effects are explained by either impairment of the blood flow and oxygen supply to or in the heart, or interference with autonomic innervation. Moreover, repeated exposure to PM_{10} may, by increasing systemic inflammation, exacerbate the vascular inflammation of atherosclerosis and promote plaque development or rupture of blood vessels. Atherogenesis is an inflammatory process, initiated via endothelial injury, which produces systemic markers of inflammation that are risk factors for myocardial and cerebral infarction. The inflammatory response to particles or the particles themselves may also impact on the neural regulation of the heart, leading to death from fatal dysrhythmia. In support of this hypothesis, studies in humans and animals have shown changes in the heart rate and heart rate variability in reponse to repeated to particle exposures.

Combustion and model nanoparticles (NPs) as well as diesel particles, which also contain NPs, may gain access to the blood following inhalation and enhance experimental thrombosis, but it is not clear whether this is an effect of pulmonary inflammation or of particles translocated to the blood. High exposures of combustion-derived NP (CDNP) by inhalation caused altered heart rates in hypertensive rats and dogs, which are interpreted as a direct effect of CDNP on the pacemaker activity of the heart. Inflammation in distal sites has also been associated with increased progression of atheromatous plaques in rabbits, ApoE-/- mice and humans.

3.11.3 Summary

The heart is a functional unit comprising the working muscle (myocardium), a system of electric stimulation and conduction, and the system of coronary arteries and veins supplying the heart muscle. Basically, all three systems can be damaged by toxic substances.

Metals cause damage to the myocardium as well as to the heart conduction system, whereas halogenated hydrocarbons produce cardiac arrhythmias, i.e. they can disturb regulation of the heart beat frequency. The effects on coronary arteries are principally due to chemicals which can either produce an acute constriction of these blood vessels, such as nitro compounds, or constrict them after chronic exposure by sclerosing (coronary sclerosis). In both cases, the consequence is an insufficient supply of oxygen to the myocardium, such as can also be induced in the blood by increased CO-haemoglobin levels. Sclerosing effects are also produced, e.g. by cigarette smoke, carbon disulfide, lead and cadmium. An association between low levels of water hardness and arteriosc1eroses needs further evaluation.

Inflammation can be seen as the main driver of the cardiovascular effects of particulate matter (PM). There is evidence of systemic inflammation following increased exposure to PM, as shown by elevated C-reactive protein, blood leukocytes, platelets, fibrinogen and increased plasma viscosity.

Further Reading

Burkhoff D, Mirsky I, Suga H: Assessment of systolic and diastolic ventricular properties via pressure-volume analysis: a guide for clinical, translational, and basic researchers. *Am J Physiol Heart Circ Physiol* **289**: 501–512, 2005.
Clerico A, Recchia FA, Passino C, Emdin M: Cardiac endocrine function is an essential component of the homeostatic regulation network: physiological and clinical implications. *Am J Physiol Heart Circ Physiol* **290**: 17–29, 2006.
Drexler H, Ulm K, Hardt R, Hubman M, Goen T, Löang E, Angerer J, Lehnert G: Carbon disulfide.IV. Cardiovascular function in workers in the viscose industry. *Int Arch Occup Environ Health* **69**: 27–31, 1996.
Frishman WH, Del Vecchio A, Sanal S, Ismail A: Cardiovascular manifestations of substance abuse: Part 2: Alcohol, amphetamines, heroin, cannabis, and caffeine. *Heart Disease* **5**, 253–271, 2003.
Gibbs CL, Chapman JB: Cardiac mechanics and energetics: chemomechanical transduction in cardiac muscle. *Am J Physiol Heart Circ Physiol* **249**: 199–206, 1985.
Leone A: Biochemical markers of cardiovascular damage from tobacco smoke. *Current Pharmaceut Design* **11**, 2199–2208, 2005.
Malpas SC: Neural influences on cardiovascular variability: possibilities and pitfalls. *Am J Physiol Heart Circ Physiol* **282**: 6–20, 2002.
Meininger GA, Davis MJ: Cellular mechanisms involved in the vascular myogenic response. *Am J Physiol Heart Circ Physiol* **263**: 647–659, 1992.
Routledge HC, Ayres JG: Air pollution and the heart. *Occup Med* **55**, 439–447, 2005.
Thornell LE, Eriksson A: Filament systems in the Purkinje fibers of the heart. *Am J Physiol Heart Circ Physiol* **241**: 291–305, 1981.
Young ME: The circadian clock within the heart: potential influence on myocardial gene expression, metabolism, and function. *Am J Physiol Heart Circ Physiol* **290**: 1–16, 2006.
Zhang X, Li S-Y, Brown RA, Ren J: Ethanol and acetaldehyde in alcoholic cardiomyopathy: From bad to ugly en route to oxidative stress. *Alcohol* **32**, 175–186, 2004.
Zhang DX, Gutterman DD: Mitochondrial reactive oxygen species-mediated signaling in endothelial cells. *Am J Physiol Heart Circ Physiol* **292**: 2023–2031, 2007.
  Cardiovascular system: http://www.innerbody.com/anim/card.html: http://www.innerbody.com/image/cardov.html.

3.12 The Endocrine System[1]

Gerlinde Schriever-Schwemmer

3.12.1 Introduction

> **In mammals, the nervous and endocrine systems are responsible for signalling. Nerves transmit signals very quickly, whereas messengers (hormones) of the endocrine system relay signals more slowly.**

Hormones, in the strict sense of the word, are produced in endocrine cells. They are transported via the circulatory system and act on target cells carrying receptors.

---

[1] This chapter is based on the chapter "Endokrines System" of the German edition by V. Strauss and B. van Ravenzwaay. (H. Greim, ed., Das Toxikologiebuch. Wiley-VCH, Weinheim, 2017)

Tissue hormones (e.g. gastrin) are messengers that are not transported via the blood, but act in the immediate vicinity of the producing cells. Neurotransmitters (e.g. acetylcholine) are a special form of tissue hormones. Mediators (e.g. growth factors and cytokines), in contrast, are messengers that are produced by many types of cells rather than by a specific glandular cell. However, the distinctions between hormones and mediators are not clear-cut (e.g. histamine).

Exogenous substances may play a role in the regulation of the hormones in the body. However, this chapter and Chapter 6.5 focus on those parts of the endocrine system relating to thyroid and sex hormones in humans and laboratory animals that are of interest to toxicology.

3.12.2 Structure and Function

> **Strictly speaking, the mammalian endocrine system is made up of the cells that produce hormones in various organs of the body and the hormones themselves. Decisive for the hormonal effect are the circulatory system as the transport medium, the target cells of various tissues with their hormone receptors, and the degrading and excretory organs of the body (in most cases liver and kidneys). The release of hormones from endocrine cells is regulated.**

Chemically, hormones are classified in the following substance groups: steroid hormones, amino acid derivatives (e.g. thyroid hormones and catecholamines), peptide hormones (e.g. thyrotropin-releasing hormones (TRH)), protein hormones (e.g. calcitonin), glycoprotein hormones (e.g. thyroid-stimulating hormone (TSH)), and eicosanoids (e.g. prostaglandins).

The release of hormones from endocrine cells is regulated via a feedback system. Thus, steroid hormone concentrations influence the release of the superordinate hormones in the hypothalamus and pituitary gland. However, endocrine cells are influenced and controlled by many other factors (other hormones, stress, daylight, season, etc.).

Hormones are transported to the target organs by the blood. Hereby, some primarily bind to proteins in an inactive form (e.g. steroid and thyroid hormones), while others are enzymatically converted to a more active form on the target cells (e.g. T4 to the more active T3 by deiodinase or testosterone to the more active dihydrotestosterone (DHT)).

Hormones are bound to receptors on the target cells. There are two groups of receptors: nuclear receptors (for steroid hormones, T3 and T4) and cell membrane-bound receptors (for protein hormones and catecholamines). A hormone may bind to several receptors (e.g. oestradiol binds to at least ERα and ERβ) or a receptor may bind to several different hormones (e.g. α1 adrenergic receptor to norepinephrines and epinephrines). Receptor density and hormone affinity vary among the different types of target cells and may change depending on the stage of development (age, pregnancy and menstrual cycle). Therefore, the hormonal effect may differ depending upon the type of cell and time. The hormonal effect of nuclear receptors is additionally influenced by the binding of co-regulators to the receptor.

Figure 3.34 shows the sites at which the hormonal balance in mammals may be affected by an exogenous substance.

> **In adults, hormonally active substances generally only have an effect as long as the substance is present in the body. In most cases, the effect is reversible. If, however, that same substance is given during certain stages of pregnancy, the offspring may be permanently damaged.**

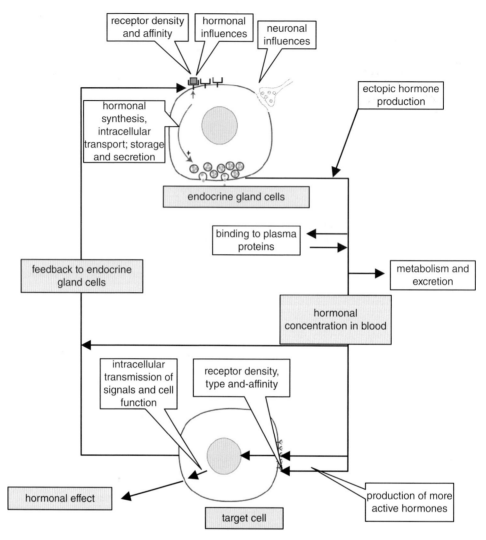

Figure 3.34 The endocrine system. Sites at which the hormonal homeostasis may be affected by an exogenous impact.

Any evaluation of endocrine active substances has to consider the differences in developmental physiology between humans and laboratory animals (primarily rats and rabbits) and the toxicokinetics of hormones and test substances (Chapters 2.1 and 2.2). The following focuses only on those aspects relevant to sex hormones and thyroid hormones.

Thyroid

Toxicologically Relevant Functions
Biosynthesis

Thyroid hormones are synthesized in the follicular epithelial cells (thyrocytes) of the thyroid. Iodide is oxidized by thyroperoxidase to form a reactive iodine molecule, which

> **binds to tyrosyl residues of thyroglobulin. Monoiodine and diiodine tyrosyl residues are combined to triiodine tyronyl and tetraiodine tyrosyl residues, leading to the formation of the two thyroid hormones triiodothyronine and tetraiodothyronine.**

Ingested iodide is absorbed into thyrocytes by a secondary active transport mechanism involving the transport protein sodium iodide symporter (NIS). Transport proteins (e.g. pendrin) bind iodine in the thyrocytes and carry it to the follicular membrane, where it is oxidized to an activated iodine molecule by thyroperoxidase (TPO).

Thyroglobulin (TG), a protein containing L-tyrosyl residues that serve as a scaffold for thyroid hormones, is synthesized in the thyrocytes. Activated iodine molecules are bound to the aromatic rings of tyrosine, leading to the formation of monoiodine and diiodine tyrosyl residues. These are then combined to tetraiodine or triiodine tyronyl residues, forming the two thyroid hormones tetraiodothyronine (T4; 90–95%) and triiodothyronine (T3; 5–10%). These thyroid hormones are stored in the follicular colloid of the thyroid in a covalently cross-linked form. The thyroid can store hormones in an amount sufficient to last for a few days in adult rats and a few months in humans, but only enough to last about one day in newborn babies.

Thyroid hormone release starts with the solubilization of covalently cross-linked thyroglobulin in the thyroid follicle, endocytosis in the thyrocytes and further degradation in the endolysosomes by means of cathepsins. The released T4 (and small amounts of T3) is secreted into the bloodstream by exocytosis, but is primarily transported there bound to proteins.

Transport proteins differ by species. In humans, 80% of T4 and 90% of T3 are transported bound to the highly affine transport protein thyroxine-binding globulin (TBG), while small fractions are transported bound to the less affine and non-specific transthyretin (= pre-albumin), albumin and lipoproteins. In the rat, TBG is only formed prenatally (as of foetal day 16) up to about 4 weeks after birth. In older rats, thyroid hormones are bound mainly to albumin.

> **As hormones are bound more loosely in rats, they can be metabolized more rapidly. Therefore, the half-life of T4 in the blood of rats is only about 1/10 and the half-life of T3 is only 1/4 of that in humans. In comparison to humans, the thyroid of the rat has to synthesize about 10 times more T4 per kilogram body weight to be able to maintain blood hormone levels. To maintain this increased rate of synthesis, TSH values in rats are also higher than those in humans.**

This difference between rats and humans is the reason why prolonged stimulation of the thyroid through increased TSH levels (e.g. by substances leading to an enhanced degradation of thyroid hormones) induces thyroid tumours in rats, but not in humans (Table 3.26).

<u>Effect on Target Cells</u> Thyroid hormones may reach the target cells by means of passive diffusion (lipid-soluble hormones), but most are actively absorbed by means of organic anionic transport proteins (OATPs) and monocarboxylate transporters (MCTs) to a concentration that is ten times higher than that in the blood. T4 is generally converted to the more active T3 by membrane-bound deiodinases. There are three types of deiodinases, which, depending on tissue development, differ with regard to the localization of deiodination at the hormone molecule and their occurrence in different tissues. The following generally applies to humans and rats:

Table 3.26 Comparison of thyroid parameters in humans and rats.

Parameter	Humans	Rat
Main binding protein of T4 and T3 in the blood	Thyroxine-binding globulin (TBG)	Albumin in adults
Half-life of T4 in the blood	5–9 days	0.5–1 day
Half-life of T3 in the blood	1 day	0.25 days
T4 production rate/kg body weight	1	10 × higher
TSH concentration* in the blood	1	5–10 × higher

From US EPA (1998), modified.
T4, thyroxine; T3, triiodothyronine; TSH, thyroid-stimulating hormone.
*Human and rat TSH levels converted according to the 2nd IRP WHO 80/558 standard (37 mIU = 7.5 μg TSH).

Deiodinase type I: converts T4 to T3 or rT3 (reverse T3 molecule that is not active as a thyroid hormone); commonly found in the thyroid, kidneys, liver, brain, skeletal muscle and placenta.
Deiodinase type II: converts T4 to T3; commonly found in the brain, pituitary gland, brown adipose tissue and placenta.
Deiodinase type III: converts T4 to rT3 and T3 to T2 (diiodothyronine); found in many tissues except in the liver, kidneys, thyroid and pituitary gland.

Intracellularly, T3 is bound to the cytosolic T3 binding protein (CTBP) and transported to the nuclear thyroid hormone receptor (TR). There are two thyroid hormone receptors with two or three isoforms: TRα1, TRα2 (normally does not bind thyroid hormones), TRβ1, TRβ2 and TRβ3. The number and ratio of receptors on the target cells differ depending on the tissue type and stage of development. T3 binds to a number of TRs with the same affinity. However, thyroid hormone analogues can bind to certain isoforms (e.g. desethylamiodarone to TRβ1). The different TRs can induce various functions in the target cell by binding to different DNA segments and by binding different co-factors.

> **Apart from regulating foetal and postnatal development, the main activity of thyroid hormones in adults is to increase the basal metabolic rate as well as gluconeogenesis, glycogenolysis, proteolysis, lipid synthesis and lipolysis.**

Thyroid hormones also interact with a series of other hormones, such as growth hormones, insulin, catecholamines and sex hormones.

Effects of T3 on target cells that are not mediated by TR are also being discussed, such as an increased uptake of glucose in cells and the influence of oxidative phosphorylation on mitochondria.

Regulation of Synthesis

> **T3 and T4 hormone levels in the blood regulate the release of thyrotropin-releasing hormone (TRH) in the hypothalamus and of thyroid-stimulating hormone (TSH = thyrotropin) in the pituitary gland via a negative feedback mechanism. TRH in turn increases the release and glycolization (important for the biological activity) of TSH.**

The release of TRH is regulated not only by the level of the thyroid hormone, but also by the presence of other hormones (e.g. dopamine, serotonin and glucocorticoids inhibit its release) or by nerve impulses (e.g. cold, stress and hunger inhibit its release). After T4 is converted to T3, this hormone triggers the release of TSH in the pituitary gland. TSH controls its own release in an ultrashort feedback loop. The possibility of TSH playing a role in the release of TRH release has not been discussed. TSH regulates iodine uptake in the thyroid and various steps of thyroid hormone synthesis and release, and also influences thyrocyte hypertrophy and hyperplasia. As a protein hormone, TSH acts on the thyrocytes via G protein-coupled membrane receptors. Iodide is not only necessary for the synthesis of thyroid hormones, but also regulates thyroid functions (e.g. an excess inhibits the sodium iodide symporter (NIS), the Wolff–Chaikoff effect).

Metabolism

> **Thyroid hormones are conjugated to glucuronic acid (by UDP-glucuronosyl transferases, UDPGTs) or sulfuric acid (by sulfotranferases, SULTs) in the microsomes of the liver and are eliminated in the intestines via the bile. Further metabolic pathways include decarboxylation of the amino acid group and oxidative deamination of amines.**

Following glucuronidation and sulfation, conjugates can be deglucuronidated and re-absorbed in the intestines to some extent. Further metabolic pathways of thyroid hormones include decarboxylation of the amino acid group to produce thyronamines (thyroxamine and triiodothyronamine) and deamination of amines to produce acetic acid derivatives (tetraiodothyroacetic acid (TETRAC) and triiodothyroacetic acid (TRIAC)). Deiodinated T3/T4 molecules can also be metabolized via the described pathway. This leads to the formation of endogenous metabolites, which also have effects themselves (e.g. 3-iodothyronamine (T1AM) which leads to a hypometabolic status at pharmacological doses). Another, at least theoretical, possibility would be ether cleavage of the T4/T3 molecule to form diiodotyrosine and monoiodotyrosine. However, this metabolic pathway hardly plays a role under physiological conditions.

> **Decarboxylases, monoaminoxidases, UDPGTs and SULTs may be induced by xenobiotic metabolism (activation of the constitutive androstane receptor (CAR) and pregnane x receptor (PXR)), thus increasing the clearance of thyroid hormones.**

Foetal Development of the Hypothalamus–Pituitary–Thyroid Axis When compared based on time of birth, the thyroid system matures earlier in humans than in rats (39 weeks after fertilization in humans and 22 days after fertilization in rats). The thyroid hormone receptors are the first to be detected in foetuses (humans: as of gestation week (GW) 10; rats: as of gestation day (GD) 14). In the human foetus, the first T4 levels are detected in GW 11. Deiodinases convert T4 to T3 in certain tissues in the foetus. As of GW 18, T3 and TSH can also be determined in the human foetus. TSH, T3 and T4 are found in the rat foetus as of GD 17. In the human foetus, the hypothalamus–pituitary–thyroid (HPT) axis begins to function during the last trimester of gestation (as of GW 27), whereas in rats it starts functioning only at birth. In both species, full maturation is achieved 4 weeks after birth.

Testing of Thyroid Function in Toxicology
Thyroid Endpoints in Regulatory In Vivo Studies

> **Regulatory in vivo toxicology studies should investigate the following endpoints when screening for thyroid function disorders: TSH, T3 and T4 levels in the blood, thyroid weights and histological examinations.**

Hormone values in the blood of rats should be determined by means of a species-specific immunoassay of the TSH protein hormone and immunoassays that can reliably detect any reduction in T3 and T4 hormone concentrations. According to OECD Test Guideline 407 (Repeated Dose 28-day Oral Toxicity Study in Rodents), variation coefficients in the control group of rats should be kept below 25% for T3 and T4 and below 35% for TSH. In addition to any variation found in the immunoassay, hormone values are influenced by the manner in which rats are handled during blood sampling (stress), the time of day the sample was taken and by the oestrus cycle in females (blood samples are preferably taken in the morning at defined intervals). As hormone concentrations are also affected by the type of anaesthesia and the type of blood sampling used in rats, the age of the rats, and the rat strain, a comparison with historical control values can only be made if testing is carried out under the same conditions.

Further Possible Thyroid Endpoints in In Vivo Studies Apart from the above-described endpoints, thyroid disorders may also be detected using other in vivo methods: more sophisticated histological methods, such as morphometric measurements in the rat brain at various stages of development, additional stainings to determine such parameters as the degree of myelination in the brain, deiodinase activities in various tissues and UDPGT activities in the liver tissue ex vivo. Mechanistic in vivo studies may be carried out to elucidate the mode of action of the substance (e.g. a perchlorate discharge assay to determine the inhibition of direct iodine uptake in the thyroid by blocking the sodium iodine symporter (NIS, perchlorate) or the increase in hormone clearance by the liver (phenobarbital) or a TRH stimulation test to determine the inhibition of pituitary or thyroid activities). Furthermore, knock-out mice are used to clarify the mode of action (e.g. Pax-/- mice that cannot synthesize thyroid hormones; mice having different thyroid hormone receptor deficiencies). Omics studies may also provide evidence of the mode of action of exogenous substances through blood or tissue samples obtained from repeated dose rat studies.

In Vitro Screens for Thyroid Function

> **A large number of in vitro test systems are available to study individual mechanisms. However, some of them have not yet been validated (overview in OECD, 2014).**

These tests can be divided into the following groups:

- Thyroid hormone synthesis and transport: the inhibition of TPO is investigated with rat, porcine and human TPO. Furthermore, iodine uptake can be tested in various cell lines (e.g. FRTL-5). Binding studies are carried out to test whether a substance cleaves thyroid hormones from the binding proteins (TBG and transthyretin), which would lead to a faster degradation of the hormones.
- Metabolism: some enzyme activity tests determine deiodinase (mainly D1), UDP-glucuronosyltransferase (UDPGT) and sulfotransferase (ST) activities in liver microsomes. Furthermore, the binding of substances to hepatic nuclear receptors (CAR and PXR) and their activation can be tested in vitro.
- Effect on target cells: the uptake of thyroid hormones in target cells is tested in cell cultures using radioactively labelled T3 and T4. The binding and activation or inhibition of various nuclear thyroid receptors (TRα1, TRα2 and TRβ1) can be tested in reporter gene assays (examples in rat cell lines, GH3 and PC12; examples in human cell lines, HEK293T or in yeasts).

Sex Hormones

Toxicologically Relevant Functions
Biosynthesis

> In male animals and in men, sex hormones are produced in the Leydig cells of the testes (testosterone). In female animals and in women, sex hormones are produced in the ovarian follicles (theca interna cells (progesterone and androgens) and granulosa cells (aromatization of androgens to oestrogens)) and in the ovarian corpora lutea (large and small lutein cells (progesterone; species-specifically oestrogen, e.g. in humans)) as well as species-specifically in the placenta during gestation (e.g. progesterone and oestrogen in humans and guinea pigs). Furthermore, oestrogens and androgens are synthesized in both sexes in the zona reticularis of the adrenal cortex.

The steroid hormone-producing cell absorbs cholesterol via hormone-sensitive low-density lipoprotein (LDL; absorption is mediated by membrane receptors). Thereafter colesterol is stored in the form of an ester. Cholesterol ester is cleaved and free cholesterol is actively transferred into the mitochondrium by the steroidogenic acute regulatory (StAR) protein. The cholesterol side-chain cleavage enzyme (P450scc) pregnenolone is produced in the mitochondrium and then transported out into the cytoplasm. In the cytoplasm, steroid hormones are synthesized further by microsomal enzymes. Figure 3.35 provides an overview of sex steroid synthesis. Steroid

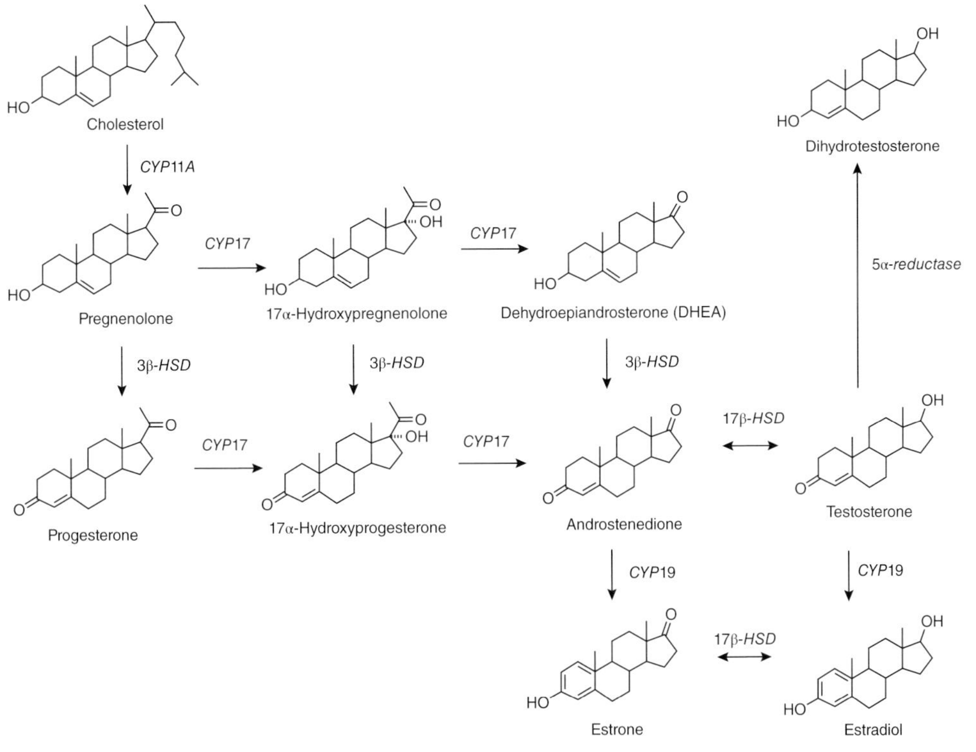

Figure 3.35 Overview of androgen and oestrogen synthesis and the enzymes involved (from Sanderson and van den Berg, 2003). 17β-HSD, 17β-hydroxysteroid dehydrogenase.

hormones are not stored intracellularly, but enter the bloodstream across the cell membrane in the form of lipid-soluble molecules. Steroid hormones are carried by the blood bound to transport proteins, mainly albumin. In foetuses, they are also bound to alpha-foetoprotein (AFP) and, in some species, to steroid hormone binding globulin (SHBG; not in adult rodents).

Effect on Target Cells

> **Lipid-soluble steroid hormones (not bound to protein) passively enter the target cells across the cell membrane and then bind to receptors in the cytoplasm.**

The steroid hormone receptors are one androgen receptor (AR) with two isoforms (ARa and ARb), two oestrogen receptors (ERα and ERβ; ERγ not in mammals) and one progesterone receptor (PR) with two isoforms (PRa and PRb). The binding of a hormone to this type I nuclear receptor induces the cleavage of heat shock proteins, the active transport of the receptor–hormone complex into the nucleus and the binding to hormone responsive elements (HREs) of the DNA together with the binding of other proteins as co-regulators. Depending on these bindings mRNA transcription is induced, which is then translated into proteins. Proteins alter cell functions. Furthermore, cell membrane-associated receptors (e.g. GPR30) were found for oestrogens; they are responsible for rapid (non-genomic) effects of the cell after hormone binding.

Regulation of Synthesis

> **Sex steroid hormones regulate the release of gonadotropin-releasing hormone (GnRH) in the hypothalamus and of gonadotropins (luteinizing hormone (LH) and follicle-stimulating hormone (FSH)) in the pituitary gland via a feedback mechanism.**

In male animals and in men, testosterone synthesized in Leydig cells (possibly after local aromatization to oestradiol or hydrogenation to the more potent dihydrotestosterone) inhibits LH release in the pituitary gland. FSH synthesis is primarily inhibited by inhibin (a protein hormone that is produced mainly in Sertoli cells in male animals). In contrast to human Leydig cells, rat Leydig cells also have prolactin and GnRH receptors. In Leydig cells, prolactin increases LH receptor density and thus the capacity for steroidogenesis. GnRH directly activates testosterone synthesis. Therefore, permanently reduced prolactin levels (e.g. through the administration of dopamine agonists) or increased GnRH concentrations (through the administration of GnRH agonists) may induce Leydig cell tumours in rats, but not in humans.

In female animals and in women, oestrogens induce a dose-dependent inhibition of gonadotropin release (more FSH than LH) by reducing the secretion of GnRH from the hypothalamus; in rats, this may also be achieved by stimulating dopamine activity. Moreover, in the pre-oestrus phase, oestrogens have a positive feedback effect on gonadotropin release once an oestrogen threshold concentration has been reached in the blood, mainly by increasing the number of GnRH receptors at the pituitary gland. In rats, increased noradrenaline activity and decreased dopamine and β-endorphin activities in the CNS enhance the positive feedback effect. Progesterone primarily inhibits the ovulation-inducing LH surge (inhibiting the release, but not the synthesis of LH in the pituitary gland). Inhibin formed in the granulosa cells of the ovarian follicle during the cycle and in the lutein cells of the corpus luteum during pregnancy works in parallel to oestrogen to inhibit FSH biosynthesis.

In both sexes, gonadotropins have a short negative feedback effect on the GnRH release in the hypothalamus, and GnRH, which regulates its own release, has an ultrashort negative feedback effect (the latter feedback mechanism is probably of greater importance).

> **In addition to the effects of gonadotropins and sex steroids on ovaries and testes, further mediators are active in gonad regulation:**
>
> 1. **intragonadal peptide hormones (e.g. insulin-like growth factor 1 (IGF-1), follistatin, angiotensin and oxytocin) (overview in Döcke, 1994)**
> 2. **metabolic hormones (e.g. thyroid hormones, glucocorticoids, insulin and growth hormones)**
> 3. **cytokines**
> 4. **specific enzymes (e.g. plasminogen activator)**
> 5. **vegetative nervous system (primarily β-adrenergic system).**
>
> **The literature on reproductive physiology provides more information on the processes involved and general differences in the oestrus cycle between humans, rodents and rabbits (e.g. Neill and Wassarman, 2005).**

Metabolism Steroid hormones are inactivated with a half-life of a few minutes mainly in the liver as well as in other tissues, such as the kidneys, lungs and gonads, and in the blood, primarily through the introduction of hydroxyl groups and oxidation. They are then conjugated with glucuronic acid and sulfuric acid and eliminated mainly via the urine and faeces as well as in breast milk.

Oestrogens can be hydroxylated at different molecular positions (Figure 3.36), which may induce the production of genotoxic degradation products (4-OH-estradiol; 16α-OH-oestrone). The direction of oestrogen degradation can be controlled through the activation of the respective CYP enzymes.

3.12.3 Foetal Development of the Hypothalamus–Pituitary–Gonad Axis

The following describes developmental aspects relevant to toxicology that can be influenced by having an effect on the hypothalamus–pituitary–gonad (HPG) axis. The literature on reproduction can be reviewed for more details (e.g., Neill and Wassarman, 2005).

> **An endocrine disruption in parental animals may initially have an effect on fertilization. In female animals, hormonally active substances may disturb follicular maturation in the ovaries and thus the oestrus cycle, while in male animals they may affect sperm maturation or libido.**

In rats and mice, the uterus has first to be prepared hormonally by oestradiol and progesterone from the ovary to ensure that the fertilized ovum can implant in the endometrium. In contrast, in guinea pigs and rabbits, only progesterone from the corpus luteum is responsible for implantation. As the embryo is located in an inverted yolk sac in rodents, there is an especially close exchange of substances between the maternal blood and the embryo during the early embryonic phase. In rabbits, the yolk sac does not completely enclose the embryo. In humans, the chorioallantoic

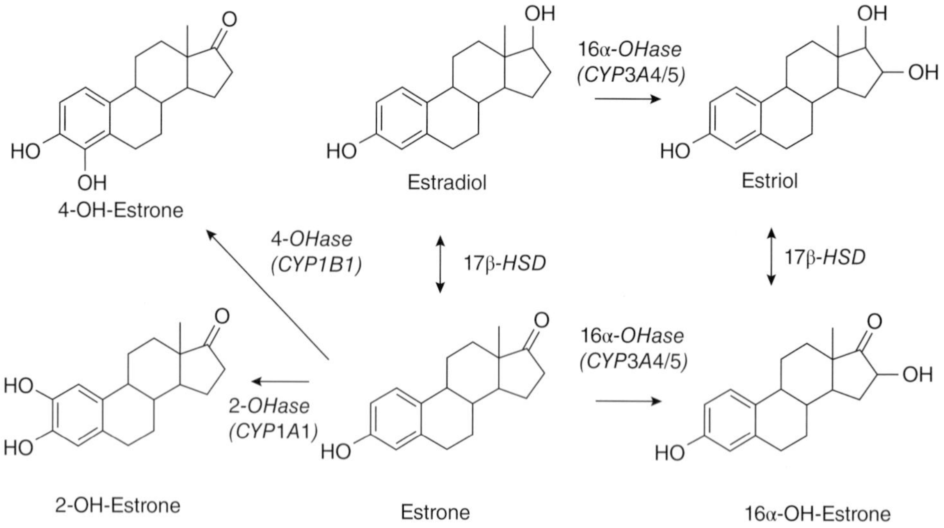

Figure 3.36 *Hydroxylations of oestradiol. 2-OH-oestrone and 4-OH- oestrone (= catechol oestrogens). They are further degraded by methylation with catechol O-methyltransferase (COMT). All hydroxylated oestrogens undergo glucuronidation or sulfation. From Sanderson and van den Berg (2003). 17β-HSD, 17β-hydroxysteroid_dehydrogenase.*

membrane allows a selected exchange of substances (see Carney et al. (2005) for details). After the placenta develops, haemomonochorial (humans and guinea pigs), haemodichorial (rabbits) and haemotrichorial placentas (mice and rats) are differentiated depending on the number of cell layers between maternal and foetal blood.

During gestation, oestradiol and progesterone are only produced by the corpora lutea. In humans and guinea pigs, the placenta synthesizes sex steroids after the first trimester of gestation. In humans, human choriogonadotropin (hCG), a glycoprotein primarily produced in the placenta during the first trimester of gestation, activates the corpus luteum function beginning on day 10 after fertilization. In mice and rats, the corpus luteum function is initially activated by prolactin from the anterior lobe of the pituitary gland of the dam during the first half of gestation and then by lactogenic hormones (polypeptide hormones similar to the growth hormone), which are produced in the foetal part of the placenta.

Androgens are also produced during gestation; they are primarily converted to oestrogens by aromatization during the second half of gestation. In rats and mice, androgens (primarily androstenedione) are formed in the placenta and aromatized in the corpora lutea. This increases oestradiol levels in the blood of the dam from mid-gestation to one or two days before birth. In humans, the adrenal glands of a pregnant woman and foetus produce an increased amount of androgens (primarily dehydroepiandrosterone sulfate) during the last two trimesters of gestation; these are converted to oestrogens in the placenta. In rodents, the corpus luteum synthesizes oestradiol and oestrone, and the placenta of humans additionally produces oestriol. For oestradiol levels to be affected by exogenous oestrogenic substances, the oestradiol levels in the maternal blood have to be several hundred times higher in humans (about 15–20 ng/ml) than in rodents (about 20–60 pg/ml).

Hormone concentrations vary in dams close to parturition because steroid hormones are synthesized at different sites in different species. In rats, mice and rabbits, a drop in the progesterone level by luteolysis of the corpora lutea, which is triggered by prostaglandin F2α synthesis from the mature placenta, is necessary for the induction of birth. In humans and guinea pigs, progesterone levels do not decrease in the maternal blood close to birth. Here, a functional progesterone block resulting from a change in the receptor isoforms at the uterus, a change in the progesterone effect by transcriptional co-regulators, locally increased progesterone metabolism or other paracrine factors at the uterus are discussed.

> **Exogenous substances may affect both the course of gestation and foetal development.**

In mammals, chromosomes determine the primary sex. However, masculinization of the male foetus primarily depends on three hormones: testosterone, insulin-like factor 3 (synthesized in foetal Leydig cells) and anti-Mullerian hormone (AMH; produced in foetal Sertoli cells). The foetus is especially sensitive to exogenous, endocrine active substances during the masculinization programming window. This programming window is between gestation day (GD) 15.5 and 19.5 in rats (total gestation generally 22 days depending on the rat strain) and between gestation week (GW) 8 and 15 in humans (total gestation 39 weeks). In foetal rat testis, testosterone synthesis and the expression of LH receptors starts between GD 14.5 and 15.5, but LH is later secreted from the pituitary gland (GD >17.5). The human foetus also synthesizes testosterone and LH receptors at the beginning of masculinization (GW 8–12). However, hCG produced by the placenta immediately induces testosterone synthesis at this site (LH is produced as of GW 16). This difference means that oestrogens and glucocorticoids may inhibit foetal testosterone synthesis during masculinization in rats, but not foetal hormone synthesis in humans. Testosterone is necessary in foetal development for such processes as the formation and growth of male sex organs (development of the Wolffian duct into epididymis, seminal duct and accessory sex glands) and for foetal brain development following local aromatization to oestrogens. Insulin-like factor-3, which is also synthesized in foetal testicular Leydig cells, is responsible for the initial, intra-abdominal stage of testis descent in the foetus and the formation of the gubernacular cord between testis and scrotum. The second stage of testicular (inguinoscrotal) descent through the inguinal region to the scrotum is mediated by androgens. The anti-Mullerian hormone, which is synthesized in Leydig cells, induces a regression of the Mullerian duct, from which the uterus, oviducts and parts of the vagina develop in female foetuses. Figure 3.37 compares the foetal development of male rats and humans.

The development and proliferation of Sertoli cells in the foetal testis determines the sperm count in the sexually mature individual. Sertoli cells proliferate during a specific period, which in rats lasts from the end of gestation up to day 15 post partum (time window similar in mice and rabbits).

In humans, there are two phases of proliferation: phase 1 from the middle of gestation up to one year after birth and phase 2 from the age of 10 to puberty. If exogenous oestrogens are given during these periods, pituitary FSH secretion is suppressed, which leads to a decreased proliferation of Sertoli cells and thus to reduced sperm production and smaller testes during sexual maturation.

3.12.4 Testing of Sexual Function in Toxicology

Sexual Function endpoints in Regulatory In Vivo Studies

> **Repeated dose studies in rats, mice and dogs include the examination of morphological changes in the sex organs of adult animals caused by endocrine active substances by gross necropsy as well as histopathological examination of the sex organs.**

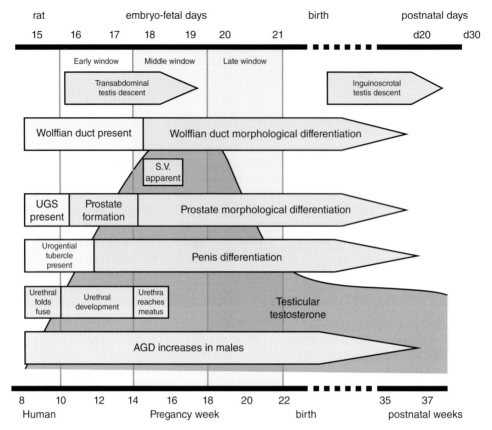

Figure 3.37 Development of male sex organs in rats as compared with humans (early, middle and late window: masculinization window; from Welsh et al. (2008), modified).

A further option is the determination of blood hormone levels. However, this option may involve large inter-individual and intra-individual variations in hormone levels. Studies would have to include a larger number of animals per group to detect the effects of the test substance on hormone levels. In females, the oestrus cycle has to be taken into account. Moreover, blood hormone levels do not reflect hormone concentrations in synthesis or target tissues.

In addition to the above-mentioned endpoints, reproduction studies investigate the mating behaviour, fertilizing ability, development and number of foetuses, parturition and development of offspring (see Chapter 2.10 for details). OECD Test Guidelines 414, 415, 416, 421 and 422 have been designed to investigate these end points. OECD Test Guideline 443 (Extended One-Generation Reproductive Toxicity Study) is a particularly complex reproductive toxicity study that involves detailed histopathological examinations and examines neurotoxicity, immunotoxicity and toxic effects on the thyroid.

In Vivo Screening Assays for Endocrine Active Substances

> **In addition to the above-mentioned regulatory studies, which are required for the toxicological evaluation of substances submitted for authorization, the following rat study designs have been developed to investigate specific issues concerning endocrine active substances.**

Uterotrophic assay (OECD Test Guideline 440): This test is carried out to investigate oestrogen receptor agonists and antagonists and aromatase inhibitors. Substances are given to prepubertal rats or pre-adult rats for three days after ovariectomy (to stop endogenous oestrogen synthesis). Oestrogenic substances increase uterine weight. Oestrogen receptor blockers are administered together with oestrogens to test the inhibitory effect on increases in uterine weight. Aromatase inhibitors are administered to prepubertal rats together with an androgen. The substance then inhibits endogenous aromatase, which converts the androgen to oestrogen.

Hershberger assay (OECD Test Guideline 441): This test is used to detect androgen receptor agonists or antagonists. Substances are given to castrate-peripubertal males (postnatal day (PND) 42) for 5 to 7 days; the weights of organs sensitive to androgens are then determined, such as the weights of the ventral prostate, seminal vesicle, Cowper's gland, levator ani plus bulbocavernosus muscles, and glans penis. Anti-androgens are administered together with androgens and the inhibition of the androgenic effect is recorded.

Pubertal development with male or female rats (OPPTS 890.1450 and OPPTS 890.1500): The development of the sex organs is investigated following administration of substances to prepubertal rats (PND 22/23) up to PND 43 (females) and PND 54 (males).

15-day intact adult male assay (EDSTAC, 1998): In this assay, the substances are given to rats 70–90 days of age for 14 days. The sex organs and sex hormones in the blood are then determined. Each group consists of 15 animals to increase the statistical power of the tests. The assay is designed to detect inhibitors of androgen receptors, testosterone synthesis and 5α-reductase.

In Vitro Screening Assays for Endocrine Active Substances

> **In vitro test systems can be used to investigate the individual endpoints of sex hormone synthesis and effects.**

Effect on Steroid Synthesis Substances are incubated with a human adenocarcinoma cell line (H295R) for 48 hours and the supernatant is examined for steroid hormones (mainly oestradiol and testosterone; OECD Test Guideline 456). As an important enzyme of steroid synthesis, aromatase activity (plus CYP450 reductase) can be determined in the microsomes of various cells (OPPTS 890.1200).

Effect of Sex Hormones on Target Cells The binding of endocrine active substances to **oestrogen receptors** (ER; OPPTS 890.1250) or androgen receptors (AR; OPPTS 890 1150) is tested by displacing radioactively labelled ligands (oestradiol or androgen) from receptors in rat uterine or prostate cells.

ER and AR transcription activation tests with human receptors and receptor-regulated reporter genes in transfected cell lines (ER: hERα-HeLA-9903 (OECD Test Guideline 455; OPPTS 890.1300) or hERα-MCF-7 (EDSTAC, 1998); AR: hAR CV-1 (EDSTAC, 1998; OECD Test Guideline 458)) are used to test both the binding capacity of substances and the activation of DNA transcription in target cells.

Tests with **yeast cells transfected with ER and AR** are an alternative (YES/YAS assays). Oestrogen/androgen agonists are incubated with the cells and receptor activation is then quantitatively indicated by a signal mediated by the reporter gene. Anti-androgens/anti-oestrogens are incubated together with an androgen/oestrogen and evidence to show the inhibition of the androgenic/oestrogenic effect.

Table 3.27 summarizes the OECD Guideline tests/studies relevant for the detecting of endocrine disrupting activity (see Chapters 2.10 and 4.1).

Table 3.27 OECD Guideline tests/studies relevant for the detecting of endocrine disrupting activity.

OECD guideline	OECD guideline number	Year of publication or update
In vitro		
Performance-based test guideline for stably transfected transactivation in vitro assays to detect oestrogen receptor agonists and antagonists	455	2016
H295R steroidogenesis assay	456	2011
Stably transfected human androgen receptor transcriptional activation assay for detection of androgenic agonist and antagonist activity of chemicals	458	2016
In vivo		
a) Screening-assays for endocrine-disrupting activity		
Uterotropic bioassay in rodents	440	2007
Hershberger bioassay in rats	441	2009
b) Reproductive/developmental studies		
Prenatal development toxicity study	414	2001
One-generation reproduction toxicity study	415	1983
Two-generation reproduction toxicity study	416	2001
Reproduction/developmental toxicity screening test	421	2016
Combined repeated dose toxicity study with the reproduction/developmental toxicity screening test	422	2016
Extended one-generation reproductive toxicity study	443	2012

3.12.5 Hazard Identification and Risk Assessment of Endocrine Disruptors

It is not enough to assess whether or not a substance possesses endocrine disrupting properties. Before the identification of endocrine disrupting activity with the test systems described above the biological plausibility has to be proven. This is done by linking the adverse effect and endocrine activity through an endocrine-disrupting mode of action, and by regarding the potency and the dose of the endocrine disruptor in humans. Until now there has been no epidemiological evidence that endocrine disruptors except drugs such as diethylstilbestrol and possibly high exposure to phytoestrogens affect the endocrine system in humans. Endocrine disruptors interact with receptors or via indirect effects like enzyme induction, which are threshold mechanisms. Endocrine disruptors can therefore be treated like most other substances of concern for human health and the environment.

3.12.6 Summary

This chapter compares the hypothalamus–pituitary–thyroid (HPT) and hypothalamus–pituitary–gonad (HPG) axes of the endocrine system of humans with those of laboratory animals used in toxicology. A description of the assays used in toxicology to investigate the effects that substances have on the HPT and HPG axes are given. In adults, endocrine effects are generally only observed as long as the substance is active, whereas permanent damage may be caused to the development of the offspring during certain stages of gestation. When carrying out a risk assessment of endocrine active substances, it is therefore important to take the similarities and differences in foetal development in laboratory animals and humans, the potency and the dose of the endocrine-disrupting substance and overall the biological plausibility into consideration.

Further Reading

Carney, E.W., Scialli, A.R., Watson R.E., DeSesso, J.M. Mechanisms regulating toxicant disposition to the embryo during early pregnancy: an interspecies comparison. *Birth Defects Research, Part C: Embryo today*, **72**, 345–360, 2005.

Döcke, F. (Ed.) Veterinärmedizinische Endokrinologie, 1994, Gustav Fischer Verlag, Jena, 3. Aufl., 1994.

ECETOC Seven Steps for Identification of endocrine Disrupting Properties, ECETOC 7SI-ED, Technical Report No 130, 2017, http://www.ecetoc.org/publication/tr130 (last access 10/2/2017).

EDSTAC: Final Report, Aug. 1998, http://www.epa.gov/scipoly/oscpendo/pubs/edspoverview/finalrpt.htm.

EFSA. Scientific Opinion on Endocrine Active Substances, *EFSA Journal* **11**(3), 3132–3216, 2013.

Neill, J.D. and Wassarman, P. (Eds.) *Knobils and Neill's Physiology of Reproduction*, 3296 pages, Academic Press, 3rd edition, 2005.

OECD. Guidelines for the Testing of Chemicals, Section 4. Health Effects, http://www.oecd-ilibrary.org/environment/oecd-guidelines-for-the-testing-of-chemicals-section-4-health-effects_20745788 (last access 10/2/2017).

OECD Draft Detailed Review Paper State of the Science on Novel in vitro and in vivo Screening and Testing Methods and Endpoints for Evaluating Endocrine Disruptors, July 2011.

OECD Series on Testing and Assessment No. 150: Guidance Document on Standardized Test Guidelines For Evaluating Chemicals For Endocrine Disruption, August 2012.

OECD Series on Testing and Assessment No. 207: New Scoping Document on In Vitro Ex Vivo Assays for Identification of Modulators of Thyroid Hormone Signalling, Part 1 and 2, 2014.

Sanderson, T., van den Berg, M. Topic 3.1: Interactions of xenobiotics with the steroid hormone biosynthesis pathway. *Pure Appl. Chem.*, **75** (11–12), 1957–1971, 2003.

Welsh, M., Saunders, P.T.K., Fisken, M., Scott, H.M., Hutchison, G.R., Smith, L.B., Sharpe, R.M. Identification in rats of a programming window for reproductive tract masculinization, disruption of which leads to hypospadias and cryptorchidism. *J. Clin. Invest.*, **118**, 1479–1490, 2008.

WHO and UNEP (Eds. A. Bergman, J.J. Heindel, S. Jobling, K.A. Kidd, R. T. Zoeller). *State of the Science of Endocrine Disrupting Chemicals – 2012*, WHO Library, 2013.

4
Methods in Toxicology

4.1 OECD Test Guidelines for Toxicity Tests *in vivo*

Rüdiger Bartsch

4.1.1 Introduction

> **Animal tests are necessary to discover possible adverse effects of chemical substances and pharmaceuticals to humans. Certain *in vivo* tests are, therefore, required by regulatory agencies and must be carried out according to internationally recognized test guidelines by qualified test laboratories.**

Tests, or bioassays, in animals are still a prerequisite for the evaluation of the toxicity of chemicals to humans. Although *in vitro* tests are a valuable tool in assessing certain effects, the behavior of a chemical in the body cannot yet be predicted reliably. Uptake via the various routes of administration, distribution and metabolism can be modeled with so-called physiologically based pharmacokinetic (PBPK) models but the complex interactions of the substance or its metabolites with cell components, e.g. proteins or enzymes, and the resulting toxicity can only be investigated in the living organism. Animal tests have come under criticism because the results supposedly cannot be scaled to humans. While it is true that some effects cannot be reliably detected in animals, e.g. combined effects of medicines, hypersensitivity, etc., most of the adverse reactions caused by chemicals can be recognized using *in vivo* tests on mammals. Therefore, animal tests are necessary and are required by regulatory agencies throughout the world to protect patients, consumers, and workers from the hazards of chemicals to which they are exposed.

In vivo tests are required for a variety of regulatory purposes:

- authorization for sale and use of pharmaceuticals, pesticides, fertilizers, food additives
- registration of work-place chemicals (REACH).

These fields demand different tests and requirements but the tests to be carried out can be principally grouped into:

- acute toxicity
- skin and eye irritation
- sensitization
- toxicity after repeated treatment
- reproductive toxicity
- mutagenicity
- carcinogenicity.

The results of the tests must be reliable and reproducible to be accepted by regulatory agencies. This is ensured by standardized study protocols for each test and by an extensive report of the study director on the conduct of the study, so that any deviation from the study guideline can be reconstructed and assessed to determine whether or not it significantly influenced the outcome of the study. Only qualified laboratories are accredited to perform tests for regulatory purposes and they must adhere to nationally and internationally accepted **good laboratory practice (GLP)** guidelines. These guidelines describe how to report and archive laboratory data and records to avoid manipulation of the results. The GLP guidelines also require standard operating procedures (SOPs), statistical procedures for data evaluation, instrumentation validation, materials certification, personnel qualification, proper animal care and independent quality assurance (QA). GLP regulations were first developed by the US Food and Drug Administration. The latest version is available online at http://www.fda.gov/downloads/ICECI/EnforcementActions/BioresearchMonitoring/UCM133730.pdf. Similar principals have been adopted by the Organization for Economic Cooperation and Development (OECD) (http://www.oecd.org/document/63/0,2340,en_2649_34381_2346175_1_1_1_37465,00.html) and the World Health Organization (WHO, http://www.who.int/tdr/publications/documents/glp-handbook.pdf).

The OECD became the successor to the Organization for European Economic Co-operation in 1961 and then consisted of 20 countries, mainly from Europe. The present 34 member countries comprise an international group for the exchange of information and policies. Common legally binding standards and guidelines are also developed by the OECD. The tests used for regulatory purposes have been standardized internationally by the OECD since the 1980s and are referred to as OECD test guidelines (OECD TGs). Therefore, tests conducted according to an OECD TG are recognized by regulatory agencies of any member country. As a result of this agreement the number of animals used for tests is reduced because each substance need only be tested once in each assay and the results will be accepted in every OECD member country.

An overview of the important OECD TGs for *in vivo* tests is given below. It is not possible to list every detail of the guidelines, instead the most important aspects are highlighted.

4.1.2 Requirements for *in vivo* Tests

> **The first prerequisite for initiating a series of tests is that the physicochemical properties and purity of the test substance have been documented. The exposure route chosen should be the same as for the anticipated (or known) human exposure. Several dose groups are used to detect a possible dose–response relationship, including a no observed adverse effect level (NOAEL) and a lowest observed adverse effect level (LOAEL). Test animals must be healthy to ensure that no disease interferes with a substance-related effect. The experiment should include an adequate number of appropriate control animals and a thorough statistical evaluation to ensure that an observed effect is not a chance finding.**

Test Substance

The test substance should be of the highest purity to avoid possible influences from impurities. Physicochemical analyses such as mass spectroscopy, NMR, gas chromatography, etc. offer measures of both identity and purity of the compound. Normally only pure substances are tested except when there is exposure to technical products, which generally are of a lesser purity.

The stability of the substance in the vehicle employed for the test has to be verified to avoid loss of the compound due to decomposition. If it is to be administered via the animal feed, the homogeneity of the preparation must be controlled. In cases of inhalation exposure the homogeneity of the test chamber atmosphere needs to be verified during exposure.

Exposure Route

The exposure route for animals in the study should be the same route by which human exposure is known or anticipated. Oral exposure via gavage, i.e. gastric intubation, has the advantage that the administered dose is known with precision. In many cases the test material is administered in the feed or in drinking water, and only an average dose can be calculated from the water or feed consumption and the body weight during the study. By adjusting the concentration according to feed and water intake and body weight changes a constant dose can also be achieved over the course of the study. Administration by gavage is less preferable due to the possibility of overwhelming metabolism of a bolus dose. Furthermore, in many cases it does not represent the human pattern of exposure. For exposure via the diet or drinking water it should be ensured that the amounts administered do not interfere with normal nutrition and water balance. For these exposure routes, reduced palatability of feed or water caused by a bad taste of the test substance can lower the feed and water consumption, which in turn can have adverse effects on body weight and functions of some organs such as the kidney.

Other exposure routes commonly used are dermal and inhalation. When dermal exposure is intended it is necessary to ensure that the applied substance cannot be ingested by the animal. Application of the test substance can be open, semi-occlusive or occlusive depending on the volatility of the test substance. An inhalation exposure for rodents can be whole body, head only or nose only. The latter is preferable if the test substance is a dust or an aerosol that can be deposited on the animals' fur and ingested by licking.

Animals

The test guidelines require that only healthy animals be used and the health of the animals is checked before and during the study. The animals used should be free of infection and should be monitored for disease during exposure to avoid confounding influences which might significantly impair interpretation of the results. Specific pathogen-free animals (SPF animals), which are certified to be free of certain known diseases, may be used to help control this variable. Change in body weight is an important, albeit non-specific, measure of toxicity. Therefore, at the beginning of a study animals are assigned to dose groups so that the average body weight is similar in each group.

Control Animals

Except for acute tests, a concurrent control group of animals of the same strain and sex is used to ensure that the observed effect is indeed caused by the chemical. The control group is handled in the same manner as the test group and only receives the vehicle (if any is used) in the highest amount used for the substance-treated animals. For substances administered via the diet which cause a reduced food intake, the use of a pair-fed control group may be considered. This control group is fed the same amount of diet as is consumed by the dosed animals. In carcinogenicity studies it is especially important to compare the results to a historical control (i.e. all control animals of the same sex and strain used for the same type of test and exposure route in the last few years in the testing laboratory) because tumor incidences can vary with time and the incidences for certain tumor locations display high background variability.

Interpretation of Results and Statistical Evaluation

At the termination of most of the studies the animals are examined for gross and microscopic pathology. Ideally, histopathological examination of the organs of animals from the study should be conducted without knowledge of the treatment group to minimize bias. The physical findings of control and treated animals are compared and evaluated by appropriate statistical methods to exclude the possibilities of chance findings. Only statistically significant results can be regarded as establishing a cause-and-effect relationship between the test substance and the generation of an adverse effect. Evaluation of the data should also include an evaluation of the biological plausibility of the findings. In some cases a statistically significant effect of a chemical is not observed despite a mechanistically plausible expectation of an effect. But the reverse is also possible, namely, an effect is statistically significant but not biologically plausible. Therefore, a high degree of expertise is required to interpret marginal or contradictive results.

4.1.3 Acute Toxicity

> **Acute toxicity tests provide information on lethality (at high doses), symptoms (at lower doses) and target organ toxicity after a single application of graded amounts of a substance via the oral, dermal or inhalation route. Efforts to reduce animal numbers used in these tests have led to the abolishment of the LD_{50} test for acute oral toxicity. New test guidelines require relatively few animals to classify substances into predefined acute toxicity classes for regulatory purposes.**

Significance of Acute Toxicity Tests

Acute toxicity tests provide information on symptoms of acute overdoses, a hint of general toxicity, and the basic mechanisms of toxicity. They are the starting point for further, more elaborate, tests, e.g. genotoxicity *in vivo*, subchronic and chronic toxicity. Since the results of LD_{50} tests on the same substance may vary, a grouping of the substance into an acute toxicity class is sufficient for classification and labeling purposes. Moreover, information on acute toxicity is important for emergency planning.

Oral Toxicity

The oral toxicity test is aimed at determining the lethality of a chemical after a single dose given by gavage. Before 2002 the LD_{50} test (OECD TG 401) was employed. Groups of rats were given the test substance in graded amounts and the mortality in each dose group was recorded up to 14 days after application. The dose causing lethality in 50% of the animals (LD_{50}) was calculated using standardized mathematical procedures.

The LD_{50} test has been replaced by three new test methods (Table 4.1): the fixed-dose procedure (OECD TG 420), the acute toxic class method (OECD TG 423), and the up-and-down procedure (OECD TG 425), each of which has been validated against the LD_{50} test and shown to produce similar toxicity rankings for various chemicals.

Table 4.1 OECD Test Guidelines for acute mammalian toxicity tests.

OECD Test Guideline	Species, number of animals per dose group	Duration/dosing regime	Observation time points
TG 420, 423, 425 (oral)	Rat, see text	Stepwise procedure	See text
TG 402 (dermal)	Rabbit, see text	Fixed-dose procedure	Mortality, up to 14 days p.a.
TG 403 (inhalation)	Rat, 10	4 h, ≥ 3 concentrations	Mortality, up to 14 days p.a.

OECD TG 420 Groups of animals (usually five female rats) are dosed in a stepwise procedure with 5, 50, 300 or 2000 mg/kg body weight (bw). The starting dose, which is expected to produce some signs of toxicity without causing severe toxic effects or mortality, is selected on the basis of a range-finding study with single animals. Further groups of animals may be dosed with higher or lower fixed doses, depending on the presence or absence of signs of toxicity or mortality. A period of 3–4 days between dosing at each dose level is recommended (if needed) to allow for the observation of delayed toxicity. This procedure continues until (1) the dose causing evident toxicity is established, (2) no more than one death is identified, (3) no effects are seen at the highest dose or (4) deaths occur at the lowest dose.

OECD TG 423 Groups of animals (usually three female rats) are dosed in a stepwise procedure with 5, 50, 300 or 2000 mg/kg bw. Absence or presence of mortality will determine whether (1) further dosing is needed, (2) the same dose is given to three additional animals or (3) the next higher or lower dose is given to three additional animals. Treatment of animals at the next dose should be delayed until the survival of the previously treated animals is certain. The method will enable a judgment with respect to classifying a substance according to one of a series of acute toxicity classes defined by fixed LD_{50} classification values.

OECD TG 425 The test consists of a single ordered dose progression, where animals (usually female rats) are dosed, one at a time, at a minimum of 48-hour intervals. The first animal receives a dose a step below the best estimate of the LD_{50}. If the animal survives, the dose for the next animal is increased by a factor of 3.2; if it dies, the dose for the next animal is decreased by a similar dose progression. Each animal is carefully observed for up to 48 hours before making a decision on whether and how much to dose the next animal. Dosing is ended when one of several stop criteria is fulfilled. The LD_{50} and a confidence interval is estimated with the method of maximum likelihood based on the status of all the animals at the end of the test.

The difference between OECD TG 420 and the other methods is that the decisive endpoint is not the mortality but rather clear signs of toxicity. The exact LD_{50} cannot be calculated with any of the replacement methods. These methods, however, allow the substance to be ranked and classified in one of the acute oral toxicity categories according to the Globally Harmonized System (GHS) for the Classification of Chemicals.

Dermal Toxicity

The OECD TG 402 for testing of dermal toxicity was updated in 2017. The rabbit is the preferred species for testing the toxicity of a chemical applied by the dermal route. The substance is applied once undiluted in case of liquids or moistened with water in case of solids under a semi-occlusive dressing to avoid ingestion of the test material. Groups of animals, of a single sex, are exposed to the test chemical in a stepwise procedure using the appropriate fixed doses. The initial dose level is selected at the concentration expected to produce clear signs of toxicity without causing severe toxic effects or mortality. Further groups of animals may be tested at higher or lower fixed doses, depending on the presence or absence of signs of toxicity or mortality. This procedure continues until the dose causing toxicity or no more than one death is identified, or when no effects are seen at the highest dose or when deaths occur at the lowest dose. Subsequently, observations of effects and deaths are made. Animals which die during the test are necropsied, and at the conclusion of the test the surviving animals are sacrificed and necropsied.

Inhalation Toxicity

Gases, vapors, and aerosols (mists or dusts) suspended in air are administered by inhalation to groups of five male and five female animals (usually rats). At least three adequately spaced concentrations are used. The exposure duration is 4 hours. For several regulatory requirements (e.g. emergency planning) the exposure time can be varied. An LC_{50} is calculated from the data. In the updated OECD TG 403 from 2009 a second method, the $c \times t$ method, is described, where animals are exposed for different durations to several concentrations. Animals can be exposed whole body, but the nose-only method is often preferred to avoid uptake via the skin or licking of the substance from the fur. In the case of aerosols it is important to characterize the particle size distribution of the test sample. Only particles with an aerodynamic diameter of less than 4 μm can reach the lower respiratory tract of rats. The particle size distribution should range from 1 to 4 μm with a geometric standard deviation of 1.5–3.0. Further information on the conduct and interpretation of studies according to the test guideline can be found in the Guidance Document on Acute Inhalation Toxicity Testing (http://www.oecd.org/officialdocuments/publicdisplaydocumentpdf/?cote=env/jm/mono(2009)28&doclanguage=en). For acute inhalation toxicity an acute toxic class method (OECD TG 436) and a fixed concentration procedure (OECD TG 433) have also been published.

Limit Test (All Exposure Routes)

If there is information that the toxicity of a compound will be low, e.g. from tests on chemicals with a similar structure, a single oral or dermal dose of 2000 mg/kg bw (5000 mg/kg bw if required by regulatory agencies) or an inhalation concentration of 5 mg/l can be tested on five animals of each sex. If these doses/concentrations do not produce mortality, a full study may not be necessary.

Examinations Necessary for All Acute Toxicity Tests

The animals will be observed for clinical signs at multiple intervals during the course of the study. Survivors will be killed and all animals examined *post mortem* for macroscopically visible signs of organ damage which may provide evidence of possible mechanisms of toxicity. Histopathology of target organs is optional.

4.1.4 Skin and Eye Irritation

These tests measure the effectiveness of a substance to elicit adverse effects when applied to the skin or eye. The species used is the rabbit.

Acute Dermal Irritation/Corrosion

> **Dermal irritation is the production of reversible damage of the skin following the application of a test substance for up to 4 hours. Dermal corrosion is the production of irreversible damage to the skin observed as visible necrosis through the epidermis and into the dermis, following the application of a test substance for up to 4 hours.**

A fixed amount of the substance is applied to the clipped skin and held in contact with a semi-occlusive dressing. Solid substances should be moistened to ensure good contact, whereas liquids are applied undiluted. Residual material will be removed from the skin without altering the existing response or the integrity of the epidermis. Erythema and edema at the application site are graded and recorded at the time points indicated in Table 4.2. If effects last beyond 72 hours animals will be observed for up to 14 days. If reversibility is seen before 14 days the experiment should be terminated at that time. This test detects dermal irritation, which is reversible damage to the skin, as well as dermal corrosion, which is visible necrosis through the epidermis. This test guideline has been updated several times, the last time in 2015, with the aim of reducing both the number of animals used and the extent of pain for the individual animal. A sequential test strategy is proposed where as much information as possible is gathered to avoid unnecessary testing. The test guideline includes reference to the Guidance Document on Integrated Approaches to Testing and Assessment (IATA) for Skin Irritation/Corrosion (OECD, 2014), proposing a modular approach for skin irritation and skin corrosion testing. *In vivo* testing should not be undertaken until all available data relevant to the potential dermal corrosivity/irritation of the test chemical have been evaluated in a weight-of-evidence analysis as presented in the IATA. Existing data covering human data, *in vivo* data, *in vitro* data, physicochemical properties and non-testing methods are addressed, and a weight-of-evidence analysis is performed. If the result is still inconclusive, additional testing should be performed, starting with *in vitro* methods and using *in vivo* tests as the last resort.

Acute Eye Irritation/Corrosion

> **Eye irritation is defined as the production of changes in the eye that are fully reversible within 21 days following the application of a test substance to the anterior surface of the eye. Eye corrosion is the production of tissue damage in the eye or serious physical decay of vision that is not fully reversible within 21 days following application of a test substance to the anterior surface of the eye.**

A fixed amount of substance is placed in the conjunctival sac of one eye of each animal used. The lids are gently held together for about 1 second. The eyes of the test animals should not be washed for at least 24 hours following instillation of the test substance, except for solids and in case of immediate corrosive or irritating effects. Local anesthetics may be used on a case-by-case basis. The eyes are examined after the time points indicated in Table 4.2 and the grades of the reaction of conjunctiva, cornea and iris are recorded according to a standardized evaluation scheme. Animals showing no ocular reactions are not examined beyond day 3. Animals with mild to moderate lesions should be observed up to 21 days to evaluate reversibility of the lesions. This test

Table 4.2 OECD Test Guidelines for in vivo *skin and eye irritation and skin sensitization.*

OECD Test Guideline	Species, number of animals	Duration/dosing regime	Observation timepoints
Acute dermal irritation/corrosion			
TG 404	Rabbit, see text	0.5 ml of liquid or 0.5 g of solid substance, 6 cm^2, 4 h	60 min, 24, 48 and 72 h after patch removal
Acute eye irritation/corrosion			
TG 405	Rabbit, see text	0.1 ml of liquid or 0.1 g of solid substance, aerosols sprayed from a distance of 10 cm	60 min, 24, 48 and 72 h after application
Skin sensitization TG 406	Guinea pig		
Maximization test	10 5 controls	**First induction:** day 0, intradermal ± FCA **Second induction:** day 6–8, epicutaneous, occlusive, 48 h ± FCA **Challenge:** day 20–22, epicutaneous, occlusive, 24 h, naïve application site	48 and 72 h after start of challenge
Buehler test	20 10 controls	**First induction:** day 0, epicutaneous, occlusive, 6 h **Second induction:** day 6–8 epicutaneous, occlusive, 6 h, on same application site **Third induction:** day 13–15 epicutaneous, occlusive, 6 h, on same application site **Challenge:** day 27–29, epicutaneous, occlusive, 24 h, naïve application site	30 and 54 h after start of challenge
Local lymph node assay TG 429	Four CBA mice	At least three concentrations Days 1, 2, 3: daily 25 µl on backside of ear Day 6: iv injection of ^{3}H-methylthymidine or ^{125}I-iododesoxyuridine, after 5 h removal of aurical lymphnodes, measurement of radioactivity in DNA	

FCA, Freunds Complete Adjuvant.

guideline was updated at last in 2017 and recommends the use of systemic analgesics and topical anesthetics to reduce pain for the individual animal. The test guideline includes reference to the Guidance Document on IATA for Serious Eye Damage and Eye Irritation (OECD, 2017). *In vivo* testing should not be undertaken until all available data relevant to the potential of serious eye damage or eye irritation of the test chemical have been evaluated in a weight-of-evidence analysis as presented in the IATA. Existing data covering human data, *in vivo* data, *in vitro* data, physico-chemical properties and non-testing methods are addressed, and a weight-of-evidence analysis is performed. If the result is still inconclusive, additional testing should be performed, starting with *in vitro* methods and using *in vivo* tests as the last resort.

4.1.5 Skin Sensitization

> **In the skin sensitization test, the immunologically mediated cutaneous reaction to a substance is determined. Usually, a positive reaction is manifested by dermal edema and erythema in the experimental animals.**

Validated structure–activity relationships and *in vitro* models may not be available. As of now, three *in vitro* models are validated to distinguish roughly between sensitizers and non-sensitizers in accordance with the GHS (OECD TG 442C, TG 442D and TG 442E). They cannot, however, replace *in vivo* tests.

The test species to be used is the guinea pig. Two tests can be performed: the maximization test of Magnusson and Kligman, in which sensitization is potentiated by injection of Freunds Complete Adjuvant, and the Buehler test, which does not use adjuvant. Other procedures may also be used but only these two are described in detail in the test guideline. Recently an alternative test, the local lymph node assay (OECD TG 429) has been developed, for which two modifications exist (OECD TG 442A and 442B). These tests use fewer animals than the guinea pig tests and are less stressful to the animals.

In both guinea pig tests, the animals will be initially treated with the test substance (induction dose) and after a certain time (induction period), during which an immune reaction may develop, they are treated again (challenge dose). At two defined time points after the start of the challenge (Table 4.2), the skin reactions are graded and evaluated against a control group, which will be handled identically, but only receives the vehicle. A concurrent positive control is not required. Instead, animals are tested every 6 months with a known sensitizer to prove that the test system is stable and sufficiently sensitive. The concentrations of the test substance have to be chosen in a pre-test such that the induction dose(s) elicits a mild to moderate skin irritation but the challenge dose does not lead to a skin irritation. A substance is considered to be a sensitizer if a positive reaction is seen in at least 30% of the animals in the maximization test or in 15% in the Buehler test. In borderline cases, additional animals can be tested.

In the local lymph node assay mice are given at least three doses (concentrations) of the test substances on the backside of the ear on three consecutive days. The doses are selected according to a preliminary test in which the local and systemic tolerance is assessed. On day 6 ^{3}H-methylthymidine or ^{125}I-iododesoxyuridine and fluorodesoxyuridine are given intravenously. The incorporation of the radioactivity into the DNA of the auricular lymph node cells is measured. Thereby, their proliferation induced by the sensitization by the substance is assessed. A substance is regarded as a skin sensitizer if it increases the proliferation at least three times over that of the negative control.

4.1.6 Toxicity after Repeated Dosing

> In bioassays, which are intended to mimic either brief or extended human exposure to chemicals, animals can be exposed daily for various time periods ranging from 2 weeks to 2 years. The oral, dermal or inhalation route may be selected and groups of animals may be exposed to a range of different doses. The main goal is to identify a NOAEL, which can be used as a starting point for calculating acceptable exposures for humans. Hematological changes, clinical biochemistry and histopathological alterations of most tissues and organs can be observed. Carcinogenicity studies can be performed in animals to detect potential oncogenic activity in humans. Because the incidence of a given tumor type within a population may be low and there are practical limits to the number of animals that can be studied in a bioassay, high doses, which might lead to earlier detection of neoplasms, have to be used to overcome the problem of low statistical power. All of these studies require careful interpretation of the effects observed in rodents and the extrapolation of the findings to humans.

Bioassays to assess the effects of chemical treatment after multiple exposures are designated as subacute (14–28 days), subchronic (90 days), or chronic (about the average lifespan of the species, i.e. for rats and mice about 2 years) studies. The test species are usually rats and mice. For dermal toxicity studies rabbits or hamsters are usually used. For chronic tests of pharmaceuticals in a non-rodent species, dogs are the preferred animals. At least three dose groups and an optional satellite group (i.e. additional animals in the highest dose group which are held untreated for an additional period after termination of the study to check for delayed effects or recovery from effects) are used. Dose selection and dose spacing are critical factors in identifying possible toxic effects of long-term treatment and in revealing a NOAEL.

Ideally the highest dose should lead to clear toxic effects without producing lethality because the death of the animal would preclude observations for the entire intended time period and would impair evaluation of the results. The lowest dose should not result in any adverse effects. Finding a NOAEL in a chronic study is one of the most important and most expensive tasks in the testing of chemicals and, therefore, proper doses have to be selected.

To ensure that the entire effective dose range is adequately encompassed, a chronic study is always preceded by a shorter-term study, the results of which are used to select appropriate doses for the chronic study. A shorter-term assay can thus be regarded as a dose-finding study for a longer-term experiment.

Data Collection During and Following Repeated Treatments with Chemicals

Prior to initiating the study all animals will be allocated to "control" or "treated" groups with equal mean body weights in each group. Animal weights should be measured and recorded throughout the study and at its termination. A record of feed and water consumption should also be kept throughout the study and animals should be monitored for clinical signs of toxicity during the course of the study.

Laboratory studies of samples periodically taken from the animals will include hematology, e.g. measurement of blood cell counts, hemoglobin, hematocrit, and clotting time, and clinical biochemistry of plasma or serum components, e.g. liver enzymes, creatinine, and blood urea nitrogen. For subacute and subchronic studies urinalysis is optional. These measurements may indicate treatment-related functional impairment of organs such as the bone marrow, kidney, or liver. In some cases it will also be deemed appropriate to measure methemoglobin, cholinesterase, and/or the levels of various hormones.

At the termination of the study survivors will be killed and all animals, including survivors and those that died during treatment, will be subjected to gross necropsy. The weights of selected organs will be determined (Table 4.3). All organs with gross lesions and additional selected organs of all animals of the control group and of the highest dose group will be examined histopathologically. These examinations should be extended to all animals of the other dose groups if treatment-related changes are observed in the highest dose group. Finally, measurements of functional neurotoxicity such as grip strength or motor and sensory activity may be measured in either a subacute or a subchronic study.

Results will be evaluated statistically and if possible the NOAEL will be identified. Study designs of tests for toxicity after repeated treatment are outlined in Table 4.3.

Subchronic Studies Ophthalmological examinations have to be carried out before the study and at its termination. The new version of OECD TG 413 allows the study director to use additional satellite groups, interim kills, and to perform broncho-alveolar lavage, neurological tests, clinical-pathological, and histopathological examinations.

Chronic Studies For pharmaceuticals both a rodent and a non-rodent species, usually the dog, are used. Exposure route and frequency are not fixed but should be chosen to simulate likely human exposure. Data from toxicokinetic studies can be used to determine exposure frequency. Hematology and urinalyses should be performed every 3 months, clinical chemistry investigations every 6 months on 10 rodents per group and on all non-rodents.

Combined Chronic/Carcinogenicity Studies The test species are usually rats, mice, or hamsters. A satellite group of 20 animals/sex at the highest dose is treated for at least 12 months to evaluate chronic toxicity. Fifty animals/sex/dose are used to study carcinogenicity over a major part of the lifetime of the species chosen. The exposure time is not fixed. Hematology and urinalyses should be performed every 3 months and clinical chemistry investigations every 6 months. For a valid negative test to be acceptable, no more than 10% of any group may be lost to autolysis, cannibalism or to laboratory-related management problems, and survival in each group has to be no less than 50% at 18 months for mice and hamsters and 24 months for rats.

Carcinogenicity Studies Rats and mice are used because of their relatively short lifespan and the wealth of information on their susceptibility to tumor induction, physiology, and pathology. Moreover, a large historical database on tumor incidence in most strains and tissues exists, which is important in view of the large variability of tumor incidence in untreated animals and among different strains. The animals are exposed after weaning for the major part of their life but the experiment is terminated prior to the natural death of the animals (Table 4.3). The incidence of spontaneous and substance-induced tumors increases in older animals, so it is necessary to terminate the study after a defined period (which, however, is not fixed in the test guideline) to avoid the impact of different lifespans on the interpretation of the data.

At least three adequately spaced doses are tested. It is necessary to use relatively large doses/concentrations to maximize the chance of finding a possible increase in tumor incidence because only relatively few animals (50 per sex/dose) can be used for practical reasons. The highest dose used is called the maximum tolerated dose (MTD). It should not cause excessive premature deaths, which would reduce the number of animals at risk. Ideally the top dose should result in no overt signs of toxicity but a decrease in body weight of no more than 10–15% compared with the control group. Careful selection of the MTD is crucial because if the MTD is not reached and no increase in tumor incidence is found, the negative results of the study might be questioned.

At 12 months, 18 months, and before termination of the study a blood smear is obtained from all animals. A differential blood count is performed on animals of the high-dose group

Table 4.3 OECD Test Guidelines for in vivo tests on toxicity after repeated exposure and carcinogenicity.

OECD Test Guideline	Number of animals per sex and dose	Duration/dosing regime	Number of organs to be weighed	Tissues/organs to be examined histopathologically
Subacute				
TG 407 (oral)	5 (high-dose satellite group optional)	7 d/week, can be reduced to 5 d/week Gavage, 1/d, or in feed or drinking water usually continuously, 28 d	9	20
TG 410 (dermal)		Clipped skin, 10% of body area, 6 h/d, 21 or 28 d, semiocclusive	4	4
TG 412 (inhalation)		Nose only, head only or whole body, 6 h/d, 14 or 28 d, MMAD 1–3 µm, GSD 1.5–3.0	4	6
Subchronic				
TG 408 (oral) – rodents	10 (high dose satellite group optional)	7 d/week (can be reduced to 5 d/week), 90 days; satellite group: at least 28 days Gavage, 1/d, or in feed or drinking water usually continuously	11	30
TG 411 (dermal)		Clipped skin, 10% of body area, 6 h/d, semiocclusive	4	27
TG 413 (inhalation)		Nose only, head only or whole body, 6 h/d, MMAD 1–3 µm, GSD 1.5–3.0	5	30
Chronic				
TG 452	20 rodents 4 non-rodents	≥12 months exposure route and frequency depending on anticipated human exposure	5	35 + respiratory tract if inhalation study
Carcinogenicity				
TG 451	50	Duration pre-defined: mice, hamsters ≥18 months, rats ≥24 months (depending on strain) or when 25% survival in lower dose groups is reached Exposure route and frequency depending on anticipated human exposure		34 + respiratory tract if inhalation study + all grossly visible tumors or lesions suspected of being tumors
Combined chronic/carcinogenicity				
TG 453	50 + 20 for satellite group + additional animals for interim sacrifices	See TG 451	5	34 + respiratory tract if inhalation study

GSD, geometric standard deviation; MMAD, mass median aerodynamic diameter.

and control animals. Clinical biochemistry and urinalysis is not required. For a valid negative test to be acceptable, no more than 10% of any group can be lost due to autolysis, cannibalism, or management problems, and survival in each group should be no less than 50% at 18 months for mice and hamsters and at 24 months for rats. If a significant difference is observed in hyperplastic, pre-neoplastic, or neoplastic lesions between the highest dose and control groups, microscopic examinations should be made on that particular organ or tissue of all animals in the study. If the survival of the high-dose group is substantially less than the control, animals of the next lower dose group should be examined microscopically.

It should be pointed out that in some cases, due to the large doses used, the metabolism of the animal is overwhelmed and toxifying or detoxifying mechanisms may not be operative. When interpreting the data it should also be recognized that a number of tumors are species-specific and, therefore, are not relevant to humans. The incidence and type of tumors found have to be evaluated by experts and the aforementioned influences of metabolism have to be taken into account in the final judgment of whether or not a substance is carcinogenic to humans.

Special Consideration for Dermal Studies The concentration of test substance in the vehicle can be reduced if skin irritation is severe. It may be necessary to restart the study with lower concentrations if skin has been badly damaged early in the study.

Limit Tests Subacute and subchronic limit tests via the oral and dermal routes with a dose of 1000 mg/kg bw can be performed for relatively non-toxic substances. The limit test applies except when human exposure indicates the need for a higher dose level to be used.

4.1.7 Reproductive Toxicity

> **Tests on reproductive toxicity assess the effects on fertility of the parental generation and on development of the offspring. In developmental toxicity studies, pregnant females receive the test substance from implantation to delivery. The soft tissues and the skeleton of the offspring are examined in detail. In fertility tests, male and female animals are exposed to the substance during the pre-mating period and the mating period. The females are further exposed during pregnancy and lactation until weaning of the offspring (one-generation study). These offspring are then further exposed till they reproduce to examine possible effects of exposure from conception to the reproductive stage (two-generation study). These tests serve to determine a NOAEL for toxic effects on fertility and development.**

Developmental Toxicity

Studies on developmental toxicity are focused on the effects of chemicals given to pregnant animals between implantation and delivery. The test animals used are mice or, preferably, rats and a non-rodent species such as the rabbit. Each test and control group should contain a sufficient number of females to result in approximately 20 animals with implantation sites at necropsy. Groups with fewer than 16 animals with implantation sites may be inappropriate. The exposure route should be the same as the anticipated human exposure. At least three dose levels and a vehicle control are used. Ideally the highest dose should induce toxicity but no mortality in the parental animals. Maternal mortality does not necessarily invalidate the study provided it does not exceed approximately 10%. The substance is administered daily from the day of implantation to the day before expected delivery, when dams are killed and the uterine content is examined. Females are weighed on day 0 (mating day), on the first day of dosing, at least every 3 days during the dosing period, and on the day of scheduled kill. Food consumption is measured on the same days. The fetuses are examined immediately after the scheduled kill. For rodents half of the fetuses should be examined for skeletal alterations and the remainder for soft-tissue alterations. Each rabbit fetus should be examined for both types of alterations. Numerical results should be evaluated using the litter as the unit for data analysis. Data from animals that did not survive to the scheduled kill should be reported. Whether or not these data are included in the data analysis should be judged on an individual basis. For some endpoints historical data are useful. NOAELs for maternal and developmental toxicity should be derived if possible. A very important point is to consider whether developmental toxicity is manifested in the presence or absence of maternal toxicity. In the latter case, the developmental effects are clearly caused by the substance whereas in the former case an observed effect may not be attributed unequivocally to the test substance because maternal toxicity might have contributed to the result. For example, inhibition of weight gain in the dam may lead to retardation of skeletal ossification in the fetus.

Fertility

One- and Two-generation Studies Reproductive toxicity is comprised of effects on both parental fertility and developmental toxicity. Effects on pregnancy and lactation may be observed on the parental generation (F0 generation) and on postnatal development in the offspring (F1 generation) (OECD TG 415). TG 415 has been deleted because it is no longer used and better alternatives are available to address current regulatory needs (OECD TG 443, see below). In a two-generation study (OECD TG 416), the F1 generation is further exposed and also mated to assess possible reproductive effects in this generation as a result of exposure from conception to the reproductive stage.

The test species usually employed are mice and rats because of their high fecundity. Males should be dosed for at least one complete spermatogenic cycle, i.e. 70 days in the rat or 56 days in the mouse. Females should be dosed for at least two complete estrus cycles. Animals are then mated and the test substance is administered to both sexes during the mating period and thereafter only to females during pregnancy and lactation. The exposure route should be identical to the route by which human exposure is anticipated. At least three dose levels and a vehicle control should be used. Ideally the highest dose should induce toxicity but no mortality in the parental animals. The number of animals per sex and dose is not exactly specified but should contain a sufficient number of animals to yield about 20 pregnant females at or near term. Note that failing to achieve this number does not necessarily invalidate the study; the results should be evaluated on a case-by-case basis. One male can be mated to one or two females. The maximum mating period is 3 weeks. Reproductive organs of pairs that fail to mate should be examined microscopically to determine the cause of the apparent infertility. Males are examined after mating. Females are allowed to litter normally and rear their progeny until weaning. All of the offspring in the litter may continue in the study or litters may be standardized on day 4 after birth to contain, as nearly as possible, four males and four females per litter. The study is terminated after weaning of the offspring and dams and pups are then examined.

In a two-generation study, the F1 animals are dosed in the same manner as the F0 animals except that dosage starts after weaning. F1 animals are mated with pups from another litter of the same dose group when the animals reach the age of 13 weeks for rats or 11 weeks for mice. Further conduct of the study is as described above for F0 animals. In certain instances, such as treatment-related alterations on litter size or equivocal effects observed in the first mating, it is recommended that the adults be mated again to produce a second litter.

The data to be collected and reported are shown in Table 4.4. TG 416 was updated in 2001 and requires a more detailed report of the data than TG 415.

Table 4.4 OECD Test Guidelines for in vivo tests on reproductive toxicity.

OECD Test Guideline	Number of animals per sex and dose	Duration/dosing regime	Parameters measured
Developmental toxicity			
Prenatal developmental toxicity study TG 414	>16	Rat, rabbit: from day of implantation to one day before cesarian section (ca. d 6 to 19 after mating for rats), daily	**Dams:** body weight, food consumption, clinical observations, weight of gravid uterine, number of corpora lutea and implantation sites. Pre- and post-implantation losses, number of dead and viable fetuses, sex and body weight of each fetus, external alterations, skeletal and visceral variations, anomalies and malformations
Toxicity to fertility			
One-generation reproduction toxicity study TG 415 (this test guideline has been deleted, see text)	>20	Males (rat 70 d, mouse 56 d) + mating period, females: 2 weeks + mating period + pregnancy + lactation until weaning, 7 d/week	**Parental animals:** clinical observations, body weight, food consumption, indices of fertility and gestation, duration of gestation **Offspring:** number of pups, sex ratio, stillbirths, live births, presence of gross anomalies, weight of the litter at birth and on days 4 and 7, thereafter weight of the pups weekly until weaning, viability index **Histopathology, parental animals:** all target organs and ovaries, uterus, cervix, vagina, testes, epididymides, seminal vesicles, prostate, coagulating gland, pituitary gland of the high-dose group and controls if not done in other repeated-dose studies
Extended one-generation reproductive toxicity study TG 443	>20	F0 generation: at least 2 weeks before mating + 2 weeks mating period Males: further dosed for 6 weeks Females: + pregnancy + lactation until weaning (ca. 10 weeks for males and females)	**Parental animals:** clinical observations, body weight, food consumption, indices of fertility and gestation, duration until gestation, duration of gestation, weight of reproductive organs, accessory sex organs, brain, liver, kidneys, heart, spleen, thymus, pituitary, thyroid, adrenals, target organs, sperm parameter

(continued)

Table 4.4 (Continued)

OECD Test Guideline	Number of animals per sex and dose	Duration/dosing regime	Parameters measured
		F1 generation: from weaning until adulthood, optional F2 generation, 7 d/week	**Offspring:** number, sex ratio, still births, live births, macroscopic anomalies, clinical observations, body temperature, activity, reaction to handling, litter weight at birth and 4 and 7 days later, thereafter weekly; anogenital distance, nipple retention in males, balano-preputial separation or vaginal opening for females/males, respectively **Cohort 1A:** 20 males and 20 females for examination of reproductive toxicity, clinical biochemistry, thyroid hormones, hematology, urinalysis, sperm parameter, organ weights as for parental animals, additionally lymph nodes **Cohort 1B:** 20 males and 20 females, weight of reproductive organs, accessory sex organs, target organs **Cohort 2A:** 10 males and 10 females for examination of functional developmental neurotoxicity at day 24 post partum, and between day 63 and day 75 post partum; after day 75 post partum histopathological examination of neurotoxicity **Cohort 2B:** 10 males and 10 females for histopathology of the brain **Cohort 3:** 10 males and 10 females for examination of immunotoxicity **Histopathology, parental animals and Cohort 1A:** all known target organs and weighted organs **Histopathology, Cohort 1B:** reproductive organs, if results from Cohort 1A inconclusive
Two-generation reproduction toxicity study TG 416	>20	F0 generation: same as in TG 415 F1 generation: after weaning, until 13 (rat) or 11 (mouse) weeks of age + mating period Females: + pregnancy + lactation until weaning of F2-animals, 7 d/week	Same as in TG 415 **Additionally:** sperm analysis in parental males, water consumption if test substance is administered in drinking water, estrous cycle length and normality in parental females, physical or behavioral abnormalities in dams and offspring **Offspring:** age and body weight at vaginal opening or balanopreputial separation, anogenital distance in F2 pups if triggered by alterations in F1 sex ratio or timing of sexual maturation, functional investigations (motor activity, sensory function, reflex ontogeny), if not included in other studies

Study	N	Dosing	Observations
Reproduction/ developmental toxicity screening test TG 421	10	Daily, 2 weeks prior to mating + during mating period Males: further dosed until 28 days of dosing Females: + pregnancy + up to day 3 post partum (dermal and inhalation: up to day 19 of gestation)	**Additionally:** ear and eye opening, tooth eruption, hair growth may be recorded Functional investigations may be omitted in groups showing otherwise clear signs of toxicity, they should not be done on pups selected for mating. Organ weights of uterus, ovaries, testes, epididymides (total and cauda), seminal vesicles with coagulating gland and their fluids, brain, liver, kidneys, spleen, pituitary, thyroid, adrenal glands and known target organs; brain, spleen and thymus from one pup/sex/litter Calculation of indices of mating, fertility, gestation, birth, viability and lactation, reporting of time-to-mating, number of implantations, post-implantation loss, and runts **Parental animals:** body weight (weekly), food consumption, clinical observations, gross necropsy **All animals:** weights of testes and epididymides Histopathology of testes, epididymides, ovaries and optionally accessory sex organs and all organs showing macroscopic lesions (all animals), number of implantations, post-implantation loss, estrous cycle length, thyroid hormone levels **Offspring:** number and sex of pups, stillbirths, live births, runts, gross anomalies, weight of litters on day 0 or 1 post partum and day 4 post partum, anogenital distance, nipple retention in males, thyroid hormone levels
Combined repeated dose toxicity study with the reproduction/developmental toxicity screening test TG 422	10	Daily, 2 weeks prior to mating + during mating period Males: further dosed until 28 days of dosing Females: + pregnancy + up to day 3 post partum (dermal and inhalation: up to day 19 of gestation)	Same as in TG 421 **Additionally:** **Parental animals:** hematology, clinical chemistry, optionally urinalysis, functional observations (each endpoint five animals/sex; except when done in a 90-day study), weights of liver, kidneys, adrenals, thymus, spleen, brain and heart (five animals/sex) Histopathology of these seven organs and of gross lesions, spinal cord, stomach, small and large intestines, thyroid, trachea, lungs, uterus, urinary bladder, lymph nodes, peripheral nerve, bone marrow (five animals/sex)

In an extended one-generation study (OECD TG 443), in addition to effects on fertility and development effects on the developing nervous and immune system of the F1 generation are also examined. Optionally, an F2 generation can be assessed for reproductive toxicity.

Note that offspring in fertility studies are exposed indirectly *in utero* via the bloodstream if the substance crosses the blood–placenta barrier, during lactation via the milk if the substance is excreted into milk, and, in case of feeding studies, via the feed as pups start to eat for themselves during the last week of lactation, before scheduled dosing begins.

Screening Tests These tests can be used to provide initial information on possible effects on reproduction and development, either at an early stage of assessing the toxicological properties of chemicals, or for chemicals of concern or chemicals for which little or no information is available. The tests do not provide complete information on all aspects of reproduction and development. OECD TG 421 is a reproductive/developmental toxicity screening test with shortened exposure of males but histopathological examination of male sex organs is made after 28 days of exposure. OECD TG 422 additionally includes a more detailed histopathological examination of organs usually evaluated in a subacute test (OECD TG 407). The optional functional neurotoxicity screen contained in OECD TG 407 is also included. Both test guidelines were updated in 2016 to encompass additional endpoints relevant for endocrine disruptors. At least three dose groups and a control are used. Initially, histopathology is only performed in high-dose and control animals but has to be extended to lower dose groups if triggered by other findings.

Differences of OECD TG 421 and OECD TG 422 to a full fertility test making these screening tests less sensitive as a full test are:

- males are dosed for a shorter period of time during the pre-mating period, which does not cover the whole spermatogenic cycle
- the number of animals and pups is only half, thereby reducing the chance of detecting effects
- the tests are terminated already on day 4 post partum and not at weaning so that delayed effects during the lactation period cannot be registered.

Limit Tests A so-called limit test (applies for testing of toxicity to fertility, and developmental toxicity as well as for both screening tests) with a single dose of at least 1000 mg/kg bw can be performed for substances showing no evidence of toxicity at this dose level in repeated-dose studies. Thereby, the number of animals used and the costs compared to a complete test with more dose groups are reduced. The limit test does not apply when human exposure indicates the need for a higher oral dose level to be used. For other types of administration, the physical and chemical properties of the test substance often may indicate the maximum attainable level of exposure (e.g. dermal application should not cause severe local toxicity).

4.1.8 Other Test Guidelines

In addition to the test guidelines outlined above, further tests can be performed in special cases:

- OECD TG 424: neurotoxicity study in rodents
- OECD TG 426: developmental neurotoxicity study
- OECD TG 440: uterotrophic bioassay in rodents, a short-term screening test for estrogenic properties
- OECD TG 441: Hershberger bioassay in rats, a short-term screening assay for (anti) androgenic properties.

4.1.9 Other Regulatory Bodies

Other regulatory bodies, such as the European Community, the US Food and Drug Administration, and the US Environmental Protection Agency, have published similar test guidelines. They are available from the websites listed as references below.

4.1.10 Summary

In vivo tests are required to discover possible adverse effects of chemical substances and pharmaceuticals to humans. They must be carried out according to internationally recognized test guidelines and under the principles of good laboratory practice to ensure validity of the data. Although *in vitro* tests offer a great deal of information on the actions of a substance, *in vivo* tests cannot be replaced due to the complexity of the organism and the various possibilities of toxification, detoxification, and reactions at the cell surface, within cells and in subcellular structures. However, efforts have been undertaken to reduce the number of animals used for testing purposes. Important *in vivo* tests concern acute toxicity, i.e. toxicity after a single application, skin and eye irritation, skin sensitization, toxicity after repeated administration, and reproductive toxicity, which comprises effects on the developing fetus and on fertility of the parent animals. For each of these tests, their most important principles and requirements according to the latest OECD TGs are described. Prerequisites are known purity of the test substance, healthy test animals of sufficient number, appropriate control animals and statistical evaluation of the results. The exposure route chosen should be the same as for the anticipated (or known) human exposure. Dose selection is of critical importance, therefore long-term studies or full-range studies are preceded by studies with fewer animals to find appropriate doses. Ideally, tests with repeated application should give a dose–response relation and a clear no observed adverse effect level from which safe doses for humans can be extrapolated. A dose that exerts some but not excessive toxicity is necessary for several tests to be valid. Possibilities and limitations of the various tests and the importance of expert judgment when interpreting the findings are highlighted.

Further Reading

European Communities, Directive 67/548/EEC Annex V Part B, http://ec.europa.eu/environment/archives/dansub/pdfs/annex5b_da.pdf.

OECD, GLP-Regulations, Organisation for Economic Cooperation and Development, Paris, OECD Series on Principles of Good Laboratory Practice (GLP) and Compliance Monitoring.

OECD, Guidelines for the Testing of Chemicals, Section 4: Health Effects, https://www.oecd-ilibrary.org/environment/oecd-guidelines-for-the-testing-of-chemicals-section-4-health-effects_20745788.

OECD, Guidance Document on Acute Inhalation Toxicity Testing, Organisation for Economic Cooperation and Development, Paris, http://www.oecd.org/officialdocuments/publicdisplaydocumentpdf/?cote=env/jm/mono(2009)28&doclanguage=en.

OECD, Guidance Document on Integrated Approach to Testing and Assessment (IATA) for Skin Irritation/Corrosion. Environmental Health and Safety Publications, Series on Testing and Assessment No. 203, Organisation for Economic Cooperation and Development, Paris, 2014, http://www.oecd.org/officialdocuments/publicdisplaydocumentpdf/?cote=env/jm/mono(2014)19&doclanguage=en.

OECD, Guidance Document on an Integrated Approach on Testing and Assessment (IATA) for Serious Eye Damage and Eye Irritation. Series on Testing and Assessment No. 263. ENV Publications, Organisation for Economic Cooperation and Development, Paris, 2017, http://www.oecd.org/officialdocuments/publicdisplaydocumentpdf/?cote=ENV/JM/MONO(2017)15&doclanguage=en.

US Environmental Protection Agency, Series 870 – Health Effects Test Guidelines, https://www.epa.gov/test-guidelines-pesticides-and-toxic-substances/series-870-health-effects-test-guidelines.

US Food and Drug Administration, Good Laboratory Practice Regulations; Final Rule, http://www.fda.gov/downloads/ICECI/EnforcementActions/BioresearchMonitoring/UCM133730.pdf.

US Food and Drug Administration, Red Book 2000, Toxicological Principles for the Safety Assessment of Food Ingredients, www.fda.gov/downloads/Food/GuidanceRegulation/UCM222779.pdf.

WHO, Handbook: Good Laboratory Practice, World Health Organization, Geneva, http://www.who.int/tdr/publications/documents/glp-handbook.pdf.

4.2 Genotoxicity

4.2A *In vitro* Tests for Genotoxicity

Hans-Jörg Martus[1]

4.2A.1 Introduction

> Genotoxicity is a term that describes events involved in genetic damage, which can be permanent mutations (gene mutations, structural chromosome mutations, genome mutations) or other effects that are associated with the formation of mutations (e.g. DNA modifications, DNA repair, or recombination). Germ cell mutations can increase the risk for offspring with heritable disease, and somatic mutations may lead to the development of cancer or other conditions in the affected individual. Genetic damage may be the result of a direct reaction of a chemical or its metabolite, or a physical agent with DNA but also the consequence of an interference with factors involved in genome integrity, such as the proteins of the spindle apparatus or of chromosome architecture.
>
> Genotoxicity tests of chemicals are conducted in order to obtain information on a compound's capability to induce mutations. Due to the clear mechanistic link, genotoxicity tests are primarily employed to identify a carcinogenic potential and, with restrictions, a risk to induce heritable mutations in germ cells. Further, genotoxicity tests in vitro are employed to clarify the molecular mechanisms underlying the genotoxic effects of chemicals.

Genetic Endpoints of in vitro Test Systems for Genotoxicity

In order to measure a biological effect, the appropriate endpoint needs to be in place. Genotoxic effects exerted by chemicals include alterations of the genetic material that are transmitted to the progeny of the treated cells. These heritable changes are termed **mutations**. However, genotoxic effects also include effects which can be detected in the treated cell itself, e.g. modifications of DNA structure, or certain responses of the cell occurring as a specific consequence of the damage to its genetic material. Thus, mutagenic effects represent a subgroup of genotoxic effects and are defined by a test system employing mutations as the genetic endpoint. Technically, gene mutations, where biochemical functions are used to measure genetic damage, or chromosome damage, where optical methods like microscopy or flow cytometry predominate, are used most frequently to identify and quantify genotoxic activity.

---

[1] Based on the chapter "*In vitro* Tests for Genotoxicity" by U. Andrae and G. Speit of the previous edition.

4.2A.2 Bacterial Test Systems

Tests for the Induction of Gene Mutations

> The Salmonella reverse mutation test, also referred to as the Ames test, is the most frequently employed *in vitro* mutagenicity test (Figure 4.1). The indicator organisms for the detection of induced mutations are various strains of *Salmonella typhimurium* and/or *Escherichia coli* which, as a consequence of mutations induced in genes coding for enzymes for the biosynthesis of certain amino acids (histidine or tryptophan, respectively), have lost the ability to grow on agar lacking this amino acid. Mutagenic chemicals induce reverse mutations in these cells, which recovers the amino acid proficiency and allows mutants to grow and form colonies on minimal agar restricted in the respective amino acid.

Since the genetic modifications introduced in enzymes of the histidine or tryptophane operons of the individual tester strains confer specific sensitivities for reverse mutations, the ability of a compound to induce mutations in specific tester strains allows some conclusions with respect to the molecular mechanisms (Table 4.5). The majority of the *Salmonella* strains have guanine–cytosine (GC) base pairs at the primary reversion site. In contrast, strain TA102 and the *E. coli* strain WP2 have an adenine–thymine (AT) base pair at the primary reversion site, which renders them more sensitive to oxidizing and cross-linking mutagens. The target gene for reversion is carried on the bacterial chromosome, with the exception of strain TA102, where it is located on a multicopy plasmid (pAQ1).

In addition to the mutations resulting in amino acid dependence, most tester strains carry additional mutations that were specifically introduced in order to increase the sensitivity of the cells to chemical or physical mutagenesis. One of these mutations (*rfa*) causes the partial loss of

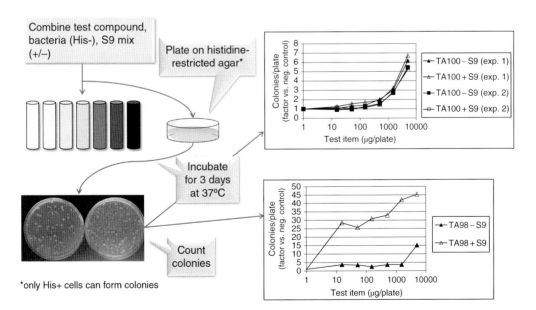

Figure 4.1 Salmonella reverse mutation (Ames) test.

Table 4.5 Properties of the most frequently used S. typhimurium and E. coli strains.

Tester strain	Repair defect	Plasmid	Base pair at primary reversion site	Type of mutation detectable
S. typhimurium				
TA1535	*uvr*B	–	GC	Base substitution
TA100	*uvr*B	pKM101	GC	Base substitution
TA1538	*uvr*B	–	GC	Frame shift
TA98	*uvr*B	pKM101	GC	Frame shift
TA1537	*uvr*B	–	GC	Frame shift
TA97	*uvr*B	pKM101	GC	Frame shift
TA102	–	pKM101, pAQ1	AT	Base substitution, frame shift
E. coli				
WP2 *uvr*A	*uvr*A	–	AT	Base substitution
WP2 *uvr*A (pKM101)	*uvr*A	pKM101	AT	Base substitution

the lipopolysaccharide layer on the outer of the two cell membranes of *S. typhimurium*. As a consequence, the uptake of large hydrophobic chemicals into the bacteria is facilitated. Further mutations (*uvr*A or *uvr*B) result in a defective repair of DNA damage, thereby enhancing the sensitivity to many chemicals. Moreover, several of the tester strains carry the plasmid pKM101, which allows the replicative bypass of DNA lesions and, thereby, increases the yield of mutations.

Table 4.6 gives an overview of the most frequently employed *S. typhimurium* and *E. coli* tester strains and the types of mutations that can be detected. For routine testing of chemicals, the use of a test battery comprising the base substitution strains TA1535 and TA100, the frame shift strains TA1537 (or TA97 or TA97a) and TA98, and strain TA102, which is particularly sensitive to oxidative mutations, has gained general acceptance and is recommended by practically all current regulatory guidance. Instead of TA102, *E. coli* strains WP2 *uvr*A or WP2 *uvr*A (pKM101) are acceptable as well.

Most commonly, the bacterial gene mutation test is performed as a so-called **plate incorporation assay**. Bacteria, S9 mix or buffer, and test material solution are mixed with soft agar and plated on top of a layer of bottom agar. In a modification of the test, the **preincubation assay**, bacteria, S9 mix or buffer and the test material are first preincubated together for a certain period (generally 20–30 minutes), subsequently mixed with the soft agar and plated. In this version test chemical, S9 and bacterial concentrations are higher for the preincubation period, which has been shown to increase the sensitivity to some chemical classes. When repeat experiments are conducted, often both methods are used for either of the repetitions. In both variants, the top agar mix is plated on the bottom agar, which contains traces of histidine or tryptophan (minimal agar), i.e. an amount that allows the bacteria to proceed through a few rounds of cell replication in order to fix any primary DNA damage into permanent mutations. Thereafter, only mutants that have regained the respective amino acid proficiency (i.e. revertants) continue growth and form colonies after 2–3 days of incubation. A background lawn of growth is formed by the non-mutants, which is used as a sensitive parameter to assess test item cytotoxicity. An increase in the number of revertant colonies over the number of spontaneously growing colonies following exposure of the cells to a test compound is taken as an indication of a mutagenic effect of the latter. Statistical methods are used to identify positive effects, but often than this a 2-fold (or higher) increase in mutant numbers (1.5-fold for strain TA102) is used as the criterion to identify a mutagenic effect. Other criteria to qualify or disprove mutagenicity are dose response and reproducibility of the effect, which means that if those are not observed, individual increases in colony numbers are generally not considered to indicate a true mutagenicity of the test compound.

In another variation of this test, sometimes termed miniscreen, six-well plates are employed instead of the 9-cm Petri dishes of the standard test. This is done in order to save test material, but otherwise the assay conditions are comparable to the standard test, with the exception that some concentrations (bacterial suspension, S9) are higher than in the standard test. Also, results obtained with both tests are widely comparable, so this test can be used as a replacement for cases where test material supply is limited. However, for situations where human safety is directly involved, such as in the support of clinical trials for drug development, the standard test is mandatory.

The widespread use of the bacterial mutagenicity assay is still mandatory, which has resulted in the assembly of a huge database especially for the *Salmonella* tester strains, is largely due to the simplicity and speed with which mutagenicity can be detected. The system is very sensitive for the detection of mutations because of the tailor-made properties of the cells and the use of very large numbers of cells (about 10^8 per treatment), which allows the detection of small increases in mutant frequency by the test compound. Furthermore, this test is relatively insensitive for artifactual effects that would resemble a mutagenic response but result from other effects, such as excessive cytotoxicity, a problem that is not infrequent in mammalian cell systems. Interestingly (and understandably), test material that releases the respective amino acid will enable the formation of colonies which are not mutants (phenocopies), but in reality such a test material will be rarely used. A disadvantage of the assay is the limited capacity of the bacteria to metabolize xenobiotics, which makes the use of exogenous the S9 mix essential. Moreover, the structure of the bacterial chromosome is different from mammalian chromosomes. As a consequence, compounds interfering with specific mammalian cell components involved in chromosome integrity, e.g. chromosomal proteins or spindle fibers, cannot be detected in bacterial test systems.

A variation of this test has been described, which is based on a liquid dilution protocol, using 96-well (or even smaller) plates instead of agar-containing Petri dishes or six-well plates. In this test, generally referred to as Ames II, bacteria are incubated in the liquid phase with test compound and S9 if needed but instead of plating the mix on agar a limited dilution protocol is employed to plate a defined number of cells in each well of multiwell plates (usually 96). Revertant growth, rather than counting colonies forming on minimal agar as in the standard test, is assessed by counting wells with cell growth, as can be easily assessed by a colorimetric reaction in the medium. Whereas this method, similar to the minscreen, is not yet accepted to directly assess human risk, it offers the advantage of automation and higher throughput due to the conductance of the assay entirely in the liquid phase, avoiding the technically challenging automatic handling of agar.

4.2A.3 Test Systems employing Mammalian Cells

Mammalian Cell Gene Mutation Tests

> **In gene mutation assays, the mutagenicity of a test compound in cells becomes apparent by mutational alteration of defined cellular functions. In contrast to the Salmonella gene mutation test, where the altered cell function is due to a reverse mutation leading to the gain of a function (i.e. amino acid synthesis proficiency), the common gene mutation assays employing mammalian cells detect forward mutations, i.e. mutations which lead to a loss of function. Gene mutation assays allow the direct identification of heritable changes of the genetic material of the cell.**

In mammalian gene mutation tests, functionally altered cells are identified by selection procedures that allow the detection and quantification of rare mutants within a large excess of non-mutated cells. In general, selection is based on the mutation-induced resistance to toxic substances that will kill the non-mutated cells, which is the opposite of the Salmonella reverse mutation test, where the gain of a function is the selection principle. The mutants survive and form colonies because the resistance is passed on to the daughter cells.

For routine testing, originally primarily Chinese hamster cell lines, e.g. V79 or CHO cells, gained acceptance for the detection of an induced resistance to guanine analogues and ouabain. In addition, the mouse lymphoma cell line L5178Y is used for the detection of an induced resistance to thymidine analogues. In addition, using the same target gene and selection system, TK6 human lymphoblastoid can also be used. These cells exhibit a stable karyotype, proliferate rapidly and have a high potential for growing in colonies, which can be counted.

In the different gene mutation assays (Table 4.6), the cells are exposed to the test compound in the absence or presence of a metabolic activation system. Subsequently, they are further cultivated in growth medium in order to allow a fixation of induced DNA damage as mutations by DNA replication. In addition, in those mutation detection systems such as the HPRT test or the TK$^{+/-}$ test where the induced resistance is due to the loss of an enzymatic activity, this post-treatment "expression time" is necessary for the dilution of the protein present in the cell, by cell division or proteolytic degradation, in order to render the cell deficient for the respective function, which forms the basis for the selection of mutants. Following the expression time, the cells are cultured for several days in the presence of the selecting agent and then the number of mutant colonies is counted. Simultaneously, the proportion of cells surviving the exposure to the chemical is determined by growing the cells in the absence of the selection agent.

HPRT Gene Mutation Test

> **The HPRT gene mutation test detects the heritable loss of the hypoxanthine guanine phosphoribosyl transferase (HPRT) activity. It primarily identifies mutations within the *hprt* gene, e.g. base substitutions, frame shifts or small deletions. Large deletions, which extend into neighboring genes that may be essential for the survival of the cell and which could be detected as chromosome aberrations, are frequently lethal and can therefore not be detected well with the HPRT test.**

HPRT converts free purine bases into the corresponding nucleoside monophosphates for the synthesis of nucleic acids. The selection of the mutants is based on the different toxicities of

Table 4.6 Genetic endpoints of common gene mutation assays employing mammalian cells.

Target gene	Genetics	Selection agent
Hypoxanthine guanine phosphoribosyl transferase (*hprt*)	Recessive, X chromosomal	6-thioguanine
Thymidine kinase (*tk*)	Recessive, autosomal, hemizygous in test cells	Trifluorothymidine
Ouabain resistance locus/Na$^+$/K$^+$-ATPase gene	Dominant	Ouabain

the synthetic purine base 6-thioguanine (6-TG) to mutated and non-mutated cells. In cells with functional HPRT, 6-TG is converted to nucleotides which are strongly cytotoxic and cause cell death. In mutated cells, the loss of the HPRT activity makes the cells resistant to TG because they use a salvage pathway circumventing the incorporation of 6-TG.

The *hprt* gene is localized on the X chromosome. Cells derived from a male animal, e.g. V79 cells, contain only one X chromosome, whereas in cells from a female animal, e.g. CHO cells, one of the two X chromosomes is naturally genetically silenced. Thus, cells from either gender are functionally hemizygous for the *hprt* gene, which is a prerequisite for the mutational principle since in both cell types a single mutation can result in the loss of the HPRT activity and the formation of the 6-TG-resistant phenotype.

$TK^{+/-}$ Gene Mutation Test

> The $TK^{+/-}$ gene mutation test detects the heritable loss of the activity of the enzyme thymidine kinase (TK) in cell lines heterozygous for the thymidine kinase gene ($tk^{+/-}$). Cell lines that have been found suitable for this test are the human lymphoblastoid cell line TK6 and the mouse lymphoma cell line L5178Y. For routine testing, the mouse lymphoma cells are usually employed, so that this test is also referred to as mouse lymphoma assay (MLA). It detects both mutations limited to the *tk* gene and large size change-mutations, which extend into regions flanking the *tk* gene and which can be identified as chromosome aberrations.

Similarly to the HPRT test, for the selection of the mutated cells use is made of the different toxicities of the thymidine analogue trifluorothymidine to mutants and wild-type cells. Following phosphorylation by thymidine kinase in the non-mutated cells, TFT is incorporated into DNA and methylated, which results in cytotoxicity and cell killing during subsequent rounds of DNA replication.

In contrast to the *hprt* gene, in normal cells the *tk* gene is located on an autosome and is therefore present in two alleles per cell. As a consequence, in order to render a cell resistant to TFT, both copies would have to be mutationally inactivated. Thus, for mutagenicity testing, specific heterozygous cell lines are employed in which one of two *tk* alleles has been inactivated whereas the other remains functionally active. In these cells, a single mutation in the tk^+ allele can be sufficient for the production of the $TK^{-/-}$ phenotype.

In this test the induction of small and large colonies is routinely recorded, and these are relatively easy to distinguish. In theory, mutants produced by a mutation within the tk^+ allele, which is not essential for growth, form rapidly growing large colonies since flanking regions, which could be important for general growth properties of the cell, are unaffected. In contrast, small colonies have been associated with larger arrangements spanning the cell's genome and impacting the cell's overall growth properties, which are also detectable in this system. However, this strict distinction has not been confirmed in all cases so that definitive conclusions on the mechanism of mutagenicity of a compound should not be drawn without further analyses.

Pig-a Gene Mutation Assay

A recently introduced assay is the Pig-a assay. It utilizes mutations in the catalytic subunit of an N-acetyl glucosamine transferase, which is involved in glycosylphosphatidylinositol (GPI) biosynthesis. In the absence of GPI anchors, GPI anchored proteins (e.g. CD59) can no longer be localized to the exterior surface of the cytoplasmic membrane, so cells without those surface markers are generated. Pig-a is an X-linked gene, highly conserved between species, so that mutations in the one active allele are sufficient to produce a phenotype. Technically, via FACS analysis cells without the surface markers are quantified among the excess of wild-type cells. Originally this test was developed for *in vivo* applications, where surface markers on, for example, erythrocyes in peripheral blood are analyzed. However, recently *in vitro* versions have been introduced that allow a quick and efficient analysis of gene mutations, e.g. by using human lymphoblastoid TK6 cells.

Tests for Chromosome Aberrations

The term "chromosome aberrations" includes both alterations in chromosome structure (**structural aberrations**) and alterations in chromosome number (**numerical aberrations**). Accordingly, agents that induce either of these effects are called clastogens or aneugens, respectively. However, whereas numerical aberrations can be seen in this test, it is not designed to measure those precisely and therefore generally is not used for that purpose.

> **The *in vitro* chromosome aberration test generally detects structural aberrations only. The first mitoses (metaphases) occurring after treatment of the cells with the test substance are scored, i.e. the analyzed cells have not yet divided. Thus, the aberration assay does not detect mutations in the narrower sense, but potential precursors of heritable chromosome aberrations, i.e. chromosome mutations.**

For the testing of a substance with regard to potential clastogenic properties, cell cultures are incubated with the test compound for a few hours. After a period equivalent to about 1.5 normal cell cycle lengths the cells are fixed and prepared for microscopic chromosome analysis. The analysis of chromosomal aberrations is performed on metaphase cells, as only in this phase of the cell cycle are the chromosomes condensed enough to be visible by light microscopy. Spindle poisons such as colcemide or related compounds are added to the cultures before fixation to accumulate cells in the metaphase.

This type of assay is specifically sensitive to agents inducing DNA double-strand breaks, which can be induced directly by an attack of the compound on DNA or formed during the processing or repair of DNA damage, usually during the S phase of the cell cycle.

In principle, chromosomal aberrations induced by chemicals can be detected in every proliferating mammalian cell. Cells used are selected on the basis of growth ability in culture, stability of the karyotype, chromosome number, chromosome diversity, and spontaneous aberration frequency. Often Chinese hamster cell lines or human lymphocytes are employed. In the case of primary cells, proliferation has to be induced by the addition of a mitogen, e.g. phytohemagglutinin.

Even though the formation of DNA double-strand breaks is an essential step in the formation of chromosome aberrations, the clastogenicity of a compound is not necessarily due to the reactivity of the compound or its metabolites with DNA. Chromosomes are complex, dynamic structures made up of DNA and proteins, and their structural integrity depends on a number of factors. Changes in chromosome structure can be induced by disturbances of DNA replication and repair, by effects on topoisomerases, depletion of cellular energy, interference with cell membrane function or triggering of apoptosis.

Micronucleus Test

> **Micronuclei are small extranuclear chromatin-containing bodies. They consist of membrane-surrounded chromosomal fragments or complete chromosomes that have not been integrated into a nucleus of one of the daughter cells during cell division. Micronuclei are considerably smaller than the main nucleus. They can be easily identified by light microscopy or flow cytometry, and automatic image analysis systems are available.**

Compounds can induce micronuclei by causing chromosome breaks or disturbances of the mitotic apparatus in proliferating cells (Figure 4.2). During mitosis, chromosomes assemble at the equatorial plate of the cell, and the chromatids are separated by the spindle fiber apparatus and distributed to the cell poles. Subsequently, cell division occurs and new daughter cell nuclei are formed. Spindle fibers attach to specific proteins of the centromeric region of chromosomes, the kinetochores, and pull the chromatids towards the cell poles. When chromosome breaks result in the formation of so-called acentric fragments, i.e. chromosome fragments without a centromeric region, these fragments remain at the equatorial plate and are not correctly

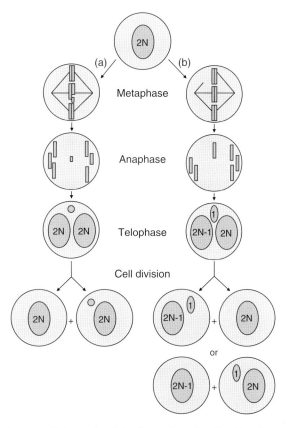

Figure 4.2 *Micronucleus test. Schematic of the formation of cells with micronuclei consisting of (a) an acentric chromosome fragment and (b) a whole chromosome.*

distributed to form the **micronucleus** in one of the daughter cells after cell division. Micronuclei containing whole chromosomes are formed when the treatment of the cells affects the function of the spindle apparatus or other functions responsible for correct cell division, so that chromosomes are lagging behind.

Since the formation of micronuclei consisting of acentric fragments and whole chromosomes involves different mechanisms, differentiation of the two types of micronuclei can yield valuable information on the mechanism of action of a chemical. The differentiation is achieved by staining the micronuclei for the presence of centromere-specific chromosome regions by either anti-kinetochore antibodies or fluorescence *in situ* hybridization (FISH) with pan-centromeric DNA probes.

Thus, the micronucleus test allows the identification of **clastogens**, which cause chromosome breaks, and of **auneugens**, e.g. spindle poisons, which disturb chromosome segregation, so that this test is particularly suitable to obtain information on the mode of action of chromosomal aberrations.

In vitro micronucleus tests are easy to perform, readily evaluated by image analysis or flow cytometry, and can be conducted in any mammalian cell line and certain primary cells, such as mitogen-stimulated freshly isolated hepatocytes or peripheral lymphocytes, which can also be obtained from human donors.

Indicator Tests

> **Indicator tests are genotoxicity assays which do not detect the induction of mutations in the progeny of the treated cells. Rather, various effects mechanistically associated with the formation of mutations are identified.**

Covalent Binding of Chemicals to DNA

> **The capability to react with DNA and form DNA adducts is a characteristic property of many genotoxic mutagens and carcinogens or their metabolites.**

In order to detect DNA binding, cells can be treated with the radioactively labelled test chemical and subsequently analyzed for the presence of radioactivity covalently bound to DNA. However, this test requires that the chemical in question is available in radioactively labelled form. Moreover, laborious control experiments are generally necessary to exclude experimental artifacts.

Immunochemical methods and the ^{32}P post-labeling technique allow sensitive detection of DNA adducts without being dependent on radioactively labelled test compounds. These are valuable tools for investigations on the mechanism of action of genotoxic chemicals and also available as *in vivo* versions suitable for the detection of organ-specific DNA binding.

Induction of DNA Strand Breaks

> **Assays for the detection of DNA strand breaks are easy to perform and highly sensitive. As a consequence of the extreme length of the DNA strands in chromosomes, even very few breaks suffice to cause a drastic reduction of the molecular mass, which can be easily detected.**

Following treatment of the cells with the test chemical, double-stranded DNA is generally exposed to an alkaline solution in order to separate the DNA strands and to uncover single-strand breaks. Subsequently, characteristic changes in the physicochemical properties of the DNA strands are determined to identify strand breaks.

The induction of DNA strand breaks can occur directly, e.g. by radical attack of reactive species at the sugar phosphate backbone of the DNA, or by the action of endonucleases during the repair of DNA damage. In addition, there are DNA lesions that are not caused by strand breaks in the cell, but are converted into breaks under the alkaline conditions employed for strand break analysis. These alkali labile sites include abasic sites formed by enzymatic or spontaneous cleavage of a base from the deoxyribose or phosphotriester, i.e. products of the alkylation of phosphoric acid groups of the DNA.

DNA strand breaks can be also induced by agents affecting the function of proteins involved in the maintenance of DNA structure. Such compounds include inhibitors of DNA polymerases and DNA repair enzymes or intercalating compounds which interfere with the activity of DNA topoisomerases. Finally, DNA strand breaks can occur as a consequence of unspecific cytotoxic effects, such as membrane damage, shifts of the intra- and extracellular distribution of ions, or inhibition of protein synthesis. Thus, for the interpretation of observations from strand break experiments it is mandatory to discriminate between unspecific or indirect actions of chemicals and specific effects originating from the reactivity of the test compound or its metabolites with the DNA, normally achieved by not testing excessively cytotoxic test compound concentrations.

Alkaline Elution

> **In principle, the isolated DNA is deposited on membrane filters and subsequently eluted by an alkaline buffer. The rate of elution through the membrane pores depends on the length of the DNA molecules, with short strands of DNA resulting from strand breaks eluting faster than intact DNA, so that the elution rate forms the basis of quantifying the extent of DNA damage in a sample.**

Following treatment with the test compound cells are collected on a membrane filter, lysed and digested proteolytically. The DNA on the filter is eluted with an alkaline buffer, fractions of the eluate are collected and the amount of DNA contained in each fraction is determined. The elution rate, i.e. the amount of DNA eluted per unit of time, is proportional to the number of DNA strand breaks in the DNA. Modifications of the experimental protocol are available which preferentially detect DNA–DNA and DNA–protein cross-links as well as DNA double-strand breaks.

The alkaline elution methodology can be used essentially with every cell type. When proliferating cells, such as cell lines, are employed, the DNA can be labeled with radioactive thymidine prior to the treatment of the cells so that the eluted DNA can be determined conveniently by measuring the acid-insoluble radioactivity in the eluate. When non-proliferating cells, such as hepatocytes, are employed, the DNA is quantified fluorimetrically.

Comet Assay (Alkaline Single-cell Gel Electrophoresis)

> **The principle of the comet assay, named after the characteristic microscopic picture of cells with DNA damage in this test, is based on the electrophoretic migration of DNA from nuclei immobilized on microscopic mini gels. It can be applied essentially in any cell type *in vivo* or *in vitro* from which a single cell suspension can be obtained (Figure 4.3).**

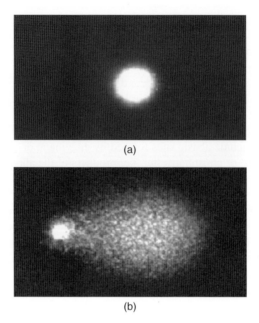

Figure 4.3 Detection of DNA strand breaks by the Comet assay (alkaline single cell gel electrophoresis): (a) control cell and (b) cell after treatment with a strand break inducing agent.

The cells are exposed to a test compound, embedded in agarose on a microscope slide, and lysed with a detergent. In the most widely used type of this assay, the alkaline Comet cells are then treated with an alkaline buffer and then subjected to electrophoresis followed by fluorescent staining to visualize cellular DNA. Migration of DNA from the immobilized nuclei depends on the extent of fragmentation so that undamaged DNA will not migrate much, in contrast to fragmented DNA, which forms the characteristic comet tail. Results are evaluated either by image analysis systems or manually, classifying cells according to the extent of DNA migration, tail length, tail vs head DNA, or similar.

Modifications of the technique employing treatment of the cells on the slides with lesion-specific endonucleases for the production of strand breaks allow the detection of specific types of DNA lesions, such as oxidative base damage.

Sister Chromatid Exchange Test

> **Sister chromatid exchanges (SCEs) represent the exchange of corresponding stretches of DNA between the two chromatids of a chromosome during DNA synthesis. The exchange reaction is based on breaks of the DNA molecules of both chromatids at genetically identical sites with subsequent rejoining of both DNA molecules at the breakage sites during replication. Thus, the exchange is "reciprocal" and does not usually lead to alterations in the DNA sequence. Therefore, while this test determines exchange of genetic material between two sister chromatids, the molecular mechanism and its relevance for the detection of genetic damage is largely nuclear, so that it is no longer used for routine applications in genetic toxicology.**

For the detection of SCE cells are grown for two cell cycles in the presence of the synthetic nucleoside 5-bromo-2′-deoxyuridine (BrdU). BrdU is incorporated into the newly synthesized

DNA strand instead of thymidine. After two rounds of DNA synthesis the chromosomes consist of two chromatids which are asymmetrically substituted with BrdU. During metaphase the two chromatids stain differentially and can therefore be discriminated. SCEs become apparent by color switches between the two chromatids. Quantitative evaluation of the experiments is performed by counting the number of color switches per metaphase.

DNA Repair Assays

> **DNA repair assays analyze the induction DNA repair in cultured mammalian cells. DNA repair is quantified by measuring DNA repair synthesis. Induction of DNA repair synthesis is an indicator that a chemical or its metabolites have induced DNA damage.**

DNA repair synthesis is a step in the process of excision repair. Both base excision repair and the nucleotide excision repair pathways enable cells to detect DNA lesions and to excise them from the damaged strand. Excision creates a gap of 1 to about 30 nucleotides in length, which is filled by repair synthesis using the undamaged strand as template. This gap-filling step leads to DNA repair synthesis. Repair synthesis can be detected and quantified by measuring the incorporation of radioactively labeled deoxyribonucleosides, such as $[^3H]$thymidine or $[^3H]$deoxycytidine, into the DNA. These nucleosides are also incorporated, in much larger amounts, by "normal" replicative DNA synthesis. Thus, the measurement of DNA repair synthesis requires a clear-cut discrimination between both kinds of DNA synthesis in the cells.

Unscheduled DNA Synthesis Test

> **The autoradiographic determination of repair synthesis makes use of the fact that replicative DNA synthesis is restricted to the S phase of the cell cycle whereas repair synthesis is independent of S phase and also occurs in other cell cycle phases. It is therefore termed unscheduled DNA synthesis (UDS).**

When cells are incubated with $[^3H]$thymidine and subsequently analyzed autoradiographically, the nuclei of cells that were in S phase during the incubation appear heavily labeled due to the large amounts of incorporated radioactivity. Nuclei of cells that were not in S phase are substantially labeled only if the test compound induced UDS. DNA repair synthesis can then be quantified by counting the number of silver grains or the area of the grains in these nuclei.

In this test, often primary cultures of rat hepatocytes are employed because of their metabolic competence. The cells are cultured with the test substance and $[^3H]$thymidine for 18–20 hours. Autoradiographs are evaluated by counting the number or the area of silver grains in the nuclei and correcting this value for the number of grains counted in a cytoplasmic area of the same size, to correct for incorporation in mitochondrial DNA. The resultant number of net grains per nucleus is an indirect measure for the incorporation of $[^3H]$thymidine into nuclei of non-S phase cells, i.e. for UDS.

The measurement of DNA repair synthesis enables the detection of the genotoxic properties of many different types of DNA damaging chemicals. Since the biological consequences of the corresponding DNA lesions can be very different, no quantitative conclusions regarding the mutagenicity of chemicals can be drawn from the results of repair tests. However, an increased repair synthesis clearly indicates a chemical reactivity of the test compound or its metabolite(s) with the DNA.

4.2A.4 Cell Transformation Assays

> In certain mammalian cells, carcinogenic chemicals induce heritable alterations in cell morphology or growth behavior, including the loss of growth control, loss of contact inhibition, the acquired ability to grow in soft agar, or the capability for unlimited cell division (immortalization). Furthermore, the clone morphology on a solid surface changes, appearing as "criss-cross" and multilayered. Many of those acquired features resemble the behavior of tumor cells, and generally this "transformation" results in a tumorigenic potential when tested *in vivo* (although this is not done in routine application). Thus, cell transformation appears to be a complex process having many aspects in common with the process of cancer development.

The close relation of the alterations observed in transformed cells to cancer development is shown by the observation that the implantation of transformed cells into experimental animals frequently results in tumor formation. Cell transformation resulting in cells capable of forming tumors is called "malignant transformation".

The molecular mechanisms responsible for the individual steps of cell transformation are largely unknown.

Strictly speaking cell transformation assays are not genotoxicity tests, and they have not gained broad acceptance in routine testing, primarily because of the uncertainties regarding the molecular mechanisms of cell transformation, the large experimental effort required for conducting them, and their sensitivity against modifications of experimental parameters.

Several cell transformation systems have been described, derived from different rodent species. The Syrian hamster embryo (SHE) assay is a primary cell system, and BALB/c 3T3, C3H10T1/2 and Bhas 42 are murine systems using established cell lines, the latter being constructed by transfection of the v-Ha-ras oncogene into the BALB/c 3T3 A31-1-1 cell line. As for most of the permanent cell lines, those have lost the majority of their metabolic activity. However, the primary SHE as well as the Bhas 42 have shown some activating capacity which facilitates analysis of compounds needing metabolic activation. Generally those cell lines are sensitive to initiating, i.e. genotoxic, as well as tumor-promoting or non-genotoxic carcinogens. As such, they could represent a true broadband assay to detect carcinogens. However, the true sensitivity and specificity is insufficiently characterized for different chemical classes. The evaluation of those assays relies on morphological parameters of altered cell growth and colony formation, a factor that introduces some degree of subjectivism. Also, the conductance of the assays requires a significant expertise and is subject to variation. For these reasons, cell transformation tests have not been introduced into routine testing but can be helpful in certain situations.

4.2A.5 Xenobiotic Metabolism

A fundamental problem of all *in vitro* test systems is that uptake, distribution and excretion cannot be simulated with precision, but are limited by the inability to reproduce in a test-tube the complex pharmacokinetics that regulate the absorption, distribution, metabolism, and excretion (ADME) of chemicals in the body. Many chemicals exert their toxicity only after metabolic activation. Differences between species, organs, and the cell types within an organ may result in a diversity of metabolic pathways for any given chemical. Since bacteria and the majority of the mammalian cells, particularly permanent cell lines, utilized in genotoxicity studies lack certain xenobiotic-metabolizing enzymes, most test systems depend on the addition of subcellular fractions of cells which contain the appropriate enzymes. Alternatively, cells which possess certain enzymatic activities, e.g. by heterologous expression, can be used.

Primary mammalian cells, i.e. cells isolated from an organ and maintained in cell culture for a limited period of time, generally reflect the xenobiotic metabolism of the respective organ reasonably well, at least during the first hours after their isolation. A disadvantage of these cells is that they generally have a very low proliferative activity, which limits their use for some genotoxicity assays that require cell division, such as tests for gene mutation. However, these cells can be used in test systems for the detection of DNA strand breaks or DNA repair as indicators of a potentially mutagenic effect of the chemical.

Alternatively, metabolic competence can be introduced into established cell lines, allowing a broad analysis of genotoxic activity but also a precise investigation of mechanisms, due to the precise knowledge of the genetic modification introduced.

Cell fractions prepared from the livers of rats or mice treated with a mixture of enzyme inductors such as polychlorinated biphenyls, e.g. Aroclor 1254, are used routinely. With this preparation, the cytochrome (CYP)-mediated activating metabolism (Phase I) is highly active. A routinely used preparation is rat liver S9, which contains microsomes and cytosol with the respective endoplasmatic and soluble enzymes, and a cofactor mix specifically designed to support CYP-mediated metabolism. The activities of the conjugating enzymes such as glucuronosyl S-transferases or sulfotransferases in the S9 mix and the resultant incubation mixture are very low due to the dilution of the cofactors. As a consequence, the proportion of activating and inactivating reactions as they occur in intact liver cells is generally shifted towards an overrepresentation of certain activating reactions, which is a desired effect if enzymatic activation is required. For that reason, for unknown molecules, experiments with or without metabolic activation are routinely conducted in parallel.

An alternative to the metabolic activation of test compounds by exogenous cell fractions is provided by co-cultivation techniques with metabolically competent cells, such as primary rat hepatocytes. Physical contact between the two cell types is usually required for an efficient transfer of the reactive metabolites formed. This approach is limited when the reactive metabolites cannot leave the activating cell because they are too reactive or electrically charged, or when they are efficiently inactivated enzymatically in the activating cell.

4.2A.6 Summary

An advantage of *in vitro* test systems to analyze genetic toxicity is the easy manipulation of environmental factors such as test compound concentrations or metabolic activation systems, and the relative ease and speed of conductance, on top of animal welfare considerations. The broad spectrum of genotoxic effects chemicals can exert in cells is reflected by the variety of *in vitro* test systems available for their detection and quantification. *In vitro* tests for genotoxicity can be performed with various test systems which differ in their capacity for the metabolism of xenobiotics and which allow the detection of a spectrum of different genetic endpoints. With the knowledge of the mechanisms underlying the various genotoxic effects, apparently contradictory results on the activity of compounds in different *in vitro* test systems can often be explained as the consequence of the properties of the compound. Thus, *in vitro* studies on the genotoxicity of chemicals are not only suited for use as fast and simple methods for the detection of genotoxic properties, but are also ideal tools for the clarification of the action mechanisms of chemicals.

Further Reading

International Conference on Harmonization of Technical Requirements for Registration of Pharmaceuticals for Human Use. ICH S2(R1), Guidance on genotoxicity testing and data interpretation for pharmaceuticals intended for human use (2011): http://www.ich.org/products/guidelines/safety/article/safety-guidelines.html (accessed 1 November 2017).

OECD guidelines for the testing of chemicals: http://www.oecd-ilibrary.org/environment/oecd-guidelines-for-the-testing-of-chemicals-section-4-health-effects_20745788 (accessed 1 November 2017).

OECD (2007) Detailed review paper on cell transformation assays for detection of chemical carcinogens. OECD environment, health and safety publications. Series on testing and assessment no. 31: http://www.oecd.org/officialdocuments/publicdisplaydocumentpdf/?cote=ENV/JM/MONO(2007)18&doclanguage=en (accessed 1 November 2017).

Pfuhler S, Fellows M, van Benthem J, Corvi R, Curren R, Dearfield K, Fowler P, Frötschl R, Elhajouji A, Le Hégarat L, Kasamatsu T, Kojima H, Ouédraogo G, Scott A, Speit G (2011). In vitro Genotoxicity Test Approaches with Better Predictivity: Summary of an IWGT workshop. *Mutation Research* **723**:101–107.

Rees BJ, Tate M, Lynch AM, Thornton CA, Jenkins GJ, Walmsley RM, Johnson GE (2017). Development of an in vitro PIG-A gene mutation assay in human cells. *Mutagenesis 1*; **32**(2):283–297.

4.2B Mutagenicity tests in vivo

Ilse-Dore Adler and Gerlinde Schriever-Schwemmer

4.2B.1 Introduction

> **Animal experiments for testing mutagenicity are divided into two groups: the analysis of effects on somatic cells and the analysis of effects on germ cells. Mutational events in somatic cells are indicative of the carcinogenic potential of the test chemical. Mutational events in germ cells imply a genetic hazard for the progeny of exposed individuals.**

Three basic classes of genetic alterations can be distinguished: (1) chromosomal aberrations, which are structural alterations to chromosomes and entail the loss or translocation of chromosomal segments, (2) gene mutations, which are changes in the genetic code, and (3) genome mutations, which involve changes in chromosome numbers.

Additionally, a number of indicator tests are available, e.g. for DNA strand breaks (comet assay, TUNEL assay), DNA repair (unscheduled DNA synthesis (UDS) test) or sister chromatid exchange (SCE test). These indicators do not represent genetic end points but point indirectly to a possible genetic alteration. They will not be discussed in this chapter. The *in vitro* genotoxicity tests described in Chapter 4.2A can also be used *in vivo* in somatic and germ cells.

The detection of chromosome mutations induced in somatic cells is mainly performed by means of the **microscopic analysis of structural or numerical chromosomal aberrations** in dividing cells, or the **observation of micronuclei in polychromatic erythrocytes of bone marrow**. Dividing cells from other tissue or organs (liver, lungs, colon, epithelial cells of the buccal mucosa, or similar tissue) can also be used.

For the induction of gene mutations in somatic cells originally only the mouse coat color spot test was established, which detects mutations in recessive coat color loci in melanoblasts of transplacentally treated fetuses.

Using the methods of molecular biology and to reduce the number of animals used in testing, transgenic mouse and rat strains (Big Blue® mice and rats, Muta™Mouse and Muta™Rat, gpt delta Mouse) were developed through the integration of microbial reporter genes.

After isolation of the integrated bacterial DNA, mutations are detected by *in vitro* methods. The newest test, the Pig-a assay, detects mutations in an endogenous X-chromosomal gene coding for anchor proteins of the cell surface of hemopoietic cells.

In germ cells, as in bone marrow cells, the induction of **chromosomal aberrations** can be directly analyzed microscopically during various stages of cell division, in male germ cells **during mitotic divisions of spermatogonia**. In early spermatid stages the micronucleus test indirectly provides information about the induction of chromosome fragments or the maldistribution of entire chromosomes in spermatocytes.

The **dominant lethal test** provides indirect evidence of the induction of chromosomal damage in different stages of spermatogenesis as most of the chromosomal damage with the loss of (DNA) information leads to embryonic death. The **heritable translocation assay** detects those forms of chromosomal alterations that are compatible with fetal survival, predominantly **reciprocal translocations**, but also Robertsonian translocations and aneuploidy of the sex chromosomes (XO, XXY and XYY). The sperm-FISH assay detects the induction of maldistribution of entire chromosomes (genome mutations) by chromosome-specific fluorescence-marked DNA samples that allow the recognition of missing or extra chromosomes in sperm.

The classical method for the analysis of induced **gene mutations in germ cells** is the **mouse specific locus test**, which detects mutations at seven recessive loci. Because large animal facilities and specific mouse strains are required for this test, in recent years the focus has been on transferring the methods of molecular biology (Big Blue® mice and rats, Muta™Mouse and Muta™Rat, gpt delta Mouse) used in somatic mutation testing to germ cells. Recently, molecular genetic technologies for detecting mutations in human families with inherited diseases (expanded single tandem repeats (ESTRs), single nucleotide polymorphisms (SNPs), copy number variations (CNVs)) have been adopted for mutation testing in animals.

Most experiments for the detection of germ cell mutagenicity are performed with male animals because germ cell divisions take place during the entire reproductive lifespan of males. The number of available sperm is unlimited and all developmental stages from spermatogonial stem cells to mature sperm are analyzable. In female animals, germ cell multiplication and some steps in the germ cell maturation process occur during prenatal development. At birth, the number of oocytes is fixed and only the last steps of germ cell maturation take place in adult females.

The analysis of **clastogenic** (chromosome breaks) and **mutagenic effects of environmental toxins** is usually performed with **rodents**. Mice of specific genetic constitution are commonly used and are essential for some of the tests, i.e. for the coat color spot test and for the specific locus test, Big Blue® or Muta™Mouse test. In order to integrate the genetic tests into a standard toxicology test battery procedure, different tests in rats were performed (micronucleus test, gene mutation test in Big Blue® rats, Muta™Rats and *gpt* delta Rats).

Commonly, animals are treated with the test substances by intraperitoneal injection (ip), gavage (po), or inhalation. Usually, the test substance is not administered in food or drinking water because of the uncertainty of the dose taken up when these routes are used. Occasionally dermal application (skin painting) is used if the induction of dermal cancer is suspected and it has to be assessed whether there is an underlying genetic mechanism. Uncommon routes of administration are intravenous, intramuscular or subcutaneous injections. In most cases short-term treatment is preferred to medium-term or long-term treatment because the first cell divisions after treatment have to be examined. However, there will be an increase in studies with long-term treatment integrated into standard toxicity testing. It is recommended that a maximal tolerable dose (MTD) is used to optimize the effect, but to avoid severe toxicity. To establish a dose–response relationship, in addition two lower doses are tested (e.g. MTD/2 and MTD/5). For non-toxic substances a single dose of 2 g/kg body weight is usually recommended.

Since 1993, workshops on genotoxicity testing (IWGTs) have been performed within the scope of the Meetings of the international Mutagenicity Society, which take place every 4 years, to follow the development and standardization of new test methods. Thereby, multinational interlaboratory test trials were initiated to complete databases and to make laboratory comparisons possible. Through the publication of the workshop initiatives, standardized new methods can find their way into legislation. At the moment, the Pig-a assay is being evaluated and an OECD Guideline is expected by 2020.

4.2B.2 Chromosomal Mutations in Somatic Cells

> **To detect the clastogenic effects of a test substance, structural chromosomal aberrations are analyzed in bone marrow cells, or any other cells undergoing mitosis.**

As a rule, the first mitotic division following treatment should be scored for chromosomal aberrations. The reason for scoring the first mitotic divisions following treatment is that certain aberration forms, e.g. chromosomal fragments, which are lost during cell division, often lead to cell death. Microscopically visible aberrations are classified as chromosome-type and chromatid-type aberrations, depending on the involvement of both or just one chromatid. Chemicals can have an S-dependent or S-independent mode of action, i.e. they may or may not require a round of DNA synthesis for the translation of the preclastogenic lesion (DNA double strand break) into a visible aberration. The experimental protocol is described in EC Directive 1986 and OECD Guideline 475 (1984, 2016). Chromosome analysis can be performed in mitotic cells in other tissues/organs as well, if suitable preliminary information, e.g. organ-specific carcinogenic effects, is available and the mechanism of action must be clarified.

In interlaboratory studies it was shown that the assessment of aberrations, especially the differentiation between achromatic lesions (gaps) and breaks, can vary widely. Gaps are defined as unstained regions of the chromatids, which are not regarded as a discontinuity of the DNA, but rather an alteration of the condensation grade of the chromatin. They have to be differentiated clearly from breaks, which are a disruption of the chromatids, indicated by the dislocation of the originating fragment. Substances which increase only the number of gaps cannot be regarded as clastogens. As a result of the development of fluorescence *in situ* hybridization (FISH) with chromosome-specific DNA samples it has become possible to recognize reciprocal translocations and other balanced aberration forms. The increase in stable chromosomal aberrations is particularly important because they can be transmitted to daughter cells and form genetically abnormal cell clones, which are considered to be the basis for the development of cancer.

To detect clastogenic effects, the micronucleus test with young (polychromatic) erythrocytes is often used as an alternative to chromosome analyses (Figure 4.4). Micronuclei are formed during cell division and can be observed in the daughter cells as a result of acentric fragments or entire misplaced chromosomes.

> **In principle, micronuclei can be scored in any proliferating tissue. However, polychromatic erythrocytes are especially suited for counting micronuclei because these cells have extruded the main nucleus during maturation and only the micronuclei remain.**

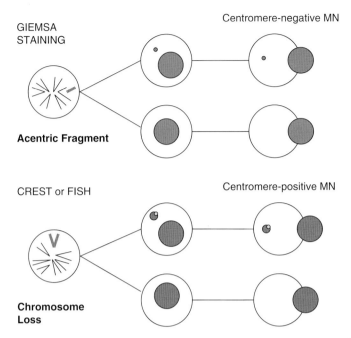

Figure 4.4 Origin of micronuclei: acentric fragments or entire chromosomes with a centromeric region which are visible by CREST antibodies or fluorescence in situ hybridization (FISH) with pan-centromeric DNA samples.

Comparative studies in several laboratories have shown that interlaboratory variation is far smaller for the results of micronucleus tests than for the results of analyses of chromosomal aberrations. Micronuclei are less prone to subjective interpretation than chromosomal aberrations. The experimental protocol is described in EC Directive Annex V (1986) and OECD Guideline 474 (1997, 2016).

Over the past few years, technical modifications to the micronucleus assay, such as preparation and staining variants, have been published which have improved the test considerably. For example, appropriate filtration and centrifugation steps can increase the number of polychromatic erythrocytes in a bone marrow preparation, which facilitates automatic scoring by image analysis or flow cytometric procedures. Furthermore, depending on the scoring method, different fluorescent stains are used to improve the discrimination of young and old erythrocytes, which differ in their RNA content. Young erythrocytes have a high RNA content, in old erythrocytes the RNA is exhausted.

> **The micronucleus assay has also been modified to recognize genomic mutations, since micronuclei are not only formed by acentric fragments but also by entire chromosomes which have lost the spindle attachment (Figure 4.4).**

To distinguish between the two possible origins of micronuclei, immunological (CREST staining, CD71 antibodies) and molecular-genetic (FISH with centromeric DNA samples) methods have been employed. If a centromere is demonstrated in the micronucleus it can be concluded

Table 4.7 Micronucleus tests in somatic cells.

Species	Tissue/cell type	Staining	Analysis
Mouse/rat	Bone marrow/polychromatic erythrocytes Blood/reticulocytes	Giemsa, AO/DAPI, PI/CD71 antibody, CREST, FISH	Microscopy, fluorescence-microscopy, image analysis, flow cytometry
Mouse/rat (adult)	Liver/hepatocytes (after partial hepatectomy or chemical-induced cell proliferation)	AO/DAPI	Fluorescence-microscopy, image analysis, several interlaboratory collaborative studies (IWGT; Morita et al., 2011)
Rat (juvenile)	Without pretreatment		
Mouse/rat	Colon/epithelial cells		
Mouse/rat	Skin/epithelial cells		
Mouse/rat (fetuses, pups)	Blood/erythrocytes, liver/hepatocytes		

AO, Acridin Orange; DAPI, 4′,6-diamidino-2-phenylindole; CREST, antibody against centromere-near heterochromatin; FISH, fluorescence *in situ* hybridization with chromosome-specific DNA samples; IWGT, International Workshop on Genotoxicity Testing.

that the micronucleus contains an entire chromosome and originated as a result of an aneugenic effect of the test substance.

Table 4.7 summarizes the micronuclei examinations in different species, tissues and staining procedures.

It was shown that reticulocytes in peripheral blood can also be used for micronucleus scoring because fluorescent staining and subsequent flow cytometry analysis allows large samples to be handled. The advantage of using peripheral blood samples is that they can be obtained from animals in ongoing toxicological studies and several samples can be obtained at appropriate intervals from the same treated animal. Thereby, the numbers of experimental animals required are reduced considerably.

> **The induction of chromosomal or genomic aberrations in somatic cells *in vivo* by a substance leads not only to clarification of its mechanism of action as a carcinogen, but also to the classification of the substance as a germ cell mutagen (Table 4.12).**

Reactive metabolites of indirect mutagens may be short-lived and therefore may not reach the bone marrow. Since the liver is the most metabolically active organ, several methods have been developed to perform micronucleus tests in rat hepatocytes. Other tissues in which mitosis takes place (such as colon and lung epithelial cells, buccal mucosa or skin cells) are suitable for micronucleus analysis.

Usually, for verification of a positive clastogenic result in an *in vitro* test, an *in vivo* chromosomal aberration test is performed. The possibility of false-positive and false-negative results in the *in vivo* micronucleus test in the bone marrow of rodents was discussed in detail at the 4th IWGT workshop. Substance-dependent failure in the physiology of the test organism (rat or mouse) may lead to micronucleus formation without being the effect of genotoxicity. These failures can be long-lasting changes in the body temperature/hypo- and hyperthermia, changes in the mitosis rate of the erythroblasts and the inhibition of protein biosynthesis. To avoid false-positive interpretations, the mechanism of action of the test substance should be taken into consideration and additionally a chromosomal aberration test could be performed. If a bone marrow test for a proven carcinogen is clearly negative, an organ-specific effect should be taken into account and a variation of the micronucleus test, as stated above, can be used to avoid a false-negative conclusion.

4.2B.3 Gene Mutations in Somatic Cells

With the mouse coat color spot test, gene mutations are detected in pigment-forming melanocytes after transplacental treatment of fetuses.

In this test, mice with a specific, genetically determined coat color are crossed. The fetuses are homozygous for non-agouti (black fur color) and heterozygous for different other recessive fur color genes and therefore their fur is black too. Pregnant females are treated with the test chemical between the 8th and 12th day of gestation. A mutation in a recessive coat color gene in embryogenic melanoblasts forms a cell clone that can be recognized in the progeny as a fur patch of a different color. Microscopic analysis of hair samples allows the recognition of the pigment alterations and, thereby, the identification of the mutated loci. A very good correlation has been found between the results of the mouse spot test and the mouse specific locus test. The experimental protocol is described in EC Directive Annex V (1986). However, mouse strains with appropriate gene constellations are nearly extinct. Therefore new data is not to be expected. There is a good database with different chemicals against which new molecular genetic tests can be calibrated. Table 4.8 gives an overview of gene mutation tests in somatic cells.

With the development of the transferal of bacterial genes to somatic cells it has been possible to develop transgenic mouse strains. With the aid of λ vectors, bacterial *Lac* Z (Muta^{TM}Mouse)

Table 4.8 Gene mutation tests in somatic cells.

Animal strain	Test system	Reporter gene	Cell type	Analysis	Acceptance
Transgenic mouse and rat	Big Blue® Muta^{TM} Mouse CPG	*Lac* Z *Lac* I *Gpt delta*	All organs	Bacterial plaques with color change, bacterial colonies with mutated plasmids	OECD 488 2013
Wild-type mouse and rat, humans	Pig-a	Endogenous	Bone marrow, spleen	Flow cytometry for CD59 and CD48-negative cells	Validation in progress Dertinger et al. (2015), Olsen et al. (2017)
Wild-type and test stock mouse	Mouse spot test	Five coat color genes	Melanoblasts	Visible as coat color spots	EC Directive 1986, OECD 484 deleted 2014

Lac Z, bacterial gene (coding β-galactosidase); *Lac* I, bacterial gene (coding lactosidase repressor protein); *Gpt delta*, bacterial gene in λEG10 phage DNA; Pig-a, X-chromosomal gene (coding an anchor protein of the cell surface); CD59 and CD48, antibodies against cell surface proteins.

or *Lac I* genes (Big Blue®) were stably integrated into the genome of the mouse or rat via fertilized oocytes. DNA is isolated from the tissue to be examined and packed *in vitro* into λ phages with which *Escherichia coli* bacteria were infected. The Lac operon regulates the metabolism of lactose through the formation of β-galactosidase and is expressed in bacteria. 5-Bromo-4-chloro-3-indolyl-b-galactopyranoside (X-gal) is used as a selection factor. Mutant bacteria clones can be detected as white clones on a blue bacterial lawn (Muta^{TM} out) or as blue clones (Big Blue®) on a white bacterial lawn (Transgenic Rodent Somatic and Germ Cell Gene Mutation Assay: OECD Guideline 488, 2013).

The *in vivo*/*in vitro* examinations of bacterial reporter genes have some disadvantages. They are highly methylated, not transcribed in mammalian cells, and therefore DNA repair mechanisms other than to endogenous DNA may operate. Moreover, these reporter genes contain excessive amounts of guanine and cytosine bases in the form of CpG dinucleotides. The mutability of the bacterial reporter genes may differ considerably from coding genes of mammalian organisms. The same also applies for other tests with transgenic rodents in which λ phage EG10 DNA has been integrated in the mammalian chromosomes. *E. coli* YG6020 colonies are infected with the isolated phages from the tissues to be examined. Thereby phage DNA is changed into a plasmid (pYG142) which contains the bacterial genes for chloramphenicol resistance and GPT. The *gpt* gene codes for cytosolic alanine aminotransferase1 (ALT1), also known as GPT (glutamate pyruvate transaminase), which plays a key role in the intermediate metabolism of glucose and amino acids. *gpt* mutants build colonies in culture medium containing chloramphenicol and 6-thioguanine. With this selection, the method recognizes gene mutations. A second selection step runs towards the *chi* complex, which is flanked by *red* and *gam* genes. In culture media with p2-lysogen of *E. coli*, only plasmids with deletions in the *chi* complex, which are insensitive against p2-lysogen, build plaques. The mutations are named lambda Spi-. This method allows the recognition of deletions of various size.

> **The induction of gene mutations in somatic cells *in vivo* by a substance leads not only to clarification of its mechanism of action as a carcinogen, but also to the classification of the substance as a germ cell mutagen (Table 4.12).**

The newest development allows the examination of mutations in an endogenous gene expressed in cells of the hematopoietic system. Mutations in the X-chromosomal Pig-a gene are known from human genetics and are the cause of the life-threatening disease paroxysmal nocturnal hemoglobinuria (PNH). This disease was so named because of the sudden and irregularly occurring (paroxysmal) dark urine (including hemoglobin) voided preferentially in the night or early morning. In this form of hemolytic anemia, the expression of membrane-protective proteins (anchor proteins, e.g. CD55 and CD59) is lacking or reduced. As a consequence, the blood's immune system (complement system) attacks and destroys these cells. These anchor proteins are ubiquitously expressed in the hematopoietic system (bone marrow, spleen), also in rodents. Already in interlaboratory collaborative *in vivo* animal studies, fluorescence-marked anti-CD59 (reticulocytes, blood) or anti-CD48 (T cells, spleen) antibodies are used for the detection and quantification of Pig-a mutations. Deficient, mutated cells can be quantified by flow cytometry. Because this test can be linked with the *in vivo* micronucleus test in reticulocytes, it is currently being considered whether it can be integrated into short-term or 28-day toxicity studies (Kimoto et al., 2016). In small blood samples, clastogenic effects and gene mutations can be analyzed in parallel. The investigation is ongoing. Recently a Pig-a test in humans has been reported

(Dertinger et al., 2015), allowing the comparison of animal test data with biological monitoring data in humans (Olsen et al., 2017).

4.2B.4 Chromosome Mutations in Germ Cells

> All mutagenicity tests in germ cells require exact knowledge of the germ cell's developmental stages, characteristics and duration. The analysis of chromosomal aberrations can be performed in mitotic male germ cells, differentiated spermatogonia. The commonly used cells are mitotically dividing spermatogonia. The method for cytogenetic analysis in germ cells is nearly the same as for that in bone marrow.

The maturation of male germ cells in mice, rats and humans is illustrated in Figure 4.5.

At the base of the testis tubules, spermatogonial stem cells divide slowly, with a cell cycle of 6–8 days, and build new spermatogonial stem cells. Through an unknown impulse, the cell cycle is shortened and the spermatogonia go through a series of mitoses in a few days. During the spermatocyte stage, the first meiotic cell division takes place, resulting in two daughter cells with half the number of chromosomes. In the second meiotic cell division, chromatids are separated and every daughter cell has a haploid set of chromosomes. In the post-meiotic stages (spermatids), the metabolism is reduced, DNA repair is stopped and DNA is condensed via the special protein protamine into a transport form. The mature sperm leave the testis tubules with an active flagellum movement and are stored in the epididymis until ejaculation. Only the duration of the various stages of spermatogenesis differs between mice, rats and humans.

Analysis of chromosomal aberrations can be performed in mitotic and meiotic divisions of the germ cells (Table 4.9). The most common test is for chromosomal aberrations in mitotic germ cell stages, the differentiated spermatogonia. The method of cytogentic analysis in germ cells is nearly the same as that in bone marrow.

The test protocol differs depending on the tissue. The duration of the cell cycle in the bone marrow of mice is 14 hours, in mitotic dividing differentiated spermatocytes it is 28 hours. Positive results in the spermatogonial test demonstrate that the chemical reaches the germ cells and has a clastogenic effect.

The different forms of aberrations can be analyzed microscopically in differentiated spermatogonia just as accurately as in bone marrow cells. The test protocol is described in EC Directive 1986 and in OECD Guideline 483 (1997, 2016).

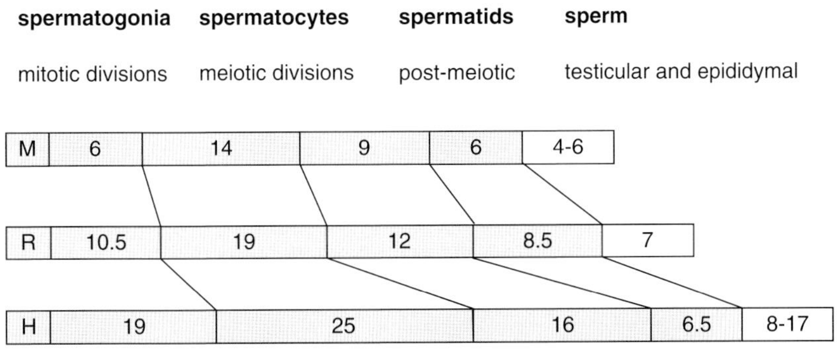

	spermatogonia	spermatocytes	spermatids	sperm
	mitotic divisions	meiotic divisions	post-meiotic	testicular and epididymal

M	6	14	9	6	4-6
R	10.5	19	12	8.5	7
H	19	25	16	6.5	8-17

Figure 4.5 Comparison of the timing of different stages of spermatogenesis in the mouse (M), rat (R) and humans (H) given in days.

Table 4.9 Chromosomal mutation tests in germ cells.

Test methods	Germ cell stages treated/analysed	End point	Analysis	Acceptance
Chromosomal analysis	Differentiated spermatogonia	Chromtid aberrations	Microscopy	EC Directive 1986 OECD 483 1997, 2016 Parry and Parry (2012)
Micronuclei	Spermatocytes/ early spermatides	Acentric chromosomal fragments, entire chromosomes	Microscopy, flow cytometry after CREST or FISH staining	Parry and Parry (2012)
Sperm-FISH	Spermatocytes/ sperm	Chromosomal maldistribution	Microscopy, flow cytometry	Parry and Parry (2012)
Dominant lethal mutations (after mating)	Entire spermatogenesis/ F_1 implantations	Unbalanced chromosomal aberrations	Macroscopy *in utero* (resorptions, dead implantations)	EC Directive 1986 OECD 478 1984, 2016
Heritable translocation test (after mating)	Entire spermatogenesis/ F_1 implantations	Reciprocal translocations	Fertility testing, evidence of cytogenetic translocation	EC Directive 1986 OECD 485 1986

Cytogenetic analyses in female germ cells are rarely performed because the mitotic divisions of oocyte propagation and the initial stages of meiosis occur during fetal development in female mammals. At birth, the numbers of oocytes are fixed and they are arrested in a specific stage of meiotic prophase, termed the dictiate stage. Oocytes are released from this meiotic arrest in small numbers during each estrus cycle of the adult female. During ovulation, oocytes are at the second meiotic division (MMII), which is only completed after entry of the sperm at the time of fertilization. Female germ cells are difficult to collect for cytogenetic analyses and are limited in number, so that their use cannot be recommended for routine cytogenetic analyses.

Chromosomal aberrations induced in post-meiotic germ cell stages, i.e. spermatids and spermatozoa, can only be analyzed during the first cleavage division after fertilization. The preparation technique for the isolation of the first cleavage division zygotes is very tedious, which prevents the routine use of this method. It is only used when scientifically justified.

> **The micronucleus test in spermatids has gained a similar importance to that in bone marrow. In addition, modern molecular genetic staining methods (CREST, FISH) can be used to distinguish between clastogenic and aneugenic effects.**

The micronucleus test in spermatids can be carried out as a flow cytometric analysis, it is time-saving in comparison with standard microscopic analysis, and the number of analyzed cells can be increased considerably.

Flow cytometry and FISH are used to detect the induction of chromosomal maldistribution (sperm-FISH assay). In sperm, autosomes and gonosomes are detected with fluorescence-marked DNA samples (chromosome 8 red, X white, Y green). With fluorescence microcopy or flow cytometry sperm with a aneuploid/abnormal number of color signals are detected. With an adequate interval between treatment and sperm sampling, distribution failure in the meiotic division can be analyzed. This method can be used for sperm samples from exposed men. It was exemplarily demonstrated that diazepam (Valium®) induces aneuploidy (change in chromosomal number) not only in mouse sperm but also in the sperm of male patients after suicide attempts. This is important information because trisomy, especially trisomy 21, is considered to be a chromosomal maldistribution during the female meiosis, which increases with the aging of the oocytes. Obviously, the exposure of the father to an aneugenic substance, such as diazepam, raises the risk of siring a child with trisomy 21. Because the risk is limited to the meiotic cells, the risk returns to that of the normal unexposed population after a time interval that corresponds to the duration of the development of spermatocytes to mature sperm. After short- or long-term exposure, exposed men should be advised to avoid conception for at least 3 months.

> **The induction of chromosome or genomic changes in germ cells *in vivo* by a substance leads to its classification as a germ cell mutagen (Table 4.12).**

As in somatic cells, false-positive cytogenetic test results can be induced in germ cells. The causes are the same. False-negative results can be judged only on the basis of the database of the tests no longer performed. An error in the test procedure is suspected.

Tests No Longer Performed

With the above-named cytogenetic methods, the transferal of chromosomal changes to the next generation cannot be demonstrated. Therefore, large-scale animal studies are necessary. However, today such studies are rarely carried out because of issues regarding animal welfare. Despite this, these tests are described here briefly because this contributes towards understanding data from older publications and because there is a large historical database which can serve as a comparison to cytogenetic tests with a limited number of animals.

Dominant lethal mutations are the result of structural chromosome aberrations in germ cells and are indirect evidence of clastogenic effects of the test substance.

Chromosomal aberrations which lead to the loss of genetic material cause embryonic death after fertilization at the time of implantation in the uterus. Dominant lethal experiments do not require animal strains with a specific genetic background. However, the litter size of the strains used should be relatively large (9–12 pups per female) and constant. After treatment, male mice or rats are mated with untreated virgin females at intervals of 4–7 days. In mice, successful mating is indicated by a vaginal plug, in rats vaginal smears have to prove mating has taken place. The pregnant females are sacrificed on day 14–16 of gestation and the contents of the uterus are inspected for live and dead implants. The dominant lethal effect is expressed as 1 minus the ratio between the number of live implants in the treated group divided by the number of live implants in the control group expressed as a percentage. These data include pre-implantation and post-implantation losses.

The increases in dead embryos over the number in concurrent controls at the various mating intervals reflect clastogenic effects at specific germ cell developmental stages, namely mature sperm, mid-early spermatids, late spermatids, spermatocytes or differentiating spermatogonia

Table 4.10 Mating scheme to sample different stages of mouse spermatogenesis after short-term exposure.

Mating interval (days)	Treated spermatogenic stages
1–7	Spermatozoa
8–14	Late spermatids
15–21	Mid-early spermatids
22–28	Spermatocytes
29–35	Spermatocytes
36–42	Differentiated spermatogonia
43 to many months	Spermatogonial stem-cells

(Table 4.10). This stage-specificity is characteristic of individual chemicals or chemical classes. Chemically related compounds show a very different stage-specificity. This depends on which physiological processes are taking place during the different stages, such as DNA replication, DNA repair, chromosome segregation (meiosis), histone modifications (chromatin status methylated), and compaction of the sperm. The stage-specific effect of a substance can be identified using a specific mating scheme. The experimental protocols are described in EC Directive Annex V 1986 and OECD Guideline 478 (1986, 2016).

> **The induction of dominant lethal mutations by a substance leads to its classification as a germ cell mutagen (Table 4.12).**

The number of live progeny with reciprocal translocations is determined in the ***heritable translocation assay***.

Unlike the dominant lethal test, the heritable translocation assay detects structural chromosome aberrations which do not lead to the loss of genetic material, e.g. reciprocal translocations, and therefore are compatible with fetal survival. Germ cells carrying reciprocal translocations can result in living offspring which generally do not have any physical malformations but display reduced fertility.

The analysis of the progeny (F1 generation) of treated parental animals can be performed in two ways. The classical method analyses only male F1 progeny and uses reduced fertility as the selection criterion for the suspected reciprocal translocation carriers. During the meiotic segregation of the translocated chromosomes and their normal homologues, 50% unbalanced gametes are formed. After fertilization, the unbalanced gametes lead to embryonic death at the time of implantation of the embryo *in utero*. The fertility of F1 translocation carriers is therefore reduced to about 50% (semi-sterility). Complete sterility is observed in XYY mice with certain types of reciprocal translocations, namely the c/t types with breaks near the centromere and telomere of the two chromosomes involved, or with several translocations. The presence of a reciprocal translocation in a semi-sterile or sterile F1 animal is confirmed by cytogenetic analysis of the first meiotic divisions in the form of translocation multivalents (Figure 4.6).

In a variation of the heritable translocation test, female F1 progeny are also included in the analysis. Thereby, no additional females are needed for the fertility tests with F1 males and the number of F1 progeny analyzed is almost doubled. Using this protocol, XO and XXY progeny

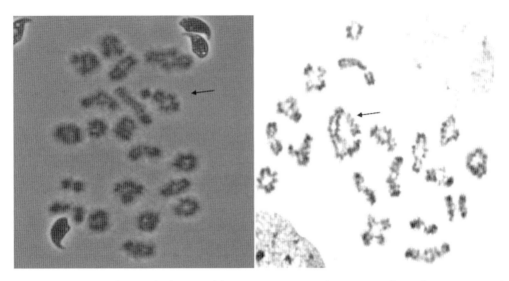

Figure 4.6 Left: Bivalent with chromatid fragment (arrow), right: reciprocal translocation (ring of four chromosomes, arrow), in mouse spermatocytes at the first meiotic metaphase.

(resulting from non-disjunction of the sex chromosomes) are achieved. The results of heritable translocation tests with cytostatics were most relevant for the quantification of the genetic risk. For cancer patients who wish to conceive after successful chemotherapy, a 6-month period of abstinence is recommended.

The experimental protocol is described in EC Directive Annex V (1986) and OECD Guideline 485 (1986).

> **The induction of heritable translocations by a substance leads to its classification as a germ cell mutagen (Table 4.12).**

4.2B.5 Gene Mutations in Germ Cells

The **specific locus test** is the classical gene mutation test in germ cells. It detects gene mutation by examining seven recessive loci which code for externally visible traits such as coat color and pattern, eye color and ear size in the offspring of exposed parental animals.

T-stock mice are homozygous for seven recessive alleles of genes which determine the coat color and pattern, eye color and ear size. The homozygous T-stock animals have a white coat, red eyes and small ears. Wild-type mice are homozygous for the dominant wild-type alleles of the seven loci and have a brown coat, brown eyes and normal sized ears. Crosses of wild-type and T-stock mice produce offspring that have the phenotype of the wild-type parent because they are heterozygous for the dominant wild-type alleles.

Commonly, wild-type mice are treated with the test chemical. A germ cell mutation in one of the wild-type alleles to a recessive allele produces a phenotypically altered mutant offspring. The mutant is homozygous for the respective allele and displays the characteristics of the respective gene, such as red eyes, light fur, short ears, white coat spots or another coat color spot.

By taking advantage of the time intervals over which spermatogonia develop into mature sperm it is possible to determine at which stage in spermatogenesis mutations are chemically induced, a procedure similar to the mating scheme of the dominant lethal test. Thus, T-stock females can be mated with chemically treated wild-type males at weekly intervals and exposed post-spermatogonial germ cells are fertilized. After 42 days, continuous mating starts and exposed spermatogonia are fertilized.

The stage-specificity during the post-spermatogonial developmental stages of male germ cells by different chemicals coincides between the specific locus test and the dominant lethal test. The research group of Liane Russell from Oak Ridge proved that the majority of the specific locus mutations induced in post-spermatogonical germ cells are deletions, whereas only a few alkylating chemicals induce gene mutations in spermatogonia.

The induction of mutations in spermatogonial stem cells is of particular importance for the assessment of the radiation-induced genetic risk. Unlike mutations in post-spermatogonial stages, mutations in stem cells are not eliminated by the removal of this cell type from the system, but are reproduced and transmitted during the entire process of spermatogenesis.

Unfortunately, there is no longer any laboratory able to conduct the specific locus test in mice because a huge animal facility and the special T-stock breed are needed. The specific locus test provides the important experimental data for the assessment of the genetic risk from a mutagen for an individual human or for the human population. The majority of cancer therapeutics with a known genetic mechanism of action have been tested. Cured patients often have a strong wish to have a child and want to know the risk of having a child with a hereditary disease. Since the development of freezing methods for sperm and oocytes, and *in vitro* fertilization (IVF), the risk can be prevented by sperm or oocyte donation before chemo or radiotherapy.

The Molecular Genetic Method for the Detection of Gene Mutations

As for somatic cells, molecular genetic methods for the detection of gene mutations have been developed for germ cells. This has been reasonably successful with transgenic mouse tests and they are internationally accepted (Table 4.10). However, only the induction of gene mutations in germ cells is tested, not their transferal to the next generation. To prove the latter has occurred, the progeny of exposed animals must be examined, e.g. in the specific locus test. Isolated experiments, including the progeny of mice exposed to radiation or ethylmethanesulfonate (ENU), have been conducted. Although these are exemplary, the effort and number of experimental animals needed is great.

An advantage of the transgenic tests is that the mutation incidence in DNA samples of different tissues can be compared. The OECD recommends the integration of the test in multiple test strategies. Furthermore, the automated evaluation procedures are of great benefit and reduce the number of animals. An instruction video describing how to do a germ cell test with transgenic rodents can be found on the Internet (O'Brien et al., 2014).

The disadvantages of the tests are that stage-specificity at the time testis tissue is generated cannot be clearly recognized, special animal strains (Big Blue®, Muta™, GPG) have to be used, and the examined genes consist of non-transcribing, exogenous DNA, which are highly methylated. Therefore, analyzed genes correspond more to non-coding endogenous DNA and the mutability may differ from that of active genes.

> **Gene mutations in germ cells caused by a substance lead to its classification as a germ cell mutagen (Table 4.12).**

Endogenous microsatellite DNA, consisting of long homologous repeats of short non-coding base pairs, can exhibit spontaneous and induced differences in length. In view of this, the expanded single tandem repeat assay (ESTR) has been developed. It can also be conducted in tissue samples of all organs from various animals and in particular in human cell samples. The spontaneous mutation incidence is relatively high and replication-dependent. The first applications focus on sperm DNA, which is extracted, diluted and amplified in multiple samples. Each sample contains approximately one ESTR molecule. This way, an infinite number of *de novo* mutations can be detected, which considerably decreases the number of experimental animals. Low doses of ionizing radiation and chemotherapeutics have been demonstrated to yield a positive result.

However, the mechanism for the induction of mutations in the ESTR loci is not yet completely understood. The mutation incidence is too high to be explained by a direct effect on these relatively short loci. It is assumed that originally mutation-induced DNA lesions somewhere else in the genome increase the mutation incidence in the ESTR loci.

This very promising method, which can be integrated into other test approaches, still requires further development, however. In Table 4.11, several methods used in human genetics and also applicable to animal tests are listed. These enabled us to compare responses in mutagenicity tests between animals and humans.

SNPs and CNVs are used in the genetic diagnosis of genetically determined but not Mendelian diseases in humans (autism, schizophrenia and others). Structural changes in sperm chromatin (SCSA) and terminal deoxynucleotide transferase-mediated dUTP nick-end labeling (TUNEL), which recognizes apoptosis and DNA fragmentation) are well-known methods in andrology and are used for sperm preparations from subfertile men before *in vitro* fertilization (IVF). These molecular techniques can be used in experimental mutagenicity testing.

Table 4.13 summarizes the OECD Guideline tests/assays for mutagenicity *in vivo*.

4.2B.6   Summary

- The classical *in vivo* mutagenicity tests, i.e. the cytogenetic translocation test, coat color test, dominant lethal test, heritable translocation test and specific locus test, were developed during the 1950s and 1960s to determine the genetic effects of ionizing radiation.
- Subsequently, these methods were used to test the mutagenicity of chemicals. The original intent of mutagenicity testing was to protect human progeny from the adverse effects of mutational events in germ cells. Today these tests are no longer used because huge animal facilities are required, which are not in accordance with animal welfare.
- With the realization that mutational events may lead to carcinogenesis, the primary interest in mutagenicity testing shifted to detecting the ability of chemicals to induce cancer. The mutagenicity data was used as a step in the strategy aimed at preventing chemically induced cancers. Tests performed *in vitro* were developed for the primary purpose of detecting potential genotoxic carcinogens as quickly and reliably as possible.
- Mutagenicity tests were adopted worldwide because a good correlation between mutagenicity and carcinogenicity was found for certain classes of chemicals. For almost two decades, the original purpose of mutagenicity testing, namely to prevent genetic burden to future generations, was largely neglected.
- The emphasis on mutagenicity testing is again undergoing reconsideration. Despite a wealth of *in vitro* mutagenicity tests with microorganisms and mammalian cells in culture, the data do not reliably predict the actual carcinogenic or mutagenic hazard of chemicals for humans because

Table 4.11 Gene mutation tests in germ cells.

Organism	Test system	Reporter gene	Cell type	Analysis	Acceptance
Wild-type and test stock mouse	Specific locus test	Five coat colors gene, one eye color gene, one ear color gene	Entire spermatogenesis exposed, mutants in F_1 progeny, progeny analyzed	Visible change in coat color, eye color or ear size	EC Directive 1986, OECD 485 1986
Transgenic mice or rats	Big Blue® Muta™Mouse GPG	Lac Z Lac I Gpt delta	Spermatogonia until spermatids exposed, sperm DNA analyzed	Flow cytometry	OECD 488 2013
Mouse, human	ESTR	Microsatellite DNA	Sperm	PCR individual DNA samples of the whole genome	Hardwick et al. (2009)
Human	SNP	Single nucleotide polymorphism	Sperm	Microarrays, aCGH	Relevant in human genetics, development of application in animal experiments Campbell and Eichler (2013)
Human	CNV	Copy number variations			
Human	SCSA, TUNEL	Structural change in sperm – chromatin, DNA breaks	Sperm	Flow cytometry	Fertility disorders, development of application in animal experiments Sharma et al. (2015)

Lac Z, bacterial gene (coding β-galactosidase); Lac I, bacterial gene (coding lactosidase repressor protein); Gpt delta, bacterial gene in λEG10 phage DNA; ESTR, expanded simple tandem repeats; PCR, polymerase chain reaction; aCGH, high-resolution array comparative genomic hybridization; SCSA, sperm chromatin structure assay; TUNEL, terminal deoxynucleotidyl transferase-mediated fluorescein-dUTP nick end labelling.

Table 4.12 *Strategies for the classification of a substance as a germ cell mutagen.*

Positive data in test system Name	CLP regulation EC 1272/2008 Category New germ cell mutagen	MAK category Germ cell mutagen
In vivo tests in somatic cells	2	3A*/3B
Cytogenetic tests in germ cells (CA, MN, ANEU)	1B	3A
Cytogenetic tests in offspring (DL, HT)	1B	2
Gene mutation tests in germ cells (Big Blue®, Muta™, GPC)	1B	3A
Gene mutation tests in offspring (SL)	1B	2
Evidence of inherited mutations in humans	1A	1

*Additional evidence that the substance or a metabolite reaches the germ cells is necessary.
CA, chromosomal aberrations; MN, micronuclei; ANEU, test with centromeric staining (MN with FISH or sperm-FISH); DL, dominant lethal test; HT, F1 translocation test; BigBlue®, Muta™, CPG, transgenic mouse or rat strains; SL, specific locus mutation.

Table 4.13 *OECD Guideline tests/studies for mutagenicity in vivo.*

OECD Guideline	Test in somatic cells	Test in germ cells	OECD Guideline	Publication or last update
Mammalian erythrocyte micronucleus test	x	–	474	2016
Mammalian bone marrow chromosomal aberration test	x	–	475	2016
Transgenic rodent somatic and germ cell gene mutation assays	x	x	488	2013
Pig-a gene mutation assay	x	–	In progress	To be completed 2020
Mouse Spot test	x	–	484	1986 (deleted 2014)
Mammalian spermatogonial chromosomal aberration test	–	x	483	2016
Specific locus test	–	x	485	1986
Rodent dominant lethal test	–	x	478	2016
Mouse heritable translocation assay	–	x	485	1986

of the complexity of metabolic processes in mammalian organisms, the multiplicity of stages in cancer development, and the biological dynamics of germ cell development. Currently, all internationally accepted criteria for categorizing carcinogens and germ-cell mutagens require data derived from *in vivo* testing.

- The newly developed and commercially available transgenic animal models, such as the Muta™Mouse (Hazelton Laboratories Corp.) or the Big Blue® mouse (Stratagene), introduced a new period of mutagenicity testing in which simple, rapid, molecular biological methods *in vitro* are combined with the complexity of *in vivo* testing. Additionally, the incidence of mutation in somatic and germ cells can be compared.
- Very promising is the Pig-a assay, particularly because it can be combined with the micronucleus test to test clastogenic, aneugenic and mutagenic effects in genes in only one test approach. The necessary database for standardization of this test combination is currently being developed (Gollapudi et al., 2015). With the interdisciplinary workshop Assessing Human Germ-Cell Mutagenesis in the Postgenome Era: A Celebration of the Legacy of William Lawson (Bill) Russell (Wyrobek et al., 2007), which took place in 2007 at the Jackson Laboratory in Bar Harbor, Maine, USA, the focus returned to the original task of mutagenicity testing. Here, learning from human genetics and adopting the well-established methods and applying them to germ cell mutagenicity testing were first suggested. The results of the discussions on the new approaches to germ cell mutagenicity testing from the 6th IWGT workshop in 2013 were published in 2015 (Yauk et al., 2015).
- To reduce the number of animals in these tests, attempts have increasingly been made to integrate mutagenicity tests (or indicator tests for genotoxicity) into the general toxicity tests. This is achievable with the new molecular biological method. Experimentally, compromises have to be made as the biology behind the genetic test systems requires certain test procedures. If the question is "Yes or No", such compromises can perhaps be acceptable, but if the question is to clarify underlying genetic mechanisms, individual test details adapted to the biological system have to be used.
- The consideration of molecular genetic methods will contribute to the improvement in rapidness and sensitivity of the tests, but it is essential that additional fundamental research is carried out to provide an insight into the way molecular changes in genes lead to effects on the phenotype of the cell and the organism (see also Chapters 2.5 and 4.6). The aim of mutagenicity research is to now discover which phenotypical effects are induced by particular molecular DNA changes, to allow a meaningful quantitative evaluation to be made of the carcinogenic and mutagenic risk.

Further Reading

Campbell CD, Eichler EE (2013) Properties and rates of germline mutations in humans. *Trends Genet.* **10**: 575–584. doi: 10.1016/j.tig.2013.04.005.

CLP-Regulation (EC) No 1272/2008 of the European Parliament and of the Council of 16 December 2008 on classification, labelling and packaging of substances and mixtures, amending and repealing Directives 67/548/EEC and 1999/45/EC, and amending Regulation (EC) No 1907/2006 (Text with EEA relevance) OJ L 353, 31.12.2008, p. 1–1355, http://data.europa.eu/eli/reg/2008/1272/oj.

Dertinger SD, Avlasevich SL, Bemis JC, Chen Y, MacGregor JT (2015) Human erythrocyte PIG-A assay: an easily monitored index of gene mutation requiring low volume blood samples. *Environ Mol Mutagen* **56**, 366–377.

EC Directive (1986) Annex V to Directive 67/548/EEC on the classification, packaging and labelling of dangerous substances: Testing Methods. Official Journal Eur. Comm., No. L225, http://ec.europa.eu/environment/archives/dansub/annex_v_table_default_en.htm.

Gollapudi BB, Lynch AM, Heflich RH, Dertinger SD, Dobrovolsky VN, Froetschl R, Horibata K, Kenyon MO, Kimoto T, Lovell DP, Stankowski LF Jr, White PA, Witt KL, Tanir JY (2015) The in vivo Pig-a assay: A report of the International Workshop on Genotoxicity Testing (IWGT) Workgroup. *Mutat Res Genet Toxicol Environ Mutagen* **783**, 23–35. doi: 10.1016/j.mrgentox.2014.09.007.

Hardwick RJ, Tretyakov MV, Dubrova YE (2009) Age-related accumulation of mutations supports a replication-dependent mechanism of spontaneous mutation at tandem repeat DNA Loci in mice. *Mol Biol Evol* **26**, 2647–2654.

Kimoto T, Horibata K, Miura D, Chikura S, Okada Y, Ukai A, Itoh S, Nakayama S, Sanada H, Koyama N, Muto S, Uno Y, Yamamoto M, Suzuki Y, Fukuda T, Goto K, Wada K, Kyoya T, Shigano M, Takasawa H, Hamada S, Adachi H, Uematsu Y, Tsutsumi E, Hori H, Kikuzuki R, Ogiwara Y, Yoshida I, Maeda A, Narumi K, Fujiishi Y, Morita T, Yamada M, Honma M. (2016) The PIGRET assay, a method for measuring Pig-a gene mutation in reticulocytes, is reliable as a short-term in vivo genotoxicity test: Summary of the MMS/JEMS-collaborative study across 16 laboratories using 24 chemicals. *Mutat Res* **811**, 3–15. doi: 10.1016/j.mrgentox.2016.10.003. Epub 2016 Oct 21.

List of MAK and BAT Values 2017, Permanent Senate Commission for the Investigation of Health Hazards of Chemical Compounds in the Work Area. Report 53, Wiley-VCH, Weinheim, p. 220–221, www.mak-collection.com.

Morita T, MacGregor JT, Hayashi M (2011) Micronucleus assays in rodent tissues other than bone marrow. Mutagenesis 26, 223–230. Reproduction, 3296 pages, Academic Press, 3rd edition (2005).

OECD Guidelines for the Testing of Chemicals, Section 4. Health Effects, http://www.oecd-ilibrary.org/environment/oecd-guidelines-for-the-testing-of-chemicals-section-4-health-effects_20745788 (last access 10 February 2017).

O'Brien JM, Beal MA, Gingerich JD et al. (2014) Transgenic rodent assay for quantifying male germ cell mutant frequency. J Vis Exp 90, e51576. doi:10.33791/51576, http://www.jove.com/video/51576/transgenic-rodent-assay-for-quantifying-male-germ-cell-mutant-frquency.

Olsen AK, Dertinger SD, Krüger CT, Eide DM, Instanes C, Brunborg G, Hartwig A, Graupner A (2017) The Pig-a Gene Mutation Assay in Mice and Human Cells: A Review. *Basic Clin Pharmacol Toxicol* **121**, Suppl 3, 78-92. doi: 10.1111/bcpt.12806.

Parry JM, Parry EM (Eds.) (2012) *Genetic Toxicology*. Princ Meth, Humana Press, c/o Springer Science + Business Media, New York, USA.

Sharma R, Agarwal A, Rohra VK, Assidi M, Abu-Elmagd M, Turki RF (2015) Effects of increased paternal age on sperm quality, reproductive outcome and associated epigenetic risks to offspring. *Reprod Biol Endocrinol* **13**, 35–45. doi: 10.1186/s12958-015-0028-x.

Wyrobek AJ, Mulvihill JJ, Wassom JS, Malling HV, Shelby MD, Lewis SE, Witt KL, Preston RJ, Perreault SD, Allen JW, Demarini DM, Woychik RP, Bishop JB (2007) Assessing human germ-cell mutagenesis in the postgenome era: a celebration of the legacy of William Lawson (Bill) Russell. *Environ Mol Mutagen* **48**(2), 71–95.

Yauk CL, Aardema MJ, Benthem Jv, Bishop JB, Dearfield KL, DeMarini DM, Dubrova YE, Honma M, Lupski JR, Marchetti F, Meistrich ML, Pacchierotti F, Stewart J, Waters MD, Douglas GR (2015) Approaches for identifying germ cell mutagens: Report of the 2013 IWGT workshop on germ cell assays. *Mutat Res Genet Toxicol Environ Mutagen* **783**, 36–54.

4.3 Assessment of the Individual Exposure to Xenobiotics (Biomonitoring)

Thomas Göen

4.3.1 Introduction

Biomonitoring in medical practice refers to the quantitative determination of substances, their metabolites and reaction products with endogenous (macro)molecules as well as strain parameters in human biological material. In toxicology, this term may also cover the monitoring of experimental animal exposure in biological material.

The areas of application of biomonitoring may be divided into the following categories:

- exposure studies with human and animal subjects to determine the toxicokinetics and metabolism of a substance
- assessment of the exposure and potential biological effects in human and animal experimental studies on the toxic effects, mechanism and potency of xenobiotics
- assessment of the exposure and burden in occupationally exposed persons (occupational medicine) or non-occupationally exposed persons (environmental medicine).

In the last two areas of application, the human health risk of exposure to xenobiotics is eventually to be assessed. Finally, biomonitoring results have to be compared with assessment values, which can partly be derived from studies of the first two areas of application.

> **Biomonitoring in medical practice refers to the quantitative determination of substances, their metabolites and reaction products with endogenous (macro)molecules as well as strain parameters in human biological material. In toxicology, this term may also cover the monitoring of experimental animal exposure in biological material.**

The tasks of human biomonitoring are to determine the human exposure and health risk, to compare the analytical data obtained with assessment values in order to identify health hazards and, if necessary, to propose suitable measures in order to reduce the exposure and thus the impact on health.

Most methods currently available deal with the quantitative determination of internal exposure by measuring the concentration of xenobiotics and their metabolites in biological material (**biological exposure monitoring**). By contrast, **biological effect monitoring** is used to detect alterations in biological material reacting sensitively and as specifically as possible to exposure, even if these alterations themselves do not have adverse effects on the organism. Parameters for monitoring biological effects include, among other, alterations in enzyme activities, products of intermediary metabolism and cytogenetic parameters. Furthermore, **biochemical effect monitoring**, as it is termed, is often given as an independent parameter category. This method basically permits the quantification of reaction products of mutagenic substances, which are covalently bonded as adducts to macromolecules such as proteins (e.g. albumin or haemoglobin) and DNA.

Biomonitoring is particularly suitable for assessing individual health risks as it

- detects the overall internal exposure level, irrespective of the route of exposure (inhalation, oral, dermal)
- enables a differentiated health risk assessment by measuring the biologically effective substance concentrations
- includes almost all individual disposition factors, which is an important factor in the toxicokinetics and toxicodynamics of the substance.

> **Biomonitoring is particularly suitable for assessing individual health risks as it detects the overall internal exposure level, irrespective of the route of exposure, enables a differentiated health risk assessment by measuring the biologically effective substance concentrations, and includes almost all individual disposition factors.**

Figure 4.7 shows the relationships between external exposure, internal exposure and toxic effects. It illustrates that individual disposition factors can play a role along the entire exposure and impact pathways. Such disposition factors can be rooted in the status of biochemical factors, especially enzyme activities, as well as in morphological or functional characteristics.

In the first step of this process, for instance, impaired or increased pulmonary function or skin lesions affect the uptake of substances. The metabolism, and in particular the balance between Phase I and Phase II metabolism, plays a crucial role in the transformation of a substance into a biologically active type. At this stage, the individual disposition is thus particularly characterized by the availability and activity of the enzymes involved in the metabolism. The general rule applies: the closer the diagnostic parameter is linked to the essential adverse effects in humans, the more the individual disposition factors are included in the test result.

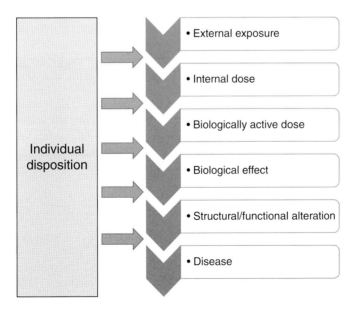

Figure 4.7 Course from external exposure to health effects and impact of individual disposition.

Occupational exposure scenarios are usually much more clearly characterized than environmental exposure scenarios. Basically, biomonitoring is particularly important or even essential for the proper assessment of individual human exposure in the following situations:

- in the case of skin contact and if the dermal uptake contributes significantly to the overall exposure
- if oral uptake must be taken into consideration
- in the case of exposure to substances with long biological half-lives
- in the case of accidental exposure
- in the case of strongly fluctuating external exposure, for example strongly fluctuating air concentrations
- in situations when the respiratory minute volume can be modified by physical activity.

However, there are also exposure situations when biomonitoring cannot be applied at all or can be applied only to a very limited extent. In particular, biomonitoring is not suitable:

- for substances that are not absorbed by the body at all or that are absorbed only to a limited extent (this applies, for example, to inert dusts, fibre dusts and insoluble substances)
- for substances for which essential hazards are related to irritating effects on skin and mucosa
- for substances with an extremely short biological half-life
- for chemical structures that are already physiologically formed in huge quantities in the human body.

4.3.2 Prerequisites for Carrying Out Biomonitoring

Some basic requirements have to be fulfilled to ensure that biomonitoring can be carried out appropriately and ultimately yields interpretable results. Fundamental requirements for carrying out biomonitoring are:

- the selection of suitable biological material, which can be obtained reasonably and practicably
- biomonitoring parameters (biomarkers) that are sensitive and specific enough to indicate exposure to a certain substance
- efficient analytical methods that allow reliable quantification of the biomarker
- availability of health-based, risk-based or descriptive values for the assessment of the results.

> **Fundamental requirements for carrying out biomonitoring are the selection of suitable biological material and biomonitoring parameters, the use of reliable analytical procedures, and the availability of assessment values for the evaluation of the results**

Biological Material for Biomonitoring

As can be seen from Figure 4.7, the assessment of the health hazards or health risk caused by chemical exposure can in principle be most reliably ensured by quantification of the substances or their metabolites in the target organs, on which the substances have a toxic effect. As a direct determination of the concentrations of the substances or metabolites in the target organs is often not acceptable for ethical reasons, the contents of the analytes in easily accessible human biological materials that are obtained by minimally invasive methods and that are at equilibrium with organ exposure are usually used for this purpose. These requirements are in particular met by blood and urine. Moreover, toxicological studies almost always involve tests in these two media, so that, as a result, the greatest amount of knowledge of toxicokinetics and toxicodynamics has been amassed with these two media. In principle, however, other human biological materials, for example alveolar air, saliva, hair, teeth, fingernails or toenails, can also be used as examination materials. For these matrices, however, knowledge of toxicokinetics and toxicodynamics as well as toxicologically founded assessment values are often lacking so the contents in these materials do not allow a health risk assessment but only indicate previous exposure.

> **Blood and urine are the most recommended biological materials because they are vital compartments of the pharmacokinetic scheme and due to their exceptional state of knowledge.**

In connection with the selection of the examination material, the question of the correct sampling time arises, even though this question can usually only be answered in due consideration of the specifications of the particular biomonitoring parameter. Basically, however, it has to be differentiated between an acute, accidental exposure and chronic exposure over an extended period of time. In the case of accidental exposure, samples should preferably be taken directly after exposure. When using highly persistent biomonitoring parameters, such as haemoglobin adducts (see below), deviations from this rule may be permissible. Moreover, a retarded uptake of the hazardous substances should be taken into account after accidental, massive dermal exposure. In the case of prolonged exposure sampling should take place at a time when the subject's internal and external exposure is in a state of equilibrium. This may vary depending on the test parameter. For example, with many volatile substances, an equilibrium between air exposure and the blood concentration is achieved only after a quarter of an hour. However, for many metabolites that are excreted in urine, the steady-state level may not be reached even after several hours. In these cases it might be useful to collect the urine sample preferably at the end of exposure to be able to detect the maximum cumulative amount. For test parameters that are determined in urine, the question arises as to whether a 24-hour urine collection holds any advantages over spontaneous urine samples. The crucial point in answering this question is on which basis the reference values were obtained that are used for the subsequent assessment. In toxicokinetic studies, the results are often based on the cumulative urinary excretion within 24 hours after exposure. By contrast, limit values which are derived from occupational medical studies, for example, are usually obtained on the basis of spontaneous urine samples, which are collected at the end of exposure and the end of the shift, respectively. The amounts of excreta which can be determined in a spontaneous urine sample are basically prone to diuresis-related fluctuations. However, in many cases the creatinine level or the specific weight of the urine sample can be used as a reference value to allow for these fluctuations.

Biomonitoring Parameters

A suitable biomonitoring parameter has to indicate the exposure to or effect caused by a substance reliably, sensitively and in particular specifically. Generally, parameters of biological effect monitoring are less substance-specific than parameters of biological exposure monitoring since even effect parameters that have been chosen because of their substance reference can be affected by various noxae. However, the substance specificity may also vary significantly within the parameter group of the biological monitoring of exposure. Of course, the determination of the unchanged substance in the biological material yields the highest specificity. However, such a biomonitoring parameter is not available for many substances due to fast metabolism. The further the metabolism has advanced, the more the metabolite's structure will differ from the structure of the substance and thus lose its specificity. This holds particularly true for the metabolism of hydrocarbon-based xenobiotics. For almost all organic substances, carbon dioxide is the ultimate metabolite, which is eventually exhaled from the lungs. However, carbon dioxide is a parameter without substance specificity and, moreover, the substance-related carbon dioxide excretion may be negligible as compared to the general carbohydrate metabolism of the organism.

> **Specificity for exposure and association with the toxicological effect are the most eminent criteria for the selection of the biomonitoring parameter. Generally, specificity decreases during advanced metabolism.**

With biological effect monitoring, only a few parameters are known to react specifically to exposure. A classic example of such monitoring is the measurement of the renal excretion of delta-aminolevulinic acid or of the erythrocyte protoporphyrin level in the case of lead exposure. Effect parameters therefore usually have to be examined and assessed in conjunction with exposure parameters.

Biochemical effect parameters are protein and DNA adducts, which can be measured at very low concentrations using modern analytical methods. Adducts are usually reaction products from highly reactive intermediates, which have been formed in the Phase I or Phase II metabolism of a substance.

Analytical Methods

Analytical methods are described in studies examining such parameters. However, these descriptions do not always give an account of the reliability of the methods used. A collection of methods for analyses in biological material is published by the Senate Commission for the Investigation of Health Hazards of Chemical Compounds in the Work Area (MAK commission) of the German Research Foundation (DFG). First, this method collection provides a comprehensive overview of the biomonitoring methods currently available. Second, these methods have been field-tested and validated by at least one independent laboratory to ensure reproducibility and reliability. Moreover, several international proficiency tests are offered that enable external quality assurance for most of the biomonitoring parameters and thus verify comparability and accuracy of the biomonitoring methods.

Assessment Values

Assessment values with different evaluation strategies (see Table 4.14) permit the appraisal of biomonitoring results. These assessment values are estimated worldwide by several national and international committees as well as some governmental institutions. The most eminent scientific committees for the evaluation of biological assessment values are the Biological Exposure Indices (BEI) Committee of the American Conference of Governmental Industrial Hygienists (ACGIH), the MAK Commission and the Scientific Committee on Occupational Exposure Limits (SCOEL) of the European Commission. Occasionally governmental agencies provide additional assessment values, particularly for questions that cannot be cleared on a scientific basis, e.g. setting of socially accepted risk limits.

Table 4.14 Categories of biological assessment values.

Labelling	Abbreviation	Appraising committee/agency
Health-based values		
Biological tolerance value	BAT	MAK Commission
Biological guidance value (Biologischer Leitwert)	BLW	MAK Commission
Biological limit value	BLV	SCOEL
Biological exposure limit	BEI	ACGIH
Risk-based values		
Equivalent to acceptance risk	–	Committee on Hazardous Substances
Equivalent to tolerable risk	–	Committee on Hazardous Substances
Descriptive assessment values		
Reference value	RV_{95}	Federal Environment Agency
Reference value for working materials (Biologischer Arbeitstoff-Referenzwert)	BAR	MAK Commission
Biological guidance value	BGV	SCOEL
Exposure equivalents for carcinogenic substances	EKA	MAK Commission

> The evaluation of biomonitoring results may be based on assessment values that are health-based, risk-based or derive from an analysis of population data.

Health-based Values The main group of assessment values comprises values that relate to the threshold concentration or dose of the relevant adverse effects (no observed adverse effect level (NOAEL) and lowest observed adverse effect level (LOAEL)). These assessment values are either directly derived from studies in which both human exposure and burden were determined by measuring the biomonitoring parameters, or based upon the relationship between external and internal exposure from air limit values, which themselves are based on the toxicological effect thresholds. In some cases the values derived from animal toxicological studies are also applicable to human assessment. In principle, there can be various toxicologically derived assessment values for one hazardous substance, which are defined for different protection goals and differ accordingly. Assessment values that are used in occupational medicine usually refer to the average person of working age. An uncertainty factor that ensures that particularly sensitive individuals are sufficiently protected is usually not applied when deriving these values. By contrast, several uncertainty factors that reflect inter-individual variability are usually applied for the derivation of no-effect levels (DNEL) under the REACH registration and biological assessment values for the general population. It should also be noted that the determination of occupational medical assessment values is based upon chronic exposure over the entire working life (assuming a total of 40 years and 40 working hours a week), whereas assessment values for environmental medical purposes consider continuous exposure throughout a lifetime (24 hours a day, 70 years). Examples of health-based assessment values are the biological tolerance values (BAT) and biological guidance values (BLW) established by the MAK Commission, the biological limit values (BLV) set by SCOEL and the biological exposure indices (BEI) published by the American Conference of

Governmental Industrial Hygienists (ACGIH). For environmental medical purposes, the Human Biomonitoring Commission of the German Federal Environment Agency establishes the human biomonitoring values HBM I and HBM II.

Risk-based Values A second group of assessment values are risk-related values. These values are particularly established for the limitation of effects for which, as in the case of genotoxic carcinogens, no threshold concentration or dose can be specified. For these effects, however, stochastic relationships between the exposure level and the cancer risk can usually be established. On the condition of socio-politically accepted or tolerated risk limits and on the basis of appropriate exposure scenarios, these relationships can be used for the derivation of relevant exposure values, which indicate the corresponding increased risk. In Germany, such a concept for the determination of occupational exposure has been developed by the Committee for Hazardous Substances (AGS) of the Federal Ministry of Labour and Social Affairs. For the concentration of some carcinogenic substances in the workplace air, "acceptance values" and "tolerance values" have been published by the AGS, which correspond to a "tolerance risk", defined as an additional cancer risk of 4:1,000, or an "acceptance risk" of initially 4:10,000 and 4:100,000 as of 2018 according to epidemiologic-toxicological data (TRGS 910). Applying these acceptance and tolerance values to biomonitoring presupposes knowledge of quantitative relationships between air exposure and resulting internal exposure. This is made possible by the exposure equivalents for carcinogenic workplace substances (EKA), which have been established by the MAK Commission for more than 30 years for workplace materials that have proved to be carcinogenic in either animal or human studies. It can be seen from the EKA tables what internal exposure would result from uptake of the substance exclusively by inhalation. On the basis of the EKA, the respective concentrations of biomonitoring parameters, which correspond to the acceptance and tolerance risks, can be derived from the air limit values.

Reference Values and Other Descriptive Assessment Values The third group of assessment values comprises descriptive values, which are taken from either a frequency distribution or a correlation. The concept of frequency distribution is applied to describe the general background level of a biomonitoring parameter. It takes into account the fact that every person is in some way, even to a limited extent, exposed to almost any substance even without a specific exposure situation. Effect parameters should be of a general specificity even with non-exposed persons. In biomonitoring, the "reference value" as it is termed is used as a descriptive assessment value. By definition, the reference value for a chemical substance in human biological material is a value that is derived from a series of measuring results according to a specified statistical method. Samples to be used for this purpose have to be collected from a defined group of the general population. The reference value is a strictly statistically defined value that is of no health relevance. It is usually derived from the 95th percentile of a group of individuals who are not occupationally exposed to the substance. However, this also means that concentrations may occur in blood and urine samples of a part of the general population that are above the specified reference value. Reference values are used for the assessment of biomonitoring results for both occupational and environmental medical purposes. In order to provide the necessary data for the environmental medical sector, the HBM Commission of the German Federal Environment Agency publishes reference values that are partly differentiated by gender and age. For some years now, the MAK commission have evaluated and published biological reference values for workplace substances (BAR) for occupational medical requirements. Moreover, SCOEL and the BEI committee recently also published biological assessment values derived from the background levels in the general population. Evaluation of reference values is primarily performed under the aspect of preventive occupational medical care and solely relates to persons of working age. By comparing biomonitoring results with reference values it can be ascertained whether an exposure to a substance exceeds the general level. Basically, it must be noted that a reference value describes the exposure situation of a reference population at the time of random sampling and that re-evaluation may be required after alteration of the general exposure situation, for example by a decrease in environmental exposure. The second concept of descriptive values comprises the aforementioned exposure equivalents for carcinogenic workplace substances (EKA), which describe, as outlined above, the relationships between the employee's external exposure (air exposure at the workplace) and the concentration of a biomarker (see above).

4.3.3 Examples of Biomonitoring of Special Substance Groups or Special Biomonitoring Parameters

Substances with a High Vapour Pressure

Substances with a high vapour pressure usually have an impact on the human organism as gas or vapour and are thus almost solely absorbed through the respiratory tract. In some workplaces, however, direct skin contact with such hazardous substances is possible; due to the low vapour pressure the exposure time on the skin is very short. In the case of constant air exposure, an equilibrium between external exposure and the concentration of the hazardous substance in blood is partly very quickly established. The status of the equilibrium is described by the blood–air partition coefficient. High blood–air partition coefficients, which imply a high uptake into the blood, are usually found for hydrophilic substances, whereas hydrophobic substances have low blood partition coefficients. Analyses of toluene and styrene show that the steady-state level is achieved within 15–30 minutes. The high volatility also implies that these substances can be very quickly eliminated through the airways at the end of exposure. Contrary to the resorption phase, compounds with a low blood–air partition coefficient can be eliminated through the airways to a greater extent than substances with a high partition coefficient. Elimination half-lives also often range from 15 to 30 minutes. Due to the great significance of the equilibrium between blood concentration and the concentration in the alveolar gas compartment, the determination of the unchanged substances both in blood and exhaled air can be used for biological monitoring of exposure. However, the short half-life of the substance in blood after the end of exposure means that sampling has to take place either directly after exposure or after a defined period of time after the end of exposure to obtain an interpretable measuring value. Examples of this biomonitoring strategy are the determination of dichloromethane or tetrachloroethene in blood, for which there are assessment values for sampling directly after exposure or 16 hours after the end of exposure.

Even though exhalation makes up a substantial proportion of elimination of substances with a high vapour pressure, some agents in this substance category are metabolized to a relevant extent. In these cases, the determination of urinary metabolites can also be used for biomonitoring of these substances. Examples include the determination of trichloroacetic acid as a parameter for trichloroethene, o-cresol for exposure to toluene or 2,5-hexanediol for exposure to n-hexane.

Volatile hydrophilic substances are only poorly eliminated through the airways. They can also be eliminated through the urinary tract without metabolism. Therefore, for such substances, the determination of the unchanged substance in urine can often be used for biomonitoring. Examples include the determination of the urinary concentrations of methanol, acetone, methyl ethyl ketone (2-butanone) and tetrahydrofuran.

Phase I and Phase II Products

The situation is completely different with regard to substances that are very reactive themselves or are easily and completely metabolized. These substances are also quickly cleared from the blood;

however they are eliminated through the airways only to a very limited extent, but particularly as metabolites in urine. Metabolites that are readily eliminated through the airways take a special position. In particular, carbon dioxide should be mentioned, which is an end product of the complete oxidation of organic substances. However, as already mentioned, it is an unspecific metabolite, which is therefore not applicable as a biomonitoring parameter. The pre-formed intermediates of most hazardous substances are readily eliminated in urine either themselves or after Phase II reactions. Examples of metabolites of hazardous substances that are eliminated in urine without Phase II reaction are ethoxyacetic acid after exposure to ethoxyethanol as well as 5-hydroxy-N-methylpyrrolidone after exposure to N-methylpyrrolidone. Hydroxyl groups, especially in phenolic structures, are often conjugated with glucuronic acid or sulphate and are thus primarily excreted as glucuronides and sulphates in urine. Metabolites that are almost completely excreted as glucuronide and sulphate conjugates are para-nitrophenol after exposure to parathion as well as bisphenol A (see ACGIH and List of MAK and BAT Values). Apart from glucuronic acid and sulphate conjugation, the glutathione S-transferase reaction is one of the most important Phase II reactions of the metabolism of xenobiotics. These are primarily addition reactions with electrophilic hazardous substances or metabolites, such as epoxides, α,β-unsaturated nitriles and ketones. However, substitution reactions, for example with halogenated hydrocarbons, can also lead to glutathione conjugation. After the split-off of the amino acids glutamic acid and glycine as well as after acetylation, the formed S-conjugated acetylcysteine (synonym: mercapturic acid) is excreted in urine. Examples of such biomonitoring parameters are S-phenyl mercapturic acid after exposure to benzene and 3-hydroxypropyl mercapturic acid after exposure to acrolein.

Biomonitoring Parameters with a High Biological Persistence

Hazardous substances that have a low vapour pressure, that are very lipophilic, readily adsorbed by cell surfaces or macromolecules and metabolize only slowly play a special role. Examples of such persistent hazardous substances, which can accumulate in the human body, are polychlorinated biphenyls (PCB) as well as the perfluorinated compounds perfluorooctanoic acid (PFOA) and perfluorooctanesulphonate (PFOS). For the representatives of the two substance groups, it was ascertained that their elimination half-lives from blood are several years. Due to the poor elimination, the biomonitoring of these substances is usually carried out in blood or in blood plasma.

Biomonitoring parameters with long half-lives are also adducts at the globin chains of haemoglobin. Along the lines of glutathione conjugation, such adduct formation occurs particularly with electrophilic substance and metabolite structures as well as substitution reactions with substances that contain functional groups that can be easily be eliminated in chemical terms, such as halogenides. Adducts are covalent, usually very stable reaction products. In addition, the erythrocyte represents a "Cave" situation, which protects the reaction product against degradation by extracellular enzymes. An adduct can thus be detected in blood until apoptosis of the erythrocyte. Exposure is usually of an order that poses no stress situation for the erythrocyte. The adduct life corresponds to the mean erythrocyte life of approximately 126 days. During this period, the adduct level can also accumulate in the case of chronic exposure. The fact that particularly the N-terminal amino acid valine is a preferred reaction site for such adducts plays a special role in biomonitoring. Such N-terminal adducts can be specifically separated from the protein structure without total hydrolysis by using a modified Edman degradation method and used for analytical determination. Examples of biomonitoring parameters on the basis of the N-terminal haemoglobin adducts are N-2-hydroxyethyl valine for exposure to ethylene oxide and N-methyl valine for exposure to dimethyl sulphate.

4.3.4 Summary

Biomonitoring enables the assessment of an organism's exposure to a substance (exposure monitoring) as well as the resulting effects on a biochemical level (effect monitoring). While in human or animal studies biomonitoring serves to determine the concentration of a substance or an intermediate in the target organ or to investigate the metabolism and toxicokinetics, human biomonitoring in routine applications serves to assess the associated health risk using the quantification of the substance exposure. To this end, suitable biological materials and biomonitoring parameters have to be selected. Knowledge of the metabolism and toxicokinetics of a substance or its metabolites is essential for the planning of biomonitoring examinations and the interpretation of the results. These are the deciding factors in whether exposure monitoring should be performed by determining the unchanged substance, a metabolite or an adduct in blood or urine and at what time of an exposure phase sampling should take place. For interpretation, biomonitoring results are compared with assessment values, which are established either on the basis of toxicological threshold concentrations or doses or in respect of certain risk levels or merely describe the basic exposure in the general population (reference values). Biomonitoring is thus an effective method to confirm or exclude substance exposure as the presumed cause of a health effect or risk.

Further Reading

American Conference of Governmental Industrial Hygienists (2016) *Threshold Limit Values for Chemical substances and Physical Agents & Biological Exposure Indices*. ACGIH, Cincinnati.

Bolt HM, Their R (2006) Biological monitoring and Biological Limit Values (BLV): The strategy of the European Union. *Toxicol. Lett.* **162**: 119–124.

Boogaard PJ (2002) Use of haemoglobin adducts in exposure monitoring and risk assessment. *J. Chrom. B* **778**: 309–322.

Boogaard PJ, Hays SM, Aylward LL (2011) Human biomonitoring as a pragmatic tool to support health risk management of chemicals – Examples under the EU REACH programme. *Regul. Toxicol. Pharmacol.* **59**: 125–132.

Committee on Hazardous Substances (2016) Risk-related concept of measures for activities involving carcinogenic hazardous substances. Technical Rules for Hazardous Substances, TRGS 910, Joint Ministerial Bulletin, GMBl 2014 No. 12, pp. 258–279, last amended: GMBl 2016 No. 31, pp. 606–609.

Deutsche Forschungsgemeinschaft (2017) List of MAK and BAT Values 2017. Commission for the Investigation of Health Hazards of Chemical Compounds in the Work Area. Report No. 53. WILEY-VCH, Weinheim.

Drexler H, Lehnert G, Hartwig A, Greim H, Henschler D (1994–2015) *The MAK-Collection for Occupational Health and Safety*. Part II. BAT Value Documentation. Vol. 1–5, Wiley-VCH, Weinheim.

Drexler H, Göen T, Schaller KH (2008) Biological tolerance values: change in a paradigm concept from assessment of a single value to use of an average. *Int. Arch. Occup. Environ. Health* **82**: 139–142.

Göen T, Angerer J, Schaller KH, Hartwig A, Greim H, Henschler D (1985–2015) *The MAK-Collection for Occupational Health and Safety. Part IV. Biomonitoring methods*. Vol. 1–13, Wiley-VCH, Weinheim.

Göen T, Schaller KH, Drexler H (2012) Biological Reference Values for Chemical Compounds in the Work Area (BAR) – An approach for the evaluation of biomonitoring data. *Int. Arch. Occup. Environ. Health* **85**: 571–578.

Greim H (2001) Endpoints and surrogates for use in population studies in toxicology. *Toxicol. Lett.* **120**: 395–403.

Greim H, Csanády G, Filser JG, Kreuzer P, Schwarz L, Wolff T, Werner S (1995) Biomarkers as tools in human health risk assessment. *Clin. Chem.* **41**: 1804–1808.

Manno M, Viau C, Cocker J, Colosio C, Lowry L, Mutti A, Nordberg M, Wang S (2010) Biomonitoring for occupational health risk assessment (BOHRA). *Toxicol. Lett.* **192**, 3–16.

Müller M, Schmiechen K, Heselmann D, Schmidt L, Göen T (2014) Human biological monitoring – A versatile tool in the aftermath of a CBRN incident. *Toxicol. Lett.* **231**: 306–314.

Mutti A (1999) Biological monitoring in occupational and environmental toxicology. *Toxicol. Lett.* **108**: 77–89.

4.4 Epidemiology

Kurt Ulm

4.4.1 Introduction

The goal of epidemiology is to study the distribution and determinants of health-related status or events in human populations, and the application of these studies to control of health problems (Last, 1988).

One of the first epidemiological studies was undertaken in the middle of the 19th century, when in 1854 John Snow in London linked the mortality due to cholera with the quality of the drinking water.

A more systematic development of epidemiological methods started in the middle of the 20th century when the research focused on chronic diseases such as cancer and coronary heart diseases. In 1949 the Framingham study began to identify risk factors for cardiovascular events. In the field of cancer, the first studies were designed to explore the relationship between smoking and lung cancer.

4.4.2 Measures to Describe the Risk

The risk of a certain disease can be described for a specific point in time (prevalence) or for a time period (incidence).

The proportion in the population to be studied can be defined by demographic factors such as age, gender, or location. There are the two ways to describe the risk of disease: prevalence and incidence.

The **prevalence** is the ratio of the number of individuals who have a specific disease at a particular point in time divided by the population at risk at that point in time.

The **incidence** describes the rate at which new cases occur. The incidence rate is given as the number of new cases in a defined period of time divided by the population at risk of experiencing the event during this period, expressed as person-time or person years at risk.

$$\text{prevalence} = \frac{\text{number of persons with specific disease at a given point in time}}{\text{persons under risk at this time}}$$

$$\text{incidence} = \frac{\text{number of new cases within a defined period}}{\text{person years at risk within this period}}$$

A high prevalence can be associated with a low incidence (e.g., diabetes) or a low prevalence with a high incidence (e.g., infectious diseases). The link is the duration of the disease.

To give an example of **prevalence** a study investigating the association between phenacetine and renal disease will be considered. 7311 women were asked about the use of phenacetine, and 623 women responded positively. 39 out of the 623 women had a renal disease, leading to a prevalence of 6.3% ($=39/623 \cdot 100\%$). Among the remaining women not taking the drug a random sample of 621 women was selected. Out of this group 19 had a renal disease, denoting a prevalence of 3.1% ($=19/621 \cdot 100\%$) among the non-users.

To calculate the incidence the number of newly diagnosed cases as well as the individual observation period of time is important. The start of the observation is either a fixed point in time or the time at which the individual entered in the study, e.g. starting the job with a specific exposure, whichever occurred later.

The end of the observation is defined by one of three events:

- the person is diagnosed with the disease
- the end of observation is reached
- the person drops out.

To give an example for the incidence a study from Denmark can be used in which the cancer incidence of workers exposed to 2,4-D, 2,4,5-T and dioxin, i.e. 2,3,7,8-TCDD, during the production of herbicides was estimated. In a cohort of 3390 workers 159 were diagnosed with cancer between 1964 and 1981. The person years at risk of the 3390 workers summarized to 49.879 years, or an average of 14.7 years per worker. The incidence rate per year is calculated as $159/49.879 = 0.0032$ per person year or 3.2 cases per 1000 person years. It is important to note that the term 1000 person years is equivalent to 100 persons observed over 10 years or 1000 persons observed over 1 year.

The mortality rate is calculated in the same way as the incidence. Within the extended follow-up of the British-doctor study, which ran from 1950 until 1978, among the smokers about 195 doctors died of lung cancer with 109.386 person years at risk, leading to a lung cancer mortality rate of 1.78 per 1000 person years ($=195/109.386 \cdot 1000$).

Comparison of the Rates

In order to investigate the association between a certain factor and the disease one has to compare either the prevalence or the incidence between exposed and unexposed or high and low exposed persons.

The difference between two groups can be described by the ratio of either the prevalence, or incidence or mortality and is denoted as risk ratio or **relative risk** (*RR*).

In the British-doctor study the mortality rate among the smokers was 1.78 per 1000 person years. Among the non-smokers the corresponding mortality rate was 0.086 per 1000 person years. The relative risk for smokers dying from lung cancer compared to non-smokers is therefore $RR = 1.78/0.086 = 20.7$. The interpretation of the relative risk is that the mortality rate for smokers dying of lung cancer is 20.7 times higher than for non-smokers.

Beside the relative risk there are some other measures to describe the difference between exposed and unexposed persons. Two of them are the **excess risk** (*ER*), which is the difference between both rates, and the **attributable risk** (*AR*), which denotes the proportion of events among exposed persons which are due to exposure.

I_1 : incidence rate in the exposed group

I_0 : incidence rate in the non-exposed group

Relative risk $(RR) = I_1/I_0$

Excess risk $(ER) = I_1 - I_0$

Attributable risk $(AR) = \dfrac{I_1 - I_0}{I_1} = \dfrac{RR - 1}{RR}$

In the British-doctor study the excess risk gives a value of 1.69 per 1000 person years (=1.78 – 0.086 per 1.000 person years). The attributable risk is $AR = 95.2\%$ (=1.78 – 0.086/1.78). The interpretation is that 95.2% of all lung cancer deaths among the smokers are due to smoking (195 · 95.2% ≈ 185 deaths).

The comparison of the rates for the exposed and unexposed persons gives an average value for the relative risk. More information can be drawn if the exposed group is divided with respect to the exposure and a dose–response relationship can be established.

In the British-doctor study the smokers are classified into certain categories with respect to the cigarettes smoked per day. In Table 4.15 the relative risks for certain categories compared to non-smokers are given. This analysis shows an increase in the risk depending on cigarette consumption.

4.4.3 Standardization

> **The exposed and unexposed groups may also be different in other factors, like age, gender, smoking, etc. In this situation an increase in the relative risk cannot immediately be attributed to the exposure. One has to adjust for the differences in the other factors. Two approaches are possible: direct and indirect standardization.**

Direct Standardization

The procedure will be explained in relation to age.

The data are divided into several age groups ($i = 1, \ldots, I$). The incidence or mortality rates are calculated within each age group separately for exposed and unexposed persons.

Afterwards the rates are summed using weights w_i for the different groups. Mostly the weights w_i are determined by the proportion of person years of that particular age group per total person years.

Importantly the same weights w_i are used for the summation of the rates for the exposed and unexposed groups, which gives the standardized rates.

Standardized incidences rate: $SI = \Sigma w_i I_i$, where I_i = incidence rate in age groups ($i = 1, \ldots, I$) and w_i = weight of age group i ($\Sigma w_i = 1$).

The ratio of the standardized incidence rates between the exposed and unexposed persons is called the relative risk adjusted for age.

Example Again the British-doctor study will be used. The smokers of 30 cigarettes or more per day are compared to the non-smokers. The unadjusted relative risk is 46.29 (see Table 4.15).

If the data are separated into three age groups, the following result is obtained (see Table 4.16).

Table 4.15 Results of the British-doctor study.

Cigarettes/day	Lung cancer	Person years	Mortality rate · 1000	RR
0	6	69.905	0.086	1
1–9	7	14.877	0.471	5.48
10–19	34	35.037	0.970	11.31
20–29	88	42.862	2.053	23.92
≥30	66	16.610	3.973	46.29

Table 4.16 Results of the British-doctor study separated into three age groups.

	Non-smokers		Smokers (≥30 cigarettes/day)		
Age group	LC*	Σt*	LC*	Σt*	w_i*
<50	0	33.679	3	5.881	0.46
50–64	3	27.380	34	8.682	0.42
≥65	3	8.846	29	2.047	0.12
Σ	6	69.905	66	16.610	1.00

*LC, lung cancer; Σt, sum of person years at risk; w_i, weight of age group i.

The standardized mortality rate among the smokers (SM_s) is given as:

$SM_s = 3/5.881 \cdot 0.46 + 34/8.682 \cdot 0.42 + 29/2.047 \cdot 012 = 3.649$ per 1000 person years

The standardized mortality rate among the non-smokers (SM_{ns}) is 0.088 per 1000 person years. Therefore, the relative risk adjusted for age is 41.29 (=3.649/0.088), which is somewhat smaller than the unadjusted relative risk of 46.29.

Another way to adjust for certain factors is to use a regression model, like the logistic regression or the Cox model, taking into account the various factors.

Indirect Standardization

The incidence or mortality rate of a certain group can be compared with a standard population, e.g. the population of a certain area or country. This approach is very common in occupational epidemiology in order to describe the risk of a certain group of workers in relation to the total population. The corresponding rates for the population are usually available. The rates for the population are given separately for age groups divided into 5-year intervals for gender and calendar time. Based on these rates the so-called expected number of events can be calculated using the observed person years at risk. The ratio of the number of observed to expected events is called the standardized incidence ratio (SIR) or standardized mortality ratio (SMR) adjusted for age, gender and calendar time.

Example The mortality rate of workers exposed to silica dust who received worker's compensation from the stone and quarry industry for contracting silicosis on the job between 1988 and 2000 has been investigated. A cohort of 440 workers was enrolled and followed up until the end of 2001. Within this period 144 workers died. Based on the mortality rates of Germany 74.35 deaths could be expected, leading to an SMR of 1.94 (=144/74.35). Therefore, nearly twice as many deaths were observed compared with the mortality rates for Germany. The calculation can

4.4.4 Types of Epidemiological Studies

> In general there are two types of epidemiological studies: observational and experimental.

Observational studies are further divided into cohort and case-control studies (Figure 4.8). Within a **cohort study**, also called a prospective study, a defined group of persons are followed up over a certain period of time. In the British-doctor study all physicians willing to participate were enrolled in this study. The whole cohort can further be divided into different subcohorts, e.g. with respect to the level of exposure, age, gender, etc. Incidence and mortality rates can only be determined within a cohort study.

The span of time between the onset of exposure and the diagnosis of a disease is called the latency period. For mesothelioma, the latency period is a least 35 years. In the design of a cohort study to investigate the health effect of a certain factor one has to keep in mind that the follow-up period has to be long enough to avoid false-negative results. The only way to shorten the follow-up period is to transfer the onset of the observation into the past. This approach is called a historical cohort study and is commonly used in occupational epidemiology.

In a **case-control study**, also called a retrospective study, persons with a defined disease (cases) are compared with disease-free persons (controls). To investigate the association between smoking and lung cancer, patients with and without lung cancer are asked about their smoking habits. This type of study is faster and more efficient than cohort studies if the disease is rare or has a long latency period. The disadvantage of this type of study is that no incidence rate can be estimated and therefore no relative risk can be calculated. Nevertheless, case-control studies permit an approximation of the relative risk.

Assume we have a cohort study with P_1 exposed persons followed up over an average of t years. If a certain number (a) out of the P_1 persons develop the disease, the incidence is given as $I_1 = a/(P_1 \cdot t)$. The incidence among the non-exposed group is expressed as $I_0 = b/(P_0 \cdot t)$. The numbers of new cases in both cohorts are denoted by **a** and **b**, and the persons without the disease

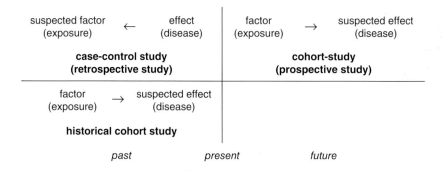

Figure 4.8 Design of a cohort study and a case-control study.

in both cohorts are called **c** and **d**, where $P_1 = a + c$ and $P_0 = b + d$. If the disease is rare the size of each cohort, P_0 and P_1, can be replaced by the persons without the disease (c and d) and both incidence rates can be approximated by:

$$I_1 = a/(P_1 \cdot t) \approx a/(c \cdot t)$$
$$I_0 = b/(P_0 \cdot t) \approx b/(d \cdot t)$$

The relative risk $RR = I_1/I_0$ can be estimated by

$$RR = I_1/I_2 \approx \frac{a/(c \cdot t)}{b/(d \cdot t)} = \frac{a/c}{b/d} = \frac{a/b}{c/d}$$

The numerator (a/b) is the ratio of the exposed and unexposed cases, called the odds of the cases (odds (K^+), where K^+ denotes the disease). The denominator (c/d) gives the odds among the non-diseased persons, the controls (odds (K^-)). The ratio of both odds is called the odds ratio (OR) and is an estimate of the relative risk. The great advantage of the odds ratio is that it can be calculated even with a random sample of the controls. If among both cohorts (exposed and unexposed) the same proportion (p) of non-diseased persons (controls) are sampled, e.g. $p = 10\%$, than the odds (K^-) is estimated by $(p \cdot c)/(p \cdot d) = c/d$. For estimating the relative risk it is necessary to get full information from all cases and only from a proportion of the controls. But it is important to note that the proportion p has to be the same in both cohorts. Random sampling is essential, otherwise one introduces a bias, which can lead to an over- or underestimation of the relative risk.

A case-control study can also be performed directly in sampling cases and controls. The main problem is the selection of the controls. The controls can either be sampled among patients with other diseases from the same hospitals (hospital controls) or from the general population (population controls). The main problem with this type of study is the selection bias. The participation rate among the hospital controls is usually much higher than among the population controls. The question is whether the hospital control is representative of the controls. A better way to perform a case-control study is within a cohort study. All cases are compared with a sample of controls. This approach is called nested case-control study.

Example In order to investigate the lung cancer risk among plutonium workers, a cohort of about 22,000 workers from a plant in the USA was identified (Brown et al., 2004). The internal lung dose was estimated for all lung cancer cases ($n = 180$) and a randomly selected group of workers without lung cancer ($n = 718$); 25 cases and 103 controls had no exposure, the remaining 165 cases and 615 controls were exposed. These figures give an *OR* of $(165/25)/(615/103) = 1.11$. The advantage of this approach is that for estimating the lung dose a relatively small sample is adequate.

In addition to observational studies, it is possible to perform an experimental study. For example, one can evaluate the effect of a certain intervention. This type of study is similar to a controlled clinical trial. Subjects are randomly assigned to the intervention or the control group. An example is the Women's Health Initiative study. Within this study the risks and benefits associated with the use of hormones in healthy post-menopausal women were investigated. About 17,000 women were enrolled. About half of them received estrogen plus progestin, whereas the other half received placebo. After a follow-up of 5 years it was shown that treatment resulted in a decrease in fractures to the hip and other fractures ($RR = 0.76$ for hormone users).

However cardiovascular diseases and cancer showed increased incidence rates ($RR = 1.29$ resp. ($RR = 1.26$)) in hormone users.

The intervention can also be a certain diet or exercise program. An example is the MRFIT study, in which several preventive measurements for reducing the risk of myocardial infarction in high-risk patients were investigated.

Beside the cohort- and case-control studies there are two other types of studies: cross-sectional and ecologic studies. Within a cross-sectional study the exposure and the disease are ascertained at the same point in time. With these data one can estimate the prevalence. However, it remains unclear whether the exposure has an influence on the risk.

Within an ecologic study groups of persons rather than individuals are considered. In this type of study the correlation between a certain factor and a specific disease can be investigated.

4.4.5 Statistics

> **To assure the significance of the values determined for incidence, mortality rate, prevalence, relative risk, or odds ratio, statistical tests and the calculation of confidence limits are necessary.**

The interpretation of a $P\%$ confidence interval is as follows: with a probability of $P\%$ (mostly $P\% = 95\%$) the "true" value of the parameter of interest is within the interval.

Based on statistical considerations it follows that the logarithm of the incidence (I) is approximately normally distributed. If the observed number of events is denoted by d and the person years risk with person year, the incidence I is estimated by $I = d/\text{person years}$. The 95% confidence interval (95% CI) for the "true" incidence is given by

$$I \cdot \exp(\pm 1.96/\sqrt{d})$$

Example For the TCDD study from Denmark:

$$d = 159, \text{ person years} = 49.879, I = 3.2 \text{ cases per 1000 person years}$$

$$95\% \; CI: \; 2.7\text{--}3.7 \text{ cases per 1000 person years}$$

Relative Risk The incidence rates in the exposed and unexposed groups are denoted by I_1 and I_0, the number of observed events by d_1 and d_0. The 95% CI around the estimate of $RR \; (= I_1/I_0)$ is given by

$$RR \cdot \exp\left\{\pm 1.96 \sqrt{1/d_0 + 1/d_1}\right\}$$

In addition to the calculation of the 95% CI a statistical test can be performed. The hypothesis of interest is whether RR is equal to 1 or different from 1 ($H_0: RR = 1$ vs. $H_1: RR \neq 1$).

The test statistics can be derived as:

$$T = \ln RR / \sqrt{1/d_1 + 1/d_0}$$

If T is between ± 1.96 the null hypothesis (H_0: $RR = 1$) cannot be rejected at a significance level of $\alpha = 5\%$. In this situation the 95% *CI* covers the value of 1, the true value of *RR* under H_0. If T is larger than 1.96 or below -1.96, H_0 is rejected and the value of 1 is outside the 95% *CI*.

Odds Ratio The calculation of the exact confidence interval for *OR* is difficult, therefore some approximations are available. One formula is called a test-based approach and uses conventional χ^2 statistics for investigating the association between a factor and a disease. The 95% *CI* is calculated as:

$$OR^{1 \pm 1.96/\sqrt{\chi^2}}$$

In the example about the association between the exposure to plutonium and the lung cancer risk, an *OR* of 1.11 was observed. The corresponding χ^2 tests for proofing the hypothesis *OR* = 1 gives a value of $\chi^2 = 0.17$. This leads to a 95% *CI* of 0.69 to 1.77. Therefore H_0 cannot be rejected and no statistical significant association between plutonium and lung cancer can be observed.

Standard Mortality Ratio The approximation of the 95% *CI* is based on the formula for *RR*. The number of events among the unexposed (= d_0) is much larger than the number of events among the exposed (= d_1) and $1/d_0$ is approximately 0.

$$SMR \cdot \exp\left\{\pm 1.96/\sqrt{d_1}\right\}$$

There are also other formulas available.

4.4.6 Meta-analysis

Meta-analysis has gained great attention since the psychologist G. Glass introduced this term in 1976. The basic idea behind this method is to combine the results from several studies dealing with similar aspects in order to increase power. This approach is mainly used in combining the results from clinical trials but it is also used in epidemiology. It produces a weighted average of the studies included.

The critical part of a meta-analysis is to identify all relevant studies. The start is usually a sensitive and comprehensive search using various databases. The next step is to identify all studies which should be included or excluded in the analysis. The criteria used have to be defined *a priori* and have to be independent from the outcome of the studies, e.g. a study has to be excluded if the exposure is not clearly described.

In order to account for differences between the studies usually a so-called random effect model is applied. One has to select the outcome measure (SMR, RR, SIR, or OR). All studies are weighted by the inverse of the variance. One can combine cohort and case control studies. In cases of heterogeneity, certain factors, if available, can be used to explain the differences between the studies. The results are usually plotted in the form of a forest plot.

As an example the outcome of a recent analysis of the bladder cancer risk among dry cleaners is presented (Figure 4.9). There are three cohort and four case-control studies available.

The cohort studies show a relative risk of 1.46, the case-control studies result in an *OR* of 1.50 and all studies together give a *RR* of 1.47 with a 95% *CI* from 1.16 to 1.85.

One main concern is related to the lack of studies not included in the meta-analysis, which is called publication bias. This means that not all studies dealing with the same problem have been published.

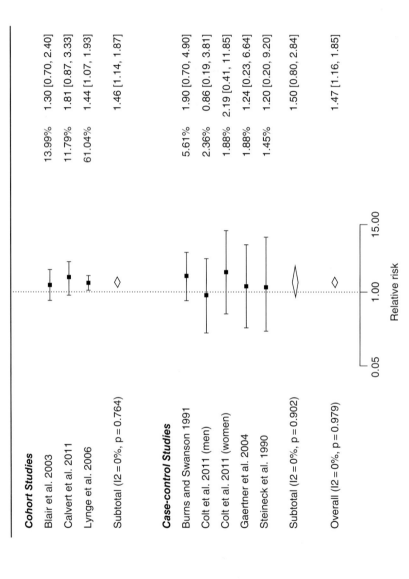

Figure 4.9 Forest plot of cohort and case-control studies included in the meta-analysis assessing the risk of bladder cancer in relation to occupation as a dry cleaner.

4.4.7 Bias, Confounding, Chance, Causality

> In epidemiological studies only statistical associations can be observed. In order to establish causality certain criteria mentioned first by Bradford Hill (1965) have to be fulfilled. Before proofing these criteria three other aspects have to be considered: bias, confounding, chance and causality.

A series of criteria which should be met when attempting to establish a "cause and effect" relationship between exposure to a certain compound and subsequent disease was postulated by Hill (1965). Prior to utilizing these criteria it is important to validate the data to insure that it cannot be impugned because of bias, confounding factors or chance.

Bias is any systematic error introduced into the process of collecting and evaluating data which leads to failure to detect the truth. Selection bias and information bias can be important sources of errors. Selection bias in case-control studies can arise from the improper selection of controls or from a participation rate of study subjects. Information bias can result from investigation of cases and controls using different methodology. The same holds for cohort studies in the event of misdiagnosis of disease or incorrect pronouncement of the cause of death. There is no way to correct for these errors.

Confounding represents a source of bias that occurs when an extraneous variable interferes with the determination of the effect of the variable under study. The extraneous variable may occur in both the control and experimental groups, and unless accounted for may lead to erroneous interpretation of the data. For example, a study is conducted to determine whether a black strain of mice metabolizes a drug more rapidly than a white strain of mice. The results do not indicate a difference. It is later determined that both strains were fed a diet containing a pesticide that induced the metabolism of the chemical in both strains. Repetition of the experiment using a non-inducing diet shows that the black mice metabolized the drug faster than the white. Thus, the results were confounded by the unknown administration of the pesticide to both species.

Chance may play an important role in the outcome of an epidemiological study in the absence of bias. For example, if a variety of endpoints are investigated within a cohort study and several of them are statistically significant, one is faced with a multiple comparison problem. If several tests are performed the type I error is inflated. The same problem occurs in case-control studies if more than one factor is considered. From the statistical point of view there are two ways to overcome this problem. One way to solve this problem is to select one primary endpoint, i.e. the disease or the factor of greatest interested. This analysis is called confirmatory. All other diseases or factors are considered as secondary endpoints. The outcomes of these diseases or factors can be used to generate new hypotheses. The other approach is to adjust the type I error by altering the number of tests (n) performed. The type I error for the single test is reduced to $\alpha^* = \alpha/n$, with α the overall type I error, e.g. $\alpha = 5\%$. The p value of a particular test has to be below α^* in order to be statistically significant at the α level.

Causality   If an association between a factor and a disease is statistically significant and bias, confounding or chance can be ruled out, the criteria for establishing causality include the following:

- *Temporality:* Taking appropriate latency periods into account, does exposure to a chemical precede the disease.

- *Consistency:* Consistency refers to the repeated observation of an association between exposure and outcome in different populations under different circumstances. This criterion can often help to distinguish between real associations and chance.
- *Dose–response relationships*: Strong support for a causal relationship can come from observing a biological gradient, i.e. a monotonic, increasing dose–response curve.
- *Strength:* A strong association is more likely to be causal than a weak association. However, the strength of the association may be modified by the level of exposure.
- *Experimental evidence:* This criterion refers to the situation in which an intervention is performed to reduce a risk and a decrease in the risk is observed.
- *Plausibility:* There should be a biologically plausible explanation connecting the exposure and the disease. Plausibility may depend on the extent to which the mechanism of action of the agent is understood.

4.4.8 Summary

The goal of epidemiology is to identify risk factors for certain diseases. Several types of epidemiological studies are available. The best-known types are cohort and the case-control studies. The association between a factor and a disease is described by the relative risk or the odds ratio. If a statistically significant association can be observed causality has to be proven. A variety of criteria are mentioned. Not all of them have to be fulfilled in order to establish causality, but the influence of bias, confounding or chance has to be ruled out.

Further Reading

Brown, S.C., Schonbeck, M.F., McClure, D., Barón, A.E., Navidi, W.C., Byers, T., Ruttenber, A.J. Lung Cancer and internal lung doses among plutonium workers at the Rocky Flates plant: A case-control study. *Am. J. Epidemiol.*, 2004, **160**, 163–172.

Doll, R., Peto, R. Cigarette smoking and bronchial carcinoma: dose and time relationships among regular smokers and lifelong non-smokers. *J. Epidemiol. Community Health*, 1978, **32**, 303–313.

Dubach, U.C., Levy, P.S., Mueller, A. Relationships between regular intake and urorenal disorders in a working female population of Switzerland. I. Initial results (1968). *Am. J. Epidemiol.*, 1971, **93**, 425.

Glass G. Primary, secondary and meta-analysis of research. *Educational Researcher* **5**, 1976, S. 3–8.

Hill, A.B. The environment and disease: association or causation? *Proc. R. Soc. Med.*, 1965, **58**, 295–300.

Last, J.M. (Ed.) *A Dictionary of Epidemiology*, Oxford University Press, New York, 1988.

Lynge, E. A follow-up study of cancer incidence among workers in manufacture of phenoxy herbicides in Denmark. *Br. J. Cancer*, 1985, **52**, 259–270.

Rothman, K.J., Greenland, S. *Modern Epidemiology*, Little Brown and Company, Boston, 1998.

Ulm, K. A simple method to calculate the confidence interval of a standardized mortality ratio (SMR). *Am. J. Epidemiol.*, 1990, **131**, 373–375.

Vlaanderen, J. et al. Tetrachloroethylene exposure and bladder cancer risk: a meta-analysis of dry-cleaning-worker studies. *Environ. Health Perspect.*, 2014, **122**(7), 661–666.

Women's Health Initiative Investigators. Risks and benefits of estrogen plus progestin in healthy post-menopausal women: principal results from the Women's Health Initiative randomized controlled trial. *J. Am. Med. Assoc.*, 2002, **288**, 2819–2825.

4.5 Alternatives to Animal Testing

Thomas Hartung

4.5.1 Introduction

> **In the last 150 years, chemists have synthesized about 85 million substances; this is the number of Chemical Abstracts on individual chemicals. More than 100,000 of these are used in products of daily consumption, in drugs, in cosmetics, in cleaners, in our food and clothing, and not least are found in our environment. Furthermore, there is an even larger number of natural substances we are in contact with and which can also represent a risk.**

When, 120 years ago, Bayer brought aspirin to the market – our oldest synthetic drug – there was no legislation as to safety assessments using animals. If at all, the producer was held liable for his products and associated problems. This did not always work out as positively as in the case of aspirin: Only a few days later, the same chemist, Felix Hoffmann (1868–1946), developed in the same company a cough syrup: heroin ... not every product is as harmless as aspirin. It is easy to understand that with each safety problem, the need for safety evaluations increases.

Only in the 1920s did the use of mice and rats in laboratories start on a large scale. Until then, it was considered odd that animals could mirror humans. However, it was soon convincing how quickly experiments with them could be done. These animals cost little, they reproduced quickly and many could be kept in small cages. This created a research hype similar to today's introduction of stem cells.

With every scandal, the toolbox of toxicology expanded in order to prevent similar incidents in the future. In the early 1930s, for example, in the USA, LashLure led to a scandal. This cosmetic was used to dye lashes permanently; unfortunately, the anilin dye in it led to severe inflammatory responses. More than 3000 reports of side-effects were collected, five women became blind and one fatality occurred. This led to the first legislation on cosmetics, which is now controlled by the USA by the Food and Drug Administration (FDA). Their employee John H. Draize (1900–1992) developed in 1944 the rabbit-eye irritation test, aka the Draize eye test, in which chemicals are put into the eye of the animal. Many people today perceive this as cruel, but over the last 70 years this test has prevented another case like LashLure. In this way, toxicology grew like a patchwork quilt. The thalidomide (Contergan) scandal of the 1960s, for example, prompted the introduction of reproductive toxicology.

4.5.2 The Birth of Doubt in Animal Experiments

Concerns about animal experiments are very old. In ancient Greece discussions on whether or not animals should be killed were recorded. In Germany in the 1920s, there were 700 animal welfare organizations.

> **In 1959, in England, Bill Russel and Rex Burch developed a concept that they called the 3Rs principle: we have to – wherever possible – reduce the number of animals (*reduce*) to achieve the same information, allieviate unnecessary pain and distress (*refine*), e.g. by using painkillers and anaesthesia, and, ultimately, substitute animal experiments (*replace*) when there is an alternative.**

This last principle in particular was largely utopic 60 years ago – cell culture or computer programs were still in their infancy and only a few could imagine that they would become successful. Over the last few decades, the commitment to the 3Rs by industry, academia and policy has led to a societal compromise with many of those citizens who would like to see the end of animal testing today rather than tomorrow. At the same time a credible investment into the abolition of animal experiments has been undertaken. In fact, the number of experiments on animals decreased substantially up to the turn of the century, when it was estimated to be two-thirds less than its peak in the mid-1970s. However, since then numbers have increased again, mainly because of new technologies for genetically modifying mice.

Beside the ethical argument for avoiding animal suffering, there are a number of other reasons why toxicology needs new (alternative) approaches that rely less on animals:

1. Most animal tests were never validated and their reproducibility is often astonishingly low, even for highly standardized and quality-assured guideline tests. For example, about 120 cancer bioassays were repeated with only 57% concurring results, and "OECD-validated" tests of the last decade, such as the uterotrophic assay for estrogenic endocrine disruptors, resulted in the infamous case of bisphenol-A with up to 10,000 times different potencies with both positive and negative findings. Similar variability was reported for acute fish toxicity. Most animal tests have simply never had their reproducibility tested.
2. The costs of individual animal safety tests are in the four- to seven-digit range (e.g. $1 million per substance for a cancer bioassay in the rat), which limits the feasible number of such tests. The comprehensive safety assessment of a single pesticide amounts to $20 million, which does not even include the costs to synthesize the 20 kg of any substance needed for such extensive testing.
3. Many animal tests take too long – a cancer bioassay can last more than 4 years from planning to results.
4. We are not 70-kg-rats – even mice and rats in general correspond to each other only 60–70% in safety tests.
5. Most toxicological methods were developed for drug safety testing, i.e. for substances with targeted biological activity, used in high doses in humans. These use scenarios cannot necessarily be translated to existing industrial chemicals or low-level contaminants in the environment.
6. New products such as biological drugs (e.g. human proteins and antibodies), cell therapies or nanoparticles often cannot be tested with traditional methods.
7. For a number of possible chemical health hazards we do not have animal tests, for example weak endocrine effects especially of the thyroid system, childhood asthma, immunotoxicity, autism, diabetes, atherosclerosis, and obesity. This does not mean that chemicals are the cause of these diseases, but their contribution can only be assessed if there are tests.
8. The combined effect of different chemicals is hard to test as traditional tests are too resource-intensive and testing all possible combinations is unaffordable.
9. Individual differences in the susceptibility of various groups of people, e.g. newborns, immunosuppressed or the elderly, cannot be tested in animals.

4.5.3 Early Successful Alternatives

A number of examples show that the 3Rs concept can succeed (Table 4.17). Since the 1920s, for example, the LD_{50} test has been used. Here, a lethal dose is determined that kills 50% of treated rats. Until 1989, 150 animals per substance were used (at each of seven test doses, 10 male and 10 female rats, and a control group). This represented an enormous animal use, especially since essentially all chemicals entering the market were tested. Based on this test, the poisons were classified and measures to protect workers and the transport of the chemicals proposed. In 1989, following intense analysis of historical data, agreement at OECD level was reached to drastically reduce the number of animals used. Since then, only groups of five animals of one sex have been used, reducing the number of animals from 150 to 45. In the 1990s, the next step was taken with a simple idea: Why treat all animals in parallel? Starting with one dose, a higher dose can then be used if the animals survive, and a lower dose if they die. At the same time, supportive data showed that groups of three animals were sufficient. In consequence, in 2002 three methods were internationally accepted that use on average 8–12 rats. From 150 to 45 to 8–12 animals – an enormous reduction. In addition, in one of these methods provisions were introduced to sacrifice the animals when signs of severe damage or likely death were observed and not to await the agonizing death of the animal. This is called *humane endpoints* and is an example of refinement, the second R.

Another example is allergic sensitization of the skin (contact dermatitis), which traditionally was done on guinea pigs. With the local lymphnode assay (LLNA), there has been, since 2001, an internationally accepted test in mice, which stops at the characteristic stage of lymphnode swelling and spares the animals the ensuing contact dermatitis. Furthermore, the test reduces animal numbers from 20 to 8–16 (this is an example of refinement and reduction).

In addition, more and more computer-based methods (*in silico* methods) have emerged which complement or sometimes also replace animal tests. A lot of progress has been made recently with read-across, an approach where chemical similarity is used to justify the assumption that substances share toxic or non-toxic properties. So far, no *in silico* method has been formally validated as an alternative to animal testing, but they are very successful, especially in combination with other testing strategy approaches.

4.5.4 The Replacement of Animal Tests is Possible

Increasingly, it has been possible to completely replace specific animal tests in toxicology – the third R for replacement. For example, human skin, which can be obtained from surgeries, can be cultured and maintained in the laboratory. Tiny skin samples are sufficient to grow square meters of artificial skin. These methods were originally developed for skin transplantations in burn patients. Very quickly, however, companies realized that these methods could also be used for testing of chemicals. In fact, it has been shown that such artificial skin is as well suited to test for skin lesions as rabbits. The respective international test guidelines have been accepted following formal validation (Table 4.17). This represents a milestone not only for the cosmetic industry, but most importantly because it represents a proof of principle that well-established animal tests can be fully replaced by alternatives in an international consensus. Similarly, a test for phototoxicity was successful, in which this chemical hazard is tested in cultures by comparing them with and without UV irridation (Table 4.17).

Another example of the successful replacement of animal tests is the quality control of drugs for pyrogenic (fever-inducing) contaminations, especially endototoxins from Gram-negative bacteria. First, alternative methods based on the coagulation of horsehoe crab blood and then new

Table 4.17 ECVAM validations and OECD test guidelines of alternative methods until 2017.

Area	Test	ECVAM validation	OECD acceptance (test guideline, TG)
Kinetics	Skin absorption	–	TG 428, 2004
	P450 induction	Ongoing	–
Eye irritation	Bovine corneal opacity and permeability (BCOP)	2007	TG 437, 2009, revised 2013
	Isolated chicken eye (ICE)	2007	TG 438, 2009, revised 2013
	Cytosensor microphysiometer (CM)	2009	–
	Fluorescein leakage (FL)	2009	TG 460, 2012
	Reconstructed human tissue (RhT)-based test methods: Skin Ethic human corneal epithelium and EpiOcular™ eye irritation test	2014	TG 492, 2015
	Low-volume eye test (LVET)	2009	TG 491, 2015
Phototoxicity	3T3-NRU in vitro phototoxicity test	1997, 1998 (UV filter)	TG 432, 2004
Skin corrosion	Transcutaneous electrical resistance test, EpiSkin™, piDerm™, SkinEthic™, EpiCS™ (EST-1000)	1998, 1998, 2000, 2006, 2009	TG 431, 2004, revised 2014
Skin irritation	EpiSkin™, EpiDerm™ SIT (EPI-200), modified EpiDerm (EpiDerm-200), SkinEthic™ reconstructed human epidermis	2007, 2009, 2008, 2009	TG 439, 2010, revised 2013
Skin sensitization	Local lymphnode assay (LLNA, refinement method)	2000	TG 429
	Reduced local lymphnode assay and non-radioactive variant	2007, – (USA only)	TG 442 a and b, 2010
	Direct peptide reactivity assay (DPRA)	2012	TG 442 c, 2015
	KeratinoSens™	2014	TG 442 d, 2015
	Human cellline activation test (h-CLAT)	2015	TG 442 e, 2016
	U937 skin sensitization test (U-SENS™)	2016	Draft OECD TG (2016)
Acute toxicity	Fixed-dose procedure (refinement method)	2007	TG 420, 2001
	Acute toxic class method (refinement method)	2007	TG 423, 2001
	Up-and-down procedure (refinement method)	2007	TG 425, 2001
	Acute toxic class method for acute inhalation toxicity	–	TG 436, 2009
	Colony-forming unit granulocyte/macrophage (CFU-GM) test	2006	–
	3T3 neutral red uptake cytotoxicity test	2011	Guidance Document 129, 2010

Endpoint	Test	Year validated	OECD TG
Carcinogenicity	Cell transformation assay	2012	TG 214, 2015
Genotoxicity	Ames test	–	TG 471, 1997
	Mammalian chromosome aberration test	–	TG 473, 1997
	Mammalian gene mutation test	–	TG 476 (under revision), 1997
	Mammalian bone marrow chromosome aberration test	–	TG 475, 1997 (revised 2014)
	Transgenic rodent gene mutation test with somatic and germ cells	–	TG 488, 2011 (revised 2013)
	Mammalian erythrocyte micronucleus test	–	TG 474, 1997 (revised 2014, 2016)
	Micronucleus test	2006	TG 487, 2010
	Comet *in vivo* assay	(Japan)	TG 489, 2014
(Sub-)chronic toxicity	One-year dog study, obsolete for pesticides	2006	n/a
Reproduction toxicity	Embryonic stem cell test	2002	–
	Micromass test for embryotoxicity	2002	–
	Rat embryo test (refinement method)	2002	–
Endocrine disruptors	Stably transfected human estrogen receptor-α transcriptional activation assay for the detection of oestrogenic agonist activity of chemicals	(OECD, US EPA)	TG 455, 2009, updated 2015 and 2016
	H295R steroidogenesis assay	(OECD, US EPA)	TG 456, 2011
	BG1Luc estrogen receptor Transactivation *in vitro* assay to detect estrogen receptor agonists and antagonists	(USA, 2012)	TG 457, 2012
	Transcriptional assay for the detection of estrogenic and anti-estrogenic compounds using MELN cells	(validation stopped 2009 after consultation with test submitter)	Ongoing
Acute fish toxicity	Upper threshold concentration step-down approach	2006	TG 203, guidance 126, 2010
	Zebrafish embryotoxicity test	2014	TG 236, 2013
Pyrogenicity	Monocyte activation tests	2006	Not applicable (European and US pharmacopoia, FDA)

Note: Not all alternative methods have undergone formal validation and some validated tests have not (yet) been accepted for regulatory guidelines. The table does not include ICCVAM (USA) and JaCVAM (Japan) validity statements and only refers to the ECVAM ones even for studies carried out by these other validation bodies. Only OECD Test Guidelines (TG) are referenced, not other regulatory uses and national test guidelines. Alternative methods for vaccine testing are not included.

methods based on the human fever reaction (monocyte activation tests) have been introduced following successful validation in 2006 and acceptance in 2010 in the USA and Europe. In fact, these tests are even superior to the rabbit assay in some aspects: they are quantitative, better reproducible, faster, cheaper and include a positive control.

Enormous expectations are linked with the integrated combination of different tests, which are called **integrated test strategies** or **integrated approaches to testing and assessment** (IATA) by the OECD. However, the ways to combine and and validate such strategies are in their infancy, with skin sensitization a forerunner.

4.5.5 Validation of Alternative Methods: Animal Welfare must not Trump Patient and Consumer Safety

The prerequisite for the use of alternative methods was that safety standards for consumers must not be lowered. Thus, the concept of formal validation was introduced. In 1991, for this purpose, in Ispra, Italy, the European Centre for the Validation of Alternative Methods (ECVAM), part of the European Commission's Joint Research Centre, was created. To date about 50 alternative methods have been validated (Table 4.17) and several others are currently in ring trials for evaluation. American (1995), Japanese (2005), Korean (2009) and Brazilian (2014) and Canadian (2017) counterparts to ECVAM followed, and currently there is discussion to create a similar validation body in China.

In order to validate alternative methods, several features have to be demonstrated:

1. **The method has to be sufficiently defined:** beside the defined protocol, this includes what can it be used for and what not?
2. **Does the method have a scientific basis that represents the mechanism in animals and humans according to our current scientific understanding?**
3. **Is the method reproducible, i.e. does the repetition of a test first in the same and then in other laboratories lead to the same results?**
4. **Are the results relevant, which in general means can the results of the traditional animal test be predicted?**

The last aspect is certainly the most critical and problematic, especially as most animal tests have never been validated or shown to be relevant for humans. Only rarely are there data from poisoning centers or clinical tests, for example, which can help here. However, we can carry out the same animal test with different animal species and assess, for example, how well mice predict rats or hamsters guinea pigs. Obviously, there is no reason to assume that any of these species should predict human effects better than they predict effects in each other. Rodents like mice and rats are much closer to each other that they are to humans. Even non-human primates are quite distant in evolutionary terms. All together, the correspondence of results between species is usually only 60–70%. What can we do? Traditionally, two approaches were taken: testing is done in two animal species and tests are made precautionary, e.g. by testing extremely high doses of the chemicals. Both approaches, however, boost false-positive results, which means we exclude many substances – often after enormous development efforts – unnecessarily from further development and use because of false allegations of toxicity. For drugs, it is often economical to show by investigative toxicology that certain animal results are not relevant for humans to allow the further development of a substance, although this prolongs the time to market and decreases the time when the drug can be sold under patent. The considerably lower profit margins for industrial chemicals does typically not allow this as, at least for new substances, it is usually not economical.

4.5.6 How Reliable are Animal Tests?

Animal tests have made the world a safer place, but it is problematic if we eliminate more and more substances because of possible but not certain problems. The example of aspirin is quite telling: aspirin would fail most of today's safety tests. Aspirin kills half of the rats (LD_{50}) at doses that are the maximally used human doses. Aspirin is irritating for eyes, skin and lungs. In some tests, aspirin is genotoxic, and though it has not induced cancer in the respective animal studies, it increased the carcinogenicity of other substances when given at the same time. Furthermore, aspirin leads to embryonic malformations in essentially all animal species tested (rats, mice, rabbits, cats, dogs and monkeys). All these are tests that are today commonly required for drugs and pesticides, and such findings would challenge the further development and marketing of a substance. Aspirin has been extremely well studied – 25,000 scientific articles are available; one trillion pills have been swallowed – and this gives the impression that its toxicological results are not very relevant for humans.

It would be very difficult to bring aspirin to the market today as a new drug, which illustrates the downside of our safety needs. This too is killing people – a drug that is stopped unnecessarily in development cannot cure patients.

> **Animals are just a model of humans and as coined by George Box in 1970: "All models are wrong, some are useful". This holds true for both animal models and alternative methods. The *art of toxicology* requires a high safety standard to be achieved by the proper combination of different information sources, including traditional and new alternative methods, and certain precautions. This includes the comparison of species with respect to substance kinetics (adsorption, distribution, metabolism and excretion) and the increasing mechanistic understanding of how a substance damages the organism. Most importantly, we need to remain aware that these are just models, which represent only a part of reality.**

Despite all this, testing in animals is not able to prevent all hazards. When substances proceed to human clinical trials after animal safety testing, 10–30% show unintended side-effects that stop further development. We are not 70-kg rats. Uncertainty remains after animal tests in both directions – there are false-positive and false-negative results. However, the alternative methods come with similar limitations – except on the ethical side. It is key that we analyze these limitations and combine methods properly.

4.5.7 The Animal Test Ban for Cosmetics in Europe as an Engine of Change

The seventh amendment of the European cosmetics legislation from 2002 introduced a testing ban for finished products in 2004 and for their ingredients in 2009 and 2013 for more complex health effects. Europe represents 50% of the world market for cosmetics and this created pressure for companies world-wide to replace animal tests, especially since the European legislation does not distinguish where the animal test was done; if the ingredient was tested after the deadline, there is a marketing ban for the respective products. This law has made the cosmetic industry an engine of change. Several of the methods in Table 4.17 are the result of these efforts. Nevertheless, at the deadline in 2013, there were no full replacements for a number of human health effects by alternatives, as confirmed by independent assessment. This prompted the development of a strategic roadmap to meet these challenges, but this still largely lacks implementation.

4.5.8 "Toxicological Ignorance": the European REACH Program as a Driver for Alternative Methods

An enormous problem is that testing of chemicals using animals is too expensive and takes too long: Testing whether or not a substance produces cancer takes at least 4 years at a cost of more than $1 million per animal species. No surprise, therefore, that in 30 years under the European Dangerous Substance Directive only 14 out of 5000 new industrial chemicals were tested for this health effect; of the more than 100,000 chemicals on the market, only about 3000 have been assessed. This has been called "toxicological ignorance". The European REACH legislation tackles the problem, but with traditional animal testing the necessary throughput cannot be achieved. More than 30,000 substances have to be registered with up to 16 animal tests depending on their production and marketing volume. There are simply not enough laboratories to suddenly test that many substances. The legislation therefore calls for alternative methods, but with registration ending in 2018, there was not much time to develop such new approaches and validate them. Notably, the USA reauthorized their Toxic Substance Control Act in 2016, which is a similar – though much smaller – program and is currently implemented.

> **REACH is the largest chemical safety testing program in history. Remarkably, one of three goals of the legislation defined in Article 1 is the promotion of alternative methods. Such alternative methods for some hazards (skin and eye irritation, skin corrosion, acute fish toxicity, genotoxicity, skin sensitization), which do not need live animals, are available; for others (systemic and chronic toxicities) there are at best strategies, how testing can be at least reduced or in the future avoided. However, we lack internationally accepted approaches for the most animal-consuming methods, such as repeated-dose and reproductive toxicity as well as carcinogenicity.**

Industry is largely employing *read-across*, i.e. the principle that similar chemicals have similar properties. In a weight-of-evidence argument typically based on structure but increasingly also similar biological properties, this fills data gaps without testing. The best practice of how to do this and how to assess the uncertainty of this approach is still under discussion, but so far 75% of REACH registrations make use of read-across. More automated read-across has become available due to the emerging large databases of historical test data and machine-learning, but these await formal validation.

In addition to the ethical discussion prompting the avoidance of animal testing, there is increasingly also a practical one: we cannot assess quickly enough and with certainty what is coming to the market. The US National Academy of Science suggested in 2007 that for this reason we have to move away from animal testing and find a new toxicology (*Toxicology for the 21st Century*). This has created an intensive discussion on how to implement this. Similar to the Human Genome Project, we talk of a Human Toxicology Project or the elucidation of the Human Toxome. The resources are still limited but this movement promises a successful way to move forward.

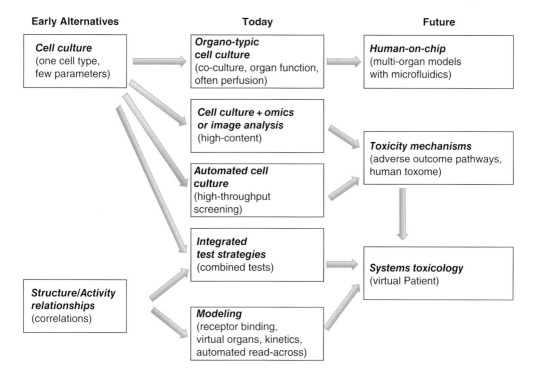

Figure 4.10 The evolution of alternative methods in toxicology.

4.5.9 Outlook

Where are the hopes and expectations with these new technologies? It has been claimed that knowledge doubles every 7 years in medicine and the life sciences. In this case, we have now 1000 times more knowledge than when most animal tests were designed. The revolution of biotechnology and informatics has brought new methods into our laboratories (Figure 4.10). Today, we have cultures for essentially all human organs and tissues. New approaches allow organo-typic cultures in three dimensions combining different cell types, often including perfusion (**organ-on-chip**). With the reproduction of some organ functionalities, we now speak of **microphysiological systems**. The combination of such organs with human-on-chip models has already begun.

At the same time, we can get more and more information from these cell cultures, for example based on gene chips, which can easily address the activity of all 27,000 human genes (transcriptomics) in a given model in response to a given toxicant. Similar toxicants lead to similar responses (signatures). This can also be studied on the level of produced proteins (proteomics) or to determine the impact on metabolism (metabolomics), i.e. using mass spectroscopy or NMR, of large numbers of enzyme proteins or metabolites. Increasingly, high-content imaging is employed, i.e. automated and computer-analyzed microscopy, which can be combined with cell functional assays and interventions with inhibitors, genetic manipulation and others.

In parallel, computer-aided methods have flourished: three- and four-dimensional receptor modeling, modeling of substance kinetics, automated read-across or the creation of virtual organs are just some examples of the many approaches in computational toxicology.

Increasingly, different *in vitro* and *in silico* approaches are now combined, as they are both fast and affordable in comparison to animal tests. These integrated testing strategies contrast strongly with the current "one-size-fits-all" animal test guidelines. Their complexity challenges validation and acceptance as well as the need to maintain flexibility for users.

We already know many pathways, how cells work and how these can be perturbed by exogenous chemicals. The term adverse outcome pathway (AOP) has been coined by OECD. Very sensitive analytical methods, robotized testing and complex measurement methods (omics and high-content imaging) allow a lot of information to be obtained and modern computers provide the capacity to analyze them. The term "systems toxicology" has been coined and describes the attempt to organize existing knowledge and large datasets by computer modeling.

The strategies in the USA and Europe differ significantly. In the USA, large programs by the EPA, NIH and FDA have started to carry out robotized tests of thousands of substances in hundreds of biological and chemical tests, and the results are made public. This broad characterization of substances is currently used predominantly for prioritization of testing. In the area of screening for endocrine disruptors there is emerging use of such data in regulatory decision-taking.

In Europe, the approach is mainly to replace traditional tests one by one; for the introduction of modern technologies into regulatory practice there is a larger gap. They are applied, if at all, for weight-of-evidence arguments, i.e. the valuing and balancing of incomplete and contradictory information. In general, the European approach is to replace animal testing, especially in cell cultures for organ-specific identification of hazards and the estimation of dose–response relationships with the no observed adverse effect level (NOAEL) via modeling and estimation of the external dose, which corresponds in the intact organism to the NOAEL level *in vitro*. This allows estimation of whether the relevant concentration can be reached in the organism at all.

Risk assessment based only on methods that do not involve animals is still utopic, but modern cell phones and the modern internet were utopic 20 years ago too, as they were then only in their infancy. In research laboratories the technologies for a new toxicology are already available. They now have to find their market in order to undergo the necessary optimization and implementation. This is then mainly an engineering challenge.

There clearly is a market. Every year, industry world-wide is spending $3 billion on animal tests and probably already several times this on *in vitro* testing. The European REACH program for existing chemcials alone will collect in only one decade data equivalent to $10 billion spent on animal testing. And this is only the start: nanoparticles, genetically modified food, cell therapies, electronic cigarettes with 10,000 flavors etc. – new products lead to challenges to control their risks. The market demand for toxicologists is high, hardly met by current education programs, though some universities have started to expand their programs.

Whether the new approaches will make consumer products safer remains to be shown. The pressure to develop something new comes from making clear the need for such technologies. The use of animals is ultimately a technology, the systematic use of an approach to solve problems. Blind faith in the relevance of animal tests makes animals a ritual sacrifice to appease the gods for the future of our products.

A realistic evaluation of the strengths and weaknesses of animal tests allows their targeted use, their pragmatic complementation and exchange, safeguarding that consumers receive safe products with acceptable animal use. It also allows producers to understand safety information gaps and toxicologists to develop the necessary new approaches. This will automatically reduce animal consumption in this field. In the end, consumer and patient safety is the primary concern, but animal welfare should not be just a secondary one.

4.5.10 Summary

Animal testing, especially in rodents, over the last century was the principal experimental approach in toxicology. Bill Russel and Rex Burch and their visionary 3Rs-principle (reduce, refine, replace) from 1959 resulted in a movement from the 1970s toward the replacement of animal tests in safety tests for predominantly animal welfare reasons. About 50 alternative methods have since been internationally validated and increasingly used. In recent years, increasing doubt about the usefulness of animal testing has increased with respect to human relevance, reproducibility, costs and duration. At the same time, new technologies have allowed a new approach based on the mechanistic understanding of toxicity. Under the slogan *Toxicology for the 21st Century*, a possibly revolutionary paradigm change in toxicology is on the way.

Further Reading

Adler, S., Basketter, D., Creton, S. et al. (2011) Alternative (non-animal) methods for cosmetics testing: current status and future prospects - 2010. *Arch. Toxicol.*, **85**, 367–485.

Basketter, D.A., Clewell, H., Kimber, I. et al. (2012) A roadmap for the development of alternative (non-animal) methods for systemic toxicity testing. *ALTEX*, **29**, 3–89.

Bouhifd, M., Andersen, M.E., Baghdikian, C. et al. (2015) The human toxome project. *ALTEX*, **32**, 112–124.

Hartung, T. (2009) Toxicology for the twenty-first century. *Nature*, **460**, 208–212.

Hartung, T. (2010) Food for thought … on alternative methods for chemical safety testing. *ALTEX*, **27**, 3–14.

Hartung, T. (2013) Look Back in anger – what clinical studies tell us about preclinical work. *ALTEX*, **30**, 275–291.

Hartung, T. (2017) Opinion versus evidence for the need to move away from animal testing. *ALTEX*, **34**, 193–200.

Hartung, T. and Hoffmann, S. (2009) Food for thought on … in silico methods in toxicology. *ALTEX*, **26**, 155–166.

Hartung, T., van Vliet, E., Jaworska, J., Bonilla, L., Skinner, N. and Thomas, R. (2012) Systems toxicology. *ALTEX*, **29**, 119–128.

Hartung, T., Luechtefeld, T., Maertens, A. and Kleensang, A. (2013) Integrated testing strategies for safety assessments. *ALTEX*, **30**, 3–18.

Leist, M., Hasiwa, M., Daneshian, M. and Hartung, T. (2012) Validation and quality control of replacement alternatives – current status and future challenges. *Toxicol. Res.*, **1**, 8, DOI: 10.1039/C2TX20011B.

Leist, M., Hasiwa, N., Rovida, C. et al. (2014) Consensus report on the future of animal-free systemic toxicity testing. *ALTEX*, **31**, 341–356.

NRC (2007) *Toxicity Testing in the 21st Century: A Vision and a Strategy*. Washington DC, USA, National Academies Press.

Patlewicz, G., Ball, N.K., Becker, R.A. et al. (2014) Read-across approaches – misconceptions, promises and challenges ahead. *ALTEX*, **31**, 387–396.

van Vliet, E. (2011) Current standing and future prospects for the technologies proposed to transform toxicity testing in the 21st century. *ALTEX*, **28**, 17–44.

4.6 Omics in Toxicology

Laura Suter-Dick

4.6.1 Introduction

In the last decade, holistic approaches or "omics" technologies have been changing the way toxicology studies are performed and safety data are generated and interpreted. When speaking about "omics" technologies, we usually mean genomics, proteomics and metabolomics. These new approaches have been propelled by technological advancements in the last 20 years, which provided the means to analyse a substantial part of specific analytes (transcripts, proteins and metabolites) in parallel at an affordable cost. This has enabled us to predict toxicity and to increase the understanding of the molecular events underlying a given toxicity. Moreover, toxicogenomics is helping scientists to integrate toxicology into earlier discovery phases by including sensitive molecular parameters that should help toxicologists recognize liabilities at lower doses (pharmacological rather than toxicological doses) or after short exposure times (acute rather than chronic exposures).

4.6.2 Concept of Toxicogenomics

> *Toxicogenomics* emerged in the late 1990s as a contraction of *toxicology* and *genomics*. The development of gene expression microarrays, also known as chips, enabled simultaneous analysis of the transcription levels of thousands of genes to assess global gene expression (called *genomics or transcriptomics*). The application of these novel technologies in toxicology gave rise to toxicogenomics as a discipline. The more recently developed next-generation sequencing technologies are now state of the art for gene expression analysis. In general, toxicogenomics is understood as the use of genomics (or transcriptomics) for the study of toxic effects, while proteomics and metabolomics are often included in the broader definition of the toxicogenomics discipline.

Originally, toxicogenomics referred to the analysis and quantification of the expression of protein coding transcripts (mRNAs). These so-called gene expression analyses were used to study adverse effects of chemicals and pharmaceuticals that perturbed biological systems. Recently, much progress has been made in the research of non-coding RNA species. The quantification of miRNAs (micro RNAs) and other non-coding RNAs is now also included under the umbrella of toxicogenomics. Further progress in the field has been achieved through advances in protein analytics and the availability of protein sequence databases. With this, the use of two-dimensional electrophoresis, protein chips, and diverse mass spectroscopy applications for peptide identification has made *proteomics* analysis in large scale possible. Thus, large numbers of newly synthesized proteins and biologically relevant post-translational modifications (e.g. phosphorlylated proteins) in a given biological sample can now be evaluated. In the field of endogenous metabolites, *metabolomics* using proton nuclear magnetic resonance (H-NMR) and/or liquid or gas chromatography coupled with mass spectrometry (LC-MS/GC-MS) makes it possible to determine all the metabolites present in a biofluid (generally urine or serum) simultaneously. Thus the holistic, global approach comprises the analysis of gene expression (*genomics or transcriptomics,* including non-coding RNA species), of proteins (*proteomics*, including specific post-translational modifications) and of metabolites (*metabolomics*) (see Figure 4.11). The parameters most commonly assessed by each of these technologies are summarized in Table 4.18.

It is also possible, but less widely applied in toxicology, to determine epigenetic modifications of genomic DNA (e.g. DNA methylation). In addition, there are other "omics" that can be used and have been reported in the literature, such as lipomics and glycomics. They refer to the determination of levels of specific types of molecules such as lipids or sugars. In toxicology, they are usually considered part of proteomics or metabolomics.

The concept of using gene and protein expression measurements, and metabolite composition for mechanistic and predictive toxicology is not completely new. However, it is the amount of information that can be gathered with relatively small effort using these tools that has transformed molecular toxicology. It was less than 50 years ago that the structure and function of DNA was deduced and by the end of 2002 the genomes of 800 organisms, including the human, had been fully sequenced. Alongside the major genome sequencing projects, protein databases now include thousands of identified proteins with their sequences and the structure of a fair number of endogenous metabolites is also available either in the public domain or in proprietary databases. Several

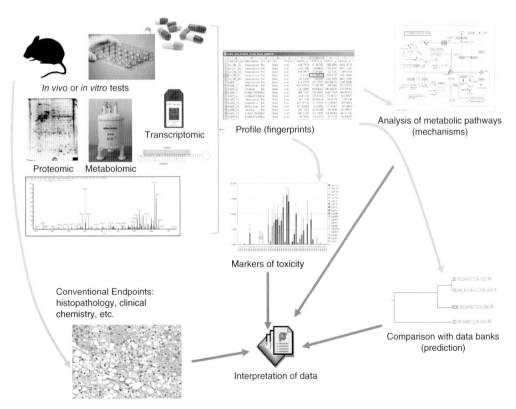

Figure 4.11 *Schematic representation of the holistic "omics" approach in toxicology. The results obtained by means of "omics" and conventional assessments generate mechanistic hypotheses. They also allow the classification of unknown substances based on the comparison of obtained data with available databases.*

Table 4.18 Summary of currently available technologies, biological matrices and analytes usually included in "omics" studies.

"Omics"	Biological material	Analytical method	Parameter	Example	Number of analytes
Genomics/transcriptomics	Tissue, cell cultures Medium/serum	Next-generation sequence microarray q-PCR	mRNA (transcripts) miRNA (intra or extracellular)	Induction of heat shock proteins after cell damage Release of miR122 as marker of hepatotoxicity	>10,000 ~1000
Proteomics	Tissue, cell culture, serum, urine	2D-PAGE, MS, MALDI, SELDI, DIGE	Protein products, including post-translational modifications	Increased expression of P450 after treatment with inducers Adduct formation	>1000
Metabolomics	Urine, serum, cell culture	NMR, LC-MS, GC-MS	Endogenous metabolites, small molecules	Reduction of intermediate products of the Krebs cycle due to mitochondrial damage	>500

publications have corroborated the use of toxicogenomics for the prediction of toxic liabilities, for the investigation of toxic mechanisms and for the identification of biomarkers of toxicity.

> **The results generated in the toxicogenomics field prove that there is a relationship between omics changes and toxicologically relevant processes. This indicates that toxicogenomics is still a relevant tool for the understanding of mechanisms of toxicity, for the prediction of potential toxic effects and also for the identification of specific biomarkers.**

miRNA

> **Micro RNAs (miRNAs) are short (21–23 nucleotides long), highly conserved, non-coding RNAs that play a key role in the regulation of protein expression. Any given miRNA generally represses the translation of several proteins encoded by their specific target genes.**

MiRNAs were first discovered in 1993 in *C. elegans*. They are key protein regulators that do not influence the transcription of the DNA but rather the fate of specific mRNAs, and count therefore as post-translational modifications. All miRNAs are produced as a primary transcript (pri-miRNA) which is spliced in the cell nucleus through the enzyme DROSHA (the double-stranded RNA-specific endoribonuclease) to give rise to a pre-miRNA (precursor miRNA). The pri-miRNA is then exported into the cytoplasm and further processed into mature miRNA through the RNase DICER. The generated miRNAs are complementary in sequence to the target genes that they regulate and can influence the fate of the mRNA in two ways: A mature miRNA can bind to the complementary sequence in the target transcripts, and the double-stranded RNA formed is recognized by the RNA-induced silencing complex (RISC) and degraded. Alternatively, the binding of miRNA to its target transcripts can also hinder the incorporation of the transcript into the ribosome and therefore hinder protein translation. Both mechanisms lead to a repression of protein synthesis of the specific target proteins. It is remarkable that one miRNA species has many (up to 100) target transcripts. This makes them powerful pleiotropic modulators with key functions in developmental biology, the development of cancers and also in toxicology. In addition to this, the interest in miRNAs has been greatly increased by the fact that they are not only present in tissues or cells, but can also be released into the bloodstream or cell culture medium. It is not yet completely clear why and how miRNAs are released from the cells and which physiological functions these extracellular miRNAs have. It is, however, of major interest that miRNAs are more or less tissue-type specific and stable in circulation. These characteristics makes them ideal biomarkers to detect specific tissue damage, such as that caused by chemical exposure. Indeed, the liver-specific mir122 has been detected after acetaminophen-induced hepatocellular injury in humans, rodents and *in vitro*.

4.6.3 Technology Platforms

This part of the chapter introduces the technological tools more widely used in the toxicogenomics field and the general principles they are based on. For the interested reader, additional specific literature is recommended in the reference section.

Genomics

> Nucleic acids have the characteristic of having a four-nucleotide code with very well-defined complementarity, which is used during the biological processes of replication and transcription. Detection technologies such as microarrays, quantitative polymerase chain reaction (qPCR) and sequencing have made use of this biochemical property and use highly specific sequences for the quantification of the genes of interest (generally called *targets*).

Methods Most RNA quantification platforms use specific complementary sequences (probes) in solution or attached to a solid surface. These probes hybridize to target transcripts that have been extracted from biological samples and allow their quantification, usually by the detection of fluorescence. Most technologies allow the simultaneous detection of many transcripts that are present in a given sample. Microarrays were one of the first platforms available to analyse many messenger RNA species. Other methods, like next-generation sequencing, have been developed more recently and are currently state of the art for global gene expression analysis. RNA sequencing is characterized by methods with high throughput that allow the sequencing of many RNAs in parallel. Both methods allow the quantification of the full transcriptome and are therefore genomic or transcriptomic tools.

Microarrays For each target gene that is to be assessed in a microarray, one or several specific probes complementary to a stretch of its sequence are represented and immobilized onto a solid support (glass or membrane). Most test systems (platforms) use 20–70-bases long probes. The high spatial density of the microarrays allows tens of thousands of probes, covering virtually all known transcripts, to be represented in one single microarray and thus be measured simultaneously. Hence, the whole *transcriptome* (transcriptome is the sum of all transcribed genes in a given tissue at a given time) for a given tissue can be assessed with one microarray experiment. For this, all transcripts (mRNA) from a given tissue or cell culture are extracted, usually amplified, labeled, and hybridized onto the microarray. Hybridized probes are measured (usually fluorimetrically), generating an output where the intensity of the signal detected for each probe is proportional to the amount of transcript (mRNA target) in the original sample. Some microarray platforms use two-color dyes for the simultaneous determination of the expression level in two samples, one being typically a baseline (control) and the other being the treated sample. In these cases, both the control and treated sample are hybridized together onto the same microarray exactly under the same hybridization conditions, thus minimizing the technical variability. However, the disadvantage of obtaining relative rather than absolute gene expression values is that they always refer to the sample that was arbitrarily selected as baseline. In all cases, microarrays provide a (semi)-quantitative measurement of absolute or relative expression level for each gene represented on the microarray. Due to the differences in the genetic code between the species and the exact complementarity required between a probe on a microarray and the target in the tissue of interest, microarrays are highly species-specific. There are many commercial providers of microarrays, including (but not limited to) Affymetrix, Applied Biosystems, Agilent, Illumina, and Codelink. Also, several laboratories in academia and industry use to employ spotting devices to spot oligonucleotides and produce their own low-cost microarrays.

Next-generation Sequencing Prior to the development of next-generation sequencing (NGS), the classic Sanger method was the most commonly used technique to determine the sequence of nucleotides. This method is very robust and, despite its low throughput, is still used for the validation of sequences of small numbers of genes. For omics applications, requiring a much higher throughput, the Sanger method has been replaced by NGS. This method relies on microfluidics systems that allow the parallelization of millions of sequencing reactions. The template DNA or RNA (for DNA and RNA sequencing, respectively) is fragmented into smaller pieces in an initial step. These fragments are then amplified either through emulsion-PCR (on bead-based instruments such as Roche/454, IonTorrent, or ABI/SOLiD) or through bridge-PCR on immobilized nucleotides (Illumina). The sequencing reactions are finally physically separated either via immobilization on picotiter or microtiter plates (Roche/454, IonTorrent) or via microfluidics (Illumina, ABI/SOLiD).

Quantitative Reverse Transcriptase Polymerase Chain Reaction In addition to the microarray and sequencing platforms quantitative reverse transcriptase polymerase chain reaction (qRT-PCR) can be employed for the detection and quantification of transcripts. This technology uses specific amplification of a transcript of interest as a means to determine its expression level. The amplified product is highly specific due to the use of specifically designed primers and can be quantified after each PCR cycle by means of a fluorescent dye. It is mandatory to use a specific primer set (two primers, forward and reverse) that will allow the specific amplification of a sequence of interest. An additional specific probe hybridizing within the amplicon can greatly increase the specificity of the reaction and is known as a TaqMan probe. After each amplification cycle, the intensity of the fluorescence, which is proportional to the amount of generated amplicon and therefore to the template in the original sample, is measured. This allows the reaction to be followed in real time (qRT-PCR is often also called real-time PCR). In general, qRT-PCR is considered to be more sensitive, specific, and accurate than most hybridization-based microarray platforms. Strictly speaking, qRT-PCR is not an "omic" application, since it was conceived for single transcripts and even now can only be applied to a limited number of genes. However, automation on 384-well plates and the use of commercially available microfluidics cards (e.g. from Applied Biosystems) allows us to measure hundreds of genes simultaneously using qRT-PCR. Moreover, qRT-PCR is currently the technology of choice for most scientists to corroborate results obtained using microarrays and therefore plays a major role in toxicogenomics.

Proteomics

> **Proteomics is the study of the proteome, defined as the sum of all proteins expressed in a given tissue at a given time. The proteome is more complex than the transcriptome, as it includes the study of post-translational modifications such as phosphorylation. In addition, the physicochemical properties of proteins are much more diverse than those of nucleic acids, particularly size, charge, concentration in a biological matrix and hydrophilicity.**

Proteins are more complex chemically than nucleic acids. First, it needs to be taken into consideration that there is no natural biochemical complementarity as with the genetic code, which precludes a simple hybridization process. Moreover, most proteins undergo extensive modifications after translation, such as cleavage of signaling peptides, phosphorylation/dephosphorylation, or glycosylation. In addition, treatment with xenobiotics can also lead to drug–protein covalent binding, a protein modification of relevance in toxicology. These post-translational modifications have a major impact on the function of the proteins. For example, the activity of many proteins is regulated through phosphorylation/dephosphorylation, which can cause dramatic biological changes, whereas the total protein content remains unchanged. In addition, proteins are present in serum and tissues with a very wide dynamic range: some proteins are highly expressed, while others are up to 10 orders of magnitude less abundant. For example, the concentration of albumin, the most abundant protein, in serum is about 10 billion times greater than that of interleukin-6. This complex situation challenges the technologies around proteomics since such a dynamic range is extremely difficult to achieve.

Methods Separation of proteins can be achieved by fractionation of cellular components (subcellular fractionation) and further by two-dimensional gel electrophoresis (2D-PAGE) or chromatographic methods. Generally, 2D-PAGE has been considered the workhorse of proteomics, although is very time-consuming and lacks sensitivity. Low-abundance proteins and highly hydrophobic proteins such as cytochrome P450 can usually not be assessed with this method. As a further development of 2D-PAGE, and similarly to the two-color dyes in microarrays, two-dimensional gel electrophoresis (2D-DIGE) includes a fluorescence labeling step of the proteins before the electrophoresis and the simultaneous separation (by 2D-PAGE) of two samples labeled each with a different color. The advantage of 2D-DIGE is that it allows relative quantification while controlling for technical variability, thus increasing the dynamic range and sensitivity of traditional 2D-PAGE.

Retentate chromatography-mass spectrometry (RC-MS), also known as surface enhanced laser desorption/ionization (SELDI) is an alternative to 2D-PAGE-based methods. SELDI uses surfaces with different physicochemical properties to separate the proteins in a complex mixture. The adsorptive chromatographic support, placed on thin metal chips, acts as a bait to adsorb proteins in the sample. This is based on the fact that different chemical surfaces present affinities to groups of proteins with specific characteristics. A mass spectrum profile is created by desorbing the proteins with the laser from the matrix-assisted laser disorption ionization (MALDI) instrument. This technology is relatively easy to use, requires a small sample volume, and allows a higher throughput than 2D-PAGE. However, SELDI has several technical drawbacks and has been reported to suffer from large experimental variability.

Identification of the proteins after 2D-PAGE is usually performed by digesting the protein spots of interest with trypsin and determining the masses of the tryptic fragments with mass spectroscopic (MS) approaches. The mass spectrograms are like fingerprints that can be compared to public databases for identification purposes, thus this process is called peptide fingerprinting. After SELDI analysis, the identification of the identified possible "markers" is rather difficult. Some researchers try to use this platform to identify "fingerprints" without identifying the actual proteins, but this is not ideal. If identification is required, the peaks of interest must be isolated to allow subsequent peptide fingerprinting or tandem-MS to be performed.

Improvements in protein labelling that makes possible the utilization of MS methods possible without a cumbersome separation of the protein species are becoming popular. The advent of MS-based tagging methods, in particular iTRAQ® Reagents, have permitted relative expression measurements of large sets of proteins with a high degree of automation.

A different approach to proteomics is the use of protein arrays, in analogy to gene expression microarrays. The most frequently used protein array is the antibody array. These arrays rely on the spotting of specific antibodies for the proteins of interest and are conceptually similar to the DNA microarrays. The antibodies fixed onto a solid support can capture through affinity binding the proteins present in the complex mixture hybridized onto the chip. The main limiting factor of antibody arrays is the need for specific, good-quality antibodies for a large number of proteins of interest.

Metabolomics

> ***Metabolomics*** **studies the collection of all endogenous metabolites present in body fluid or tissue. While transcriptomic and proteomic analyses provide partial information regarding the biological processes occurring in a cell, metabolic profiling gives an overall picture of the cellular physiology. Moreover, changes in the metabolic profile reflect the reaction of the system to external stimuli, such as feeding or exposure to chemicals. These changes in the metabolic profile can be investigated in a cell, tissue or organ.**

The *metabolome* is the final downstream product of the genome and is defined as the total quantitative collection of small molecular weight compounds (metabolites) present in a cell or organism which participate in metabolic reactions required for growth, maintenance and normal function. Therefore, *metabolomics* refers to the study of the collection of all metabolites in a biological organism, which are the end products of the activity of expressed genes and proteins and their interactions. Therefore, metabolomics measurements give an overall picture of the physiology of an individual and its reaction to a given treatment, which can be measured in body fluids such as serum and urine.

The applications of metabolomics include the characterization and prediction of drug toxicity diagnosis and monitoring of clinical disease, evaluation of therapeutic intervention and understanding the effects of genetic modifications. Metabolomics has often been used to assess the effects of nutrients and food additives as well. These are achieved by measuring the metabolite profiles, generally in biofluids (urine, serum, plasma), although metabolomics measurements can also be applied to tissues. The study of the metabolome faces similar challenges to the study of the proteome, since it cannot rely on simple chemical hybridization to separate, quantify and identify metabolites of interest. Metabolites are even more chemically diverse than proteins and are also present in a wide dynamic range of concentrations, with differences in concentration of up to nine orders of magnitude. It is a goal of researchers in this field to develop technologies that enable large-scale high-throughput screening without losing sensitivity. However, there is currently no technology that fulfils all these requirements and investigators are forced to evaluate the advantages and disadvantages of each platform for each experiment.

Methods Metabolomics uses technologies such as H-NMR or a chromatographic separation (LC or GC) coupled to an MS determination. NMR spectroscopy is the only detection technique which does not rely on separation of the analytes and is not-destructive, e.g. it allows the sample to be recovered for further analyses. It requires relatively little sample processing and the measurement times are short, allowing for relatively high throughput. All kinds of small molecules can be measured simultaneously in a complex mixture. However, it is relatively insensitive compared to MS-based techniques and the processing of the NMR spectra requires very specialized technical expertise and is relatively time-consuming. Additionally, the identification of metabolites of interest poses a challenge. The other technologies commonly used for metabolic profiling are chromatographic separation followed by MS (LC-MS or GC-MS). Complex samples are separated by LC or GC and the masses of the components of each fraction are determined by MS. This method is more sensitive than NMR, but requires additional sample processing before successful chromatographic separation can be achieved. Moreover, the sample cannot be recovered for subsequent analyses.

4.6.4 Bioinformatics and Biostatistics

> The technologies listed above allow the study of the whole transcriptome, proteome, and metabolome. Therefore, "omics" experiments produce large amounts of data. Due to cost and throughput constraints, these enormous data sets are generally generated using few biological replicates (three to five at most) and tens of thousands of parameters (genes, proteins, and metabolites). This represent a real challenge for the data analysis and requires new concepts for data processing, data analysis, and data visualization.

From the statistical point of view, "omics" technologies pose an enormous challenge. The huge amount of data does not usually fulfil prerequisites of commonly used parametric statistics such as normal distribution and independence of variables. Variables are highly correlated, since groups of genes/proteins might be co-regulated and different NMR signals or MS peaks represent one and the same metabolite/protein. Several strategies are being used to analyze this kind of highly multivariate data. Commonly used univariate statistical tests include analysis of variance, pairwise comparisons (i.e. Student's t-test) and determination of false discovery rates, usually performed by means of specialized software. Tools for multivariate analysis include cluster analysis, principal component analysis (PCA), significance analysis of micorarrays (SAM), and self-organizing maps. These unsupervised methods are used to determine if gene expression patterns allow the discrimination of natural subpopulations that might bear a biological meaning and experimental relevance, such as treated/untreated or healthy/diseased. These methods will also recognize differences in profiles of groups with less experimental meaning, such as male/female. In addition to the unsupervised analyses performed for each experiment, the use of toxicogenomics also covers the field of predictive toxicology, relying on mathematical predictive models that generate so-called *signatures* or *fingerprints* that enable the classification of unknown compounds. For this purpose, in genomics and metabolomics, scientists have produced databases with model compounds known to cause a certain type of toxicity, to which the compounds under investigation are compared. For this type of application, complex supervised multivariate analysis methods (e.g. support vector machines, discriminant analysis, partial least squares, neuronal networks) are required. These supervised methods necessitate a so-called training set (results obtained from the model compounds) that is used for the generation of the predictive model. In general the generation of such predictive models requires several steps: (i) training with data from a database, (ii) cross-validation with additional samples to ensure its appropriate performance, and (iii) classification of unknown samples. Once the model has been generated and cross-validated, test compounds (new compounds under investigation) can be classified based on the gene expression results or the metabolite profiles. The efficiency of the models depends highly on the size and the quality of the database, the appropriate training set and a robust cross-validation to diminish the risk of "overfitting". Despite several successful examples of these predictive characteristics, toxicity prediction based solely on "omics" data is currently rarely used.

An additional point that needs to be addressed when evaluating complex data sets is their appropriate standardization. This applies to experimental conditions and bioinformatics and biostatistical tools alike. Although many significant results have been derived from "omics" studies, the lack of standards limits the exchange and independent verification of the data across groups. Therefore, there is a need to develop standards that are acceptable within the scientific community. Again, most progress in this area can be reported from the genomics environment, where the *Minimum Information About a Microarray Experiment* (MIAME) has been defined and is widely accepted. In addition to robust statistical models and standardized data, it is good practice to corroborate key findings from "omics" studies by means of alternative techniques (such as PCR, Western blot, etc.).

4.6.5 Applications of Toxicogenomics

> **The main goals of the use of toxicogenomics are *predictive* and *mechanistic investigations*. An additional feature of "omics" analysis is that it can serve the identification and validation of novel biomarkers. The ideal biological system would be either an *in vitro* system or a short-term animal study. As with any *in vitro* assay, cell cultures can be easily manipulated but their biological relevance is often questioned as they do not address the whole organism. Consequently, *in vivo* investigations are preferred for predictive toxicogenomic and metabolomic studies.**

It is the task of predictive toxicogenomics that gene, protein and metabolite profiles enable the segregation of compounds with a toxic liability from others. Mechanistic investigations aim to unveil the molecular processes underlying an observed or suspected toxicity. Thus, the goal of mechanistic investigations is to understand the reasons for a given toxicity, identifying the affected cellular pathways. Both predictive toxicogenomics based on signatures and mechanistic approaches based on specific pathways can lead to the discovery of novel and specific *safety biomarkers*, an additional application of these technologies (Figure 4.12).

Predictive Toxicogenomics

> **Predictive toxicogenomic studies usually compare the gene expression patterns elicited by chemicals with unknown toxic potential to the profiles of model compounds with known toxicity, with the goal of predicting organ toxicity. To this end, "omics" profiles generated after exposure to a substance under scrutiny are compared to databases containing profiles of toxicants and non-toxic compounds. The similarity to one or the other group indicates a potential for toxicity at a defined target organ.**

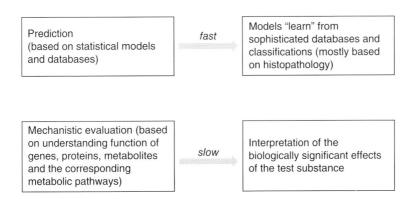

Figure 4.12 Typical application of "omics" data in toxicology.

In particular in the pharmaceutical industry, it was expected that the predictive approach would yield faster, more accurate and possibly more sensitive results than conventional toxicology assessment in the animal. As several toxicologically relevant issues from medicines were only detected after launching, there were great hopes placed in toxicogenomics to help avoid such failures in the future. This situation led to substantial investments in the pharmaceutical industry in toxicogenomics efforts. Although *in vitro* test systems sound ideal to fulfil these purposes, advances in the field of *in vitro* toxicogenomics are still limited. This is due to various reasons. As with any *in vitro* assay, it is difficult to select a biologically suitable cell system that provides information relevant to the whole organism. Despite this, the use of primary hepatocytes (mainly rodent, but also human), hepatic cell lines, and kidney cells have shown that prediction of toxicity by means of gene expression *in vitro* is feasible.

However, a substantial amount of gene expression and metabolite profiling data has been generated in animal models (mostly rat) with known toxicants, mainly hepatotoxicants and nephrotoxicants; but also with genotoxic and cardiotoxic compounds. The results show that gene expression and metabolite composition analysis can provide information to allow classification of compounds according to their mechanism of toxicity as well as identifying cellular pathways related to the toxic event. In any case, for the use of toxicogenomics as a predictive tool, the prior knowledge of gene expression patterns related to toxicity is absolutely necessary. Consequently, this approach is highly dependent on the availability of reference gene expression databases and robust software with appropriate algorithms for the comparison of complex fingerprints.

Figure 4.13 represents an illustrative example of how predictive toxicogenomics can be used in practice. In this case, the hepatotoxicity database content was used to build a model with gene expression data using support vector machines (SVM), a supervised analysis tool. The generated model allows vehicle-treated animals (controls) and animals without histopathological findings (low-dose and non-responders) to be discriminated from animals that had been exposed to a hepatotoxicant and displayed liver necrosis. Furthermore, the use of the database in conjunction with

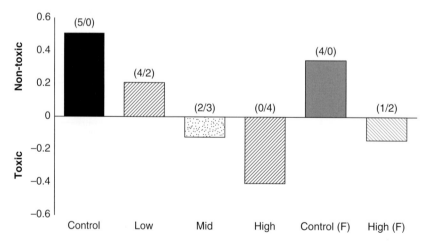

Figure 4.13 Example of the classification of a compound (identified as toxic) based on gene expression data. Livers from male animals treated with vehicle (control) or three different doses of the test substance (low, medium, high). Results obtained with female animals are identified by an (F). Gene expression data were generated using Affymetrix microarrays. The ordinate gives the classification as toxic or non-toxic in arbitrary units. The numbers in brackets indicate the incidence of a unique classification to the "toxic" group.

a model including several classes of toxicity allows us also to distinguish several classes of hepatotoxicities, among others, compounds that directly cause hepatocellular damage (apoptosis or necrosis). In the selected example, the low-dose group was exposed to a subtoxic dose and could not be segregated from the controls, whereas the animals showing liver damage were assigned correctly to the toxicity category (so-called "direct acting compounds"). Furthermore, a clear distinction of doses, time, and extent of liver injury can be obtained using gene expression analysis.

In vitro prediction models have proved less robust and showed a much lower prediction rate than *in vivo* toxicogenomics. However, several interesting results have been achieved and documented. Marc Fielden reported that the use of rodent and human hepatocytes led to the successful detection of non-genotoxic carcinogens after relatively short *in vitro* studies. This represents major progress in comparison to the required life-long rodent studies.

Both *in vivo* and *in vitro* results provide ample evidence that omics profiles can be predictive of potential side effects. However, histopathology and other validated biomarkers are still the gold standard for the detection of organ damage. Nowadays it is more common to use "omic"s technologies for the identification of cellular and metabolic pathways that are affected by a substance.

Mechanistic Approach

> **Mechanistic analyses usually focus on genes, proteins or metabolites of mechanistic interest and investigate the cellular and molecular mechanisms related to the exposure to a (toxic) compound. This type of analysis investigates changes on specific molecules or pathways that are known to be associated with the effect of a xenobioticum.**

The information obtained from mechanistic analyses is extremely useful to deepen knowledge of mechanisms of toxicity for compounds or compound classes. However, it relies on the availability of accurate functional information on the affected genes, proteins, or metabolites. Moreover, it allows scientists to introduce knowledge-driven bias into the data, overemphasizing particular findings and disregarding perhaps important information. Hence, it is vital that the generated mechanistic hypothesis are corroborated by appropriately designed follow-up experiments. These subsequent experiments aim to clarify two main sources of inaccuracy. On the one hand, it needs to be elucidated whether the parameters of interest could be technical artefacts (false signals). On the other hand, it should be assessed whether a change in the expression level of a gene or protein bears biological relevance and is actually causally associated with the toxic event under scrutiny. For the former, a technological validation needs to be performed, generally measuring the analyte(s) of interest with an alternative platform (such as qPCR or Western blot). This is particularly important for "omics" applications, since when analyzing thousands of parameters, a small percentage of statistically expected false positives translate into a considerable number of data points that will wrongly be identified as significantly changed. For biological validation, the investigation of the hypothesized mechanism in another biological system, *in vivo* or *in vitro*, or the use of molecular biology tools such as gene knock-ins (transfection) or knock-downs (RNAi), will help the investigator realize if the pathway of interest is related to the studied phenomenon or not.

Mechanistic explanations based on "omics" data have been reported with metabolomics, proteomics and transcritpomics. For example, metabolomics analyses show that changes in the intermediates of the citrate cycle are usually related to mitochondrial toxicity. In an investigative study, the investigators observed depletion in tricarboxylic acid cycle intermediates, and the appearance of medium chain dicarboxylic acids. These findings led to the mechanistic hypothesis of defective metabolism of fatty acids in the mitochondria; which was confirmed by subsequent *in vitro* experiments. In addition, genomic data have often been used for the identification of a given molecular mechanism of toxicity. Fibrates, for example, are hypolipidemic drugs that rely on the activation of the nuclear receptor PPARα. Gene expression profiling of livers of rats exposed to fenofibrate show the up-regulation of the fatty acid metabolic pathway, confirming a known pharmacological effect.

It has been shown that the use of genomics data could help clarify differences between rodents and primates after exposure to several fibrates. These drugs cause hepatic peroxisome proliferation, hypertrophy, hyperplasia, and eventually hepatocarcinogenesis in rodents. However, primates are relatively refractory to these effects, although the mechanisms for the species differences are not clearly understood. Hepatic gene expression profiles in the monkey led to the following findings. Genes related to the pharmacological effect of the compounds were strongly up-regulated, but to a lesser extent to that reported for rodents. Also, a number of key regulatory genes known to be involved in proliferation, DNA damage, oxidative stress, proliferation, and apoptosis were either unchanged or regulated in the opposite way in primates in comparison to rodents. Thus, it can be concluded that the molecular mechanisms triggered in the liver of primates are qualitatively and quantitatively different from those in the rodent, which might be the reason for the species specificity. These results arose originally from gene expression analysis but were later further substantiated by proteomics and metabolomics data. This example shows that it is of great advantage to combine results from several omics technologies, together with conventional endpoints, to draw accurate conclusions at a cellular, organ, and organism level. These sort of results add to the weight of evidence on the human relevance of toxicity findings in animals and can therefore contribute to the risk assessment process of molecules.

An excellent example of the use of several "omics" technologies are the results obtained in a EU framework 6 project called PredTox, where the consortium identified liver toxicity and markers for cholestasis. An example of the outcome of this multiomics effort is depicted in Figure 4.14. A hallmark in the field of the use of "omics" for the identification of biomarkers is the work performed with nephrotoxicants several years ago. Putative biomarkers identified by means of gene expression analysis were corroborated, validated and submitted to the health authorities in the USA (FDA) and Europe (EMA) in a process called "qualification", which led to the acceptance of these markers as indicators for kidney damage in a specific experimental context.

4.6.6 Summary

Toxicogenomics, as a relatively new field in toxicology, has faced many challenges in the last 20 years. In the meantime, this discipline has evolved, matured and found a place in toxicity assessments. The "omics" family has made major progress in terms of technology, data analysis, use, and general acceptance within the scientific community. However, the original expectation that "omics" would lead to the abolishment of animal experimentation or that all side effects in patients could be predicted have proven over-optimistic. In spite of this, toxicogenomics plays an important role in modern toxicology. The generated data, the technological advances and the knowledge gained in molecular pathways underlying toxicity manifestations have indeed changed the way in which we define toxicity. Currently, toxicogenomics is not seen as an additional discipline in toxicology but rather as an additional endpoint or technology that helps elucidate the effects of tests substances and identify useful biomarkers. As the employed technologies are relatively novel, health regulators maintain a given scepticism with regards to the interpretation of the results. A major hurdle in this area is the appropriate standardization and validation of the data. Regulatory agencies in the USA and elsewhere are currently experimentally involved and active in the generation, storage, analysis and interpretation of "omics" data. This is a clear sign that these types of data are here to stay.

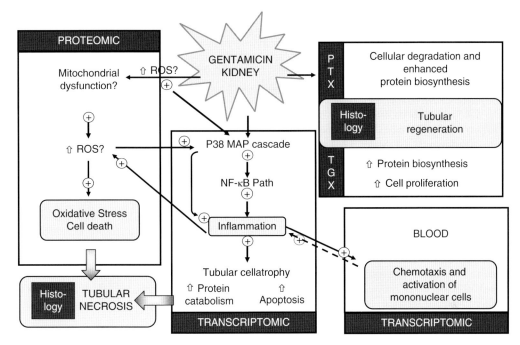

Figure 4.14 Scheme representing the interpretation of "omics" data after administration of the nephrotoxicant gentamicin to rats. Kidney tissue was analyzed using proteomics and transcriptomics, as well as histopathology. Blood from the same animal was submitted to proteomic and transcriptomic analysis. ROS, reactive oxygen species; PTX, proteomics; TGX, genomics.

Further Reading

Boitier E, A Amberg, V Barbié, A Blichenberg, A Brandenburg, H Gmuender, A Gruhler, D McCarthy, K Meyer and B Riefke. (2011). A comparative integrated transcript analysis and functional characterization of differential mechanisms for induction of liver hypertrophy in the rat. *Toxicol Appl Pharmacol* **252**:85–96.

Buness A, A Roth, A Herrmann, O Schmitz, H Kamp, K Busch and L Suter. (2014). Identification of metabolites, clinical chemistry markers and transcripts associated with hepatotoxicity. *PloS One* **9**:e97249.

Cariello NF, EH Romach, HM Colton, H Ni, L Yoon, JG Falls, W Casey, D Creech, SP Anderson, GR Benavides, DJ Hoivik, R Brown and RT Miller. (2005). Gene expression profiling of the PPAR-alpha agonist ciprofibrate in the cynomolgus monkey liver. *Toxicol Sci* **88**:250–264.

Dieterle F, F Sistare, F Goodsaid, et al. (2010). Renal biomarker qualification submission: a dialog between the FDA-EMEA and Predictive Safety Testing Consortium. *Nat Biotechnol* **28**:455–462.

Dunn, WB, NJ Bailey and HE Johnson. (2005). Measuring the metabolome: current analytical technologies. *Analyst*, **130**, 606–625.

Ellinger-Ziegelbauer H, B Stuart, B Wahle, W Bomann and H-J Ahr. (2004). Characteristic expression profiles induced by genotoxic carcinogens in rat liver. *Toxicol Sci* **77**:19–34.

Eriksson L, H Antti, J Gottfries, E Holmes, E Johansson, F Lindgren, I Long, T Lundstedt, J Trygg and S Wold. (2004). Using chemometrics for navigating in the large data sets of genomics, proteomics, and metabonomics (gpm). *Anal Bioanal Chem* **380**:419–429.

Fielden MR, R Brennan and J Gollub. (2007). A gene expression biomarker provides early prediction and mechanistic assessment of hepatic tumor induction by nongenotoxic chemicals. *Toxicol Sci* **99**:90–100.

Fielden MR, A Adai, T Robert et al. on behalf of the Predictive Safety Testing Consortium, Carcinogenicity Working Group. (2011). Development and evaluation of a genomic signature for the prediction and mechanistic assessment of nongenotoxic hepatocarcinogens in the rat. *Toxicol Sci* **124**:54–74.

Heslop JA, C Rowe, J Walsh, R Sison-Young, R Jenkins, L Kamalian, R Kia, D Hay, RP Jones, HZ Malik, S Fenwick, AE Chadwick, J Mills, NR Kitteringham, CE Goldring, B Kevin Park. (2017). Mechanistic evaluation of primary human hepatocyte culture using global proteomic analysis reveals a selective dedifferentiation profile. *Arch Toxicol* **91**(1):439–452.

Luhe A, L Suter, S Ruepp, T Singer, T Weiser and S Albertini. (2005). Toxicogenomics in the pharmaceutical industry: hollow promises or real benefit? *Mutat Res* **575**:102–115.

Mortishire-Smith RJ, GL Skiles, JW Lawrence, S Spence, AW Nicholls, BA Johnson and JK Nicholson. (2004). Use of metabonomics to identify impaired fatty acid metabolism as the mechanism of a drug-induced toxicity. *Chem Res Toxicol* **17**:165–173.

Nicholson JK, J Connelly, JC Lindon and E Holmes. (2002). Metabonomics: a platform for studying drug toxicity and gene function. *Nat Rev Drug Discov* **1**:153–161.

Starkey Lewis PJ, J Dear, V Platt, KJ Simpson, DG Craig, DJ Antoine, NS French, N Dhaun, DJ Webb, EM Costello, JP Neoptolemos, J Moggs, CE Goldring and BK Park. (2011). Circulating microRNAs as potential markers of human drug-induced liver injury. *Hepatology* **54**:1767–1776.

Steiner G, L Suter, F Boess, R Gasser, MC de Vera, S Albertini and S Ruepp. (2004). Discriminating different classes of toxicants by transcript profiling. *Environ Health Perspect* **112**:1236–1248.

Suter L, S Schroeder, K Meyer, J-C Gautier, A Amberg, M Wendt, H Gmuender, A Mally, E Boitier and H Ellinger-Ziegelbauer. (2011). EU framework 6 project: predictive toxicology (PredTox) – overview and outcome. *Toxicol Appl Pharmacol* **252**:73–84.

Taylor EL and TW Gant. (2008). Emerging fundamental roles for non-coding RNA species in toxicology. *Toxicology* **246**:34–39.

Wetmore BA and BA Merrick. (2004). Toxicoproteomics: proteomics applied to toxicology and pathology. *Toxicol Pathol* **32**:619–642.

Yadav NK, P Shukla, A Omer, S Pareek, AK Srivastava, FW Bansode and RK Singh. (2014). Next generation sequencing: potential and application in drug discovery. *Sci World J* **802437**.

4.7 Introduction to the Statistical Analysis of Experimental Data

György Csanády

4.7.1 Introduction

> **Measurements carried out under the same experimental conditions result in observations that are not exactly identical due to random variability. This variability might easily mislead data interpretation guided solely by the common sense. Applied statistical methods are useful to describe and analyse the random variability and thus can help researchers to reach a scientifically and statistically valid conclusion about the collected data. Statistical procedures should be applied to the planning phase of the experiments, to summarize and analyse the data, and to draw inferences from the data.**

This chapter is intended to assist experimentally working scientists by discussing:

- summarizing data by using **descriptive statistics**
- how uncertainty can be characterized by **error propagation** if variability of collected data is propagating in subsequent calculations

- how **probability distributions** can be used to describe variability
- how data collected from a sample of a population can be used to generate a **statistical inference** about the population using **estimation procedures** and **hypothesis tests**
- the use of **regression analysis** to describe a linear relationship between two variables
- determination of the median lethal dose using **probit analysis**
- experimental design
- the use of computers in statistical calculations.

In any toxicological experiment there are several variables and factors, such as species, strain, sex and age of the animals, season, diet, dose, time, experience of the experimenter, etc., which may affect the outcome. Interpretation of experimental results is facilitated when all but one of the factors is held constant and the effect of a single **independent variable** is studied. In toxicology the dose is among the most frequently employed independent variable. Thus, in an experimental situation the effect of varying the dose leads to observations on the **dependent variable** collected in the form of data. (A single piece of information is termed a datum.) Data can be either **qualitative,** e.g. tumor or inflammation present, or **quantitative**, e.g. body weight, concentration of a metabolite in blood.

If the experiment is carried out twice under the same experimental conditions the two outcomes will not be exactly identical. The difference between the two outcomes is governed in a well-conducted experiment only by chance. Variability is the reason why in toxicological studies it is essential to make comparisons between "treated" and "control" groups, for which identical untreated animals are used. Some of the variability can be attributed to known factors. However, the major share of the variability is considered unpredictable and is called **random variability** composed of several factors, such as biological variability, sampling variation, and experimental error. Determining this variability is the object of statistics. Statistical procedures can help to describe and characterize this variability in order to facilitate an accurate interpretation of experimental results.

The entirety of all possible outcomes of a given experiment conducted under identical conditions is called the **population**. The population – a theoretical concept – might refer to an infinite number of experiments, which could never be carried out experimentally. The outcome of a set of measurements results in data, which from a statistical point of view correspond to a sample independently drawn by chance (Figure 4.15). It is assumed that the sample is large enough to be representative of the population. After collecting the data, statistical procedures are applied with the aim of drawing an inference, i.e. a tentative conclusion, about the population itself. It is somewhat surprising that limited data can lead to any conclusion about a large, often infinite population. The reason for this peculiarity lies in the fact that underlying assumptions govern the behavior of the population. Such assumptions must be known in order to perform a valid statistical analysis. For many statistical procedures it is assumed that the variability present in the observed data follows **Gaussian** or **normal distribution** (see below). However, biological data show often another distribution called **lognormal**.

Example

In a carcinogenicity study Sprague–Dawley rats are treated with a chemical. It is impossible to study all existing Sprague–Dawley rats which make up the theoretical rat population. Instead, a given number, usually 50 rats, which comprise the sample are treated per dose group for about two years and the incidence of tumors is observed. The underlying assumption is that the finite sample investigated represents the infinite population (Figure 4.15). We are not interested in the

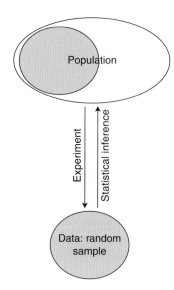

Figure 4.15 *Relationship between population and sample. The sample consisting of a set of observations embodied in the data obtained in an experiment is considered to be a random realization of the statistical population. The sample is analyzed with the aim of drawing statistical inferences about the population itself.*

fate of any particular rat but in the carcinogenicity of the tested chemical in the rat population. For this purpose the tumor incidences obtained in a treated group and a control group are compared.

Comment

Bias is a term used to describe **systematic errors** resulting in inaccuracies which lead to misinterpretation of the results. For example, use of a non-random sample of the population can lead to a constant deviation which cannot be revealed by statistical analysis per se. **Systematic errors** can only be identified by using different experimental methods or by participating in **interlaboratory tests**. **Randomization** might help to minimize bias (see Chapter 4.7.8).

4.7.2 Descriptive Statistics

> **Large data sets or even a single set of similar numbers cannot be easily comprehended. Therefore, they should be organized and summarized in such a way that the main characteristics of the data are preserved. Descriptive statistics provide methods to present and summarize measured data by calculating specific sample characteristics determined solely by the type of data (Table 4.19).**

Data collected in an experiment are either **qualitative (categorical)** or **quantitative (numerical)**. Many toxicological observations are qualitative and can be ordered into dichotomous categories, e.g. mortality data, fetal abnormalities, histopathology data and clinical signs. There are also classifications requiring more than two categories, e.g. blood groups. These characteristics are given in the **nominal scale**, which does not allow the establishment of any hierarchical order.

Table 4.19 *The type of data collected determines the type of graphical representation as well as the sample characteristics.*

	Qualitative		Quantitative	
	Nominal scale	Ordinal scale	Discrete	Continuous
Examples	Gender, blood groups	Severity of lesions (+, ++ and +++)	Some haematology data, SCE, size of litter	Body and organ weight, concentration
Graphical representation	Bar/pie diagram	Bar/pie diagram	Histogram Box-and-whisker plot	Scatter plot Box-and-whisker plot
Sample characteristics	Frequency, mode	Frequency, mode, median	Mean/median Standard deviation/centiles	Mean/median Standard deviation/centiles

Mode, most frequent value; median, the mid point in an ordered series of numbers; SCE, sister chromatid exchange.

Other classifications, e.g. severity of a lesion (+, ++, +++), are established by a hierarchical order according to a predefined **ordinal scale**. The magnitude of the effect determines the ranking, although differences between ranks might not have an arithmetical meaning.

Other experiments result in **quantitative data**, which can be either discrete or continuous. **Discrete data** represent numerical values expressed as an integer, since they represent counts of events such as radioactivity measurements, some hematology data (cell counts), number of tumors or size of litter. **Continuous data** are characterized by continuous quantities. The readings may have any values between certain limits. Such data include, for example, body and organ weights, food and water consumption, and concentrations of toxic agents in the blood.

For continuous variables full precision is retained during data analysis, except for final rounding. The number of significant digits presented should correspond to the precision of the measurement. For example, a single value of 9 usually implies that the true number is between 8.5 and 9.5. Therefore, the precision of this value is 11% if it was measured to the nearest digit (Table 4.20). To present results from most toxicological studies two significant digits are sufficient since the accuracy of the measured data is generally not below 1%. Statistical descriptors, such as mean values, can be given by one more significant figure.

Raw data are often represented in figures and tables. Information from a large body of data can be condensed by calculating their characteristic measures. Data can be summarized by reporting

Table 4.20 *Number of significant digits of a datum in relation to its accuracy.*

Datum	Number of significant digits	Limit	Accuracy %
1	1	0.5–1.5	100 (100 · 1/1)
9	1	8.5–9.5	11 (100 · 1/9)
1.1	2	1.05–1.15	9.1 (100 · 0.1/1.1)
9.9	2	9.85–9.95	1.1 (100 · 0.1/9.9)
1.11	3	1.105–1.115	0.90 (100 · 0.01/1.11)
9.99	3	9.985–9.995	0.10 (0100 · 0.01/9.99)

a measure for the **central tendency,** for the **variability**, together with the number of observations. There are several measures for the central tendency:

- the **arithmetic mean** or average equals the sum of observations (X_i) divided by the number of observations (N): $\overline{X} = \frac{1}{N} \sum_{i=1}^{N} X_i$

- the **geometric mean** is the Nth root from the product of all observations: $\sqrt[N]{\prod_{i=1}^{N} X_i}$

- the **median** divides the ordered data set into two equal groups
- the **mode** is the most common value.

The most useful measure for the central tendency depends on the type of data. The mode describes nominal scale data, giving the most prevalent sample characteristics. For ordinal scale data the median is often considered to be useful. Mean values are appropriate for quantitative data. Outliers within the data confound the mean whereas the median and mode are less sensitive. Arithmetic mean, median and mode are very similar (theoretically indistinguishable) if the data are collected from a symmetrical distribution.

The variability or spread observed in the data can also be described by different measures termed the **range**, the **centiles**, the **standard deviation**, and the **coefficient of variation**.

The **range** is defined by the difference between the largest and the smallest value. It is very sensitive to outliers since it takes into account only the two most extreme values and does not reflect the majority of data. The **centile** or **percentile** gives the value below which a given percentage of the values lie. The data must be arranged in an order, e.g. $x_1 \leq x_2 \ldots \leq x_N$, before centiles can be calculated. The **median** is defined as the 50th **centile** because 50% of the values are below and 50% are above this value. The difference between the 25th and 75th centiles (**quartiles**) gives the **interquartile range**, which is also a measure for the spread in the data.

The **standard deviation** ($S\hat{D}E = \sqrt{\frac{\sum_{i=1}^{N}(X_i - \overline{X})^2}{N-1}}$) is related to the average distance of the data from their mean value. It is especially useful to characterize the variation in the data when the underlying distribution is known. For example, the majority (68%) of the data lie within one standard deviation of the mean ($\overline{X} \pm SD$) if the data follow normal distribution (see below). The share of data increases to 95% if the range is extended to about two times the standard deviation ($\overline{X} \pm 1.96\ SD$). The precise two times range of standard deviation around the mean covers exactly 95.5% of the data. However, if the distribution is unknown then at least 75% of the data will be in the range of $\overline{X} \pm 2\ SD$. The **coefficient of variation** ($CV = SD/\overline{X}$) is a measure of the relative standard deviation and is widely used to compare the variability among different samples and for estimating error propagation (see below). For example, the CV of chemical analysis is usually below 0.05 whereas that of biological experiments may often exceed 0.4.

The **box-and-whisker plot** can be helpful to demonstrate the distribution of a large data set by showing the median and the quartiles of data in a box plot together with the extreme values indicated by the whiskers (Figure 4.16). The median is centered between the upper and lower quartiles if the data come from a symmetrical distribution. Asymmetric portions between the median and the quartiles as well as between the quartiles and the whiskers are indicative of a skewed distribution.

The majority of the toxicological data follow a **unimodal** distribution having a single peak with tails on either side. The most common unimodal distribution is symmetrical, with equal tails on either side, and can be depicted by a symmetrical box-and-whisker plot corresponding to a

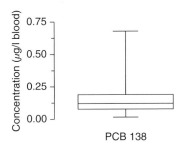

Figure 4.16 Box-and-whisker plot of a polychlorinated biphenyl (PCB138) concentration measured in the blood of children visiting a contaminated school (age 6–16 years; single determinations in N = 218 individuals).

bell-shaped curve (see below). A given sample from such a distribution is summarized by its mean and standard deviation.

A **skewed** or asymmetric distribution is characterized by asymmetric tails. Such a distribution is often encountered if the data are positive and the twofold standard deviation is larger than the corresponding mean value. In such a case the median or the geometric mean and the range between the 5th and 95th centiles can be used as measures for central tendency and variation.

It is also possible that data arise from distributions having more than one peak. For example a **bimodal** distribution might be observed if the data show a corresponding cluster, e.g. they include different sexes or age groups. In such cases, appropriate categorization of the data could result in unimodal distributions.

Example

The polychlorinated biphenyl concentrations plotted in Figure 4.16 show a skewed distribution since the median is closer to the lower than to the upper quartile. Furthermore, the range between the lower quartile and lowest values is several times smaller than the range of the upper quartile and the largest data point. Consequently, the data are summarized using the median for central tendency, i.e. 0.122 µmol/l. The variability is given by 5th and 95th centiles, corresponding to the values of 0.056 and 0.357 µmol/l.

Comment

Before applying any statistical methods to the raw data they should be carefully checked for plausibility and consistency since any error in the data, e.g. due to the measurement itself or data transfer, may have serious implications on the outcome of most statistical procedures.

4.7.3 Error Propagation

> **Continuous raw data measured in an experiment are characterized by their mean and standard deviation. In many cases such mean values are used in further calculations. Since the mean values incorporated in the calculation are affected with uncertainty represented by their standard deviations the result will also exhibit a definite uncertainty.**

To calculate this particular uncertainty in the result the law of propagation of error is used. If means of two independent variables (X, Y) are added or subtracted $(Z = \overline{X} + \overline{Y}, \text{ or } Z = \overline{X} - \overline{Y})$

then the squares of the resulting standard deviation equal the sum of the squares of the two standard deviations ($SD_Z^2 = SD_X^2 + SD_Y^2$). If the two means are multiplied or divided ($Z = \overline{X} \cdot \overline{Y}$, or $Z = \overline{X}/\overline{Y}$) then the squares of coefficient of variations are added ($CV_Z^2 = CV_X^2 + CV_Y^2$). Thereafter, the standard deviation of Z can be calculated using the value of CV_Z^2. For more complicated cases formulas can be derived or found in the literature.

Comment

To use the above formulas it is assumed that the standard deviations are small with respect to their means (<20%) and the two data sets are uncorrelated. By subtracting two numbers of the same magnitude the result could be statistically non-significant (see below) due to an increase in the standard deviation of the result.

4.7.4 Probability Distribution

> A probability distribution is a mathematical function describing the probability associated with all possible values of the random variable within the population.

Large data sets with more than 50 observations ($N > 50$) can also be plotted by means of a **frequency histogram** depicting the frequency of observations in ordered classes. To obtain an acceptable visualization, the x axis is divided into about 7–20 **classes** covering the full range of the data. The number of classes can be approximated as the square root of the number of observations if $N < 500$. Observations falling within a given class are grouped together. In the **relative frequency histogram** (Figure 4.17A) a bar represents the relative frequency of the observations falling within a given class, i.e. the number of observations within the interval divided by the total number of observations. Therefore, the total area of all bars equals one.

The **cumulative frequency histogram** is obtained by adding the bars of the frequency histogram together cumulatively. Each class contains the number of observations that fall within the class plus all that fall below it. The **relative cumulative frequency histogram** is obtained if

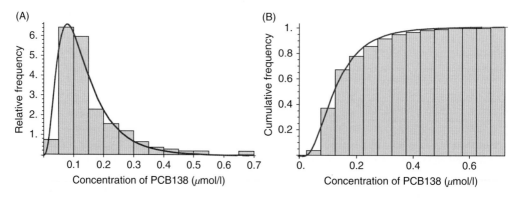

Figure 4.17 *Relative frequency histogram (A) and cumulative relative frequency histogram (B) of the concentration of a polychlorinated biphenyl (PCB138) measured in the blood of children attending a contaminated school (age 6–16 years; single determinations in N = 218 individuals). The solid grey bars correspond to the probability density function (A) and cumulative distribution function (B).*

the class values are divided by the total number of observations (Figure 4.17B). Therefore, this histogram approaches the value of one at its right boundary. The histogram would become less dispersed, i.e. the edges would disappear, and a continuous smooth function would appear as the number of data points approach infinity and the class width nears zero. These smooth functions depicted by the solid lines are referred to as the **probability density function** (Figure 4.17A) and **probability distribution function** (Figure 4.17B). The latter is also called **cumulative distribution function**. In general, the probability density function is specified by a mathematical equation and the cumulative distribution function can be calculated by integrating the density function. The area under a probability density function equals one as does the area under the relative frequency histogram.

Example

The relative frequency histogram of the PCB138 concentration (Figure 4.17A) shows that the data are not symmetrically distributed. Moving from the mode which displays a maximum of about 0.09 µmol/l to the left side, values rapidly decrease to zero whereas on the right side the decrease takes place more slowly and values up to 0.7 µmol/l are reached. The majority of the data points (70%) are located at the right side of the mode. Such a distribution is called "skewed right" and displays a longer tail on the right side.

The **normal** or **Gaussian distribution** plays a central role in statistics. The Gaussian distribution is often assumed to describe the random variation in the data especially when the variability resulting from several random sources acts additively. Such probability density functions,

$$f(x) = \frac{1}{\sigma\sqrt{2\pi}} e^{-\frac{(x-\mu)^2}{2\sigma^2}}$$

are shown in Figure 4.18. The two parameters identified by the Greek letters μ and σ^2 describe the location of the curve on the x axis (**central tendency, expected value**) and the spread of the curve (**variance**), respectively. There are two curves with the same expected value of 10 (Figure 4.18A) located at the same place on the x axis. One of these two curves has a fourfold larger variance, resulting in a flattened, wide shape. All possible normal distributions can be transformed ($z = \frac{x-\mu}{\sigma}$) to a single function located at zero ($z = 0$) and having a unit variance ($\sigma = 1$). This function is called the **standard normal distribution** ($f(z) = \frac{1}{\sqrt{2\pi}} e^{-\frac{z^2}{2}}$) and is depicted in Figure 4.18B. The area under the distribution curve between any two abscissae gives the probability that the x value falls between the specified limits. Tabulated values of the area under the standard normal distribution can be found in many statistical tables.

Comment

The expected value of a distribution, referred to as E(x), indicates its average or central value and is calculated as a probability-weighted average value:

$$E(x) = \int_{-\infty}^{\infty} x \cdot f(x) dx$$

Applying this definition on the Gaussian distribution results in E(x) = μ. Note the analogy between the arithmetic mean and the expected value. It can be shown that the arithmetic mean of a sample approximates the population mean. Therefore, the sample mean value is an estimator

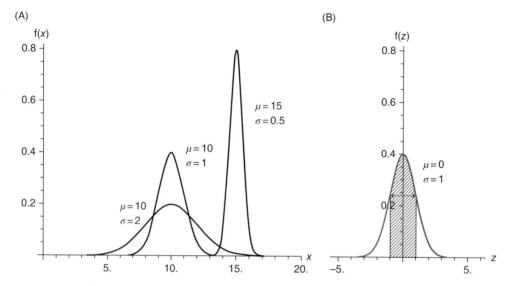

Figure 4.18 Probability density functions of normal distributions (A) with different parameters (μ, σ). The standard normal distribution is depicted in B. The shaded area has a value of 0.68, representing the probability that x falls within one standard deviation of the mean.

of the unknown population mean, if the sample is large enough, sufficiently representative and unbiased.

The variance of a distribution, given as V(x), indicates the spread of the curve. It is calculated as the expected value of the square of the deviation of X from its own expected value:

$$V(x) = E((X - E(X))^2) = \int_{-\infty}^{\infty} (x - E(X))^2 \cdot f(x)dx$$

The probability density function is the weighting function similar to the expected value. Applying this definition to the Gaussian distribution results in $V(x) = \sigma^2$. Due to the analogy between the standard deviation and σ, it can be shown that the square of the standard deviation of a sample approximates the population variance.

Many measurements show a "skewed right" distribution as exemplified for the **lognormal distribution** in Figure 4.17. Such a distribution becomes apparent if the variation results from multiplicative effects. Biological mechanisms are known for inducing lognormal distributions. Many biological variables, such as alveolar ventilation or blood flow through tissue, concentrations of several exo- and endogenous substances in biological material and latency periods of diseases, follow such a distribution. Skewed data such as the blood concentrations of polychlorinated biphenyls can often be transformed to normal distribution just by using the logarithm of the data or plotting them on a logarithmic scale. Consequently, lognormal distributions can be described by the log transformed variable resulting in a normal distribution, which is characterized by its expected value and variation expressed on a log scale. The skewed distribution (Figure 4.17A) becomes a symmetrical bell-shaped curve (Figure 4.19A) following the logarithmic transformation by stretching out the lower end and compressing the upper end of the original distribution. Statistical analysis can be carried out on the log-transformed data before converting the results back to the original scale.

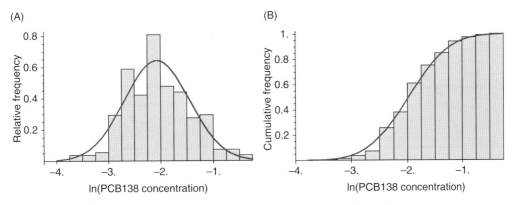

Figure 4.19 *The relative frequency histogram (A) and cumulative relative frequency histogram (B) obtained after a logarithmic transformation of the measured concentrations of a polychlorinated biphenyl PCB138 (see Figure 4.17). The grey bars correspond to the probability density function (A) and cumulative distribution function (5B).*

Example

On a natural logarithmic scale, the mean value and standard deviation of the log PCB138 data amount to –2.072 and 0.621, respectively. By taking the exponent of these values of 0.126 and 1.86 are obtained. Thereafter the range covering 95.5% of the data is calculated on the log scale corresponding to –3.31 (–2.072 – 2 · 0.621) and –0.83 (–2.072 + 2 · 0.621). These values represent the PCB138 concentrations of 0.036 (0.126/(1.86^2) or e$^{-3.31}$) and 0.436 (0.126 · (1.86^2) or e$^{-0.83}$). The interval is calculated in a multiplicative manner ($\overline{X}_{\log}/SD^2_{\log}$ and $\overline{X}_{\log} \cdot SD^2_{\log}$) in contrast to the additive manner demonstrated for the normal distribution ($\overline{X}$-2 · SD and $\overline{X}$ + 2 · SD). The index log indicates that both the mean and the standard deviation were obtained using the log transformed data that followed the "back-transformation" $\left(\overline{X}_{\log} = \mathrm{Exp}\left(\dfrac{\sum \ln(X_i)}{N}\right)\right)$. Note that the mean of the log transformed data ($\overline{X}_{\log}$) corresponds to the geometric mean on the original scale.

> **A visual approach to evaluate whether a given data set follows a normal distribution is based on plotting the data on normal probability paper. This graph paper is the equivalent of logarithmic paper on which exponential functions are depicted by straight lines.**

On **normal probability paper**, a normal distribution yields a straight line since the ordinate is scaled in percentages according to the inverse normal distribution. It covers the ranges 1–99% or 0.1–99.9%, but never includes the adjacent values 0 or 100%. Normally distributed data will appear close to a straight line in the probability plot where the percentiles of 16, 50 and 84% correspond to the abscissae of $\overline{X}$-SD, $\overline{X}$ and $\overline{X}$+SD. A departure of the data points from the theoretical line in the middle region indicates a deviation from normality, especially when the departure follows a given pattern, e.g. a hump, or a divergence. In contrast, a deviation of the data from the line towards its ends is less important since it can result from outliers or sampling effects. The PCB138 data demonstrate a significant deviation from the theoretical line in the probability plot (Figure 4.20A). However, after plotting the data on a logarithmic scaled *x* axis they show a nearly normal distribution, which is fitted reasonably well by a straight line (Figure 4.20B). Probability

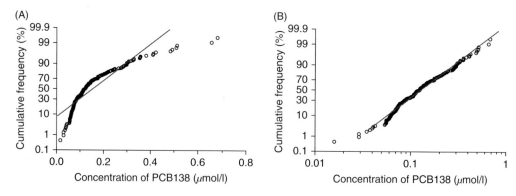

Figure 4.20 Cumulative percentages of the PCB138 data, taken from Figure 4.17, depicted on normal (A) and lognormal (B) probability paper.

paper can be used to test various distributions such as **lognormal, exponential** or **Weibull**. They are available commercially or can be downloaded (www.weibull.com/GPaper).

There are statistical tests to investigate formally whether a random sample stems from the normal distribution. These include the **Shapiro–Wilk test** if $N < 50$, the **D'Agostino–Pearson normality test** if $N \geq 50$, and the **Kolmogorov–Smirnov test**. The first two tests are considered to be more robust and have more power than the Kolmogorov–Smirnov test, although any of these tests might fail if the sample size is too small ($N < 10$).

In statistical theory, the most important distribution is the normal distribution. There are other relevant distributions which can be applied to different fields of toxicological research. For example, the **binomial distribution** can be useful to analyze results obtained by flow cytometry whereas the **Poisson distribution** can be applied to mutagenicity assays. Time to tumor data and survival data can be interpreted using the **Weibull distribution**.

For statistical analysis the **t distribution** which compares means, the **F distribution**, which compares variances, and the **Chi (χ^2) squared distribution**, which compares distributions, are of importance. The areas under their probability density functions can be found in statistical books or tables.

Comment

If the random variable is discrete then the probability distribution is also discrete. For example, the lung cancer deaths in a sample of a given population depict a discrete distribution.

4.7.5 Inferential Statistics

> **The major challenge for many toxicologists is to draw reliable conclusions on the basis of measured or published data. Statistical inference can provide a major tool to support the assessment process, which relies on objective procedures. Inferential statistics generalize about the population on the basis of a collected sample. There are two main methods used for this purpose: estimation and hypothesis testing. Estimation procedures are used to estimate a given parameter of a distribution, such as point estimates for expected value or variance, based on the sample and to construct its confidence interval. Hypothesis testing is used when a comparison has to be made.**

Estimation

> The aim of the estimation is to calculate an unknown population parameter, such as the expected value of the underlying distribution, using the mean value calculated from collected sample (a point estimate). Thereafter, around this single value a confidence interval is constructed enclosing the value of the population parameter with a predefined confidence level.

The mean value of a given sample ($\overline{X}$) represents a single value known as a **point estimate** for the expected value of the normally distributed population (μ) since by increasing the size of the sample the sample mean approximates the expected value. By drawing repeated samples from the same population the calculated means of the samples differ from each other and show a scatter around the true population mean. Its true value remains unknown unless an exhaustive sampling is carried out, which is impossible in most cases. Therefore, it is necessary to provide the point estimate with a measure of confidence. This measure can be derived since the distribution of the sample mean is normal if the population from which the data were taken underlies a normal distribution. The variation of the random variable corresponding to the sample mean is characterized by the **standard error of the mean** (SE), which describes the expected extent of the variation among the means of different samples drawn from the same population. Its value is calculated by dividing σ by $\sqrt{N}$, where σ^2 is the variance of the population from which the sample including N measured data was drawn. The standard error of the mean is a measure of the uncertainty of a single sample mean related to the population mean. For example, the mean calculated from 68% of the samples will lie within the interval of $\overline{X}$ – SE and $\overline{X}$ + SE where $\overline{X}$ – SE is called the lower limit (L_L) and $\overline{X}$ + SE is the upper limit (L_U). By increasing the probability from 68% to 95% the interval becomes less precise and almost two times ($\overline{X} \pm 1.96\mathrm{SE}$) wider. Such a symmetric range defined by its two limits (L_L and L_U) is called the **confidence interval** (CI) in which the population mean can be found with a specified level of confidence. The corresponding probability is called the **confidence level** ($1 - \alpha$) and is associated with the complementary probability called the **significance level** (α). The significance level is assigned before the experiments are carried out, having usual values of 0.05 or 0.01. The multiplier of the standard error can be obtained by looking up in a statistical table the $100(1 - \alpha/2)$ quantile of the standard normal distribution if σ is known. In many cases the squared standard deviation of the sample is used as a surrogate for the unknown variance of the population and SE is obtained as the ratio of SD and $\sqrt{N}$. For such samples, the critical values for calculating the confidence interval are taken from the ***t*-distribution** ($100(1 - \alpha/2)$ quantile) at the specified level of significance α. The *t*-distribution is also a symmetrical distribution, having one single parameter called the degree of freedom. For estimating the confidence level for the mean by taking a sample with size N, the degree of freedom equals $N - 1$. The *t*-distribution approaches the normal distribution if the sample size becomes large enough. The deviation from the normal distribution becomes smaller than 2.5% if the sample size is larger than 50 ($N > 50$).

Example Ten 5-week-old mice were purchased with the following body weights: 22.3, 21.8, 22.7, 21.6, 19.8, 20.4, 22.6, 21.9, 20.8, 22.4 g. The mean value and the standard deviations of this sample are $\overline{X} = 21.6$ and SD = 0.988, respectively. The 95% confidence interval of the mean can be calculated to be 21.6 ± 0.706 ($21.6 \pm 2.26 \cdot 0.988/\sqrt{10}$) by selecting the critical *t*-value of 2.26 that can be taken from a statistical table given for the *t*-distribution for degrees of freedom = 9

498 Toxicology and Risk Assessment

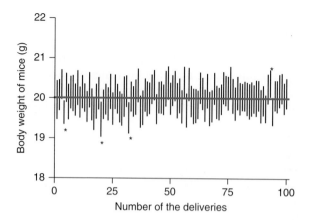

Figure 4.21 The confidence intervals of the body weight calculated for 100 different deliveries containing 100 mice each of similar age by selecting a 0.95 confidence level (grey vertical lines). The solid black line shows the theoretical population mean of 20 g. In fact, there are four deliveries (marked with an asterisk) for which the calculated confidence interval does not include the population mean. The confidence intervals calculated for the major share of deliveries (96%) enclose the population mean.

and $\alpha = 0.05$ (two-sided test) or $\alpha = 0.025$ (one-sided test). Note that the critical t-value is only slightly larger than the corresponding value obtained from the standard normal distribution (1.96). This deviation becomes more obvious if the number of degrees of freedom is smaller.

The probabilistic concept of the confidence interval is demonstrated in Figure 4.21, indicating that the confidence interval calculated from a single sample at a significance level of 0.05 is likely to contain the true population mean 95% of the cases.

The mean ($\overline{X}$) of a normally distributed sample represents only a single value estimate or a point estimate for μ whereas the confidence interval provides an interval estimate by quantifying the degree of confidence that the population mean lies within the calculated confidence interval. Therefore, a confidence interval is considered to be superior compared to a single point estimate.

Comments

About the distribution of the sample mean, interesting findings can be derived from the central limit theorem. If the sample size is large enough ($N > 50$) then the mean of repeatedly obtained samples (sampling distribution) is nearly normally distributed and independent from the distribution of the sample.

Confidence intervals can be constructed for other population parameters, e.g. the median, the variance, or slopes obtained by linear regression.

The confidence interval is a concept similar to the reference interval (95%) used in medicine usually defining the difference between the upper (97.5 centile) and the lower limit (2.5 centile) of a test value when measured in a population of apparently healthy individuals. The International Federation of Clinical Chemistry and Laboratory Medicine recommends reference interval measurements in a least 120 individuals. Upper limits (L_U) rather than estimates of expected values or confidence intervals are used in quantitative cancer risk assessment.

Hypothesis Testing (Significance Test)

> Knowledge is often generated by comparing treated and control groups. To draw an objective decision is possible only with the aid of statistical hypothesis testing. For this purpose, the probability is calculated that the difference observed between the two groups can be explained by chance alone. A distinct characteristic of statistical decision making is that uncertainty cannot be eliminated since decisions must be taken on the basis of limited samples.

Many toxicological studies are carried out to test potential adverse effects of a given chemical by formulating a specific research hypothesis, e.g. styrene is a carcinogen. For this purpose the outcome of a treatment-related effect. i.e. tumor incidence, is measured and a comparison between two groups called "treated" and "control" is made. The only way to confirm a research hypothesis is to reject its negated form. Therefore, the negated research hypothesis is formulated as the statistical **null hypothesis** (H_0) stating in general that the effect variable between the two groups does not differ or, in other words, the treatment has no effect. In contrast, the **alternative hypothesis** (H_1) describes the situation if the null hypothesis is invalid, corresponding to the research hypothesis.

One of the most commonly used tests is the **_t_-test**, which can be used in two different situations. Using this test either a known population mean can be compared with a sample mean or means of two different samples can be compared, provided that the sample(s) are taken from normally distributed populations. The use of t-test will be demonstrated on the following example.

Example

A sample of ten 5-week-old mice was obtained. The mean value and the standard deviation of the body weight are $\overline{X} = 21.63$ and $SD = 0.988$, respectively. The historical records of the breeding laboratory indicate that the mean body weight of such mice in the past has been 22 g. Is the weight of the animals in the delivered sample different from that of the historical controls?

The null hypothesis is that the ten mice have the same mean body weight as the historical population (H_0: $\overline{X} = \mu$). The alternative hypothesis is that they are different (H_1: $\overline{X} > \mu$ and $\overline{X} < \mu$).

By performing the t-test a probability (p) is calculated under the condition that the null hypothesis is true. In the above case the test statistics are given by the difference between the mean weights of the ten mice (sample mean) and the weights of the historical population (μ) in relation to the standard error:

$\left(t = \frac{|\text{sample mean} - \mu|}{\text{standard error of observed mean}} \right)$. For this example the value of 1.2 $\left(\frac{|21.63 - 22|}{0.988/\sqrt{10}} \right)$ is obtained, indicating that the population mean is 1.2 standard errors below or above the sample mean. The outcome of the t-test is the probability p, which corresponds to the grey area in Figure 4.22 and represents the area under the t-distribution over the limits between –infinity and –1.2 and between 1.2 and +infinity. The value of p gives the likelihood that a sample having at least such a large or even larger difference between measured and population mean values can be obtained by chance. In this example p has a value of 0.27. This rather large value indicates that in at least one in four cases a sample can be obtained in which the sample mean shows at least such a large departure from the historical control value. Because such a sample can be obtained frequently, the null hypothesis seems to be very likely. Therefore the null hypothesis cannot be rejected and the conclusion is that the sample mean is not significantly different from that of the historical population.

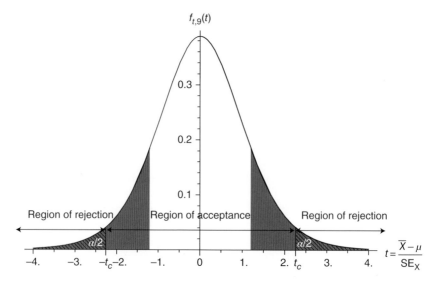

Figure 4.22 The regions of acceptance and rejection for the hypothesis test H_0: $\overline{X} = \mu$ with H_1: $\overline{X} > \mu$ and $\overline{X} < \mu$ by selecting a significance level of $\alpha = 0.05$ as demonstrated using the mouse data set. The area below the curve colored in grey is 0.27. The area below the shaded curve is 0.05. The symbols $-t_c$ and t_c give the upper and lower limits of the region of rejection (see text for details).

In contrast, a very small p value, e.g. 0.001, would indicate that the sample collected is highly improbable if indeed H_0 is true. In other words it would be almost impossible (once in 1000 cases) to observe the data just by chance if the null hypothesis is true. Therefore, the null hypothesis should be rejected in favor of the most likely alternative hypothesis.

It is obvious how to deal with p values above 0.2 and below 0.001. The grey area between these probabilities (0.001 and 0.2) is covered solely by convention. For this purpose an arbitrary threshold value or cut-off point of 0.05 (α) is defined, although another cut-off value (e.g. 0.01) might be considered as well. It is important that cut-off values are defined in advance before data collection takes place. When the p value is below the cut-off point ($p \leq \alpha$) the sample is called significantly different from the population tested, otherwise it is called not significantly different. Nowadays, statistical programs give the actual p value, which should be included in the publication. For this purpose two significant digits are often sufficient for citing p values. Alternatively, critical values for the test statistics (t_c) can be obtained from statistical tables for a given cut-off point. Its value can be used to construct the upper and lower limits of the region of rejection (Figure 4.22). If the absolute value of the test statistics exceeds the critical value (t_c) then the null hypothesis is rejected. For the example given above, statistical tables give a value of $t_c = 2.26$ by selecting $\alpha = 0.05$ and the degrees of freedom (9). Consequently, the null hypothesis cannot be rejected, indicating no statistically significant difference between the sample mean and the population mean ($\mu = 22$).

> **There is a close link between estimation and hypothesis testing since both methods are based on similar theoretical grounds. Therefore, the hypothesis test given above can be converted to calculate the confidence interval at the defined significance level (α).**

The region of acceptance is essentially the same as the corresponding confidence interval. Consequently, the null hypothesis cannot be rejected (i.e. it is accepted) at the defined significance level if the confidence interval of a sample mean includes the hypothetical mean value (μ) given in the null hypothesis. For the body weight example given above, the 95% confidence interval of the mean is calculated to be 21.6 ± 0.706, enclosing the hypothetical mean value of 22. Consequently, no statistically significant difference can be reported between the sample mean (21.6) and the hypothetical mean ($\mu = 22$).

Confidence intervals are considered to be superior compared to a single p value since they present a numerical estimate of the uncertainty. In contrast the p value does not indicate the size of any difference. It solely gives the probability that the observed difference had arisen by chance alone.

The most suitable statistical test depends on the nature of the hypothesis and the kind of variable measured. The result of any hypothesis test is given by a p value, which can be interpreted as discussed above. An overview of the different tests is given in Table 4.21.

Comment

The research hypothesis should be clearly defined in advance to be able to unequivocally formulate the statistical null hypothesis. The null hypothesis can be formulated by an equality stating that there are no differences between the two groups. In this case there are two alternative hypotheses stating that deviations may occur in both directions (colored areas in Figure 4.22). For such a bidirectional hypothesis test two-tailed or two-sided p values are used. The formulation of the null hypothesis by an inequality leads to a unidirectional test, which is performed on a single tail of the distribution. As a consequence the resulting significance is greater in the one-tailed test than in the two-tailed, i.e. a one-tailed test is more likely to detect an effect.

To make a comparison between treated and control groups they should be similar with respect to known sources of variability, e.g. test conditions, season, source of chemicals, etc. The statistical test can either result in a rejection of the null hypothesis or it can fail to reject. Failing to reject the null hypothesis is not the same as verifying that it is correct since statistically non-significant results can be due to several circumstances: too small a difference in the effect variable measured between the groups, too large a variation in the samples, or limited **power** resulting from a small sample size. The sample size necessary to reach a more definitive answer can be calculated in advance by selecting the power of the test (see below).

Parametric statistical tests are based on the assumption of normality and homogeneity of variance. It is further assumed that the sample is randomly drawn from the population and the sample size is large enough to be representative. Some of the tests, such as the *t*-test, are considered to be robust with regard to these assumptions. Statistical tests developed for continuous data cannot be used with ordinal and nominal variables. **Non-parametric tests** should be applied if no assumption can be made about the underlying population, if it is known to be strongly skewed or if measurements were done on a nominal scale. Non-parametric methods may lack power compared with those that are parametric, especially if the sample size is small.

For comparison of multiple group means the **analysis of variance** (**ANOVA**) should be used. For this purpose the variation between the groups is compared to that within the groups. A multiple pairwise comparison based on the *t*-test is not correct since the cumulative probability of the type I error increases rapidly, resulting in multiples of the preselected significance level (α). Typically, the type I error is $n \cdot \alpha$, if n pairwise comparisons are to be made. Consequently, the chance of a false-positive result will increase.

Table 4.21 Overview of statistical methods based on the data type and on distributional assumption.

Aim of the method	Continuous data from normal distribution	Ranks, scores or data from non-normal distribution	Dichotomous or binominal data
Comparison of the mean of one group to a reference value	t-test	Wilcoxon test Kolmogorov–Smirnov test	χ^2-test
Comparison of the means of two unpaired groups	Unpaired t-test	Mann–Whitney test Kolmogorov–Smirnov test	Fisher exact probability test
Comparison of the means of two paired groups	Paired t-test*	Wilcoxon matched pairs signed rank sum test	McNemar sign test
Comparison of the means of three or more groups	Analysis of variance (ANOVA)	Friedman test Kruskal–Wallis test	χ^2-test
Linear association between two variables	Pearson correlation	Spearman rank correlation Kendall's coefficient of rank correlation	Contingency coefficient
Linear relationship between two variables	Simple linear regression	Grizzle–Starmer–Koch linear model	Simple logistic regression
Linear relationship among several variables	Multiple linear regression	Grizzle–Starmer–Koch linear model	Multiple logistic regression
Comparison of the distribution of data with a theoretical distribution	Shapiro–Wilk test D'Agostino–Pearson test Kolmogorov–Smirnov test	χ^2-test Kolmogorov–Smirnov test	

*The distribution of differences should be normal.

> The hypothesis test may result in a false decision since the ultimate truth about the hypothesis is unknown. Furthermore, there is still a non-zero chance that the limited random samples taken might lead to a wrong decision.

Two types of errors are associated with hypothesis testing (Table 4.22). A **type I error** (α **error**) occurs when one would claim a significant difference and reject the null hypothesis although in reality it is really true, i.e. the finding is false positive. A **type II error** (β **error**) addresses the possibility that although the null hypothesis is false one would fail to reject it and therefore generate a false-negative finding. In other words, the β error gives the probability of not detecting an existing difference. The size of an α error has to be determined in advance, e.g. 0.05, whereas that of a β error depends on the α error, the difference in the effect studied, its variation and the sample size. The value of β is used to calculate the **power** of a test ($1 - \beta$) which is the probability of detecting an existing difference and rejecting a false null hypothesis. Therefore, the power is considered as a measure of the quality of the statistical test. A higher power is more likely to result in a test which detects statistically significant results.

In each study enough experiments should be included to ensure that the outcome is not only scientifically but also statistically significant. The necessary sample size can be estimated in advance by **power analysis**. Qualitatively, a two-sided test, a more stringent significant level, and a larger power require a larger sample size. Larger sample sizes are also needed to get reliable results when the average effect difference is small and the measurement is variable. For two-sided problems, comparing a given difference (d) between two groups, a simple formula $N = 1 + 15.7 \cdot \left(\frac{SD}{d}\right)^2$ can be derived by assuming the difference (d) follows a normal distribution, the same variation in both groups and by fixing $\alpha = 0.05$ and $\beta = 0.2$ (usually $\beta = 4 \cdot \alpha$). For more general cases, the probable number of samples can be calculated by using either large and expensive full-featured statistical programs or small free programs (e.g. G*POWER (www.gpower.hhu.de/) or DSTPLAN, which runs under Mac OS or Windows. Such estimates help to design studies with optimal sample sizes. If fewer experiments yield sufficient data, excessive experiments waste resources. If too few experiments are carried out the resulting data may not be adequate to reach statistically significant conclusions. In a study with human individuals or animals, the size of the study should comply with obligatory ethical considerations as well.

Comment

The magnitude of the acceptable type I and type II errors depends on the aim of the research. For example, in a carcinogenicity study the α error leading to the claim that a chemical is a carcinogenic when it is not is not the primary concern. In contrast, the reduction of the β error which argues that the chemical is not carcinogenic when it is, is of utmost importance.

Table 4.22 An overview of the possible test outcomes related to a hypothesis test.

	Test outcome	
	H_0 is rejected	Fail to reject H_0
H_0 is true	False positive (α error)	Correct decision ($1 - \alpha$)
H_0 is false	Correct decision ($1 - \beta$)	False negative (β error)

Example

How many Sprague–Dawley rats are necessary in an experiment in which one wishes to measure at least a 15% exposure-related change in hepatic cytochrome P450 2E1 (CYP 2E1) content? The hepatic CYP 2E1 content in control rats is expected to be 9.7 ± 0.9 nmol/mg protein. The treatment might induce or reduce the CYP 2E1 content. Therefore, the data will be evaluated using a two-sided statistical test. By assuming normal distribution and equal variances in both groups and by fixing $\iota = 0.05$ and $\beta = 0.2$, the number of animals in the control group as well as in the exposed group should equal

$$7\left(1 + 15.7 \cdot \left(\frac{0.9}{0.15 \cdot 9.7}\right)^2\right)$$

After carrying out the experiments in 14 rats, the statistical test should have a power of 0.8 to detect a 15% difference between the group means (9.7 in the control group, either less than 8.2 or greater than 11.2 in the treatment group) if the standard deviation in both groups is 0.9.

4.7.6 Regression Analysis

> **The ultimate aim of many investigations is to derive and characterize dependencies between measured variables. To establish functional relationships among variables regression analysis is a widely used statistical technique. The following investigates the simplest linear relationship between two continuous variables.**

Propylene oxide (PO) concentrations measured in the blood (C_{blood}; **effect variable** or **dependent variable**) of rats exposed to constant atmospheric concentrations (C_{air}; **independent variable** or **predictor variable**) of PO are expected on a theoretical basis to follow a linear relationship, at least over a certain range of exposure concentrations. Furthermore, no PO can be found in the blood of control, non-exposed rats since the compound is not endogenously formed. Therefore, the theoretical relationship can be described in the form of a straight line which passes through the origin ($C_{blood} = m \cdot C_{air}$), and has a slope equal to m. The "best" value of this slope is determined using the data set shown in Figure 4.23A. The criterion of "best" depends on the statistical approach used. For example, the **least-squares method** minimizes the summed deviations between measured (C_{blood}) and predicted ($m \cdot C_{air}$) blood concentrations. Thus, the regression line, depicted by the straight line in Figure 4.23A, is fitted to the data by minimizing the vertical departure of the regression line ($\varepsilon_i = C_{blood,i} - m \cdot C_{air,i}$) from the measured data points. The smaller the residuals become the closer the line fits the data. The value of the sum of squares $\left(\sum_{i=1}^{N}\varepsilon_i^2 = \sum_{i=1}^{N}(C_{blood,i} - m \cdot C_{air,i})^2\right)$ can be considered as a measure of the quality of the fitted line. Graphically, the sum of the squared residuals can be plotted versus the slope depicting a function with a distinct minimum at about 0.058 (Figure 4.23B). Alternatively, statistical programs can be used to perform the necessary calculations, resulting in an estimate for the slope and its standard error ($C_{blood} = (0.058 \pm 0.0011) \cdot C_{air}$).

Comment

In order to carry out a valid regression analysis it is expected that the observations are independent and the dependent variable is normally distributed with a constant variability (homoscedasticity).

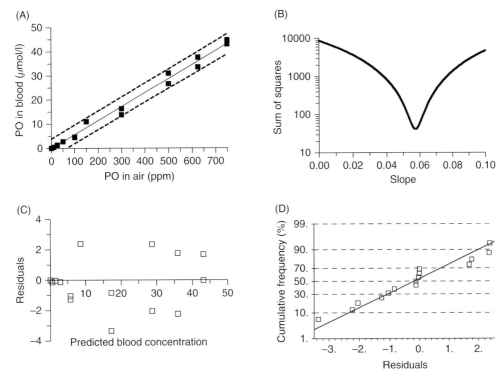

Figure 4.23 Blood concentrations (C_{blood}) of propylene oxide (PO) measured in rats exposed to constant atmospheric concentrations (C_{air}) of PO (A). Symbols represent measured values. The solid line was obtained by linear regression ($C_{blood} = 0.058 \cdot C_{air}$). The dashed lines represent the prediction bands calculated at a confidence level of 0.95. The plot of the sum of the squared residuals versus the slope (B) indicates a minimum at the slope value of 0.058. The plot of residuals versus the predicted value of the blood concentrations (C) depicts no specific pattern. The normal probability plot of the residuals (9D) demonstrates that residuals point closely to a straight line.

The **weighted least-squares method** can be considered if these assumptions do not hold. Furthermore, it is assumed that the independent variable is error free. If the latter assumption does not apply, special regression techniques, such as the **Bartlett regression**, should be used. Finally, a linear relationship is expected between the two variables. The fulfillment of these assumptions can be revealed by a graphic analysis of residuals. The normal probability plot can be used to demonstrate whether the residuals are normally distributed. Departure from linearity, outliers, independence and constant variance become evident by plotting the residuals versus predicted variable or independent variable. These plots should be evenly scattered and should not show any specific pattern such as lines having a non-zero slope, arch or shape of a funnel.

Regression analysis can also be conducted for observational studies when the assumptions given above are fulfilled. However, the interpretation of the regression parameters. i.e. slope or intercept, is often more complex.

The justification for the error-free independent variable is given by the fact that the variability of the air measurements can be neglected when compared to the variability of the blood determinations since the biological variability is expected to be several times larger than the variability resulting from the analytical procedure. Therefore, C_{air} might be considered practically error free when compared to C_{blood}. The necessary assumptions seem to be fulfilled since the scatter of residuals versus predicted variables appears to be random (Figure 4.23C) and also the residuals point close to a straight line in the normal probability plot (Figure 4.23D).

The standard error can be used to calculate the confidence interval of the slope (0.058 ± 0.0022) at a confidence level of 0.95 under the assumption that the blood concentration follows normal distribution. In addition, a hypothesis test stating that the slope equals zero (H_0: $m = 0$ and H_1: **m** $\neq$ 0) can be performed if necessary. For the example given above the slope differs significantly from zero and the null hypothesis should be rejected at $p < 0.0001$.

The regression line given in Figure 4.23A represents predicted values describing the average PO blood levels for a given exposure concentration. The expected confidence interval for the mean of many blood concentrations measured at an identical exposure concentration with a specified confidence level can also be calculated. By repeating the calculation over the entire range of exposure concentrations, upper and lower **confidence bands** for the regression line are obtained. Confidence bands contain the true regression line at a given confidence level.

A confidence interval can also be constructed describing the uncertainties for the blood concentration of a single new observation at a given exposure concentration. Repeating the calculation for a range of exposure concentrations with a confidence level of 0.95, **prediction bands** are obtained in which a measured blood concentration will be with 95% probability (dashed lines in Figure 4.23A). For example, it is expected with 95% confidence that rats exposed to PO concentrations of 400 ppm will exhibit blood concentrations between 19 and 26.9 µmol/l. The prediction band accounting for a single observation is expected to be broader than the confidence band related to an average response.

Comment

The regression line through the origin should be used when the regression line is expected on a theoretical basis to pass through the origin. Equations obtained by regression are often used to predict the value of the dependent variable. The predicted data may be invalid if extrapolation is carried out, i.e. the value of the independent variable is beyond its measured range. If the scatter plot of the data indicates a non-linear relationship then non-linear regression should be used. Computer programs provide effective algorithms to deal with non-linear regression. A linearization of the relationship by transforming the dependent and independent variables is, in general, not advised. The simple linear regression analysis can be extended to more than one independent variable (multiple or multivariate regression).

4.7.7 Probit Analysis

> **Probit analysis is used to determine the quantiles of dose-response curves such as the LD_{50} by linearizing the cumulative normal distribution.**

Dose–response curves observed in biological assays are depicted by S-shaped sigmoid curves similar to a cumulative distribution function (see Figure 4.19B). In fact, the sigmoid curve resembles the cumulative normal distribution if the dose is given on a logarithmic scale and the tolerance levels (threshold below which the dose is tolerated or there are no response results) in the population are normally distributed. Sigmoid curves such as obtained in an acute toxicity experiment are typically investigated by **probit analysis** in order to calculate the median lethal concentration

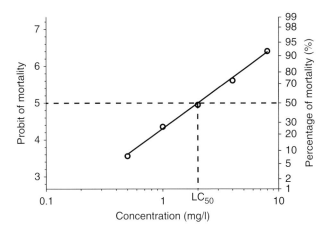

Figure 4.24 *Toxicity of an insecticide to the codling moth (*Cydia pomonella*). Percentage of mortality and probit of mortality as a function of the insecticide concentration. The median lethal concentration estimated graphically corresponds to about 2 mg/l. Circles: simulated data. Solid line: obtained by visual fit.*

(LC_{50}) or median lethal dose (LD_{50}). For this purpose, the S-shaped curve is transformed to a straight line by changing the scales of the dose and the observed frequencies, e.g. the portion of animals affected, using the logarithmic and **probit transformations**, respectively. This transformation is similar to that shown in Figure 4.20B. However, the ordinate is scaled linearly using the so-called **probit**, which is a normally distributed random variable with a mean of 5 and a variance of 1. The ordinate values are positive and linearly scaled in the range between 2 and 8. For example, if 50% of the investigated population are responding, the corresponding probit value on the ordinate is 5. The frequencies of 16% and 84% correspond to the probit values of 4 and 6 (Figure 4.24). To transform the experimentally observed incidences to probit values, suitable tables can be used. Thereafter, the resulting linear relationship is analyzed and the median effective dose is calculated by appropriate regression techniques. For a rough estimation of the median lethal dose, a graphical method using **log-probit paper** or **lognormal probability paper** can be considered (Figure 4.24).

4.7.8 Experimental Designs

> **Statistical design of experiments insures that the necessary amount of the right type of data will be generated in order to reach valid and objective conclusions with an acceptable degree of precision.**

Experiments are performed to confirm or refute hypotheses. Carefully planned experiments are required to enable the researcher to reach an unequivocal conclusion. The ultimate aim of good experimental design is to rule out any alternative causes or confounders which might explain the results, and minimize bias. For this purpose the goal of the experiment should be clearly defined by a research hypothesis before carrying out any experimental work. The further planning of the experimental procedure has to be based on three major principles: **replication**, **randomization** and **stratification (blocking)**.

Appropriate **replication** of the experiment is required to obtain a statistically significant result. The number of necessary repetitions required to ensure significant results can be estimated using the principles of **power analysis**. In general, **replication** reduces variability in the result, which in turn leads to a narrow confidence level and statistical significance.

Randomization is regularly used in experiments to minimize an eventual bias by creating homogenous groups. A completely randomized design is achieved, if subjects are assigned to treatment or control groups completely at random. If the experimental subjects are not comparable with respect to age, sex, treatment schedule, etc. simple randomization will not account for heterogeneity.

Stratification is a technique used to divide heterogeneous experimental subjects into homogeneous blocks (strata) according to some of their properties, such as gender. The resulting **blocking** balances the treatments across the variability represented by the blocks. In a carcinogenicity study each of the treatment and control groups might contain two different blocks assigned to male and female animals.

Using these principles a detailed experimental plan can be established by listing the number of experimental subjects, how they will be selected and assigned to groups, what specific treatments will be applied, in which order these treatments will be carried out, which observations will be taken and which statistical procedures will be used to evaluate the results.

Different types of experimental designs can be realized according to the research objective: **comparative design** (to compare the effect(s) of one independent variable) or **factorial design** (to compare the effect(s) of two or more independent variables or to identify important factors affecting the outcome). A detailed description of such designs can be found in books on advanced statistics. The quality of an experimental design is measured in terms of cost related to experimental efforts and in the accuracy of the conclusions that can be drawn. Finally, it should be taken into account that the most carefully planned experiments will lead only to trivial results if a reasonable research hypothesis is lacking.

Comment

To confirm a hypothesis, it is often required to demonstrate that the investigated phenomenon occurs exclusively under the conditions of a certain treatment. Therefore, it is a good practice to include both a positive and a negative control.

4.7.9 Statistical Software

> **Nowadays, statistical analysis is carried out using computers with the aid of a given statistical software package.**

There are large differences among the packages with respect to price, user-friendliness, graphical design, accuracy and complexity (for a review see http://www.statsci.org/statcomp.html). The major generalists are **SPSS** from SPSS Inc., USA (Windows), **SAS** from SAS Institute Inc., USA (Windows and UNIX), **Statistica** from StatSoft Inc., USA (Windows), **STATA** form StataCorp LP, USA (Macintosh OS X, Windows and Unix) and **JMP** from SAS Institute Inc. USA (Macintosh OS 9, Windows and Linux). Also, software designed for general mathematical or technical computing, e.g. **Mathematica** from Wolfram Research Inc., USA (Linux, Macintosh OS X, Windows and UNIX) or **Matlab** from The Math Works Inc., USA (Linux, Macintosh

OS X, Windows and UNIX), can be considered. There are many other products, such as Prism from Graphpad Software Inc., USA (Macintosh OS X, Windows), which cost less yet but offer limited scope and functionality. The public domain software **R Foundation for Statistical Computing** has been developed within the framework of the Free Software Foundation and can be used on Linux, Macintosh OS X as well as on Windows computers (http://www.r-project.org). The choice of software is often determined by existing installations, convenience or monetary constraints, and by the complexity of the data analysis. In any event, a certain amount of time must always be invested in order to use a given software package efficiently.

The major disadvantage of most statistical software is that it allows the user to perform any test, even one incompatible with the collected data. This might easily lead to a wrong conclusion. Therefore, at least a basic understanding of the statistical tests is indispensable before using any statistical software package. It must be emphasized that data collected in a well-done study can be analyzed in many ways leading to scientifically and statistically significant results, but even the most sophisticated statistical method cannot compensate for shortcomings in the experiments or study design.

4.7.10 Summary

Repeated measurements result in data that are not identical because of random variability. This variability may lead to erroneous interpretation. Statistical methods are indispensable for analysis and description of the random variability and to enable scientifically reliable conclusions.

Statistics should be an integral part in designing an experiment, in particular to obtain a sufficient number of data for the required degree of precision and to minimize bias. Considering the corresponding sample characteristics, data should be presented by means of descriptive statistical methods. Inferential statistics are used to confirm or refute scientific hypotheses.

The present chapter gives an introduction to basic principles of statistical reasoning, including error propagation, linear regression and probit analysis.

Further Reading

Online References

The Engineering Statistics Handbook (NIST/SEMATECH, 2004; http://www.itl.nist.gov/div898/handbook/index2.htm) provides an overview and in-depth details of several statistical methods. The main emphasis lies in the exploratory data analysis. There are many technical but no biological examples.

The online edition of Statistics at Square One (T.D.V. Swinscow) is a popular introduction to medical statistics (http://bmj.bmjjournals.com/collections/statsbk).

Printed References

The reader might refer to books intended for an introductory treatise with practical examples (e.g. Weisbrot, 1985; Miller and Miller, 1988; Gad, 1999).

For more exhaustive coverage it is recommended that readers consult in-depth statistical books either with less mathematical exposition (e.g. Altman, 1996) or with thorough details (e.g. Sokal and Rohlf, 1981).

There are books covering one or more aspects of statistical analysis, e.g. with major emphasis on experimental design (e.g. Box et al., 1978; Montgomery, 2005), on regression techniques (e.g. Kleinbaum and Kupper, 1978), on analysis of quantal response data (Morgan, 1992) and how to report statistical results properly (Lang and Secic, 1997).

Computational References

Straightforward introductions to the public domain software R, which provides a powerful environment for the statistical analysis on multiple computing platforms:
David Rossiter (2003–2006) Introduction to the R Project for Statistical Computing for use at ITC. http://www.itc.nl/personal/rossiter/teach/R/RIntro_ITC.pdf.
Vincent Zoonekynd (2005) Statistics with R. http://zoonek2.free.fr/UNIX/48_R/all.html.

Publications

Altman D (1996) Practical Statistics for Medical Research. Chapman & Hall, London, Weinheim, New York, Tokyo, Melbourne, Madras.
Box G, Hunter W and Hunter J (1978) Statistics for Experimenters. John Wiley & Sons, New York, Chichester, Brisbane, Toronto, Singapore.
Gad S (1999) Statistics and Experimental design for Toxicologists. CRC Press, Boca Raton, London, New York, Washington, D.C.
Kleinbaum D and Kupper L (1978) Applied Regression Analysis and Other Multivariable Methods. Duxbury Press, Boston, Massachusetts.
Lang T and Secic M (1997) How to Report Statistics in Medicine. American College of Physicians, Philadelphia.
Miller J and Miller J (1988) Statistics for analytical chemistry. Horwood Limited, Chichester, Ellis.
Montgomery D (2005) Design and analysis of experiments. Wiley and Sons, New York.
Morgan B (1992) The analysis of quantal response data. Chapman and Hall, London.
Sokal R and Rohlf F (1981) Biometry. W.H. Freeman and Company, San Francisco.
Weisbrot I (1985) Statistics for the clinical laboratory. Lippincott Company, Philadelphia.

4.8 Mathematical Models for Risk Extrapolation

Jürgen Timm

> This chapter describes the use of mathematical models to extrapolate the risk of low-dose carcinogenic exposure starting with high-dose exposure data in experimental animals or occupancy studies.

4.8.1 Introduction

Carcinogenicity is a broad field of research. The results are important for diagnostics, therapies and prophylaxis. Assessing the risk of cancer due to chemical exposition of human populations with typically low exposure is an important goal, and it is also a field of regulation policies. Scientific research tries to detect the mechanisms of carcinogenicity. Regulation praxis, however, needs easy-to-handle upper bonds of risk estimates in the framework of precaution. Such boundaries may be derived by risk extrapolation as usually no empirical data exist for relevant doses near zero. The results of extrapolation are sensitive to model assumptions, thus they are often controversially discussed. The following limitations of these methods should be kept in mind.

Limits of Empirical Methods

There are two main sources of data available: animal experiments and epidemiological studies.

Methods for the determination of carcinogenicity in animals have been standardized. Provided that relevant sources of interference have been eliminated, one can effectively assess the relationship between dose d of an agent (or mixture) and its probability P of inducing tumors. The precision of this assessment mainly depends on the number of animals used. Since many agents must undergo transportation processes via body compartments and metabolic activation to induce carcinogenic effects, the application of animal data to humans is complicated by the need to thoroughly understand mode of action, metabolism, repair mechanism and kinetics in both animals and humans.

Epidemiological carcinogenicity studies face the problem of separating the effect of the agent in question from the effects of a wide range of potentially harmful chemicals to which the population is exposed. The additional or **excess risk** (ER), caused by an additive **exposure** d can be estimated by comparing the risk P_e of the exposed group with the risk P_0 of an appropriate control group providing an estimate of the **relative risk** (RR) and the excess risk by the following equations:

$$\text{RR}(d) = P_e/P_0 \tag{4.1}$$

$$\text{ER}(d) = P_e - P_0 \tag{4.2}$$

These estimates depend heavily on the right choice of controls and corrections for confounders. Moreover, in many cases adequate information about the amount and duration of exposure is missing. This is clearly available in animal experiments. While effects in animal experiments may be interpreted as causal, the effects of epidemiologic studies yield only associations.

Limits of Risk Levels to be Detected

The precision of experimental animal results is highly dependent on the number of animals used. With standard designs and usual numbers of animals risks down to 10% are accessible. Assessment of risks below 10% needs relevant higher case numbers. Considering ethical aspects as well as technical and financial limitations excludes experiments to detect very low risks. If a spontaneous tumor rate is observed, demonstrating the additional effect of small doses is difficult.

The epidemiological method proved to be successful in estimating cancer risk in the occupational setting or in cigarette smokers where exposures tend to be relatively high. Relative risks down to about 1.1 seem to be accessible by sophisticated study design and sufficient number of cases.

Limits of Perspectives

Different perspectives or goals interfere with the process of risk assessment. There are at least two general perspectives to look at carcinogenic risks caused by small doses of carcinogens: the scientific and the regulatory point of view. Scientists try to understand the carcinogenic process in full detail and base their knowledge on empirical data. They are concerned with each individual chemical and its specific mode of action (MOA), often resulting in highly sophisticated **dose–response relations** (DRRs), which are communicated to the scientific community. The real risks and not the extrapolations are the main scientific focus.

The regulatory authorities try to find easy-to-handle estimates of risks in a framework of precaution. In order to protect the population or some sensitive subgroups they prefer generalized concepts that are even applicable where data are lacking, and are appropriate for risk management and communication to the public. Extrapolation of risk data to relevant low doses of everyday contamination is a main focus of regulation processes.

Conclusion

> **Risk evaluation involves mathematical models fitting the measured data. Extrapolation should be as simple as possible and as complex as necessary.**

Neither animal experiments nor epidemiology will, in general, provide direct risk determination for environmental contaminants inducing cancer rates of 1:100,000 or less. It is therefore necessary to extrapolate the DRR from high to low exposure and risk. One should allow some overestimation of risk as a matter of precaution. Mathematical methods solving this problem require a quantitative DRR and an extrapolation function from accessible high-dose risks to risk at very low doses. The mathematics should not be excessively challenging, but should provide a reasonable measure of risks.

With respect to the higher regulatory interest in extrapolation we will consider regulatory aspects of risk extrapolation and its development in more detail than the scientific aspects of DRRs. Thus we start with the most common extrapolation in regulatory frameworks, the linear extrapolation.

4.8.2 Basic Approach of Linear Extrapolation

Extrapolation with Straight Lines

The main principle of linear extrapolation is demonstrated in Figure 4.25: the (real) DRR curve is extrapolated beyond the data range by some straight lines (g_1, g_2 and g_3). The figure refers to the absolute risk, but the results can also be expressed as the relative risk using Equation 4.1.

Linear extrapolation from empirical data (dotted line g_1) tends to underestimate risks at the lower dose range because most DRRs are sigmoid in shape.

The linear extrapolation $P_{\text{lin}}(d)$ (straight line g_2) involves extending the line from some starting point (**point of departure, POD**) of the empirical range to the origin of the coordinates. The straight lines through origin are marked only by their slopes, i.e. by the risk $P_{\text{lin}}(1)$ at the unity point ($d = 1$ unit of the dose scale), called the **unit risk** (UR). Carcinogens may be compared by (linear extrapolated) URs as measures. Conversely the dose d_y associated with a specific risk P_y ($Y = 1\%$, 5% or 10%, for example) is known as the Y **benchmark dose** (BD or BD(Y)). The linear extrapolation also yields an approximation of BD(Y). The relation between both estimates is given by $Y = \text{UR} \cdot \text{BD}(Y)$. Benchmark doses are commonly preferred for interpretation and communication of results as they are defined by (meaningful) standardized risks instead of (arbitrary) dose units. From the mathematical point of view, however, the slope (unit risk) of the straight line is the key parameter.

Straight Lines as Upper Boundary Estimates

Note that UR and BD are not the true parameters of the dose–response curve but linear extrapolations. The real risk $P(d)$ is smaller than the extrapolated risk between zero and the observed dose range for sigmoid dose–response curves (the "standard case" of chemical carcinogens), including DRRs with a threshold parameter. Under this assumption the relation to reality is given by:

$$P_{\text{lin}}(d) = \text{UR}d > P(d) \text{ and } \text{BD}(Y) = Y/\text{UR} < d_Y \qquad (4.3)$$

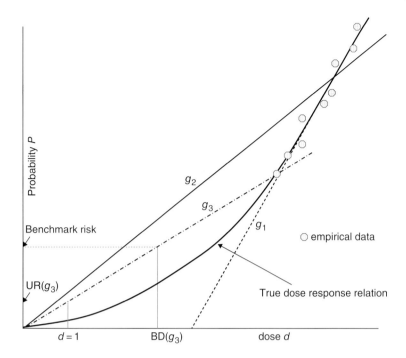

Figure 4.25 Main principles of linear extrapolation.

This relation between reality and extrapolation yields an overestimation of risk and an underestimation of benchmark doses.

The use of linear extrapolation near the origin need not be abandoned due to this overestimation. It may serve as a rule of precaution and it still makes sense to have a simple and understandable method performing an upper boundary of risks near the origin. As long as there is great uncertainty about the shape of the dose–response curve or in the absence of empirical data it may serve as a practical default method. This is the reason for its use by regulatory authorities.

Consistency with Empirical Facts

There is a danger of model oversimplification to the extent that it does not correspond with the empiric data anymore. Therefore the linear model has to be cross-checked with biological knowledge empirical data for each single substance. The regulatory need for a conservative estimate (i.e. risk overestimation) and the scientific knowledge have to be kept in an acceptable relation.

A first step to minimize this problem is to use a POD, left to the empirical range, but consistent with the empirical facts, for example on a 95% confidence limit of a function fitted to the data. Approaching zero contamination will cause problems by increasing uncertainty, however. The widespread praxis is to use a confidence limit for an appropriate BD for the POD.

If more information is available more sophisticated approaches are necessary and possible. Based on theory and empirical results about the MOA and the biological base of the toxicological process the DRR can be estimated more and more reliably.

The curvature of the DRR near zero is the main criterion for a decision about linear extrapolation. The absolute and relative overestimation of precautious extrapolations is an important tool for this purpose (see Chapter 4.6).

4.8.3 Some Special Methods of Linear Extrapolation

There are a number of different proposals for how to derive a linear extrapolation. Three examples are presented here.

Linear Extrapolation from Single Data Points

WHO Method   The simplest method of linear extrapolation was used by the World Health Organization (WHO) when drawing up the air quality guidelines for Europe. Here the individual records, which could be derived from different studies, are evaluated, and unreliable results are eliminated methodically. Using the average of the remaining data points as the POD one draws a straight line to the origin. The UR was defined by "the additional lifetime cancer risk in a hypothetical population in which all individuals are exposed continuously from birth throughout their lifetimes to a concentration of 1 microgram/cubic meter of the agent in the air they breathe".

ICRP Method   The International Commission for Ray Protection (ICRP) used a slightly different method, incorporating only the two data points with the lowest dose of all data.

Then, by using a linear regression analysis, the straight line through the origin can be estimated, which gives the best fit to those two data points. This method is widely used for the assessment of low radioactive contamination.

Extrapolation with Function Adjustment (EPA Method)

A different extrapolation was introduced by the Environmental Protection Agency (EPA) in 2005 to fit a function f starting at the origin and reaching the empirical data. The function f has to be collected from a suitable function class in accordance to data and knowledge. The next steps are to use an effect level, for example $P = 10\%$, calculate BD(P) and its lower 95% confidence limit LBD(P), use <LD(P), P> as the POD and draw a straight line from the POD to the origin as conservative linear approximation of risk near the zero. The EPA has made consistent evaluations of many carcinogens using this method. This approach has two disadvantages: it is quite sensitive to the choice of f and to mistakes in single records.

The EPA recommends this method as a default in case of insufficient knowledge or if there is some evidence that the dose–response function has a linear component, i.e. the tangent slope at origin is greater than zero.

Evaluation of Linear Extrapolation Methods

The three extrapolation methods have different advantages and disadvantages. As the mathematical effort can be kept small, the ICRP and WHO methods have the great advantage of being easily understood by non-experts. In coherence with environment carcinogens the ICRP method has the disadvantage of being based on extreme data points of generally poor precision compared with the remaining data set, which is neglected and therefore prey to high random fluctuations.

This aspect is solved in a much more stable way by the WHO method, especially when pre-results have been averaged. However, it is questionable whether this method uses the available information sufficiently, and whether it is in contradiction with empirical results. It should be checked to ascertain whether the approach is sufficiently correlated with the empirical results for each mathematical model used (see Chapter 4.6 for details).

The EPA method uses the whole data and more existing information about the DRR modeled by *f*, but is bound to the specific choice of allowed functions with linear components.

4.8.4 Consideration of Time Aspects

The sections above consider a risk extrapolation independent of time. This approach is effective for toxicological situations with acute contamination, but it represents the real relation only incompletely in the case of chronic diseases like cancer, as it considers neither the variation in the amount of contamination over time nor the quite different cancer risks in various age groups. Within the regulative scope of precaution one is interested in the risk (excess risk, ER) of cancer dependent on time, which is the focus of the following paragraphs.

Mathematical Description of Risk and Survival

> A more precise mathematical description regards the risk as a function of time, that is, it introduces risk changes in the course of time into the concept.

The following survey shows some definitions and formulas of great importance for this concept. More details can be found in the relevant literature:

- $L_0(t)$ basic **survival function**, probability to survive at least until time t (in an uncontaminated population)
- $s_0(t)$ **basic hazard function**, the border probability of dying by a relevant cancer at time t, if one has been living up to then in an uncontaminated population (with background dose d_0)
- $d(t)$ biological **effective dose** (internal dose of ultimate carcinogen) at a time t
- $s(d,t)$ **hazard function** of contaminated population with $d = d(t)$
- $L_e(t)$ **survival function** in the contaminated population.

Simplification in Environment Carcinogenesis

Survival Function Concentrating on a special carcinogen the general mortality of the population is hardly changed by typical environmental carcinogens, as they contaminate at low dose levels and result in rare cases of cancer only. Therefore one can approximate the model by assuming that contaminated and non-contaminated overall survival functions are equal, i.e.

$$L_e(t) = L_0(t) \quad (4.4)$$

Integrated over the entire life span the risks in question are given by

$$P_0 = \int_0^\infty s_0(t) \cdot L(t) dt \quad (4.5)$$

for the uncontaminated (basic) risk and

$$P(d) = \int_0^\infty s(d,t) \cdot L(t) dt \quad (4.6)$$

for the risk of the contaminated population.

Proportionality Assumption Usually one assumes that the hazard function may be split into the basic (spontaneous) hazard $s_0(t)$, depending only on time t (resp. age) and a dose factor. In order to simplify later formulas this factor will be presented in the specific form $1 + f(d)$, such that the general hazard s is given by

$$s(d,t) = s_0(t)\,[1 + f(d)] \tag{4.7}$$

Equation 4.7 is cited as the **proportional hazard assumption**. Note that in combination with Equations 4.5 and 4.6 the specific formulation in Equation 4.7 implies that $f(0) = 0$ and $P(d) > 0$ if $d > 0$.

Using Equation 4.7 the lifetime excess risk may be written as

$$ER = P(d) - P_0 = \int_0^\infty f(d(t)) \cdot s_0(t) \cdot L(t)\,dt \tag{4.8}$$

From Integral to Sum In reality the precise functions are unknown but there is some information about mean d_i, s_{0i} and L_i for distinct time intervals i ($i = 1, 2, 3, \ldots, N$) of length Δt through the life span. This is clearly the case in animal experiments. Epidemiologic studies may use $s_0(t)$ and $L(t)$ derived from life tables. The integral may be approximated by these interval data as

$$ER = \sum_{i=1}^{N} f(d_i) \cdot s_{0i} \cdot L_i \cdot \Delta t \quad \text{in case of } 5-\text{year life tables } \Delta t = 5 \tag{4.9}$$

Dose Specification

> **Before practical use of these formulas one has to consider the dose over time, including the conversion of high-dose contamination to lifelong low contamination.**

Actual or Cumulated Dose? The carcinogenic effect is not only caused by the actual concentration of the contamination (*actual-dose principle*) as earlier exposures may lead to cancer incidence later on. In some cases the cumulated overall dose (*cumulated-dose principle*) seems to be the causal factor. Varying doses and their impact on carcinogenesis is a broad field of scientific research. As a result neither of these simple concepts (actual or cumulated) is acceptable from a scientific point of view in most cases. A general mathematical concept consists of construction of the effective dose at time t using a function $w(\tau)$ weighting the density of earlier contaminations $\delta(\tau)$ by

$$d(t) = \int_0^t w(\tau) \cdot \delta(\tau)\,d\tau \tag{4.10}$$

Note that "$w = 1$" yields the cumulative and "$w = 0$ except $w = 1$ in the last time unit" the actual concept.

In a regulatory context one normally uses the cumulated dose, forming an upper dose limit. The WHO simplified the calculation by arguing that cancer mortality is mainly concentrated in the final time periods. Thus one may approximate ER by replacing the time-dependent doses in Equation 4.9 by a constant dose near to the overall cumulated dose (see Equation 4.11). The failure will be great in the first time intervals, but the hazard there is very small, such that the impact of this replacement may be tolerable.

$$ER = \sum_{i=1}^{N} f[d_N]] \cdot s_{0i} \cdot L_i \cdot \Delta t = f(d_N) \cdot \sum_{i=1}^{N} s_{0i} \cdot L_i \cdot \Delta t \approx f(d_N) \cdot P_0 \tag{4.11}$$

Constant Exposition For a constantly contaminated population with actual dose δ per year the average cumulated dose during a 5-year interval i is given by

$$d_i = \delta \cdot (i - 0.5) \cdot 5 \qquad (4.12)$$

and

$$ER = \sum_{i=1}^{N} f[\delta \cdot (i - 0.5) \cdot 5] \cdot s_{0i} \cdot L_i \cdot 5 \qquad (4.13)$$

Following the WHO arguments the relevant cumulated dose is almost the life-long cumulated dose for 70-year-olds (age group 70–75). Taking this approximation, the effect will be overestimated and the result stays on the safe side:

$$ER \approx f(\delta_{70}) \cdot P_0 \qquad (4.14)$$

Dose Conversion for Different Time Slots of Exposure

Using working studies for the estimation of environmental risks one has to transform the dose of short time interval exposure at work to lifelong exposure. Usually two factors of conversion are introduced, for example:

$g_1 = 8/24$ for the difference between the duration of the working day and the calendar day

$g_2 = 240/365$ for the different numbers of working days, and the days of the calendar year.

If T_{work} denotes the years of contamination in the workplace studied and T is the expected lifetime of the population in question, then the conversion of constant working-place dose δ into constant lifetime dose δ' is given by

$$\delta' = \delta \cdot g_1 \cdot g_2 \cdot (T_{work}/T) \qquad (4.15)$$

The situation is far more complex if an extrapolation from animal experiments has to be calculated.

Determining the Dose-effect Function **f(d)**

The dose-effect function f, introduced in Equation 4.7, has to be further specified. We demonstrate the method for a typical situation with an epidemiologic workplace study as source of information. Analyzing certain subgroups j ($j = 0, 1, 2, 3, \ldots, k$) with cumulated dose d_j or constant actual dose δ_j such a study reports associated cancer rates P_j, excess rates ER_j or relative risks RR_j. Let $j = 0$ denote the control group with zero exposition with respect to the carcinogen in question. Applying the approximation and cumulative dose concepts above one gets for $j = 1, 2, 3, \ldots, k$:

$$f(d_j) \approx ER_j/P_0 \quad \text{or} \quad f(d_j) \approx 1 - RR_j \qquad (4.16)$$

The function f as a whole may then be estimated by connection, or inter- or extrapolation of these point estimates for dose group j. A superior approach is fitting f out of a convenient, empirical or theoretically derived function class to the data. Unless the deviation from empirical facts is too great the linear approximations discussed above may be directly applied to $f(d)$ or to the excess risk given by $ER(d) = f(d)P_0$ as described there.

Non-linear Extrapolation

Non-linear extrapolations may be discussed using the Taylor expansion of the DRR function f at $d = 0$:

$$f(d) = a_1 d + a_2 d^2 + a_3 d^3 + \ldots \tag{4.17}$$

In this formula a_1 is the slope of the tangent at the origin. Note that a_0 is omitted as $f(0) = 0$, hence $a_0 = 0$. Further $a_1 \geq 0$ as the risk is a number between 0 and 1.

As long as $a_1 > 0$, i.e. f has a positive linear component, the relative overestimation by linear extrapolations will converge to a finite constant, the quotient of the estimated UR and a_1 at zero. The absolute overestimation is limited by the extrapolated risk at the POD. From this there are concrete criteria to decide if the precautious overestimation is acceptable.

If $a_1 = 0$ then the DRR becomes asymptotic to the dose axis for $d \to 0$. In this case a linear extrapolation may be too conservative because the relative failure made by a linear extrapolation tends to infinity approaching $d = 0$.

Consider first the "normal" case $a_i > 0$ for the smallest non-zero coefficient in Equation 4.17, i.e. f is convex (left curved) near zero. This holds for all sigmoid functions as well as for DRRs with a threshold parameter d_0 that is statistically distributed (e.g. probit models). For threshold models with fixed $f = 0$ for $d < d_0$ a right shift of the origin to d_0 may be applied and d interpreted as the dose exceeding d_0, while the risk is zero beneath d_0.

A possible solution in this situation is to construct a 95% confidence limit function $Lf(d)$ of $f(d)$ which is forced to pass the origin as conservative non-linear extrapolation. Another possibility is to define an acceptable risk P_a, for example 1:100,000 or 1:1,000,000, determine the corresponding dose d_a by solving $f(d_a) = P_a$ and construct a 95% confidence interval and its lower bound Ld_a. Expositions below Ld_a may then be handled as acceptable. These are only two of the possible derivations of conservative (non-linear) extrapolations in this case.

A different approach is needed if hormesis appears, i.e. the DRR is negative near zero (probability smaller than P_0). In that case the first non-zero coefficient a_i in Equation 4.17 will be negative but there exists a cutting point d_0 with $f(d_0) = 0$. From a toxicological or a regulative (defense of risks for the population) point of view f might be approximated by a threshold model for $d > d_0$ and extrapolation done as above. If the positive effects of very low expositions should be taken into account, more information about the U-shaped function is needed.

A rare case is a DRR curve with right curvature near zero. For such DRRs it is not possible to find an upper risk limit by a straight line from a POD on the curve or its surrounding to origin. However, one can still find a straight line serving as the upper limit of the effects by using a confidence limit of the gradient at the origin. The relative overestimation will then tend to a constant near zero but the absolute overestimation may be high for a dose $d > 0$. The only exception is a tangent angle of 90°, where no extrapolation through origin but a straight line from a POD parallel to the dose axis may serve as the upper limit for small doses, but again the relative overestimation will tend to infinity if d approaches zero. Both limitations of straight lines as upper boundaries for right curved DRRs are omitted if a confidence bound for the entire function through the origin is constructed.

4.8.5 Models of Carcinogenesis

Principally, one can extend these derivations to any kind of dose–effect relationship. Empirical data may be fitted using an appropriate function class. Conservative extrapolations (linear or not) may be derived following the rules above based on the Taylor approximation near the origin, but

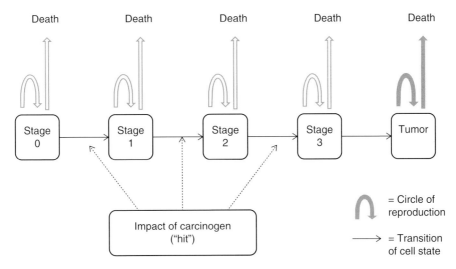

Figure 4.26 *Multihit multistage model. Reproduction and death are balanced until in the last state reproduction outnumbers deaths developing a tumor. Example with three hits and four stages.*

this will be considerably dependent on the choice of the respective function used. This is not satisfactory and leads to the demand that this approach must not be arbitrarily set up, but should be based on a simple and fundamentally plausible biological model.

Multihit Multistage Models

Such a model has been introduced for carcinogenic processes in the early works of Armitage and Doll, assuming that the cells have to pass a number of stages before they mutate to malignant tumor cells. The transition between certain stages can be triggered by small doses of harmful chemical substances, i.e. in this family of models the stage transition probabilities are functions of those doses. The reactions between carcinogen and cells causing these transitions from stage to stage are called hits. Figure 4.26 shows a graphical representation of the model. Several hits may be necessary in order to induce a single-stage transition and the probability of this determines the probability of stage transition.

Depending on how many stages and hits are considered, one arrives at some simplifying assumptions for formulas for the expected risk, at dose d and time t, which contain both as special cases.

For the simpler case discussed here the following equations hold:

$$P(t,d) = 1 - \exp[-a \cdot d \cdot t] \quad \text{one hit, one stage} \quad (4.18)$$

$$P(t,d) = 1 - \exp[-a \cdot d^b \cdot t] \quad \text{multihit, one stage} \quad (4.19)$$

$$P(t,d) = 1 - \exp[-a \cdot d \cdot t^k] \quad \text{one hit, multistage} \quad (4.20)$$

$$P(t,d) = 1 - \exp[-a \cdot d^b \cdot t^k] \quad \text{multihit, multistage} \quad (4.21)$$

The most general formula of this sequence (Equation 4.21), with b hits and k stages, is suitable to describe many relevant experiments in cancer research. It yields a Weibull-type distribution function.

Expansions and Alternative Models An additional term modeling a basic risk and a time to realize that a cell is in state "tumor" (tumor is visible) may be added.

A more complex result is derived if one considers not only stages in balance between death and replication, but also a last stage with an increased replication rate ("tumor"). In this set up the replication rates vary and are not balancing deaths. Thus there may be stages with increased clonal expansion due to exposition in between. These models are defined as multistage clonal expansion models. The problem is that the calculation of risks becomes challenging and there are no analytical solutions to access the risk by numerical approximation only.

If sufficient biological information is available it may serve as a starting point for more complicated model considerations, leading to different biologically based cancer-risk models.

We concentrate here on the simpler model given by Equation 4.21 as the main principles have been demonstrated already with this approach and many experiments have been successfully analyzed by it (see next subsection).

Empirical Evidence for the Multihit Multistage Model An implication of Equation 4.21 is that for a given risk P the following equation holds:

$$d^b t^k = -\log(1-P)/a = \text{constant} \qquad (4.22)$$

The Druckrey law is a special case of Equation 4.22. It states that the time points t and respective dose d for $P = 50\%$ follow this rule. Thus it is an early example to verify the consequence of Equation 4.21. Many animal experiments with carcinogens demonstrate the validity of this law within the limits of their precision. Figure 4.27 presents an example of this.

In order to further check Equation 4.21 graphically one can use a double-logarithmic scale for age-specific tumor incidence and age, where Equation 4.24 predicts a bundle of parallel straight lines for various doses. Figure 4.28 shows this principal for a typical animal experimental setting (with different doses of benzopyrene). In human populations one can find similar parallel lines, as Armitage and Doll have pointed out.

Extrapolation to Small Doses What does a transition to smaller doses at fixed observation times implies for this model? One easily recognizes that all models are approximated by the following

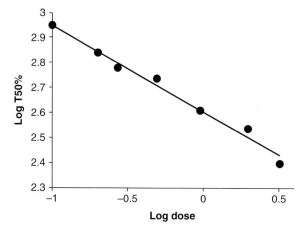

Figure 4.27 Example for Druckrey's law (taken from Duckrey et al., 1962) showing tumor-induction times up to 50% for tumor-bearing animals versus dose of 4-(dimethylamino)azobenzolandrats.

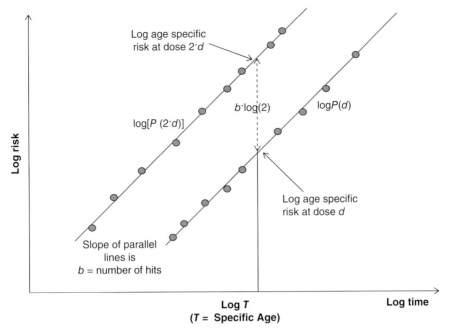

Figure 4.28 *Graphical check of the multihit multistage model by parallel lines in double logarithmic scaling. Many mouse skin experiments have been analyzed this way in concordance with the model.*

formula near to zero (for fixed $t = T$):

$$P(d) = ad^b \qquad (4.23)$$

This result is important as there are two cases:

For $b = 1$ the model collapses to a one-hit model with possibly many stages. The linear extrapolations discussed above yield acceptable overestimations for regulatory use in this case (and only in this case). In some publications these models are called linear no threshold (LNT) models, although linearity is only given near zero.

For $b > 1$ a non-linear extrapolation (see Chapter 4.6) should be used.

4.8.6 Assumptions and Limits of Extrapolation in Mathematical Models

Finally, the underlying assumptions and limits of extrapolation methods should be discussed. We will limit ourselves here to the four most important requirements.

Linearity

Linearity surely is the most often mentioned and discussed assumption. We have mentioned above that a non-linear DRR function can be approximated by a straight line at the origin if the slope of its tangent at the origin is positive. In this case the linear extrapolation is acceptable because absolute as well as relative deviation of extrapolation is limited.

For (near-zero) left curved DRRs with zero tangent angle a linear extrapolation from the POD to the origin may be used as a matter of precaution as the DRR stays sublinear near the origin.

The absolute overestimation is still limited but the relative overestimation tends to infinity if the dose approaches zero. Thus linear extrapolation is less acceptable in this case.

Right curved DRRs allow conservative linear extrapolations with limited relative overestimation as long as the slope of the tangent at the origin is smaller than 90°. However, one has to be careful as in this case the absolute risk difference may be great. Non-linear confidence limits may be the better choice.

Other shapes of the DRR, although relatively rare, cause greater problems and special extrapolations, mostly not linear, are the method of choice.

Summing up, the linearity of the DRR is not postulated for linear extrapolations in the framework of precaution. Positive linear components of the DRR function are sufficient for acceptable linear extrapolations. Linear upper bounds may be constructed in other cases but they suffer from essential problems and have to be discussed for each single carcinogen.

Additivity

The additivity of risks is often postulated to construct upper limits of risk estimates for mixtures of carcinogens by the equation

$$P(d_1, d_2, \ldots, dk) = P(d_1) + P(d_2) + \ldots + P(d_k) \tag{4.24}$$

At the extremely low dose level for environment toxic substances this practically leads to an independent effect of small single contaminations, a situation which surely exists near the origin in a mathematical sense. The question is whether the field of relevant doses as a whole is close enough to zero to ensure that this condition works. This seems to be the case with most mixtures of carcinogens in the environment evaluated up to now. Of course this does not mean that the interaction with massive other contamination, such as smoking, stress, or certain working factors, can be neglected. As far as such other sources are concerned, complementary reconsiderations are necessary. The behavior of mixtures is a broad field of research.

Proportionality

The proportionality assumption in Equation 4.7 yields that the relative risks are constant for various doses over all observation times and age groups. The Cox model widely used in epidemiology requires this assumption. When transferring results from occupational studies to environment situations, one has to consider that for work contamination this assumption is often not true. Typically, one rather finds a temporal course of the relative risk between case and control group computed on the basis of the same age structure, as shown in Figure 4.29. One has to pay attention to a sufficiently extended observation time for the work study, so that the relative risk, which is connected to the contamination at work, can be correctly evaluated.

Accumulation of the Dose

The usual transformation of job-related data to life-long contamination, typical for environmental risk evaluation, assumes that the effect relates only to the overall dose and not to the time pattern of the contamination. This condition has to be seen critically, for theoretical reasons, and will probably be quite rare. Time-dependent risk evaluation for cohort studies with working exposure shows an increase of risk during exposure, which may be associated with the cumulative exposure. In general, further evaluation of these cohorts does not present a stable level of risk after exposure but a decrease, which cannot be explained by the constant cumulated dose alone. The decrease may be steep (type 1 in Figure 4.29) resulting in a risk at control-group level, or slowly flattening out and remaining above that level (type 2 in Figure 4.29).

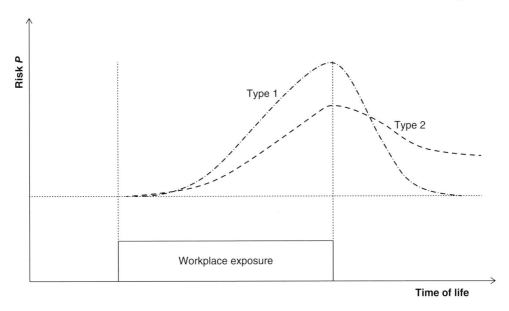

Figure 4.29 Evaluation of occupational studies show different risk curves over time (see type 1 and type 2).

The problem of induction time, which for environmental cancer surely may last for decades, shows that the overall dose cannot be decisive alone. This problem deserves greater attention, considering the risk evaluation. In doing so, the impact of toxicokinetics and metabolism as well as repair mechanisms have to be considered. By this the weight functions $w(t)$ of Equation 4.10 may be specified in detail. The results received so far are too incomplete to solve this problem for many of the carcinogens, so there is a need for more research on this concept. First results indicate, however, that the evaluation on this basis of the total dose leads to upper limits that should be corrected at least by using the concept of induction times.

4.8.7 Summary

The risk resulting from exposition to toxic substances, especially carcinogenic materials, can be evaluated by epidemiological studies or animal experiments. For different reasons both deliver information about effects of higher doses but fail to result in exact data for low-dose effects in human populations, therefore one has to extrapolate the DRR near zero by mathematical methods.

In the framework of precaution regulative authorities are interested in easy-to-handle extrapolations as upper limits for possible risks. Different approaches for linear extrapolations fulfilling this demand are presented here in detail. Because of its simple mathematical structure, the concept of straight lines through the origin is well suited as a base for risk management, but the limits and assumptions behind this method have to be taken into account. In many cases this tends to overestimate the risk considerably. Scientific results about the mechanism of cancer, the mode of action and the function classes to be fitted to experimental data allow better (non-linear) extrapolations if enough evidence is reached. If relevant uncertainty remains, some linear extrapolations (e.g. the EPA method) may be applied in regulative settings.

In either case a biologically based cancer risk assessment should be the goal of further research, and results within the framework of this research program will provide a better fit of mathematical models (some of them newly derived) and practical use of more sophisticated procedures based on such models.

Further Reading

Alison M (ed.), 2007. *Cancer Handbook*, 2nd edition, John Wiley & Sons, West Sussex.
Armitage P, Doll R, 1961. A two-stage theory of carcinogenesis in the relation to age distribution of human cancer, *Br. J. Cancer*, **11**, 161–169.
Calabrese EJ, Staudemayer JW, Stanek EJ, Hoffmann GR, 2006. Hormesis outperforms threshold model in NCI anti-tumor drug screening data. *Toxicol, Sci.*, **94**, 368–378.
Cederlöf R, Doll R, Fowler B, Friberg L, Nelson N, Vouk, V (ed.), 1978. *risk assessment methodology and epidemiological evidence*. Environ. Health Prospect., **22**, 1–12.
Cogliano VJ, Luebeck, EG, Zapponi GA (ed.), 1999. *Perspectives on biologically based cancer risk assessment*. NATO, Challenges of a Modern Society, volume **23**, Springer Science & Business Media, LLC.
Cox DR, 1972. Regression models and life tables. *J. R. Statist. Soc. B*, **34**, 187–220.
Davidson JWF, Parker JC, Beliles RP, 1986. biological basis for extrapolation across mammalian species. *Regul. Toxicol. Pharmacol.*, **6**, 211–237.
EPA (Environmental Protection Agency), 1984. The Carcinogen Assessment Group's Final Risk Assessment on Arsenic, EPA-600/8-83-02IF.
EPA (Environmental Protection Agency), 1995. *The use of the benchmark dose approach in health risk assessment. Office of Research and Development*, Washington DC, US EPA 630/R-94/007.
EPA (Environmental Protection Agency), 2005. *Guidelines for carcinogen risk assessment, Office of Research and Development*, Washington DC, US EPA 630/P-037/001F.
Hsu H-C, Stedeford T, 2010. *Cancer Risk Assessment*, Hsu H-C, Stedeford T (eds.), John Wiley & Sons Hoboken, New Jersey.
IRCP (International Commission on Radiological Protection), 2007. Report No 103.
Lee P, O'Neill J, 1971.The effect of time and dose applied on tumor incidence rate in benzopyrene skin painting experiments. *Br. J. Cancer*, **25**, 759–770.
Malarkey DE, Maronpot, RR, 2005. Carcinogenesis, in *Encyclopedia of Toxicology*, 2nd edition, Wexler, P. (ed.), Elsevier, New York.
Meza R., Jeon J, Moolgavar SH, 2010. Quantitative cancer risk assessment of nongenotoxic carcinogens, in *Cancer Risk Assessment*, Hsu H-C, Stedeford T (eds.), John Wiley & Sons Hoboken, New Jersey.
Portier CJ, Sherman CD, Kopp-Schneider A, 2000. Multistage, stochastic models of the cancer process: a general theory for calculating tumor incidence. *Stochastic Environ. Res. Risk Assess.*, **14**, 173–179.
Rechard RP, 1999. Historical relationship between performance assessment for radioactive waste disposal and other types of risk assessment. *Risk Anal.*, **19**, 763–807.
WHO, 1987. Air Quality Guidelines for Europe, WHO Regional Publications. European Series, 23. 494 Toxicology and Risk Assessment: A Comprehensive Introduction.

5
Regulatory Toxicology

5.1 Regulations on Chemical Substances in the European Union

Werner Lilienblum and Klaus-Michael Wollin

5.1.1 Introduction

Until the middle of the last century, each country in Europe had its own legislation on prevention and managing technical dangers, including those from chemical substances. Increase in transboundary trade in Europe and worldwide required more and more harmonized rules on trade in an increasing common market in Europe. After a predecessor, the European Coal and Steel Community (ECSC), established in 1952, the Treaty of Rome entered into force in 1958 setting up the *European Economic Community (EEC)* and the *European Atomic Energy Community (Euratom)*. Whereas the Euratom Treaty essentially remained unchanged over decades, there were several successor or amendment treaties modifying structures of the Community, extending the number of political areas beyond the Common Market, and specifying the goals on harmonization over a period of half a century. In the context of this chapter, only the Treaty of Maastricht on a European Union (TEU), which established the *European Community (EC)*, and the Treaty of Lisbon on the *European Union (EU)* need mentioning. The latter is also termed the Treaty on the Functioning of the European Union (TFEU). These treaties entered into force in November 1993 and December 2009, respectively, and were amended later. During this time, the number of Member States that joined the Community and the Euratom Treaty increased from 6 to 28. The European institutions were established. Those relevant for European legislation and for the purpose of this chapter are shown in Table 5.1. The EU is based on the rule of law: Everything it does is founded on treaties voluntarily and democratically agreed by its Member States. The EU is also governed by the principle of representative democracy, with citizens directly represented at EU level in the *European Parliament* and Member States represented in the *European Council*.

EU legislation has replaced, amended or otherwise modified the national law of the Member States on a voluntary basis in many areas for the purpose of harmonization and achieving common law among Member States where required according to the objects and aims of the treaties. It confers rights and obligations on the authorities in each Member State as well as individuals and businesses. The authorities in each Member State are responsible for implementing EU legislation in national law and enforcing it correctly, and they must guarantee citizens' rights under these laws. EU case law is made up of judgments from the *Court of Justice of the European Union*, which interprets EU legislation.

Toxicology and Risk Assessment: A Comprehensive Introduction, Second Edition. Edited by Helmut Greim and Robert Snyder.
© 2019 John Wiley & Sons Ltd. Published 2019 by John Wiley & Sons Ltd.

Table 5.1 Main players in European legislation.

Court of Justice of the European Union (CJEU)	since 2009*
European Commission (COM)	since 1958
European Parliament (EP)	since 1962**
European Council (EC)	since 1974

*With predecessors established in 1952 and 1958, respectively.
**First direct elections in 1979.

5.1.2 Current Legislation in the EU

The EU has legal personality and as such has its own legal order. EU law is divided into primary and secondary legislation. Primary legislation covers the TEU and TFEU treaties (consolidated versions in 2016) and general legal principles as ground rules for all EU actions. The treaties lay down in which specific areas they confer the EU exclusive competence, for instance customs union and competition rules necessary for the functioning of the internal market. Rules and areas with shared competence between the EU and Member States apply among others in the following principal areas: agriculture and fisheries, environment, consumer protection, transport, and common safety concerns in public health matters. As can be seen in *EU Legislation on Chemical Substances and Their Uses* and *Legislation on Chemical Substances in the Environment* (Chapters 5.1.5 and 5.1.6) shared competence in the legislation on chemical substances in these areas may vary depending on the issue and is often delicately balanced between EU institutions and Member States. The working environment is another important field where chemicals are widely used in certain industry sectors. However, besides declarations of intent, the EU has less competence because it only "shall support and complement the activities of the Member States" (Articles 153 ff. TFEU).

EU secondary legislation is derived from the principles and objectives set out in the treaties. A set of legal instruments with different legal implications has been established, taking into account case by case different necessities of harmonization on one hand, but on the other hand not interfering in the domestic systems of law any more than necessary. The instruments of secondary legislation are binding legislative acts, namely regulations, directives, and decisions, and non-legislative acts issued by the European institutions, such as recommendations, communications, and opinions that are not legally binding (Table 5.2). *Directives* are binding upon any or all of the Member States to whom they are addressed, but national authorities must adopt a transposing act or 'national implementing measure' to transpose directives and bring national law into line with their objectives.

Areas that are not regulated by binding EU legislation remain, of course, under the national legislation of the Member States. This also applies for gaps in EU legislation where either no necessity exists or no compromise can be reached. For instance, protection from ionizing radiation by radionuclides is in general covered within the Euratom treaty, mainly by directives, and mainly requires additional national protection rules. Also, whereas an EU law for the protection of groundwater exists, other sectors of soil use and protection are subject to the legislation of the Member States (see *Soil Protection*, this chapter). Gaps also exist in the area of consumer protection. In Germany, for instance, commodities that are often in contact with the body, such as textiles etc., do not fall under EU law but are regulated by national law.

Legal acts are planned and adopted taking into account the principles of subsidiarity and proportionality, both defined in Article 5 TEU and described in Protocol No. 2 to TFEU. The *principle of subsidiarity* aims to ensure that decisions are taken as closely as possible to the citizen and that constant checks are made to verify that action at EU level is justified in light of the possibilities available at national, regional or local level. Specifically, it is the principle whereby the EU does not take action unless it is more effective than action taken at national, regional or local level. It is

Table 5.2 Features of European legal acts.

Type of legal act	Addressees	Legal effects
Regulation	All Member States, natural and legal persons	Directly applicable and binding in their entirety
Directive	All or specific Member States	Binding with respect to the intended result
		Directly applicable only under particular circumstances
Decision	Not specified	Directly applicable and binding in their entirety
	All or specific Member States, specific natural or legal persons	
Recommendation	All or specific Member States, other EU bodies, individuals	Not binding
AVIS	All or specific Member States, other EU bodies	Not binding

Source: Borchard KD (2010) The ABC of European Union law (modified) (Borchardt, 2010).

closely bound up with the *principle of proportionality*, which requires that any action by the EU should not go beyond what is necessary to achieve the objectives of the treaties. Both principles also apply in the Member States at the national and subnational level. Both principles play an important role in the legislation, for example, on chemical substances when the roles, duties and responsibilities of authorities in the Member States or at EU level have to be defined or when adequate risk management actions have to be taken.

Three different EU institutions are involved in the development of a new law or modification of an existing legislation, the *European Commission*, the *European Council* and the *European Parliament* (Table 5.1). Whereas the *European Commission* implements EU legislation and is the sentinel of EU treaties and existing EU legislation, it is also responsible for proposals of new or changes of existing legislation. However, if a policy area is not cited in a treaty, the Commission cannot propose a law in that area. After a proposal of the Commission has been discussed and agreed by both Parliament and Council, the new or modified law is approved by Parliament and Council. Because different actors take part in the development of a new law or modification of an existing law (Commission, Council, Parliament and many stakeholders, political and lobby groups in the Member States), such changes often take several years until agreement and adoption, if any.

As an example, the first chemicals law at the European level, Directive 67/548/EEC, was adopted in 1967. Over decades until 2007, it underwent many updates and other changes, mainly adaptions to technical progress (ATP) due to, for instance, increased scientific knowledge about properties of the chemical substances. Directive 67/548/EEC was repealed in 2007 and replaced by Regulation 1907/2006/EC.

How to Find and Deal with EU Legislation

Any EU law is based on the respective treaty and identified by the year of issue and a number. Hence, the acronyms /EEC and /EC of directives and regulations go back to the political and legal predecessors of the European Union mentioned above, i.e. the European Economic Community (EEC, 1967–1993) and the European Community (EC, 1993–2009). From 2010 on, the acronym /EU is used. Directives and regulations can easily be identified by their notations. Directives are numbered by the year of issue followed by consecutive numbering, for instance 67/548/EEC or 2001/83/EC. The notation of regulations starts with a number followed by the year of issue, for instance 1907/2006/EC. A different official notation is, for example, Regulation (EC) No 1907/2006, but the former is shorter and will be used throughout this chapter.

All EU laws, recommendations, and communications are available at the website http://eur-lex.europa.eu/. A complicated feature of EU laws is that any changes or corrective actions are issued as separate law documents which reference to the original directive or regulation. Thus, over some time, a rag rug of law documents may be generated. For instance, the Chemicals Regulation 1907/2006/EC underwent 17 changes and amendments and 5 corrigenda within six years until 2012. From time to time, scattered law documents published in different issues of the *Official Journal of the European Union* are combined, indicating the changes in one easy-to-read document that is published by the European Commission. However, such consolidated texts are only intended for use as documentation tools, and the institutions do not assume any liability for the contents. For such consolidated legislation documents see http://eur-lex.europa.eu/collection/eu-law/consleg.html. For legal purposes one should refer to the texts published in the *Official Journal* that are accessible at the website Eur-Lex via the link above.

5.1.3 Risk Issues and Some Definitions in Terms of Chemical Substances

Risk Assessment, Risk Management, and Risk Communication in the EU

Risk governance refers to the institutions, rules conventions, processes and mechanisms by which decisions about risks are taken and implemented. Risk governance goes beyond traditional *risk analysis*, which mainly consists of risk assessment and risk management. It promotes the involvement and participation of various stakeholders as well as considerations of the broader legal, political, economic and social contexts in which a risk is evaluated and managed (Renn, 2008). Key principles governing the use of scientific knowledge on chemical risks as basis of policy are the separation between scientific advice and decision-making, transparency, and the open exchange of information with stakeholders. Such principles underlie the *three-pillar approach to risk governance* that the EU has adopted. The approach, which is also common in other OECD countries, distinguishes risk assessment, risk management, and risk communication on an institutional basis. *Risk assessment* covers the scientific aspects of identification and characterization of hazards, evaluation of the probability of certain events and of exposure to these events, and characterization of risk. *Risk management* covers the identification and assessment of the policy options to face risk, and the choice of the best option such as registration, authorization, restriction or ban of chemical substances from a societal view. *Risk communication* consists of a dialogue between EU institutions and stakeholders or society at large, and intervenes at both of the previous stages. In particular, it helps in checking the consistency of messages delivered by risk assessors and managers, who operate in separate and autonomous spheres.

In the EU, therefore, scientific risk assessment of chemical substances and risk management are considered as separate procedures and are assigned to different institutions. Independently acting scientific bodies such as the European Medicines Agency (EMA), the European Food Safety Authority (EFSA), the European Chemicals Agency (ECHA) and some scientific committees are charged with risk assessment issues and scientific advice. They publish their reports and conclusions in a transparent manner whereas risk management in the EU has to remain primarily in the responsibility of the European Commission or the authorities of the Member States.

Hazard, Risk Types and the Precautionary Principle

Hazard, better termed the *hazard potential* of a chemical substance, depends on the inherent properties of the substance, either physical properties such as inflammability or chemical/toxicological

Table 5.3 Classification of health hazards according to Regulation 1272/2008/EC, Part 3 of Annex I.

Hazard class	Hazard class
Acute toxicity	Carcinogenicity
Skin corrosion/irritation	Reproductive toxicity
Serious eye damage/eye irritation	Specific target organ toxicity — single exposure
Respiratory or skin sensitization	Specific target organ toxicity — repeated exposure
Germ cell mutagenicity	Aspiration hazard

properties in contact with biological matrices such as corrosion on skin or any other acute or chronic effects. In the EU, according to the REACH regulations, hazards from chemical substances, e.g. to human health, are defined according to the classification categories given in Table 5.3.

Risk is often defined from a technical perspective as the product of probability of the occurrence of an event and the level of damage, which also includes adverse effects such as health damage and deaths. In toxicological terms, according to WHO (2004), "risk is the probability of an adverse effect in an organism, system, or (sub)population caused under specified circumstances by exposure to an agent". Correspondingly, the risk to human health by exposure to a chemical substance can be defined as the probability and severity of adverse effect(s) depending on (a) the chemical's inherent toxicological properties, (b) the substance dose(s) (i.e., exposure characteristics), (c) the number of the persons exposed, and (d) the proportion of potential individual susceptibilities towards the substance. Consequently, risk regulation requires that known risks should be excluded according to best knowledge and adequate measures such as limiting exposures below toxicity thresholds. *Remaining risks* due to uncertainties or gaps of knowledge should be minimized, for instance in case of mutagenic or genotoxic carcinogens, when no threshold of adverse effect can be derived. For details of the risk assessment concept and corresponding procedures see Chapter 2.1.

A dilemma of risk regulation is that *zero exposure* of humans to chemical substances is often not possible for practical reasons or is in conflict with important socio-economic considerations. Therefore, the concept of *acceptable risk* has been developed. This concept requires the setting of limits for the exposure to toxic chemicals, i.e. acceptable limit values or other measures, that can be considered to have no or a negligible impact on the incidence of adverse effects in an exposed population. Setting the acceptable risk and adequate measures is a risk regulation issue case by case: According to WHO (2004) the acceptability of a risk depends on scientific data, social, economic, and political factors, and the perceived benefits arising from exposure to an agent. Sometimes, in case of complex risk assessment situations and scientific uncertainties, potential *remaining risks*, although considered presumably small and in general not measurable, cannot be excluded and are inherently also part of the accepted risk.

The *precautionary principle* has been developed for special cases, when scientific uncertainties prevent a full risk or cost-benefit analysis but severe or irreversible adverse effects for a part of the population or for the environment cannot be excluded. Meaning and impact of the precautionary principle have been clarified in a Communication of the European Commission in 2000 by use of several criteria (COM, 2000), which have essentially been confirmed by the European Court. According to the Communication of the Commission, the precautionary principle can be invoked when a major risk cannot be determined or excluded by a scientific and objective evaluation with sufficient certainty. The cause of concern may be a technical process or product, or a phenomenon such as climate change. Measures based on the precautionary principle have to comply with the principles of non-discrimination and proportionality, and should be provisional until the time when more comprehensive information and data concerning the risk are accessible and can be analysed.

Applying the precautionary principle aims at ensuring a higher level of environmental protection through preventative decision-taking in the case of risk (as originally set out in Article 191 TFEU). However, in the practice of EU legislation, the scope of this principle is far wider and also covers legislation on protection of human health. One application of the precautionary principle is the principle of minimization, as found in several legal acts. For instance, the well-known *ALARA principle* (as low as reasonably achievable) is often used, for instance in case of minimization of human exposure to genotoxic carcinogens or ionizing radiation by radionuclides. Therefore, in situations of scientific uncertainty, taking into account the application principle of proportionality as another important criterion, decisions of risk regulators may be justified when adopting protective regulatory measures at the expense of, for example, enterprises that produce or emit toxic chemical substances.

Definitions: Chemical Substances and Mixtures

The clear characterization of the essential term *substance* is a prerequisite for the identification and description of a chemical compound's identity. In the common chemical understanding, as specified by the International Union of Pure and Applied Chemistry (IUPAC), a *chemical substance* is a "matter of constant composition best characterized by the entities (molecules, formula units, atoms) it is composed of. Physical properties such as density, refractive index, electric conductivity, melting point etc. characterize the chemical substance." Furthermore, IUPAC defines a *mixture* as a "portion of matter consisting of two or more chemical substances called constituents" (IUPAC, 2014). The applicable EU chemicals law is based on IUPAC's definitions. Pursuant to Regulation 1907/2006/EC concerning the Registration, Evaluation, Authorization and Restriction of Chemicals (REACH), "*substance* means a chemical element and its compounds in the natural state or obtained by any manufacturing process, including any additive necessary to preserve its stability and any impurity deriving from the process used, but excluding any solvent which may be separated without affecting the stability of the substance or changing its composition" whereas *preparation* means a "mixture or solution composed of two or more substances". An *article* means per definition an object which during production is given a special shape, surface or design which determines its function to a greater degree than does its chemical composition (Regulation 1907/2006/EC). Nonetheless REACH regulates articles indirectly in Article 7 via the registration and notification of hazardous substances in articles, if specific conditions are fulfilled. These are (a) the hazardous substance is present in those articles in quantities totaling over one metric ton per producer or importer per year and (b) the hazardous substance is present in those articles above a concentration of 0.1% weight by weight (w/w). It should be noted that the legal term "article" according to REACH is used in a wider regulatory context than the similar legal term "product", which is applicable alone for consumer products offered on sale on the market (Consumer Products, Chapter 5.1.5, and Chapter 5.3).

Classification of Health Hazards

The *health hazard* of a substance or mixture is legally defined in Article 3 of the Regulation 1272/2008/EC on classification, labelling and packaging of substances and mixtures (CLP regulation). A substance or a mixture fulfilling the criteria relating to health hazards as laid down in Part 3 of Annex I is hazardous and shall be classified in relation to the respective hazard classes provided for in Annex I (Table 5.3).

Where hazard classes are differentiated on the basis of the route of exposure or the nature and the severity of the health effect, the substance or mixture shall be classified in accordance with such differentiation into hazard categories/subcategories. It is important to note that the hazard class of a substance only describes the inherent property of a substance but not a risk per se. Risk assessment and hazard identification by use of the classification categories in Table 5.3 are different procedures with different objectives. Hazard identification and classification describes an inherent toxic potential of a substance or mixture at a sufficiently high dose and is a means for risk communication, i.e. warning of unintended high exposure. This is done by labelling according to the CLP Regulation (see Chemicals, Chapter 5.1.5, and Chapter 5.3). In contrast, in the risk assessment process, the objective is to determine a dose range that is considered safe for human exposure irrespective the hazard class that is relevant at a much higher dose range.

5.1.4 International Co-operation and Harmonization Supported and Implemented by the EU

Co-operation with International Institutions

The role of the EU at the international level can be considered as another source of EU law. The EU concludes agreements in international law with non-member countries ("third countries") and with international organizations. Against the background of the global production and use of chemicals, in the past and at present, enormous worldwide efforts of several stakeholders have been made to assess and manage the risks associated with exposures to hazardous chemicals. In addition to the programs mainly of international organizations, some networks have been established in order to harmonize the different activities more effectively, avoiding duplication and thereby optimizing the use of assessment resources. Major programs, funds and specialized agencies are part of the United Nations (UN) under the platform of the *Economic and Social Council* (ECOSOC), which co-ordinates regional commissions such as the *Economic Commission for Europe* (UNECE), specialized agencies as, for example, WHO, FAO, ILO, and IMO, and programs such as the *United Nations Environment Programme* (UNEP).

The *World Health Organization* (WHO) is active in many areas of health protection and hence also initiates and shares a multiplicity of programs on chemical safety. To support global efforts, the *WHO Chemical Risk Assessment Network*, a network of institutions involved in chemical risk assessment activities, was recently initiated (WHO, 2013; http://www.who.int/ipcs/network/about/en/). Important WHO projects within the network are recommendations and hazard assessments (Environmental Health Criteria (EHCs) documents, Concise International Chemical Assessment Documents (CICADs) via ICPS), the assessment of pesticides via the Joint FAO/WHO Meeting on Pesticide Residues (JMPR), and the *Strategic Approach to International Chemicals Management* (SAICM).

WHO's main contribution to chemical safety is the *International Programme on Chemical Safety* (IPCS). Through IPCS, WHO is establishing scientifically sound methodologies for the evaluation of risks to human health from exposure to chemicals and facilitating aid in risk management. The main areas of ICPS's work are the assessing the public health impacts of chemicals, providing tools for assessing chemical risks, poisons information, prevention and management, chemical incidents and emergencies, strengthening capacity building, and joining chemical risk assessment networks. The *International Agency for Research on Cancer* (IARC) is the specialized cancer agency of the WHO.

The strategic objectives of the *Food and Agriculture Organization* (FAO) are global activities of food supply (eradication of hunger, food insecurity and malnutrition) and the sustainable management and utilization of natural resources, including land, water, air, climate and genetic resources. The *Joint Meeting on Pesticide Residues* (JMPR) is an expert ad hoc body administered jointly by FAO and WHO with the purpose of harmonizing risk assessment on pesticide residues. JMPR comprises the WHO Core Assessment Group and the FAO Panel of Experts on Pesticide Residues in Food and the Environment. The WHO Core Assessment Group is responsible for reviewing pesticide toxicological data and estimating *acceptable daily intakes* (ADIs) and *acute reference doses* (ARfDs), and examines other toxicological criteria. The FAO Panel is responsible for reviewing pesticide residue data and estimating *maximum residue levels* (MRLs), supervised trials median residue values (STMRs), and highest residues (HRs) in food and feed. The Codex Alimentarius Commission adopts the recommended MRLs as Codex MRLs (CXLs) for pesticides. Another international expert scientific committee administered by FAO and WHO is the *Joint FAO/WHO Expert Committee on Food Additives* (JECFA). JECFA undertakes the risk assessment and safety evaluation of food additives, processing aids, flavoring agents; residues of veterinary drugs in animal products, contaminants and natural toxins. Its activities are related to exposure assessment, specifications and analytical methods, residue definition, and MRL proposals (veterinary drugs).

The *International Labour Office* (ILO) has the mandate for chemicals safety at the workplace. ILO develops policies and devises programs on codes of practice and guides in the field of chemical safety (e.g., safety in the use of chemicals at work, prevention of major industrial accidents, safety and health in the use of chemicals at work, safety in the use of mineral and synthetic fibers, and safety and health in the use of agrochemicals). The International Chemical Safety Cards (ICSC) project is a common undertaking between the WHO and ILO, with the cooperation of the European Commission. The *Globally Harmonized System for Classification and Labelling of Chemicals* (GHS) is a project initiated by ILO.

The *International Maritime Organization* (IMO) has created a comprehensive shipping regulatory framework, addressing safety and environmental concerns, legal matters, technical cooperation, security, and efficiency. IMO provides global regulations for the transport of packaged hazardous chemicals as well as liquid, solid and gaseous chemicals as bulk goods. It also recommends guide values for discharging of chemicals into the sea and deals with global permission of active substances in biocides for ballast water treatment.

The *United Nations Economic Commission for Europe* (UNECE) is one of five regional commissions of the United Nations and includes more than 50 Member States in Europe, North America and Asia. By negotiating on international legal instruments, development of regulations and norms, exchange and application of best economic and technical practices, UNECE sets out norms, standards and conventions to facilitate international co-operation within and outside the region. Since 1979, four international legally binding conventions have been developed within UNECE on transboundary air pollution, environmental impact assessment, industrial accidents across borders, and international watercourses and lakes. These treaties are important elements of a common European legal framework on the protection of the environment (see *Legislation on Chemical Substances in the Environment*, this chapter). Another important area in the context of this chapter is the harmonization of legal and technical rules on the transport of dangerous goods (TDG; see *Transport of Dangerous Goods*, this chapter).

The *United Nations Environment Programme* (UNEP) will unfold activities on a global or regional level aimed at protecting the environment. UNEP promotes chemical safety by providing policy advice through its Chemicals and Waste Subprogramme, technical guidance and capacity building to developing countries and those with economies in transition, including activities on chemicals related to the implementation of the *Strategic Approach to International Chemicals Management* (see below). UNEP aims to support institutional strengthening at the national level for implementation of the *Basel Convention on the Control of Transboundary Movements of*

Hazardous Wastes and their Disposal, the *Rotterdam Convention on the Prior Informed Consent Procedure for Certain Hazardous Chemicals and Pesticides in International Trade* and the *Stockholm Convention on Persistent Organic Pollutants*, the *Minamata Convention on Mercury* and the *Strategic Approach to International Chemicals Management*.

The *Organisation for Economic Co-operation and Development* (OECD) is an intergovernmental organization aiming at promoting policies that will improve the economic and social well-being of people around the world. OECD also develops environmental and health policy tools and sets international standards for the safety of chemicals. Chemical safety and biosafety activities within OECD's Environment, Health and Safety Programme consider topics such as the testing of chemicals and their assessment, good laboratory practice (see *Good Laboratory Practice* and *OECD Test Guidelines*, this chapter, and Chapter 4.1), the risk management of chemicals, chemical accident prevention, preparedness and response, the pollutant release and transfer register, concepts on the safety of manufactured nanomaterials, and agricultural pesticides and biocides. OECD shares information on chemicals via its eChemPortal and has developed the *International Uniform Chemical Information Database* (IUCLID), an electronic tool for data submission, evaluation and exchange.

The manifold international activities on chemical safety increasingly require co-ordination. The *Inter-Organization Programme for the Sound Management of Chemicals* (IOMC), established in 1995, aims to strengthen cooperation and increasing coordination in the area of chemical safety. The participating nine organizations of IOMC, which each carry out their own programs with specific priorities, are FAO, ILO, the United Nations Development Programme (UNDP), UNEP, the United Nations Industrial Development Organization (UNIDO), the United Nations Institute for Training and Research (UNITAR), WHO, the World Bank, and OECD. The IOMC is the pre-eminent mechanism for initiating, facilitating and coordinating international action to achieve the *World Summit on Sustainable Development* (WSSD) 2020 goal. This means that by 2020 chemicals are used and produced in ways that lead to the minimization of significant adverse effects on human health and the environment.

The *Strategic Approach to International Chemicals Management* (SAICM) represents a policy framework to foster the sound management of chemicals. SAICM supports the achievement of the WSSD 2020 goal ensuring that chemicals are produced and used in ways that minimize significant adverse impacts on the environment and human health.

The work of the *International Commission on Radiological Protection* (ICRP) helps to prevent cancer and other diseases and effects associated with exposure to ionizing radiation, and to protect the environment. ICRP has developed, maintained, and elaborated the *International System of Radiological Protection* used world-wide as the common basis for radiological protection standards, legislation, guidelines, programs, and practice. ICRP has published more than 100 reports on all aspects of radiological protection. The *International Atomic Energy Association* (IAEA), located in Vienna, is responsible for international co-operation, scientific exchange and advice on nuclear safety and protection towards ionizing radiation whereas regulation of nuclear safety is a national responsibility.

Harmonization: Some Examples

Harmonized and internationally accepted test guidelines are an essential prerequisite for the identification of possible health-damaging effects of substances. The development of alternative testing methods in terms of the *3R principles* is a challenge of increasing importance. The type and extent of the necessary toxicological tests for the determination of the potential health risks arising from a substance are specified in the relevant legal context. The data requirements can be comprehensive, such as the placing on the market of pesticides or the approval of medicinal products. The testing of chemicals is labor-intensive and expensive. To avoid duplication in safety testing and government assessments, the OECD Council adopted a decision in 1981 on *Mutual Acceptance of Data* (MAD), stating that test data generated in any member country in accordance with OECD Test Guidelines and *Principles of Good Laboratory Practice* (GLP) shall be accepted in other member countries for assessment purposes and other uses relating to the protection of human health and the environment. A further Council Act was adopted in 1989 describing the framework of national and international compliance monitoring procedures to provide assurance that the data are indeed developed in compliance with the Principles of GLP.

OECD Test Guidelines The OECD is the key institution in the process of international coordination of chemical safety testing guidelines. The general objective of the OECD Test Guidelines Program is the development and updating of the test guidelines as well as policies that cover all health-related endpoints, thereby reflecting the current state of knowledge and regulatory needs by regular updates. Moreover, the matter of animal welfare is strengthened. In addition to OECD's Principles of Good Laboratory Practice, the OECD's test guidelines are the essential basis for the MAD system. The OECD test guidelines are primarily used in regulatory safety testing and subsequent chemical notification or registration. Furthermore, OECD publishes explanatory guidance documents, reviews, and test strategies, which are, however, not part of the MAD system. The OECD Guidelines for the Testing of Chemicals have been developed in order to (a) enhance the validity and international acceptance of test data, (b) make the best use of available resources in both governments and industry, (c) avoid the unnecessary use of laboratory animals, and (d) minimize non-tariff trade barriers.

The OECD Test Guidelines, a collection of about 150 of the most relevant internationally agreed testing methods, cover non-clinical safety testing of chemicals in a broad sense regarding physical-chemical properties, effects on biota (eco-toxicity), environmental fate (degradation/accumulation), health effects (toxicity), and other test guidelines (e.g., test guidelines on pesticide residues chemistry and metabolism studies). The OECD test guidelines and their updates are taken up by the EU in the Regulation 440/2008/EC and respective ATP regulations (adaptions to technical progress).

Good Laboratory Practice Good Laboratory Practice (GLP) is a quality assurance and quality management concept covering the organizational process and the conditions under which laboratory studies and field studies are planned, performed, monitored, recorded, reported, and archived. However, GLP principles do not concern the scientific and methodological approach, which is generally covered by testing guidelines. OECD's Principles of Good Laboratory Practice have been developed to promote the quality and validity of test data from non-clinical safety and environmental studies used for determining the safety of chemicals and to prevent fraudulent practices. First published in 1981 and subsequently updated and amended, the Series on Principles of Good Laboratory Practice and Compliance Monitoring comprises presently (as of October 2016) 17 Guidance Documents, which describe and explain the principles in more detail.

The EU has adopted OECD's GLP principles and the revised OECD Guides for Compliance Monitoring Procedures for GLP as annexes in its two core GLP Directives, both last revised in 2004. The Directive 2004/9/EC of 11 February 2004 on the inspection and verification of good laboratory practice (GLP) lays down the obligation of EU Member States to designate the authorities responsible for GLP inspections in their territory. Furthermore, it comprises reporting and MAD requirements. Directive 2004/10/EC requires EU Member States to take all measures necessary to ensure that laboratories carrying out non-clinical safety studies and environmental studies on chemicals comply with the OECD Principles of GLP. When submitting data to the receiving authorities, the application of GLP is an obligation required by a variety of different EU substance and product-specific legal acts.

International co-operation and Harmonization of Requirements for Pharmaceuticals for Human and Veterinary Use The *International Council for Harmonization of Technical Requirements for Pharmaceuticals for Human Use* (ICH) makes recommendations to achieve greater harmonization in the interpretation and application of scientific and technical guidelines and requirements for pharmaceutical product registration and approval. Amongst others, the ICH provides guidelines that are harmonized between Europe, Japan and the USA on the safety of medicines. ICH's safety guidelines cover all relevant areas, including biotechnological products, non-clinical evaluation for anticancer pharmaceuticals, photo safety evaluation, and non-clinical safety testing. At the EU level, the EMA adopts the scientific and technical guidelines of ICH. *The Veterinary International Conference on Harmonization (VICH)* is the equivalent international forum for veterinary medicines. Regulatory co-operation and exchange of information with international regulators is also assured through the *International Pharmaceutical Regulators' Forum (IPRF)*.

International Co-operation on Ionizing Radiation and Nuclear Power The *International Atomic Energy Association* (IAEA) is required by its statute to promote international co-operation on nuclear power whereas regulating safety is a national responsibility. International co-operation is facilitated by international safety-related conventions, codes of conduct and safety standards. The IAEA safety standards constitute a tool for contracting parties to assess their performance under international conventions on, for example, nuclear accidents, nuclear safety, and radioactive waste management. The *United Nations Scientific Committee on the Effects of Atomic Radiation* (UNSCEAR) compiles, assesses and disseminates information on the health effects of radiation and on levels of exposure to radiation from different sources. Its findings and the recommendations of international expert bodies, notably the *International Commission on Radiological Protection* (ICRP), are taken into account in developing the IAEA safety standards, which are issued in the IAEA Safety Standard Series (Safety Fundamentals, Safety Requirements, and Safety Guides).

Chemical Weapons Convention In this context, last but not the least, the *Chemical Weapons Convention* (CWC) should be mentioned, which is a treaty under international law between the Organisation for the Prohibition of Chemical Weapons (OPCW) located in The Haag/Netherlands and OPCW Member States. The convention's main objectives are a worldwide ban, non-proliferation and disarmament of chemical weapons, including their elimination and destruction of facilities for their production. The convention, which came into force in 1997, distinguishes three classes of controlled chemicals that can either be used as weapons themselves or can be misused in the manufacture of chemical weapons. An OPCW notification and inspection regime allows for the control of production, uses and consumption, and trade of chemicals which might be converted to chemical weapons. Whereas the EU is not a signatory, all Member States of the EU have ratified the convention, most of them in 1994. Meanwhile 192 OPCW Member States, which represent about 98% of the global population and landmass, as well as 98% of the worldwide chemical industry (status June 2016) have joined the convention.

5.1.5   EU Legislation on Chemical Substances and Their Uses

Chemicals

Adopted in 1967, the chemicals directive 67/548/EC was one of the first European laws covering chemicals substances. During the next two decades, the legislation was adapted/modified and a distinction was made between "existing" and "new" substances in terms of registration requirements. Around 100,000 substances registered as existing on the European market before 1981 by definition were taken up in the European Inventory of Existing Commercial Chemical

Table 5.4 EU legislation on the safety of chemical substances.

Legislation area	Scope of the law	EU Directive (DIR) or Regulation (REG)
Chemical substances in general	REACH; establishment of ECHA	REG 1907/2006/EC
	Classification, labelling and packaging of chemicals	REG 1272/2008/EC
Import and export of hazardous substances	Import and export based on prior informed consent (PIC) (Rotterdam Convention)	REG 649/2012/EU
Ban of substances	Ban or limitation of production and use of persistent organic pollutants according to the Stockholm Convention (2001)	REG 850/2004/EC
	Export ban and safe storage of metallic mercury and certain mercury compounds	REG 1102/2008/EC
GLP	Harmonization of legislation on the application of the principles of GLP	DIR 2004/10/EC
	Inspection and verification of GLP; guidance on national monitoring programs on GLP	DIR 2004/9/EC
Guidelines for safety testing of substances	Guidelines on tests required to generate information on intrinsic properties of chemical substances	REG 440/2008/EC

REACH, registration, evaluation, authorization of chemicals; ECHA, European Chemicals Agency; GLP, good laboratory practice.

Substances (EINECS) and more or less received a preservation of their status quo regarding data and testing requirements. On the other hand, new substances intended to be brought onto the market with a production volume of 1 ton per year or more had to be registered together with a set of testing data with requirements depending on the production volume per year. Mainly due to various problems with the existing substances, the chemicals legislation of the EU was completely revised. For the current basic chemicals legislation see Table 5.4.

The previous legislation on chemicals was largely repealed by REACH Regulation 1907/2006/EC and the ECHA was established, which has been responsible for chemicals registration, evaluation, and authorization procedures since 2008. For a more detailed review on REACH Regulation, the reader is referred to Chapter 5.3.

Providing adequate information on hazards is an important principle of protection from hazards and risks from chemicals to human health and the environment. This information system is based on the classification, labelling and packaging (CLP) of a substance or mixture according to the set of hazard categories given in Table 5.3. The information on CLP characteristics of a substance or mixture is the backbone for risk communication and protection measures in the production and supply chain with varying exposure conditions at work and for consumers. Meanwhile, the *CLP Regulation 1272/2008/EC* fully replaced the remaining parts of the former Chemicals Directive 67/548/EWG and the Directive 1999/45/EC for dangerous mixtures/preparations. The CLP Regulation is based on the *Globally Harmonised System of Classification and Labelling of Chemicals* (GHS), which was developed by the United Nations with the objectives of international harmonization, improvement of transparency, and comparability of classification and labelling issues. According to the switch to GHS, there was also a change in the symbols, codes and in part also the criteria of classification compared with the previous system.

Import and Export of Dangerous Chemicals The *Prior Informed Consent Regulation* (PIC, Regulation 649/2012/EU) administers the import and export of certain hazardous chemicals and places obligations on companies that wish to export these chemicals to non-EU countries. It aims to promote shared responsibility and co-operation in the international trade of hazardous chemicals by providing developing countries with information on how to safely store, transport, use, and dispose of hazardous chemicals. The PIC Regulation entered into operation in 2014 and implements, within the EU, the *Rotterdam Convention* of 2001 on the prior informed consent procedure for certain hazardous chemicals and pesticides in international trade. Banned or severely restricted chemicals are listed in Annex I, including industrial chemicals, pesticides, and biocides. The export of such chemicals is subject to two types of requirement: export notification and explicit consent. Chemicals that are banned for export are listed in Annex V. Furthermore, all chemicals when exported must comply regarding their packaging and labelling with relevant EU legislation. The Regulation also applies to the export of articles if the article is a finished product containing or including a chemical the use of which has been banned or severely restricted within the EU in that particular product. For legislation on the transport of dangerous goods see *Transport of Dangerous Goods*, this chapter.

Special Groups of Dangerous Substances Certain groups of dangerous substances are covered by special chemicals legislation. One example is the POP Regulation 850/2004/EC covering the group of *persistent organic pollutants (POPs)*. Taking into account, in particular, the precautionary principle, the POP Regulation is to protect human health and the environment from the release of POPs by prohibiting, phasing out or restricting the production, placing on the market and use of substances that are subject to the *Stockholm Convention on Persistent Organic Pollutants*. These highly halogenated substances persist in the environment and bio-accumulate through the food web. POPs are transported via air and water across international boundaries far from their sources. This group of priority pollutants consists of halogenated pesticides (such as DDT), industrial chemicals (such as polychlorinated biphenyls, PCBs) and unintentional by-products of industrial processes (such as dioxins and furans). Meanwhile, additional substances regarded as POPs, including wastes containing POP substances, have been added to the annexes of Regulation 850/2004/EC by means of amending EU regulations.

Similarly, according to Regulation 1102/2008/EC the ban on exports of *mercury and certain mercury compounds* from the EU since 2011 will contribute to reducing the global mercury supply and, indirectly, to limiting the emissions of this heavy metal, which is extremely toxic to human health and the environment. Member States must guarantee the safe storage of this metal when it is used in or produced by certain industrial activities. In 2016, a proposal to repeal the Regulation was published by the European Commission.

Medicinal Products for Human or Veterinary Use

"Medicinal product" or "medicine" is the regulatory term for a drug or pharmaceutical in European legislation. Medicinal products are substances that are used to treat diseases, to relieve complaints, or to prevent such diseases or complaints in the first place. This definition applies regardless of whether the medicinal product is administered to humans or animals. The substances can act both within or on the body. The regulation of medical devices does not fall within the scope of the European regulatory system for medicines and is covered in *Medical Devices*, this chapter. The European regulatory system for medicines is based on Directive 2001/83/EC and Regulation 726/2004/EC, and consists of a network between the regulatory authorities for medicines in the EU Member States (and also Iceland, Liechtenstein and Norway), the European Commission and the EMA. To protect public health and ensure the availability of high-quality, safe and effective medicines for European citizens, all medicines must be authorized before they can be placed on the market in the EU. The regulatory system offers different routes for such an authorization (EMA, 2016).

The *centralized procedure* allows the marketing of a medicine on the basis of a single EU-wide assessment and marketing authorization, which is valid throughout the EU. Pharmaceutical companies submit a single authorization application to the EMA. The Agency's Committee for Medicinal Products for Human Use (CHMP) or Committee for Medicinal Products for Veterinary Use (CVMP) then carries out a scientific assessment of the application and gives a recommendation to the European Commission for a decision on whether or not granting a marketing authorization. Once granted by the European Commission, the centralized marketing authorization is valid in all EU Member States. The use of the centrally authorized procedure is compulsory for most innovative medicines, including medicines for rare diseases (orphan drugs). The centralized procedure is also obligatory for veterinary medicines intended primarily for use as performance enhancers in order to promote the growth of treated animals or to increase yields from treated animals.

The majority of medicines authorized in the EU, however, do not fall within the scope of the centralized procedure but are authorized by national competent authorities (NCAs) in the Member States. When a company wants to authorize a medicine in several Member States, it can use one of the following procedures:

- *decentralized procedure*: companies can apply for the simultaneous authorization of a medicine in more than one EU Member State if it has not yet been authorized in any EU country and does not fall within the scope of the centralized procedure;
- *mutual-recognition procedure*: companies that have a medicine authorized in one EU Member State can apply for this authorization to be recognized in other EU countries. This process allows Member States to rely on each other's scientific assessments.

Rules and requirements applicable to medicinal products in the EU are the same, irrespective of the authorization route for a medicine. To achieve transparency in the system and the regulatory decisions, a European Public Assessment Report (EPAR) is published for every human or veterinary medicine that has been granted or refused a marketing authorization following an assessment by the EMA. For a medicine that is authorized by a Member State, details on the assessment of the medicine are also available in a Public Assessment Report.

In addition to the centralized procedure, specific regulations apply for the authorization of innovative medicinal products developed by means of certain biotechnological processes, medicines for rare (orphan) diseases, advanced therapy medicinal products, and new active substances for the treatment of most immunological diseases, cancer, neurodegenerative disorders, diabetes, and viral diseases (Table 5.5). Specific rules hold to facilitate the development and accessibility of medicinal products for use in the pediatric population. Traditional herbal medicines may be authorized through a simplified registration procedure.

The EMA as an independent body is responsible for the scientific evaluation of innovative and high-technology medicines developed by pharmaceutical companies for use in the EU whereas the European Commission or the competent authorities of the Member States are responsible for risk management actions such as authorization or withdrawal (institutional separation of scientific evaluation and risk management; see Chapter 5.1.3). Besides the already mentioned scientific committees CHMP and CVMP, the EMA has other scientific committees that carry out its scientific assessments (Table 5.6).

In the work of the EMA, external experts participate as members of its scientific committees, working parties, etc. or as members of the national assessment teams that evaluate medicines. The

Table 5.5 *EU legislation on the safety of medicinal products.*

Legislation area	Scope of the law	EU Directive (DIR) or Regulation (REG)
General legislation on medicinal products for human or veterinary use	Production, distribution and use of medicinal products	DIR 2001/83/EC
	Production, distribution and use of veterinary medicinal products	DIR 2001/82/EC
	Authorization and supervision of medicinal products for human and veterinary use; establishing of the EMA	REG 726/2004/EC
Special legislation on medicinal products for human use	Conduct of clinical trials, GCP	REG 536/2014/EU
	Principles and guidelines of GMP	DIR 2003/94/EC
	Specific rules for advanced therapy medicinal products	REG 1394/2007/EC
	Medicinal products for pediatric use	REG 1901/2006/EC
	Medicinal products for rare (orphan) diseases	REG 141/2000/EC
	Traditional herbal medicinal products	DIR 2001/83/EC

EMA, European Medicines Agency; GCP, good clinical practice; GMP, good manufacturing practice.

Table 5.6 *Specialized EMA committees for medicinal products.*

Pharmacovigilance Risk Assessment Committee (PRAC)
Committee for Orphan Medicinal Products (COMP)
Committee on Herbal Medicinal Products (HMPC)
Committee for Advanced Therapies (CAT)
Paediatric Committee (PDCO)

EMA prepares scientific guidelines in co-operation with experts from its scientific committees and working groups. These guidelines reflect the most recent developments in biomedical science. They are available to guide the development programs of all medicine developers. The authorization and oversight of a clinical trial is the responsibility of the Member State in which the trial is taking place. The European Clinical Trials Database (EudraCT) tracks which clinical trials have been authorized in the EU.

Manufacturers, importers and distributors of medicines in the EU must be licensed before they can carry out those activities. The regulatory authorities of each Member State are responsible for granting licenses for such activities taking place within their respective territories. All manufacturing and importing licenses are entered into EudraGMDP, the publicly available European database operated by the EMA. The network between the regulatory authorities for medicines in the EU Member States, the Commission and the EMA not only enables exchange of information on the regulation of medicine, for example regarding the reporting of side effects of medicines, but also the oversight of clinical trials and the conduct of inspections of medicines' manufacturers and compliance with good clinical practice (GCP), good manufacturing practice (GMP), good distribution practice (GDP), and good pharmacovigilance practice (GVP). Manufacturers listed in the application of a medicine to be marketed in the EU are inspected by a national competent authority. Inspection outcomes can be accessed by all Member States and are made publicly available across the EU through EudraGMDP. Equivalence between Member States' inspectorates is ensured and maintained by common legislation, common GMP and common procedures for inspections. Compliance with GMP and other controls of active pharmaceutical ingredients and medicines when imported is a prerequisite for their marketing in the EU. This also includes inspections of manufacturers located outside the EU unless a mutual recognition agreement (MRA) is in place between the EU and the country of manufacture.

The European Commission and the EMA work in close cooperation with partner organizations around the world, such as the WHO. One of the main forums for multilateral international cooperation is the International Council for Harmonization of Technical Requirements for Pharmaceuticals for Human Use (ICH; see also Chapter 5.1.4), which brings together medicines regulatory authorities and pharmaceutical industries from around the world. The Veterinary International Conference on Harmonization (VICH) is the equivalent forum for veterinary medicines. Both are dedicated to international harmonization in safety, quality and efficacy as the main criteria for approving and authorizing new medicines.

Medical Devices

Medical devices cover a broad range of products from plasters and pregnancy tests, to hip implants and state-of-the-art pacemakers, X-ray machines and genetic tests. Most of the medical devices on the market are mainly effective in a technical or physical way. In the context of this chapter, they are of interest as far as contact of substances with the human body may pose health risks, such as harmful substances that may cause skin sensitization or other types of allergies. Yet other risks may be posed if substances are systemically available, either intentionally by release from the device into the body or unintentionally/inevitably from the surface of the device such as abrasion or attrition from implants (e.g., particles or ions from toxic metals).

The regulation of medical devices does not fall within the scope of the European regulatory system for medicines/medicinal products but rather differs considerably and is in some way similar to the general regulation of consumer products in the EU (see Consumer Products, Chapter 5.1.5). Unlike medicinal products, medical devices and in vitro diagnostic medical devices are not subject to pre-market authorisation. Instead, they undergo a *conformity assessment* to establish whether they meet the applicable standards. Depending on the risk posed by a product given by a risk classification, the assessment may involve a so-called "*notified body*", e.g. a laboratory, a national standards authority or the like, which performs the conformity assessment. However, the rules which determine the scope of controls carried out by these bodies currently vary from one Member State to another. Medical devices should, as a general rule, bear the *CE marking* to indicate their conformity. Rules of GMP and GCP also apply for medical devices (see Chapter 3.2 and Table 5.6). The recent scandals related to metal-on-metal artificial hips and faulty silicone breast implants have strengthened the case for modernising the current directives for medical devices (Table 5.7).

For these and other reasons, the current medical device directives were revised and replaced by two regulations, the Medical Device Regulation 2017/745/EU (MDR) and the In Vitro Diagnostic Medical Device Regulation 2017/746/EU (IVDR). These new regulations place a special focus on *pre-market conformity*, with requirements on safety and quality standards, and contain a series of important improvements to modernize the current system. Among them are the introduction of a new risk classification system for in vitro diagnostic medical devices, stricter ex-ante control for high-risk devices, reinforcement of the criteria for designation and processes for oversight of *notified bodies*, reinforcement of the rules on clinical evidence, *post-market oversight* by manufacturers, and *traceability* of all medical devices throughout the supply chain. The latter should be achieved inter alia by the establishment of a Unique Device Identification (UDI) system and the creation of a European database on medical devices (Eudamed). Both regulations will apply with transitional periods of several years, the MDR in 2020, the IVDR in 2022.

Table 5.7 *EU legislation on the safety of medical devices.*

Legislation status	Scope of the law	EU Directive (DIR) or Regulation (REG)
Current Directives	In vitro diagnostic medical devices	DIR 98/79/EC
	Active implantable medical devices	DIR 90/385/EEC
	Other medical devices	DIR 93/42/EEC
	Directive amending DIR 90/385/EEC and DIR 93/42/EEC	DIR 2007/47/EC
Regulations in force since May 2017 but which apply with transitional periods	Medical devices (including active implantable medical devices)	REG 2017/745/EU
	In vitro diagnostic medical devices	REG 2017/746/EU

Plant Protection Products and Biocidal Products

From a regulatory perspective, the term *pesticide* includes, amongst others, herbicides, fungicides, insecticides, acaricides, nematicides, molluscicides, rodenticides, growth regulators, repellents, and rodenticides. Pesticides legislation in the EU distinguishes between *plant protection products (PPPs)* and *biocidal products* depending on the intended use. Whereas biocidal products are widely used to control harmful organisms (see below), the use of PPPs is restricted to crops. As crops can be affected by pests and pathogens, agriculture is implementing a wide range of measures to prevent this, including the use of PPPs. However, the use of PPPs can lead to pesticides residues of, in particular, active substances or their metabolites in crops and foodstuffs. For further details on the legislation of PPPs see Chapter 6.10.

Biocidal products are covered by Regulation 528/2012/EU, which concerns the availability of biocidal products on the market and use, laying down rules for: (a) the establishment at EU level of a list of active substances which may be used in biocidal products, (b) the authorization of biocidal products, (c) the mutual recognition of authorizations within the EU, (d) the making available on the market and the use of biocidal products within one or more Member States or the EU, and (e) the placing on the market of treated articles. Any active substance contained in a biocidal product must be approved in advance. Regulation 528/2012/EU distinguishes between an "existing active substance" (a substance which was on the market on 14 May 2000 as an active substance of a biocidal product for purposes other than scientific or product and process-orientated research and development), a "new active substance" (substance which was not on the market on 14 May 2000 as an active substance of a biocidal product), and a "substance of concern" (any substance, other than the active substance, which has an inherent capacity to cause an adverse effect on human health or the environment and is present or is produced in a biocidal product in sufficient concentration to present risks of such an effect). The distinction between existing active substances which were on the market in biocidal products on the transposition date set in Directive 98/8/EC and new active substances which were not yet on the market in biocidal products on that date has practical consequences. During the ongoing review of existing active substances, EU Member States should continue to allow biocidal products containing such substances to be made available on the market according to their national rules until a decision is taken on approval of those active substances. Active substances need to be assessed and approved before they can be used in biocidal products in the EU. The ECHA coordinates the assessment that is done by a Member State and is followed by a peer review involving all EU countries.

Biocidal Product Regulation 528/2012/EU was amended by Regulation 334/2014/EU amending Regulation 528/2012/EU concerning the making available on the market and use of biocidal products, with regard to certain conditions for access to the market.

Food

European food law is highly complex and closely linked to neighboring legal areas, such as consumer goods, cosmetics, feedstuffs, animal diseases, and veterinary checks. The general principles and requirements of food law and procedures in matters of food safety have been laid down in Regulation 178/2002/EC (General Food Law Regulation). The General Food Law Regulation is aiming at ensuring the quality of food for human or animal consumption. It forms a horizontal framework reinforcing EU and national measures relating to food and feed, and establishes the common principles of risk analysis. Key points are that no foods dangerous to health or unfit for consumption may be put on sale, considering the following factors:

- the normal conditions under which the food is used by the consumer;
- information provided to the consumer;
- the short- and long-term effects on health;
- cumulative toxic effects;
- individual sensitivities of certain consumer groups, e.g. children.

EU food legislation applies at all stages of the food chain, from production, processing, transport and distribution to supply. It is based on divided responsibilities for a sound risk assessment, risk management, and risk communication as the three inter-related components of risk analysis. Regulation (EC) No 178/2002 creates the main procedures and tools for the management of emergencies and crises as well as a rapid alert system for the notification of a direct or indirect risk to human health deriving from food or feed (RASFF). In addition, the General Food Law regulation provides the following measures on food safety: (a) the establishment of the Standing Committee on Plants, Animals, Food and Feed (PAFF Committee), (b) the adoption of emergency measures, and (c) the establishment of a general plan for crisis management. The PAFF Committee delivers opinions on draft measures that the European Commission intends to adopt. If food or feed (including those imported from a non-EU country) presents an emerging serious and uncontainable risk to human health, animal health or the environment, the Commission can put in place protective measures, such as suspending the placing on the market or use of products originating from the EU or suspending imports of products originating from non-EU countries based on an opinion from the PAFF Committee. In some cases, where incidents related to food or feed that pose potential serious risks to human health cannot be managed properly within routine procedures, the Commission, EFSA and the affected Member States should follow the general crisis-management plan as adopted by Decision 2004/478/EC.

Scientific and technical evaluations are undertaken by the EFSA, which is responsible for scientific advice and support only but is an independent body (institutional separation of scientific risk assessment and risk management; see Chapter 5.1.3).

The precautionary principle (Article 7 of the General Food Law Regulation) takes effect in specific situations where (a) reasonable grounds for concern that an unacceptable level of risk to health exists and (b) the available supporting data are not sufficiently complete to make a comprehensive risk assessment (http://ec.europa.eu/food/safety/general_food_law/principles_en).

The horizontal provisions apply to the common fields of labeling and nutrition, biological safety, chemical safety, food improvement agents, novel food, and animal feed as well as food and feed safety alerts (RASFF), and food waste. Table 5.8 shows an overview of the horizontal legislation selecting regulations and directives that focus on chemical and toxicological aspects of food and feed. The vertical food legislation is directly related to defined foodstuffs or product groups.

Drinking water is ingested directly or indirectly like other foods, thereby contributing to the overall exposure of a consumer to ingested compounds, including chemical and microbiological contaminants. The quality of water intended for human consumption is regulated separately by the Council Directives 80/778/EEC and 98/83/EC.

The European food legislation pertaining to chemical safety is divided into the following areas (cf. Table 5.8): contaminants, residues of veterinary medicines, hormones in meat, pesticide residues, and food contact materials. Contaminants, i.e. substances that have not been intentionally added to food, may be present in food as a result of its production, packaging, or transport/holding, or also might result from environmental contamination. The EU has established maximum levels for the following contaminants:

- mycotoxins (aflatoxins, ochratoxin A, fusarium-toxins, patulin, citrinin);
- metals (cadmium, lead, mercury, inorganic tin, arsenic);
- dioxins and polychlorinated biphenyls (PCBs);
- polycyclic aromatic hydrocarbons (PAHs);
- 3-MCPD (3-monochloropropane-1,2-diol);
- melamine;
- erucic acid ((Z)-docos-13-enoic acid);
- nitrates.

Veterinary medicines from treating food-producing animals may leave residues in the food from these animals. In this case, the levels of veterinary medicines residues in food should not harm the consumer. The EMA provides scientific advice for assessing maximum residue limits for residues of veterinary medicinal products marketed in the EU. Member States must implement residue monitoring plans to detect illegal use or misuse of authorized veterinary medicines in food-producing animals and the reasons for residue violations should be investigated. Non-EU countries exporting to the EU must implement a residue monitoring plan comparable to that in the EU, which guarantees an equivalent level of food safety. Member States must implement national residue monitoring plans and inform the Commission annually of the measures that have been taken for non-compliant results in food of animal origin.

In 1981, the EU banned the use of hormonally active substances in animals. The legal instrument in force to control hormones in meat is Directive 96/22/EC as amended by Directive 2003/74/EC. Examples for hormones used as growth promoters are estradiol 17ß, testosterone, progesterone, zeranol, trenbolone acetate, and melengestrol acetate. Because the United States and Canada contested the prohibition of the use of hormones as growth promoters in food-producing animals and in 1997 a panel of the World Trade Organization (WTO) ruled that the EU measure was not in line with the Agreement on the Application of Sanitary and Phytosanitary Measures (SPS), the EU amended Directive 96/22/EC by adoption of Directive 2003/74/EC and thus implemented its international obligations in the WTO context.

In the EU legal context, the traces that pesticides leave in treated products are called "residues" (see *Plant Protection Products and Biocidal Products*, this chapter). The key obligation is that the levels of residues found in food must be safe for consumers and must be as low as possible. A maximum residue level (MRL) is defined as the highest level of a pesticide residue that is legally tolerated in or on food or feed when pesticides are applied correctly under the conditions of good agricultural practice. The European Commission sets MRLs for all food and animal feed which can be found for all crops and all pesticides in the MRL database on the

Table 5.8 Overview of the regulated areas of food (selection of regulations, directives, and recommendations).

Area		Horizontal legislation
Labelling and nutrition	Food supplements	Directive 2002/46/EC on the approximation of the laws of the Member States relating to food supplements
	Addition of vitamins and minerals	Regulation 1925/2006/EC on the addition of vitamins and minerals and of certain other substances to foods
	Natural mineral waters	Commission Directive 2003/40/EC establishing the list, concentration limits and labeling requirements for the constituents of natural mineral waters
	Food for specific groups	Regulation 609/2013/EU on food intended for infants and young children, food for special medical purposes, and total diet replacement for weight control
Chemical safety	Contaminants	Regulation 315/93/EEC laying down Community procedures for contaminants in food
		Regulation 1881/2006/EC setting maximum levels for certain contaminants in foodstuffs
		Directive 96/22/EC concerning the prohibition of the use in stockfarming of certain substances having a hormonal or thyrostatic action and of beta-agonists
		Directive 96/23/EC on measures to monitor certain substances and residues thereof in live animals and animal products
	Residues of veterinary medicines	Directive 2001/82/EC on the Community code relating to veterinary medicinal products
		Regulation 396/2005/EC on maximum residue levels of pesticides in or on food and feed of plant and animal origin
		Commission Regulation (EC) No 1881/2006 setting maximum levels for certain contaminants in foodstuffs
		Regulation 470/2009/EC on Community procedures for the establishment of residue limits of pharmacologically active substances in foodstuffs of animal origin
		Regulation 37/2010/EU on pharmacologically active substances and their classification regarding maximum residue limits in foodstuffs of animal origin
	Hormones in meat	Directive 2003/74/EC concerning the prohibition on the use in stockfarming of certain substances having a hormonal or thyrostatic action and of beta-agonists

	Pesticide residues	Regulation 396/2005/EC on maximum residue levels of pesticides in or on food and feed of plant and animal origin
	Food contact materials	Regulation 1935/2004/EC on materials and articles intended to come into contact with food
	Extraction solvents	Directive 2009/32/EC on extraction solvents used in the production of foodstuffs and food ingredients
Food improvement agents	Additives	Regulation 1333/2008/EC on food additives
	Enzymes	Regulation 1332/2008/EC on food enzymes
	Flavourings	Regulation 1334/2008/EC on flavourings and certain food ingredients with flavouring properties for use in and on foods
Novel food		Regulation 2015/2283/EU on novel foods
Animal feed	Feed additives	Regulation 1831/2003/EC on additives for use in animal nutrition
	Medicated feed	Directive 90/167/EEC on preparation, placing on the market and use of medicated feedstuffs Commission proposal on the manufacture, placing on the market and use of medicated feed (COM(2014) 556 final)
	Undesirable substances	Directive 2002/32/EC on undesirable substances in animal feed Commission Recommendation 2006/583/EC on the prevention and reduction of fusarium toxins in cereals and cereal products Commission Recommendation 2013/165/EU on the presence of T-2 and HT-2 toxin in cereals and cereal products

Commission website (http://ec.europa.eu/food/plant/pesticides/eu-pesticides-database/public/?event=homepage&language=EN).

Materials and articles which come into contact with food during its production, processing, storage, preparation, and serving are called food contact materials (FCMs). This includes direct or indirect contact. FCMs do not cover fixed public or private water supply equipment. The safety of FCMs is evaluated by the EFSA. The safety of FCMs is tested by the business operators placing them on the market and by the competent authorities of the EU Member States during official controls. Scientific knowledge and technical competence on testing methods is maintained by the European Reference Laboratory for Food Contact Materials (EURL-FCM).

Consumer Products

The EU aims to ensure a high level of consumer safety when the public buy goods on sale in Europe. The General Product Safety Directive (GPSD) 2001/95/EC encompassing manufactured non-food consumer products requires firms to ensure that items on sale are safe and to take corrective action when that is found not to be the case. However, in the context of EU regulation of chemicals, the focus of this section primarily lies on chemical risks by consumer products. A product is considered safe if it meets specific national requirements or EU standards. If no such requirements or standards exist, the safety assessment should be based on Commission guidelines, best practice in the sector concerned, state of the art and technology, and reasonable consumer safety expectations. Products placed on the EU market must bear information enabling them to be traced, such as the manufacturer's identity and a product reference. Where necessary for safe use, products must be accompanied by warnings and information about any inherent risks. National enforcement authorities have powers to monitor product safety and take appropriate action against unsafe items.

In practice, the GPSD applies in the absence of other EU legislation, national standards, Commission recommendations or codes of practice relating to the safety of products. It also complements sector-specific EU legislation. Specific vertical EU legislation exists for the safety of toys, electrical and electronic goods, cosmetics, chemicals and other specific product groups (see Table 5.9). Chemicals in articles are widely covered by the REACH Regulation (see also Chapter 5.3). Where no specific EU legislation exists, e.g. commodities with intensive contact with the human body, such as textiles, it is up to the Member States to implement respective standards and to provide monitoring programs.

The GPSD introduces an EU rapid alert system for dangerous non-food products (separate arrangements are in place for food, medicinal products, and medical devices). Commission Decision 2010/15/EU lays down guidelines for the management of the *Community Rapid Exchange of Information System (RAPEX)*. This enables national authorities to share information promptly on any measures taken to withdraw such products from sale. In 2015, products bearing chemical risks attained a maximum of 25% of all notifications in the system. Textiles, toys, cosmetics, tattoo inks, and liquids for e-cigarettes are examples of such notified products. Similar to other areas of legislation in this chapter, under certain conditions the European Commission may take rapid EU-wide measures if a specific product poses a serious risk. For instance, the Commission may adopt a formal temporary decision requiring the Member States to ban the marketing of a product posing a serious risk, to recall it from consumers or to withdraw it from the market. A decision of this kind is normally only valid for up to one year, but it may be renewed and result in permanent legislation, such as restricting or banning the substance under the REACH Regulation. An example of a substance regulated in this way is dimethylfumarate, a powerful antimold chemical that was widely used in everyday consumer products such as sofas and shoes. The substance can provoke allergic reactions causing skin itching, irritation, redness, burns and rheumatic pain.

Table 5.9 *EU legislation on consumer products.*

Legislation area	Scope of the law	EU Directive (DIR) or Regulation (REG)
Products in general	General principles and procedures ensuring safety of products	DIR 2001/95/EC
	Accreditation and market surveillance of products	REG 765/2008/EC
	Framework for the marketing of products	Decision 768/2008/EC
	Guidelines for the management of the Community Rapid Exchange of Information System (RAPEX)	Decision 2010/15/EU
	Rules on European standardization of products and services	REG 1025/2012/EU
Specific product groups	Framework for cosmetics ensuring protection of human health	REG 1223/2009/EC
	Criteria for the justification of claims used in relation to cosmetic products	REG 655/2013/EU
	Safety of toys	DIR 2009/48/EC
	Products that look like foodstuffs but are not edible	DIR 87/357/EEC
	Restriction of certain hazardous substances in electrical and electronic equipment (RoHS-2 Directive)	DIR 2011/65/EU
	Construction products, harmonized conditions for marketing	REG 305/2011/EU

To avoid or minimize possible health risks by chemical substances in or released by consumer products, the *special product legislation* on toys and electronic devices, respectively, covers, among other safety issues, substance groups such as toxic metallic elements (lead, cadmium, mercury etc.) and their compounds, certain halogen-organic compounds and phthalate esters that may be released from the surfaces of the products. Another example is Directive 87/357/EEC on dangerous imitations, which prohibits the marketing, import and manufacture of products which, appearing to be other than they are, endanger the health or safety of consumers products, particularly if they look like foodstuffs but are not edible. Construction products must be designed and built in a safe way regarding potential emissions of toxic gases, dangerous substances, volatile organic compounds or dangerous particles into indoor air or the environment.

A proposal of the European Commission from 2013 for a revision of the GPSD aims to improve its functioning and to ensure consistency with developments in EU legislation as regards market surveillance, obligations of economic operators, and standardization. The proposal is currently under discussion in the European Council and Parliament.

Cosmetics

As mentioned in the previous section, legislation on cosmetics is part of the special legislation of products. Regulation 1223/2009/EC on cosmetic products is the main regulatory framework for finished cosmetic products when placed on the EU market. Cosmetic products are substances or mixtures intended to be placed in contact with external parts of the human body (skin, hair, lips, nails) or with the teeth and the mucous membranes of the oral cavity with the purpose of cleaning, perfuming, protecting or keeping them in good condition or changing their appearance. Generally, they may be distinguished between rinse-off and leave-on products. Cosmetic products should be safe under normal or reasonably foreseeable conditions of use. Depending on the intended use, cosmetic products may be composed of a wide variety of substances and mixtures.

Before placing a cosmetic product on the market, the manufacturer or importer has to fulfil a set of legal obligations in order to achieve a high degree of safety and product quality. To establish clear responsibilities, each cosmetic product should be linked to a legal or natural *responsible person* established in the company or within the EU. Product information should include a *cosmetic product safety report* (Annex I of the Regulation) documenting inter alia normal and reasonably foreseeable use, composition including impurities, properties and manufacture, microbiological quality of the product and that a safety assessment has been conducted. Correspondingly, a *product information file* (PIF) should be sent to the competent authority in the Member State where the file is kept. Cosmetic products placed on the market should be produced according to GMP. Among other documentation, traceability of a cosmetic product throughout the whole supply chain should be ensured. For transparency for the consumers, the ingredients used in a cosmetic product should be indicated on its packaging. In case ingredients consist of or contain manufactured nanomaterials in certain amounts, their names should be followed by the word "nano" in brackets. Cosmetic products should be notified via an internet portal at the European Commission so that confidential information about the frame formulation can be made available, in particular to poison control centers and assimilated entities to allow for rapid and appropriate medical treatment in the event of necessity.

Whereas no restrictions are necessary for many substances and mixtures used such as many oils, waxes, etc., the Regulation contains various annexes for prohibited or restricted substances. Prohibited substances are listed in Annex II (around 1400 entries). Annex III contains substances with use and concentration restrictions or warning labels (around 300 entries). For certain cosmetic ingredients or uses, in addition to the manufacturer's safety assessment, safety evaluation by the Scientific Committee on Consumer Safety (SCCS) and finally authorization is required. Besides Annex III substances (or candidates), this concerns skin and hair colorants (Annex IV), preservatives (Annex V), active UV-filter substances for sunscreens (Annex VI), manufactured nanomaterials, and substances classified as carcinogenic, mutagenic or toxic for reproduction (CMR), category 2 pursuant to the CLP Regulation 1272/2008/EC. The use of CMR substances classified as category 1A or 1B in cosmetic products is generally prohibited unless in the exceptional case that these substances comply with food safety requirements, inter alia as a result of their natural occurrence in food, moreover that no suitable alternative substances exist, and the SCCS considers the use of the substance as safe. An example for such an authorization is the category 1A substance formaldehyde in nail hardeners because inhalation and systemic exposure by use is extremely low in this particular case.

In the framework of this Regulation, all animal studies for testing of cosmetic ingredients have been strictly forbidden since 2013. This implies the development of suitable alternative methods according to the Reduce, Refine, Replace (3R) principles. However, such methods have still to be developed for several toxicological endpoints, not only in the cosmetics area (see Chapter 4.5).

Work Protection

Directive 89/391/EEC on the safety and health of workers at work represents the legal basis of EU activities regarding the field of protection of labour. It contains general principles concerning the prevention of occupational risks, the protection of safety and health, the elimination of risk and accident factors, the informing, consultation, balanced participation and training of workers and their representatives, as well as guidelines for the implementation of the general principles. The provisions of Directive 89/391/EEC apply to all risks, and in particular to those arising from the use of chemical, physical and biological agents at work. The duties of employers are comprehensively regulated. They are obliged to keep themselves informed of the latest advances in technology and scientific findings concerning workplace design, account being taken of the inherent dangers in their enterprise, and to inform workers' representatives to be able to guarantee a better level of protection of workers' health and safety. The employer shall implement measures on the basis of the general principles of prevention as follows: (a) avoiding risks, (b) evaluating the risks which cannot be avoided, (c) combating the risks at source, (d) adapting the work to the individual, especially as regards the design of workplaces, the choice of work equipment, and the choice of working and production methods, (e) adapting to technical progress, (f) replacing dangerous environmental factors in the workplace by non-dangerous or less dangerous ones, (g) developing a coherent overall prevention policy, (h) giving collective protective measures priority over individual protective measures, and (i) giving appropriate instructions to the workers.

Adverse health effects that can be caused by working with and exposure to dangerous substances range from mild eye and skin irritations/diseases, allergies, respiratory diseases to severe effects such as reproductive problems and teratogenic defects and cancer.

Taking into account that cancer is estimated to account for more than half of work-related deaths in the EU, in 2016 changes to the Carcinogens and Mutagens Directive (2004/37/EC) were proposed to limit exposure to 13 cancer-causing chemicals in the workplace, including respirable crystalline silica (RCS), by new or amended limit values. In January 2017, the European Commission adopted workplace limit values for a further seven carcinogenic chemicals. Introducing these limit values will lead to fewer cases of occupational cancer and improve the legal protection of exposed workers. By reducing the differences between Member States in terms of workers' health protection, this proposal will encourage more cross-border employment because workers can be reassured that minimum standards and levels of protection of their health will be guaranteed in all Member States.

The Scientific Committee on Occupational Exposure Limit Values (SCOEL) mandate is to advise the European Commission on occupational exposure limits for chemicals in the workplace in evaluating the latest available scientific data and in proposing occupational exposure limits (OELs) for the protection of workers from chemical risks. SCOEL's approach is documented in its Methodology for the Derivation of Occupational Exposure Limits: Key Documentation (2013, http://efcc.eu/document.aspx?di=1225&fn=2013).

As laid down in Commission Decision 2014/113/EU, SCOEL develops scientific recommendations for the Commission which are used to support regulatory proposals on occupational exposure limit values (OELVs) for chemicals in the workplace. SCOEL recommends *indicative occupational exposure limit values* (IOELVs), *binding occupational exposure limit values* (BOELVs), and *biological limit values* (BLVs) as the three main types of OELs. The OELs can be supplemented by further notations as *eight-hour time-weighted average* (TWA – 8 h) and *short-term exposure limits* (STEL) (Wollin and Illing, 2014). Community IOELVs are health-based, non-binding values derived from the most recent toxicological data available and taking into account the availability of measurement techniques for chemical analysis. They set threshold levels of exposure below which, in general, no adverse effects are expected for any given substance after short-term or per working day exposure over a working lifetime. IOELVs are European objectives to assist employers in determining and assessing risks. Socio-economic and technical feasibility factors are not taken into account when establishing IOELVs. For any chemical agent for which an indicative OEL value is established at EU level, Member States must establish a national exposure limit value, taking into account the Community indicative limit value, determining its nature in accordance with national legislation and practice. Moreover, due to the huge number of chemicals in workplaces and as there is only a limited number of agreed OELVs, Member States can set occupational limit values or recommended values for avoiding or limiting health risks for workers by chemical exposures.

Table 5.10 Protection from ionizing radiation by radioactive materials.

Legislation area	Scope of the law	EU Directive (DIR) or Regulation (REG)
Basic safety standards	Protection from exposure to ionizing radiation	DIR 2013/59/Euratom
Nuclear safety	Framework for the safety of nuclear installations	DIR 2009/71/Euratom
	Reinforcement of safety measures and accident prevention	DIR 2014/87/Euratom
Radioactive waste	Management of spent fuel and radioactive waste	DIR 2011/70/Euratom
	Supervision and control of shipments of spent fuel and radioactive waste	DIR 2006/117/Euratom
Cross-border shipment	Shipment of radioactive substances or materials between Member States	REG 1493/93/ Euratom

Protection from Ionizing Radiation by Radioactive Materials

The general regulation basis for radioactive substances and materials is the Euratom treaty (see *International Co-operation on Ionizing Radiation and Nuclear Power*, this chapter). Only major legislation is described in this section (Table 5.10). Taking into account the up-to-date recommendations of the *International Commission on Radiological Protection* (ICRP), Directive 2013/59/EURATOM is the current law laying down improved procedures and basic safety standards for protection of humans and the environment against the dangers arising from exposure to ionizing radiation. The provisions of the Directive comprise management systems to monitor or control exposures of humans in workplaces, in the medical area, and existing exposures of the general public to naturally occurring radionuclides such as radon. Emergency situations are also covered by the Directive, in particular with improvements in respect to emergency preparedness and response as a consequence to the Fukushima accident in 2011. As minimum standards, dose limitations for workers and dose constraints such as reference levels for existing exposures and emergencies are given. Within the given framework of the Directive, Member States may set more stringent exposure levels where appropriate and should bring into force the laws and administrative provisions necessary to comply with this Directive until February 2018.

The *safety of nuclear installations* is governed by national legislation and international conventions such as the recommendations of the *International Atomic Energy Association (IAEA)* (see *International Co-operation on Ionizing Radiation and Nuclear Power*, this chapter). In the EU, Directive 2009/71/Euratom establishes a Community framework for ensuring nuclear safety from possible dangers by nuclear installations. In the light of the Fukushima accident, the Directive was amended in 2014 by Directive 2014/87/Euratom, which has to be transposed into the legislation of the Member States by 2017. A consolidated version was published in August 2014. The amended Directive reinforces the provisions of the existing Directive such as emphasising accident prevention and the avoidance of significant radioactive releases, regular safety reassessments of nuclear installations, enhancing accident management and on-site emergency preparedness and response arrangements and procedures, and increasing transparency on nuclear safety matters, including involvement of the public.

Shipments of radioactive substances/sources between Member States within the Euratom community are governed by Regulation 1493/93/Euratom. The Regulation requires that such shipments are controlled and documented if the quantities and concentrations exceed certain levels. In particular, it requires the source holder (exporter) to obtain a prior written declaration from the importer to the effect that the importer has complied with national requirements for the safe storage, use and disposal of the source being received. Additional provisions apply to the export of sealed sources. Source holders must provide quarterly summaries of all the shipments they made to a Member State during that period to the relevant competent authority in that Member State. The Regulation does not affect transport requirements. Nuclear materials are not covered by this Regulation.

In the EU, national legislation on the *management of spent fuel and radioactive waste* is supplemented by Directive 2011/70/Euratom, which establishes a Community framework for the responsible and safe management of radioactive waste and aims to ensure a high level of safety, avoiding undue burdens on future generations and enhancing transparency. Whereas each Member State remains free to define its nuclear fuel cycle policy, the Directive reaffirms the ultimate responsibility of Member States for the management of the spent fuel and radioactive waste generated in them, including establishing and maintaining national policies and frameworks, and ensuring the needed resources and transparency. Whatever option is chosen, the disposal of high-level waste, separated at reprocessing, or of spent fuel regarded as waste should be considered. Prime responsibility of the licence holder for the safety of spent fuel and radioactive waste management under the supervision of its national competent regulatory authority is also reaffirmed. Member States are obliged to establish and implement national programmes for the management of spent fuel and/or radioactive waste from generation to disposal and to notify to the Commission their national programmes and any subsequent significant changes.

Directive 2006/117/Euratom repealing Directive 92/3/Euratom aims to reinforce the Community system of supervision and control of shipments of radioactive waste and spent fuel between Member States and into and out of the Community. The Directive now applies not just to shipments of radioactive waste but also to shipments of spent fuel, whether they are destined for final disposal or for reprocessing. Limits for quantity or concentration of radioactive waste at which Directive 2006/117 applies are taken over from the repealed Directive 92/3/Euratom. The Directive sets out the various formalities which must be undertaken by the "holder" of radioactive waste or spent fuel, who is defined as any natural or legal person who is responsible under the applicable national law for such materials and plans to carry out a shipment to a consignee.

Transport of Dangerous Goods

Dangerous goods are transported by road, rail, water or airplane, and often across borders. To prevent or minimize risks for human health and the environment, international conventions on the transport of dangerous goods (TDG) have been established. For the *Rotterdam Convention* and PIC Regulation 649/2012/EU on the import and export of dangerous chemicals and pesticides, the reader is referred to Chapter 5.1.4. Dangerous goods legislation in the EU is constituted by global and European conventions and EU Directives, which are endorsed into the national law of the Member States. Recommendations covering the TDG, including hazardous wastes and substances, are issued and regularly revised by the Committee of Experts on the Transport of Dangerous Goods, which has its secretariat at the *United Nations Economic Committee for Europe* (UNECE; see Chapter 5.1.4). For instance, the Committee has drawn up the classification (grouping) of dangerous goods, by type of risk involved, for all modes of transport, i.e. rail, road, inland waterways, sea, and air. These recommendations serve as the basis for EU and national legislation as well as for international instruments covering the TDG by sea, air, rail, road and inland waterways all over the world. Amongst these, those listed below have also been developed and are regularly updated by the UNECE. For instance, as a comprehensive rule system for international TDG on roads, the *European Agreement concerning the International Carriage of Dangerous Goods by Road* (ADR) entered into force around 50 years ago, with the most recent consolidated revision in 2017. Apart from some excessively dangerous goods, other dangerous goods may be carried internationally in road vehicles subject to compliance with the conditions on their packaging and labeling and the conditions on the construction, equipment and operation of the vehicle carrying the goods in question. A corresponding *European Agreement concerning the International Carriage of Dangerous Goods by Inland Waterway* (ADN) entered into force in 2008. Regulations also exist concerning the *International Carriage of Dangerous Goods by Rail* (RID). On the basis of ADR, RID and ADN, the EU has set up a common regime by Directive 2008/68/EC covering all aspects of the TDG including inland transports by road, rail and water. More specialized EU legislation on the TDG by road refers to the type-approval requirements for the general safety of motor vehicles and their equipment (Regulation 661/2009/EC), uniform procedures for checks of the TDG (Directive 95/50/EC), and transportable pressure equipment (Directive 2010/35/EU).

The carriage of dangerous goods and marine pollutants in sea-going ships is regulated by the *International Convention for the Safety of the Life at Sea* (SOLAS) and the *International Convention for the Prevention of Pollution from Ships* (MARPOL), respectively. Relevant parts of both SOLAS and MARPOL have been worked out in great detail and are included in the International Maritime Dangerous Goods (IMDG) Code, thus making this Code the legal instrument for maritime TDG and marine pollutants. Since 2004, the IMDG Code has become a mandatory requirement.

The International Civil Aviation Organization's (ICAO) Technical Instructions are an internationally agreed set of provisions governing the requirements for transporting dangerous goods by air. The International Air Transport Association (IATA) publishes Dangerous Goods Regulations in accordance with the ICAO technical instructions.

5.1.6 Legislation on Chemical Substances in the Environment

General Aspects

The EU environmental *acquis* is highly complex und is estimated to amount to over 500 Directives, Regulations and Decisions. A turning point in EU environmental policy was attained in 1987, when following an Inter-Governmental Conference (IGC) in 1986 to amend the Treaty, a new Environment Title (Articles 130r–t) was introduced by the 1987 Single European Act. The early environmental law of the EU was characterized by a strictly environmental compartment-related view. Subsequently, a cross-sectoral analysis was increasingly implemented.

The EU is pursuing the following basic principles with its environmental regulation. EU environmental policy is aimed at a *high level of protection* (Articles 114 and 191 TFEU). This principle does not necessarily impose reaching the highest possible level of protection, but rather prohibits the adoption of environmental policy measures with a low level of protection. The *precautionary principle* (Article 191 TFEU) requires that preventive measures shall be taken. These measures must be taken as soon as the credible proof has been provided that a certain action could harm the environment, even if the causal relation between the action and the negative impact is not scientifically proven (Chapter 5.1.3). This principle applies for both environmental and health issues. The *avoidance principle* (Article 191 TFEU) comprises a preventive approach to environmental issues. Measures that avoid environmental damage from the outset should take precedence over measures to restore the already damaged environment. Measures should primarily address environmental damage at their source (Article 191 TFEU). This means that the EU should focus on problem areas where pollution is generated (*principle of origin*). According to the *"polluter pays" principle* (Article 191 TFEU), the polluter must pay

Table 5.11 Overview of EU legislation on chemical substances in the environment.

Legislation area	Scope of the law	EU Directive
Ambient air	Definition and limit value setting for harmful air pollutants, development of standardized methods and criteria for assessing air quality, and information of the public on air quality	DIR 96/62/EC; DIR 2008/50/EC
Surface water	Water Framework Directive: maintaining and improving the quality of the aquatic environment	DIR 2000/60/EC
Ground water	Specific measures to prevent and control groundwater pollution	DIR 2006/118/EC
Waste	Providing for a general framework for the handling of waste and sets the basic waste management requirements for the EU	DIR 2008/98/EC
Soil	Preventing soil contamination	Soil Thematic Strategy (COM (2006) 231)

for environmental damage. The corresponding Directive 2004/35/EC ensures that environmental damage is prevented or remedied and the operator who caused it is held responsible. Environmental damages to be avoided include damage to water resources, natural habitats, animals and plants as well as contamination of land which causes significant harm to human health. In general, integrating environmental concerns into other EU policy areas has become an important concept in European politics (the *Cardiff process*). An overview of EU legislation is given in Table 5.11.

Ambient Air

With the Air Quality Framework Directive 96/62/EC, the EU has created the legal framework for future air quality development. The objectives were, in particular, the definition and limit value setting for harmful air pollutants, the development of standardized methods and criteria for assessing air quality, public information on air quality by means of alarm levels at high concentrations of, for example, groundlevel ozone and, in general, the maintenance of good air quality and its improvement. The recent Directive 2008/50/EC on ambient air quality and cleaner air for Europe has combined the original Framework Directive 96/62/EC and three of its subsidiary directives (except for the fourth daughter directive) as well as obligations for EU data exchange. The ambient air quality standards according to 2008/50/EC are given in Table 5.12. Under EU law, a *limit value* is legally binding from the date it comes into force subject to any exceedances permitted by the legislation. In contrast, a *target value* is to be attained as far as possible by the attainment date and so it is less strict than a limit value. Member States must ensure that limit values are met. Where limit values are not met, Member States must take all necessary measures and have to prepare an *air quality plan* to address the problems identified.

In addition to Directive 2008/50/EC, which addresses ambient air quality standards and objectives, a multitude of provisions exist to prevent atmospheric pollution from *emissions* of industrial sources, land motor vehicles or specific chemicals (e.g., petrol vapors, heavy metals, and substances depleting the ozone layer).

Water

Directive 2000/60/EC of 23 October 2000 establishing a framework for Community action in the field of water policy (Water Framework Directive; WFD) unifies the legal framework for

Table 5.12 EU air quality standards (http://ec.europa.eu/environment/air/quality/standards.htm).

Pollutant	Concentration	Averaging period	Legal nature	Permitted number of exceedences each year
Sulfur dioxide (SO_2)	350 µg/m³	1 hour	Limit value	24
	125 µg/m³	24 hours	Limit value	3
Nitrogen dioxide (NO_2)	200 µg/m³	1 hour	Limit value	18
	40 µg/m³	1 year	Limit value	n/a*
PM10**	50 µg/m³	24 hours	Limit value	35
	40 µg/m³	1 year	Limit value	n/a
Lead (Pb)	0.5 µg/m³	1 year	Limit value	n/a
Carbon monoxide (CO)	10 mg/m³	Maximum daily 8 hour mean	Limit value	n/a
Benzene	5 µg/m³	1 year	Limit value	n/a
Ozone	120 µg/m³	Maximum daily 8 hour mean	Target value	25 days averaged over 3 years
Arsenic (As)	6 ng/m³	1 year	Target value	n/a
Cadmium (Cd)	5 ng/m³	1 year	Target value	n/a
Nickel (Ni)	20 ng/m³	1 year	Target value	n/a
Polycyclic aromatic hydrocarbons (PAHs)	1 ng/m³ (expressed as concentration of Benzo(a)pyrene)	1 year	Target value	n/a
Fine particles (PM2.5)***	25 µg/m³	1 year	Limit value	n/a

*n/a, not applicable.
**Particulate matter <10 µm.
***Particulate matter <2.5 µm.

water policy within the EU. The WFD aims to maintain and improve the status of the aquatic environment (i.e., surface waters, ground waters, transitional waters, and coastal waters) in the EU. It therefore contains measures for the achievement of environmental protection in surface water, groundwater, and water protection areas. The approach to water management requires that objectives of water management are based on the overall ecology of these aquatic systems, taking into account biological, chemical and hydromorphological criteria. The WFD pursues a combined approach using control of pollution at the source through the setting of emission limit values and setting of aquatic environmental quality standards.

The WFD demands that all waters achieve "good ecological status", "good ecological potential" or "good status" (groundwater) or that "high status" waters are maintained. A good surface water chemical status requires meeting the environmental objectives for surface waters established in Article 4(1) (a) and defined by the environmental quality standards established in Annex IX. Good groundwater chemical status meets all the conditions set out in Annex V. The requirements to ensure a good chemical status both of surface water and ground water are quite broadly expressed and comprise only a few parameters (oxygen content, pH value, conductivity, nitrate, and ammonium).

Decision 2455/2001/EC established for the first time a comprehensive list of 33 priority substances in the field of water policy. The priority list was derived from an initial survey of other lists, including those in Directive 76/464/EEC, the OSPAR and HELCOM Conventions and further sources in combination with monitoring data on environmental contamination in surface waters and sediments at that time. The list of priority substances aimed to tackle eco-toxicological effects, bio-accumulation and health impacts. Directive 2008/105/EC on environmental quality standards in the field of water policy, in which Annex II replaced Annex X of the WFD, was adopted in 2008. Directive 2008/105/EC establishes environmental quality standards for surface waters (annual average concentrations and maximum allowable concentration (MAC)) for the 33 priority substances in order to achieve the environmental objectives of the WFD.

The groundwater Directive 2006/118/EC complements the WFD. This Directive establishes specific measures in order to prevent and control groundwater pollution. The measures include (a) establishing criteria for the assessment of good groundwater chemical status and (b) criteria for the identification and reversal of significant and sustained upward pollution trends and for the definition of starting points for trend reversals. Groundwater quality standards as laid down in Annex I pertain to the pollutants nitrates (50 mg/l) and active substances in pesticides, including their relevant metabolites, degradation and reaction products (0.1 μg/l and 0.5 μg/l, the sum of all individual pesticides detected and quantified, respectively). According to Annex II, EU Member States should establish threshold values for all those pollutants and indicators of pollution which characterize bodies of groundwater as being at risk of failing to achieve good groundwater chemical status. The determination of threshold values should include the origins of the pollutants, their possible natural occurrence, their toxicology and dispersion tendency, their persistence and their bioaccumulation potential. The minimum list of pollutants and their indicators comprises (a) substances or ions or indicators which may occur both naturally and/or are of anthropogenic origin (arsenic, cadmium, lead, mercury, ammonium, chloride, sulfate), and (b) synthetic substances (trichloroethylene, tetrachloroethylene), and parameters indicative of saline or other intrusions (conductivity). EU Member States should amend the list of threshold values when new information on pollutants, groups of pollutants, or indicators of pollution indicates that a threshold value should be set for an additional substance, that an existing threshold value should be amended, or that a threshold value previously removed from the list should be re-inserted in order to protect human health and the environment.

Waste

European waste law is characterized by a large number of legal acts (http://ec.europa.eu/environment/waste/hazardous_index.htm). The revised Directive 2008/98/EC waste (Waste Framework Directive) provides for a general framework for the handling of waste and sets the basic waste management requirements for the EU. Its scope is to lay down measures to protect human health and the environment by preventing or reducing adverse impacts of the generation and management of waste and by reducing overall impacts of resource use and improving the efficiency of such use. Directive 2008/98/EC defines waste as "any substance or object which the holder discards or intends or is required to discard". The introduced concepts of "by-product" and "end-of-waste" enable a distinction between waste and non-waste. The categorization into hazardous and non-hazardous waste is based on the Community legislation on chemicals, which ensures the application of similar principles over their whole lifecycle. Annex III (Properties of waste which render it hazardous) provides an overview of toxicological characteristics which mark a waste as hazardous. Directive 2008/98/EC establishes – as a key concept – a waste hierarchy as follows: (a) waste prevention, (b) preparing for re-use, (c) recycling or other recovery, e.g. energy recovery, and (d) disposal. The competent authorities of the Member States are obliged to establish waste management plans that will set out an analysis of the current waste management situation in the geographical entity concerned, as well as the measures to be taken to improve environmentally sound preparation for re-use, recycling, recovery and disposal of waste and an evaluation of how the plan will support the implementation of the objectives and provisions of Directive 2008/98/EC. Furthermore, the Member States must establish waste prevention programs. Such programs should be integrated into either the waste management plans or other environmental policy programs, as appropriate, or function as separate programs. The recent proposal by the European Commission for amending Directive 2008/98/EC on waste (COM (2015) 595 final) pursues new ambitious waste management policy objectives.

Soil Protection

To date, the EU does not have a soil protection law on its own and the key term "contaminated soil" is not defined in legal acts at Community level. Nevertheless, aspects of soil or soil protection are objectives in many other EU environmental policies in areas such as agriculture, water, waste, chemicals, and prevention of industrial pollution that indirectly contribute to the protection of soils. A relatively great closeness exists, e.g. in relation to the provisions of Directive 2008/98/EC on waste. In 2006, the Commission adopted a Soil Thematic Strategy (COM (2006) 231) including a proposal for a Soil Framework Directive (COM (2006) 232). Taking note that the proposal for a Soil Framework Directive has been pending for almost eight years without a qualified majority in the Council in its favor, the Commission withdrew this proposal on 30 April 2014. Recently, aspects of EU soil protection were targeted in the Decision 1386/2013/EU on a General Union Environment Action Programme to 2020 "Living well, within the limits of our planet".

5.1.7 Summary

The chapter provides an overview of general aspects of EU legislation and its institutions, the principles underlying legislation and risk governance, and the special legislation on chemicals and their uses including environmental issues. The first two sections are dedicated to a general description of EU law in terms of the EU treaties and secondary legislation as necessary for the understanding of the rules and laws on chemical substances and radionuclides for a reader unfamiliar with EU law. After a short introduction on the development of the EU, an overview is provided on the roles of the EU institutions, the rules of their co-operation, and the allocation of tasks and competences between the EU institutions and the Member States, which may vary depending on the law considered. Some general principles of EU legislation are explained, such as the *principle of subsidiarity* and the *principle of proportionality* as well as the different kinds of EU law with different addressees and different legal binding force, respectively, i.e. Regulations, Directives, Decisions, Recommendations etc. In Chapter 5.1.3, issues of risk governance, i.e. risk assessment, risk management, hazard, risk types and the *precautionary principle*, risk communication and some basic terms such as hazard, chemical substance and mixture, are described. The general part of this chapter includes international co-operation of the EU with international institutions on the control of dangerous substances and radionuclides. Legislation on chemicals and their uses is described in Chapter 5.1.4 with subsections on special laws on chemicals, medicinal products, medical devices, plant protection and biocidal products, various aspects of food, consumer products and cosmetics, work protection, ionizing radiation by radioactive materials, and finally transport of dangerous goods. In the last section, legislation on chemicals in the environment is covered.

Further Reading

Borchardt KD (2010) The ABC of European Union law (ed. European Union). ISBN 978-92-78-40525-0. doi:10.2830/13717. http://bookshop.europa.eu/en/the-abc-of-european-union-law-pbOA8107147/.
COM (2000) Communication from the Commission on the precautionary principle. COM (2000), 1 final. http://eur-lex.europa.eu/legal-content/EN/TXT/PDF/?uri=CELEX:52000DC0001&from=EN http://eur-lex.europa.eu/legal-content/EN/TXT/?uri=celex:52000DC0001.
EMA (2016) The European regulatory system for medicines. A consistent approach to medicines regulation across the European Union. http://www.ema.europa.eu/docs/en_GB/document_library/Leaflet/2014/08/WC500171674.pdf#.
Lilienblum W, Dekant W, Foth H, Gebel T, Hengstler JG, Kahl R, Kramer PJ, Schweinfurth H, Wollin KM (2008) Alternative methods to safety studies in experimental animals: Role in the risk assessment of chemicals under the new European Chemicals Legislation (REACH). *Arch. Toxicol.* **82** (4), 211–236.
Renn O (2008) *Risk Governance: Coping with Uncertainty in a Complex World*. Earthscan, London.
WHO (2004) IPCS Risk Assessment Terminology. Part1: IPCS/OECD Key Generic Terms used in Chemical Hazard/Risk Assessment; Part 2: IPCS Glossary of Key Exposure Assessment Terminology. Harmonized Project Document No. 1. http://www.who.int/ipcs/publications/methods/harmonization/en/index.html.
Wollin K-M, Illing HP (2014) Limit value setting in different areas of regulatory toxicology. In: Reichl, F.-X., Schwenk, M. (Eds.), *Regulatory Toxicology*. Springer, Berlin Heidelberg, pp. 649–659.

General Information

For EU institutions and legislation see Borchardt (2010).
Access to European Union law: http://eur-lex.europa.eu/.
Website of the European Commission: https://europa.eu/european-union/index_en.
Milieu (2011) Considerations on the application of the Precautionary Principle in the chemicals sector. Final Report. http://ec.europa.eu/environment/chemicals/reach/pdf/publications/final_report_pp.pdf.
Organisation for Economic Co-operation and Development: http://www.oecd.org/.
World Health Organization: http://www.who.int/en/.

5.2 Regulations Regarding Chemicals and Radionuclides in the Environment, Workplace, Consumer Products, Foods, and Pharmaceuticals in the United States

Dennis J. Paustenbach

5.2.1 Introduction

The United States was among the first countries to establish a myriad of legislative and regulatory initiatives intended to control the release of chemicals to the ambient and workplace environment. Beginning in about 1970, the first of many of the modern era initiatives was promulgated. In that year, Congress approved forming the Occupational Safety and Health Administration (OSHA) and the Environmental Protection Agency (EPA). In 1976, the Consumer Products Safety Commission (CPSC) was formed to protect persons against the possible hazards posed by these products. By 1975, the impact of the Environmental Revolution, which began with Rachel Carson in 1962, was well recognized.

Advancements in science related to synthesizing organic chemicals and mining from the 1920s played a major role in bringing about the need for these agencies and regulatory oversight (e.g.,

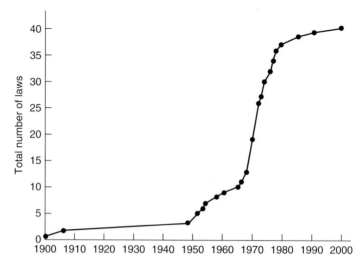

Figure 5.1 Major legislative initiatives in the United States which deal with the manufacture, use, transportation, sale, or disposal of hazardous materials. Reprinted from Paustenbach. Copyright (2002), with permission from John Wiley & Sons.

it has been said that the number of new chemicals synthesized between 1940 and 2018 was more than 300,000, although only about 1000 are produced in large quantities. From 1970 to the present, literally thousands of regulations were introduced in the United States and elsewhere in order to reduce exposure of humans and wildlife to these molecules. The significant changes in the regulatory environment in the United States regarding industrial chemicals are illustrated in Figure 5.1. This chapter presents a brief description of the various major initiatives that dictate the manufacture, use, and disposal of various chemicals in the United States. Table 5.13 summarizes these major regulations. It is, of course, noteworthy that the number of new major initiatives has been small since 2000. Many believe that no additional ones have been needed because existing regulatory agencies appear to have provided adequate public safety.

5.2.2 Occupational Health Regulations

The development of early health guidelines on toxic substances in the workplace, circa 1920 to 1970, was a fragmented process involving various scientific or professional bodies. Early guidelines, or occupational exposure limits (OELs), for example, were written by a combination of the National Research Council (NRC), an independent council of scientists, the American National Standards Institute (ANSI), a conglomerate of government, industry, and private individuals that develops voluntary standards, the National Safety Council (NSC), and the American Conference of Governmental Industrial Hygienists (ACGIH). Established in 1941, the ACGIH comprises hygienists and toxicologists employed in government, academia or consulting who have the objective of protecting workers. For the past 70 years it has been considered the premier group for setting OELs.

The US Occupational Safety and Health Administration (OSHA), together with the National Institute for Occupational Safety and Health (NIOSH), are responsible for developing and enforcing regulations on workplace exposures. OELs are identified by these two organizations. OELs can be recommended exposure limits (RELs) prepared by NIOSH, which are based solely

Table 5.13 Summary of major US regulations on occupational and environmental health.

Law	Date	Agency	Scope
Occupational Health			
Occupational Safety and Health Act	1970	OSHA	Develop worker safety guidelines, permissible exposure limits (PELs)
Food, Drugs, Cosmetics			
Food, Drug, Cosmetic Act	1938	FDA	Protect foods from harmful additives and pesticides, regulate drugs and medical devices, including veterinary drugs
Food Quality Protection Act	1996	EPA	Lowered allowable of residual pesticides on crops
Environment			
National Environmental Policy Act	1969	EPA	General environmental protection, requires environmental impact statements (EIS)
Clean Air Act	1970	EPA	Addresses public exposure to airborne contaminants; develop National Ambient Air Quality Standards (NAAQS)
Clean Water Act	1977	EPA	Addresses water pollution; restricts pollution discharge into water and streams
Comprehensive Environmental Response, Compensation, and Liability Act (CERCLA)	1980	EPA	Clean up of abandoned contaminated sites; establishes financial responsibilities for clean up
Toxic Substances Control Act	1976	EPA	Develop tracking system for hazardous chemicals
Resource Conservation and Recovery Act	1976	EPA	Regulate chemicals 'cradle-to-grave' from manufacture, transportation, treatment, storage and disposal, also regulates non-hazardous waste
Federal Insecticide Fungicide Rodenticide Act	1972	EPA	Regulate use of pesticides through licensing and registration
Consumer Products			
Consumer Product Safety Act	1972	CPSC	Regulate and ensure safety of consumer products
Radionuclides			
Atomic Energy Act	1946	AEC, NRC (1975)	Assure safe handling and management of radioactive materials and facilities

on health considerations and are used to prepare proposed regulations, and permissible exposure limits (PELs), which are promulgated by OSHA and are usually higher (less protective) than RELs because they need to account for technical and economic considerations. In contrast, threshold limit values (TLVs) are guidelines established by the ACGIH and are probably the most widely recognized and accepted of the OELs. However, they are not regulations and are, therefore, not enforceable.

Occupational Safety and Health Administration

> **Established in 1970, OSHA is responsible for developing and enforcing workplace safety and health regulations. The agency's mission is to ensure the safety and health of America's workers by setting and enforcing standards, providing training, outreach, and education, establishing partnerships, and encouraging continual improvement in workplace safety and health.**

On December 29, 1970, President Richard M. Nixon signed the Occupational Safety and Health (OSH) Act. This legislation created both OSHA and NIOSH. It was the first federal act which mandated that every state and nearly every business enterprise meet certain minimal standards with respect to safety and health. Prior to this time, regulations regarding the healthfulness of the workplace were promulgated by each state or regional governmental organization. OSHA officers had responsibility for conducting inspections of the workplace and they could issue fines, as well as recommend criminal penalties for company officials who failed to comply with the regulations.

The OSH Act was intended to ensure worker and workplace safety. This regulation focused on exposure to toxic chemicals, excessive noise levels, nonionizing radiation, mechanical dangers, safety hazards (falls, electrocution, slips, and ergonomic stressors), heat or cold stress, and unsanitary conditions.

The goal of the Act was to assure safe and healthful working conditions for working men and women by:

- authorizing enforcement of the standards developed under the Act
- assisting and encouraging the States in their efforts to ensure safe and healthful working conditions
- providing for research, information, education, and training in the field of occupational safety and health.

It is generally recognized that OSHA has improved the well-being of workers since its inception in 1970. Unfortunately, due to legal requirements placed upon them, the Agency has been unable to pass many regulations during its almost 50-year history. Thus, the TLVs have been more widely recognized as the best OELs because they are routinely updated as more information from toxicology and epidemiology studies becomes more widely available.

National Institute for Occupational Safety and Health

> **Formed through the OSH Act of 1970, the NIOSH is the federal agency responsible for conducting research and making recommendations for the prevention of work-related injury and illness.**

NIOSH is an agency established to help ensure safe and healthful working conditions for working men and women by providing research, information, education, and training in the field of occupational safety and health. NIOSH is part of the Centers for Disease Control and Prevention (CDC) in the Department of Health and Human Services (DHHS). It is generally considered the organization which focuses on conducting research which will inform the OSHA regarding the kind of standards that need to be promulgated in order to protect workers. It was intentionally placed in a separate division of the government to OSHA (which is in the Department of Labor) in an attempt to ensure objectivity in its recommendations.

Information pertaining to the responsibilities of NIOSH are found in Section 22 of the Occupational Safety and Health Act of 1970 (29 CFR § 671). NIOSH is authorized to:

- develop recommendations for occupational safety and health standards
- perform all functions of the Secretary of Health and Human Services under Sections 20 and 21 of the Act
  - conduct Research on Worker Safety and Health
  - conduct Training and Employee Education
- develop information on safe levels of exposure to toxic materials and harmful physical agents and substances
- conduct research on new safety and health problems
- conduct on-site investigations (health hazard evaluations) to determine the toxicity of materials used in workplaces (42 CFR Parts 85 and 85a)
- fund research by other agencies or private organizations through grants, contracts, and other arrangements.

The Federal Mine Safety and Health Amendments Act of 1977 delegated additional authority to NIOSH for coal mine health research.

The mine health and safety law authorized NIOSH to:

- develop recommendations for mine health standards for the Mine Safety and Health Administration (MSHA)
- administer a medical surveillance program for miners, including chest X-rays to detect pneumoconiosis (black lung disease) in coal miners
- conduct on-site investigations in mines similar to those authorized for general industry under the OSH Act
- test and certify personal protective equipment and hazard-measurement instruments.

In order to help promote a safe and healthful workplace, NIOSH has issued literally hundreds of guidance documents for the public, as well as for occupational health professionals. For most of the past 50 years these documents have been relied upon by agencies and professionals around the world in their attempts to identify the appropriate workplace regulations or standards.

5.2.3 Food and Drug Regulations

Amid public outcry due to health hazards from unsanitary food processing and inappropriate food additives, the Pure Food and Drugs Act was adopted in 1906. The act, administered by the Food and Drug Administration (FDA), has undergone a number of amendments, but remains the primary regulation for food and drug safety in the United States today.

Food and Drug Administration

> **The FDA is responsible for protecting public health by ensuring the safety, efficacy, and security of human and veterinary drugs, biological products, medical devices, the food supply, cosmetics, and products that emit radiation.**

Food and drugs are regulated by the FDA. The history of the US FDA dates back to 1862, when President Lincoln appointed a chemist, Charles M. Wetherill, to serve in the new Department of Agriculture (USDA). This was the beginning of the Bureau of Chemistry, the predecessor of the

FDA. In 1927, the Bureau of Chemistry was reorganized into two separate entities. Regulatory functions were located in the Food, Drug, and Insecticide Administration, and nonregulatory research was located in the Bureau of Chemistry and Soils. In 1930, the name of the Food, Drug, and Insecticide Administration was shortened to the Food and Drug Administration.

The FDA regulates all food and food-related products except commercially processed egg products, and meat and poultry products, which are regulated by the USDA's Food Safety and Inspection Service (FSIS). Fruits, vegetables, and other plants are regulated by the Department's Animal and Plant Health Inspection Service (APHIS) to prevent intrusion of plant diseases and pests into the United States. The Agricultural Marketing Service (AMS) of the USDA is responsible for the voluntary grading of fruits and vegetables.

The FDA is responsible for advancing public health by helping to:

- facilitate innovations that make medicines and foods more effective, safer, and more affordable
- inform the public with accurate, science-based information on medicines and foods.

Over the years, the FDA has grown to become one of the most respected agencies in the world. They routinely issue guidance and regulations which protect hundreds of millions of individuals who ingest foods or take pharmaceuticals.

Food, Drug, and Cosmetic Act On 30 June 1906, President Theodore Roosevelt signed the Food and Drugs Act, also known simply as the Wiley Act. This legislation, which the Bureau of Chemistry was charged to administer, prohibited the interstate transport of unlawful food and drugs under penalty of seizure of the questionable products and/or prosecution of the responsible parties.

The basis of the law rested on the regulation of product labeling rather than pre-market approval. Drugs or pharmaceuticals, defined in accordance with the standards of strength, quality, and purity in the *United States Pharmacopoeia* and the *National Formulary,* could not be sold in any other condition unless the specific variations from the applicable standards were plainly stated on the label. The law prohibited the addition of any ingredients that would substitute for the food, conceal damage, pose a health hazard, or constitute a filthy or decomposed substance.

The Food, Drug, and Cosmetic Act (FDCA) was signed on June 25, 1938 by President Franklin Delano Roosevelt. This new law brought cosmetics and medical devices under control, and it required that drugs be labeled with adequate directions for safe use. Moreover, it mandated pre-market approval of all new drugs, such that a manufacturer would have to prove to the FDA that a drug was safe before it could be sold. It prohibited false therapeutic claims for drugs, although a separate law granted the Federal Trade Commission (FTC) jurisdiction over drug advertising. The Act also corrected abuses in food packaging and quality, which had become commonplace in the United States. It also mandated legally enforceable food standards. Tolerances, or the safe concentration of certain chemical contaminants, for certain crops, foods, or drugs, were addressed. The law formally authorized factory inspections, and it added injunctions to the enforcement tools at the agency's disposal.

Cosmetics and medical devices, for which the Post Office Department and the FTC had limited oversight prior to 1938, also came under FDA authority after 1938. While pre-market approval did not apply to devices, in every other sense the new law equated them to drugs for regulatory purposes. As the FDA had to deal with both increasing fraudulent medical devices and the expansion of medical technology in and after World War II, Congress passed the 1962 drug amendments for medical devices.

The 1938 Act required colors to be certified as harmless and suitable by the FDA for their use in cosmetics. Further amendments to the Federal FDCA included a series of laws addressing food additives in 1958 and color additives in 1960. These laws gave the FDA tighter control over the growing list of chemicals entering the food supply and gave manufacturers the responsibility for establishing the safety of their products. The laws included a provision, known as the Delaney Clause, which established that no food or color additive could be deemed safe – or given FDA approval – if found to cause cancer in humans or animals above *de minimus* concentrations.

In 1962, the Delaney Clause was modified to permit the FDA to approve the use of carcinogenic compounds in food-producing animals if certain conditions were met. This modification to the Delaney Clause is known as the diethylstilbestrol (DES) proviso, named for a hormone approved in 1954 to promote growth in cattle and sheep. DES had also been approved much earlier in humans. It was thought to prevent miscarriages in women, but the hormone later was linked to vaginal cancers in the daughters of women who were treated with the drug during pregnancy. Under the DES proviso, the FDA could approve a carcinogen for food animal use if the concentration of any residue remaining in the edible tissues was so low that it presented an insignificant risk of cancer to consumers.

The 1960 color amendments strengthened the safety requirement for color additives, necessitating additional testing for many existing cosmetics to meet the new safety standard. The FDA attempted to interpret the new law as applying to every ingredient of color-imparting products, such as lipstick and rouge, but the courts rebuffed this proposal.

Another agency responsibility, veterinary medicine, had been stipulated since the 1906 Act; foods included animal feed and drugs included veterinary pharmaceuticals. Likewise, animal drugs were included in the provisions for new drugs under the 1938 law and the 1962 drug amendments. However, the Food Additives Amendment of 1958 had an impact too, since drugs used in animal feed were also considered additives, and thus subject to the provisions of the food additive petition process.

5.2.4 Environmental Regulations

Federal environmental regulations in the United States began with the adoption of the National Environmental Policy Act (NEPA) in 1969. This Act led to the creation of a Council on Environmental Quality (CEQ) within the Executive office of the President, which was responsible for developing and coordinating the nation's environmental programs and policies. The nation's primary environmental regulatory agency is the EPA, which was established in 1970.

Environmental Protection Agency

> **The US EPA was established on December 2, 1970. The EPA's mission is to protect human health and to safeguard the natural environment: air, water, and land. The agency leads the nation's environmental science, research, education, and assessment efforts.**

In July of 1970, the White House and Congress worked together to establish the EPA in response to the growing public demand for cleaner water, air, and land. Prior to the establishment of the EPA, the federal government was not structured to make a coordinated effort to minimize exposure to pollutants that harm human health and the environment. Since its inception, the US

EPA has been generally recognized globally as the premier agency for conducting research and promulgating regulations which protect the public from harm due to contaminated air, water, soil, sediments, and certain consumer products. In recent years, agencies in the EU have moved to be their equal in many areas.

In 2016, the US EPA employed nearly 30,000 people across the country, including the headquarters in Washington, DC, 10 regional offices, and more than a dozen laboratories. The EPA works to:

- *develop and enforce regulations*: The EPA works to develop and enforce regulations that implement environmental laws enacted by Congress. The EPA is responsible for research and setting national standards for a variety of environmental programs, and delegates to States and Tribes the responsibility for issuing permits and for monitoring and enforcing compliance. Where national standards are not met, the EPA can issue sanctions and take other steps to assist the States and Tribes in reaching the desired levels of environmental quality.
- *offer financial assistance:* In recent years, between 40% and 50% of the EPA's enacted budgets have provided direct support through grants to state environmental programs. EPA grants to States, nonprofits, and educational institutions support high-quality research that will improve the scientific basis for decisions on national environmental issues and help the EPA achieve its goals by:
  - providing research grants and graduate fellowships
  - supporting environmental education projects that enhance the public's awareness, knowledge, and skills to make informed decisions that affect environmental quality
  - offering information for state and local governments and small businesses on financing environmental services and projects
  - providing other financial assistance through state environmental pro- grams.
- *perform environmental research:* At laboratories located throughout the nation, the agency works to assess environmental conditions and to identify, understand, and solve current and future environmental problems; integrate the work of scientific partners such as nations, private sector organizations, academia and other agencies; and provide leadership in addressing emerging environmental issues and in advancing the science and technology of risk assessment and risk management.
- *sponsor voluntary partnerships and programs:* The Agency works through its headquarters and regional offices with over 10,000 industries, businesses, nonprofit organizations, and state and local governments, on over 40 voluntary pollution-prevention programs and energy-conservation efforts. Partners set voluntary pollution-management goals; examples include conserving water and energy, minimizing greenhouse gases, slashing toxic emissions, re-using solid waste, controlling indoor air pollution, and getting a handle on pesticide risks. In return, the EPA provides incentives like vital public recognition and access to emerging information.
- *further environmental education:* the EPA advances educational efforts to develop an environmentally conscious and responsible public, and to inspire personal responsibility in caring for the environment.
- *publish information:* Through written materials and its website, the EPA informs the public about its activities.

Clean Air Act Adopted in 1970, the Clean Air Act (CAA) is the comprehensive Federal law that regulates air emissions from area, stationary, and mobile sources. This law authorizes the US EPA to establish National Ambient Air Quality Standards (NAAQS) to protect public health and the environment.

The goal of the Act was to set and achieve NAAQS in every state by 1975. The setting of maximum pollutant standards was coupled with directing the states to develop state implementation plans (SIPs) applicable to appropriate industrial sources in the state. The Act was amended in 1977 primarily to set new goals (dates) for achieving attainment of NAAQS since many areas of the country had failed to meet the deadlines. Significant amendments were made to the Act in 1990 to meet unaddressed or insufficiently addressed problems such as acid rain, ground-level ozone, stratospheric ozone depletion, and air toxics. Many have credited these changes for dramatically improving air quality in the United States and for significantly reduced air-pollution-related disease (such as asthma, heart attacks, etc.).

The CAA is often considered among the most important of all environmental regulations, not only in the United States, but also in providing guidance to dozens of countries around the world. It currently regulates more than 100 chemicals that may be emitted into the ambient air.

Clean Water Act Growing public awareness and concern for controlling water pollution led to enactment of the Federal Water Pollution Control Act Amendments of 1972. As amended in 1977, this law became commonly known as the Clean Water Act (CWA). Some have said that when the Cuyahoga river caught fire, Americans and their elected officials decided to pursue aggressive legislative initiatives to bring about a significant improvement in drinking water quality and river quality.

The Act established the basic structure for regulating discharges of pollutants into the waters of the United States. It gave the EPA the authority to implement pollution control programs such as setting wastewater standards for industry. The CWA also continued requirements to set water quality standards for all contaminants in surface waters. The Act made it unlawful for any person to discharge any pollutant from a point source into navigable waters, unless a permit was obtained under its provisions. It also funded the construction of sewage-treatment plants under the construction grants program and recognized the need for planning to address the critical problems posed by nonpoint-source pollution.

Subsequent enactments modified some of the earlier CWA provisions. Revisions in 1981 streamlined the municipal construction grants process, improving the capabilities of treatment plants built under the program. Changes in 1987 phased out the construction grants program, replacing it with the State Water Pollution Control Revolving Fund, more commonly known as the Clean Water State Revolving Fund. This new funding strategy addressed water quality needs by building on EPA–State partnerships.

Comprehensive Environmental Response, Compensation, and Liability Act The Comprehensive Environmental Response, Compensation, and Liability Act (CERCLA), commonly known as the Superfund, was enacted by Congress on December 11, 1980. Some believe that Congress was convinced that such a law was necessary after it was discovered in Love Canal (New York) that significant waste had been buried near homes and a community. Although the liabilities associated with this site have been debated, there is little doubt that this incident urged Congress to act.

This law created a tax on the chemical and petroleum industries and provided broad Federal authority to respond directly to releases or threatened releases of hazardous substances that may endanger public health or the environment. During the initial five years after passage (about 1986), $1.6 billion was collected and the tax went to a trust fund for cleaning up abandoned or uncontrolled hazardous waste sites. The Superfund or CERCLA:

- established prohibitions and requirements concerning closed and abandoned hazardous waste sites
- provided for liability of individuals responsible for releases of hazardous waste at these sites

- established a trust fund to provide for clean-up when no responsible party could be identified.

The law authorizes two kinds of response actions:

- short-term removals, where actions may be taken to address releases or threatened releases requiring prompt response
- long-term remedial response actions that permanently and significantly reduce the dangers associated with releases or threats of releases of hazardous substances that are serious, but not immediately life-threatening. These actions can be conducted only at sites listed on the EPA's National Priorities List (NPL).

CERCLA also enabled the revision of the National Contingency Plan (NCP). The NCP provided the guidelines and procedures needed to respond to releases and threatened releases of hazardous substances, pollutants, or contaminants. The NCP also established the NPL. CERCLA was amended by the Superfund Amendments and Reauthorization Act (SARA) on October 17, 1986. SARA reflected EPA's experience in administering the complex Superfund program during its first six years and made several important changes and additions to the program.

The reauthorization of SARA:

- stressed the importance of permanent remedies and innovative treatment technologies in cleaning up hazardous waste sites
- required Superfund actions to consider the standards and requirements found in other State and Federal environmental laws and regulations
- provided new enforcement authorities and settlement tools
- increased State involvement in every phase of the Superfund program
- increased the focus on human health problems posed by hazardous waste sites
- encouraged greater citizen participation in making decisions on how sites should be cleaned up
- increased the size of the trust fund to $8.5 billion.

SARA also required the EPA to revise the Hazard-Ranking System (HRS) to ensure that it accurately assessed the relative degree of risk to human health and the environment posed by uncontrolled hazardous waste sites that may be placed on the NPL. Superfund has not been reauthorized since 1986 even though it has been brought before the US Congress on a number of occasions.

Toxic Substances Control Act The Toxic Substances Control Act (TSCA) (pronounced "tosca") was enacted by Congress in 1976 to give the EPA the ability to track the 75,000 industrial chemicals currently produced or imported into the United States. The EPA repeatedly screens these chemicals and can require reporting or testing of those that may pose an environmental or human-health hazard. The EPA can ban the manufacture and import of those chemicals that pose an unreasonable risk.

Also, the EPA has mechanisms in place to track the thousands of new chemicals that industry develops each year with either unknown or dangerous characteristics. The EPA can control these chemicals as necessary to protect human health and the environment.

On June 22, 2016, the Frank Lautenberg Chemical Safety for the 21st Century Act, which amends TSCA, was signed into law. The new law, which received bipartisan support, included many improvements:

- a mandatory requirement for the EPA to evaluate existing chemicals with clear and enforceable deadlines
- new risk-based safety standards

- increased public transparency for chemical information
- consistent source of funding for the EPA to carry out the responsibilities under the new law.

Many of the requirements of the 2016 amendment put the societal expectations of US citizens in line with the actions in Europe that were evidenced by the passage of the REACH program, which has been implemented in the EU for at least 10 years.

Fundamentally the TSCA initiative of 2016 called for much more toxicology testing, the conduct of hundreds of exposure assessments, and an equal number of risk assessments. Among other objectives, one was to diminish the introduction of new chemicals into commerce unless there was a considerable understanding of the possible human and ecological health risks.

Resource Conservation and Recovery Act The Resource Conservation and Recovery Act (RCRA, pronounced "rick-rah"), was adopted in 1976 and gave the EPA the authority to control hazardous waste from cradle to grave. The intent was to regulate all chemicals from the time of manufacture, transportation, treatment, storage, and disposal of hazardous waste. RCRA also set out a framework for the management of nonhazardous wastes.

The 1986 amendments to RCRA enabled the EPA to address environmental problems that could result from underground tanks storing petroleum and other hazardous substances. RCRA focuses only on active and future facilities and does not address abandoned or historical sites (see CERCLA).

The Federal Hazardous and Solid Waste Amendments (HSWA, pronounced "hiss-wa") are the 1984 amendments to RCRA that required the phasing out of the land disposal of hazardous waste. Some of the other mandates of this strict law include increased enforcement authority for the EPA, more stringent hazardous waste management standards, and a comprehensive underground storage tank program.

Federal Insecticide, Fungicide, and Rodenticide Act The primary focus of the Federal Insecticide, Fungicide, and Rodenticide Act (FIFRA) was to provide federal control of pesticide distribution, sale, and use. The EPA was given authority under FIFRA not only to study the consequences of pesticide usage but also to require users (farmers, utility companies, and others) to register when purchasing pesticides.

Through later amendments to the law, users also must take exams for certification as applicators of pesticides. All pesticides used in the United States must be registered (licensed) by the EPA. Registration ensures that pesticides will be properly labeled and that if in accordance with specifications will not cause unreasonable harm to the environment.

Some key elements of FIFRA include:

- it is a product-licensing statute; pesticide products must obtain an EPA registration before manufacture, transport, and sale
- registration is based on a risk/benefit standard
- strong authority to require data and authority to issue data call-ins
- the ability to regulate pesticide use through labeling, packaging, composition, and disposal
- emergency exemption authority: permits approval of unregistered uses of registered products on a time-limited basis
- the ability to suspend or cancel a product's registration: appeals process, adjudicatory functions, etc.

Food Quality Protection Act With the enactment of the Food Quality Protection Act (FQPA) of 1996, Congress presented the EPA with the enormous challenge of implementing the most comprehensive and historic overhaul of the nation's pesticide and food safety laws in decades.

The FQPA amended both the FIFRA and the Federal FDCA by fundamentally changing the way the EPA regulates pesticides.

Some of the major requirements include stricter safety standards, especially for infants and children, and a complete reassessment of all existing pesticide tolerances.

In 1996, Congress unanimously passed landmark pesticide food safety legislation supported by the administration and a broad coalition of environmental, public health, agricultural, and industry groups. President Clinton promptly signed the bill on August 3, 1996, and FQPA became law. One of the key provisions of this Act was that the residual concentrations of a large fraction of pesticides on crops was to be substantially lowered; often by at least 10- to 100-fold.

The EPA regulates pesticides under two major Federal statutes. Under FIFRA, EPA registers pesticides for use in the United States and prescribes labeling and other regulatory requirements to prevent unreasonable adverse effects on human health or the environment. For the FDCA, EPA establishes tolerances (maximum legally permissible levels) for pesticide residues in food. FQPA streamlined these two regulations into one law that:

- mandates a single, health-based standard for all pesticides in all foods
- provides special protections for infants and children
- expedites approval of safer pesticides
- creates incentives for the development and maintenance of effective crop protection tools for American farmers
- requires periodic re-evaluation of pesticide registrations and tolerances.

5.2.5 Consumer Product Regulations

Consumer Product Safety Commission

> **The US Consumer Product Safety Commission (CPSC), created in 1972 under the Consumer Product Safety Act, is charged with protecting the public from unreasonable risks of serious injury or death from more than 15,000 types of consumer products under the agency's jurisdiction.**

It has been reported that deaths, injuries, and property damage from consumer product incidents cost the United States more than $700 billion annually (this estimate seems high in the author's opinion).

The goal of the CPSC is to protect consumers and families from products that pose a fire, electrical, chemical, or mechanical hazard or can injure children. The CPSC's work to ensure the safety of consumer products, such as toys, cribs, power tools, cigarette lighters, and household chemicals, contributed significantly to the 30% decline in the rate of deaths and injuries associated with consumer products over the past 30 years.

The CPSC is involved in:

- developing voluntary standards with industry
- issuing and enforcing mandatory standards or banning consumer products if no feasible standard would adequately protect the public
- obtaining the recall of products or arranging for their repair
- conducting research on potential product hazards
- informing and educating consumers through the media, State and local governments, private organizations, and by responding to consumer inquiries, including local and national media coverage, publication of numerous booklets and product alerts, a website, a telephone hotline,

the National Injury Information Clearinghouse, CPSC's Public Information Center, and responses to Freedom of Information Act (FOIA) requests.

Consumer Product Safety Act The Consumer Product Safety Act (CPSA), enacted in 1972, is CPSC's umbrella statute. It established the agency, defined its basic authority, and provided that when the CPSC finds an unreasonable risk of injury associated with a consumer product it can develop a standard to reduce or eliminate the risk. The CPSA also provides the authority to ban a product if there is no feasible standard, and it gives CPSC authority to pursue recalls for products that present a substantial product hazard. (Generally excluded from CPSC's jurisdiction are food, drugs, cosmetics, medical devices, tobacco products, firearms and ammunition, motor vehicles, pesticides, aircraft, boats and fixed-site amusement rides.)

5.2.6 Radionuclides Regulations

Nuclear Regulatory Commission

> **Created in 1974, the Nuclear Regulatory Commission (NRC)'s primary mission is to protect public health and safety, and the environment from the effects of radiation from nuclear reactors, materials, and waste facilities.**

The NRC regulates nuclear materials and facilities to promote the common defense and security. The commission carries out its mission by conducting the following activities:

- commission direction-setting and policymaking: policy formulation, rulemaking, and adjudication oversight activities performed by NRC's five-member Commission
- radiation protection: providing information about radiation and ensuring protection of the public and radiation workers
- regulation: rulemaking, oversight, licensing and certification
- emergency preparedness and response: integration of the NRC emergency and preparedness programs and response to a wide spectrum of radiological emergencies
- nuclear security and safeguards: regulating licensees' accounting systems for special nuclear and source materials, and security programs and contingency plans for dealing with threats, thefts, and sabotage relating to special nuclear material, high-level radioactive wastes, nuclear facilities, and other radioactive materials and activities that the NRC regulates
- public affairs: interactions with the media and the public
- congressional affairs: interactions with Congress
- State and Tribal programs: cooperative activities and interactions with Federal, State, and local governments, interstate organizations, and Indian Tribes
- international programs: cooperative activities with other governments and the international nuclear regulatory community and licensing for nuclear imports and exports.

The NRC is headed by a five-member Commission. The President designates one member to serve as Chairman and official spokesperson. The Commission as a whole formulates policies and regulations governing nuclear reactor and materials safety, issues orders to licensees, and adjudicates legal matters brought before it. The Executive Director for Operations (EDO) carries out the policies and decisions of the Commission and directs the activities of the program offices.

The offices reporting to the EDO ensure that the commercial use of nuclear materials in the United States is safely conducted. As part of the regulatory process, the four regional offices conduct inspection, enforcement, and emergency response programs for licensees within their borders.

Atomic Energy Act The Atomic Energy Act of 1946, which was legislated when tensions with the Soviet Union were developing into the cold war, acknowledged the potential peaceful benefits of atomic power. The Act established the five-member Atomic Energy Commission (AEC) to manage the nation's atomic energy programs.

In 1954, Congress passed new legislation that for the first time permitted the wide use of atomic energy for peaceful purposes. The 1954 Atomic Energy Act redefined the atomic energy program by ending the government monopoly on technical data and making the growth of a private commercial nuclear industry an urgent national goal. It instructed the agency to prepare regulations that would protect public health and safety from radiation hazards. Thus, the 1954 Act assigned the AEC three major roles: to continue its weapons program, to promote the private use of atomic energy for peaceful applications, and to protect public health and safety from the hazards of commercial nuclear power.

In 1974, Congress divided the AEC into the Energy Research and Development Administration and the NRC. The Energy Reorganization Act, coupled with the 1954 Atomic Energy Act, constituted the statutory basis for the NRC. The new agency inherited a mixed legacy from its predecessor, marked both by 20 years of conscientious regulation and by unresolved safety questions, substantial anti-nuclear activism, and growing public doubts about nuclear power.

By 1974, the AEC's regulatory programs had come under such strong attack that Congress decided to abolish the agency. Supporters and critics of nuclear power agreed that the promotional and regulatory duties of the AEC should be assigned to different agencies. The Energy Reorganization Act of 1974 created the NRC and it began operations on January 19, 1975.

The NRC (like the AEC before it) focused its attention on several broad issues that were essential to protecting public health and safety. In many ways, the NRC carried on the legacy inherited from the AEC. It performed the same licensing and rulemaking functions that the regulatory staff had discharged for two decades. It also assumed some new administrative and regulatory duties. The NRC, unlike the AEC's regulatory staff, was the final arbiter of regulatory issues; its judgment on safety questions was less susceptible to being overridden by developmental priorities.

The AEC and the NRC published standards that were intended to provide an ample margin of safety from radiation that was generated by the activities of its licensees. The radiation standards embodied available scientific information and the judgment of leading authorities in the field.

Although reactor safety issues received the majority of public notice, the NRC also devotes substantial resources to a variety of complex questions in the area of nuclear materials safety and safeguards, such as the protection of nuclear materials from theft or diversion and the safety of depositories for the disposition of high-level and low-level radioactive waste.

5.2.7 Governmental Agencies on Human Health

Agency for Toxic Substances and Disease Registry

> **In 1980, Congress created the Agency for Toxic Substances and Disease Registry (ATSDR) to implement the health-related sections of laws that protect the public from hazardous wastes and environmental spills of hazardous substances.**

As mentioned previously, the CERCLA provided the Congressional mandate to remove or clean up abandoned and inactive hazardous waste sites and to provide federal assistance in toxic emergencies. As the lead agency within the Public Health Service for implementing the health-related provisions of CERCLA, ATSDR is charged under the Superfund Act to assess the presence and nature of health hazards at specific Superfund sites, to help prevent or reduce further exposure and the illnesses that result from such exposures, and to expand the knowledge base about health effects from exposure to hazardous substances. ATSDR is part of the CDC.

In 1984, amendments to the Resource Conservation and Recovery Act (RCRA) of 1976, which provides for the management of legitimate hazardous waste storage or destruction facilities, authorize ATSDR to conduct public health assessments at these site when requested by the EPA, States, or individuals. ATSDR was also authorized to assist the EPA in determining which substances should be regulated and the levels at which substances may pose a threat to human health. With the passage of the SARA in 1986, ATSDR received additional responsibilities in environmental public health. This Act broadened ATSDR's responsibilities in the areas of public health assessments, establishment and maintenance of toxicologic databases, information dissemination, and medical education.

5.2.8 Centers for Disease Control

> **The CDC, established in 1946, is one of the 13 major operating components of the Department of Health and Human Services (HHS), which is the principal agency in the United States government for protecting the health and safety of all Americans and for providing essential human services.**

Since the CDC was founded in 1946 to help control malaria, it has remained at the forefront of public health efforts to prevent and control infectious and chronic diseases, injuries, workplace hazards, disabilities, and environmental health threats. Today, the CDC is globally recognized for conducting research and investigations and for its action-oriented approach. The CDC applies research and findings to improve people's daily lives and responds to health emergencies, something that distinguishes the CDC from its peer agencies. The CDC seeks to accomplish its mission by working with partners throughout the nation and the world to:

- monitor health
- detect and investigate health problems
- conduct research to enhance prevention
- develop and advocate sound public health policies
- implement prevention strategies
- promote healthy behaviors
- foster safe and healthful environments
- provide leadership and training.

These functions are the backbone of the CDC's mission. Each of the CDC's component organizations undertakes these activities in conducting its specific programs. The steps needed to accomplish this mission are also based on scientific excellence, requiring well-trained public health practitioners and leaders dedicated to high standards of quality and ethical practice.

5.2.9 Litigation is Nearly as Effective as Regulation in the United States

It is fairly well known across the globe that the United States is a litigious society. The benefits and disadvantages of such a culture have been debated for decades. Of significance to occupational and environmental health professionals is that the pressures placed upon corporations is sufficiently strong that the threat of legal action serves as a powerful "backstop" in the event that a regulatory agency fails to pass new laws or have aggressive enforcement.

It would not be an exaggeration to say that many large corporations in the United States often set internal standards which are more stringent than the current regulatory expectations in an attempt to go a bit beyond the minimum expected in part, to ward off the likelihood of claims that might bring them to the courtroom. Most firms have found that it is important to understand not only the specific expectations of a regulatory agency, and to comply, but also to have a sense as to what the citizens expect. This "reading of the stars" is difficult but it has become part of staying in business in the United States for firms who handle large quantities of chemicals.

5.2.10 Summary

The regulation of chemicals in the United States has increased steadily over the years. This chapter presents a brief overview of major environmental, human health, and occupational safety regulations in the United States. The evolution of new technologies and scientific knowledge plays an important role in how these regulations are created and enforced. It is anticipated that the overall effectiveness of regulation in protecting human health and the environment will continue to improve in the years to come.

Further Reading

Ashford, N.A. (2000) Chapter 9: Government Regulation of Occupational Health and Safety. In: Levy, B.S. and Wegman, D.H. (eds.) *Occupational Health. Recognizing and Preventing Work-Related Disease and Injury*, 4th Ed. Philadelphia, PA: Lippincott Williams & Wilkins. pp. 177–199.

ATSDR (February 2013) *ATSDR Congressional Background and Mandates*. Retrieved September 7, 2017 from: https://www.atsdr.cdc.gov/about/congress.html.

Bingham, E. and Grimsley, L.F. (2001) Chapter 8: Regulations and Guidelines in the Workplace. In: Bingham, E., Cohrrsen, B., and Powell, C.H. (eds.) *Patty's Toxicology, Volume 1*, 5th ed. New York: John Wiley & Sons. pp. 321–352.

CDC (April 2014) Mission, Role, and Pledge. Retrieved September 7, 2017 from: https://www.cdc.gov/about/organization/mission.htm.

CDC – NIOSH (June 2016) About NIOSH. Retrieved September 7, 2017 from: https://www.cdc.gov/niosh/about/default.html.

CPSC (n.d.) CPSC Overview. Retrieved September 7, 2017 from: https://www.cpsc.gov/PageFiles/164687/CPSCOverviewEn.pdf.

DEFRA (n.d.) The Government's Chemicals Strategy: Sustainable Production and Use of Chemicals - A Strategic Approach. Retrieved September 7, 2017 from: http://webarchive.nationalarchives.gov.uk/20110318161936/http://www.defra.gov.uk/environment/quality/chemicals/documents/chemicals-strategy0904.pdf.

FDA (May 2009) Milestones in US Food and Drug Law History. Retrieved September 7, 2017 from: https://www.fda.gov/AboutFDA/WhatWeDo/History/Milestones/default.htm.

Mazuzan, G.T. and Walker, S.J. (1997) *Controlling the Atom: the Beginnings of Nuclear Regulation, 1946–1962*. Washington, D.C.: US Nuclear Regulatory Commission.

OSHA (n.d.) OSHA's Mission. Retrieved September 7, 2017 from: https://www.osha.gov/about.html.

UNEP (September 2017) The Hazardous Chemicals and Wastes Conventions. Retrieved September 7, 2017 from: http://www.pic.int/Portals/5/ResourceKit/A_General%20information/d.3Convention%20brochure/UNEP_threeConventions_engV4.pdf.

US Congress (June 2016) Frank R. Lautenberg Chemical Safety for the 21st Century Act. Public Law 114-118. 130 Stat. 448. June 22, 2016. Retrieved September 7, 2017 from: https://www.gpo.gov/fdsys/pkg/PLAW-114publ182/pdf/PLAW-114publ182.pdf.

USEPA (August 2017) Laws and Regulations. Retrieved September 7, 2017 from: https://www.epa.gov/laws-regulations.

Weiner, J.B, Rogers, M.D., Hammit, J.K., and Sand, P.H. (2011) *The Reality of Precaution: Comparing Risk Regulations in the United States and Europe*. Resources for the Future. Washington, DC.

5.3 The Concept of REACH

Jörg Lebsanft

5.3.1 Introduction

> **Substances as defined by the REACH regulation include natural substances as well as substances produced by chemical synthesis. They are the main ingredients of physical goods, such as washing and cleaning agents, cosmetics, toys, construction products, textiles, furniture, electrical and sports equipment, vehicles and many other products. The REACH chemicals legislation is the regulatory basis of the European Union (EU) for determining the risks posed by substances to human health and the environment.**

During production of substances or products, their use, disposal or recycling, substances may be released and lead to exposure of humans or the environment. The magnitude of the risk to human health or the environment depends on the level of exposure and the properties of the substance. Since in many cases the available information was insufficient for risk assessment, a comprehensive revision of chemicals legislation was initiated at the beginning of the last decade in the EU which finally resulted in the adoption of the REACH regulation in 2006. The provisions of the REACH regulation are closely linked with the legislation on occupational health and waste disposal. They are complemented by the regulation for classification, labelling and packaging (CLP Regulation) and product-specific legislation (e.g. the regulations on plant-protection products, biocides and cosmetics).

The expectations for such regulations are high. On the one hand, a high level of protection of human health and the environment should be ensured. On the other hand, the regulations should burden industry as little as possible and encourage innovation in order to avoid the competitiveness of EU products on the global market being negatively affected. Trade barriers for imported products should also be avoided or at least be minimized as far as possible. Challenges also exist regarding the methods for identifying the hazardous properties of the substances. Toxicological and eco-toxicological testing should, whenever possible, not involve the use of vertebrate animals. If this condition cannot be met, the number of animals used and their suffering must be minimized. Furthermore, the tests should be carried out according to internationally agreed test guidelines and meet internationally recognized quality standards (good laboratory practice) in order to ensure acceptance of test results in other parts of the world. Finally, compliance with the regulations should be easy to monitor by authorities because effective and rigorous enforcement discourages free-riders and thus contributes to a level playing field for competitors. Increasingly shorter product cycles and the fact that many products consist of components that are manufactured in non-EU countries with different legal requirements create a particular challenge for the authorities. Often substantial effort is required to identify the substances contained in and released from a product.

This chapter describes the basic elements of the REACH regulation. Professionals who assess and register substances need to undertake a more in-depth examination of this complex legislation. The European Chemicals Agency (ECHA) provides comprehensive information that can be downloaded from its website. Some readers may be inspired to study REACH and the related legislation more intensively. They may discover a highly diverse and exciting subject at the interface between science, law and politics. The REACH regulation is considered a milestone of EU environmental and consumer protection rules. The legislative process leading to the adoption of REACH was accompanied by controversies of unprecedented intensity.

5.3.2 Historical Development

> **REACH emerged from previous EU legislation which is briefly described below in order to facilitate understanding of the REACH concept.**

At the end of the 1970s, EU legislation was adopted stipulating that substances must be tested prior to their marketing to identify the hazardous properties. Since this provision came into force on September 18, 1981, all substances placed on the market for the first time have been affected by this directive, which is known as the Dangerous Substances Directive. In contrast, substances that were already on the market at this time (so-called Existing Substances) were exempted. The high degree of pragmatism inherent to this rule is typical for EU chemicals legislation. At that time, neither industry nor authorities would have been able to deal with the vast number of about 100,000 Existing Substances.

More than 10 years later, in 1992, specific provisions for Existing Substances were issued (Existing Substances Regulation). However, as the regulation did not require systematic testing and risk assessment of substances by industry, their impact was limited (Strategy for a future Chemicals Policy – White Paper). In the late 1990s, the deficits of the regulation triggered a discussion on the need for a fundamental reform of EU chemicals legislation. Though rightly criticized, the Existing Substances Regulation nevertheless made an important contribution to the advancement of chemicals legislation. The risk assessment methodology developed, standardized and applied by authorities under this regulation has become the binding standard for industry under the REACH regulation.

5.3.3 Substances, Mixtures and Articles

> **According to chemicals legislation substances are chemicals which are produced by chemical synthesis, as well as substances of natural origin, such as plant extracts and minerals. Typically, they are mixtures rather than chemically pure compounds.**

A substance as defined by the REACH regulation is the product of a manufacturing process which may be a chemical reaction, a distillation or an extraction. While, from the chemist's point of view, it may seem more convincing to define substances as chemically pure compounds, such an approach would have caused undesirable implications. Most substances arising from manufacturing processes and placed on the market are in fact mixtures (in the chemical sense). Costs for the chemical producers or importers would have been far higher if the obligations specified below referred to individual chemical compounds. In this case, manufacturers would have been obliged to separate the individual components of substances from each other and test and register them individually. However, the REACH definition also has drawbacks, which are described in more detail in *Joint Registration of Same Substances* (this chapter) and Section 5.3.7.

It may be confusing that the regulation also uses and defines "mixtures" and distinguishes them from substances. According to REACH, a "mixture" is produced by mixing two or more substances or dissolving substances where the substances do not chemically react with each other or with the solvent. Typical mixtures are detergents or cleaning agents. It therefore depends on the production process rather than on the chemical composition whether a "substance" or a "mixture" is obtained.

There are significant legal differences regarding the obligations for substances and mixtures. Only substances require systematic testing and registration. Extension of the registration duty to mixtures would have led to a much higher administrative workload for industry and authorities. On the other hand, exempting mixtures from registration implies that authorities have far less information on these, although mixtures are used more frequently in daily life than substances. The safety concept of REACH is based on the principle that the information gained from testing of substances is used for the assessment of the hazards of both substances and mixtures. If the assessment shows that a substance or mixture has hazardous properties (e.g. toxic, carcinogenic, flammable), the substance must be classified accordingly and the information must be printed on the packaging (labelling). These obligations are laid down in the Classification, Labelling and Packaging of Substances and Mixtures Regulation, which is based on the globally harmonized system (GHS) for classification and labelling. Hazardous substances contained in a mixture trigger classification and labelling of the mixture if they exceed the concentration limits defined in the CLP regulation.

> **Most of the products currently on the market are neither substances nor mixtures but articles. REACH defines articles as objects whose function is determined to a greater degree by their special shape, surface or design than by their chemical composition.**

Typical articles are furniture and textiles or even highly complex products such as computers or automobiles. For articles, neither a general obligation for registration nor classification and labelling exist. As they are usually chemically inhomogeneous, the classification and labelling rules for mixtures which are based on the concentrations of the substances in the mixture would not provide satisfactory results for most articles. The lack of specific provisions for classification and labelling entails a general lack of information about the hazardous substances in articles. Moreover, even if the identity and content of all hazardous substances in an article have been identified, it is challenging to assess the risk to human health and the environment due to a general shortage of standardized methods for the determination of the release of substances from products. For example, to determine the exposure of children to phthalates from teething rings, special chewing studies had to be carried out. The development of adequate methods to assess the chemical safety of products still represents an enormous challenge.

5.3.4 The Main Elements of REACH

> **The acronym REACH is derived from three of the four main elements of REACH, namely Registration, Evaluation and Authorization of CHemicals. The fourth element is called "restriction" and designates a partial or total restriction of the manufacturing or use of a substance. It should be noted that the term "chemicals" is only used in the title of the regulation, while the legal text uses the terms "substance" or "mixture", as described in the previous section.**

Evaluation, authorization and restriction of substances may be considered as tools of the authorities to address potential deficiencies or risks. For example, if the available information is insufficient for risk assessment, authorities may oblige registrants by "evaluation" to generate the necessary information and provide it to the authorities. If there is a suspicion that certain uses of a substance may cause relevant health or environmental risks, such uses may be banned by means of "restriction". For restrictions, the burden of proof lies with the authorities. Substances with properties of very high concern may be subjected to authorization, which implies that users must demonstrate safe use and thus reverses the burden of proof.

Registration

The Submission of Dossiers to ECHA

> **Manufacturers and importers are required to register their substances with ECHA. ECHA examines the completeness of the dossiers and assigns a registration number to complete dossiers. This number entitles the manufacturer or importer to produce or import the substance. Registrants are required to update their registration dossier if uses of their substance change or if relevant new knowledge about the hazardous properties becomes available. Registration dossiers are "living documents" and subject to frequent adaptations.**

Registration dossiers contain information about the identity of the substance, its physical, chemical, toxic and eco-toxic properties. The chemical safety report, which is part of the registration dossier, includes a description of the use conditions, an assessment of exposure and risk, and information on the risk management measures applied or recommended by the registrant.

A large part of the data contained in the registration dossiers is made available to the public by ECHA. When using this information it should be taken into account that most of the information has not been validated by authorities. At the level of registration, ECHA only examines whether the registration dossier contains information for all parameters as provided for by the regulation. A systematic scientific examination of the relevance and validity of the submitted information would be associated with a very high administrative burden.

The obligation to register only applies to substances manufactured or imported in quantities of 1 ton or more per year. In order to enable registrants and ECHA to cope with the vast number of Existing Substances, deadlines for registration have been defined depending on the annual tonnage. For example, the last day for registration of phase-in substances produced or imported at a volume of more than 1000 tons per year was November 30, 2010. Substances with an annual tonnage between 100 and 1000 tons had to be registered by May 31, 2013 and for substances between 1 and 100 tons the latest registration date is May 31, 2018. The name "phase-in" substances is derived from this gradual, tonnage-dependent "phase-in" process and is used by REACH instead of the term Existing Substances. The deadlines do not apply to "non-phase-in" substances, previously termed New Substances. These substances have to be registered before the annual volume exceeds 1 ton.

Joint Registration of Same Substances

> **A significant proportion of substances subject to registration are produced by more than one manufacturer. REACH requires registrants to submit registration dossiers of same substances jointly.**

For a joint registration of same substances manufacturers and importers organize themselves in substance information exchange fora (SIEF) in order to exchange and assess the existing data on the intrinsic properties. The obligation for joint registration only applies to those parts of the registration dossier which are related to the hazard assessment. The chemical safety report, which includes use-specific exposure and risk assessments, may be submitted separately. The duty to share data and register jointly was predominantly introduced for animal welfare reasons to avoid repetition of testing on vertebrate animals. As chemical safety reports may contain confidential business information on the uses of a substance, the reports were exempted from the duty of joint registration.

In practice, the assessment of sameness is not trivial. Differences regarding, for example, the production processes often lead to different compositions of substances. It is often difficult to decide whether substances with different compositions should be considered as same substances or trigger separate testing and registration, in particular for a group of substances called UVCB substances (substances of unknown or variable composition, complex reaction products or biological materials). The information on substance identity provided by the registrants has turned out to be often insufficient. Without a precise description of the composition of the substances, however, an assessment of sameness is impossible. The term UVCB substance was introduced decades ago when analytical methods were less well developed. With current methodology, the composition of most UVCB substances can be determined.

Testing and Assessment Strategy

> **The annexes of the REACH regulation contain detailed provisions on the required physical, chemical, toxicological and eco-toxicological studies. The extent of the mandatory testing depends on the annual tonnage of a substance.**

For the lowest tonnage range (1–10 tons per year), testing for acute toxicity, skin and eye irritation/corrosion, and mutagenicity is required. For substances between 10 and 100 tons, additional testing for mutagenicity and a short-term repeated dose toxicity study (exposure duration 28 days) should be carried out. Substances above 100 tons require a subchronic toxicity (90 days) and a developmental toxicity study and substances above 1000 tons an additional reproductive toxicity study referred to as a Extended One-Generation Reproductive Toxicity Study (EOGRTS). Registrants may waive any of these tests if the conditions laid down in Annex XI of the REACH regulation are met. According to this annex, the standard tests are not required if registrants can demonstrate by a weight of evidence approach that the existing data overall provide at least equivalent information as compared to the standard test. Furthermore, testing may be replaced by a prediction of toxicity using qualitative or quantitative structure–activity relationship (QSAR) models or read-across. Experience has shown that registrants make extensive use of the options to waive testing, particularly regarding expensive toxicological tests (The use of alternatives to testing on animals for the REACH Regulation – Second report under article 117.3 of the REACH Regulation). The Annex XI provisions aim to avoid unnecessary animal testing and costs. On the other hand, it is often a matter of controversy between experts whether an adequate prediction of toxicity is possible without testing. In practice, application of Annex XI provisions often complicates administrative processes, leads to delays in decision-making and obtaining the required information (see section on *Evaluation*, this chapter) and increases the likelihood that relevant toxic effects of a substance remain undetected.

> **The risk assessment is a part of the chemical safety report and contains information on the derivation of the derived no-effect level (DNEL), an assessment of use-specific exposure of consumers and workers as well as a comparison of the DNEL with the estimated exposures.**

A use is considered safe (adequately controlled) if exposure during the entire lifecycle of the substance is below the relevant DNEL. If the substance is used as component of a mixture or an article, exposure and risk from such use must also be covered. For the assessment of environmental risks, corresponding provisions apply.

DNELs are derived from existing human or animal studies by applying assessment factors to the observed adverse effect thresholds. The assessment factors are used to compensate for differences regarding the exposure between the reference study and the assessed use and to account for differences of sensitivity between the tested animal species and humans (interspecies variability) and within the same species (intraspecies variability). The assessment of exposure is carried out on the basis of so-called exposure scenarios, which are representative for the use conditions (operational conditions), including the risk management measures. For many professional and industrial standard processes, exposure scenarios have been developed and algorithms derived to facilitate the estimation of the level of exposure. ECHA has published comprehensive and detailed guidance documents on the methodology for risk assessment.

Evaluation

> **The term "evaluation" is used for administrative processes which aim to find out whether registration dossiers comply with REACH information requirements and whether the submitted information is sufficient for risk assessment. There are three different evaluation processes: (1) the examination of testing proposals, (2) the compliance check, and (3) substances evaluation.**

Before carrying out costly and complex tests, in particular long-term animal studies, registrants must obtain ECHA's permission. In the REACH process **examination of testing proposals** ECHA decides on testing proposals submitted by registrants. The agency examines the need of the proposed test as well as the study design (e.g. animal species, route of application). Avoidance of unnecessary animal tests and costs to registrants has motivated the establishment of this evaluation process.

In the **compliance check**, ECHA assesses the compliance of information submitted by registrants with the legal requirements, focusing on the quality and adequacy of the data. This REACH process complements the initial cursory completeness check carried out at the registration stage. REACH provides for a minimum of 5% of the registration dossiers to be checked for compliance by ECHA. It is at the agency's discretion to decide which dossiers are selected and which elements of the dossiers the compliance check is focused on. The EU legislator has entrusted ECHA with a wide margin of discretion when assessing registration dossiers. The increased flexibility was inspired by the failure to use resources efficiently for chemicals safety under the previous Existing Substances Regulation. Under this EU regulation, assessment could only be completed for few substances due to the enormous resources required to meet the high standards regarding the comprehensiveness and depth of the assessment.

Substance evaluation may be carried out if there is a concern that a use of a substance presents a relevant risk to human health or the environment. In contrast to the other evaluation processes, the authorities of EU Member States carry out the technical examination of the dossiers. Under this process, registrants may be required to generate and submit information required to verify or dismiss a concern. In contrast to the previous two evaluation processes, registrants may also be obliged to submit information that is not part of the REACH standard information requirements.

The decision-making procedure is common to all three evaluation processes. After ECHA or a Member State authority has prepared a draft decision specifying the required information, registrants and, subsequently, Member State authorities may provide comments on the draft. If no Member State objects, the procedure is finished and the decision is taken accordingly by ECHA. Otherwise, ECHA submits the draft decision to the Member State Committee, which is composed of representatives of EU Member States. If the Committee fails to reach unanimous agreement, the draft decision is forwarded to the European Commission for decision-making.

Authorization

> **Authorization is the most complex and demanding element of REACH and may be applied to regulate substances of very high concern (SVHC). SVHC include carcinogenic, mutagenic and reprotoxic substances (CMR), substances which are persistent, bioaccumulative and toxic (PBT) and substances which are very persistent and very bioaccumulative (vPvB).**

REACH does not stipulate that all substances with SVHC properties are subject to authorization. Only uses of substances listed in Annex XIV of the REACH regulation require authorization. Substances with hazardous properties other than those mentioned above may also be included into this annex, provided Member State authorities have shown that the use of such substance gives rise to an equivalent level of concern as compared to CMR, PBT or vPvB substances.

The REACH authorization is composed of three distinct phases. In the first phase, an EU Member State or the European Commission proposes to include a substance meeting at least one of the SVHC criteria in an ECHA inventory called the candidate list. Inclusion in this list does not (already) entail that uses of the substances require authorization. It (only) triggers limited communication requirements, which are not further described here. In the second phase the substance is subjected to the actual authorization. This phase is initiated by ECHA, which selects a substance from the candidate list and proposes its inclusion into Annex XIV. The decision on the inclusion is taken by the European Commission. After a date referred to as the sunset date, further use of the substance is prohibited unless users of the substance have obtained authorization. The third phase consists of the submission of applications for authorization of specific uses by manufacturers, importers or (downstream) users of the substance. ECHA carries out a technical/scientific assessment of the application dossier and, based on the result of the assessment, the European Commission decides whether or not an authorization should be granted.

Restriction

> **The restriction process is employed to prohibit production, marketing or use of substances. It can also be used to issue mandatory risk management measures.**

EU Member States and the European Commission are entitled to initiate the restriction process by submitting a dossier to ECHA. The dossier must include an assessment of the risks to human health or the environment to justify the proposed restrictions and a description of the proposed risk management measures. ECHA assesses the scientific/technical validity of the assessment and analyses the socio-economic impact of the restriction measure. The restriction process is concluded with a decision by the European Commission.

5.3.5 Allocation of Responsibilities and Administration of REACH

On reading the previous sections, readers may have wondered why the regulator has created administrative processes of such high complexity. Why have not all tasks and competencies been allotted to ECHA instead of being distributed to different authorities, including Member State authorities (substance evaluation and compilation of restriction dossiers) and the European Commission (decision-making)? Knowing the answers to this question helps in understanding the REACH concept.

When designing REACH, a key objective was to make the administrative procedures robust towards corruption and manipulation, and to ensure that all steps of the process and their outcome can be followed by the public. A further intention was to promote that decisions are taken on the basis of assessments of high scientific/technical quality. It was envisaged that the implied high administrative burden would be rewarded by increased public acceptance of decisions.

Status and Function of ECHA

> **ECHA is an independent EU institution. Its main task is to carry out an examination of dossiers for restriction and authorization and adopt conclusions, referred to as opinion. It carries out public consultations on submitted dossiers, thus allowing everybody to feed in additional information and submit comments. Furthermore, ECHA is responsible for carrying out administrative tasks under registration and evaluation.**

ECHA comprises several committees and a secretariat managed by an executive director. The scientific-technical dossier assessment is carried out by two committees, namely the Committee for Risk Assessment (RAC) and the Committee for Socio-Economic Assessment (SEAC). These committees are composed of independent experts from EU Member States. The RAC examines the submitted risk assessment and the SEAC analyses the socio-economic impact of restrictions or uses applied for under authorization. The opinions are published on ECHA's website. The independence is safeguarded through the organizational set-up of ECHA (independency from political instructions) as well as by the personal independency of RAC and SEAC members. Also, members of RAC and SEAC stake their reputation on the quality and independency of their opinions. The publication of the names of committee members and the accessibility of the opinions to external experts provide an incentive for the members to conduct a thorough analysis.

Whereas RAC and SEAC act as independent reviewers, the ECHA secretariat carries out administrative tasks referred to in the section on registration (receipt and completeness check of registration dossiers) and on evaluation (examination of testing proposals and compliance check). The assignment of these tasks reflects the intention of the regulator to centralize the responsibility for ensuring the quality of the registration dossiers. Member States, however, retain considerable influence on these two evaluation processes. The ECHA secretariat must obtain approval from all Member States for all evaluation decisions.

Role of the European Commission

> **Decision-making on restrictions and authorizations under REACH implies arbitration between conflicting public interests and exertion of political discretion. For this type of decision, responsibility for decision-making cannot be delegated to (committees of) independent experts but, rather, decisions must be taken by institutions having sufficient democratic legitimacy. For this reason the decision-making power could not be delegated to ECHA but instead had to remain with the European Commission.**

The arbitration of conflicting interests in the context of REACH includes weighing risks to human health or the environment against socio-economic impact where the level of the accepted risk depends, among others, on the socio-economic benefits. For example, most workers are willing to accept a limited cancer risk from exposure to carcinogenic substances for keeping their job. However, weighing health or environmental risks against socio-economic benefits may become politically highly delicate in cases where, in contrast to the example, different persons are affected, i.e. the exposed persons are not among the beneficiaries. In practice, the margin of discretion of the European Commission is limited by the published opinions of RAC and SEAC, which function as reference points regarding the nature and level of risk and benefit. Readers interested in EU policy-making processes are referred to the White Paper on European Governance.

The Commission is only involved in the evaluation processes if Member States cannot reach unanimous agreement. As the decision-making process at Commission level is rather burdensome, this procedural set-up saves time and resources. Through evaluation decisions additional information is requested from the registrants which is needed for either compliance of dossiers or clarifying a concern. From the formal point of view, decisions are legitimized through approval by Member States.

Role of Member States

> **Member States are responsible for investigating indications of human health or environmental risks and assessing the level of risk. Indications may be received from a variety of sources, for example from occupational health checks or studies, from doctors, emergency health response centers, enforcement authorities, individual persons or the media.**

Over the last few decades, the competence for deciding on risk management measures has been progressively transferred from Member States to EU level. In particular in the area of tradeable products, any national measure may lead to trade barriers, thus disturbing the European single market. The regulatory competence of the Member States has been replaced by the right of initiative. Every EU Member State is entitled to propose a restriction of the production or use of a hazardous substance, either as such or as component of products. In the course of the restriction process it is assessed at EU level whether the concern of the Member State is justified. In case the Member State believes that the available information is insufficient to decide on the need or scope of a restriction measure, it may conduct a substance evaluation first in order to force registrants of the substance to generate the information needed.

At first glance it may not be obvious why time- and resource-consuming EU assessments are required instead of using Member State assessments as the basis for EU-wide restrictions. The case of the wood preservative pentachlorophenol (PCP) illustrates the importance of an assessment at EU level. In the 1980s many residents of houses with indoor wood treated with wood preservatives containing PCP complained of fatigue, insomnia, diminished appetite, forgetfulness and other unspecific symptoms. As a result of these complaints, all uses of PCP were banned in Germany in 1989 while production and use of the substance continued in France. In order to avoid creation of a trade barrier for PCP and PCP-treated products, the European Commission proposed to restrict the use of PCP. After controversial discussions among EU Member States, PCP was restricted in 1991. However, the EU restriction did not cover all uses of PCP as proposed by Germany. EU discussions on PCP continued until a complete ban of PCP was finally agreed in 1999. The establishment of an EU process comprising an independent scientific risk assessment was intended to facilitate and accelerate agreement between Member States. It is noteworthy to add that a causal relationship between the exposure to PCP and the reported symptoms of impaired health has never been conclusively demonstrated.

5.3.6 Downstream Users

> **Downstream users (defined as natural or legal persons who use a substance either on its own or in a mixture) are obliged to assess whether their uses are covered by the exposure scenarios drawn up by registrants, or, in other words, whether their described use conditions and risk management measures match with the assumptions of the registrant(s).**

As referred to in *Registration* (this chapter), registrants are obliged to assess the risks to human health and the environment for all identified uses during the whole lifecycle of the substance. However, a substance may be contained in various upstream products and may have changed owner several times until it finally reaches the end product. In many cases, manufacturers or importers do not know the users, the technical function of the substance or the use conditions. By selling a substance, they lose control of it. For this reason, the use conditions described in the chemical safety reports may differ from the actual downstream uses. In order to close this gap, the regulator has established obligations for downstream users and provisions regarding the exchange of information in the supply chain.

Downstream users are required to check if their use conditions match with the exposure scenarios of the registrant of the substance. In case of deviations they may inform the registrant. Registrants are generally obliged to prepare exposure scenarios for all reported uses and include them in their chemical safety reports unless they consider the use is not safe. In the latter case, registrants are obliged to provide ECHA and relevant downstream users with the reasons for their concern. Provided downstream users, nevertheless, wish to adhere to the use, they have to carry out a chemical safety assessment themselves and inform ECHA. The obligation to perform an own assessment also applies to downstream users who do not want to disclose their uses or use conditions to registrants (e.g. for reasons of confidentiality).

The provisions regarding the information exchange in the supply chain intend to make sure that all actors in the supply chain receive the information relevant for safe use of the substance. By extending the duties to downstream users, the number of affected companies was multiplied. While it is obvious that safe use of chemicals cannot be reached without involving all actors, it should be acknowledged that the REACH provisions constitute a considerable challenge to small and medium-sized enterprises. On the other hand, many actors support the notion that REACH has increased the awareness of companies regarding chemical risks and that the learning process is not yet finished.

5.3.7 Outlook

Experience shows that an adequate level of safety cannot be reached without a legal framework laying down specific obligations for producers of substances and products and for all other actors in the supply chain. In today's competitive market environment, economic incentives are insufficient to trigger comprehensive testing and risk assessment. The REACH regulation was a milestone in the development of EU chemicals legislation and is used as international benchmark. However, there is still ongoing criticism and the expectations of some stakeholders have not been met. Advancement of REACH is essential and desirable. Considering the divergence of interests and the numerous conflicting objectives, legislation on the safe use of chemicals is likely to remain a matter of controversy.

5.3.8 Summary

REACH regulates substances produced and imported in the EU and establishes a complex administrative framework for the evaluation, authorization and restriction of substances. It obliges producers and importers to submit registration dossiers to ECHA for substances produced or imported at an annual volume of 1 ton or more per year. The registration dossier must contain information on intrinsic properties, uses and exposure to the substance. The minimum information requirements depend on the annual tonnage of the substance. Administrative processes called "evaluation" are used to ensure compliance of registration dossiers with REACH and to obtain additional information in case of a concern. Substances with intrinsic properties of very high concern may be subjected to authorization. This implies that they must not be used unless permission was granted. REACH allocates distinct roles, rights and responsibilities to the European Commission, ECHA and authorities of the Member States. Complex administrative processes aim to secure the rights of affected parties and independence regarding the technical and scientific assessment of risks and socioeconomic impact. REACH also affects downstream users of substances. They are obliged to exchange relevant information in the supply chain and carry out their own risk assessments for uses not covered by registration dossiers.

Further Reading

CLP regulation: Regulation (EC) No 1272/2008 of the European Parliament and of the Council of 16 December 2008 on classification, labelling and packaging of substances and mixtures, amending and repealing Directives 67/548/EEC and 1999/45/EC, and amending Regulation (EC) No 1907/2006.

Guidance on REACH; European Chemicals Agency, http://echa.europa.eu/guidance-documents/guidance-on-reach (07/01/2017).

REACH regulation: Regulation (EC) No 1907/2006 of the European Parliament and of the Council of 18 December 2006 concerning the Registration, Evaluation, Authorization and Restriction of Chemicals (REACH), establishing a European Chemicals Agency, amending Directive 1999/45/EC and repealing Council Regulation (EEC) No 793/93 and Commission Regulation (EC) No 1488/94 as well as Council Directive 76/769/EEC and Commission Directives 91/155/EEC, 93/67/EEC, 93/105/EC and 2000/21/EC.

Strategy for a future Chemicals Policy – White Paper; Commission of the European Communities, COM(2001) 88 final.

The use of alternatives to testing on animals for the REACH Regulation – Second report under article 117.3 of the REACH Regulation; European Chemicals Agency, ECHA-14-A-07-EN (2014).

6
Specific Toxicology

6.1 Persistent Halogenated Aromatic Hydrocarbons[1]

Heidrun Greim and Karl K. Rozman

6.1.1 Introduction

(Poly)halogenated aromatic hydrocarbons (PHAHs) are highly persistent organic chemicals that, as a result of their structural similarities, have common toxicological properties. The well-known representatives of this class of substances are the polychlorinated dibenzodioxins (PCDDs) and dibenzofurans (PCDFs), polychlorinated biphenyls (PCBs) and some pesticides such as *p,p'*-dichlorodiphenyltrichloroethane (DDT) and hexachlorobenzene (HCB). These compounds are regarded toxicologically as representatives of the whole substance class. Brominated and mixed halogenated congeners have similar physical and biological properties and lead to comparable effects. PHAHs are ubiquitous environmental pollutants of mainly anthropogenic origin. Many representatives are extremely persistent and cause damage to both the environment and to humans. They were listed together with other persistent organic pollutants (POPs) as the so-called "Dirty Dozen" (see Table 6.1) and will be or have already been banned by the Stockholm Convention of 2001, which was signed by 122 countries, including the USA. It came into effect on 17 May 2004, after ratification by 50 countries.

> **In Europe, the Stockholm Convention was incorporated in EC Regulation No. 850/2004 of the European Parliament and of the Council of 29 April 2004 on persistent organic pollutants and amended Directive 79/117/EEC. It was also adopted into national law in Switzerland by means of an edict. In the USA most of the persistent PHAHs listed in the Stockholm Convention were categorised as hazardous air pollutants in the Clean Air Act (CAA) and as priority toxic pollutants in the Clean Water Act (CWA).**

---

[1] This chapter is an adapted and updated version of the original chapter "Persistent Polyhalogenated Aromatic Hydrocarbons" of the 1st edition of this book published in 2008.

Table 6.1 The "Dirty Dozen" according to the Stockholm Convention: almost half of the compounds are PHAHs (in bold type).

Compound	Use	Structure
Aldrin	Insecticide	Chlorinated 1,4:5,8-dimethanonaphthalene derivative
Chlordane	Insecticide	Chlorinated 4,7-methanoindene derivative
DDT	**Insecticide**	**Chlorinated diphenylethane derivative**
Dieldrin	Insecticide	Chlorinated 1,4:5,8-dimethanonaphthalene derivative
Endrin	Insecticide, rodenticide	Chlorinated 1,4:5,8-dimethanonaphthalene derivative
HCB	**Fungicide, industrial compound and by-product**	**Perchlorinated benzene**
Heptachlor	Insecticide	Chlorinated 4,7-methanoindene derivative
Mirex	Insecticide	Perchlorinated cyclobuta[cd]pentalene
PCBs	**Industrial compounds**	**Chlorinated biphenyls**
PCDDs	**Product of combustion, industrial by-products**	**Chlorinated dibenzo-*p*-dioxins**
PCDFs	**Product of combustion, industrial by-products**	**Chlorinated dibenzofurans**
Toxaphene	Insecticide	Chlorinated norbornane derivatives

6.1.2 Polychlorinated Dibenzodioxins and Dibenzofurans

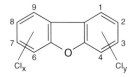

Polychlorinated dibenzodioxins   Polychlorinated dibenzofurans

> **Polychlorinated dibenzodioxins and dibenzofurans (PCDDs and PCDFs) are a group of chlorinated organic compounds that comprises 75 dioxin and 205 furan congeners. The congeners differ in the number and position of the chlorine atoms. The PCDDs/PCDFs are planar molecules. In addition to the chlorinated substances there are also brominated and both brominated and chlorinated congeners. Although only a few comparative studies have been carried out to date, it can be assumed that the brominated compounds are very similar to the chlorinated compounds as regards the toxicological potency, toxicokinetics and metabolism. Homologues with halogen substituents in the 2,3,7,8 positions are the biologically and toxicologically most effective compounds of the substance class.**

PCDDs and PCDFs are unwanted by-products of almost exclusively anthropogenic origin. They are formed during practically all combustion processes in the presence of inorganically or organically bound chlorine. In 1872, Merz and Weith reported the first proven synthesis of PCDDs. The high biological potency of this substance class was not discovered, however, until 1957, when Sandermann carried out investigations with the fungicide pentachlorophenol. He discovered PCDDs as the by-product of a process for manufacturing plywood, using PCP as a wood preservative. The completely chlorinated congener octachlorodibenzo-p-dioxin (OCDD) showed no activity towards termites and mould, while tetrachlorodiphenylene dioxide (later referred to as 2,3,7,8-tetrachlorodibenzo-p-dioxin, 2,3,7,8-TCDD) was highly effective. Researchers involved in this discovery soon developed symptoms of exposure, above all chloracne, which led to the end of Sandermann's research (Sandermann, 1984). Kimmig and Schulz linked these symptoms to identical clinical signs in workers at plants manufacturing trichlorophenoxyacetic acid (2,4,5-T). It was found that dioxins are formed as by-products during the industrial synthesis of this chlorinated pesticide as well as other chloro-organic syntheses, such as the production of trichlorobenzene or of polychlorinated biphenyls. PCDDs were found as contaminants in the defoliant Agent Orange, which was used during the Vietnam War in Operation Ranch Hand. The concentration of 2,3,7,8-TCDD was found to be 0.1–47 μg in 1 g Agent Orange. As a result of the public awareness and the health problems of the veterans, research on dioxins started around 1970. Reports of industrial accidents with the release of PCDDs date back as early as 1949, but a major accident in a 2,4,5-T-producing plant in Seveso, Italy, in 1976 became the most infamous episode.

The main source of dioxins and furans in the environment are thermal processes such as the production of metals and mineral products, combustion processes such as waste incineration, heating and electricity plants, open combustion processes and the organochlorine industry. Each production condition generates a characteristic congener cluster.

During the various combustion processes characteristic mixtures of PCDD and PCDF congeners are formed. Samples from the environment can therefore be linked with sources of emission. PCDDs and PCDFs are not produced commercially, except for small amounts for scientific research. The thermal formation of PCDDs is a combination of de novo synthesis and their generation from precursors in a temperature range of 250–350°C, as found in smelters or outdated waste-combustion plants. PCDDs/PCDFs are also generated during natural processes, including forest fires, biodegradation and biosynthesis, and geothermal activities. In addition, congener clusters are found worldwide in historical and prehistorical sediments (Australia, Germany, China, Japan, USA). The composition of these mixtures is not related to any pattern generated by anthropogenic processes, which indicates truly natural formation. The United Nations Environmental Programme (UNEP) published a comprehensive inventory of contributions to dioxin release, which allows countries to estimate their total emissions according to their domestic economy.

The environmental fate of 2,3,7,8-TCDD is determined by its persistence. Investigations after the accidental release in Seveso showed a half-life in soil of about 9–12 months. Studies at various different test sites showed the half-life to be dependent of the climate, with values of 6.3 months in Florida and 11 months in Utah. The number of bacteria capable of anaerobic, reductive dehalogenation in soil or sediment is limited and still under investigation for purposes of developing large-scale decontamination methods. The degradation of 2,3,7,8-TCDD in the gaseous phase is rapid due to its susceptibility to photolysis by ultraviolet light ($\lambda = 290$). Its theoretical half-life under photolytic conditions in the vapour phase is 1 hour. On the other hand, its half-life in the air under the conditions of radical reactions with OH is about 8.3 days. The persistence of 2,3,7,8-TCDD in surface waters depends on the extinction coefficient of the respective body of water and on the seasonal changes in ultraviolet radiation. It varies between 21 hours in summer and 118 hours in winter.

6.1.3 Polychlorinated Biphenyls

co-planar

non-planar

> **Polychlorinated biphenyls (PCBs) are industrial chemicals that are no longer used in developed countries. The substance class comprises 209 congeners, the toxicological potencies of which differ by the degree as well as by the pattern of the chlorination. There are co-planar and non-planar PCBs with differential toxicity profiles.**

PCBs possess favourable technical properties, which have led to their widespread application in industry. These properties include a very high dielectric coefficient and high boiling points. PCBs are virtually non-combustible, heat-resistant, chemically stable and show very low acute toxicity. Their viscosity varies with the degree of chlorination. For this reason PCBs were used in electric transformers, capacitors, and heat exchangers, and as lubricants, flame retardants, plasticisers and additives in printing inks and lacquers. An estimated cumulative total of 1.2 million tonnes were produced worldwide, half of which was manufactured by the USA. Since 1978 the production and new use of PCBs has been banned there under the Toxic Substances Control Act (TSCA), and since the mid-1980s most countries have followed suit. Various interim arrangements were introduced in the EU for the eradication of products containing PCBs. In accordance with the Stockholm Convention, the use of systems containing PCBs must be terminated by 2028.

PCBs are cost-effectively produced by the chlorination of the biphenyl parent compound. Technical-grade mixtures are marketed according to their chlorine content (30–60% w/v), often indicated by their commercial trade name (e.g. Aroclor 1242 contains 42% chlorine). The IUPAC nomenclature of individual congeners is based on the ascending numerical order of chlorination: the PCB congeners are assigned numbers from 1 to 209.

PCBs are found ubiquitously in the environment and accumulate above all in the fatty tissue of animals and thus also in the food chain. The congener cluster in individual samples differs greatly from that in the technical-grade mixtures, among other things as a result of the different species-specific half-lives. When monitoring the PCB concentrations in food or in environmental samples, for practical reasons usually only six selected congeners are quantitatively determined, the so-called indicator congeners (see Table 6.2), which are found in almost all samples and in the largest amounts.

The biological effects of the PCBs depend very much on their pattern of chlorination and thus on their conformation. *Meta*-, *para*- and mono-*ortho*-substituted congeners are co-planar and show a toxicological profile similar to that of the PCDDs, but at much lower potency. The

Table 6.2 Indicator congeners for analysing PCB concentrations.

Name	Congener number
2,4,4'-Trichlorobiphenyl	28
2,2',5,5'-Tetrachlorobiphenyl	52
2,2',4,5,5'-Pentachlorobiphenyl	101
2,2',3,4,4',5'-Hexachlorobiphenyl	138
2,2',4,4',5,5'-Hexachlorobiphenyl	153
2,2',3,4,4',5,5'-Heptachlorobiphenyl	180

benzene rings rotate freely around the central C–C bond. Di-*ortho*-chlorinated PCBs show a barrier of rotation of about 80 kJ/mol, which still allows racemization at room temperature. Higher chlorination in the *ortho* positions, however, greatly increases the rotation barrier since only cisoid transition states are possible. The energy required for the C–C rotation is 180 kJ/mol for tri-*ortho*-chlorinated congeners and 246 kJ/mol for tetra-*ortho*-chlorinated congeners. As a result, racemization is prevented in these compounds, even at higher temperatures. For this reason, a planar transition state cannot be attained in these compounds under physiological or environmental conditions. It has been reported that some (+) and (−) enantiomers differ in their toxic potency. Kinetic studies have shown slight differences in disposition, providing a possible explanation for different potencies. Tri- and tetra-*ortho*-congeners elicit a toxicological profile different from those of the planar congeners. They are also neurotoxic, possibly by altering calcium signalling.

6.1.4 Dichlorodiphenyltrichloroethane

> **The insecticidal effect of dichlorodiphenyltrichloroethane (DDT) was first discovered in 1939 by Paul Müller. The application of DDT considerably reduced the outbreaks of vector-borne diseases such as typhus and malaria. In 1948 Paul Müller was awarded the Nobel Prize for medicine for his contribution to preventive medicine. DDT was banned in the USA in 1972 because of its persistence in the environment.**

As a member of the so-called "Dirty Dozen", DDT was also banned by the Stockholm Convention. However, exemptions were granted to 25 countries for vector control until cost-efficient, environmentally safe and locally available alternatives were developed. The cumulative world-wide production of DDT is estimated to exceed a total of 2 million tonnes. The highest level of production in the USA took place in the early 1960s, with 85,000 tonnes in 1962, with major applications in agriculture (cultivation of cotton, peanuts and soya beans). During the same period Rachel Carson's book *Silent Spring* was published, which awakened a growing environmental awareness in the USA. The author linked the excessive use of pesticides to a reduction in bird populations, leading to the silence in the woods reflected in the title. Later DDT and its metabolites were found to be endocrine disruptors. In 1980 the wastewater pond of a pesticide-manufacturing facility containing DDT residues overflowed into Lake Apopka in

Table 6.3 The effect of DDT on malaria infections in Ceylon (Sri Lanka).

Year	1948	1963	1964	1969
Number of malaria cases	2,800,000 (before DDT usage)	17 (cessation of DDT spraying)	150	2,500,000

Florida and caused toxic effects on reproduction and development in alligators. By 1968, several US states had banned the use of DDT, followed in 1972 by a USA-wide ban on most uses and in 1989 for all uses. Nevertheless, production in the USA remained at the level of 1 tonne per day until the 1990s. Today, DDT is neither produced there, nor is it imported or exported. In most western industrial countries DDT was banned in the 1970s.

From a historical point of view, the role of DDT in preventive medicine by far exceeds its economic importance in agriculture. DDT has proved very efficient in controlling malaria, even in chronically afflicted regions. This disease, which is transmitted by mosquitos, was almost driven to extinction in Ceylon (Sri Lanka) by areal and local spraying of DDT. After cessation of application in 1963 the infection rates soon returned to previous levels (Table 6.3), indicating the effectiveness of DDT.

DDT is a contact poison, irritant and repellent for insects. The combination of these properties enhances its efficiency in controlling mosquitos and lice by deterring them from entering sprayed areas as well as by killing them. However, gradually resistance to DDT started to develop, for example in Anopheles, requiring increasing doses to control the mosquito populations.

Its structurally analogue mitotane (see Figure 6.1) is used as a chemotherapeutic agent in the treatment of adrenocortical carcinomas. The mechanism of action is unclear, but mitotane appears to act as an endocrine disruptor, and to inhibit certain functions in the adrenal cortex and thus steroidogenesis. It causes atrophy of the adrenal cortex, which leads to the "starvation" and subsequent destruction of the tumour.

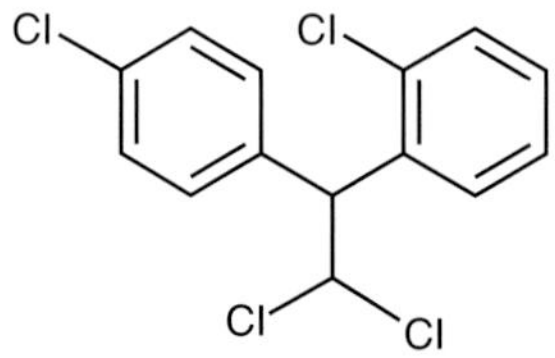

Figure 6.1 Mitotane, a DDT analogue used as a drug in the treatment of adrenocortical carcinoma.

6.1.5 Hexachlorobenzene

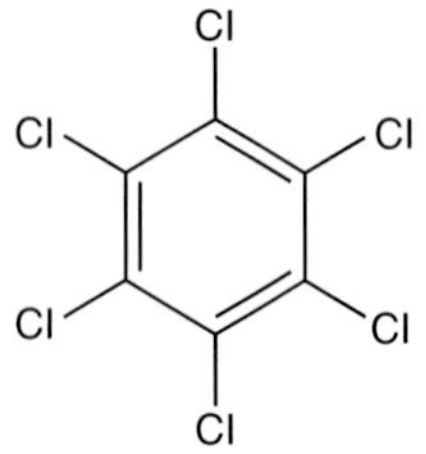

> **Hexachlorobenzene (HCB) is a potent fungicide, an industrial intermediate, and is generated as a by-product in chloro-organic syntheses, for example of carbon tetrachloride, chlorophenols or vinyl chloride.**

HCB is synthesised commercially by the direct chlorination of benzene in the presence of a Friedel–Crafts catalyst, but most HCB in the atmosphere is generated as a by-product of the chloro-organic industry. Its primary use was in the treatment of agricultural seed grain, such as wheat and sorghum, as well as onions, but it was used temporarily also as a flame retardant in plastics and in the production of pyrotechnics. The fungicidal properties of HCB led to its widespread use in the 1950s and 1960s. Its use as a fungicide was phased out in the USA, starting in the mid-1960s. Its registration was voluntarily cancelled in 1984. Today, in the USA and Europe, HCB is neither produced nor imported or exported as a pesticide. However, it can be manufactured and used for chemical intermediates. Estimates of the amount of HCB generated as an impurity range from 68 to 690 tonnes per year in the mid-1990s in the USA alone.

6.1.6 Physico-chemical Properties

> **PHAHs share common characteristics, such as high lipophilicity, low vapour pressure, a high melting point and slow biodegradation, leading to biomagnification and environmental persistence.**

The hydrophobicity of a compound increases with the degree of chlorination. This effect is amplified in the PHAHs that lack further functional groups. Besides creating steric hindrance for enzymatic reactions, a high degree of halogenation also causes increased molecular weight, with elevated melting points and very low vapour pressures (Table 6.4). These properties are the predominant reason for the persistence of PHAHs in the environment.

Owing to their (at least partial) planarity as a result of the attached aromatic rings, PHAHs easily intercalate in micro-layered minerals, such as clay, leading to geoaccumulation. Their Henry's constants indicate low volatility from an aqueous solution into the gaseous phase, quite independent of the environmental conditions. Therefore, their vertical mobility in the soil and sediment is very limited. Horizontal mobility occurs by erosion only. As a consequence, PHAHs show high compartment persistence in the environment. Ecotoxicological evaluations need to consider

Table 6.4 Physico-chemical properties of PHAHs at 20–25°C.

	2,3,7,8-TCDD	Aroclor 1242	DDT	HCB
Molecular weight	321.97	266.5	354.49	284.78
Melting point (°C)	305	No data	109	230
Boiling point (°C)	No data	325–366	Disintegrates	322 (sublimated)
Water solubility	7.9 ng/l	0.1-0.34 mg/l	25 µg/l	6 µg/l
$pK_{O/W}$	6.79	5.6	6.91	3.59–6.08
Vapour pressure (mm Hg)	1.5×10^{-9}	4.06×10^{-4}	1.6×10^{-7}	1.089×10^{-5}
Henry's constant (atm m^3/mol)	1.62×10^{-5}	5.2×10^{-4}	8.3×10^{-6}	6.8×10^{-4} to 1.3×10^{-3}

the resulting low bioavailability for these compounds from the adsorbed states. The high $pK_{O/W}$ values of the PHAHs reflect their high lipophilicity causing bioaccumulation (i.e. their uptake from the surrounding medium and food), particularly in aquatic animals and, as a consequence, biomagnification (i.e. their uptake from food only) throughout the food chain.

These substances are slowly degraded via chemical and biochemical processes. However, they are subject to photolysis. The dehalogenation of higher chlorinated PCDDs/PCDFs and PCBs can lead to the formation of more toxic congeners, increasing the total toxicity of a mixture. In aqueous media OCDD is preferentially dechlorinated in the *peri* positions (C1, C4, C6, C9) yielding the more potent heptachloro congener. It has also been shown that the ultraviolet irradiation of PHAHs results in the formation of other classes of PHAHs that were previously not present. The irradiation of water samples containing PCP yielded the formation of PCDDs and PCDFs in similar ratios. Therefore, the degradation of PHAHs, depending on the substrates and conditions, can lead to an increase or decrease in the total toxicity.

6.1.7 Toxicity

Chlorinated Dibenzodioxins, Dibenzofurans and Biphenyls

> **The toxicity profiles of PCDDs, PCDFs and co-planar PCBs, in the following referred to as dioxin toxicity, are qualitatively very similar. However, amongst the congeners potency can differ by orders of magnitude. This relative potency is expressed in terms of toxic equivalency factors (TEFs), normalized to 2,3,7,8-TCDD, which is the most potent member of the substance class.**

In the accident during the production of 2,4,5-T in Seveso, the most potent dioxin congener, 2,3,7,8-TCDD, was released. The estimates for the amount released range from 300 g to 3.4 kg; most widely accepted is an estimate of 1.3 kg, which contaminated an area of about 2.5 square kilometres and affected around 37,000 people. Chloracne was the main symptom observed.

The first reports of mass poisoning with PCBs are from 1968 from the island Kyushu in Japan (Yusho). PCBs were used in the heat-exchanging system of a rice oil manufacturing plant. Owing to a leak in a pipe, PCBs entered the final product and were consumed by approximately 1800 people. In 1979 a similar incident occurred in Taiwan (Yusheng), where likewise around 2000 people were affected. It was later shown that as a result of the repeated heating of the PCBs in the heat exchanger, traces of PCDFs were formed in the product. Therefore, the symptoms observed in the affected population, such as chloracne and skin pigmentation, are considered a combination of PCB and PCDF toxicity.

The most toxic congeners of PCDDs, PCDFs and PCBs are chlorinated in the 2,3,7,8 and 3,3'4,4' positions, respectively. This is probably due to two main factors. First, this chlorination pattern is required for the compound to interact with the target site. Secondly, it renders these compounds largely resistant to metabolism. For this reason they persist long enough to reach the target site at concentrations high enough to elicit an effect. The potency of these PHAHs is ranked in terms of their TEFs (see Table 6.5). The respective numbers were determined based on a combination of the acute effects *in vitro* and *in vivo*. The evaluation of the data by different groups of experts led to slightly different TEF values between agencies (bold in Table 6.5). This may result in diverging risk assessments of mixtures.

It has been unequivocally demonstrated that the high-dose acute toxic effects as well as carcinogenic effects of PCDDs, PCDFs and co-planar PCBs are additive. Also the effects of medium

Table 6.5 TEF values of PCDDs, PCDFs and PCBs according to NATO and WHO determinations.

Congener	NATO	WHO
PCDDs		
2,3,7,8-TCDD	1	1
1,2,3,7,8-PeCDD	**0.5**	**1**
1,2,3,4,7,8-HxCDD	0.1	0.1
1,2,3,7,8,9-HxCDD	0.1	0.1
1,2,3,6,7,8-HxCDD	0.1	0.1
1,2,3,4,6,7,8-HpCDD	**0.01**	**0.001**
1,2,3,4,6,7,8,9-OCDD	**0.001**	**0.0003**
PCDFs		
2,3,7,8-TCDF	0.1	0.1
1,2,3,7,8-PeCDF	**0.05**	**0.03**
2,3,4,7,8-PeCDF	**0.5**	**0.3**
1,2,3,4,7,8-HxCDF	0.1	0.1
1,2,3,7,8,9-HxCDF	0.1	0.1
1,2,3,6,7,8-HxCDF	0.1	0.1
2,3,4,6,7,8-HxCDF	0.1	0.1
1,2,3,4,6,7,8-HpCDF	0.01	0.01
1,2,3,4,7,8,9-HpCDF	0.01	0.01
1,2,3,4,6,7,8,9-OCDF	**0.001**	**0.0003**
PCBs		
IUPAC No.	Structure	
77	3,3',4,4'-TCB	0.0001
81	3,4,4',5-TCB	0.0003
105	2,3,3',4,4'-PeCB	0.00003
114	2,3,4,4',5-PeCB	0.00003
118	2,3',4,4',5-PeCB	0.00003
123	2',3,4,4',5-PeCB	0.00003
126	3,3',4,4',5-PeCB	0.1
156	2,3,3',4,4',5-HxCB	0.00003
157	2,3,3',4,4',5'-HxCB	0.00003
167	2,3',4,4',5,5'-HxCB	0.00003
169	3,3',4,4',5,5'-HxCB	0.03
189	2,3,3',4,4',5,5'-HpCB	0.00003

Pe, penta; Hx, hexa; Hp, hepta; O, octa.

doses have been shown to be additive for reproductive end points. A mixture of 2,3,7,8-TCDD, 1,2,3,7,8-PeCDD, 1,2,3,4,7,8-HxCDD, 2,3,4,7,8-PeCDF and 3,3',4,4',5-PeCB (PCB 126) inhibited ovulation in an entirely additive manner. This additivity allows the calculation of the total toxicity of mixtures based on the concentrations of its individual components and the respective TEFs (see Equation 6.1). The sum of these products is expressed as the toxicity equivalent (TEQ).

Formula to calculate the TEQ for a mixture by applying the TEF concept:

$$\sum_{k=1}^{n}(c_k \cdot \text{TEF}_k) \tag{6.1}$$

where c is the concentration, TEF is the toxicity equivalency factor, and k and n are specific congeners.

The TEF of the most potent congener, 2,3,7,8-TCDD, is defined as 1. The TEQ represents the theoretical amount of 2,3,7,8-TCDD eliciting a response identical to that of a given mixture. For this reason the TEQ is also referred to as the 2,3,7,8-TCDD equivalent.

> **The toxicity of dioxins shows marked species differences, with the guinea pig being the most sensitive and the hamster being the most resistant of the animal models not genetically manipulated.**

The effects of the PCDDs and PCDFs as well as of the PCBs are thought to occur by the same or a similar mechanism. For this reason, most studies have focused on the most potent representative of this substance class, 2,3,7,8-TCDD. It serves as the standard model compound for toxicological studies. The LD_{50} of 2,3,7,8-TCDD varies considerably among the various species (see Table 6.6). The most sensitive mammal is the guinea pig, with an LD_{50} of 0.7–2.0 µg/kg body weight. It is well over 1000 times more sensitive than the hamster, the most resistant species, with an LD_{50} of >5051 µg/kg body weight. The LD_{50} also varies greatly among the strains, as most prominently revealed by two rat strains. The LD_{50} in the sensitive Long Evans rat is 10 µg/kg body weight and differs by a factor of 1000 from the resistant Han/Wistar rat (LD_{50} >9600 µg/kg body weight). These differences are observed also in mice, but to a lesser degree. C57BL mice were found to be about 30 times more sensitive than DBA mice.

The explanation for some of the intra-species and inter-strain differences has turned out to be quite complex. Nevertheless. a striking general correlation has been identified between the total body fat (TBF) content of a species and the acute toxicity of 2,3,7,8-TCDD (see Equation 6.2).

Formula to calculate the approximate LD_{50} of 2,3,7,8-TCDD in mammals:

$$LD_{50} = 6.03 \times 10^{-4} (TBF)^{5.30} \qquad (6.2)$$

where TBF is the total body fat (% of body weight).

A correlation between the toxicity of highly lipophilic compounds and the TBF appears intuitively correct considering the disposition of such compounds. The larger the compartment that serves as sink or pool (peripheral compartment) for a compound is, the lower is its concentration in the circulation and, thus, at the target site (assuming that the target site is not part of the peripheral compartment).

The above empirical formula can be used to calculate a hypothetical acutely lethal dose for humans. The calculated LD_{50} of 6230 µg/kg body weight for a body weight of 70 kg and 21%

Table 6.6 LD_{50} values for 2,3,7,8-TCDD in different species.

Species	LD_{50} of 2,3,7,8-TCDD (µg/kg body weight)	Total body fat (% of body weight)
Guinea pig (Dunkin Hartley)	0.7–2.0	4.5
Rat (Sprague–Dawley)	45	8.9
Rhesus monkey	50	10.3
Mouse (C57 BL/6)	100	7.9
Rabbit (New Zealand White)	115	10.1
Dog (Beagle)	1000	13.8
Hamster (Golden Syrian)	5051	17.3

TBF supports the common notion that adult humans are not as sensitive to the acute toxicity of dioxins as are some of the animal models. In addition, these calculations allow an estimate of the LD_{50} values across ages, using the respective body fat contents. This leads to the tentative conclusion that newborn babies (13.6% TBF) are about 10 times more sensitive to the acute toxicity of 2,3,7,8-TCDD than adult humans.

> **The effects of PCDDs, PCDFs and PCBs are quite divers and well characterised in experimental animals. They include a so-called wasting syndrome and carcinogenicity (lungs, liver) at high doses, liver injury, immunosuppression, reproductive effects and lowered serum insulin-like growth factor-1 (IGF-1) levels at medium doses and effects on thyroid hormones, thymic atrophy, enzyme induction at low doses and chloracne.**

Chloracne The term 'chloracne' comes from a form of acne first observed more than 100 ago after occupational exposure to octachloronaphthalene. The term 'haloacne' would be more correct, since it was previously known that bromine and, in some cases, iodine compounds led to disturbance of the structure and function of the sebaceous glands.

In the early stages of acne, thickening of the excretory ducts of the sebaceous glands occurs up to total occlusion. In the later stages, the sebaceous glands fill with horny cells, and disfiguring pustules and abscesses develop based on secondary bacterial infections (see also Chapter 3.6). In humans, chloracne is the most sensitive symptom of elevated 2,3,7,8-TCDD body burdens. It occurs approximately two weeks after the exposure, manifested by swollen skin follicles, particularly on the face. Three to five weeks post exposure they turn into comedones. In rabbits, the initial symptom of chloracne is dermatitis. This occurs two to four weeks after the administration of 2,3,7,8-TCDD, followed by swollen follicles and cysts several days later. In susceptible humans (young girls) symptoms may occur at 2,3,7,8-TCDD-concentrations of 800 ppt, based on the serum lipid content. In adolescents, the differential diagnosis of acne-like skin conditions is notoriously difficult. Most individuals do not show symptoms below 11,000 ppt. The highest reported 2,3,7,8-TCDD level in humans was 144,000 ppt blood fat in a 30-year-old woman, which corresponds to a dose of 25 µg 2,3,7,8-TCDD/kg body weight. The symptoms comprised severe chloracne, nausea, vomiting and gastrointestinal pain as well as the cessation of menstruation. The hypothalamus–pituitary axis seemed to be unaffected.

In 2004, the case of the Ukrainian president Viktor Yushchenko hit the headlines. He suffered dioxin poisoning after the uptake of a very large amount of 2,3,7,8-TCDD and among the symptoms produced chloracne. Initially, a concentration of 108,000 ppt 2,3,7,8-TCDD in serum was determined, which 3.5 years later had decreased to 20,500 ppt.

In contrast to the halogenated aromatics, the polycyclic aromatic hydrocarbons, which are also supposed to cause acne, are not known in detail. Various mineral oil and coal tar products, which come into intense contact with the skin in occupational settings as cutting, drawing and lubricating oils or drilling oil emulsions and semi-synthetic, water-mixed cooling agents, cause a more or less severe irritation of the sebaceous glands, which is called oil acne. Even the use of oil-based cosmetics can cause a mild form of acne. The underlying origin mechanism of the oil acne is not yet known.

Carcinogenicity

> **Epidemiological studies could not link exposure to 2,3,7,8-TCDD clearly to increased cancer mortality. Some studies associated occupational exposure to PCBs with hepatic,**

> biliary and intestinal cancers as well as with skin melanomas. Except for liver and biliary tract cancer, confirmatory evidence in animal models is lacking.

Investigations after accidental exposure suggest increased risks of cancer of the digestive tract and the respiratory tract in smokers exposed to 2,3,7,8-TCDD. The association supports the notion that 2,3,7,8-TCDD is a carcinogen even though it is not a mutagen. Workers exposed to high concentrations of 2,3,7,8-TCDD over prolonged periods of time showed significantly elevated cancer incidences ≥ 20 years after initial exposure. However, when considering the total cohort, no increased mortality rate was found in comparison with the general population. Another cohort of 5100 workers revealed, on the other hand, a decreased mortality rate from strokes and gastrointestinal diseases. In summary, the epidemiological data suggest that 2,3,7,8-TCDD might have a weak tumorigenic effect at higher doses, in particular when the total incidence for all types of tumour is taken into consideration. The evidence is strongest for lung tumours, soft-tissue sarcomas and non-Hodgkin's lymphomas. High dioxin doses cause tumours also in experiments with rats, mice and hamsters, including in the lungs and liver. A lack of evidence for genotoxicity *in vitro* and *in vivo* suggests that 2,3,7,8-TCDD, like PCBs and several other chlorinated hydrocarbons, exerts its carcinogenic effect through an epigenetic mechanism, for example by a promoting effect.

Immunosuppression

> In many, if not in most species, 2,3,7,8-TCDD has immunosuppressive effects at doses that are much lower than those causing acute toxicity.

Doses of 1 µg 2,3,7,8-TCDD/kg body weight and week caused an increased susceptibility to Salmonella infections in mice. Immunosuppression was observed in various lymphoid organs, such as the thymus, spleen and lymph nodes, in different species exposed to a wide range of doses. Rats showed a decrease in cell-mediated immunity after a dose of 40 µg 2,3,7,8-TCDD/kg body weight, whereas 10 µg 2,3,7,8-TCDD/kg body weight caused an increase. The most prominent effect was thymus atrophy with impaired differentiation of T lymphocytes. The antibody-mediated humoral immunity was likewise reduced. There are contradictory reports regarding effects on the immune system in humans. Epidemiological studies in the Seveso cohorts could not establish an association.

Reproductive Toxicity

> In animals, dioxins have been shown to affect reproduction and development, to be foetotoxic and embryotoxic as well as teratogenic. In humans, the cessation of menstruation has been reported.

Studies in animal models demonstrated effects of 2,3,7,8-TCDD on fertility and on pre- and post-parturitional growth and development. At higher doses, these effects included miscarriages and stillbirths. Morphological aberrations in the ovaries and uterus of rats were seen after doses of 1 µg/kg body weight and day for 13 weeks. The ED_{50} for the inhibition of ovulation in immature rats was determined to be 3–10 µg 2,3,7,8-TCDD/kg body weight. A no observed adverse effect level (NOAEL) of 0.03 µg 2,3,7,8-TCDD/kg body weight and day was reported for rats when administered from days 6 to 15 of gestation. In mice, teratogenic effects, such as cleft palates, hydronephrosis and thymic hyperplasia, were observed below maternal toxic doses. The same effects were reported also in rats and hamsters, but at doses that caused significant toxicity in the dams. A delayed onset of puberty in rats was observed for the offspring when pregnant dams were dosed with 0.8 µg TEQ/kg body weight on day 15 of gestation. Studies in primates likewise revealed reduced fertility, lowered birth weights and increased prenatal mortality, often accompanied by considerable toxicity in the pregnant animals.

Epidemiological studies in the Seveso cohorts did not reveal an increase in miscarriages or teratogenic effects, but did find an increase in the number of female offspring, which was assumed to be linked with the paternal side. This could not be confirmed in the cohorts from Operation Ranch Hand, who applied defoliants such as Agent Orange in the Vietnam War, which leaves a considerable uncertainty regarding the reliability of the findings in Seveso.

Further Diseases Other alleged effects associated with human dioxin exposure include hyperkeratosis, hyperpigmentation, hirsutism, liver damage, elevated blood fat content and cholesterol levels, gastrointestinal effects such as diarrhoea, cardiovascular effects, headaches, peripheral neuropathy, reduced sensory performance, loss of libido and psychiatric changes. Some of these effects are also age-related and therefore it is difficult to assess if dioxins are indeed contributory or not. This demonstrates once more that the lack of information on dose or dose rates and exposure time in epidemiological studies often leads to equivocal conclusions.

2,3,7,8-TCDD has been shown to reduce hormone levels, such as thyroid hormones, insulin or IGF-1, causing endocrine effects and also the dysregulation of intermediary metabolism.

It has been determined that 2,3,7,8-TCDD causes a decrease in the serum levels of total thyroxin (TT4) in rats within four days after the administration of 1 µg 2,3,7,8-TCDD/kg body weight and day. The concentration of total triiodothyronine (TT3) was, however, not affected. Both hormones were lowered in a dose-dependent manner in C57BL and DBA mice, with an LOAEL of 0.1 µg 2,3,7,8-TCDD/kg body weight and 100 µg 2,3,7,8-TCDD/kg body weight and day, respectively. Sprague–Dawley rats undergo a transient hypoinsulinaemia at doses of 25 µg 2,3,7,8-TCDD/kg body weight, together with insulin-hypersensitivity. In addition, a decrease in IGF-1 signal transmission within eight days was reported in rats maintained at steady state after a loading dose of 3.2 µg 2,3,7,8-TCDD/kg body weight. Similar doses were shown to decrease ovulation, but to prolong the lifespan of experimental animals as well as to reduced cancer rates below those in the control animals. Figure 6.2 illustrates that these effects could be mediated by decreased IGF-1 signalling.

Evaluation by National and International Organisations The International Agency for Research on Cancer (IARC) has classified 2,3,7,8-TCDD, 2,3,4,7,8-PeCDF and 3,3',4,4',5-pentachlorobiphenyl as human carcinogens (Group 1). An ACGIH TLV value is not established. The Permanent Senate Commission of the Deutsche Forschungsgemeinschaft (DFG) for the Investigation of Health Hazards of Chemical Compounds in the Work Area has likewise classified 2,3,7,8-TCDD as a human carcinogen in Category 4, for threshold carcinogens for which an occupational exposure level (OEL) can be derived. The OEL is very low at 1.0×10^{-8} mg/m^3 for the inhalable fraction. In addition, 2,3,7,8-TCDD is classified in Pregnancy Risk Group C, which comprises substances for which prenatal toxicity is unlikely when the OEl value is observed.

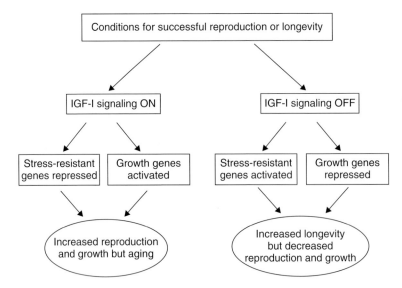

Figure 6.2 *Arking's theory of IGF-1 signalling and associated effects. Reprinted from Geyer et al. Copyright (1990/2002), with permission from Elsevier.*

The Permanent Senate Commission has likewise classified the chlorinated biphenyls in Category 4 with an OEL of 0.003 mg/m³ for the inhalable fraction. In addition, the chlorinated biphenyls are classified in Pregnancy Risk Group B, which comprises substances for which prenatal toxicity cannot be excluded at the level of the OEL.

For dioxins and dioxin-like PCBs the EU Scientific Committee on Food derived a health-based tolerable weekly intake of 14 pg/kg body weight for the general population, which corresponds with the recommendation of the Joint Expert Committee on Food Additives of the WHO and the UN Food and Agriculture Organization, which recommends 70 pg/kg body weight per month. The US EPA, on the other hand, derived an oral reference dose of 0.7 pg/kg body weight and day.

Example Evaluation: Chlorinated Dibenzodioxins and Dibenzofurans in Hens' Eggs In the mid 1999s, increased PCDD/PCDF concentrations were determined above all in poultry products. Hens' eggs were found to contain up to 1000 pg TEQs/g fat. To calculate the additional internal exposure from eating a contaminated hen's egg it is assumed that a hen's egg weighs 60 g, has a fat content of 10% and the contamination is thus 1000 pg/g fat. An egg therefore contains 6000 pg or 6 ng TEQs.

For a body weight of 70 kg, this results in the uptake of 86 pg TEQs/kg body weight. As described above, the tolerable daily amount is 0.7 or 2 pg/kg body weight. The increase in the internal exposure (body fat) can also be estimated from this. Assuming a TEQ concentration in body fat of 30 ng/kg, a body weight of 70 kg and 20 kg fat, the TEQ amount for an adult (30 years old) is 600 ng. An egg with 6 ng increases the internal exposure from 600 to 606 ng or from 30 to 30.3 ng/kg body weight. If an egg is eaten every day for 30 days, the exposure increases to 39 ng/kg body weight, and after consumption for 90 days to 57 ng/kg body weight. The usual TEQ exposure is, depending on age, 10–80 ng/kg body fat.

It must be remembered when taking children into consideration that they have a smaller relative amount of fat and the dioxin exposure is age-dependent; it increases with age. This means that the occasional consumption of eggs contaminated with PCDDs/PCDFs causes an only insignificant increase in the internal exposure and thus has no consequences on health.

DDT

> **DDT and certain metabolites are substances with endocrine effects that can cause hormonal disturbances in wild animals.**

Rachel Carson's book *Silent Spring* triggered a large number of studies of the ecotoxic effects of DDT and other pesticides. There were reports that the DDT metabolites p,p'-dichlorodiphenyldichloroethylene (DDE) and p,p'-dichlorodiphenyldichloroethane (DDD) caused eggshell thinning in birds of prey, resulting in the breaking of eggs under the weight of the breeding parents. This led to decreased reproduction, for example in bald eagles, which recovered after the ban on DDT. In 1980 the wastewater pond of a pesticide-manufacturing facility containing DDT residues overflowed into Lake Apopka in Florida, USA. In the female alligators from that lake the plasma levels of 17β-estradiol in serum were two times higher than in control animals, while male animals had significantly reduced testosterone levels. As a consequence of these endocrine effects, the testes of the male alligators showed irregular structures, and phalli were abnormally small.

The human toxicity of DDT is very low, only one lethal intoxication has ever been reported. In rats and mice DDT leads to liver tumours. The substance has tumour-promoting properties, but is not genotoxic.

The IARC classified DDT as 'probably carcinogenic to humans, Group 2A' on the basis of positive results in animal experiments and limited evidence from epidemiological studies.

HCB

Between 1955 and 1959 several episodes of poisoning with HCB were reported in Turkey, as grain treated with HCB was used for baking bread as a result of bad harvests. Epidemiological studies yielded an association with porphyria cutanea tarda, a heme imbalance, in these populations exposed to HCB.

An estimated 4000 people were affected by the so-called 'black sore' syndrome, which comprised dermal sensitivity to the sun, skin rashes with blisters, scarring, changed pigmentation and hirsutism. As a result of the induction of the activity of cytochrome P450 (CYP), HCB also leads to hepatomegaly. The substance causes liver tumours in rats, mice and hamsters, the occurrence of which was found to be dose-dependent. In rats and mice, HCB was furthermore found to be a potent teratogen causing kidney and palate defects; in rats neurotoxic effects on development were also observed.

The IARC classified HCB as 'possibly carcinogenic to humans, Group 2B' as a result of positive findings in animal experiments. The ACGIH TLV value is 0.002 mg/m^3 and the substance is classified as carcinogen group A3 (confirmed animal carcinogen with unknown relevance to humans). The Permanent Senate Commission of the DFG for the Investigation of Health Hazards of Chemical Compounds in the Work Area classified HCB in Category 4 for carcinogenic substances for which a MAK or BAT value can be derived. In this case there is only a BAT value of 150 µg/l plasma/serum.

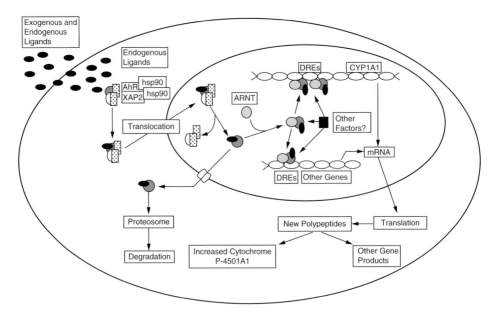

Figure 6.3 *AhR signalling pathway. Reprinted from Denison and Nagy. Copyright (2003) with permission from Annual Reviews.*

6.1.8 Mechanisms of Action

The Aryl Hydrocarbon Receptor

> **Much research has been conducted to elucidate the mechanism of dioxin toxicity, most of it related to the aryl hydrocarbon receptor (AhR). However, it has also been argued that one single mechanism is unlikely to explain a complex toxicity profile such as that displayed by the dioxins.**

Cytochrome P450 plays a major role in Phase I metabolism. Compounds can influence the transcription of these proteins via promotors. As first described by Poland in 1976, 2,3,7,8-TCDD induces CYP1A1 activity by interacting with its gene through a receptor-mediated mechanism. The cytoplasmic AhR and its nuclear partner, the AhR nuclear translocator (ARNT), play key roles in this signal transduction (see Figure 6.3). Although the AhR is not 2,3,7,8-TCDD-specific, 2,3,7,8-TCDD is the most potent AhR-agonist known. Therefore 2,3,7,8-TCDD is used as a model compound for AhR-binding kinetics and the study of downstream effects. The best-known AhR signalling pathway is initiated by an agonist binding to the receptor. In this process, the proteins associated with the AhR are released. The AhR/agonist complex translocates into the cell nucleus and binds to the ARNT. This heterotrimer attaches to the so-called dioxin response element (DRE), a core-heptanucleotide sequence in the DNA, and acts as a transcription factor. Mediated through translation, the AhR signalling cascade elicites a biological response such as the induction of CYP1A1 *in vitro* and *in vivo*.

Although CYP1A1 metabolizes many exogenous compounds, an endogenous substrate for the AhR has yet to be identified. In the rat, the highest tissue concentrations of the AhR are found in the thymus, lungs, liver and kidneys. Some researchers attribute most or even all of the toxic effects of the PCDDs to AhR interaction. It has, however, also been argued that the AhR might not be the universal key to the acute toxicity, as the ED_{50} for the CYP1A1 induction in rats (Sprague–Dawley 0.2–0.3 µg 2,3,7,8-TCDD/kg body weight) is about two orders of magnitude lower that the LD_{50} (43 µg 2,3,7,8-TCDD/kg body weight). In addition, calculations of the protein content in the liver yield a saturation of the AhR at 1.27 ng 2,3,7,8-TCDD/g tissue, whereas sublethal doses of 2.5 µg 2,3,7,8-TCDD/rat cause 100-fold higher liver concentrations (117 ng/g). That could be explained, however, by the fact that no further increase in the dose–response curve is possible when the receptor is saturated. The investigation of sensitive and resistant rat strains likewise yields controversial results. The specific binding affinity of the AhR to 2,3,7,8-TCDD does not differ between sensitive (Long Evans, 20 fmol/mg cytosolic protein) and resistant (Han/Wistar, 23 fmol/mg cytosolic protein) rat strains. Furthermore, different effects are produced at different 2,3,7,8-TCDD doses in sensitive and resistant mice (Table 6.7). While C57BL mice show a decrease in the thyroid hormones at 0.1 µg 2,3,7,8-TCDD/kg body weight, the same effect is only elicited in DBA mice by almost 1000 times higher doses. The ED_{50} for the CYP1A1 induction, however, differed only by a factor of 15.

Studies in AhR-knockout mice showed at least 10 times higher resistance to 2,3,7,8-TCDD as compared with the wild-type mice. However, 2,3,7,8-TCDD treated knockout mice displayed scattered necrosis of the liver and lymphocytic infiltration of the lungs, which were effects also observed in 2,3,7,8-TCDD-treated wild-type mice. Although less severe the very presence of these effects suggests an AhR-independent component in the manifestation of the toxicity.

Further studies with AhR-knockout as well as wild-type mice will have to be conducted to address the contradictions and to fully reveal if and to what extent this receptor modulates or mediates dioxin toxicity.

Phosphoenolpyruvate Carboxykinase

> **The hallmark of dioxin toxicity is the so-called wasting syndrome in animal models: reduced feed intake combined with decreased body weights and the derailment of intermediary metabolism. This eventually leads to a lethal hypoglycaemia, the apparent cause of the acutely lethal effect of 2,3,7,8-TCDD.**

Overall, the energy metabolism is directly related to the respiratory quotient. In 2,3,7,8-TCDD-treated animals, this quotient is reduced, as compared with pair-fed controls when the wasting

Table 6.7 Enzyme induction and acute toxicity of 2,3,7,8-TCDD in sensitive (C57 BL) and resistant (DBA) mouse strains.

Mouse strain	ED_{50} induction of CYP1A1 (liver) (µg/kg)	LD_{50} (µg/kg body weight)	ED_{50} reduced serum glucose, activity of PEPCK and glucose-6-phosphatase (µg/kg)	Effects of reduced T3/T4 levels in serum (µg/kg)
C57 BL	1.1	100	100	0.1
DBA	16	>3000	1000	97.5
Difference between the strains	15-fold	>30-fold	10-fold	975-fold

syndrome has progressed beyond the stage of glycogen depletion. This has led to the supposition that intermediary metabolism is impaired in animals treated with 2,3,7,8-TCDD. In further studies gluconeogenesis was confirmed as the target of that alteration. It has been shown in rats that a key enzyme of hepatic gluconeogenesis, phosphoenolpyruvate carboxykinase (PEPCK), is inhibited by 2,3,7,8-TCDD in a dose-dependent manner, leading to a decreased activity to the extent of 44% of pair-fed controls. This reduction of the liver PEPCK activity was found in the exact same dose range as acute toxicity occurs. In addition, hepatic tryptophan-2,3-dioxygenase in rats was likewise dose-dependently reduced and, consequently, plasma tryptophan level was dose-dependently increased, which is compatible with the feed intake reduction observed. Subsequent studies in mouse strains with different sensitivity to 2,3,7,8-TCDD (C57BL/6J and DBA/2J) revealed a plateau in the decreasing PEPCK activity in the liver, coinciding with the onset of acute toxicity in the same dose range. At lethal doses, the PEPCK activity was reduced by 80% in the liver of the mice. However, hepatic tryptophan-2,3-dioxygenase activity and plasma tryptophan were unchanged in mice, which is compatible with a lack of reduced feed intake and a lack of reduced body weight in mice. Still, lethal hypoglycaemia ensued, apparently by the more pronounced inhibition of PEPCK in mice than in rats.

Studies of glucose homeostasis in the most 2,3,7,8-TCDD-resistant and most sensitive species, the hamster and the guinea pig, respectively, do not support the notion of hepatic PEPCK being the mediator of the lethal 2,3,7,8-TCDD toxicity in all species.

A decreased PEPCK activity, but unaffected levels of liver glycogen, were observed in hamsters at doses that did not induce a body weight loss. Additionally, in hamsters a dose-dependent decrease in serum free fatty acids was also detected. While guinea pigs showed a dose-dependent decrease in the liver glycogen levels, they revealed a dose-dependent increase in serum free fatty acids. No inhibition of PEPCK activity was seen in liver cytosol of guinea pigs. It is noteworthy that in guinea pigs most of the hepatic PEPCK is located in the mitochondria, whereas in both the rats and mice PEPCK is mainly found in the cytosol.

This leads to the conclusion that lethality in rats, mice, hamsters and guinea pigs after the administration of 2,3,7,8-TCDD is due to severe impairment of intermediary metabolism. However, a uniform mechanism of action for all species is unlikely to be discovered because of the very different regulation of intermediary metabolism between hibernators (hamsters), herbivores (guinea pigs) and omnivores (rats, mice).

Stimulation of the Central Nervous System

> **DDT is a central nervous system (CNS) stimulator by interfering with sodium/potassium conductance in the neurons.**

DDT is not toxic to humans up to an estimated acute dose of 10 mg/kg body weight. Chlorinated insecticides such as DDT are CNS stimulants. In mammals, high doses cause neurotoxicity by interfering with sodium/potassium conductance in the cells. This results in repetitive neural firing, causing tremor and seizures, which can be treated with anticonvulsants.

Uroporphyrinogen Decarboxylase

> **HCB is porphyrinogenic by inhibiting uroporphyrinogen decarboxylase (UROD) in the liver, leading to the accumulation of porphyrins.**

The toxicity of HCB is thought to be related to the inhibition of UROD, a key enzyme in haem synthesis. This inhibition has been reported to precede porphyrinogenic symptoms. Specifically, HCB blocks the conversion of the first cyclic tetrapyrrol, uroporphyrinogen III, leading to increased re-oxidation and thus to the accumulation of uroporphyrins in the liver. Since a direct inhibition of UROD could not be demonstrated *in vitro*, an indirect effect is assumed. A role for metabolites is unlikely considering the very slow rate of metabolites of HCB. Lipid peroxidation as a result of CYP induction has been suggested as a possible mechanism of action. Lipid peroxidation and mitochondrial dysfunction are known to alter potassium membrane permeability, which has been shown to increase the hepatic porphyrin levels. Furthermore, iron exacerbated the effects of HCB. Therefore, altered potassium and/or iron homeostasis is thought to be the mechanism of HCB toxicity.

6.1.9 Metabolism

> **Biotransformation is the slowest and hence the rate-limiting step in PCDD elimination. Therefore, non-biliary intestinal elimination by desquamating enterocytes and by redistribution of the PHAHs into faecal fat becomes the main route of excretion.**

PHAHs can be ring-hydroxylated, which requires vicinal unsubstituted aryl positions. Lacking these, biotransformation occurs on a very slow time-scale, leading to long half-lives of the different congeners. Therefore, PCDDs and PCDFs chlorinated in 2,3,7,8 positions are poor substrates for both oxidation and reductive dehalogenation. The same holds true for PBC with *meta*- and *para*-substituents. One of the few low-yielded metabolites of 2,3,7,8-TCDD is 2,3,7-trichloro-8,9-dihydroxydibenzo-*p*-dioxin. These enzymatic reactions can also involve a chlorine-switch from a lateral position towards the central ring, as shown by the identification of the minor 2,3,7,8-TCDD metabolite 2-hydroxy-1,3,7,8-TCDD. Other metabolic pathways are the oxygen bridge cleavage in PCDDs and PCDFs, leading to the formation of tetrachlorodihydroxydiphenylether and hydroxylated PCBs, respectively. Perchlorinated PHAHs have to undergo reductive dehalogenation in order for further metabolism to occur. Epoxide intermediates as well as hydroxylated metabolites, once formed in low yield, are readily biotransformed by Phase II metabolism. The derivatised PCB-glutathione adduct 2,2',5,5'-tetrachloro-4,4-bis(methylsulfonyl)biphenyl was reported to accumulate in lungs after intraperitoneal administration to rats. Furthermore, methylsulfone derivatives of PCBs were identified in the liver and adipose tissue after chronic exposure, suggested to be due to protein binding. However, the metabolites are generally less toxic than the parent compounds and much more rapidly eliminated due to the increased hydrophilicity and additional Phase II metabolism.

Overall, biotransformation plays a minor role in the disposition of PHAHs.

6.1.10 Enzyme Induction

> **PHAHs are potent enzyme inducers *in vitro* and *in vivo*. Enzyme induction is one of the most sensitive effects of 2,3,7,8-TCDD and related compounds.**

It was demonstrated in *in vitro* studies that prior exposure of hepatocytes to 2,3,7,8-TCDD increases the rate of its metabolism 3.2-fold for subsequent exposures. However, an auto-induction of 2,3,7,8-TCDD-metabolism *in vivo* could not be confirmed. The previously described AhR signalling pathway is the best-known route of enzyme induction by co-planar PHAHs. It predominantly leads to the induction of CYP1A1 and CYP1A2 by a factor of 50 to 100 times compared with the levels in controls. The activity of NAD(P)H-quinone reductase was likewise induced by 2,3,7,8-TCDD, but to a far lesser degree. In addition, it was reported that PCBs induce the activity of the phenobarbital group (CYP2B, CYP2C, CYP3A) as well as CYP2A1, epoxide hydrolase, DT-diaphorase (NAD(P)H-quinone oxidoreductase) and aldehyde dehydrogenase activity. PCB atropisomers play differential induction patterns of enzymes by (+) and (−) enantiomers, including the patterns observed by the phenobarbital-type inducers. PHAHs induce also Phase II enzymes, but fewer studies have been conducted in this field. It was demonstrated that 2,3,7,8-TCDD induces UROD-GT (uridine diphosphate glucuronosyltransferase) around 25-fold, whereas the activities of glutathione S-transferase (GST), sulfotransferase and N-acetyltransferase remained essentially unaffected.

6.1.11 Kinetics

> **After an initial distribution phase, PHAHs are deposited mainly in adipose tissue and in the liver. Owing to their persistence they have very limited elimination, and thus long elimination half-lives. Because of these very long half-lives the rate-limiting steps in the toxicity of this class of compounds are driven by their kinetics.**

The bioavailability of 2,3,7,8-TCDD after a single per os administration ranges from about 50% in guinea pigs to 70–85% in the rat, and is also dependent on the choice of vehicle. Human data are available only on one self-administered dose. It was determined as >86% derived from faecal excretion of [1,6-^{3}H]-TCDD. The faeces are also the main elimination pathway in the rat. Urinary excretion plays a minor role at only 5–13% of the dose. Following absorption, the initial distribution of 2,3,7,8-TCDD depends on physiological parameters such as the perfusion rate and relative size of the tissues. The final distribution of 2,3,7,8-TCDD follows the affinity of the compound to the liver (5% of the dose/g tissue) and to white adipose tissue (1% of the dose/g tissue) within the first 24 hours. The half-life of 2,3,7,8-TCDD in the rat is 11 days for the parent compound in serum and 21 days in the liver. In guinea pigs, 2,3,7,8-TCDD has a half-life of around 30 days, whereas the mean half-life of 2,3,7,8-TCDD in humans is 7.8 years. The higher the degree of chlorination, the longer the biological half-life of a PHAH as compared with its congeners.

In humans DDT has an elimination half-life of about 5 years; its metabolite DDE is even more persistent. DDT is excreted mainly with the urine. The fractions excreted via biliary or lactational pathways increase with increasing DDT doses.

HCB shows a half-life of 24 days in rats, 32 days in rabbits and about one year or more in humans. Much less HCB distributes into the liver than 2,3,7,8-TCDD because HCB, unlike 2,3,7,8-TCDD, does not have a high affinity to CYP1A2, which is the main and most abundant binding protein of PHAHs in the liver.

The kinetics of persistent xenobiotics in rats and humans can be calculated by an empirical formula (Equation 6.3), which also allows the calculation of elimination half-lives of very persistent compounds, such as OCDD.

Formula for the extrapolation of elimination half-lives of persistent xenobiotics from rats to humans:

$$t_{1/2(\text{humans})} = 17.78 \cdot \left(t_{\frac{1}{2(\text{rat})}}\right)^{1.34} \tag{6.3}$$

where $t_{1/2}$ is the elimination half-life.

The difference in the half-lives between rats and humans not only determines the species-dependent persistence of these chemicals in the body, but also the steady-state concentration during chronic exposure (Equation 6.4).

Formula for the calculation of steady-state concentrations during chronic exposure:

$$c_{\text{SS}} = \frac{1.44 \times t_{\frac{1}{2}} \times f \times \text{dose}}{V_d \times \tau} \tag{6.4}$$

where c_{SS} is the steady-state concentration, f is the fraction absorbed, V_d is the volume of distribution and τ is the dosing interval.

Small species differences in the volume of distribution and in the fraction absorbed only have a minor impact on this equation. The differences in the half-lives, however, are orders of magnitude larger and directly proportional to the steady-state levels achieved. The half-life in rats is approximately 140 times shorter than in humans. It takes 6.64 elimination half-lives to reach 99% of the steady-state during chronic exposure. Therefore, rats reach it considerably faster than humans (see Equations 6.5–6.7). Furthermore, at identical daily dose rates, humans will eventually reach a steady state about 140 times higher than rats, which represents the safety factor needed when extrapolating the data obtained in rats to humans.

Calculation of eliminated fractions:

$$C_p = C_0 \times e^{-kt} \tag{6.5}$$

$$C_p = 0.5 \times C_0 \rightarrow t_{\frac{1}{2}} = \frac{\ln 2}{k} \tag{6.6}$$

$$C_p = 0.01 \times C_0 \rightarrow t_{1/99} = \frac{\ln 100}{k} = 6.64 \times t_{1/2} \tag{6.7}$$

where C_p is the concentration in plasma, C_0 is the concentration in plasma at time $t = 0$, k is the elimination rate constant, t is time, $t_{1/2}$ is the elimination half-life and $t_{1/99}$ is the time to reach 99% elimination.

The long period of time to steady state also affects the bioconcentration factors (BCFs). This is the reason why exposure studies with PHAHs in aquatic species must be extended for at least 6.64 half-lives. In the 1970s, the lack of an appropriate study design led to the assumption that highly chlorinated PCDDs would not bioconcentrate. This has been clearly refuted both theoretically and experimentally.

The lipophilicity of PHAHs leads to high concentrations in breast milk. With the onset of lactation, lipophilic compounds are redistributed into this newly formed lipophilic compartment. This results in high concentrations, especially for the first pregnancy, when the body burden of PHAHs after decades of accumulation is available for redistribution into the milk fat. In later pregnancies the levels in breast milk are significantly lower, since the body burden has been reduced by the previous lactational elimination. Because of more frequent lactations, the PHAH levels in cows' milk are considerably lower than in breast milk.

6.1.12 Summary

The production, emission and use of persistent PHAHs peaked in the middle of the last century with an increase in synthetic chemistry and an upswing in agriculture and technology. Today, most of these materials are being phased out or have already been banned. Unintentional emissions have been reduced and the clean-up of heavily contaminated sites is underway in developed countries.

These facts could lead to the wrong assumption that PHAHs belong to our past and merit little further attention. However, these compounds are highly persistent and are ubiquitously present in the environment and the human food chain at the present time and in the foreseeable future. The effects of PHAHs on humans, animals and the environment have been investigated for decades. Yet studies of the toxicity of these compounds applying classical and modern technologies still yield new findings, providing new and improved bases for risk assessment and for the improvement of directives in regulatory agencies worldwide. PHAHs serve also as model compounds to study mechanisms of toxicity, which might facilitate the development of new drugs.

Further Reading

ATSDR (1995) *Toxicological Profile for Polybrominated Biphenyls, US Department of Health & Human Services, Agency for Toxic Substances and Disease Registry*, Atlanta, Georgia.

ATSDR (1996) *Toxicological Profile for Hexachlorobenzene, US Department of Health & Human Services, Agency for Toxic Substances and Disease Registry*, Atlanta, Georgia.

ATSDR (1997) *Toxicological Profile for Polychlorinated Biphenyls (Update), US Department of Health & Human Services, Agency for Toxic Substances and Disease Registry*, Atlanta, Georgia.

ATSDR (2002) *Toxicological Profile for DDT, DDE, and DDD (Update), US Department of Health & Human Services, Agency for Toxic Substances and Disease Registry*, Atlanta, Georgia.

Denison, M.S. and Nagy, S.R. (2003) Activation of the aryl hydrocarbon receptor by structurally diverse exogenous and endogenous chemicals. *Annu. Rev. Pharmacol. Toxicol.*, **43**, 309–334.

EFSA (2015) Scientific statement on the health-based guidance values for dioxins and dioxin-like PCBs. *EFSA Journal*, **13** (5), 4124. doi:10.2903.

Golden, R. and Kimbrough, R.D. (2009) Weight of evidence evaluation of potential human cancer risks from exposure to polychlorinated biphenyls: an update based on studies published since 2003. *Crit. Rev. Toxicol.*, **39**, 299–331.

IARC (2012) 2,3,7,8-Tetrachlorodibenzo-*para*-dioxin, 2,3,4,7,8-pentachlorodibenzofuran, and 3,3',4,4', 5-pentachlorobiphenyl. In: *Chemical Agents and Related Occupations, IARC Monographs on the Evaluation of Carcinogenic Risk of Chemicals to Humans*. Vol. **100F**, Lyon, France, International Agency for Research on Cancer, pp. 339–378.

Robertson, L.W. and Hansen, L.G. (2001) *PCBs – Recent Advances in Environmental Toxicology and Health Effects*, The University Press of Kentucky, Lexington, Kentucky.

Rozman, K.K. (1989) A critical view on the mechanism(s) of toxicity of 2,3,7,8-tetrachlorodibenzo-p-dioxin. Implications for human safety assessment. *Dermatosen*, **37**, 81–92.

Rozman, K.K., Roth, W.L. and Doull, J. (2006) Influences of dynamics, kinetics, and exposure on toxicity in the lung. In: *Toxicology of the Lung*, 4th edition (ed. D.E. Gardner). CRC Press, Boca Raton, Florida, pp. 195–229.

Saurat, J.H., Kaya, G., Saxer-Sekulic, N., Pardo, B., Becker, M., Fontao, L., Mottu, F., Carraux, P., Pham, X.C., Barde, C., Fontao, F., Zennegg, M., Schmid, P., Schaad, O., Descombes, P. and Sorg, O. (2012) The cutaneous lesions of dioxin exposure: lessons from the poisoning of Victor Yushchenko. *Toxicol. Sci.*, **125**, 310–317.

Serdar, B., LeBlanc, W.G., Norris, J.M. and Dickinson, L.M. (2014) potential health effects of polychlorinated biphenyls (PCBs) and selected organochlorine pesticides (OCPs) on immune cell and blood biochemistry measures: a cross-sectional assessment of the NHANES 2003-2004 data. *Environ. Health*, **13**, 114. doi:10.1186/1476-069X-13-114.

Sorg, O. (2014) AhR signalling and dioxin toxicity. *Toxicol. Lett.*, **230**, 225–233.

6.2 Metals

Andrea Hartwig and Gunnar Jahnke

6.2.1 General Aspects

Metal ions can be both essential and toxic; frequently both properties are closely linked. Metals and their compounds are ubiquitously present in the environment, but anthropogenic sources contribute significantly to human exposure. Some metals, including calcium, magnesium, zinc, cobalt, nickel, copper and iron, are essential components of biological systems. They mediate oxygen transport and metabolism, catalyze electron transfer reactions, are involved in signal transduction and stabilize the structure of macromolecules. For some other metals or semi-metals, such as arsenic, lead, cadmium and mercury, no essential functions are known.

From a toxicological point of view, metals and their compounds have some special properties. Thus, toxic and even carcinogenic effects are not limited to non-essential metals, but are also observed in the case of essential metals.

> **Essential and toxic effects of metals and their compounds are often closely interrelated, for example in the case of the transition metals iron and copper. While one of their essential biological functions is catalyzing one-electron transitions, this ability can lead to the formation of reactive oxygen species (ROS) with the consequence of damage to cellular macromolecules. Therefore, the precise regulation of the metal ion concentrations in tissues and cells is necessary in order to prevent toxic effects. This is achieved by a strict control of intake, transport, intracellular storage and excretion.**

Toxic effects of essential metals arise when their homeostatic control is overridden by either excessively high uptake or non-physiological uptake pathways. Thus the optimal supply with essential trace elements is regulated by the control of absorption via the gastrointestinal tract and excretion via kidney or liver. However, this control does not exist in the case of inhalative exposure and resorption via the lungs. Besides the generation of ROS, one other principle in metal toxicity consists of the competition of toxic metal ions with essential metal ions. These interactions occur at the level of uptake and intracellular functions. For example, different divalent metal ions are absorbed by the same transporter (divalent metal transporter 1, DMT1). In addition, so-called zinc finger structures, a common motif found in DNA binding proteins, such as DNA repair proteins and transcription factors, are sensitive target structures for toxic metal ions. Possible consequences are reduced bioavailability of essential metal ions as well as disturbances in signal transduction and in the structure and function of macromolecules.

Metal atoms can form ionic, covalent and coordinative bonds. Ligands containing oxygen, nitrogen or sulfur are preferably bound. Thus, many important biological molecules, including proteins and nucleic acids, are the preferred target structures for the interaction with metals.

Toxic effects are not only dependent on the particular metals, but are also decisively determined by the respective metal species; the oxidation stage and solubility have an especially pronounced impact on uptake and ultimately bioavailability.

While absorption and distribution depend mainly on the chemical form of the metal (poorly soluble, poorly absorbable metal phosphates, lipid-soluble or readily absorbable organometallic compounds), the toxicity of a metal is decisively influenced by the concentration, the (bio)chemical complex, and the binding partner inside the cell. Concerning acute toxicity, local

Table 6.8 Specific metal-binding proteins.

Protein	Molecular weight (Daltons)	Occurrence	Metal	Function
Calmodulin	14,000	Ubiquitous	Ca	Activates several enzymes
Ferritin	470,000	Liver, spleen and bone	Fe	Storage protein
Transferrin	90,000	Plasma, extracellular space	Fe	Transport protein
Ceruloplasmin	132,000	Plasma	Cu	Transport protein
Metallothionein	6,500	Ubiquitous	Zn, Cu, Ag, Hg, Cd	Metal homeostasis, storage protein, in part detoxification

effects are often found at the site of application or absorption. Chronic toxicity of metals is often tissue specific, and in some cases, specific metal-binding proteins may be particularly important. Such metal-binding proteins have in most cases no enzymatic activity, but are rather temporary storage or transport forms for essential metals. Examples of such proteins are the calcium-binding **calmodulin**, the iron-binding **ferritin** and the glycoprotein **transferrin**, the transport form for copper **ceruloplasmin** and **metallothionein** (Table 6.8).

Of particular importance with respect to both essential metal homeostasis and metal toxicity is **metallothionein**. It binds mainly zinc and copper (up to 7 Zn or 12 Cu atoms per molecule) under physiological conditions, but it can also bind up to 7 atoms of cadmium, mercury, cobalt and nickel and up to 18 atoms of silver. It is an ubiquitous protein with a molecular weight of 6500 Da and a cysteine content of 30%. All cysteine sulfur atoms are involved in metal binding in two clusters. Due to its high affinity to metals and its metal-inducible synthesis, metallothionein plays a special role in the detoxification and resistance towards toxic metals. Accordingly, for example, humans with elevated cadmium exposure, such as occupationally exposed workers or smokers, show higher cadmium and metallothionein levels in the kidneys. However, this leads to highly prolonged half-lifes within this organ; also, it must be borne in mind that metallothionein not only can bind metal ions, but can also release them again, e.g. under conditions of oxidative stress. Both aspects contribute to the chronic toxicity of the respective metal ions.

6.2.2 The Importance of Bioavailability

> **The metal species as well as the particle size of metal-containing dusts often determine the bioavailability and thus their toxicity. One key aspect of metal toxicology is therefore the risk assessment of different metal species. One example is chromium. While chromium(VI) compounds are carcinogenic, this is not established for chromium(III) compounds. This can be attributed to differences in bioavailability. Water-soluble chromium(VI) compounds are absorbed via anion transporters and are reduced intracellularly to chromium(III) via various intermediate steps. Ultimately, chromium(III) species cause DNA damage and mutations. For soluble chromium(III) compounds, however, the cell membrane is almost impermeable.**

One further question concerns the toxicological assessment of poorly water-soluble, particulate compounds when compared to readily water-soluble compounds. One well-studied example is nickel. Both water-soluble and particulate nickel compounds are carcinogenic in humans. In animal experiments, compounds of medium solubility and average toxicity, such as nickel sulfide (NiS) and nickel subsulfide (αNi_3S_2), are among the strongest known carcinogens. Decisive are, above all, the solubility in extracellular fluids, the uptake of the compounds into the cells of the target organs, as well as the subsequent intracellular release of nickel ions as the ultimate damaging agent. Soluble nickel compounds are absorbed into the cells via ion channels. Crystalline particles, which are largely insoluble in water, are phagocytized and enter the lysosomes, dissolve gradually there due to the acidic pH and release high amounts of nickel ions intracellularly. Even though the toxicity is mediated in all cases by nickel ions, intracellular concentrations are much higher in the case of particulate compounds. Additionally, the far longer retention time of particulate nickel oxide and subsulfide in experimental animals most likely contributes to the strong carcinogenicity. Similar effects are observed with metal-containing nanomaterials. For example, copper-containing nanoparticles show more pronounced cytotoxicity and genotoxicity when compared to water-soluble or microscale compounds of the same chemical composition. Thus, investigations in cell culture systems showed a very rapid release of large amounts of copper ions from the lysosomes after endocytosis, which lead to significantly higher intracellular concentrations of copper in the cytoplasm as well as in the nucleus. Overall, the cellular copper homeostasis is bypassed, leading to pronounced increases in redox reactions and ROS formation (the Trojan horse effect).

6.2.3 Acute and Chronic Toxicity as well as Carcinogenicity

> **Acute metal-induced health problems are rare nowadays. Chronic effects include both carcinogenic and non-carcinogenic effects.**

Acute Toxicity

Acute metal poisoning is nowadays usually limited to accidental exposure; in particular, cadmium, lead, mercury and arsenic may be important. In this case, the ability of toxic metal ions to form complexes with organic molecules is used therapeutically for accelerated metal excretion. If the ligands are organic molecules with more than one group capable of coordination, these usually very stable complexes are called **chelates**. Thus, some chelating agents such as dimercaprol (BAL), the calcium disodium salt of ethylenediaminetetraacetic acid (EDTA), the hydroxamic acid derivative desferoxamine, 2,3-dimercaptopropanesulfonate (DMPS), and β,β-dimethylcysteine (D-penicillamine) are used in the treatment of metal poisoning (Table 6.9). Basic prerequisites for the therapeutic use of chelating agents as **antidotes** are:

- high affinity to toxic and low affinity to essential metal ions
- low toxicity of chelating agent and chelate
- no metabolism of chelating agent and chelate
- substantial excretion of chelate in urine or bile
- stability of chelate at physiological pH and in acidic urine.

Chelating ligands generally do not react specifically with the metal to be mobilized. This can lead to undesirable excretion and a life-threatening loss of essential metal ions such as calcium, zinc, or copper. Therefore, it needs to be emphasized that the diagnostic application of complexing

Table 6.9 Chelating agents as antidotes in metal poisoning.

Chelator	Metal
Dimercaprol (BAL)[a]	As, inorganic Hg, Pb (with EDTA)
EDTA[b]	Pb
Desferoxamine	Fe
DMPS[c]	Methyl-Hg, inorg. Hg, Pb, Cd, Cu, Ni
D-Penicillamine[d]	Cu, (Pb)

[a] British Anti-Lewisite (BAL); originally used to detoxify the warfare agent Lewisite.
[b] Calcium disodium salt of ethylenediaminetetraacetic acid.
[c] 2,3-Dimercaptopropanesulfonate (Dimaval®).
[d] β,β-Dimethylcysteine.

agents, such as DMPS (Dimaval®), is not justified for the verification of the body burden with metals such as cadmium, mercury or lead. At least for lead and mercury there is an equilibrium between blood, tissue and urine, so that the body burden can be assessed by the measurement of the metals in the blood.

Chronic Organ Toxicity

In contrast to the diminishing importance of acute metal toxicity, the recognition of chronic health effects after long-term exposure to even comparatively low concentrations of toxic metal compounds has considerably increased during recent years. Adverse chronic health effects include damage to the central nervous system by lead, manganese, mercury and aluminum, damage to the immune system and kidney damage caused by cadmium. Furthermore, for some metals and their compounds lung damage after inhalation exposure has been identified as a critical endpoint.

Carcinogenicity

The question of a potentially carcinogenic effect of metal compounds is of special interest. Thus, for example, chromates as well as nickel, cadmium and arsenic compounds were identified as carcinogens in epidemiological studies and/or in experimental animals. The assessments of individual metals and their compounds regarding their carcinogenicity for humans by the International Agency for Research on Cancer (IARC) and the Permanent Senate Commission of the DFG for the Investigation of Health Hazards of Chemical Compounds in the Work Area (MAK Commission) are summarized in Table 6.10.

With respect to metal carcinogenicity, the direct interaction of metal ions with DNA constituents is usually of minor importance. One exception are chromates. They are reduced intracellularly to chromium(III), and together with reducing agents such as ascorbate, potentially mutagenic ternary Cr-DNA adducts are formed. For most other metal compounds, rather indirect mechanisms have been demonstrated, such as the increased formation of ROS, the inactivation of DNA repair processes, changes in gene expression, and interactions with signal transduction processes. Thus, for some metal compounds elevated levels of oxidative DNA damage have been detected in cellular test systems, but these are caused by indirect mechanisms. Examples are the catalysis of Fenton-like reactions with H_2O_2 by transition metal ions and thus the generation of very reactive hydroxyl radicals, as well as the interference with the cellular response towards ROS, such as the inactivation of ROS-detoxifying enzymes. One other common mechanism in metal carcinogenicity is interference with DNA repair systems. Metal compounds, such as those of nickel, cadmium, arsenic, cobalt, and antimony, have been shown to inhibit DNA repair

Table 6.10 Classifications of selected carcinogenic metals and their compounds.

Substance	IARC classification	MAK category
Antimony and its inorganic compounds	n.e.	2 (except for stibine)
Antimony trioxide (Sb_2O_3)	2B	2
Antimony trisulfide (Sb_2S_3)	3	2
Arsenic and inorganic arsenic compounds	1	1
Beryllium and its inorganic compounds	1	1
Lead (metal)	n.e.	2
Lead compounds, inorganic	2A	2
Butyltin compounds	n.e.	4
Cadmium and its inorganic compounds	1	1
Chromium (metal)	3	n.e.
Chromium(III) compounds	3	–
Chromium(VI) compounds	1	1
Cobalt and cobalt compounds	2B	2
Hard metal containing tungsten carbide and cobalt	2A	1
Nickel (metal)	2B	1
Nickel compounds	1	1
Mercury and inorganic mercury compounds	3	3B
Mercury, organic compounds	2B (methylmercury compounds only)	3B
Rhodium and its inorganic compounds	n.e.	3B
Selenium and its inorganic compounds	3	3B
Uranium and its hardly soluble inorganic compounds	n.e.	2
Uranium, soluble inorganic compounds	n.e.	3B
Vanadium and its inorganic compounds	n.e.	2
Vanadium pentoxide (V_2O_5)	2B	2

IARC, International Agency for Research on Cancer; MAK, Permanent Senate Commission for the Investigation of Health Hazards of Chemical Compounds in the Work Area; –, not classified; n.e., not evaluated.
IARC classifications: Group 1, Carcinogenic to humans; Group 2A, Probably carcinogenic to humans; Group 2B, Possibly carcinogenic to humans; Group 3, Not classifiable as to its carcinogenicity to humans.
MAK categories: Category 1, Substances that cause cancer in humans and can be assumed to contribute to cancer risk; Category 2, Substances that are considered to be carcinogenic for humans because sufficient data from long-term animal studies or limited evidence from animal studies substantiated by evidence from epidemiological studies indicate that they can contribute to cancer risk; Category 3B, Substances that cause concern that they could be carcinogenic for humans but cannot be assessed conclusively because of lack of data; Category 4, Substances that cause cancer in humans or animals or that are considered to be carcinogenic for humans and for which a MAK value can be derived. A non-genotoxic mode of action is of prime importance and genotoxic effects play no or at most a minor part provided the MAK and BAT values are observed. Under these conditions no contribution to human cancer risk is expected.

processes in very low concentrations, thereby decreasing the protection against endogenous and environmental mutagens. In addition, for some metals, epigenetic effects, i.e. changes in the DNA methylation patterns, have been observed, which may lead to altered gene expression patterns. In this context, the activation of growth-related genes (oncogenes) or the inactivation of tumor suppressor genes is particularly critical in the context of cancer development.

6.2.4 Toxicology of Selected Metal Compounds

Arsenic

> **Of arsenic compounds, the most critical effect of toxicological concern is the carcinogenicity, which cannot be excluded even under environmental exposure conditions.**

Occurrence and Exposure Arsenic, which belongs to the semimetals, occurs in the oxidation states +5, +3, 0 and –3. From a toxicological point of view, organic and inorganic arsenic compounds need to be distinguished. Both natural and anthropogenic sources are relevant for human exposure, including the burning of fossil fuels. Furthermore, with respect to the general population and depending on geological conditions, drinking water may have a significant impact on arsenic exposure. Here, arsenic is mostly present in inorganic form as arsenate (+5), and under reducing conditions also as arsenite (+3). The concentration of arsenic in the groundwater is usually less than 10 µg/l, but in some areas of the world, such as India or Bangladesh, concentrations of more than 3000 µg/l can be reached. Rice often contains inorganic arsenic in concentrations of 0.1–0.4 mg/kg dry mass, in part clearly above, so that significant exposure results also from rice products, especially in children. The majority of the dietary intake of organic arsenic comes from fish and seafood, where arsenic is predominantly found as arsenobetaine and contains average concentrations of 0.1–1.8 mg arsenic/kg. In algae, arsenic is predominantly present as arsenosugars, the concentrations usually being in the range of 2–50 mg arsenic/kg dry mass. Occupational exposure to arsenic occurs in metal production and processing. Arsenic and arsenic compounds are used, for example, in semiconductors as gallium arsenide, in wood preservatives and in alloys. In the past, arsenic-containing pesticides were used.

Toxicokinetics Arsenic is distributed into all tissues after absorption. In experimental animals, the highest levels were found in the skin, kidneys, liver, and lung. Excretion occurs in most species, including humans, mainly via the urine in the form of arsenite, arsenate, and the methylated metabolites monomethylarsonic acid (MMA(V)) and dimethylarsinic acid (DMA(V)).

Inorganic arsenic in the form of arsenite or arsenate is absorbed well after inhalation with approximately 30–90%. After oral intake, resorption is about 45–75%. In the case of sparingly soluble arsenic compounds, the absorption is markedly lower, both after inhalation and after oral administration. In the case of arsenobetaine, an almost complete absorption and unchanged excretion is found after oral ingestion. In contrast, in most mammalian species, including humans, inorganic arsenic compounds undergo an extensive biotransformation in the liver. First, arsenate is reduced to arsenite and subsequently tri- and pentavalent mono- and dimethylated arsenic species are obtained in the form of monomethylarsonous acid (MMA(III)), dimethylarsinous acid (DMA(III)), MMA(V) and DMA(V). While this methylation was originally considered as a detoxification process, research results of recent years show that the trivalent methylated metabolites in particular exert increased toxicity and also genotoxicity when compared to arsenite.

Toxicity Inorganic arsenic compounds, especially arsenic trioxide (As_2O_3), were a popular murder poison. While 0.1 g of orally taken arsenic trioxide is fatal, small daily doses of 2 mg consumed by so-called arsenic eaters have a postulated performance-boosting effect (mountain farmers, forestry workers, but also horses), aiming to increasingly tolerate arsenic without acute toxic effects. This may be due to an increased formation of metallothionein; however, from today's point of view, chronic toxicity including carcinogenicity is expected at these exposure levels (see below).

Acute toxic effects of inorganic arsenic and its compounds include gastrointestinal, cardiovascular, renal, and neurotoxic effects. Trivalent inorganic arsenic compounds are usually more toxic than the pentavalent compounds and organic arsenic compounds usually have a markedly lower acute toxicity. Nowadays, more attention is paid to the chronic toxicity of arsenic. Besides exposure at the workplace, chronic exposure towards elevated levels of inorganic arsenic via drinking water has been associated with pronounced toxicity. Adverse health effects include skin changes and blood flow disorders ("blackfoot disease"), cardiovascular diseases, neurotoxicity as well as developmental toxicity. With respect to carcinogenicity, epidemiological studies provide clear evidence for an increased lung tumor risk after inhalative exposure to arsenic. Also, after oral exposure (via drinking water) to inorganic arsenic, increased risks of skin, lung, and urinary bladder tumors were found. As mechanisms of the carcinogenic effect of inorganic arsenic compounds and their methylated metabolites, the induction of oxidative stress, the impairment of DNA repair processes as well as epigenetic mechanisms have been described. The toxicity of organic arsenic compounds is still largely unknown. Even though the acute toxicity is lower when compared to inorganic arsenic, there is still a need for further research, especially with regard to chronic effects.

Limit Values and Classifications Arsenic and its inorganic compounds have been classified as carcinogenic in humans by the IARC and the MAK Commission (carcinogenicity category 1). A drinking water limit of 10 μg arsenic/l was established by the World Health Organization (WHO) and the US Environmental Protection Agency (EPA). Based on dose–response relationships from epidemiological studies, the European Food Safety Authority (EFSA) has identified lower confidence limits for an additional risk of 1% ($BMDL_{01}$) after oral intake of 0.3–8 μg inorganic arsenic/kg body weight and day for lung, skin and urinary bladder cancer, as well as for skin damage by means of benchmark calculations. Since the estimated average dietary exposure of consumers to inorganic arsenic is already in this range, a possible risk for the general population cannot be excluded. Therefore, a tolerable daily intake (TDI) value could not be derived.

Cadmium

Cadmium is considered to be a lung carcinogen on inhalative exposure. Additionally, kidney and bone toxicity occur at low exposure conditions.

Occurrence and Exposure Cadmium is a naturally occurring element of the earth's crust, but its distribution in the environment is also due to industrial and agricultural sources. Nowadays, cadmium is used mainly in nickel–cadmium batteries, but also in solder, in pigments, as stabilizers in PVC, and in the electroplating industry. The use of cadmium in some applications, such as pigments, has significantly declined in recent years due to toxicity and legal requirements. The most important sources of soil contamination are the use of cadmium-containing phosphate fertilizers and sewage sludge. Due to its ubiquitous occurrence in soil, cadmium is present in all foods of plant origin as well as in foods of animal origin. Particularly rich in cadmium are fungi, oil seeds and cocoa beans, where mean levels are about 0.2 mg/kg, with maximum values of more than 2 mg/kg. The content in foods of animal origin is dependent on the cadmium content in feed. Particularly high levels of 0.2 mg/kg are found in kidney and liver. In the case of non-smokers, dietary intake is the predominant source of exposure for the general population, while inhalation contributes significantly to the cadmium exposure of smokers.

Toxicokinetics Oral absorption of cadmium or inorganic cadmium compounds is in the range of 1–10%. After inhalation, absorption of up to 25% is reported. Skin absorption is low, at <1%. After absorption, cadmium is bound to metallothionein in the liver and enters the kidneys via the blood as cadmium-metallothionein. Because of its low molecular mass, the complex is filtered through the glomerular membranes and reabsorbed in the proximal tubule. Upon chronic exposure, approximately 50% of the metal is found in the kidneys, 15% in the liver, and 20% in the muscles. Due to the binding to metallothionein, cadmium has a long biological half-life of about 10–30 years. The very slow excretion takes place in approximately equal parts via urine and feces.

As a consequence of the long half-life, cadmium concentrations increase with increasing age in most tissues and achieve maximum levels at the age of about 50 years, especially in the kidney cortex, the target organ for chronic toxicity. Because of the effective barrier of the placenta for cadmium, newborns have very low cadmium body burdens.

Toxicity Acute inhalation exposure to cadmium may lead to respiratory problems and in severe cases to pulmonary edema, inflammation, and death. Acute effects after oral exposure include mainly gastrointestinal symptoms. Target organs after repeated oral exposure are kidneys and bones, and in case of repeated inhalation exposure the lungs are also affected. A particularly serious case of environmental cadmium poisoning was the occurrence of the so-called "Itai-Itai" disease in Japan. The symptoms were severe pain in the back and legs triggered by osteomalacia and osteoporosis. The underlying reason was massive renal damage, which led to disturbances in calcium, phosphorus and vitamin D metabolism in the bones. Renal dysfunctions occur at comparatively low concentrations; beginning effects are observed close to environmental exposure conditions. People with an above-average consumption of cadmium-containing food, people with calcium deficiency, iron deficiency, vitamin D deficiency, smokers and indivuduals with occupational exposure can be particularly affected.

After inhalation exposure, cadmium is considered as carcinogenic to humans (lung and kidney, in animal experiments lung and prostate). Several inhalation studies with rats relevant to the exposure conditions in the workplace show that both water-soluble cadmium chloride and cadmium sulfate as well as water-insoluble compounds such as cadmium sulfide and cadmium oxide are carcinogenic at low concentrations.

Cadmium ions do not directly interact with DNA. Relevant mechanisms of cadmium-induced carcinogenicity include the induction of oxidative stress due to interference with antioxidative enzymes, the inhibition of DNA repair systems, the altered regulation of cell proliferation, the inactivation of tumor suppressor proteins as well as changes in the DNA methylation pattern and thus gene expression.

Limit Values and Classifications Cadmium and its compounds have been classified by the IARC as well as by the MAK Commission as carcinogenic in humans (Category 1). A tolerable weekly intake (TWI) for cadmium of 2.5 µg/kg body weight was derived from the kidney toxicity in humans from the EFSA.

Lead

> **Exposure to lead compounds has dropped significantly since the ban on leaded fuels. At low exposure, developmental neurotoxicity is the most sensitive adverse health effect.**

Occurrence and Exposure Lead, which is ubiquitously present in the environment, can occur in the oxidation states 0, +2 and +4. An important source for the entry of lead into the environment was the use of tetraethyl lead as an fuel additive. After the prohibition of leaded fuels, the concentrations of lead in the air, in food and subsequently in the blood of the general population have declined significantly. Lead concentrations in the air in industrialized areas are 0.03–0.1 µg/m^3, while in rural areas they are significantly lower. Typical lead concentrations in foodstuffs are between 10 and 200 µg/kg, whereby foods of plant origin generally contain less lead than those of animal origin. Higher contents are found in offal, mussels, mushrooms and spices, some of which can be well above 1 mg/kg. The biggest contribution to the lead exposure from food in the European population comes from cereal products, vegetables and drinking water. Other sources of exposure include, in addition to occupational exposure, lead-containing drinking water pipes and, especially for children, the ingestion of lead-containing paint. The main current use of lead is in the production of batteries and accumulators; in many other areas the use of lead is declining.

Toxicokinetics Inorganic lead compounds can be absorbed after inhalation, oral, and, to a lesser extent, dermal exposure. In children, an oral absorption of water-soluble lead compounds of 40–50% was found, whereas in adults it was about 3–10%. In addition to age, the absorption after oral intake is affected by the nutritional status, the composition of the food, the iron and calcium status, and other factors. In the case of organic lead compounds, a good dermal absorption was found in animal experiments. After absorption, lead accumulates mainly in the bones, where in adults up to 94% of the body burden is located. Comparatively high concentrations are also found in the liver and kidneys. In the blood, lead has a half-life of about 30 days, while in the bones it was found to be several years. Lead is predominantly excreted via urine and feces.

Toxicity Acute lead poisoning is rare today; the symptoms range from vomiting and intestinal colic to kidney failure. Chronic effects include, in particular, disturbed blood synthesis, (developmental) neurotoxic effects, renal toxicity, and cardiovascular disorders.

The **hematological effects** are based on the inhibition of the following enzymes of the heme synthesis (Figure 6.4):

1. δ-aminolevulinic acid dehydratase, leading to an increase in δ-aminolevulinic acid in blood and urine
2. coproporphyrinogen III oxidase
3. ferrochelatase and therefore inhibition of the insertion of ferrous iron into protoporphyrin IX.

The inhibition of the enzymes results in hemoglobin deficiency with the consequence of anemia. The changes in the enzyme activity, in particular the activity of the δ-aminolevulinic acid

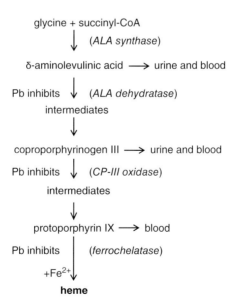

Figure 6.4 *Impact of lead on heme synthesis. Inhibition of ALA dehydratase, coproporphyrinogen III oxidase, and ferrochelatase results in increased levels of δ-aminolevulinic acid, coproporphyrinogen III, and protoporphyin IX in blood and urine as well as anemia.*

dehydratase in the peripheral blood, as well as the excretion of δ-aminolevulinic acid with the urine correlate well with the blood lead levels and are therefore used as sensitive biochemical markers for the current lead exposure.

The **neurotoxicity** of lead, e.g. the effects on the peripheral and central nervous system, are of great importance with regard to the health effects of chronic exposure.

Peripheral lead neuropathy is caused by functional disorders of motor nerve fibers and leads to paralysis of the upper extremities in blood levels of 500–700 µg Pb/l. Even the smooth muscles of the colon can be paralyzed and leads to painful colic with persistent constipation. Central nervous disorders after chronic lead exposure (lead encephalopathy) manifest themselves with fatigue, headache, dizziness, and tremor. Subclinical changes in the central and peripheral nervous system are already described at far lower blood lead levels.

One of the most critical effects of lead is neurotoxicity in children, which leads to persistent intelligence deficits, i.e. a lowering of the intelligence quotient (IQ). Thus, epidemiological studies revealed that in children under the age of 10 years blood levels of less than 100 µg Pb/l lead to impairments of the IQ. No NOAEL has yet been derived for this effect.

Epidemiological studies provide indications for a possible **carcinogenic effect** of lead and lead compounds. In animal experiments tumors were found in kidneys, adrenals, testes, prostate, lung, liver, pituitary, thyroid, and breast glands as well as leukemias, sarcomas of the hematopoietic system, and cerebral gliomas. The mechanisms of carcinogenicity have not yet been fully elucidated. Indirect genotoxic effects, for example by a disruption of DNA repair mechanisms, are suspected to contribute to tumor formation. In the case of the organic lead compounds such as tetramethyl- and tetraethyllead, neurotoxic effects are predominant.

Limit Values and Classifications The IARC classified inorganic lead compounds in Group 2A, the MAK Commission classified metallic lead and lead compounds in Category 2. The provisional tolerable weekly intake (PTWI) established by the WHO in 1999 is 25 µg/kg body weight. In Europe, maximum levels for lead in individual foodstuffs have been set, for example 20 µg/kg for milk, 50 µg/kg for fruit juice, 100 µg/kg for meat, fruit and vegetables, 200 µg/kg for cereals, 300 µg/kg for leafy vegetables, and 1500 µg/kg for mussels (Commission Regulation (EC) No 1881/2006). On the basis of results on developmental neurotoxicity in children as well as cardiovascular effects and renal toxicity in adults, the EFSA has evaluated lower confidence limits for an additional risk of 1 or 10% ($BMDL_{01}$ or $BMDL_{10}$) of 0.5 ($BMDL_{01}$), 1.5 ($BMDL_{01}$) and 0.6 ($BMDL_{10}$) µg/kg body weight and day (benchmark calculations), respectively. The previous PTWI of 25 µg/kg body weight was considered to be no longer appropriate.

Mercury

> **The toxic effect of mercury depends on the actual species. In particular, organic mercury compounds and metallic mercury are highly neurotoxic.**

Occurrence and Exposure Mercury is released into the environment from both natural and anthropogenic sources. Mercury occurs in elementary (Hg^0) as well as inorganic and organic species. Mercury compounds occur in the oxidation states +1 and +2. Elemental mercury is the only metal that is liquid at room temperature, with a relatively high vapor pressure. Inorganic mercury can be transformed into organic mercury compounds by aquatic microorganisms and therefore accumulates in the food chain. Here, methylmercury compounds are the most abundant organic mercury compounds. For the occupationally unexposed population, the diet and in particular the consumption of fish and other marine organisms is the most important source of exposure to mercury. In addition, mercury may be released from amalgam fillings in teeth. In the workplace, exposure to mercury vapor is of particular relevance. The most important applications of mercury are use as cathode material in the chloralkali process, thermometers, manometers, mercury vapor lamps, energy saving lamps and special batteries, in metal production, and in amalgams in dentistry. Organic and inorganic mercury compounds have been used as fungicides and insecticides, as seed, wood and animal hair pickling. In many areas the use of mercury is declining or is already discontinued or prohibited.

Toxicokinetics Vaporous elementary mercury is absorbed in the lung to approximately 80%. In the gastrointestinal tract, on the other hand, there is virtually no absorption of metallic mercury, so that the swallowing of small amounts, e.g. from broken thermometers, does not lead to intoxication. The absorption of inorganic mercury compounds after oral administration is in the range of 2–38% in animal experiments and depends on the actual species. Methylmercury compounds are absorbed after oral uptake in humans and in animal experiments to about 80%. Data for the inhalative absorption of these compounds are not available. After exposure to inorganic mercury compounds, an accumulation takes place mainly in the kidneys but also in the liver. In contrast to inorganic mercury compounds, metallic mercury as well as methylmercury compounds can pass the placenta and the blood–brain barrier by virtue of its high lipophilicity. The excretion of metallic mercury is by exhalation, but also, like the excretion of inorganic mercury, via urine and feces. Methylmercury is excreted as Hg(+2) with a half-life of 70–80 days in humans, mainly via the feces.

Toxicity Acute inhalative exposure to high mercury vapor concentrations is associated with lung toxicity. Chronic human exposure to mercury vapor affects the central nervous system, with characteristic forms of tremor, as well as psychological and neurological changes. Acute poisoning of a large number of people with more than 450 deaths due to exposure towards organic mercury compounds occurred in Minamata (Japan) around 1960 and in Iraq (1971/72). The consumption of methylmercury-contaminated fish in the Minamata region caused a variety of neurological symptoms such as paresthesia, visual and hearing disorders, indistinct language, uncertain gait, and loss of memory. The reason for the poisoning in Iraq was the use of methylmercury-treated seeds as bread cereals.

The neuronal development of fetuses and newborns is particularly sensitive to even low concentrations of organic mercury compounds. The high neurotoxicity of inhaled elemental mercury and organic mercury compounds is related to the high lipophilicity. Comparatively high concentrations of both species enter the bloodstream and can also cross the blood–brain barrier.

> **Once inside the cell, metallic mercury and alkylmercury compounds are metabolized or oxidized to Hg^{2+}, which exerts a high affinity to SH groups, thereby, for example, damaging the microtubules of neuronal cells.**

Inorganic mercury compounds cross biological membranes to a lesser extent, which affects the mercury distribution in the body and the patterns of damage. Target organs for the oral intake of inorganic mercury compounds are the gastrointestinal tract and the kidneys. Mercury(I) compounds disproportionate to Hg^{2+} and metallic mercury in the plasma and exhibit corresponding toxic effects. Mercury and inorganic mercury compounds have a sensitizing effect on the skin. For both organic and inorganic mercury compounds there is a suspicion of a carcinogenic effect due to kidney tumors in male mice as well as clastogenic effects *in vitro* and *in vivo*. Discussed mechanisms include interactions with proteins of the spindle apparatus, inhibition of DNA repair processes or other enzymes involved in DNA processing as well as the formation of ROS.

Limit Values and Classifications EFSA has set a TWI for methylmercury of 1.3 µg/kg body weight based on studies on developmental neurotoxicity in humans and for inorganic mercury of 4 µg/kg body weight based on renal toxicity in rats.

Mercury and its inorganic and organic compounds are classified by the MAK Commission in carcinogenicity category 3B. The MAK value for mercury and inorganic mercury compounds is 0.02 mg/m^3, derived from the correlation to mercury concentrations in the urine at which no nephrotoxic and neurotoxic effects occur. A MAK value for organic mercury compounds is not available.

Chromium

> **Chromium(VI) but not chromium(III) compounds are carcinogenic. While Cr(VI) is able to cross biological membranes, they are largely impermeable to Cr(III). Within the cell, however, Cr(VI) is reduced to Cr(III), which is ultimately involved in the formation of DNA damage.**

Occurrence and Exposure Chromium is a transition metal and occurs in the oxidation states 0 to +6, among which 0, +3 and +6 represent the most stable forms. Naturally occurring chromium is present as Cr(III). In food, chromium is predominantly found as Cr(III), but in drinking water, relevant amounts can be present as Cr(VI). The release of chromium into the environment is mainly due to anthropogenic sources. For the general population, food is the main chromium source. The highest chromium levels of up to 200 µg chromium/kg are found in meat, fish, fats, and oils as well as in bread, nuts, and cereals. Commercially, chromium is used mainly in the metal industry, for example for the production and application of stainless steels. Cr(III) and Cr(VI) are further used as pigments, for example the light-resistant and insoluble red and yellow pigments used for traffic signs and safety vests, but also in the production of leather (Cr(III)) and as a wood preservative (Cr(VI)).

Toxicokinetics The absorption of chromium compounds after inhalative exposure depends on the particle size, the solubility, and the oxidation state. Cr(VI) compounds usually are more efficiently absorbed from the lung than Cr(III) compounds. The absorption of chromium compounds after oral uptake is also dependent on the solubility and the oxidation state; available studies indicate higher absorption of Cr(VI) when compared to Cr(III). Even though the absorption of Cr(III) from the diet is low, at up to 2.5%, it covers the essential requirements. Also the dermal penetration is relevant for Cr(VI), but very small for Cr(III). The distribution of absorbed chromium occurs in almost all tissues, but highest concentrations are reached in kidneys and liver. The excretion occurs primarily with the urine.

Essential and Toxic Effects As an essential trace element, chromium is required for carbohydrate, fat and protein metabolism by its impact on insulin activity; however, the exact mechanism of action is still unclear. An oligopeptide complex with four Cr(III) ions, also referred to as chromodulin, is postulated to be involved in chromium transport in the body.

Concerning toxic effects, a clear distinction must be made between Cr(VI) and Cr(III) compounds. While Cr(VI) compounds are carcinogenic in humans and in animal experiments, no carcinogenicity is anticipated for Cr(III) and its toxicity is significantly lower, presumably due to differences in bioavailability. While Cr(VI) compounds are absorbed into cells via anion channels and lead to damage to cellular macromolecules, including DNA, after intracellular reduction to Cr(III), biological membranes are largely impermeable for Cr(III) compounds. With regard to dietary exposure, Cr(III) prevails in food where oxidation to Cr(VI) is not likely, therefore toxic effects are unlikely through a normal diet. In the case of occupational exposure to Cr(VI) compounds, however, increased tumor frequencies in the lung and nasal sinuses were observed, e.g. in chromate production, chromate pigment production, and electroplating. The mechanisms of chromate-induced carcinogenesis are comparatively well understood. Thus, after the uptake via ion channels, Cr(VI) is reduced intracellularly via different intermediate steps to Cr(III). In the course of this reduction by ascorbate and/or glutathione, reactive intermediates are formed depending on the reducing agent, which subsequently lead to different types of DNA damage, such as DNA adducts, oxidative DNA base damage, DNA strand breaks, and DNA-protein cross-linking. Under physiological conditions, ternary Cr(III) ascorbate DNA adducts with a high mutagenic potential have been identified, which also lead to repair-deficient cell populations. A further mechanism of action is the induction of aneuploidy, which can lead to cell transformations. Furthermore, chromium compounds, in particular Cr(VI), have contact-sensitizing properties which have led to increased contact dermatitis, especially in cement workers. The introduction of chromate-poor cement has led to a reduction in chromate sensitization.

Limit Values and Classifications Chromium(VI) compounds are classified as carcinogenic in humans by both the IARC and the MAK Commission (Category 1). To maintain the essential functions, the estimated values for adequate intake are 30–100 µg Cr/day, with no evidence of insufficient supply. The EFSA has derived a TDI of 0.3 mg/kg body weight and day for Cr(III). The starting point for the derivation was the highest dose in a chronic oral toxicity study in rats, where no effects were observed.

Copper

> **Copper is an essential trace element with high toxicity under overload conditions due to the increased formation of ROS.**

Occurrence and Exposure Copper is a transition metal which can be present in the oxidation states 0, +1 and +2. In biological systems, Cu(I) and Cu(II) are prevalent, and respective redox reactions are required for its essential functions in several enzymes. Comparatively high copper contents are found, for example, in grain products, innards (liver and kidneys), fish, shellfish, legumes, nuts, cocoa, and chocolate. Relevant exposures can also occur via drinking water, in particular when delivered via copper pipes at acidic pH. Relevant exposure occurs also at the workplace. Here, copper is used in many technical areas in metallic form or as a component of alloys such as brass or bronze. In addition, copper compounds are used in fertilizers, wood preservatives, and fungicides.

Toxicokinetics The human body has a very well-regulated system of copper homeostasis, which includes regulation of gastrointestinal intake, protein binding during transport and intracellular storage as well as excretion via the bile. The absorbed fraction from the stomach and the duodenum amounts to 35–70%. After absorption, copper is bound to proteins and low molecular weight ligands and transported to the liver. Here, it is partially stored, incorporated into copper-dependent enzymes and released into the plasma bound to ceruloplasmin. The total copper content in the body of an adult is estimated to be 50–150 mg, of which approximately 40%

is located in the musculature, 20% in the skeleton, 15% in the liver, 10% in the brain, and 6% in the blood. The excretion of copper occurs mainly via the feces.

Essential and Toxic Effects Copper is a constituent of many metalloproteins, where it is involved in electron transfer reactions due to its redox activity. Copper-containing enzymes are of essential importance for the cellular energy metabolism (respiratory chain) as well as for the synthesis of melanin, of connective tissue, and of neuroactive peptide hormones such as catecholamines. In addition, the copper-containing enzymes ceruloplasmin and hephaestin are directly involved in iron metabolism due to their ability to oxidize iron. In the nervous system, copper is also required for myelin formation. However, under overload conditions, copper exerts pronounced toxicity, mediated by the increased formation of ROS generated via Fenton-type reaction, and with the consequence of lipid peroxidation as well as DNA and protein damage. In order to enable essential functions and to prevent toxic reactions to a large extent, the body has a well-regulated system of copper homeostasis. If the capacity of this homeostasis is exceeded, this can lead to serious health damage. This becomes particularly evident in the case of genetic metabolic disorders such as Menkes syndrome, which disrupts gastrointestinal copper absorption, and Wilson syndrome, in which copper excretion from the liver is disturbed due to the diminished activity of the Wilson's ATPase, which transfers copper to ceruloplasmin and excretes copper via the bile. In the latter case, severe liver damage occurs due to copper overload as well as damage to the central nervous system and the eyes. Infants are particularly susceptible to increased copper intake, since the biliary copper excretion is not fully developed within the first years of life. This may lead to premature liver cirrhosis when infant food is prepared with highly copper-containing water (e.g., German childhood cirrhosis).

Limit Values and Classifications The daily requirement of the essential trace element copper for children from 7 years, adolescents and adults is estimated to be 1.0–1.5 mg/day. This requirement is met in Europe, and consequently there is no risk of an undersupply. To prevent an oversupply, a tolerable upper intake level (UL) of 5 mg/day and for children between 1 and 4 mg/day depending on age was set by the European Commission's Scientific Committee on Food. This UL was based on a NOAEL of 10 mg/day, which showed no evidence of hepatic dysfunction after 12 weeks of administration to healthy volunteers. Concerning inhalative exposure, the MAK Commission set a MAK value of 0.01 mg/m^3 for the respirable fraction of copper and its inorganic compounds based on inhalation studies in rats and mice, to prevent lung toxicity.

Iron

> **Iron is an essential trace element that can provoke oxidative stress if the homoeostatic control is exceeded.**

Occurrence and Exposure Iron is the fourth most abundant element and the most common transition metal, and occurs primarily in nature in the form of oxidic and sulfidic ores. From these ores, iron can be obtained by reduction. In addition to the use of iron itself and in steel production, synthetic iron oxides are used, inter alia, as pigments in paints, lacquers, inks, and coatings.

In the organism, iron is present in the oxidation states +2 and +3 and functions as a reducing or oxidizing agent. Iron-containing proteins include those involved in oxygen transport and storage, hemoglobin and myoglobin, as well as iron–sulfur proteins, for example within the respiratory chain. Iron-rich foods of animal origin are pig liver, liver sausage and beef, and with respect of foods of plant origin some vegetable varieties and cereal products provide iron for essential functions.

Toxicokinetics The resorption of iron after oral uptake occurs in particular in the duodenum and upper jejunum. It is influenced by several factors and is also regulated as a function of physiological needs. The bioavailability depends on the iron species, thus heme iron possesses a comparatively high bioavailability with approximately 15–35% due to a specific heme receptor, while only about 1–15% is absorbed in case of non-heme iron, derived mainly from plants. The latter is usually found as trivalent iron which is reduced to Fe(II) either by vitamin C or the membrane-specific ferric reductase in order to be absorbed into the intestinal cells in a gradient-dependent co-transport (with H^+) from the DMT1 transporter. Since other divalent ions such as cadmium can also be absorbed via the same transporter, with increased cadmium exposure, an impairment of the iron absorption may occur. In the enterocytes, iron is oxidized to Fe(III) and stored in ferritin or bound to transferrin and transported via the blood to other tissues. The excretion of iron from enterocytes into the blood is mediated via ferroportin, which in turn is regulated by hepcidin. Thus, at high iron levels, the hepcidin synthesis is upregulated and thus the release of the iron is inhibited. At low iron levels, synthesis of hepcidin is inhibited and thus iron is made available. Iron remaining in the intestinal cells is excreted with desquamated cells via the feces. Transferrin contains two iron binding sites; it is taken up via receptor-mediated endocytosis and within the acidic environment of the lysosomes, iron is released from transferrin, reduced to iron(II), and transported out of the lysosomes. Subsequently, transferrin together with its receptor is recycled via exocytosis.

Approximately 70% of the total iron in the organism is present in hemoglobin and approximately 10% each in myoglobin and ferritin; the rest is located in the cytochromes of the respiratory chain, the endoplasmic reticulum (Phase I enzymes), and small amounts bound to transferrin. High tissue concentrations are found mainly in liver, spleen and bone marrow. Notably, the iron body content is controlled exclusively by absorption; no regulation occurs at the level of iron excretion.

Essential and Toxic Effects Iron is the most common trace element of the human organism. It is essential for humans, animals, and plants. The biologically most important forms are Fe(II) and Fe(III). Since the Fe(III) hydroxide present in aqueous solutions is sparingly soluble, organisms use iron-binding proteins and chelators to make iron bioavailable. As a component of enzymes, iron is, for example, involved in the transport of oxygen from the lung into the target tissue via hemoglobin, in the storage of oxygen in the muscles via myoglobin, and in the electron transport chain of oxidative phosphorylation via iron-sulfur proteins and cytochromes as well as in the metabolism of xenobiotics.

> **Toxic effects of iron depend, like the essential functions, on the redox activity of iron. Elevated levels of "free" or loosely bound iron can lead to damage to proteins, nucleic acids, lipids and cellular membranes via the formation of ROS (Fenton reaction, Haber–Weiss reaction) and promote the growth of cancer cells. In order to facilitate essential and prevent toxic reactions, organisms have evolved a complex system of iron homeostasis, in which iron is transported in the blood bound to transferrin, taken up into and excreted from cells in a controlled manner, and stored intracellularly in ferritin. If the storage capacity is exceeded, serious health damage may occur.**

This is particularly evident in the case of hereditary haemochromatosis, in which the regulation of intestinal iron released from the mucosal cells is disturbed and the accumulation of iron in important organs occurs. A major cause is the disturbed regulation of hepcidin synthesis, due to a mutation in the HFE gene. Respective homozygous carriers of this mutation exert severe organ toxicity, including liver cirrhosis.

Acute iron poisoning in non-genetically predisposed persons occurs predominantly when an overdose of iron-containing medication is ingested, for example accidental intake of iron tablets by children. Acute toxic effects are seen at doses between 20 and 60 mg iron/kg body weight; doses above 180 mg iron/kg body weight can be fatal. Consequences of acute iron intoxication are bloody vomiting, bloody diarrhea, heart failure, liver necrosis with organ failure, coagulation disorders, hypoglycaemia, lethargy, coma, and convulsions. Similarly, chronic overdosed intake of iron preparations (150–1200 mg/day) may lead to liver cirrhosis, diabetes mellitus, and heart failure. There are indications for an increased risk of cancer for patients with hemochromatosis, and a link between high iron stores and cancer has been discussed. The epidemiological data on increased cancer risk in exposed workers are, however, contradictory. Even though epidemiological studies have revealed an increased risk of lung cancer in workers exposed to iron oxide, in all cases there was simultaneous exposure to other potentially carcinogenic metal compounds and/or polycyclic aromatic hydrocarbons. Mechanistically, an increased tumor risk can be explained by persistent oxidative stress, including accelerated levels of oxidative DNA damage when the controlled uptake and storage of iron is exceeded.

Limit Values and Classifications The recommended intake is 15 mg/day for women before menopause, otherwise for adults 10 mg/day. A daily intake of 30 mg of iron is recommended during pregnancy. Even though the higher intake recommendations for women are not fully achieved, there is no undersupply of iron in the industrialized nations, in contrast to the so-called developing countries. The upper limits derived by different bodies with regard to iron as food supplements and for iron-enriched foodstuffs vary. Since the range between optimal uptake levels and potential oversupply is quite narrow, for reasons of preventive health protection the German BfR, for example, recommends not to use iron in food supplements as well as not to add iron to conventional foods.

Manganese

> **Manganese is an essential trace element, but is neurotoxic upon inhalation at fairly low concentrations.**

Occurrence and Exposure Manganese occurs as a transition element in a variety of minerals in the oxidation states –3 to +7. In biological systems, it is predominantly present as Mn(II) and Mn(III). Regarding non-occupationally exposed persons, manganese is mainly taken up via food. Particularly high values are found in cereals, rice and nuts.

Industrially, manganese is used for the reduction and desulfurization of iron and steel as well as a component of alloys. It is also used in the production of pigments, in corrosion protection, in some countries as an additive to fuels, in drinking water treatment, in data processing technology, and as an additive to feed and fertilizers. Significant occupational exposure occurs during welding, cutting, and grinding of manganese-containing steels.

Toxicokinetics After oral uptake, absorption in the gastrointestinal tract in humans is around 3–5% and is homeostatically controlled. Quantitative data on the systemic availability of manganese after inhalation are not available. Depending on the particle size it can be assumed that inhaled manganese is absorbed in the nasal mucosa, in the lungs, and in the gastrointestinal tract. Some data suggest that ultrafine manganese oxide particles may also translocate directly via the olfactory nerve to the olfactory bulb into the nervous system. The excretion of manganese occurs mainly with the bile and feces.

Essential and Toxic Effects Manganese is a trace element and cofactor for various metabolic enzymes, such as the pyruvate carboxylase, as well as the superoxide dismutase within the antioxidative defense system. Deficiencies in humans are very rare. The main chronic toxic effect of manganese, which may occur after occupational exposure via inhalation, but also after excessive oral intake, is neurotoxicity evoked by damage to the central nervous system. The "manganism" is characterized by tremor, muscle stiffness, slowed movements, and gait disturbances, but also by neuropsychological disturbances and personality changes. The underlying mechanism of manganese-induced neurotoxicity has not yet been fully elucidated; however, manganese has been found to induce oxidative stress, exacerbate mitochondrial dysfunction, dysregulate autophagy, and promote apoptosis, ultimately enhancing neurodegeneration. Chronic inhalation can also lead to damage to the respiratory tract (manganese pneumonia).

Limit Values and Classifications For manganese, there are no reliable data available on the required intake, therefore only estimates for adequate intake are provided. They range from 2.0 to 5.0 mg/day for adolescents and adults. Up to now, no upper limit for oral intake has been established by the European Commission's Scientific Committee on Food (SCF) or EFSA due to a lack of human data and a lack of an NOAEL in animal studies with respect to neurotoxicity. The occurrence of preclinical neurotoxic effects was considered by the MAK Commission as the most sensitive endpoint for the derivation of workplace exposure limits for manganese. From a series of epidemiological studies at work, MAK values were derived of 0.2 mg/m^3 for the inhalable and 0.02 mg/m^3 for the respirable fraction.

Nickel

Nickel compounds are carcinogenic, sensitizing, but also toxic to the lungs after inhalation.

Occurrence and Exposure The transition metal nickel can be present in various oxidation states from (I) to (IV), with the oxidation state (II) being the most important in biological systems. Nickel is used mainly as a component of stainless steels and other alloys in plating, in the production of batteries, in pigments, and as a catalyst. Given the manifold applications, occupational exposure towards nickel is widely distributed, mainly via inhalation and/or skin contact.

Nickel intake of the non-occupationally exposed population occurs mainly via food; particularly nickel-rich foods are those of plant origin such as nuts, chocolate and cocoa powder, legumes, and cereals.

Toxicokinetics After inhalation exposure, approximately 20–35% of the nickel deposited in the lung is absorbed into the blood, depending on the solubility of the nickel compound. Thus, after exposure to soluble nickel compounds, such as nickel chloride and sulfate, higher nickel concentrations were found in the urine of workers than after exposure to poorly soluble compounds (nickel oxide, nickel subsulfide).

The absorption from the gastrointestinal tract in humans is between 1 and 40%, depending on the solubility of the nickel compound and the food matrix. Nickel can also be absorbed through the skin. After absorption, nickel ions bind to serum proteins and are distributed in the organism. Excretion of absorbed nickel occurs with the urine.

Essential and Toxic Effects Even though nickel is considered as a trace element, nickel-dependent functions have not yet been identified in humans. In animal experiments, however, nickel deficiency causes growth limitations and disorders of glucose metabolism and methionine synthesis. Inorganic nickel compounds are carcinogenic to humans. Elevated pulmonary and nasal cancer incidences were observed in several epidemiological studies, with both water-insoluble and water-soluble nickel compounds being carcinogenic. Nickel compounds are also carcinogenic in experimental animals; the carcinogenic potential, however, depends significantly on the respective species. Compounds of medium water solubility and average toxicity such as nickel sulfide (NiS) and nickel subsulfide (αNi_3S_2) are among the strongest known carcinogens, while both water-soluble and sparingly soluble nickel compounds show weaker effects. Differences in the uptake into the cells appear to be decisive: while soluble nickel compounds can only slowly pass cell membranes via metal transporters, largely water-insoluble, crystalline particles are phagocytized and dissolve in the lysosomes due to the acidic pH, thus releasing high amounts of nickel ions inside the cell. The ultimate damaging agent is the Ni^{2+} ion at all solubility levels. Metallic nickel was carcinogenic in experimental animals only after intratracheal instillation, but not after inhalation. Nickel-induced carcinogenicity appears to be mainly due to indirect mechanisms of action. Nickel compounds are not or only weakly mutagenic, but show distinct co-mutagenic properties in combination with other DNA-damaging agents due to the inactivation of DNA repair systems. In addition, elevated levels of oxidative DNA damage have been observed, predominantly due to an inhibition of their repair. Finally, altered gene expression of oncogenes and tumor suppressor genes have been attributed to changes in the DNA methylation pattern. In addition to carcinogenicity, all nickel compounds, including metallic nickel, show pronounced lung toxicity at low concentrations in animal inhalation studies. After excess oral intake, hepatic and renal damage occurred in animal experiments. Nickel poisoning by the consumption of food is not known. Nickel is the most commonly diagnosed cause of allergic contact dermatitis worldwide. After skin contact with nickel-containing jewelry and everyday items such as coins and trouser buttons, contact eczema can occur as a result of an allergic reaction of type IV in the epidermis, which can worsen with the consumption of nickel rich foods.

Limit Values and Classifications An adequate intake of 25–30 µg nickel/day has been estimated; since the actual intake is far higher, an undersupply is not to be expected. A TDI of 2.8 µg nickel/kg body weight was derived by the EFSA.

The Scientific Committee for Occupational Exposure Limits (SCOEL) of the European Commission has evaluated nickel compounds as a carcinogen with a practical threshold because of the indirect genotoxic mechanism of action. Based on data for lung toxicity from animal studies, an occupational exposure limit of 0.005 mg/m^3 was set for the respirable fraction of metallic nickel and for poorly soluble nickel compounds. In addition, a limit value of 0.01 mg/m^3 for the inhalable fraction of soluble and poorly soluble nickel compounds was derived to prevent from carcinogenicity.

6.2.5 Summary

Metals occur everywhere in nature. Since, unlike most organic chemicals, they are not degraded or destroyed, their anthropogenic use leads to a permanent enrichment in ecological circuits. The general population is mainly exposed via food. With the exception of arsenic and perhaps cadmium, dietary exposure is usually not associated with adverse health effects; nevertheless, toxic levels may be reached by an oversupply of dietary supplements. Workplace exposure occurs predominantly via inhalation, which can lead to organ toxicity such as chronic inflammation in the lung and/or carcinogenicity. Regarding molecular mechanisms of carcinogenicity, with the exception of chromium(VI), direct interactions with DNA are of limited if any relevance. Rather, indirect genotoxic mechanisms are assumed, such as increased formation of ROS, inactivation of DNA repair processes, changes in gene expression, and signal transduction. As underlying interactions, the binding to or the oxidation of SH groups often is very important, rendering, for example, zinc-binding structures in DNA repair proteins, transcription factors, and tumor suppressor proteins highly susceptible molecular targets.

Further Reading

ATSDR *(Agency for Toxic Substances and Disease Registry) Toxicological profiles*. US Department of Health and Human Services, Public Health Service, Agency for Toxic Substances and Disease Registry, Atlanta, GA, USA, http://www.atsdr.cdc.gov/toxprofiles.

Beyersmann D, Hartwig A (2008) Carcinogenic metal compounds: recent insight into molecular and cellular mechanisms, *Arch Toxicol* **82**: 493–512.

EFSA, Scientific Opinions, http://www.efsa.europa.eu/de/publications/efsajournal.

EFSA (2006) Tolerable Upper Intake Levels for Vitamins and Minerals by the Scientific Panel on Dietetic products, nutrition and allergies (NDA) and Scientific Committee on Food (SCF), http://www.efsa.europa.eu/sites/default/files/assets/ndatolerableuil.pdf.

Hartwig A (2013) Metal interaction with redox regulation: an integrating concept in metal carcinogenesis? *Free Radic Biol Med* **55**: 63–72.

Henschler D, Greim H, Hartwig A (Editors) *MAK Value Documentations*, Wiley-VCH, Weinheim, Germany, http://onlinelibrary.wiley.com/book/10.1002/3527600418/topics.

IARC, IARC monographs on the evaluation of the carcinogenic risk of chemicals to humans, IARC, Lyon, France, http://monographs.iarc.fr/ENG/Monographs/PDFs/index.php.

6.3 Toxicology of Fibers and Particles

Paul J.A. Borm

6.3.1 Introduction

Historically, workers in the mining, sanding, wood, construction and agricultural industries were exposed to high concentrations of particles. Massive exposure to inhalable and respirable particles over a period of years at more than 5 mg/m^3 is known to cause chronic obstructive pulmonary diseases (COPD) including bronchitis, emphysema, fibrosis, and pneumoconiosis, and lung cancers. In coal mines dust and asbestos induce pneumoconiosis. Reducing particle exposure decreases, but does not eliminate these disorders. Particle-induced lung diseases recognized more recently include allergic and non-allergic asthma caused by a wide variety of occupational allergens. During the past decade a series of epidemiological studies have shown that ambient exposure to particles which display a mass mean diameter of 10 μm (PM_{10}) is related to both respiratory and cardiovascular mortality and morbidity. Observations made on the workplace environment and experimental studies have led to the appreciation that particle components such as transition metals and ultrafine particles play a role in the induction of pulmonary oxidative stress and inflammation. These processes are considered crucial to many of the acute effects of ambient particle exposure. Furthermore, it is currently thought that ambient particulate matter (PM) with a mean diameter of 2.5 or 10 μm ($PM_{2.5}$, PM_{10}) can be carcinogenic, but it is not clear which components or characteristics of PM are responsible.

6.3.2 Particle Toxicology: Basic Concepts

> To understand particle effects, one can apply the "D concept", i.e. the Dose, Deposition, Dimension, Durability and Defense (Table 6.11). This paradigm builds a conceptual understanding of the adverse effects of particles and fibers.

Particles are a special case in toxicology. They are widespread in many different occupational and environmental situations, as well as in consumer products such as cosmetics, paints, etc. An overview of particles, their most important applications and/or exposure situations and effects is given in Table 6.12. There are several characteristics which distinguish the toxicological properties of particles and soluble chemicals. First, if a particle is insoluble only its surface will interact with the biological environment. For example, a fly-ash particle with a hazardous bulk composition of heavy metals and quartz may be totally inert, since it is covered by an inert amorphous glass layer. This cover does not allow leaching or contact of the potential toxic constituents with the biological environment. Second, particles do not distribute homogenously throughout tissue compartments or even cells, in contrast to lipophilic agents such as solvents or drugs. Therefore, it is difficult to define the effective dose within cells. A single 5-µm particle in a cell may exert different actions in the same cell, caused by the fact that one side of the particle interacts with the cell membrane and the other part with mitochondria or the nucleus. Third, particles may carry other agents such as absorbed polycyclic aromatic hydrocarbons (PAHs), gases such as SO_2, bacteria, or proteins into the lung, thereby changing the deposition and bioavailability of the absorbed contaminants.

Particle Deposition

> Particle size largely determines where particles are deposited in the respiratory tract. The major fraction of fine (<2.5 µm) and ultrafine (<100 nm) particles is deposited on the fragile epithelial structures of the gas exchange region. Larger particles are deposited in the nasopharyngeal region (>30 µm) and tracheobronchial region (10 µm < diameter < 30 µm). Particle deposition is further determined by shape and density.

Table 6.11 Processes and particle parameters that play a major role in determining the toxic response upon particle inhalation.

Dose	Cumulative dose for chronic effects; can be based on particle or fiber mass, fiber or particle number, or particle surface dose
	Bulk composition is not equal to surface
Dimension	Size (diameter, length)
Deposition	Dependent on dimension but also on airway properties (hot spots)
Durability	Biopersistence dependent on defense as well as particle properties (dissolution)
Defense	Mucociliary clearance, macrophage clearance, inflammatory cells
	If macrophage clearance is saturated, overload occurs; dose increases exponentially with time

Table 6.12 Various sources of particles, exemplary exposure situations, and the current knowledge of their toxicological properties and effects.

Particle type	Source	Example(s)	Who is exposed?	Effects
Crystalline silica	Quarrying	Coal mine dust, quartz flours, Kieselguhr	Miners	Silicosis, lung cancer
Coal mine dust	Coal mining		Miners	Pneumoconiosis
Asbestos	Mining, insulation	Crocidololite, chrysotile	Insulators, shipyard workers	Asbestosis, mesothelioma, lung cancer
Organic dusts	Agriculture	Grain, cotton, flour	Bakers, cotton workers	Asthma, COPD, allergic alveolitis
PM_{10}, $PM_{2.5}$, TSP[a]	Traffic, industry	Fly-ashes, diesel, sea salt, road dust	Everyone	Increase mortality from cardiovascular and respiratory causes
Man-made mineral fibers	Industry	Rockwool, ceramics	Occupational	
Organic synthetic fibers	Industry	Aramid, polyethylene	Occupational	
NP (<100 nm)				
Combustion-derived NP	Combustion	Diesel soot	Everyone	Pulmonary and CV effects in humans
Bulk manufactured NP	Combustion	Carbon black	Occupational	Fibrosis and lung cancer in rats
	Synthesis	Amorphous silica, TiO_2	Occupational	
	Geological/synthesis	Alumina	Occupational	
Engineered NP	Nanotubes	Combustion	Occupational	Lung damage rats
	Q-dots	Sytnthesis	Patients	Hepatic damage

[a] Definitions used in environmental particle control.
COPD, chronic obstructive pulmonary disease; PM_{10}, particulate matter with diameter smaller than 10 µm; $PM_{2.5}$, particulate matter with diameter smaller than 2.5 µm; TSP, total suspened particles; NP, nanoparticles; CV, cardiovascular.

The cumulative dose at a specific pulmonary site determines the adverse effects of particles. The deposited dose is dependent on the inhaled concentration and the dimensions of the particle. Smaller particles increase the probability of particle deposition in the respiratory tract. A major fraction of the fine (<2.5 µm) and ultrafine (<100 nm) particles will be deposited onto the fragile epithelial structures of the gas exchange region. Larger particles will be deposited in the nasopharyngeal region (>30 µm) and tracheobronchial region (10 µm < diameter < 30 µm). As illustrated in Figure 6.5 the particle size largely determines where particles will be deposited in the respiratory tract. The terminology given in Figure 6.5 includes the inhalable and respirable fractions, which are important definitions in the standard setting for occupational and environmental exposure limits.

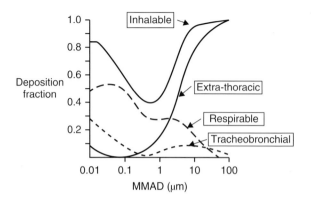

Figure 6.5 The relationship between particle size and the efficiency of deposition in various anatomical compartments of the lungs. The fractions depicted in the deposition curves (e.g. respirable fraction) are conventions used for design of sampling devices and setting of occupational standard limits for different particles. MMAD, mass median aerodynamic diameter.

The processes driving deposition are impaction of heavier particles, sedimentation in alveoli, interception of fibers, and, rarely, electrostatic interactions with airway wall constituents. Small ultrafine particles have little mass and behave by diffusion. This causes most of them to be deposited in the nasal compartment and the alveoli.

Shape and Density

The size for spherical particles is simply defined by the diameter. For non-spherical particles, the term "diameter" does not appear to be strictly applicable because it does not properly describe the geometry of a flake of material or a fiber. Also, particles of identical shape can be composed of different chemical compounds and, therefore, have different densities. To provide a simple means of categorizing particles and fibers of different shape and density the term "aerodynamic diameter" has been introduced, which is defined as the diameter of a spherical particle having a density of 1 gm/cm^3 that has the same inertial properties. Figure 6.6 explains the concept of aerodynamic diameter. Particle density affects the motion of a particle through a fluid and is taken into account in Equation 6.8. The Stokes diameter for a particle is the diameter of a sphere that has the same density and settling velocity. It is based on the aerodynamic drag force caused by the difference in velocity of the particle and the surrounding fluid. For smooth, spherical particles, the Stokes diameter is identical to the physical or actual diameter. The aerodynamic diameter for all particles greater than 0.5 μm can be approximated using the following equation:

$$d_{pa} = d_{ps} \sqrt{\rho_p} \tag{6.8}$$

where d_{pa} is the aerodynamic particle diameter in μm, d_{ps} is the Stokes diameter in μm, and ρ_p is the particle density in gm/cm^3.

Most dosimetry models and calculations assume a uniform deposition at the bronchial or alveolar surfaces and, therefore, assume a similar target dose for all epithelial cells within the respiratory tract. However, analysis of particles in human lung tissue as well as mathematical modeling show that local particle deposition occurs at the bronchial airway bifurcations independent of the

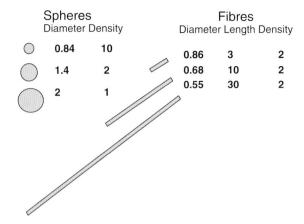

Figure 6.6 *Illustration of the concept of aerodynamic diameter by a set of spherical and fiber-shaped particles all with the same aerodynamic diameter of 2 µm. A small particle with a high density virtually behaves as a bigger particle. On the other hand, fiber length does not really affect aerodynamic diameter. This latter explains why long fibers pose a risk to the lower airways.*

particle size. Moreover, major differences in particle deposition can be found between lung lobes. In particular, cells located in the vicinity of the dividing spur may receive local doses that are a few hundred times higher than the average dose for the total airway. It is no coincidence that it is at these so-called *hotspots* located at the broncho-alveolar bifurcations that tumors related to exposure to inhaled asbestos or other particles is usually seen.

Particle Clearance and Lung Overload The lung has potent defense systems to remove particles. These include mucociliary clearance in the upper airways and macrophage clearance in the lower airways and the alveoli (Figure 6.7). Particle transport by macrophages from the alveolar region towards the larynx is rather slow in humans, even under normal conditions, and eliminates about a third of the deposited particles in the peripheral lung only. This implies that the other two-thirds accumulate in the lower lungs without significant clearance unless the particles are biodegraded and cleared by other mechanisms. In the course of inhalation studies high concentrations of particles, e.g. 1–3 mg or 0.1 µl per g of lung in rats, may exceed the macrophage clearance capacity and overburden the lung, a phenomenon known as lung overload.

> **Lung overload results when deposition exceeds elimination. It is the consequence of exposure that results in a lung burden of particles that is greater than the steady-state burden predicted from the deposition rates and clearance kinetics of particles inhaled during exposure. The overload concept has important implications for hazard assessment as well as for setting occupational standard values for particles when based on the outcomes of animal studies.**

The cause of particle overload is impaired macrophage clearance function. Volumetric overloading of macrophages starts at 6% of normal alveolar macrophage (AM) volume and this was originally used to develop to the overload concept, as illustrated in Figure 6.8. Nowadays the surface dose is suggested to be a better indicator of biological responses occurring at overload, especially in explaining the effects of nanoparticles.

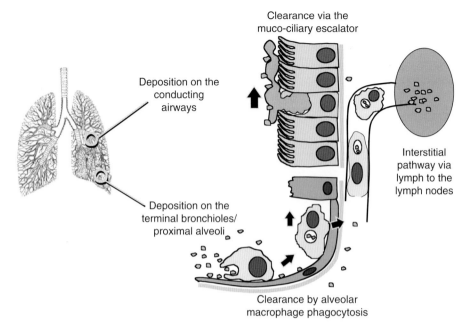

Figure 6.7 Clearance pathways as host defenses in the respiratory tract are associated with the anatomical compartments in the lung. Larger particles (>10 μM) predominantly deposit in the tracheo-broncial area, where the airways are covered with ciliated epithelial cells and particles are transported upwards through the mucociliary escalator. Smaller particles that penetrate into the broncho-alveolar area can only be removed by phagocytosis from alveolar macrophages. This clearance is much slower than in the upper respiratory tract.

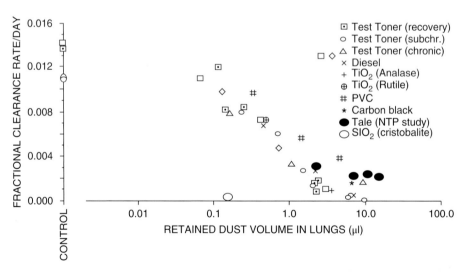

Figure 6.8 The clearance rate of deposited particles in the lung is dependent on the total volume of retained particles in the lung. The graph is adapted from Morrow (1992) and shows the relation between clearance rate of various so-called poorly soluble, low-toxicity particles (PSP) in the lung in relation particle volume (in μl) per gram of lung. Regardless of particle type, it is shown that beyond 0.1 μl saturation of clearance occurs. This limit is the basis for many exposure limits of particles.

At particle overload, macrophage clearance function is impaired and at this point particle accumulation starts and inflammatory cell influx increases sharply. It needs to be emphasized that this concept specifically applies to low-toxicity poorly soluble particles (PSP). Other more toxic particles, such as crystalline silica, man-made fibers and toxic metal particles, affect AM-mediated clearance as well, but at much lower lung burdens since they can actively damage AM. Nanoparticles have also been shown to impair phagocytosis to a much greater extent than larger particles when evaluated on an equal mass-basis (see Section 6.3.4). Thus, every impairment of AM-mediated particle clearance should not be viewed as a particle overload.

Particle Translocation

> **Particle translocation has only recently been recognized as a mechanism to explain possible systemic effects of inhaled ambient particles.**

Basically, inhaled particles can translocate via two different routes:

- Passage through the pulmonary epithelium and endothelium to enter the systemic circulation. This may occur through endocytosis, transcytosis or transport by immune cells.
- Passage to the central nervous system, more specifically the olfactory bulb or the cranial bulb, through the olfactory epithelium in the nose, or the trigeminal nerves in the upper airways.

Both routes seem to be limited to particle sizes under 100 or even 50 nm. Particle properties such as surface charge seem to be crucial in translocation through the lung barriers, while for transport along olfactory nerves size seems to be the main determinant. More details are discussed in *Nanoparticles*.

Inflammation and Oxidative Stress

> **One of the crucial properties of particles is their ability to generate oxidants such as reactive oxygen species (ROS) and reactive nitrogen species (RNS). This factor is generally thought to be involved in many pathological outcomes of particle exposure, including fibrosis and the proliferative effects of particles.**

One of the unifying paradigms in particle toxicology is the linkage between the generation of ROS and RNS, and the subsequent inflammatory response. The ability of particles to induce different endpoints in both lung and systemic organs seems to be associated with their ability to generate oxidants that overwhelms the endogenous antioxidant defense mechanisms and is, therefore, called oxidative stress.

Particle exposure can induce oxidative stress by two mechanisms: particles may have acellular oxidant generating properties themselves or they may stimulate cellular oxidant generation. Furthermore, these actions may be subdivided into primary particle driven and secondary inflammation driven formation of oxidants. The inflammation driven formation of oxidants is described here, while the particle driven formation is discussed under *Surface Reactivity*, this chapter.

Within the lung, various cell types, including vascular endothelial cells and lung epithelial cells, can endogenously generate ROS and RNS. For example, epithelial lung cells generate intracellular ROS upon exposure to fly ash, diesel exhaust particles (DEP), particulate matter (PM), and quartz particles, processes in which mitochondrial respiration and activation of NAD(P)H-like enzyme systems are involved. However, the pool of **inflammatory phagocytes** constitutes the most significant and important cellular ROS/RNS generating system in the lung. During *in vivo* particle exposure, increased levels of ROS/RNS can be found in pulmonary tissue, and are associated with the influx of inflammatory phagocytes. Consequently, environmental particles and fibers such as asbestos, crystalline silica, heavy-metal-containing dusts, oil fly ash, coal fly ash and ambient PM induce ROS production by neutrophils and macrophages *in vitro*. Several particle characteristics result in the activation of the phagocytic oxidative burst. For mineral dusts such as crystalline silica, it has been shown that ROS release from inflammatory cells is related to the physical dimensions and the surface-based radical-generating properties of the particles. In chemically complex particles such as PM or fly ash (Table 6.12), the presence of metals is related to ROS release. Finally, organic substances adsorbed on the particle surface of PM have been related to the oxidative burst in neutrophils.

Genotoxic RNS such as nitric oxide (NO) and peroxynitrite (ONOO) can be generated in phagocytes by virtue of their inducible nitric oxide synthase (iNOS) activity. Although neutrophils are considered to be the most potent phagocytes with respect to particle-related ROS generation, the major source of RNS in the lung is the alveolar macrophage (Figure 6.9).

> **Apart from producing ROS, inflammatory phagocytes produce growth factors such as cytokines and chemokines, which regulate cell migration, tissue remodeling, cell proliferation, and repair of damage. These mechanisms are induced by quartz, carbon black, and particulate matter.**

Particles activate NF-kB and other signaling pathways and cause increased release of various pro-inflammatory mediators such as IL-8, TNF-α, IL-6 and anti-inflammatory mediators as IL-10 and TGF-α. Such acute pro-inflammatory effects explain the exacerbations in COPD and asthma as well as cardiovascular events following air pollution episodes. Chronic inflammation helps to understand, detect, and treat long-term sequelae of particle exposure, including fibrosis,

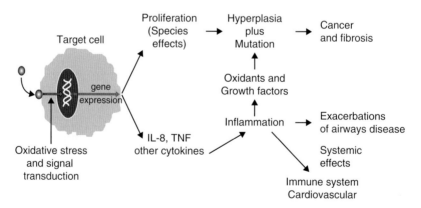

Figure 6.9 The pathways and mediators leading to the acute and chronic pathology induced by particle or fiber inhalation. Particles induce oxidative stress in target cells, which is amplified by an inflammatory response. Both oxidants and pro-inflammatory mediators may contribute to acute effects of particles (lower part of figure) as well as the chronic effects.

Table 6.13 Comparison of different indices for biodurability of carcinogenic and non-carcinogenic fibers.

Fiber	Carcinogen	$t_{1/2}$ (days)[a]	K_i[b]	K_d (ng/cm^2.hour)[c]	KNB[d]
RCF1, 2 and 3	+++	>60	>90	<6	<2
E-glass					
Crocidolite					
MMVF-21	−	30–60	<5	6–30	>18
MMVF-11	−	15–30	25	30–300	>18
Soluble fibers	−	<15	>30	>300	>18

[a] $t_{1/2}$ is the half-life (days) after inhalation or intratracheal instillation.
[b] K_i is an index calculated based on KNB + BaO + B$_2$O$_3$ − 2Al$_2$O$_3$;
[c] K_d is the measured dissolution rate in an acellular system.
[d] KNB is an index calculated as the sum of percentual composition of Na$_2$O + CaO + K$_2$O + MgO.

emphysema, and lung cancer. TNF-α release from phagocytes is related to the risk of developing pneumoconiosis. Polymorphisms in the TNF-α gene have been associated with elevated risk of silicosis. Treatment with TNF-soluble receptor is now being applied as a treatment for pulmonary fibrosis.

6.3.3 Particle Properties

The relationship between the types of particles to which workers are exposed and the lung diseases that they manifest is complex, for example pneumoconiosis can be caused by different types of particles. Defining the dose of a particle requires information on the durability of the particle and its surface area. The locus of action, which is usually the sight of deposition, is determined by factors that include the dimensions and shape of the particle and its inherent surface activity.

Durability

> **If a particle is not soluble or not degradable in the lung it has a high durability and there will be rapid local accumulation upon sustained exposure. Since durability, or bio-persistence, is a major determinant of fiber pathogenicity, it is now incorporated into test protocols to characterize and classify new fibers.**

Durability or bio-persistence is a major determinant of fiber pathogenicity. In Table 6.13 a comparison of different indices of bio-persistence or dissolution are given for various fibers. This shows that carcinogenic fibers have a high durability, as illustrated by long a half-life (>60 days), and a slow dissolution rate (<6 ng/cm^2 hour), which is reflected in the chemical fiber indices K_i and KNB. Since *in vivo* and *in vitro* bio-persistence assays are not always congruent, a tiered approach combining both simple and rapid bench-screening tests for *in vitro* dissolution and breakdown after intratracheal instillation in animal models is recommended. In addition, based on chemical composition a prediction can be made of dissolution rates of specific fiber classes (e.g. glass fibers) and correlations between dissolution rates and *in vivo* carcinogenicity allow the screening out of potentially harmful fibers.

Dimensions: Size and Shape

> **Fiber dimensions can have profound effects on the ability of the body to defend against particles and, thereby, on cumulative dose. Long (>20 μm) fibers are not taken up by**

> alveolar macrophages and, therefore, have longer half-lives in the lung than shorter fibers of the same material, consequently resulting in a higher cumulative dose at similar inhaled fiber number or mass. Inhaled fibers longer than 20 µm also show interception at the bronchiolar bifurcations due to their length and may achieve high local doses by this process.

For long fibers, which are not easily cleared by the immune system, durability (bio-persistence) is the main determinant of whether or not the fiber is carcinogenic in animal studies. Fiber bio-persistence *in vivo* and fiber dissolution *in vitro* are now used as screening methods for new synthetic fibers to select out potential durable and pathogenic products (see *Durability* (this chapter) and Table 6.13). The intent is to develop new fibers that are rapidly broken down and dissolve under biological conditions. It is recognized that after phagocytosis by AM short fibers are trapped in an acidic (pH 4.5) environment. Fibers that are highly stable under normal conditions become biodegradable upon deposition in the lung.

Surface Area

> A larger surface area adsorbs more material, which is carried by inhaled particles to their deposition sites.
> Smaller particles of a large surface area dissolve faster than larger particles.
> Acute (inflammation) and chronic (lung cancer) responses to particle treatment are better correlated to surface area dose than to mass dose or volumetric dose.

Surface area is a particle property that has to be considered apart from surface reactivity. Equations for calculating the volume and surface area of spheres are provided below, where r is the diameter of the sphere:

$$\text{Surface area of a sphere, } S = 4\pi r^2 \quad (6.9)$$

$$\text{Volume of a sphere, } V = 4\pi r^3/3 \quad (6.10)$$

The surface of the airborne particles adsorbs materials such as PAHs from the air and/or gases (see *Absorbed and Soluble Components*, this chapter). The adsorbed material could be toxic even if the particles are not. Alternatively, the adsorbed material could interfere with the normal chemical activity of the particles.

The larger the surface area, the more contaminating material reaches targets in the lung. On the other hand, stronger binding of adsorbed material on larger surfaces, mediated by van der Waals forces, diminishes the release of adsorbed components from the particle surface. An example is provided by PAHs absorbed on carbon blacks that only become biovailable from highly contaminated versions of small surfaced carbon black particles.

The time for dissolution is approximately proportional to the square of the particle size, which means much more rapid dissolution for smaller particles. This can be understood by looking at the surface/volume ratio of a sphere, which is inversely proportional to the diameter. For a particle of 6 µm the S/V ratio is 0.5 while for a 100 nm particle it is 30. The concentration at the site of deposition will be determined by fluid flow in the vicinity of the particle and by chemical reactions at the site. The faster the dissolution or reaction, the shorter the time before the material has been totally dissolved or removed.

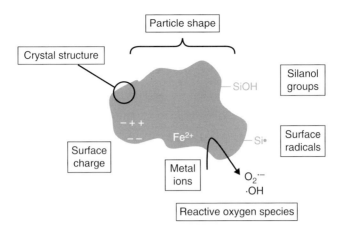

Figure 6.10 Illustration of the different properties that play a role in the reactivity of a particle, using a crystalline silica particle as an example. Reproduced from Schins & Borm, in press in Occupational Lung Disease: An international perspective, 2nd edition (Banks, Parker, White & Lindale), Common cellular and molecular mechanisms in the development and progression of pneumoconiosis.

Empirical observations that show that both acute (inflammation) and chronic (lung cancer) responses to particle treatment are better correlated to surface area dose than to mass dose or volumetric dose. This concept has been mainly developed with PSP and considering its relevance for dose–response curves, it is expected to impact on standard setting, compliance measurements, and hazard identification. For more detailed discussions on this issue see Oberdörster (2001), Borm et al. (2004), and Tran et al. (2000).

Surface Reactivity

Differences in particle toxicity can be related to their surface reactivity and ability to induce oxidative stress either directly or by inducing inflammation.

As already indicated above, one of the crucial concepts in particle toxicity is the ability to induce oxidants. Various physicochemical properties of particles that play a role in oxidant generation have been characterized. Particles such as crystalline silica generate oxidants as a function of their reactive particle surfaces (Figure 6.10). Selective blockade of the active quartz particle surface with polymers (PVNO) or metals (Al) is associated with a reduction in its ROS-generating capacities and toxicity. On the other hand, freshly ground dusts, in which the reactive particle surface is increased, not only contain more surface-bound reactive groups but are also more potent in generating free radicals in aqueous suspensions. Among the various existing types of particles one discriminates between those with inherent toxic activity, such as hard-metal dusts, welding fumes, or quartz dust, and those that, due to their material properties, have a much lower inherent particle reactivity. The latter are commonly referred to as poorly soluble particles (PSP) of low acute toxicity, or granular bio-durable particles (GBP) without known specific toxicity.

Absorbed and Soluble Components

The ultimate oxidant-generating capacity of particles is determined by their ability to adsorb various chemicals, including metals that also may enhance ROS generation. The adsorbed material could be toxic even if the particles are not. Alternatively, the adsorbed material could interfere with the normal chemical activity of the particles.

Absorbed transition metals are of particular importance because they are involved in the generation of ROS via Haber–Weiss reactions. Indeed, in transition metal-containing particles like coal fly ashes and PM, the generation of ROS can be easily modified by the use of metal chelators or by removing the soluble particle fraction. Additionally, organic constituents associated with the particle surface may also contribute to oxidant formation. Semiquinone radicals produced during their metabolic activation can undergo redox cycling, leading to the formation of ROS. Polycyclic aromatic hydrocarbons are present in many particle samples as a result of incomplete combustion. Particles may carry, hold and release PAHs into cellular compartments that normally would not be reached. Also the kinetics of particle-associated PAHs and soluble PAHs have been demonstrated to differ substantially and in some conditions particle-associated PAHs can be considered a slow-release deposit. On the other hand, larger surfaces usually exhibit stronger binding of components through van der Waals forces, which diminishes the ability of absorbed components to be released from the particle surface. An example of this is provided by PAHs absorbed on carbon blacks that only become biovailable from low-surface, highly contaminated versions of carbon black.

6.3.4 Nanoparticles: A Special Case?

> **Nanotechnologies are expected to bring a fundamental change in manufacturing, mainly by producing new materials with highly added value. Anticipated applications and sectors are electronics, the automotive sector, food packaging, drug delivery, and imaging. Engineered nanoparticles (diameter <100 nm) are an important tool to realize these new applications and products.**

The reason why nanoparticles (NP) are attractive for many purposes is because of their unique features, such as their high surface to mass ratio, their unique (quantum) properties, and their ability to adsorb and carry other compounds. Up to now many of these special purpose engineered NP, such as carbon nanotubes, quantum dots, dendrimers and polymer carriers, have been produced in small quantities, but production and applications are anticipated to increase steeply. Although chemically equivalent to larger particles of the same material, NP have a surface that is usually more reactive. Studies with fine and ultrafine TiO_2 particles have highlighted that a material of low toxicity in the form of fine particles could be toxic in the form of ultrafine particles. Later studies have demonstrated that pulmonary inflammation, usually measured as the number of neutrophilic granulocytes (PMN) in bronchoalveolar lavage (BAL), is related to the surface area of particles, although at similar surface areas some NP seem to be more inflammatory than others. Although the **deposition** of NP in the lungs follows largely the same distribution as fine particles, the underlying mechanisms are different. NP have a size dimension that makes them less subject to gravity and turbidometric forces and, therefore, their deposition occurs mostly by diffusion. In addition, their size causes them to interact more with other potential targets (brain, heart, endothelial tissue) than conventional fine particles. Secondly, NP may circumvent endogenous **defense** because, due to their small size, they may not be recognized by macrophages. In addition, NP can be **translocated** to body compartments away from their deposition sites in the respiratory system. NP may cross the epithelial and endothelial layer and they have access to cells in the epithelium, the interstitium, and the vascular wall. Rapid translocation toward the liver of more than 50% of 26 nm NP occurred within 24 hours in a rat model after inhalation or instillation. In contrast to such information from animal studies a rapid but no more than 3–5% uptake of radiolabeled carbonaceous NP into the bloodstream within minutes of exposure and subsequent uptake in the liver has been found in humans. NP can also be transported along the olfactory nerves into the brain after crossing the olfactory epithelium or via uptake along trigeminal nerves in the upper respiratory tract. It remains to be determined whether this uptake can lead to changes in the central nervous system. Experimental animal studies suggest that uptake leads to activation of pro-inflammatory mediators such as TNF and COX-2. No human data are available at the moment, although circumstantial evidence for a relation between ambient particles in the brain and Alzheimer-like pathology has been observed in brain sections. Interestingly, the uptake and potential effects of NP in the brain have not been reported for fine particles and this effect seems to be limited to NP.

> **Most toxic mechanisms of NP are probably qualitatively not different from cell–particle interactions for fibers and fine particulates. Quantitatively, however, NP can induce more inflammation at considerably lower gravimetric lung burdens than their larger sized analogues. Other mechanisms, such as translocation to the brain, are substantially different from their fine analogues.**

Similar to fine-sized particles, cells in contact with NP such as macrophages and neutrophilic granulocytes are activated and produce ROS. Within hours, cytokines and chemokines are synthesized and secreted into the affected area. These mediators interact with specific receptors on the surfaces of many cell types and result in activation of local cells as well as those in the blood and other tissues. As a result, cells are attracted from the bloodstream and enter the fluid-filled interstitial spaces, where they can attack the foreign material. Consequently, particle-induced cell activation in the airways results in inflammation. Epithelial and nerve cells may also contribute to airway inflammation by producing pharmacologically active compounds such as capsacein. In this *neurogenic inflammation*, stimulation of sensory nerve endings releases neurotransmitters that may affect many types of white blood cells in the lung, as well as epithelial and smooth muscle cells.

For hazard characterization and classification of newly engineered nanomaterials several crucial questions need to be answered:

- Which effects are specific for nanomaterials and which effects are merely stronger?
- Can we extrapolate available data and concepts generated with combustion-derived particles to newly engineered materials?
- Are current testing procedures specific enough to detect the effects of nanomaterials?
- Is our current regulatory system ready to handle and communicate the risks of nanomaterials?

6.3.5 Special Particle Effects

Carcinogenicity

> **Our current knowledge of particle-induced lung tumors in experimental animals can be summarized by stating that all inhaled particles, fibrous and non-fibrous, are likely to induce lung tumors in rats, provided that these particles are inhaled chronically or instilled intratracheally at sufficiently high doses, are respirable to the rat, and are highly durable.**

The retained lung burden leading to lung tumors in the rat can differ for different particles and, apart from dose, greatly depends on particle properties such as surface area and chemistry, cytotoxicity, and size/dimensions. The gravimetric dose needed for the onset of particle overload and risk for subsequent neoplastic events is 1 mg/g lung tissue or 200–300 cm^2 surface burden of PSP. The surface dose where lung tumors start to develop after both inhalation and instillation of PSP lies between 0.2 and 0.3 m^2 per lung, which conforms to 5–15 mg of PSP of average surface area (20–40 m^2/g). It is now generally accepted that the continued presence of non-toxic particle material in the lungs, upon impairment of AM clearance, leads to a chronic inflammatory response, fibrosis, and tumor genesis in the rat. As discussed earlier, the overall pattern is one of chronic inflammation, which occurs upon saturation of lung clearance by overloading of macrophages. At this point, particle accumulation starts and inflammatory cell influx increases sharply.

The influx of neutrophils and associated DNA damage and proliferation are responsible for the mutagenicity and the lung tumors after chronic particle exposure to PSP are due to their mutagenesis. However, several studies have generated data that deviate from this paradigm. A number of studies question the validity of the inflammation paradigm. In a rat study, HPRT-based mutations in lung epithelial cells upon exposure to crystalline and amorphous silica were not increased in rats exposed to amorphous silica compared to controls, although the inflammation was similar to that induced by crystalline silica. Similarly, depletion of circulating neutrophils in rats by injection of antineutrophil serum before short-term inhalation of quartz particles (3 days, 100 mg/m^3) did not affect acute lung damage by quartz. It remains to be investigated whether the findings in these models using toxic quartz can be reproduced with PSP.

A comparison of lung tumors at similar gravimetric dose (30 mg) of several NP (Table 6.14) shows that the number of lung tumors induced by three different insoluble ultrafines is proportional to their surface area. DEP with the lowest surface area (34 m^2/g) induced 22 tumors in 46 animals, while the carbon black (CB, 300 m^2/g) induced 40 tumors in a total of 45 animals. Importantly, the high-surface amorphous silica induced few lung neoplasms, which might be due to its high solubility, i.e. a low durability *in vivo*. However one has to caution against oversimplification and not rely solely on surface area, since apart from their small size, NP for commercial applications often have chemically different surfaces. Evidence for the relevance of the chemical surface in PSP has also come from studies with surface-modified TiO$_2$. Surface modifications resulting in an enhanced hydrophobicity of TiO$_2$ have generally been found to lead to an amelioration of the inflammatory response, although initial studies showed an increased toxicity of surface-modified TiO$_2$. No data are available that connect this ameliorated response to chronic outcomes such as carcinogenicity.

Table 6.14 Prevalence of lung tumours in rats 129 weeks after intratracheal administration of a 30 mg dose of different poorly soluble, low-toxicity particles.

Particle	Size (nm)	Surface area (m^2/g)	Lung tumours Benign	Lung tumours Malign
Control			0/91	0/91
TiO$_2$	30	50	16/45	24/45
Carbon black	14	300	32/48	16/48
Diesel	ND	34	21/46	1/46
Aerosil	14	200	3/39	0/39

Female Wistar rats were treated with multiple intratracheal injections (6 mg) at weekly intervals to reach the final cumulative dose (30 mg). Control animals remained untreated and received no injections. Tumours were evaluated by histopathological scoring of two HE sections per lung lobe.

Although some of the particles listed in Table 6.12 have been characterized as confirmed human carcinogens (Group I by IARC), PSP such as coal mine dust, pigmentary TiO_2, and CB were not associated with an increase in lung cancer in exposed workers. Thus, hazard assessment using rat studies raises the question of whether particles that induce tumors in this bioassay should be labeled as possible or even probable human carcinogens.

Cardiovascular Effects

> **Epidemiological studies indicate that the major cause of increased deaths upon ambient particle increase is among patients with cardiovascular diseases. In addition, experimental animal studies with combustion NP show that high exposures to diesel soot NP or other surrogate NP causes observable cardiovascular effects.**

Combustion and model NPs can gain access to the blood following inhalation or instillation and can enhance experimental thrombosis. Diesel particles instilled into hamster lungs also enhance thrombosis but it is not clear whether this is an effect of pulmonary inflammation or particles translocated to the blood. High exposures of combustion-derived NP (CDNP) by inhalation caused altered heart rate in hypertensive rats and dogs and are interpreted as a direct effect of CDNP on the pacemaker activity of the heart. Inflammation in distal sites has also been associated with increased progression of atheromatous plaques in rabbits, ApoE-/- mice, and humans. In summary, the cardiovascular effects of inhaled particles (PM) are explained by theories that involve either impairment of the blood flow to or in the heart or interference with autonomic innervation (Figure 6.11).

> **Inflammation can be seen as the main driver of the cardiovascular effects. There is evidence of systemic inflammation following increases in PM, as shown by elevated C-reactive protein, blood leukocytes, platelets, fibrinogen, and increased plasma viscosity.**

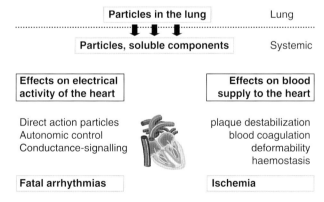

Figure 6.11 Iillustration of the main mechanisms that may be involved in the cardiovascular effects of inhaled ambient particulate matter. It is suggested that particles cause an exacerbated inflammation in the lung, which may lead to translocation of particles, soluble components, and/or inflammatory mediators. This may affect electrical innervation of the heart and/or blood flow to the heart, leading to arrhythmias or ischemic heart damage.

Atherosclerosis is the underlying cause of acute coronary syndrome, the main cause of cardiovascular morbidity and mortality. Atherogenesis is an inflammatory process, initiated via endothelial injury, which produces systemic markers of inflammation that are risk factors for myocardial and cerebral infarction. Repeated exposure to PM_{10} may, by increasing systemic inflammation, exacerbate the vascular inflammation of atherosclerosis and promote plaque development or rupture of blood vessels. The inflammatory response to particles or the particles themselves may also impact on the neural regulation of the heart, leading to death from fatal dysrhythmia. In support of this hypothesis, studies in humans and animals have shown changes in heart rate and heart rate variability in repose to particle exposures.

6.3.6 Summary

Massive exposure to inhalable and respirable particles over a period of years at more than 5 mg/m^3 is known to cause both chronic obstructive pulmonary diseases, including bronchitis, emphysema, fibrosis, and pneumoconiosis, and lung cancers. Epidemiological studies have shown that ambient exposure to particles of 10 μm (PM_{10}) is related to both respiratory and cardiovascular mortality and morbidity. It is currently thought that ambient particulate matter with a mean diameter of 2.5 or 10 μm ($PM_{2.5}$ or PM_{10}) can be carcinogenic, but it is not clear which components or characteristics of PM are responsible.

To understand particle effects, one can apply the "D concept" (Dose, Deposition, Dimension, Durability, and Defense). This paradigm builds a conceptual understanding of the adverse effects of particles and fibers. Fiber dimensions determine the ability of the body to defend against particles and on persistence. Long (>20 μm) fibers are not taken up by alveolar macrophages and, therefore, have longer half-lives in the lung than shorter fibers and result in a higher cumulative dose at similar inhaled fiber number or mass. Inhaled fibers longer than 20 μm also show interception at the bronchiolar bifurcations due to their length and may achieve high local doses by this process.

If a particle is not soluble or not degradable in the lung it has a high durability and there will be rapid local accumulation upon sustained exposure. Particle size largely determines where particles are deposited in the respiratory tract. Particle deposition is further determined by shape and density. Small ultrafine particles have little mass and behave by diffusion. Most of them are deposited in the nasal compartment and the alveoli. Particle translocation has only recently been recognized as a mechanism to explain possible systemic effects of inhaled ambient particles.

One of the crucial properties of particles is their ability to generate ROS and RNS. This factor is involved in many pathological outcomes of particle exposure, including fibrosis, proliferative effects, and carcinogenicity. Apart from producing ROS, inflammatory phagocytes produce growth factors such as cytokines and chemokines that regulate cell migration, tissue remodeling, cell proliferation, and repair of damage.

Differences in particle toxicity can be related to their surface reactivity, their ability to absorb various chemicals, including metals that enhance ROS generation, and inflammation.

Most toxic mechanisms of nanoparticles are probably qualitatively not different from cell–particle interactions for fibers and fine particulates. Quantitatively, however, nanoparticles can induce more inflammation at considerably lower gravimetric lung burdens than their larger sized analogues. Other mechanisms, such as translocation to the brain or the vascular system, are substantially different from their fine analogues.

Further Reading

Borm PJA, Fowler P, Kirkland D. An updated review of the genotoxicity of respirable crystalline silica. *Particle and Fibre Toxicology* 2018, **15**: 23. https://doi.org/10.1186/s12989-018-0259-z

Borm PJA, Kreyling W. Toxicological hazards of inhaled nanoparticles- potential implications for drug delivery. *J Nanosci Nanotechnol.* 2004, **4**: 521–531.

Borm,PJA, Schins RPF, Albrecht CA. Inhaled particles and lung cancer. Part B: Paradigms and risk asessment. *Int J Cancer.* 2004, **110**(1): 3–14.

Dockery DW, Pope CA 3rd, Xu X et al. An association between air pollution and mortality in six US Cties. *N Engl J Med.* 1993, **329**(24): 1753–1759. *First epidemiological study highlighting the potential effect of ambient particles on mortality.*

Donaldson K, Tran L, Jimenez LA, Duffin R, Newby DE, Mills N, MacNee W, Stone V. Combustion-derived nanoparticles: a review of their toxicology following inhalation exposure. *Part Fiber Toxicol.* 2005, **2**: 10.

Greim H, Borm P, Schins R, Donaldson K, Driscoll K, Hartwig A, Kuempel E, Oberdorster G, Speit G. Toxicity of Fibers and Particulates – Report of a Workshop held in Munich, Germany, October 26–27, 2000. *Inhal Toxicol.* 2001, **13**: 737–754.

Knaapen A, Borm PJA, Albrecht C, Schins RPF. Inhaled particles and lung cancer. Part A: Mechanisms. *Int J Cancer.* 2004, May 10, **109**(6): 799–809. Review.

Mills NL, Amin N, Robinson SD, Anand A, Davies J, Patel D, de la Fuente JM, Cassee FR, Boon NA, Macnee W, Millar AM, Donaldson K, Newby DE. Do inhaled carbon nanoparticles translocate directly into the circulation in humans? *Am J Respir Crit Care Med.* 2006, Feb 15, **173**(4): 426–431.

Nemmar A, Hoylaerts MF, Hoet PH, Dinsdale D, Smith T, Xu H, Vermylen J, Nemery B. Ultrafine particles affect experimental thrombosis in an in vivo hamster model. *Am J Respir Crit Care Med.* 2002, Oct 1, **166**(7): 998–1004.

Nel A, Xia T, Madler L, Li N. Toxic potential of materials at the nanolevel. Science. 2006, **311**(5761): 622–627. *Short review with the conceptual thinking of toxicologists on nanoparticle adverse effects.*

Oberdorster G, Sharp Z, Atudorei V, Elder A, Gelein R, Kreyling W, Cox C. Translocation of inhaled ultrafine particles to the brain. Inhal Toxicol. 2004, Jun,**16**(6–7): 437–445. *Key publication for the future and has set the stage for further research on effects of particles in the brain.*

Oberdorster G, Oberdorster E, Oberdorster J. Nanotoxicology: an emerging discipline evolving from studies of ultrafine particles. Environ Health Perspect. 2005, Jul, **113**(7): 823–839. *Comprehensive review on mechanisms and actions of inhaled nanoparticles by the founder of this field.*

Pope CA 3rd, Burnett RT, Thurston GD, Thun MJ, Calle EE, Krewski D, Godleski JJ. Cardiovascular mortality and long-term exposure to particulate air pollution: epidemiological evidence of general pathophysiological pathways of disease. *Circulation.* 2004, **109**(1): 71–77.

Tran CL, Buchanan D, Cullen RT, Searl A, Jones AD, Donaldson K. Inhalation of poorly soluble particles. II. Influence of particle surface area on inflammation and clearance. Inhal Toxicol. 2000, **12**: 1113–1126. *Surface area and not mass drives the inflammatory response to particles.*

6.4 Principles of Nanomaterial Toxicology

Thomas Gebel

6.4.1 Introduction

Nanoscaled particulates possess specific physico-chemical properties. For instance, nanomaterials possess quite high specific surface areas related to mass due to small

> **primary particle size. Smaller nanoparticles may possess novel chemical properties compared to the standard material, making them interesting for technological and pharmaceutical applications.**

Nanoscaled particulates possess specific physico-chemical properties which make them interesting in certain technological and pharmaceutical applications. In particular, smaller nanoparticles may possess novel chemical properties compared to the standard material. In nanotechnology, nanoscale is generally defined as particulate structure smaller than 100 nm in one of three dimensions. This size threshold was established by the International Organization for Standardization (ISO). Novel physico-chemical and chemical properties may become apparent at sizes smaller than about 30 nm. ISO applied a safety factor of roughly three to arrive at 100 nm as the upper cut-off in their definition. It is relevant to note in this context that marketed nanomaterials mostly do not consist of primary particles with identical sizes but show a wide-ranging particle size distribution. As a consequence, not all particles contained in marketed nanomaterials may comply with the definition of nanosize smaller than 100 nm in one dimension. The established definition for nanomaterial size in pharmaceutical applications is not limited to 100 nm for primary particle size, but may be larger, i.e. up to 1000 nm. The term nanomaterial generally covers manufactured or technically engineered materials. Nanomaterials have to be discriminated from particles generated by processes like combustion or human activities like welding. Dusts generated by these processes also contain nanoscaled particles. Usually the term ultrafine particles or dusts is used in this case. This chapter deals with manufactured or technically engineered nanomaterials. With respect to environmental fine and ultrafine particulate matter and human health effects, an extensive body of mainly epidemiological literature is available. It is unclear whether findings from these data may be cross-read to nanomaterials due to the chemical complexity of environmental particulate matter.

> **Due to small size and basic physical rules, the primary particles contained in nanomaterials have a strong tendency to adhere to each other. In other words, they agglomerate to larger structures, i.e. secondary particles in air and biological milieu. The tendency to agglomerate is higher the smaller the primary particles are. The occurrence of free single primary particles is rare.**

Due to small size and basic physical rules, the primary particles contained in nanomaterials have a strong tendency to adhere to each other. In other words, they agglomerate to larger structures in air and biological milieu. Alternatively, and due to production process (e.g. at higher temperatures they appear in processes like sintering) they even may be bound more strongly to each other in larger structures called aggregates. The appearance of free isolated primary particles is rare and the exception rather than the rule. There are only very few exceptions to this agglomeration rule. Particulate material freshly generated by, for example, combustion may contain higher portions of isolated primary particles. However, agglomeration occurs within seconds. In low concentrations (low µg/m^3 range depending on primary particle size and material density) particles may be found to a higher extent disagglomerated, e.g. in smaller agglomerates. Silver or gold nanomaterials may appear to higher portions as primary particles or very small agglomerates. Surface coating with specific chemicals may reduce or prevent the tendency to agglomerate.

Some nanomaterials have been used for decades in various technological applications. For instance, carbon black is used in car tires and lends them their black color. Amorphous silicon dioxide is used as flow agent additive in powdered foods. Nanosized titanium dioxide is used as a UV-absorbent in sunscreens. The use of nanopharmaceuticals is currently less important as it is still in its infancy. The main aim is to develop carrier systems for bioactive compounds aiming at a targeted delivery at the site of desired action.

6.4.2 Toxicology

> **The crucial question in nanotoxicology is how physico-chemical properties changed by nanoscale may influence toxicokinetics and toxicodynamics. Thus, the focus is on materials remaining nanosized in biological systems for longer durations, i.e. materials which possess higher biodurability. A biosoluble nanomaterial rapidly loses its nano-specific properties and putative effects are caused by the substances the material consists of.**

It was assumed that the novel technological properties of nanomaterials might lead to novel toxicological properties. This could be associated with unexpected negative impacts on human health on exposure to such materials. With increasing market relevance and the increasing number of novel materials developed, nanotechnology has increasingly been accompanied by safety research. The aim is to obtain toxicological information early to inform risk assessment and avoid any undesirable consequences on human health.

> **Nanomaterials are heterogenous with respect to chemical identity, which poses an immense challenge for risk assessment. Thus, in practice it is not feasible to carry out a thorough risk assessment for each marketed material.**

Nanomaterials often consist of inorganic metal salts but may also be of organic origin, e.g. particulate polystyrene. Surface properties of nanomaterials may vary due to production process, for example. For instance, amorphous silicon dioxide is produced by either pyrogenic processes or chemical precipitation. This has an influence on the surface hydroxyl group density, which is lower for pyrogenic silica. Nanomaterials may be produced with various surface coatings, which increases their heterogeneity. The design of nanomaterials in pharmaceutical applications is often even more complicated with respect to chemical identity and design. For instance, such materials may consist of an inorganic core and a chemically complex organic shell. The reason for this complexity is to achieve the desired behavior of these materials in the body to control resorption, distribution, drug transport, targeted drug release, and disintegration or elimination.

The early phase of safety research in nanotoxicology was characterized by intensive scientific experimentation and study that aimed to identify novel effects of nanomaterials. One major drawback was that many of the effects reported were not interpretable with respect to their significance for human risk assessment. This was because these studies were not performed according to standard study protocols established in regulatory toxicology but focused on basic research. Another drawback was that the standard analogous material in microsized form (the bulk material) was mostly not studied in parallel. Thus, it was unclear whether the reported findings were specific and unique for the tested nanomaterial. It became evident that sample preparation and characterization were neither standardized nor transparent, introducing a further uncertainty into how study results should be interpreted. Sample preparation and particle generation partly led to rather artificial test samples as the aim was to generate highly dispersed single primary particles and/or very small agglomerates. This is rather different to real-life exposure situations, in which more or less larger agglomerates predominate.

The crucial question to address in nanotoxicology is how physico-chemical properties that are altered because of nanosize may impact toxicokinetics and toxicodynamics. Thus, the focus is on materials keeping their nanostructure in biological systems, i.e. materials with a higher biodurability. A biosoluble material will quickly lose its specific nano properties, and its effects will be mediated by the chemical substances the material is composed of. This chapter will focus on nano-specific issues of toxicity.

Nanotoxicokinetics

> **Dermal absorption and transdermal transport of nanomaterials are reported to be generally very low, as is enteral resorption. This is generally also true for the inhalation route.**

There is no evidence so far that materials consisting of nanosized primary particles show a markedly different higher uptake and distribution in the body compared to other particulate materials. Nanomaterials are generally taken up by phagocytosing immune system cells in larger portions. This explains the higher particle levels in lung-associated lymph nodes and in the spleen. The systemic distribution of particulate material may be higher during particle-mediated inflammation in the deep airways. This leads to tissue damage and a consequential enhancement of systemic particle transfer. In intravenous application and distribution with blood flow, nanomaterials are trapped in larger quantities in the liver and spleen. The distribution to other organs, e.g. the brain, is generally extremely low. Following a maximized bioavailability by intraperitoneal or intravenous application it may be slightly higher, but it is generally far below 1% of the applied dose.

There are some experimental studies published that study the uptake of nanomaterials via olfactory tissue. It was found that there may be a certain uptake of particulate material in low amounts. Such uptake was also reported previously for mercury vapor, metal ions, and polio viruses. From a quantitative perspective, this route of uptake does not seem to be relevant for risk assessment and it is neither restricted nor specific for nanomaterials.

> **Particles entering the body are generally covered with body material such as proteins, lipids, and others. This so-called corona may have an impact on the distribution and cellular uptake of these particles. This may especially be the case if the particles present protein structures with a targeting function. Besides particle corona, agglomerate/aggregate size influences the uptake of particulates into cells. It is not a general rule that the smaller the particle or the agglomerate, the better it passes barriers or enters cells.**

Cellular uptake is characterized by different mechanisms and is not restricted to nanosize. A prominent example are nickel compounds of lower solubility, such as nickel sulfide or nickel subsulfide. Microscaled particulate nickel was shown to be taken up by cells to a far higher level than dissolved ionic nickel. It is not a general rule that the smaller the particle or the agglomerate, the better it passes barriers or enters cells. Particles of sizes up to 100–150 nm may be taken up into cells via clathrin-mediated endocytosis, particles of sizes up to 200–500 nm may enter cells via caveolae-dependent endocytosis, and particles of sizes from 500 to 5000 nm may be taken up by (macro-)pinocytosis. Predicting cellular uptake based on size only is not possible. Further properties such as shape, charge, agglomeration/aggregation, protein and cell membrane binding, and the protein corona may affect uptake. The general conclusion is that particles may enter cells, normally in rather low numbers.

There are several studies available which have investigated whether aggregated or agglomerated nanomaterials disaggregate or disagglomerate after uptake in the body. The existing data indicate that disaggregation or disagglomeration of nanomaterials in the body does not appear to be significant.

After dermal, oral, and inhalation exposure particles are systemically distributed, generally at a low level. In the few studies which have investigated in parallel the systemic availability of the same compound in both the nanomaterial and the micromaterial form it was found that micromaterial particles were resorbed and distributed to a similar or slightly lower extent. The elimination half-lives of resorbed biodurable particulate matter are generally rather high. Thus, if continuously exposed, particle body burden may increase constantly during the lifetime. On the other hand, few experimental animal studies have indicated that systemically distributed particle material may level off at low concentrations. This could at least partly be explained by the fact that particulate material may also be actively excreted from cells. Long-term studies are lacking but are currently underway to assess whether low-dose accumulation may lead to systemic particle levels with the consequence of adverse effects. So far there is no indication from epidemiological or experimental studies so far that this issue is relevant.

Nanotoxicodynamics

> **Dermal and oral exposures are not the pathways of major relevance with respect to hazard and risk assessment as the uptake rates of nanoparticulates are very low. Inhalation is the prominent route of exposure to be taken into account. Nanomaterials can be categorized according to different basic modes of toxic action of fibrous particles, granular particles, and other particles mediating effects via specifically toxic components (Figure 6.12).**

Fibrous Nanoparticles The principle of fibre toxicity, also called the fibre paradigm, applies to certain nanomaterials. Rigid respirable nanoscaled fibrous structures with high biodurability and a high aspect ratio are capable of inducing tumours similar to asbestos. It is well known that after inhalation, asbestos fibres cause lung cancer and mesothelioma, a type of cancer that develops from the mesothelium, the protective lining covering several internal organs. Carcinogenesis induced by asbestos is known to depend on specific fibre properties: they need to be respirable, sufficiently biopersistent and exhibit a specific 'architecture'. The relevant parameters for this fibre-specific effect are the so-called WHO fibre dimensions (length >5 µm, diameter <3 µm, aspect ratio >3:1). Fibres cannot be phagocytosed by alveolar macrophages in case they are too long and rigid (Figure 6.13a,c). This mode of action is transferable to rigid fibrous nanomaterials. Currently, some high aspect ratio nanomaterials are known to act by an asbestos-like mode of action, such as certain multi-walled carbon nanotubes. Note that this mode of action is not restricted to nanofibers but to respirable fibres.

By contrast, nanotube materials that are too short or are not rigid are not expected to exhibit an asbestos-like action (Figure 6.13b) and indeed the available data do not indicate such a mode

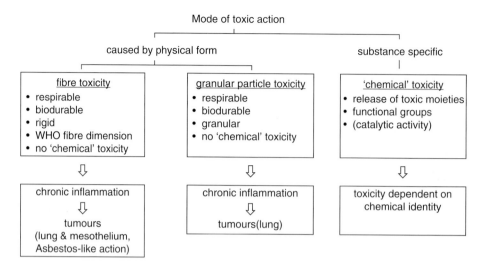

Figure 6.12 Basic nanomaterial categorization according to type of mode of toxic action.

of action. Induction of lung tumours and mesothelioma are the assumed relevant toxic effects for biopersistent fibres of WHO dimensions. Other toxic properties are not relevant. The relevant dose metric for risk assessment is the particle number concentration of WHO fibres. It should also be noted whether fibres adhere or clump together since the crucial mode of toxic action can only appear if the fibres in the critical dimension are available as free fibres. Fibrous dusts with diameters of more than 3 µm cannot reach the deep respiratory tract (the alveoli, i.e. they are not respirable) and thus cannot cause asbestos-like toxicity. Biopersistent rigid fibres of WHO dimensions are highly potent carcinogens.

Granular Nanoparticles

The toxicology of another type of nanomaterials can also be described by a common mode of action. These particles are characterized as respirable granular and biopersistent but not fibrous according to the WHO dimension (Figure 6.13d). If biopersistent respirable fibres do not meet the WHO fibre definition in that they are too short and/or they are not sufficiently rigid, they may fall into the category of granular biodurable particles. Their toxicity is not mediated by specific substances contained in the particles, nor by specific functional chemical surface groups, nor by specific surface-related toxicity, as is the case for, for example, crystalline silica. These particles have been termed poorly soluble particles (PSP), respirable granular biodurable particles without known significant specific toxicity (GBP), or poorly soluble, low-toxicity particles (PSLT). In the following, the abbreviation GBP is used as it is the most comprehensive description. Some high-volume production nanomaterials and non-nanomaterials can be characterized as GBP. For instance, carbon black can be assigned to this category of dusts. For all GBP materials, inhalation of respirable dust reaching the bronchi and alveoli is the exposure scenario that is critical with respect to health hazards. For this exposure pathway, chronic inflammation and carcinogenesis in the lung are the crucial effects to be considered. Lung carcinogenicity was shown in the rat for nanosized carbon black and nanosized titanium dioxide and was considered to be relevant for extrapolation to humans. For instance, the International Agency for Research on Cancer (IARC) classified carbon black as a Group 2B carcinogen (possibly carcinogenic to humans) on the basis of sufficient evidence in experimental animals for inhalation carcinogenicity.

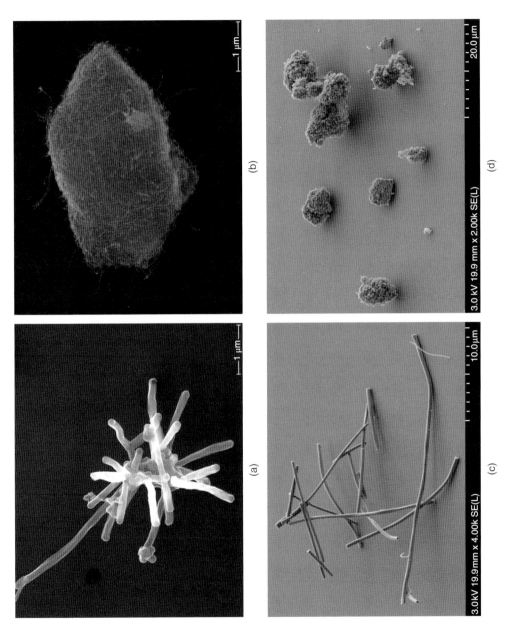

Figure 6.13 Examples of various nanomaterials (source BAuA). (a) Multi-walled carbon nanotubes. This material has to be considered as being rigid. (b) Multi-walled carbon nanotubes. This material has to be considered as not being rigid. (c) Fibrous nanosized titanium dioxide. (d) Granular nanosized titanium dioxide.

> **The toxic potency of GBP nanomaterials is higher than for GBP microsized materials for both relevant toxicological endpoints when comparing cumulative mass concentration exposure. The carcinogenic potency was estimated to be higher for GBP nanomaterials by a factor of 2–3.**

The potency of lung inflammation is higher for GBP nanomaterials by roughly half an order of magnitude. The carcinogenic potency was estimated to be slightly higher for GBP nanomaterials by a factor of 2–3 in comparison with the respective bulk materials when referring to the dose metrics mass concentration. This difference in potency is relatively low when taking into account the fact that the potency of carcinogens may vary over many orders of magnitude. When the comparison for carcinogenic potency was performed based on the dose metrics cumulative surface area concentration, the difference in potency was even smaller, if present at all. This could indicate that the higher specific surface area might be one reason for the slightly higher carcinogenic potency of GBP nanomaterials. An analogous hypothesis has been generated for acute inflammation that could be explained by the adsorption of lung surfactant. However, the data do not generally support the finding that surface area concentration is the only crucial parameter explaining mechanism of action after longer-term exposure. Another possibility is that the agglomerate volume is the crucial factor accounting for the long-term inflammatory potency of GBP materials in the lung. When alveolar macrophages are more highly loaded with particles, they react by mediating inflammatory processes in their vicinity. Subsequently this may damage the alveolar tissue close to the macrophages. The hypothesis is that the particle agglomerate volume determines macrophage loading and thus determines macrophage-mediated inflammation. GBP agglomerates are assumed to remain stable and not to disintegrate within the macrophages. GBP nanomaterials in general possess a lower agglomerate density, i.e. they are more 'fluffy' than GBP micromaterials. In other words, there is more void space in GBP nanomaterial agglomerates. For such lower agglomerate densities, the macrophages are loaded to a critical volume level, leading to inflammation at lower mass concentrations. As a consequence, alveolar macrophages react by initiating oxidative stress and an inflammatory response at lower mass concentrations in case they have to deal with GBP nanomaterials. According to current knowledge, no further health hazards other than lung inflammation and carcinogenicity seem to be relevant for GBP nanosized or GBP microsized materials. The available studies do not indicate any clear evidence for systemic toxicity of GBP nanomaterials.

6.4.3 Summary

The toxicity of more or less biodurable nanomaterials may be mediated by the specific toxicological properties of their chemical components. A nanomaterial may release toxicants, usually toxic ions. For instance, cadmium-based quantum dots may release cadmium ions or silver nanoparticles may release silver ions. A particulate material may also contain chemically reactive groups on its surface that cause toxicity, which is well known for crystalline silica. Likewise, the surface of nanomaterials may have catalytic properties that might lead to specific toxic effects.

Such catalytic activity could also become toxicologically relevant and should be considered. However, a technically relevant catalytic activity is not necessarily relevant for toxicity. For instance, the catalytic activity of gold nanoparticles has as yet not been shown to be associated with a changed profile of toxicity. Taken together, for a chemically mediated toxic action, the chemical identity of the nanomaterial is critical for the evaluation of health hazards. The nanoparticle may modify the toxic potency of its integrated chemicals, usually due to altered kinetic behaviour, but it will not lead to qualitatively different properties. In comparison with the respective micromaterial, the specific surface area of a nanomaterial is higher, allowing more chemical functional groups to be available on its surface. Therefore, reactivity and toxicity, as well as the rate of release of a toxic moiety, may be increased due to a larger specific surface area.

Further Reading

Donaldson K, Poland CA. Nanotoxicity: challenging the myth of nano-specific toxicity. *Curr Opin Biotechnol.* 2013; **24**(4):724–734.

Gebel T. Small difference in carcinogenic potency between GBP nanomaterials and GBP micromaterials. *Arch Toxicol.* 2012; **86**(7):995–1007.

Gebel T, Foth H, Damm G, Freyberger A, Kramer PJ, Lilienblum W, Röhl C, Schupp T, Weiss C, Wollin KM, Hengstler JG. Manufactured nanomaterials: categorization and approaches to hazard assessment. *Arch Toxicol.* 2014; **88**(12):2191–2211.

Krug HF. Nanosafety research – -are we on the right track? *Angew Chem Int Ed Engl.* 2014; **53**(46):12304–12319.

Landsiedel R, Fabian E, Ma-Hock L, van Ravenzwaay B, Wohlleben W, Wiench K, Oesch F. Toxico-/biokinetics of nanomaterials. *Arch Toxicol.* 2012; **86**(7):1021–1060.

Moreno-Horn M, Gebel T. Granular biodurable nanomaterials: No convincing evidence for systemic toxicity. *Crit Rev Toxicol.* 2014; **44**(10):849–875.

Rittinghausen S, Hackbarth A, Creutzenberg O, Ernst H, Heinrich U, Leonhardt A, Schaudien D. The carcinogenic effect of various multi-walled carbon nanotubes (MWCNTs) after intraperitoneal injection in rats. *Part Fibre Toxicol.* 2014; **11**(1):59.

6.5 Endocrine Active Compounds

Volker Strauss and Bennard van Ravenzwaay

6.5.1 Introduction

Endocrine active compounds (this chapter focuses on the hypothalamic–pituitary–gonadal (HPG) and hypothalamic–pituitary–thyroidal (HPT) axes) are used as drugs in humans (Table 6.15). Side effects are assessed for registration in preclinical as well as clinical studies. In the framework of a risk-benefit assessment a therapeutic dose of the drug is defined for several patient groups. Administering drugs to healthy humans may result in hormone dysregulation. For example, anastrozole, an aromatase inhibitor, is used to treat hormone-sensitive breast cancer in post-menopausal women. It is not permitted to administer this drug to premenopausal, pregnant or breast-feeding women, or children.

> Endocrine active compounds occur in nature and are synthesized as drugs and anthropogenic chemicals. Humans and animals may absorb these compounds via the oral, dermal or inhalation route.

An **adverse** effect in toxicology is defined as a change in morphology, physiology, growth, development, reproduction or life span of an organism, system or (sub)population that results in an impairment of functional capacity, an impairment of the capacity to compensate for additional stress, or an increase in susceptibility to other influences (IPCS, 2004).

Table 6.15 Drugs and their mechanisms influencing the HPG and HPT axes.

Mechanism	Example of drug	Additional effect
HPG axis		
Steroid synthesis inhibitor	Ketoconazole	Antifungal (ergosterol synthesis inhibitor)
Estrogen receptor agonist	Ethinylestradiol	
Estrogen receptor antagonist	Mepitiostane	Anabolic
Selective estrogen receptor modulator	Tamoxifen	Aromatase inhibitor
Aromatase inhibitor	Anastrozole	
Progesterone receptor agonist	Medroxyprogesterone	
Progesterone receptor antagonist	Mifepristone	Glucocorticoid receptor antagonist
Selective progesterone receptor modulator	Asoprisnil	
Androgen receptor agonist	Methyltestosterone	
Androgen receptor antagonist	Flutamide	
Selective androgen receptor modulator	Andarine	
5α-Reductase inhibitor	Finasteride	
FSH/LH receptor agonist	Menotropin	
FSH/LH receptor antagonist	Not known	
GnRH receptor agonist	Buserelin	
GnRH receptor antagonist	Ganirelix	
HPT axis		
T4 substitution	Levothyroxine	
T3 substitution	Liothyronine	
Thyroid hormone analog	Triiodothyro-acetic acid	
Thyroid hormone antimetabolite	Dibromotyrosine	
Thyroid peroxidase inhibition	Propylthiouracil	Deiodinase inhibition
Sodium-iodide symporter Inhibitor	Sodium perchlorate	
Deiodinase inhibitor	Sodium ipodate	

FSH, follicle-stimulating hormone; LH, luteotropic hormone; GnRH, gonadotropine releasing hormone.

Endocrine active compounds leading to adverse effects are **endocrine disruptors**. They are defined as exogenous compounds that alter the function of the endocrine system and cause adverse health effects in an intact organism, or its progeny, or (sub)populations. Endocrine disruption is not regarded as a toxic endpoint per se, but as a functional alteration potentially leading to an adverse effect.

The risk of adverse effects of endocrine active compounds is assessed the same way as with other toxic endpoints (see Chapter 2.1). Preclinical screening tests (e.g., *in vitro* tests, the Hershberger assay and uterothropic test) and mechanistic studies are performed in order to identify the effect on the endocrine system (hazard identification). In human health risk assessment, other factors must be considered, such as adverse effects in reproductive toxicity studies (e.g., multigeneration studies), toxicokinetic data in animals and humans as well as exposure data in different subpopulations (children, adults, pregnant women etc.).

6.5.2 Thyroid Hormone Affecting Compounds

Thyroid Gland

> The following mechanisms for dysregulations of the HPT axis are discussed: inhibition of the iodine uptake in the thyroid gland, inhibition of thyroid peroxidase, displacement of thyroid hormones from binding proteins in blood, increased degradation of thyroid hormones in the liver, and thyroid hormone receptor agonism.

Reference compounds for these mechanisms are described in Figure 6.14.

Iodine Uptake Inhibition in the Thyroid Gland **Perchlorate** inhibits the sodium–iodide symporter (NIS) in the thyroid gland. It is found in nature and also produced in anthropogenic processes: vegetables and fruits can be contaminated by organic fertilizers containing perchlorate or by water treated with biocides containing chlorine. Because of its ubiquitous occurrence, a basic perchlorate level has been determined in humans: in Europe, the exposure in adults is up to 0.26 µg/kg bodyweight/day (bw/d) and in children up to 0.75 µg/kg bw/d (95th percentile). In the USA, an uptake of perchlorate in infants via breast milk was estimated to be up to 6.5 µg/kg bw/d. The EFSA set the tolerable daily intake (TDI) at 0.3 µg/kg bw/d perchlorate on the basis of a study in which the reduction in iodine uptake after perchlorate administration was measured in test persons, this being the most sensitive parameter. It was assumed that a chronic 50% reduction of the iodine uptake leads to multinodular goiter. In this 14-day study with the maximum dose at 500 µg/kg bw/d, no thyroid hormone changes occurred. Regarding the reduced iodine uptake, the NOAEL was 7 µg/kg bw/d. Studies with rats are not relevant because of the species-specific higher thyroid hormone turnover, resulting in a greater sensitivity towards a reduced iodine uptake. In children and breast-fed infants the intake of perchlorate may exceed the TDI in some cases, although epidemiologic reports about effects in these target groups are not currently available.

Thyroid Peroxidase Inhibition

Ethylene bis-dithiocarbamate (fungicides: Maneb, Mancozeb und Metiram) is quickly metabolized in mammals as well as in the environment to ethylene thiourea (ETU). ETU inhibits thyroid peroxidase, resulting in hypothyroidism. To date the US Environmental Protection Agency has classified the risk of acute, subacute and chronic toxicity as well as carcinogenicity as being minimal while the endocrine effects have yet to be assessed.

Displacement of Thyroid Hormones from Binding Proteins in Blood

Pentachlorophenol (PCP) was used in the past as a pesticide, disinfectant, and ingredient in anti-fouling paint. Since the early 1980s, the general public in the USA are no longer permitted to purchase or use PCP. Technically produced PCP (84–90%) contains other polychlorinated phenols and dioxins which are more toxic than PCP. PCP displaces T4 from the transporter molecule transthyretin *in vitro* whereas the displacement of T4 from thyroxine binding globulin (TBG) is not as marked. *In vivo* a single PCP dose of 0.56 mmol/kg bw is able to reduce the serum T4 level. Additionally, an *in vitro* antiandrogen effect (positive YAS assay) is assumed. In a two-generation study with rats a dose of >10 mg/kg bw/d resulted in reproductive toxicity effects (reduced fertility

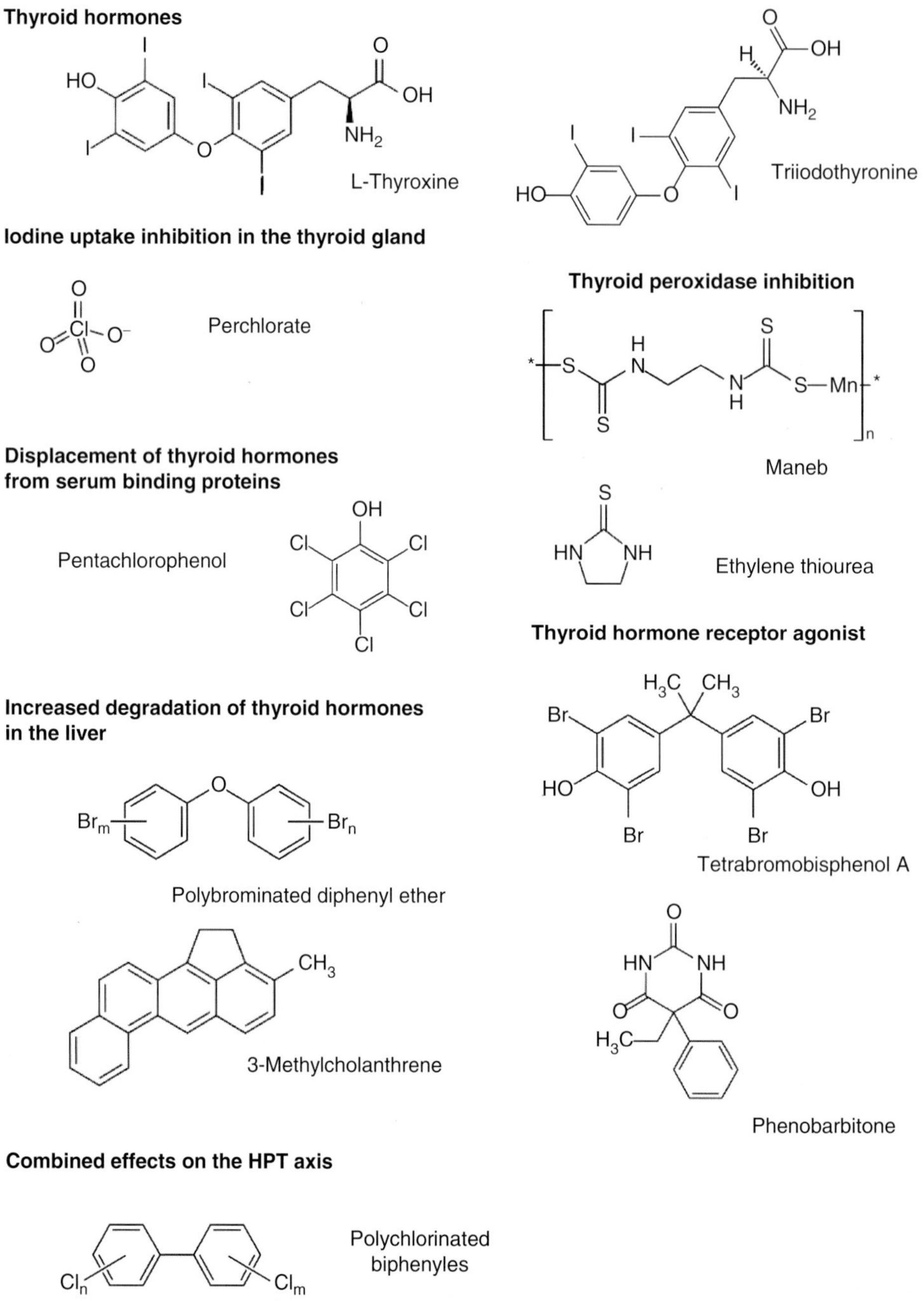

Thyroid hormones

L-Thyroxine

Triiodothyronine

Iodine uptake inhibition in the thyroid gland

Perchlorate

Thyroid peroxidase inhibition

Maneb

Displacement of thyroid hormones from serum binding proteins

Pentachlorophenol

Ethylene thiourea

Thyroid hormone receptor agonist

Tetrabromobisphenol A

Increased degradation of thyroid hormones in the liver

Polybrominated diphenyl ether

3-Methylcholanthrene

Phenobarbitone

Combined effects on the HPT axis

Polychlorinated biphenyles

Figure 6.14 Compound structures with an effect on the hypothalamic–pituitary–thyroidal (HPT) axis.

index, reduced offspring per litter, reduced sperm counts, retarded vaginal opening in offspring). However, the most sensitive endpoint was liver toxicity in a canine dermal chronic study, with a LOAEL of 1.5 mg/kg bw/d. Using this study, a reference dose (RfD) of 5 µg/kg/d for the chronic dermal and oral intake with a safety factor of 300 was calculated. In 1991, the US Food and Drug Administration identified the oral route only as being relevant because PCP was found solely in food. In the USA, oral intake of PCP is estimated to be 2.4% of the RfD in children and <0.5% of the RfD in adults.

Increased Degradation of Thyroid Hormones in the Liver

An increased stimulation of thyroid hormone degradation in liver cells may result in induction of thyroid neoplasms in rats when compounds are administered in long-term studies. For the risk assessment of thyroid tumors in humans, two reference compounds – phenobarbitone and 3-methylcholanthrene – with different modes of action are used. Phenobarbitone has been administered as an antiepileptic drug for a long time without any increased rates of thyroid tumors being observed in patients. In contrast, 3-methylcholanthrene is known from animal studies to be carcinogenic (GHS classification 1B). Administered compounds may activate nuclear receptors (e.g., constitutive androstane receptor (CAR), pregnane-X-receptor (PXR), arylhydrocarbon receptor (AhR) and peroxisomes proliferator activated receptor α (PPARα)) in liver cells, which induce cytochrome P450 and Phase II enzymes such as UDP-glucuronosyltransferase, also resulting in increased degradation of thyroid hormones.

> **Phenobarbitone activates CAR and PXR followed by an induction of CYP2B enzymes, whereas 3-methylcholanthrene induces CYP1A enzymes via the AhR. CYP2B inducers do not cause thyroid neoplasms in humans.**

Polybrominated diphenyl ethers (PBDEs) were used in the past as flame retardants in plastics and textiles. Molecular structures with four to eight bromine atoms persist in the environment, hence the production of penta- and octa-BDEs was banned worldwide in 2009 in accordance with the Stockholm Convention. Although the application of deca-BDEs is also restricted, their use is still allowed as flame retardants in textiles due to the lower risk of bioaccumulation. Reproductive, neuro- and liver toxicity depends on the PBDE congener. PBDEs decrease serum thyroid hormone levels as a result of an enhanced metabolism induced by CAR- and PXR-triggered mechanisms. Apart from this, displacement of thyroid hormones from binding proteins and suppression of the thyroid-hormone-receptor-dependent gene expression are being discussed. This thyroid hormone dysregulation results in developmental neurotoxic effects, with altered locomotion activity in mice being the most sensitive endpoint.

> **The half-life of PBDEs in rodents is different to that in humans. Therefore, an internal dose at the benchmark-dose level (BMDL) of the most sensitive parameter in rodent studies is estimated. For risk assessment, the measured exposure in humans (95th percentile) is compared with the internal dose at the BMDL in a margin of exposure (MOE) model.**

This reflects the "worst-case-scenario" because the longest measured half-life in humans is taken into account. A lower MOE means a higher health risk. As an example, the MOE of the

PBDE congener BDE-47 for adults is between 90 and 38, and for children is between 27 and 11. A MOE <2.5 is assessed as a health risk for all PBDE congeners. This cut-off is reached in breast-fed infants for two congeners (BDE-99 and BDE-153). This assessment is very conservative because the half-life for PBDEs in humans is about 10 years and an internal dose level is only reached after three to four half-lives.

Thyroid Hormone Receptor Agonist

Tetrabromobisphenol A (TBBPA) is used worldwide as a flame retardant. In 2001 the production rate was assessed to be 150,000 t/a. In *in vitro* studies TBBPA has an agonist effect on estrogen receptors alpha (ERα) and an antagonist effect on progesterone receptors. In addition, it is a thyroid hormone receptor agonist in GH3 cells. In a one-generation rat study, F1 generation rats had lower serum T4 levels (BMDL: 16 mg/kg bw/d; non-dose-dependent higher testis and pituitary gland weights in the F1 generation). In another postnatal reproductive toxicity study, nephrotoxic effects were found in the progeny (NOAEL 40 mg/kg bw/d). In additional rat studies (subchronic and two-generation studies) no adverse effects were observed up to 1000 mg/kg bw/d. The endocrine effects of TBBPA in *in vitro* studies could not be confirmed *in vivo*. Contaminated food is the main source of exposure in humans. The daily intake of TBBPA is <20 ng/kg bw/d in adults and 0.3–257 ng/kg bw/d in breast-fed infants. Therefore, even with a very conservative risk assessment (NOAEL 16 mg/kg bw/d) the safety factor for the estimated exposure in breast-fed infants is >6000 and in adults is >6,000,000.

Combined Effects on the HPT Axis

Polychlorinated biphenyls (PCBs) were used worldwide as dielectrics and coolants in electrical apparatus and as plasticizers. Because they are persistent, they accumulate in the environment. PCBs result in neuro-, immune and reproductive toxicity and are carcinogenic in animal studies, hence production and use of these substances were banned worldwide in 2001 in accordance with the Stockholm Convention. Coplanar as well as non-coplanar PCBs and their hydroxylated metabolites decrease T4 plasma levels without affecting T3 and TSH levels. The mechanisms under consideration are a decreased effect of TSH on the thyroid gland, induction of T4-UDP-glucuronosyltransferase in liver cells and displacement of T4 from binding proteins. Pre- as well as postnatal administration of PCBs in rats interferes with brain development in the progeny with cognitive and motor control disturbances. Additional mechanisms for these effects may be an induction of type II deiodinase and increased expression of the thyroid hormone responsive element in the fetal brain.

6.5.3 Sex Hormones

> **Compounds affecting the HPG axis may mainly act as estrogen receptors (ERs) agonists, androgen receptor antagonists, aromatase inhibitors or a combination thereof.**

Reference compounds for these mechanisms are described in Figure 6.15.

Estrogen Receptor Agonists

The insecticide **methoxychlor (MXC)** is a pro-estrogen because its metabolite HPTE (2,2-bis(p-hydroxyphenyl)-1,1,1-trichloroethane) shows estrogen activity by binding to the

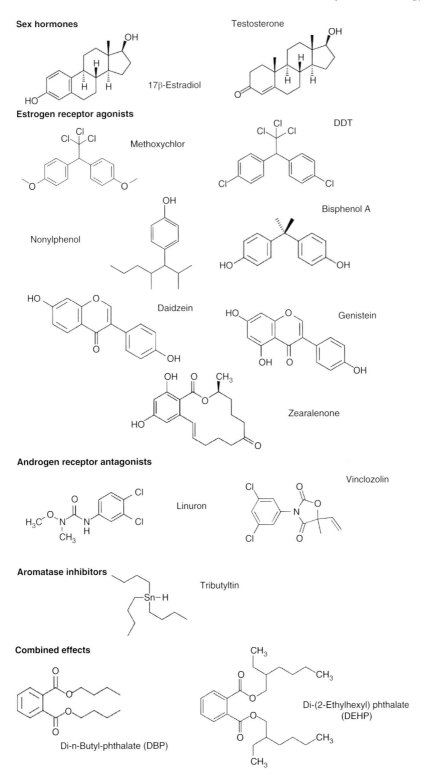

Figure 6.15 *Compound structures with effects on the hypothalamus–pituitary–gonadal (HPG) axis.*

ERα receptor. Reproductive toxicity studies with MXC administration, starting on different gestation days, resulted in pubertal developmental defects (premature vaginal opening, but retarded preputial separation in males). The uterotrophic assay is positive. Additionally, HPTE blocks androgen receptors and inhibits testosterone synthesis by decreasing the expression of the cholesterol-side-chain-cleaving enzyme (CYP450scc) as a consequence of binding to the ERα. These mechanisms explain the fertility defects of adult rats caused by MXC and HPTE beginning at a dose of 50 mg/kg bw/d, i.e. a decreasing fertility index, litter size and viability index. As a result of these endocrine disruptor effects along with bioaccumulation, the use of MXC has been banned since 2002 in the EU and since 2004 in the USA.

DDT (dichlorodiphenyltrichloroethane) was used worldwide as an insecticide in the 1970s. Because of its accumulation in fatty tissue and long persistence in the environment as well as endocrine activity in wildlife, it has been banned from agricultural use. In accordance with the Stockholm Convention 2004, DDT usage is restricted to vector control of human diseases like malaria. At the end of the 1980s the mean DDT concentration in fatty tissue of humans in North America and Europe was 1 mg/kg. DDT is a mixture of isomers: about 80% p,p'-DDT which acts as an insecticide and about 20% o,p'-DDT with a mainly estrogen effect by binding to the ER. Additionally, the metabolites p,p'-DDE and o,p'-DDE show endocrine effects, for example p,p'-DDE blocks the androgen receptor. Moreover, DDT is classified as potentially carcinogenic (GHS category 2: in animal studies, e.g. liver, pancreas, testis and mamma tumors).

Bisphenol A (BPA) and nonylphenol (NP) are basic high-volume chemicals. BPA is used as an antioxidant in epoxy resins and in plasticizers during the synthesis of plastics such as polysulfones, polyether ketones and polycarbonates. Additionally, it is used as a polymerization inhibitor in polyvinyl chloride (PVC) and as a color developer in thermal paper. NP is used in manufacturing antioxidants, lubricating oil additives, laundry and dish detergents and solubilizers. Both compounds bind weakly to ER and are weakly positive in the uterotrophic assay. They are metabolized quickly in the body into inactive metabolites (half-life of BPA in rats by oral gavage: 14–28 hours). In multigeneration studies with NP in rats, the NOAEL is 10 mg/kg bw/d with kidney damage and premature vaginal opening (estrogenic effect) in the progeny with higher doses. With BPA doses in rats of <50 mg/kg bw/d (NOAEL) no estrogenic effect but a slightly lower body weight occurred.

> The ingestion of BPA with food contaminated by packaging material represents the main uptake route in humans. According to an EU estimation, children may absorb 14 µg BPA/kg bw/d and adults about 1.4 µg/kg bw/d with food. A comparison with the NOAEL (50 mg/kg bw/d) from a multigeneration study in rats reveals a high safety factor of 3570 for children and 35,700 for adults.

Phytoestrogens belong to several chemical groups, such as flavonoids, isoflavones, lignans, stilbenes and steroids. They bind to both estrogen receptors ERα and ERβ with agonistic effects, and have a positive reaction in the uterotrophic assay. Coumestrol is the most potent phytoestrogen (100–1000-fold less active than 17β-estradiol) followed by the soybean isoflavones **genistein** and **daidzein**. The suggestion has been raised that isoflavones affect thyroid hormone levels as well, as positive results have been reported in *in vitro* studies (inhibition of thyroid peroxidase and sulfotransferase; displacement of T4 from transthyretin). Because genistein and daidzein are derived from soy beans, human exposure depends on eating habits, i.e. omnivores about 0.5 mg/day and vegetarians 5–50 mg/day. Infants with a soybean-rich diet have a high isoflavone intake of 4.5–8 mg/kg bw/d, whereas the mean daily intake in adults is 0.1–1 mg/kg bw/d.

The effect of **isoflavonoids** on hormone-sensitive breast and prostate tumors has been under detailed investigation, as it was observed epidemiologically that both types of tumor are rarer in Asian people on a soybean-rich diet than in Europeans (plasma levels of isoflavonoids in Asians 870 nmol/l; in Europeans 10 nmol/l). Regarding breast cancer, it was found in *in vitro* studies that low levels of genistein (<10 µmol/l) promote tumor growth, whereas higher levels (>10 µmol/l) inhibit growth. In postmenopausal women with breast cancer, a parallel dosage of tamoxifen (aromatase inhibitor) and isoflavonoids is contraindicated because of contrasting mechanisms. Long-term, daily administration of isoflavones in humans (150 mg isoflavones/day) results in endometrium hyperplasia in the uterus.

The mycotoxin **zearalenone** is an estrogen receptor agonist which is produced by several Fusarium species. Cereal-rich food may contain zearalenone. Pigs are the most sensitive species to zearalenone with a NOAEL for reproductive toxicity (infertility and abortion) of 40 µg/kg bw/d. In humans with normal eating habits a mean exposure to zearalenone is estimated to be 0.05–0.1 µg/kg bw/d. Children tend to have higher uptakes compared to adults: in unfavorable events the uptake may reach 1 µg/kg bw/d. Depending on the assumed safety factor a daily intake of 0.1–0.5 µg/kg bw/d zearalenone is considered to be low risk for humans. However, even in cases where this intake has been exceeded, no effects were observed.

Only a few **estrogen receptor antagonists** were observed in the estrogen receptor transactivation assay (e.g., the pyrethroids cyhalothrin and deltamethrin). **Androgen receptor agonists** were not found using the androgen receptor transactivation assay.

Androgen Receptor Antagonists

The fungicide **vinclozolin** and one of its metabolites (M2 = 3,5-DCA = 3′,5′-dichloro-2-hydroxy-2-methylbut-3-enanilide) show androgen receptor antagonistic effects (positive Hershberger assay). In two-generation studies male rats exhibited the following changes: increased plasma LH and testosterone values, and reduced anogenital distance, nipple retention, hypospadias and cryptorchidism. Female progeny showed virilization at higher dosage. In the carcinogenicity study, male rats developed Leydig cell tumors. The NOAEL in rat studies with the most sensitive effects was 1.2 mg/kg bw/d. Considering a safety factor of 1000, a tolerable daily intake (TDI) of 0.0012 mg/kg bw/d was determined. Since 2004, vinclozolin use has been restricted to special crops, like canola, therefore human exposure is estimated to be normally <1% of the TDI.

A second androgen receptor antagonist resulting in a positive Hershberger assay is the herbicide **linuron**. In a two-generation study with rats, in the male progeny changes of sex hormone levels as well as testis and epididymis weights occurred at a dose above 45 mg/kg bw/d. For the calculation of the reference dose, the most sensitive parameters were used (i.e., changes in hematology in a chronic dog study) with a NOEL of 770 µg/kg bw/d. Based on a safety factor of 100, a TDI of 7.7 µg/kg bw/d was determined. Between 1986 and 1991 in the USA, human exposure to linuron was measured at 0.0024 µg/kg bw/d for children and 0.0005 µg/kg bw/d for adults.

Aromatase Inhibitors

Tributyltin had been used as a biocide in antifouling paints, reducing the growth of organisms on the bottom of ship hulls. Because of the long half-life in ocean sediments (about two years) and consequent bioaccumulation in marine organisms, its use has been banned since 2008 by the International Maritime Organization. It was observed that in marine snails (Neogastropoda) tributyltin leads to a masculinization of female snails. Tributyltin is known to be an aromatase inhibitor, blocking the enzyme forming estrogens from androgens. However, a second mechanism of tributyltin is assumed because selective aromatase inhibitors are much less potent in inducing the masculinization of female snails.

In two-generation studies with rats, tributyltin reduced the pup weights, increased the anogenital distance, and delayed the vaginal opening in the female progeny. However, the most sensitive endpoint of tributyltin relates to immunotoxicity: it inhibits 11β-hydrosteroid dehydrogenase type 2, which oxidizes active cortisol to inactive cortisone in peripheral tissue, thus influencing the immune system. In special immunotoxicity studies in Wistar rats (reduced resistance to *T. spiralis* infections) a NOAEL of 0.025 mg/kg bw/d was observed. Based on a safety factor of 100, a TDI of 0.25 µg/kg bw/d was established for humans. Calculated intakes of tributyltin and other organotins in a human subpopulation eating mainly fish (Norway) were 0.018–0.17 µg/kg bw/d. Therefore, it cannot be excluded that eating fish from extremely contaminated locations (in the proximity of harbors) may result in uptakes above the TDI cut-off. Although some reports described acute intoxication symptoms in humans (neuronal, pulmonary and dermal toxicity), no reports about symptoms with chronic exposure above the TDI were found.

Compounds with Combined Effects

Phthalates are used as plasticizers in PVC, cellulose nitrate, and synthetic rubber. For toxicity risk assessment, it is important to differentiate between high molecular weight (e.g., di-isononyl phthalate (DINP), di-isodecyl phthalate (DIDP)) and low molecular weight (e.g., di-(2-ethylhexyl) phthalate (**DEHP**), di-*n*-butyl phthalate (DBP)) phthalates. The latter are regarded as endocrine disruptors: in adult rats, high doses result in testis toxicity and infertility in males. In prenatal development, synthesis of testosterone and the insulin-like factor 3 (insl3) in Leydig cells is reduced. Additionally, developmental defects in Sertoli cells, forming great multinucleated gonocytes, have been observed. It is important to note that when taken orally, intestinal lipases hydrolyze phthalates to more toxic monoesters. For humans, many exposure routes for DEHP are possible: oral uptake with food and beverage, inhalation of indoor air, dermal contact with plastics and in hospitalized patients the use of medical devices, like catheters. The general intake of DEHP in the Danish population was estimated to be 4.5 µg/kg bw/d for adults and 26 µg/kg bw/d for children. In a multigeneration study with rats, the NOAEL of DEHP regarding testis toxicity in the progeny was found to be 5 mg/kg bw/d (most sensitive parameter: atrophy of the seminiferous tubules in the testis). On the basis of a safety factor of 100, a TDI of 50 µg/kg bw/d was calculated.

> **For patients in intravenous contact with plastic medical devices (e.g., during blood transfusion) a TDI of 600 µg/kg bw/d DEHP was calculated. This cut-off value can be reached in acute emergency cases. In long-term blood transfusions, the exposure to DEHP is 10-fold lower compared to the TDI. In newborns in intensive care, the DEHP exposure may even rise to 3000 µg/kg bw/d when PVC-rich medical devices are used in the longer term. In this case, replacing the device with one made of a different material would be necessary. However, adverse effects in newborns with an exposure above the TDI have not been reported to date. Due to ethical limitations, detailed investigations would prove to be difficult.**

6.5.4 Low-dose, Non-monotonic Dose-effect Relation and Additive Effects

Serum hormone levels and their effects on target cells depend on the developmental stage, physiological status (e.g., sexual cycle phase) and exogen influence factors (e.g., time of day, stress). The concentrations of various hormones are related to each other and are linked to the nerve system in a network of feedback mechanisms. Depending on the sensitivity of the parameters used, changes in the hormone system and/or the target organs may be observed even if low doses are administered. It is important for the assessment that any changes are related to the compound administration and not to other secondary factors (e.g., postnatal prostate weight may be influenced by the fetal position *in utero* during intrauterine compound administration). If results from repeated studies differ significantly, the statistical power of parameters has to be taken into account. This may result in a more plausible negative result compared to a positive effect. The assessment of an adverse versus an adaptive effect is mentioned in the introduction to this chapter.

> **A low dose is defined as a dose below the NOAEL of regulatory studies (NTP, 2001).**

The question remains whether a low dose of endocrine active compounds has to follow a monotone dose–effect curve or if a non-monotonic, for example U-shaped, effect curve may be possible (i.e., occurrence of an effect at a low and a high dose, but no effect at the mid dose). Assuming the latter option, a NOAEL may not be correctly determined. However, it must be excluded that the effect is a sum of several single endpoints. For example, in an *in vitro* cell-based receptor binding assay a high, cytotoxic dose will result in a decrease of the receptor-binding signal (inverse U-shaped effect curve). There may be an increased risk that *in vivo* assay effects are influenced by several parameters, not all of them having a treatment-related cause.

> **It is necessary to define key events of a mechanistic adverse outcome pathway. Dose–effect curves can be established solely for the single key events (OECD, 2013).**

A second subject under discussion is whether an additive effect can occur when low doses of different compounds with the same molecular target are administered in parallel, but there is no effect when the compounds are administered individually. This additive effect has been shown in *in vitro* assays (e.g., estrogen receptor agonists in the receptor transactivation assay). However, the applied concentrations extrapolated to *in vivo* studies result in doses far above the NOAELs. Furthermore, the adaptation capacity of a whole organism and toxicokinetic aspects were not considered.

> **Although additive effects are described for higher effect doses of endocrine active compounds in animal studies, the extrapolation to additive effects of low doses below the NOAEL is questionable.**

6.5.5 Summary

In this chapter risk assessments for humans regarding typical endocrine active compounds are discussed. It is known that the most sensitive endpoint for these compounds is not in every case the endocrine effect, but rather other endpoints, such as the immunotoxic effect with tributyltin, hematologic effects with linuron or liver toxic effects with pentachlorophenol. Some chemicals have been banned because of endocrine effects in wildlife (e.g., DDT, tributyltin). The human risk assessment is based on the NOAEL values determined in animal studies and safety factor and human exposure calculations. A margin of exposure model was applied for the risk assessment of polybrominated diphenyl ethers, as the half-lives differ between rats and humans: an internal dose of the most sensitive parameter at BMDL was assessed in rodents and this dose was compared with the human exposure.

For some compounds, like tetrabromobisphenol A and bisphenol A, human exposure is far below the TDI. With other compounds, some cases exist where human exposure may exceed the TDI (e.g., DEHP, zearalenone). However, there are no reports on adverse effects in humans in such cases. With isoflavonoids and phenobarbitone epidemiologic studies are available which indicate adverse effects only when high doses are administered.

Finally, in this chapter the endocrine effects of low doses, additive low-dose effects and seemingly non-monotonic dose–effect curves are discussed. Here it is important to emphasize that key events in adverse outcome pathways must be regarded individually.

Further Reading

ECETOC Workshop and Technical Reports 2013 http://www.ecetoc.org/publications.

EFSA. Scientific Opinion on the risk assessment of several compounds (e.g., perchlorate, polybrominated diphenyl ethers (PBDEs), zearalenone, tetrabromobisphenol A (TBBPA), bis(2-ethylhexyl)phthalate (DEHP)), EFSA Journal, webpage: http://www.efsa.europa.eu/en/efsajournal/all.htm.

FAO/WHO. Safety evaluation of certain contaminants in food prepared by the Seventy-second meeting of the Joint Food and Agriculture Organization/World Health Organization Expert Committee on Food Additives.

IPCS Risk Assessment Terminology, Harmonization Project Document No. 1, ISBN 9241562676, 117 pages, 2004.

NTP. NTP-CERHR Monograph on the Potential Human Reproductive and Developmental Effects of Bisphenol A. Sept (2008). NIH Publ No. 08-5995. Center for the Evaluation of Risks to Human Reproduction.

NTP. National Toxicology Program's report of the endocrine disruptors low dose peer review. NTP Office of Liaison and Scientific Review NIEHS, NIH, 487 pages, 2001.

OECD Series on Testing and Assessment No 184: Guidance Document on Developing and Assessing Adverse Outcome Pathways, 45 pages, April 2013.

Owens, J.W. and Chaney, J.G. Weighing the results of differing 'low dose' studies of the mouse prostate by Nagel, Cagen, and Ashby: Quantification of experimental power and statistical results. *Regulatory Toxicology and Pharmacology*, **43**, 194–202, 2005.

US EPA. Reregistration Eligibility Decision (RED) for several compounds (e.g., Vinclozolin, Maneb, Pentachlorophenol) webpage: http://www.epa.gov/opp00001/reregistration/status.htm.

Welshons, W.V., Thayer, K.A., Judy, M., Taylor, J.A., Curran, E.M., vom Saal, F.S. Large effects from small exposures: I. Mechanisms for endocrine-disrupting chemicals with estrogenic activity. *Environmental Health Perspectives*, **111**, 994–1006, 2003

WHO Food Additives Series 63, 685-762. 2011 Available at: http://www.inchem.org/documents/jecfa/jecmono/v63je01.pdf.

WHO and UNEP (Eds. A. Bergman. J.J. Heindel, S. Jobling, K.A. Kidd, R. T. Zoeller). State of the Science of Endocrine Disrupting Chemicals – 2012, WHO Library, 2013.

Witorsch, R.J. Low-dose in utero effects of xenoestrogens in mice and their relevance to humans: an analytical review of the literature, *Food and chemical Toxicology*, **40**, 905–912, 2002.

6.6 Assessment of Xenoestrogens and Xenoantiandrogens[2]

Helmut Greim

6.6.1 Introduction

During the last decade endocrine active compounds (EACs), i.e. compounds capable of interfering with the endocrine system, have become a widely discussed issue and topic of active research in toxicology and related disciplines. An area of concern is that these agents are widely used commercial chemicals and exposure to them may result in detrimental effects in humans and animals. Chemicals which act as estrogen mimics or function as antiandrogens are known as *xenoestrogens* and *xenoantiandrogens*. They are the best studied group of EACs and illustrate the issue of endocrine-related modes of action and potential toxicities of commercial chemicals. An endocrine activity is not necessarily associated with the induction of adverse effects, EACs that induce negative health effects (via their endocrine activity) are considered as "endocrine disruptors". The following WHO definition has been generally accepted to define the endocrine issue.

> **An endocrine disruptor is an exogenous substance or mixture that alters function(s) of the endocrine system and consequently causes adverse health effects in an intact organism or its progeny, or (sub)populations.**

The essential point for toxicologists is that endocrine disruption is inherently described as a mode of action. These endocrine modes of action are among several that may potentially lead to adverse effects on reproduction, growth and development. An underappreciated implication is to distinguish assays identifying modes of action (typically referred to as screens) from the classical assays used by toxicologists to characterize adverse effects, i.e. reproductive hazards and dose responses, such as multigenerational assays.

Most EACs studied in laboratory experimental animals affect the target tissues of sex hormones; other organs may also be affected, such as the thyroid, immune and neuroendocrine system, for instance by chemicals covered in Chapter 6.5. The effects vary from subtle changes in the physiology to permanently altered sexual differentiation. The spectrum of changes is well known from reprotoxicity studies with potent hormones developed as drugs.

> **Exposure to EACs is considered to be particularly critical during prenatal and postnatal phases and may later result in a permanent change of function or sensitivity to hormonal signals.**
>
> **Exposure to EACs in adulthood may be compensated for by homeostatic mechanisms or result in toxicities that are reversible when exposure ends.**

An instructive case is that of **diethylstilbestrol** (DES), an estrogen whose potency is equivalent to ethinylestradiol (EE) and that readily passes the placenta. In the 1950s and 1960s, DES was

---

[2] This chapter is based on the chapter "Xenoestrogens and Xenoandrogens" by G.H. Degen and J.W. Owens of the 1st edition of this book published in 2008.

prescribed to a very large number of pregnant women at massive doses. These doses ranged from 5 mg per day up to 125 mg per day or approximately 2 mg/kg body weight/day (bw/d), which compares to a normal contraceptive dose of 1 μg EE/kg bw/d for birth control.

DES use was found to be associated with an increased risk of breast cancer in the "DES mothers", who received high doses. In their male offspring exposed to DES *in utero*, urogenital tract abnormalities (cryptorchidism, hypospadias, epidydimal cysts, and hypoplastic testes) were found more frequently than in non-exposed males. In grown-up "DES sons", average sperm counts were only slightly lower than in controls. Although sperm quality was clearly lower, this has been attributed to a higher incidence of hypoplastic testes in this group. However, there was no indication for reduced fertility when these individuals were studied again at the age of 38–41. Earlier concerns of a possibly increased incidence of testicular tumors (untreated cryptorchidism is a known risk factor) have not been confirmed in more recent epidemiological studies.

DES increases the incidence of a very rare tumor in young women exposed *in utero* during critical phases of development. The risk of this cancer (clear cell adenocarcinoma of the vagina) was estimated to be in the order of 1 per 1000 in the "DES daughters". Other, non-malignant reproductive tract abnormalities were observed much more frequently in this group.

Recent studies compared breast cancer incidence in women with or without prenatal DES exposure (where cohorts have been followed since the 1970s). The results indicate that women with prenatal exposure to DES have an increased risk of breast cancer after the age of 40 years. Apparently, *in utero* exposure to DES significantly increases postmenopausal, but not premenopausal breast cancer incidence.

Work with DES in experimental animals has greatly improved our knowledge on the spectrum of effects to be expected in males and females from perinatal EAC exposure, and even predicted some outcomes observed later in clinical studies on DES-exposed humans. DES studies also provide important data on **dose–effect relationships** in rodents and in humans. Epidemiological data gave some insight because of marked differences in DES dosing schedules used at various clinical centers. In cohorts of males prenatally exposed to comparatively low maternal doses (estimated mean total of 1.4 g at the Majo Clinic) there were no indications for adverse consequences such as those described in the high-dose DES cohorts (estimated mean total dose of 11.6 g at the University of Chicago). The human data are consistent with the existence of maternal dose levels below which adverse non-cancer effects may not occur. The extensive rodent DES reproductive/developmental toxicity database is also consistent with this view: DES effects on fertility and the occurrence of genital tract abnormalities are consistent with the observations of dose levels below which adverse non-cancer effects may not occur, i.e. dose–response thresholds.

> **Many well-conducted studies on the reproductive/developmental toxicity of various xenoestrogens confirm the view that adverse effects are related to the dose and hormonal potency of the compound in question.**

6.6.2 Modes of Action and Testing

The fundamental hormonal signaling system is based on a feedback system involving the hypothalamic/pituitary/gonad axis, and this is schematically depicted in Figure 6.16. As the steroid hormones act via an intracellular, nuclear receptor (see Chapter 2.6), the feedback system can be modulated in two basic ways: (1) agonists or antagonists of the respective estrogen and androgen receptors and (2) interference with steroid biosynthesis and metabolism, particularly important terminal steps such as the conversion of testosterone to estrogen by aromatase and the conversion of testosterone to the more potent dihydrotestosterone by 5α-reductase.

The set of *in vitro* and *in vivo* tools developed by research pharmacologists are now being applied to address questions about weakly potent commercial chemicals. This section is devoted to briefly illustrating several of these tools and their potential application.

Screening for Modes of Actions **in vitro**

Data from *in silico* (structure-activity relationships) and *in vitro* studies can provide a very rapid orientation on whether compounds may interact with elements of the endocrine system and the (relative) hormonal potency of chemicals, e.g. an estrogenic activity in relation to estradiol, and do so while avoiding the use of laboratory animals. However, there are also limitations:

(i) *in silico* systems do not incorporate and cell-based *in vitro* systems often lack enzymes required for the conversion of pro-hormones to biologically active metabolites (e.g., methoxychlor)
(ii) neither do they reflect other important aspects of the toxicokinetics (bioavailability, tissue distribution) of a given compound that also determine their potency *in vivo*
(iii) cell culture systems may simulate the responses of one but not those of another target organ, therefore *in silico* systems and *in vitro* assays are used for rapid prioritization, thereby avoiding using significant numbers of animals. However, it is inappropriate to use *in silico* to identify substances as potential endocrine disruptors, and *in vitro* systems should be used with caution.

Direct Receptor Binding A significant number of xenobiotics bind to estrogen receptors (ER, more precisely the subtypes ERα and ERβ). A smaller number interact with androgen receptors (AR), mostly acting as antagonists. While endogenous steroid hormones act in nM or sub-nM concentrations, the binding affinity of environmental chemicals is often several orders of magnitude lower than that of the endogenous ligands. Thus, efficient receptor occupancy by xenobiotics requires μM to mM intracellular concentrations of the free ligand. Accordingly, higher xenobiotic doses are needed to elicit hormonal effects *in vitro* or in bioassays *in vivo*.

Advantages and limitations of existing test methods are briefly discussed below. Compounds and metabolites that bind directly to sex hormone receptors (Figure 6.16, case A) are rather easy to detect and this may explain the higher number of suspect compounds by comparison to EACs operating by indirect modes of actions.

Steroid Biosynthesis Chemicals may affect the biosynthesis of sex hormones (Figure 6.16, case B) and/or its regulation by the hypothalamic–pituitary–gonadal axis. With several enzymes catalyzing the many steps in steroid biosynthesis from cholesterol, the challenge is to pinpoint the site of interference. The recently developed adenocarcinoma cell line H295R possesses an intact steroid synthesis pathway and therefore can identify modulation of the entire steroidogenesis pathway from cholesterol recruitment by the STAR protein through the terminal conversion of androgens to estrogens. This is accomplished by analytical determination of the concentration of the final and intermediate products in response to doses of the putative disruptor.

Steroid and Thyroid Hormone Metabolism The transport and degradation of steroid and thyroid hormones (Figure 6.16, cases C and D) may be altered by xenobiotics, e.g. by inducers or inhibitors of hepatic enzymes. A large number of drugs and natural food ingredients are known to modulate expression or activity of Phase I/II enzymes. These indirect and difficult to demonstrate mechanisms reinforce the need for rather stringent definitions of (potential) endocrine disruptors. This is now widely adopted (see the definition at the start of this chapter), and the following

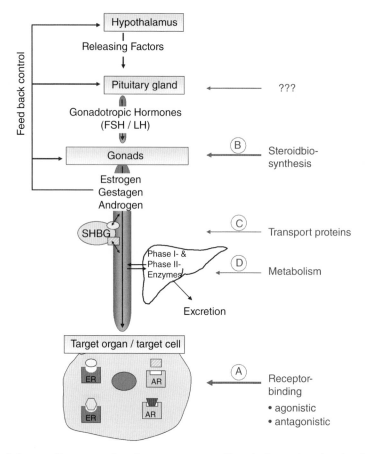

Figure 6.16 *Scheme of hormone signaling components. Chemicals can interfere by different mechanisms (A to D) with the sex hormone system (Chapter 6.6.2).*

statement acknowledge the higher rank of *in vivo* studies for a risk characterization of EACs. Hepatocyte cell lines employing toxicogenomic and proteomic techniques are under development to identify the genes and metabolic pathways that can be induced by both natural and anthropogenic compounds. The capability of these enzymes to metabolize steroids and thyroid hormones can then be assessed.

Screening for Modes of Actions *in vivo*

The current regulatory strategy to identify potential EACs is to adapt and utilize existing bioassays as they are rapid, use a minimum number of animals, have proven to be highly specific for drugs, and often correlate to assays for adverse reproductive effects. The advantages are that they incorporate toxicokinetics, but due to their short periods of administration, the MTD in subchronic and chronic studies can be exceeded.

Estrogens and Antiestrogens The uterotrophic bioassay has been employed for over 70 years to identify estrogens and was adapted in the 1970s to identify antiestrogens. Uterine growth is a direct target for estrogens, and the uteri of ER knock-out mice do not respond to estrogens. The bioassay has been validated by the OECD using several weak, commercial estrogenic compounds using small group sizes ($n = 6$). The blotted uterine weight is equally sensitive to histological markers such as epithelial cell height of the uterus, and recent toxicogenomic studies have confirmed the estrogenic nature of the weight increase. For estrogens, with the short estrous cycle of rodents, after three consecutive days of administration, the uteri from the putative test substance groups is compared to the uteri from vehicle controls on day 4. For antiestrogens, a stimulating dose of EE and the putative antiestrogen are coadministered to assess the inhibition of uterine growth.

Androgens and Antiandrogens The Hershberger bioassay has also been employed for over 70 years to identify androgens and was adapted in the late 1960s to identify anitandrogens. Male accessory reproductive tissues and muscles such as the levator ani are direct targets for androgens, and the tissues of AR knock-out mice as well as humans with genetic defects in the AR (androgen insensitivity syndrome) have little or no response to androgens. The bioassay has been validated by the OECD for regulatory use using several weak, commercial antiandrogen compounds, again using small group sizes ($n = 6$). Androgens are administered to castrate males after a period of tissue regression, while antiandrogens are coadministered with a stimulating dose of testosterone propionate.

Aromatase The uterotrophic bioassay has been modified by the pharmaceutical industry for over 20 years to identify aromatase inhibitors. The immature rodent is administered a potent, aromatizable androgen, and upon conversion to estrogen rapid uterine growth is observed. To study aromatase inhibitors, the putative inhibitor is coadministered with the androgen,and, if effective, the increase in uterine weight is inhibited.

5α-Reductase The antiandrogen protocol of the Hershberger bioassay has shown to present a feasible *in vivo* model for 5α-reductase inhibition, using finasteride as a model compound during the OECD validation program. Testosterone propionate was coadministered, and decreases in tissue weight gains were elicited by the finasteride.

General Steroidogenesis A protocol based upon the 15-day intact male assay has been developed that presents an *in vivo* model for steroidogenic inhibitors as well as other modes of action. The premise of the assay is to combine a battery of tissues and circulating levels of several serum hormones to produce a profile characteristic of endocrine modes of action. The assay requires a larger group size ($n = 15$) to provide sufficient power for the hormone analyzes, a detailed protocol to minimize hormonal variation, and a laboratory with expertise in sampling animals and running a battery of hormonal assays. The assay has been demonstrated against pharmaceutical compounds, and awaits formal validation with weakly potent compounds. Classical assays like 28- or 90-day repeat-dose studies in rodents can also contribute to the evaluation of such effects, but as normally only weights and histology of hormone producing glands and their target tissues are investigated, little information on the exact mode of action is observed.

Thyroid Toxicity Thyroid toxicity is an example where classical assays are already in place and sufficient to evaluate this endocrine mode of action, such as thyroid weight changes and histopathological evaluations after 28- or 90-day repeat-dose studies in rodents. These can be supplemented with hormone assays where animal handling, serum sampling and hormonal analyses are carefully done to minimize T3 and T4 variations. Alternatively, the 15-day intact male assay can be used. The rat is a sensitive model due to the lack of a specific, high affinity serum carrier for thyroid hormones, and should be sensitive for vertebrates in general. However, the precise mode of action is not identified, e.g. synporter inhibition, peroxidase inhibition, or increased metabolic removal of circulating hormone.

Steroid and Thyroid Hormone Metabolism The liver is a primary target for the induction of metabolic pathways that increase the excretion of steroid and thyroid hormones. Classical observations of increased liver weight and histopathology are now being supplemented with identification of individual messenger RNA and protein quantities/activities for specific metabolic enzymes, while toxicogenomic and proteomic techniques are under development to elucidate complete pathways of metabolism. Metabolic screens of serum, bile and feces also provide the capability to assess increased hormonal turnover when the potential for increased hormone metabolism has been identified.

Wildlife Mode of action assays have been also developed in certain wildlife models. The egg lipoprotein, vitellogenin (VTG), is being validated in fish as a marker for estrogens, antiestrogens, and aromatase inhibitors. VTG production in the liver and its circulating levels are under estrogen control. The external features of sexually dimorphic species such as the nuptial fat pad and sexual tubercles in the fathead minnow are under the control of androgens. They are now being explored as suitable parameters as well as androgen-specific proteins, such as the production of spigin protein in the kidney of the three-spined stickleback.

6.6.3 A Weight of Evidence Approach and Future Improvements

Data for the interaction of a chemical with components of the endocrine system and any adverse health outcome will come from a variety of sources, and thus evaluation will require a weight of evidence approach. The weight of evidence approach has several components, many of which were noted by Hill (1965), such as plausibility, reproducibility, coherence, consistency, and relevance. A set of decision rules is also needed, such as the toxicological relevance of *in vivo* screens should outweigh or supersede *in vitro* findings. These *in vivo* screens then may provide a basis for estimating whether exposures (1) may be sufficient to pose a threat to wildlife or humans or (2) large margins of safety may exist between doses that could lead to endocrine activity in screens and actual exposures. In this latter case, in order to use a screen in this context, comparative data between the screen LOEL and those observed in definitive studies will be needed. Thus, the entire knowledge set of hazard identification, characterization of endocrine modes of action responses, and exposure should be considered together to identify those compounds requiring characterization for adverse effects.

In regards to identifying specific modes of action, the promise of the application of toxicogenomic and metabolic techniques is worthy of note. All of the above screens have limitations and a certain low rate of false positives has been observed in the uterotrophic and Hershberger validation programs. The inherent advantage to toxicogenomics is the specificity of a gene profile where the annotation of individual genes supports a specific mode of action. Work has already been conducted by several laboratories showing that a specific profile of uterine and ovarian genes can be identified that correlate with uterotrophic action. Work has also been recently published to elucidate similar profiles for androgens and antiandrogens.

Definitive tests are exemplified by the multi-generation reproductive and development assays which identify adverse effects in both adult sexes and adverse effects in offspring elicited in the absence of maternal toxicity. These definitive tests are the basis for both classification and labeling as well as the NOAELs used to establish regulatory allowable daily intakes (ADIs) or margins of exposure or safety (MOS). A number of endpoints are associated with endocrine action, such as premature vaginal opening in the case of estrogens and delayed preputial separation, the appearance of specific patterns of male reproductive tract malformations, and nipple/areola retention in the case of antiandrogens. However, such endpoints have not traditionally been commonly employed, so these data may be absent from many studies. Even with these endpoints, definitive assays for adverse effects are not specific for modes of action. Thus, clearly defined criteria to accept a causal linkage between positive findings in mechanistic assays and the appearance of particular endpoints in definitive assays is needed. Two criteria would be (1) a similarity of the dose–response characteristics (similar LOELs and LOAELs) and (2) that the putative endocrine-related effect observed in the definitive study was indeed the clear primary effect in the definitive assay. Again, a weight of evidence approach will be needed combining both screening and definitive data to arrive at the conclusion that a compound fulfills the Weybridge definition of a true endocrine disruptor that adverse effects are the result of an endocrine mode of action, as illustrated in the box below.

> **In assessing whether a compound is indeed an endocrine disruptor, it is important to distinguish between the responses observed in *in vitro* and *in vivo* screens, and the observation in a definitive assay of an *adverse* effect. Screening assays such as the uterotrophic bioassay generate alerts to *potential endocrine disruptor*. These and other data can be utilized to prioritize substances for definitive tests. These assays, such as one- or two-generation rodent assays or lifecycle assays in non-mammalian species, determine the toxicological consequences and characterize the hazard. Together, these data indicate whether an adverse effect has occurred that is due to an endocrine mode of action and thus fulfills the Weybridge definition of an *endocrine disruptor*.**

6.6.4 Limited Evidence for Endocrine Disruption

A number of compounds in Table 6.16 have been suggested to be endocrine disruptors based upon limited evidence or uncertain interpretations of data. Examples include lindane, mirex, heptachlor, and toxaphene. To date, these compounds do not have the structural alerts for well-known modes of action such as estrogen and androgen receptor binding and have not been positive in well-performed *in vitro* and *in vivo* screens. This points to the need for validated regulatory assays, a tiered framework for a battery of assays, and the need for a weight of evidence approach to arrive at conclusions that substances are potential endocrine disruptors. Subsequently, there is the need for a clearly defined set of endpoints in definitive assays that would lead to a clear conclusion that adverse effects have occurred through an endocrine mode of action.

Other compounds present additional difficulties. The polychlorinated and polybrominated biphenyls, the polybrominated diphenyl ethers, and the polychlorinated dibenzodioxins and furans are complex chemical isomer families. Several isomers are potent binders of the Ah receptor, which appears to cross talk with the ER, so that the compounds may have antiestrogenic properties. Other compounds may displace thyroid hormones from circulating serum carrier proteins, accelerating metabolism. In these cases, the action is indirect, and the mode of action and relationship to adverse effects are difficult to establish with sufficient toxicological certainty to establish a weight of evidence. As a result, weight of the evidence reviews by the International Program for Chemical Safety have concluded that the data are insufficient to say these compounds fulfill the Weybridge definitions for being endocrine disruptors.

6.6.5 Summary

Chemicals of different origin which act to mimic estrogens or act as antiandrogens may be detrimental to reproduction and development in humans and animals. The evidence for a causal link

Table 6.16 Environmental chemicals of natural origin and of anthropogenic origin.

Substances by class	In vitro screen[a]	In vivo screen[b]	Definitive data
Mycoestrogens natural chemicals			
Zearalenone [E]	+	+	+ Domestic animals
(α-,β-)Zearalenol [E]	+	+	+ Domestic animals
Phytohormones natural chemicals			
Coumestrol [E]	+	+	+ Domestic animals/reproductive studies[c]
Daidzein [E]	+	+	+ Short-term developmental study[c]
Equol [E]	+	+	+ Domestic animals
Genistein [E]	+	+	+ Domestic animals/reproductive studies
Glycitein [E]	+	+	No study available
Glycyrrhetinic acid [GC]	NA	NA	Reproductive studies not available; + human case reports [GC]
Resveratrol [E]	+	Mixed	No study available
ß-Sitosterol [E]	+	+	Studies inadequate for determining estrogenic and androgenic activity
Current and banned pesticides, anthropogenic			
Aldrin [E]	–/+	–	– Reproductive study[d]
Atrazine [E, AE, AA]	–	–	– Reproductive study[e]
Chlordecon/kepone	–	–	Studies inadequate for determining estrogenic and androgenic activity
o,p'-DDT, o,p'-DDE [E]	+	+	+ Short-term developmental study
p,p'-DDT, p,p'-DDE [AA]	+	+	+ Short-term developmental study
Dieldrin (major metabolite of aldrin) [E]	–/+	–	– Multi-generation repro study[d]
Endosulfan [E]	–	–	– Reproductive study[d]
Fenarimol [AA]	+	+	+ Reproductive study
Heptachlor	–	–	– Multi-generation repro study[d]
β- and γ-hexachlorohexane (Lindan)	–	–	– Multi-generation repro study[d]
HPTE [E, AA]	+	+	+ Metabolite – see methoxychlor
Linuron [AA]	+	+	+ Reproductive study
Methoxychlor [E, AA]	+	+	+ Reproductive study
Mirex	–	–	Studies inadequate for determining estrogenic and androgenic activity
Procymidone [AA]	+	+	+ Multi-generation repro study
Toxaphene [E]	–/+	–	Studies inadequate for determining estrogenic and androgenic activity
Vinclozolin (M1 and M2) [AA]	+	+	+ Multi-generation repro study
Natural steroids in meat and food products			
Equilin, Equilenin [E]	+	+	No study available
Estradiol [E]	+	+	+ Multi-generation repro study
Estrone, Estriol [E]	+	+	+ Reproductive study
Progesterone [G]	+	+	+ Human volunteers and + reproductive studies
Testosterone [A]	+	+	+ Reproductive study
PBTs (some banned, some under consideration for ban) anthropogenic			
PBDEs	–	–	– Reproductive studies[f]
PCBs	–	–	– Reproductive studies[f]
Hydroxy PCBs [E]	+[g]	Mixed	– Reproductive studies[f]
PCDDs and PCDFs [AE]	–	–	+ Reproductive study, cancer study

Table 6.16 (continued)

Substances by class	In vitro screen[a]	In vivo screen[b]	Definitive data
TCDD [AE]	−	−	+ Reproductive study, cancer study
Tributyltin [A, GC]	−	−	+ Environmental populations/+/− multi-generation repro study[h]
Commercial substances anthropogenic			
Amsonic acid (DAS) [E]	+	+	No study available
Bisphenol A [E]	+	+	− Multi-generation repro study[d]
Butylbenzylphthalate [AA]	−	−	+ Multi-generation repro study[i]
Di-N-butylphthalate [AA]	−	−	+ Reproductive study[i]
Diethylhexylphthalate [AA]	−	−	+ Multi-generation repro study[i]
Nonylphenol [E]	+	+	+ Multi-generation repro study
Octylphenol [E]	+	+	− Multi-generation repro study
Organosiloxanes (D4, D5) [E]	+	+	− Multi-generation repro study
Pharmaceuticals anthropogenic			
Anastrozole [AE]	+	+	+ Multiple studies
Cyproterone acetate [AA, G]	+	+	+ Multiple studies
Ethinylestradiol [E]	+	+	+ Multiple studies
Finasteride [AA]	+	+	+ Multiple studies
Flutamide [AA]	+	+	+ Multiple studies
Tamoxifen [AE]	+	+	+ Multiple studies
Trenbolone [A]	+	+	+ Multiple studies

Agents and metabolites with hormonal activities other than estrogenicity [E] are marked as anti-/androgenic [AA, A], antiestrogenic [AE], gestagenic [G], or glucocorticoid-like [GC] and + positive, − negative, or +/− equivocal.
NA, not available.
[a] *in vitro* assay: estrogen or androgen receptor binding or reporter gene cell-based assays.
[b] *in vivo* assay: uterotrophic or Hershberger assays.
[c] Observation of estrogenic-related (e.g., accelerated vaginal opening in reproductive study or for antiestrogenic, decrease in mammary cancer) or androgenic-related (e.g., phallic clitoris and masculinization in female offspring and, for antiandrogenic, malformations after *in utero* exposure during gestational days 14–18) findings.
[d] No observation of estrogenic or androgenic-related findings.
[e] Findings indicate action as dopamine antagonist.
[f] Evidence suggests some congeners may displace T3 and T4 from carrier proteins, resulting in accelerated metabolism leading to thyroid toxicity.
[g] Only a subset of the family of congeners.
[h] Some changes in anogenital distance suggesting possible androgenic activity, but other endpoints not affected.
[i] Evidence suggests no direct antiandrogenic action at the AR receptor, but action by interference with testosterone synthesis.

between exposures to such agents and adverse health effects in humans is limited to prenatal exposures to pharmacological doses of the potent drug diethylstilbestrol. Endocrine disruption in wildlife observed in areas with high levels of pollution has been linked to persistent and bioaccumulating chemicals such as DDT and TBT.

The question as to what extent xenoestrogens and xenoantiandrogens can indeed exert adverse effects on humans remains somewhat controversial. Some uncertainties remain regarding the role of combination effects, the existence of practical thresholds, and the complex regulation of the endocrine system, but it is recognized now that:

(i) risk assessments for synthetic chemicals with hormonal activity should also take into account naturally occurring compounds which may act as endocrine modulators
(ii) adverse effects of xenoestrogens and xenoantiandrogens are related to time of exposure, dose and potency of the compound in question, regardless of its mode of action (direct receptor binding or indirect effects on the hormone system) and origin.

A toxicological evaluation of the hazards, mode of actions, and risks from endocrine active compounds is feasible within the existing regulatory framework. This requires data from animal studies on dose–effect relationships for relevant endocrine endpoints and reliable information on human exposure, as well as data on the kinetics to aid in route-to-route and species extrapolation. The database for judgments on the biological relevance of human exposure to endocrine active chemicals with foods and other consumer products has improved. Margins of safety assessed case by case (e.g. bisphenol A, zearalenone) or when appropriate by a group approach (e.g. phthalate esters) cover a wide range, thereby allowing a focus on "suspects" that deserve further investigation.

The more difficult challenge may be conducting a weight of evidence review that ascertains that a substance can indeed act in the intact animal by an endocrine mode of action (potential EDC), that the endocrine-related effects are observed in a definitive tier test, and finally that the endocrine-related effects are the primary effects that are causally related to the mode of action (a true EDC).

Further Reading

Bolt HM, Janning P, Michna H, Degen GH (2001) Comparative assessment of endocrine modulators with oestrogenic activity. I. Definition of a hygiene-based margin of safety (HBMOS) for xenooestrogens against the background of European developments. *Arch Toxicol* **74**:649–662.

Committee on Toxicity of Chemicals on Food/Consumer Products and the Environment. COT Report-Phytoestrogens and Health. London: Food Standards Agency; 2003; available at http://www.food.gov.uk/multimedia/webpage/phytoreportworddocs.

Damstra T, Barlow S, Bergman A, Kavlock R, Van Der Kraak G, eds. (2002) Global assessment of the state-of-the-science of endocrine disruptors. International Programme on Chemical Safety; WHO/IPCS/EDC/02.2; http://ehp.niehs.gov/who/.

David RM (2006) Proposed mode of action for in utero effects of some phthalate esters on the developing male reproductive tract. *Toxicol Pathol* **34**: 209–219.

European Commission, Scientific Committee on Food. Opinion of the Scientific Committee on Food on Bisphenol A. SCF/CS/PM3936 Final 3 May 2002; EC, Brussel, Belgium; http://europa.eu.int/comm/food/fS/sc/scf/index_en.html.

European Commission, Scientific Committee on Consumer Safety. Memorandum on Endocrine Disruptors. SCCS/1544/14. Adopted on 16 December 2014.

European Food Safety Authority (2013) Scientific Opinion on the hazard assessment of endocrine disruptors: scientific criteria for identification of endocrine disruptors and appropriateness of existing test methods for assessing effects mediated by these substances on human health and the environment. *EFSA Journal* **11**(3):3132; http://www.efsa.europa.eu/en/efsajournal/doc/3132.pdf.

European Food Standards Agency (2004) Opinion of the Scientific Panel on Contaminants in the Food Chain on a request from the Commission related to Zearalenone as undesirable substance in animal feed. *The EFSA Journal* **89**:1–35; http://www.efsa.eu.int/.

Golden RJ, Nollet KL, Titus-Ernsthoff L, Kaufman RH, Mittendorf R, et al. (1998) Environmental endocrine modulators and human health: an assessment of the biological evidence. *Crit Rev Toxicol* **28**:109–227.

Hill AB (1965) The environment and disease: Association or causation? *Proc Roy Soc Med* **58**:295–300.

Kelce WR, Wilson E (2001) Antiandrogenic effects of environmental endocrine disruptors. In: Metzler M (Ed.) Endocrine Disruptors (Part I) Springer-Verlag, Berlin Heidelberg, pp. 40–61.

Kennel PF, Pallen CT, Bars RG (2004) Evaluation of the rodent Hershberger assay using three reference endocrine disruptors (androgens and antiandrogens). *Reprod Toxicol* **18**:63–73.

Mueller SO (2004) Xenoestrogens: mechanisms of action and detection methods. *Anal Bioanal Chem* **378**:582–587.

National Toxicology Program (NTP), Center for the Evaluation of Risks to Human Reproduction (CERHR) (2005) *NTP-CERHR Expert Panel update on the reproductive and developmental toxicity of di(2-ethylhexyl)phthalate*; http://cerhr.niehs.nih.gov/reports/index.html.

Newbold R (2002) Effects of perinatal estrogen exposure on fertility and cancer in mice. In: Metzler M (Ed.) *Endocrine Disruptors (Part II)* Springer-Verlag,, Berlin Heidelberg, pp. 171–186.

Palmer JR, Wise LA, Hatch EE, Troisi R, et al. (2006) Prenatal diethylstilbestrol exposure and risk of breast cancer. *Cancer Epidemiol Biomarkers Prev* **15**:1509–1514.

Sharpe RM, Irvine DS (2004) How strong is the evidence of a link between environmental chemicals and adverse effects on human reproductive health? *Brit Med J* **328**:447–451.

Tinwell H, Ashby J (2004) Sensitivity of the immature rat uterotrophic assay to mixtures of estrogens. *Environm Health Perspect* **112**:575–582.

Vos JG, Dybing E, Greim HA, Ladefoged O, Lambre C, Tarazona JV, Brandt I, Vethaak AD (2000) Health effects of endocrine-disrupting chemicals on wildlife, with special reference to the European situation. *Crit Rev Toxicol* **30**:71–133.

6.7 Solvents

Wolfgang Dekant

6.7.1 Introduction

Many chemicals are used individually and in the form of mixtures as solvents in industry and technology. Chemicals with typical solvent properties, but with very low boiling points, are also used as blowing gases for the application of foams and as refrigerants in air-conditioning systems. Propellants for gasoline and diesel engines also consist mainly of chemcials with typical properties of solvents; gasoline is also partly used as a solvent ("white spirit"). When selecting a solvent for a particular process, safety-related problems such as flammability or formation of explosive mixtures with air, toxicity, degradability in the environment, and physicochemical parameters such as volatility, solubility and evaporation temperature play a role. Main application areas of technical solvents with possible exposure outside the workplace are applications for the dilution of paints, varnishes and adhesives, and as cleaning agents. Major applications with workplace exposure are metal degreasing and textile cleaning as well as the use of solvents as reaction media for chemical syntheses.

> **Due to the volatility of many solvents required for the applications, inhalation is the most important exposure pathway for humans. Certain solvents can also be absorbed through the skin when large areas of the skin are exposed. The result may be toxic effects after skin contact. Ethanol also exhibits typical physicochemical and toxicological properties of a solvent and is also used as such. The absorption of large quantities of alcoholic drinks within a short time therefore also leads to the typical toxic effects of an acute solvent intoxication.**

Solvent intoxications show typical symptoms which are often very specific and, like all toxic effects, are decisively influenced by exposure duration and exposure level:

- After short-term inhalation exposure to high concentrations, as with all lipophilic chemicals, the efficient absorption from the lung into the blood and the rapid transport of solvents dissolvedin the blood into well-perfused tissues such as the central nervous system and the brain rapidly results in high concentrations in these tissues. The specific membrane changes triggered by high concentrations of lipohilic solvents in the brain lead to effects which are very similar to the narcotic effects of inhalation narcotics. Some previously widely used solvents such as chloroform, diethyl ether and trichloroethene have therefore also been used as inhalation anesthetics. The currently used inhalation narcotics are specifically developed compounds without importance as solvents. The narcotic effect of lipophilic compounds is a direct physical effect; chemically inert gases such as the lipophilic xenon also induce marcosis when inhaled in sufficient quantities. The potency of the narcotic effect of specific solventsvaries widely.
- Because of psychoactive and mood-enhancing ("high") effects, solvents are also abused as drugs. Abuse is facilitated by the easy availability of solvent-containing products and usually takes place by inhalation; many solvent-containing products, but also gases such as butane from gas stove cartridges, are used for purposes of "flushing". For some of the substances used, the difference between doses that induce psychoactive effects or toxicity is low, and lethal intoxications are reported after acute exposures. For long-term abuse, typical and chemical-specific symptoms of the intoxication with the respective solvents are observed.
- After subchronic exposure to solvents, mainly after occupational exposures, decreased concentration, fatigue, insomnia and other nonspecific symptoms are reported. These are sometimes attributed to "solvent encephalopathy", but their presence is controversial and usually cannot be reconciled with the known mechanisms of action of the toxic effects of solvents. As triggering factors for the psychical changes, multiple exposures to very high concentrations with acute-toxic effects are discussed.
- Chemical-specific effects play an important role in chronic solvent poisoning. For many solvents, chemcial-specific damage to the peripheral nervous system, liver or kidneys is observed after long-term exposure. The potency and the affected target organs are different from solvent to solvent, and often the effects on target organs are highly specific. These organ-specific effects are based on metabolic activation reactions, which are well-characterized for many chemicals used as solvents.
- In the case of skin or mucous membrane contact, many solvents lead to a rapid degreasing of the outer skin, partly coupled with a strong irritation of mucous membranes.

6.7.2 Toxicology of Selected Solvents

Halogenated Aliphatic Hydrocarbons

> **The introduction of halogen atoms to aliphatics and olefins reduces their chemical reactivity. The solubility properties of such substances for fat and oils are also improved by the presence of halogen atoms. Technically, large amounts of halogenated hydrocarbons – mainly fluorine and chlorine compounds – are still used as solvents. Because of the high lipid solubility, all compounds from this group have narcotic effects after short-term intake of high doses; other effects of acute intoxications differ from substance to substance.**

The extent to which the starting compounds undergo bioactivation reactions is different and critical for the acute toxicity of substances such as chloroform and carbon tetrachloride to the liver. For other compounds such as tri- and perchlorethene, bioactivation is not important for acute effects, and narcotic effects are responsible for acute toxicity. Many halogenated solvents show characteristic toxic effects after prolonged exposure, and metabolic activation reactions are responsible for the induction of these effects.

Trichloroethene and Perchloroethene Trichloroethene and perchlorethene are used in many processes and remain important technical solvents due to their low acute toxicity and lack of flammability. Perchlorethene is mainly used in chemical cleaning and trichloroethene for the degreasing of metal parts in the metal-processing industry.

In cases of acute oral uptake of high doses or inhalation of high concentrations, a narcotic effect is the main concern in the case of tri- and perchlorethene; lethality after acute exposures is usually due to respiratory paralysis. Because of the low acute toxicity and good narcotic properties, trichloroethene was used as an inhalation anesthetic until the 1950s. This use has been discontinued since trichloroethene can decompose by contact with basic components of the anesthetic system and form the highly toxic dichloroethyne. There were cases where dichloroethyne was formed in submarines after trichloroethene was used for purification purposes and the air passed through alkaline filters for absorption of carbon dioxide. Dermal contact with both tri- and perchlorethene leads to the degreasing of the skin without any further toxic consequences.

Both after ingestion and after inhalation, trichlorethylene and perchlorethene are rapidly absorbed into the systemic circulation and, due to high lipophilicity, distribute into body fat with the possibility of storage after prolonged and high exposure. The extent of the metabolic conversion of trichloroethene is dependent on the exposure level and the route of exposure; even at higher doses, a considerable fraction of trichloroethene absorbed is biotransformed and excreted as polar metabolites with the urine. In contrast, excretion of absorbed perchlorethene occurs mainly by slow exhalation due to the high chemical stability and the higher boiling point compared to trichlorethylene. The main pathway of the biotransformation of both haloolefins is metabolic oxidation by cytochrome P450 enzymes. These enzymes catalyze the conversion of trichloroethene to chloral, which is further converted to trichloroacetic acid and trichloroethanol. In the oxidative metabolism of perchlorethene, trichloroacetic acid (Figure 6.17) is the first stable metabolite. The second metabolic pathway via glutathione conjugation is more pronounced with perchlorethene (Figure 6.17). With trichloroethene, this pathway plays only a very minor role in the biotransformation. In contrast to most other substances where coupling with glutathione is a detoxification, the formation of glutathione S-conjugates for tri- and perchlorethene is an activation reaction. The degradation of the initially formed glutathione S-conjugates produces reactive metabolites in the kidney, whose interaction with biological macromolecules is responsible for the renal toxicity effects of long-term exposure to very high doses of tri- and perchlorethene in rats.

Chlorinated Methanes

Chlorinated derivatives of methane are widely used because of their favorable solvent properties, low combustibility and low cost.

Tetrachloromethane Tetrachloromethane (carbon tetrachloride) has often been used as a solvent due to its inexpensive manufacture. However, due to its high toxicity, the use of tetrachloromethane has been severely restricted. Ttetrachloromethane is a by-product of some technical syntheses.

Acute poisoning by tetrachloromethane is characterized by pronounced liver damage; ingestion of only a few milliliters of tetrachloromethane can lead to liver failure in humans. Narcotic effects are only slightly pronounced in acute tetrachloromethane poisoning or do not occur at all. The first

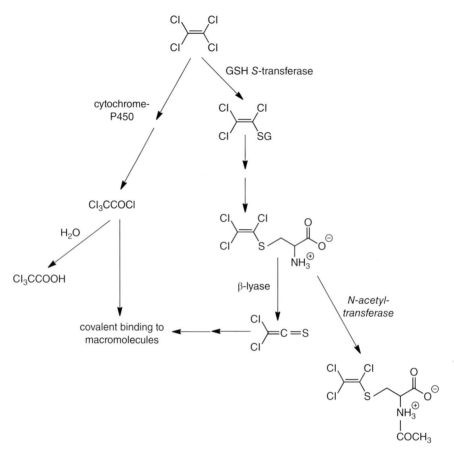

Figure 6.17 *Biotransformation of perchlorethene in the rat. Perchlorethene is metabolised by a cytochrome P450-catalyzed oxidation; the second metabolic pathway via glutathione conjugation plays only a minor role. Activation of the conjugates formed by glutathione conjugation in the kidney was detected only in the rat, but not in humans.*

sign of liver injury after oral intake of tetrachloromethane is an increase in the activities of liver enzymes in the plasma. As a result of the injury, the liver size increases due to the introduction of fat and becomes pressure-sensitive. In addition to liver damage, kidney damage also occurs after high doses of tetrachloromethane and renal failure has been described as the cause of death after acute poisoning. Chronic exposure to tetrachloromethane leads to the formation of liver tumors in rats. These tumors develop in regions of the liver that are massively altered by the local toxicity.

> **Both the acute and chronic toxicity of tetrachloromethane is due to bioactivation reactions. Tetrachloromethane is reduced by cytochrome P450 in a one-electron transfer reaction, resulting in the trichloromethyl radical (Figure 6.18). In the reaction of this free radical with unsaturated fatty acids in membrane lipids, hydrogen atoms are abstracted and lipid peroxidation is initiated (Figure 6.19).**

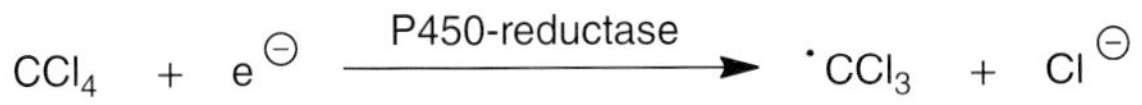

$$CCl_4 + e^{\ominus} \xrightarrow{\text{P450-reductase}} {}^{\bullet}CCl_3 + Cl^{\ominus}$$

Figure 6.18 Bioactivation of tetrachloroethane by a cytochrome P450-catalyzed reduction.

The consequence of the lipid peroxidation is a destruction of the membranes of organelles in the cell and the cell membrane. Alcohol abuse increases the liver toxicity of tetrachloromethane. The cytochrome P450 isoenzyme CYP2E1 is induced by alcohol, which in turn catalyzes the reductive activation to the trichloromethyl radical (Figure 6.18).

Chloroform Chloroform (trichloromethane) has been used since about 1860 as an inhalation anesthetic. Because of many incidents during the use of chloroform for anethesia, this application has been abandoned and its use as a solvent is severely restricted. However, chloroform is a significant contaminant in chlorinated drinking water and is formed from natural ingredients (humic acids) of groundwater in the process of chlorination. However, the amounts of chloroform absorbed by consumption of chlorinated drinking water are below the threshold for toxicity.

Acute exposure to high chloroform concentrations causes narcotic effects. Chloroform is rapidly absorbed into the organism because of its high lipophilicity. As with tetrachloromethane, bioactivation reactions are responsible for the toxic effects after both short-term and long-term administration of high doses. After oral administration, chloroform produced toxic changes in the liver and kidneys in rats and mice. In addition, renal tumors are seen in male rats, and liver tumors in mice. Tumor induction occurs only after administration of doses that trigger massive cytotoxic effects on liver and kidneys. Cytochrome P450-dependent oxidations are responsible for both liver and kidney damage (Figure 6.20). Phosgene is formed as a reactive metabolite. This electrophile binds to cellular macromolecules and leads to the initiation of toxic effects. The final product of the metabolic conversion of chloroform is carbon dioxide, which is produced by hydrolysis of phosgene.

Dichloromethane Due to low toxicity, low combustibility and high volatility, dichloromethane is widely used as a solvent. In contrast to tetrachloromethane, the acute toxicity of dichloromethane is low; after short-term exposure to high concentrations, anesthesia and

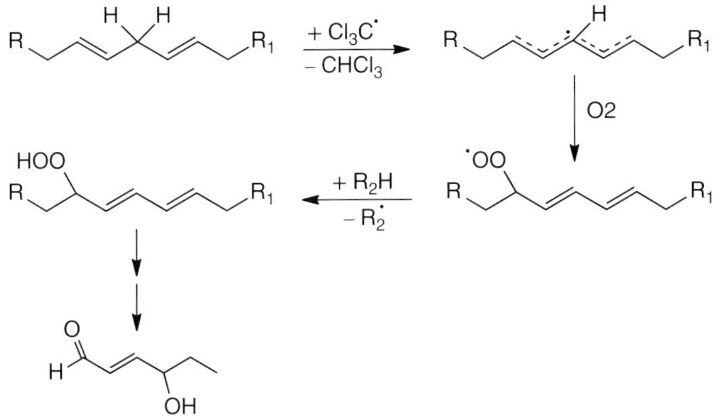

Figure 6.19 The attack of the trichloromethyl radical on unsaturated fatty acids of biological membranes and the initiation of lipid peroxidation. The decomposition of the unsaturated fatty acids produces α, β-unsaturated carbonyl compounds such as 4-hydroxy-2-trans-hexenal.

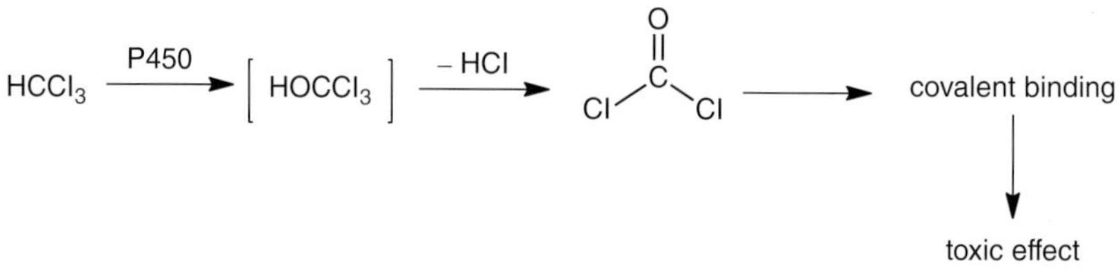

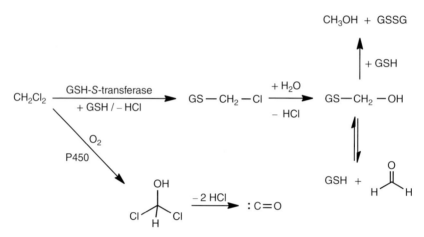

Figure 6.20 Biotransformation of chloroform via cytochrome P450: bioactivation to phosgene.

Figure 6.21 Biotransformation of dichloromethane via cytochrome P450 and glutathione-S-transferase catalyzed conversion of dichloromethane to reactive intermediates.

the formation of carboxyhemoglobin are the main effects. Damage to the liver and other organs is rarely observed in acute intoxications. Long-term inhalation of dichloromethane resulted in the formation of liver and lung tumors in mice, but not in hamsters and rats. In contrast, administration of dichloromethane in drinking water did not produce tumors in mice. Dichloromethane is mutagenic in bacteria but does not show genotoxicity in mammalian cells.

Absorbed dichloromethane is rapidly eliminated by exhalation and biotransformation and shows no tendency to accumulate. The biotransformation of dichloromethane in mammals takes place by two pathways, and the final products of the reaction sequences are carbon monoxide and carbon dioxide. Carbon monoxide is formed from dichloromethane by a cytochrome P450-catalyzed oxidation with formyl chloride as a short-lived intermediate (Figure 6.21). A second metabolic pathway can take place after saturation of the cytochrome P-450 mediated oxidation. In this pathway, carbon dioxide is produced from dichloromethane via a glutathione-dependent mechanism. This pathway is initiated by a glutathione-S transferase-catalyzed reaction of dichloromethane with glutathione followed by hydrolysis of the S-(chloromethyl)-glutathione formed to give formaldehyde (Figure 6.21). Formaldehyde is further oxidized with formic acid as an intermediate to carbon dioxide. Pharmacokinetic studies have shown that the glutathione-dependent pathway is responsible for the carcinogenicity of dichloromethane in the mouse. The toxic metabolite is the electrophilic S (chloromethyl) glutathione. In humans, this bioactivation pathway has no significance after exposure to low dichloromethane concentrations.

6.7.3 Hydrocarbons

n-Hexane

n-Hexane was used as a solvent for paints and adhesives and for extractions because of favorable solvent properties. Short-term exposure to concentrations up to 1000 ppm does not lead to observable toxic effects, and anesthesia occurs only after exposure to very high concentrations. In the case of long-term and high exposure in the workplace or when n-hexane-containing products are misused as recreational drugs, n-hexane produces a neuropathy that manifests itself through numbness in the fingers and toes. In less severe cases, this may be the only symptom; effects are reversible at this stage. If exposure to n-hexane continues, irreversible effects on contact and temperature sensitivity as well as muscle weakness in the hands and feet may occur. First symptoms of poisoning may occur only after several months of n-hexane exposure. Pathologically, a degeneration of the peripheral nerves can be observed as a result of chronic exposure in humans.

> **The neurotoxicity of n-hexane after prolonged exposure is also due to metabolic activation reactions. N-hexane is hydroxylated by cytochrome P450 enzymes, producing 2-hexanol, 2,5-hexanediol and other isomeric hexanols and hexanediols. These can be oxidized further to the corresponding ketones. 2,5-Hexanedione is considered to be the metabolite responsible for the neurotoxicity of n-hexane.**

2,5-hexanedione can react chemically with the ε-amino group of the amino acid lysine in proteins. Pyrrols are formed by ring closure of the initial reaction product in proteins, leading to protein cross-links through further oxidation (Figure 6.22). Such cross-links in the nerve system (axons) can affect the axonal transport of proteins and the transmission of electrical impulses. This results in the above-described toxic effects. Because of the biotransformation, the toxicity of n-hexane is strongly dependent on the presence of other hydrocarbons that compete for the cytochrome P450 enzymes. For example, simultaneous exposure to toluene and n-hexane significantly reduces the neurotoxicity of n-hexane.

Gasoline and Kerosene

Gasoline and kerosene are mixtures of saturated and partially unsaturated hydrocarbons containing low concentrations of aromatic compounds. The gasoline used in engines and for textile cleaning consists mainly of isomers of hexane, heptane and octane. Despite beiing readily available and widely used, acute intoxications are rarely observed due to the low toxicity of gasoline. After exposure to high concentrations (>2000 ppm), narcotic effects occur. Commercially available gasoline often contains methanol, ethanol and other additives, such as methyl tert-butyl ether (MTBE) in concentrations of up to few percent in order to improve the combustion behavior. MTBE shows typical solvent symptoms in acute intoxications, and the toxicity of MTBE after long-term administration is very low.

Diesel fuel and fuel for oil lamps consists of longer-chain hydrocarbons with higher viscosity, which have only a low toxicity. Despite the low toxicity, intoxcations are caused by the oral intake of petrol and fuels for oil lamps in children, with a possible fatal outcome. Swallowed oils irritate the gastric mucosa and cause vomiting. During vomiting, oil droplets may enter the airways and lead to massive inflammation in the lungs and gasoline pneumonia, which is difficult to treat.

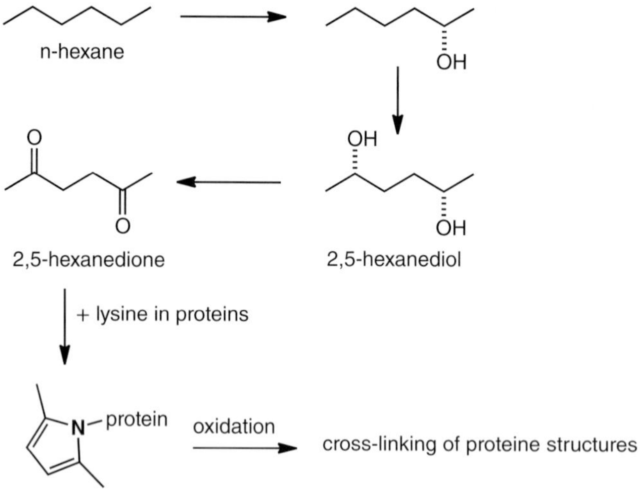

Figure 6.22 Conversion of n-hexane to 2,5-hexanedione and reaction of the dione with the amino acid lysine in proteins. The auto-oxidation of the pyrroles formed leads to cross-linking of protein structures.

Benzene

Benzene was formerly used as a solvent in the rubber industry, for the application of adhesives in shoe manufacture, and in the production of paints and plastics. Nowadays, benzene is still present in unleaded gasoline (up to 2%). Benzene is a final product of many combustion processes and reaches the environment from these sources. For non-smokers, time spent in buildings and vehicles can be a significant source of benzene exposure due to the exhaust air from heating systems or the evaporation of gasoline. Tobacco smoke also contributes to the exposure of non-smokers to benzene indoors. In the environment (air and also surface water) measurable concentrations of benzene are present and inhalation is most important exposure pathway for humans. However, indoor sources of benzene have a much higher contribution to overall benzene exposures compared to outdoor sources due to the longer time spent in buildings. For smokers, cigarette consumption is by far the most important benzene source. Smokers have up to 10-fold higher exposure to benzene compared to non-smokers.

Acute exposure to very high doses of benzene causes anesthesia and often fatal cardiac arrhythmias. The target organ for benzene toxicity after long-term exposure is the hematopoetic system. In humans, changes in the blood count were observed during chronic benzene exposures at air concentrations above 1 ppm, and increased incidences of leukemia were observed at air concentrations above 30 ppm benzene over many years. In animal models, benzene produces changes in hematopoesis, but no leukemias. This is explained by the fact that no continuous but intermittent exposure is necessary for the formation of leukemias. Among the early indicators of chronic benzene exposure in humans is a reduction in the number of leukocytes. Absorption via the lung is the most important pathway for humans due to the high vapor pressure of benzene. Absorbed benzene is rapidly distributed and metabolically converted into polar metabolites. At low concentrations of benzene in the air, about 50% of the inhaled benzene is biotransformed (Figure 6.23) and the remaining 50% is cleared from the body by exhalation. In the first step of biotransformation, benzene oxide, which is in equilibrium with the corresponding oxepine, is formed from benzene in a cytochrome P450-catalyzed reaction. All

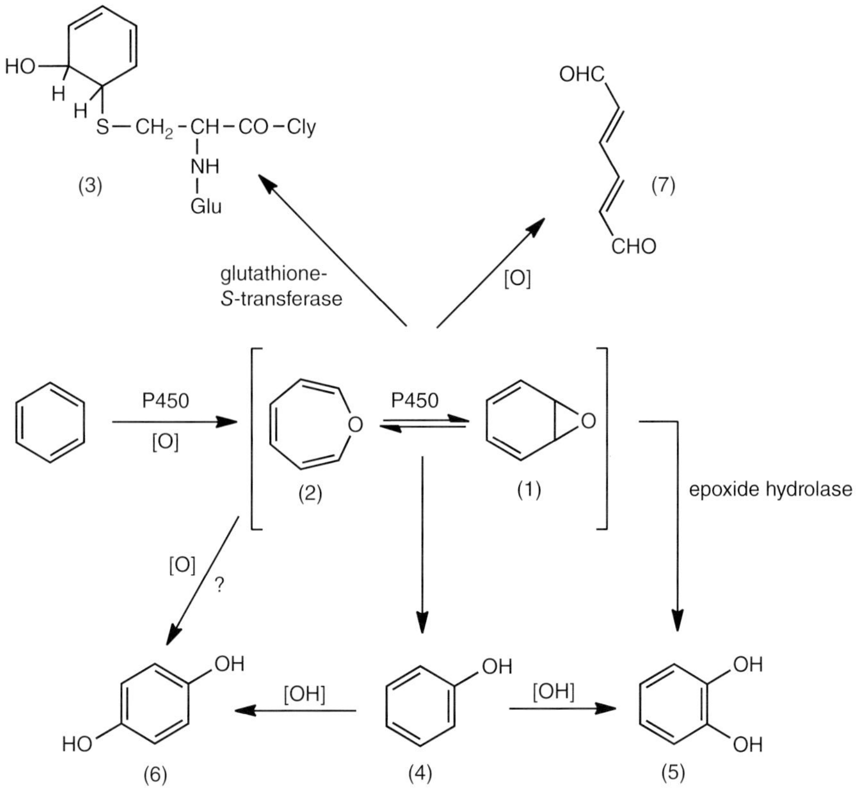

Figure 6.23 Metabolic activation of benzene to hematotoxic metabolites, where: (1) = benzene epoxide, (2) = oxepine, (3) = glutathione conjugate, (4) = phenol, (5) = catechol, (6) = hydroquinone, and (7) = Muconaldehyde.

known metabolites of benzene are derived from benzene oxide. The epoxide can be converted into a glutathione S-conjugate by glutathione-S-transferases under ring opening. The hydrolysis of benzene oxide leads to the formation of phenol, which is a major metabolite of benzene in the urine in humans. Catechol is formed by an epoxide hydrolase-catalyzed reaction from benzene oxide; hydroquinone (6, Figure 6.23) is formed by further oxidation of the metabolite phenol. Muconaldehyde (7, Figure 6.23) is formed under ring opening by a still unknown mechanism, as the end product of this metabolic pathway muconic acid is excreted with the urine.

Benzene metabolites bind covalently to proteins and the metabolite responsible for the covalent binding is benzoquinone. The bone marrow as the target organ of the carcinogenic effect of benzene in humans has only a limited capacity for the metabolism of benzene. It is therefore assumed that benzene metabolites formed in the liver, such as hydroquinone and catechol, are distributed with blood to the bone marrow. There, hydroquinone and catechol undergo bioactivation by myeloperoxidases to form electrophilic quinones, which can trigger the specific effects of benzene on the hematopoetic system.

Toluene

Toluene (methylbenzene) is far less toxic than benzene. Inhalation of higher concentrations produces the typical picture of inhalation anesthesia. Changes in the hematopoesis in humans have not been observed after toluene exposures. The different effects of toluene and benzene on the hematopoetic system can be attributed to different biotransformation pathways. Toluene is oxidized by cytochrome P450 enzymes mainly at the exocyclic methyl group with benzyl alcohol and benzaldehyde as intermedates to form benzoic acid, not resulting in electrophilic intermediates. Oxidations on the aromatic ring, which might lead to the formation of hematotoxic quinones, play only a very minor role in the biotransformation of toluene. Toluene is neither mutagenic nor carcinogenic.

6.7.4 Aliphatic Alcohols

All aliphatic alcohols have narcotic effects after intake of higher doses, and the extent of narcotic effects is dependent on lipid solubility and toxicokinetics. Regarding toxic effects, primary alcohols differ significantly in the mechanisms responsible for the initiation of toxicity. In the case of ethanol, the acute toxic effect is triggered by parent ethanol. The ethanol metabolite acetaldehyde also plays an important role in chronic effects. In the case of methanol and propanol, the accumulation of the metabolites formic acid (methanol) and acetone (isopropanol) is decisive for the acute toxicity because of slow excretion. Again, the parent alcohol is important for toxic effects when chain length increases.

> **All primary alcohols are oxidized mainly by alcohol dehydrogenase, rates of oxidation are highest in ethanol and decrease with increasing chain length.**

Methanol

Methanol is used as a solvent and as a fuel additive in gasoline. The most frequent cause of methanol poisoning in humans is the deliberate addition of methanol to alcoholic beverages (due to similar taste and smell). In humans, methanol intoxications occur mostly after oral uptake and have a characteristic time course. Shortly after the ingestion of toxic doses, symptoms often are not observed, only after high and repeated doses an excitement state similar to that induced by ethanol is observed. This disappears after a few hours due to rapid oxidation of methanol. After a symptom-free interval of 12–24 hours, metabolic acidosis with forced breathing, increased blood pressure, headaches, dizziness and vomiting occurs. The metabolic acidosis can persist for one to four days depending on the methanol dose. As a second consequence of an acute methanol poisoning, often parallel with the acidosis, light sensitivity, eye pain and vision impairments are observed. These manifest themselves in a limitation of vision and blurry images. After two to six days, symptoms usually improve. However, irreversible degeneration of the optic nerve can lead to blindness following the accumulation of large amounts of methanol; in acute poisoning, fatal doses (death by central respiratory paralysis) in the range of 1 ml methanol/kg body weight are described. The rodents frequently used for toxicological testing are much less sensitive to the toxic effects of methanol; in rodents, even a pronounced methanol poisoning does not lead to blindness. The effects observed in humans are attributable to the metabolic oxidation of methanol to formic acid as a toxic metabolite. In humans and in mammals, methanol is slowly but completely absorbed and is distributed in body water because of its low fat solubility. The first step of the oxidation of methanol, catalyzed by alcohol dehydrogenase (ADH), leads to the formation of formaldehyde as a short-lived intermediate (Figure 6.24). Formaldehyde formed is rapidly

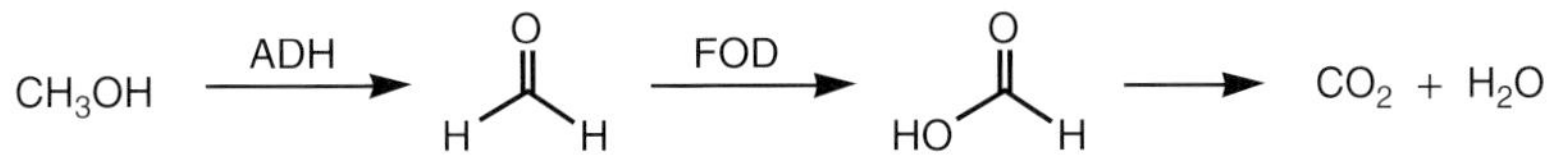

$$CH_3OH \xrightarrow{ADH} \underset{H}{\overset{O}{\underset{|}{H-C-H}}} \xrightarrow{FOD} \underset{HO}{\overset{O}{\underset{|}{H-C-H}}} \longrightarrow CO_2 + H_2O$$

Figure 6.24 *Metabolic activation of methanol. ADH, alcohol dehydrogenase; FOD, formaldehyde dehydrogenase.*

oxidized to formic acid by formaldehyde dehydrogenase (FOD). In rodents, formic acid is also rapidly oxidized to carbon dioxide. Therefore, toxic concentrations of formic acid in the blood of rodents cannot be reached. In humans, the conversion of formic acid to carbon dioxide is inefficient and formic acid excretion with the urine is slow. This leads to an accumulation of formic acid in the blood with the consequences of acidosis and effects on the optic nerve. The effect on the optic nerve is probably mediated by a disturbance of the function of the mitochondria in the optic nerve.

Visual disturbances after chronic methanol exposure have been described. However when the metabolic degradation of the formed formic acid is faster than its rate of metabolic formation from methanol, toxic concentrations of formic acid cannot occur. Methanol is neither genotoxic nor carcinogenic For the treatment of methanol poisoning, ethanol or 4-methylpyrazole are used to inhibit alcohol dehydrogenase to prevent methanol oxidation. During therapy, a blood ethanol level of about 0.1% is maintained for an extended period of time by infusion of alcohol. Ethanol has a higher affinity for alcohol dehydrogenase than methanol, thereby inhibiting its oxidation. Under these circumstances, unchanged methanol is excreted by exhalation and with urine. In the context of the treatment of poisoning, the correction of metabolic acidosis by sodium bicarbonate is also necessary. The oxidation of formic acid to carbon dioxide can be accelerated by the administration of folate, but at high doses of methanol and the appearance of effects on the eye, hemodialysis is necessary for the rapid reduction of methanol blood levels.

Ethanol

Because of its low toxicity, ethanol is also used as a solvent. At the present workplace exposure limits, elevated blood alcohol levels do not occur by inhalation. On the other hand, the use of ethanol as recreational drug may have toxic consequences. Ethanol is produced during the fermentation of mono-, di- and polysaccharides from plants. By fermentation, beverages with ethanol contents of 4–18% (beer and wine) can be obtained and a higher concentration of ethanol in beverages can be achieved (usually usually 40%) by distillation. Ethanol-containing beverages are a socially acceptable method of intake of a potentially toxic chemical in many parts of the world. The annual consumption of ethanol in Germany is about 12 liters/person, converted to pure ethyl alcohol.

Toxicokinetics of Ethanol After oral intake, ethanol is completely absorbed within a period of a few minutes to two hours depending on the filling of the gastrointestinal tract. Carbonated drinks such as champagne accelerate the resorption process. Ethanol distributes rapidly into the body water (distribution volume of 0.6–0.7 l/kg body weight) and maximal blood concentration is reached within one to two hours.

Ethanol can pass from the systemic circulation into the placenta and into the mother's milk. Because of the rapid equilibration, the blood alcohol level is representative of the concentrations of ethanol in the central nervous system. In contrast to almost all other chemcials, the rate of elimination of ethanol is not dependent on its blood concentration and is constant over the entire elimination period (85–100 mg/kg body weight/hour).

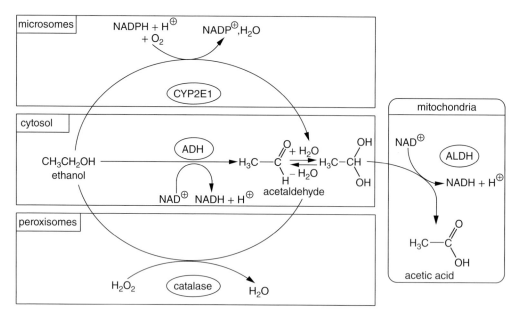

Figure 6.25 *Biotransformation of alcohol to acetaldehyde by alcohol dehydrogenase (ADH), cytochrome P450 (CYP2E1) and catalase. Oxidation of acetaldehyde to acetic acid by aldehyde dehydrogenase (ALDH).*

> **Because of this elimination kinetics, blood ethanol concentrations can easily be calculated back from a measured blood alcohol concentration at a specific time point, taking into account the constant degradation rate. The biotransformation of ethanol by alcohol dehydrogenase (ADH) leads to a reduction in the blood concentration of ethanol after the end of the intake in the range of about 0.015% per hour. The reason for linear elimination is a saturation of ADH even at very low blood concentrations of ethanol.**

Only a small portion of an oral dose of ethanol is exhaled through the lungs (2–3%) or excreted unchanged with the urine (1–2%). Elimination is almost as fast for alcohol-dependent people as in the general population.

Ethanol is mainly oxidized by ADH in the liver to acetaldehyde. The acetaldehyde formed is further oxidized to acetic acid by aldehyde dehydrogenase (plasma half-life of 1.7 minutes). The acetic acid obtained (plasma half-life of 6.4 minutes) is consumed predominantly in the tricarboxylic acid cycle to give carbon dioxide and water.

Ethanol can also be oxidized at high blood concentrations by cytochrome P450 2E1 and peroxisomal catalase to acetaldehyde (Figure 6.25).

Acute and Chronic Effects of Consumption of Alcoholic Beverages Causes of acute poisoning are predominantly excessive drinking. Acute poisoning with ethanol has very characteristic symptoms with pronounced dose dependence (Table 6.17). With blood levels above 0.2%, signs of anesthesia are predominant. This limit is noticeably shifted in the case of individuals with a history of chronic abuse of ethanol.

Other symptoms of acute intoxication include nausea and vomiting, hyperventilation, hypoglycemia (hypoglycemia), and decrease in body temperature. Low alcohol levels lead to a slight

Table 6.17 *Acute central nervous alcohol effects in unaccustomed persons.*

Ethanol concentrations in blood (%)	Effects
0.03	Impaired coordination detectable
0.04	Decreased visual performance, increased response time to stimuli
>0.05	Euphoria, legal limit for operating motor vehicles in several countries in the EU
>0.08	Visual impairment (nystagmus), legal limit for operating motor vehicles in many states in the USA
>0.1	Poor judgement, labile mood, ataxia
>0.14	Marked ataxia, nausea and vomiting
>0.2	Stage one anethesia, memory lapse, slurred speech, severe confusion
>0.4–0.5	Lowest limit for lethal intoxication, respiratory depression, acidosis, coma

rise in blood pressure, higher doses result in a shift of blood flow to the periphery with a resulting drop in blood pressure. As a consequence of a massive intoxication, central respiratory depression can occur. Alcohol has a diuretic effect. The increased formation and release of heat increases the basic metabolic turnover, which results in hypoglycaemia. Dizziness and nausea are caused by local irritation of the gastric mucosa, but also by direct effects on the central nervous system.

Intensive care is necessary in cases of severe poisoning, especially in the case of mixed intoxications with drugs. Characteristic of a massive alcohol poisoning are hypoglycemia and hypothermia, therefore the infusion of 10% glucose and covering the patient with blankets is recommended. Due to the lack of adsorption of ethanol to charcoal, activated charcoal therapy is not effective. Gastric lavage is also ineffective since ethanol is rapidly absorbed from the gastrointestinal tract. In the case of life-threatening concentrations (depending on the condition of the patient from 0.3% to 0.6%), ethanol concentrations in blood can be reduced by hemodialysis.

> **Targets of chronic abuse of high doses of ethanol are primarily the liver, followed by the nervous system and the cardiovascular system.**

Liver injury is the most frequent and serious consequence of chronic alcohol abuse. The mechanism of liver damage by ethanol is complex. Long-term uptake of high doses of ethanol induces a cascade of pathophysiological events in the liver. These include accumulation of triglycerides and protein adducts of acetaldehyde, lipid peroxidation, and excretion of cytokines (regulatory peptides that can alter cell growth, differentiation, and immune responses). The first stage of chronic ethanol poisoning is fatty liver, which is due to the accumulation of lipids formed from the fatty acids that are biosynthetisized from the excess of acetic acid present as ethanol metabolite. Fat accumulation in the liver is initially reversible, but can lead to a stationary fatty liver. Fatty liver hepatitis is often formed with massive proliferation of the connective tissue and destruction of the normal liver architecture. As a result, progressive liver cirrhosis develops. The critical dose and duration of the development of a fatty liver is about 40–80 g (pure) ethanol/day for men over a period of between 5 and 10 years, thresholds for hepatitis and development of liver cirrhosis are measured with daily ethanol intake of 80–160 g/day over the duration of more than 10 years, and women are more sensitive than men. The chronic consumption of high levels of alcohol is clearly associated with the development of liver diseases, but less than 20% of alcohol addicts develop liver diseases. Therefore, other factors such as type of consumed alcohol (beer, wine, liquor) as well as interindividual differences in alcohol consumption, diet and immune response play a role. Alcohol withdrawal can slow or stop the development of a liver cirrhosis even in advanced stages of the fatty liver.

Alcohol-dependent persons often have a chronic tremor that is more pronounced in the morning than in the evening. The tremor probably results from the increased catecholamine stimulation of β-adrenergic receptors.

In addition to diabetes mellitus, chronic alcohol abuse is the most frequent cause of nerve degeneration (peripheral polyneuropathy). Typical symptoms are paresthesia and pain in the feet and hands, as well as diminished reflexes, muscular atrophies, and mild symptoms of paralysis.

Withdrawal symptoms (tremor, tachycardia, anxiety) and, in rare cases, delirium tremens are induced by the abrupt discontinuation of a long-lasting and excessive alcohol consumption; however, delirium tremens can also occur after a pronounced alcohol drinking excess. The clinical symptoms of withdrawal include anxiety, tremor, tremor, visual and acoustic hallucinations, and disorientation. Numerous vegetative symptoms such as mydriasis, perspiration, facial redness, acceleration of breathing, and severe fluctuations in blood pressure also occur. In contrast to the blood pressure reducing effect of a single low dose of ethanol, ethanol abuse leads to an increase in blood pressure and an increased incidence of cardiovascular diseases. Chronic alcohol abuse also affects fertility in both sexes. Effects on brain structures as well as the direct damage of the testes in men are discussed as causative changes. Maternal alcohol abuse is also by far the most frequent cause of damage to the embryo by exogenous factors (alcohol embryopathy). In more than half of the children with alcohol embryopathy, there are also intrauterine growth retardation, microcephalus with characteristic head and face malformations, as well as development retardations. Hypoglycemia and malnutrition occurring in alcoholic pregnant females may be involved in the development of alcohol embryopathy.

Glycols

Diols such as ethylene glycol (1,2-ethane diol) and propylene glycol (1,2-propane diol) are used for many technical applications. 1,2-Ethane diol is used as an additive to the cooling water for combustion engines in vehicles for lowering the freezing temperature (antifreeze) and, in some cases, also as a solubilizer in cosmetics. Ethylene glycol is a liquid with a sweet taste that induces similar effects to ethanol. Because of the sweet taste, 1,2-ethane diol was also used in small quantities to increase the sweetness of wine. 1,2-Ethane diol has a characteristic acute toxicity to the kidney due to its oxidative biotransformation (Figure 6.26); renal damage is also observed after chronic uptake. As an end product of the oxidation of 1,2-ethane diol, oxalate is formed, which reacts with Ca^{2+} ions to give calcium oxalate, which is poorly soluble in water. During the concentration of the urine in the renal tubules, calcium oxalate present can precipitate and impair the renal function by the formation of crystals. However, the actually toxic intermediate in the metabolism of 1,2-ethanediol appears to be glyoxylic acid (CHO–COOH), which has a direct

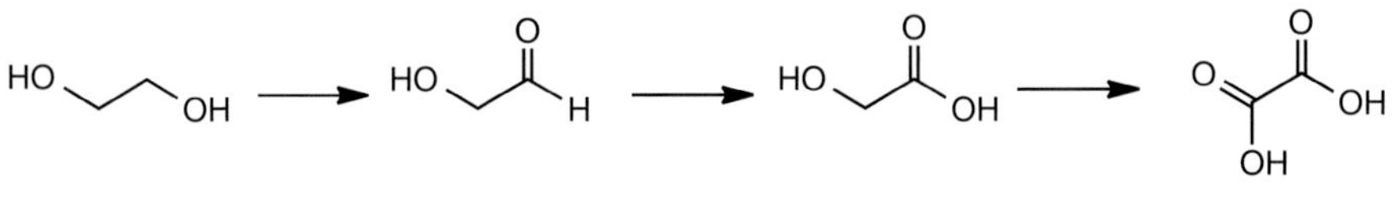

Figure 6.26 *Biotransformation and activation of 1,2-ethanediol.*

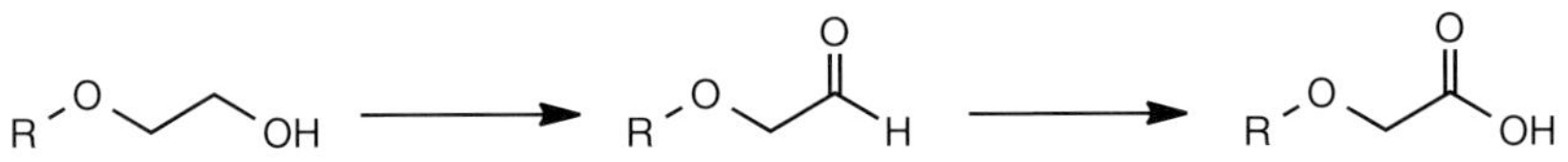

Figure 6.27 Bioactivation of glycol ethers.

toxic effect on renal tubules. Pronounced poisoning with 1,2-ethane diol leads to uraemic coma and an oral uptake of 100–200 ml can lead to life-threatening intoxication. Rapid initiation of dialysis to maintain the renal function is therapeutic. During treatment, inhibitors of alcohol dehydrogenase such as ethanol or 4-methylpyrazole can be used as an antidotes to prevent oxidation of 1,2-ethane diol.

1,2-propane diol has a very low potential for toxicity and is therefore often used as a solvent for chemcials that are difficult to dissolve in water. 1,2-propanediol is also metabolically oxidized, but the metabolites lactic acid and pyruvic acid are not toxic.

Glycol ethers are used as solvents for paints, for the solution of plastic films, and in adhesives. For these highly bioavailable substances, liver and kidney lesions have been observed in humans at very high exposures. Long-term exposure of test animals against high doses of 2-methoxyethanol, 2-ethoxyethanol and 1-methoxy-2-propanol resulted in toxic effects on the testes with sterility as a result; exposure during pregnancy induced malformations and increased death rates of the embryos. These toxic effects are also based on bioactivation reactions, in which 2-methoxyethanol is converted into 2-methoxyacetic acid as a toxic metabolite (Figure 6.27). This toxic metabolite appears to affect the function of mitochondria, and the specific effects on the male reproductive organs can be explained by the high energy requirement of the testes.

6.7.5 Summary

All solvents, although with varying intensity, lead to alcohol-like effects on the central nervous system, therefore many of these chemicals have been used as narcotic agents. These effects are caused by the parent chemicals. In the case of long-term exposures, the toxic changes are mainly due to metabolites formed with different organ specificities. Therefore, depending on the parent chemical, very different and chemical-specific signs and symptoms of intoxications can occur. These, like all toxic effects, are decisively influenced by exposure duration and exposure level.

Most solvents are irritant to mucous membranes and very high air concentrations can lead to toxic pulmonary irritation.

Solvents degrease the skin and remove the bactericidal protective coat of fat. Repeated skin contact therefore leads to cracked skin and bacterial inflammation.

Solvent intoxications show typical symptoms which are often very specific and which, like all toxic effects, are decisively influenced by exposure duration and exposure level.

Further Reading

Literature on solvents can be found in the following online databases, which are constantly updated:

The MAK Collection for Occupational Health and Safety http://onlinelibrary.wiley.com/book/10.1002/3527600418

US Environmental Protection Agency – Integrated Risk Information System (IRIS) http://www.epa.gov/iris

Agency for Toxic Substances and Disease Registry (ATDSR) http://www.atsdr.cdc.gov/

6.8 Noxious Gases

Kai Kehe and Horst Thiermann

6.8.1 Introduction

According to their mechanism of action, airborne toxicants may be classified as pulmonary irritants which act locally or as systemic poisons. In general, pulmonary irritants cause irritation of the airways and damage leading to alveolitis and toxic lung edema, while systemic poisons cause asphyxia, hypoxia, and central respiratory arrest.

In order to warrant optimal gas exchange only a very thin membrane, consisting of alveolar epithelia, capillary endothelia, and the basal membrane in between, separates respiratory air from the blood in the lung. As gaseous compounds reach the alveolar space they pass practically more or less unhindered into the blood and are rapidly distributed in the organism. The total exchange surface in the lung approximates the size of a tennis court. Thus, even relatively polar noxious gases such as hydrogen cyanide and hydrogen sulfide, which hardly penetrate the skin, rapidly enter the circulation.

6.8.2 Airborne Systemic Poisons

Carbon Monoxide

Carbon monoxide (CO, MW 28.01; 1 ppm = 1.16 mg/m^3) is a colorless, odorless gas with a relative density of 0.96 (air = 1). CO exhibits about 200–300 times higher affinity than oxygen (O_2) for the divalent iron atom of hemoglobin (Hb). Relatively low concentrations of CO in air lead to substantial binding and formation of carboxyhemoglobin (COHb), thereby reducing the availability of hemoglobin for O_2 transport to the tissues. Exposure to 50 ppm CO in air results in approximately 5% COHb concentration in the blood. These values have been described in smokers and are harmless for healthy persons. COHb at concentrations higher than 10% results in fatigue, weakness and/or headache. Individuals with pre-existing cardiovascular diseases are much more prone to CO toxicity.

Sources

> **Incomplete combustion in inadequately ventilated areas frequently leads to severe poisoning with CO.**

Combustion of carbon-containing material always produces CO to a certain extent. The yield increases with decreasing availability of O_2. Severe CO poisoning in people is, therefore, almost completely limited to exposure in confined places. Faulty heating equipment and the use of charcoal grills or open kerosene or gas heaters in closed spaces are frequent sources of CO exposure and poisoning. The most frequent cause of CO poisoning is automobile exhaust fumes. While exposure in the car or in closed garages is often life threatening, exposure in poorly ventilated high traffic places, such as underground passages, may cause some mild symptoms but is, in general, not a source of exposure leading to serious poisoning.

Methylene chloride, an ingredient of paint and varnish removers, is metabolized to CO in the organism. Its use in poorly ventilated places for longer periods may lead to CO poisoning.

Small amounts of CO are endogenously formed from Hb degradation in the body, leading to physiological concentrations in the blood of non-smokers of 0.4–0.7%. Higher amounts (8% HbCO) were observed during hemolysis. Smoking one pack of cigarettes per day results in 4–6% COHb and 2–3 packs in 7–9%. Natural sources of CO in the atmosphere are the oxidation of methane, forest fires and marine microorganisms. The air usually contains less then 0.1 ppm CO.

The maximum allowable working place concentration of CO in Germany (MAK) is 30 ppm. In the USA the OSHA PEL is 50 ppm; the ACGIH TLV is 25 ppm.

Mechanism of Action The greater affinity of CO than oxygen for hemoglobin can result in substantial occupation of oxygen binding sites by CO at relatively low CO concentrations, resulting in hypoxia.

> **The binding of CO to hemoglobin is reversible: CO + Hb <=> COHb.**

The physiological impact of CO on the ability of hemoglobin to bind or release oxygen depends upon the source of hemoglobin and the biological system under observation. CO is a competitive antagonist of O_2 at the divalent iron atom of Hb, exhibiting 200–300 times greater affinity. The effect may relate to the oxygen requirements of specific organs. Significant differences can be encountered between normal adults and the fetus *in utero*.

CO toxicity results from the reduction in Hb available for O_2 transport. Elevated levels of COHb can lead to tissue hypoxia. Organs and tissues with high metabolic activity, such as the brain (nervous tissue) and heart (cardiac muscle), are most vulnerable. Thus, for the same proportion of binding to Hb, when compared to O_2, a 200–300 times lower concentration of CO in the air can suffice to produce toxicity. At 21% O_2 in the air, arterial blood is completely saturated. In contrast, only 0.1% (1000 ppm) CO in the air is sufficient to produce 50% COHb.

COHb does not bind and transport O_2. Moreover, in presence of COHb the ability of Hb to release O_2 in the periphery decreases, causing a further aggravation of hypoxia.

It has been suggested that the increased release of nitric oxide, destructive enzymes and excitatory amino acids from leukocytes, platelets and damaged vascular endothelia are contributory factors in aggravating hypoxia. CO is also bound to the divalent iron atom of myoglobin and other cytochromes, but the relevance for poisoning is not clear because of the low affinity of myoglobin for CO.

CO can be of special concern in pregnant women. The fetus has a 10–15% higher burden of COHb because of the higher affinity of CO for fetal hemoglobin. In addition, fetal blood exhibits lower oxygen tension compared to that of the mother. Fetal Hb also exhibits a decreased O_2 release capacity in the periphery, which further aggravates the hypoxia. Thus, newborns are more susceptible to CO toxicity than adults.

Pharmacokinetics

> **Inhaled CO readily enters the circulation and binds to hemoglobin.**

The velocity of CO uptake through the lungs is determined by the diffusion of the gas through the alveolar membrane. The subsequent binding to Hb is relatively slow and the time required to reach maximum binding at a concentration of 1000 ml/m^3 is 8–10 hours. The time may be reduced by a few hours by increasing the minute volume of respiration. About 85% of CO will be bound to Hb, the remainder being distributed to myoglobin and other protein hemoproteins.

The elimination of CO, which depends upon the concentration of O_2 in air and the minute volume of ventilation, is slow. The half-life of COHb when breathing ambient air, i.e. air containing about 20% oxygen, is 3–4 hours. However, the half life of CO elimination can be reduced to 60–80 minutes with normobaric oxygen and to 15–20 minutes at hyperbaric conditions (100% O_2, 2.5 atm).

Therapeutic Principles

> **Free airways and plenty of oxygen are decisive for effective recovery.**

Vomiting and aspiration are frequent in CO poisoning. Clear airways and removal of the victim from the contaminated area are necessary for recovery. Oxygen accelerates the elimination of COHb. Oxygen can be supplied either as 100% O_2 or better still as carbogen (95% O_2, 5% CO_2) to stimulate the respiratory center and enhance respiratory minute volume hyperbaric oxygen (HBO) is the most effective therapy, but it is dependent on the availability of a HBO facility. Further treatment of severe metabolic acidosis may be required.

Effects

> **Symptoms of mild to moderate CO poisoning are not very specific, leading frequently to misdiagnosis.**

Table 6.18 shows a list of symptoms of observed with increasing COHb concentrations. Signs of mild to moderate poisoning might also be caused by other diseases such as pneumonia, cardiac infarction, viral infections, cholecystitis, or epilepsy. As a result CO poisoning might be misdiagnosed or overseen. Part of this problem may be due to the fact that CO is odorless and hence poses a danger for those accidentally exposed as well as rescuers. Several fire brigades are therefore equipped with CO detectors. In patients with pre-existing cardiovascular diseases and limited tissue oxygenation even a small increase in COHb, e.g. to 5%, might induce signs of angina pectoris.

Chronic and Delayed Effects

> **Chronic and delayed effects, usually focusing on neuropsychiatric complaints, are frequent after CO poisoning.**

Table 6.18 Signs and symptoms of CO poisoning in relation to COHb concentration in the blood.

Signs and symptoms	Blood COHb concentration (%)
No obvious signs and symptoms, but decreased psychomotor performance and decreased exercise tolerance	5
Headache, dizziness, malaise	10–20
Nausea, vomiting, throbbing headache, visual disturbances	20–30
Shortness of breath, headache, palpitation, tachycardia, dizziness, syncope	30–40
Confusion, coma, collapse	40–50
Coma, respiratory failure, cardiovascular depression, seizures, death	50–80

Apathy, disorientation, amnesia, hypokinesia, mutism, irritability, urinary and fecal incontinence, gait disturbance, and increased muscle tone are among the most common delayed effects of CO poisoning and may not become apparent for 2–40 days after exposure.

Cyanide

Sources

> Industry, combustion, pesticides, therapeutics, and plants are sources of cyanide exposure leading to poisoning.

Hydrocyanic acid (HCN; MW 27.03; 1 ppm = 1.12 mg/m^3) has a relative density of 0.9 (air = 1). HCN and the cyanide group (CN$^-$)-containing compounds (cyanides) are commercially used in electroplating, ore extraction, metal processing, and the synthesis of plastics, pesticides, and drugs among others. Combustion of organic compounds containing carbon and nitrogen (e.g., polyurethane or polyacrylonitrile) may lead to the release of cyanides.

Hydrolytic enzymes from plants and human gut β-glucosidase can hydrolyze cyanogenic glycosides from plants such as bitter almond, apple seeds, plum, peach, and apricot seeds, cassava beans and roots. HCN in the organism originates from food and from vitamin B$_{12}$ metabolism. In the blood cyanide is concentrated in erythrocytes. Normal plasma levels are 4 ng/ml in non-smokers and 6 ng/ml in smokers. These values may be expressed as 15 and 40 ng/ml, respectively, in whole blood. Sodium nitroprusside is an intravenous antihypertensive medication used in emergency medicine, which is metabolized to cyanide. The maximum allowable working place concentration in Germany is 1.9 ppm. In the USA the Occupational Safety And Health Administration (OSHA) and permissible exposure limit (PEL) for general industry relates to skin exposure and is 11 mg/m^3 (10 ppm); the American Conference of Governmental Industrial Hygienists (ACGIH) and National Institute for Occupational Safety and Health (NIOSH) both recommend an exposure limit (REL) and a short-term exposure limit (STEL) of 4.7 ppm.

Mechanism of Action

> The high affinity of the cyanide ion for the trivalent iron in cytochromes is the cause of cytochrome a-a$_3$ inhibition and reduction of mitochondrial oxidative metabolism.

The high affinity of CN$^-$ for oxidized iron in cytochromes of the respiratory chain blocks the last step of oxidative phosphorylation and consequently causes cellular ATP depletion. The binding to cytochrome iron is reversible.

Toxicokinetics

> Rapid uptake of hydrocyanic acid occurs after inhalation whereas absorption through the skin is toxicologically not relevant.

Gaseous HCN rapidly enters the bloodstream and is distributed throughout the body. Mucous membranes are also rapidly penetrated, whereas penetration via the skin is slow. After oral uptake of cyanide salts, the acid is liberated by hydrochloric acid in the gastric juice and rapidly absorbed.

Cyanogenic glucosides such as amygdalin, prunasin, dhurrin and linamarin yield HCN under the influence of β-glucosidase or linase in the gut or by hydrolysis, which is then absorbed.

HCN is detoxified predominantly by the enzyme rhodanese to thiocyanate in presence of thiosulfate:

$$CN^- + Na_2S_2O_3 \xrightarrow{rhodanese} SCN^- + Na_2SO_3$$

In humans the enzyme metabolizes about 1 mg cyanide/kg body weight per hour. Cyanide is also eliminated by binding to hydroxocobalamine with formation of cyanocobalamine (vitamin B_{12}). Small amounts are also exhaled and excreted by sweat, causing a bitter almond odor.

Intoxications

> **Loss of consciousness with seizures, metabolic acidosis, and respiratory failure in the absence of cyanosis indicate acute cyanide poisoning. Chronic cyanide poisoning is an unusual event.**

Breathing about 20 ppm of HCN for a few hours may result in headache. Death may follow exposure to 100 ppm after one hour and at 300 ppm after about 10 minutes. The oral lethal dose of KCN and NaCN in humans is between 1 and 5 mg/kg. The symptoms are the result of a deficiency of cellular energy. Tissues with high energy demand show highest sensitivity. CNS symptoms occur first, but are not very specific, for example dizziness, nausea, dyspnea, tachypnea, restlessness, and anxiety. Severe poisoning is associated with seizures, loss of consciousness, respiratory failure, and dilated fixed pupils. The cardiovascular system is much less sensitive. In very severely poisoned individuals cyanosis might be absent initially despite respiratory arrest because of good saturation of the blood with oxygen through tachypnea and decreased intracellular oxygen consumption.

Continuous exposure to cyanide can cause headache, dizziness, nausea, vomiting, and bitter almond taste. Vitamin B_{12} and folate metabolism as well as thyroid function may be mildly disturbed. In Nigeria, a tropical ataxic neuropathy with enhanced thiocyanate levels was linked to chronic cyanide exposure from cassava consumption.

Therapeutic Principals

> **The aim of antidote use is rapid elimination of cyanide ions from the cells by binding in the blood and transformation to less toxic metabolites to be excreted in the urine.**

Cyanide exhibits a high affinity for oxidized iron (Fe^{3+}) in cytochromes. Conversion of some hemoglobin to methemoglobin draws cyanide from tissue cytochromes to erythrocytes, enabling recovery of oxidative metabolism in the tissues. Transformation of 30% Hb to methemoglobin results in enough capacity to bind about six times the LD_{50} of cyanide ion. This can be achieved by an intravenous (iv) injection of **4-dimethylaminophenol** (4-DMAP), 3 mg/kg.

In some countries, **sodium nitrite** or **amyl nitrite** are more commonly used for methemoglobin formation in cyanide poisoning. The dose at which either is used is limited because of strong vasodilatation and reduction in blood pressure. Up to 12 mg/kg $NaNO_2$ can be given iv, which appears sufficient for formation of about 15% methemoglobin. Amyl nitrite, a very volatile liquid, is in some textbooks recommended to be used by vapor inhalation in cyanide poisoning. However, inhalation of amyl nitrate is practically ineffective because of rapid exhalation of the vapor.

The use of these methemoglobin-forming drugs should also be combined with sodium thiosulfate. A complimentary mechanism for detoxifying cyanide is through the enzyme rhodanese, a normal body constituent which mediates the reaction between cyanide and thiosulfate to yield the detoxification product thiocyanate. For maximum effectiveness $Na_2S_2O_3$ (100 mg/kg, iv) must be administered. In contrast to CN, thiocyanate is much less toxic and more rapidly excreted in the urine.

Hydroxocobalamine, a vitamin B_{12} congener, detoxifies cyanide by a reaction that yields cyanocobalamine, which is eliminated from the body by excretion in the urine. Cyanide exposure resulting from breathing smoke released by a fire may be accompanied by CO inhalation, which can impair tissue oxygenation. Unlike methemoglobin formation by nitrite, the hydroxocobalamine method of trapping cyanide does not interfere with tissue oxygenation. The International Programme on Chemical Safety/Commission of the European Communities suggests that large doses of hydroxocobalamine are required. It was estimated that about 1406 mg are required to inactivate 1 mmol of cyanide, i.e. 65 mg of KCN. In most countries commercial preparations of the antidote can only be obtained in ampoules containing 1–2 mg, although a 4 g preparation is available in France. A solution of hydroxocobalamine in 5% dextrose, which may or may not contain thiosulfate, can then be administered intravenously. Some organ toxicity or allergic response to treatment may be encountered.

Hydrogen Sulfide

Sources

> **Hydrogen sulfide occurs in industry, agriculture, sewage and the environment.**

Industrial sources of hydrogen sulfide (H_2S, MW 34.08, 1 ppm = 1.41 mg/m^3), which yields a smell commonly described as that of rotten eggs, include facilities for the production of viscose and rayon, paper, mineral oil, vulcanized rubber, and illuminating gas, and from heavy water production in nuclear plants. Waste from the production of cellulose, sugar, and glue and from tanneries contains the gas. Other sources include places where fermentation and rotting of fish, manure, and raw sewage occurs. Natural sources are volcanoes and sulfur springs. Small amounts of H_2S are also formed in the gut of living organisms. H_2S has a relative density of 1.19 (air = 1) and accumulates at the ground. The maximum allowable working place concentration in Germany is 5 ppm. In the USA the OSHA PEL is 10 ppm, ACGIH TVL is 1 ppm (TWA) and 5 ppm (STEL), and NIOSH REL is 10 ppm/10 minutes.

Mechanism of Action

> **Hydrogen sulfide, much like cyanide, is thought to inhibit cytochrome a-a$_3$ and thereby oxidative metabolism.**

H_2S exhibits a somewhat higher affinity for oxidized iron (Fe^{3+}) in cytochrome oxidase than cyanide. Consequently, oxidative metabolism is inhibited and metabolic acidosis may occur.

Pharmacokinetics

> **Rapid pulmonary but negligible dermal absorption, rapid binding, and metabolism in the tissues characterize hydrogen sulfide poisoning.**

H_2S rapidly penetrates from the air into the blood and is distributed to the tissues. Dermal uptake is negligible. The gas is rapidly eliminated by oxidation to sulfate, methylation, and synthesis of SH-containing proteins. The probability of accumulation is low and very little is eliminated unchanged.

Effects

> **Sudden collapse and a smell of rotten eggs is indicative of poisoning. Chronic poisoning and delayed effects are uncommon.**

Table 6.19 shows that perception of the smell occurs much earlier than irritation of eyes or respiratory epithelia. At concentrations causing mucous membrane irritation olfactory perception rapidly becomes paralyzed.

Therapeutic Principals

> **Rapid rescue with 100% oxygen respiration for the patient and airway protection for the rescuer are important measures.**

Although H_2S binds to the oxidized iron of cytochrome oxidase, similar to cyanide, methemoglobin formation is an ineffective tool in mobilizing the poison from the tissue because of the rapid decay of the sulfmethemoglobin complex and the rapid elimination of sulfide. Furthermore, the H_2S–cytochrome oxidase complex seems to be very unstable. Patients in general recover very rapidly after termination of exposure.

6.8.3 Respiratory Tract Irritants

> **The primary localization of airway damage by irritant gases is largely determined by their water solubility.**

Properties

An important protective factor which may determine the extent of damage by irritant gases is the initial warning provided by irritation and pain. Detection of painful stimuli motivates people to move away from the area of exposure. The mucous membranes of the eye, nose, throat and larynx

Table 6.19 Effects of hydrogen sulfide relative to the concentration in air.

Concentration (ppm)	Effect
0.02–0.1	Odor threshold
10–50	Irritation threshold (eyes, respiratory tract)
100–200	Loss of smell after 3–5 minutes
>250	Pulmonary edema after several hours of exposure
>500	Systemic poisoning with unconsciousness and respiratory paralysis after 0.5–1 hour exposure
>700	Respiratory failure and death within minutes

are very sensitive. In deeper lung areas perception of irritation and pain is absent. Therefore, water-soluble irritant gases have a higher warning capacity and cause a faster active termination of exposure. The sensory perception of exposure to lipid-soluble gases is much lower and may be negligible, especially at low concentrations capable of causing damage.

The solubility of a gas in water largely determines the depth of the lung at which irritant gases exert their toxicity. Thus, highly water-soluble gases tend to be absorbed into the mucous layers higher in the lung than less water-soluble gases. Gases that impact on the deeper lung areas are those which are relatively non-polar. In the absence of upper airway irritation these gases may reach deep into the lung, undetected, in sufficient concentration to produce severe damage such as lung edema.

Ammonia, hydrogen chloride, hydrogen fluoride, formaldehyde, and acrolein are examples of water-soluble gases. Moderate water solubility characterizes sulfur dioxide, chlorine, and bromine, whereas oxygen, ozone, nitrous oxides, and phosgene are poorly soluble in water.

Exposure to gases with high water solubility causes immediately irritation of eyes, nose, throat, and larynx, and at higher concentrations erosions of the above mucous membranes and laryngospasm. Characteristics of damage by gases with moderate water solubility are symptoms of bronchial damage such as coughing, expectoration, bronchospasm, and subsequent bronchitis and bronchopneumonia. Gases displaying poor water solubility exert their effects predominantly in the alveolus, where they target alveolar epithelia and capillary endothelia. As a consequence of the damage, an inflammatory reaction is initiated with swelling of the alveolar membrane leading to inhibition of gas exchange. Alveolar exudation is a result of damage to alveolar epithelia and the basal membrane. Nevertheless, prolonged or high dose exposure to water-soluble irritant gases may also cause deep lung damage. Thus, in accidental exposure to ammonia, a highly water-soluble gas, victims who are not able to escape the room where the gas is being emitted may develop lung edema.

Therapeutic Principles

Remove victims from the contaminated area and provide immediate support of airway and breathing. After exposure to gases of high and moderate water solubility, mucous membranes of eyes, nasopharynx and, if necessary, exposed skin should be decontaminated with a physiological saline solution or tap water.

Sympathomimetics may be used to treat cases of larynx edema and bronchospasm. Tracheostomy might become necessary.

Pulmonary edema is a consequence of alveolar damage which often results in mortality. Several hours may pass until alveolar membrane damage results in exudation of fluid into the alveolus, resulting in detectable disturbances of gas exchange. The development of a life-threatening lung edema can be visualized by chest x-ray several hours after intake and even before symptoms occur. Clinical symptoms such as coughing, exudation, cyanosis, and dyspnea can appear up to 24 hours after exposure to phosgene. In order to prevent development of a toxic lung edema after high dose exposure, early administration of inhalative glucocorticoids may be considered. Among suggested treatments, which may or may not be effective, are oxygenation, bronchial lavage, and administration of high doses of glucocorticoids. In every case, maintenance of sufficient oxygenation, e.g. by early intubation and artificial ventilation with positive endexpiratory pressure (PEEP), is important.

Chronic Effects

Highly water-soluble respiratory irritants may cause necrotic damage to the mucous membranes of the eyes and upper respiratory tract and scar formation. These lesions are slow to heal. Chronic bronchitis can follow extensive acute damage, leading to pulmonary fibrosis. Extensive necrosis of small bronchi and alveoli can lead to lung fibrosis. Increased lung vascular resistance can lead to increased workload for the right heart.

6.8.4 Irritant Gases

Hydrogen Chloride

> **Hydrogen chloride gas and aqueous hydrochloric acid are widely used in industry and in most chemical, pharmaceutical, and medical laboratories.**

Hydrogen chloride (HCl, MW 36.46) is a highly water-soluble, colorless gas, with a pungent odor. It is heavier than air (density 1.27; air 1.0). HCl is widely used in industry, e.g. pharmaceutical manufacturing, synthesis of organic compounds, and vinyl chloride production. Combustion of polyvinyl chloride, chlorinated acrylics, and flame-retardant materials produces hydrogen chloride.

The maximum allowable working place concentration in Germany is 2 ppm. In the USA the PEL value is 5 ppm (ceiling) and the ACGIH TVL is 2 ppm (ceiling).

Effects HCl gas and hydrochloric acid are very strong irritants causing coagulation necrosis at all surfaces they come into contact with.

Hydrogen Fluoride

> **Hydrogen fluoride and its aqueous solution are among the most corrosive compounds known.**

Hydrogen fluoride is a colorless highly corrosive gas, which is heavier then air (HF; density 1.27; air 1.0). Commercial use includes mineral oil processing, aluminum manufacture, separation of uranium isotopes, glass and enamel etching, and production of fluorinated resins and paint.

The maximum allowable working place concentration in Germany is 1 ppm. In the USA the OSHA PEL is 3 ppm and the ACGIH TLV is 0.5 ppm with a ceiling value of 2 ppm.

Effects HF causes enzyme inhibitions and cellular damage leading to necrosis. In contrast to other halogens HF binds calcium to form insoluble calcium fluoride. The deprivation of calcium exacerbates tissue damage, which can be ameliorated by calcium replacement.

Formaldehyde

> **Formaldehyde is a gas with pungent odor and broad commercial use, and is a suspected carcinogen.**

Formaldehyde (HCHO, MW 30.03) is a reactive, colorless, combustible gas with a pungent odor (synonyms: formalin, methyl aldehyde, methylene oxide). Commercial use of HCHO includes the production of phenol- and urea-based polymers. It is used for medical purposes as a disinfectant, antiseptic, deodorant, tissue fixative, and embalming agent. Because of suspected carcinogenicity, permissible exposure in Germany is limited to 0.0.3 ppm by technical regulations for hazardous materials (TRGS). In the USA the OSHA PEL is 0.75 ppm and the ACGIH TVL is 0.3 ppm (ceiling).

Effects HCHO is a mucous membrane and skin irritant. It is metabolized to formic acid in the body. HCHO forms various adducts and cross-links with proteins, RNA and DNA. Carcinogenic effects were observed in the nose of mice and rats exposed to 5 ppm for their lifetime. In all cases local necrosis were observed prior to tumor appearance in affected areas. Cytotoxicity stimulates cell proliferation and results in tumor promotion. In the absence of cytotoxicity no carcinogenicity was observed, therefore doses at which no cytotoxicity was observed are considered not to be carcinogenic.

Sulfur Dioxide

> **Sulfur dioxide is an important environmental toxicant that is a major cause of acid rain and originates mainly from combustion of fossil fuels**

Sulfur dioxide (SO_2) a highly water-soluble, colorless, and non-combustible gas with a pungent odor that is 2.3 times heavier than air. Commercial applications include preservation of fruit and vegetables via fumigation, disinfection of breweries, and bleaching of textiles, straw, wicker-work, gelatin, glue, and beet-sugar. Other sources of SO_2 include smelting, fossil fuel combustion, paper manufacturing, and fabrication of rubber.

The maximum allowable working place concentration in Germany is 1 ppm. In the USA the PEL value is 5 ppm and the ACGIH TLV and NIOSHSTEL values amount to 5 ppm (10 ppm ceiling) and 0.5 ppm, respectively.

Effects In aqueous media such as mucus, SO_2 is found as sulfurous acid, which is capable of damaging the epithelia of the airways and inhibiting mucocilliary transport. The odor threshold is about 1 ppm. Airway irritation and, eventually, nosebleeds occur at about 10 ppm, and eye irritation at 20 ppm or higher. The maximum tolerable concentration is reached at 50–100 ppm. Concentrations higher than 400 ppm represent an immediate danger to life. Patients having hyperreactive airways indicative of bronchial asthma are particularly sensitive to SO_2. It is assumed that the severity of pulmonary diseases is increased by sulfur dioxide, leading to increased mortality from cardiovascular and pulmonary causes.

Chlorine

> **Chlorine is, after fluorine, the second most reactive and toxic gas.**

Chlorine is a gas with a pungent odor. It is 2.49 times heavier than air, yellow-green in color and very reactive. Commercial use includes the production of vinyl- and polyvinyl chloride, solvents, and bleaching materials. Chlorine is widely used in water purification and paper bleaching. It forms explosive mixtures with hydrogen. During World War I chlorine was the first gas used as a chemical warfare agent and its release resulted in many casualties.

The maximum allowable working place concentration in Germany is 0.5 ppm of air. In the USA the OSHA PEL value is 1 ppm, the ACGIH TLV is 0.5 ppm and the STEL is 1 ppm.

Effects The odor threshold of chlorine is about 0.2 ppm. Mucous membrane irritation occurs at about 1 ppm. At 30 ppm dyspnea, chest pain, nausea, vomiting, and coughing are observed. Inhalation of 40 ppm can cause pulmonary edema. Lethality occurs after 30 minutes at 400 ppm.

In aqueous media (mucus) chlorine forms hydrochloric (HCl) and hypochlorous (HOCl) acid. The latter decomposes to chloric acid ($HClO_3$) and oxygen free radicals. Effects on biological tissues include necrosis and chlorination, and oxidation of biologic materials. HOCl formed by leukocytes is an important mediator of inflammation.

Isocyanates

The highly reactive NCO group of isocyanates is responsible for their broad use as precursors in the chemical industry.

Isocyanates are widely used for the production of plastics, polyurethane foam, lacquers, adhesives, and fibers. The most important representatives of this class are toluene diisocyanate (TDI), methylene bisphenyl diisocyanate (MDI), hexamethylene diisocyanate (HDI), and 1,5-naphthalene diisocyanate (NDI).

For the assessment of an exposure towards isocyanates it was decided to determine the total reactive isocyanate group concentration (TRIG). Basing on the maximum allowable working place concentration in Germany a so-called exposure guide value (ELW) was derived which amounts to 0.018 mg/m^3 NCO.

Effects In December 1994, a severe accident in which approximately 27 tons of methyl isocyanate was released at a pesticide plant in Bhopal, India, resulted in more than 3300 deaths. Isocyanates cause irritation of mucous membranes in the eye, nose and throat. At higher doses pulmonary symptoms (cough, dyspnea, choking sensation), gastrointestinal symptoms (nausea, vomiting, abdominal pain), skin inflammation, and neurologic symptoms may develop. In the occupational setting sensitization is frequently, resulting in bronchial asthma. Although less frequent, skin reactions may also be seen.

Whereas the di-isocyanates are strong sensitizers the mono-isocyanates are not.

Nitrogen Oxides

Among the various nitrogen oxides, the toxicologically most important is nitrogen dioxide.

The nitrogen oxides include nitrous oxide (N_2O, "laughing gas"), nitric oxide (NO), nitrogen dioxide (NO_2) and the commercial product nitrogen tetroxide (N_2O_4), the latter usually made up of a mixture of N_2O and N_2O_4. NO is rapidly oxidized in air to N_2O. NO and N_2O_4 are colorless gases. NO_2 is slightly yellow to reddish-brown. The commercial N_2O_4/NO_2 mixture is 1.61 times heavier than air and, while not combustible, can promote the combustion of carbon, phosphorus, and sulfur. Water solubility is low and the odor of NO_2 is faint and chlorine-like.

The maximum allowable working place concentration of NO and NO_2 in Germany is 0.5 ppm. For NO_2 (nitrous oxide) there is no OSCHA PEL, the ACGIH TVL is 50 ppm, and the NIOSH REL TWA is 25 ppm (over time exposed). For NO (nitric oxide) the OSHA PEL is 25 ppm, the ACGIH TVL is 25, and the NIOSH REL is 25 ppm.

Occupational exposure can be observed in manufacture of dyes, fertilizers, or lacquers. Welding, glass blowing, and food bleaching are also known sources. Silo filler's disease is a consequence of NO, NO_2 and CO_2 inhalation in silos originating from decomposing plants in the presence of high nitrate concentration. NO_x in the atmosphere originates from air nitrogen oxidation and combustion of fossil fuels (power plants, home heating, and automobiles).

Effects NO is not an irritant, but is a physiologically important mediator. It causes vasodilatation and methemoglobinemia. It represents only a small fraction of NO_X fumes and its contribution to toxicity is negligible. The effect of nitrogen dioxide is typical of poor water soluble irritant gases. Initial irritation of eyes, nose and throat is mild, but after a latent period of 3–30 hours lung edema accompanied by fever, dyspnea, coughing, hemoptysis, wheezing, rales, and cyanosis may develop, leading to respiratory failure. 50% of patients surviving lung edema develop bronchiolitis obliterans two to three weeks later.

In the aqueous medium of the mucus nitric acid formed from nitrogen oxides may damage the epithelium. The reaction of nitrogen dioxide with olefinic binding sites in fatty acids can result in radical formation and subsequent lipid peroxidation of capillary endothelia; 100 ppm causes airway damage after 30–60 minutes and 200 ppm may be fatal.

Phosgene

Phosgene was responsible for 80% of fatalities caused by chemical agents in World War I.

Phosgene ($Cl_2C=O$, MW 98.92) is a colorless gas, heavier than air, with a musty hay-like odor. It is used for production of isocyanates, dyes, insecticides, and pharmaceuticals. Thermal decomposition of chlorinated hydrocarbons can lead to phosgene release and potential poisoning.

The maximum allowable working place concentration in Germany is 0.1 ppm. In the USA both the OSCHA PEL and the ACGIH TLV are 0.1 ppm.

Effects Phosgene has poor warning properties. Only mild irritation of the eyes and throat occur, even at concentrations causing pulmonary edema. The molecular mechanism of damage is unknown.

In order to prevent development of a toxic lung edema administration of inhalative glucocorticoids with high receptor affinity should be considered. This approach is being discussed scientifically at present. In any case, the patient should be decontaminated as fast as possible and set at rest. After verified exposure hospitalization for 24 hours is recommended.

Ozone

In the stratosphere ozone provides protection from excessive ultraviolet radiation. In the troposphere this irritant oxidizing gas is an important pollutant formed in light by the interaction of oxygen with nitrogen oxides and volatile organic compounds (VOCs).

Ozone (O_3, MW 48) is a bluish, explosive gas with low water solubility. The odor is pleasant at low concentrations (up to 2 ppm) and pungent at higher concentrations. Occupational sources of exposure are electric arc welding, mercury vapor lamps, photocopy machines, water purification processes, bleaching, and the synthesis of organic compounds.

In the stratosphere the ozone concentration is about 10 ppm and in the troposphere 1 ppm. Ozone is the most important pollutant in photochemical smog, where it comprises about 90% of the oxidants.

In the USA the OSCHA PEL value is 0.1 ppm. The ACGIH TLV values are based on work effort and range from 0.05 ppm for heavy work to 0.1 ppm for light work.

Effects The damage caused by ozone is based on direct oxidation of various molecules in the tissues (–SH, –NH_2, –OH, phenol groups, etc.), formation of free radicals, diene conjugation,

Table 6.20 Toxic gases in fires.

Gas	Source	Effect*
Carbon monoxide	Organic materials at oxygen deficiency	S
Cyanide	Polyacrylate, polyurethane, nylon, wool, silk	S
Hydrogen sulfide	Rubber	S, I
Sulfur dioxide	Rubber, sulfur-containing materials	I
Hydrogen chloride	Polyvinylchloride	I
Phosgene	Chlorinated hydrocarbons	I
Isocyanate	Polyurethanes	I
Nitrogen oxides	Cellulose nitrate, wool, silk	I
Acrolein	Polyolefines and cellulose pyrolysis	I

*S, systemic effects; I, irritation (local effects).

and lipid peroxidation. Increased release of cytokines in asthmatics exposed to ozone appears to exacerbate the disease.

Because of somewhat higher water solubility, compared to phosgene and nitrogen oxides, the irritation of eyes and upper airways is more pronounced. High concentrations of ozone can cause lung edema. At low concentrations ozone causes initially increased reactivity of the bronchi, which disappears after prolonged exposure, as the airways adapt to oxidative stress.

Smoke Inhalation

Up to 80% of fire-related deaths are due to poisoning and not to thermal injuries.

Smoke is the volatilized product of combustion consisting of various gases and particulate matter. The gaseous phase can contain irritant and systemically poisonous gases. Particulate matter can transport toxic materials, with smaller particles reaching the lower respiratory tract. Hot smoke can cause burns of the upper airways, but the damage usually does not reach areas deeper than the larynx. Damage to deep lung areas develops from exposure to toxic materials.

Fires in closed or poorly ventilated places bear an especially high risk of poisoning. In most cases carbon monoxide is the main toxicant, but depending on the combusted material a large number of other systemically or locally poisonous gases can be formed (Table 6.20).

Effects Release of toxic gases in fires can lead to systemic poisoning (mainly asphyxiation) and lung irritation (Table 6.20). While systemic gases have to be taken up and distributed in the circulation to the tissues, locally acting gases primarily cause irritation and damage to respiratory airway epithelia.

6.8.5 Summary

Acute poisoning by airborne toxicants is based on either systemic effects or local damage of airway epithelia. Systemic poisoning is caused by gases such as carbon monoxide, hydrogen cyanide or hydrogen sulfide. These compounds interfere with oxygen availability or its utilization in the tissues, thereby reducing oxidative metabolism in the mitochondria and the availability of energy for metabolic processes. High-energy-demanding tissues such as the central nervous system are most sensitive, leading to respiratory and circulatory failure in severe poisoning.

Locally acting irritant gases affect the epithelia of the respiratory tract, leading to irritation at lower doses and necrosis at higher concentrations. Extensive damage may lead to fibrosis of the lung. Gases displaying high water solubility, such as ammonia and formaldehyde, primarily affect the upper respiratory airways such as the nasopharynx, the larynx, and the eyes. This area has a very dense sensory innervation resulting in early warning in the form of irritation and pain. In contrast, smaller amounts of gases with low water solubility, such as phosgene and nitrogen oxides, are dissolved in the mucus and reach the epithelia of the upper respiratory tract and eyes. The irritation and the warning effect is not prominent and higher concentrations reach and damage the lower parts of the lung, i.e. small bronchi and alveoli, therefore lung edema is more characteristic of poisoning by gases with low water solubility.

Further Reading

Brent J, Burkhart K, Dargan P, Hatten B, Megarbane B, Palmer R, White J (2017) *Critical Care Toxicology Diagnosis and Management of the Critically Poisoned Patient*. Springer International Publishing.

Brandt B, Assenmacher-Maiworm, Hahn JU (2013) Messung und Beurteilung von Isocyanaten an Arbeitsplätzen unter Beachtung der TRSG 430, Gefahrstoffe-Reinerhaltung der Luft 73: Nr 5 – Mai.

Burton LL, Hilal-Dandan R, Knollmann BC (2018) *Goodmans & Gilmans The Pharmacological Basis of Therapeutics*. McGraw-Hill, New York.

Zilker T (2008) *Klinische Toxikologie für die Notfall- und Intensivmedizin*. UNI-Med Verlag, Bremen.

6.9 Fragrance Materials

Anne Marie Api

6.9.1 Introduction

Smell is one of the chemical senses and the sense of smell has always been an essential part of "living" and "evolving" from the day life began. In humans, the sense of smell is unique in having powerful effects on human emotions. When we smell a particular fragrance, we unconsciously connect that smell to a portion of our memory. Smell can evoke feelings and bring back memories that we forgot we had. The ancient Egyptians were responsible for the inception of perfumes.

> **Fragrance materials are used in a wide variety of consumer products, including both personal care and household products. Fragrance mixtures (also called fragrance compounds or fragrance oils) are formulations consisting of specific combinations of individual materials or mixtures. Consumer exposure to fragrance materials ranges from skin contact to inhalation.**

6.9.2 Evaluation of Toxicity

With the implementation of the European regulation on the Registration, Evaluation, Authorization and Restriction of Chemicals (REACH) substantial additional data are becoming available for chemicals, including fragrance materials. While it is theoretically possible to evaluate each and every chemical, there are ethical and practical considerations, such as the aim to minimize animal use and testing laboratory capacity, that drive the need to use data for one or more materials to support related chemicals that do not have sufficient data.

> **Key toxicological endpoints include genotoxicity, repeated dose toxicity, developmental and reproduction toxicity, skin sensitization, phototoxicity/photoallergy, local inhalation effects, and environmental considerations.**

Safety assessments of materials used in fragrances should be carried out by evaluating the available data for relevant toxicological endpoints for local and systemic effects, including (but not limited to) genotoxicity/carcinogenicity, reproductive and developmental toxicity, repeated dose toxicity, skin sensitization, respiratory toxicity, phototoxicity, and environmental effects. These data are put into context with the expected exposure from various fragrance products via the dermal route (including both leave-on and rinse-off applications) and from inhalation exposure. Oral exposure is also relevant for fragrance materials that are used in oral care products, such as toothpaste and mouthwash.

Any safety assessment must consider both the human and the environmental impact of a material. As such, the environmental assessment is an integral part of a safety assessment process. The processes for assessing human health and environmental safety, while not identical, are complementary in their design following a tiered screening approach to set safety assessment priorities.

> **There are several important considerations for the safety assessment process: exposure, threshold of toxicological concern (TTC), read-across, and adequacy of data.**

6.9.3 A Tiered Approach to the Risk Assessment of Fragrances

This approach employs a tiered testing strategy and provides an overall profile for the material of interest and operates for each of the different endpoints through a series of steps.

First one must prioritize materials for review by evaluating volume of use, exposure, and chemical structure. Fragrance materials are prioritized by key human health and environmental endpoints and are evaluated by using an initial risk prioritization based on exposure and the data available. Prioritization of assessments is more heavily weighted on direct consumer exposure than on volume of use. Although high volume of use suggests the potential for high human exposure, there are instances where high-volume fragrance materials are used in products that result in relatively low human exposure. Conversely there may be lower volume materials that, in part due to their scent characteristics, are used in products with relatively high exposure potential. In addition, other factors to be considered in prioritization of assessments include existing data of concern and/or the need for additional information on one or more toxicological endpoints under review and/or regulatory requirements.

Step 1: Data

Step 1 in all endpoint assessments examines the acceptability of any existing data. The focus is on studies carried out according to established methods and latest guidelines and that are considered reliable based on accepted evaluation criteria such as Klimisch scores. The quality of the test material for all studies evaluated should be defined and correspond to fragrance materials in commerce. For studies which do not necessarily meet accepted guidelines, a weight of evidence approach is considered.

Step 2: Read-across

Read-across is an important technique to estimate missing data for a single or limited number of chemicals using an analogue approach. In principle, read-across utilizes common endpoint information, including physicochemical properties and toxicity, for one (or more) chemical(s) to make a prediction on the same endpoint for another chemical. It may be performed in a qualitative or quantitative manner. This process can help to avoid the need to carry out specific tests on every substance for every endpoint. The criteria for providing sufficient information via read-across will be specific to each analogous set of chemicals and may be specific to each endpoint.

Step 3: Exposure Assessment and Threshold of Toxicological Concern

> **Hazard and exposure evaluations can be developed almost simultaneously since low exposures may permit use of the TTC approach.**

The exposure and risk assessment of any fragrance material should be an iterative process that incorporates the available hazard data for the key toxicological endpoints coupled with the exposure assessment. Exposure is an essential part of the safety assessment process and is required in order to conduct a safety assessment. Fragrances are used in a wide variety of products, including decorative cosmetics, fine fragrances, shampoos, toilet soaps, and other toiletries as well as in other consumer products such as household cleaners, detergents, oral and air care products. For example, two types of exposure data on fragrance materials may be used. The first is volume of use data, which is provided by the International Fragrance Association (IFRA) approximately every five years through a comprehensive survey of IFRA and Research Institute for Fragrance Materials (RIFM) member companies that manufacture fragrances. The second method is an aggregate exposure model using deterministic and probabilistic exposure data to describe real-life consumer exposure to a specific fragrance material. This model called the Creme RiFM Aggregate Exposure Model, will address exposure from all routes, including that from inhaled products.

The TTC approach is based on the concept that reasonable assurance of safety can be given, even in the absence of chemical-specific toxicity data, provided the exposure is sufficiently low, i.e. that an exposure level can be defined below which there is no significant risk to human health. The TTC is based on the Threshold of Regulation, FDA's priority-based assessment of food additives, which was expanded to include consideration of the chemical structure in conjunction with toxicity data. These analyses originally focused on systemic exposure following oral administration. More recently, the TTC approach was extended to consider systemic exposure following topical application of cosmetic products, including the use of default skin penetration values. In 2012, a joint opinion from the European Scientific Committees (Scientific Committee on Consumer Safety (SCCS), the Scientific Committee on Health and Environmental Risks (SCHER), and the Scientific Committee on Emerging and Newly Identified Health Risks (SCENIHR) considered the TTC approach, in general, scientifically acceptable for human health risk assessment of systemic toxic effects caused by chemicals present at very low levels, as based on sound exposure information. In 2016 another review of the TTC approach and the development of a new TTC decision tree was provided by the European Food Safety Authority and the World Health Organization. These experts concluded that the TTC approach is based on scientific risk assessment principles and fit for purpose as a screening tool, to assess low dose chemical exposures, and to identify those for which further data are necessary to assess the human health risk. The expert group made recommendations to improve and expand the TTC concept, and proposed a tiered approach (revised decision tree) considering the current state of the science and available toxicological databases. A more recent publication by Yang et al. in 2017 provided a new dataset of cosmetics-related chemicals for the TTC approach. These researchers compiled a dataset comprising 552 chemicals. Data were integrated and curated to create a database of no/lowest observed adverse effect level (NOAEL/LOAEL) values, from which the final COSMOS TTC dataset was developed. The 966 substances in the federated database provide corresponding TTC values that are broadly similar to those of the original Munro dataset.

The TTC concept has also been applied to evaluating potential skin sensitizers. The dermal sensitization threshold (DST) establishes a level below which there is no appreciable risk for the induction of sensitization and is based on a probabilistic analysis of potency data for a diverse range of known chemical allergens. There has also been the suggestion that TTC can be applied to inhalation exposure and risk assessment. With respect to inhalation exposure another important consideration is the potential for site of contact (local) effects in all parts of the respiratory tract.

Step 4: Generate Data

Additional testing may be recommended as determined on a case-by-case basis. Other testing deemed necessary to complete the safety assessment could include *in silico*, *in vitro* or *in vivo* studies. If *in vitro* assays are available, these assays are always considered before *in vivo* tests. Testing an analogue material or metabolite may also be considered. Another consideration is to forego testing and consider risk management measures which would limit the use of the material to a low level and consider applying the TTC approach.

6.9.4 Summary

Fragrance materials are used in a wide variety of consumer products including both personal care and household products. Fragrance mixtures (also called fragrance compounds or fragrance oils) are formulations consisting of specific combinations of individual materials or mixtures. Consumer exposure results from dermal application primarily from cosmetics and body care products, including both leave-on and rinse-off applications, and from inhalation exposure. Oral exposure is also relevant for fragrance materials that are used in oral care products, such as toothpaste and mouthwash. Although safety assessment comprises the array of toxicological studies, including local and systemic effects, genotoxicity/carcinogenicity, reproductive and developmental toxicity, and repeated dose toxicity, the most relevant effect is skin sensitization. To prevent sensitization the sensitizing agents need to be identified and concentrations in the different products are reduced to levels that avoid both induction of sensitization and elicitation after use of a product which contains the specific agents.

Further Reading

Api, A.M., et al., 2015. Criteria for the Research Institute for Fragrance Materials, Inc. (RIFM) safety evaluation process for fragrance ingredients. *Food Chem Toxicol.* **82** Suppl, S1–S19.

Bickers, D.R., et al., 2003. The safety assessment of fragrance materials. *Regul Toxicol Pharmacol.* **37**, 218–273.

Carthew, P., et al., 2009. Exposure based waiving: the application of the toxicological threshold of concern (TTC) to inhalation exposure for aerosol ingredients in consumer products. *Food Chem Toxicol.* **47**, 1287–95.

Comiskey, D., et al., 2015. Novel database for exposure to fragrance ingredients in cosmetics and personal care products. *Regul Toxicol Pharmacol.* **72**, 660–72.

Comiskey, D., et al., 2017. Integrating habits and practices data for soaps, cosmetics and air care products into an existing aggregate exposure model. *Regul Toxicol Pharmacol.* **88**, 144–156.

European Food Safety Authority and World Health Organization, 2016. Review of the Threshold of Toxicological Concern (TTC) approach and development of new TTC decision tree. *EFSA Supporting Publication 2016*; **13**(3):EN-1006, 50 pp. doi:10.2903/sp.efsa.2016.EN-1006/2017.

Escher, S. et al., 2010. Evaluation of inhalation TTC values with the database Rep Dose. *Regul Toxicol Pharmacol.* **58**, 259–274.

Klimisch, H.J., et al., 1997. A systematic approach for evaluating the quality of experimental toxicological and ecotoxicological data. *Regul Toxicol Pharmacol.* **25**, 1–5.

Kroes, R., et al., 2007. Application of the threshold of toxicological concern (TTC) to the safety evaluation of cosmetic ingredients. *Food Chem Toxicol.* **45**, 2533–2562.

Roberts, D.W., et al., 2015. Principles for identification of high potency category chemicals for which the dermal sensitisation threshold (DST) approach should not be applied. *Regul Toxicol Pharmacol.* **72**, 683–693.

Safford, B., et al., 2015. Use of an aggregate exposure model to estimate consumer exposure to fragrance ingredients in personal care and cosmetic products. *Regul Toxicol Pharmacol.* **72**, 673–682.

Safford, B., et al., 2017. Application of the expanded Creme RIFM consumer exposure model to fragrance ingredients in cosmetic, personal care and air care products. *Regul Toxicol Pharmacol.* **86**, 148–156.

Salvito, D. et al., 2002. A framework for prioritizing fragrance materials for aquatic risk assessment. *Environ Toxicol Chem.* **21**, 1301–1308.

SCCS, SCHER, SCENIHR, Joint Opinion on the Use of the Threshold of Toxicological Concern (TTC) Approach for Human Safety Assessment of Chemical Substances with focus on Cosmetics and Consumer Products, 8 June 2012. Available: http://ec.europa.eu/health/scientific_committees/consumer_safety/docs/sccs_o_092.pdf/2017.

Schultz, T.W., et al., 2015. A strategy for structuring and reporting a read-across prediction of toxicity. *Regul Toxicol Pharmacol.* **72**, 586–601.

Yang, C., et al., 2017. Thresholds of toxicological concern for cosmetics-related substances: New database, thresholds, and enrichment of chemical space. *Food Chem Toxicol.* **109**, 170–193.

6.10 Pesticides

Roland Alfred Solecki and Vera Ritz

6.10.1 Introduction

> **The terms "biocides" and "pesticides" are regularly interchanged and often confused with "plant protection products". To clarify this, pesticides include both biocides and plant protection products, where the former includes substances for the protection of human or animal health, or goods excluding direct applications on food and feed, and the latter includes substances for the protection of plants or plant products, including food and feed. According to Directive 2009/128/EC a pesticide means a plant protection product as defined in Regulation (EC) No. 1107/2009 or a biocidal product as defined in Directive 98/8/EC (which was replaced by the biocides regulation (EU) No. 528/2012).**

One of most important measures to protect crops and crop commodities against pests including weeds and to raise the efficacy of agricultural production is the use of plant protection products (PPPs). These measures are not only beneficial but can also bear risks and hazards to

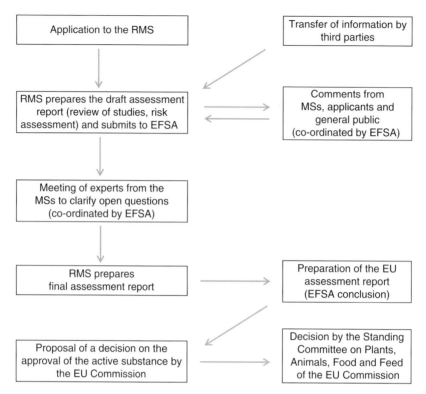

Figure 6.28 *EU approval process of active substances in plant protection products. MS, Member State; RMS, Rapporteur Member State; EFSA, European Food Safety Authority.*

humans, animals and the environment, particularly if they are put on the market without prior risk assessment and authorisation, and are used improperly. Therefore, all pesticidal active substances have to pass an approval process on a European level according to the EU PPPs regulation (EC) No. 1107/2009, which is re-evaluated every 10 years (see Figure 6.28). For detailed information on the regulation of biocides see Chapter 5.1.

According to Regulation (EC) No. 1107/2009 PPPs are intended for one of the following uses:

(a) **protecting plants or plant products against all harmful organisms or preventing the action of such organisms, unless the main purpose of these products is considered to be for reasons of hygiene rather than for the protection of plants or plant products;**
(b) **influencing the life processes of plants, such as substances influencing their growth, other than as a nutrient;**
(c) **preserving plant products, in so far as such substances or products are not subject to special Community provisions on preservatives;**
(d) **destroying undesired plants or parts of plants, except algae unless the products are applied on soil or water to protect plants;**
(e) **checking or preventing undesired growth of plants, except algae.**

Specific Toxicology 705

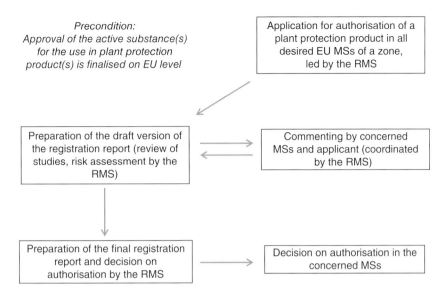

Figure 6.29 Zonal authorisation procedure for plant protection products in the EU. MS, Member State; RMS, Rapporteur Member State.

For authorisation of a PPP in a zonal process in the EU only approved active substances are eligible. For zonal authorisations one of the EU Member States is nominated as the Rapporteur Member State who evaluates and issues the initial authorisation. After giving the other Member States in the same zone the opportunity to comment on the assessment report they are legally obliged to recognise mutually the initial authorisation (see Figure 6.29).

In non-agricultural areas some of the same pesticidal active substances are applied as in biocidal products that are used in disinfectants to protect from infectious diseases, as pest control agents to protect humans and animals from vector-borne diseases or as preservatives to protect building materials and industrial goods from deterioration from microbes and insects. As for PPPs, for biocidal products there is a European approval procedure for all active substances according to the biocidal products Regulation (EU) No. 528/2012.

> **According to Regulation (EC) No. 528/2012 biocidal products are necessary for the control of organisms that are harmful to human or animal health and for the control of organisms that cause damage to natural or manufactured materials.**

The subject of this chapter is the toxicology of pesticidal active substances applied mainly in PPPs. It is possible to allocate those active substances systematically to target pests (see Table 6.21).

Herbicides are the most commonly used chemical PPP with a market share of 40% in Germany in 2014, followed by inert gases for storage protection with a market share of 26% and fungicides with 24%. Insecticides and acaricides have a low market share (2%) but can have significant toxicological effects.

Since the 1960s there has been a considerable change regarding the evaluation of PPPs. It was realised that the "wonder pesticides" applied in huge amounts can have ecotoxicological

Table 6.21 Use categories of pesticidal active substances with respect to target organisms.

Target organism	Pesticidal category
Bacteria	Bactericide
Fungi (yeast, moulds)	Fungicide
Algae	Algicide
Weeds	Herbicide
Insects	Insecticide
Acarids (mites, ticks)	Acaricide
Nematodes	Nematicide
Gastropods (slugs, snails)	Molluscicides
Rodents	Rodenticide

effects in birds as well as carcinogenic effects in humans. This triggered a more intense assessment of the toxicological side effects on humans and animals, and raised also awareness of the toxicological evaluation of other chemicals. This changed the regulation of chemical substances completely as legally binding testing requirements and regulatory assessment principles evolved. Innovative methods based on molecular biology, *in silico* methods as well as the validation of experimental alternatives to the classical animal study, the evolution of adverse outcome pathways, and other methods to investigate mechanisms of action has lead to an accelerating development that will greatly influence the regulatory practice of pesticides toxicology in the 21st century.

Many assessments and regulatory approaches of the PPP area are comparable for pesticidal active substances in biocidal products. The main groups of biocidal products and the individual product types are summarised in Table 6.22.

6.10.2 Toxicity of Selected Pesticidal Active Substances

The core tasks of human health risk assessment of pesticidal active substances include the comprehensive evaluation of inherent toxicological properties and the dose–response relationship, the derivation of toxicological limit values, classification and labelling, and the exposure assessment of humans, companion animals and livestock. This results in the deduction of risk mitigation measures and the derivation of maximum residue levels aiming to avoid an adverse outcome on the health of humans and animals.

In the following paragraphs general aspects of the common toxicology of selected pesticidal active substances and chemical substance groups out of the classes of herbicides, fungicides, insecticides and other application groups will be summarised.

More detailed information and assessments as well as up-to-date limit values and maximum residue levels for each substance can be found on the web pages of the European Commission (Pesticides Database[**]), the European Food Safety Authority (EFSA) and the World Health Organization/Food and Agriculture Organization (WHO/FAO) Joint Meeting on Pesticide Residues (JMPR[***]). Further sources of information are given in the *Further Reading* section.

[**] http://ec.europa.eu/food/plant/pesticides/eu-pesticides-database/public/?event=homepage&language=EN.
[***] http://apps.who.int/pesticide-residues-jmpr-database/Home/Search.

Table 6.22 Biocidal main groups and their product types (a detailed description is given in biocides regulation (EU) No. 528/2012).

Main Group 1: Disinfectants
Product type 1: Human hygiene
Product type 2: Disinfectants and algaecides not intended for direct application to humans or animals
Product type 3: Veterinary hygiene
Product type 4: Food and feed area
Product type 5: Drinking water
Main Group 2: Preservatives
Product type 6: Preservatives for products during storage
Product type 7: Film preservatives
Product type 8: Wood preservatives
Product type 9: Fibre, leather, rubber and polymerised materials preservatives
Product type 10: Construction material preservatives
Product type 11: Preservatives for liquid-cooling and processing systems
Product type 12: Slimicides
Product type 13: Working or cutting fluid preservatives
Main Group 3: Pest control
Product type 14: Rodenticides
Product type 15: Avicides
Product type 16: Molluscicides, vermicides and products to control other invertebrates
Product type 17: Piscicides
Product type 18: Insecticides, acaricides and products to control other arthropods
Product type 19: Repellents and attractants
Product type 20: Control of other vertebrates
Main Group 4: Other biocidal products
Product type 21: Antifouling products
Product type 22: Embalming and taxidermist fluids

Herbicides

> **Herbicides are used to prevent or selectively minimise the growth of weeds to enhance the crop yield.**

They are the most extensively used group of pesticides in agriculture. However, regarding residues in food or feed they are of minor importance.

For weed control, a wide spectrum of substance groups is applied. These substance groups comprise a wide variety of single compounds. In specific crop cultures, specific substance groups dominate the ranking, e.g. the organo phosphorus compounds (29%), amides and anilines (21%) and triazines (13%) followed by urea-based herbicides and phenoxy herbicides.

Glyphosate The organophosphorus compound glyphosate is worldwide one of the most extensively applied substances in PPPs. This phosphonate is an inhibitor of the plant enzyme 5-enol pyruvyl shikimate 3-phosphate (EPSP) synthase essential for forming the amino acids phenyalanine, tyrosine and tryptophan in plants.

In mammals, glyphosate is absorbed to 20–30% from the gastrointestinal tract and eliminated nearly completely as glyphosate and the main metabolite aminomethylphosphonic acid (AMPA) within seven days. In animal studies glyphosate showed a very low acute toxicity and was not skin irritating or sensitising. However, the substance revealed eye-irritating properties in animal studies.

After repeated administration changes in liver and salivary glands as well as irritation of mucous membranes in the gastrointestinal tract and the urinary bladder were observed.

Exposure of pregnant rabbits to very high doses of glyphosate associated with poisoning symptoms in dams caused single cases of anomalies in foetuses.

Carcinogenicity tests in rats and mice showed elevated but non-consistent incidences of single tumours in different organs at much higher dose levels compared to the low human exposure to this pesticide. Therefore, no risk regarding carcinogenic or reprotoxic effects was expected by the by the EFSA[****] and the JMPR[#] for humans, although the International Agency for Research on Cancer (IARC) assessed glyphosate as "probably carcinogenic to humans". Following an evidence-based assessment of data from epidemiological studies and animal studies, the European Chemicals Agency (ECHA) announced in 2017 that a hazard classification of glyphosate as carcinogenic, mutagenic and reprotoxic based on the CLP Regulation is not justified.[##]

Glyphosat-containing products are formulated as aqueous solutions as well as in combination with co-formulants that act as a wetting agent. Specific enhancers like polyethoxylated (POE) tallowamines reveal a higher toxicity than glyphosate itself, mainly caused by irritating properties. Reported poisoning cases after ingestion of high amounts of glyphosate-containing formulations in humans erroneously or in suicide attempts are due to the effects of those substances.

Chloro Acetaniline Herbicides The group of chloro acetaniline herbicides comprises substances such as acetochlor, alachlor, butachlor, dimethachlor, metazachlor, metolachlor and propachlor. These compounds act as selective herbicides before crop emergence and influence different metabolic processes, e.g. protein biosynthesis and cell division in weeds.

Metolachlor and S-Metolachlor In the EU, metolachlor has not been approved since 2002, but S-metolachlor is approved instead. In mammals, resorption and distribution of the substance after oral administration is rapid (most within 48 hours) and efficient (70–93%) with a uniform distribution and a slight preference for well-perfused organs.

The highest concentrations of the substance are found in erythrocytes, where the substance is bound to rat but not to human haemoglobin. Metolachlor and S-metolachlor have shown a low acute toxicity in rats, are not irritating to skin or eyes but are skin sensitisers. After subchronic and chronic administration metolachlor is a liver toxin in rats, mice and dogs. In carcinogenicity studies with S-metolachlor emergence of eosinophilic foci and neoplastic nodules were observed. Mechanistic studies showed that metolachlor is, like phenobarbital, a potent inductor of xenobiotic-metabolising enzymes (CYP2B) in the liver. In multi-generation studies, a lower body weight gain in progeny was observed at parental toxic doses of metolachlor whereas in developmental toxicity studies delayed ossification in foetuses was observed at maternally toxic doses. S-metolachlor did not reveal any developmental effects, even at maternally toxic doses.

Triazine Herbicides The group of triazine herbicides comprises simazine, atrazine, cyanazine and terbuthylazine. The herbicidal mode of action of the triazines is mainly due to inhibition of photosynthesis. From the toxicological point of view, triazines predominantly affect the endocrine system, resulting in dysfunction of reproduction and fertility as well as mammary gland tumours in rat studies. While simazine, atrazine and cyanazine are not approved in Europe

---

[****] http://onlinelibrary.wiley.com/doi/10.2903/j.efsa.2015.4302/abstract.
[#] http://www.who.int/entity/foodsafety/jmprsummary2016.pdf?ua=1.
[##] http://www.bfr.bund.de/cm/349/echa-classifies-glyphosate-as-non-carcinogenic-non-mutagenic-and-non-reprotoxic.pdf.

anymore terbuthylazine is authorised in many herbicides. High acute toxicity of terbuthylazine results in classification as harmful if swallowed, the substance is a slight eye and skin irritant and it is sensitising to skin. After repeated administration of terbutylazine, reduced body weight and effects on the red blood cell count were observed. There is limited evidence of a carcinogenic effect because in carcinogenicity studies a higher incidence of mammary gland tumours could be demonstrated.

Urea Herbicides As for triazine herbicides, the herbicidal mode of action of urea herbicides is mainly due to inhibition of photosynthesis. To demonstrate the toxicology of the group, diuron is representative for the urea herbicides: It has shown rapid and nearly complete absorption, is widely distributed with highest residues in blood, organs that produce or contain blood, excretory organs and ovaries, and is excreted within 48 hours, mainly via urine. After subchronic administration, haemolytic anemia including generation of met- and sulfhaemoglobin was observed in dogs, rats and mice which eventually resulted in a compensatory elevated haemotopoesis and haemosiderosis as well as damage to the liver and the urogenital tract. Decreased fetal weight, skeletal alterations and delayed ossification at maternally toxic doses were described in rats.

Phenoxy Herbicides Phenoxy herbicides are applied as free acids, esters or alkaline or ammonium salts and are chemical analogues of auxins which act as phytohormones and affect the growth of weeds, resulting in plant death. In mammals comparable hormones to auxins do not exist. 2,4,5-T was principally contaminated with the dioxin TCDD and therefore banned in most countries. 2,4-DB and 2,4-D are representative approved substances for the phenoxy herbicides. 2,4-D is rapidly and almost completely absorbed after oral administration and widely distributed, with highest residues in liver, kidney and brain. It is poorly metabolised and rapidly eliminated, mainly via urine. Higher levels of the substance are found in the kidneys and liver. In acute poisoning cases mainly local irritating effects as well as dysfunction of the central nervous system are observed, while short-term and long-term exposure can result in kidney degeneration as well as liver and thyroid effects. Dogs were found to have a reduced capacity for urinary excretion of weak organic acids and therefore the dog is not considered the most relevant species to extrapolate 2,4-D toxicity to humans. Reproductive effects and offspring toxicity were noted in the presence of excessive parental toxicity in multi-generation studies and fetotoxicity was observed in the presence of maternal toxicity in developmental studies.

Bipyridylium Herbicides Bipyridylium herbicides also affect photosynthesis by acting as electron acceptors preventing the reduction of NADP+. This results in the generation of reactive oxygen species, which are strong cellular toxins and rapidly lead to plant death.

The most important substances from this group are diquat and paraquat, which are contact fast-acting herbicides used on the aerial parts of weeds. Due to its high acute toxicity, paraquat is not approved in Europe anymore. Diquat shows poor oral absorption and is excreted within 96 hours. There is some evidence for accumulation in the eye lens. It is acutely toxic, an eye and skin irritant and sensitising to skin. After repeated exposure with rats and dogs, critical effects were observed in the eyes (cataracts, both species) and epididymis (decreased weight, dog). Kidney effects were observed only in mice.

6.10.3 Fungicides

> **Fungicides are organic or inorganic substances inhibiting the growth of fungi or killing them.**

The most commonly used fungicidal substances are inorganic fungicides, carbamates and dithiocarbamates as well as the azoles, i.e. imidazoles and triazoles. The market share of the azoles is nearly a quarter of the fungicide market. Further important substance groups are the dicarboximides and strobilurines. The different chemical classes of fungicides differ in their toxicological properties.

Azoles

The fungicidal mode of action of azoles is inhibition of CYP51 and thus blocking of the ergosterole synthesis of fungi. Besides this fungicidal mode of action, other cytochrome P450 enzymes of mammals are inhibited. Some azoles have a pharmaceutical use as antimycotics. There is no evidence that the use of azoles as PPPs is a cause of the development of resistance of human pathogenic fungi. Furthermore, liver toxicity, endocrine disrupting effects and developmental and reproductive toxicity are observed in animal studies of many azole fungicides. Thus, several azoles, such as cyproconazole, difenoconazole, epoxiconazole, propiconazole, fenbuconazole, tebuconazole and tetraconazole, are known liver carcinogens via induction of enzymes involved in xenobiotics metabolism and liver toxicity. Studies on tumour promotion and induction of drug-metabolising enzymes showed, for example, that propiconazole is a promoter of proliferative changes and causes induction of hepatic enzymes. Increased incidences of thyroid tumours by diniconazole and fenbuconazole in rats could also be observed due to liver enzyme induction and subsequently increased turnover of thyroid hormones. An inhibition of aromatase as well as steroid-11-hydroxylase and steroid-21-hydroxylase are assumed to be the cause of decreased oestrogen and corticosterole as well as aldosterole synthesis leading to tumours of the ovaries and adrenal cortex. Developmental and reproductive effects are triggered by some azoles, such as epiconazole and propiconazole, which consist of prenatal mortality, growth retardation, prolonged pregnancy, craniofascial malformations (e.g. cleft palate), and increased visceral and skeletal variations at dose levels causing maternal toxicity.

Dicarboximides

Dicarboximides inhibit the spore germination and the growth of hyphes of fungi.

Antiandrogenic effects of the dicarboximides procymidon and vinclozolin observed in mammals are due to a competitive antagonism at androgen receptors while in contrast iprodion causes inhibition of the androgen synthesis. There is limited evidence of a carcinogenic effect of iprodione because chronic effects observed in animal studies are increased incidence in liver cell carcinoma, luteomas, and Leydig cell tumours, but no teratogenic effects have been described.

Vinclozolin and procimidone are not authorised anymore. Vinclozolin caused decreased fertility or infertility and malformations and feminisation of male offspring. Reduced anogenital distance, increased testicular weight, decreased prostate weight in the offspring as well as reduced anogenital distance, hypospadia, testicular atrophy, and undescended testes in rat fetuses were observed as reproductive effects of procimidone hypospadia, based on antiandrogenic activity and hypersecretion of testosterone in the rat.

Strobilurins

Strobilurins are synthetic analogues of strobilurin A, a natural compound produced by fungi. They act via inhibition of germination and spore development by interference with the mitochondrial respiration. The most relevant active substances are azoxystrobin, kresoxim-methyl, pyraclostrobin and trifloxistrobin. The prevailing toxic effect of strobilurins observed in animal studies is liver toxicity. Chronic studies in rats and mice revealed liver cell tumours in rats with kresoxim-methyl.

6.10.4 Insecticides

> **Insecticides are applied to kill, harm, repel or mitigate one or more species of insects which can destroy food and feed or any other products, or which can transfer any of animal or human diseases. They are the second most extensively used group of pesticides in agriculture and can be classified in two major groups: systemic insecticides and contact insecticides.**

Some insecticides disrupt the nervous system of insects, whereas others may damage their exoskeleton, or repel or control them by some other means.

Long before the introduction of synthetic insecticides, materials of natural origin provided means for controlling insects affecting the human population both directly and indirectly. Insecticides of natural origin are obtained from animals, plants, bacteria, or certain minerals. The body of scientific literature documenting the effects of natural pesticides continues to expand, yet only a handful of botanicals are currently used in agriculture in the industrialised world. More recently, the immense potential of bacteria and other microorganisms for the production of biologically active insecticides was realised and many new pesticides commercialised since the middle of the 20th century are of microbial origin. As Rachel Carson published in her book *Silent Spring* in 1962 a vision of adverse effects of incecticidal substances such as DDT and other chlorinated hydrocarbons, a second harmful face of insecicides became obvious, which was related mainly to ecotoxicological effects against bird populations and cancer effects in humans. In the following years, the use of bioaccumulating and persistent substances was strongly restricted, including the total ban on the use of DDT in agriculture.

Organophosphates

> **The insecticidal action and toxicological effects of most organophosphates (OPs) are based on the inhibition of acetylcholinesterase, which metabolises the neurotransmitter acetylcholine in choline and acetate. The cholinesterase inhibition results in the accumulation of acetylcholine at the nicotinerge receptors of the parasympathic and sympathic ganglia and the motoric neuromuscular end-plates, leading to muscular rigidity, tremor, subsultus, tonic-clonic seizures, speech disorders, paresthesia, impaired consciousness and respiratory paralysis.**

OPs are applied as PPPs mainly as contact and systemic insecticides. They are also used as biocidal products and veterinary drugs against ecto- and enteroparasites. The poisonousness of OPs is quite different and ranges from very toxic with paraoxone (oral LD_{50} < 2 mg/kg bw) to the relatively slightly toxic malathion (oral LD_{50} of 2800 mg/kg bw). Both the insecticidal action as well as the toxicological effects of most OPs are based on an inhibition of acetylcholinesterase, which metabolises the neurotransmitter acetylcholine in choline and acetate. Inhibition of the acetylcholinesterase is reversible in the beginning but due to conformational changes of the OP–acetylcholinesterase complex a covalent binding leads to irreversible inhibition ("aging") within minutes to days depending on the substance. The cholinesterase inhibition results in an accumulation of acetylcholine at the nicotinerge receptors of the parasympathic and sympathic ganglia, and the motoric neuromuscular end-plates. The following nicotinergic effects may be observed: muscular rigidity, tremor, subsultus, tonic-clonic seizures, speech disorders, paresthesia, impaired consciousness and respiratory paralysis. Furthermore, some OPs can induce a delayed neuropathy characterised by paralysis and paresthesia of extremities. Hens are very sensitive to this syndrome and were therefore used for testing the potential of different OPs to induce this effect.

After repeated exposure to OPs, an adaptation of the organism is responsible for less pronounced neurotoxic symptoms at comparable dose levels. This effect is based on a decline of muscarinergic acetylcholine receptors.

In the case of an acute OP poisoning oximes, which may be applied alone or together with atropine, can be effective antidotes if given before most of the acetylcholinesterase is irreversibly inhibited ("aged").

Carbamates

Carbamates are N-substituted esters of the carbamic acid with the main effect of choline esterase inhibition. Although their general efficacy and toxicity are comparable to those of OPs, there are relevant differences between both chemical classes, especially in relation to the binding to aetylcholinesterase. Carbamates have shown a substantially larger margin between doses leading first to poisoning symptoms and then to mortality. The reversibility of this intoxication is also higher based on a more effective reactivation of the acetylcholinesterase compared to OPs.

Pyrethrins and Pyrethroids

> **Extracts derived from chrysanthemum flowers of the genus *Chrysanthemus* have been used as insecticides for a long time. The insecticidal neurotoxic activity of these extracts is due to a mixture of three naturally occurring, closely related insecticidal esters of chrysanthemic acid (pyrethrins I) and three closely related esters of pyrethric acid (pyrethrins II).**

Selection of varieties of chrysanthemum rich in pyrethrins and extraction techniques have improved over the years. Pyrethrum extract contains pyrethrins and other phytochemicals, including triglyceride oils, terpenoids, and carotinoid plant colours. Pyrethrum induces a toxic effect in insects in the nervous system by binding to sodium channels, which are responsible for nerve signal transmission along the nerve cells by permitting the flux of sodium ions. This results in hyperexcitation and loss of function of the nerve cell. The shutdown of the insect nervous system and insect death are most often the consequences of insect exposure to pyrethrins. In acute toxicity studies in mammals, pyrethrins have been shown to be harmful if swallowed, harmful if inhaled and are also classified as harmful in contact with skin. They do not induce skin or eye irritation, or skin sensitisation. In oral short-term toxicity studies, the target organ is the liver in all species and adverse effects are observed in haematological and clinical chemistry values in rats and dogs. The dog is the most sensitive species.

Synthesized chemical analogues of pyrethrins called pyrethroids are the most abundant group of insecticides used in households and public health. Their advantage is a higher stability and a higher purity compared to the natural pyrethrins without being persistant in the environment like OPs. Based on their toxicity profile, which is dependent on their chemical structure, two types of pyrethroids can be distinguished: type I pyrethoids which cause a tremor (T) syndrome, e.g. permethrin, tetramethrin and allethrin, and type II pyrethroids causing a choreoathetosis (CS) syndrome, e.g. cypermethrin, fenvalerate and cyfluthrin. The T syndrome is characterized in animal studies by aggressive behaviour, hyperexcitation, ataxia, tremor and convulsions, while effects comprising the CS syndrome are hypersensitivity, profound salivation, choreoathetosis and clonic seizures.

6.10.5 Substances of Biological Origin

> **Beyond pyrethrins, other insecticides of biological origin such as further compounds derived from plant extracts (rotenone, azadirachtin, quassin, anabasin), fermentation products from soil microorganisms (avermectins and spinosins) and baculoviruses are more or less widely used as insecticides in PPPs or biocides but may also play a role in public health to prevent the spread of infectious or parasitic diseases.**

Substances of biological origin have been touted for a long time as attractive alternatives to synthetic chemical insecticides for pest management because these substances reputedly pose little threat to the environment or human health. One benefit of plant insecticides is that many of them are readily biodegradable. However, insecticides of natural origin have certain disadvantages in that they are often mixtures of active and inactive components and the active ingredient content may be low, depending on origin, harvest, storage conditions and manufacturing process.

Some naturally occurring substances can also control pests by non-toxic mechanisms, such as insect sex pheromones that interfere with mating, as well as various scented plant extracts that attract insect pests to traps. Other plant oils are complex mixtures of substances made by lemon, orange, and anise, which are used as pesticides to repel and to kill certain insects.

Rotenone

Rotenone is an example of a biologically active isoflavonoid. Rotenone is relatively harmless to plants, highly toxic to fish and many insects, moderately toxic to mammals, and leaves no harmful residue on vegetable crops. It can be applied as a spray on fruits and row crops, even several times before harvest time because the chemical residues do not linger. Rotenone dusts and sprays have been used for years to control aphids, certain beetles and caterpillars on plants, as well as fleas and lice on animals. Rotenone is a mitochondrial poison and its insecticidal activity is based on inhibition of mitochondrial oxidation. The critical effects in insects are exerted by blocking of electron transport in mitochondria from complex I to ubiquinone, thus inhibiting oxidation linked to $NADH_2$, which results in nerve conduction blockade. Acute poisoning in animals is characterised by an initial respiratory stimulation followed by respiratory depression, ataxia, convulsions, and death by respiratory arrest. The acute toxicity appears to vary considerably between species, with oral LD_{50} values ranging from 25 to 3000 mg/kg bw. Chronic exposure to rotenone in rats has reproduced the anatomical, neurochemical, behavioural and neuropathological features of Parkinson's disease.

Neem Tree Extracts

The neem tree contains in its roots, bark, leaves, fruit, flowers, and seed kernels a number of substances which possess, alone or in combination, remarkable insecticidal, "antifeedant" or insect-repellent properties that are successfully used in plant protection and for protection of stored food but also in veterinary medicine and public health.

Neem seed oil or extracts from kernels or leaves have been used for centuries in traditional Indian medicine for treatment of a wide range of diseases including febrile illnesses, hepatitis, respiratory diseases, gastrointestinal disturbances and helminthoses, or more generally as "health strengtheners". Azadirachtin and the neem kernel extracts are very toxic to aquatic organisms and have shown reprotoxic effects in mammals. The acute oral and dermal toxicity of neem extracts in rats is very low, with LD_{50} values greater than 5000 mg/kg bw for oral or greater than 2000 mg/kg bw for dermal application. Proper risk assessment is therefore needed before PPP authorisation is granted.

Active Compounds from Microorganisms

Some insecticidal active substances may be produced by microorganisms, such as the avermectins discovered in the 1970s, which are derived from the soil bacterium *Streptomyces avermitilis*. These complex macrocyclic disaccharides have shown biological activity against insects, nematodes and arthropods by potentation of GABAergic neuronal and neuromuscular transmission. The hyperpolarisation of neuronal membranes mediates paralysis in arthropods and nematodes.

Spinosad Spinosad is a macrolide antibiotic with insecticidal activity that is naturally produced under aerial fermentation conditions by the actinomycete bacterium *Saccharopolyspora spinosa*. Spinosad is a mixture of numerous spinosyns, but nearly all of the insecticidal activity of spinosad is produced by two closely related compounds, spinosyns A and D. The spinosyns have a structure consisting of a large complex hydrophobic ring, a basic amine group and two sugar moieties. Insects exposed to spinosad exhibit classical symptoms of neurotoxicity, including lack of coordination, prostration, tremors and other involuntary muscle contractions leading to paralysis and death. Although the mode of action of spinosad is not fully understood, it appears to affect nicotinic and gamma-aminobutyric acid (GABA) receptor function. Spinosad is of low acute toxicity after oral or dermal administration and by inhalation. The main effect associated with repeated exposure to spinosad in all test species was observed histologically as cellular vacuolation, inflammatory changes including necrosis, histiocytosis, and regenerative and degenerative changes in a wide range of tissues.

Viruses

Baculoviruses, a host-specific family of viruses infecting exclusively arthropods predominantly of the insect order *Lepidoptera*, are considered as low-risk substances as there is no scientific evidence that baculoviruses have any negative effect on non-target animals and humans.

6.10.6 Insect Growth Regulators

> **Althought most insecticides are neurotoxic, there are some substances acting against specific biological characteristics of insects. These are known as insect growth regulators and include chitin synthesis inhibitors, ecdysone agonists and juvenile hormone analogues.**

Dibenzuron

Diflubenzuron, an acaricide and insecticide, is a chitin synthesis inhibitor and inhibits the growth of many leaf-eating larvae, mosquito larvae, aquatic midges, rust mites, boll weevils, and flies, and appears also to have an ovicidal action in disturbing chitin storage in the cuticle. Diflubenzuron was not irritating or sensitising and of very low acute oral, dermal and inhalative toxicity in rats. In medium- and long-term studies on mice, rats and dogs the most sensitive endpoint was increased concentrations of methaemoglobin and sulf-haemoglobin in males and females.

Tebufenozide

Tebufenozide, a nonsteroidal ecdysteroid agonist, can cause a variety of molting and behavioural effects in certain orders of insects and some larval Crustaceae and mimics the activity of natural ecdysteroids by inducing a premature and lethal larval molt. Tebufenozide is of minimal mammalian toxicity, was not irritant nor did it have sensitisation potential. In medium- and long-term studies on mice, rats and dogs, hemolytic anemia with compensatory increased hematopoesis and increased methemoglobin content was observed.

6.10.7 Other Pesticidal Active Substances

Rodenticides

> **Rodenticides are used to control rodent pests, i.e. rats and mice, as vectors for human and animal pathogens but also as storage pests. To kill rodents successfully and not only drive them off it takes advantage of the physiological characteristics of rodents.**

Fumigation with gases like phosphine as well as anticoagulant-containing baits can be employed to control rodents. The most important group of anticoagulants are coumarin derivatives, which need to be taken up repeatedly in small doses to be efficacious. These substances inhibit blood coagulation if taken up repeatedly and they increase the permeability of blood vessels. The resulting internal and external hemorrhages lead to the death of the animal within a few days. Coumarin derivatives are vitamin K antagonists and act mainly in the liver, where they replace vitamin K in the prothrombin formation. This results in insufficient prothrombin levels for the transformation of thrombin, which is necessary for fibrin formation from fibrinogen.

Warfarin Warfarin is a first-generation anticoagulant that causes extensive resistance development in rats and mice. Warfarin is also used as a pharmaceutical for the prophylaxis and therapy of thromboembolic disorders in humans. Therefore, substantial data from clinical use are available for this substance which were used for read-across for the toxicological evaluation of second-generation anticoagulants, which are much more potent. The clinical administration of warfarin revealed placental hemorrhages in pregnant women which resulted in early death of the embryo, therefore warfarin was classified as toxic for developmental toxicity in category 1A. In rats, embryotoxic effects like an hemorrhagic syndrome and malformations of hind leg and brain were observed after substitution of vitamin K which were comparable to the effects observed in humans (warfarin embryopathy). For animal welfare reasons, no teratogenicity studies for second-generation anticoagulants were performed in rats. Instead, the results from warfarin were used for read-across regarding developmental toxicity, resulting in classification of all anticoagulants in category 1.

Molluscicides

Molluscicides are mainly used to control slugs and snails. There are two forms of chemical molluscicides: oxidising and non-oxidising. Usually, oral poisons like metaldehyde and ferric III phosphate in bait formulations are applied.

Metaldehyde Metaldehyde is a polymerised acetaldehyde that is resorbed quickly after ingestion and is catabolised via acetaldehyde to acetyl CoA. After entering the citrate cycle 80–90% of ingested metaldehyde is exhaled as carbon dioxide but part of it is used for anabolic reactions. In molluscs, metaldehyde causes strong mucus production and subsequently, via deregulation of the fluid balance, the death of the animals. Metaldehyde is acutely toxic to mammals, therefore respective classification is necessary. Symptoms observed after acute poisoning by metaldehyde are abnormal behaviour, increased salivation and bleeding. After subchronic exposure, the target organs in rodents are the liver and central nervous system. There is no evidence for carcinogenicity, mutagenicity or developmental toxicity. At parentally toxic doses there are indications of impairment of reproduction, like decreased body weight gain of the offspring.

Observations in humans after acute intoxication revealed gastrointestinal and central nervous disorders.

Ferric III phosphate Ferric III phosphate is a natural constituent of the soil, animals, plants and the human body. The formulations applied as PPPs are ingested by molluscs. Afterwards, the active substance causes cellular changes in the head and midgut of the animals. The molluscs stop feeding and die within a few days. It is presumed that high concentrations of ferric ions in the gastrointestinal tract damage the mucosa and lead to a translocation of ferric ions into the bloodstream. The resorption is increased because the ferric ions cannot be adequately bound. In mammals, target organs for toxicity of high doses of ferric ions are the liver, central nervous system, cardiovascular system and gastrointestinal tract.

6.10.8 Regulatory Toxicology of Pesticidal Active Substances

In the EU, pesticides, i.e. PPPs and biocidal products, cannot be placed on the market without prior authorisation. The EU legislation for both PPPs and biocides stipulates a two-tier procedure to assess the impact on human health. In the first stage, the safety of each active substance used within a pesticidal product is evaluated in a peer-review program between Member States and a European Authority, which leads to approval or non-approval at the European level by the European Commission (Figure 6.28). In the second stage, the Member States of the EU evaluate and authorise the pesticide at the national level with mutual recognition in zones with comparable agricultural conditions for PPPs or in other Member States for biocidal products (Figure 6.29). Regulations (EC) No. 1107/2009 (PPPs) and (EU) No. 528/2012 (biocidal products) constitute the legal base for these procedures and allow the harmonised and efficient approval of pesticides within the EU based on mutual principles and decision criteria. Although this chapter is focused on PPPs, the regulatory principles are comparable for biocidal products. Further details on the regulatory toxicology of biocides can be found in *Regulatory Toxicology in the European Union* (Ritz & Solecki).

For the authorisation of PPPs, the EU is divided into three zones. Within these zones, authorisations should be transferred more easily to other Member States by means of mutual recognition procedures. PPPs contain one or several active substances which are chemical substances or microorganisms with a general or a specific action against harmful organisms. Moreover, PPPs may also contain safeners and synergists as well as co-formulants. The European legislation therefore clearly requires taking interactions within the formulated product as well as direct and indirect cumulative and synergistic effects through different exposure pathways into account. Further tests on human and animal health are only required if the toxicity of the PPP cannot be predicted by the toxicological evaluation of the active substance and co-formulants or if the PPP is more acutely toxic than the active substance.

For a pesticidal active substance to be approved, the health of all groups of people who come into contact with the product or its residues must be ensured if the product is used correctly and in line with its intended purpose. In order to achieve this, applicants have to address all data requirements by submitting original studies that have been carried out according to harmonised OECD or EU test methods as well as according to GLP principles. These are developed based on the current state of the art in science and technology and are adapted if necessary. The health risk assessments carried out by the authorities is also considering the published literature to these compounds.

Recent progress in analytical and biomedical techniques has provided regulatory toxicology with a wide range of new methods and possibilities in which "omics" and complex cell cultures enable a comprehensive analysis of molecular toxic effects. By application of modern molecular techniques, more mechanistic information should be gained in future to support standard toxicity studies and to contribute to a reduction and refinement of animal experiments required for certain regulatory purposes. However, research activities more relevant for regulatory purposes regarding new methodologies for alternative testing are considered necessary.

6.10.9 Toxicological Endpoints

For pesticidal active substances hazard identification and assessment of the dose–response relationship are the two main steps required to identify quantitative and qualitative aspects of a potential risk associated with the use of pesticidal active substances in a weight-of-evidence approach. For active substances in PPPs the toxicological and metabolism studies are required in Regulation (EU) No. 283/201[*]. Data requirements for toxicological studies on preparations are required in Regulation (EU) No. 284/201[**]. Data requirements for toxicological studies on biocidal products and their active substances required in Regulation (EU) No. 528/201[***].

These data requirements aim to identify the toxicological profiles, including toxicological targets, severity of toxicity, interspecies differences, no observed adverse effect levels (NOAELs) and lowest observed adverse effect levels (LOAELs) or Benchmark doses (e.g. BMDL) for the most relevant observed toxicological effects of all pesticidal active substances to derive the following health-based limit values:

- ADI, acceptable daily intake: the amount of a substance that consumers can ingest on a daily basis over a lifetime without any discernible health risk
- ARfD, acute reference dose: an estimate of the amount of a substance in food and/or drinking water, normally expressed on a bodyweight basis, that can be ingested in a period of 24 hours or less without appreciable health risk to the consumer, on the basis of all the known facts at the time of evaluation
- AOEL, acceptable operator exposure level: the exposure limit for users of PPPs and uninvolved third parties. These are people who come into direct contact with the PPP by chance either during or shortly after its application.

Safety margins accounting for uncertainties in the extrapolation from toxicity data from animal studies to the exposed human population have to be applied for the translation of critical NOAELs into regulatory health based limit values. A safety margin of at least 100 should be used, taking into account the type and severity of effects and the vulnerability of specific groups of the population (Figure 6.30).

---

[*] http://eur-lex.europa.eu/LexUriServ/LexUriServ.do?uri=OJ:L:2013:093:0001:0084:EN:PDF
[**] http://eur-lex.europa.eu/LexUriServ/LexUriServ.do?uri=OJ:L:2013:093:0085:0152:EN:PDF
[***] https://eur-lex.europa.eu/legal-content/EN/TXT/PDF/?uri=CELEX:32012R0528&from=de

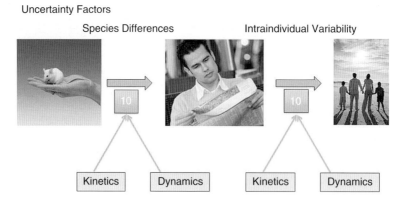

Figure 6.30 *Parameters that define the safety margins for determination of health-based limit values.*

For all approved pesticidal active substances the health-based limit values are published by the EU Commision at[****].

6.10.10 Classification and Cut-off Criteria

For the harmonised classification and labelling of active substances used in PPPs as well as in biocidal products, the CLP Regulation (Regulation (EC) No. 1272/2008) ensures that potential hazards caused by pesticides are clearly communicated to users and consumers through standard statements and pictograms on labels.

Several hazard-based exclusion criteria for active substances concerning human health have been established and these can lead to a direct refusal of the approval. With regard to human health protection, active substances classified as mutagenic (Category 1A or 1B) will in principle not be approved. Substances which are known or presumed to cause cancer (carcinogenic, Category 1A or 1B), have adverse effects on sexual function and fertility in adults and developmental toxicity in the offspring (reproductive toxicity, Category 1A or 1B), or have endocrine disrupting properties according to the Regulation (EC) No. 2018/605[*****] that may cause adverse effects in humans will not be approved unless the exposure to humans is negligible.

If these exclusion criteria regarding human health are not met, the risk assessment of the active substances will be continued. If they are met, it has to be checked whether the criteria for negligible exposure apply (Figure 6.31).

6.10.11 Human Health Risk Assessment

For the final assessment of health risks that may be caused by PPPs, the exposure levels that various groups of people could be exposed to are compared with the health-based limit values, including an assessment of uncertainty. As long as the exposure level is not above the calculated limit values, there is no unacceptable health risk for users, uninvolved third parties or consumers.

[****] http://ec.europa.eu/food/plant/pesticides/eu-pesticides-database-redirect/index_en.htm
[*****] http://eur-lex.europa.eu/legal-content/EN/TXT/PDF/?uri=CELEX:32018R0605&from=EN

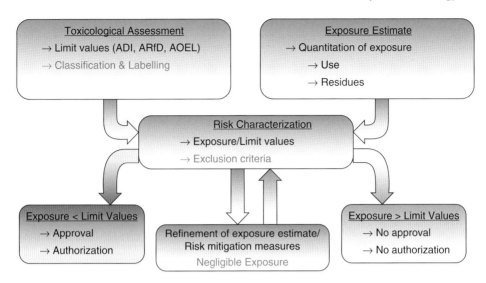

Figure 6.31 Classification and cut-off criteria.

Consumer Protection

Even when authorised PPPs are used properly and in line with their intended purposes, residues can remain in the harvested crops and in processed plant commodities used as food and animal feed. The residues that can be ingested by humans must be determined from the level of residue in a food and from the amount of this food normally consumed. The concentration of residues in food is derived from studies in which the PPP was used under real conditions (residue monitoring studies). Consumption data are available from so-called consumption surveys, such as the EFSA Pesticide Residue Intake Model (PRIMo). The determination of acceptable quantities of an active substance in food items follows the ALARA (as low as reasonably achievable) principle. The fact that different groups of people can be exposed in very different ways with regard to intake path, duration and level of exposure is taken into account. To ensure that levels of residues in food are not harmful for consumer health, either through lifelong daily food intake or short-term consumption of large portions of food, a comprehensive risk assessment as part of the authorisation procedure must be performed and maximum residue levels on the basis of this risk assessment are established. Maximum residue levels are the highest concentrations of the active substances contained in PPPs and their degradation products permitted in foods and animal feeds. They are not determined solely on the basis of the health risk assessment but also take into account good agricultural practice and the technically feasible sensitivity of analytical methods. Compliance with a maximum residue level is the decisive factor when determining whether a food is fit for sale or must be removed from the market.

The authorisation for the use of a PPP on plants or plant products to be used as feed or food is only possible if maximum residue levels for the affected agricultural products have been set or modified in accordance with Regulation (EC) No. 396/2005. Thereby it is ensured that the resulting residues exceed neither the ADI or the ARfD, and the PPP has no immediate or delayed harmful effect on human health, including that of vulnerable groups, or animal health, directly or through drinking water, food or feed. Consumers are normally exposed to a complex pattern of multiple residues with low individual levels which are consumed in a broad range of food items. The methods for cumulative exposure assessment in food differ by their degree of complexity. Methods used range from simple deterministic calculation models to laborious probabilistic models.

Safe Use

During the application of PPPs or follow-up work, users, farmers, workers, residents or bystanders may be exposed to a specific PPP. The maximum expected exposure for each of the potential routes of exposure should therefore be assessed to determine whether the health of all persons who come into contact with the product during its application can be assured when the product is used correctly and in line with its intended purpose. For safe use of pesticides, it is generally necessary to assign obligatory instructions for operators to comply with risk mitigation measures. The hazard and risk-based allocation of safety instructions to operators handling pesticides is based on the intrinsic toxicological properties of the products combined with qualitative and quantitative exposure aspects, e.g. for different formulation types and application scenarios. The toxicological properties of PPPs are evaluated on the basis of the uniform principles as published in the Regulation (EU) No. 546/2011.

For the final risk characterisation, exposure is compared with the harmful effects of the product or the active substances and other ingredients it contains. If the determined exposure levels do not exceed the relevant limit values, further considerations are not necessary. If the relevant limit values are exceeded, the exposure assessment is refined using specific parameters to reflect more realistic and precise assumptions or additional risk mitigation measures can be specified to reduce exposure. Such measures may involve the wearing of protective glasses, protective gloves or a protective suit when handling the product. The use of warning signs or minimum waiting times before re-entering treated areas may also be required and set as a condition of authorisation. If a potential risk of uninvolved individuals during application of PPPs cannot be excluded based on these assessments, the use of special, drift-reducing devices, for example, may be required. Established exposure models are used for exposure assessment.

Cumulative risk assessment with respect to application safety considers different substances during the same exposure period. Operators and other unintentionally involved people may be simultaneously exposed to multiple substances if the PPP contains more than one active substance.

6.10.12 Summary

The terms "biocides" and "pesticides" are regularly interchanged, and often confused with "plant protection products". According to Directive 2009/128/EC pesticides include both biocides and PPPs. This chapters describes general aspects of the common toxicology of selected pesticidal active substances and chemical substance groups out of the classes of herbicides, fungicides, insecticides and other application groups. More detailed information and assessments as well as up-to-date limit values and maximum residue levels for each substance can be found on the web pages of the European Commission (Pesticides Database).

Principles and data requirements for the health risk assessment of pesticides in the EU are harmonised based on EU Regulations. The data submitted by the applicants to the authorities have to meet strict standards regarding the methodological performance of studies (OECD test methods) and quality assurance (GLP requirement). The authorities responsible for the evaluation of the applications take their decisions on standardised criteria that are subject to permanent adaptation to technical and scientific progress. The health risk is determined by comparison of hazard and exposure, including consideration of uncertainty. According to the exclusion criteria, pesticidal active substances that are carcinogenic, mutagenic or toxic to reproduction and/or fertility or endocrine disruptors are subject to a hazard-based assessment approach. In future, robust health risk assessments for pesticides will require a higher degree of harmonisation and relevant research. Future risk assessment of pesticides will have a stronger focus on alternative methods and new assessment fundamentals.

Further Reading

BVL, online – Registration reports: http://www.bvl.bund.de/DE/04_Pflanzenschutzmittel/01_Aufgaben/02_ZulassungPSM/02_Zulassungsberichte/psm_zulassungsberichte_node.html.

European Union (2009) Regulation (EC) No 1107/2009 of the European Parliament and of the Council of 21 October 2009 concerning the placing of plant protection products on the market and repealing Council Directives 79/117/EEC and 91/414/EEC.

European Union (2009) Directive 2009/128/EC of the European Parliament and of the Council of 21 October 2009 establishing a framework for Community action to achieve the sustainable use of pesticides.

European Union (2012) Regulation (EU) No. 528/2012 of the European Parliament and of the Council of 22 May 2012 concerning the making available on the market and use of biocidal products.

European Union (2013) Commission Regulation (EU) No. 283/2013 of 1 March 2013 setting out the data requirements for active substances, in accordance with Regulation (EC) No. 1107/2009 of the European Parliament and of the Council concerning the placing of plant protection products on the market.

European Union (2013) Commission Regulation (EU) No. 284/2013 of 1 March 2013 setting out the data requirements for plant protection products, in accordance with Regulation (EC) No. 1107/2009 of the European Parliament and of the Council concerning the placing of plant protection products on the market.

EU, online – EU Pesticides database: http://ec.europa.eu/food/plant/pesticides/eu-pesticides-database-redirect/index_en.htm

Hamilton D, Yoshida M, Wolterink G and Solecki R. (2017). *Evaluation of Pesticide Residues by FAO/WHO JMPR Food Safety Assessment of pesticide residues.* In: A. Ambrus and D. Hamilton (eds). Imperial College Press, London, pp. 113–196.

Ritz, V. and Solecki R. (2018). Regulatory Toxicology of Pesticides: Concepts. In Marrs & Woodward (ed.) Regulatory Toxicology in the European Union, © Royal Society of Chemistry 2018 (ISBN: 978-1782620662) pp 402–438.

Solecki R. (2004). Toxicology of miscellaneous insecticides. In: T. Marrs and B. Ballantyne (ed.), *Pesticide Toxicology and International Regulation (Current toxicology Series)*. John Wiley & Sons Ltd. The Atrium, Southern Gate, Chichester, West Sussex, pp. 195–189.

Solecki R. (2007). Insektizide. In: H. Dunkelberg, T. Gebel, and A. Hartwig (eds), *Handbuch der Lebensmitteltoxikologie – Belastungen, Wirkungen, Lebensmittelsicherheit*, Hygiene. Wiley-VCH Verlag, Weinheim, pp. 1427–1487.

Solecki R and Pfeil R. (2013). Pflanzenschutzmittel und Biozidprodukte. In: H. Marquardt, S. Schäfer, and H. Barth (eds), *Toxikologie*, 3rd edn. Wissenschaftliche Verlagsgesellschaft mbH Stuttgart, pp. 695–740.

Solecki R, Davies L, Dellarco V, Dewhurst I, Raaij M and A Tritscher. (2005). Guidance on setting of acute reference dose (ARfD) for pesticides. *Food Chem Toxicol* **43**(11): 1569–1593.

Solecki R, Kortenkamp A, Bergman Å, Chahoud I, Degen GH, Dietrich D, Greim H, Håkansson H, Hass U, Husoy T, Jacobs M, Jobling S, Mantovani A, Marx-Stoelting P, Piersma A, Slama R, Stahlmann R, van den Berg M, Zoeller RT and Boobis AR (2017). Scientific principles for the identification of endocrine-disrupting chemicals: a consensus statement. *Arch Toxicol* **91**(2): 1001–1006.

Solecki R, Schumacher DM, Pfeil R, Bhula R and MacLachlan DJ (2017). OECD Guidance documents and test guidelines. *Food Safety Assessment of pesticide residues.* In: A. Ambrus and D. Hamilton (eds). UK, Imperial College Press, London, pp. 13–36.

6.11 Polycyclic Aromatic Hydrocarbons

Heidrun Greim and Hermann Bolt

6.11.1 Introduction

> **Polycyclic aromatic hydrocarbons (PAHs) are a large class of many hundreds of organic compounds consisting of two or more fused aromatic rings of carbon and hydrogen atoms. PAHs are formed ubiquitously by combustion and pyrolysis processes of organic materials.**

The term PAHs (polycyclic aromatic hydrocarbons) refers to unsubstituted non-heterocyclic PAHs, including alkyl-substituted derivatives. The general terms polycyclic aromatic compounds, polycyclic organic matter or polynuclear aromatic compounds include not only PAHs, but also derivatives in which hydrogen atoms are replaced by other atoms or functional groups, e.g. chlorine, alkyl, nitro and amino groups, and/or heterocyclic analogues, in which one or more carbon atoms in the rings are replaced by nitrogen.

PAHs are constituents of tar, tar vapours, coke oven emissions, emissions and soot from the combustion of wood, coal, oil, gas, fuels, tobacco etc. They are released to the environment through natural sources like volcanoes or forest fires or to a much greater extent by man-made sources. In air, due to their low volatility, most PAHs occur adsorbed to particulates. Human exposure takes place via inhalation, dermal absorption or oral uptake, for example via grilled or smoked foodstuffs.

Single PAHs do not occur isolated, but as components of complex mixtures that contain many different compounds. Only some of the huge number of PAHs have been investigated toxicologically. An important marker for exposure to PAHs has been for a long time benzo[a]pyrene. It may serve as a surrogate for health effects by PAH mixtures since it is one of the most extensively studied compounds of this class of substances. Benzo[a]pyrene is considered to be one of the strongest genotoxic carcinogens. Other well-known important representatives of the substance class are anthanthrene, chrysene, benzo[b]fluoranthene, benzo[j]fluoranthene, benzo[k]fluoranthene, benzo[a]anthracene, dibenzo[a,l]pyrene, naphthalene, phenanthrene and pyrene (see Figure 6.32).

In addition, some of the alkylated representatives, such as methylpyrene, or the nitrated PAHs are of importance with respect to the carcinogenicity of PAH mixtures.

6.11.2 Physico-chemical Properties

> **PAHs in general are non-volatile, lipophilic substances occurring in air adsorbed to particles.**

Most PAHs are non-volatile compounds (see Table 6.23). Present in air, they are adsorbed to particulate matter. PAHs are highly lipophilic substances and have very poor aqueous solubility. They are readily absorbed through the skin. Some physico-chemical properties of PAH are listed in Table 6.23.

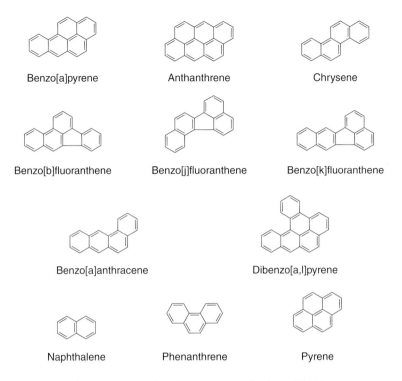

Figure 6.32 Molecular structures of selected PAHs.

6.11.3 Occurrence in the Environment and in the Workplace

Even though most PAHs are released into the atmosphere, considerable amounts are found in water due to distribution processes between water and air, water and sediment, and water and biota. For instance, PAHs can enter soil systems and the aquatic environment via wastewater effluents from the coke and petroleum refining industries, accidental oil spills and leakages, rainwater runoff from highways and roadways, or from intentional disposal in the past. The low aqueous solubility of PAHs and high octanol–water partition coefficient (K_{OW}) often result in accumulation in soils and sediments to levels several orders of magnitude above aqueous concentrations. Their affinity for organic matter in sediment, soil, and biota is high, and PAHs therefore accumulate in organisms, water, and sediments. In Daphnia, accumulation of PAHs from water is correlated with their K_{OW}. In organisms that actively metabolise these chemicals, absorbed concentrations are not correlated with the partition coefficient. Biomagnification is not observed. PAHs undergo photodegradation, microbial degradation, and metabolism in higher organisms. Hydrolysis plays essentially no role in their degradation. PAHs are photooxidised in air and water in the presence of, for example, OH·, NO_3· or O_3· radicals. Reactions with O_3· may lead to mutagenic nitro derivatives.

The production and use of coal tar and coal tar-derived products are major sources of occupational exposure to PAH. Crude coal tar is a by-product of coke production and was formerly also a by-product of gas works. Crude coal tar is usually distilled, and blends of distillation fractions are used for various purposes, such as wood conservation, paints, road tars, and roofing materials.

Table 6.23 Physico-chemical properties of selected PAHs (from Hartwig, 2012).

Name	CAS No.	Molecular formula	Molecular mass (g/mol)	Melting point (°C)	Boiling point (°C)	Vapour pressure (°C)	log K_{ow} (Pa at 25°C)
Anthanthrene	191-26-4	$C_{22}H_{12}$	276.3	264	547		
Benzo[a]anthracene	56-55-3	$C_{18}H_{12}$	228.3	160.7	400	2.8×10^{-5}	5.61
Benzo[b]fluoranthene	205-99-2	$C_{20}H_{12}$	252.3	168.3	481	6.7×10^{-5}	6.12
Benzo[j]fluoranthene	205-82-3	$C_{20}H_{12}$	252.3	165.4	480	2.0×10^{-6}	6.12
Benzo[k]fluoranthene	207-08-9	$C_{20}H_{12}$	252.3	215.7	480	1.3×10^{-8}	6.84
Benzo[b]naphtho-[2,1-d]-thiophene	239-35-0	$C_{16}H_{10}S$	234.3	185–188	160–180 (3 torr)	–	–
Benzo[a]pyrene	50-32-8	$C_{20}H_{12}$	252.3	178.1	496	7.3×10^{-7}	6.50
Chrysene	218-01-9	$C_{18}H_{12}$	228.3	253.8	448	8.4×10^{-5}	5.91
Cyclopenta[cd]pyrene	27208-37-3	$C_{18}H_{10}$	226.3	170	439	–	–
Dibenzo[a,h]anthracene	53-70-3	$C_{22}H_{14}$	278.4	266.6	524	1.3×10^{-8}	6.50
Dibenzo[a,e]pyrene	192-65-4	$C_{24}H_{14}$	302.4	244.4	592	–	–
Dibenzo[a,h]pyrene	189-64-0	$C_{24}H_{14}$	302.4	317	596	–	–
Dibenzo[a,i]pyrene	189-55-9	$C_{24}H_{14}$	302.4	282	594	3.2×10^{-10}	7.3
Dibenzo[a,l]pyrene	191-30-0	$C_{24}H_{14}$	302.4	162.4	595	–	–
Indeno[1,2,3-cd]pyrene	193-39-5	$C_{22}H_{12}$	276.3	163.6	536	1.3×10^{-8}	6.58
Naphthalene	91-20-3	$C_{10}H_{8}$	128.2	81	217.9	10.4	3.4
Phenanthrene	85-01-8	$C_{14}H_{10}$	178.2	100.5	340	1.6×10^{-2}	4.6
Pyrene	129-00-0	$C_{16}H_{10}$	202.3	150.4	393	6.0×10^{-4}	5.18
1-Methylpyrene*	2381-21-7	$C_{17}H_{12}$	216.3	70–71	410		

*Included as a representative of alkylated PAH.

The most extensively studied PAH, as a surrogate for total PAH exposure, is benzo[a]pyrene. It is released from a great variety of different PAH sources. In the 1930s, benzo[a]pyrene was identified as the predominant carcinogenic compound in coal tar. Since then, exposure assessment as well as research on health effects have largely focused on this particular compound. In addition, various national and international authorities have recommended and used benzo[a]pyrene as an indicator for total PAH exposure.

6.11.4 Toxicity

Toxicokinetics: Absorption, Distribution, Metabolism, Excretion

Exposure to PAHs can occur through the airways, the skin and the digestive tract; absorption into the circulation is possible through all three routes. Metabolic activation of lipophilic PAHs occurs primarily in the liver, but also in many other tissues, including the epithelial barriers. The ready skin penetration of individual PAHs has been demonstrated in humans and animal studies. It can be assumed that there is no substantial difference in the penetration behaviours of different PAHs. In animal experiments PAHs have been shown to be rapidly transported to all other tissues via blood and lymph after inhalation, oral and dermal administration. Adipose tissues may be a depot from which the PAHs are released again.

> PAHs are oxidized to epoxides, phenols and dihydrodiols, mainly by cytochrome P450 (CYP) 1A1, and then conjugated with glucuronide, sulphate or glutathione to yield water-soluble metabolites. Biotransformation does not always mean deactivation. In some cases DNA-reactive intermediates are formed, which can initiate cancer.

The biotransformation of PAHs starts when they are absorbed through the epithelia of the lungs, skin, or gastrointestinal tract. The longer the retention time in the epithelium, e.g. of the respiratory tract, the more PAHs will be metabolised. The metabolism of PAHs is complex. In a first step, PAHs are oxidised to form epoxides, phenols and dihydrophenols (so-called Phase I metabolites, see Chapter 2.3), which produce DNA-reactive metabolites. In a second step, these metabolites are conjugated with glutathione, sulphate or glucuronic acid to form much more polar and water-soluble metabolites (so-called Phase II metabolites, see Chapter 2.3). Most PAHs metabolised in this way are deactivated. However, some of them are activated to DNA-binding species, which can initiate cancer (see *Carcinogenicity*, this chapter).

Most information on the carcinogenic bioactivation of PAHs has been obtained from *in vivo* and *in vitro* studies with benzo[a]pyrene as model compound. Other carcinogenic PAH compounds are metabolised in a comparable way. In the first step, benzo[a]pyrene is oxidised by CYP1A1, resulting in epoxide and phenol groups at several sites in its ring structure. These epoxides may be hydrated by epoxide hydrolase to dihydrodiols or spontaneously rearrange to phenols. Quinone structures can also be formed. The epoxides may be conjugated with glutathione, while the phenols are conjugated with glucuronide or sulphate. Most conjugates are detoxification products. After the initial formation of the (+)benzo[a]pyrene-7,8-epoxide and its subsequent dihydrodiol product (by epoxide hydrolase), it is activated to the ultimate reactive intermediate (+)*anti*-benzo[a]pyrene-7,8-dhydrodiol-9,10-epoxide. This is an extremely reactive species that covalently interacts with cellular DNA. The DNA adducts formed may lead to mutations. An additional complexity of PAH metabolism is stereoselectivity. The location and stereochemical configuration of the epoxide ring within the PAH ring system results in marked differences in chemical reactivity towards DNA.

Bioactivation pathways other than the diol-epoxide mechanism may also play a role. These include radical-cation, quinone and benzylic oxidation mechanisms, which may occur simultaneously for various PAH components.

The extent to which PAHs will express carcinogenic effects in human individuals are thought to be partly genetically controlled. Although similar enzyme systems are involved in PAH metabolism, the inducibility and activity of relevant enzymes may differ between species, within species and in different organs.

The phenols and dihydrodiols formed in the metabolism of PAH are excreted in urine as water-soluble sulphates and glucuronides. It was demonstrated, at least for rodents, that arene oxides are also metabolised to glutathione conjugates by the catalytic effect of glutathione S-transferase; these are finally excreted, after cleavage of two amino acid residues (glycine and glutamic acid) and subsequent acetylation, as mercapturic acids.

Cigarette smokers excrete most inhaled PAHs within 24 hours. For pyrene, a half-life of 4–12 hours is reported in volunteers after oral and dermal exposure.

Biological Monitoring Biological monitoring of exposure to PAHs is a difficult issue, since PAHs always occur as complex mixtures of variable composition. Furthermore, the individual compounds exert various adverse effects and their potency may vary. In addition, biotransformation is highly complex and specific for individual compounds, and it results in metabolites with potentially different effects, e.g. different carcinogenicity. Several groups of PAH metabolites have been considered as biomarkers of exposure, e.g. 1-hydroxypyrene, 3-hydroxybenzo[a]pyrene and hydroxylated phenanthrenes. For the most promising biomarkers specified analytical techniques to support implementation of biomonitoring approaches are available.

At present, it appears that 1-hydroxypyrene is a preferred biomarker for human exposures to PAH, despite several disadvantages. This marker does not represent adequately the internal exposure to carcinogenic PAHs in quantitative terms, as pyrene is not carcinogenic. It also does not well represent the variability of PAH mixtures.

Acute Toxicity

The acute toxicity of PAH is relatively low. Accidental intoxications in humans are reported only with naphthalene, mainly with mothballs in children. Some of the children were found to have haemolytic anaemia. Reported acutely lethal oral doses are 5–15 g for adults and 2 g for a 6-year-old child.

Local Toxicity and Irritancy

Mixtures of carcinogenic PAHs are described in the older literature as causing skin disorders in humans and in animals. However, information on specific effects in humans or animals of individual PAHs is scarce.

Naphthalene is irritating to human, rabbit and mouse skin, as well as to human and rabbit eyes. Pyrene is a mild dermal irritant in animals. Benzo[a]pyrene and dibenzo[a,l]pyrene in mice have been reported to cause dermal irritation and erythema, respectively.

However, more important is the carcinogenic effect of PAHs after dermal application (see *Carcinogenicity*, this chapter).

Sensitisation and Immunotoxicity

Anthracene increases the sensitivity of the human skin to sunlight. A positive reaction to naphthalene (patch test) was observed in a dermatitis patient.

Benzo[a]pyrene has been proven to be sensitising in guinea pigs and mice, and in the local lymph node assay. For dibenzo[a,l]pyrene contact sensitisation was also observed, whereas anthracene and phenanthrene were negative in this respect.

Pyrene increases photosensitivity and suppresses the immune system.

PAHs are generally thought to be immunotoxic, but the mechanism of this type of toxicity is not clear.

Toxicity after Repeated Exposure

Apart from carcinogenicity, information on the toxicity of single PAH compounds is very limited. A typical acute systemic effect after accidental dermal, oral or inhalation exposure in humans to naphthalene is acute haemolytic anaemia. Lens opacity, corneal ulcerations and cataracts have been described after inhalative exposure to naphthalene in the workplace. In addition, after dermal application, anthracene, fluoranthene and phenanthrene may induce specific skin reactions.

In subchronic oral studies with rats, naphthalene led to liver damage and signs of an inflammation of the kidneys, and at higher doses to the formation of cataracts. In mice, naphthalene caused weight reduction of the thymus and spleen, and in dogs diarrhoea and anaemia.

Pyrene, orally applied for 13 weeks, led to reduced kidney weight and nephropathy in mice.

Genotoxicity

> Genotoxicity induced by PAHs was found in a huge number of test systems with bacteria, mammalian and human cells *in vitro* as well as in test systems *in vivo*, either with single PAH or with PAH mixtures.
> Genotoxic action by PAH generally requires metabolic activation.

Benzo[a]pyrene has been extensively studied in various *in vitro* and *in vivo* genotoxicity tests. It showed positive results in many endpoints, such as DNA repair and mutation in bacteria, mutations in *Drosophila melanogaster*, DNA binding in various species, DNA repair, sister chromatid exchange, chromosomal aberration, point mutation and transformation in mammalian cells, sperm abnormalities and *in vivo* somatic mutations at specific loci *in vivo*.

Other representatives of PAHs, as well as PAH mixtures, predominantly yielded positive results in different genotoxicity endpoints *in vitro* and *in vivo* (see also classification as germ cell mutagen in Table 6.24). As exemptions phenanthrene, pyrene, or naphthalene are to be be mentioned.

Data compilations can be found in detail in WHO (1998), IARC (2010) and Hartwig (2012).

Table 6.24 Classifications of selected PAHs as (germ cell) mutagen or carcinogen (from Hartwig, 2012).

Compound	Germ cell mutagen MAK*	Carcinogen MAK**	Genotoxicity (WHO, 1998)	Carcinogenicity (WHO, 1998)
Anthanthrene	No data	2	(Positive)***	Positive
Benzo[a]anthracene	3B	2	Positive	Positive
Benzo[b]fluoranthene	3A	2	Positive	Positive
Benzo[j]fluoranthene	3B	2	Positive	Positive
Benzo[k]fluoranthene	3B	2	Positive	Positive
Benzo[b]naphtho[2,1-d]thiophene	3B	2		
Benzo[a]pyrene	2	2	Positive	Positive
Chrysene	No data	2	Positive	Positive
Cyclopenta[cd]pyrene	3B	2	Positive	Positive
Dibenzo[a,h]anthracene	3A	2	Positive	Positive
Dibenzo[a,e]pyrene	3B	2	Positive	Positive
Dibenzo[a,h]pyrene	3B	2	(Positive)	Positive
Dibenzo[a,I]pyrene	3B	2	Positive	Positive
Dibenzo[a,l]pyrene	3B	2	(Positive)	Positive
Indeno[1,2,3-cd]pyrene	No data	2	Positive	Positive
Naphthalene	3B	2	Negative	Questionable
Phenanthrene	–	–	Questionable	Questionable
Pyrene	–	–	Questionable	Questionable
1-Methylpyren	No data	2	Not given	Not given

*Germ cell mutagenicity: Category 2, shown to increase the mutant frequency in the progeny of exposed mammals; Category 3A, shown to induce genetic damage in germ cells of humans or animals, or to produce mutagenic effects in somatic cells of mammals *in vivo* and to reach the germ cells in an active form; Category 3B, suspected of being germ cell mutagens because of their genotoxic effects in mammalian somatic cells *in vivo* (or, if no *in vivo* data, but clearly mutagenic *in vitro* and structurally related to known *in vivo* mutagens); –, no classification.
**Carcinogenicity: Category 2, animal carcinogen; –, no classification.
***Small database.

Carcinogenicity

> **PAH mixtures from different working processes have been shown to lead to significantly increased cancer incidences for the lungs, bladder and skin in typical industries with high PAH exposure. Evidence of cancer in organs such as kidneys, larynx, oesophagus and stomach is less consistent.**

In 1775, Sir Percival Pott reported extraordinarily high incidences of scrotal carcinoma in London chimney sweeps, which he regarded as induced by exposure to soot and ash. More than 100 years later, in 1892, cancer of the skin and inner organs in chimney sweeps, as well as in workers in the coal tar and pitch processing industries, was found. In 1915, Yamagiwa and Ichikawa experimentally demonstrated for the first time the carcinogenic effect of coal tar on rabbit skin. A large number of studies on mouse skin followed that showed this effect for numerous coal tar products containing PAH mixtures. In 1932 neat benzo[a]pyrene had been applied to mouse skin and numerous PAHs were investigated later in this particular test system.

Since then, numerous human and animal studies have been published on the carcinogenic effects of PAHs by various routes of exposure, as complex mixtures or as single compounds in animal studies. These studies revealed that PAHs act mainly as local carcinogens, e.g. lung cancer by inhalation or skin cancer by dermal exposure. Some publications also report on risks of systemic cancer, such as bladder cancer in humans after occupational exposure to PAHs. However, in none of these studies could the presence of specific bladder carcinogens like 2-naphthylamine be ruled out.

In the literature a vast amount of epidemiological data is presented associating lung cancer with work-related PAH exposure. However, interpretation and comparison of these data is hampered due to differences in study design, i.e. case control versus cohort studies, differences in exposure measurements, not taking into account lifestyle factors, unawareness of co-exposure and incomplete data presentation. Nevertheless the majority of the epidemiological data associate airborne PAH exposures with increased lung cancer risk. In addition, skin cancer has been reported to be positively associated with dermal PAH exposure, but not with inhalation exposure.

The carcinogenicity of single PAH- and complex PAH-containing mixtures has much been investigated in experimental animals. With respect to single PAH, most of the studies were performed with benzo[a]pyrene and the majority with dermal exposure. Various complex PAH-containing mixtures have also been investigated for carcinogenic properties in experimental animals. In most of these studies the animals were dermally exposed, e.g. to extracts, tars or condensates from tobacco smoke, diesel and gasoline engine exhaust, carbon black, coal tar and coal gasification-derived products. Overall, these mixtures caused dermal tumours after repeated dermal exposure, mainly of benign origin.

Exposure via the respiratory tract was applied in a much smaller number of studies. Female Wistar rats and female NMRI/BR mice showed lung tumours. Tumours in organs other than the lung were not studied.

In a few carcinogenicity studies lung tumours were reported after chronic feeding of PAH-containing mixtures. Animal data are also published on systemic cancer after oral, intraperitoneal and intrarectal administration of single PAH, but the quality of these studies was insufficient for a final conclusion.

Detailed compilations of epidemiological studies and animal experiments can be found in, for example, ATSDR (1995), WHO (1998), IARC (2010), Hartwig (2012) and SCOEL (2016).

To date benzo[a]pyrene is the only individual PAH that has been officially classified as a confirmed human carcinogen (IARC Group 1). Others, like cyclopenta[cd]pyrene, dibenzo[a,h]anthracene and dibenzo[a,l]pyrene, are considered as probably carcinogenic to humans (IARC Group 2A).

The classification of single PAHs and PAH mixtures by different national and international organisations is described in *Evaluation by National and International Organisations* (this chapter; see also Table 6.24).

Carcinogenicity Risk Assessment Since the lung is considered the main target organ for inhalative PAH exposure, numerous studies have been carried out to examine the relative lung cancer risk imposed by PAHs. Results of meta-analyses have been published, which included a substantial number of cohorts. In one meta-analysis, 39 different cohorts with coal-derived PAH sources from various industries, e.g. coking plants, gas production from coal, aluminium smelters, asphalt use or tar distillation, were analysed for PAH exposure and lung cancer cases. In this review quantitative dose–response relationships were considered, and the risk estimates were specified as unit relative risks, i.e. as relative risks related to 100 µg/m^3 benzo[a]pyrene years. The overall mean unit relative risk was calculated at 1.20 (95% CI: 1.11–1.29, $p < 0.001$), which means an increase of 20% at a cumulative exposure of 100 µg/m^3 benzo[a]pyrene years.

Reproductive Toxicity

There are hardly any data on the reproductive toxicity of PAHs. Two reports of transplacental poisoning in humans are available for naphthalene. Naphthalene, benzo[a]anthracene and dibenzo[a,h]anthracene induced vaginal and intraplacental haemorrhages in rats after subcutaneous treatment. These compounds, as well as chrysene, led to the death of foetuses in the F1 generation in rats and mice and to the induction of benzo[a]pyrene hydroxylase in the rat liver. Benzo[a]pyrene had a severe embryotoxic effect in mice and rats and impaired fertility in mice.

6.11.5 Mechanisms of Action

> **CYP metabolism of a number of PAHs leads to carcinogenicity and mutagenicity. PAHs have different toxicity profiles; some are more toxic than others. The mechanism of toxicity often involves adduct formation with macromolecules following biotransformation.**

PAHs as such are relatively inert hydrophobic chemical substances, which can be metabolised to highly reactive *syn*- or *anti*-dihydrodiol epoxides in the mammalian organism. The diol epoxides react with DNA, preferably with the N^6 position of guanine and the N^2 position of adenine, yielding cis and trans adducts for the different *syn* and *anti* enantiomers. Depending on their structure and reactivity, the activated PAHs can form further adducts. Covalent binding to DNA may be associated with only marginal or even major changes in conformation.

Many PAHs are investigated for their carcinogenicity. It has been observed that compounds with a so-called "bay region", an indentation formed by an angular benzene ring (see Figure 6.33), are severe carcinogens. However, compounds with a so-called "fjord region" (see Figure 6.33) have been proved to be even more carcinogenic, whereas the so called "K-region", the region with the most pronounced double bindings of the structure, is of minor importance for the DNA binding reactivity. The structural characteristics of PAHs thus influence both their metabolic activation as well as the stereochemistry after binding to DNA.

As an example, a higher carcinogenic potency has been reported for dibenzo[a,l]pyrene compared to benzo[a]pyrene (see Figure 6.33).

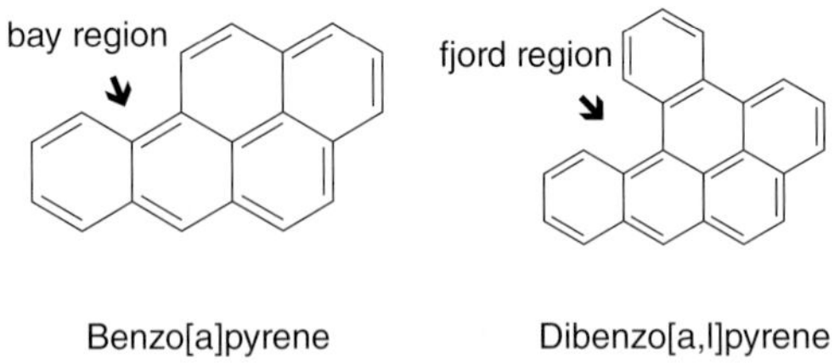

Figure 6.33 Reactive sites for PAH metabolism.

Despite this, there is no conclusive structure-response regularity to predict the carcinogenic potency of a single PAH. The binding frequency is only a rough standard for the carcinogenicity. The type of DNA binding is of greater influence because it decides whether programmed cell death occurs or DNA repair, whether DNA repair occurs rapidly enough relative to the cell cycle, whether the DNA barrier can be overcome during DNA replication, and if so, for example, at the cost of which copy errors. Frequent findings are mutations in the H-*ras*, K-*ras* and *p53* oncogenes.

PAH also may lead to gene activation. The induction of Phase I metabolising enzymes like CYP1A1 as well as Phase II conjugating enzymes, e.g. glutathione S-transferase or UDP-glucuronyl transferase, have been shown. They lead to an increase in PAH metabolism and thus to higher activation. Increased gene expression, e.g. of the proto-oncogene *c-myc*, or of the transferrin receptor gene has been described as well.

6.11.6 Evaluation by National and International Organisations

The International Agency for Research on Cancer (IARC) has classified occupational exposure during coal gasification, coke production, coal-tar distillation, paving and roofing with coal-tar pitch, aluminium production and as a chimney sweep as carcinogenic to humans (Group 1).

Occupational exposure during carbon electrode manufacture and creosotes were evaluated as probably carcinogenic to humans (Group 2A).

Occupational exposure during calcium carbide production was evaluated as not classifiable to its carcinogenicity to humans (Group 3).

Regarding single PAHs, benzo[a]pyrene is regarded as being carcinogenic to humans (Group 1), cyclopenta[cd]pyrene, dibenz[a,h]anthracene and dibenzo[a,l]pyrene are seen as probably carcinogenic to humans (Group 2A), and benz[j]aceanthrylene, benz[a]anthracene, benz[b]fluoranthene, benz[j]fluoranthene, benz[k]fluoranthene, benzo[c]phenanthrene, chrysene, dibenzo[a,h]pyrene, dibenzo[a,i]pyrene, indeno[1,2,3-cd]pyrene and methylchrysene were rated as possibly carcinogenic to humans (Group 2B).

Many other single PAHs were not classifiable as to their carcinogenicity to humans (Group 3), partly due to lack of data.

The Permanent Senate Commission of the Deutsche Forschungsgemeinschaft for the Investigation of Health Hazards of Chemical Compounds in the Work Area (MAK Commission) has classified 19 selected PAHs for their germ cell mutagenicity and carcinogenicity (see Table 6.24).

Excerpts from the evaluation of the available genotoxicity and carcinogenicity tests by the World Health Organization (WHO) in 1998 for 33 PAHs studied are shown in Table 6.24.

The EC Scientific Committee on Food recommended in 2002 that exposures to PAH should be "as low as reasonably achievable". The maximum daily intake of benzo[a]pyrene from food is estimated to be approximately 6 ng/kg body weight and day, which is about five to six orders of magnitude lower than the doses inducing tumours in experimental animals.

6.11.7 Toxicity Equivalency Factors for PAHs

> **Monitoring all single components of a complex PAH mixture is almost impossible. It has therefore been suggested that a few key component(s) should be determined and the carcinogenic or genotoxic potency of the whole mixture derived by means of toxicity equivalency factors.**

To monitor all PAHs of a given exposure mixture might be desirable, but is practically almost impossible due to the complex and varying composition of these mixtures and the large number of candidate components. It is therefore proposed that a few representatives of the substance class are measured, with well-established practice being to use only benzo[*a*]pyrene as key indicator component.

Another proposal, suggested by the US Environmental Protection Agency in 1984, was to use 16 distinct PAHs together.

The EU Scientific Committee on Food in 2002 concluded that benzo[a]pyrene may be used as a marker of occurrence and effect of carcinogenic PAHs in food, based on examinations of PAH profiles in food. A conservative assessment was that the carcinogenic potency of total PAHs in food would be 10 times that contributed by benzo[a]pyrene alone. In contrast, the EFSA Panel on Contaminants in the Food Chain concluded in 2008 that benzo[a]pyrene is not a suitable indicator for the occurrence of PAHs in food, but instead suggested monitoring of eight, or at least four, PAHs as suitable indicators.

In order to provide reliable estimates of the carcinogenicity of PAH mixtures the carcinogenic and mutagenic potencies of single PAHs have been ranked compared to benzo[a]pyrene. For this ranking, a toxicity equivalence factor approach was used.

Proposals made by different authors or organisations can be found in Hartwig (2012).

These rankings are based on comparative studies in which benzo[a]pyrene and other PAHs were investigated by the same protocol and within the same time frame. Overall, these comparisons show that the genotoxic and carcinogenic potency increases with the number of rings. Carcinogenic three- or four-ring PAHs are less potent than five- and six-ring PAHs. Dibenz[a,h]anthracene appears to be at least equipotent to benzo[a]pyrene, whereas other five-ring PAHs tested, such as benzofluoranthenes and benzo[e]pyrene, are less potent. A carcinogenic potency higher than benzo[a]pyrene is reported for dibenzo[a,l]pyrene, dibenzo[a,i]pyrene, dibenzo[a,h]pyrene, dibenz[a,h]anthracene and 5-methylchrysene.

In essence, estimated cancer risks of complex PAH mixtures with benzo[a]pyrene used as the sole exposure indicator may be over- or underestimated.

Nevertheless, for practical reasons benzo[a]pyrene alone may still serve as an acceptable genotoxicity or carcinogenicity indicator for PAH mixtures at present (SCOEL, 2016).

To obtain an improved estimation of the cancer risk of a given PAH mixture, the German MAK Commission, for example, suggested monitoring of 19 selected PAHs applying the derived equivalency factors, as shown in Table 6.25.

Table 6.25 Relevant PAH analytes suggested for routine determination and their toxicity equivalency factors (from Hartwig, 2012).

Compound	Equivalency factors
Anthanthrene	0.1
Benzo[a]anthracene	0.1
Benzo[b]fluoranthene	0.1
Benzo[j]fluoranthene	0.1
Benzo[k]fluoranthene	0.1
Benzo[b]naphtho[2,1-d]-thiophene	0.01
Benzo[a]pyrene	1
Chrysene	0.01
Cyclopenta[cd]pyrene	0.1
Dibenzo[a,h]-anthracene	1
Dibenzo[a,l]pyrene	10
Dibenzo[a,e]pyrene	1
Dibenzo[a,h]pyrene	10
Dibenzo[a,i]pyrene	10
Indeno[1,2,3-cd]-pyrene	0.1
Naphthalene	0.001
Phenanthrene	0.001
Pyrene	0.001
1-Methylpyrene*	0.1

*Equivalency factor was attributed with the restriction that only one dose level was tested in newborn mice, expressing relatively few sulphotransferases.

6.11.8 Summary

PAHs are neither produced nor used as such, but are ubiquitously formed during combustion and pyrolysis processes of organic materials. Human exposure takes place via inhalation, dermal absorption or oral uptake via food. The most extensively studied representative of this class of substances is benzo[a]pyrene, one of the strongest genotoxic carcinogens known, which significantly contributes to the carcinogenic potency of PAH mixtures. After metabolism within the body, PAHs exert various toxic effects. The most critical toxicological endpoint of PAHs is carcinogenicity. The carcinogenic effect of individual PAHs has been demonstrated in animal studies, mainly resulting in local tumours after inhalation or dermal exposure. Genotoxic effects of PAHs were observed in numerous animal studies *in vivo*, as well as in mammalian (including human) cells *in vitro*. The pathways leading to genotoxicity and carcinogenicity have been investigated thoroughly. Carcinogenic PAHs, such as benzo[a]pyrene, may undergo biotransformation to highly reactive dihydrodiol epoxides. These diol epoxides react with double-strand and single-strand DNA, preferably with the N^6 position of guanine and the N^2 position of adenine.

Although it might be desirable to monitor total PAHs or a selection of PAHs, as proposed by different national and international bodies, considering the vast and consistent amount of data presented for benzo[a] pyrene and the fact that benzo[a]pyrene is considered to be one of the most potent carcinogens of the substance class, for practical reasons most available studies have used benzo[a]pyrene as a indicator substance for overall PAH exposure.

Further Reading

Armstrong, B., Hutchinson, E., Unwin, J. and T. Fletcher (2004) Lung cancer risk after exposure to polycyclic aromatic hydrocarbons: a review and meta-analysis. *Environmental Health Perspectives* **112**, 970–978.

ATSDR (1995) *Toxicological Profile for Polycyclic Aromatic Hydrocarbons. Agency for Toxic Substances and Disease Registry*, US Department of Health and Human Services, TP 69, August 1995, Atlanta, GA.

DECOS (2006) Benzo[a]pyrene and PAH from coal-derived sources. Health-based calculated occupational cancer risk values of benzo[a]pyrene and unsubstituted non-heterocyclic polycyclic aromatic hydrocarbons from coal-derived sources. No. 2006/01 OSH, Dutch Expert Committee on Occupational Standards, Health Council of the Netherlands, The Hague, 21 February 2006.

EC (2002) Opinion of the Scientific Committee on Food on the risks to human health of polycyclic aromatic hydrocarbons in food. Scientific Committee on Food SCF/CS/CNTM/PAH/29 Final, European Commission, 4 December 2002, https://ec.europa.eu/food/sites/food/files/safety/docs/sci-com_scf_out153_en.pdf.

Hartwig, A. (2012) Polycyclic aromatic hydrocarbons (PAH). *The MAK-Collection Part I: Occupational Toxicants*, Vol. **27**, 1–230. Wiley-VCH, Weinheim.

IARC (1973) *Certain polycyclic aromatic hydrocarbons and heterocyclic compounds*. In: IARC Monographs on the Evaluation of the Carcinogenic Risk of Chemicals to Humans, Vol. **3**. International Agency for Research on Cancer, Lyon.

IARC (2010) *Some non-heterocyclic polycyclic aromatic hydrocarbons and some related exposures*. In: IARC Monographs on the Evaluation of Carcinogenic Risks to Humans, Vol. **92**. International Agency for Research on Cancer, Lyon.

NTP (2016) *Report on Carcinogens, 14th edition*. National Toxicology Program, US Department of Health and Human Services, Public Health Service, Research Triangle Park, NC.

SCOEL (2016) Polycyclic aromatic hydrocarbon mixtures containing benzo(a)pyrene. Recommendation from the Scientific Committee on Occupational Exposure Limits, SCOEL/REC/404. Scientific Committee on Occupational Exposure Limits, European Commission, Brussels.

WHO (1998) *Selected non-heterocyclic polycyclic aromatic hydrocarbons*. IPCS Environmental Heath Criteria 202, World Health Organization, Geneva.

6.12 Diesel Engine Emissions

Heidrun Greim

6.12.1 Introduction

> **Diesel engine emissions contain several hundred chemical compounds. They are emitted both in gaseous form and as particles.**

Diesel engine emissions are formed during the combustion of diesel fuels in engines. Exposure of the general population occurs depending on the proximity to emissions, for example from street traffic. In addition, in workplaces, for example in mining or in the railway, construction and transport industries, there is increased exposure.

> In 2012, on the basis of evidence of a relationship between exposure to diesel exhaust fumes and an increased lung cancer risk, the International Agency for Research on Cancer (IARC) classified diesel engine emissions as carcinogenic for humans (Group 1). There is also thought to be a relationship between exposure to diesel fumes and inflammatory effects on the lungs. In addition, exposure to diesel fumes has been suggested as the cause of cardiovascular effects and the intensification of asthma and allergic diseases.

The carcinogenic effects of "old diesel" are attributed mainly to carbon particles with polycyclic aromatic hydrocarbons (PAHs) adsorbed to them, while the effects of "new diesel" are determined predominantly by the level of nitrogen oxides, therefore carcinogenic effects are no longer expected.

6.12.2 Contents of Diesel Engine Emissions

The amount and exact composition of the emissions from diesel engines depend, for example, on the kind of engine, the composition and physical properties of the fuel, and the methods of exhaust-gas purification.

The main components of the **gas phase** of diesel engine emissions include nitrogen, carbon dioxide (CO_2), oxygen and water vapour. Nitrogen oxides (NO_x) and carbon monoxide (CO) are formed as products of incomplete combustion. In addition, there are smaller amounts of sulphur dioxide (SO_2) and various organic compounds (alkanes, alkenes, carbonyl compounds, carbonic acids, aromatics and their (nitrated) derivatives). The **particle phase** consists mainly of elementary carbon and adsorbed organic compounds such as PAHs, oxidized PAHs, nitro-PAHs, sulphates, nitrates, metals and other trace elements.

The particles vary in size from 0.03 to 2.5 μm, and therefore also include so-called ultrafine particles with a diameter of less than 100 nm.

Over the last 20 years, diesel engine emissions have been subject to ever more stringent regulations in the European Union and worldwide, which has led to considerable further development of diesel technology. As a result, the emissions and the composition of exhaust fumes have changed greatly. Thus, a considerable influence on the effects of exposure to diesel fumes on health is also expected.

> While the effects of "old diesel" are attributed predominantly to carbon particles and the carcinogenic PAHs and nitro-PAHs adsorbed to them, "new diesel" contains hardly any elementary carbon; instead, nitrogen oxides are more important. This is reflected also in the toxicity observed after long-term exposure in animal experiments.

When we talk of diesel engine emissions, usually diesel fuels derived from crude oil are meant. They belong to the middle distillates of crude oil. In addition, there are alternatives such as "biodiesel", which is produced from vegetable oil or animal fats, and synthetic fuels from coal, gas or biomass.

6.12.3 Toxicokinetics

> Diesel engine emissions consist of particles and gaseous components. The inhalation, deposition, clearance and retention of diesel engine particles takes place according to the same principles as for other barely soluble, granular, biopersistent dust particles.

Particles

Inhaled particles are deposited in the respiratory tract in amounts of about 7% in the alveolar region and 4% in the tracheobronchial region in humans, according to model calculations. With increased breathing the amount in the alveolar region increases. In the rat, 11% is deposited in the alveolar region and 4% in the tracheobronchial region.

The particles are removed from the tracheobronchial region by means of mucociliary clearance and are transported into the gastrointestinal tract. Clearance in the alveoli takes place by means of phagocytosis via the macrophages and subsequent transport in the alveolar and tracheobronchial lumen and then likewise mucociliary clearance in the airways. The rate of clearance from the alveolar region is very much slower than from the tracheobronchial region. The retention half-time for particles in the rat is 60–100 days at concentrations of up to 1 mg/lung and increases at higher levels of exposure. In humans it is even longer and is given as several hundred days.

The particles in diesel engine emissions consist of a nucleus of elementary carbon onto which organic substances, in particular PAHs, nitro-PAHs, oxidized PAHs, and to a smaller extent also sulphates, nitrates, metals or other trace elements, are adsorbed. Studies in exposed workers demonstrated the absorption, distribution, metabolism and elimination of metabolites of PAHs. In exposed groups of people, 1-hydroxypyrene in urine and haemoglobin adducts of nitro-PAHs or lower-molecular alkenes such as hydroxyethylvaline and hydroxypropylvaline were observed.

As a result of the numerous different components, there are no studies available for the metabolism of diesel engine emissions.

> Emissions from diesel engines using newer technologies contain markedly fewer particles than "old diesel". In "new diesel" the gaseous components, in particular nitrogen oxides, play the greater role.

Gaseous Components

The gaseous components of diesel engine emissions, such as NO_2, NO or CO, are absorbed into the respiratory tract by breathing, with NO_2 and NO absorbed in amounts of up to 90%.

NO_2 and NO are absorbed into the systemic circulation, NO_2 mainly in the form of nitrite or nitrate. NO reacts with haemoglobin to form methaemoglobin. Both substances are eliminated with the urine as nitrate. CO also enters the blood, especially from the lower respiratory tract, where it binds to haemoglobin, to form carboxyhaemoglobin, or also with other proteins containing haem. CO is eliminated mainly via exhalation.

6.12.4 Toxicity

> **The target organ of the effects of diesel engine emissions is the lung. Even after short-term exposure, signs of inflammation occur and after longer-term exposure impairments in lung function are observed. In studies with rats, lung tumours were observed. Lung cancer is attributed to diesel engine emissions in many epidemiological studies in humans, which, however, is controversial.**

Acute Toxicity

The acute effects of exposure to diesel engine emissions are irritation and inflammation of the eyes, upper airways and lungs, and exposure leads to respiratory and neurophysiological symptoms, as well as haematological, immunological and cardiovascular effects.

In controlled studies with volunteers, the first signs of an inflammatory reaction in the lungs (inflammation markers in the bronchioalveolar lavage) were reported after exposure to about 100 µg/m^3 for two hours.

The mortality observed in experiments with animals after high-level exposure to diesel fumes is attributed to the effects of carbon monoxide and nitrogen dioxide.

Chronic Toxicity

Longer-term exposure to diesel engine emissions in workplaces with high-level exposure, for example in mining, railway or tunnel workers, has been associated with the development of chronic obstructive pulmonary disease (COPD) or impairments in lung function. In some studies the increased incidence of cardiovascular diseases or of asthma is attributed to diesel engine emissions, although other air pollution factors may also have an influence here.

High concentrations of (old) diesel engine emissions of 1000 µg elementary carbon particles/m^3 and above lead in humans and experimental animals, especially rats, to the accumulation of particles in the macrophages, changes in the lung cell population, fibrotic effects and metaplasia of the squamous epithelium, accompanied by the impairment of lung clearance. A NOAEL cannot be given for this; first effects were reported at 100–200 µg/m^3.

In rats, exposure to the emissions from diesel engines using new technologies ("new diesel") causes relatively slight changes in lung function and histological changes which are reminiscent of the picture after exposure to nitrogen dioxide (NO_2). Mice were found to be less sensitive in parallel investigations. NO_2 is also the main component in "new diesel", which is used as a parameter as the concentrations of carbon particles here are extremely small.

Irritation, Sensitization and Immunotoxicity

Diesel engine emissions can cause acute irritation of the eyes and mucous membranes of the respiratory tract.

Diesel engine emissions themselves are not allergenic, but there is evidence that the allergenic effects of known allergens and of asthma-like symptoms are intensified by exposure to them. In addition, diesel engine emissions were shown in various studies to have effects on the immune systems of humans and experimental animals.

Developmental and Reproductive Toxicity

There are no studies available of possible effects on reproduction and development in humans as the action of particles is the main focus of the toxic effects of "old diesel". As regards developmental and reproductive toxicity, "old diesel" is not considered to be critical. Experiments with animals have confirmed this. In several epidemiological studies of air pollution, an association with premature births, low birth weights and increased mortality in newborn babies was found, but in such studies the exact influencing factors are, however, difficult to determine. In most of the studies with rats, mice and rabbits, neither teratogenic nor embryotoxic or foetotoxic effects occurred at concentrations of up to 12,000 µg/m^3, neither was impairment of female or male fertility observed.

Studies with "new diesel" are not available, but in view of the composition of the emissions such effects are not to be expected.

Genotoxicity

> **Studies of the genotoxicity of diesel engine emissions *in vitro* and *in vivo* yielded positive findings that were attributed to genotoxic components such as PAHs and nitroaromatics. Also secondary genotoxicity resulting from the formation of reactive oxygen species (ROS) occurs, as with other particles, at high concentrations.**

Some of the studies with exposed groups of people yielded genotoxic effects, others did not. The expression of genes found in connection with oxidative stress and inflammation in blood lymphocytes, and in cells involved in inflammatory processes in the bronchioalveolar lavage, was increased. Diesel engine emissions caused "bulky" DNA adducts, DNA damage and micronuclei.

In mammalian cells, gene expression profiles likewise showed that diesel engine emissions, particle suspensions and particle extracts caused the increased expression of genes connected with oxidative stress, inflammation, DNA damage, antioxidative reactions, cell cycle, cell transformation and apoptosis.

DNA damage, gene mutations, DNA strand breaks, chromosomal aberrations and morphological cell transformations were demonstrated in various test systems *in vitro*, for example in cell lines and primary cells from humans and rodents, and gene mutations in bacteria, and *in vivo* in rats and mice. The effects *in vivo* were seen after inhalation, intratracheal instillation and oral administration of entire diesel engine emissions and in particle form and after the topical application and intraperitoneal injection of organic extracts.

"New diesel", on the other hand, has not been found to be genotoxic in the investigations (comet assay, micronucleus test) carried out so far.

Carcinogenicity

> **Diesel engine emissions are carcinogenic in experiments with rats, but not in mice and hamsters.**

In rats, after exposure to "old diesel" engine emissions with a particle content of ≥ 2200 µg/m^3 for 104 or 130 weeks, an increase in tumours was observed in the lungs; tumours did not occur in other organs. Investigations have shown that the PAH content alone cannot be responsible for the carcinogenic effects. Also, after exposure to filtered, and therefore particle-free diesel engine emissions or at concentrations of ≤ 800 µg/m^3 no carcinogenic effects were observed.

Emissions from "new diesel" engines with 13 μg particles/m^3 and 3.6 ppm NO$_2$ likewise did not lead to tumours in the rat lung.

> **In 2012, the International Agency for Research on Cancer (IARC) classified diesel engine emissions on the basis of evidence of a relationship between exposure to diesel fumes and an increased risk of lung cancer as carcinogenic for humans (Group 1).**

Most epidemiological studies describe an increased lung cancer risk for people who were exposed at the workplace for a long period to high-level diesel engine emissions. The weakness of all these studies is the demonstration of a dose–response relationship. The exposure data cannot be determined in most studies, or only incompletely. Often the duration of exposure is given as a surrogate for the determination of the exposure level. Another difficulty is that the term "diesel engine exhaust fumes" is not precisely defined. The emissions depend very much on the technology of the diesel engines, with the old engines being successively replaced with more modern engines and often different technologies are used in parallel. There are no epidemiological data which fulfil present-day requirements available yet for the effects of diesel engines using new technology because the latency period between exposure and the observation of possible effects is not yet long enough.

The carcinogenic potential of diesel engine emissions was investigated in numerous epidemiological studies mainly in workers exposed to high levels, for example in mining, tunnel construction, railway construction, lorry traffic and shipping. The results are inconsistent.

A clear dose–response relationship could not be demonstrated to date even in the available well-documented case-control studies or in extensive pooled or in meta-analyses.

In addition to the lungs, the bladder, larynx, pancreas and colon have also been suggested as target organs of the carcinogenic effects of diesel engine emissions; leukaemia is also mentioned in this context. However, the associations here are much weaker.

6.12.5 Mechanisms of the Carcinogenic Effects of Diesel Engine Emissions

> **The mechanisms of the carcinogenic effects of diesel engine emissions have not been completely clarified. In addition to the genotoxic effects of some components, such as PAHs or nitroaromatics, in particular the action (as particles) of the elementary carbon nucleus is held responsible, especially for the lung tumours observed in rats.**

One hypothesis suggests that the inflammatory-proliferative effects of carbon nucleus particles form the basis for effects on lung cells. Another hypothesis assumes that predominantly soluble and insoluble inorganic and organic substances adsorbed on the carbon particles, for example PAHs or nitro-PAHs, induce the carcinogenic effects. Studies have shown, however, that the PAHs amount to only a small part of the diesel fume particle mass and cannot explain the increased tumour incidences. Presumably, the adsorbed substances are metabolized to mutagenic substances, which lead to the DNA adducts in the lung cells of rats and monkeys observed after exposure to diesel fumes.

Genotoxicity

Diesel engine emissions contain genotoxic components, such as PAHs and oxidized or nitro-PAHs, which bind directly to DNA and can cause DNA damage. The amount of these alone, however, cannot explain the observed tumours. Direct genotoxic effects cannot be excluded, but chronic inflammation, which leads to oxidative stress and ROS, in other words secondary genotoxicity, seems, together with the increased cell proliferation, to play the greater role. Therefore, a non-linear dose–response relationship for carcinogenicity is also probable.

Carcinogenicity

Several factors are probably responsible for the carcinogenicity of diesel engine emissions. In studies with rats, the carcinogenic effects of diesel engine emissions were induced mainly by the particle fraction. Although the gaseous fraction contains small amounts of carcinogenic substances and yielded positive results in bacterial mutagenicity tests, filtered, and therefore particle-free, diesel engine emissions were not found to be carcinogenic. Species differences in the deposition and clearance of particles in the lungs may be responsible for the particular sensitivity of the rat as regards tumours induced by substances in dust form (so-called "overload", see below).

Acute and Chronic Toxicity: Irritative-inflammatory Effects on the Airways

The inflammatory effects of the particles from diesel engine emissions are attributed to the formation of ROS. The emissions contain substances such as PAHs, quinones and transition metals, which can form ROS as a result of redox reactions within and outside the lung cells. ROS can, however, also be formed by alveolar macrophages during particle phagocytosis. At low concentrations, oxidative stress leads to the activation of antioxidants and to detoxifying enzymes such as haemoxygenase-1 and glutathione S-transferase, which protect the cell against oxidative damage. At higher concentrations, these protective mechanisms are overloaded (so-called "overload") and inflammatory and cytotoxic effects occur.

The inflammatory effects caused by diesel particles are mediated by redox-sensitive mitogen activated protein (MAP) kinase and transcription factor nuclear factor kappa B (NF-κB) cascades, which are responsible for the production of inflammatory cytokines, chemokines, and adhesion molecules. Together with the adhesion molecules, cytokines and chemokines, including tumour necrosis factor alpha (TNF-α) as well as interleukines (IL), are involved in recruiting and activating inflammatory cells in the lung. Large quantities of ROS are produced by activated leukocytes, causing further oxidative damage to the surrounding cells. The inflammatory cascade includes the activation of phospholipase A2, leading to an increase in local vasodilatation and vasopermeability, enhancing accumulation of inflammatory cells. Oxidative stress inside a cell, besides the inflammatory effects, may lead to cytotoxicity through pro-apoptotic factors released by mitochondria.

One of the components of the gas phase is NO_2, a strong irritant, which reacts with antioxidants, lipids and proteins on the surface of cell membranes in the lower respiratory tract and thus produces further reactive substances such as peroxides, which in turn lead to oxidative stress and inflammatory reactions. NO_2 in addition diminishes the activity of the cilia in the airways, which impairs mucociliary clearance.

Cardiovascular Effects

Three mechanisms have been suggested for the cardiovascular effects caused by diesel engine emissions and other fine dust particles. Firstly, inflammation mediators such as cytokines or activated inflammatory cells can enter the systemic circulation from the lungs and likewise lead to inflammatory processes there. The inflammation mediators can damage the vascular system directly or via the increased production of coagulation factors in the liver. Alternatively, it has been suggested that the particles disturb the equilibrium of the autonomous nervous system or the heart rhythm via interaction with receptors in the lungs or the nerves. A third possibility might be that the particles enter the systemic circulation via translocation from the lungs and act directly on the vascular system. However, the amount of particles detected in the vascular system is very low compared to that in the lungs. At the molecular level, the ROS and oxidative stress in various stages, for example the intensification of systemic inflammation, the stimulation of vasoconstriction and atherosclerotic plaques, play a central role.

In addition, among the components of the gas phase, CO is known to bind to haemoglobin and other proteins containing haem, which, at high concentrations, can lead to hypoxia, cardiac dysfunction and ischemia. NO is a pulmonary vasodilator and can cause methaemoglobin, but at concentrations that do not occur in the context of diesel engine emissions.

Immunotoxicity

Various mechanisms have been suggested for the intensification of asthma and allergic diseases as a result of diesel engine emissions. In addition to oxidative stress and inflammatory mechanisms, which can contribute to pulmonary inflammation in association with allergic asthma, there is evidence that diesel particles together with allergens stimulate the TH2-type immune response and increase the production of allergen-specific IgE and IgG. Oxidative stress plays a role here both in inflammatory and adjuvant effects.

As a result of its effects on inflammatory processes in the lungs and on lung permeability, NO_2 is known to also intensify asthma.

6.12.6 The Exposure of Humans to Diesel Engine Emissions

The exposure of the general population to the components of diesel engine emissions is determined by the proximity to traffic emissions and depends on the extent and kind of traffic.

In workplaces there might additionally be particular high exposure levels, for example as a result of street vehicles such as cars, lorries and buses, but also other vehicles such as diggers, fork-lift trucks, tractors, harvesters or military vehicles. Exposure can also take place in shipping, or as a result of locomotives or stationary equipment such as compressors, pumps, building machinery, generators etc. in industry and agriculture. Miners and workers in (tunnel) construction, warehouse workers, mechanics, vehicle drivers, shipping and railway workers, but also workers in forestry and agriculture, in waste disposal or environmental sanitation are exposed to particularly high levels.

The consumption of diesel fuel in 2010 in Europe was almost 300 million tonnes, and it is increasing steadily.

6.12.7 Evaluation by International Organizations

The International Agency for Research on Cancer (IARC) classified diesel engine emissions in 2012 as carcinogenic in humans (Group 1).

The World Health Organization (WHO) regarded diesel fumes as early as 1996 as "probably carcinogenic to humans". For endpoints other than cancer, a so-called guidance value of 5.6 µg/m^3 (or of 2.3 µg/m^3 with a different calculation procedure for extrapolation from experiments with animals to humans) was derived for lifelong exposure for the general population.

The Permanent Senate Commission for the Investigation of Hazardous Substances at the Workplace of the German Research Foundation (MAK Commission) in Germany classified diesel engine emissions in 1987 in Carcinogen Category 2 (carcinogenic in animal experiments), whereas the epidemiological studies have been considered inconclusive regarding human carcinogenicity. This evaluation was confirmed in 2008, taking into consideration new epidemiological data expressly for exposure to "old diesel".

The US National Toxicology Program (NTP) regarded exposure to diesel engine emissions in 2001 and 2011 as "reasonably anticipated to be carcinogenic to humans".

According to the US Environmental Protection Agency (EPA) in 2002, diesel engine exhaust fumes are "likely to be carcinogenic to humans by inhalation". In addition they represent a non-cancer hazard with long-term exposure, lead to sensory irritation, effects on the airways and neurophysiological symptoms. Evidence has also been seen of immunological effects. A reference concentration (RfC) of 5 µg/m^3 for the lifelong inhalation of diesel particles was derived for inflammatory effects on the lungs.

Risk estimations for lung cancer, based on studies with rats, are in the range of 1.6×10^{-5} to 7.1×10^{-5} for diesel exhaust particle concentrations of 1 µg/m^3.

6.12.8 Summary

Diesel engine emissions are formed during the combustion of diesel fuels in engines. They contain several hundred chemical compounds, which are emitted partly in gas form, partly as particles. They are suspected of causing cancer at least in experimental animals. The carcinogenic effects are attributed to the carbon particles with PAHs bound to them, which are emitted mainly from engines using older technologies. The effects of emissions from more modern engines, on the other hand, are determined predominantly by the level of nitrogen oxides. The main effects of short-term exposure are irritation and inflammatory reactions in the lungs. In addition, cardiovascular effects and the intensification of asthma and allergic diseases have been suggested.

Further Reading

Attfield, M.D., Schleiff, P.L., Lubin J.H., Blair, A., Stewart, P.A., Vermeulen, R., Coble, J.B., Silverman, D.T. (2012) The diesel exhaust in miners study: a cohort mortality study with emphasis on lung cancer. *J. Natl. Cancer. Inst.*, **104**, 1–15.

European Commission (2012) Commission Regulation (EU) No 459/2012 of 29 May 2012 amending Regulation (EC) No 715/2007 of the European Parliament and of the Council and Commission Regulation (EC) No 692/2008 as regards emissions from light passenger and commercial vehicles (Euro 6). Official Journal of the European Union L142/16-24.

Greim, H. (ed.) (2014) *Diesel engine emissions, Supplement 2008, The MAK-Collection Part I, MAK Value Documentations*, Wiley-VCH, Weinheim.

IARC (1989) Diesel and gasoline engine exhausts. In: Diesel and gasoline engine exhausts and some nitroarenes. *IARC Monographs on the Evaluation of Carcinogenic Risk of Chemicals to Humans*, vol. **46**. International Agency for Research on Cancer, Lyon, pp. 41–185.

IARC (2013) Diesel and gasoline engine exhausts. In: *Diesel and gasoline engine exhausts and some nitroarenes, IARC Monographs on the Evaluation of Carcinogenic Risk of Chemicals to Humans*, vol. **105**. International Agency for Research on Cancer, Lyon, pp. 39–486.

Möhner, M., Kersten, N., Gellissen, J. (2013) Diesel motor exhaust and lung cancer mortality: reanalysis of a cohort study in potash miners. *Eur. J. Epidemiol.*, **28**, 159–168.

NEG (2016) Diesel engine exhaust, No. 149. Taxell, P., Santonen, T., The Nordic Expert Group for Criteria Documentation of Health Risks from Chemicals and the Dutch Expert Committee on Occupational Safety, Arbete och Hälsa 49(6).

Silverman, D.T., Samanic, C.M., Lubin, J.H., Blair, A.E., Stewart, P.A., Vermeulen, R., Coble, J.B., Rothman, N., Schleiff, P.L., Travis, W.D., Ziegler, R.G., Wacholder, S., Attfield, M.D. (2012) The diesel exhaust in miners study: a nested case-control study of lung cancer and diesel exhaust. *J. Natl. Cancer Inst.*, **104**, 1–14.

US EPA (2002) *Health Assessment Document for Diesel Engine Exhaust*. EPA/600/8-90/057F. US Environmental Protection Agency, Washington, DC.

US EPA (2003) *IRIS, Integrated Risk Information System*. Diesel engine exhaust. Last revised 02/28/2003. Washington, DC. http://www.epa.gov/iris/subst/0642.htm.

Vermeulen, R., Silverman, D.T., Garshick, E., Vlaanderen, J., Portengen, L., Steenland, K. (2014) Exposure-response estimates for diesel engine exhaust and lung cancer mortality based on data from three occupational cohorts. 10.1289/ehp.1306880.

Wikipedia Emission Standard. https://en.wikipedia.org/wiki/Emission_standard.

WHO (1996) *Diesel Fuel and Exhaust Emissions. Environmental Health Criteria 171, International Programme on Chemical Safety*, World Health Organization, Geneva.

6.13 Animal and Plant Toxins

Thomas Zielker

6.13.1 Introduction

> **There is an enormous variety of naturally occurring toxins. The chemical structures of these toxins range from simple organic compounds to very complex proteins. Animal toxins consist mostly of a mixture of polypeptides and different digestive enzymes. In plants alkaloids and glycosides are responsible for the toxic effects.**

The morbidity of intoxications due to natural toxins differs from country to country and continent to continent according to the habitat of the toxic plants and animals. Nevertheless, as a consequence of travelling and the habit of keeping exotic animals in aquaria and terrariums, intoxications with animal toxins can happen around the world via animals that are not indigenous to that particular place. Moreover, snakes, scorpions and spiders can transmigrate from continent to continent through modern transportation. Mushroom and plant intoxications occur mostly by oral intake, but dermal toxic reactions are possible following the contact with some plants.

A description of all known toxins can fill whole books. In this chapter only those intoxications are described which can lead to severe illness or even death.

6.13.2 Animal Toxins

Intoxications caused by animal toxins happen by bites, stings, through dermal transmission or by oral intake of seaborne food that is contaminated with a toxin. Bites happen through snakes and spiders. Stings happen through hymenoptera, scorpions and fish.

Snakes

> Snake bites from poisonous animals are the most common cause of severe morbidity and mortality among poisonous animals. In some areas of the world snake bites are among the top ten fatal diseases. The Elapidae and Viperidae families are those that cause the most morbidity and mortality.

There are four families of snakes that only contain either exclusively venomous or some toxic species. The families with only venomous species include the Elapidae, the Viperidae and the Atractaspididae. This last family is found only in Africa and the Middle East. Envenomations with the toxins of these snakes are usually mild. The family Colubridae includes mostly non-venomous snakes. A few species of this family have fangs with venom glands, others have a toxic saliva which leads to an inoculation of the venom after a bite with ordinary teeth.

The Viperidae can be divided in two subfamilies, Viperinae and Crotalinae. The species vipera berus (common in Europe) belongs to the Viperinae; the Crotalinae, which are also call pit vipers, includes the North American rattlesnakes. The Elapidae consists, among others, of the cobras and the kraits in Asia, the mambas in Africa, the tiger snake and taipan in Australia, and the coral snakes in the USA. Though the poisons of many snakes in the Elapidae family are neurotoxic, there are few species in this family, which produce only local tissue damage (African tree cobra). Most venoms of the Viperinae are more or less toxic, producing mild to extreme local damage, but there are some that are neurotoxic (e.g. the Middle East horned viper).

The four families of snakes have all different sorts of fangs. The Elapids have hollow front fangs, which are short because they are permanently erect, whereas the Viperinae and Crotalinae have fangs which fold back on hinges and can therefore be long. The Atractaspids have a small maxilla but still bear large hollow front fangs, which are erect and stick out by the side of the mouth as so-called side-striking fangs.

The poisonous Colubrids have grooved fangs at the back of their mouths (rear fang snakes), which are permanently erect, or have no fangs and poisonous saliva.

Venom Variability There is significant intraspecies and even intraindividual venom variability in snake venom, which depends on the geographic range and even the season. The different groups of toxins and their various effects will be described.

Neurotoxins

> There are two different kinds of neurotoxins in snake venom: presynapticly and postsynapticly acting venoms. The former are phospholipases, the latter small polypeptides.

The *presynaptic neurotoxins* belong to the phospholipase A_2 family of enzymes. Their target is the terminal axon of the neuromuscular junction. The early effect is a release of the neurotransmitter, the later a destruction of the axonal structure leading to a disruption of the synaptic vesicles with a complete stop of transmitter release.

Signs and symptoms in a bitten patient are a flaccid paralysis starting with the cranial nerves, ptosis coming first, followed by opthalomoplegia, dysarthria, loss of airway protection, paralysis of the respiratory muscles and the diaphragm, and finally limb paralysis with loss of deep tendon reflexes.

Postsynaptic neurotoxins are polypeptides with about 60–70 amino acids. Their relatively small-sized proteins contain four (neurotoxin I) or five (neurotoxin II) disulfide bonds.

These bonds fold the peptide to three loops. One of the loops (B) probably attaches to the acetylcholine-binding site of the acetylcholine receptor. The receptor is a ligand-gated channel protein, allowing ions passing through when activated. The postsynaptic snake neurotoxins attach to the two subunits (a glycoprotein) of the receptor and block the formation of the ion channel.

The onset of symptoms commences one hour after the bite. Whereas the effect produced by the postsynaptic toxins can be blocked by a specific antivenom, the damage done to the presynaptic structures cannot be influenced. If the presynaptic effect is dominant it usually takes weeks to resolve. This means that long-term respiratory therapy may be necessary even when antivenom treatment was applied.

Among the Elapids, mambas have a special venom containing two neurotoxic groups: dendrotoxins and fasciculins. Their action is different from the action of the already mentioned snake neurotoxins. *Dendrotoxins* affect the potassium channels in the terminal axon membrane, leading to a continuous release of acetylcholine and an over-stimulation of the muscle endplate. *Fasciculins* act as cholinesterase inhibitors in the junctional cleft, reducing the degradation of acetylcholine. Together the two actions lead to a surplus of acetylcholine similar to that in organophosphate poisoning. The over-stimulation of the endplate leads to fasciculation and myocloni of the peripheral muscles with the consequence of being unable to use the peripheral musculature, leading to respiratory paralysis.

The action of these two groups of toxins may be very fast, leading to symptoms within less than one hour. The fasciculins have a similar structure to the postsynaptic neurotoxins of other Elapids. They have an amino-acid chain consisting of 57–60 amino acids cross-linked by three disulfide bonds. Dendrotoxins have a molecular weight of around 7000. They have considerable sequence homology with beta-bungarotoxin venom of the krait, which has a presynaptic action and is also a sodium channel blocker.

Myotoxins

Snake venoms that cause damage to the musculature are phospholipase A_2 enzymes, which destroy skeletal muscles.

The myotoxins destroy muscle cells, sparing the basement membrane. This allows regeneration of the muscle cells in due course (one month). As muscles are destroyed a release of myoglobin occurs; potassium as a diagnostic indicator for the damage creatine kinase (CK) is released. Myoglobin damages the tubulus cells of the kidney leading to secondary kidney failure, which can be treated by hemodialysis. It is questionable whether antivenom can do any good if myolysis has already ensued.

Hematotoxins

The mechanism by which a mixture of enzymes in snake venom damage blood homeostasis is twofold: on a cellular level the functioning and the amount of platelets are affected, on a humoral level the blood coagulation is deranged. The major venom constituents are procoagulants, anticoagulants, platelet aggregation inhibitors or promoters, as well as hemorrhagins that cause local damage to vessel walls.

A mixture of enzymes interact with hemostasis. The strange effect of these components is that hypercoagulopathy with thrombosis and bleeding can happen simultaneously. The cause of the tendency to bleed lies in the procoagulant activation of Factor V, Factor X, Factor IX and prothrombin (Factor II) thus using up fibrinogen.

In addition to the mechanism of using up fibrinogen, there is a toxin with a direct fibrinolytic action promoting fibrinogen degradation. Therefore, in many snake poisoning cases, especially by Crotalinae, a coagulopathy with no measurable fibrinogen is present. As the platelets are affected by platelet aggregation inducers, the laboratory testing (low platelets, no fibrinogen, high PTT and PT) may point to a disseminated intravascular clotting (DIC) which in fact is not happening.

Nearly all snake venoms contains hemorrhagins, which are zinc metalloproteinases that cause local damage of vessel walls leading to local bleeding and ecchymosis. Bleeding is enhanced by substances with anticoagulant action such as protein C, thrombin inhibitor and phospholipase A_2. The chemistry of all these diverse toxins reaches from small to large molecules that resemble the prothrombinase complex. Although after a snake bite, especially from Viperidae, severe coagulopathy is notable, spontaneous bleeding is rare. Difficult-to-control hemorrhage may happen after medical interventions such as puncture from an intravenous line, surgical debridement or surgical cuts for treatment of swelling and compartment syndrome. In these circumstances the substitution of clotting factors alone cannot stop the bleeding so use of an antivenom is essential.

There is one venomous snake that has a particular nephrotoxin, the Russell's viper, which leads to acute renal failure that is not secondary to shock or myolysis.

Local Damage

Most snake bites – with the exception of some Elapids (e.g. Krait) – create more or less local damage (blisters, necrosis) around the bite due to the venoms' content of enzymes, kinins, leukotrienes, histamine, phospholipases, collagenases and metallic ions.

Generally, within minutes after the bite, the region around the fang marks starts to swell and becomes painful. Within hours ecchymosis, blisters and tissue necrosis develops and the swelling spreads over the whole limb, sometimes involving adjacent areas of the trunk. This can lead to compartment syndrome, in that the swelling gets so bad that the circulation in the affected limb is impaired. Without surgical intervention this could lead to the loss of the limb.

The toxic components act at varying times. The venoms may remain local and be fixed to tissue but can reach the circulation as well and thereby cause systemic toxicity.

Systemic Symptoms Usually the venom is injected subcutaneously during the snakebite. It spreads through lymphatic and superficial venous vessels and reaches the circulatory system slowly, inducing systemic signs. The rare intravascular envenomation produces systemic symptoms within minutes. Direct intravenous venom injection seldom happens but may account for the majority of fatalities by snakes, which are thought not to be toxic. In some cases the venom of the snake can lead to overt anaphylaxis with a fatal or near fatal reaction. Less severe systemic signs are weakness, nausea, malaise and anxiety. More severe signs manifest as abdominal pain, vomiting, diarrhea, dyspnea, profuse sweating, salivation, metallic taste and confusion. The most severe symptom is circulatory collapse with hypotension and tachycardia.

Management of Snakebites

First Aid

> **The "traditional" first-aid methods, like incision or sucking the wound, are ineffective. Instead, it is effective to immobilise the bitten limb and the patient to reduce the distribution of the venom.**

The only approved first aid for a snake bite is to immobilise the bitten limb and keep the victim still. For bites of many elapids the Australian pressure immobilisation method is effective. It is able to retard venom transport in the lympathics by immobilising the limb on a splint. The limb is wrapped with an elastic bandage just hard enough to stop the transport of lymph and compress the superficial veins. Though this method has proven to be effective for the extremely toxic Australian elapids, immobilisation of the bitten limb on a splint makes sense for all snake bites to reduce the transport of the venom through the muscle pump.

Hospital Treatment

> **Hospital treatment requires careful medical history of the event, description of symptoms and swelling to judge the progression, neurological and haematological investigations, antivenom therapy and if necessary anticoagulant therapy and surgical intervention.**

First of all a medical history has to be taken, not forgetting to ask for the current tetanus immunisation status and for known allergies. A careful examination, with photographs taken of the extremity that was bitten, has to be made. The diameter of the limb has to be measured to be able to compare with later measurements and thus judge the progression. The extent of the edema and skin discoloration can be marked by a pen to help recognise the development of the envenomation. A physical examination records the vital signs, and cardiorespiratory and neurological status. The disposition for bleeding can be clinically seen by searching for blood in urine, blood in stool and gingival petechiae.

The baseline laboratory tests cover a complete blood count, including the platelets, electrolytes, urine analysis, BUN, glucose, INR, PTT and fibrinogen. If normal, these tests have to be repeated within six hours. If there are any signs of circulatory depression, IV fluid replacement and, when this is not sufficient, vasopressors, have to be administered.

Antivenom Therapy The treatment with antivenom depends to some extent on the toxicity of the snake involved. Whereas for the European viper antivenom is only indicated if there are systemic signs and/or if the edema is progressing quickly, an antivenom treatment for crotaline or elapid envenomations is more often necessary.

Besides antivenoms derived from horses, ovine-derived F(ab) fragments have been developed that have fewer side effects than horse serum. CroFab is available for the most common American croatline and ViperaTab for the most common European vipers.

Treatment of Coagulopathy Crotalinae, far more than Viperinae or Elapidae, envenomation leads to severe coagulopathy. Coagulopathy can only be restored if sufficient antivenom is applied, as all blood products are neutralized by circulating snake venom. Bleeding calls for a large amount of antivenom before fibrinogen and clotting factors are given. The same holds true for platelet transfusion as platelets are destroyed by the venom as long as not enough antivenom is present.

Surgical Intervention   Surgical intervention should be reduced to a minimum as most snake bites, even with enormous swellings, can achieve full recovery. There are two reasons for surgical treatment. First, wound debridement around the bite if necrosis of the tissue has developed and hemorrhagic blebs and blisters have become large enough for resection. This is usually occurs 3–6 days after envenomation. Second, a compartment syndrome has to be decompressed by fasciotomy. Fasciotomy has to be instituted only if there is found by direct measurement to be a high tissue pressure, if there is no more blood flow through the limb and if the neuromuscular transmission is interrupted.

Scorpions

> **Scorpion stings exhibit a high morbidity and mortality in India, the Middle East, North Africa, Brazil, Mexico, southern states of the USA and in Central- and South Africa. Scorpion venoms are both cardiotoxic and neurotoxic as a consequence of increasing neuronal sodium influx by interfering with the sodium channel. The initial symptoms are similar to poisoning by organophosphates.**

Scorpions are the next most poisonous animals after snakes. Scorpion stings are common in various regions but not in Europe. In Mexico it has been estimated that 1000 people die as a result of 200,000 scorpion stings per year. In Saudi Arabia the fatality rate for scorpion stings is about 1.5%.

High mortality rates of up to 25% have been reported from India in patients who received only supportive care. The most toxic species are scorpions of the Buthus, Leiurus, Antroctonus, Tityus, Centruroides and Parabuthus families. As a rule the thinner the claws and bigger the stingers are, the more toxic is the scorpion.

Venoms of the Scorpions   Scorpion venoms are neurotoxic by influencing the vegetative and central nervous systems. Similar to snake venom the constituents are a mixture of mucopolysaccharides, hyaluronidases, serotonin, histamine, protease inhibitors, histamine releasers and amino acids. The local necrotising reactions are far less than in snake bites with the exception of stings of Hemiscorpius lepturus. Scorpion venoms increase neuronal sodium influx by interfering with the sodium channel. This leads to an increased conductance at the voltage-gated calcium channel in presynaptic nerve fibres with an increased release of neurotransmitters including acetylcholine. Similar to poisoning by organophosphates, first a heavy sympathetic stimulation with the release of epinephrine and norepinephrine ensues. This is followed by a parasympathetic reaction with low cardiac output. Salivation, abdominal pain nausea and vomiting are common signs of this vagal reaction. Pancreatitis can occur.

The nicotinergic neuromuscular transmission is manifested as tongue and muscle fasciculations, and involuntary movements of skeletal muscles. The catecholamine over-stimulation leads to hypertension, creating lung congestion or even pulmonary edema. Myocardial ischemia and even infarction can occur. Central nervous system (CNS) effects may be due to hypertensive encephalopathy as well as to a direct central mechanism of toxicity. Agitation, hyperthermia, seizures and even coma may be the signs and symptoms of this development. Whereas most scorpion species exhibit both cardiotoxicity and neurotoxicity there are some scorpions of the Centruroides family that are merely neurotoxic.

Clinical Presentation of Scorpion Stings Scorpions sting humans most often in the extremities after hiding in clothing. Children are more often stung than adults and show more serious reactions than adults, especially if they are under the age of 10. The severity grade of scorpion envenomation depends on the scorpion's species, age and size. Very often local reactions are the only symptoms. Paresthesia and pain at the sting site are usual. In more severe cases these symptoms spread out over the limb involved and can even reach the peri-oral area. Systemic signs and symptoms comprise stimulation, followed by depression of the CNS, hypertension, lung edema, dysrhythmia and rarely shock. Cranial nerve palsy and neuromuscular dysfunction, e.g. involuntary movement, restlessness and shaking, can occur. Parasympathetic stimulation leads to gastrointestinal over-stimulation and may cause priapismus. In the circulatory system the sympathomimetic effects prevail. This is reflected by tachycardia and hypertension; bradycardia and hypotension can happen but are rare. In cases with fatal outcome, cardiac and respiratory failures are the predominant symptoms. A dilated cardiomyopathy with sequelae is found in few cases. Scorpion stings may produce some laboratory abnormalities. Hyperglycemia, leucocytosis and CRP elevation are commonly seen. The later are due to an inflammatory reaction caused by interleukin release. There maybe a transient elevation of CKMB and Troponin T due to myocardial ischemia which can be diagnosed by ischemic signs in the ECG. It was possible to demonstrate cardiac dysfunction in echocardiography showing a decreased ejection function.

Therapy of Scorpion Stings A light constricting bandage proximal of the wound to prevent lymphatic spread and application of ice at the sting site slowing venom absorption and relieving pain is generally recommended. In cases of systemic signs and symptoms an antivenom should be used. Nevertheless, the use of antivenom is still controversial as supportive care has improved over the years and the death rate has come down in the last 40 years

Spiders

> **All spider venoms are complex mixtures consisting of toxic peptides and proteins, but most spiders are unable to bite humans to penetrate the skin and therefore bites do not lead to an envenomation. For humans the most toxic spider is the Australian funnel-web spider.**

Taxonomy divides spiders into the orders Mygalomorphae and Araneomorphae. Within the Mygalonomorphae is the Australian funnel-web spider (*Atrox robustus*), the most toxic spider for humans. All the other spiders which are toxic for humans belong to the Araneomorphae, namely *Lactrodectus* spp. (widow spiders), *Phoneutria* spp. (banana spiders) and *Loxosceles* spp. (recluse spiders).

Spider venoms are, like snake and scorpion venoms, complex mixtures consisting of toxic peptides and proteins. There is great divergence from one spider venom to another. Some venoms are very potent neurotoxins, others create an endogenous catecholamine storm, some contain necrotoxins which can destroy large skin areas of the human body.

Australian Funnel-web Spiders *Atrox robustus*, the Australian funnel-web spider, has caused the most severe bites and fatalities of all spider bites.

The mouse spiders (Missulena) have venom similar to the Australian funnel-web spider, but only a few severe human envenomings occur. The venom of these spiders has a neurotoxic component that is potentially lethal for humans but does little harm to other mammals. The venom of the male spiders is more toxic than that of the females. The neurotoxic component is robustoxin. It is a protein of molecular weight 4854 Da with 42 amino acid residues and four disulfide bridges. The LD_{50} for mice is 0.16 mg/kg.

Clinical Signs and Symptoms   Since the spider is large, and after the bite it hangs on, the bite is usually recognized. Bite marks are present and the bite is painful.

Similar to some scorpion stings perioral paresthesias and tongue fascicultations develop. This is followed by a sympathomimetic phase as the catecholamines are forced out of their repositories, including tachycardia, hypertension, piloerection, cardiac arrhythmias and pulmonary edema, which is partially due to left ventricular congestions and partially of neurogenic nature. As in scorpion stings, muscarine-cholinergic over-stimulation can ensue with hypersalivation, hyperlacrimation, increased sweating – first around the bite and later generalized – with nausea, vomiting and abdominal pain. The nicotinic-cholinergic hyperactivity leads to muscle fasciculation and involuntary movements. As soon as catecholamine depletion becomes manifest, the second life-threatening phase is reached in which excitatory effects come to an end, secretion stops and hypertension resolves only to progress into terminal circulatory failure apnoea and cardiac arrest.

Treatment of Spider Bites by Australian Funnel-web Spiders   There are bites by the Australian funnel-web spiders that cannot be survived without the use of intravenous antivenom treatment. An antivenom is available on the Australian market against the male *A. robustus* venom. It is a rabbit IgG with the name FWSAV.

Latrodectus Spiders

> **Latrodectus spiders are found around the world, usually between 50° N and 45° S. They are very rarely found in Middle Europe. 2500 cases of lactrodectus bites are reported to US poison centres per year. In Australia 5000–10,000 bites occur annually. The venoms contain high-molecular-weight proteins that cause release of transmitters in various synapses.**

There are about 50 species of the genus Lactrodectus, of which about nine are of medical importance: *L. mactans*, *L. hesperus*, *L. bishopi* and *L. variolus* (North America), *L. hasseltii* (red back spider, Australia), *L. Katipo* (New Zealand), *L. indistinctus* (South Africa), *L. geometricus* (brown widow, world-wide) and *L. tredecimguttatus*.

The male spiders have a small biting apparatus and therefore cannot envenomate humans, whereas the female can. The Lactrodectus venoms contain high-molecular-weight proteins. Some of the proteins have been cloned and sequenced. The α-Latrotoxin (A-LTX) is produced by all Lactrodectus species, it has a molecular weight of 150,000 Da and seems to be the constituent of the venom that is responsible for human envenomation.

A-LTX causes massive release of transmitters in various synapses and is effective not only at cholinergic but also at noradrenergic and glutaminergic sites due to a depletion of the synaptic vesicles. The main action of Lactrodectus venom is mediated by the binding of A-LTX to the presynaptic Ca^{2+}-independent receptor (CIRL). CIRL is believed to be coupled to phospholipase C and interfers with phosphoinositide metabolism, which is responsible for secretion.

There are several CIRLs with different affinities to A-LTX. CIRLs 1 and 3 are high-affinity receptors in neuronal tissue and in many other organs such as the kidney, spleen, ovary, heart, lung and brain. CIRL belongs to the family of 7-transmembrane domain G-protein-coupled receptors. The nervous system is the first target for α-LTX but other tissues can be involved due to the presence of CIRLs.

Signs and Symptoms of Lactrodectism

> **Within 60 minutes after a bite local pain, local sweating and piloerection may occur. Only a quarter of the bitten victims experience systemic toxicity such as muscle pain, which in severe cases may become generalized and in very severe cases require treatment with analgetics.**

A pain similar to a pinprick is what the victim feels if bitten by a *Lactrodectus* spp. A pair of red spots with a halo can be seen at the site. Very often the bite remains unnoticed. The local reaction at the bite site is uneventful until 60 minutes after the bite. Within that time a pathognomonic triad appears: local pain, local sweating and piloerection. It may take hours before the bite fully develops.

Muscle pain starts around the bite and spreads over the extremity involved. Lymph tenderness of the regional nodes evolves. In severe cases the pain becomes generalized. The pain is felt in the neck, chest, abdomen, lower back and thighs. The pain can become very severe and mimic acute peritonitis. Severe diaphoresis is present. Dysphoria, restlessness, horror and fear of death torment the patient. The victims show – if a fully developed latrodectism ensues – a facia lactrodectismica which means trismus, flushing, grimacing and blepharo-conjunctivitis. Symptoms peak within 12 hours after the bite and resolve within 24 hours. Symptoms may last for a week, although this is rare, and muscle weakness may be present over months. Laboratory test exhibit leucocytosis, hyperglycemia, and elevation of CK and LDH, which are of course non-specific.

Management of Lactrodectism   The treatment of a patient who has been bitten by a spider of the Iatrodectus species depends on the severity of symptoms. It is most important to treat patients with analgesics to relieve pain. Since muscle cramping plays a prominent role in lactrodectism the use of benzodiazepines has been recommended. The indication for the use of antivenom is given if hypertension or pain is not managed by this treatment.

Loxosceles Spiders

> **Loxosceles venom is one of the most potent natural toxins, containing necrotizing and hemolysing substances. Most bites occur in South and Central America.**

Loxosceles spiders are found throughout the world. *L. reclusa*, the brown recluse spider, lives in the southern states of USA. *L. intermedia* and *L. laeta* are found in South Africa. *L. rufescens* resides in the Mediterranean regions of Europe and Africa. Very few bites happen in Europe, some in North America and a lot in South and Central America.

So far, 11 major components of the venom have been identified. The most important subcomponents are hyaluronidase, deoxyribonuclease, ribonuclease, alkaline phophatase, lipase and, last but not least, sphingomyelinase-D. Hyaluronidase works as a spreading factor that allows the venom to penetrate tissue. Sphingomyelinase-D (MW 32,000) is the necrotizing and hemolysing substance. Sphingomyelinase-D stimulates platelets to release serotonin and interacts with the membrane of cells, which contain sphingomyelin. This leads to a release of hemoglobin, choline and N-acylsphingosine phosphate from red blood cells. The resulting hemoglobinemia can cause kidney failure. The other substances trigger a cascade of reactions by releasing inflammatory mediators such as thromboxanes, leukotriens, prostaglandins and the activation of neutrophil granulocytes. The sequelae of this release is vessel thrombosis, tissue ischemia and skin necrosis.

Complete hemolysis is induced by the activation of a metalloprotinase, leading to a cleavage if glycophorins from the erythrocyte surface, mediating complement activation. The hemolytic process is maintained by transfer of the toxins from one erythrocyte to another, attacking even newly synthesized erythrocytes. Cytokinin activation by the toxic proteins of the spider venom can evoke an endotoxin shock resembling a septic shock.

A typical histopathology develops around the lesion of the bite. It starts as perivasculitis with a polymorphonuclear inflammation, followed by local hemorrhage and edema with epidermal necrosis and ulcerations. A panniculitis with subcutaneous fat necrosis is the next step in this development, together with arterial wall necrosis. The ulcerations become eschar-covered. Slowly resolution of the lesion leads to scarring of the dermis and subcutis. Skin transplantation is sometimes needed after several weeks.

Symptoms of Loxosceles Spider Bites Loxosceles bites can be classified in three severity grades:

1. A bite with little venom injected leads to a small erythematous papule associated with a localized urticarial reaction.
2. A cytotoxic reaction from a painless bite that evolutes to blistering, local bleeding and ulcerations after 2–8 hours. Within 3 days a violaceous necrosis develops, surrounded by ischemic blanching of the skin with an outer erythema (blue, white, red: French flag). The central blister becomes necrotic within 3–4 days with eschar formation between 5 and 7 days. After several weeks the ulceration starts to heal by secondary intention.
3. A systemic manifestation with fever, chills, general edema, vomiting, arthralgia, petechial bleeding, rhabdomyolysis, DIC and renal failure.

Treatment of Loxoceles Bites Treatment is according to severity grading. Grade 1 doesn't need treatment, apart from tetanus vaccination if necessary, and observation for several days. Grade 2 needs general wound care with immobilisation of the extremity that is involved. Stimulation of natural healing by wound debridement and delayed primary wound closure is a useful surgical measure.

Antibiotics should be used to treat cutaneous or systemic infections.

For systemic signs and symptoms (Grade 3) there is an antivenin available but no clinical studies about its side effects or effectiveness are available.

6.13.3 Plant Toxins

Introduction

Plant chemistry is very variable and one plant can contain multiple compounds that act and interact together. This makes a grouping of the ingredients of plants necessary to get an idea of the underlying chemistry and action.

The occurrence of plant poisoning varies from country to country. In Europe there are many cases involving children who eat berries or other parts of a plant in summer leading to diarrhea or vomiting; whereas in Sri Lanka there are deadly plant poisonings from yellow oleander occurring when people commit suicide. A classification of plant toxins into five groups allows differentiation among the most toxic substances.

> **The major plant toxins are alkaloids, glycosides, phenols and phenylpropanoids, lectins, and terpenes.**

Classification of Plant Toxins

Alkaloids Alkaloids are bases containing at least one nitrogen and their structure is heterocyclic. The best-known alkaloids belong to the *tropane* group and consist of atropine, hyoscyamine, scopolamine, apoatropine and mandragonine (belladonna alkaloids). Most illegal drugs stem from alkaloids, e.g. codeine, morphine, cocaine, psilocibine and LSD. Some natural stimulants, like nicotine and caffeine (theine), are also alkaloids. The most severe intoxications with alkaloids happen with aconitine, colchicine and strychnine.

Glycosides Glycosides consist of a sugar or sugar derivative (glycone) and a non-sugar moiety called aglycone. The aglycone allows further classification into *saponin glycosides*, *cyanogenic glycosides*, *anthraquinone glycosides, atractylosides* and *salicines*. The best-known glycosides are the cardioactive *saponins*, such as digoxin and digitoxin. Intoxication with naturally occurring plant toxins of this group are poisonings with oleander (nerium oleander), lily of the valley (convallaria majalis) and squill (urginea maritima). Not all saponins glycosides are cardiotoxic. There are, for example, saponin glycosides in holly berries (*Ilex*), which induce gastrointestinal symptoms, and glycyrrhizin in licorice (*Glycyrrhiza glabra*), which inhibits 11 β-hydroxysteroide dehydrogenase producing pseudohyperaldosteronism. *Cyanogenic glycosides* are amygdalin, cyacosin, linamorin and sambunigrin, which can under certain circumstances lead to acute or, more likely, chronic cyanide poisoning. *Anthraquinone glycosides* are laxatives. *Atractyloside* is hepatoxic and *salicin* is hydrolyzed to salicylic acid.

Phenols and Phenylpropanoids Phenylpropanoids have a phenyl ring with a propane side chain. There is no nitrogen in the molecule. This group includes the *coumarins* (with a lactone side chain), the *flavonoids* (built around a flavan 2,3-dihydro-2 phenylbenzopyran nucleus), the *lignans* (linked double phenylpropanoids like podophyllin), the *lignins* (polymers of lignans) and *tanins* (with phenylhydroxylgroups that can condense with proteins).

Lectins Lectins are glycoproteins with different carbohydrate ligands, e.g. galactosamines. The most toxic lectins are the *toxalbumines*, e.g. ricin and abrin. Ricin has an A chain and a B chain with 267 and 262 amino acids, respectivel,y linked by a disulfide bridge. The sugars are lactose, galactose and N-acetylgalactosamine.

Terpenes Terpenes are plant constituents of the essential oils. The basic structure of terpenes is the hydrocarbon isoprene ($CH_2=CH-[CH_3]-CH=CH_2$). The number of the isoprene units is used for the classification into monoterpene (two isoprenes), sesquiterpene (three isoprenes) diterpene (four isoprenes), triterpene (six isoprenes) and tetraterpene (eight isoprenes). There may be many functional groups attached, such as alcohols, phenols, ketones and esters (terpenoids). Toxic terpenes and terpenoids are ginkgolides, kava lactones, thujone, anisatin, ptaquiloside and gossypol.

Alkaloids

Belladonna Alkaloids

> **Atropine, the best known representative of these alkaloids, is a competitive muscarinic receptor antagonist. It competes with acetylcholine and other muscarinic agonists for a common binding site on the muscarinic receptor.**

Belladonna alkaloids show a strong anticholinergic effect by competitive inhibition at the muscarin receptor. The receptor is activated by acetylcholine or blocked by atropine. There are five different muscarinic receptors located in different places in the body: M_1–M_5. All known muscarinic receptors belong to the family of G protein-coupled receptors. M_1 receptors are found in the cerebral cortex, the hippocampus and in ganglia, in the gastrointestinal tract and in parietal cells of the stomach. M_2 receptors are present in the conducting tissue of the heart and in presynaptic terminals. M_3 receptors exist in exocrine glands like lacrimal and sweat glands and in smooth muscles, including the endothelium of vessels. They are also seen in lymphocytes. M_4 receptors dwell in the cerebral cortex, amygdala and hypocampus. Outside the brain they are in the keratinocytes of the skin. M_5 receptors can be detected in keratinocytes and lymphocytes.

Clinical Signs of Poisoning by Belladonna Alkaloids The anticholinergic syndrome caused by belladonna alkaloids includes peripheral and central effects. Peripheral muscarinic receptors differ in their sensitivity to antagonists. First, excretoric glands are antagonized, decreasing salivary and bronchial secretion and sweating. This is followed by mydriasis, loss of accommodation and tachycardia. Finally, urinary sphincter, intestinal and gastric motion are influenced, leading to atony of these organs.

Atropine causes peripheral anticholinergic signs at lower doses; at higher doses central anticholinergic effects are added. They start with stupor, followed by delirium with restlessness and hallucinations. These hallucinations are usually visual, less often auditory. Seizures are common signs of anticholinergic poisoning. The inhibition of sweating, which reduces the ability to dissipate heat, leads to hyperthermia. Hyperthermia together with increased muscle activity can create rhabdomyolysis with kidney failure.

A very high dose of atropine can result in immediate respiratory failure with a deep coma and shock, death may ensue within a short time.

The cardiac effects of belladonna alkaloids may consist of abnormal conduction, such as bundle-branch block and atrioventricular dissociation besides sinus tachyarrhythmia. Rarely ventricular fibrillation can be induced. Treatment includes sedation with benzodiazepines and the use of physostigmine. In severe poisoning deep sedation has to be performed, which makes mechanical ventilation mandatory.

Aconitines Aconitines include three (19 diterpenoid ester) alkaloids: acontine, mesaconitine and hypoaconitine. These alkaloids influence the sodium channel in the conductive tissue of the heart as well as in central and peripheral nerves. The sodium influx through these channels is increased, increasing inotropy and at the same time delaying repolarization and promoting premature excitation.

Signs of cardiotoxicity are bradycardia and later on ventricular dysrhythmia. Symptoms can occur 5 minutes to hours after ingestion. Early signs of peripheral neurotoxicity are paresthesias around the mouth and of the oral mucosa followed in severe cases by progressive skeletal muscle weakness. Vegetative symptoms consist of nausea, vomiting, diarrhea and hypersalivation. CNS involvement is manifested by seizures. There is no real antidote for aconitine. Atropine may be useful to treat bradycardia and hypersalivation. For ventricular dysrhythmias sodium channel blockade by Lidocaine, flecainide or, probably safest, amiodarone is recommended.

Strychnine Strychnine alkaloid is found in vomit button (*Strychnos nux-vomica*). It was isolated in 1818 and is a colourless crystaline powder with an extremely bitter taste.

In the past, strychnine was used in many medical preparations as a digestive stimulant, as an analeptic and as an antidote for barbiturate overdose. It caused many fatal poisonings in the 1920s and 1930s, and has become a very rare cause of intoxication in the present western world. It is still a common poisoning in Asia, occurring from unintentional overdose of a Chinese herbal medicine.

The lethal dose of strychnine is around 50–100 mg but much higher doses have been survived with proper treatment.

Strychnine is well absorbed. Protein binding is minimal. It is quickly distributed to peripheral tissue and has a volume of distribution of about 13 l/kg. Strychnine is metabolized by hepatic P450 microsomes. The main metabolite is strychnine-N-oxide. The half-life amounts to 10–16 hours.

Mechanism of Toxicity of Strychnine

Strychnine competitively inhibits the binding of glycine to the α-subunit of a glycinergic chloride channel. Although glycine receptors are found in the whole nervous system, the predominant target of strychnine is the spinal cord.

The inhibition of the glycinergic receptor by strychnine interrupts the control of the neurons in the ventral horn of the spinal cord on muscle tone. The result is an increase in impulse transmission to the muscles manifested by generalized muscular contractions.

This is a similar action to that seen in tetanus, where a blockade of glycine release is due to the toxin of *Clostridium tetani* (tetanospasmin).

Clinical Signs and Symptoms of Strychnine Poisoning   After ingestion of strychnine, symptoms commerce within 1 hour. The characteristic signs of poisoning are generalized involuntary muscle contractions. Each period of musculature contraction lasts for 1–2 minutes. Due to different strengths of various muscle groups flexion and extension are asymmetric, leading to opisthotonus, flexion of the upper limbs and extension of the lower limbs. Other symptoms are hyper-reflexia with convulsions, nystagmus and hyperthermia up to 43°C as consequence of muscular activity. Loss of consciousness can follow due to severe metabolic acidosis as lactate is produced in surplus by muscle contraction.

Late fatality is a consequence of multi-organ failure induced by hyperpyrexia plus hypotension.

Treatment of Strychnine Poisoning   The primary aim of treatment is to stop the muscular hyper-reactivity preventing metabolic acidosis and respiratory impairment by administration of benzodiazepines. Hyperthermia requires aggressive cooling with ice water, cooling blankets and fanning. In severe cases hemodialysis is the treatment of choice as it resolves metabolic acidosis and allows rigorous cooling.

Colchicine   The name colchicine can be traced back to the king of Colchis, a kingdom in Asia Minor. His imprisoned daughter Medea killed her children to take revenge. Colchicine is found mainly in two plants, *Colchicum autumnale* (autumn crocus) and *Gloriosa superba* (glory lily). Colchicine is readily absorbed in the jejunum and ileum. Due to a strong first-pass metabolism the bioavailability is about 25%.

Plasma elimination half-lives follow a two-compartment model with a distribution half-life between 90 and 100 minutes and a delayed terminal elimination half-life of 2–30 hours.

Colchicine can be found in leucocytes or urine even 10 days post exposure. Protein binding is circa 50%. Colchicine is found in blood cells in concentrations 5–10 times higher than in plasma. Peak plasma concentration occurs between 1 and 3 hours after ingestion. Toxic effects are only found if plasma concentrations are above 3 μg/l.

Pathophysiology of Colchicine Poisoning

Colchicine inhibits microtubulin formation by binding to tubulin, resulting in its conformational change, which affects microtubule growth. The impaired formation of the

> microtubule spindle in metaphase leads to cellular dysfunction and cell death. Colchicine competitively inhibits the γ-aminobutyric acid A (GABA$_A$) receptors in the brain and impairs leukocyte mobility and adhesiveness.

The best-known pathomechanism is the inhibition of microtubule formation, which is responsible for cellular mitosis. Microtubules also take part in the maintenance of cellular structures and transport within cells. Colchicine binds to subunits of the tubules and alters tubulin secondary structure. These conformational changes in the tubules prevent binding of the next tubule subunit and thereby interrupt microtubule growth. These alterations result in a lack of conformation of the microtubule spindles in the metaphase of mitosis, leading to cellular dysfunction and cell death.

The effects of colchicines are dose dependent and reversible. High doses disintegrate microtubules, low concentrations prevent new formation of microtubules. Colchicine also inhibits intracellular granule transport, which is mediated by microtubules. As shown from animal experiments colchicine might also be able to inhibit DNA synthesis.

Colchicine can furthermore influence the γ-aminobutyric acid A (GABA$_A$) receptors in the brain by competitive inhibition, which might be partially responsible for the neurological symptoms seen in colchicine poisoning.

As colchicine accumulates in leucocytes it exhibits an inhibitory effect on leukocyte mobility and adhesiveness. It reduces expression of adhesion molecules on endothelial cells and inhibits cytokine production in polymorphonuclear leucocytes. These latter mechanisms are at therapeutic doses responsible for the anti-inflammatory action of colchicine in the treatment of gout or familial Mediterranean fever (FMF).

Symptoms of Colchicine Poisoning   There are three phases of colchicine poisoning. Within the hours following ingestion there is gastrointestinal disturbance, within a week organ dysfunction follows, with possible death from cardiac, lung failure or septicemia due to bone marrow suppression. Myopathy, neuropathy and a combined myoneuropathy can result in respiratory insufficiency. During recovery sequelae like alopecia and myoneuropathy are seen after a week to several months.

Treatment of Colchicine Poisoning   Besides cardiovascular monitoring, treatment with catecholamines is indicated if lowered blood pressure is not responsive to fluids. In such a case multi-organ failure is imminent and early endotracheal intubation and artificial respiration become necessary. In cases of very likely infection a calculated treatment with antibiotics is warranted. Hemodialysis is needed if kidney failure occurs.

Glycosides

Cardioactive Steroids   The best-known cardiac glycosides are digoxin and digitoxin because they were used for centuries to treat cardiac insufficiency. Digitoxin stems from the purple foxglove (*Digitalis pupurea*) whereas digoxin and digitoxin are found in the Grecian and woolly foxglove. Similar digitaloides are contained in oleander (*Nerium oleander*) as oleandrin, in yellow oleander (*Thevetia peruviana*) as thevetin A plus B, in lily of the valley (*Convallaria majalis*) as convallotoxin, convallarin and convallananin, in squill (*Urginea maritima*) as scillaren and scillarenin, and finally in Christmas rose (*Helleborus niger*) as helleborin.

The most toxic of them is yellow oleander: four seeds can kill an adult. The next most toxic is oleander, with high morbidity in children; again the seeds have the highest concentration of glycosides.

The leaves and flowers of lily of the valley have a low glycoside concentration. The red fruits taste very bitter and the seeds with the highest toxin concentrations are too hard to be bitten open. Poisoning with foxglove is very rare, probably due to the bitter taste of these plants. Even after the ingestion of six leaves in a suicide attempt only a mild intoxication was seen. There are no known severe cases of Christmas rose poisonings.

Pathophysiology of Cardiac Glycosides

> **The toxicity of cardiac glycosides is due to the same mechanism as their therapeutic action. They bind to specific subunits of the sodium–potassium ATPase in all sorts of muscles, resulting there in inhibition of the Na^+–K^+ pump. Potassium is no longer pumped into the muscle cells and in severe poisoning hyperkaliemia can be seen. Since sodium and calcium are exchanged more calcium is taken up intracellularily, resulting in increased contractility. Excessive increase in intracellular calcium concentration results in elevation of the resting potential, producing myocardial sensitization and predisposing to dysrhythmia. Depolarization is delayed.**
>
> **Within the CNS cardiac glycosides interact with the vegetative nerve system. At therapeutic levels the vagal tone is increased, at toxic levels the sympathetic system prevails, which can cause an increase in automaticity and life-threatening dysrhythmias.**

Symptoms of Cardiac Glycoside Poisoning   The non-cardiac manifestation of cardiac glycoside poisoning includes gastrointestinal, neurological and visual symptoms. The cardiac toxicity is manifested by conduction abnormalities and dysrhythmia. The disturbed conduction includes bradycardia, sinuatrial block and I–III degrees of atrio-ventricular block. Patients usually die in ventricular fibrillation not responsive to defibrillation.

Laboratory Findings in Cardiac Glycoside Poisoning   Serum digitoxin and digoxin glycoside concentration can be measured by different immunoassays. If cardiac glycosides from plants are ingested the immunoassay gives positive results due to a cross-reactivity within both immunoassays. No non-toxic or toxic levels are defined for these glycosides. Any measurable value after a plant ingestion means that glycosides were within the plant. Therapeutic levels of digoxin or digitoxin found after such plant ingestion indicate a severe intoxication by these naturally occurring cardiac glycosides.

Treatment of Cardiac Glycoside Poisoning   For gastrointestinal decontamination a single dose of 10 g of activated charcoal can be administered within 2 hours after ingestion. Gastric lavage is not advised as it can trigger dysrhythmias.

Since digoxin-specific Fab fragments have become available, all previous treatments such as the administration of potassium, phenytoin, lidocaine or an external pacemaker, are no longer necessary.

For I–II degree AV-block atropine can still be used. Whereas for digitoxin or digoxin poisoning the dosage of digitalis antidote is well defined if the amount taken in or the serum levels are known, the dosing for plant digitaloides intoxications has to be pragmatic.

Cyanogenic Glycosides   Cyanogenic glycosides are found in about 2500 plants. The best known is amygdalin in bitter almonds (*Prunus* spp.) but they are also found in the seed kernels of apricots, peaches, pears, apples and plums. Other examples are prunosin, limarin, dhurin, sambunigrin and cyacosin. The chemical structure of these compounds consists of mono- or disaccarides with a benzaldehyde-cyanhydrin attached. On digestion the saccharides are split of by β-glucosidase and the benzaldehyde is freed. Cyanide is produced by further hydrolysis.

Eating the fruits or leaves can lead to cyanide poisoning. Treatment is very seldom needed as the natural detoxification is faster than the liberation of cyanide. Treatment of cyanide poisoning with sodium thiosulfate or hydroxocobalamin is theoretically possible (for treatment of cyanide poisoning see Chapter 6.8). A chronic sort of poisoning by cyanogenic glycosides is quite common in East Africa by the vegetable cassava (*Manihot esculenta*), which contains limarin. The disease caused by the ingestion of cassava is called konzo. The manihot is usually dried by oven or sun exposure. Mass poisonings can happen when, due to food shortages, there is insufficient drying of this traditional food. The disorder shows in a symmetric spastic paraparesis and hypothyroidism with the development of a goitre.

Phenols and Phenylpropanoids

From this group of plant toxins only podophyllin is discussed here.

Podophyllin or podophyllotoxin is a phenylpropanoid, which is the main toxic compound in the rhizome and roots of mayappel (*Podophyllum pelatum*) and wild mandrake (*Podophyllum emodi*). Podophyllin in purified form is still used as ointment for the treatment of anogenital warts. It is also used to treat oral hairy leucoplakia.

Pathophysiology of Podophyllin Poisoning

> **Podophyllin, similar to colchicines, binds reversibly to tubulin at the same site as colchicine. This leads to mitotic arrest. Since the mitotic spinals cannot be formed without microtubles the metaphase comes to a halt and the chromosomes are clumped. Potophyllotoxin inhibits nucleoside transport into the cell nucleus. Disruption of the microtubles hampers all sorts of intracellular transport mechanisms.**

Signs and Symptoms Intoxications are reported after ingestion of mandrake root, herbal remedies containing podophyllin and from topical application of podophyllin. The symptoms start within several hours with nausea, vomiting, abdominal pain and diarrhea.

Compared to colchicine poisoning podophyllin exhibits more CNS symptomatology and less delayed hematologic toxicity.

Patients rapidly develop confusion and even coma. Auditory and visual hallucinations are found early in the course of the poisoning. Paresthesia, loss of tendon reflexes and pyramidal tract signs can be seen. The cranial nerves are included in the neuropathology with diplopia, nystagmus, dysconjugate gaze and facial nerve paralysis.

As in colchicine poisoning a peripheral sensorimotoric axonopathy with slow, or no, recovery can develop. After an initial leucocytosis a pancytopenia with its nadir around day 7 can develop. As a complication leucopenia infection can develop. Treatment is the same as described for colchicine poisoning.

Lectins

Lectins are glycoproteins. The most important are the toxalbumins such as ricin and abrin. They are so toxic that they even might be used as biological weapons. Abrin is found in prayer beans (*Abrus precatorius*) and ricin in castor beans (*Ricinus communis*). Further plants containing toxalbumines are the black locust tree (*Robinia pseudacacia*) and the European mistletoe (*Visum album*).

Abrin and Ricin

Pathophysiology of Toxalbumin Poisoning Ricin and abrin are the two main toxalbumines. They consist of an A chain and a B chain bound by a disulfide linkage. The molecular weights are about 65,000 Da. Cleaving the disulfide bond eliminates toxicity. The B chain binds to a galactose-containing receptor in the cell wall and both chains are internalized by the cells thereafter.

> **Toxalbumins disrupt DNA synthesis, and thereby protein synthesis, at the 60 S ribosomal subunit. An agglutinin in the *R. communis* plant can stimulate hemagglutination and intravascular haemolysis by binding to glycoproteins on red blood cell surfaces. The hemolysis can lead to kidney failure. Abrin and ricin can cause multi-organ failure. They can damage the endothelium of the alveolar space, leading to pulmonary edema. Liver failure can result from direct hepatotoxicity. Direct toxin contact with gastrointestinal mucosa results in ulceration, fluid loss and bleeding. Damage to all parenchymatous organs were seen, for instance the pancreas, spleen, liver and kidney.**

Symptoms of Toxalbumin Poisoning Oral ingestion of abrin or ricin creates a biphasic toxidrome. In the first stage patients, usually 2 hours after ingestion, have nausea, vomiting, and diarrhea with blood in faeces. Corrosive burns can be found in the throat and stomach. Thereafter more systemic symptoms develop, like circulatory problems with signs of hypovolemic shock. Multi-organ involvement follows, partially through direct toxic effect on the different organs and partially due to the shock itself. Liver, kidney, heart and lungs are involved. Cardiac dysrhythmia and epileptic convulsions can be seen. The second phase develops over 3–4 days.

Treatment of Toxalbumin Poisoning Early administration of activated charcoal may be helpful. When diarrhea has started fluid resuscitation has to be commenced with infusion of electrolytes and fluid. Daily, or even twice daily, laboratory testing for liver enzymes BUN, creatinine, clotting parameters and blood count are necessary. If after volume replacement the blood pressure stays low, catecholamine infusion has to be given for circulatory support. Pulmonary edema and prolonged shock call for intubation and ventilation. Massive hemolysis requires transfusion and alkalinization to keep up urinary output. If these measures are not successful, hemodialysis may be required.

Terpens and Terpenoids

This group of plant toxins, though structurally similar, have quite different effects. Ginkgloides from *Ginkgo biloba* have an antiplatelet aggregation effect. Kava lactones found in kava kava (*Piper methysticum*) cause CNS effects and may be hepatotoxic.

Ptaquilosides found in bracken fern (*Pteridium aqulinum*) lead to hemorrhage following a severe thrombocytopenia.

Kava lactones, thujone and antisatin (*Illicum* spp.) are terpens or terpenoids that have an effect at $GABA_A$ and $GABA_B$ receptors. They are all associated with seizures. Thujone – the main ingredient in absinthe, an extract from wormwood (*Artemisia absinthium*) – creates CNS excitation followed by depression and seizure. Thujone has a similar structure to tetrahydrocannabinol (THC), which might explain its psychoactive effects. Chronic intake of thujone produced a clinical picture of absinthism with hallucinations, cognitive impairment and personality change.

Thujone has α and β isomers. The toxic one is the α-stereoisomer, which antagonizes the $GABA_A$ receptor at the picrotoxin site of the chloride channel, leading to neuroexcitation and seizure.

Treatment of Terpens and Terpenoids Poisoning The treatment is symptomatic. In case of hemorrhage blood transfusion or platelet transfusion has to be performed. For hepatic failure in kava kava poisoning liver transplantation has to be undertaken in some cases. The acute CNS effects such as excitation, hallucination and seizures can be treated with benzodiazepines dosed as necessary.

6.13.4 Summary

Animal Toxins

Snake bites, spider bites and scorpion stings can cause severe morbidity and even mortality in humans. Four families of snakes are responsible: Viperidae, Elapidae, Colubridae and Atractaspididae. The most toxic snakes belong to the Viperidae, especially the subfamily of Crotalinae (rattlesnakes) and the Elapidae.

All these snake families produce neurotoxins, myotoxins, hematotoxins and necrotoxins. Most elapid bites exhibit neurotoxicity, most crotaline bites show coagulopathy. Most snake toxins can produce mild to severe tissue damage and systemic symptoms. The neurotoxins act presynapticly and postsynapticly. The myotoxins destroy muscle cells, sparing the basement membrane. The hematotoxins induce coagulopathy, thrombocytopenia and hemolysis. Systemic symptoms are gastrointestinal and cardiovascular. Antivenom therapy is available but well-tolerated antivenoms are not on the market for all poisonous snakes.

Scorpions are neurotoxic and induce little local tissue damage. Their toxins stimulate acetylcholine release, leading to cholinergic crisis. The most toxic spiders are the Australian funnel-web spiders, the widow spiders and the recluse spiders. The funnel-web spider's venom leads to a catecholamine storm, and the widow spiders cause a massive release of different neurotransmitters like norepinephrine, acetylcholine and glutamine. They induce pain that can generalize, sweating and piloerection, dysphoria and restlessness. Recluse spiders are necrotizing and lead to extended skin necroses. They can create disseminated intravascular clotting and hemolysis. For some scorpions and spiders antivenoms are available but their effectiveness is not well studied.

Plant Toxins

Plant toxins can be classified by either their chemistry or their action. The toxic ingredients belong to five different chemical groups: alkaloids, glycosides, phenols, lectins and terpenes. Their actions can include peripheral and central nervous system, cardiovascular or gastrointestinal toxicity. The most important toxic alkaloids are atropine, aconitine, colchicine and strychnine.

Atropine shows a strong anticholinergic effect by competitive inhibition at most muscarinic receptors.

Aconitine enhances the influx of sodium at the sodium channel in peripheral, central and conductive nerve tissue.

Strychnine is a competitive inhibitor of glycine.

Colchicine interrupts the formation of microtubulin, blocking mitosis in the metaphase.

The relevant toxic glycosides are digitalis or digitaloids, and cyanogenic glycosides.

Digitalis and digitaloids are cardiotoxic by binding to the sodium potassium ATPase, especially at the myocardium.

The cyanogenic glycosides can interact with cytochrome a_3 in the respiratory chain, blocking oxygen consumption. Chronic poisoning leads to a peripheral neuropathy.

Podophyllin is a toxic phenylpropanoid leading to mitotic arrest.

The most toxic lectins are ricin and abrin. They disrupt protein synthesis at the 60 S ribosomal subunit.

Terpens interfere with the $GABA_A$ receptor, inducing excitation and CNS depression.

Further Reading

Abroug F, ElAfrous S, Nouria S et al. Serotherapy in scorpion envenomation. A randomised controlled trial. *Lancet* 1999; **354**: 906–909.

Brent J, Wallace KL, Burkhart KK, Phillips SD, Donovan JW Eds. *Critical Care Toxicology*, Philadelphia: Elsevier Mosby, 2000.

Chang LW, Dyer RS, Eds. *Handbook of Neurotoxicology*, New York: Marcel Dekker Inc., 1995.

Escouba P, Diochot S, Corzo G. Structure and pharmacology of spider venom neurotoxins, *Biochimie* 2000; **82**: 897–907.

Flomenbaum NE, Goldfrank LR, Hofmann RS, Howland MA, Lewin NA, Nelson LS Eds. *Goldfrank's Toxicologic Emergencies*, New York: Mc Graw-Hill, 2006.

Ford MD, Delaney KA, Ling LJ, Erickson T, Eds. *Clinical Toxicology*, Philadelphia: WB Saunders Company, 2001.

Gold B S, Barish RA. Venomous snakebites: Current concepts in diagnosis, treatment and management. *Emerg Med Clin North Am* 1992, **10**: 249–267.

Jyaniwura TT. Snake venom constituents: Biochemistry and toxicology, Parts I + II. *Vet. Hum Toxicol* 1991: 468–480.

Winkel KD, Hawdon GM, Levick N. Pressure immobilization for neurotoxic snake bites. *Ann Emerg Med* 1999; **34**: 294–295.

Glossary of Important Terms in Toxicology

Absorption, distribution, metabolism and elimination (ADME) Determination of the fate of a compound in an organism.

Abundance (ecology) Number of organisms, relative to an area or room unit.

Acceptable daily intake (ADI) The dose of a pollutant that does not lead to health problems with lifelong daily intake. The ADI value is calculated by dividing the NO(A)EL by a safety factor; synonymous with TDI. *See also* no observed (adverse) effect level, tolerable daily intake.

Acute toxicity test Single-dose toxicity test at an observation time of up to 14 days after exposure. *See also* lethal dose.

Ames test *See* Salmonella mutagenicity test.

Aneugenic Inducing an abnormal number of chromosomes (numerical chromosome aberrations).

Antidote Substance that can counteract a form of poisoning.

Apoptosis Programmed cell death; a controlled, energy (ATP)-dependent process.

Application Supply of substances into the organism by one of these methods:
  p.o. per os (via the gastrointestinal tract)
  s.c. subcutaneously (under the skin)
  i.m. intramuscularly (into the skeletal muscle)
  i.v. intravenously (into the bloodstream)
  i.p. intraperitoneally (into the abdominal cavity)

dermal via the skin.

Bioaccumulation Accumulation of substances from the environment in humans, animals, and plants via water, air, soil or food.

Bioactivation Enzymatic conversion of a foreign substance into a toxic or carcinogenic metabolite.

Bioavailability The amount of substance delivered to the organism.

Biocenosis Cohabitation of the plant and animal organisms living in a habitat, which interact with each other and with their abiotic environment.

Biological exposure index (BEI) Limit values for biological materials used to protect workers against occupational injuries. BEIs are defined by the American Conference on Governmental Industrial Hygienists (ACGIH).

Biological guidance value (BGV) The upper concentration of a substance or a metabolite in any appropriate biological medium corresponding to a certain percentile (generally 90th or 95th percentile) in a defined reference population. If background levels cannot be detected, the BGV may be equivalent to the detection limit of the biomonitoring method, which is specified in the document. Defined by the Scientific Committee on Occupational Exposure Limits (SCOEL).

Biological limit value (BLV) Health-based exposure concentration that generally does not affect the health of an employee adversely. Defined by the Scientific Committee on Occupational Exposure Limits (SCOEL).

Biological tolerance (BAT) value The maximum permissible quantity of a substance or its metabolites in the human body or a biological indicator at which no health effects are considered to occur. BAT values are defined by the MAK Commission of the Deutsche Forschungsgemeinschaft (DFG, German Research Foundation).

Biomonitoring (biological monitoring) Measurement of the internal body burden of foreign substances in individual persons (internal individual exposure control).

Biotope Abiotic, i.e. a habitat characterized by physical and chemical properties in the ecosystem.

Biotransformation Enzymatic conversion of foreign substances in the organism.

Cancer risk factors Substances and substance mixtures that are involved in the development of cancer in humans or in carcinogenicity tests. *See also* carcinogenicity test.

Carcinogen Substance that leads to the formation of tumors in animals or humans.

Carcinogenesis Process of uncontrolled growth of tissues (neoplasia) as a result of exposure to carcinogenic substances. The phases of initiation, promotion and progression can be differentiated. *See also* carcinogenic, initiation, promotion, progression.

Carcinogenic Inducing cancer.

Carcinogenicity test Chronic toxicity test *in vivo* that is designed to detect the carcinogenic effects of substances. The test substance is administered to experimental animals five times a week for 18–24 months and the occurrence of tumors is examined.

Chronic toxicity test Repeated administration of a test substance with a test duration of more than 3 months.

Clastogenic Inducing chromosome breakages (structural chromosome aberrations).

Clearance Elimination of a substance from the organism or removal of particles from the lower respiratory tract.

Cocarcinogens Substances that are effective in the first phase of carcinogenesis favoring the development of DNA damage in initiated cells without being carcinogenic itself. *See also* carcinogenesis.

Cohort A group of people characterized by a common feature (occupation, place of residence, exposure).

Combination effect Enhancement or weakening of an effect that can occur if several simultaneously present chemicals have comparable mechanisms of action and thus act on certain organs or biochemical functions of the organism.

Complete carcinogen Substance that causes tumors in animal experiments with chronic administration without additional provocation by a tumor-promoting agent. According to the multistage concept of carcinogenesis, complete carcinogens have both initiating and promoting properties. *See also* carcinogenesis.

Confidence interval Variability of an estimate such as odds ratio, relative risk, or standardized mortality risk.
Confounder Disruptive factor that can influence the result of, for example, an epidemiological investigation.
Destruents (ecology) Organisms that feed on the remains of consumers and producers (dead organic substances).
Dose Amount of a substance administered to an organism, usually given in mg/kg body weight.
Draize test Toxicity test *in vivo* for the detection of skin and mucous membrane irritation and corrosive effects of chemicals.
Effective concentration (EC_{50}) Concentration of an active substance causing the expected effect in 50% of exposed individuals.
Effective dose (ED_{50}) Dose of an active substance causing the expected effect in 50% of exposed individuals.
Embryo Designation for the developing organism in its mother's womb until the completion of organ formation; in humans approximately until the third month of pregnancy, in rats until days 15–16. *See also* fetus.
Embryotoxicity Damage to the embryo by chemicals leading to death, developmental delay, disturbance of organ function or malformations. *See also* fetotoxicity.
Endocytosis Incorporation of material into the cell by invagination and subsequent constriction of the cell wall.
Endogenous Caused by factors inside or produced within the organism.
Environmental hygiene Science of the research of environmental impacts on the population or on population groups and detection and elimination of unfavorable factors and the use of beneficial factors.
Environmental medicine In contrast to environmental hygiene, this concerns the individual in terms of clinical medicine. Where environmental damage is suspected, an anamnesis (survey) of the load conditions is performed and the necessary therapeutic measures are initiated.
Environmental toxicology Branch of toxicology that identifies and quantifies health risks from environmental chemicals.
Epidemiology Medical discipline that investigates the incidence of diseases and the physiological variables and social consequences of disease in human populations as well as the factors that influence them.
European Chemicals Agency (ECHA) The ECHA is the driving force among regulatory authorities in implementing EU chemicals legislation for the benefit of human health and the environment as well as for innovation and competitiveness. The ECHA helps companies to comply with the legislation, advances the safe use of chemicals, provides information on chemicals, and addresses chemicals of concern. The ECHA is located in Helsinki, Finland.
Existing chemicals Substances that have been notified by manufacturers in the European Community are listed in the European Inventory of Existing Commercial Chemical Substances (EINECS). The EINECS was drawn up by the European Commission in the application of Article 13 of Directive 67/548/EEC, as amended by Directive 79/831/EEC, and in accordance with the detailed provisions of Commission Decision 81/437/EEC.
Exogenous Caused by factors from outside the organism.
Extrathoracic region Upper respiratory tract, nose, mouth, throat, pharynx, and larynx.
Fetotoxicity Harmful effect on the fetus. *See also* embryotoxicity.
Fetus Designation for the developing organism in its mother's womb after the completion of organ formation; in humans approximately from the third month of pregnancy, in rats from days 15–16 to birth at days 21–22. *See also* embryo.

First-pass effect When substances are absorbed from the digestive tract and transported to the liver via the portal vein, where they are metabolized during the first passage through the liver.

Food chain Relation between food producers and food consumers (e.g., fodder plant – herbivore – carnivore). Important for toxicology because persistent lipophilic pollutants may accumulate in the food chain. *See also* bioaccumulation.

Function (ecology) Processes that take place in ecosystems, such as energy flow and material cycles.

Genotoxicity General term for toxic effects on the genetic material of cells (e.g., DNA damage and damage of the mitotic apparatus).

Globally Harmonized System (GHS) System of classification, labeling, and packaging of chemicals.

Guide(line) values (reference values) Values for concentrations of pollutants in soil, air, water, and food that are derived from comparative measurements in burdened and unloaded media, and that can provide orientation for interpretation of a measured value.

Half-life, biological Period of time after administration of a substance within which half of the original dose is eliminated by degradation or excretion.

Health damage Any temporary or permanent adverse effects caused by chemicals, radiation, accidents or lifestyle.

Health risk Probability of the occurrence of a particular health impairment in a population exposed to a harmful factor. The health risk depends on the level and duration of exposure to a pollutant and its potency.

Homeostasis Maintaining the balance of a physiological condition.

Hyperplasia Enlargement of a tissue by increasing the number of cells.

Hypertrophy Enlargement of a tissue by increasing the cell volume.

Hypoxia Deficiency of the oxygen supply of a tissue or organism.

Immunosuppression Weakening or suppression of the immune system, which may result in impaired immune response to infections.

Immunotoxicity Harmful consequence of the toxic effect of a chemical on the immune system.

Incidence Frequency of a new disease of a defined population during a certain period of time.

Initiation Phase of damage to DNA that can lead to mutations and formation of tumors.

Initiator Substance that can cause irreversible damage to DNA and thereby trigger (initiate) the formation of tumors.

Intervention value A value suggested in the context of contamination regulations for residues or contaminants in food up to the respective limit values. If exceeded, appropriate measures should be taken to reduce the entry of the substance into the environment and thus the contamination of the food.

Intoxication Poisoning.

Lethal concentration (LC_{50}) Concentration of an active substance in the surrounding atmosphere, or in water-living organisms in the water, that leads to death in 50% of the exposed individuals.

Lethal dose (LD_{50}) Dose of an active substance that causes death in 50% of the treated animals within 1–2 weeks.

Limit values Concentrations of pollutants in the environment that protect humans and the environment from the harmful effects of chemicals or radiation. Toxicologically justified limit values are, for example, the maximum permitted quantities of plant protection products in food or the maximum allowed workplace concentrations.

Lowest observed (adverse) effect concentration (LO(A)EC) Lowest concentration of a substance at which an (adverse) effect is observed. *See also* no observed (adverse) effect concentration.

Lowest observed (adverse) effect level (LO(A)EL) Lowest dose of a substance at which an (adverse) effect is observed. *See also* no observed (adverse) effect level.

Margin of exposure (MoE) Distance between a point of departure such as the experimentally determined weak toxic effect (LO(A)EL or benchmark dose) and human exposure.

Maximum immission concentration (MIK value) Maximum permissible concentration of pollutant as gas, vapor or suspended matter in the atmosphere. MIK values are defined to protect the health of the population.

Maximum tolerated dose (MTD) The dose in animal experiments that triggers weak toxic effects such as a slight reduction in body weight.

Maximum workplace concentration (MAK value) Maximum permissible concentration of a substance as gas, vapor or suspended matter in the workplace air that does not affect the health of the employee after long-term exposure (8 hours per day, 5 days per week). The MAK values are proposed by the MAK Commission of the Deutsche Forschungsgemeinschaft (DFG; German Research Foundation).

Mechanism of action (toxicological) Molecular, biochemical or biophysical effect underlying toxicity, such as saturation or inhibition of detoxifying reactions, interaction with receptors, cytotoxicity, formation of reactive metabolites, etc.

Mesofauna, meiofauna (ecology) Very small (0.2–2 mm) animals living predominantly in or on the ground.

Mutagenicity Capacity of a substance to induce mutations.

Mutagenicity testing To clarify the question of whether a substance is mutagenic or not, numerous *in vitro* and *in vivo* test systems are available, e.g. for bacteria, mammalian cells or laboratory animals. *See also* Salmonella mutagenicity test.

Mutation Persistent alteration of the genetic material of a cell, which is transmitted to the daughter cells.

Necrosis Uncontrolled cell death triggered by external factors or disease, in contrast to apoptosis. *See also* apoptosis.

Neoplasia Abnormal growth of tissue, usually synonymous with tumors.

No observed (adverse) effect concentration (NO(A)EC) Highest concentration of a substance at which no (adverse) effect is observed. *See also* lowest observed (adverse) effect concentration.

No observed (adverse) effect level (NO(A)EL) Highest dose of a substance at which no (adverse) effect is observed. *See also* lowest observed (adverse) effect level.

Noxa Something that exerts a harmful effect on the body, e.g. physical (noise, vibration, radiation) or chemical (pollutants); apoptotic regulator.

Overload Overloading of the lungs by a large number of particles, which leads to an impairment of their clearance by uptake into macrophages and their removal via the respiratory tract.

Persistence Resistance of a substance against degradation in the environment or in the organism.

Phase I enzymes Enzymes that insert functional groups into a molecule.

Phase II enzymes Enzymes that couple a water-soluble substrate such as glucuronic acid or glutathione to molecules with functional groups.

Physiology-based pharmacokinetic (PBPK) modeling A mathematical modeling technique for predicting the absorption, distribution, metabolism, and excretion (ADME) of synthetic or natural chemical substances in humans and animal species.

Point of departure (POD) Starting point, e.g. NO(A)EL or LO(A)EL, for a risk assessment.

Prevalence Number of patients with a defined disease relative to a defined population.

Progression Phase of increasing growth autonomy and malignancy in the development of a tumor. *See also* carcinogenesis.

Promoters Substances that accelerate the development of tumors from pre-damaged (initiated) cells, but are themselves unable to trigger tumorigenesis.

Promotion Phase of the proliferation of initiated cells in carcinogenesis. *See also* carcinogenesis.

Pulmonary area Lower respiratory area of the lung that includes bronchioles, alveolar ducts, and alveoli.

Registration, Evaluation, Authorization and Restriction of Chemicals (REACH) EU Regulation from 2007 that addresses the potential impacts of chemical substances on both human health and the environment, and their regulation.

Relative risk (RR) In statistics and epidemiology this is the ratio of the probability of an event occurring, e.g. developing a disease, in an exposed group to the probability of the event occurring in a non-exposed group.

Retention Failure to eliminate a substance from the body, e.g. particles in the respiratory tract that are not removed by clearance. *See also* clearance.

Risk assessment (risk evaluation) Assessment of a risk based on hazardous properties, dose, and exposure.

Risk characterization (risk description) Quantitative or qualitative determination of possible health hazards due to chemicals or radiation as a result of intensity of effect, duration of exposure, and level of exposure or ingested dose.

Safety factor To take into account the fact that humans may be more sensitive than the most sensitive experimental animal species and that differences in sensitivity also exist within the population, the highest dose without effect in animal experiments is usually divided by a safety factor of 100 to set an acceptable exposure limit for the general population, such as ADI. *See also* acceptable daily intake.

Salmonella mutagenicity test Method for testing the genotoxic (mutagenic) efficacy of substances identifying mutations in bacterial cultures (*Salmonella typhimurium*).

Short-term exposure limit (STEL) Short-term exposure at the workplace, which may be achieved a maximum of four times per shift for a maximum of 15 minutes each.

Skin sensitization Reaction of the immune system to the impact of sensitizing agents on the skin. If sensitization has occurred, the immunological reaction at re-exposure will result in skin changes such as redness, edema or scabbing.

Structure (ecotoxicology) Organisms of a biotope.

Subacute (subchronic) toxicity test Toxicity tests with repeated administration of the test substance for 28 (90) days.

Syncancerogenesis Interaction of several carcinogenic substances, in contrast to co-cancerogenesis, in which the action of a carcinogen is intensified by the simultaneous action of a non-carcinogenic substance.

Synergistic effect If chemicals influence their respective effects on the organism, this is called synergism if the total effect of the chemicals is higher than the individual effects (additive or supra-additive).

T25 Tumorigenic dose leading to 25% additional incidence.

Teratogenic substances Substances that cause a disturbance of the germinal development (teratogens). The best-known example of a teratogen is thalidomide (Contergan), which caused severe malformations of internal organs, but especially of the extremities.

Teratogenicity The property of a substance during the development of the germ from the fertilized egg to the embryo causing disorders; in contrast to mutagenic effects that affect the germ cells.

Threshold concentration, threshold value Dose or concentration in the range of the effect threshold, i.e. between the just observable effect (LO(A)EL) and the not observable effect (NO(A)EL). *See also* lowest observed (adverse) effect level, no observed (adverse) effect level.

Threshold limit value (TLV) Occupational exposure limit for the protection of workers from damage to health in the United States. It is defined by the American Conference of Governmental Industrial Hygienists (ACGIH).

Threshold of effect This term is misleading. It is used as a no-effect concentration, although it refers only to the change in the course of the dose–response curve. *See also* no observed (adverse) effect level, lowest observed (adverse) effect level.

Threshold of toxicological concern (TTC) Tolerable concentration of a chemical with a very low likelihood of showing adverse effects.

Tolerable daily intake (TDI) Dose of a pollutant that, based on the current state of knowledge, does not cause any health disturbances during lifelong daily intake. The TDI value is calculated by dividing the NO(A)EL by a safety factor; synonymous with ADI value. *See also* no observed (adverse) effect level, safety factor, acceptable daily intake.

Topoisomerases Enzymes that reduce supercoiling in DNA by breaking and rejoining one or both strands of the DNA molecule prior to DNA replication.

Toxicity Adverse or harmful effects of a substance.

Toxicity categories (acute toxicity) Allocation of substances based on acute toxicity by the oral, dermal or inhalation route according to numeric cut-off criteria (e.g., LD_{50} in mg/kg bodyweight) according to the Globally Harmonized System of Classification, Labelling and Packaging of Chemicals.

Toxicity test Tests in which experimental animals are treated with different doses of a substance and the effects are investigated time- and dose-dependently. They are used to characterize the effect of a substance after short-term or chronic exposure.

Toxicodynamics Changes in the organism under the influence of a pollutant.

Toxicogenetics Branch of toxicology that detects genetically determined individual differences in the effect of pollutants on an organism.

Toxicokinetics Branch of toxicology that quantifies the fate of a pollutant in an organism, i.e. absorption, distribution, metabolism, and elimination (ADME).

Toxicology Teaching of the harmful effects of chemicals on living organisms. It describes the nature of the effects depending on the dose as well as the cellular, biochemical, and molecular mechanisms of action and the kinetics.

Toxin (natural) Toxic substance, e.g. in plants (phytotoxin) or fungi (mycotoxin).

Tracheobronchial area Middle part of the respiratory tract with trachea and bronchial area without respiratory bronchioles.

Trophic levels (ecology; nutritional levels) Interdependent nutrition groups, interconnected by energy flow, forming, for example, the food chain green plants, herbivores, carnivores.

Unit risk Estimated cancer risk resulting from the lifetime exposure to 1 $\mu g/m^3$ of a carcinogenic substance (used by the US Environmental Protection Agency).

Index

abnormal mitotic spindles   180
abrin   758
absorption, distribution, metabolism, and elimination (ADME)   10
acceptable daily intake (ADI)   13, 16, 29
acceptable exposure level (AEL)   342
acceptable risk   529
acetamidofluorene   86
acetaminophen (APAP)   242
aconitines   753
acrylamide (ACR)   277
acrylonitrile   128–130
action potential   273
activated glucuronic acid   74, 75
active substance, pesticides
  molluscicides   715
  regulatory toxicology of   716–717
  rodenticides   715–716
  target organisms   706
  toxicity   707–709
  toxicological effects   717
acute damage   369
acute methanol poisoning   360
acute toxicity
  diesel engine emissions   736, 739
  metals   609
  OECD TGs for *in vivo* tests
    dermal toxicity   393–394
    examinations   394
    inhalation toxicity   394
    limit test   394
    oral toxicity test   392–393
    significance of   392
  PAHs   726

adaptive immune system   332
adenosine 5′-phosphosulfate (APS)   76
adenosine triphosphate (ATP)   271
adverse effects, endocrine active compounds   649–650
adverse outcome pathway (AOP)   26, 469
aerodynamic diameter   628
aflatoxin B_1 (AFB1)   246, 247
Agency for Toxic Substances and Disease Registry (ATSDR)   570–571
AhR nuclear translocator (ARNT)   600
airborne systemic poisons
  carbon monoxide   686–689
  cyanide   689–691
  hydrogen sulfide   691–692
ALARA *see* as low as reasonably achievable (ALARA)
alcohol dehydrogenase (ADH)   71
  alcohol biotransformation   682
  ethanol   681–684
  methanol   680–681
aldehyde dehydrogenase   71–72
aldo-keto reductases (AKR)   72
aliphatic alcohol, solvents
  ethanol
    acute, chronic effects   682–684
    toxicokinetics   681–682
  glycols   684–685
  methanol   680–681
alkaline single-cell gel electrophoresis   417–418
alkaloids toxin
  aconitines   753
  belladonna alkaloids   752
  colchicine   754–755
  strychnine   753–754

Toxicology and Risk Assessment: A Comprehensive Introduction, Second Edition. Edited by Helmut Greim and Robert Snyder.
© 2019 John Wiley & Sons Ltd. Published 2019 by John Wiley & Sons Ltd.

allergic contact dermatitis (ACD)   302–303, 341
α–Latrotoxin (A-LTX)   749
alveolar macrophage (AM)   250, 629
Amanita muscaria   278
American Conference of Governmental Industrial Hygienists (ACGIH)   18, 444–446
2-amino-1-methyl-6-phenylimidazo-[4,5-b]pyridine (PhIP)   134
aminomethylphosphonic acid (AMPA)   707
amyl nitrite   690
anaplerosis   105
androgen receptor (AR)   380, 385
  agonists   657
  antagonists   657
  modes of actions   662
androgens   665
aneuploidy   174
animal testing
  aspirin   467
  ban for cosmetics, in Europe   467–468
  cell cultures   468
  computer-aided methods   468
  consumer safety   466–467
  ECVAM validations   463–465
  experiments   461–462
  IATA   466
  in vitro and *in silico* approaches   469
  OECD test guidelines   463–465
  organo-typic cultures   468
  reduce, refine, replace (3Rs-principle)   463
  toxicological ignorance   468
animal toxins
  scorpions   747–748
  snakes
    bites   743
    first aid   746
    hematotoxins   744–745
    hospital treatment   746
    local damage   745
    myotoxins   744
    neurotoxins   743–744
    symptoms   745
  spiders
    Australian funnel-web   748–749
    latrodectus   749–750
    loxosceles   750–751
antiandrogens *see also* endocrine active compounds (EACs)
  diethylstilbestrol   661
  limited evidence   667
  modes of action
    in vitro   663

  in vivo   664–666
  weight of evidence   666–667
  xenoantiandrogens   661–662
  xenoestrogens   661–662
antidotes   609
antiestrogens   664–665
antivenom therapy   746
apoptosis
  DNA repair   178
  extrinsic pathway of   179
  intrinsic pathway of   178–179
  p53 protein   178
  premature senescence   177
Arking's theory of IGF-1   598
aromatase inhibitors   657–658, 665
aromatic amines   161
arsenic metal   309, 611–613
arteriosclerosis   370
aryl hydrocarbon receptor (AhR)   600–601
as low as reasonably achievable (ALARA)   27, 32, 530
astrocytes   271
ataxia telangiectasia (AT)   174
ataxia-telangiectasia mutated (ATM)   190
Atomic Energy Act   570
Atomic Energy Commission (AEC)   570
ATRad-3 related (ATR) protein   190
ATR-interacting protein (ATRIP)   170, 190
atropine   226
α_{2u}-globuline-mediated effects   316
Australian funnel-web spiders venoms   748–749
autophagy   179
axonopathy   275–277
azole fungicides   710

barbiturates   83, 84
base excision repair (BER) system   165–167, 188–190
Bcl-2 family   178
behavioral neurotoxicology
  exposure assessment   283–284
  neurobehavioral effects in humans
    lead   293–294
    mercury   294–295
    solvents   292–293
  neurobehavioral methods
    batteries   291, 292
    cognitive functions   286–289
    evaluation of   291
    executive functions   286
    personality characteristics   289, 290
    personality or emotional changes   287

psychomotor functions 286–289
  sensory functions 291
  symptom measurements 289, 290
 neuromorphological method 284–285
 neuropathological methods 284–285
 neurophysiological methods 285
 sensory evoked potentials 285
belladonna alkaloids 752
benchmark dose (BMD) 24
benchmark-dose level (BMDL) 653
Benton test 288
β-lyase-mediated carcinogenicity 316
β-N-acetyl-glucosaminidase (NAG) 257
bile salt export pump (BSEP) 240
bioaccumulation potential 215
bioavailability 41
biocidal products, EU legislation 541–542
bioconcentration 215
biological exposure indices (BEI) 444, 445
biological guidance values (BLW) 445
biological limit values (BLV) 445
biological membrane
  active transport 37
  convective transport 37
  facilitated diffusion 37
  passive diffusion 36, 37
  pinocytosis 37
biological origin substances
  microorganisms active compounds 714
  neem tree extracts 713–714
  rotenone 713
  viruses 714
biological tolerance values (BAT) 445
biomagnification 215
biomonitoring
  analytical methods 444
  application of 440
  biochemical effect 440
  biological assessment values
    descriptive values 446–447
    health-based values 445–446
    reference values 446–447
    risk-based values 446
  biological effect 440
  biological material for 441–443
  definition 440
  of effects 5
  of exposure 5
  external exposure 441
  fundamental requirements 442
  health risks 441
  internal exposure 441
  occupational exposure 442
  parameters
    high biological persistence 448
    high vapour pressure substances 447
    Phase I and Phase II products 447–448
    protein and DNA adducts 444
  toxic effects 441
biotransformation of xenobiotics
  barbiturates 83, 84
  bioactivation reactions 87–90
  hepatic first-pass effect 81
  inhibition of biotransforming
    enzymes 84–85
  metabolic enzymes induction 82–84
  Phase II reactions
    acetylation 78
    amino acid conjugation 77–78
    functionalization 62, 63
    glucuronidation 74–76
    glutathione S-transferases 78–80
    methylation 80
    sulfatation 76–77
  Phase I reactions
    ADH 71
    carbonyl-reducing enzymes 72
    cytochrome P450 enzymes 63–69
    epoxid hydrolases 73–74
    esterases and amidases 70–71
    FMO 68–69
    functionalization 62, 63
    prostaglandin synthase 69–70
    reductive reactions 72–73
  reactive intermediates
    detoxifications 90–91
    DNA interactions 92–96
    lipids interactions 92
    proteins interactions 91–92
  species differences in 86–87
  TCDD 62, 83
  toxic effects 87, 88
bipyridylium herbicides 709
bisphenol A (BPA) 656
bladder cancer 130–133, 315, 318–319
blood-brain barrier (BBB) 268–269
bloom syndrome 174
Botulinum toxin (Botox) 279
Bowman's capsule 312
brainstem 268
brain stem auditory-evoked responses 285
bromism 307
bronchoalveolar lavage fluid (BALF) 256, 257

cadmium metal   613–614
Ca^{2+}-independent receptor (CIRL)   749
calmodulin   608
carbamate insecticides   712
carbon monoxide (CO) poisoning
   chronic, delayed effects   688–689
   effects   688
   hypoxia   275, 279
   mechanism of action   687
   pharmacokinetics   687–688
   signs and symptoms   688
   sources   686–687
   therapeutic principles   688
carbon tetrachloride (CCl$_4$)   242, 673
carbonyl-reducing enzymes   72
carboxyhemoglobin (COHb)   686
carcinogen   595–596
   diesel engine emissions   737–738
   lead effects of   616
   metals classification of   611
   PAHs   728–729
   particle effects   637–639
   vs. non-carcinogen fibers   633
carcinogenic aromatic amines   130
carcinogenicity
   animal experiments   511
   assumptions and limits of extrapolation
      accumulation of dose   522–523
      additivity   522
      linearity   521–522
      proportionality   522
   dose-response relations (DRRs)   511
   epidemiological studies   511
   linear extrapolation
      biological knowledge empirical data   513
      EPA method   514
      extrapolation with straight lines   512
      ICRP method   514
      straight lines, upper boundary estimates   512–513
      WHO method   514
   multihit multistage models
      empirical evidence for   520
      expansions and alternative models   520
      extrapolation to small doses   520–521
      graphical representation of   519
   of time aspects
      dose-effect function   517
      dose specification   516–517
      environment carcinogenesis   515–516
      non-linear extrapolation   518
      risk and survival   515

carcinogenic natural products   161
Carcinogenic Risk Assessment   186
cardiac glycoside poisoning   755–756
cardiomyopathy   368
cardiovascular effects
   diesel engine emissions   740
   special particle effects   638–640
cardiovascular system
   structure and function
      coronary blood vessels   366
      excitation and conduction system   365
      myocardium   364, 365
      vascular system   366
   toxic damage to
      coronary arteries and blood vessels   369–371
      excitation and conduction system   368–369
      myocardium   367–368
caretakers   173
caspases   178
catalase   163
cataract formation   360
cationic amphiphilic drugs (CAD)   260
celiac disease   227–228
cell nucleus   105
cellular defence mechanisms   194
cellular membranes   107–108
Centers for Disease Control (CDC)   571
central nervous system (CNS)   267–269, 602
cerebellum   268
ceruloplasmin   608
chelates   609
chemical carcinogenesis
   cancer development
      aneuploidy   180
      apoptosis   177–179
      carcinogen-induced immortalization of cells   179–180
      chromosomal instability   174–175
      chromosome segregation   180
      colon cancer   172–173
      initiation   171–172
      mutator phenotype   173–174
      neoangiogenesis   181
      oncogenes   175–176
      progression   172
      promotion   172
      tumor suppressor genes   176–177
      two-hit hypothesis   173
   DNA damage   160
      aberrant non-duplex DNA forms   161
      antioxidant mechanisms   163

cross-links 161
damage to DNA bases 161
definition 160–161
endogenous 162
exogenous 161
halting cell cycle progression 170–171
lesions, DNA backbone 161
prevention and repair of 162
tolerance 169–170
DNA repair
base excision repair 165–167
damage reversal mechanisms 164–165
definition 163
DSBs 168–169
MMR 168
nucleotide excision repair 167–168
non-genotoxic mechanisms
definition 181
epigenetics 181–182
tumor promoters 183–184
risk assessment 184–185
tumor initiation 159
tumor progression 160
tumor promotion 159–160
chemically-induced cell death 115
chemical substances, defined 530
chemical substances regulations, in EU
EU legislation 525–528
hazard potential 528–529
precautionary principle 529–530
risk assessment 528
risk communication 528
risk governance 528
risk management 528
risk types 529
Chemical Weapons Convention (CWC) 535
chloracne 306–307, 595
chlorinated methanes 673–674
chlorine 695–696
chloro acetaniline herbicides 708
chloroform 675
chloroquine 361
cholestasis 240
choreoathetosis 712
choroid 354
chromium metal 618–619
chromosomal aberrations 423
chromosomal instability (CIN) 180
amplifications 175
deletions 175
inversions 175
translocations 174–175

chromosome aberrations 414
chronic damage 369
chronic obstructive pulmonary diseases (COPD) 625, 736
chronic organ toxicity 610
chronic toxic encephalopathy 292
chronic toxicity 736, 739
cigarette smoking 370
citric acid cycle 105
clara cells 250
classification and labeling (C&L) of chemicals 17–18
classification, labelling and packaging (CLP) 15, 536
clastogenic 423
Clean Air Act (CAA) 564–565, 585
Clean Water Act (CWA) 565, 585
clonal selection 174
coal tar 309
cocarcinogens 172
Codex Alimentarius Commission 532
colchicine 754–755
colon cancer 172–173
colony-forming units (CFU) 321, 322
colorectal cancer 133–135
combustion-derived NP (CDNP) 371, 639
comet assay 417–418
Commission Regulation (EC) No. 440/2008 15
Committee for Risk Assessment (RAC) 580
Committee for Socio-Economic Assessment (SEAC) 580
compartment models
one-compartment model
apparent volume of distribution 49
area under the concentration 52–54
Bateman function 51–52
continuous administration 54
definition 48
extravascular administration 51–52
intravascular administration, elimination, total clearance, and half-life 50, 51
repeated administration 54–56
saturation kinetics 55–57
physiologically-based toxicokinetic models 59–61
two-compartment model 56–59
competitive inhibition 44
Comprehensive Environmental Response, Compensation, and Liability Act (CERCLA) 565–566
consolidated legislation documents 528
constitutive androstane receptor (CAR) 83, 377

Consumer Product Safety Act (CPSA) 569
Consumer Product Safety Commission (CPSC) 568–569
contact dermatitis 463
convective transport 37
copper metal 619–620
coronary blood vessels
  structure and function 366
  toxic damage to
    cadmium 371
    carbon disulfide 369
    carbon monoxide 370
    cigarette smoking 370
    fibers and particles 371
    lead 370
    nitrate esters 369–370
    water hardness 370
cosmetic products 547–548
cross-linking and alkylating agents 161
cyanide (CN^-) poisoning
  hypoxia 279–280
  intoxications 690
  mechanism of action 689
  sources 689
  therapeutic principals 690–691
  toxicokinetics 689–690
cyanogenic glycosides 756–757
cycasin 228
CYP1A2 133, 134
CYP2E1-mediated metabolism 128, 130
cytochrome c 178
cytochrome P450 (CYP) 63–69, 83, 85, 224
cytochrome P450-2E1 82
cytochrome P450 enzyme (1A1) 87
cytoskeleton 105
cytotoxicity 315–316
  cell
    cytoplasm components 105–106
    membrane system 103–104
    structural elements 102, 104
  cell death
    apoptosis 115, 117, 118
    calcium homeostasis 118–120
    DED 118
    necrosis 115, 117, 118
  cell functions
    animal and plant poisons 99–102
    chemicals 99, 102
  cellular targets of toxins
    ATP formation inhibition 112–114
    cellular membranes 107–108
    critical enzymes inhibition, signal transduction 114, 115
    lipid peroxidation 108–110
    redox systems disturbance 110–112
    synthesis of macromolecules 114
    transporters inhibition 114–117
  toxin targets 98–99

daidzein 656
Dangerous Substances Directive 574
DDT *see* dichlorodiphenyltrichloroethane (DDT)
death-associated proteins (DAPs) 179
death effector domain (DED) 118
death-inducing signal complex (DISC) 179
death receptors 117
degradation of fatty acids 105
Delaney Clause 563
delayed-type hypersensitivity (DTH) reactions 337
demyelination 276, 277
dendrites 269
dendrotoxins 744
deoxyribonucleic acid (DNA) 106
derivation of no-effect levels (DNEL) 445
derived minimal effect levels (DMELs) 29–31
derived no effect level (DNEL) 16, 29–31
dermal irritation 8
dermal sensitization threshold (DST) 702
descriptive statistics 488–491
detoxifications 90–91
diacylglycerole (DG) 172
dicarboximides fungicides 710
dichlorodiphenyltrichloroethane (DDT) 108
  central nervous system 602–603
  definition of 589
  effect of 590
  estrogen receptor antagonists 654–656
  metabolites 599
dichloromethane 675–676
diesel engine emissions
  carcinogenic effects 738–740
  gas phase of 734
  humans exposure of 740
  international organizations 740–741
  mechanisms of 738–740
  new diesel 734
  old diesel 734
  toxicity 737
    acute 736
    carcinogenicity 737–738
    chronic 736
    development/reproductive 737

genotoxicity 737
irritation/immunotoxicity 736
toxicokinetics 735
diesel engine exhaust fumes 738
di-(2-ethylhexyl) phthalate (DEHP) 658
diethylstilbestrol (DES) 661
diflubenzuron 714–715
Digit Symbol Test (DST) 287
2,3-dimercaptopropanesulfonate (DMPS) 609
4-dimethylaminophenol (4-DMAP) 690
dioxin response element (DRE) 600
diphtheria toxin 277
Dirty Dozen 585–586, 589
disorders of movement 280
disseminated intravascular clotting (DIC) 745
dissimilar joint action 149–150, 157
distal axonopathy 276
DNA damage response (DDR) system 188
DNA polymerase I (Pol I) 170
DNA polymerases 173
DNA repair
  apoptosis 178
  cellular reaction to DNA damage
    apoptosis and necrosis 192
    ATM and ATR 190–191
    base excision repair system 188–190
    DNA adducts 191–192
    epigenetic mechanisms 192
    intercellular communication, gap junctions 192–193
  chemical carcinogenesis
    base excision repair 165–167
    damage reversal mechanisms 164–165
    definition 163
    DSBs 168–169
    MMR 168
    nucleotide excision repair 167–168
  genes 173
DNA-strand breaks 111
dominant lethal test 423
domoic acid 278
dose-effect relation 659, 662
dose-response relations (DRRs)
  linear extrapolation 512–514
  non-linear extrapolations 518, 521
double-strand breaks (DSBs) 168–169, 190
D-penicillamine 609
drug-induced ocular cicatricial pemphigoid 358
dying-back neuropathy 276

EACs *see* endocrine active compounds (EACs)
ecotoxicology

cascading effects 212
damping effects 212
definition 209–210
dose/concentration-effect relationships 211
exposure assessments 214–215
fast-track and short-cut techniques 217–218
LOAEL 211
NOAEL 211
protection targets 210–211
risk assessment 216–217
test systems
  freshwater systems 212–213
  marine systems 213
  micro-, meso/macrocosm studies 214
  terrestrial systems 213, 214
EC regulation
  biocides 707
  PPPs 703–705
  toxicology 716
electron transport chain of mitochondria 105, 106
electroretinograms 285
EMA committees, for medicinal products 538, 539
encephalopathy 280
endocrine active compounds (EACs)
  adverse effects 649–650
  dose–effect 659
  low dose 659
  perchlorate 651
  sex hormones
    androgen receptor antagonists 657
    aromatase inhibitors 657–658
    compounds with combined effects 658
    estrogen receptor 654–657
  thyroid hormones
    in blood 651–653
    in liver 653
    PBDEs 653–654
    TBBPA 654
    thyroid gland 651
    thyroid peroxidase 651
endocrine disruptors 199
  compounds 667
  definition 661
  potential 661
endocrine modes of action
  in vitro studies
    direct receptor binding 663
    disadvantages 662
    steroid biosynthesis 663
    thyroid hormone metabolism 663–664

endocrine modes of action (*continued*)
in vivo
androgens and antiandrogens 665
5α-reductase 665
aromatase 665
estrogens and antiestrogens 664–665
steroid and thyroid hormone metabolism 666
steroidogenesis 665
thyroid toxicity 665
wildlife 666
endocrine system
foetal development, HPG axis 381–383
hazard identification 386
hormones 372
HPG axis 381–383
risk assessment 386
risk assessment of 386
sexual function in
endocrine active substances 384–386
regulatory *in vivo* studies 383–384
structure and function
hormonal balance in mammals 373, 374
sex hormones 379–381
steroid hormone 373
thyroid (*see* thyroid hormones)
tissue hormones 373
endoplasmic reticulum (ER) 105, 385
end replication problem 179
5-enol pyruvyl shikimate 3-phosphate (EPSP) 707
enterohepatic circulation 10, 46
Environmental Protection Agency (EPA) 14, 514, 563
CAA 564–565
CERCLA 565–566
CWA 565
FIFRA 567
FQPA 567–568
RCRA 567
TSCA 566–567
environment carcinogenesis 515–516
enzyme induction, PHAHs 603–604
EPA *see* Environmental Protection Agency (EPA)
epidemiology
bias 459
British-doctor study 451–452
causality 459–460
chance 459
confounding 459
direct standardization 452–453
experimental study 455–456
history of 450–451
indirect standardization 453–454
meta-analysis 457–458
observational study 454–455
risk of disease 450–451
statistics 456–457
epigenetic mechanisms for carcinogenesis 181–182
epoxid hydrolases 73–74
error propagation 491–492
erythrocytes
carbon monoxide poisoning 325
erythropoiesis 323–324
hemoglobin function 324
hemolytic anemias 324–325
Escherichia coli 409, 410
estrogen receptor 654–657
1,2-ethane diol *see* ethylene glycol
ethylene bis-dithiocarbamate 651
ethylenediaminetetraacetic acid (EDTA) 609
ethylene glycol 684–685
ethylene thiourea (ETU) 651
EU Chemicals Agency (ECHA) 201
EU laws 527, 528
EU legislation 544–545
biocidal products 541–542
on chemical substances in environment
air quality standards 553, 554
avoidance principle 552
high level of protection 552
polluter pays principle 552–553
precautionary principle 552
soil protection 556
waste management 555–556
Water Framework Directive 553–555
on consumer products 546, 547
cosmetic products 547–548
dangerous chemicals, import and export of 537
dangerous substances, groups of 537
drinking water 543
EU laws 527, 528
European legal acts, features of 526, 527
food 542–543
medicinal products, for human/veterinary use 537–540
plant protection products 541
primary legislation 526
protection from ionizing radiation, by radioactive materials 550–551
safety of
chemical substances 535–536

medical devices   540, 541
  medicinal products   538, 539
 secondary legislation   526
 transport of dangerous goods   551–552
Euratom Treaty   525
European Centre for the Validation of Alternative Methods (ECVAM)   466
European Chemicals Agency (ECHA)   15, 29, 573, 708
European Commission   527
European Commission Pesticides Database   706
European Council   527
European Food Safety Authority (EFSA)   186, 706
European legal acts, features of   526, 527
European Medicines Agency (EMA)   15
European Parliament   527
EUROQUEST   289
EU Scientific Committee on Food in 2002   731
EU Technical Guidance Document on Risk Assessment   5
excitation and conduction system
 structure and function   365
 toxic damage to
  barium   369
  cadmium   368
  halogenated hydrocarbons   368
  mercury   368–369
excitatory amino acids   278
excretion   36
 intestinal tract   46
 kidneys   44–46
 lung   46–47
 mammary glands   47
 skin   47
executioner caspases   178
existing active substance, EU legislation   541
exogenous DNA damaging agents   161
expanded single tandem repeat assay (ESTR)   435
exposure guide value (ELW)   696
Extended One-Generation Reproductive Toxicity Study (EOGRTS)   577
eye
 anterior segment
  acidic and alkaline corrosion   301
  ciliary body   353
  conjunctiva   352
  cornea   353
  corneal limbus   352–353
  iris and chamber   353
  irritation and contact dermatitis   301–302

 lens   353–354
 metabolism   300
 ocular absorption   299
 structure   297
 tear film   352
 toxicology of
lids   352
OECD TGs for *in vivo* tests   395–397
posterior segment   354
routes of delivery of xenobiotics
 blood flow   355
 lacrimal drainage   355
 ocular absorption   354–355
toxicology of
 acids   356–357
 alkalis   356
 cataract formation   360, 361
 chemical injuries   355–356
 conjunctival irritation   357
 contact allergy   357–358
 cornea   358–360
 inflammation   357
 mucocutaneous disease   358
 retinal toxicity (*see* retinal toxicity)
 toxic optic neuropathy   363

Fas-associated death domain protein (FADD/Mort1)   179
fasciculins   744
fatty acid hydroperoxides (FS-OOH)   163
fatty liver/steatosis   241
Federal Insecticide, Fungicide, and Rodenticide Act (FIFRA)   567
ferric III phosphate   716
ferritin   608
fiber dimensions   633–634
fibre toxicity   645
fibrosis   238–240
Fick's first law   41
first-order kinetics   44
first-pass effect   40, 41
5α–reductase   665
flash-evoked potentials   285
flavin adenine dinucleotide (FAD)   68
flavin-dependent monooxygenases (FMOs)   68, 69
follicle-stimulating hormone (FSH)   197, 380
Food and Agriculture Organization (FAO)   531, 532, 706
Food and Drug Administration (FDA)   561–562
food contact materials (FCMs)   546
Food, Drug, and Cosmetic Act (FDCA)   562–563

food-induced allergy 227
Food Quality Protection Act (FQPA) 567–568
formaldehyde (HCHO) 694–695
formaldehyde dehydrogenase (FOD) 681
fractional factorial design 152
fragrance materials
  chemical senses 699
  risk assessment of 700–702
  toxicity 699–700
Friedel–Crafts catalyst 591
full factorial design 152
fungicides 709–710

gamma-aminobutyric acid (GABA) 273
gastrointestinal (GI) tract
  definition 221
  function 222–223
  microbiota 223
  structure 221–222
  tissue repair 223
  toxicology
    cytotoxic effects 226–227
    disturbance of function 226
    experimental methods 229–230
    immune reactions 227–228
    toxic damage 230, 231
  xenobiotics in
    absorption of 223–224
    biotransformation of 224–225
    enterohepatic circulation of 225–226
Gatekeeper genes 174
General Food Law regulation 542
General Product Safety Directive (GPSD) 546, 547
general steroidogenesis 665
Genetic Susceptibility to Environmental Carcinogens (GSEC) 136
genistein 657
genome-wide association studies (GWASs) 137, 138
genotoxic carcinogens, threshold effects 25, 27
  cellular reaction to DNA damage
    DNA repair 188–193
    mechanism 187, 188
    development of cancer 187
    dose-dependent reactions 193
genotoxicity
  definition 408
  diesel engine emissions 737, 739
  in vitro test systems
    bacterial test systems 409–411
    cell transformation assays 420

    genetic endpoints of 408
    mammalian cells (*see* mammalian cells)
    xenobiotic metabolism 420–421
  in vivo mutagenicity tests
    germ cells 429–437
    somatic cells 424–429
  PAHs 727
germ cells
  chromosome mutations in 429–433
  gene mutations in 433–437
GI tract *see* gastrointestinal (GI) tract
global genomic repair (GGR) 167
Globally Harmonised System of Classification and Labelling of Chemicals 536
Globally Harmonized System (GHS) 17, 340
gluconeogenesis 105
glucose-6-phophate dehydrogenase 122, 163
glucose-6-phosphate dehydrogenase deficiency (G6PD) 325
glucuronidation 74–76
glutamic acid 278
glutathione (GSH) 111, 112
glutathione peroxidases 163
glutathione reductase 163
glutathione S-transferase (GST) 78–80, 129, 163
glycols 684–685
glycosides 755–757
glycosylphosphatidylinositol (GPI) biosynthesis 413
glyphosate herbicides 707–708
gonadotropin-releasing hormone (GnRH) 197, 380
good laboratory practice (GLP) 14, 15, 390, 534
G-protein-coupled receptors (GPCR) 144–145
granular bio-durable particles (GBP) 635, 646
granular nanoparticles 646–648
granulocytes 326–327
gut-associated lymphoid tissue (GALT) 227

Haber–Weiss reaction 255
halogen acne 306–307
halogenated aliphatic hydrocarbons 672–673
halting cell cycle progression 170–171
haplotype 125–126
Hashimoto thyroiditis 346
hazard identification
  acute toxicity studies 8
  carcinogenicity 9–10
  chronic studies 8
  genotoxicity 9
  institutions 6, 7

irritation   8–9
phototoxicity   9
qualitative description   22
reproduction and development   10
toxicokinetics   10–11
Hazardous and SolidWaste Amendments (HSWA)   567
hazard potential, of chemical substances   528–529
headache   369
health-based limit values   717
health hazards, classification of   529–531
hematopoietic system (bone marrow and blood)
    bone marrow niche   323
    CFU   320, 321
    components of   320, 321
    depression of bone marrow function   329
    leucocytes   326–328
    leucocytosis and leukemias   329–330
    platelets/thrombocytes   328
    toxicological features, circulating blood cells
        methemoglobinemia   325–326
        red blood cells   323–325
hematotoxins   744–745
hepatocyte growth factor (HGF)   235
hepatocytes   232
hepatotoxicants   482
herbicides
    bipyridylium herbicides   709
    chloro acetaniline herbicides   708
    glyphosate   707–708
    phenoxy herbicides   709
    triazine herbicides   708–709
    urea herbicides   709
hereditary non-polyposis colon cancer (HNPCC)   173
heritable translocation assay   423
heterocyclic amine metabolism   134, 135
hexachlorobenzene (HCB)   346, 590–591, 599
hexachloro butadiene (HCBD)   317
hexachlorophene   277
2,5-Hexanedione (HD)   277
hormone responsive elements (HREs)   380
hormones   372
human health risk assessment   718–720
    feasibility and achievability issues   32
    importance of   32
    OELs   30–32
    precautionary principle   31–32
    principles of
        animal data   26
        application of assessment factors   29

exposure assessment   28
    genotoxic carcinogens   27
    hazard characterisation   22, 24–25
    human data   25–26
    non-genotoxic carcinogens   28
    qualitative and quantitative method   22
    risk characterisation   28–29
    in silico tools   26
    terminology   22–24
    in vitro models   26
human repeat insult patch test (HRIPT)   342
hydrocarbons
    partition coefficients for   36, 37
    solvents for
        benzene   678–679
        gasoline and kerosene   677
        n-Hexane   677
        toluene   679–680
hydrochloroquine   361
hydrogen chloride (HCl)   694
hydrogen fluoride (HF)   694
hydrogen sulfide (H_2S) poisoning
    effects   692
    mechanism of action   691
    pharmacokinetics   691–692
    sources   691
    therapeutic principals   692
hydroxocobalamine   691
hypothalamic–pituitary–gonadal (HPG) axis
    compound structures, effects   655
    foetal development of   381–383
    human drugs   649–650
hypothalamus-pituitary-thyroid (HPT) axis
    foetal development of   377
    human drugs   649–650
hypoxanthine guanine phosphoribosyl transferase (HPRT) activity   411–412
hypoxia   279

immune system
    adaptive   332
    advantage   332
    antigen recognition   333–334
    B lymphocytes   336
    chemical immunosuppression   348–350
    chemical-induced autoimmunity   345–347
    elicitation thresholds   340–342
    human studies   340
    immunologic tolerance   336–337
    immunostimulation by chemicals   346–348
    innate   332
    neutrophils   332–333

immune system (*continued*)
   risk assessment of
      oral and parenteral sensitization   345
      respiratory tract sensitization   343, 344
      skin sensitization   342–343
      sensitization and allergy   337–339
      T lymphocytes   334–335
      in vitro studies   339–340
immunosuppression   596–597
immunotoxicity, diesel engine emissions   740
inferential statistics   496–504
inflammatory phagocytes   632
initiator caspases   178
innate immune system   332
inorganic lead   277
insect growth regulators   714–715
insecticides   711–712
insulin-like growth factor-1 (IGF-1)   598
Integrated Approaches to Testing and Assessment (IATA)   466
   for Serious Eye Damage and Eye Irritation   397
   for Skin Irritation/corrosion   395
International Agency for Research on Cancer (IARC)   531, 597, 730, 734, 740
International Atomic Energy Association (IAEA)   533, 535
International Chemical Safety Cards (ICSC) project   532
International Civil Aviation Organization (ICAO) Technical Instructions   552
International Commission for Ray Protection (ICRP)   514
International Commission on Radiological Protection (ICRP)   533
International Conference on Harmonisation of Technical Requirements for Registration of Pharmaceuticals for Human Use (ICH)   15, 535, 540
International Conference on Harmonization (ICH)   199
international co-operation, EU role   531–533
International Council of Chemical Associations (ICCA) program   14
International Fragrance Association (IFRA)   701
International Labour Office (ILO)   532
International Maritime Dangerous Goods (IMDG) Code   552
International Maritime Organization (IMO)   532
International Organization for Standardization (ISO)   642

International Programme on Chemical Safety (IPCS)   531
International System of Radiological Protection   533
International Uniform Chemical Information Database (IUCLID)   533
Inter-Organization Programme for the Sound Management of Chemicals (IOMC)   533
intoxications   690
ionized organic substances   36
iron metal   620–622
irritant gases
   chlorine   695–696
   formaldehyde   694–695
   hydrogen chloride   694
   hydrogen fluoride   694
   isocyanates   696
   nitrogen oxides   696–697
   ozone   697–698
   phosgene   697
   smoke inhalation   698
   sulfur dioxide   695
ischemia   280
isocyanates   696
isoflavonoids   657

Joint FAO/WHO Expert Committee on Food Additives (JECFA)   532
Joint Meeting on Pesticide Residues (JMPR)   532, 706

kainic acid   278
kerosene   677
kidneys
   blood filtration   312–313
   carcinogenic effects   318
   phase I and phase II enzymes   311
   toxicology
      glomerulus   314
      heavy metals   316–317
      hydrocarbons   315–316
      intoxication   313–314
      polyvalent alcohols   317
      tubular system   314
   tumors of   318
kinetics   604–605
Kupffer cells   232

lactate dehydrogenase (LDH)   257
Langerhans cells   340
latrodectus spiders venoms   749–750

lead metal   614–616
lectins   757
leucocytes
    granulocytes   326–327
    lymphocytes   327
    monocyte/macrophage system   328
lifetime weighted average exposure (LWAE)   283–284
Li Fraumeni syndrome   174
ligand-receptor interactions
    biological consequences of
        agonistic effects   143
        antagonistic effects   143–144
    law of mass action   141
    membrane-bound hormone   141
    neurotransmitter receptors   141
linuron   657
lipid peroxidation   108–110
lipid-soluble steroid hormones   380
lipopolysaccharides (LPS)   232
liver
    acetaminophen   242
    bile formation   234–235
    cell death   237–238
    cell types   232
    cholestasis   240
    cirrhosis   238
    concentration dependent *vs.* idiosyncratic hepatotoxicity   245–246
    endogenous metabolism   234–235
    fibrosis   238–240
    hepatotoxic compounds
        allyl alcohol   244
        APAP   242–244
        CCl_4 242–244
        ethanol   244–245
    increased liver weight   246–247
    inflammation   241–242
    interspecies differences   247
    in vitro systems   247
    liver tumors   246
    microstructure   232, 233
    myths of   247–248
local lymph node assay (LLNA)   9, 339
loss of hearing   280
loss of heterozygosity (LOH)   173
low acute toxicity   635
lowest observed adverse effect level (LOAEL)
    in behavioural toxicology   284
    biological assessment values   445–446
    ecotoxicology   211
    POD   25, 29

risk assessment process   6
lowest observed effect concentrations (LOECs)   211, 342
loxosceles spiders venoms   750–751
lung cancer   262–263
luteinizing hormone (LH)   380
Lyell Syndrome   303
lymphocytes   327
lysosomes   105

major histocompatibility complex (MHC)   228, 333–334
malignant transformation   420
mammalian cells
    alkaline elution   417
    chromosome aberrations   414
    comet assay   417–418
    DNA binding   416
    DNA repair assays   419
    DNA strand breaks Induction   416–417
    gene mutation tests   411–412
    HPRT gene mutation test   412–413
    indicator tests   416
    micronucleus test   415–416
    SCEs   418–419
    TK+/−gene mutation test   413–414
    UDS   419
mammalian reproductive cycle   196
manganese metal   622–623
margin of exposure (MOE)   6, 653
margin of safety (MOS)   216, 666
maximal tolerated dose   9, 10
maximum residue level (MRL)   543
mechanism of action
    aryl hydrocarbon receptor   600–601
    carbon monoxide   687
    central nervous system   603
    cyanide   689
    hydrogen sulfide   691
    PAHs   729–730
    PEPCK   601–602
    UROD   602–603
membrane-bound sulfotransferases   76
Mendelian diseases   136
mercury and mercury compounds, exports of   537
mercury metal   616–618
messenger-ribonucleic acid (mRNA)   106
metabolites
    DDT   599
    saturation kinetics   43–44
    toxicokinetics   43–44

metaldehyde 716
metallothionein 608
metals
  acute toxicity 609–610
  aspects 607–608
  bioavailability importance 608–609
  carcinogenicity 610–611
  chronic organ toxicity 610
  proteins 607
  toxic effects of 607
  toxicology of
    arsenic 611–613
    cadmium 613–614
    chromium 618–619
    copper 619–620
    iron 621–622
    lead 615–616
    manganese 622–623
    mercury 616–618
    nickel 623–624
methoxychlor (MXC) 654–656
methylation 80
methylbenzene *see* toluene
methylguanine-DNA methyltransferase (MGMT) 163
methyl mercury (MeHg) 274
methyl methanesulfonate (MMS) 184
1-methyl 4-phenyl 4-propionoxypiperidine (MPPP) 274–275
1-methyl-4-phenyl-1,2,3,6-tetrahydropyridine (MPTP) 274–276
methyl tert-butyl ether (MTBE) 677
Michaelis-Menten equilibrium 141
Michaelis-Menten kinetics 43, 44
microcystins (MCs) 114
microglia 271
micro RNAs (miRNAs) 475
microsatellite instability (MIN) 180
microsomal epoxide hydrolases 73–74
microsomes 63
midbrain 268
minimal risk level (MRL) 29
Minimum Information About a Microarray Experiment (MIAME) 480
miRNAs 475
mismatch repair (MMR) 168
mitochondria 105, 178
mitogen activated protein (MAP) 739
mitotic spindle checkpoints 180
mixture, defined 530
mixtures and combinations of chemicals
  complex mixture 149
  cumulative exposure 149
  joint action
    direct chemical-chemical interaction 150
    dissimilar 149–150
    similar 150
    toxicokinetic interaction 150
  polycyclic aromatic hydrocarbons 157
  safety evaluation
    dissimilar joint action 157
    flow chart for 154
    individual constituents 155–156
    interaction 157
    number of fractions 155
    similar joint action 156–157
    single entity mixture 154–155
    toxicity study design of mixtures/combinations 151–152
  simple mixture 148–149
  specified combination 149
molluscicides 715
monoamine oxidase (MAO) 224
morphine 226
mucociliary clearance 629–631
muscimol 278–279
mutator phenotype 173–174
myocardium 364, 365
  structure and function 364, 365
  toxic damage to
    antimony 367
    arsenic 367
    arsines 367
    cadmium 367
    cobalt 367
    ethanol 367
    lead 367
myotoxins 744

N-acetylation 78
N-acetyl-p-benzoquinone imine (NAPQI) 244
nanomaterials toxicology
  granular nanoparticles 646–648
  mode of toxic action 646
  nanotoxicodynamics 645–646
  nanotoxicokinetics 644–645
  types of 647
nanoparticles (NPs)
  deposition of 636
  fibrous 645–646
  granular 646–648
  toxic mechanisms of 637
nanotoxicodynamics 645–646
nanotoxicokinetics 644–645

narcotic effect   107
nasopharyngeal (NP) region   249–250
National Adult Reading Test (NART)   287
National Institute for Occupational Safety and
      Health (NIOSH)   560–561
NAT2 polymorphism   131
natural toxins   99–101
neoangiogenesis   181
nephrotoxicants   482
nervous system
    neurotoxins
      axonopathy   275–277
      clinical signs and symptoms   279–281
      demyelination   276, 277
      neuropathy   274–276
      neurotransmission effects   277–279
    structure and function of
      cells of   269–273
      central nervous system   267–269
      peripheral nervous system   265–267
      transmission of neuronal information
            264–265
Neurobehavioral Evaluation System (NES)   291
neurogenic inflammation   637
neurons   269, 270
neuropathy   274, 275
    MeHg   274
    MPTP   274–275
    TMT   276
neurotoxicity of lead   616
neurotoxins   744–745
neurotransmission   272–273
neurotransmitters   273
neutrophils   332–333
new active substance, biocidal product   541
n-Hexane   677
nickel metal   623–624
nicotinamide adenosine dinucleotide (NAD)   105
nitrilotriacetic acid (NTA)   317
nitrogen oxides (NO_2)   696–697
nitrosamine formation, stomach   224, 225
N-nitroso compounds   161
NOAEL *see* no observed adverse effect level
      (NOAEL)
no-expected sensitization induction level (NESIL)
      342
non-alcoholic steatohepatitis (NASH)   238
non-genotoxic carcinogens   27, 28
non-genotoxic mechanisms   181–184
non-homologous end joining (NHEJ)   163, 169
non-steroidal anti-inflammatory drugs (NSAIDs)
      227

nonylphenol (NP)   656
no observed adverse effect level (NOAEL)   596,
      659
    ADI   13, 14
    animal test   469
    in behavioural toxicology   284
    biological assessment values   445–446
    ecotoxicology   211
    non-genotoxic carcinogens   18
    OECD TGs for *in vivo* tests   398–400
    pesticidal active substances   716–717
    POD   29
    threshold   24
    uncertainty analysis   16
no observed effect concentrations (NOECs)   211,
      342
no observed genotoxic effect (NOGEL)   193
noxious gases
    airborne systemic poisons
      carbon monoxide   686–689
      cyanide   689–691
      hydrogen sulfide   691–692
    irritant gases
      chlorine   695–696
      formaldehyde   694–695
      hydrogen chloride   694
      hydrogen fluoride   694
      isocyanates   696
      nitrogen oxides   696–697
      ozone   697–698
      phosgene   697
      smoke inhalation   698
      sulfur dioxide   695
    respiratory tract irritants   692–694
Nuclear Regulatory Commission (NRC)   569
nucleotide excision repair (NER)   163, 167–168

occupational exposure limits (OELs)
    acceptability of risk   30–31
    margins of safety   30–31
    SCOEL   30
Occupational Safety and Health (OSH) Act   560
Occupational Safety and Health Administration
      (OSHA)   14
octachlorodibenzo-*p*-dioxin (OCDD)   587
ocular absorption   354–355
oestrogen receptors   385
oestrogens   381, 382
okadaic acid (OA)   114
oligodendrocytes   271
oocyte maturation   197
optic nerve   354

oral sensitization 345
oral toxicity test 392–393
Organisation for the Prohibition of Chemical Weapons (OPCW) 535
Organization for Economic Co-operation and Development (OECD) 199–201, 337, 340, 533
Organization for Economic Cooperation and Development test guidelines (OECD TG) 385, 386, 534
  animal testing 463–465
  in vivo tests
    acute toxicity 392–394
    control animals 391–392
    exposure route 391
    eye irritation 395–397
    healthy animals 391
    OECD TG 424 406
    OECD TG 426 406
    OECD TG 440 406
    reproductive toxicity 401–406
    skin irritation 395, 396
    skin sensitization 397
    statistical evaluation 392
    test substance 391
    toxicity after repeated dosing 398–401
  OECD TG 401 392
  OECD TG 407 378
  OECD TG 420 392, 393
  OECD TG 423 392, 393
  OECD TG 425 392, 393
  OECD TG 433 394
  OECD TG 436 394
  OECD TG 440 385, 406
  OECD TG 441 385, 406
organophosphate derivatives (OPDs) 278
organophosphates (Ops) 226, 361, 711–712
organ toxicology
  behavioral neurotoxicology (*see* behavioral neurotoxicology)
  cardiovascular system 364–371
  endocrine system
    hazard identification 386
    hormones 372
    HPG axis 381–383
    risk assessment 386
    sexual function 383–386
    structure and function 373–381
    tissue hormones 373
  eye 355–363
  gastrointestinal tract
    cancer 228–230
    cytotoxic effects 226–227
    definition 221
    disturbance of function 226
    experimental methods 229–230
    function 222–223
    immune reactions 227–228
    microbiota 223
    structure 221–222
    tissue repair 223
    toxic damage 230, 231
    xenobiotics in 223–226
  kidneys 311–319
  liver
    acetaminophen 242
    bile formation 234–235
    cell death 237–238
    cell types 232
    cholestasis 240
    cirrhosis 238
    concentration dependent *vs.* idiosyncratic hepatotoxicity 245–246
    endogenous metabolism 234–235
    fibrosis 238–240
    hepatotoxic compounds 242–245
    increased liver weight 246–247
    inflammation 241–242
    interspecies differences 247
    liver tumors 246
    microstructure 232, 233
    myths of 247–248
    porphyria 241
    regeneration 235–237
    steatosis 241
    in vitro systems 247
    xenobiotic metabolism 234–236
  nervous system (*see* nervous system)
  respiratory system
    allergy and asthma 261–262
    anatomy 249
    epithelia and cell types 252, 254
    function 252–254
    lung cancer 262–263
    nasopharyngeal (NP) region 249–250
    protective systems 254–256
    pulmonary (P) region 250–252
    target for toxicity 256–261
    toxic effects, inhaled materials 263
    tracheobronchial (TB) 250
  skin (*see* skin)
oxidative phosphorylation 105
oxidative stress 110, 631–633
ozone (O_3) 697–698

PAHs *see* polycyclic aromatic hydrocarbons (PAHs)
painter's syndrome   292
parasympathetic nervous system   266
parenteral sensitization   345
particle deposition   626–628
particle surface with polymers (PVNO)   635
particle toxicology
   deposition   626–628
   and effects   627
   inflammation   631–633
   lung overload   629–631
   mucociliary clearance   629–631
   nanoparticles   636–637
   oxidative stress   631–633
   properties   627
     absorbed components   635–636
     bio-persistence   633
     dimensions   633–634
     soluble components   635–636
     surface area   634–635
     surface reactivity   635
   shape and density   628–629
   special particle effects
     carcinogenicity   637–639
     cardiovascular effects   639–640
   translocation   631
particulate matter (PM)   371
passive diffusion   36, 37, 47
pathogen-associated molecular patterns (PAMP)   227, 332
pattern-recognition receptors (PRR)   332
pentachlorophenol (PCP)   581–582, 651–653
PEPCK *see* phosphoenolpyruvate carboxykinase (PEPCK)
perchlorate   651
perchlorethene   673
perfusion limited model   60
peripheral distal axonopathy   276
peripheral nervous system (PNS)   265–267, 269, 272
peroxidase   69
peroxisomes   105
persistent, bioaccumulative, and toxicity (PBT) criteria   218
persistent organic pollutants (POPs)   537, 585
Pesticide Residue Intake Model (PRIMo)   719
pesticides
   active substances
     molluscicides   715
     regulatory toxicology of   716–717
     rodenticides   715

   target organisms   706
   toxicity   707–709
   biological origin substances   713–714
   classification   718
   fungicides   709–710
   human health risk assessment   718–720
   insect growth regulators   714–715
   insecticides   711–713
   plant protection products   710
   toxicological effects   717
phagocytic cells   332
PHAHs *see* polyhalogenated aromatic hydrocarbons (PHAHs)
pharmacokinetics
   carbon monoxide   687–688
   hydrogen sulfide   691–692
phenobarbital   172
phenols   757
phenothiazines   362–363
phenoxy herbicides   709
phenylpropanoids   757
phosgene ($COCl_2$)   697
3′-phosphoadenosine-5′-phosphosulfate (PAPS)   76, 77
phosphoenolpyruvate carboxykinase (PEPCK)   601–602
physiologically based pharmacokinetic (PBPK) models   389
physiologically-based toxicokinetic models   59–61
phytoestrogens   656
Pig-a assay   413, 423, 424
plant protection products   541
plant protection products (PPPs)
   active substances   704
   consumer protection   719–720
   EC regulation   703–705
   herbicides   705
   regulatory toxicology   716
plant toxins
   alkaloids   752–755
   glycosides   755–757
   lectins   757
   phenols and phenylpropanoids   757
   terpens and terpenoids   758–759
plate incorporation assay   410
podophyllin/podophyllotoxin   757
point of departure (POD)   24, 25
polybrominated diphenyl ethers (PBDEs)   653–654
polychlorinated biphenyls (PCBs)   588–589, 654

polychlorinated dibenzodioxins (PCDDs) 47
   biotransformation 603
   PHAHs 586–588
   toxicity 592
polychlorinated dibenzofurans (PCDF) 47
   PHAHs 586–588
   toxicity 592
polycyclic aromatic hydrocarbons (PAHs) 161
   biological monitoring 725–726
   classifications of 727
   environment occurrence 723–724
   mechanisms of action 729–730
   molecular structures of 723
   national and international organisations 730–731
   particles 635
   physico-chemical properties 722
   skin cancer 309
   toxicity
      acute toxicity 726
      after repeated exposure 726
      carcinogenicity 727–728
      equivalency factors for 731–732
      genotoxicity 727
      local toxicity and irritancy 726
      reproductive 729
      sensitisation and immunotoxicity 726
polyethoxylated (POE) 708
polyhalogenated aromatic hydrocarbons (PHAHs)
   DDT 589–590, 599
   Dirty Dozen 585
   enzyme induction 603–604
   hexachlorobenzene 590–591, 599
   kinetics 604–605
   mechanisms of action
      aryl hydrocarbon receptor 600–601
      central nervous system 602
      PEPCK 601–602
      UROD 602–603
   metabolism 603
   PCDDs/PCDFs 586–588
   physico-chemical properties 591–592
   polychlorinated biphenyls 588–589
   toxicity
      carcinogenicity 595–596
      chloracne 595
      chlorinated dibenzodioxins 592–595
      immunosuppression 595
      reproductive toxicity 596–597
polymorphism 123
polyvinyl chloride (PVC) 656
poorly soluble particles (PSP) 631, 635, 646

porphyria 241
positive endexpiratory pressure (PEEP) 693
postsynaptic neurotoxins 743–744
PPPs *see* plant protection products (PPPs)
pre-and postnatal toxicology
   congenital human malformations 205–207
   drug effects principles, pregnancy 201–204
   drug metabolism in pregnancy 204
   embryonic and fetal toxicology 204–205
   endocrine disrupters 208
   lactation 207–208
predicted no-effect concentrations (PNECs) 216
predictive toxicogenomic 481–483
PredTox 484
Pregnan X receptor (PXR) 83, 377
preincubation assay 410
premature cataract formation 359
presynaptic neurotoxins 743
primary carcinogens 172
principle of proportionality 527
principle of subsidiarity 526–527
Prior Informed Consent (PIC) Regulation 537
probability distribution 492–496
probit analysis 506–507
proliferation-dependent nuclear antigen (PCNA) 166
promoters 172
1,2-propane diol *see* propylene glycol
propionibacterium acnes 307
propylene glycol 684–685
prostaglandin G2 (PGG2) 69
prostaglandin H2 (PGH2) 69
prostaglandin synthase 69, 70
protein kinase C (PKC) 172
protein phosphatases (PPPs) 114
proto-oncogenes 172, 176
pulmonary circulation 366
pulmonary (P) region 250–252
purified cytochrome P450 enzymes 65
pyrethrins 712–713
pyrethroids 712–713

quantitative reverse transcriptase polymerase chain reaction (qRT-PCR) 477
quantitative structure activity relationships ((Q)SARs) 26, 27

REACH *see* Registration, Evaluation, Authorization and Restriction of Chemicals (REACH)
REACH Registration process 29
reactive nitrogen species (RNS) 162, 631

reactive oxygen species (ROS)   607, 631, 737
receptor-mediated mechanisms
  ligand-receptor interactions   141–144
  Michaelis-Menten equilibrium   141
  signal transduction
    GPCR   144–145
    ion channels   145–146
    nuclear receptors   146–147
    RTK   146
receptors tyrosine kinases (RTK)   146
reciprocal translocations   423
recommended exposure levels (RELs)   29
red blood cells *see* erythrocytes
reference concentration (RfC)   29
reference dose (RfD)   29, 653
Registration, Evaluation, Authorization and Restriction of Chemicals (REACH)   14, 15, 31, 445, 468, 573, 583, 699
  authorization   579
  downstream users   582
  ECHA, status and function of   580
  European Commission, role of   581
  evaluation process
    compliance check   578
    decision-making procedure   579
    examination of testing proposals   578
    substance evaluation   579
  historical development   574
  joint registration of same substances   576–577
  Member States, role of   581–582
  registration dossiers   576
  Regulation (EC) No 1907/2006   29
  restriction process   579–580
  safety concept   575
  substances, mixtures and articles   574–575
  testing and assessment strategy   577–578
regression analysis   504–506
Regulation (EC) No. 1223/2009   15
renal clearance   45
repeated open application tests (ROATs)   341
replication protein A (RPA)   191
reproductive toxicity   596–597
  developmental toxicity   401–402
  fertility   401–406
reproductive toxicology
  characteristics of   196–197
  definition   195
  female and male fertility, adverse effects
    cleavage divisions and implantation   198–199
    fertilization   198
    OECD   199
    oocyte maturation   197
    spermatogenesis   197–198
    international test methods   199–200
  pre-and postnatal toxicology
    congenital human malformations   205–207
    drug effects principles, pregnancy   201–204
    drug metabolism in pregnancy   204
    embryonic and fetal toxicology   204–205
    endocrine disrupters   208
    lactation   207–208
Research Institute for Fragrance Materials (RIFM)   701
Resource Conservation and Recovery Act (RCRA)   567, 571
respiratory system
  allergy and asthma   261–262
  anatomy   249
  function   252–254
  lung cancer   262–263
  protective systems   254–256
  structure
    nasopharyngeal (NP) region   249–250
    pulmonary (P) region   250–252
    tracheobronchial (TB)   250
  target for toxicity
    BALF   256, 257
    gases   260
    particles   257–260
    substance-specific pulmonary toxicity   260–261
  toxic effects, inhaled materials   263
respiratory tract irritants   692–694
respiratory tract sensitization   343, 344
retina   354, 359–360
retinal toxicity   360
  chloroquine   361
  clofazimine and methanol ingestion   363
  desferrioxamine   363
  hydrochloroquine   361
  ibuprofen   363
  indomethacin   363
  isoretinoin   362
  lead   360
  methanol   360–361
  organophosphates   361
  phenothiazines   362–363
  sildenafil   362
  solvent   361
  tamoxifen   362
  vigabrantin   362
rheumatoid arthritis   346
ricin   758

Rio Conference on Environment and Development 16
risk assessment process
   acute toxicity studies 8
   ADI 13
   carcinogenicity 9–10
   chronic studies 8
   dose-response curves 3, 6
   exposure assessment 4–5
   genotoxicity 9
   hazard identification 2–3
   institutions 6, 7
   irritation 8–9
   LOAEL 6
   NOAEL 6
   phototoxicity 9
   reproduction and development 10
   reversible and irreversible effects 6
   toxicokinetics 10–11
   toxic potency 3
risk, defined 529
risk management measures (RMM) 22
RNA-induced silencing complex (RISC) 475
rodenticides 715
rough endoplasmic reticulum (rER) 105

Saccharopolyspora spinosa 714
Salmonella gene mutation test 411
Salmonella typhimurium 409, 410
saturnine angina pectoris 370
Schwann cells 271, 272
Scientific Committee on Consumer Safety (SCCS) 701
Scientific Committee on Emerging and Newly Identified Health Risks (SCENIHR) 701
Scientific Committee on Health and Environmental Risks (SCHER) 701
Scientific Committee on Occupational Exposure Limits (SCOEL) 18, 30, 444, 445, 549
sclera 354
SCOEL, *see* Scientific Committee on Occupational Exposure Limits (SCOEL)
scorpions venoms 747–748
scrotal squamous cell cancer 309
secondary carcinogens 172
second cancer 330
serious adverse effects 280
sex hormones
   androgen receptor antagonists 657
   androgens 665
   aromatase inhibitors 657–658
   biosynthesis 379–380
   biosynthesis of 663
   compounds with combined effects 657–658
   estrogen receptor 654–657
   estrogens 664–665
   metabolism 381, 382
   regulation of synthesis 380–381
   target cells effect 380
short-chain dehydrogenase/reductases (SDR) 72
sildenafil 362
similar joint action 150, 156–157
single nucleotide polymorphism (SNP) 125
sinusoidal endothelial cells 240
sister chromatid exchanges (SCEs) 418–419
skin
   function
      dermal absorption 298–299
      excretory function 299–300
      metabolism 300
   OECD TGs for *in vivo* tests 395
   sensitizers 341
   structure 296–297
   toxicology of
      acidic and alkaline corrosion 301
      allergic contact dermatitis 302–303
      chloracne 306–307
      disturbances of pigmentation 307–309
      hair loss 306
      irritation and contact dermatitis 301–302
      photoallergy 304, 305
      phototoxic reactions 304
      toxic epidermal necrolysis 303
      tumors 309
      urticaria 305
skin sensitization 397
slow acetylators 122
small ubiquitin-related modifier (SUMO) 192
smoke inhalation 698
snakes venoms 743
   bites 743
   first aid 746
   hematotoxins 744–745
   hospital treatment 746
   local damage 745
   myotoxins 744
   neurotoxins 743–744
   symptoms 745
sodium iodide symporter (NIS) 375, 651
sodium nitrite 690
solar retinopathy 360
solvent-induced psycho organic syndrome 292
solvents

aliphatic alcohols
    ethanol   681–684
    glycols   684–685
    methanol   680–681
  hydrocarbons
    benzene   678–679
    gasoline and kerosene   677
    n-Hexane   677
    toluene   679–680
  intoxications   671
  toxicology
    chlorinated methanes   673–674
    chloroform   675
    dichloromethane   675–676
    halogenated aliphatic hydrocarbons   672–673
    perchlorethene   673
    tetrachloromethane   673–674
    trichloroethene   673
somatic cells
  chromosomal mutations in   424–427
  gene mutations   427–429
space of Dissé   232
species sensitivity distribution (SSD)   216, 217
specific estrogen receptor modulators (SERM)   146
spermatogenesis   197–198
spiders venoms
  Australian funnel-web   748–749
  latrodectus   749–750
  loxosceles   750–751
spinosad   714
statistical analysis
  descriptive statistics   488–491
  error propagation   491–492
  experimental designs   507–508
  inferential statistics   496–504
  population and sample relationship   487–488
  probability distribution   492–496
  probit analysis   506–507
  regression analysis   504–506
  Sprague-Dawley rats   487
  statistical software   508–509
  systematic errors   488
steatosis   241
stellate cell   232, 239
stem cell model   322
steroid biosynthesis   663
steroidogenic acute regulatory (StAR) protein   379
steroid synthesis   385
Stevens–Johnson syndrome (SJS)   303, 358
stimulation index (SI)   339

Stockholm Convention of 2001   585–586
Strategic Approach to International Chemicals Management (SAICM)   533
Streptomyces avermitilis   714
strobilurins fungicides   710
structural changes in sperm chromatin (SCSA)   435
structure-activity relationships   663
strychnine alkaloid   279, 753–754
substance, defined   530
substance of concern   541
Substances of Very High Concern (SVHC)   201
sulfotransferases   76–77
sulfur dioxide (SO_2)   695
Superfund Amendments and Reauthorization Act (SARA)   566
superoxide dismutase (SOD)   256
Swedish Performance Evaluation System (SPES)   291
Switching Attention   287
Symbol Digit Substitution Test (SDST)   287, 288
sympathetic nervous system   266
Syrian hamster embryo (SHE) assay   420
systemic circulation   366
systemic lupus erythematosus (SLE)   346

tamoxifen   362
2,3,7,8-TCDD *see* 2,3,7,8-tetrachlorodibenzo-*p*-dioxin (2,3,7,8-TCDD)
T-cell receptor (TCR)   333
TDI *see* tolerable daily intake (TDI)
tebufenozide   715
technical regulations for hazardous materials (TRGS)   694–695
telomerase   172, 179–180
terpenoids   758–759
terpens   758–759
Test Digit Span   288
tetrabromobisphenol A (TBBPA)   654
2,3,7,8-tetrachlorodibenzo-*p*-dioxin (2,3,7,8-TCDD)   83, 151
  acute toxicity of   601
  biotransformation reactions   62
  enzyme induction   603–604
  LD_{50} values for   594
  metabolism   602
  PCDDs/PCDFs   587
  TEFs   592
  total body fat   594
tetrachloromethane   673–674
tetradecanoyl phorbol acetate (TPA)   172

tetraiodothyroacetic acid (TETRAC)   377
Tetrodotoxin (Tedox)   279
thalidomide   461
6-thioguanine (6-TG)   413
thorotrast   246
three-pillar approach to risk governance   528
threshold of toxicological concern (TTC)   17
thyroglobulin (TG)   375
thyroid gland   651
thyroid hormones
   biosynthesis   374–375
   in blood   651–653
   HPT axis   377
   in liver   653
   metabolism   377, 663–664
   PBDEs   653–654
   receptor   654
   regulation of synthesis   376–377
   target cells effect   375–376
   thyroid gland   651
   thyroid peroxidase   651
   in vitro screens   378
   in vivo toxicology studies   377–378
thyroid peroxidase inhibition   651
thyroid-stimulating hormone (TSH)   376, 377
thyroid toxicity   665
thyroperoxidase (TPO)   375, 378
thyrotropin-releasing hormone (TRH)   376
thyroxine binding globulin (TBG)   651
tissue hormones   373
tolerable concentration (TC)   29
tolerable daily intake (TDI)   29, 613, 651, 657
toluene   679–680
toxic epidermal necrolysis (TEN)   303
toxic equivalency factor (TEF) method   156, 592, 731–732
toxic gases, fires   698
toxicity
   active substance, pesticides   707–709
   acute (*see* acute toxicity)
   arsenic   612–613
   cadmium   614
   cardiac glycosides   756–757
   chromium   618–619
   copper   619–620
   diesel engine emissions   737
     acute   736
     carcinogenicity   737–738
     chronic   736
     development/reproductive   737
     genotoxicity   737
     irritation/immunotoxicity   736

   fragrance materials   699–700
   iron   621–622
   lead   615–616
   manganese   622–623
   mercury   617–618
   nickel   623–624
   PAHs
     acute toxicity   726
     after repeated exposure   726
     carcinogenicity   727–728
     equivalency factors for   731–732
     genotoxicity   727
     local toxicity and irritancy   726
     reproductive   729
     sensitisation and immunotoxicity   726
   PCDDs/PCDFs   592
   pesticides active substance   707–709
   PHAHs
     carcinogenicity   595–596
     chloracne   595
     chlorinated dibenzodioxins   592–595
     immunosuppression   596
     reproductive toxicity   596–597
toxicity equivalent (TEQ)   593, 598
toxicity exposure ratio (TER)   216
toxicodynamics (TD)   215
toxicogenetics
   favism   122
   genotyping   124–127
   phenotyping   125–127
   polymorphic xenobiotic-metabolising enzymes
     acrylonitrile   128–130
     bladder carcinogens   130–133
     colorectal cancer   133–135
     CYP2C9*2   127
     NAT2   128
   slow acetylators   122
   study numbers and effect size   135–138
   toxicogenomics   123
toxicogenomics   123
   bioinformatics   479–480
   biostatistics   479–480
   concept of   472–473, 475
   genomics
     methods   476
     microarrays   476
     next-generation sequencing   476–477
     qRT-PCR   477
   mechanistic analyses   483–485
   metabolomics   479
   predictive   481–483
   proteomics   477–478

toxicokinetics 10–11
  absorption
    definition 35
    gastrointestinal tract 40–41
    intraperitoneal injection 42
    intravascular 42
    oral cavity 40–41
    respiratory tract 38–40
    skin 41
    subcutaneous/intramuscular injection 42
  arsenic 611–613
  biological membrane (*see* biological membrane)
  cadmium 614
  chromium 618
  compartment models (*see* compartment models)
  copper 619–620
  cyanide 689–690
  definition and purpose 34–35
  diesel engine emissions 735
  distribution 42–43
  elimination
    excretion 44–47
    metabolites 43–44
  ethanol solvents 681–682
  iron 621
  lead 615
  manganese 623–624
  mercury 617
  nickel 623–624
  PAHs 724–725
toxicokinetics (TK) 215
toxicological concern (TTC) 700–701
toxicological evaluation
  animal experiments 15
  C&L of chemicals 17–18
  evaluation of mixtures 15–16
  existing chemicals 14
  hazard identification and risk assessment
    acute toxicity studies 8
    ADI 13
    carcinogenicity 9–10
    chronic studies 8
    genotoxicity 9
    institutions 6, 7
    irritation 8–9
    phototoxicity 9
    reproduction and development 10
    toxicokinetics 10–11
  issues 13–14
  mode and/or mechanism of action 10–11
  precautionary principle 16
  sensitization and photosensitization 9

  test guidelines 14–15
  TTC concept 17
  uncertainty analysis 16
Toxic Substances Control Act (TSCA) 468, 566–567, 588
tracheobronchial (TB) 250
transcription-coupled repair (TCR) 167
transferrin 608
translesions synthesis 170
transport of dangerous goods (TDG) 551–552
triazine herbicides 708–709
tributyltin 657–658
trichloroethene 673 *see also* chloroform
trichloromethyl radical (CCl_3) 108
triiodothyroacetic acid (TRIAC) 377
tumor necrosis factor (TNF) receptor 118
tumor suppressor genes 172
  ras proto-oncogenes 176–177
  TP53 177
type II cell death 179
type II pneumocytes 250
type I pneumocytes 250
tyramine 224, 225

UDP-glucuronic acid 74, 75
UDP-glucuronyltransferases 75, 76
United Nations Economic Commission for Europe (UNECE) 532
United Nations Environment Programme (UNEP) 532–533
United Nations Scientific Committee on the Effects of Atomic Radiation (UNSCEAR) 535
United States regulations 557
  ATSDR 570–571
  CDC 571
  CPSA 569
  CPSC 568–569
  environmental regulations 563–568
  Food and Drug Administration 561–562
  litigation 571–572
  on occupational and environmental health 558, 559
  OSHA 558–560
  radionuclides regulations 569–570
unscheduled DNA synthesis (UDS) 419
urea herbicides 709
urinary tract 313, 317–319
  bladder tumors 315, 318–319
  collection and excretion 313
  toxicology of 317–318

uroporphyrinogen decarboxylase (UROD) 602–603
urticaria 305
US Environmental Protection Agency (EPA) 186, 741
US National Toxicology Program (NTP) 741
uterotrophic assay 385

very low density lipoprotein (VLDL) formation 241
vessel cooption 181
Veterinary International Conference on Harmonization (VICH) 535, 540
vigabrantin 362
vinclozolin 657
viral oncogenes 175
visual system 280
vitellogenin (VTG) 666
vitreous body, eye 354

warfarin 715
Wechsler Intelligence Scale 287
Werner syndrome 174
white blood cells 326–328
WHO Core Assessment Group 532
whole-mixture study 152
wildlife toxicology *see* ecotoxicology
Wiley Act 562
Wnt factors 235
World Health Organization (WHO) 514, 531, 706, 741

xenoantiandrogens *see* antiandrogens
xenoestrogens *see* antiandrogens
Xeroderma pigmentosum 173, 309

zearalenone 657
zero-order kinetics 44
zonation of liver lobules 235, 236